The Universe Revealed

Chris Impey
PROFESSOR OF ASTRONOMY, UNIVERSITY OF ARIZONA

William K. Hartmann
SENIOR SCIENTIST, PLANETARY SCIENCE INSTITUTE

Pacific Grove • Albany • Belmont • Boston • Cincinnati • Johannesburg • London • Madrid
Melbourne • Mexico City • New York • Scottsdale • Singapore • Tokyo • Toronto

Sponsoring Editor: Gary Carlson
Developmental Editors: Heather Dutton and Patricia Longoria
Assistant Editor: Marie Carigma-Sambilay
Marketing Team: Steve Catalano, Dena Donnelly, and Christina DeVeto
Editorial Assistant: Larisa Lieberman
Production Editor: Mary Anne Shahidi
Editorial Project Management: GTS Publishing Services
Manuscript Editors: Patti Scott and Patsy Dilernia
Permissions Editor: Mary Kay Hancharick

Interior Design: Paul Uhl
Cover Design: Roy R. Neuhaus
Cover Photo: Tom Bullock
Interior Illustrations: GTS Graphics
Photo Researcher: Judy Mason
Print Buyer: Vena Dyer
Composition: GTS Graphics
Cover Printing: Phoenix Color
Printing and Binding: R. R. Donnelley-Roanoke

COPYRIGHT © 2000 by Brooks/Cole
A division of Thomson Learning
The Thomson Learning logo is a trademark used herein under license.

For more information about this or any other Brooks/Cole Products, contact:

BROOKS/COLE
511 Forest Lodge Road
Pacific Grove, CA 93950 USA
www.brookscole.com
1-800-423-0563 (Thomson Learning Academic Resource Center)

All rights reserved. No part of this work may be reproduced, transcribed or used in any form or by any means—graphic, electronic, or mechanical, including photocopying, recording, taping, Web distribution, or information storage and/or retrieval systems—without the prior written permission of the publisher.

For permission to use material from this work, contact us by
Web: www.thomsonrights.com
fax: 1-800-730-2215
phone: 1-800-730-2214

Printed in the United States of America
10 9 8 7 6 5 4 3 2

Library of Congress Cataloging-in-Publication Data
Impey, Chris
   The universe revealed / Chris Impey, William K. Hartmann.
      p. cm.
   Includes bibliographical references and index.
   ISBN 0-534-24894-2
   1. Astronomy. I. Hartmann, William K. II. Title.
QB43.2.I47 1999
520—dc21   99-28561
                                                         CIP

BRIEF CONTENTS

1. How Science Works
2. The First Discoveries About Earth and Sky
3. The Copernican Revolution
4. Matter and Energy in the Universe
5. The Earth-Moon System
6. The Terrestrial Planets
7. The Giant Planets and Their Moons
8. Interplanetary Bodies
9. How Planetary Systems Form
10. Detecting Radiation from Space
11. Our Sun: The Nearest Star
12. Properties of Stars
13. Star Birth and Death
14. The Milky Way
15. Galaxies
16. The Expanding Universe
17. Cosmology
18. Life in the Universe

DETAILED CONTENTS

Preface xvii

CHAPTER 1 — How Science Works 2

LIFE ON MARS? 3
Themes in Studying Science 4
The Scientific Method 5
- Evidence 7
- Measurements 7
- Logic and Mathematics 8
- SCIENCE TOOLBOX: *Estimation* 9
- Observations and Errors 12

Building a Theory 14
- SCIENCE TOOLBOX: *Errors and Statistics* 15

Other Systems of Knowledge 17
Summary 19
Important Concepts 19
How Do We Know? 19
Problems 20
Projects 21
Reading List 21

CHAPTER 2 — The First Discoveries About Earth and Sky 22

STONEHENGE AND ITS PURPOSE 23
Cycles in the Sky 24
- What You Can See in a Year 24
- Navigation 24
- Keeping Track of the Seasons 26
- Calendars and Time 28
- SCIENCE TOOLBOX: *Dividing Time* 31
- Ancient Observatories 32
- Systems of Counting and Measurement 34

Real Astrology: Ancient Origins of a Modern Superstition 35
The Birth of Science 36
- Early Greek Thinkers 36
- Angles, Sizes, and Distances 37
- Phases of the Moon and Eclipses 37
- SCIENCE TOOLBOX: *The Small-Angle Equation* 38

v

    The Legacy of Aristotle 41
    Earth and the Seasons 42
       SCIENCE TOOLBOX: *Aristarchus and a Sun-Centered Cosmology* 43
Astronomy in Other Cultures 47
    Ancient Arab Science: Carrying the Flame 47
    Astronomy in India: A Hidden Influence 47
    Astronomy in China: An Independent Worldview 48
    Native American Astronomy: Science Cut Short 48
Summary 49
Important Concepts 49
How Do We Know? 49
Problems 50
Projects 51
Reading List 52

CHAPTER 3 — The Copernican Revolution 52

GALILEO'S TRIAL 53
Moving Beyond the Greek Model 54
The Five Key Players in the Copernican Revolution 56
    Copernicus and His Simple Idea 56
    Brahe's Careful Observations 59
    Kepler and Elliptical Orbits 61
       SCIENCE TOOLBOX: *Kepler's Laws of Planetary Motion* 62
    Galileo and the Revelations of the Telescope 64
    Newton's Great Synthesis 66
    A Universal Law of Gravity 68
       SCIENCE TOOLBOX: *The Inverse Square Law of Gravity* 69
The Plurality of Worlds 71
The Layout of the Solar System 71
    The Scale of the Solar System 72
    The Planets and Their Spacings 73
Exploration of the Solar System 74
    Dreams of Escaping Earth 74
    Orbits 74
A Brief History of Space Exploration 75
       SCIENCE TOOLBOX: *Periodic Processes* 77
Summary 80
Important Concepts 80
How Do We Know? 81
Problems 81
Projects 82
Reading List 83

CHAPTER 4 — Matter and Energy in the Universe 84

RUTHERFORD AND THE STRUCTURE OF THE ATOM 85
Basic Structure of Matter 86
    Early Greek Ideas 86
    Dalton and the Theory of Atoms 86
    Rutherford and the Atomic Nucleus 87
    Structure of Atoms and Molecules 89

Nature of Energy 91
  Different Forms of Energy 91
  Energy and Temperature 92
    SCIENCE TOOLBOX: *Potential and Kinetic Energy* 94
  Transformation and Conservation of Energy 95
    SCIENCE TOOLBOX: *Velocities of Atoms and Molecules in a Gas* 96
States of Matter in the Universe 98
Heat Transfer and the Concept of Equilibrium 99
  Thermal Equilibrium 99
  Three Modes of Heat Transfer 100
A Closer Look at Radiation 101
  Wien's Law and Thermal Radiation 103
    SCIENCE TOOLBOX: *Using Wien's Law* 105
  Radiation from Planets and Stars 106
  Internal Heat in Planets and Stars 107
Summary 107
Important Concepts 107
How Do We Know? 107
Problems 108
Projects 109
Reading List 109

CHAPTER 5 The Earth-Moon System 110

BUFFON AND THE AGE OF EARTH 111
How Old Is the Earth-Moon System? 112
  First Estimates of the Earth's Age 112
  Measuring Rock Ages by Radioactivity 113
  Measurements of Earth's and the Moon's Age 115
    SCIENCE TOOLBOX: *Working with Half-Lives* 115
Internal Structure of Earth and the Moon 116
  The Engine That Drives the Planet 116
  Layering of Earth and the Moon—Core, Mantle, and Crust 116
  Three Basic Rock Types in Earth's Crust 118
  Physical Layering of Earth and the Moon 118
Influences That Shape a Planet 119
  The Changing Earth and the Static Moon 119
  Plate Tectonics: The Hidden Sculptor of Earth's Surface 120
  Geological Processes in Earth's Evolution 121
  Impact Cratering: Sculptor of Geologically Dead Worlds 123
Geological Time Scale 125
  Sudden Changes During Geological History 126
  The Catastrophe 65 Million Years Ago—A Giant Impact 127
  The Catastrophe 250 Million Years Ago—Volcanic Eruptions? 130
  The Cosmic Connection 131
Earth's Atmosphere, Oceans, and Environment 131
  Origin and History of Earth's Atmosphere and Oceans 131
    SCIENCE TOOLBOX: *Random Processes* 132
  The Ozone Problem 134
  The CO_2 Increase and Global Climate Change 135
  Translating Scientific Research to Public Policy 137
Explorations of the Earth-Moon System 138
  Tides and Their Effects 138

SCIENCE TOOLBOX: *Tidal Forces* 140
Ice on the Moon? 141
The Origin of the Moon 141
Summary 142
Important Concepts 143
How Do We Know? 143
Problems 144
Projects 144
Reading List 145

CHAPTER 6

The Terrestrial Planets 146

LANDING ON THE SURFACE OF MARS 147
Terrestrial Planets 147
Mercury 149
A Day on Mercury 149
Mercury, Dr. Einstein, and Mr. Spock 150
A Surprising Polar Cap on Mercury 150
Venus 151
Venus' Infernal Atmosphere 151
The CO_2 Greenhouse Effect: Cause of Venus' High Temperature 151
The Volcanic Landscape of Venus 152
SCIENCE TOOLBOX: *Venus and the Greenhouse Effect* 153
Surface Ages and Internal Activity of Venus 154
Mars 155
The Mythic Mars—An Example of Evolution in Scientific Ideas 155
The Real Mars—A Modern View 158
Martian Mystery Number 1: Why Did the Climate Change? 161
Martian Mystery Number 2: Life on Mars? 162
Future Exploration of Mars 164
The Two Small Moons of Mars 164
Comparative Planetology: Rules of Planetary Evolution 165
Older Surfaces Have More Impact Craters 166
SCIENCE TOOLBOX: *Counting Craters on the Moon* 167
Bigger Planets Have Greater Internal Heat and More Geological Activity 168
Bigger Planets Have Stronger Gravity and Therefore Retain More Atmosphere 169
CO_2 is the "Normal" Composition for the Atmosphere of a Terrestrial Planet 171
Free Oxygen Means Something Strange Is Going On 171
Summary 172
Important Concepts 172
How Do We Know? 172
Problems 173
Projects 173
Reading List 174

CHAPTER 7

The Giant Planets and Their Moons 175

DISCOVERY OF VOLCANOES ON IO 176
Jupiter, Saturn, Uranus, and Neptune 177
A Variety of Icy Worlds 178
General Properties of the Atmospheres 178
The Nature of the Clouds 179
Internal Structure of Giant Planets 180

    Discoveries of Uranus and Neptune 182
    SCIENCE TOOLBOX: *Gas Laws* 183
    Infrared Thermal Radiation of Giant Planets 184
    Life on Giant Planets? 185
Comparative Planetology: Why Giant Planets Are Giant 186
Ring Systems of the Giant Planets 187
    Four Distinctive Ring Systems 187
    Structures Within Ring Systems 188
    Roche's Limit 189
    Where Did the Ring Particles Come From? 189
Satellite Systems of the Giant Planets 189
    SCIENCE TOOLBOX: *Resonance and Harmonics* 190
    SCIENCE TOOLBOX: *Roche's Limit and the Outer Edge of Ring Systems* 192
    The Largest Moon in the Solar System: Jupiter's Ganymede 193
    Jupiter's Icy Moon: Europa 194
    Jupiter's Volcanic Moon: Io 195
    Saturn's "Missing Link" Moon: Enceladus 196
    Saturn's Smoggy Moon: Titan 196
    Uranus' Fractured Moon: Miranda 197
    Neptune's Smoking Moon: Triton 197
Pluto: Ninth Planet or Interplanetary Body? 199
Summary 200
Important Concepts 201
How Do We Know? 201
Problems 201
Projects 202
Reading List 202

CHAPTER 8 Interplanetary Bodies 203

THE TUNGUSKA EVENT 204
Interplanetary Bodies: Many Types 205
Comets 207
    Comet Names and Orbits 208
    The Comet Nucleus: A Dirty Iceberg in Space 211
    The Oort Cloud and the Kuiper Belt 213
    SCIENCE TOOLBOX: *Kepler's Laws and Comet Orbits* 214
    The Life Story of Comets 215
Meteors and Meteor Showers 216
Asteroids 217
    Discovery of Asteroids 218
    SCIENCE TOOLBOX: *Gravitational Perturbations* 221
    Asteroids' Moons and Compound Shapes 222
    Rocky, Metallic, and Carbonaceous Asteroids 223
Meteorites 224
    The Scientific Discovery of Meteorites 224
    Meteorite Impacts on Earth 224
    Types and Origin of Meteorites 225
Interplanetary Threat or Opportunity? 226
    The Interplanetary Threat 227
    Observing a Major Impact on Jupiter 229
    SCIENCE TOOLBOX: *Probability and Impacts* 230
    The Interplanetary Opportunity 230

Summary 231
Important Concepts 232
How Do We Know? 232
Problems 233
Projects 234
Reading List 234

CHAPTER 9

How Planetary Systems Form 235

VICTOR SAFRONOV AND THE THEORY OF PLANET FORMATION 236
Archaeology of the Solar System 237
The Protosun 238
Early Contraction and Formation of the Disk 238
SCIENCE TOOLBOX: *Momentum and Angular Momentum* 240
Helmholtz Contraction of the Protosun 241
The Solar Nebula 242
Condensation of Dust in the Solar Nebula 242
Meteorites as Direct Evidence 243
A Nearby Presolar Explosion? 245
From Planetesimals to Planets 245
The Process of Accretion 246
Origins of Satellites 249
Gradual Evolution Plus a Few Catastrophes 250
Planets Orbiting Other Suns 251
Direct Detection 251
SCIENCE TOOLBOX: *Chaos and Determinism* 252
Indirect Detection 254
The Payoff: Discoveries of Extrasolar Planets 256
SCIENCE TOOLBOX: *The Doppler Effect and Planet Detection* 257
Summary 260
Important Concepts 261
How Do We Know? 261
Problems 262
Projects 263
Reading List 263

CHAPTER 10

Detecting Radiation from Space 264

BEYOND THE VISIBLE SPECTRUM 265
Radiation and the Universe 266
The Nature of Light 266
The Electromagnetic Spectrum 267
Characteristics of Radiation 267
Waves and Particles 269
How Radiation Travels 272
Thermal Spectra 272
Radiation and the Structure of the Atom 273
SCIENCE TOOLBOX: *Properties of Electromagnetic Radiation* 274
Spectral Lines 277
Emission Lines 277
SCIENCE TOOLBOX: *Uncertainty and the Quantum World* 278
Molecules and Emission Bands 279

Absorption Lines 281
   Molecules and Absorption Bands 281
Astronomical Uses of Radiation 281
Telescopes and Detectors 281
   Optical Telescopes 282
   Optical Detectors 285
   Image Processing 285
   SCIENCE TOOLBOX: *Digital Information* 286
   Radio Telescopes 288
   Telescopes in Space 288
   Interferometry 289
   SCIENCE TOOLBOX: *Collecting Area and Resolution of Telescopes* 290
   New Frontiers 292
Summary 293
Important Concepts 293
How Do We Know? 293
Problems 294
Projects 295
Reading List 295

CHAPTER 11

Our Sun: The Nearest Star 296

KELVIN AND THE AGE OF THE SUN 297
Understanding the Nearest Star 298
The Sun's Properties 298
   Rotation 299
   Studying the Sun's Spectrum 299
   Composition 301
Energy from the Atomic Nucleus 302
   The Conversion of Mass to Energy 303
   Energy from Nuclear Fission 303
   SCIENCE TOOLBOX: *Mass-Energy Conversion* 304
   Energy from Nuclear Fusion 304
Solar Energy from Nuclear Reactions 305
The Sun's Interior 306
   How Energy Gets from the Sun's Core to Its Surface 307
   SCIENCE TOOLBOX: *Collisions and Opacity* 308
   Solar Neutrinos 309
   Solar Oscillations 310
The Sun's Atmosphere 310
   Chromosphere and Corona 310
   SCIENCE TOOLBOX: *Testing a Hypothesis* 311
   Sunspots 313
   The Solar Cycle 315
   Solar Wind 316
Effects of the Sun on Earth 317
   Auroras 317
   The Sun and Climate Change on Earth 317
   Solar Energy and Other Cosmic Fuels 318
Summary 319
Important Concepts 319
How Do We Know? 319
Problems 320

Projects 321
Reading List 321

CHAPTER 12 Properties of Stars 322

FINGERPRINTING STARS 323
The Nature of Stars 324
  Names of Stars 325
Distances to Stars 325
  Apparent Brightness 326
  Absolute Brightness 328
  Measuring Distance Directly 329
Observed Properties of Stars 329
  SCIENCE TOOLBOX: *Parallax and Stellar Distances* 330
  Spectra of Stars 331
  Temperature 332
  Chemical Coposition 334
  Stellar Motion 334
Fundamental Properties of Stars 335
  Luminosity 335
  Diameter 335
  Mass 335
  SCIENCE TOOLBOX: *The Stefan-Boltzmann Law* 336
  Hydrostatic Equilibrium 337
Classifying Stars 337
  The Hertzsprung-Russell Diagram 338
  Nearby Stars 340
  Prominent Stars 340
  Non-Main-Sequence Stars 340
  Stellar Radius and the H-R Diagram 341
  SCIENCE TOOLBOX: *Comparing Giants and Dwarfs* 343
The Evolution of Stars 346
  Stellar Mass and the H-R Diagram 346
  Explaining the Main Sequence 346
  The Rate of Stellar Evolution 347
  Determining the Ages of Stars 348
Summary 349
Important Concepts 349
How Do We Know? 349
Problems 350
Projects 351
Reading List 351

CHAPTER 13 Star Birth and Death 352

SUPERNOVA 1987A 353
Understanding Star Birth and Death 354
Star Formation 355
  Molecular Clouds 355
  Toward a Theory of Star Formation 357
  Protostars and Pre-Main-Sequence Stars 358
  Mass Limits, Large and Small 359
  The Transition from Planet to Star: Brown Dwarfs 360

Young Clusters and Associated Young Stars 361
The Main-Sequence Stage 362
   Nuclear Reactions in Low-Mass Main-Sequence Stars 364
   Nuclear Reactions in High-Mass Main-Sequence Stars 364
   Main-Sequence Lifetimes 365
Evolved Stars 365
   SCIENCE TOOLBOX: *Lifetimes of Stars* 366
   Red Giants 367
   The Creation of Heavy Elements 368
   Variable Stars 368
   Mass Loss in Evolved Stars 370
The Death of Stars 370
   The Laws of Thermodynamics 371
   White Dwarfs 373
   SCIENCE TOOLBOX: *Understanding Entropy and Time* 374
   Supernovae 375
   SN 1987A 377
   Neutron Stars and Pulsars 378
   Einstein's Theories of Relativity 380
   SCIENCE TOOLBOX: *Properties of Black Holes* 386
Summary 387
Important Concepts 389
How Do We Know? 389
Problems 389
Projects 390
Reading List 390

CHAPTER 14   The Milky Way 391

HERSCHEL SCANS THE SKIES 392
The Distribution of Stars in Space 393
Stellar Companions 396
   Types of Binary Systems 396
   How Many Stars Are Binary or Multiple? 398
   Mass Transfer in Binaries 399
   SCIENCE TOOLBOX: *Binaries and Stellar Mass* 400
   Novas and Supernovas 401
   Exotic Binary Systems 401
   How Do Binary and Multiple Stars Form? 402
The Environment of Stars 403
   Composition of the Interstellar Medium 404
   Intersellar Medium and Starlight 405
   Structure of the Interstellar Medium 409
   SCIENCE TOOLBOX: *Dust Extinction and Reddening* 411
Groups and Clusters of Stars 412
   Open Clusters and Associations 412
   Globular Star Clusters 412
   Distances to Stars and Groups of Stars 415
   The Ages of Groups of Stars 417
The Layout of the Milky Way 420
   SCIENCE TOOLBOX: *Isotropy and Anisotropy* 422
Summary 424
Important Concepts 424

How Do We Know? 425
Problems 425
Projects 426
Reading List 426

CHAPTER 15 Galaxies 428

HUBBLE AND THE NATURE OF GALAXIES 429
The Milky Way Galaxy 430
   Components of the Milky Way 430
   Mapping the Rotating Disk 432
   The Mass of the Galaxy 436
   SCIENCE TOOLBOX: *Weighing a Galaxy* 437
   The Galactic Center 438
   Stellar Populations 440
   The Formation of the Milky Way 442
Discovering the Distances to Galaxies 443
Properties of Galaxies 445
   A Classification System 445
   The Local Group 447
   SCIENCE TOOLBOX: *Light Travel Time* 451
   Size and Luminosity 452
   Colors, Stellar Populations, and Mass 452
   The Evidence for Dark Matter 455
   SCIENCE TOOLBOX: *The Gravity of Many Bodies* 457
How Galaxies Evolve and Interact 459
The Puzzle of Galaxy Formation 462
Summary 464
Important Concepts 464
How Do We Know? 464
Problems 465
Projects 466
Reading List 466

CHAPTER 16 The Expanding Universe 467

THE DISCOVERY OF QUASARS 468
The Redshifts of Galaxies 469
Expansion of the Universe 470
   The Nature of the Redshift 470
   The Hubble Relation 473
   Distances to Galaxies Revisited 474
   SCIENCE TOOLBOX: *Relating Redshift and Distance* 475
   The Hubble Constant 473
   SCIENCE TOOLBOX: *Size and Age of the Universe* 477
Large-Scale Structure 479
   Clusters of Galaxies 479
   Superclusters and Voids 482
   SCIENCE TOOLBOX: *Galaxy Clustering* 483
   Dark Matter Revisited 486
   The Most Distant Galaxies 489
Active Galaxies and Quasars 491

Black Holes in Nearby Galaxies 491
Active Galactic Nuclei 492
Quasars and the Nature of the Redshift 494
Properties of Quasars 496
SCIENCE TOOLBOX: *Gravitational Lensing* 498
Understanding the Power Source 499
Quasars and Probes of the Universe 501

Summary 503
Important Concepts 503
How Do We Know? 504
Problems 504
Projects 505
Reading List 505

CHAPTER 17 Cosmology 506

THE RADIATION OF THE BIG BANG 507
Early Cosmologies 508
Relativity and Curved Space 510
The Big Bang Model 513
The Cosmological Principle 514
The Expansion of the Universe 515
Cosmic Nucleosynthesis 517
Cosmic Background Radiation 517

Measuring Cosmological Parameters 520
The Current Expansion Rate 520
The Curvature of Space 520
SCIENCE TOOLBOX: *The Critical Density of the Universe* 524
The Mean Density 524
The Age of the Universe 526
The Fate of the Universe 526
The Status of the Big Bang Model 528

The Early Universe 528
Limitations of Space and Time 528
The Evolution of Structure 531
Particles and Radiation 533
SCIENCE TOOLBOX: *Particles and Radiation in the Early Universe* 535
Inflation and the Very Early Universe 536
Symmetry and the Mass of the Universe 540
The Forces of Nature 541
The Limits to Knowledge 543

Summary 544
Important Concepts 545
How Do We Know? 545
Problems 545
Projects 546
Reading List 546

CHAPTER 18 Life in the Universe 547

CANALS ON MARS 548
Are We Alone? 549

The Nature of Life 550
Sites for Life 552
The Origin of Life on Earth 554
    The Production of Complex Molecules 554
        SCIENCE TOOLBOX: *Life as Digital Information* 555
    Complex Molecules in Space 557
    From Molecules to Cells 558
    Earth's Earliest Life-Forms 559
Evolution, Intelligence, and Technology 561
    From Cells to Intelligence 562
Natural Selection 563
    Is Intelligence Inevitable? 565
    The Influence of the Cosmic Environment 566
    Culture and Technology 566
The Search for Extraterrestrial Intelligence 567
    Is There Alien Life in the Solar System? 569
    Intelligent Life Among the Stars 569
        SCIENCE TOOLBOX: *The Drake Equation* 570
    Where Are They? 571
    The Best Way to Communicate 575
    How to Recognize a Message 577
    SETI Past, Present, and Future 577
The Anthropic Principle 580
Summary 580
Important Concepts 581
How Do We Know? 581
Problems 582
Projects 582
Reading List 583

Appendices 585

Appendix A: Scientific and Mathematical Techniques 585
A-1 Scientific Notation and Logarithms 585
A-2 Algebra and Working with Dimensions 587
A-3 Units and the Metric System 589
A-4 Precision and Measurement Errors 593
A-5 Angular Measurement and Geometry 595
A-6 Observing the Night Sky 597
A-7 Ways of Representing Data 605

Appendix B: Astronomical Data 609
B-1 Planetary Data 609
B-2 The Nearest Stars 611
B-3 The Brightest Stars 612
B-4 Local Group Galaxies 613

Glossary 614

Index 622

PREFACE

Welcome to the Universe

This book will introduce you to the physical nature of the universe. Our intellectual journey takes us from the bounds of Earth, to worlds far from our own, to the centers of stars, to galaxies scattered through space, and to the universe as it was in the distant past. It is remarkable that the same set of physical laws that govern our life on Earth can also describe these strange and unfamiliar places. These physical laws come alive through examples of the exciting, difficult and often inspiring process of scientific discovery. The quest to understand our place in the universe is one of the greatest human adventures.

Astronomy is the oldest science, but it is also tremendously active right now. Discoveries are being made by large new telescopes on high mountaintops and by sophisticated new observatories in space. Not a week goes by without an exciting new research result. Many of these projects address profound questions about our place in the universe: Do planets exist around other stars? What is the fate of stars like the Sun? How big is the universe? Will the universe expand forever? In this book, you will learn the answers to these questions. You will also see how scientists have systematically revealed how the universe works.

Scientists strive to understand the universe based on observations and evidence, not opinions. Therefore, this book contains facts—because they represent the basic information of science, and it contains terminology—because scientists need a precise language for communication. You will also learn how we make measurements and how we establish and test theories. The first chapter covers how science works; we will then illustrate these basic ideas with many examples throughout the book. Most of the material is rooted in physical science, but we use material from the fields of life science and technology to broaden the material and show how the methods of science apply to a wide range of knowledge. Our society depends on science and technology, so an understanding of how science works and why it is important is essential for every citizen.

Astronomy is a wonderful adventure of discovery. Earth spins through space in a universe that is vast and ancient. It is remarkable that humans have been able to reach out with their minds and telescopes and find out so much about our cosmic setting. This book is written for you. Welcome to the universe.

Themes

To help you grasp basic scientific principles, this book is built around three central themes. One theme examines humanity's relationship to Earth and the larger universe. Another explores the idea that the universe is governed by a small set of mathematical and physical rules, of which the law of gravity is just one example. The theme of science as a human process to uncover the rules of the universe is also presented. You will see exactly how science works to reveal these rules. While all three themes do not appear with equal prominence in every chapter, they serve as threads that connect the material throughout the book. Each theme is developed in a variety of settings across the universe. We sum up the themes below.

THE UNIVERSE IS BIG AND YOU ARE NOT We are dwarfed by the universe we live in. The earliest civilizations believed that Earth was the center of a small universe. We now know that planet Earth orbits an average star, one of billions in the Milky Way galaxy, and that the Milky Way is one of billions of galaxies flung through space. The first step in displacing Earth from the center of the universe was the Copernican revolution. We will follow the Copernican idea throughout the book and try to get a true sense of the scale of the universe.

THE UNIVERSE FOLLOWS RULES The universe is governed by a small set of simple physical rules. Think of the differences between the atmosphere of a planet, the fiery core of a star, and the entire universe just after the big bang. The gas in each of these situations is described by physical laws. The same law of gravity describes the way an apple falls from your hand and the way two galaxies orbit each other in distant space. The radio waves that let you listen to music in the car and the X rays in your doctor's office have the same fundamental nature. Scientists have found a small set of physical rules that explain the diverse behavior of the natural world. We will present these rules as we tour the universe.

KNOW THE RULES As a citizen of the universe, you should know the rules. Science is a method of learning the rules by making observations, conducting experiments, devising theories, and then testing them. Science is built on a foundation of evidence and repeatable observations. The method of science is the most effective way humans have devised of acquiring knowledge; it has revealed a universe of surprising beauty and grandeur. We will explore the way that astronomers have come to know the rules. In effect, we will answer the question: How do we know what we know?

What You Will Find in This Book

You will be studying the universe starting from Earth and working outward. This book therefore moves from the planets to stars to galaxies to cosmology. The first three chapters deal with the development of the scientific method, the roots of Greek science, and the work of Copernicus, Galileo, and Newton. These chapters will give you a view of the emergence of science and the often bold application of physical laws beyond Earth. Next we present the basic ideas of radiation, heat, and the physics of Isaac Newton. We describe Earth and the planets of the solar system in the succeeding chapters. A chapter on 20th-century physics and the relationship between matter and energy is followed by the discussion of stars, galaxies, and modern cosmology. We end with a chapter that brings together material from earlier in the book to consider the nature of living organisms and the possibilities of life elsewhere in the universe.

We have aimed for a direct, conversational style of writing and hope you will find this book easy to read. You will encounter stories of many of the scientists who have helped us understand how the universe works. You will also encounter simple mathematics. One of the most remarkable aspects of science is the fact that natural phenomena are so well described by mathematics. Most of the math is separated into a feature called a Science Toolbox, but you will also find simple calculations sprinkled through the text. These calculations are designed to get you comfortable with estimation—the ability to do a rough multiplication or division and see the order of magnitude of a quantity.

When dealing with unfamiliar ideas, we often use analogies in this book. Remember that science is a continuing process of discovery—we do not yet know all the answers. Use your imagination as you read this book. The universe has surprised us in the past, and it will surprise us in the future.

How to Use This Book

This book has a number of features that are designed to organize the material and help you learn. You will find these features in each chapter. Use them to understand the material better.

- *What to watch for in this chapter.* This brief summary, at the front of each chapter, describes the ideas and topics you will encounter. Use this summary to help you anticipate the most important material in each chapter.
- *Prereading questions on the themes of the book.* Three themes provide links among the many topics of astronomy. The first theme is our role in the universe. The second theme is how the universe works. The third theme is how we acquire knowledge. Prereading questions at the beginning of each chapter will help you keep the themes in mind.
- *Opening Story.* Science is a continuing story of discovery. To set the stage for a new topic, we begin each chapter with the story of a major scientific breakthrough. The goal is to acquaint you with some of the major figures in the history of science and to show how science works in practice. We also want to put some flesh on the bare bones of the scientific method by revealing the twists and turns of scientific discovery.
- *Science Toolboxes.* Each chapter has between two and four Science Toolboxes. These boxes present mathematical techniques used in astronomy and other branches of science. They are useful demonstrations of the "nuts and bolts" of the scientific method. Equations are expressed in words in the text, but a Science Toolbox is the place to look for worked examples and applications of simple mathematics. Your instructor may omit this level of math; the text is designed to be self-contained without the boxed material.
- *Summary.* There is a summary after the material in each chapter. Use the summary to make sure you understand the most important aspects of each topic. Also use it as a study aid for quizzes, tests, and exams.
- *Important Concepts.* A list at the end of each chapter corresponds to boldfaced words or phrases in the chapter. These are the most important pieces of terminology in astronomy. You should not think of them as just words to be memorized; each one is a concept that relates to some aspect of the material you have just read. Try to define each term and use it in a sentence, to make sure you understand the concept.
- *How Do We Know?* This set of questions (and answers) at the end of each chapter is designed to engage you in thinking about how astronomers learn about the universe. For example, the chapter on the Sun asks how we know the physical conditions in the solar core, and the chapter on light asks how we detect invisible forms of radiation. These questions are key reinforcements of the scientific method.
- *Problems and Projects.* Most of the problems at the end of each chapter are conceptual. Problems marked with an asterisk require the application of simple mathematics. There are no problems that test terminology or ask you to simply restate material from the chapter. Even if your instructor does not assign these questions, it is worth trying them to make sure you understand the material. Many projects require naked-eye observa-

tions of the night sky, or the use of small telescopes or binoculars. Some of the projects are group activities.

- *Reading List.* This book is self-contained. However, if you want to learn more about a topic, consult the reading list for resources. You will find books and magazine articles written at a popular or nontechnical level. The magazines *Astronomy* and *Sky and Telescope* are the best sources to learn what is going on in the night sky from month to month. Several times a year, *Discover* magazine and *Scientific American* carry astronomy articles.
- *Appendices.* Two types of material follow the main contents of the book. One set of seven appendices covers Scientific and Mathematical Techniques. Here you will find important background information that is useful in the study of any scientific subject, such as discussions of units, errors, and ways of representing large and small numbers. Appendix A-6 contains star charts and tips to help you find your way around the night sky. Another set of four appendices presents basic Astronomical Data.
- *Glossary.* Here you will find an alphabetical listing of all the boldfaced words or phrases that appear in the book. Each term has a one- or two-sentence definition; these terms are the same as the Important Concepts that are listed at the end of each chapter. For searching through the material of the book in general, consult the Index.
- *Art Program.* Color images can be powerful aids to learning. You will find this book filled with images from the Hubble Space Telescope and other major facilities. These are not just pretty pictures; each astronomical image illustrates an idea from the text. Read the figure captions carefully because they often contain information needed to understand the image. We use diagrams to explain key concepts in astronomy. This text has a unique feature: astronomical art that presents scientifically plausible views of places no humans have visited.

Using the World Wide Web

The World Wide Web is an enormous source of information on any subject. There is a lot of excellent astronomy material on the Web; the difficulty is finding what you want in the ocean of material. This book comes with access to a World Wide Web site that contains links to the best sources of astronomy on the Internet. Explore these sites or others that your instructor might refer you to. However, remember that your textbook does two things that the World Wide Web cannot do. We have organized the material and developed the ideas in a systematic and logical manner. We have also maintained a strong narrative as we tell the story of scientific discovery.

Included with each new copy of this book is a four-month subscription to InfoTrac, an online library of journals, magazines, and other periodicals. It is an invaluable research tool.

Study Tips

Probably you are a non–science major taking a college science course for the first time. Perhaps you are approaching the subject of astronomy with a feeling of nervous anticipation—excited to be taking a journey through the universe, but worried about the math or the technical nature of the subject. There is so much information in this book! But don't worry, because this book was written to help you succeed in learning about astronomy.

Try to keep up with the material. You instructor will assign chapters according to a syllabus; try to read these chapters before you see the material in class. Do not try to remember everything or take copious notes. Just read a chapter like a story, and note ideas you find difficult to understand. After you have seen the material in class, review the chapter, paying particular attention to those topics or concepts that were stressed by your instructor. Try to write down the main ideas of the chapter in your own words. Pay particular attention to the chapter summary, which tells you what you should have learned.

Use your instructor. You may be in a large lecture class where individual attention is rare. Your instructor is the primary source of information in the course, but amazingly few students go to office hours. Go to your instructor if you have problems or misconceptions, or if you just want to review the material. Use your fellow students. Get together for study groups to study for exams or to review difficult concepts. Don't fall into the trap of rote memorization of key words—challenge each other to define a concept or give examples. Most importantly, if you are struggling in any way, get help. There is a lot of material in an astronomy course, and it is easy to fall behind.

Why study astronomy? You may never take a science course again. You may be taking astronomy to satisfy a science requirement. We live in a technological society, and the ability to think clearly and rationally about science has never been more valuable. In this book you will learn modes of logical thought that can help you in other aspects of your life. Primarily, a knowledge of astronomy is part of your intellectual heritage. There has been no greater human adventure than our quest to reach beyond our tiny home planet and understand the vastness of the universe. Enjoy the journey!

Acknowledgments

We acknowledge our colleagues for many ideas and suggestions and for unstintingly sharing the results of their research. We also thank our families for their patience during the long gestation of this book. This book depended on the talents and prodigious efforts of a number of people at Brooks/Cole and beyond—Gary Carlson, Heather Dutton, Mary Anne Shahidi, Pat Longoria, Tanya Nigh, Heather Stratton, Katy Spining, and Bob Richardson.

We thank the following reviewers for their helpful comments and suggestions: Karen Bjorkman, Universitiy of Toledo; Dennis Dawson, Western Connecticut State University; James Dull, Husson College (ME); Christopher

Godfrey, Missouri Western State College; Albert D. Grauer, University Arkansas–Little Rock; David J. Griffiths, Oregon State University; Fred Jaquin, Onandaga Community College (NY); Darrel James Johnson, Missouri Western State College; Burton Jones, University of California at Santa Cruz; Steve Kipp, Mankato State University; Dave Kriegler, University of Nebraska–Omaha; Lynn Lindsay, Snow College (UT); Michael C. Lopresto, Henry Ford Community College (MI); Peter R. McCullough, University of Illinois; J. Scott Miller, University of Louisville; J. Ward Moody, Brigham Young University; Dwight Russell, University of Texas–El Paso; and Dave Vesper, Indiana State University.

Finally, we thank our past and present students for inspiring us to attempt to explain the universe to a larger audience.

Chris Impey and William Hartmann
Tucson, Arizona

About the Authors

Chris Impey is the Deputy Department Head and a Professor of Astronomy at the University of Arizona. His research interests center on cosmology and the study of diffuse galaxies, active galaxies, and quasars. He has over 100 research publications in the professional literature, and has had 14 projects approved for observations with the Hubble Space Telescope. His work to develop interactive teaching materials for the Web is supported by a major grant from the National Science Foundation. He is a former Associate Director of the NASA Arizona Space Grant Program, which supports a variety of education and outreach activities. He has won six University awards for undergraduate teaching, and is coauthor of the Brooks/Cole textbook *Astronomy: the Cosmic Journey*.

William K. Hartmann is known internationally as a planetary astronomer, writer, and painter. He is a Senior Scientist at the Planetary Science Institute in Tuscon. His research has involved the origin and evolution of planets and studies of the surfaces of Mars, the moon, asteroids, and comets. Asteroid 3341 is named after him in recognition of this work, and in 1998 he was named first recipient of the Carl Sagan medal of the American Astronomical Society for communicating planetary science to the public. He has authored three other astronomy books for Wadsworth/Brooks-Cole, several popular astronomy books, and most recently *Mars Underground*, a novel about Mars.

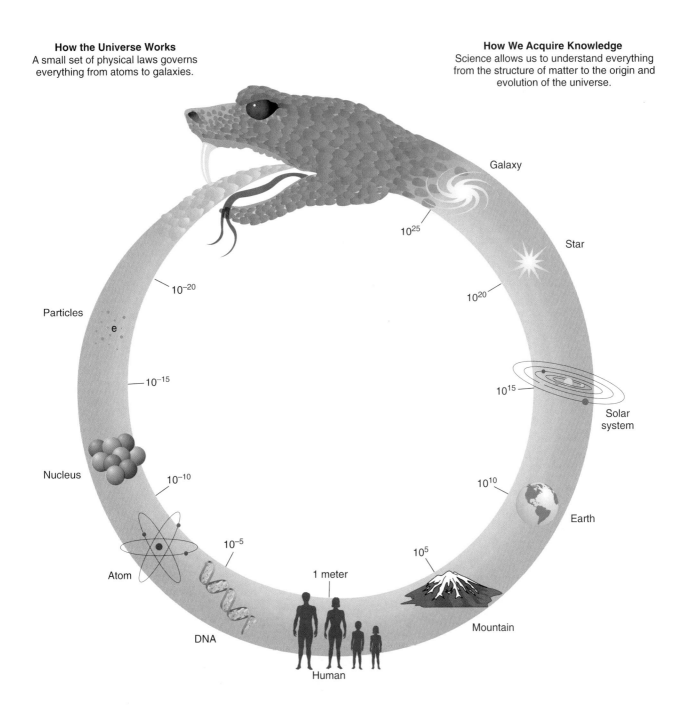

The ancient image of a snake swallowing its tail is called **orobouros.** As we study the fantastically small and the unimaginably large, we learn more about the unity of nature. The properties of fundamental particles tell us how stars shine and what the universe was like when it was young. We are poised to understand the universe of which we are a tiny part. (Based on a concept first presented at a research symposium by Sheldon Glashow, Nobel Prize winner in physics.)

CHAPTER 1

How Science Works

CONTENTS
- Life on Mars?
- Themes in Studying Science
- The Scientific Method
- Building a Theory
- Other Systems of Knowledge
- Summary
- Important Concepts
- How Do We Know?
- Problems
- Projects
- Reading List

Our home, Earth, set against a backdrop of stars and dust. Humans have always wondered whether they are alone in the universe. Astronomers have recently begun to discover and explore sites where life may exist beyond Earth.
(SOURCE: Painting by WKH.)

WHAT TO WATCH FOR IN CHAPTER 1

If you awoke one day to discover that you had been put on a strange island, as Robinson Crusoe was, your first project, after getting food and water, probably would be to find out where you were. If you spotted other distant islands around you, you would be interested in finding out what was on them. Did they have other resources? Did other people live there? This is just what happened to all of us. We were born onto an island in space: the planet Earth. We are all passengers on this chunk of rock, sailing on a cosmic voyage through trackless seas. In the distance we see other planets and stars around us. What are they like? Where did they come from? Do they follow the same laws of nature that we experience on Earth? How and when did Earth itself form? Are we alone, or do we have companions in the universe?

Questions like these are the basis of the three themes of this book. As a prelude to our exploration of the universe, we address the theme, How do we know what we know? In the last 2500 years, and especially in the last few centuries, we have invented a powerful system for learning about the world, called the *scientific method*. It is not perfect, but it has a better track record for creating knowledge and making predictions than any other system. This chapter explains the basic workings of the scientific method. Even though science is complex, the same methods and procedures are used in almost every scientific advancement. The key idea is that science relies on evidence to test ideas. This chapter discusses observations, their errors, and the logic that is the basis of all scientific arguments. You will learn the difference between a hypothesis and a theory and will find out that the scientific method is never a guarantee of truth. Despite its limitations, science can be used to build reliable theories that accurately describe the universe.

PREREADING QUESTIONS ON THE THEMES OF THE BOOK

OUR ROLE IN THE UNIVERSE
What is the role of humans in the universe?

HOW THE UNIVERSE WORKS
What are the physical rules that govern the universe?

HOW WE ACQUIRE KNOWLEDGE
How does science allow us to understand the universe?

LIFE ON MARS?

In early August 1996, rumors of an exciting discovery were flying among astronomers and biologists. On August 7, a respected group of scientists announced that they had found evidence that life existed on Mars 3.6 billion years ago, in the form of microscopic organisms. The story created a media sensation and also served as a striking example of how modern science works. The evidence seemed straightforward: Scientists found traces of ancient microbes inside a rock that they believed had been blasted off Mars by an impact from space. The rock had traveled for millions of years through space, falling to Earth thousands of years ago. Inside the rock they also found traces of organic material—compounds often produced by microbial life, as shown in Figure 1–1.

Think of the many scientific questions involved in this discovery. How did scientists identify the rock as Martian? How did they detect organic concentrations? How did they figure out that the organic material was Martian, rather than being due to contamination during the time the rock was on Earth? The answers involve intertwined threads of modern scientific work in dozens of different laboratories by chemists, biologists, geologists, and astronomers. For example, researchers found that the gas trapped in the rocks had the same composition as that measured in the Martian atmosphere with space probes years earlier. This was strong evidence that this rock and several others like it came from Mars.

In some respects, however, the evidence for ancient Martian life seemed less

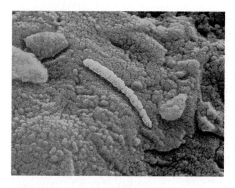

FIGURE 1-1
Evidence of past life on Mars? Scientists identified microscopic structures, such as the tubular forms in the center, inside a meteorite discovered on Antarctic. Note the scale on the image—200 nanometers (nm) [2×10^{-7} meters (m)]—is 10,000 times narrower than a human hair. Some researchers interpret the structures as fossil microbes. Whether they really represent ancient life on Mars is still open to debate.
(SOURCE: NASA photograph of meteorite ALH84001, courtesy Everett Gibson.)

compelling. The rock did not contain any actual fossils of microbes or direct imprints of complex organisms. Biologists pointed out that the proposed "fossil life-forms" were smaller than any known microbes on Earth—only one-thousandth the diameter of a human hair—and questioned whether objects this small could contain enough genetic material to qualify as life. Some chemists thought that the chemical evidence could be explained without requiring a living organism. Others doubted whether any form of life could have survived the extremes of temperature to which the rock had been subjected. The issue is still unresolved, and most planetary scientists remain skeptical that traces of extraterrestrial life have been discovered. This is one of the hottest topics in science. Later in this book we will explore the many lines of evidence that converge in this discussion of possible ancient life on Mars.

Three larger philosophical questions underlie the interpretive details of the chemistry, biology, geology, and astronomy. Why did the announcement capture the public imagination so completely? How could such life have once thrived on the red planet, which today is a frozen desert? And why were different scientists able to view the same evidence and come to quite different conclusions? These are major questions, not just about the Martian rock, but about how science works.

Themes in Studying Science

In fact, these questions reflect the three major themes of this book. The answer to the first question—Why did the public react so strongly to the possible discovery of life on another planet?—is that humans have been wondering for several centuries if life on Earth is unique. What is our role in the universe? Could there be other worlds that have environments like that of Earth and life-forms like ours? Is Earth the apex, or pinnacle, of all creation and the "capital of the universe"? Is it a unique cosmic accident or just one of many planets scattered among the galaxies that support life? These concerns are part of a broad intellectual movement spanning several centuries.

Until the 1500s, most people thought that Earth was the center of the universe, with the Sun, planets, stars, and other bodies moving around it. But then Polish scientist Nicolaus Copernicus hypothesized that the Sun, not Earth, is at the center of our planetary system. Within generations, this was proved correct, and humanity learned that the Sun was just one ordinary star among millions of stars. This discovery became known as the *Copernican revolution*. Copernicus showed for the first time that humans were not at the center of creation, and he spurred people to think about their relationship with the universe. Another aspect of this revolution came in the 1800s, when English scientist Charles Darwin and others showed that humans are just one among many species that have evolved on Earth. In fact, biologists revealed a multimillion-year procession of species, with most becoming extinct. The next milestone in the changing worldview that started with the Copernican revolution came in the 1920s, when astronomers showed that the Sun is not at the center of our gigantic disk-shaped galaxy, but at its outskirts. Moreover, our galaxy is only one of millions of visible galaxies. Once again, these findings displaced human beings from our central role in the universe.

The question of whether life ever existed on Mars can thus be seen as a current chapter of this long adventure of discovering where we fit in the universe. As we will see later in the book, planets have been found around nearby stars. With perhaps a solar system for every star and so many stars to choose from, the number of potential sites for life is enormous. Is it possible that life on Earth is just one of many cases of life scattered throughout the universe, like flowers in a meadow? Or is life so complex and unlikely that it has originated just once in the universe? It is a big question, and we do not yet know the answer. However, after centuries of speculation, we may be on the verge of a breakthrough.

The second question—How could life have appeared on Mars?—contains the essence of a second broad theme. This book will explain the physical nature of our cosmic environment. We are really asking whether the rules that govern our terrestrial experience also apply across vast reaches of space. What are the planets and stars really like? Where do elements come from? How is matter organized? How does it interact with radiation? How do carbon atoms combine with other atoms to create complicated molecules that can reproduce themselves? Do these processes occur on other planets? To answer these questions, this book reviews some basic concepts of physics and chemistry; later chapters will tour the universe of planets, stars, and galaxies. You will learn the remarkably small set of physical rules that govern the behavior of everything in the universe.

The third question—How do we know that the results of science are correct?—addresses a third theme of the book. How do we *know* something? Humans have come to believe that there are physical explanations for the diverse phenomena of the natural world. For example, we know that solar eclipses are caused when the Moon passes in front of the Sun. We have rejected the notion that eclipses express the displeasure of Sun gods, as some ancient cultures believed. If we encounter a phenomenon we do not understand, we can study it by a logical, systematic method. This method has yielded physical ideas that describe everything from the reactions going on at the center of a star to the conditions just after the creation of the universe.

The main subject of this book is **astronomy,** which is the study of all the kinds of matter and radiation beyond Earth. Astronomy is an excellent vehicle for studying science because of the sheer diversity of the universe. The primary backdrop for astronomy is **physics,** because we need to understand how the material of the universe interacts and behaves. Matter and radiation in all its forms will be encountered in every chapter. And while the interaction of matter and radiation might sound a little forbidding, in astronomy it leads to sights as beautiful as the northern lights and the rings of Saturn. In the first part of the book, we will also encounter **geology** when we consider the composition of the planets. We also require **chemistry** for our study of the way that elements combine in rocks and in space. The existence of life on Earth leads us to consider **biology** and the exciting possibility of how life might have evolved in different environments. The language of science is **mathematics,** and we will see in each Science Toolbox how it provides elegant shorthand for conveying many of the ideas in this book.

For a single image that encapsulates the three themes of this book, look at Figure 1–2. The digital image shows a swarm of remote galaxies as revealed by the Hubble Space Telescope. This is the opposite end of the cosmic scale from the microscopic structures shown in Figure 1–1! Each fuzzy blob contains many millions of stars. The existence of so many stars beyond the Sun leads us to wonder about our relationship to the larger universe around us. The stars in those faint galaxies are many trillions of miles away, yet each operates according to the laws of physics that apply to the Sun and the nearby stars we see in the night sky. The image also makes us wonder how we "know" in astronomy. After all, in terms of direct exploration, humans have made only the small jump to the Moon, and our spacecraft emissaries have barely left the solar system. Yet with large telescopes, we have studied radiation that has traveled for billions of years through space to reach us from distant, ancient galaxies.

In this chapter we focus on the third theme: how we know. We study the method that scientists use to probe, observe, and interpret the universe. Interestingly, many of the physical environments discussed in this book—extremes of temperature, pressure, and density—cannot be duplicated in any laboratory. To understand how astronomers have built up a reliable base of knowledge about conditions so remote

FIGURE 1-2
A universe of galaxies without end? This picture from the NASA Hubble Space Telescope is the deepest image of the sky ever made. The region of sky is only 1 arc minute across, or one-thirtieth the diameter of the full moon. This angle is the size of the head of a nail held at arm's length. Yet it contains hundreds of fuzzy images, each of which is a system of stars like our own Milky Way galaxy. The existence of so many galaxies, each with billions of stars, gives us a new perspective on our place in the universe. (SOURCE: Courtesy of Space Telescope Science Institute, NASA.)

from ordinary experience, we will consider the basics of the scientific method, including its tools of measurement, logic, and mathematics. Next, we will consider what a hypothesis is and how a scientific theory is established. Finally, we will contrast the scientific method with other belief systems. Not all systems of knowledge are equal; science provides the most reliable way of learning about the universe we live in. We will revisit and develop many of the elements of this chapter later in the book.

The Scientific Method

Since the time of the ancient Greeks, people who study the natural world have developed a system for establishing knowledge, called the **scientific method.** The scientific method requires, as a minimum, the following: terminology that is precisely defined, measurements that are quantitative and repeatable, and assertions that are backed up by evidence.

There are several essential steps in the scientific method. The first step is to develop curiosity about something and then gather evidence about it, usually in the form of observations or data. The evidence might be physical, such as rocks brought back from the Moon; or it might be readings from instruments, such as measures of light focused by a telescope. Statements made with no evidence to support them are called *speculation;* they might be true, or they might not. Without supporting evidence, there is no way to prove or disprove them.

The second essential step in the scientific method is to analyze the data, which usually involves a process of pattern recognition. For example, astronomers might analyze the nightly positions of a planet in the sky and recognize regularity in the motion; or they might find a similarity in the chemical composition of stars in different parts of our galaxy. The idea of searching for patterns in nature is at the heart of science. Discoveries start with a playful and curious mind at work.

We can use a deck of cards as an analogy for this process of discovery. Look at the first two sequences of cards in Figure 1–3a and b. The patterns are obvious. The first is just a sequence of cards of increasing value, and the second is an alternating sequence of red and black cards. Note that the value of the cards is irrelevant in the second example—the pattern lies in the colors. Now look at the last two sequences in Figure 1–3c and d. (Cover up the caption so you cannot read the answer!) Can you figure out the rules that govern these patterns? This is the situation of the scientist, who does not know the answer and is trying to figure out the rules. You can think of the cards as the "data." What kinds of data are required to see the patterns? Is all the information relevant? How complicated is the rule that determines the pattern? Could more than one rule describe the pattern? The example of a deck of cards is simple, even trivial, but it can illuminate some of the features of the scientific method.

The third step of the scientific method is the development of an explanation for the results of the analysis. Such an explanation is called a **hypothesis.** Often it is called a *working hypothesis* to emphasize that it is only a tentative proposal. An essential aspect of a scientific hypothesis is that it must be testable. That is, there must be some further observation or experiment that is capable of affirming or disproving the hypothesis. Scientists propose different hypotheses. The wrong ones are weeded out by such experiments. There is always more than one possible explanation for any set of data, which is one of the reasons why scientists argue so much! The last element of the scientific method is therefore the critical evaluation of hypotheses through testing.

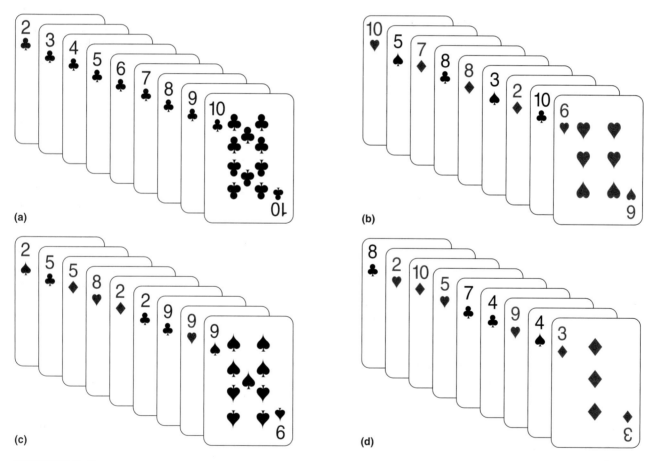

FIGURE 1–3
Patterns in a sequence of cards as an analogy for the discovery of patterns in nature. **(a)** A simple progression of cards in numerical order. **(b)** An alternating sequence of red and black cards. The number values show no pattern. **(c)** Each successive card matches either the color or the value of the card before it. **(d)** An even-numbered card is followed by any red card, and an odd-numbered card is followed by any black card. Notice that the last two patterns are quite subtle, even though a simple rule governs each one.

Often, an idea that starts out as one researcher's working hypothesis survives many tests and becomes widely accepted. A hypothesis that has been tested repeatedly and successfully is usually called a **theory,** indicating that it is stronger than a mere working hypothesis. The term *theory* is usually reserved for a hypothesis that covers a wide range of phenomena. For example, you might hypothesize that people who watch a lot of television are overweight. You might have evidence to support your hypothesis, but this insight will never have the authority of a theory. Why? Because many people who watch a lot of television are not overweight, and there are many reasons why a person might be overweight. A theory should be broad in scope and application. An example of a modern robust theory is Einstein's theory of relativity.

The scientific method can never *guarantee* truth, but it can be used to draw conclusions with a much higher degree of reliability than other systems of knowledge can. Stronger evidence yields more reliable conclusions. This statement is not an idle boast, but a conclusion based on a long history of scientific advances. The scientific method is more reliable in part because of the pains taken to gather evidence.

Evidence

There is no science without **evidence.** This means that when a scientist makes an assertion, he or she must back up the assertion with some observational or empirical data. These observations can be examined and checked by other scientists (or anyone else who is interested). This method differs from prescientific styles of learning about nature, in which an assertion might be made on the strength of mysticism, moral authority, political ideology, or the old standby "Everybody knows that. . . ." For example, for centuries "everybody knew" that stones could not fall out of the sky. Rumors that stones had fallen out of the sky were written off as superstition. Then, in 1803, a shower of meteorites fell on a town in France. French physicist Jean Baptiste Biot visited the town, heard numerous consistent eyewitness accounts, and collected actual specimens of the stones, which were unlike any known terrestrial rocks. The convincing body of evidence helped Biot prove to doubters that stones do indeed fall from space. We call them *meteorites*.

Contrast that story with unidentified flying objects (UFOs). The idea that some UFOs are extraterrestrial visitors permeates the popular culture. Do alien spaceships visit Earth? That is a legitimate hypothesis, but how can we test it? It is difficult to investigate UFOs scientifically because there is so little physical evidence. No one has come forward with fragments of alien material that can be tested in laboratories. Many "UFO photographs" are fakes; others show identifiable natural phenomena. It is true that absence of evidence is not the same as evidence of absence. However, no real science can be done on this subject without tangible evidence. People are free to propose that many UFO sightings actually involve alien spaceships, but the reports are not believable without better evidence to support the hypothesis. Scientists do not ignore the UFO issue because they believe that alien spaceships are impossible (although many think that they are extremely unlikely), but because there is little or no physical evidence to work with.

What type of evidence do we consider in astronomy? There is very little direct evidence. Astronauts have returned a few hundred pounds of Moon rocks to Earth, and occasionally a piece of debris from a more distant part of the solar system falls to Earth as a meteorite. As we learned at the beginning of this chapter, we have even been lucky enough to find a "free" piece of Martian rock! Spacecraft have viewed all the major planets of the solar system. However, our knowledge about most of the universe has been learned by pointing telescopes at the skies and interpreting the radiation that is gathered. While many of these telescopes collect and detect visible light, in the last few decades astronomers have devised technologies to record radiation that the eye cannot see. Many of the images in this book are derived from "invisible" radiation, which we discuss in Chapters 4 and 10.

Science thus depends on the idea that we can extend our senses with instruments. For example, physicists deal daily with electric and magnetic fields and subatomic particles, none of which register with our five senses. Astronomers routinely detect many types of radiation from celestial objects, including radio waves, X rays, and gamma rays. In this book, you will find these invisible types of radiation converted to a map, so we can "see" the object. Telescopes are used to gather light from sources that are billions of times too faint for the eye to see. All this information is the evidence of astronomy.

Measurements

Scientists present their evidence in the form of a **measurement.** People make many statements in everyday life. Some are qualitative and others are quantitative. A friend might say, "This music is great!" or "It was colder yesterday than today," or "There are only four forces of nature." There is no reliable way to quantify the first statement, even though your friend may feel it is true. However, the next two statements can be quantified, and they can be subjected to actual measurements that will either support or refute the assertions. In science, we try to deal *only* with statements that can be quantified. Otherwise, we have no way to compare results.

We use measurements to arrive at precise and quantitative statements. A measurement must have two components: a number and a **unit,** which specifies the type of the quantity that is being counted. Notice that numbers alone are not that useful. Units must be in a system that is well defined. The statement "It was cold yesterday" can be quantified by saying that "It was 15 degrees yesterday," but you still need to know whether this is on the Celsius or Fahrenheit system of units. The statement "The stock market fell 50 points" is quantified, but to understand it, you need to know what a point on the Dow Jones average actually represents.

Astronomers need to handle very large and very small numbers. **Scientific notation** (also called *exponential notation*) is a useful shorthand for writing numbers of any size. For example, instead of saying that the nearest star is about

40,000,000,000,000 kilometers (abbreviated as km) away, we can say it is 4×10^{13} km away. (The exponent 13 represents the number of zeros following the significant figure 4.) Appendix A-1 has a summary and examples of this important way of representing large and small numbers.

The system of units in physical science is amazingly simple. All the diverse measurements we can make in the physical world—speed, force, temperature, electric charge, energy, and so on—are derived from only three fundamental properties: **mass, length,** and **time.** To these properties we attach the familiar units in the metric system: kilograms to measure mass, meters to measure length, and seconds to measure time. Almost every other type of measurement is just some combination of these units (see Appendix A-2). For example, area is just width multiplied by length, and momentum is just mass multiplied by velocity—which is the same as mass multiplied by length divided by time. Even concepts such as energy or temperature can be expressed as combinations of the same three quantities. The metric system of units is reviewed in Appendix A-3, along with other units that are convenient for astronomers. A simple system of units helps scientists make sense of a complicated world.

Why do scientists generally use units in the metric system? The metric system was first put into widespread use after the French Revolution of 1789. The architects of the French Revolution wanted to make a break with the culture defined by royalty and hereditary power and to usher in an Age of Reason. As part of this sweeping set of social changes, they introduced a set of units based on a decimal counting system. The metric system was designed to replace the English system in which 1 gallon (gal) is 8 pints (pt); 1 foot (ft) is 12 inches (in.); 1 pound (lb) weight is 16 ounces (oz); 1 pound Sterling is 20 shillings; and so on. These units have their origins in medieval European history! At a time when few people were literate or numerate, it was easier to have measurements that could be readily subdivided. The numbers 8, 12, 16, and 20 all have three or more factors, while 10 only has two factors.

However, the metric system is much simpler. For example, 1 meter (m) [which is roughly 1 yard (yd)] is 100 centimeters (cm), and 1 km is 1000 m. Thomas Jefferson was very impressed by the metric system and pushed for it to be adopted in the United States. He would be very disappointed if he knew that 200 years later his country was the world's last holdout against adopting the metric system. It is ironic that the world's most technologically advanced society still clings to inches, miles, acres, gallons, pints, pounds, and even horsepower ratings!

In this book, we will deal with temperature extremes from the coldness of intergalactic space at 3 degrees above absolute cold to the shock wave of a supernova at 1 billion (10^9) degrees. We will study objects that range in size from microscopic cosmic dust grains (10^{-6} m) to the distance that light has traveled in the age of the universe (10^{23} km). Our discussions will range from the sparseness of the space between galaxies at a density of 10^{-19} kilogram per cubic meter (kg/m^3), to the density inside a black hole at 10^{18} kg/m^3. The ratio of these last measurements is a factor of 10^{37}, or 10 followed by 37 zeros! With such factors, it is not surprising that astronomers use scientific notation.

Notice the number of *significant figures,* or nonzero digits, in a scientific measurement. For example, look at the numbers 13,000 km and 12,756 km. In scientific notation, we write these numbers 1.3×10^4 km and 1.2756×10^4 km. The first number implies quite a rough measurement; changing the least significant figure would give you 12,000 or 14,000. But the second number implies a very fine measurement; changing the least significant figure would give you 12,755 or 12,757. So the number of significant figures is a measure of **precision.** This idea is explained in greater detail in Appendix A-4.

However, precise numbers are not always required in science and are sometimes not even possible to measure. You can understand most of the material in this book without having to deal with very precise numbers. In the example just given, it is more important to remember that Earth is roughly 13,000 km across than to try to memorize that it is exactly 12,756 km across. Few numbers in astronomy are known with a precision of more than three significant figures. Knowing an approximate value allows you to estimate many effects without making extensive calculations or looking things up in books. See the following Science Toolbox for examples of the basic scientific skill of **estimation.** Estimation saves time and allows scientists to distinguish promising hypotheses from foolish ones.

For an example of estimation, let us consider a deck of cards again. We will use this simple example or analogy in a number of places throughout the book. Everyone knows that there are 52 cards in a standard deck, but suppose you are given a large pile of cards. How could you estimate the number of cards in the pile? Clearly, the height of the pile divided by the thickness of a single card gives the number of cards. But these are two quite different types of measurement. You can measure the height of the deck with a ruler marked off in millimeters. It is not possible to measure the thickness of a single card with a ruler (can you explain why?); a more precise instrument, called a *micrometer,* would be needed. And yet the result clearly depends on the more difficult measurement. Scientists often face this situation—finding the right tool for a measurement and combining measurements of different precision.

Logic and Mathematics

Scientists must make several important assumptions to do their work. These assumptions sound reasonable, but they are hard to prove once and for all. For instance, we assume that *causality* holds true in the universe. In other words, we believe that all events have causes. This sounds very reasonable. The behavior of the universe would be capricious if causality did not exist—imagine objects moving for no reason or time flowing backward! Scientists also assume that the laws of nature that we measure on Earth hold everywhere in the universe. Since we have not traveled beyond the solar system, this is a major assumption. Although unproven, these assumptions are crucial to the scientific method. They form a backdrop against which we can

SCIENCE TOOLBOX

Estimation

Many important calculations in science can be done by estimation. Estimation gives us a quick way to gain insight by using rough or approximate numbers. This is especially appropriate in astronomy, where many measurements have low precision and few important numbers are known to more than two or three significant figures. In estimation, we are satisfied with a precision of a factor of 2 or even a factor of 10, which is known as an *order of magnitude*. Estimation is perhaps the best way to come to grips with the enormous range of scales in the universe.

To start with a frivolous example, how many pieces of paper would you have to pile up to reach the Moon? A ream of 500 pages of writing paper is roughly 40 millimeters (mm) thick. Although we might be able to make a more precise measurement of 39 mm, or 39.6 mm, in estimation we make do with the round number 40. The thickness of a single sheet is therefore

$$\frac{40}{500} = 0.080 \text{ mm}$$

The distance to the Moon is 384,000 km. Expressed in millimeters, this is 384,000 times 1,000,000, or 3.84×10^{11} mm. So the number of pieces of paper required to reach the Moon is the distance divided by the thickness of a single page:

$$\frac{3.84 \times 10^{11}}{0.080} = 4.8 \times 10^{12}$$

This number, nearly 5 trillion, is about the same order of magnitude as the sum of all the pages in all the world's books. Or imagine that the pieces of paper are instead dollar bills. The bank balance of most people would be a pile smaller than the height of a person. However, the worth of the richest person in the world would be a pile that reached several times around Earth, and the United States budget would be a pile that reached nearly halfway to the Moon!

You could also easily estimate how many times Earth would fit inside the largest planet in the solar system, Jupiter. Earth has a diameter of roughly 13,000 km, and Jupiter has a diameter of roughly 140,000 km. The volume of Jupiter, in cubic kilometers, is

$$\frac{4}{3} \pi \left(\frac{D}{2}\right)^3$$
$$= \frac{4}{3} \times 3.14 \times (70,000)^3$$
$$= 1.4 \times 10^{15} \text{ km}^3$$

If we imagine packing Jupiter with many versions of Earth, such that they just touch one another like marbles in a jar, then each Earth will take up a space equal to a cube that just nestles around a sphere. (This is an approximation; marbles in a jar will actually pack more tightly than this. Try it!) The volume of each cube surrounding Earth is D^3, or $(13,000)^3 = 2.2 \times 10^{12}$ km^3. So the number of times Earth will fit into Jupiter is given by

$$\frac{1.4 \times 10^{15}}{2.2 \times 10^{12}} \approx 600$$

Scientists use a wavy equals sign ($\approx$) or another similar symbol ($\sim$ or $\cong$) to mean approximately equals or roughly equals. Jupiter dwarfs Earth, and we will see later that the largest storms on Jupiter are even bigger than Earth.

The last example will give us a sense of the vast distance between stars. The Pioneer 11 spacecraft left the solar system several years ago and is traveling at about 110,000 kilometers per hour (km/h). How long will it take to reach the distance of the nearest stars? Alpha Centauri is 1.3 parsecs (pc), or 3.9×10^{13} km, away (the parsec distance unit will be introduced later in the book). So the number of hours that Pioneer 11 will take to reach the distance of the nearest star is

$$\frac{3.9 \times 10^{13}}{110,000} = 3.5 \times 10^8 \text{ h}$$

We can convert this to $(3.5 \times 10^8) / (24 \times 365) \approx 40,000$ years. This is a sobering reflection on the capabilities of today's spacecraft. We will need new technologies before we can explore the stars.

Scientists try to combine only numbers that have similar precision (see Appendix A-4 for a discussion of precision). Why? Because the result is governed by the number with the lowest precision, or the least number of significant figures. In other words, combining a great measurement with a lousy measurement will give you a lousy result. We can see this in our first example. Suppose that we know the distance to the Moon with a precision of eight significant figures or an accuracy of about 1 cm. (In fact we do, using radar measurements!) Yet our measurement of the thickness of a piece of paper is only good enough to have two significant figures. This means that our estimate of the number of pieces of paper to reach the Moon should only be quoted to two significant figures—it is limited by our least precise measurement.

The way to get good at estimation is to practice. Try it! Just remember to express all the quantities you are combining in the same units. Mixing meters and kilometers or grams and kilograms is the easiest way to make a mistake that will throw your answer way off. If you are using a calculator, be careful to enter very large or very small numbers correctly in scientific notation (see Appendix A-1 for tips). And if you want to estimate to only one or two significant figures, you do not even need a calculator! This is what scientists mean by a "back of the envelope" calculation.

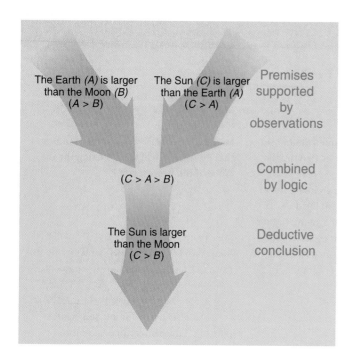

FIGURE 1-4
A schematic view of deduction. Pieces of evidence are combined logically to reach a conclusion. Arithmetic is an example of a system of deduction. The use of symbols in place of words (A = Earth, B = moon, C = sun) shows that we can use algebra for deduction, too.

interpret the observational evidence of astronomy. They also lead us down the path of logical inquiry, while our ancestors were often content not to seek causes at all, or to believe in supernatural causes.

We make a further assumption that *logic* leads to valid conclusions about the natural world. The basis of every scientific argument is logic. By combining their ideas and observations logically, scientists can draw conclusions and create secure knowledge. The scientific method uses two forms of logical proof that were originally developed by philosophers. One of these methods is **deduction**. A deductive argument starts with statements or premises and then draws a conclusion for a particular case. For example, by combining the premises that "The Sun is larger than Earth" and "Earth is larger than the Moon," we can deduce that "The Sun must be larger than the Moon." See Figure 1-4 for a schematic view of deduction.

In deductive reasoning, if the initial premises are correct, the conclusion must be correct. However, a deductive argument is like a piece of machinery that blindly gives an output for a given set of inputs. If the premises are wrong, the conclusion will be wrong. We have to be careful about premises as well as conclusions. For example, if we start with the premise that "Pigs have wings," we could waste a lot of time trying to investigate why a particular pig does not have wings. The deductive method is valuable only if we demand evidence to back up our premises. Figure 1-5 shows two examples where deduction can fail.

Many branches of mathematics such as arithmetic and algebra are deductive (basic algebra is reviewed in Appendix A-2). They can be used to illustrate one great advantage and one great disadvantage of this type of logic. The advantage is that, with accurate premises, deduction yields certain conclusions. Following the premises of basic arithmetic, we can say that $2 + 2 = 4$. This is not just true occasionally, or true every day except Thursdays; it is always true. The disadvantage of deductive logic is that the conclusion of a deductive argument contains no more information than is contained in the premises. If we say that $2x^3 + 7 = 61$, where x is unknown, we can follow the rules of algebra to deduce that $x = 3$ (see Appendix A-2). The solution to a complicated algebra problem may seem like a wonderful discovery, but it contains no more information than the original equations. Pure deduction is very powerful but it cannot always create new knowledge.

Induction is another logical tool of the scientific method. An inductive argument starts with specific observations (not broad premises) and then infers a general conclusion that is widely applicable. For example, an inductive argument might start with the observation "Not one of the 100 pigs I have seen has wings." Generalizing from this observation, we could hypothesize that no pigs anywhere have wings. Of course, we might be wrong; a mutant pig somewhere might have wings. But the conclusion seems very reliable. To take a less frivolous example, many medical studies have shown that smoking is correlated with the occurence of lung cancer. From these particular studies, doctors (and the United States government) draw the general conclusion that smoking is hazardous to your health.

We pay a price with inductive arguments, however: The conclusion of an inductive argument is never absolutely guaranteed. In practice, careful use of induction can lead to very good hypotheses. Three hundred years ago, Isaac New-

FIGURE 1-5
Examples of the failure of deduction. In the first example, one of the premises is false because the presence of four legs is a necessary but not sufficient requirement to prove the conclusion. In the second example, both premises are false, but they can be logically combined to produce a correct conclusion. Thus, to reach a valid conclusion by using deductive reasoning, all the premises and the conclusion must be carefully examined.

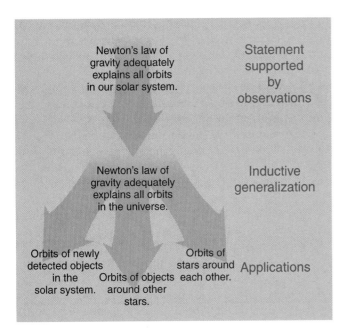

FIGURE 1-6
A schematic view of induction. A theory or model that has been tested in a limited way is generalized to apply to a much larger set of situations.

FIGURE 1-7
An example of the failure of induction. In the first instance, a limited amount of data is used to draw an inductive conclusion. In the second instance, a larger amount of data shows the first conclusion to be false.

ton jumped from the observation that "Every planetary orbit so far tested fits the universal law of gravity" to the inductive conclusion that "All astronomical orbits follow this law of gravity." So far as we know, his conclusion is correct (although minor corrections have been made to allow for the relativity effects discovered by Albert Einstein). The power of induction is that it allows us to generalize an argument and reach very broad conclusions. We have taken the evidence that the planets follow Newtonian orbits and made a hypothesis that these orbits apply far beyond the solar system. Figure 1–6 shows the schematic view of an inductive argument.

Good inductive arguments can be very reliable, but they also can be flawed if the data are too limited. For example, what about the observation that "All people examined so far are right-handed," leading to the inductive conclusion that "All people are right-handed"? Obviously, whoever conducted that piece of science did not have a large enough sample of people. Figure 1–7 shows the dangers of induction. You can understand why science is such a data-hungry enterprise: We need large data sets to be sure our inductive logic is reliable. As attractive as it would be, no logical system has been devised that yields certainty *and* creates new knowledge. Scientists use both deduction and induction in their work.

Let us look at two astronomical examples, which we will consider in greater detail later in the book. There is evidence to support the statement that "Most of the mass of the Milky Way galaxy is invisible dark matter." If we hypothesize that "The Milky Way is typical of all stellar systems," then we could conclude that "Most of the mass of the universe is dark matter." It is a bold conclusion, but it raises several questions. How good is the evidence for dark matter in the Milky Way? Is the Milky Way really typical of all galaxies? Is there evidence for dark matter in any galaxies other than the Milky Way?

The second example relates to the opening story of this chapter. We know that life has formed at least once in the solar system—on Earth. The jury is still out on the evidence from Mars. If we hypothesize that "Most stars form planetary systems as a natural by-product of their formation," we might conclude that "Given the large number of stars, life in the universe is common." Notice the danger—we are generalizing a conclusion on the likelihood of life throughout the universe based on only one example. How sure are we that planets have actually formed around other stars? If they have, how many are suitable sites for life? What about evidence for life itself? As long as we use logical arguments and insist on evidence at every step, the scientific method can lead us to the answers.

We have seen that the language of science is mathematics, a rigorous and deductive system for replacing concrete objects by numbers and symbols that represent them. Most of the physical principles that we will see in this book can be expressed simply in mathematics. Whenever you see such a principle, we will express it in words, and then in some cases we will discuss it abstractly, in symbols.

Mathematics is the wondrous system developed—we believe uniquely—by humans that makes full use of symbols and abstract thought. Until a baby is a couple of months old, an object such as a toy or a rattle that is removed from direct view is forgotten. Each time it reappears, it is new to the baby's world. Then, as the baby's brain grows, something remarkable happens. Babies become able to hold the idea of an object in their heads, so that the abstract idea of it is retained even after the object is removed from view. In that simple transition, the infant's universe expands immeasurably. A practical example of abstraction is money. In the United States, the dollar bill has become an abstract symbol for a certain amount of value or labor, backed only by the confidence we have that we all agree to use it and that the federal government will honor it. If you seek an example of something abstract, look no farther than your pocket.

Our description of the scientific method raises a serious philosophical issue. Cultures dating back to the ancient Greeks have proposed that the universe is ruled by numerical relations. The Greek word *cosmos* means more than just the universe and everything in it. It also means order and harmony, reflecting the ancient Greek ideal of a universe governed by understandable laws. For example, Pythagoras made the amazing discovery that to sound a musical note on a vibrating string that is just one octave below another note, you need a string exactly twice as long! The human brain can hear that these two notes are somehow related; but it took Greek science to discover that the relation was a precise numerical relation in the lengths of the string, not 2.7 or 1.94 or some other number, but precisely 2. In this century, Einstein and other great thinkers have posed the question, Why is the universe so well described by mathematics? Nobody knows why, but the application of mathematics to the natural world gives science its great explanatory and predictive power.

Observations and Errors

We have seen that the core of the scientific method is evidence—observational data in some form. Scientists make explanations of their data called *hypotheses,* and they combine the information according to the rules of logic. We have also seen that these logical tools have limitations. But what can we say about the limitations of the observations themselves?

Every measurement has an associated **observational error,** which is the uncertainty in the measurement. In the 18th century, the great German mathematician Karl Friedrich Gauss worked out the theory of observational errors. Scientists rely heavily on his ideas in all their work with measurements. They can understand a particular result only if they know the degree of error involved.

For example, let us revisit the deck of cards. Suppose you try to slide a card into a standard deck exactly halfway down the pile. By guessing, you would be unlikely to place it halfway through the deck. If you tried the experiment 10 times, for example, you might find that you had placed it 24, 33, 28, 27, 23, 27, 24, 32, 26, and 31 cards down into the pile. There is a large error associated with guessing. But if you measured the height of the deck with a millimeter ruler, for example, you could insert the card halfway down much more accurately. This experiment might yield results of 25, 28, 24, 27, 27, 26, 28, 26, 25 and 27 cards down into the pile. The second set of results has less error, due to careful measurement. Figure 1–8 shows how this experiment might turn out if you did it 40 times. The histogram of positions in the deck is narrower when a more accurate measurement is made.

To take an astronomical example, how can we identify the *exact* position of a star in the night sky? Look at the distribution of points in Figure 1–9. Each point represents an actual measurement of the position of a star. There are many measurements, and none of them are identical. Where is the star exactly? We cannot say! However, we can do two things. We can take an **average** of all the measurements as the best estimate of the star's position (the average is also referred to as the *mean value;* it is the cross in Figure 1–9). We can take the spread in the measurements as the **standard error** in that estimate, which is also referred to as the *uncertainty,* or *standard deviation.* In a way, the concept of "error" is a bit misleading, since no mistake was made in the measurement. There is just a limit to our certainty in the result of *any* measurement.

Where does observational error come from? Usually it just reflects the normal limitations of the measuring equipment. Suppose that a ruler is marked off with millimeters as the smallest unit. We could make a single, quick measurement with an accuracy of the smallest unit on the ruler—a millimeter. If we measure the width of a piece of paper, we might come up with the result 217 mm. Notice the number of significant figures—the digits that carry meaningful information. We could have quoted the measurement with one significant figure, as 200 mm, but that is needlessly rough. If we used two significant figures, we would say the width is about 220 mm, rounding off the last figure. We could also quote the width as 217.84 mm—to five significant figures—but that is unrealistically precise. It makes most sense to quote the result with three significant figures, which reflects the **accuracy** of the measurement. *Accuracy* is defined as the amount by which a measurement deviates from the true value.

Scientists make a clear and important distinction between precision and accuracy (see Appendix A-4). We have seen that a single measurement can be quoted with a wide range of precision, but the accuracy is set by the nature of the measuring device. For example, you can easily set most calculators to display 6 or 8 or 10 decimal places, and the results of any calculation you do will be shown with that precision. But that does not mean that every calculation you do is that accurate. Imagine you used the same ruler to measure the length of the block where your house is located. You might add all the numbers as you put the ruler down end to end and get 56,794 mm. It sounds very precise, but is your measurement really accurate to 1 mm? A better indication of the precision of the measurement would be 56.8 m. Scientists always try to quote measurements with a precision that matches the accuracy.

Recall that a scientific measurement has two components: a number and a unit. As we have seen above, the number itself has two pieces: the best estimate and the standard error. In the example just mentioned, the full measurement would be written as 217 ± 1 mm. The symbol ± (also written as +/−) is called *plus or minus*. It means that while the true value might well be 218 mm or 216 mm, it is unlikely to be 120 or 250 or even 210 mm. For details on how much a single measurement might deviate from the true value, see Appendix A-4. On a graph, the best estimate is drawn as a

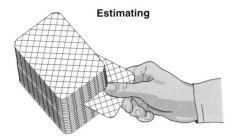
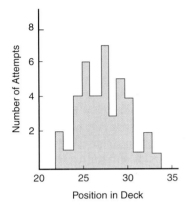
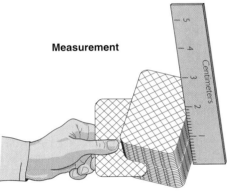
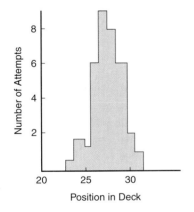

FIGURE 1-8
Using a deck of cards to demonstrate measurement error. **(a)** In this case the card is placed in the middle of the deck by estimation, or "guessing." **(b)** If the card is placed 40 separate times, this a possible histogram of the card position in the deck in each case. There is a large spread in card position, corresponding to a large error in placement. **(c)** In this case a millimeter rule is used to carefully measure halfway down the deck and insert the card. **(d)** With 40 separate measurements, this is a possible histogram of card positions in each case. The distribution is more tightly centered on the center of the deck—position 27.

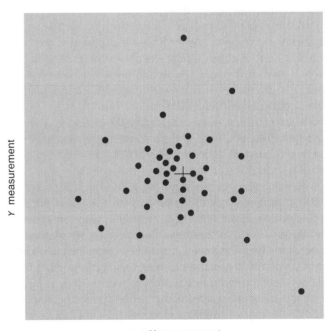

FIGURE 1-9
This represents an experiment in which the position of a star in the sky is measured repeatedly. No single measurement gives the "exact" answer. The center of the distribution—the average of all the individual X and Y coordinates given by the cross—is the best estimate of the star's true position. The probable error can be estimated from the scatter in the distribution. Ultimately, all measurements of nature should be quoted as a best estimate with an uncertainty, but in ordinary life the uncertainty of most "factual data" is small enough that we ignore it.

THE SCIENTIFIC METHOD 13

point or as some other symbol, and the standard error or uncertainty is drawn as an "error bar." These examples with a ruler are mundane. But, as we will see later, cosmologists use the same ideas in measuring the size and age of the universe!

When making measurements of your own, you will be relieved to know that there are ways to reduce observational errors. We can use more precise measuring equipment such as a micrometer marked off in tenths of millimeters. Scientists often make progress in this way, but it is not always possible to improve on existing technology. The other way to reduce observational error or improve accuracy is to take more than one measurement. As more measurements are made, the uncertainty of the average goes down, as described in the Science Toolbox on the next page. This is reflected in the way the result is stated, with a larger number of significant figures and with a smaller error bar.

Building a Theory

Let us recap what we have learned so far about the scientific method. Science is based on observations of the natural world. To qualify as scientific evidence, observations must be quantitative and repeatable. Scientists then look for patterns in the data and might form a hypothesis to explain those patterns. The tools of this process are logic and mathematics. Now let us look at how we build scientific theories to increase our knowledge about the universe.

It is important to recognize the distinction between a hypothesis and a theory. A hypothesis is a proposed explanation for a set of observations. The observations need not be very extensive, and the hypothesis need not cover a wide range of physical phenomena. Hypotheses are the building blocks of science. However, a theory must explain a large number of observations and unify them under a single coherent idea. A theory is like a completed building. Like a well-built building, a good theory will be an elegant logical construction that stands the test of time.

You will also see the word *model* used in this book and other places, referring to the progress of science. In general, you can treat *model* as being synonymous with *hypothesis*. A model is a mathematical representation of a set of observations. Just like a toy model, it may be a simplified and idealized version of the real world. When scientists "model" their data—the word *model* is used here as a verb—they are using a hypothesis to make a mathematical description of their data and are seeing how well that mathematical description fits.

For example, Johannes Kepler studied the data describing the orbits of the planets in the solar system. He found that the orbits were not circular, as had been previously thought, and hypothesized that the orbits were elliptical. Isaac Newton built on Kepler's hypothesis to build his theory of gravity. Newton's law of gravity neatly encompassed all three of the rules of planetary motion discovered by Kepler. Newton's theory explained not only the planetary orbits but also the orbits of comets and asteroids. In fact, it explained the motions of *any* two orbiting objects. A theory is broader and more generally applicable than a hypothesis. In biology, the theory of natural selection explains the rich variety of species—flowers and fungi and flamingos—in terms of evolutionary adaptation to a constantly changing environment. A successful scientific theory shows us the unity in nature.

Beyond the level of a theory, there are a few ideas that are referred to as *laws of nature*. What makes a theory strong enough to be a law of nature? There is no rigid rule, but in general a law of nature is a physical idea that has stood the test of time. In other words, it is a theory that has been successfully tested over and over. Newton's laws of motion match the motions of all mechanical objects and are used in the design of all machines. The laws of thermodynamics were devised over 200 years ago, and these descriptions of heat and energy formed the basis of the industrial revolution. Such ideas have been tested so thoroughly that scientists think it is very unlikely that these ideas could be wrong.

Scientists must always be aware of the limitations of their theories. Every set of data may have more than one possible explanation. This is why there are rival hypotheses and why scientists do not always agree with one another. We have seen that even the most accurate observation has some error or uncertainty attached to it. If there are no perfect measurements, then there can be no perfect tests of a theory. Also, the data that scientists work with are always limited in some way. New observations may be made which disagree with a theory. With rival explanations to choose from, and with limited and imperfect data to work with, how do we know if we have a "good" theory?

A good theory must be *testable*. In the 18th century, scientists proposed that migrating birds followed invisible elastic threads when they traveled long distances. This is not a very fruitful hypothesis. If the threads are undetectable, the idea cannot be tested. More recently, researchers proposed that birds navigate by a mixture of visual clues and sensing of magnetic fields. This idea was supported by the fact that migrating birds (and other animals) have high concentrations of the mineral magnetite in their brains. Magnetite allows birds to orient themselves by detecting variations in the earth's magnetic field. Through testing of the flight patterns of birds under different conditions, this hypothesis became an accepted theory. A scientific theory must be testable—that is one of the features that distinguishes science from other approaches to knowledge.

A good theory must also make reliable *predictions*. Many people believe that UFOs are alien spacecraft that the U.S. government is keeping secret from the public. The government certainly keeps many military issues secret, but there is no evidence that they include captured aliens. This idea is not very useful since it assumes the current evidence is secret, and it makes no testable predictions of when future sightings might occur. On the other hand, scientists believe that at least some UFO sightings represent well-understood astronomical phenomena. This leads to the confirmed prediction that UFO sightings peak around these events, such as meteor showers and times when Venus is particularly bright. In the same way, people used to think comets were supernatural omens until Halley and Newton used theory of gravity to accurately

SCIENCE TOOLBOX

Errors and Statistics

We can reduce observational error by using more accurate equipment and by making multiple independent measurements. Let us return to the example of measuring paper. Suppose we use a ruler marked off in millimeters to measure the width of a piece of paper. If we do this once, we might get the answer 217 mm. How accurate or reliable is this single measurement? We might think that it is accurate to the size of the smallest division on the ruler, 1mm. But the truth is that we do not really know.

Now suppose we get 10 people to measure the paper once each, with the following results: 217, 216, 216, 216, 215, 217, 216, 217, 218, and 216 mm. The best estimate is given by the average of the measurements

$$\text{Average} = \frac{217+216+216+216+215+217+216+217+218+216}{10} = 216.4 \text{ mm}$$

In the case of multiple measurements, we can calculate an error for the combined measurement. The error, also called the *standard error* or the *standard deviation,* is the scatter of the individual measurements about the average. The standard error is calculated from the differences between each individual measurement and the average. Mathematically, we square the differences and add them, dividing by the number of measurements (minus 1). Finally we take the square root of the result:

$$\text{Standard error} = \sqrt{\frac{(217-216.4)^2 + (216-216.4)^2 + \cdots + (216-216.4)^2}{N-1}} = 0.8 \text{ mm}$$

In this calculation, the dots ··· are used as a space-saving notation; we must form the sum of the squares of all 10 differences. Most calculators will do this manipulation for you automatically, but it is instructive to see it written out. Why do we divide by the number of measurements minus 1, rather than by the number of measurements? The answer is that error cannot be determined for a single measurement. Mathematically, when $N = 1$, we divide by 0 in the equation above and the error is infinitely large. Notice that when the number of measurements is very large, the difference between N and $N - 1$ is not very important.

What have we gained by making more than one measurement? We know that our combined estimate is more reliable, because it is not as sensitive to a mistake that we might make with a single measurement (for example, one person misreading the scale on the ruler). We also know what the true error of our best estimate is, given by the scatter in the individual readings.

Previously, all we could do for a single measurement was to estimate the error to be the size of the smallest division on the ruler.

Single measurement ± estimated error
= 217 ± 1 mm

The combination of 10 independent measurements gives a more reliable best estimate and a standard error

Average ± standard error
= 216.4 ± 0.8 mm

The accuracy of our combined measurement is actually smaller than the division on the rule. Multiple measurements have increased the accuracy. This in turn allows us to quote the result with a higher precision—four significant figures instead of three.

To return to the example of the measurement of a star's position, we can define the position with greater accuracy by using more and more independent measurements. We can never say with complete certainty where the star really is. But we can combine data to yield tighter and tighter estimates of its likely position. Now we are dealing with two dimensions, so the best estimate of the position is the average of the X positions and the average of the Y positions of the individual measurements. The scatter in positions gives the standard error. Figure 1–9 showed the scatter in the individual measurements, with the most probable position or the average plotted as a cross.

Figure 1–A shows a sequence of situations where the number of measurements increases from 1 to 3 to 10 to 30. The best estimate of the position is given in each case by the average of the measurements. The average does change as the number of measurements increases. However, the average changes less and less as the number of measurements increases because a single outlying point has less and less impact.

It is less clear how we should define the scatter in the measurements. If we chose an error that accounted for all the measurements, then we would use the dashed circles in Figure 1–A. This is not a good idea because (unlike the average) it would be very sensitive to a single outlying point. In Gauss' theory, and in the mathematical form given above, the standard error encompasses two-thirds of the measurements. These are the solid circles in Figure 1–A. We should remember that the star has a ⅓ probability of being outside the circle defined by the standard error (see Appendix A-4).

You can see that the solid circles in Figure 1–A shrink as the number of measurements increases. The standard error gets smaller, and the combined measurement gets more accurate. With many measurements, you minimize the effect of wayward readings and get a more reliable average. The standard error is inversely proportional to the square root of the number of measurements. In mathematical terms, this means

$$\text{Standard error} \propto \frac{1}{\sqrt{N}}$$

(continued)

SCIENCE TOOLBOX

Errors and Statistics (continued)

The math symbol ∝ means "proportional to." In both the examples of star position and ruler measurement we see why good science is so hard–reducing the standard error by a factor of 3 requires 9 times as many measurements! We tend to think of science in terms of dramatic discoveries, but most scientific progress is the whittling away at errors to match physical models of the world more and more accurately.

FIGURE 1-A
The change in the average and standard error as the number of measurements increases. In this case the measurements are of a star's position, but these principles apply to the random errors in any measurement. **(a)** With a single measurement, no error can be estimated. **(b)** With three measurements, the best estimate of the position is given by the averages of the X and Y coordinates, and the standard error is drawn as a solid circle. The dashed circle encompasses all the measurements. **(c)** With 10 measurements, the mean position becomes more reliable, and the standard error is smaller. **(d)** With 30 measurements, the mean position does not change much, but the standard error continues to get smaller. Notice that the dashed circle that encloses all the measurements is large. The standard error is not as sensitive to single outlying points, but there is a 1-in-3 chance that the true value may lie beyond the circle defined by the standard error.

The previous discussion refers to *random* errors. A more dangerous uncertainty results from *systematic* errors. These are errors that are due to some flaw in the measuring equipment or to a mistake in the way the measurement is made. Imagine if the ruler you had used were printed with an incorrect millimeter scale. In this case, you would have combined many measurements in the belief that you were achieving an accurate result when, in fact, *all* the measurements were flawed.

Systematic errors are insidious, because there is no simple experiment you can do with your measurements to reveal them. You would have to compare your ruler to someone else's ruler. We will encounter examples of systematic errors later in the book.

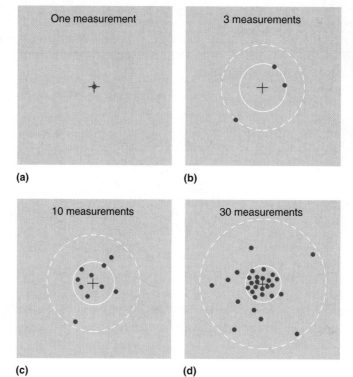

predict the return of Halley's comet. We can see in this one example the power of the scientific method. The success of the predictions moved comets from the realm of superstition into the realm of the natural world.

A good theory should also explain a wide range of phenomena. Sometimes a theory is overturned when new data show it to be a poor description of the natural world. Other times, we replace an old theory with a better one that explains more observations. At the beginning of the 20th century, scientists found that the simple view of atoms as hard billiard balls was inadequate. The quantum theory provided a better description of nature on the subatomic scale. At about the same time, Einstein came up with his theory of relativity, which described situations of strong gravity better than Newton's theory did. The older theories are not useless since we can understand the behavior of a gas perfectly well without using quantum theory and the Apollo spacecraft were sent accurately to the Moon using only Newton's law of gravity. But the new theories provide a more complete view of nature.

These are the bare bones of the scientific method. In practice, science can be very complex, but here you have the guidelines that motivate the whole enterprise. The flowchart

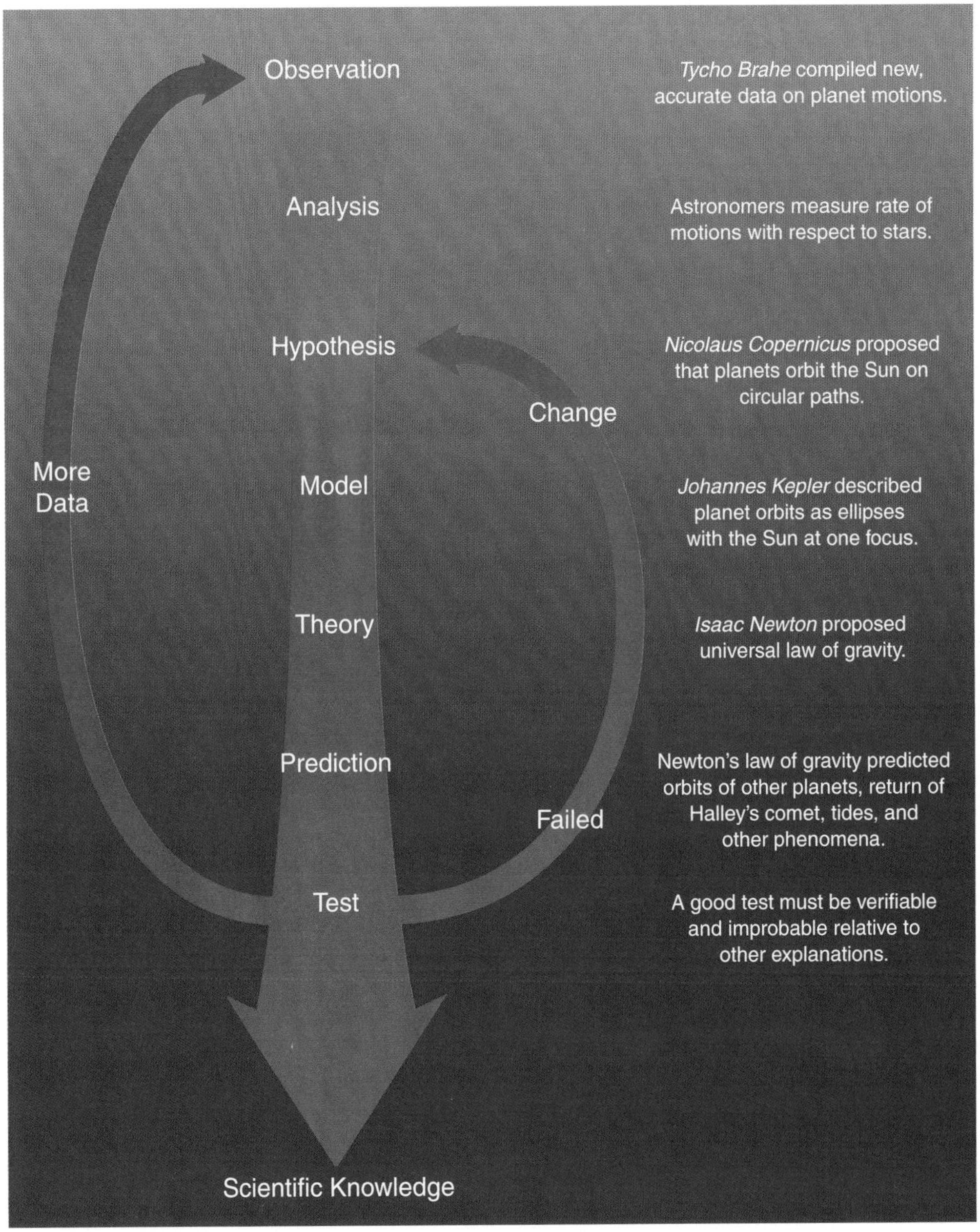

FIGURE 1–10
An example of scientific progress. Accurate observations of the planets' orbits lead to data that are good enough to reject the hypothesis that the orbits are circular. A careful study shows that the best mathematical model for the orbits is an ellipse. This type of orbit emerges naturally out of a theory of gravity proposed by Newton. The theory explains existing data well and predicts phenomena that are subsequently observed.

Observation	Tycho Brahe compiled new, accurate data on planet motions.
Analysis	Astronomers measure rate of motions with respect to stars.
Hypothesis	Nicolaus Copernicus proposed that planets orbit the Sun on circular paths.
Model	Johannes Kepler described planet orbits as ellipses with the Sun at one focus.
Theory	Isaac Newton proposed universal law of gravity.
Prediction	Newton's law of gravity predicted orbits of other planets, return of Halley's comet, tides, and other phenomena.
Test	A good test must be verifiable and improbable relative to other explanations.
Scientific Knowledge	

(Loop arrows: More Data → Observation; Failed → Change → Hypothesis)

in Figure 1–10 gives an example of how science progresses. Science is a logical and systematic endeavor, but we should not ignore the role of sudden insight, luck, and serendipity in scientific discovery. You will see many examples of the human side of science—stories of dedication and inspiration—later in this book.

Other Systems of Knowledge

What is so special about the scientific method? There is no doubt that this system of creating knowledge has transformed the human condition in only a few hundred years. If you doubt this, imagine living in a time before cars or telephones or antibiotics. All the innovations of technology start with pure scientific research, conducted according to the principles described in this chapter. When Benjamin Franklin was asked what was the use of the electricity he had just learned to harness, he replied, "What use is a newborn baby?"

In everyday life we encounter other **systems of knowledge** besides the scientific method. One ancient system is called *appeal to authority,* which means citing the opinion of an authoritative person or book rather than using evidence and logic to prove a point. This system was especially prevalent in the Middle Ages, when scholars believed that ancient Greek, Egyptian, and Arab civilizations represented a golden age of knowledge and that there had been little progress since then. Therefore, instead of trusting their own observations and ideas, scholars cited old authorities such as

Aristotle and Plato. To preserve ancient bodies of knowledge, medieval officials attacked anyone who contradicted the old authorities. Appeal to authority is very different from citing a reference when you write a paper on some topic. A citation should not be used to claim that your argument is correct because some "great person" agrees with you, but rather to let your reader check the evidence from original sources.

Appeal to authority is antiscientific because it is not based on prediction, creative experimentation, and observation. Galileo, the first person to observe the universe with a telescope, remarked that the humble reasoning of one person is worth more than the opinion of a thousand authorities. Historically, the system of appeal to authority has led to terrible repercussions for those who challenged the traditional wisdom. In ancient times, Socrates was put to death for encouraging his students to question traditional ideas. In the Middle Ages, the philosopher and mystic Giordano Bruno was burned at the stake by the Roman Catholic Church for suggesting heretical ideas about the existence of other worlds and life-forms. Galileo was shown the instruments of torture by the Inquisition and sentenced to house arrest for presenting his evidence that planets move around the Sun (see Chapter 3). In each case, a pattern was repeated: Authorities feared that the roots of their culture would be endangered if bold thinkers encouraged debate about new ideas.

One damaging aspect of the appeal to authority is secrecy. In the Middle Ages, secret societies passed on mystical symbols and supposed ancient knowledge only to people who were seen as being worthy to receive it—others could not examine the evidence. This is one reason why the scientific method was such a revolution. In the 1600s, the scientists who founded England's Royal Society debated whether to keep new discoveries quiet. They voted and decided to publish all scientific results. That debate started our modern tradition of open publication so scientists around the world can verify hypotheses by conducting their own experiments. The drive to share scientific data has spurred the most recent phase of the information revolution with the creation of the Internet and the World Wide Web.

Another belief system is *superstition,* which is based on irrational beliefs about the natural world. Most superstitions have their origins in ancient folklore. They have no basis in logical or rational thought. Instead, they appeal to unseen and undetectable influences, which makes them immune to scientific analysis. One of the most common types of superstition assumes that patterns in nature can predict or influence human destiny. For example, ancient Romans and other Mediterranean people often sacrificed animals, tore out their intestines, and "interpreted" patterns of the entrails as predicting the future. This practice is called *divination.* Reading patterns in tea leaves or patterns in the lines of your palm is a similar idea. The most widespread superstition of this type is *astrology*—the belief that the patterns of planets and stars in the sky when you were born somehow affect your personality and the events that will happen to you during the rest of your life.

Palmistry and astrology are certainly complicated belief systems, but that fact alone does not give them the power to explain and predict. Why do people believe in these ideas? Part of the reason is a phenomenon called *confirmation bias,* which is well known to psychologists. When you read a horoscope in a newspaper or magazine, you are more likely to remember the parts of the description that fit you than the larger number of parts that are vague or do not fit at all. Similarly, you are more likely to remember the occasion when you broke a mirror and something bad happened than the other times when nothing bad happened. Many superstitions keep a surprising grip on our modern world. Try to notice how many apartment or office buildings have no 13th floor, or how many of your friends and acquaintances "knock on wood" for luck.

Still another system of determining truth dominates our modern legal and political system, and so influences much of our lives. This is the *advocacy system,* in which each participant or team advocates one position in an argument. You may have experienced this in a school debate class, where you were assigned to defend one position regardless of whether you personally supported it. It is the basis of our legal system. In court cases, the primary goal is not to present all the evidence, as in science, but to win the argument. The assumption is that the truth will emerge from this vigorous contest between the two sides. In practice, it can be a troubling system. For example, the side with the most money can frequently afford to develop the best case. In addition, evidence with a clear bearing on the case may be removed from consideration by the court for procedural reasons.

The advocacy system is also the basis of marketing and advertising. In an advertisement, you hear only one side of the case. You are unlikely to hear an advertisement whose goal is to evaluate fairly the pros and cons of each competing product! That is reserved for some consumer magazines, which use scientific tests and compare products. We should also notice that in the advocacy system the advocate can be as important as the argument. Think of the eloquent trial lawyer or the sports hero pitching a product. In a scientific debate there may be good speakers and bad speakers, but the outcome is primarily determined by the evidence.

There is an unfortunate idea gaining ground in our culture: The idea that method of science is just one way among many for knowing the world. This is called *relativism,* which holds that knowledge gained from science or a psychic or an astrologer or an old superstition is equally valid. We must acknowledge the limits of science. Not every phenomenon has a scientific explanation, and science alone is not sufficient to run human affairs—try, for example, to imagine systems of law or economics or medicine that had logic but no compassion. The scientific method is not perfect. Overly cautious scientists sometimes delay the acceptance of correct new hypotheses. Fraud exists, but it is rare because an important result will always be checked. Despite these flaws, knowledge in the past century has progressed at a breathtaking rate. As a system for revealing the way nature works, science is without equal.

Summary

This book is built around three themes. The themes provide a focus and are useful for revealing connections among the wide variety of ideas and phenomena in this book. One theme is the role of humans in the universe. So far, science has indicated that we are not in a central or unique position, but are just part of a vast system. Put simply, the universe is big, and you are not! Another theme follows from the fact that a small number of physical rules govern the behavior of everything from Earth's atmosphere to the whole universe during the big bang. In effect, this book will teach you the rules. The final theme explores the idea that science is a superior method for understanding the natural world. We will answer the question, how do we know what we know?

What qualities make science a powerful means for learning about the nature of the universe and our place in it? This chapter has presented many aspects of science and compared them to other systems for obtaining knowledge. To come back to our opening story about the hypothesis of fossil microbes in the rock from Mars, only the scientific method could have identified the rock as being from Mars. In other belief systems, such as astrology or superstition, Mars is a mystical body that influences humans, not a world with its own geology and perhaps biology. Currently the scientific method is being used to answer one of the most profound questions we can ask: Has life ever existed beyond Earth?

All ideas in science are based on evidence and are subject to review and revision. We must communicate our evidence in the form of a quantitative measurement, which consists of a number, a system of units, and an error or uncertainty in the number. Using such measurements, scientists make a working hypothesis based on observations and then try to verify or disprove it with more observations. Ideas that survive this skeptical approach become known as theories. A good theory is testable, makes predictions, and explains a wide range of phenomena. Scientific ideas can be expressed either in words or in equations, and you will see both forms throughout this book. The tools for manipulating these ideas are logic and mathematics.

Good scientists know that they have no lock on the truth, and they recognize that any measurement is uncertain to some degree. If we are still unsure whether life once existed on Mars, it is because the evidence is inadequate. This provides strong motivation for the planned set of space missions to return new rocks to Earth. Humans have invented many systems of thought over the centuries. These systems have their attractions, but they all have limitations for transmitting a coherent body of knowledge and for revealing the truth. The unique character of science is that it progresses toward a more accurate body of knowledge by gathering new data, testing hypotheses, and revising theories that do not match observations. In this book, we will explore science, the system that has served us so well and the one that we have used to unlock many of the secrets of the universe.

Important Concepts

You should be able to define these concepts and use them in a sentence.

- astronomy
- physics
- mass
- length
- geology
- chemistry
- biology
- mathematics
- scientific method
- hypothesis
- theory
- evidence
- measurement
- unit
- scientific notation
- time
- precision
- estimation
- deduction
- induction
- observational error
- average
- standard error
- accuracy
- system of knowledge

How Do We Know?

These questions and answers show how the scientific method is used to learn about the universe.

Q How do scientists create knowledge?

A Knowledge in science begins with observation. We make measurements of the universe by gathering information. In astronomy, this usually requires a telescope since most objects are so remote. We look for patterns in the data and form a hypothesis to explain the patterns. A good hypothesis makes accurate predictions, which allows us to test the hypothesis. When a hypothesis is used to make a successful prediction, we learn something new about the universe. But a hypothesis can be wrong. If it does not explain the data, then we must go back to the drawing board and make a new hypothesis.

Q What makes a good measurement?

A A good measurement is quantitative and repeatable, and it must have a known degree of accuracy. You might say, "The table looks about 1 m long," but this is an estimate and might be very inaccurate. If, instead, you use a tape measure to determine the length of a table, you might come up with a good measurement of 1.238 ± 0.008 m. Note that the measurement includes a number, an error or uncertainty, and a type of unit. Since the smallest division on the tape measure is 1 mm (0.001 m), the answer has a precision of four significant figures. Then you might measure the length a number of times and might find that the spread in the answers is 8 mm (0.008 m), or an error of 0.008 m. Finally, you could ask someone else to verify your result by measuring it with a different tape measure.

Q What makes a good theory?

A A good theory is supported by evidence, is testable, and makes verifiable predictions. We also expect that a good theory will be applicable in a wide range of situations. Newton's law of gravity is an excellent example. It explains the motions of the planets and makes verifiable predictions about others stars and planets. It is also a very accurate theory—National Aeronautics and Space Administration (NASA) engineers use Newton's law to send space probes tens of millions of miles with an accuracy of less than a thousand miles. Newton's theory gives an excellent description of most regions of the universe. It is only superseded by Einstein's theory of relativity in situations of extremely strong gravity.

Q How reliable are scientific conclusions?

A Scientific conclusions are never 100 percent reliable, for two reasons. First, observations are never perfect; they always have errors. The point is to know how big the probable error is. Second, the scientific method involves induction—the generalization of an idea to a wide range of situations—and can never guarantee the absolute truth of the conclusions. In practice, scientific conclusions can be very reliable, but each case must be looked at carefully. For example, there is a big difference between the reliability of the statement "Some meteorites found on Earth originated on Mars" and the statement "There was life on Mars several billion years ago." The first statement is based on detailed chemical and geological similarities between a few meteorites found on Earth and rocks measured on the surface of Mars by the Viking and Mars Pathfinder spacecraft. The second statement is based on limited indirect evidence of long-dead organisms in a single Mars rock and is far less reliable.

Q What causes observational error?

A Most observational error is caused by limitations in the measuring device. For example, think of the difference in the error between guessing the weight of a book, measuring it on the bathroom scale, and having it measured at the Post Office on an electronic balance. In astronomy, some errors are caused not only by limitations in telescopes and instruments, but also by the limited amount of light that can be gathered from distant objects. There are only two ways to reduce the error. One is to collect more data, preferably by using different scientists to make independent measurements. The other is to build a more accurate measuring device.

Q Are there questions that science cannot answer?

A Yes. There are questions—such as "Are there Earth-like planets and civilizations beyond the solar system?"—that science cannot yet answer because the data do not exist. There are questions that go beyond the limits of current theories, such as "What is it like inside a black hole?" or "What happened before the big bang?" We know more about the universe today than ever before; but scientific knowledge is not complete, and it is not infallible. There are also some areas of human activity that are so complex that it is impossible to model them. For example, it is unlikely that the scientific method will ever lead to a theory of human behavior or a theory of why the stock market rises and falls.

Problems

Use these problems to test your understanding of the information and concepts in this chapter. The * indicates a more advanced or mathematical problem.

1. Evaluate the scientific reliability of each statement below, and decide how reliable each argument is. **(a)** Einstein said that mass can be converted to energy, so it must be true. **(b)** A modern-day cult stores food under pyramids because its members believe, without any tests, that pyramids have supernatural powers to preserve things. **(c)** At a trial, a lawyer pays expert witnesses to testify in favor of her side of the case and moves to have evidence favoring the other side thrown out. **(d)** At a medical conference, doctors discuss the effectiveness of a certain drug. Seven out of ten medical researchers on the panel report positive tests that the drug works. One finds negative results, and two find that their patients' improvement is no better than random, within the error bars. **(e)** The drug manufacturer in case (d) states in its advertising that the product is "proven effective" because a majority of doctors find that the drug works.

2. Discuss the difference(s) between a working hypothesis and a theory.

3. Ten scientists want to measure a particular quantity, but one of them claims to have measuring equipment that is much more accurate than the equipment of the others. Which strategy would produce the "best" measurement, and why? **(a)** Let only the scientist who claims to have the most accurate equipment make the measurement. **(b)** Let the scientist who claims to have the most accurate equipment and one of the others make measurements, and then compare the two answers. **(c)** Let all 10 make the measurement and then gather to compare and discuss the results. **(d)** Let only the nine scientists with ordinary equipment make measurements, and take an average of their nine answers.

4. What is the basis for our belief in causality, the idea that every action has a cause? How would it affect the value of the scientific method if this were *not* true and events in the everyday world happened spontaneously without any prior cause?

5. Evaluate the reliability of these various types of evidence for identifying UFOs as alien spacecraft. **(a)** A series of photographs published in a weekly newsmagazine. **(b)** A sighting by a close relative of yours, who saw something believed to be a metallic disk. **(c)** A similar report by a member of the U.S. President's Cabinet. **(d)** A piece of very dense metal found near scorch marks in a wheat field somewhere in the Midwest. **(e)** A report from a friend in the military who says he heard rumors from several officers of frozen aliens being kept at an Air Force base in Nevada. **(f)** Hundreds of independent visual reports and a series of 56 photographs taken in sequence on the same evening in different towns along a line from Oregon to Nevada, showing a formation of three brilliant objects moving toward the southeast at a calculated rate of 650 miles per hour (mi/h). The objects are so small or high that best photographs show them only as blurry triangular shapes. **(g)** A detailed account of an abduction by aliens, recounted on a daytime TV talk show.

What more would you want to know in each situation to evaluate the evidence? Can you think of a hypothesis other than alien spaceships that might account for each report? Give examples of the minimum evidence you would demand before believing the hypothesis that some UFOs are alien spacecraft.

6. Critique the following statements from the perspective of the scientific method, explaining what, if anything, is wrong with each one or what you would like to know in order to clarify each one. **(a)** "Nine out of ten scientists believe in the big bang model for the creation of the universe, so I believe it, too." **(b)** "Nine out of ten scientific reports published in an international medical journal find that a certain food additive causes cancer in mammals and advise against using it, so for the time being I'm going to try to avoid using it, although I recognize that future tests may give different answers." **(c)** "My favorite talk show host says that a certain food additive causes cancer, so I'm going to stop using it." **(d)** "Team, I want you to go out there and give 110 percent." **(e)** "The cake probably weighs 2.3461 kilograms (kg)." *[just an estimate]* **(f)** "Wind is caused by trees wav-

ing their branches." **(g)** "Lung cancer is caused by smoking." **(h)** "There is no chance that Los Angeles will suffer a major destructive earthquake within the next 10 years."

7. How would you design a scientific experiment to test the existence of telepathy (the ability to transmit thoughts without verbal communication)?

*8. To how many significant figures do you know your weight? Your height? Your bank account (or the amount of change in your pocket) at any given time? To how many significant figures do you think we know the population of the planet at any particular time? Discuss the sources of uncertainty in each case. Why do modern scientists often need to measure more significant figures than in these cases to make progress?

*9. Conduct a simple experiment using estimation. **(a)** Find some spherical beads or marbles of one size and put them in a box. Measure the size of one of the spheres. **(b)** Discuss a method for estimating the number in the box without counting them. Try to think first of a simple method, and then try to think of ways you might refine it by allowing for more details of the situation. *Hint:* Discuss whether the spheres fill all the space. Do they stack in regular order? **(c)** Make the estimate by as many methods as you can and then count the marbles. Compare the results. By what percentage were you in error? If you had more than one method, was the error less for your more sophisticated method(s)? **(d)** Now use the techniques you have learned to make a more accurate estimate of the number of times the Earth would fit inside Jupiter, as done in the Science Toolbox earlier in the chapter.

10. Here are some quantitative "measurements" of various bits of information. Arrange them in a list, in order from least accurate to most accurate. **(a)** A book is measured to have a length of 124 mm, using a meter stick marked off in units of 1 mm. **(b)** Someone's birth month is known, but not the actual day of birth. **(c)** A magnetic field is measured to have a strength of 560 ± 17 gauss (G). **(d)** The number of grains of sand on a beach is estimated as 10^{12}. **(e)** An aerial photograph of an open-air concert leads to a crowd size estimate of 40,000 to 60,000 people.

*11. Carry out the experiment shown in Figure 1–8. Try 10 times to place the card as near as you can to the middle of the deck by eye. Then try 10 times to measure the height of the deck with a millimeter ruler and place the card carefully halfway down the deck. Plot the distributions of your two sets of measurements. What is the mean card position in each case? What is the standard error in each case? How much more accurate is the experiment that used the ruler?

Projects

Activities to carry out either individually or in groups.

1. For a given day or week, pick the birthday of one member of the class (or the teacher) and have class members obtain as many horoscopes as possible for that specific birth date. (These can be cut from newspapers in the city where your college is, newspapers mailed by friends in other cities, magazines, entertainment and arts weeklies, or other sources.) Compare the predictions and general tone of all the horoscopes. Do different astrologers make specific predictions? Do they make the same specific predictions? Do different astrologers focus on the same general themes and do the tones agree (is it a happy period, sad, financially risky, a time for romance, etc.)? Try this for several different birthdays and discuss whether the horoscopes have any predictive power.

2. Carry out the same assignment as in Project 1, but assign some members of the class to write fake horoscopes for unspecified dates. The fake horoscopes should be made up without any astrological technique but in the same style as the ones written by astrologers. Mix them up and read them aloud in class. See if the class can tell the "real" ones from the fake ones. Assign extra points if the authors of the fakes fool the class. Then discuss whether there is any effective way to tell whether astrologers use consistent rules or "real science," or whether their predictions might just as well be made up at random.

3. Flip a coin 100 times and make a graph that shows how the cumulative fraction of heads changes as you flip the coin more times. In other words, graph the evolving percentage of tosses that came up heads. Given that the "correct result" should be 50 percent, how many observations (tosses) does it take in your experiment before you can measure the "answer" with an accuracy of 10 percent (that is, the percentage of heads stays between 40 percent and 60 percent)? Do you ever reach a point in your experiment when you get the right answer with an accuracy of 1 percent?

Reading List

Ferris, T. 1988. *Coming of Age in the Milky Way*. New York: William Morrow and Company.

Frazier, K., ed. 1991. *The Hundredth Monkey: A Skeptical Inquirer Collection*. Buffalo, N.Y.: Prometheus Books.

Giere, R. N. 1984. *Understanding Scientific Reasoning*. New York: Holt, Rinehart and Winston.

Goldsmith, D. 1991. *The Astronomers*. New York: St. Martin's Press.

Hazen, R. M., and Trefil, J. 1991. *Science Matters*. New York: Doubleday.

Kraus, L. M. 1993. *Fear of Physics: A Guide to the Perplexed*. New York: Basic Books.

Morrison, P., and Morrison, P. 1987. *The Ring of Truth: An Inquiry into How We Know What We Know*. New York: Random House.

Sagan, C. 1980. *Cosmos*. New York: Ballantine Books.

Sagan, C. 1997. *The Demon-Haunted World: Science as a Candle in the Dark*. New York: Ballantine Books.

Strauss, S. 1995. *The Sizesaurus: A Witty Compendium of Measurements*. New York: Kodansha International.

Tyson, N. D. 1994. *Universe Down to Earth*. New York: Columbia University Press.

CHAPTER 2

The First Discoveries About Earth and Sky

CONTENTS

- Stonehenge and Its Purpose
- Cycles in the Sky
- Real Astrology: Ancient Origins of a Modern Superstition
- The Birth of Science
- Astronomy in Other Cultures
- Summary
- Important Concepts
- How Do We Know?
- Problems
- Projects
- Reading List

Stonehenge, a prehistoric monument in England, has astronomical features built into its design and was apparently associated with observations of the date of summer solstice.
(SOURCE: Photograph by WKH)

WHAT TO WATCH FOR IN CHAPTER 2

This chapter describes how we first learned of our place in the universe. The first part of the chapter asks you to imagine yourself in prehistoric times, when people were still trying to understand properties of the sky, patterns of stars, and seasonal changes. We describe the changes in the sky that would be visible over the course of a year. For humans through the ages, the sky has been a map, a calendar, and a clock. Next, we describe some of the ancient observatories that are scattered around the world, and we show how astronomy has influenced systems of measurement.

The second part of the chapter covers the birth of science about 2500 years ago in the Mediterranean region. The ancient Greeks began using logic and geometric principles to analyze what they saw in nature. We will see the astonishing sophistication of some Greek naturalists, who figured out that Earth was round and measured its approximate size. One thinker even deduced that the Moon was much closer than the Sun. A final, but this time incorrect, legacy of the Greek era—an idea that lasted 2000 years—was an Earth-centered cosmology. The final part of the chapter considers the achievements of astronomers in other cultures.

PREREADING QUESTIONS ON THE THEMES OF THE BOOK

OUR ROLE IN THE UNIVERSE
Why were most ancient cultures convinced that Earth was the center of the universe?

HOW THE UNIVERSE WORKS
Is it possible to measure and predict the cycles of the sky without understanding the physical nature of celestial objects?

HOW WE ACQUIRE KNOWLEDGE
How did the Greeks use geometry to measure the sizes and distances of astronomical objects?

STONEHENGE AND ITS PURPOSE

Mysterious rings of upright stones dot the hills and dales of England. The most famous is a monument called *Stonehenge*, which many people know only as a ring of giant stones standing silently in an ancient field. Few people realize that other prehistoric structures abound in the nearby landscape. For example, a broad "avenue" lined by earthen banks leads out of Stonehenge across the plains for about ½ km (⅓ mi) toward the northeast. What was the purpose of such enigmatic features? In the Middle Ages, people believed that King Arthur's magician, Merlin, built them. Others attributed them to ancient wise men called Druids. In the 1720s, a lone scholar named William Stuckley visited many of the ancient monuments and made careful drawings of these prehistoric wonders. Stuckley was the first person to record a curious fact. The avenue leading from the center of Stonehenge points exactly toward the spot on the horizon where the Sun rises on the longest day of summer.

What could this alignment mean? More than a century later, in the 1890s, English astronomer Norman Lockyer noticed that other ancient monuments were oriented according to astronomical principles as well. Lockyer rediscovered Stuckley's work and then synthesized it with his own observations to conclude that Stonehenge was a prehistoric astronomical temple designed specifically for ceremonies on the longest day of summer. According to Lockyer, some cultures from various parts of the world who lived from 3000 B.C. to 1000 B.C. knew of subtle astronomical cycles. For example, they knew how to construct calendars accurate enough to reliably plant crops from year to year. They knew of the variations in the motions of the planets among the stars. Some cultures were even able to discover that the constellations had subtly changed their rising and setting positions relative to the horizon over the centuries.

Archaeologists and other scientists of the day ridiculed Lockyer's ideas, arguing that ancient people could not have gained such extensive knowledge. Today, most scientists agree with Lockyer. Ancient architects in various cultures sometimes used astronomical knowledge to locate and orient buildings and monuments. Precisely arranged standing stones have been found throughout Europe and the Americas.

We should avoid two tendencies when speculating about ancient cultures. The first is to assume that such people were unsophisticated because they had primitive technology. Remember that the brain function and language skills of humans have not changed for many thousands of years—the people who built Stonehenge were just like us. The second danger comes when we ascribe modern motives to ancient cultures. Stonehenge was not an observatory in the modern sense of the word. It almost certainly had a ceremonial and spiritual function as well. We can only make educated guesses at the intentions of the builders since they left no written language.

The building of Stonehenge was a terrific undertaking! The circular embankment and solstice avenue were built around 2500 B.C. During the time that followed from 2100 B.C. to 1500 B.C., workers dragged the huge stones from quarries as far as 380 km (240 mi) away and erected them to form the famous structure in the center of the ring. Archaeologists think that Stonehenge's purpose gradually shifted from precise observation to ceremony; by medieval times, its original function had been forgotten. Why would ancient people go to such enormous trouble to build this monument? Why would they take such an interest in the sky and astronomy? As we will see, the sky serves as a map, a clock, and a calendar. The "cycles of the sky" have been important to every culture throughout history.

Cycles in the Sky

Imagine that it is 40,000 years ago and that you are a nomadic hunter-gatherer. Much of your life is spent in the search for food and shelter. You are familiar with the slanting path of the Sun as it crosses the sky. You know the steady shift of the length of day and night throughout the year. The slow change of the climate gives you clues to the appearance of ripening berries and the migration of herds of animals. The Sun and the stars give you the tools for navigation. At night, by the safety of a fire, you and your tribe weave stories around the changing shapes of the Moon and the patterns of the stars.

We have lost touch with the sky. Most people are only vaguely aware of the cycles of the Sun, Moon, and stars. Our lives are regulated by watches, and by artificial heat and light. Most of us live in urban areas where the stars are barely visible. However, for much of history this information was essential for human survival. Let us see what someone at northern latitudes would observe over the course of a year.

What You Can See in a Year

The Sun rises in the east and sets in the west, traveling on a slanting path across the sky. In summer, the Sun rises north of due east, and the day is longer than the night. In winter, the Sun rises south of due east, and the day is shorter than the night. The Sun rides higher in the sky in summer than in winter. Using a distant horizon as a marker, you see that the shifting path of the Sun repeats during each cycle of the seasons. You celebrate the return of the Sun to its higher trajectory because it means that winter is easing its icy grip.

The Moon also orbits the sky along the same path as the Sun, but it appears to move at a different rate. When it is close to the Sun, the Moon appears as a crescent with the lighted side pointing toward the Sun. When it is on the opposite side of the sky to the Sun, it is a fully lighted disk. The changing face of the Moon also follows a regular pattern. The Moon seems to be a mysterious object—it is remote, yet the level of the oceans responds to it.

At night the stars also rise in the east and set in the west. Over the course of a night, they rotate about a fixed point in the northern sky. As the seasons progress, different groups of stars are seen above the horizon at sunset. The stars migrate completely around the sky during one cycle of the seasons. You note that stars twinkle but they do not change their colors. The colors of different stars vary from a dull red to a brilliant blue-white. Your imagination is captured by the delicate band of light that arcs across the winter sky.

You also notice five bright points of light that move from night to night among the fixed star patterns. You can locate these "wanderers" in the strip of sky traversed by the Sun and Moon; they do not twinkle. They move in the same direction as the Sun and the Moon with respect to the stars. They also have the peculiar habit of occasionally reversing the direction of their nightly motion for a few weeks or few months at a time. You know the sky offers other surprises. Every so often the full Moon is darkened to a blood red color and then slowly brightens again. You have been told of a rare but fearsome event when the midday Sun darkens inexplicably. You can vividly imagine the disquiet of animals as they respond to the false night and the sudden chilling of the air.

These are the cycles of the sky. They are summarized in Table 2–1. You need no telescope or advanced timekeeping to measure them. This chapter deals with the slow march toward understanding these phenomena. We will see our three themes reflected in the material. There is a natural reluctance of humans throughout history to accept that Earth might not be the center and main feature of the universe. Until a few centuries ago, people thought that the celestial objects occupied an "ethereal realm" where the laws were different from those on Earth. We will see the emergence of the idea that the same physical laws apply to the untouchable objects of space as to mundane terrestrial objects. We will also see the application of logic and mathematics to the universe. This marks astronomy as "the first science."

Navigation

How could you use the sky to find your way? The first orienteering skills were discovered over 10,000 years ago. If you

TABLE 2-1
What You Can Observe in the Sky Over a Year

1. In summer, the Sun rises north of due east, and the day is longer than the night. In winter, the Sun rises south of due east, and the day is shorter than the night.
2. The Sun, Moon, and planets traverse the same strip of sky from east to west.
3. Stars rise in the east and set in the west. They all appear to slowly rotate about a fixed point in the northern sky.
4. The pattern of stars in the constellations does not change from year to year.
5. Any particular star rises and sets slightly earlier each night. The constellations migrate through the sky completely in one cycle of the seasons.
6. Some of the planets move irregularly among the constellations, occasionally reversing their direction of motion.
7. The Moon changes its phase on a regular cycle. A full moon is high in the sky around midnight, and a new moon is high in the sky around midday.
8. Lunar eclipses are more frequent and last longer than solar eclipses. Neither occurs every month.

were out hunting during the day, you would watch the path of the Sun carefully, because you would have to turn around when it reached its highest point in the sky to avoid being out at night. If you live in the north, the Sun indicates the direction south at its highest point in the sky. The Sun can act as a clock. If you lived in a hot climate, you might need to travel at night and use the stars to guide you. Ocean voyagers must also use the stars to navigate at night. The star patterns can be a map.

Most people are vaguely aware that they can find their way at night by the stars. They also know that there is a North Star, called *Polaris,* which is always in the north. But there are many other features of the sky that are familiar and useful. Night walks or camping trips offer a great opportunity to learn your way around the sky. Knowing some basic sky terminology is a useful first step. The point directly overhead is called the **zenith,** illustrated in Figure 2–1. An imaginary line overhead, from the due south point on the horizon through the zenith, to the due north point on the horizon, is called the **meridian.** As the figure shows, the Sun, Moon, and all the stars and planets rise on the east side of the meridian, cross the meridian, and set on the west side of the meridian. The ancient term A.M. (from the Latin phrase *ante meridiem,* or before the meridian) refers to the first half of the daylight period, before the Sun crosses the meridian. The term P.M. (*post meridiem*) refers to the second half of the day, after the Sun crosses the meridian. The basis for describing the positions of astronomical objects is angular measurement (see Appendix A-5).

It seemed obvious to ancient people that the Sun, the Moon, and the stars all orbited around Earth. There was no reason to suppose that Earth was moving—after all, we do not feel any motion! We now know that Earth spins once per day. The rising and setting motions of the stars are due to Earth's rotation. The line through the center of Earth that connects the North Pole and the South Pole is Earth's rotation axis. For navigation, the most interesting star is the North Star or Polaris. Because Earth's North Pole happens to be aimed at it, it neither rises nor sets but sits at the same spot above the northern horizon all night. All the other stars in the sky move slowly in circles around it. This motion is imperceptible from minute to minute, but you can see it over a period of hours or with a long-exposure photograph. Polaris is a fairly bright star and thus serves as a beacon, pointing north. Study Figure 2–2, which shows timed exposures looking north and south. You can see that Earth is like a spinning top, with the stars wheeling around two fixed points in the sky.

Polaris reveals your latitude. Figure 2–3 demonstrates that the angle of Polaris above the northern horizon (measured in degrees) equals your latitude (the number of degrees you are north of the equator). For example, in most of the United States, the North Star is about 30° to 45° above the northern horizon. However, at the North Pole, Polaris is directly overhead, and the stars move in concentric circles parallel to the horizon. Anywhere on the equator, Polaris is low on the northern horizon, and stars rise straight out of the east and set straight into the west. Earth's South Pole does not point at any bright star, so navigation south of the equator uses other star patterns.

Ancient navigators sighted the North Star to determine their latitude, even in the middle of the ocean or in a trackless wilderness. For example, ancient Polynesian navigators sailing from Tahiti to Hawaii could sail north until the North Star confirmed the latitude of Hawaii, then head west until they came to Hawaii. Once you have memorized the main features of the night sky, star patterns can be used for navigation, too. You can find sky maps of bright stars in Appendix A-6 near the back of this book. It is too hot to travel during the day in the desert regions of the Middle East. For this reason, and because the Arabs preserved in translation the original names from ancient cultures, many of the bright stars have Arabic names (Aldebaran, Mizar, Alcor, Deneb, and others).

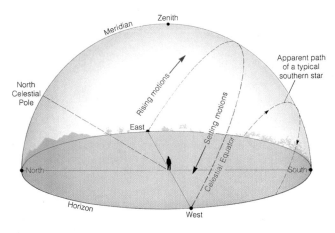

FIGURE 2–1
General properties of the sky as defined by Earth's daily rotation. The sky is shown as seen by observers at mid-northern latitudes, such as the United States. The north celestial pole marks the position of the North Star. Ancient people thought of the stars as rotating overhead on a celestial sphere.

FIGURE 2-2
A camera making a timed exposure at night toward the northern or southern sky will reveal the circular (~~clockwise~~) tracks of the stars, caused by Earth's rotation. **(a)** Northern sky. Polaris is the bright star near the center of the circular star traces. Because this photograph is made at latitude 20°N, stars within 20° of the north celestial pole never set. **(b)** Southern sky. In this photograph from the same location, the stars make rainbowlike (~~counterclockwise~~) *Clockwise* arcs as the Earth turns during the night. The stars rotate around the south celestial pole, a fixed point 20° below the southern horizon. Stars within 20° of the south celestial pole never rise. (SOURCE: Both photographs by WHK from Mauna Kea Observatory, in Hawaii, illuminated by moonlight. Stationary 35-mm camera with 15-mm wide-angle fish-eye lens; 20-min exposure on commercially available ISO 1000 film.)

Keeping Track of the Seasons

Ancient people had to be aware of the cycles of seasons. It was a matter of life and death. For most of human history, we have been wanderers. But around 7000 to 10,000 years ago, tribes learned how to domesticate animals and cultivate simple crops. The birth of agriculture created a new need to keep track of time. To eat, people had to know when to plant, when rains would come, when the birds would migrate through their region, and when they would have to store food for the lean winters.

We can track the seasons using either the stars or the Sun. The stars all have a fixed pattern relative to one another from year to year, like the pattern of major cities on a map of Earth. Many ancient cultures selected specific groupings of stars or **constellations** to represent certain animals or mythological figures. When you see the constellations—for example, Ursa Major, the great bear; or Orion, the hunter; or Pisces, the fish—it is not obvious what the patterns are supposed to represent. You have to use your imagination! The constellations are not literal drawings on the sky, but were used as a memory aid for navigation or to pay homage to gods and myths. If you learn a few major constellations, they can help you locate the North Star and other directions in the sky, as shown by Figure 2–4.

FIGURE 2-3
How to determine your latitude astronomically. The altitude of the North Star above the horizon, measured in degrees, always equals your latitude. Angle L is your latitude, the angle that your point on Earth makes with the equator. Angle θ is the elevation of the north celestial pole above the horizon. By simple geometry, these angles are always equal. Try to draw this diagram for an observer at the equator ($L = \theta = 0°$) and at the North Pole ($L = \theta = 90°$). An observer at the equator sees Polaris on the northern horizon, and an observer at the North Pole sees Polaris directly overhead.

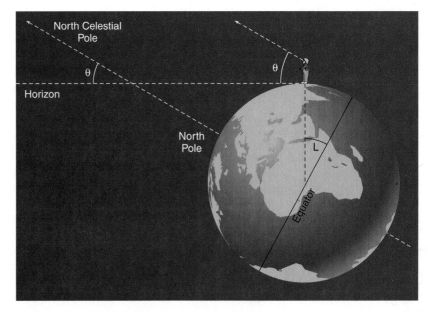

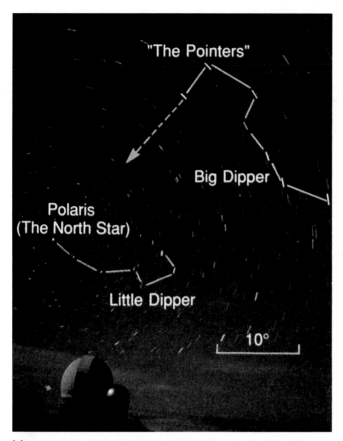

(a)

(b)

FIGURE 2-4
Notable constellations of the northern sky. **(a)** Winter sky. The constellations of the greater and lesser bears, Ursa Major and Ursa Minor, contain two asterisms, or small groups of stars, popularly known as the Big Dipper and Little Dipper. The North Star is at the end of the Little Dipper's handle, and it can be found best by using "the pointers"—the two end stars in the Big Dipper's bowl. The Big Dipper's stars are brighter and easier to find than the Little Dipper's. This 10 min exposure with a stationary camera shows that the star images trace small arcs as Earth rotates through about $2\frac{1}{2}°$; they would be complete circles if we could follow them for 24 h. **(b)** Summer sky. Scorpius is rising in the east in this wide-angle photograph, which has a field of view of 80° from top to bottom. The camera has tracked the star motions for 25 min in this exposure. The stars outline a scorpion with two curving claws at top and a hooked tail at bottom. A double star marks the end of the tail, and the noticeably red star Antares marks the heart of the scorpion. The Milky Way crosses the lower half of the picture. (SOURCE: Both photographs by WKH from Mauna Kea Observatory; using a 35-mm camera with wide-angle 24-mm lens at f 2.8; on commercially available film rated at ISO 1600.)

Many constellations are extremely old. Clues in the constellation patterns, such as the names of mythical creatures, suggest that many constellations on our sky maps were designated around 2600 B.C. by Mediterranean seafarers. A few, such as Ursa Major (which contains the asterism called the Big Dipper) were known throughout Asia and also by Native Americans. We believe this constellation was first named in Asia before 10,000 B.C., and the name was carried to America by the first Americans who traversed the land bridge that crossed the Bering Strait around that time. Ursa Major is one of the oldest human cultural artifacts.

The path traveled by the Sun with respect to the fixed stars is called the **ecliptic**. The Moon and the planets are always found near this path, and this strip of sky has always had special significance. Ancient people called this strip of sky the **zodiac**—circle of animals—and divided it into the 12 constellations that we recognize as the star signs.

The constellations on view depend on your position on the spinning globe of Earth. Figure 2–5 shows the geometry that defines what fraction of the sky you could see over a year. Observers at the poles only ever see one-half the stars in the sky. Observers at the equator see all the stars each year. Any particular star rises and sets a little earlier every night. The entire set of constellations passes through the night sky once during each cycle of the seasons. This results from Earth's annual motion around the Sun, as shown in Figure 2–6. During each season, different constellations appear in the evening sky. For example, Cygnus, the swan, rides high in the summer, and Orion, the hunter, appears in the late fall. After a full year, the pattern of constellations in the evening sky repeats its cycle. Constellations return each season, and it is fun to greet them, like old friends coming around each year.

Constellations have been used for keeping track of the seasons long before we had a scientific explanation for their motions. Hesiod was one of the first poets whose words were written down. He lived in Greece about 2500 years ago and said

> "When great Orion rises, set your slaves
> to winnowing Demeter's holy grain
> upon the windy, well-worn threshing floor.
> Then give your slaves a rest; unyoke your team.
> But when Orion and the Dog Star move
> into the mid-sky, and Arcturus sees
> The rosy-fingered dawn, then Perseus, pluck
> The clustered grapes, and bring your harvest home."

Astronomy is not only the oldest science but it is woven into the fabric of our culture in many ways. We can see this clearly when we look at the development of calendars and the way we divide up time.

Calendars and Time

In midnorthern latitudes, such as in the United States, Canada, or Europe, the Sun rides much higher above the horizon in the summer than in the winter. In the summer, it rises in the northeast, crosses the meridian nearly overhead, and sets in the northwest. But in the winter, it rises in the southeast, crosses the meridian low in the south, and then sets in the southwest. You can mark the passage of the seasons by tracking the position of sunrise or sunset on the horizon, as shown in Figure 2–7. A distant feature such as a rugged mountain range makes a simple but effective measuring device.

A **calendar** is a means of counting the days in a year. Ancient people did this by counting the days until the sunrise (or sunset) position moved back to its extreme northerly (or southerly) position after a cycle of the seasons—called a *solar year*. The earliest records we have show a count of 360 days in the year. The Egyptians had revised the count to 365

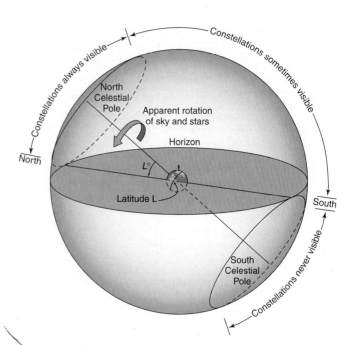

FIGURE 2–5
For an observer at latitude L in the Northern Hemisphere, no stars within an angle of L degrees ever set. Stars in the middle zone rise and set, and those in a southern zone within an angle L of the south celestial pole are never seen. Try to draw this diagram for an observer at the equator ($L = 0°$) and the North Pole ($L = 90°$) to convince yourself that someone at the equator has every star visible and someone at the North Pole has only half the stars visible.

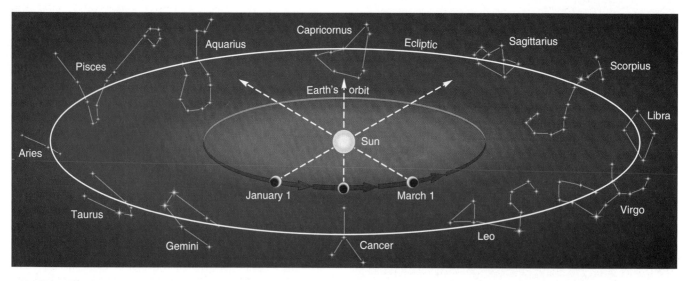

FIGURE 2–6
The changing constellations of the night sky. As the Earth moves around its orbit, we see different constellations in our night sky. In January, Gemini is in the night sky and the Sun blocks our view of Sagittarius. Three months later, Leo is in the night sky and the Sun blocks our view of Aquarius. The ecliptic is the projection of the Earth's orbit on the celestial sphere. As seen from the Earth, the ecliptic is the apparent path of the Sun in the sky. Constellations that lie on this path are the star signs of the zodiac.

FIGURE 2-7
The solstice principle. Views looking due west at sunset on about the 22nd of each month show the movement of the sunset position. On winter solstice (top, December) the Sun is as far south as it can get at this latitude in Tucson, Arizona, and it sets in the southwest. By summer solstice (June) the Sun sets in the northwest. (SOURCE: Photographs by WKH, Gayle Hartmann, Kelly Rehm, and Amy Hartmann.)

days, and they added a leap year—a year of 366 days inserted every fourth year—for an average calendar 365 1/4 days long. By 2700 B.C., the Babylonians had refined this to a calendar of 365.26 days which was accurate to about 30 minutes (min) in 1 year. This is an impressive calendar. Nearly 5000 years ago, the length of the year was known to an accuracy of better than 1 part in 10,000!

Ancient observers divided the year into seasons, using four special dates that we still recognize. Winter solstice is the first day of winter, around December 22. On this day the Sun rises and sets farthest to the south. This day has the shortest period of daylight of any day in the year in the northern hemisphere. The pre-Christian pagan cultures of England and France began the year on this date, to celebrate the return of the Sun toward the northern sky. Since the seasons vary smoothly throughout the year, it is quite arbitrary when we choose to begin the calendar. Spring equinox is the first day of spring, around March 21. On this day the Sun rises due east and sets due west. Day and night are equal in length on this day (in the word equinox, *equi-* means equal and *-nox* means night). Other pagan cultures, such as those that worshipped the goddess Maia, began the year on this date because it marked the beginning of the cycle of new growth. Summer solstice is the first day of summer, around June 22. On this day the Sun rises and sets farthest to the north. It has the longest daylight period of the year. In many ancient calendars, it marked a day of celebration, when the days were long and the weather pleasant. Autumn equinox is the first day of fall, around September 22. On this day the Sun rises due east and sets due west, and day and night are again equal.

Earlier cultures also marked the midpoints of the calendar between the solstices and the equinoxes. These dates are February 1, May 1, August 1, and November 1. In Ireland, these festivals are all still celebrated and are known by their Gaelic names: *Imbolic, Beltane, Lughnasa,* and *Samhain.* May Day was originally a fertility festival in the pagan world, and it is still a folk festival in England. And of course the time around November 1 is celebrated in many parts of the world, as All Saints' Day in England, as the Day of the Dead in Mexico, and as Halloween in the United States.

You might think that knowledge of solstices and sunrise positions is useless knowledge from the ancient past. However, it has applications in the present, especially in an environmentally conscious age. Consider the layout of windows in your home or apartment. Windows facing northeast or northwest allow sunlight to enter on summer mornings or afternoons, when you are likely not to want extra heat in the house. Draperies or shades on these windows in the summer will block sunlight and save on energy bills. Windows facing southeast or southwest, on the other hand, will let in the low-angled light of the winter Sun, giving a free input of extra heat. Figure 2-8 shows a house specifically designed to take advantage of these ideas. In this example, windows are placed to take advantage of the midwinter morning and afternoon Sun on the southeast and southwest side; and trees are planted to shade the northeast, east, west, and northwest sides of the house during the summer mornings

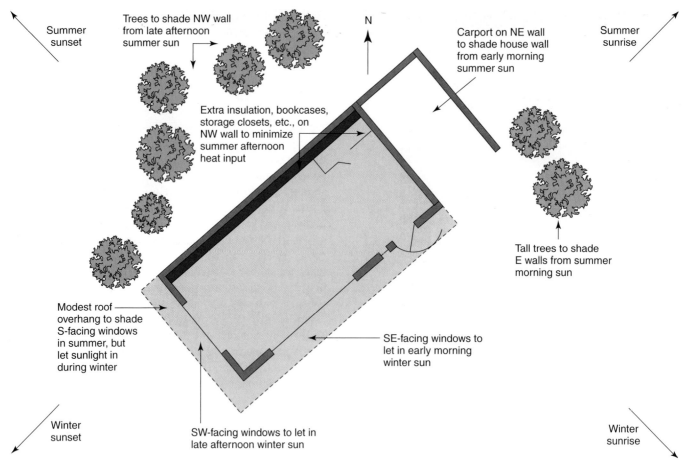

FIGURE 2-8
The solstice principle as applied to environmental architecture and landscaping. East-northeast and west-northwest walls are shaded to reduce heat from the summer morning and afternoon sun. Southeast, southern, and southwest windows allow sunshine to enter and heat the house in winter. This design would be appropriate for a mid-northern latitude such as the United States, but the same ideas can be used to design an energy-efficient house anywhere.

and afternoons. Roof overhangs shade southern exposures from the high summer midday Sun, while letting the low winter midday Sun warm the interior and exterior of the home.

All the other major divisions of time have astronomical origins. The illumination cycle of the Moon gives rise to the month. Every $29\frac{1}{2}$ days there is a full Moon or a new Moon, and this cycle divides approximately 12 times into the solar year (but not exactly; see the Science Toolbox on "Dividing Time"). We have found carved animal bones and other artifacts dating back 20,000 to 30,000 years in France and other parts of Europe. Some of these objects have numbers of notches that indicate that the cave dwellers were using them to count months. These portable calendar sticks are among the oldest human relics.

There is a difference between time kept by using the Sun and Moon and time kept by using the stars. The time taken for the Sun to pass through the meridian on successive days is a *solar day*. The time taken for a star to pass through the meridian on successive nights is a *sidereal day*. Since every star rises and sets a little bit earlier every day, a solar day is about 4 min longer than a sidereal day. You can see this by remembering that all the constellations migrate through the night sky every 12 months. Each month, a particular constellation or star appears to shift $1/12$ of the way around the sky, and so it rises and set 2 hours earlier. And every day any particular star rises and sets about 2/30 hours or 4 minutes earlier. Similarly, the time taken for the same phase of the moon to recur is the Moon's *synodic period*—29.5 days. This is longer than the time taken for the Moon to pass the same place among the stars—the Moon's *sidereal period* is 27.3 days. These differences occur because the stars that would be seen in the direction of the Sun shift gradually as the Earth spins and as the Moon orbits the Earth.

Once ancient people recognized the fixed pattern of the constellations, they discovered that five bright "stars" were different from all the others. These "stars" moved from week to week relative to the pattern of the fixed stars. They became known as *planets,* from the Greek word for wanderer. The planets had unusual attributes. For example, they were not found any place in the sky but always in the strip of sky occupied by the Sun and Moon. Some planets were only seen close to the Sun, others were seen far from the

SCIENCE TOOLBOX

Dividing Time

The cycles of the sky have been used to divide time throughout human history. A calendar marks a solar year—the time it takes for Earth to return to the same place in its orbit of the Sun. The natural division of time is the day—the time it takes for the Sun to return to its highest position in the sky, or cross the meridian. Modern accurate measurements show that

1 solar year = 365.242199 days

Note the number of significant figures. If you convert to seconds, you will see that we know the duration of Earth's orbit with an accuracy of better than 1 s! The Moon has also been used to mark time. The best measurements show that

1 lunar month = 29.53059 days

This is the time between two occurrences of the same lunar phase, also called a *synodic month*. We should not be surprised that these are not whole numbers. It would be entirely fortuitous if three objects in space moved in such a way that a lunar orbit took an exact number of Earth rotations or if an Earth orbit took an exact number of Earth rotations. Nor does a lunar month divide evenly into a solar year (365.24 / 29.53 = 12.37), so counting months will not give a good calendar to keep track of the seasons.

We have seen that the Babylonians had a very accurate calendar of 365.26 days. This number is quoted to five significant figures and differs by only about 30 min from the true astronomical number (you should review the discussion of accuracy and significant figures in Chapter 1 and Appendix A-4). Unfortunately, we inherited our calendar from the Romans, who had lost the knowledge of the Babylonian calendar. We will see that our calendar has some very unusual features.

The earliest Roman calendar, from the 7th century B.C., had only 304 days and began in mid-March. This was when the snows had melted enough for the Roman soldiers to go off on conquests. The first month was March, named after Mars, the god of war. This is also the reason that the last four months of the year—September, October, November, and December—are taken from the Latin for 7, 8, 9, and 10. Our original calendar only had 10 months! The early months of the year were named after popular Roman gods. By the 6th century B.C., two early months had been added to give a year of 354 days. You can see that after only 3 years this calendar would have slipped by (365.24 − 354) × 3 = 34 days, or more than a month with respect to the seasons. This makes it a very poor calendar.

At the height of the Roman Empire, Julius Caesar reformed the calendar. He added days to all the months to make a total of 365, and he added a leap year to make an average count of 365.25 days in the year. The months neatly alternated 30 and 31 days long, except February, which had a short count. The Romans were very superstitious, and February was chosen as the "bad luck" month; no Roman would travel or entertain during that month. Caesar also took the next unnamed month for himself. So we have July, after Julius. The Emperor Augustus followed Caesar, and his ego dictated that he grab the next available month for himself. So we have August, after Augustus. However, in the alternating sequence his month was shorter than Caesar's month, so he arbitrarily added a day to it, stealing a day from unlucky February. Then to preserve the same sum of days, he had to add and subtract 1 day from the rest of the months. That is why you have to use a little trick to remember the number of days in each month!

The Julian calendar was quite accurate. It would only lose 365.25 − 365.24 = 0.01 day per year. This error is systematic—always in the same direction—unlike the random measurement errors discussed in Chapter 1. As a consequence, this tiny error steadily accumulated as the centuries passed, and by the 16th century the calendar had slipped by 10 days. In 1582, Pope Gregory moved the date forward 10 days by decree. He also added the extra rule that leap years should be skipped in century years unless the year was divisible by 400—in other words, 2000 is a leap year, but 1900 was not. By this little trick, the Gregorian calendar has an average year that is 365.2422 days long. This is within one-thousandth of a day of the astronomical number, so it will be many thousands of years before the Gregorian calendar needs adjustment. Interestingly, the American colonies and England waited another 180 years to make the switch, as an act of independence by these primarily Protestant countries against the "imposition" of a Catholic calendar.

Religion also figures prominently in the distinction between a *solar* and a *lunar* calendar. A lunar calendar will very quickly shift with respect to a solar calendar. Twelve lunar months give a year that is 12 × 29.53 = 354 days long. Islamic countries follow the lunar calendar Islamic festivals such as Ramadan and the pilgrimage to Mecca therefore advance through our calendar by 11 days every year. Every 365 / 11 = 33 years they shift through an entire cycle of our seasons. This simple but deep-rooted difference in calendars may be one reason why Arab and Western cultures have difficulty in communicating. If biblical texts are followed explicitly, the major Christian holiday of Easter is tied to a lunar calendar and so cannot always fall on a Sunday. The Eastern Orthodox Church split from the Catholic Church over this issue in 325, and the Orthodox Jewish calendar is also lunar.

Seasonal changes are not as noticeable in the Middle East, so there is less reason to key the calendar to the Sun. Also, agriculture in the major Middle East civilizations is usually governed by irrigation from major rivers and is less sensitive to seasonal changes. All Islamic countries have a crescent moon in their flags. This is an indication that festivals and holidays depend on observations of the new moon.

Sun. Some planets would even reverse their direction of motion in the sky when viewed from week to week. Our names for the planets come from the Roman gods Mercury, Venus, Mars, Jupiter, and Saturn. (The other planets—Uranus, Neptune, and Pluto—were not discovered until the invention of the telescope, and Earth was not yet recognized as one of the planets.)

Every culture has felt the need to create a chunk of time such as a week. No one knows why. Ancient Egyptians used 10 days, the Babylonians used 7 days, the Assyrians used 5 days, and some West African tribes have used 4 days in the week. Our calendar is based on Roman tradition, which named the 7 days of the week after the seven "moving" objects in the sky—the Sun, the Moon, and five planets. Some of these names are obvious: Saturn-day, Sun-day, Moon-day. The connections to other planets are clearer in languages that are derived from Latin, such as Spanish, Italian, and French. Tuesday is Mars-day, for example (*Mardi* in French, and *Martes* in Spanish). Wednesday is Mercury-day (*Mercredi* in French, and *Miercoles* in Spanish), Thursday is Jupiter-day (*Jeudi* in French, and *Jueves* in Spanish), and Friday is Venus-day (*Vendredi* in French, and *Viernes* in Spanish). So what happened to the names of these four days in English? They were named after gods from the Anglo-Saxon culture of a thousand years ago. Tuesday comes from Tiw, the Norse god of war. Wednesday comes from Woden, the supreme deity. Thursday comes from Thor, the god of thunder. And Friday comes from Frigg, the wife of Woden and goddess of love and beauty.

Even the division of the day into hours has an astronomical origin. Egyptian astronomers used a sequence of bright stars across the sky for timekeeping at night. Twelve timekeeping stars were visible during the critical midsummer period when the Nile would flood, so the night (and later the day) was divided into 12 hours (h). Timekeeping was very primitive until the last 250 years. The Greeks used sundials, and the Romans perfected the water clock, where water could drip at a regulated rate though a small hole in a hard stone or jewel. The sand hourglass dates from 8th-century Europe. Nobody could divide time into units smaller than hours until the pendulum clock was invented in the 17th century. We can thank the Babylonians of 5000 years ago for the choice of units that divide the hour into 60 min and the minute into 60 seconds (s).

Many other features of our calendar and timekeeping spring from the pagan cult of Sun worship. Stonehenge and other great prehistoric structures were built to measure and celebrate the motions of the Sun. Many pagan traditions were borrowed by the early Christian calendar that we still follow. The year starts on January 1. This copies the pagan cultures which began their calendar when they could detect the Sun beginning to move farther north in its rising and setting position. Our rest day of Sunday follows the pagan day of worship of the Sun. Why do clocks move clockwise? In northern Europe clocks were designed to mimic the arcing motion of the Sun from left to right across the southern sky—refer back to Figure 2–2a. Our habits of timekeeping are a rich brew of astronomical ideas taken from earlier cultures.

Ancient Observatories

We now have a context for understanding enigmatic structures such as Stonehenge. Keeping a calendar was a vital job for any ancient culture. We can imagine that an enormous effort would be put into measuring and marking the changing seasons. In the absence of true understanding, we can also imagine that these structures would have a ceremonial purpose. Today, most astronomers and archaeologists agree that Stonehenge was built as a marker or ceremonial site, dedicated to observing the date of summer solstice. The avenue points from the center of Stonehenge to the position of sunrise on June 21, as shown in Figure 2–9. Someone standing in the center of Stonehenge could see the Sun rise over a big stone, called the *heel stone,* on the day of solstice. (The name *heel stone* may have come in ancient times from the Greek root *helios* for Sun.) Still more astronomical purposes have been suggested for Stonehenge, including lunar eclipse predicting, but these are controversial.

The investigation of Stonehenge's timekeeping functions helped to create a new field of astronomy called **archaeoastronomy**—the study of astronomical practices in ancient societies. Astronomical temples, which were used to help calibrate the calendar for agricultural and ceremonial purposes, have been found in many parts of the world. As we consider some examples, we can see the aspects of culture that are woven into the stone monuments.

The long span of human observations of the sky allows subtle motions to be discovered. The astronomer Hipparchus studied two centuries of star maps and realized that the direction of the north celestial pole has moved slowly across the sky. A spinning gyroscope that is not pointed straight up will wobble—its rotation axis sweeps out a circle. This conical motion is called **precession,** and the Earth precesses with a long 26,000-year cycle. From day to day or year to year, this motion is far too small to detect. The north celestial pole has not always pointed at Polaris, and ancient cultures were aware of this steady shift. In one legend, the precession was ascribed to a great whirlpool in the Mediterranean sea, which slowly twisted the heavens.

The Great Pyramid in Egypt—constructed around 2650 B.C. from over 2 million limestone blocks weighing 2 tons each—is aligned almost perfectly with the north-south axis. Two air shafts point directly from the pharaoh's tomb to the brightest star in Orion's belt. The Egyptians identified Orion with the underworld god of rebirth, and it is likely that the air shafts were intended for the passage of the pharaoh's soul on its journey to an afterlife in the heavens. One air shaft does not point at any bright star now, but if we account for 4500 years of precession, it used to point at the bright star Thuban.

The ancient city of Chichen Itza rises out of the dense rainforest in the Yucutan peninsula of Mexico. Several buildings dating from around the year 1000 incorporate astronomical alignments, most notably of Venus since it was im-

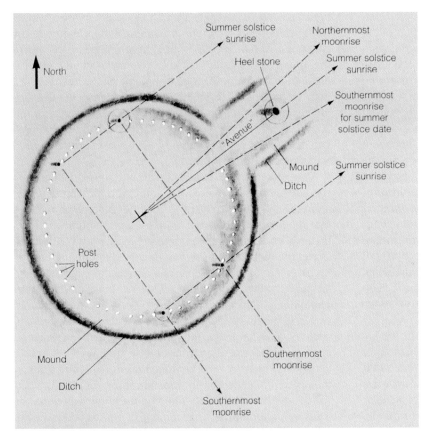

FIGURE 2-9
The solstice principle at Stonehenge. **(a)** An aerial plan shows the original Stonehenge construction around 2500 B.C., with an outer ditch, a mound, and a ring of posts. Stonehenge was designed so that an observer at the center could mark specific horizon points. A radial "avenue" and distant stone called the *heel stone* marked the position of summer solstice sunrise. Selected moonrises could also be observed and timed. **(b)** An aerial photograph shows modern Stonehenge. The central ring of giant stones was added around 2000 B.C. for ceremonial purposes; the curving path is a 20th-century addition. The avenue leads out of the right of the picture, and the heel stone is near the road.

(SOURCE: George Gerster, Photo Researchers, Inc.)

portant in the Mayan religion. The Mayan calendar was based on the cycles of both Venus and the Sun. On the morning of summer solstice, the rising Sun casts a shadow on the corner of the pyramid at Chichen Itza that climbs up the structure as a snake would.

In Arizona, the Hohokam Native Americans built a large ceremonial structure called Casa Grande in 1350. The setting Sun aligns perfectly with an unusual circular, west-facing window only on the longest day of the year. Farther north, at Chaco Canyon, the Anasazi tribe created a structure on a high promontory where light is admitted through a space between rock slabs. The gap lets in a "dagger of light" that projects onto a spiral pattern carved on the opposite wall. The placement of the Sun dagger marks both the solstices and the equinoxes. These apparitions—snaking shadows, projected beams, and light daggers—are a vivid demonstration of the diverse ways that cultures have marked astronomical time.

Not everyone uses the sky in the same way. We know that at our northern latitude, the stars move on slanting paths in the sky, and they all move in slow circles around Polaris. However, near the equator the stars rise straight out of the east and set directly in the west (review Figure 2–2). Polaris may be low on the horizon or even below the horizon. In the tropics, which is the zone between 23.5° latitude north and 23.5° latitude south, there are 2 days each year when the Sun is directly overhead at the zenith. The Incas of Peru and the Javanese of Indonesia both chose to fix their calendars around these notable days, when a tree or a vertical stick casts no shadow. Meanwhile, at far northern latitudes above the Arctic Circle, the Sun spends large chunks of the year below the horizon and is of no use for timekeeping. The Inuit and other tribes have always used the tides to keep track of time, which is a kind of lunar calendar!

Systems of Counting and Measurement

All our counting and arithmetic are based on the **decimal system.** Yet many familiar quantities are divided in ways that do not use factors of 10. Think of the 360° in a circle, or the 60 min in 1 h, or the 24 h in 1 day, or the 16 oz in 1 lb, or the 12 in. in 1 foot, or the 8 pt in 1 gal. Where do these units come from? Most ancient people could not read or write, so they preferred systems that made it easy to divide a number (of cows or coins) in many different ways.

The earliest cultures often used systems of measurement based on 60—this is called the *sexagesimal* system. Why use 60? It is the lowest number that can be divided into 2, 3, 4, 5, or 6 equal parts (see Appendix A-2). During the period from 3000 B.C. to 2000 B.C., the Babylonians invented the divisions of angle and time that we still use today. The fact that the unit of angular measurement is a degree is almost certainly based on astronomy. Each day Earth moves about 1° in its orbit around the Sun. To take the viewpoint of ancient people, if you view the sky each night at the same time of night, you find that the stars have shifted position by about 1° compared to the previous night. Over several weeks, this shift in the star positions is quite noticeable. The approximation that the year has 360 days may have reinforced the choice of 360° to define a circle.

As you can see in Appendix A-5, degrees are subdivided into minutes of arc and seconds of arc just as hours are divided into minutes and seconds. You might wonder why the Babylonians gave us our smallest units of time, when there were not even clocks or watches 5000 years ago. Remember that they used the stars to keep accurate time, and they knew the length of a year to within a fraction of an hour. Since Earth is rotating, angles and time are related. With one complete rotation every 24 h, Earth rotates 1° in about $(24 \times 60)/360 = 4$ min. So any star rises and sets 4 min earlier each night.

Astronomy starts with **angular measurement.** We should distinguish between linear measure and angular measure. Linear measure gives the actual length of something in inches, or meters, or miles. Angular measure, by contrast, gives the angle covered by an object. It measures the apparent separation between two objects at a particular distance from the observer, in units such as degrees. The verb *subtend* refers to the angle covered by such an object; for example the Moon subtends $1/2°$. To get a sense of this angle, extend your little finger to arm's length. Your fingernail at this distance subtends $1/2°$. If you are unsure of these ideas, review Appendix A-5.

We will encounter a wide range of angular measures in this book. The human eye is an amazing optical system with an angular range of roughly 70°. Cameras come with lenses that range from wide-angle to telephoto. The difference between such lenses is merely the angular width of the view, sometimes called the *field of view* of that lens. Figure 2–10 illustrates this concept. A normal lens usually covers an angular field of view of about 40° to 50°, whereas a typical telephoto lens may show about 10° to 20°. At the other extreme, the finest feature that the eye can see subtends an angle of only 3 minutes (or 3′) of arc, and many telescopes can measure features as small as 1 second (or 1″) of arc. To get a sense of these small angles, look again at the stone at the center of Figure 2–10d, which is about the size of a person. Three minutes of arc would be the angle subtended by a penny on top of the stone, and 1″ of arc would be the angle subtended by the head of a pin.

Ancient people found it useful to think of the stars as imprinted on a *celestial sphere* that rotates overhead. The location of every object in the sky can be specified by two angles. Suppose you are looking at a star in the night sky. You can measure the angle of the star above the horizon, called the **altitude.** You can then measure the angle of that point on the horizon from due north or any other reference point, called the **azimuth.** These two angles uniquely give the position of an object on the celestial sphere. In a similar way, any point on Earth's surface can be defined by two angles—longitude and latitude. Of course, the altitude and azimuth of a star will vary with time and with position on Earth. So astronomers use a different pair of angles which are referenced to the celestial equator, which is the circle on the sky that is at every point 90° away from the pole star Polaris. See the explanation and the star maps in Appendix A-6 near the end of the book, which show how astronomers measure positions on the sky and how you can use the stars to find your way.

The early counting systems like the 60-based sexagesimal system are certainly convenient for dividing things up. Many of our familiar units of measurements derive from multiples of small numbers—inches, ounces, pints, and so on. However, these systems become very clumsy when we are dealing with large numbers. For example, the old Roman number for the decimal year 1988 is MCMLXXXVIII (still used in movie credits and on dollar bills). Imagine writing large astronomical numbers or doing your taxes in this way! Yet we have tended to stick with familiar systems, like ounces and pounds, even when they are not the best or the easiest to use.

The decimal system was perfected by the Arabs, who

FIGURE 2-10
These views toward the northeast from the center of Stonehenge illustrate angular measurement using photography with different lenses. **(a)** This view uses an ultra-wide-angle fish-eye lens to subtend a horizontal angle of 120°. **(b)** A standard wide-angle lens subtends 65°. **(c)** A normal lens subtends 40°; this is the approximate field of view of most typical snapshots, postcard views, and paintings. **(d)** A telephoto lens subtends only 15°. (SOURCE: Photographs by WKH with a 35-mm camera. Curvature at the edge of the wide-angle view is a distortion common with fish-eye lenses.)

brought it from India around the year 750. Note that it requires the invention of a zero to stand for an empty place in a number. *Zero* is an Arabic word. The decimal system leads naturally to the **metric system** of measurement, where the units are related by powers of 10. For example, metric length units are kilometers and meters and millimeters, and metric weight units are kilograms and grams (see Appendix A-3). As an experiment after the Revolution of 1789, the French even developed a short-lived decimal system of time. However, you can see that our modern world is still littered with archaic units of measurement.

Real Astrology: Ancient Origins of a Modern Superstition

The night sky was a source of mystery to ancient people. They could observe regularities in the motions of celestial objects, but they had no understanding of their true nature. Ancient cultures thought that the objects of the night sky exerted a special power over humans. This belief system is called **astrology,** the idea that configurations of planets and stars in the sky cause or influence human events on Earth. As we have seen in Chapter 1, astrology is just one example of a wide set of ancient beliefs that patterns in nature cause or influence human events. In a prescientific era—when there are no rational explanations for natural events—it is perhaps inevitable that humans will ascribe special powers to the things they do not understand.

Astrology is a superstition that has no basis in scientific fact. Its roots date back to the earliest astronomical observing. The 12 constellations around the ecliptic give us the star signs. There were originally 13, but Ophiuchus was dropped. Astrology has been with us for 4000 years, apart from a brief period in medieval times when it was banned by the Catholic Church as a form of magic (and a threat to the Catholic Church's power). In Babylonian times, personal prediction was made only for kings. By the time of the Romans, astrology extended forecasting to rich people. Since then, astrology has trickled down to everyone, and we now find it routinely in newspapers and magazines. The predictions of astrology are tied to the star sign constellation and planets that were in the same location in the sky as the Sun at the time of your birth. In addition, personality traits are supposed to correlate with star sign. Poll after poll shows that the belief in astrology is still widespread.

There is no scientific basis for astrology, and its statistical claims can be refuted (e.g., the book you are holding has more gravitational influence over you than the planets). Here are several examples of scientific studies that refute the claims of astrology. Physicist John McGervey looked at the birth dates and professions of 23,000 people and found no tendency for professions to cluster among the star signs as astrologers predict. Psychologist Bernard Silverman looked at the star signs of nearly 2500 married and divorcing couples. The "compatibility" as judged by star sign had no connection with which couples separated and which stayed together. Astronomers Culver and Ianna tracked the predictions of astrologers for celebrities over a 5-year period. Only 10% of the specific predictions came to pass—about what was achieved by the educated guessing of veteran reporters.

If astrology does not work, why is it so popular? Perhaps many people consider it a harmless diversion. Perhaps people are looking for guidance in life and are not too fussy about the source of that guidance. The Australian researcher Geoffrey Dean altered horoscopes so that the phrases were the opposite of the real ones, yet his subjects declared them accurate. French mathematician Michel Gauquelin sent the horoscope of one of the worst mass murderers in French history to 150 people—141 of them said that the description fit them!

Having seen how science works, we are now in a position to judge issues that claim to be based in science but in fact are not. The basis of the scientific method is hypothesis testing. A good hypothesis must predict aspects of nature; and if the hypothesis is wrong, it must be rejected or modified. The physicist Richard Feynman said, "We are trying to prove ourselves wrong as quickly as possible, because only in that way can we find progress." A scientific hypothesis must first be *testable*. That is, it must be subject to experiment, or it must be compared with repeated observations of the real world. Second, it must be *refutable*. That is, a hypothesis must make specific enough predictions that observations can be made that could prove it wrong.

Astrology is an example of a hypothesis that is treated as true (usually with commercial intent) but that has no consistent body of supporting evidence. This is often called *pseudoscience*. It may appear to be backed by the trappings of real science, such as quotations of evidence, but the evidence is often hearsay, the references are often to other poorly researched commercial books, and the work is rarely reviewed by professional researchers before publication. As in many areas of life, you should have a healthy skepticism when faced with the extraordinary claims of pseudoscience. Pseudoscience misrepresents real scientific discovery and contributes to anti-intellectual attitudes that exchange mysticism and magic for exploration and discovery. The exchange is a poor one because, as we intend to show, real discoveries about the universe are more exciting than the erroneous claims of pseudoscience.

The Birth of Science

Ancient cultures such as the Babylonians and the Egyptians could recognize patterns in the sky. They could predict astronomical events. They had accurate calendars. However, they made no progress in answering fundamental questions about their universe. How far away are the planets, the Sun, and the Moon? Why do they shine? What are their true sizes? Progress was made by a remarkable group of thinkers who lived in the 3rd to 6th centuries B.C. on what is now the coast of Greece and Turkey. This era marks the birth of science.

The breakthrough was made by a set of philosophers who developed the first scientific ideas. They applied logic and rigorous thought to many areas of human activity. Astronomy was just one example. They also developed new tools in mathematics to carry them forward. When Plato founded the world's first university in an olive grove outside Athens, he elevated mathematics to high intellectual status. The inscription above the main entrance read "Let Only Geometers Enter." The Greeks used geometry to move beyond the impression of the stars and planets as points of light on a fixed backdrop. They reached out to the third dimension and formed startling ideas about the true size of the universe.

The Greek philosophers also speculated about the microscopic world. Democritus imagined the process of dividing a grain of sand in half again and again. He supposed that this could not go on forever and that there must be a tiny indivisible unit of matter, much smaller than the eye could see. This is the idea of an atom. Empedocles imagined that nature was composed of only four primordial substances—earth, water, fire, and air. He proposed that all the materials in the world are made of combinations of these four substances. This is the idea of elements. We will learn more about the structure of matter in Chapter 4. The Greeks did not have the tools to understand the microscopic world of the atom; but in their applications of logic and their search for simplicity they came up with some strikingly modern ideas.

Early Greek Thinkers

In 584 B.C. two Greek tribes were engaged in a bloody battle on the coast of Asia Minor. The poet Hesiod recorded the scene with vivid descriptions of burnished shields and flashing swords and carnage. Suddenly, the sky darkened, and the air chilled with a total eclipse of the Sun. The soldiers wandered dazed and confused from the battlefield, believing they had witnessed an omen from the gods. Not far away, a man called Thales was not at all surprised. He had used Egyptian eclipse records to predict the exact date of the eclipse.

Thales was a statesman, geometer, and astronomer who lived in Miletus in what is now Turkey from 634 to 556 B.C. He is the subject of perhaps the first anecdote about absent-minded scientists. A story is told of a servant girl who saw him fall into a well and chastised him for being so preoccupied with the heavens that he failed to notice what was under his feet. Thales believed that everything in the material world derived from water. While this might seem naïve, it is an important step in the development of science to suppose that all things have a source and to suppose that the source of all things is one thing. This is a basic scientific idea—the diversity of the natural world conceals an underlying simplicity.

The ideas of Pythagoras are perfect examples of the union between mathematics and cosmology. He lived from 583 to 510 B.C. Pythagoras left no writings, so we know of his work only through the writing of his followers. By observing the phases of the Moon, he realized that the Moon is a sphere and proposed the then-unusual idea that Earth is a sphere, too. Pythagoras placed spherical Earth at the center of the universe. He developed many ideas in mathematics, among them the famous *Pythagorean theorem* of right-angled triangles. Since a sphere is the most symmetric and perfect shape, it was natural to Pythagoras that it would describe Earth and the orbits of the Sun, the planets, and the stars. Pythagoras established a school in southern Italy, and some of his students proposed the idea that Earth, the planets, the Sun, and the stars all move around some distant, central "fire." Even though some of these early ideas are incorrect, we can recognize the bold and adventurous thinking of these early scientists.

In the 5th century B.C., Anaxagoras deduced the true causes of eclipses. He noticed the curved shadow of Earth on the Moon during a lunar eclipse and realized that this observation supports the idea that Earth is round. A sphere is the only solid shape that casts a circular shadow regardless of the direction of illumination. He studied a meteorite that had fallen out of the sky in 467 B.C., which indicated a connection between the celestial and terrestial realms. He also calculated that the Sun is an incandescent "stone" even larger than Greece. This idea got him in trouble because people of the day viewed the Sun as a god. He was charged with impiety and banished for teaching heretical ideas.

Angles, Sizes, and Distances

Why did it take so long to figure out the sizes and distances of celestial objects? Part of the reason is psychological. Humans have tended to regard themselves as the pinnacle of creation and the center of the universe. So there is resistance to the notion that the sky may contain objects much larger than Earth. One of the themes of this book is the slow but steady realization that Earth is only a tiny part of a vast universe. The other reason is practical. As the Greeks knew, it is almost impossible to measure the linear size of a distant unfamiliar object directly by eye. We can really specify only its angular size.

Linear size and linear distance are difficult to judge. People sometimes report an unfamiliar object in the sky and say something like "It looked as big as a dinner plate," but this statement conveys almost no useful information. When a bright meteor or fireball is seen in the sky, astronomers often receive reports that "It looked close; it landed just over the hill." These reports are almost always wrong. Fireballs are typically in the upper atmosphere, 60 to 100 mi from the observers. To use angular measurement and say the fireball looked "$1/2°$ across, the same size as the Moon" would be correct; but to unconsciously convert to linear measurement and say "It was as big as an airplane" is likely to be wrong if the object itself is unfamiliar.

The ancient Greeks invented trigonometry, which allows quantitative relationships between angles, linear sizes, and linear distances. Through simple mathematical relations, we can calculate distances of remote objects whose sizes are known (or sizes, if the distances are known). We can see the application of these important ideas in the following Science Toolbox.

Phases of the Moon and Eclipses

Armed with the geometric tools of the Greeks, we can understand some of the simple cycles of the sky. Figure 2–11 shows the changing illumination of the Moon as seen in the Northern Hemisphere. The starting point is the idea that the Moon shines by the reflected light of the Sun. Just like Earth, the Moon is always half illuminated by the Sun. However, the fraction of that half-lighted surface that we see depends on the relative positions of Earth, the Sun, and the Moon. When the Moon is near the Sun in the sky, the lighted side faces the Sun and we see only the dark half. This is a *new Moon*. When the Moon is opposite the Sun in the sky, the sunlit side faces us and we see a *full moon*. About a week after new moon, the Moon stands about 90° away from the Sun in the sky. This phase is called *first quarter*. About a week after full moon, the Moon stands once again about 90° away from the Sun, and the phase is called *third quarter*. Remember that the phases refer to the portion of the Moon that we see; the Moon is always half-bathed in sunlight. Figure 2–11 shows the **phases of the Moon.**

Most people will be mystified if you ask them where to look for a full moon or a new moon. If you understand the geometry in Figure 2–12, the answer is obvious. A new moon must rise and set with the Sun. A full moon must rise

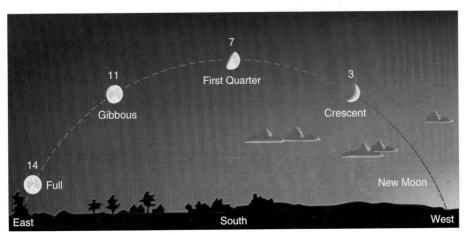

FIGURE 2–11
Wide-angle view of the sky, looking south, showing the changing illumination of the Moon as observed by someone at sunset in the Northern Hemisphere. The numbers indicate the days into the lunar cycle, starting with the day of new Moon.

SCIENCE TOOLBOX

The Small-Angle Equation

Angles and linear measures can be combined in an extremely useful and simple equation, called the *small-angle equation*. This equation involves the angular size of an object, its linear size, and its distance. If any two of these quantities are known, the third can be calculated. Let us refer to the angular size by the symbol α, expressed in seconds of arc. Let us refer to the diameter of the object as d and its distance as D. Then the small-angle equation is

$$\frac{\alpha}{206{,}265} = \frac{d}{D}$$

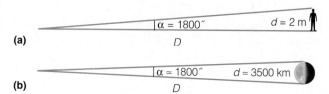

FIGURE 2–A
Relationships in the small-angle equation. (a) If your friend is 2 m tall and subtends an angle of ½°, or 1800″, his or her distance D is 230 m. (b) The same geometry can be used to measure interplanetary distances, such as the distance to the Moon. The ancient Greeks pioneered these applications. (Note that angles are not to scale; they really are much smaller.)

The number 206,265 is called a *constant of proportionality*; it stays the same in all applications of the equation. The number 206,265 is actually the number of seconds of arc in an angle of 57.3°, which is a special angle called a *radian*. A radian is defined as an angle subtending one radius of a circle, laid along the circumference of the circle. Since the circumference of a circle is $2\pi r$, a radian is $360° / (2\pi) = 57.3°$, or about one-sixth of a full circle. It is an important angle with many applications in geometry. For more background on angular measurement, see Appendix A-5.

Let us take an example. Suppose a friend who is 2 m tall is standing across a field from you, where he or she subtends an angle of ½°, or 1800″, as shown in Figure 2–A*a*. How far away is your friend? We want to solve the equation for D. Rearranging the equation, we have

$$D = \frac{206{,}265\, d}{\alpha}$$

Using metric units, we have

$$D = \frac{2.1 \times 10^5 \times 2}{1.8 \times 10^3} = 2.3 \times 10^2 \text{ m}$$
$$= 230 \text{ m}$$

If your friend is 2 m tall and subtends an angle of ½° (or 1800″), distance D is 230 m. In other words, your friend is about ⅙ mi away. Notice we are rounding all our estimates to two significant figures because the angle measurement is not likely to be very accurate.

As the Greeks realized, the small-angle equation can be used to investigate astronomical distances. They could not measure the Moon's diameter accurately, but they knew its angular size α, which is also roughly ½°, or 1800″. If we use the modern knowledge that the Moon is about 3500 km in diameter, we can estimate its distance just as we did for the friend's distance above (look at Figure 2–A*b*). In metric units, d would be 3.5×10^6 m. The equation would read

$$D = \frac{2.1 \times 10^5 \times 3.5 \times 10^6}{1.8 \times 10^3}$$
$$\approx 4 \times 10^8 \text{ m} \approx 4 \times 10^5 \text{ km}$$

This is about 400,000 km. Notice once again the symbol $\approx$, meaning "approximately equal to." It is useful whenever approximate values such as 1° are involved. In other words, the measurement of angular size is approximate, so the resulting estimate of distance must also be approximate.

To take another example, how big are the smallest craters that we can see on the Moon with a backyard telescope? To solve this, we start with the information that a good backyard telescope can resolve or make out angular detail as small as 1 second of arc, or 1″. The Moon is 384,000 km away. So we are asking how big an object is that subtends 1″ at that distance. Rearranging the small-angle equation again gives

$$d = \frac{\alpha D}{206{,}265}$$

Plugging in the Moon's distance and the angle of 1″, we have

$$d = \frac{1 \times 3.8 \times 10^5}{2.1 \times 10^5} = 1.8 \text{ km}$$

We can see features as small as 1 mi across on the Moon with a telescope. Recall from the text that the eye can only make out angular scales of about 3′. So when you stare at the Moon, the smallest features you can see are $3 \times 60 \times 1.8 \approx 320$ km or about 200 mi across.

Finally, if the Sun has a diameter of about 1.4 million km and is 150 million km away (also called one *astronomical unit*, or 1 AU; see Appendix A-2), what angle does the Sun subtend in the sky? Solving the small-angle equation for α gives

$$\alpha = \frac{d}{D} \times 206{,}265$$

Substituting the values for diameter d and distance D, we get

$$\alpha = \frac{1.4 \times 10^6}{1.5 \times 10^8}\left(2.1 \times 10^5\right) = 1960''$$

Since there are 3600″ in 1°, this angle is $1960 / 3600 = 0.54°$. The Sun and the Moon differ greatly in size and in distance from Earth. Due to the coincidence that they both subtend the same angle of ½°, the spectacular sight of an eclipse is possible.

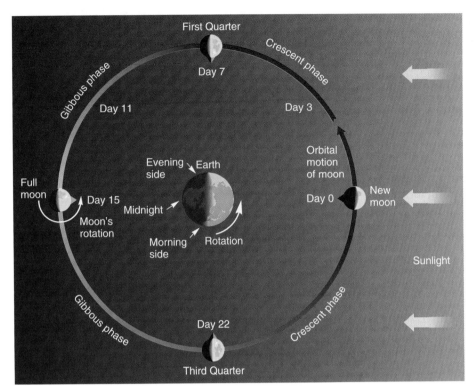

FIGURE 2-12
The Moon's motions around Earth. The Moon always keeps the same face pointed at Earth, as indicated by the small triangle. The Moon completes one rotation during one complete revolution around Earth. A full cycle of phases takes 29.5 days. The phases seen by an earthbound observer are indicated at different points in the orbit. First quarter is seen in the evening sky; third or last quarter is seen in the early morning sky.

around sunset and set around sunrise. A first quarter moon rises around midday and is high in the sky around sunset, and a third (or last) quarter moon rises around midnight and is high in the sky around sunrise. We know that the Moon takes about 29.5 days to go through a cycle of phases, so it must move 360 / 29.5 or about 12° across the sky from one day to the next. It rises and sets 24 / 29.5 h or about 50 min earlier each day. These patterns were well known to all ancient people. (You can relate the real photographs in Figure 2–13 to the schematic view in Figure 2–12.)

You can notice something interesting about our explanation for the phases of the Moon—it makes no assumption about whether Earth or the Sun is stationary at the center of the universe. This explanation requires only that the Sun and

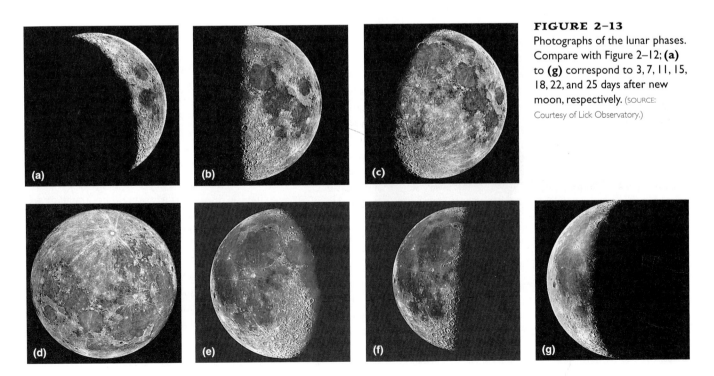

FIGURE 2-13
Photographs of the lunar phases. Compare with Figure 2–12; **(a)** to **(g)** correspond to 3, 7, 11, 15, 18, 22, and 25 days after new moon, respectively. (SOURCE: Courtesy of Lick Observatory.)

THE BIRTH OF SCIENCE

the Moon travel on a similar path through the sky (the ecliptic) and that the Moon periodically pass between the Sun and the Earth. A model in which the Sun and the Moon travel in concentric circles around Earth can do this. In other words, the phases of the Moon provide no evidence for or against the idea of an Earth-centered universe.

For thousands of years people have observed another phenomenon that is rarer and apparently less predictable than the phases of the Moon or the annual return of constellations and seasons. Every generation or so they saw the Sun blotted out (or largely blotted out) by a **solar eclipse,** and every year or so they saw the Moon blotted out (or partially blotted out) by a **lunar eclipse.** The sudden disappearance of the Sun and darkening of the sky could be a terrifying spectacle (Figure 2–14). We have heard about Greek warriors who were terrified by a solar eclipse, and Christopher Columbus used prior knowledge of an eclipse in 1504 to gain control over the inhabitants of the island of Hispaniola. Mark Twain borrowed this idea in his book *A Connecticut Yankee in King Arthur's Court.* A lunar eclipse is less spectacular but is also a beautiful sight (Figure 2–15).

Eclipses occur due to an amazing coincidence: The Sun and the Moon subtend the same $1/2°$ angle in the sky! As we saw in the last Science Toolbox, angular size alone is no indicator of true physical size. Two totally different celestial objects just happen to have the relative distance that gives them the same angular size. As shown in Figure 2–16, we know today that a solar eclipse is caused when the Moon passes between Earth and the Sun, so that the Moon's shadow falls on Earth. A lunar eclipse is caused when Earth passes between the Moon and the Sun, so that Earth's shadow falls on the Moon.

FIGURE 2–15
A nearly total lunar eclipse. Sunset-colored reddish light scatters and refracts through Earth's atmosphere, coloring the Moon so that it is not in total shadow. The bright crescent on the upper left is part of the Moon still lit by the Sun's rays. An astronaut on the Moon would see the extraordinary sight of this red light coming from a sunset that encircled Earth. (SOURCE: Celestron International.)

We can now answer some basic questions about eclipses. Use Figure 2–16 to help you. Why does a lunar eclipse happen only at full moon? As the figure shows, the Moon must be opposite the Sun, as seen from Earth, and that is when it appears fully illuminated. Why does a solar eclipse happen only at new Moon? Because the Moon must pass directly in front of the Sun and its illuminated half faces toward the Sun. Why are lunar eclipses more common than solar eclipses? The Greeks had observed Earth's round shadow during a lunar eclipse. They also observed that Earth's shadow was much bigger than the Moon. They correctly inferred that the Earth itself is bigger than the Moon. Therefore we are more likely to see the Moon pass through Earth's large shadow than to see Earth pass through the Moon's small shadow.

A full explanation of eclipses is more subtle. For example, why do we not see a lunar and a solar eclipse every month? The answer is that the Sun and the Moon do not travel on exactly the same apparent path through the sky. The Moon's path is tilted by about 5° from the ecliptic, which means that eclipses can only occur during the two times each year when the line in the Moon's orbital plane that intersects the ecliptic points at the Sun (Figure 2–17). You can see from this figure that if the Moon traveled entirely in the ecliptic, solar and lunar eclipses would occur every month. The Moon's shadow is small and tapers nearly to a point, so that only a small area on Earth, kilometers wide, is fully darkened by a total eclipse of the Sun. Thus, while a lunar eclipse can be

FIGURE 2–14
The March 1970 total solar eclipse as seen from the village of Atatlan, Mexico. The sky has darkened dramatically, giving a general appearance of dusk. (SOURCE: Photograph by WKH.)

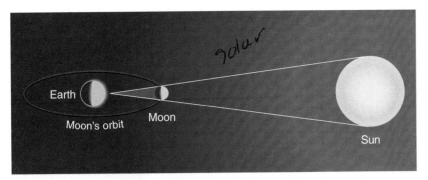

(a)

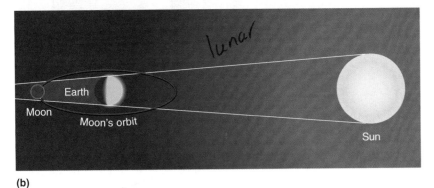

(b)

FIGURE 2-16
Causes of solar and lunar eclipses. The geometry is shown schematically, not to scale. (a) A solar eclipse happens when the Moon's shadow falls on Earth; people in the shadow's core see the Sun completely blotted out. People just off the shadow's core see a partial solar eclipse. A lunar eclipse happens when the Moon passes wholly or partially through Earth's shadow.

seen by inhabitants all over the night side of Earth, a total solar eclipse is seen only by a tiny fraction of people on the daytime side of Earth. Slight changes in the Earth–Sun distance mean that an eclipse is not always complete; sometimes a ring of light passes around the edge of the Moon in an *annular* eclipse (Figure 2–18). A solar eclipse only lasts for a few minutes as the Moon's motion sweeps the shadow across the surface at close to 2400 mi/h.

The Legacy of Aristotle

We have seen that the earliest Greek thinkers developed the tools of geometry, allowing them to distinguish between apparent size and true size. Let us continue our narrative and see how these tools were used to determine Earth's place in the universe. Aristotle (384–322 B.C.) was the most famous and influential Greek philosopher. He founded a school at Lyceum, near Athens, with a library, zoo, and lavish research equipment bought by his one-time pupil Alexander the Great, who ruled Greece and conquered much of the Mediterranean world. Aristotle applied his prodigious brain to many subjects. He developed the rules of logic that are the basis of the scientific method. He wrote books on botany, anatomy, economics, politics, and meteorology. He measured star positions from his own observatory. He is also responsible for a cosmological model that lasted for 2000 years, even though it proved to be wrong!

Aristotle and his colleagues made few new observations. In fact, they were painfully aware of the limitations of the human senses. They believed that the universe could be understood by the power of reason alone. This is truly audacious. We can see how far it is from the old idea that humans are subject to capricious and supernatural forces. The goal of these philosophers was to unify the diversity of the natural world with a few elegant ideas.

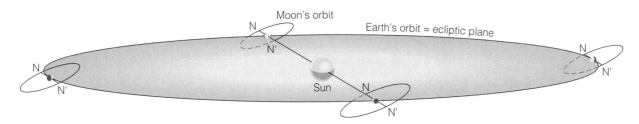

FIGURE 2-17
The plane of the Moon's orbit is tilted by 5° with respect to the ecliptic. The Moon only passes through the ecliptic at points *N* and *N'* in its orbit. As Earth moves around the Sun, there are only two periods each year when the line joining *N* and *N'* also aligns with the Sun so that eclipses occur.

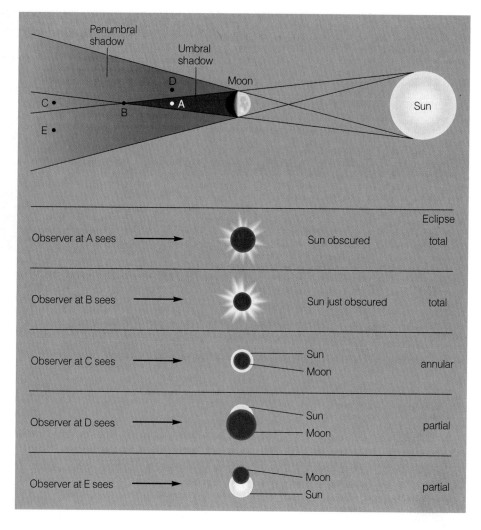

FIGURE 2-18
Geometry of different types of solar eclipse (not shown to true scale), caused by the Moon's shadow falling on Earth. Umbral shadow is totally dark; penumbral shadow is partially dark. Observers at different positions see different kinds of eclipses, and Earth's rotation causes an observer to sweep in and out of the Moon's shadow. Other differences are caused by slight variations in the distances between Earth, the Moon, and the Sun. By chance, the tip of the umbra (point B) lies very close to Earth's surface.

Aristotle's cosmology had several essential features. Earth was a sphere. Aristotle followed Pythagoras in believing that a sphere was the most perfect shape. He was also aware of the powerful evidence provided by the shape of Earth's shadow during a lunar eclipse. Earth was stationary. To Aristotle this was just common sense, since we do not feel any motion of Earth and objects fall straight down when dropped. We call this a **geocentric cosmology,** or Earth-centered cosmology, where all the other celestial bodies travel around the Earth in circular orbits. Aristotle borrowed the idea of crystalline spheres from Eudoxus. The Sun, the Moon, and each of the planets have a crystalline sphere, nested like a set of Russian dolls. The outermost sphere carried all the stars.

An alternative view came from Aristarchus (310–250 B.C.), who lived on the island of Samos off the coast of present-day Turkey. Living in the time just after Aristotle, he boldly proposed that the Earth and the planets orbited the Sun. This is a **heliocentric cosmology.** Few of his writings survive, but we have descriptions from other Greek authors. The following Science Toolbox describes the logic underlying his argument. Think of the irony that Aristarchus was measuring the relative distances of a spherical Earth, Moon, and Sun, yet 1000 years later, many Dark Age citizens of Europe still thought that Earth was flat!

This raises an important point. Why, in the history of science, do the correct ideas not always prevail? Usually it has to do with a lack of compelling evidence. Aristarchus' followers could not *prove* that his hypothesis of an orbiting Earth was correct. Aristotle's followers could not *prove* that Earth stood still. Aristotle argued that if Earth were really rushing though space, we should be able to detect its motion. This was considered a strong argument. An equally strong argument was the fact that heavenly bodies all rise in the east and set in the west—that is, they seem to go around the Earth.

We should also recognize a powerful psychological reason for favoring a geocentric cosmology. Humans are intelligent enough to consider our place in the universe. The Greek philosophers were convinced that humans were the pinnacle of creation and therefore must be at the center of the universe. It was unthinkable for them to relegate Earth to just another object flying through space. There is natural resistance to displacing Earth in its importance in the scheme of things. The discovery of Earth's true place in a vast universe is a story that will unfold throughout this book.

Earth and the Seasons

Another dramatic application of geometric reasoning was the advance made by Eratosthenes around 200 B.C. He was

SCIENCE TOOLBOX

Aristarchus and a Sun-Centered Cosmology

Let us see how Aristarchus used simple geometric ideas to deduce that Earth is larger than the Moon and that the Sun is larger than Earth. They are all based on the Greek understanding that we see the Moon in reflected sunlight. By observing the lighting and geometric relations between bodies in the sky, we can deduce their positions in three-dimensional space. This led Aristarchus to the idea of a Sun-centered, or heliocentric, cosmology. There are three steps to this conclusion.

If the Moon is a sphere illuminated by the Sun, then we can think about how that illumination should look to us for different arrangements of Earth, Sun, and Moon. Consider the situation on a day when the Moon is located 90° from the Sun; for example, when the Sun is setting and the Moon is just crossing the meridian. In Figure 2–B(a), you can see that if the Moon were much farther from Earth than the Sun is, then the Moon would look nearly fully illuminated. But in Figure 2–B(b), you can see that if the Moon were much closer to Earth than the Sun is, then the Moon would look half illuminated, which is true.

Now imagine a lunar eclipse in progress. Figure 2–C(a) shows the geometry. Figure 2–C(b) shows an actual

(continued)

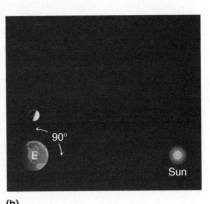

(a)   (b)

FIGURE 2–B
Example of Aristarchus' geometric methods for proving the Moon is closer than the Sun. **(a)** If the Moon is farther than the Sun, then the Moon will appear to be nearly fully illuminated when the Moon is 90° from the Sun. **(b)** If the Moon is closer than the Sun, then the Moon will be more nearly half-lighted when the Moon is 90° away from the Sun, as is actually observed. Therefore the Sun is considerably farther away than the Moon.

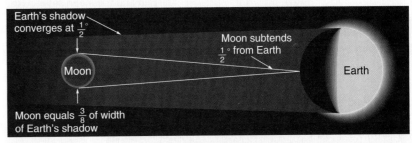

(a)

FIGURE 2–C
(a) The geometry of a lunar eclipse, not drawn to true scale. Earth casts a shadow due to the Sun's illumination off to the right. If the Moon fits within Earth's shadow, it must be substantially smaller than Earth since the Sun's rays reach Earth nearly parallel. This fact plus the small-angle equation can be used to deduce the relative sizes of Earth and the Moon. **(b)** Triple exposure taken before, during, and after the total lunar eclipse of August 16, 1989. A 4-in. refractor was used, with 30-s exposures. You can see the curved shadow and relate the geometry to the diagram above. (SOURCE: Photograph by Jim Rouse.)

(b)

SCIENCE TOOLBOX

Aristarchus and a Sun–Centered Cosmology
(continued)

time-lapse photograph. Since the Sun is so far away from Earth, it casts a shadow with nearly parallel edges (they actually converge by $1/2°$ since that is the angular size of the Sun in the sky). Aristarchus knew that the Moon looked about one-third as big as Earth's shadow and concluded that the Moon was one-third as big as Earth. The actual fraction is one-quarter; it is difficult to measure it accurately by eye. We also know that the Moon subtends an angle of $1/2°$ (1800″) as seen from Earth. Using the small-angle equation from the last Science Toolbox gives

$$D = \frac{206{,}265 d}{\alpha} = \frac{2.1 \times 10^5 \times d}{1.8 \times 10^3} = 120 d$$

So the distance to the Moon is more than 100 times larger than the Moon's diameter. It is also $120(1/3) = 40$ times farther away than Earth's diameter.

Aristarchus then looked at the relationship between the first- and last-quarter phases of the Moon. Look at Figure 2–Da. You can see that if the Sun were enormously far away compared to the Moon, then the Sun's rays would come in parallel and there would be the same amount of time between a first- and last-quarter Moon as between a last- and first-quarter Moon. Now look at Figure 2–Db. You can see that if the Sun is somewhat closer, then the angle between the Sun and Moon at first and third quarters is slightly less than 90°. Aristarchus deduced from this triangle that the Sun is 20 times farther from Earth than the Moon is. This is a difficult timing measurement with the simple timekeeping devices that were available to Aristarchus; the true ratio is a factor of 390. Nevertheless the logic is sound.

The Sun and the Moon subtend the same angle in the sky. The small-angle equation tells us that if the Sun is 20 times farther away than the Moon, then it is also 20 times larger than the Moon. We can combine the two previous results:

$$\frac{D_{sun}}{D_{Earth}} = \frac{D_{sun}}{D_{moon}} \times \frac{D_{moon}}{D_{Earth}} = 20(1/3) = 7$$

So the Sun is larger than Earth (the true ratio is larger than 7 because Aristarchus' factor of 20 is not the true value of 390). Aristarchus went further, surmising that it was more reasonable for the smaller body—Earth—to go around the larger body—the Sun—than the other way round. He likened the situation to that of a hammer thrower. It is reasonable for someone to swing a small object, but it is impossible for someone to swing an object much larger than himself or herself! Aristarchus correctly put the Sun in the middle of his system with the smaller, spherical Earth going around it.

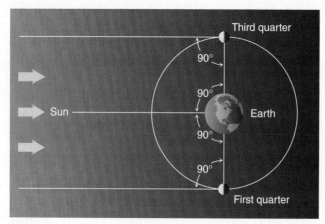
(a)

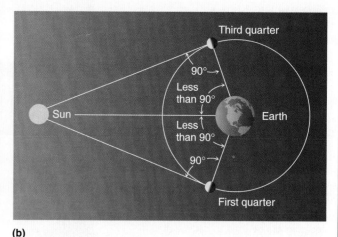
(b)

FIGURE 2–D
Geometric argument for figuring out how much farther away the Sun is than the Moon. **(a)** If the Sun is infinitely far away, the Sun's rays come in parallel and the time between third- and first-quarter Moon equals the time between first and third quarters. **(b)** If the Sun is a large but measurable factor farther away than the Moon, the time between third and first quarters is less than the time between first and third quarters and the angle between the Sun and Moon at first and third quarters is less than 90°.

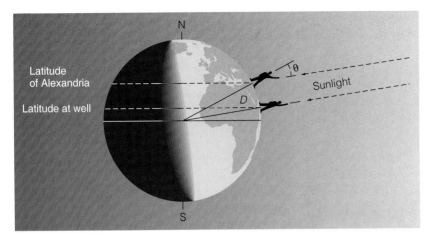

FIGURE 2-19
Eratosthenes' method for deducing the size of Earth. On a day when the Sun was directly overhead at one point, Eratosthenes measured the angle (given by the Greek letter θ) from the zenith at another point, a known distance D away. Since the angle was one-fiftieth of a circle, he knew that D was one-fiftieth of Earth's circumference and used that relationship to estimate Earth's diameter.

a researcher and librarian at the great library in Alexandria, Egypt, where books containing much of the knowledge of the ancient world were stored. He completed a catalog of the 675 brightest stars, but his most famous achievement was to use geometry to measure the size of planet Earth. Told that the Sun at summer solstice shone straight down a well to the south, near Aswan, he noted that the Sun's direction was off vertical by about 6° or one-fiftieth of a circle on the same date in Alexandria. These angles could also be measured from the lengths of shadows at the two locations. As shown in Figure 2–19, Eratosthenes realized that this difference was due to the curvature of Earth, and that the north-south distance from Alexandria to Aswan must be one-fiftieth of the circumference of Earth. Measuring the distance and multiplying by 50, he got an estimate of Earth's size that was within 20 percent of the correct answer. Figure 2–20 shows the same effect in modern photographs at different latitudes. At a time when few people traveled more than 50 mi in their lifetimes and the first circumnavigation of Earth was more than 1000 years away, most educated Greeks knew the size and shape of the planet they lived on!

Eratosthenes also wondered about the change in the apparent position of the Sun from season to season. How does the changing path of the Sun relate to the seasons? How does the path of the Sun vary when seen from different places on Earth's surface? Eratosthenes measured the difference between the summer and winter noontime elevation of the Sun (its angle above the horizon in degrees). He used this information to deduce the fact that Earth's equator is tilted by 23.5° with respect to the direction of the Sun at summer and winter solstice. A geocentric cosmology can account for this variation just as well as a heliocentric model.

Study Figure 2–21, which shows the **cause of the seasons**. Earth orbits the Sun with its equator tilted by 23.5° with respect to the plane of the Earth–Sun orbit (the path we see as the ecliptic in the sky). Earth moves around the Sun and always keeps its north polar axis tipped and pointed at the North Star. You can see that on one side of the Sun, the tilt of Earth's Northern Hemisphere is toward the Sun. Someone in the Northern Hemisphere will see the Sun travel high in the sky and the sunlight will shine nearly straight down at midday. The result is a strong heating effect. Look at Figure 2–22. This is summer in the north, while the Southern Hemisphere will have winter. Six months later, on the other side of the Sun, the tilt of the Northern Hemisphere is away from the Sun. A northern observer will see the Sun travel low in the south at midday. This is winter in the north, while the Southern Hemisphere has more direct sunlight and is in summer.

You can see a second consequence of Earth's tilted axis in Figure 2–21. When the Northern Hemisphere is tipped toward the Sun, northern latitudes receive more than 12 h of sunlight each day and southern latitudes receive less than 12 h. When the Northern Hemisphere is tipped away from the Sun, northern latitudes receive less than 12 h sunlight each day and

(a)   (b)

FIGURE 2-20
Two midday photographs taken at different latitudes on the same day demonstrate Eratosthenes' method for proving Earth is round and measuring its size. **(a)** Noontime view from Anchorage, Alaska on Feb. 20. **(b)** Noontime view from Tucson, Arizona on Feb. 20, taken with the same camera and lens. The vertical height of the photographs covers an angle of about 62°. The difference in angular elevation of the Sun between the two photographs equals the angular difference in latitude between the two cities. (SOURCE: Photographs by WKH.)

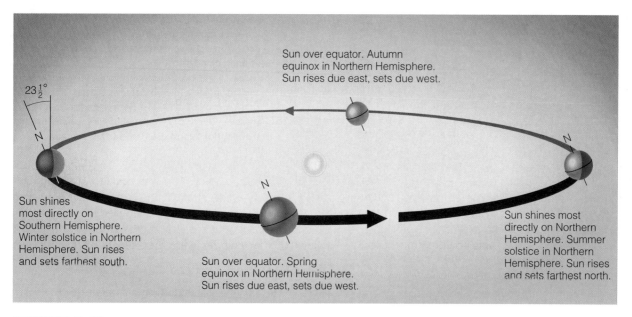

FIGURE 2-21
The cause of the seasons. Earth's axis is tipped at 23.5° to the plane of its orbit, always pointing in a constant direction (toward the North Star) as Earth moves around the Sun. Therefore the Sun shines more on the Northern Hemisphere during one season (northern summer) and more on the Southern Hemisphere 6 months later (northern winter). Note that this orbit looks elongated in this figure because it is a projection—Earth would come in and out of the plane of the page in true three dimensions. The seasons are *not* caused by changes in Earth's distance from the Sun!

southern latitudes receive more than 12 h. Spring equinox and autumn equinox are times of equal day and night everywhere. Regions between 23.5°N and 23.5°S latitude can view the Sun directly overhead. These are the tropics. You can also see from the figure that these are parts of the Earth's orbit when regions north of 67.5°N and south of 67.5°S latitudes are in continuous darkness or continuous light. These are polar regions—the "land of the midnight sun."

Many people believe that the seasons occur because Earth is closer to the Sun in summer. Not true. If it were true, summer would occur at the same time in the Northern and Southern Hemispheres. Earth stays at nearly the same distance from the Sun (in fact, Earth is slightly closer to the Sun during winter in the Northern Hemisphere—proof that our seasons are not caused by the change in solar distance). The seasons are caused by the tilt of the planet's axis; and if the axis were not tilted, there would be no seasons. You can see this by imagining Earth with no tilt in Figure 2-21. Every part of Earth would get 12 h of day and 12 h of night all year round, and the Sun would reach the same elevation in the sky all year round. The tilt certainly makes our lives and our climate more interesting!

FIGURE 2-22
The strong heating effect caused by angle of illumination. **(a)** The light from a flashlight beam pointing straight down is not spread out very much, like the situation at northern latitudes around the time of summer solstice. **(b)** With a slanting angle of the flashlight beam, the same amount of light is spread over a much larger area, causing less heating effect at the ground. This is typical of the situation at northern latitudes near winter solstice. Look at Figure 2-21 and see if you can understand why the difference between the Sun's noontime elevation in winter and summer is 47°— twice the tilt angle.

(a) (b)

Astronomy in Other Cultures

After the 2nd century B.C. the Greek civilization went into decline and was eclipsed by the Roman Empire. Claudius Ptolemy was another scholar at the Alexandrian library (Figure 2–23). Around the year 140 he wrote a 130-volume encyclopedia that synthesized the teachings of Greek scholars. In this work, Ptolemy extended the existing star catalogs to 1022 stars. He described a geocentric model of the solar system, with the Sun, Moon, and planets all moving around Earth in circular paths. Ptolemy's method of tracking planets, using these assumptions, gave fairly good predictions of the planets' positions. This model was accepted for a thousand years. As we will see in the next chapter, improving and correcting Ptolemy's system was a dangerous enterprise that cost some people their lives during the Renaissance.

Astronomy and science in general did not advance substantially under the Romans, who were more interested in such practical matters as agriculture, engineering, and governance than they were in studies of nature and the universe. With the fall of Rome in 410, maintaining the repository of knowledge in Alexandria became more difficult. Remember that this was centuries before printing; many books existed in only a few handwritten copies. Among the last guardians of the library was the first known woman astronomer, Hypatia. Widely admired for her learning and eloquence, she corresponded with leading scholars, wrote a commentary on Ptolemy's work, and invented navigation devices. But during riots that plagued Alexandria's decline, she was murdered by a mob. Over the next century, the library buildings were burned several times, and the best collection of Greek books was lost or scattered.

This marked a sad point in history. Much general knowledge was lost. The scientific way of looking at the world, gained among the ancient Mediterranean cultures, languished and was forgotten. Only through circuitous and painful routes was some of this knowledge reintroduced centuries later into Europe. During the long Dark Ages that followed, other cultures were the keepers of the flame of astronomy.

Ancient Arab Science: Carrying the Flame

Arab scholars kept alive some of the Greek knowledge and preserved a few copies of Greek books. A century after the time of the Muslim prophet Muhammad, around 760, Islamic leaders in the new capital of Baghdad began to sponsor translation and distribution of surviving Greek texts. This started an Islamic flowering of science. Ptolemy's work came down to us in an Arabic translation called *al-Magiste*, or "The Greatest." Islamic scholars refined the measurement of Earth's circumference in 820, coming within 4 percent of the actual circumference. Arab naturalists also advanced the field of optics. They recognized that the eye is an optical device and studied the ways in which light was reflected. No new cosmological models were developed because the Koran forbids pictorial representations of the heavens.

We still use Arabic names for many mathematical and chemical terms and for many stars, as indicated by the Arab prefix, *al-*. Examples of such terms are algebra, algorithm, alkali, alcohol, and the stars Algol, Alcor, and many others. By 1000, the Islamic empire had spread to Spain, carrying the old Greek knowledge and the new Arab discoveries into Europe. Spanish astronomers published astronomical tables using a longitude system with 0° in Cordoba (rather than Greenwich, England, as in the modern longitude system, introduced when Britannia ruled the waves).

Astronomy in India: A Hidden Influence

India was another center of ancient knowledge. Astronomical practices in India date back to about 1500 B.C. The first known astronomy text in India, which appeared in about 600 B.C., described planetary motions and eclipses. By this time India had contact with the Mesopotamian and Greek worlds, and influences probably traveled both ways. For example, Indian texts dating from around 450 used Greek computational methods and referred to the longitudes of both Alexandria and Benares, a major Indian astronomical center. As we have mentioned earlier, Indian mathematicians invented the number system that uses zero, which is the basis of our decimal system.

FIGURE 2–23
Ptolemy summarized the Greek knowledge of astronomy, with emphasis on the geocentric model, in a 130-volume encyclopedia called the *Almagest*. (SOURCE: The Granger Collection.)

Unfortunately, invaders in the 12th century destroyed most of the records of this fertile early period of Indian science. When we judge the scientific contributions of other cultures, we must remember that the historical record is often obscured by conquest (climate can be important, too; many artifacts from Mesopotamia and Egypt survived because the Middle East is so dry). The great astronomical observatory at Benares was destroyed in 1194, and various university libraries of Buddhist and other ancient literature were burned in religious wars. A massive observatory—one of the world's five major observatories by the 1700s—was reestablished at Benares in later centuries (and then damaged yet again by a new wave of invaders).

Astronomy in China: An Independent Worldview

According to legend, Chinese astronomers were predicting eclipses before 2000 B.C.; scholars estimate the time as being closer to 1000 B.C. In any case, Chinese astronomy may have flourished somewhat earlier than Greek astronomy. Ancient Chinese observations include records of Halley's comet and fireballs and comprehensive lists of the "guest stars"—exploding stars that we will encounter in a later chapter. Thanks to an unbroken span of court astronomers dating back 2000 years, these records are extremely useful and are still consulted by modern astronomers.

Chinese astronomers were willing to accept that Earth might be in motion without that motion's being apparent. One court astronomer is quoted as saying, "The Earth is constantly in motion, never stopping, but men do not know it; they are like people sitting in a huge boat with the windows closed; the boat moves but those inside feel nothing." Contrast this with Aristotle's view during the same era, which put a stationary Earth at the universe's center. The perceptive views of Chinese astronomers might have been the key to realizing that the Earth was actually in motion. Unfortunately, the Chinese ideas had little influence outside of Asia until after the Renaissance, since China traditionally avoided contact with the Western world.

Native American Astronomy: Science Cut Short

The highest levels of astronomical knowledge were reached in Mesoamerica around the year 400 by the Mayans. They developed a written language based on glyphs or pictures, used a complex and accurate calendar, recorded positions of planets, and predicted eclipses. Mayan astronomy was well organized and state-supported. Rulers used astronomical knowledge to decide when to wage war, perform sacred rituals, marry, or ascend the throne. This was a form of astrology. One inscription records a conference of astronomer-priests at Copan, Honduras (probably on May 12, 485), to discuss the calendar system. The Mayans passed on much of their astronomical observations and other knowledge to later cultures. Much of the Mayan knowledge was lost, however, after Spanish priests burned most of the Mayan manuscripts in 1562 because they believed them to be sacrilegious. One of three priceless Mayan manuscripts that still survive is a record of solar eclipses, the motions of Venus, and other astronomical data. Earlier in this chapter we discussed some of the major observatories of Mesoamerica.

The Mayans devised a unique calendar, which was still in use when the Europeans arrived (and was more accurate than the calendar the conquerors brought with them). The

(a)

(b)

FIGURE 2-24
At the latitude of the Mayan site of Edzna, the Sun passes overhead on July 26. Ancient Mayans called this their New Year's Day. **(a)** In the courtyard of this temple, they erected a stone device to measure the event. **(b)** Close-up view of the device. When the Sun is at the zenith at noon on July 26, the shadow of the top piece covers the shaft. On other dates, as in this photograph, part of the shaft is in sunlight. (SOURCE: Photographs courtesy V. H. Malstrom, Dartmouth College.)

Mayans celebrated the beginning of the new year on July 26! What could have led them to this choice? The Sun can pass directly overhead at noon in the tropics, but not in Europe. Perhaps because dense jungle obscured the horizons, zenith observations were particularly important to the Mayans. Near the latitude of a major Mayan observatory, Edzna, in the Yucatan peninsula, the Sun passes through the zenith at noon on July 26. Recent archaeological studies show that Edzna was a major city of some 20,000 people in the first few centuries A.D. In the courtyard in front of the main five-story pyramid, a cleverly designed stone pedestal (Figure 2–24) allowed priests to measure the important "New Year's Day" when the Sun passed through the zenith. Probably it was in or near this prehistoric city that early Mayan astronomers first selected, and then commemorated, July 26 as their New Year's Day.

Mayan astronomy is perhaps the most fascinating example of incipient Native American science. Since it survived until historic times and produced written records of complex planetary observations, astronomical conferences, eclipse predictions, and calendars, it is frustrating and sad that so much of it was lost. We may never know how the Mayan priests visualized the cosmos.

Summary

This chapter helps us understand the question raised in the opening story: Why would ancient peoples go to the trouble to build astronomical monuments such as Stonehenge? They were aware of cycles in the sky. The Sun, the Moon, and the stars all moved in predictable ways. Cultures all over the world were able to use these patterns to regulate their lives. For agricultural purposes, they needed to establish calendars, and they used the solstices and equinoxes as seasonal indicators. For navigating across land or sea, they needed to map the star patterns, or constellations. As early as 2600 B.C., astronomical observations led to the construction of astronomically aligned temples in many parts of the world. These were not observatories in the modern sense, but buildings created around a rich brew of astronomy, religion, and mythology.

In the prescientific era, people put astronomy to practical use without any real understanding of the celestial objects. They used stars to navigate on ocean voyages of thousands of miles. They created calendars of great accuracy. In its simplest form, astronomy yields only the angles of objects on the sky and gives no idea of their size or distance. These early cultures thought of the sky as a supernatural realm. This explains the widespread belief in astrology—a superstition that holds that astronomical events can influence human affairs.

We witness the birth of science in Greece around 500 B.C. to 150 A.D. The Greek philosophers made two extraordinary contributions: They applied logic and reason to understanding the universe, and they applied the principles of geometry to measure the true sizes and distances of objects in space. The Greeks established that Earth and the Moon were spherical worlds in space, and they understood the illumination that leads to the phases of the Moon and eclipses. They used the geometry of lunar and solar eclipses and other observations to estimate the shapes and relative distances of the Moon and Sun. They discovered how to make a direct measurement of Earth's circumference. Aristotle developed a model where spherical, stationary Earth stood at the center of the universe: a geocentric cosmology. The bold idea of a Sun-centered, or heliocentric, cosmology was also created at this time, but there was not enough evidence to support it and the geocentric picture was chosen by Ptolemy in his summary of Greek astronomy. After the fall of Greek civilization, astronomy made few advances in Europe for over 1000 years.

Other cultures sought to understand the nature of the universe by observing the sky. Arab astronomers revitalized Greek astronomy; they also made advances in optics and passed on the names we use for many of the bright stars in the sky. Astronomers in India built large observatories, and Indian mathematicians gave us the decimal system that is the basis of our counting and our metric system of measurement. Astronomers in China provided an almost unbroken record of 2000 years of celestial events. Astronomical monuments were built throughout Mesoamerica, and many of them show a sophisticated awareness of the cycles of the sky. Each culture developed astronomy with a unique flavor as the people tried to understand their place in the universe.

Important Concepts

You should be able to define these concepts and use them in a sentence.

zenith	altitude
meridian	azimuth
constellations	metric system
ecliptic	astrology
zodiac	phases of the Moon
calendar	solar eclipse
archaeoastronomy	lunar eclipse
precession	geocentric cosmology
decimal system	heliocentric cosmology
angular measurement	cause of the seasons

How Do We Know?

These questions and answers show how the scientific method is used to learn about the universe.

Q How did ancient people prove that the world is round?
A Early Greek naturalists noted that Earth's shadow on the Moon during a lunar eclipse is round. The only solid shape that always casts a circular shadow is a sphere. Early seafarers had also noticed that a tall ship would start to disappear over the horizon when it was far away (although this observation can be interpreted as the bending of light in our atmosphere). They would also have seen that different stars became visible as they traveled farther south.

Eratosthenes conducted an experiment in which he compared the shadows cast by the Sun at the same time at different latitudes. By comparing the angles, not only did he show that the world is round, but also he was able to calculate the circumference of Earth with surprising accuracy.

Q How do we know that some prehistoric monuments had an astronomical purpose?

A It is difficult to be absolutely certain that any prehistoric monument was used for astronomy. After all, these ancient cultures left nothing in writing, so we cannot be sure of their intentions. However, in the case of Stonehenge and a few other monuments, the evidence is very good. When major elements of a structure align with important astronomical directions—sunrise on the longest day of the year or the largest excursions of Venus from the direction of the Sun, for example—we can be confident that the monument had an astronomical function. However, we can make mistakes when we push this kind of thinking too far. If we measure a large number of alignments for a monument and also consider a large number of astronomical orientations, then some matches will occur purely by chance. Also, we must be sure that the monument was not altered by later cultures. Archeoastronomy is a complex discipline. Claims of "ancient observatories" require a lot of information about the culture and traditions of the builders.

Q How did astrology contribute to early astronomical knowledge?

A Prehistoric people thought that the sky was a separate realm and that the objects in the sky had supernatural powers. With this belief they were motivated to look for patterns in astronomical events. This spurred them to make systematic observations of the cycles of the sky. The accumulated records of the positions of planets in the sky, for example, eventually led to the understanding of the orbital arrangement and orbital motions of planets.

Q Why do scientists reject astrology?

A The belief that objects in the sky control our destiny or behavior has no scientific basis. Attempts to test astrology have been made by taking a large number of astrologers' predictions and seeing if they are accurate; such studies conclude that astrologers' predictions have no more than random accuracy. Furthermore, the basic principle of astrology—that patterns in nature control human events—is the same principle as many other ancient superstitions that are ignored in modern times. Finally, no physical theory has been proposed to explain why planets' positions could control human events or personalities.

Q How could the Greeks determine the sizes of distant objects in the sky?

A They used logic and the principles of geometry. They also started with an awareness of how celestial objects are illuminated. For example, they hypothesized that the Moon did not create its own light but reflected the light of the Sun. Starting with this, you can deduce a lot. Since the lighted part of the Moon usually has a curved edge, the Moon must be a sphere. Since the shadow of Earth is curved during a lunar eclipse, Earth must be a sphere. Since the Moon can block the light of the Sun, it must be closer than the Sun. Since the Moon does not fill Earth's shadow, the Moon must be smaller than Earth. And so on. The use of geometry allowed the Greeks to go beyond measuring angles in the sky and to begin to understand the positions of celestial objects in three dimensions.

Problems

Use these problems to test your understanding of the information and concepts in this chapter. The * indicates a more advanced or mathematical problem.

1. What is your latitude? What is the elevation of the North Star; in other words, how many degrees is it above the horizon as seen from your location?

2. Why would you expect more people to have seen a lunar eclipse than a solar eclipse? Ask a number of your friends and relatives—is it true? Why would scientists use jet planes to study solar eclipses?

3. Suppose you are standing facing north at night. Describe the apparent direction of movement of each of the following, as a consequence of Earth's rotation. **(a)** A star just above the eastern horizon. **(b)** A star just above the western horizon. **(c)** The North Star. **(d)** A star just below the North Star.

4. **(a)** Suppose you come across a set of large and ancient standing stones in a field in a remote part of Ireland. If you could observe over the course of a year, what simple measurements could you make to decide whether the stones had an astronomical function? What simple equipment would you need to make these measurements? **(b)** Discuss how local destructive events, such as the pillaging of the Alexandrian library or the destruction of the observatory and library at Benares, can have long-term bad effects on world history.

5. The stars can be used to keep time. **(a)** If we see the full cycle of constellations in a year, by what amount of time does a particular star rise or set earlier each night? **(b)** If Vega sets at midnight tonight, at what time will it set 2 months from now? **(c)** If Altair rises at 1 A.M. tonight, how long before it will rise at midnight?

6. We have described in this book what is seen in the sky by people living at a mid-northern latitude. Write an imaginary journal describing what you can see by careful naked-eye observation over the course of a year **(a)** in northern Finland above the Arctic Circle and **(b)** in a small village in Nigeria near the equator. Describe the behavior of the Sun, Moon, stars, and planets.

7. The angular height of the wide-angle photographs in Figure 2–20 is about 62°. **(a)** Use this fact to measure the difference in angular elevation of the Sun between Anchorage, Alaska, and Tucson, Arizona. **(b)** Use this to estimate the difference in latitude between Anchorage and Tucson. **(c)** Consult an atlas and determine the actual difference. Compare your results. Estimate the uncertainty in your measurements.

*8. Look up your latitude in an atlas. **(a)** How high is the celestial equator above your southern horizon? **(b)** Does this result depend on the season or time of day?

*9. Consider the advantages of environmental architecture. **(a)** Describe qualitatively how a single window awning on a south-facing window could completely shade the window at noon on the date of summer solstice but let light and heat into the window at noon during winter solstice. **(b)** Suppose you live at latitude of 40°N. Show a cross-section sketch of such a window and awning, and label (in degrees) the angle that a line from the window bottom to the edge of the awning makes with the horizontal. Label the same angle from the top of the window to the edge of the awning.

*10. **(a)** The Moon is measured to be about 384,000 km away from Earth. Using the fact that the Moon subtends roughly $1/2°$, estimate the diameter of the Moon, using the small-angle equation from the earlier Science Toolbox. Compare the number you calculate with the size of Earth. **(b)** If a kite at an altitude of 200 m just covers the Sun, what is the size of the kite?

*11. Suppose a cloud overhead in the sky is at an elevation of 10,000 m above the ground, and it has an angular size of roughly 5° by 10°. Estimate the actual dimensions of the cloud in meters.

Projects

Activities to carry out either individually or in groups.

1. If the class has access to a planetarium nearby, arrange for a class visit and a demonstration that shows the following: **(a)** Position of the North Star. **(b)** Rising and setting motions of stars and their motions around the North Star. **(c)** Prominent constellations. **(d)** Motion of the Sun from day to day and week to week, relative to the stars.

2. You will need a distant, strong light source such as the Sun or a single lightbulb, a small white ball to represent the Moon, and your eye to represent an observer on Earth. Now show that the crescent phases of the Moon prove that its orbit takes it between Earth and the Sun. In other words, show that you can only see a crescent phase on the Moon when it passes in the general area between Earth and the Sun. (This was one of the techniques used by the ancient Greeks to figure out the relative positions of bodies in the sky.)

3. If class members travel to different states during vacation, arrange for observations of the elevation of the North Star, in degrees above the horizon. Compare the different results. Is the North Star higher in northern locations?

4. With a camera and sensitive color film (ISO rating of 400 or more is best), make black-and-white or color time exposures of the night sky. Try different exposures such as 1 min, 5 min, and 10 min. You will need a tripod. Make one series including the North Star, another series of the eastern horizon, and another of the western horizon. Explain the patterns made by stars recorded in the images.

5. Choose a location where you have a reasonably clear view of the western horizon. Pick a spot where you can return to exactly the same place later. Make a careful and neat sketch of the horizon detail to scale. Mark the exact position of the Sun at sunset. *Be very careful never to stare directly at the Sun!* Repeat this observation once a week for five weeks, weather permitting. Does the Sun set at the same point each week? Does it set at the same time each week? What would you expect to see over the course of an entire year?

6. Each member of the class should ask five friends or relatives to explain in their own words **(a)** the phases of the Moon and **(b)** the cause of the seasons. Record each answer and evaluate each one to decide if it is correct. Combine the results for the entire class to get a larger statistical sample. What fraction of people got each question right? What fraction said they had no idea? What fraction thought they knew but gave an explanation that was wrong? How often do you see the classic misconceptions that the phases of the Moon are caused by Earth's shadow, and that the seasons are caused by the changing Earth–Sun distance?

Reading List

Aveni, A. 1989. *Empires of Time: Calendars, Clocks, and Cultures.* New York: Basic Books.

Bronowski, J. 1973. *The Ascent of Man.* Boston: Little, Brown.

Carlson, J. B. 1975. "Lodestone Compass: Chinese or Olmec Primacy?" *Science,* vol. 189, p. 753.

Gingerich, O. 1984. "The Origin of the Zodiac." *American Scientist,* vol. 55, p. 88.

Gingerich, O. 1986. "Islamic Astronomy." *Scientific American,* vol. 254, p. 74.

Gingerich, O. 1992. "Astronomy in the Age of Columbus." *Scientific American,* vol. 267, p. 100.

Hadingham, E. 1984. *Early Man and the Cosmos.* New York: Walker and Company.

Krisciunis, K. 1988. *The Alexandrian Museum, in Astronomical Centers of the World.* Cambridge, England: Cambridge University Press.

Krupp, E. C. 1991. *Beyond the Blue Horizon: Myths and Legends of the Sun, Moon, Stars and Planets.* New York: HarperCollins.

Lewis, D. 1973. *We the Navigators.* Honolulu: University of Hawaii Press.

Pannekoek, A. 1961. *A History of Astronomy.* London: Allen and Unwin.

Schaefer, B.E. 1989. "Dating the Crucifixion." *Sky and Telescope,* April, p. 374.

CHAPTER 3

The Copernican Revolution

CONTENTS
- Galileo's Trial
- Moving Beyond the Greek Model
- The Five Key Players in the Copernican Revolution
- The Plurality of Worlds
- The Layout of the Solar System
- Exploration of the Solar System
- A Brief History of Space Exploration
- Summary
- Important Concepts
- How Do We Know?
- Problems
- Projects
- Reading List

Galileo presents his astronomical discoveries to officials of the Catholic Church. The telescope provided evidence to support the Copernican hypothesis, but Galileo was destined to pay a heavy price for supporting this view. (SOURCE: Copyright National Geographic Society.)

WHAT TO WATCH FOR IN CHAPTER 3

In terms of exploring the first theme of this book, this chapter is one of the most important. It tells the story of a profound change, called the Copernican revolution. Before this revolution, virtually everyone pictured our planet Earth as unique at the center of the universe. They thought that planets and other celestial bodies were part of a separate ethereal realm that obeyed different laws. After the Copernican revolution, people were forced to confront the idea that Earth was just one of many worlds, replaced by the Sun at the center of the solar system. It is important to learn how and when this change happened, and who the key players were.

We will also see how new observations led to a refinement in our understanding of the solar system. Instead of circles, the planetary orbits are better described by ellipses. There is a simple relationship between the time it takes a planet to go around the sun and its distance from the Sun. Isaac Newton synthesized these observations brilliantly in a single law of gravity. He also removed the distinction between motions on Earth and motions in space—both are governed by the same universal laws of physics. The gravity law can help you understand everything from your own weight to the orbit of the Moon and the trajectories of spacecraft. Modern application of these laws has allowed direct exploration of the universe around us by spacecraft.

PREREADING QUESTIONS ON THE THEMES OF THIS BOOK

OUR ROLE IN THE UNIVERSE
How did the Copernican revolution affect our idea of Earth's place in the universe?

HOW THE UNIVERSE WORKS
Why do we believe that Newton's law of gravity is universal?

HOW WE ACQUIRE KNOWLEDGE
How can social or political forces impede the acceptance of new scientific theories?

GALILEO'S TRIAL

On the first day of October in 1632, the dreaded Inquisitor of Florence, Italy, knocked on the door of the famous astronomer Galileo Galilei and served him with a summons to appear before the Inquisition in Rome within 30 days. The noted scientist was being forced to answer charges that he had promoted heresy in his latest book. Galileo was 68 years old, and these charges were extremely serious. Anyone found guilty of heresy could be sentenced to death.

What was Galileo's crime, according to the Catholic Church? Galileo had taken a stand against traditional views about the nature of the universe. He presented startling evidence that our planet is not at the center of the universe. Galileo's own work, plus that of other scholars, had clearly convinced him that Earth and other planets moved around the Sun. These ideas seemed strange at the time. When people looked out from Earth, it certainly seemed that Earth was stationary and at the center of things, as if all bodies in the sky moved around them. This episode illustrates two major themes that we stress in this book. First, the slow and difficult progress of Galileo's ideas shows how science works, sometimes in the face of opposition. Second, this new thinking was part of a famous revolution that affected the way all humans thought about their place in the universe.

The radical new idea that Earth is not a unique central body, but one of a number of bodies that orbit the Sun, is called the Copernican revolution. This revolution in our sense of our place in the universe is named after the Polish scientist Nicolaus Copernicus, who had suggested in the 1500s that Earth was not at the center of the solar system. The Copernican revolution is one of the most important ideas in history, because it indicates that we are part of a larger cosmic environment, not the masters of nature living in the capital of the universe.

Galileo had come under fire from the Catholic Church for publicly advocating the ideas of Copernicus. He had published his initial conclusions supporting Copernicus in 1610. Six years later, a Catholic cardinal visited Galileo and advised him that Copernicus' view went against the teachings of the Roman Catholic Church. The official directed that Galileo not "hold or defend" these views, although he might discuss them as hypotheses. Galileo was shaken by this encounter, but the pressure did not stop him from continuing his scientific work.

Early in 1632, Galileo had published his masterwork, *A Dialogue on the Great World Systems,* in which two fictional characters debate whether Earth is central. This is the book that got Galileo in trouble with the Catholic Church, because he presented the Copernican idea in strong and plain language and he used the rhetorical device of having the less clever character unsuccessfully defend the Catholic Church's position.

After he received the summons to appear before the Inquisition, Galileo surrendered himself in Rome. The trial began in 1633. The surviving transcript shows the dilemma Galileo faced. At the beginning, Galileo was sure he could clear himself. One of the first questions from the Inquisitors concerned what had happened during the Cardinal's visit in 1616. Galileo described the visit and presented as evidence a 1615 letter from the Cardinal, commending him for "speaking [only] hypothetically and not with certainty" about these issues. A second letter from the Cardinal, written in 1616, gave the order that "the Copernican opinion may neither be held nor defended, as it is opposed to Holy Scripture." Based on these letters, Galileo argued that his book debating the sides of the argument was within the spirit of the instructions he had been given.

Not so, argued the Inquisitors. Galileo had clearly leaned toward Copernicus. Worse yet, the Inquisitors produced their copy of the Cardinal's instruction, which contained an added phrase that Galileo was not to "teach that opinion in any way whatsoever." Galileo brandished his letter and denied that he had ever been told this last phrase. Some modern scholars believe that the Inquisitors' document was a forgery, written by them after 1616. In any case, only one thing would satisfy the Inquisition. Galileo had to renounce his belief in the Copernican idea and deny that Earth moved around the Sun. The Inquisitors showed Galileo—by then an old man of 70—the instruments of torture that might be used on him if he refused to cooperate. There was also a real possibility of his being burned at the stake, which had happened to the mystic and astronomer Giordano Bruno 33 years earlier.

Galileo faced an impossible moral dilemma. Should he risk death to defend his scientific observations in front of the secret Inquisition court, or should he recite a confession that would satisfy the judges and live to fight another day? He gave his answer. Even under the threat of death, he told them, he would never say that he was not a good Catholic or that he had tried to deceive anyone. However, under duress he would be willing to say that he did not believe the new Copernican idea. In the end, he trusted that copies of his book would get out and that the scientific evidence would speak for itself. He begged the judges to "take into account my pitiable state of bodily illness, to which, at the age of 70 years, I have been reduced by ten months of constant mental anxiety." The trial came to a climax on June 22, 1633, when Galileo was summoned to kneel before the judges to hear his sentence and to recite a confession of error. The judges read a lengthy condemnation that included the Catholic Church's strong opposition to the Copernican revolution. Galileo was sentenced to house arrest for life. The Inquisitors ordered his book banned. He had to repeat his confession in public, saying that he would "abandon the false opinion that the Sun is the center . . . and that the Earth is not the center and moves," and vowing to "abjure, curse, and detest the aforementioned errors and heresies. . . ."

There are several postscripts to this story. As Galileo had hoped, his book was published elsewhere in Europe, and within a few decades his ideas were widely accepted. The Catholic Church could not stop the spread of new scientific ideas. In 1757, Galileo's book was removed from a list of books banned by the Catholic Church. In 1983, Pope John Paul II took up Galileo's case, and in 1992, he formally proclaimed that the Catholic Church had erred in condemning Galileo, and that science is a valid realm of knowledge "which reason can discover by its own power." This story perhaps makes clearer something that puzzles many people today—why academics, scientists, and intellectuals are so concerned about maintaining the freedom to pursue and express controversial views. The real point of Galileo's story is not the conflict between science and the Catholic Church; it is the fact that entrenched ideology is the enemy of human understanding. As shown by Italy in the 1600s, or Germany in the 1930s, or the Soviet Union in most of this century—and even by episodes in the United States—it is very easy for an overzealous society to silence the voices of people who express unpopular ideas. History shows that unpopular ideas, like Galileo's, may turn out to be true!

Moving Beyond the Greek Model

Scientists of the 1500s and 1600s inherited a model of the universe whose basic features had been defined by Aristotle 2000 years earlier. The idea was simple. Earth was stationary at the center, and the Sun, Moon, and other planets all moved around Earth. Each object was fixed to a spinning crystalline sphere. The stars were all fixed to an outermost sphere and were also carried around Earth on circular orbits. Rest was the natural state for any object on Earth and a mysterious power was required to keep the celestial bodies in motion. Medieval people pictured the whole universe as a set of concentric spherical shells centered on Earth, as shown in the diagram in Figure 3–1. If you study the diagram carefully, you can see *Terra immobilis* marked in the center, surrounded by shells of water, air, and fire, with those surrounded in turn by shells that carried the Moon, Sun, planets, and finally the distant stars. As we saw in Chapter 2, this cozy arrangement fit with the powerful idea that humans were at the center of creation.

However, Ptolemy's most successful realization of the Greek model was very complex. Motion seen from the center of a circular orbit is uniform. Yet it was known that the planets do not move among the stars at a constant rate. To account for this, Ptolemy was forced to hypothesize that the center of the motion was displaced from Earth, like the eccentric motion of a wheel when the hub is not at the center. Also, it was known that some planets can reverse their

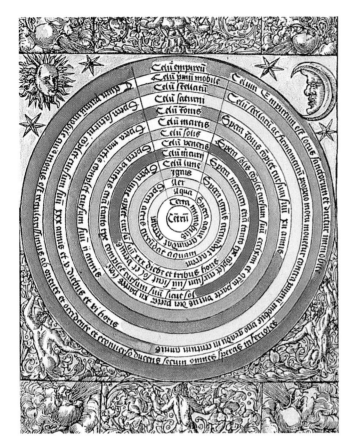

FIGURE 3-1
A medieval conception of the universe, dating from 1537, shows the persistence of Ptolemy's theory of a central, stationary Earth. The central sphere is labeled *Terra immobilis*, or immovable Earth. Around it are shells of water, air, and fire, then shells carrying the Moon, Sun, planets, and stars. The term *seventh heaven* comes from the idea that the seventh of the outer spheres was the finest and purest. (SOURCE: The Granger Collection, New York.)

steady eastward motions among the stars—a phenomenon called **retrograde motion.** The Ptolemaic model therefore required the planets not only to move in circles around Earth, but also to move along smaller circles, called *epicycles,* around imaginary points along the main circular orbits. This strange scheme is shown in Figure 3-2. Mercury and Venus are never seen far from the Sun, so they have a special status in Ptolemy's model. Their epicycle centers must lie on the line connecting Earth and the Sun. The Greeks had used geometry to estimate the distance to the stars as at least 1 million miles. Therefore the outermost crystalline sphere had to be whirring around at over 1 million mph!

How good was the Ptolemaic model? Note that we do not call it a theory because it has no physical explanation for how and why the planets move as they do. Ptolemy himself never claimed that it represented reality, only that it provided a convenient mathematical description to predict the planet positions. Initially the predictions were accurate to 1 or 2 arc minutes (recall from Chapter 2 that this is about as good as the resolution of the human eye). But the eccentric motions adopted by Ptolemy were just approximations to the true motions of the planets, and over the centuries the errors began to accumulate. As we saw in the last chapter, a small error in a calendar will also accumulate into a serious problem over a span of centuries. By the 13th century, the predictions of the model could be off by as much as 1° or 2°, several times the angular diameter of the Moon. Astronomers had to make increasingly complicated adjustments to the model in order to get correct answers. They even had to add tiny epicycles onto the larger epicycles. In 1252, Spain's King Alfonso X funded a special almanac of predicted planetary positions. Watching his astronomers laboriously calculate motions of epicycles upon epicycles, he commented that had he been present at the creation, he could have suggested a simpler arrangement.

This idea of looking for a simpler arrangement has become a key element in the scientific method. As early as 1340, the English scholar William of Ockham proposed the famous idea that among competing theories, the best theory is usually the simplest theory—that is, the one with the fewest assumptions or the fewest quantities that have to be combined to make a prediction. We did not discuss this in Chapter 1, and it would be difficult to do so because simplicity is an aesthetic judgment. Are the simplest and most elegant theories always correct? Or is the belief that the universe is simple merely a human conceit? You can judge for yourself as the book progresses.

It is appropriate that our story has taken us to 13th-century Spain because this was where the Renaissance was incubated. The Renaissance (or "rebirth" in French) was a broad cultural awakening that included science and politics as well as art. Spain at the time was an exhilarating melting pot of Christian, Jewish, and Muslim cultures. As Alfonso was commissioning new astronomical almanacs, schools of translators were rediscovering Aristotle and other Greek philosophers. They were also translating the work of Arab mathematicians, including Alhazen who invented the science of optics—the geometric study of how light travels and reflects. In the 15th century, Italian artists used this new science to put depth into painting by the use of perspective. Perspective makes art come alive by giving it the sense of a third dimension. In much the same way, the Greeks were able to use geometry to create an awareness of depth in space. Science and art were also united in turning to nature rather than established wisdom as the source of their inspiration. Leonardo da Vinci used the same optical principles to understand the shadowing of eclipses as he did to draw the shadows and contours of the human body. The Renaissance spread across Europe. Exciting new ideas are contagious!

The scientists of the Renaissance once again began to ask profound questions about the universe. How do we know that Earth is the center of the universe? What is the best way to understand the motions of the stars and planets? The **Copernican revolution** still influences our whole conception of the role of the human species in the cosmos, and even our religious and philosophical ideas. Also it illustrates how science works. The whole modern idea of searching for other planets near other stars, or asking if there are alien

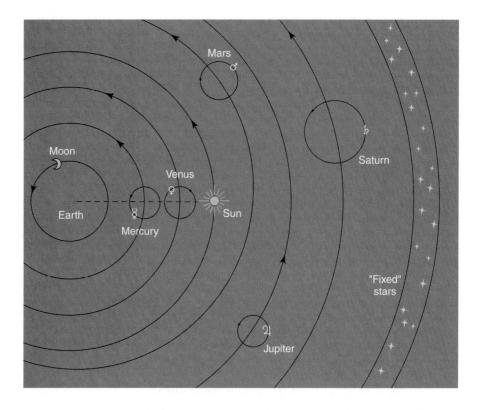

FIGURE 3-2
The solar system as conceived by Ptolemaic astronomers between about 100 and 1500. Ptolemaic astronomers made rough predictions of planetary motions, using a theory in which the Sun and other planets moved in circular orbits around a stationary, central Earth. Superimposed on the larger orbits were smaller circular motions called epicycles, introduced to try to make the theory more accurate.

civilizations, or picturing a starship "Enterprise" flying from world to world, comes from what we might call Copernican ideas, developed in the 1500s and 1600s. Before that, reasonable people thought that there was only one world in the whole universe—Earth, located at the center. To understand where we are today, we have to understand how the old idea was overthrown by the scientific method.

By the 1500s, the best scientists had noted that older predictions of planetary positions were now in error by a degree or so. Like King Alfonso X, they began to wonder if there might not be a simpler theory that could give better results than the patchwork Ptolemaic assembly of epicycles. One observer sought a new model that he said might be more "pleasing to the mind." This man was Nicolaus Copernicus, father of the Copernican revolution.

The Five Key Players in the Copernican Revolution

The Copernican revolution took a century and a half, from roughly 1540 to 1690. It involved five very famous scientists: Copernicus, Tycho Brahe, Johannes Kepler, Galileo Galilei, and Isaac Newton.

Copernicus and His Simple Idea

Nicolaus Copernicus (Figure 3–3) started the drive to visualize the Sun, not Earth, as the center of the solar system. He was born on February 14, 1473, the son of a Polish merchant. While being educated at a university in Italy, he became excited by the burgeoning scientific thought in that country. At age 24 he made his first astronomical observations. A few years later, he obtained a position as a clerical official in the Catholic Church. This post gave him the time and economic security to continue his astronomical studies. At age 31 he observed a rare conjunction, or passage of planets close to

FIGURE 3-3
Nicolaus Copernicus (1473–1543) pioneered the Sun-centered, or heliocentric, view of the solar system. He holds a model with the central Sun, circled by Earth and the Moon. This reluctant revolutionary only published his most important work on his deathbed. (SOURCE: Corbis–Bettmann.)

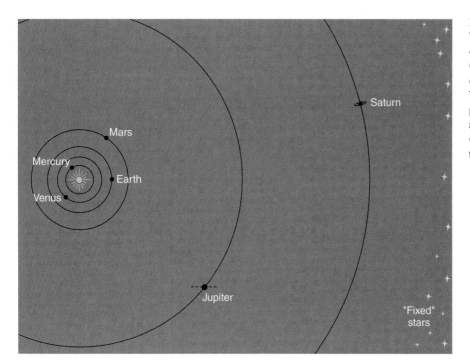

FIGURE 3-4
The solar system as it might have been conceived around 1700, at the end of the Copernican revolution. The diagram shows the orbits of the 10 known planets to true scale. The view is correct except that the outermost planets (Uranus, Neptune, and Pluto) and the asteroids had not yet been discovered. Compare with Figure 3–2 to see the change from the Ptolemaic view.

each other, as seen in the sky. The conjunction brought all five known planets as well as the Moon into the constellation of Cancer. He found that their positions departed by several degrees from an earlier set of Ptolemaic predictions.

Copernicus made few new observations. However, he spent a long time studying different models for the arrangement of the solar system. He concluded that the prediction of planetary positions would be simpler if we imagined that the Sun were at the center and Earth were one of the Sun's orbiting planets. In 1512 Copernicus circulated a short commentary containing the essence of his new hypothesis: The Sun is the center of the solar system, the planets move around it, and the stars are immeasurably more distant. This commentary was only distributed in handwritten form to a few of Copernicus' acquaintances. Copernicus continued his studies but, fearing controversy with the Catholic Church, delayed publication for many years. Finally, encouraged by visiting colleagues, including some in the clergy, he allowed the written commentary to be more widely circulated. News of Copernicus' work spread rapidly.

Late in his life, in 1543, Copernicus prepared a synthesis of all his work, called *On the Revolutions of the Celestial Spheres*. In this book he laid out and explained his evidence for the solar system's arrangement: Planet positions in the sky could be explained if one assumed that Earth and other planets move around the Sun. Only 400 copies of this book were printed, and only a small part of it deals with the heliocentric hypothesis. Yet the modern meaning of the word *revolution*—sudden political and social upheaval—originates with the title of Copernicus' book.

Turmoil was ensured because Catholic Church officials and most intellectuals held that Earth was at the center. The printer of the book, a Lutheran minister, had tried to defuse the situation by inserting a preface stating that the new theory need not be accepted as physical reality but could be seen merely as a convenient model for calculating planetary positions. This was philosophically a valid way of looking at the situation. Already Copernicus had come under fire from Protestant fundamentalists: In 1539 Martin Luther had called him "that fool [who would] reverse the entire art of astronomy. Joshua bade the Sun and not the Earth to stand still." In a world of strong dogmas, tampering with established ideas is dangerous. In the 1530s Michael Servetus had been criticized for his writings on astrology and astronomy; in 1553 he was burned at the stake as a heretic for professing a mysterious theology that offended both Protestants and Catholics. Both Protestants and Catholics suppressed heretical ideas. John Calvin masterminded Servetus' execution, although, in a fit of moderation, he recommended beheading instead of burning. Servetus, a man of wide learning and varied interests, had improved geographic data on the Holy Land and discovered blood circulation in the lungs. Copernicus was aware that he, too, had rattled the hornet's nest.

Despite the furor over his book, Copernicus was not able to *prove* that the heliocentric model was correct. He followed Greek tradition in assuming that the planet orbits must be perfect circles. Because of this, his model did not predict the positions of the planets any more accurately than Ptolemy's. Yet we have argued in Chapter 1 that the hallmark of a good theory is its ability to accurately explain observations. So why, then, did scientists come to favor the heliocentric idea?

Placing the Sun at the center brings a certain symmetry and simplicity to the model of the solar system. Figure 3–4 shows the heliocentric cosmology as accepted by astronomers by the end of the Copernican revolution. In Ptolemy's model, Mercury and Venus are special because they revolve around empty points between Earth and the

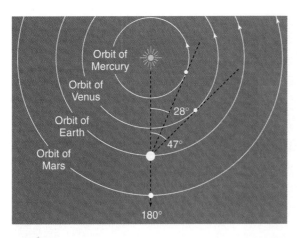

FIGURE 3-5
Mercury and Venus always appear close to the Sun, and therefore they only can be seen clearly in the dawn or evening sky. In the heliocentric model, Mercury and Venus are interior to Earth's orbit, which explains why they never stray far from the Sun. Exterior planets such as Mars can be seen at any angle from the Sun, including opposition at 180° from the Sun.

Sun (refer to Figure 3-2). Copernicus has all the planets orbiting the Sun in the same sense. He simply explains the fact that Mercury and Venus always appear close to the Sun (Figure 3-5). In Ptolemy's model the retrograde motions of some planets are explained with the artificial device of epicyclic motion. The Copernican model accounts for this naturally with the different speeds of planets in their orbits. Figure 3-6 shows that as Earth "overtakes" Mars on its interior orbit, Mars appears to temporarily reverse its motion with respect to the distant stars.

Furthermore, in Ptolemy's model there were many different combinations of epicycle size and motion that could roughly fit the planet motions. This seemed to Copernicus to be arbitrary and unsatisfactory, like a puzzle with no single solution. In the heliocentric model, the relative spacing of the planets is fixed uniquely by their apparent motions. There is regularity of the motions in that the planets closest to the Sun orbit the fastest. Interior planets are always seen near the Sun. Exterior planets are seen at any angle to the Sun and can sometimes perform retrograde motion. Copernicus also knew of the work of Aristarchus. It made sense to put the largest object, the Sun, at the center of all motions.

Objections were raised to the heliocentric model. If Earth is moving, why do we not feel the motion? Copernicus had no simple answer to this, but he pointed out that the annual movement of the Sun in the sky could equally well be explained by Earth's moving annually around the Sun with a tilted axis. The apparent motion of the celestial sphere could equally well be explained by the daily rotation of Earth. While some critics complained that it was implausible for the equator of Earth to rotate at 1000 mph, the geocentric model required the celestial sphere to rotate 1000 times faster! The last major objection was the lack of any seasonal change in the angles and brightness of stars. In a geocentric model, the stars orbit Earth at a fixed distance and so never change their brightness or angular separation on the sky. However, in a heliocentric model, Earth must change its distance from each part of the celestial sphere as the seasons pass. Yet no star appeared to brighten and dim, and no constellation appeared to change its size over the course of a year. Defenders of the heliocentric view were forced to hypothesize that the stars were so far away that these changes would be undetectable.

How far away did stars have to be in the Copernican model? To understand this, we must introduce the idea of **parallax.** Parallax is the shift in angle that occurs when a nearby object is seen against a distant backdrop from two different perspectives. This is a familiar idea. Imagine driving in a car with a distant mountain range on the horizon.

FIGURE 3-6
A schematic view showing why Mars on occasion changes its direction of motion with respect to the stars—a phenomenon called *retrograde motion*. Positions of Earth and Mars are shown for six dates. Note that Mars moves more slowly than Earth and at a slight angle to Earth's orbit. As Earth overtakes Mars, Mars' apparent path on the starry backdrop changes speed and direction. It is very difficult to explain this in a Ptolemaic model.

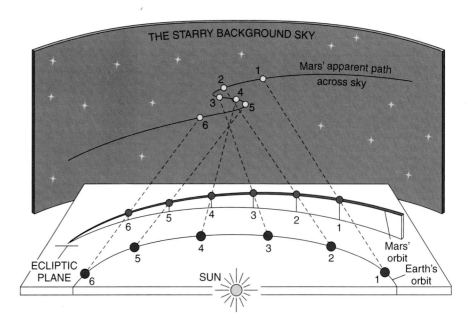

Seen by left eye   Seen by right eye

(a)

(b)

FIGURE 3-7
Parallax is the change in perspective when a nearby object is viewed against a more distant backdrop from two different positions. (a) Hold you thumb out at arm's length, and view a distant backdrop with one eye and then the other. The small shift in perspective is a parallax angle; we use binocular vision and our brains to estimate distances using this angle. (b) A tree across a river shows a shift in angle with respect to a distant mountain range when viewed from position A and then B. By simple geometry, if you know the distance from A to B and the parallax angle α, you can calculate the distance to the tree.

A nearby tree appears to shift more quickly than a distant one as seen against the horizon—this is a shift in parallax angle. Hold a finger out at arm's length and view it with one eye and then the other (Figure 3–7a). The slight change in perspective from one eye to the other is a parallax shift. This is the way we get depth perception from our binocular vision. Figure 3–7b shows a practical use of the effect. If you know the distance between the viewing points and the parallax angle, then simple geometry gives you the distance to the nearby object. If the angle α is small, you can use the small-angle equation from Chapter 2.

The same idea applies to stars. In the geocentric model, we might expect to see a difference in the angle between two stars on the celestial sphere when observations are made at different times or from different positions on Earth's surface. You can see from the geometry in Figure 3–8a that angle β must be slightly smaller than angle α. But no difference is seen, which means that the stars must be very far away compared to the size of Earth. By the time the Copernican idea was accepted, astronomers believed that stars were scattered through space rather than fixed to a crystalline sphere. In the heliocentric model, a nearby star should show a parallax shift with respect to more distant stars as Earth moves in its orbit of the Sun (Figure 3–8b). No shift had ever been observed. We can use the small-angle equation from the Science Toolbox in Chapter 2 to show how far away the stars had to be. The limit of angular observation was about 1 arc minute (1′) or 60 arc seconds (60″). So $d/D < \alpha / 206{,}265 < 0.003$, and therefore $D/d > 3300$. The stars had to be at least 3300 times farther away than the diameter of the Earth-Sun orbit for parallax to be unobservable! Many people were uncomfortable with the idea of such an immense universe.

Copernicus himself missed the height of the violent debate. The first copies of his book were reportedly delivered to him on the day of his death in 1543, at age 70. Aided by the invention of the printing press 100 years earlier, the heliocentric idea was soon being discussed in centers of learning all over Europe. The Copernican revolution was underway.

Brahe's Careful Observations

Tycho Brahe (Figure 3–9) took the next step towards confirming Copernicus' hypothesis. Tycho (as he is known) was a colorful character who was perhaps the greatest observer of the pretelescopic era. His life was marked by swift changes of fortune. Born to a penniless family, he was raised by a foster father who could not have children of his own. His foster father died of pneumonia after saving the king of Denmark from drowning in an accident, and in gratitude the king made Tycho a rich man. Tycho wore a silver nose to cover a dueling mutilation—suffered in an argument over who was a better mathematician! At age 16, he noticed the errors in the planet positions predicted by the old Ptolemaic system. This led to his lifelong interest in recording the positions of planets.

Tycho made observations that challenged key elements of Aristotle's conception of the universe. In 1572, when he was only 25 years old, he observed an exploding star that temporarily brightened and then faded. This exciting observation refuted the ancient belief that the stars were forever

FIGURE 3-8
(a) In the geocentric model, the angle between two stars should depend on the time and position of the observation from Earth. Since there is no detectable difference between angles α and β, the stars must be very far away compared to the size of Earth ($D \gg d$). This in turn implies that the celestial sphere must be moving very fast to complete an orbit of stationary Earth every 24 h. **(b)** In the heliocentric model, a nearby star shows a parallax shift when viewed against more distant stars. The nearby star lines up against different distant stars as Earth moves in its orbit of the Sun. Knowing the diameter of Earth's orbit and the parallax angle, the distance to the nearby star can be measured. The parallax angle in this diagram is exaggerated. In practice, the triangle is long and skinny, and the parallax angle is small and difficult to measure.

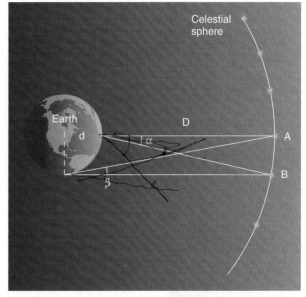

(a)

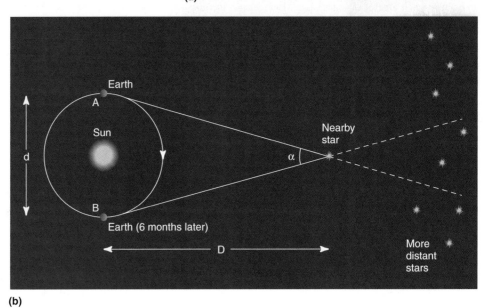

(b)

unchanging, mounted on their starry crystalline sphere, as diagrammed in Figure 3–2. Five years later he tracked the motion of a bright comet from night to night and compared his data with those of observers in other parts of Europe. The comet always formed the same pattern with respect to the stars as seen by any observer. Tycho was able to show that it was much farther away than the Moon and not an atmospheric phenomenon as Aristotle had taught. Moreover, the path of the comet required that it traverse the orbits of several planets, calling into question the whole idea of crystalline spheres. By demonstrating that no star or planet shows an angular shift compared to the pattern of stars as our position shifts with the rotation of Earth, Tycho proved that stars and planets were many times farther away than the Moon. Interestingly, Tycho did not accept Copernicus' simple model, but preferred a more complicated model of his own.

With funds from the king of Denmark, he built the first modern European observatory, named Uraniborg or Sky Castle, on an island near Copenhagen. The telescope had not yet been invented, but Tycho made the most accurate set of astronomical observations ever, thanks to his fine experimental technique. He used fittings made of rigid metal rather than wood. He sighted objects with a precise angular scale, using a large rotating device like a giant protractor (see Chapter 1 to be clear on the distinction between precision and accuracy). He made multiple independent measurements as a way of improving accuracy. Or he combined independent measurements made by himself and his assistants. He improved on the accuracy of previous star and planet positions by a factor or 3 or 4—his best measurements were accurate to 1 to 2 minutes of arc. Tycho toiled for 20 years, making observations on every clear night. This formidable set of data would be crucial in understanding the orbits of the planets.

FIGURE 3-9
Tycho Brahe (1546–1601) set up an observatory to track planet positions with greater accuracy than ever before.
(SOURCE: The Granger Collection.)

Brahe's 20-year accumulation of measured planetary positions. He first studied the orbit of Mars, whose movements in the sky had plagued astronomical theorists since Ptolemy. He found something astonishing. After all the centuries of debate about circular orbits and circular epicycles, the orbit that best fit the observed positions of Mars was not a circle at all, but an ellipse! Ellipses, as discussed in the accompanying Science Toolbox, are figures that can range from circular to highly elongated loops. Eventually, Kepler found that all planets move around the Sun in elliptical paths. Any ellipse has two special points, called *foci,* shown in Figure 3–11 on page 64. As an ellipse becomes more like a circle, the two foci move closer together. In a circle they merge to a single point at the center. Kepler made the surprising discovery that in each planet's **elliptical orbit,** the Sun lies not at the center of the ellipse, but at one focus of the ellipse. Although all planets' orbits are elliptical, most are so slightly elongated that they look almost circular.

Kepler achieved his breakthrough by giving up one of the most cherished Greek ideas—the perfect symmetry of a circular orbit. He did not do this lightly. It took him 8 years of analysis and hundreds of pages of calculations to be sure of his conclusion. However, his insight allowed him to dramatically simplify the description of planetary orbits. Within the Ptolemaic model, the accurate data of Tycho could only be matched using circular orbits by adding many epicycles. Ironically, the Copernican model also needed epicycles to fit

Then Tycho's luck took a turn for the worse. His royal patron King Frederick died, and the young new king was less inclined to tolerate Tycho's troublesome personality. Tycho became an itinerant astronomer, taking a series of posts around Europe. In 1600, he hired a 30-year-old mathematician named Johannes Kepler. Tycho's death was as colorful as his life. One evening, while dining at the house of a nobleman, Tycho drank copiously and suffered a burst bladder. Before he died he named Kepler as his successor. And so the brilliant young mathematician inherited the mass of Tycho's observations and began analyzing the planetary positions.

Kepler and Elliptical Orbits

Deeply religious and a believer in astrology, Johannes Kepler (Figure 3–10) was sure that planetary motions would turn out to be governed by hidden regularities. Kepler was greatly influenced by Pythagoras, who had discovered that two musical notes an octave apart are produced by vibrating strings with lengths in a ratio of 2 to 1. Kepler hoped that the ratios of the planet's distances would be simple mathematical ratios like those vibrations that produced harmonious sounds in music. They are not.

However, Kepler continued his studies, using Tycho

FIGURE 3-10
Johannes Kepler (1571–1630) analyzed Tycho Brahe's observations and concluded that planets follow elliptical orbits around the Sun. He was also inspired by Greek ideas about the "harmony of the spheres." (SOURCE: Corbis–Bettmann.)

SCIENCE TOOLBOX

Kepler's Laws of Planetary Motion

Kepler's discovery that planets move in elliptical orbits is pleasing because ellipses are such simple figures compared to the complex motions required in Ptolemy's theories of cycles and epicycles. Figure 3–11, on page 64, shows some examples of ellipses and a simple method by which you can draw one. Take a piece of string about 6 to 10 in long and tie it in a loop. Put two thumbtacks in a piece of cardboard. Loop the string around the tacks, put a pencil against the string, and pull it taut. Then trace a loop all the way around the tacks. This will be an ellipse.

Each tack occupies a special point in the ellipse, called a *focus* (plural: *foci*). Kepler's first law of planetary motion says that each planet orbits the Sun on an elliptical path with the Sun at one focus. What lies at the other focus? Nothing! If you put the two foci closer together, as in Figure 3–11b, the ellipse becomes more like a circle. We can continue this progression until the two foci meet at the same point. In Figure 3–11c, the ellipse has become a circle. In other words, a circle is only a special case of an ellipse. You will recall that the Greeks had favored circular orbits because a circle is the simplest and most symmetric shape. Now we have another way to think about orbits.

There is a whole family of ellipses—just think of circles that get successively more squashed—of which the circle is just one example. The ellipse is the more general figure. The generality of Kepler's description is the hallmark of a good scientific idea.

A circle is described by just one number: the radius. An ellipse is described by two numbers. The widest diameter of an ellipse is called the *major axis,* and one-half of this distance is the *semimajor axis* (given by the symbol a), or the distance from the center of the ellipse to one end. The semi-major axis is equal to the mean distance of a planet from the Sun. It is therefore analogous to the radius of a circular orbit. The word used to describe the flattening or noncircularity of an ellipse is *eccentricity* (denoted by the symbol e). Mathematically, it is the ratio of the distance between the foci to the major axis ($2a$). A circle has zero eccentricity. Most planets are in nearly circular orbits with very low eccentricity. Very flattened or elongated orbits have high eccentricity. Many comets move in eccentric orbits, but still have elliptical orbits with the Sun at one focus.

Elliptical motion offers a good example of the progress of science as more accurate data are gathered. If you only had measurements of the motions of planets accurate to 10 percent, you would find that the orbits of most planets were consistent with circular motion. But departures from uniform motion are seen when the accuracy reaches 1 percent. The effects are quite subtle, which is why Tycho's excellent data were required to make the breakthrough. For example, Earth moves in its elliptical orbit from 1.017 times the mean Earth-Sun distance to 0.983 times the mean Earth-Sun distance. This is a total variation of only 3.4 percent. In other words, the ellipse of Earth's orbit has a distance between the foci that is only 3.4 percent of the semimajor axis—a very slightly squashed circle.

If you measured the orbits with even greater accuracy, say 0.1 percent, you would find that there are were tiny discrepancies. In other words, although planets move in nearly elliptical orbits, they do not move exactly in elliptical orbits. The reason is that they are pulled slightly out of their ellipses by the gravitational tug of other planets. Scientists realized this in the 1700s. Also, at a higher level of precision it is not quite correct to say that the Sun lies exactly at a focus. Rather, the focus of the orbit is the *center of mass* of the Sun and the planet. The center of

(*continued*)

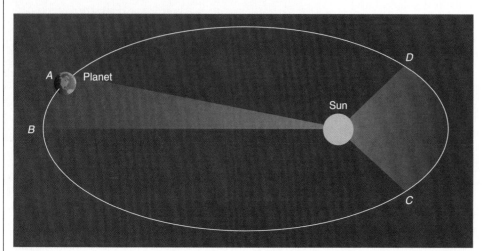

FIGURE 3-A
Kepler's second law of planetary motion. The orbit sweeps out an ellipse where an imaginary line connecting the planet to the Sun sweeps out equal areas in equal time intervals. The time taken to move from A to B equals the time taken to move from C to D. In other words, planets travel faster when they are close to the Sun and more slowly when they are far from the Sun. The true planet orbits are much closer to circles, and the speed only changes by a small percentage along the orbit.

SCIENCE TOOLBOX

Kepler's Laws of Planetary Motion (continued)

mass is defined as follows: If you took an imaginary balancing beam and put the Sun at one end and a planet at the other, the center of mass would be the balance point between the two. The Sun is so massive that the center of mass for each planet would be inside the Sun, but not quite at the center. These comments show that data of higher accuracy can lead to a whole new level of understanding.

Figure 3–A shows Kepler's second law of planetary motion. This states that a planet speeds up when it is near the Sun and slows down when it is far away. This occurs in such a way that the line joining the planet to the center of the ellipse sweeps out equal areas in equal intervals of time. Kepler's third law states that the square of the orbital period is proportional to the cube of the semimajor axis:

$$P^2 \propto a^3$$

We can set the constant of proportionality to 1 if we use units of years for the period and astronomical units (1 AU = mean Earth-Sun distance) for the semimajor axis. In this case

$$(P_{years})^2 = (a_{AU})^3$$

Mars' average distance from the Sun is about 50 percent larger than Earth's, or 1.52 AU. Therefore the period squared is equal to $(1.52)^3 = 3.52$, and the period is $\sqrt{3.52} = 1.87$ years. The Martian year is nearly twice as long as Earth's year. Figure 3–B shows Kepler's third law and the steady increase in orbital period with distance from the Sun. Notice that in these calculations we only use three significant figures because there are departures from Kepler's laws at a level below 0.1 percent.

Let us see another couple of applications of this important relation. Say we use geometry to show that Jupiter is 5.2 times the distance of Earth from the Sun. What is Jupiter's orbital period? Following the previous example, the period squared is equal to $(5.2)^3 = 140.6$, and the period is $\sqrt{140.6} = 11.9$ years. The Jovian year is nearly 12 times as long as Earth's year. Now imagine that with careful observation we were able to show that Mercury took 89 days to orbit the Sun—measured as the time it took for Mercury to move between successive appearances at its maximum angular distance from the Sun on the sky. Converting to years, we have that the semimajor axis cubed is equal to $(89 / 365)^2 = 0.0595$, so the semimajor axis is $(0.0595)^{1/3} = 0.39$ AU. Mercury lies at just over one-third the Earth-Sun distance.

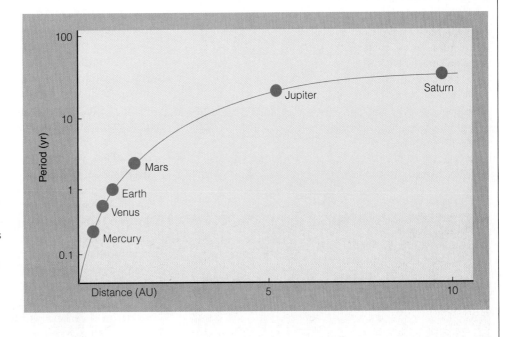

FIGURE 3–B
Kepler's third law of planetary motion. The period of a planet's orbit increases with increasing distance from the Sun. Planets discovered after the invention of the telescope obey this law, too. Notice that according to Kepler's law, planets can have properties that place them *anywhere* on this curve. The planets of the solar system have a particular pattern in their spacing from the Sun.

the data equally well. The problem is that the motion of the planets is not uniform, but a circular orbit produces uniform motion regardless of whether Earth or the Sun is at the center. Kepler avoided the contortion of epicycles by having a single ellipse to describe the orbit of each planet. His results did not fit the geocentric theory of Ptolemy, but they fit beautifully the heliocentric hypothesis of Copernicus.

Kepler summarized his results in two books, published in 1609 and 1619. They can be distilled to three famous "laws," or rules, which have come to be called **Kepler's laws of planetary motion.** The first law is the statement we already described, that all planets' orbits are ellipses with the Sun at one focus. The other laws satisfied Kepler's ambition to find patterns in the orbits of the planets. They each involve a mathematical relation between the distance of each planet from the Sun, and the rate of its motion around the Sun. The second law shows that planets move along their orbits fastest when they are nearest the Sun and most slowly when they are farthest from the Sun. If you imagine a line connecting the planet to the Sun, the line sweeps out equal areas in equal intervals of time. The third law states that there is a simple scaling between the mean distance of a planet from the Sun and the time it takes to orbit the Sun—see the Science Toolbox for more details. Kepler's laws apply to any planetary motion. Stated succinctly,

1. Planets travel in elliptical orbits with the Sun at one focus.
2. Planets travel faster when they are closer to the Sun, such that a line connecting a planet to the Sun sweeps out equal areas in equal intervals of time.
3. The square of a planet's orbital period is proportional to the cube of the semimajor axis.

Kepler's work greatly advanced Copernicus' idea that all planets moved in orbits around the Sun, rather than around Earth. The single shape of an ellipse swept away all the complications of Ptolemy's system. The simple relationships that described planets' orbits made it clear that the Sun governed the motions. What was still needed to establish the Copernican hypothesis, however, was direct observational evidence.

Galileo and the Revelations of the Telescope

Galileo Galilei provided this observational evidence. Galileo, known by his first name, like Tycho, is shown in Figure 3–12. He worked at the same time as Kepler was working on his orbital theory. As early as 1597, Galileo wrote to Kepler, "Like you, I accepted the Copernican position several years ago . . . I have not dared until now to bring [my ideas] into the open." As we saw in the opening story, Galileo's scientific views put him on a collision course with the Catholic Church.

Galileo was born in Pisa, Italy, and although he had to leave school because of financial difficulties, he returned to complete his studies in mathematics. He became a professor at the University of Padua, where he stayed for 18 years. Around 1609, probably in Holland, makers of eyeglasses discovered that they could combine lenses to magnify. This discovery led to the invention of the telescope. Galileo's brilliant idea was to turn the newly invented telescope skyward for a systematic study of the heavens. His first telescope had a magnification of only a factor of 3 and could be duplicated today for a dollar's worth of materials. Even his best telescope, with a magnification of a factor of 30, was inferior to good-quality modern binoculars. But his discoveries forever changed the way people thought of the universe.

Galileo observed mountains and plains on the Moon, proving that it was not a polished sphere or supernatural orb of the "perfect" celestial realm, but a geological world like Earth. He detected sunspots on the surface of the Sun and tracked their motion, deducing that the Sun rotated once every 4 weeks. The Sun was therefore not perfect either, and if it could rotate, why not Earth? Most important

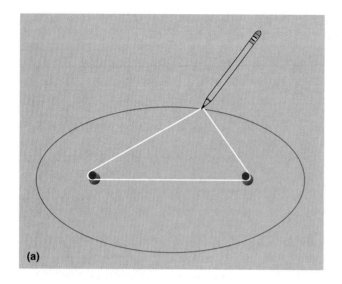
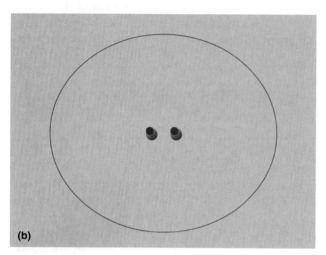
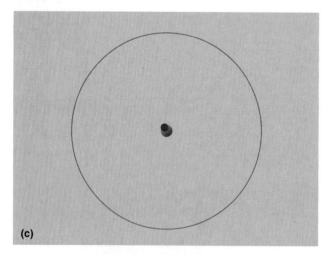

FIGURE 3–11
The properties of ellipses. **(a)** An ellipse can be drawn by looping a string around two tacks. The position of each tack is called a *focus* of the ellipse. In the case of planet orbits, Kepler's first law of planetary motion states that the orbits are ellipses with the Sun at one focus of each orbit. **(b)** When the two foci are close together, the ellipse is more like a circle. **(c)** A circular orbit is an extreme case of an ellipse where the two foci coincide; planets' orbits are nearly circular with the Sun at the center.

FIGURE 3-12
Galileo Galilei (1564–1642) was the first to use a telescope to establish the nature of the planets. He provided critical new evidence to support the Copernican hypothesis, and he established an experimental method that is typical of modern science.
(SOURCE: Corbis–Bettmann.)

were his discovery and subsequent observations of four moons moving around the planet Jupiter. This observation was crucial. It showed, once and for all, that not all bodies move around Earth.

Another important observation for the debate about the Copernican model involved the phases of Venus. Look at Figure 3–2. In the Ptolemaic model, Venus is always between Earth and the Sun, so it will always show a crescent phase since most of the lighted side faces toward the Sun. Now look at Figure 3–4. In the Copernican model, Venus can sometimes lie between Earth and the Sun, when it will have a new or crescent phase. But there are parts of the orbit where it lies on the opposite side of the Sun from Earth, when we see a full phase as the lighted side faces back at us. Figure 3–13 shows the distinction between the geocentric and heliocentric predictions. By studying the illumination of Venus over a period of months, Galileo showed that Venus went through a full cycle of phases. Not only that, but the full phase occurred when the angular size of Venus was smallest, as the Copernican model would predict (Figure 3–14). This was compelling evidence that the heliocentric view was correct.

We can see perhaps why Galileo's debate with the Catholic Church turned ugly (part of the reason was undoubtedly Galileo's cantankerous personality, since he was never one to shy away from an argument or a confrontation). In the first years after Copernicus, the heliocentric model was a mathematical abstraction—a hypothesis that presented no real threat to the established order. Then Galileo used his telescope to dethrone Earth from its special place at the center of the cosmos. It was just one of a number of worlds scattered through space. Galileo lectured and published in Italian rather than Latin, which was the traditional scholarly language. In this way he reached many people. He presented the Copernican model as an established fact.

During his long period of house arrest, Galileo completed his other major work on mechanics, or the science of motion. Aristotle had taught that rest was the natural state of any object, and that a heavier object would fall faster than a light one. Galileo showed that both of these ideas were wrong. He compared the motions of balls of different mass rolling down inclined planes, using his pulse as a timing

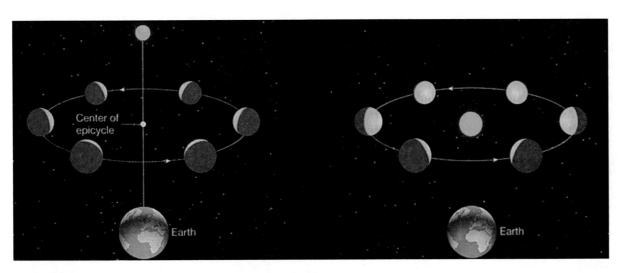

FIGURE 3-13
The phases of Venus. The changing illumination of Venus is a critical test of the Copernican hypothesis. **(a)** In the geocentric cosmology of Figure 3–2, Venus lies between the Sun and Earth at all points in its orbit. The result is that the illuminated face of Venus would point toward the Sun and we would see very little change in phases. **(b)** In the heliocentric cosmology of Figure 3–4, Venus moves from being between the Sun and Earth to being on the far side of the Sun from Earth. In this case, Venus would go through a full cycle of phases, exactly as Galileo observed.

FIGURE 3-14
Phases of Venus as seen through a telescope. Compare to Figure 3-13 to see where in Venus' orbit each photo is taken. Notice how the planet's apparent size changes with phase. This is consistent with the heliocentric model, which predicts a large change in the relative distance of Venus from Earth as it orbits the Sun. (SOURCE: Courtesy of Lowell Observatory.)

device. The inclined plane was just a device to slow the motion down enough that he could measure it—he never actually dropped objects from the Leaning Tower of Pisa. Galileo realized that friction on a surface, or air resistance, would affect the motion of a falling object. If these effects were minimized, a ball rolling on a flat surface would keep rolling. Uniform motion was just as natural as rest. In addition, objects of any material would fall toward Earth in the same way. The **acceleration,** or rate of change of velocity, does not depend on the weight or composition of the object.

Many of Galileo's ideas in mechanics were relevant to the motions of objects in space. He came up with the concept of **inertia,** the resistance of any object to a change in its motion. If a ball is dropped from the mast of a tall ship, the ball will not fall behind the ship but will continue with the forward motion of the ship and land at the base of the mast. If the atmosphere is being carried with the rotation of Earth, then we might not feel our motion in space, Galileo supposed. Similarly, someone confined to the cabin of a smoothly sailing ship could do experiments and never be aware of their motion. Galileo was convinced by these arguments that Earth could be in motion without our being aware of it. In this way yet another of the objections to the heliocentric model was overcome.

Why is Galileo important in the history of science? He was a pioneer of observational astronomy. He completed the Copernican revolution with his telescopic observations. In a broad sense, he was the first modern scientist. Rather than accept the established wisdom on any subject, he preferred to conduct his own experiments—to read the "book of nature."

Newton's Great Synthesis

The next phase of the Copernican revolution provides an excellent example of how science works. In the 1600s, scientists hammered out the method by which modern scientists approach a problem. (See the summary flowchart in Figure 1-10.) Kepler deduced that elliptical orbits were a good description or model of the motions of the planets. He also found patterns in the motions and spacing of the planets that supported the Copernican hypothesis. Astronomers now needed a more complete physical theory to explain and predict the observed phenomenon. To be successful, the theory would have to start with a few universal principles and show that Kepler's elliptical orbits were a consequence of these principles. If such a theory could be developed, scientists believed they would be able to understand other phenomena in nature as well.

The man who achieved this synthesis—the man usually considered the greatest and most creative physicist who ever lived—was Isaac Newton (Figure 3-15). Newton revolutionized the fields of physics and astronomy. He was born in the south of England in 1643, just after the death of Galileo. He was a lonely and moody boy, often preoccupied by his own thoughts. Between the ages of 23 and 25, while attending Cambridge, he single-handedly developed calculus, discovered the principle of gravitational attraction and certain properties of light, and improved the design of the telescope. Newton once said that he made his discoveries "by always thinking about them without ceasing," a trait that no doubt contributed to his reputation for absentmindedness.

Newton thought deeply about the way objects move and came up with three laws of motion that describe the mechanics of objects in the everyday world. The first stemmed from Galileo's ideas about inertia: A body at rest stays at rest or moves at constant speed in a straight line unless an unbalanced force acts on it. There are several important ideas here. The first is that uniform motion is just as natural as no motion, which is a major departure from Greek physics. The second is that an object changes its motion because a **force** is acting—this is the way we define the concept of a force. For example, you know that a rolling ball always comes to a halt. Newton realized that it slows down because the force of friction is acting. In a situation with less friction, such as a hockey puck sliding across ice, the motion would be nearly uniform. What do we mean by "unbalanced force"? When you sit on a chair, you do not move because the chair pushes up on you with the same force that gravity pulls down on you. If it did not, the chair would break and your stationary state would suddenly change!

FIGURE 3-15
Isaac Newton (1642–1727) discovered the laws of gravity and motion and showed why planets' orbits must be ellipses. He is shown here using a prism to make his discovery that sunlight is a mixture of all colors. (SOURCE: The Granger Collection, New York.)

onto the pier is countered by a force that sends the boat away from the pier. Sometimes the reaction force is concealed by the very different masses of the two objects. When you jump into the air, you do exert a downward force on the entire planet Earth, but the mass and inertia of Earth are so large that you cannot perceptibly change its motion in this way.

Newton's understanding of the way forces act in the universe was so broad that the unit of force is named after him. As with most of the other physical concepts in this book, force can be described in terms of the three fundamental units we listed in Chapter 1: mass, length, and time. Appendix A-2 shows how many complex parameters can be reduced to combinations of these three units. **Newton's laws of motion** are the basis of the subject of *mechanics,* the study of the way objects move (Figure 3–16). We summarize them as follows:

1. An object stays at rest or in constant motion unless an unbalanced force acts on it.
2. An object responds to a force with an acceleration that is proportional to the force and in the direction of the force, and inversely proportional to the mass.
3. For every force on an object, there is an equal and opposite reaction force.

Newton's laws of motion have a profound implication for how science works. We saw in Chapter 1 that science depends on the assumption of causality—that every effect has an identifiable cause. Aristotle was forced to explain the different motions of a falling feather and a falling rock in terms of their different "natures." Newton removed all this

The second law mathematically relates a force to the change in motion it causes. For every force, there is a corresponding acceleration, which is proportional to the force and is in the direction of the force, and inversely proportional to the mass of the body. In other words, force equals mass times acceleration. In equation form, $F = ma$. The key concept here is **mass,** which is a measure of inertia or the resistance of any object to a change in its motion. If you push hard on a shopping cart, it gives little resistance to the force, so you can change its motion or accelerate it. A similar push on a car (with the brake off) would produce little change in its motion. For any object a larger force produces a larger acceleration. You can also think of mass as the amount of material in any object.

The third law is sometimes called the *principle of action and reaction.* It states that for every force (sometimes called an *action*) there is an equal and opposite force (called a *reaction*). We do not see single isolated forces in nature. There is always an opposite-directed force as well. Take the case of rocket propulsion. Rocket fuel creates a force which accelerates the rocket forward, but an equal force ejects exhaust gas at very high speed backward. If you foolishly punch a wall, the force you exert on the wall is countered by an equal force on your hand that will cause much pain. The same principle occurs in the recoil of a gun. Or consider the fact that when you step off a small boat onto a pier, the force that gets you

FIGURE 3-16
Newton's laws of motion apply to objects in the everyday world. When the brakes failed, the train tried to stay in motion (first law). When it plowed through the wall, the force of Earth's gravity accelerated it downward (second law). When it hit the ground, the ground exerted an upward force equal to the weight of the train (third law). (SOURCE: H. Roger Viollet, The Liaison Agency)

unexplained mystery by identifying causes for motions of all kinds. A rolling ball stops because the force of friction is acting. The bird keeps flying because its beating wings exert an upward force on the air. Equally, nature is not capricious. Fallen objects do not spontaneously right themselves. Newton's laws establish that the universe is a rational place where events have causes.

Newton's work also had a practical consequence. The systematic understanding of motions and forces led clever inventors to try to harness those forces in useful machines. The spinning "jenny" and the steam engine were the first examples in a sequence of innovations that would transform the economic landscape of England and eventually the world. Within a hundred years of Newton's work, the industrial revolution was underway.

A Universal Law of Gravity

Newton's most brilliant step was to unite motions on Earth and in the sky. He knew that a force must cause an object such as an apple to fall to Earth. He also knew that the Moon must have a force acting on it to make it travel in a curved path as it orbits Earth. Could these be the same force? (It is not known if the story of Newton's getting his insight from watching a falling apple is true, but there is an apple orchard outside his childhood home in England!) Newton's **universal law of gravitation** is one of the most important discoveries in the history of science. Newton expressed it mathematically (see the Science Toolbox). In words, we can write: Every particle in the universe attracts every other particle with a force proportional to the product of their masses and inversely proportional to the square of the distance between them. We therefore define gravity as a universal attractive force; it is a property of every object in the universe. This expression of gravity means that if you could double the mass of the Sun, the Sun's gravitational attraction to Earth would double; but if you doubled the distance between them, the force on Earth would decrease by a factor of 4 (the square of 2). Similarly, if you tripled the Sun's mass, the force would triple; but if you tripled the distance, the force would decrease by a factor of 9.

Gravity is a force that follows an **inverse square law.** That is, the strength diminishes with the square of the distance from the object. Newton was able to show that if gravity followed an inverse square law, then the planet orbits that are predicted must be ellipses with the Sun at one focus. Here at last was the elegant explanation for Kepler's work. Newton's gravity was used to successfully predict the return of Halley's comet on its highly elliptical orbit of the Sun—a stunning confirmation of the theory. The statement of gravity given above applies to two single particles. How, then, can we calculate the force of gravity on a person, when the attractions due to all the different parts of Earth must be added together? Newton invented the calculus to solve this problem—with the result that a spherical object such as a planet behaves as if all its mass were concentrated at the center. Newton did many of these important calculations and then buried them on his desk, as his voracious mind found new problems to study. Spurred on by Edmund Halley, the famous astronomer, Newton finally published his work in a masterful book called *Principia*, in 1687.

We can see how Newton's gravity illustrates his laws of motion. Gravity is the force that keeps the Moon in its curving motion around Earth, or Earth in its curving motion around the Sun. In the absence of gravity, Earth would just fly off straight through space, just as a stone whirled overhead on a string would fly if the string broke. In the vacuum of space, there is no friction or air resistance, so the solar system can maintain its motions for a very long time. Also, gravity is a mutual force that illustrates Newton's third law of motion. Earth exerts a gravitational force on you, but you exert an equal gravitational force on Earth! Gravity has a long reach. It declines in strength as the distance increases but never becomes zero. Using this logic, Newton was sure that gravity was a universal force.

It is important to keep in mind the difference between mass and **weight.** People tend to use the words interchangeably but they are quite different concepts. Mass is a fundamental property of an object or a particle. It is the amount of "stuff" or the number of atoms in something. Mass is measured in units of kilograms in this book. Mass determines the force of gravity. Weight depends on your location in space. On Earth's surface, the acceleration due to gravity is $9.8 m/s^2$—usually given the symbol g (to distinguish it from the universal constant G). This is an increase in speed of 9.8 m/s for every 1 s of falling. On the smaller and less massive Moon, the acceleration due to gravity is only $1.6 m/s^2$. Your mass on Earth and the Moon would be the same—the number of atoms in your body does not depend on your location. However, your weight on the Moon would be $1.6/9.8 \approx 1/6$ of your weight on Earth. In orbit around Earth, you would be weightless (Figure 3–17). This is because you and the spacecraft experience the *same* gravity force, and there is no difference between the force on you and the spacecraft. You would also be weightless in deep space, far from the gravity of any star or planet. In all these situations, your mass is the same. But your weight depends on the local gravity force.

Newton the man was a bundle of contradictions. He could be humble when thinking of his predecessors. He said, "I feel as if I have been turning over pretty pebbles on a beach, while the vast ocean of knowledge lay undiscovered before me." Yet he could be brutal with colleagues and rivals. He also said, "If I have seen any further than others, it is because I have stood on the shoulders of giants." While this second quote has been taken as Newton's homage to his scientific predecessors, it is in fact a reference to physicist Robert Hooke, with whom Newton was in a dispute. Hooke was a dwarf and a hunchback. The man who gave birth to the rational and mechanistic view of a "clockwork" universe spent much of his effort on alchemy. After he died, his estate was found to contain thousands of pages of detailed analysis of the Bible.

Newton the scientist is easier to judge. He used one simple law to understand a huge variety of seemingly unrelated effects: your weight, an apple falling from a tree, the Moon's moving around Earth, the arcing path of a comet, or the planets moving around the Sun. The English writer Alexander

SCIENCE TOOLBOX

The Inverse Square Law of Gravity

Newton's universal law of gravitation is a common type of law called an *inverse square law*. The force of gravity due to some object depends on the inverse square of the distance. This dependence is typical of many forces that emanate from a point—in fact, it is a property of the three-dimensional space we inhabit.

As shown in Figure 3–C*a*, we can imagine lines of gravity force emerging from an object in all directions. The concentration of the lines of force is then a measure of the strength of the gravity. The surface area of any sphere surrounding the object is $4\pi D^2$, where D is the distance from the object. You can see from the figure that the lines of force spread out more and more as we move away from the object. This decreasing concentration corresponds to the decreasing force of gravity. The total number of lines of force does not increase moving away from the object, but they are spread over an area that increases proportionally to D^2. Therefore the concentration or number of lines of force in any particular area decreases with the inverse square of the distance, that is, proportionally to $1/D^2$.

Electrical forces often obey the same relationship. You can also understand how the inverse square law would apply to the radiation from a lightbulb, reaching surfaces at different distances (Figure 3–C*b*). Imagine rays of light moving out from a light source. If at a certain distance the light emitted in a certain direction is spread over 1 unit, then at twice that distance it is spread over 4 units. The intensity of the light at twice the distance is thus one-fourth as much. In the same way, the intensity at 3 times the initial distance is only one-ninth as much. So light, like gravity, obeys an inverse square law.

Let us follow Isaac Newton's reasoning as he used the inverse square law of gravity to connect the falling apple with the orbit of the Moon. We will use simple geometry to get the answer roughly—calculus is required for a more accurate result. The distance from the Moon to the center of Earth is 60 times the radius of Earth. So by the inverse square law, Earth's gravity at the distance of the Moon should be 60^2 times less than at Earth's surface. The acceleration of the Moon that causes it to deviate from a straight path and curve around Earth is $9.8/60^2 = 0.0027$ meters per second per second (m/s^2). In each second of its orbit, the moon falls $0.0027/2 = 0.0014$ m, or only 1.4 mm toward Earth.

Now how far does the Moon travel in its orbit in 1 s? This is just the circumference of the orbit divided by the orbit time in seconds. The orbit time is 27.3 days, or $27.3 \times 24 \times 3600 = 2.36 \times 10^6$ s. (Note that the time for the Moon to complete a

(continued)

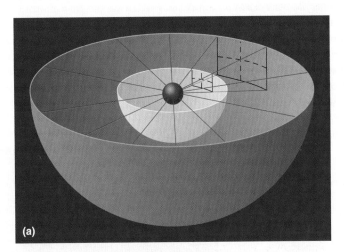

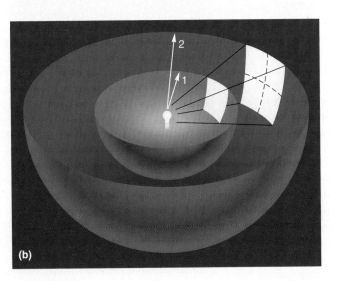

FIGURE 3–C
The inverse square law. **(a)** The force of gravity decreases according to the square of the distance from the mass. The figure shows imaginary lines of force emerging from an object. The thinning out of the lines of force indicates the diminishing strength of gravity with distance. **(b)** The intensity of light decreases according to the inverse square of the distance from a light source. At twice the distance the light is spread over 4 times the area. At 3 times the distance, it is spread over 9 times the area.

SCIENCE TOOLBOX
The Inverse Square Law of Gravity (continued)

cycle of phases—29.5 days—is longer than the orbit time because Earth moves in its own orbit in 1 month.) So in 1 s the Moon travels $2\pi (384,000) / (2.36 \times 10^6)$ = 1.02 km. For such a small piece of the orbit we can approximate the curved path as a straight line and use the small-angle equation from the Science Toolbox in Chapter 2. Figure 3–D shows the situation, where the tiny angle has been exaggerated to be visible. The small-angle equation gives $\alpha / 206,265 = d / D$, so α = 206,265(0.0014 / 1020) = 0.3 arc second. This is the angle by which the Moon deviates in its orbit each second due to the gravity of Earth. The angle by which it rotates in its orbit is also shown in Figure 3–D. The small-angle equation gives $\alpha / 206,265 = d / D$, so α = 206,265(1.02 / 384,000) = 0.5 arc second, a similar number. So we can see that an inverse square law of gravity describes the Moon's orbit.

It is actually easier to write the law of gravitation in mathematical terms than to spell it out in words. If we have two masses M_A and M_B, separated by a distance R, then Newton's law of gravity gives the force between mass A and mass B of

$$F = \frac{GM_A M_B}{R^2}$$

The number G is the gravitational constant, a fundamental constant of nature. If we measure mass in kilograms, distance in meters, and force in the normal units of newtons (N), $G = 6.67 \times 10^{-11}$ N $\times$ m^2 $\times$ kg^2. The constant G is a tiny number—gravity is actually a very weak force. It is only the presence of enormous amounts of matter that gives gravity a sizable force.

For example, the mass of Earth, 6×10^{24} kg, gives us a downward acceleration of 9.8 meters per second per second (m/s^2). But everything with mass attracts everything else with mass, so what about other objects? Suppose two ocean liners of mass 10,000 tons (10^7 kg) were sitting in the water separated by 100 m. The force between them would be (6.67 $\times$ $10^{-11} \times 10^7 \times 10^7$) / 100^2 = 0.67 N. From Newton's second law of motion, acceleration = force / mass, the acceleration on each ship would be 6.7×10^{-8} m/s^2. This is clearly too tiny to measure—the gravity of astronomical objects clearly overwhelms the gravity of everyday objects!

If we want to compare the gravity due to two different objects, we can take ratios and do not need to use the gravitational constant. To compare the relative gravitational force of objects B and C on object A, we divide:

$$\frac{F_{B\text{ on }A}}{F_{C\text{ on }A}} = \frac{M_B}{M_C}\left(\frac{R_{C\text{ to }A}}{R_{B\text{ to }A}}\right)^2$$

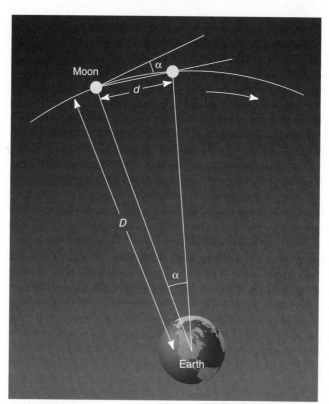

FIGURE 3–D
A direct demonstration that the force of gravity unites the motion of objects on Earth and the orbit of the Moon. The acceleration of the Moon in its orbit is related to the acceleration of a falling object on Earth by the inverse square law. The tiny angle α—exaggerated in this figure so as to be visible—is the amount by which the Moon is deflected toward Earth each second by gravity. Simple geometry shows that this is just the angle required for the Moon to travel in an orbit of Earth.

The gravitational constant and the mass of A cancel. We are pinned to Earth by its gravity. Let us see how the gravity of other objects compares. Using the equation above where B is the Sun and C is Earth, we see that the relative force is $(2 \times 10^{30} / 6 \times 10^{24}) \times (6400 / 1.5 \times 10^8)^2 = 6 \times 10^{-4}$. So the Sun exerts less than 0.1 percent as much gravity on you as Earth does. What about two people sitting in a room? If they each have mass of 50 kg and are 1 m apart, the relative force is $(50 / 6 \times 10^{24}) \times (6400 / 0.001)^2 = 2 \times 10^{-9}$. This is one-billionth of the force that keeps us down to Earth. Whatever else might attract two people, gravity has very little part in it!

FIGURE 3-17
Mass and weight are not the same thing. These astronauts are weightless because they are "falling around" Earth along with their spacecraft. With no net force acting, they float in their surroundings. However, they have the same number of atoms as they did on Earth's surface; their mass has not changed. (SOURCE: NASA.)

Pope got it right when he penned this clever verse upon Newton's death:

> Nature and Nature's laws lay hid in night;
> God said, "Let Newton be!" and all was light.

The Plurality of Worlds

Until the Copernican revolution and the invention of the telescope, Earth seemed the unique center of everything. The Copernican revolution displaced Earth forever from its central role. The excitement of this revolution at the time is hard to overemphasize. It seemed for the first time that humans could understand the "machinery" that runs the universe. Johannes Kepler recalled the moment around 1604 when he studied the observations of Mars' position from earlier years. He was the first person in history to realize that planets moved not in circles around Earth but in ellipses around the Sun. "I awoke as if from sleep, a new light broke on me," he said.

A light broke on a lot of other people, too. We have accounts of messengers running through the streets of European cities to homes of famous scientists, shouting in front of their houses about the latest discoveries and what they might mean. New knowledge can cause turmoil, too. The conflict between scientists and the Catholic Church arose because the heliocentric model broke the cozy pact between humans and God that placed the Earth at the center of creation. The revelations of the telescope did not sit comfortably with some people. Early telescopes were not perfect, but the images they produced became more and more reliable. Yet we hear of people who were unwilling to believe the evidence of their own eyes because it did not accord with their preconceived notions.

When Kepler read Galileo's descriptions of stars and planets as viewed through a telescope, he "awoke" even further. He realized that Giordano Bruno might be right after all, in speculating that stars and planets were not supernatural bodies, but worlds. Galileo's work showed that planets were spherical worlds like Earth, and the Moon could even be seen to have mountains and valleys. Kepler also cited Bruno and went further to conclude that stars are distant luminous objects like the Sun, whereas planets are moons or Earths. There might be many worlds. Perhaps some were even habitable! The hypothesis that Earth is not the only world in the universe is called the **plurality of worlds.**

Scientists also speculated freely about the size of the universe. In the Copernican model, the stars had to be very distant because no change in their brightness or relative positions was observed as Earth orbited the Sun. More and more stars became visible as the light grasp of the telescope improved. Bruno speculated that there was "not one, but countless suns, not a single earth, but a thousand, I say, an infinity of worlds." Such extravagant and heretical thinking cost him his life. Newton used logic to conclude that the universe was infinite. Gravity is a universal and attractive force, so if there were an edge to the distribution of stars, then stars near the edge would be pulled by gravity toward the center of the distribution. Newton reasoned that the universe could not be stable unless it was infinite.

These ideas spread rapidly among intellectuals after the Renaissance. Just to take one example, we can paraphrase the questions raised by an English scholar named Robert Burton (1577–1610):

> . . . who dwells in these vast bodies, these Earths and worlds?
> Rational creatures? Have they souls to be saved?
> Do they inhabit a better part of the Universe than we do?
> Are we or they the Lords of the Universe?

Those were pretty heavy-duty questions for the 17th century, and they are unanswered today. The Copernican revolution was the first major step in our awareness of our true place in the universe.

The Layout of the Solar System

The properties of elliptical orbits discovered by Kepler can be derived from Newton's laws of motion and his law of gravity. In other words, once Newton figured out his laws of gravity and motion, he could *predict* Kepler's laws of planetary orbits. An important exercise in more advanced astronomy courses is to derive all three of Kepler's laws from Newton's laws. If Newton's laws are true, the Copernican theory and Kepler's laws also have to be true. Newton therefore tidied up the miscellaneous observations of preceding centuries and completed the Copernican revolution.

The gravity law proposed by Newton represents an impressive example of the power of the scientific method. It was developed to explain the motions of only eight astronomical bodies—the Sun, Earth, the Moon, and the five visible planets. Yet by calling it a *universal* law of gravity, we are making the presumption that it applies everywhere in the universe. As you will recognize from the discussion in Chapter 1, this is an audacious use of induction! Soon afterward it was used to explain the orbits of the moons of Jupiter and to predict the return of Halley's comet. Later astronomical observations have shown that Newton's laws also apply in other parts of the universe. For example, we will see later in the book that they correctly predict cosmic movements ranging from the motions of pairs of stars that orbit each other in remote regions of our galaxy, to the motions of pairs of galaxies that orbit each other in remote parts of the universe.

The Scale of the Solar System

By the time of Newton's death in 1727, at age 84, the solar system was conceived essentially as we know it today. It lacked only the discovery of the three outer planets—Uranus, Neptune, and Pluto. The true layout of the solar system is shown in Figure 3–18. Surrounding the Sun are the orbits of nine known planets. (Many astronomers today would prefer to say eight major planets, because the last "planet," Pluto, is smaller than our own Moon and is now regarded as only the largest of a swarm of bodies in the region of Neptune and beyond. We will discuss this further in Chapter 7.)

The four small planets close to the Sun—Mercury, Venus, Earth, and Mars—are called *terrestrial planets* because they are relatively close to Earth in size, and because they all have a rocky and metallic composition, like Earth. Beyond Mars is a belt of thousands of asteroids—small rocky and metallic bodies, most of which are less than 100 km (60 mi) across. The biggest asteroid is 1000 km (600 mi) across. Beyond the asteroid belt is a colder region of the solar system, where four gas giant planets are composed mostly of ices and high-pressure forms of hydrogen and hydrogen compounds.

In addition to these planets, the solar system contains dozens of moons. Galileo discovered only the biggest ones, and many others have been sighted since. Some of these moons are distinct worlds in their own right; two of them are

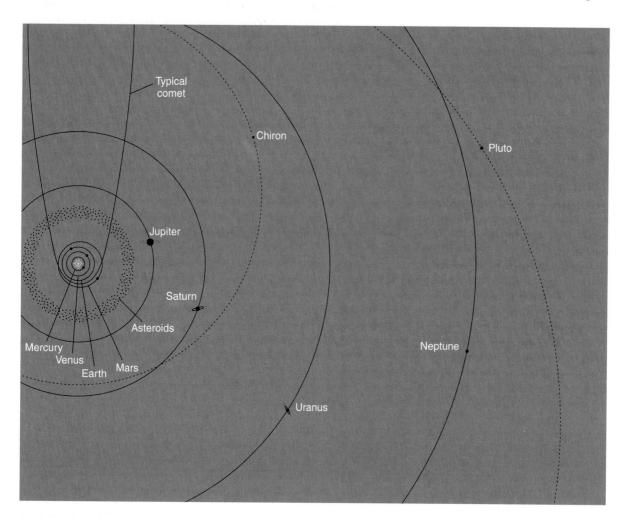

FIGURE 3–18
The arrangement of the solar system, as it is known today, shown approximately to scale. Orbits of the eight largest planets and a typical comet are shown as solid lines. The positions of typical asteroids are shown. Pluto and Chiron (dotted orbits) are just two of a growing number of modest-size bodies on the outskirts of the solar system, whose orbits are noticeably elliptical. The comet has an even more elliptical orbit.

larger than the planets Mercury and Pluto! As we will see later, these moons have distinctive features, including unique atmospheres, erupting volcanoes, and even ice-spewing geysers. Furthermore, as we show in Figure 3-18, the solar system includes additional interplanetary bodies that loop through the system on elliptical orbits that pass among the planets. These bodies range up to a few hundred kilometers in size and include additional rocky asteroids and comets, which are bodies dominated by ice.

How do we know the scale of the solar system? Kepler's laws give the *relative* distances to all the planets, or how far one planet is from the Sun compared to another planet. But they do not give the *absolute* distance in units such as kilometers. The Greeks had used geometry to try to estimate the distance from Earth to the Sun; their crude estimate was 1200 Earth radii, or about 8 million km. Once the telescope had been invented, another approach was to measure the angular size of a planet. A modest-size telescope shows the angular size of Mars to be about 30 arc seconds at its closest approach to Earth. If you assume that Mars is a world the same size as Earth, the small-angle equation (see Chapter 2) yields the distance $D = 206,265 \, (d/\alpha) = 80$ million km. This relies on a complete guess that Mars and Earth are the same size, so it is clearly not a very reliable number.

The best idea is to turn the triangle around. With the diameter of Earth as the base, a parallax angle can be used to measure the distance of Mars compared to the much more distant stars. Remember that this is a very long and skinny triangle—Mars is thousands of times farther away than the size of Earth—and so the parallax angle is small and difficult to measure. Venus is the better choice for the experiment, because twice every 110 years it passes directly in front of the Sun as seen from Earth. In astronomy this is called a *transit*. The Sun provides a perfect backdrop for observing the small dark disk of Venus; and with observations from two widely separated sites, the parallax angle can be measured and the distance to Venus calculated.

But there are two problems. First, Earth is spinning. The two observations have to be made at the same time, and an error in time measurement on a spinning planet is equivalent both to an error in distance and to an error in the measurement of parallax. The angle of stars can be used for navigation in latitude (recall our discussion of Polaris in Chapter 2). But timekeeping is required for navigation in longitude. Earth spins at 15° per hour, so a clock that is off by 30 min after a long sea voyage would give a navigational error of 7.5°/360° times the circumference of Earth, or 1000 km! Poor clocks were the reason for most shipwrecks and for the famously poor navigation of Christopher Columbus. This problem was solved in the middle of the 18th century, when the British government realized that an accurate clock was the key to mastery of the seas. In response to a competition for a prize of £20,000, John Harrison produced a clock accurate to 5 s over 80 days—an outstanding timekeeping device.

The second problem was that you had to be lucky. Edmund Halley knew that a good observation of the Venus transit would fix the scale of the solar system, but he did not live to see one (he did not live to see the return of his comet either!). Amazingly, nobody succeeded in observing the twin transits in 1631 and 1639. Eager astronomers had to wait until the next pair, in 1761 and 1769. Even then, adventurers endured arduous sea voyages only to be thwarted by pirates or typhoons or clouds. Captain James Cook made the first successful observation of a Venus transit in 1769 during a voyage to Tahiti. He measured the distance from Earth to the Sun to within 10 percent of its modern value of 150 million km. The distances to all the other planets can then be calculated from their orbit periods and Kepler's third law.

Astronomers define the mean distance from Earth to the Sun as an **astronomical unit,** abbreviated as AU. Distances to other solar system bodies are quoted in multiples of astronomical units. This is our first useful unit of measurement that is outside the metric system.

The Planets and Their Spacings

Memorization is not an important part of learning science, but some facts are so basic you should commit them to memory. You should know the names of the planets in order outward from the Sun. Generations of students have made up mnemonic devices, which are sentences whose words begin with the letters of the planets. An old (but somewhat sexist) mnemonic sentence for the planets was "Men Very Early Made Jars Stand Upright Nicely." After Pluto was discovered, some students added "Period" to the end of the sentence. In an earlier text, we challenged students to come up with a better sentence. One winner (from the University of Hawaii) was "My Very Erotic Mate Joyfully Satisfies Unusual Needs Passionately." Many students lose track of the sequence around Saturn, Uranus, or Neptune, and so it is helpful to remember that the SUN (standing for Saturn, Uranus, Neptune, in that order) is also part of the system.

Once you have learned all the planets you should also be aware that they are not evenly spaced in their distance from the Sun. In 1772, the German astronomer Johann Titius discovered a curious relationship governing the planetary distances. It was popularized by his colleague Johann Bode, and it is usually known as *Bode's rule*. Bode and Titius noticed that each planet is about 1.5 to 2 times as far from the Sun as the previous planet. This is called a *geometric progression,* where each number is a fixed factor larger than the previous number (as opposed to an *arithmetic progression,* where each number increases by the same additive constant). The discovery of Uranus in 1781 also fit the progression because it is about twice as far from the Sun as Saturn. Astronomers then noticed that Bode's rule also predicted a planet between Mars and Jupiter. German and Italian observers, nicknamed the "celestial police," set out to find the missing planet, but instead discovered Ceres and several other large asteroids at just the right distance! Today we know of thousands of asteroids between the orbits of Mars and Jupiter.

Why should we be surprised that the planets have an apparent pattern in their spacing from the Sun? Newton's law of gravity describes the force between two objects in space. This in turn gives the relationship between the distance from the Sun and the orbital period of a planet. But the gravity law allows for planets at *any* distance from the Sun. In other words, the spacing of the planets is not built into Newton's laws. So we must look for a new physical explanation for this roughly geometric progression. We now believe that this spacing is controlled by the process of planet formation, which we will discuss in Chapter 9.

Exploration of the Solar System

Dreams of Escaping Earth

The idea of space exploration is another extension of the Copernican revolution. Most people do not realize that our ability to fly spaceships to the Moon or Mars is a direct result of Newton's work in the 1600s! The dream of such flights is even older. The first known description of a flight to the Moon is from—did you guess it?—an ancient Greek writer. Lucian of Samoset in the year 190 had one of the characters in his story don eagle wings and fly to the Moon, in order to learn how the stars came to be "scattered carelessly up and down the universe." But the actual scientific basis of space travel came from Newton. How fast do you have to go to escape Earth's gravity and fly into space? How long would it take to fly to the Moon, once you reach that speed? What orbit would require the least energy and fuel to reach Mars? Newton could have answered these questions using his simple laws of gravity and motion. And students in introductory physics classes can learn how to answer them, too.

In fact, Newton realized that a satellite could be launched into space. In his 1687 masterwork, *Principia*, he drew a picture similar to Figure 3–19 and described a "thought experiment" to see how it could be done. Scientists sometimes use thought experiments to help them in their work. A thought experiment is a hypothetical experiment—a physically reasonable idea that is difficult to carry out in practice. For example, one of Newton's laws of motion states that an object in uniform motion (constant speed) will continue in uniform motion. Yet we know that a projectile is always subject to air resistance. And a rolling object is always subject to friction. So Newton could not realize his ideal of a uniformly moving object. Yet he was confident that if friction or air resistance could be eliminated, uniform motion would be the result.

Imagine a mountain high enough that it projects above Earth's atmosphere, said Newton. Now imagine a cannon pointing out from the mountaintop, parallel to Earth's surface. If you fired a cannonball at modest speed, the ball would fall near the foot of the mountain (Figure 3–19). At a higher speed, the ball would fall farther away. At a high enough speed, the ball would be falling toward Earth by the force of gravity at exactly the same rate that the curved surface of Earth was falling away. In this case it would continue to travel all the way round Earth. This is just how we define an orbit (Newton's idea is theoretical—in practice, friction in the atmosphere would cause a rapid decay of the orbit.).

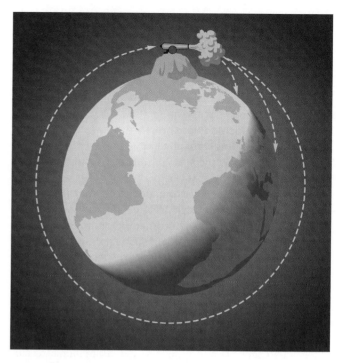

FIGURE 3-19
A cannon on a high mountain could put its projectile into orbit if it could achieve a high enough velocity. Newton published a similar figure in *Principia*.

Note that each path obeys Kepler's laws: Even the paths that hit Earth's surface are segments of ellipses around the center of Earth. The orbit is the special case of an ellipse that is a circle. Thus, launching a satellite into orbit around Earth is in fact a 17th-century idea of Isaac Newton!

Orbits

Let us examine how Newton's laws can be used to calculate the speed that an object must reach to go into a circular orbit around a planet. The answer depends on the mass of the planet and the distance from the planet. The more massive the planet, the faster the speed. The higher above the surface, the lower the speed. In the region close to Earth, above the atmosphere, the answer is 7.8 km/s. In familiar units, this is 17,500 mi/h—which is why it takes a big rocket to launch a satellite! This speed is the minimum needed to keep an object in space near Earth and is called the **circular velocity** (or orbital speed). The same idea applies to any object in orbit around a larger object. We have already seen in the last Science Toolbox that the circular velocity of the Moon around Earth is 1 km/s. Earth orbits the Sun at an average circular velocity of 30 km/s (Earth's orbit is an ellipse, but it is so close to circular that this is a good approximation). The orbital speed decreases for planets farther from the Sun. Pluto has an average circular velocity of about 5 km/s. Moving out from the Sun, planets have lower and lower orbital velocities. Figure 3–20 shows the variation with distance.

To launch a satellite, all we have to do is to raise it above Earth's atmosphere with a rocket and then accelerate it until it reaches a speed of 7.8 km/s. This reaches what is called a *low-Earth orbit*, at an altitude of about 200 km and an orbital time of 90 min. For the cost of a lot more energy, we can send a satellite to a special orbit called a *geosynchronous* orbit. At an altitude of about 36,000 km, the orbit time is just under 24 h (a sidereal day) and a satellite will rotate with Earth and appear to hang above a fixed point in the sky. The many satellite TV dishes you see are all pointed at satellites in geosynchronous orbits. Satellite launching is big business. Hundreds of satellites will be launched in the next few years to enable global telecommunications. These will be launched into low-Earth orbit by conventional rockets. However, recent advances in technology have led to cannons that can accelerate a projectile to close to the circular velocity, and electronics robust enough and compact enough to survive the forces of such a launch. Amazingly, we may soon see Newton's thought experiment realized! If we increase the speed of the satellite, it goes into an elliptical orbit that extends higher above Earth. If we keep increasing the speed enough, we reach a curve which is related to an ellipse, called a *parabola*; a projectile on a parabola never curves back, but keeps traveling farther and farther from Earth. Eventually it would escape from the vicinity of Earth, and go into orbit around the Sun.

The speed required to escape completely into space from any given point is called the **escape velocity.** Newton's laws show that the escape velocity from any given point near any planet or star is always the $\sqrt{2}$ times the circular velocity. For a point on Earth's surface or just above the atmosphere, it is about $\sqrt{2} \times 7.8 = 11.0$ km/s. We can help the process by launching toward the east from the equator. Earth's spin of 0.5 km/s acts as a slingshot to reduce the requirement on the rocket. If a rocket fired its engines long enough to reach that speed, it would escape from Earth orbit and fly off into a path around the Sun, among the planets. An object that reaches the escape velocity of Earth is still held by the gravity of the Sun. To escape from the solar system altogether requires a speed $\sqrt{2}$ times the circular velocity of Earth in its orbit or $\sqrt{2} \times 30 = 42$ km/s. Four NASA space probes—*Pioneer 10*, *Pioneer 11*, *Voyager 1*, and *Voyager 2*—have left the solar system, never to return.

An orbit is a kind of motion that repeats regularly. Earth or any other planet passes any particular point in its orbit after a fixed interval of time, called the *period* of the orbit. Orbits are just one example of a broad class of physical phenomena, called **periodic processes.** Waves and vibrations are other examples. We can use an ordered deck of cards as an analogy for a periodic sequence. Figure 3–21 shows a deck of cards in which the suits are separated and the cards within a suit are arranged in numerical order. The cards of any particular value are arranged periodically. For example, the jacks would be at positions 10, 23, 36, and 49 in the deck. The spacing is uniform, in a cycle or *period* of 13 cards. The same period applies for any other value of card. Waves and vibrations are all around us (Figure 3–22). We explore the simple characteristics of repeating phenomena in the Science Toolbox on page 77.

A Brief History of Space Exploration

Humans have been dreaming of space for thousands of years. Newton provided the physical laws that form the basis of travel beyond Earth, but for several hundred years the technology did not exist to implement those ideas. The space age is less than 50 years old. During the cold war period, both the United States and the Soviet Union developed plans to launch the first artificial satellites into space for military purposes. Space was a new arena for conquest. The Soviet Union won this race in 1957, launching the first successful artificial satellite, a small science probe called

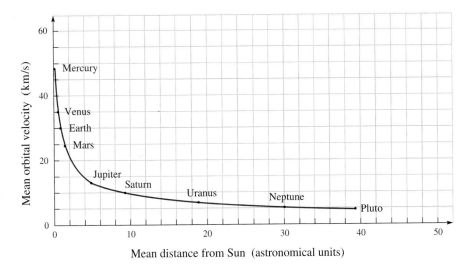

FIGURE 3–20
The circular velocity of an orbit depends on the distance from the source of the gravity. The orbital speeds of the planets go down with distance from the Sun, according to the inverse square root of the distance. At Earth's distance, the circular velocity is 30 km/s.

Sputnik I. This unexpected success shocked the United States into action. Meanwhile, the Soviet Union launched the first person into space in 1961, when popular Soviet pilot Yuri Gagarin made the first trip around the world in space. The Soviets also crash-landed the first human-made object on the Moon in this period and launched probes with mixed success toward Mars and Venus.

In response to the Soviet achievements, the United States made ambitious plans to improve U.S. science education and research funding. President John F. Kennedy made space exploration a centerpiece of the U.S. program to win the cold war competition. The National Aeronautics and Space Administration (NASA) was founded to be a leading force in new technology and science. Kennedy therefore committed the United States to land humans on the Moon by the end of the 1960s, galvanizing the U.S. space effort. The Apollo program was the result. We now know that the Soviets had secretly started their own Moon landing program at the same time as Apollo, building an enormous rocket booster and a lunar

Apollo 8 (Dec. 68)
Apollo 11 (July 69)

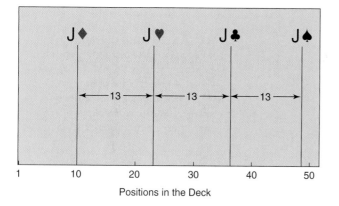

FIGURE 3-21
An ordered deck of cards is a simple analogy for a periodic sequence. Any particular value of card appears every 13 cards; this spacing is analogous to the period.

FIGURE 3-22
The everyday world is full of periodic phenomena such as waves and vibrations. In this picture, water waves, sound waves, and vibrations on a string are represented. Light is also a wave, as we will see in Chapter 10. (SOURCE: Vic Bider, PhotoEdit.)

landing module. They abandoned this program after tests of their large Moon rocket failed in dramatic explosions. In 1968, the astronauts of *Apollo 8* were the first humans to leave Earth's gravity, in a scouting expedition for a lunar landing site. A year later, Neil Armstrong and Buzz Aldrin were the first humans to set foot on another world.

Public interest in the space program peaked at this time. Six landings over the next 3 years placed 12 Apollo astronauts at six different lunar sites (Figure 3-23). However, public appetite for the expensive Apollo program dimmed, and it was abandoned in 1972. Meanwhile, several nations sent robotic probes to other planets, giving our first close-up views of conditions on their surfaces. A Soviet probe returned the first soil samples from the Moon. Several Soviet probes survived descent through the hot atmosphere of Venus, soft-landed, and took the first photographs of its surface. The Soviets also crash-landed probes on Mars, and later U.S. spacecraft landed on Mars and took the first photographs of its surface. A pair of U.S. spacecraft explored all four giant planets and their satellites, and the European Space Agency, a consortium of European nations, launched the first probe to take close-up images of the nucleus of a comet. The result of all this activity is that we have made fly-bys and taken close-up pictures of every large solar system object except Pluto. These voyages are analogous to the first global explorations of planet Earth.

Meanwhile, human space flight also moved forward. The U.S. developed the Space Shuttle system for delivering satellites and scientific experiments to orbit. The Space Shuttle was originally designed to be a "space truck," with weekly launches and eventually even paying passengers. But technical problems and the catastrophic loss of the *Challenger* slowed the Space Shuttle program. Nevertheless, it remains the best way for the United States to get a large payload into orbit. Meanwhile, the Soviets concentrated on building a space station

SCIENCE TOOLBOX

Periodic Processes

We have seen that orbits are repeatable motions in time and space: Earth passes through any particular point in its orbit every 365.25 days, and it follows the same path through space every year. Every 89 days Mercury reaches its maximum elongation or angular separation from the Sun. Every 75 years Halley's comet returns from the depths of space to brighten our skies. These cyclic motions are repeatable and predictable. Spinning motions are also cyclic—every 24 h the Sun rises due to Earth's rotation.

Orbits are an example of a wide range of physical phenomena called periodic processes. Let us look at some other examples. Oscillations or vibrations occur when an object moves backward and forward or from side to side in a regular way. A pendulum is a good example—a weight freely suspended by a string will swing from side to side repeatedly. The time it takes to finish a cycle of motion depends only on the length of the string. Galileo discovered this interesting fact while he was a young student in Pisa. During a service at the cathedral, he noticed that a swinging altar lamp had a period that did not depend on how wildly or gently it was swinging. This periodic motion of a pendulum became an essential part of clocks for hundreds of years.

Mechanical objects can vibrate in a regular way. If you hold a wooden meter rule (or a thin metal ruler) so that it is projecting off the edge of a table and flick the end, you will set up a vibration or a rapid periodic motion. Usually this motion is a blur; but if the ruler is long enough, you can see that several cycles of motion occur every second. Mechanical vibrations can be very rapid. Most musicians use a tuning fork that vibrates 440 times per second! Now we can see the connection to yet another periodic process—waves. The rapid vibration of the arms of the tuning fork sets up a periodic variation in the density of air that we perceive as a sound wave. A vibration of 440 times per second corresponds to A above middle C on the musical scale.

All these apparently diverse phenomena—orbits, rotations, oscillations, vibrations, and waves—share common features. They all follow a cycle, that is, a motion or behavior that repeats. The time it takes to complete a cycle is the *period*, which is usually labeled T. The number of cycles per second is the *frequency*, labeled f. These two quantities are simply related:

$$f = \frac{1}{T}$$

The unit of frequency is the hertz (abbreviated as Hz), after the German physicist Heinrich Hertz. Figure 3–E shows examples of periodic motion, ranging from very long periods or low frequencies to very short periods or high frequencies. In the case of orbits we use a closely related quantity called the *angular frequency*, denoted by the Greek letter omega (ω). Each orbit, an object sweeps through an angle of 360° or 2π radians (rad). So angular frequency is the number of radians moved through per second:

$$\omega = 2\pi f = \frac{2\pi}{T}$$

You might have noticed an interesting feature of the examples in Figure 3–E. Progressing toward smaller physical systems, the period decreases and the frequency increases. Smaller objects rotate (or oscillate or vibrate) faster! This behavior is obvious

(continued)

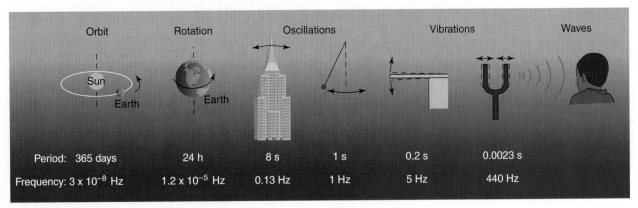

FIGURE 3–E
A variety of periodic phenomena. The period is the time of one complete orbit or rotation or oscillation or vibration. The frequency is how many occur each second. Notice that larger-scale motions correspond to longer periods and lower frequencies.

SCIENCE TOOLBOX

Periodic Processes (continued)

within the solar system, where planets with smaller orbits have shorter periods. You can see it for yourself in the example of the ruler vibrating off the edge of a table. As the ruler vibrates, slide it so that more or less is projecting off the table and watch how the frequency of vibration changes. The same behavior is also familiar from the world of vibrations and sounds. Short piano wires have more rapid vibrations than long piano wires, and they lead to higher-frequency (or higher-pitched) sounds. Also, the highest-frequency sounds come from the smallest brass or woodwind instruments—think of a trumpet compared to a tuba or a flute compared to a bassoon.

There is another striking similarity between the different periodic processes. Their motions in one dimension are described by a *sinusoidal* variation. The motion or displacement from the center position is given by

$$X = X_{max} \cos(\omega t) = X_{max} \cos(2\pi f t)$$

Figure 3–F shows this cyclic variation with time. The maximum amount of motion is given by the quantity X_{max}. Positive and negative positions just correspond to motions to the left or right of the central or resting position, so the total deflection is $2X_{max}$. For the case of the Earth-Sun orbit, X is the apparent separation of Earth from the Sun when the orbit is viewed in the ecliptic plane. For the case of rotating Earth, X is the apparent separation of any point on the surface from the rotation axis, when the rotation is viewed in the plane of the equator. In all the other examples, X is the displacement of the object from the central or resting position. For sound waves, X is the variation in air density as the wave passes any point. It is remarkable that such a wide range of phenomena can be described by the same simple equation.

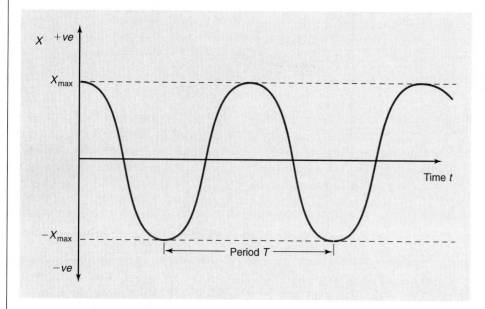

FIGURE 3–F
A wave is a sinusoidal variation. The wavelength is the distance between successive peaks or troughs, and the frequency is the number of peaks or troughs that occur each second.

(Figure 3–24). Russian cosmonauts on the *Mir* space station established world records for living more than 1 year at a time in orbit. This knowledge of the effects of extended weightlessness on humans is essential if we are ever to send people to Mars.

The space age was born out of the competition (and sometimes paranoia) of the cold war. The superpowers raced to lead the world in science and technology as a means to bolster their positions. Since the end of the cold war, budgets for space exploration as well as broader scientific research have been cut. Space exploration is expensive. In an era of tight funding, international cooperation is required to make progress. The U.S. Space Shuttle docked on the Russian space station, and several U.S. astronauts stayed aboard for extended periods. The new international space station draws on both Russian and U.S. technology. Similarly, current plans for robotic probes to Mars involve plans coordinated among the United States, Russia, Europe, and Japan. There has been

FIGURE 3-23
Apollo lunar landing module on its way back to the command module and Earth, after blasting off from the Moon. These space expeditions were bold and expensive; no nation has plans to return to the Moon anytime soon. (SOURCE: NASA.)

a long debate over the relative merits of human exploration versus robotic spacecraft. With the miniaturization of electronics, robotic probes will always be cheaper. However, many researchers believe humans will be needed during geologic exploration of other planets' surfaces. Human involvement also expands human experience.

The high costs of the space program should not obscure its enormous economic benefits. Satellites have transformed global telecommunications. The view from Earth orbit allows us to predict long-term weather more accurately, which increases the yields from agriculture. Satellites produce data to help us understand the complex workings of the oceans and the atmosphere, so that we can be better caretakers of planet Earth. Ironically, the space program is becoming a victim of its own success. There is so much debris in low-Earth orbits that satellites are endangered. Astronomy suffers because of light pollution and radio interference.

What lies ahead in space exploration? Visionaries have always argued that our future lies in space. In space we will continue the exploration that showed us the shape and size of our own planet. In space we will live and work. These visionaries foresee more astronomical telescopes in orbit above our cloudy atmosphere. They foresee space-borne technology to protect Earth from large asteroid impacts that occur every few centuries. They foresee the development of giant solar collectors in space that could collect the free, 24-h/day solar energy and beam it to Earth to reduce our dependence on fossil fuels.

There are thus pragmatic and economic reasons to stay active in space. The ability to launch satellites cheaply will transform global communications. In the future it may be economically feasible to mine rare metals from asteroids. Over the past few decades the space program has transformed from a vehicle for military technology to a means for governments to conduct high-level scientific research. In the future, the commercialization of space will lead to many new possibilities. Companies are already thinking of new propulsion systems—remember that the basis of today's chemical rockets has not changed for 70 years. Space might even become a place for tourism and entertainment. This sounds fanciful until you recall that many space probes cost less than an expensive Hollywood movie, and with a cost of around $10,000 per kilogram to launch into orbit, many rich people could afford to pay for a joyride in space.

To take a long-term view, our ability to understand the solar system and explore space has allowed us—in only the last four centuries—to cast off ancient false views of our planet, expand our finite-world horizons, and realize that we live in a much larger, cosmic environment.

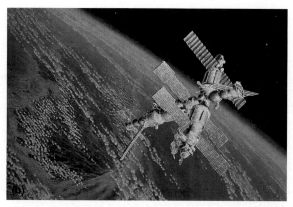

FIGURE 3-24
Complementary systems for developing human capability to work in space. **(a)** The U.S. Space Shuttle system allows delivery of astronauts and supplies but can stay in space for only about 1 to 2 weeks at a time. (SOURCE: NASA.) **(b)** The Russian space station *Mir* offered the first long-term orbital station, powered by large solar panels. The first module was launched in 1986 and the station has been expanded since. Individual astronauts have lived in it for more than a year at a time. (SOURCE: Painting by Russian space artist Andrei Sokolov.)

Summary

The Copernican revolution was a relatively brief period of dramatic change in human perception, in which we went from thinking about Earth as the fixed center of the universe to realizing that Earth is just one of a number of planets moving around the Sun. This change in thinking did not happen overnight, and there was no single observation that overturned the Greek model of the universe. Earlier we presented the principles of how science works. The Copernican revolution shows that the progress of science has many twists and turns and even setbacks. It can be dangerous to pursue a new and unpopular idea.

It was entirely natural to Greeks that Earth was stationary at the center of the universe. But Copernicus showed that the complexity and the arbitrariness of Ptolemy's model could be reduced if the Sun were at the center of the planets' orbits. The elegance and economy of Copernicus' model were strong points in its favor. Tycho Brahe then produced a new and accurate set of data on the motions of the planets. With these data in Kepler's hands, it was clear that the orbits were better described by ellipses with the Sun at one focus than by circles with Earth at the center. Kepler also came up with simple relationships that described the orbits in terms of their distance from the Sun. Galileo made the definitive break with ancient physics. His telescope observations showed that the Copernican model was correct, and that the planets were worlds like Earth, with their own moons and their own geology.

This led to another new idea: the plurality of worlds. This is the notion that there are many planets, that some of them may have properties similar to those of Earth, that there may be planets around other stars, and that some planets may even have alien life-forms. These possibilities are strikingly modern—they also relegate Earth from the center of creation to just one of many astronomical objects.

Isaac Newton transformed the ideas of the Copernican revolution into mathematical laws. These laws of gravity and motion allow us to predict planet and spacecraft motions. Newton's universal law of gravitation gives a mathematical expression for the force of gravity as an inverse square relation, where the force of gravity drops rapidly with distance from the attracting body. Newton showed that a falling rock, the Moon in its orbit, and a comet are all governed by the same type of force. By applying Newton's laws of gravity and motion, we can calculate the speeds needed to travel into interplanetary space, launch large telescopes into orbit, and send probes to the Moon and planets. These probes, in turn, have led to tremendous advances in our understanding of our cosmic environment and our place in it.

Important Concepts

You should be able to define these concepts and use them in a sentence.

retrograde motion	Newton's laws of motion
Copernican revolution	universal law of gravitation
parallax	inverse square law
elliptical orbit	weight
Kepler's laws of planetary motion	plurality of worlds
	astronomical unit
acceleration	circular velocity
inertia	escape velocity
force	periodic processes
mass	

How Do We Know?

These questions and answers show how the scientific method is used to learn about the universe.

Q How do we know that Earth goes around the Sun, rather than being in the center of the solar system?

A This simple question lies at the heart of the Copernican revolution. Since people argued about it for over 2000 years, you can imagine that it has no simple answer! It is certainly true that a heliocentric model with circular orbits is not a closer match to the planetary motions than a geocentric model. The original argument for a heliocentric universe was the fact that it was more simple and elegant than the cumbersome Ptolemaic model. With Earth at the center of the solar system, Mercury and Venus must be in special positions, and there are many possible combinations of distance and epicyclic motion that can account for the behavior of the other planets. In the Sun-centered Copernican system, the planets all orbit in the same sense, and there is a simple relationship between the distance from the Sun and the speed of the orbital motion. The Copernican model accounts for retrograde motion without having to use epicycles. The evidence for the heliocentric model became much stronger once Kepler had demonstrated that planet orbits were ellipses—a geocentric model is a much poorer fit to the data.

Q How do we know today that Earth really moves? Is motion not just relative? Why can we not say that the Sun is moving around us?

A Greek astronomers were able to use geometry to show that the Sun is much larger than Earth. It made more sense to Aristarchus for the small body to go around the larger one. For example, you can swing a light object around you without moving yourself, but you cannot swing an object around you that is heavier than you. Galileo added the idea that you could be moving without *feeling* the motion—just as a passenger in a smoothly moving car or train might not always sense its movement. Newton's laws explained that when two objects orbit each other under the influence of gravity, they actually both move around the center of mass of the system (the balance point if the two masses were connected by a slender rod). This center of mass lies closer to the more massive object. In the case of Earth and the Sun, this relation means that Earth moves essentially around the Sun, while the Sun hardly moves at all. Later, we will describe the technique by which modern astronomers can detect Earth's circular motion around the Sun, relative to all the stars. In other words, we see motion toward certain stars in January when Earth is in one part of its orbit, then toward stars in the other direction in July when Earth is halfway around the Sun and moving in the other direction. If Earth were sitting motionless in the center of the solar system, as Ptolemaic astronomers thought, this motion would not be detected. Similarly, if you could observe from the Sun itself, you would not see this annual motion. Relative to a framework set by the other stars in the sky, Earth is moving around the Sun, and not the other way around.

Q How do we know that Newton's law of gravity applies throughout the universe?

A There are many examples where we have observed the law of gravity working in remote places. Engineers can navigate a spacecraft all the way to Mars or Neptune with positional precision as good as a few hundred meters, using only Newton's law. Astronomers have observed the motions of so-called double stars, in which one star orbits around another. Using measurements of the masses and separation distances in these pairs, we can employ Newton's law of gravity to predict the orbital velocities. The observations show that Newton's laws work wherever we look. (One very important exception was discovered by Albert Einstein. In rare, high-gravity situations, Newton's law must be modified slightly. This is a good example of the continuing refinement of physical knowledge over the centuries.) Strictly, we cannot prove that Newton's law applies *everywhere* in the universe because we have not studied every part of the universe. But we can say that Newtonian gravity has passed many observational tests and is one of the most secure and well-founded theories in science.

Q How did we learn the size of the solar system?

A It all starts with a measurement of the size of Earth. We have seen that Greek astronomers used the simultaneous observation of shadows cast at different places on Earth's surface to measure the size of Earth. They also use simple geometry to deduce that the Sun was much farther away than the Moon—at a distance of 600 Earth diameters. These arguments were logically correct, but the measurements were very crude, so the results were inaccurate. After the telescope was invented, astronomers used geometry to measure the distance to the nearest planets at their closest approach to Earth. The best application of this technique required people to be widely separated on Earth's surface while measuring the transit of Venus across the face of the Sun. This rare event only occurred twice every 110 years, and the measurement was not reliable until there were accurate clocks to allow the determination of longitude. Eventually this procedure led to discovery of the distance to Venus. Given the distance to one planet, Kepler's laws give the distances to all the other planets.

Q How do we know there are only nine planets?

A We do not. In fact, searches continue for more bodies beyond Pluto. These searches have probably ruled out planets as large as Earth, but they have recently turned up some bodies more than one-tenth the size of Pluto. Perhaps more "Plutos" exist at greater distances. Because Pluto itself is only one-half the size of our own Moon, and because searches have turned up other small bodies in that region, many scientists feel that if Pluto had been discovered today, it would not have been classified as a full-fledged planet. Some questions are not really questions of physical science, but rather questions of classification or terminology. It is important to distinguish between these types of questions, in order to keep from having arguments over semantics.

Problems

Use these problems to test your understanding of the information and concepts in this chapter. The * indicates a more advanced or mathematical problem.

1. Galileo discovered that the planet Venus changes apparent size and shows phases. When it appears largest, it is close to us and shows a crescent-lighted phase. When it is small, it is far away and shows a fully lighted phase. **(a)** Explain

why these facts disprove the Ptolemaic system, which restricts Venus to positions between the Sun and Earth. **(b)** Explain why these facts are consistent with the Copernican system, in which Venus goes around the Sun in an orbit closer to the Sun than Earth's orbit.

2. Which planet can come closest to Earth? (*Hint:* Use Kepler's third law.)

3. Which planet can come closer to Jupiter, Earth or Uranus? (*Hint:* Use Kepler's third law.)

4. Kepler believed that the motions of the planets were governed by mystical relations between numbers. Following this belief, he showed in his third law that the orbital period always followed a certain relationship to the distance from the Sun. Newton believed that these motions were governed by more fundamental properties of matter, and he discovered laws of gravity and motion that allow us to derive Kepler's laws and many other phenomena. Why is Newton's approach considered to be of greater scientific importance?

5. Because Earth rotates, the equatorial regions move eastward at about 1600 km/h per hour (1000 mi/h). **(a)** Use this fact to explain why it is easiest to launch a satellite into orbit from a base near the equator in an eastward launch direction. **(b)** Jules Verne wrote a science fiction book in the 1800s about Americans launching a flight to the Moon. In his book, the Americans launched their flight from a site in Florida, amazingly close to the Kennedy Space Center where the Apollo rockets were launched. Why was this not just luck, but a reasonable prediction for Jules Verne to make?

6. The section of this chapter about space travel says that the escape velocity is always the $\sqrt{2}$ times the circular velocity. The square root of two is about 1.4. Suppose a space shuttle is in circular orbit around Earth. If you fired the engines to accelerate it in the direction of its motion, how much velocity would you have to add for the shuttle to escape from Earth into interplanetary space around Earth?

*7. One of the early arguments against the heliocentric model was the fact that the stars did not change their relative positions as Earth moved in its orbit. If Earth actually moves around the Sun, the absence of this parallax shift can be used to put a lower bound on the distance to the stars. **(a)** If we imagine the Greek model in which all the stars are fixed on a sphere, then two stars in a particular constellation will subtend a larger angle when Earth is nearest those stars than when it is farthest from those stars. Suppose that the smallest angular shift we can detect without a telescope is $1/2°$. If no shift is seen, use the small-angle equation from Chapter 2 to calculate the minimum distance to the stars compared to the Earth-Sun distance. **(b)** Consider the experiment where people on opposite sides of Earth observe the same star at the same time. If the parallax shift is less than 1 arc minute, what is the minimum distance to the stars in terms of Earth diameter? **(c)** Now suppose the parallax shift is less than 1 arc minute as measured from the opposite sides of Earth's orbit. How much farther away are the stars than you calculated in part (b)?

*8. The distance from Earth to the Sun (called the astronomical unit or AU) is 150 million km (1.5×10^8 km). The number of seconds in 1 year is about π times 10^7 seconds. **(a)** Use these two facts to prove that the velocity of Earth in its near-circular orbit around the Sun is approximately 30 km/s. (*Hint:* The velocity would be the distance traveled, or circumference of the orbit, divided by the time.) **(b)** Using the methods in Problem 6, figure out how fast an object at our distance from the Sun would have to travel, with respect to the Sun, in order to escape from the solar system into interstellar space.

*9. The Moon has about $1/81$ of the mass of Earth, and about one fourth the radius or diameter. Using these facts and Newton's universal law of gravity (in the Science Toolbox), calculate your weight on the Moon. (*Hint:* Note that you do not need to use the actual mass of Earth or the Moon, or the gravitational constant. Instead, write one version of Newton's law of gravitation for the Moon, and then write a version for Earth. Divide the first equation by the second, getting the force of attraction on the Moon divided by that on Earth, or F_M/F_E. This gives the ratio of a person's weight on the Moon to his or her weight on Earth. If you have trouble following these directions, ask your instructor.)

*10. Suppose a car manufacturer enlarges a certain car model and adds features, to the point where the new model weighs twice as much (has twice as much mass) as the old model. To save gas, the manufacturer keeps the same engine, which is able to exert the same forward force on the new model as the old model. Using Newton's second law of motion, describe how the acceleration of the new model will compare with that of the old model. What would be the difference in performance of a lunar "rover" vehicle, driven on Earth and on the Moon?

Projects

Activities to carry out either individually or in groups.

1. On a piece of paper, make a scale drawing of the orbits of the planets, based on Bode's rule or a table of actual orbital radii. Why is it hard to do this? If Earth were the size of a basketball, how far away would the Sun be? How far away would the edge of the solar system be? Repeat these calculations for the case where Earth is the size of a grain of sand—0.3 mm in diameter.

2. If a planetarium is nearby, arrange for a class visit and observe a demonstration of planetary motions across the sky from night to night, such as charted by Tycho Brahe. Alternatively, go out every three or four nights for a period of 3 or 4 weeks and follow the motions of Mars or Jupiter, if they are available in the night sky. Depending on its position in its orbit, Venus might also be favorably placed away from the Sun for this observation. Each time you make an observation, draw the position of the planet compared to nearby stars, and try to include other constellations in your sketch. Draw compass directions and a horizon on each sketch, and note the time and date of your observation. How much did the planet move over the course of your observation? What was the angular rate of the motion? If you knew the orbital velocity and period of the planet's orbit of the Sun, could you derive a distance from your observations?

3. With a telescope or binoculars of at least 2-in diameter, examine Jupiter and the four bright satellites discovered by Galileo. Sketch their positions from night to night or hour to hour. (On a given night, one or more of the satellites may be obscured by Jupiter itself, or Jupiter's shadow.) Note that this repeats Galileo's proof that not all bodies move around Earth. Do all the moons orbit in the same

plane? From the motion you observe, could you derive their orbital periods?

4. Observe the Moon with binoculars as it passes above or below a star. Remembering that the Moon is about 3480 km in diameter, observe how many times the span of its own diameter moves in 1 hour, and use that to calculate the orbital velocity of the Moon.

Reading List

Anonymous, 1992. "Vindication for Galileo." *Science,* vol. 258, p. 1303.

Beardsley, T. 1996. "Science in the Sky." *Scientific American,* vol. 274, p. 64.

de Santillana, G. 1962. *The Crime of Galileo.* New York: Time Books.

Gingerich, O. 1973. "Copernicus and Tycho." *Scientific American,* December, p. 86.

Gingerich, O. 1992. *The Great Copernicus Chase.* Cambridge, England: Cambridge University Press.

Hargrove, E. C., ed. 1986. *Beyond Spaceship Earth: Environmental Ethics and the Solar System.* San Francisco: Sierra Club Books.

Hartmann, W. K., Miller, R., and Lee, P. 1984. *Out of the Cradle.* New York: Workman Publishing Co.

Lerner, L. S., and Gosselin, E. 1973. "Giordano Bruno." *Scientific American,* April, p. 86.

Lerner, L. S., and Gosselin, E. 1986. "Galileo and the Specter of Bruno." *Scientific American,* vol. 255, p. 126.

Lewis, J., Matthews, M., and Guerrieri, M. 1993. *Resources of Near-Earth Space.* Tucson: University of Arizona Press.

CHAPTER 4

Matter and Energy in the Universe

CONTENTS

- Rutherford and the Structure of the Atom
- Basic Structure of Matter
- Nature of Energy
- States of Matter in the Universe
- Heat Transfer and the Concept of Equilibrium
- A Closer Look at Radiation
- Summary
- Important Concepts
- How Do We Know?
- Problems
- Projects
- Reading List

Atoms and energy. This sphere contains 0.15 kg of radioactive plutonium oxide encased in graphite. As the plutonium atoms decay, they release high-energy particles that heat up the carbon atoms in the casing. The sphere emits 100 W of radiant energy and glows with a temperature of 1000°C. (SOURCE: EG & G Mound Applied Technology.)

WHAT TO WATCH FOR IN CHAPTER 4

This chapter concentrates on some very basic physical properties of the universe, and gives enough physical principles to discuss the nature of our Earth and the other planets in the following chapters. The chapter describes the atomic and molecular structure of matter, and you should especially notice how temperature is related to the motions of atoms and molecules. The rest of the chapter describes how energy interacts with matter and flows through it. Two important ideas are that energy can transform from one form to another, and that in a closed system, the total energy is conserved. We also introduce the form of energy flow known as radiation, and we give a simplified introduction to the spectrum—in terms of ultraviolet, visible, and infrared radiation. The qualities of the solar spectrum, and our atmosphere's interaction with radiation, control many features of our Earth's environment and our knowledge of the universe.

PREREADING QUESTIONS ON THE THEMES OF THIS BOOK

OUR ROLE IN THE UNIVERSE
In what ways do humans experience matter and radiation?

HOW THE UNIVERSE WORKS
What are the principles that explain how energy can change forms and travel through matter?

HOW WE ACQUIRE KNOWLEDGE
How did scientists determine that matter is made of microscopic units called atoms?

RUTHERFORD AND THE STRUCTURE OF THE ATOM

Ernest Rutherford was puzzled. It was 1911, and the experimental physicist was using the newly discovered phenomenon of radioactivity to probe the structure of matter. He placed a radioactive source to send a stream of tiny particles toward a thin sheet of gold foil. Then he used a Geiger counter to detect the particles and see what happened to them as they hit the foil. Rutherford was doing a pioneering experiment to measure the structure of the atom. The ancient Greeks had come up with the idea of atoms—tiny, indivisible units of matter far too small for the eye to see. Rutherford was using a new tool to probe this invisible world.

Why was Rutherford puzzled? Previous ideas of atoms led Rutherford to think that atoms were packed tightly together like ball bearings in a jar. He expected that most of the tiny projectiles would plough into the foil and deposit their energy there. Yet when he placed the Geiger counter directly behind the foil, he found that most of the particles passed through undeviated or deflected by only small angles. The particles acted as if the foil were not there! Rutherford was forced to conclude that the gold atoms were made of mostly empty space.

The result of his next experiment was even more surprising. This time he placed the Geiger counter on the same side of the foil as the radioactive source. Expecting to see nothing, he was amazed when a small fraction of the incoming particles—about 1 in 1000—bounced right back in the direction of motion. At the time Rutherford said, "It was quite the most incredible event that ever happened to me in my life. It was almost as incredible as if you had fired a 15-inch shell at a piece of tissue paper and it came back and hit you." This indicated that the tiny particles were bouncing off something. The two observations together showed that most of the mass of the atom was contained in a small dense core or nucleus. A particle that hit the nucleus could bounce back. He had gained new insight into the structure of the atom.

This was the start of a golden age in atomic physics. Rutherford and his colleagues at the University of Manchester did many elegant experiments to probe atoms. This was in the days before enormous atom smashers. Their apparatus was built using ingenuity and the kind of components you might find in a hardware store. Rutherford was proud of his ability to do great experiments with modest equipment; he once proclaimed he could do research even at the North Pole. Rutherford won the Nobel Prize in physics for his work and became the director of the Cavendish Laboratory at the University of Cambridge. He ended his life as Lord Rutherford.

Rutherford was born to a poor family in New Zealand, relying on scholarships to get to college and travel overseas. He was a big bear of a man with a booming voice and an intense manner. He would sweep through the Cavendish Laboratory every evening to send people home—not to relax and spend time with their families, but to think more deeply about their experiments! Rutherford was undoubtedly a tough boss, but his enthusiasm for new ideas and discoveries in physics meant that students flocked to work with him. He liked clear thinking and simple explanations and once said that a theory was no good unless it could be explained to a bartender.

Basic Structure of Matter

Early Greek Ideas

To understand the physical nature of the universe, we need to explore the basic blocks of matter itself. As we learned in Chapter 2, early Greek thinkers turned their attention to the very small as well as the very large. Over 2000 years ago, they speculated that the incredible variety of materials in the world might conceal a basic simplicity. Democritus thought it unlikely that we could divide and subdivide matter endlessly. On purely logical grounds he supposed that there was a fundamental indivisible unit of matter, which he called an *atom*.

Democritus supposed that these tiny indivisible atoms were moving constantly in space. He also made a distinction between primary and secondary properties of matter. Secondary properties such as color, smell, and taste result from combinations of atoms. Atoms themselves have no color, odor or flavor. Empedocles went further. He proposed that nature contained only four substances—earth, water, fire, and air. All the diverse materials of the world are made of different combinations of these four substances, which he called *elements*.

In ancient times, the existence of atoms and elements was mere speculation. The Greeks had no evidence. Yet we can see that some of their ideas were strikingly modern. They were following an important goal of the scientific method: to find simple organizing principles to explain the diversity of nature. Who came up with the evidence that atoms actually exist?

Dalton and the Theory of Atoms

Historians of science usually credit a rather unusual hero: a self-educated English scientist who worked in the early 1800s, around the time that Beethoven composed symphonies and Napoleon ruled much of Europe. John Dalton was born into a family of Quaker weavers in 1766. By the age of 12, he was teaching other youngsters in a Quaker school in his village. He went on to learn mathematics, astronomy, foreign languages, and observational methods from John Gough, a blind philosopher who lived nearby. He published a theory about the *aurora borealis*, or northern lights, at the age of 22. Dalton correctly inferred that the aurora was caused in part by Earth's magnetism. He later published a less successful theory about the causes of color blindness, which he suffered from. Dalton never attended a university.

We have seen in Chapter 1 that science often boils down to seeing patterns in observational data. Dalton's particular skill was to be able to synthesize the available information. He could move quickly from partially complete data to brilliant insights about the "big picture," much as some people can see a few pieces from an incomplete jigsaw puzzle and deduce the entire picture. These sudden moments of insight are the rarest and most pleasurable feelings a scientist can have. Dalton also followed the example of Galileo in not trusting the opinions of previous authorities on a subject. He wrote, "Having been in my progress so often misled by taking for granted the results of others, I have determined to write [only] what I can attest by my own experience." Dalton did his own experiments, and he did his own thinking.

Dalton reached his theory of atoms by experimenting with gases. Using homemade laboratory equipment, he studied how they mixed in the air, got absorbed into water, or combined. He discovered that air is not a uniform medium, but rather a mixture of gases including nitrogen and oxygen. A single gas is a chemical **element.** Dalton also found that gases combine only in certain proportions. A combination of gases is a chemical **compound.** For example, hydrogen and oxygen combine explosively to form steam, but only in proportions of 1 part hydrogen to 8 parts oxygen, by weight. Dalton figured out the proportions for other compounds, too. The specific proportions intrigued him. "I am nearly persuaded that the circumstance depends upon the weight and number of the ultimate particles of the . . . gases . . . ," he wrote in 1803.

By writing about the "ultimate particles" of each element, Dalton was giving new life to the old Greek idea of atoms. An **atom** is therefore the particle of each element that cannot be subdivided by ordinary chemical means. Dalton inferred that the fixed proportions meant that the compounds were being formed by fixed numbers of atoms of each element, with different elements having atoms of different weights. A **molecule** is a linked set of two or more atoms that correspond to a unique compound. Dalton's measurements of water could be explained if oxygen atoms have 16 times the weight of hydrogen atoms, and if molecules of water vapor (steam) each have two atoms of hydrogen and one of oxygen. Thus, Dalton was the first to describe the composition of water as

H₂O, meaning it has two hydrogen atoms and one atom of oxygen. No known process could break down hydrogen and oxygen so they were considered fundamental.

The same ideas applied to all materials—solid, liquid, and gaseous. For example, chemists knew that most materials can be broken down into simpler chemicals. Burning wood produces carbon dioxide, water, and carbon in the form of charcoal. Water can be broken down with an electric current into hydrogen and oxygen. However, carbon cannot be broken down any further. It is an element. We can summarize Dalton's insight by saying that an atom is the microscopic, indivisible counterpart of a chemical element, and a molecule is the fundamental counterpart of a chemical compound. Figure 4–1 shows this schematically.

Dalton published the results of his elegant experiments in 1810. These ideas, which are known as the *atomic theory*, form the basis for modern chemistry. The atomic theory includes the following key ideas:

1. All elements are composed of atoms.
2. Each element has unique chemical properties, and each element is composed of atoms of a unique weight.
3. Compounds are formed when atoms are linked in certain proportions into molecules. Each compound has unique chemical properties.
4. Elements and compounds combine (and separate) in a processes called **chemical reactions.**
5. Elements—and their microscopic constituents, atoms—are fundamental and cannot be broken down into different materials.
6. One element cannot be changed into a different element by chemical reactions.

Dalton's work met with some resistance. After all, no one had ever seen or isolated an individual atom! They were apparently far too small to be seen with a microscope. The atomic theory was a fatal blow to alchemy—the ancient idea that elements could be transformed from one to another. Today we think of alchemy as a nonscientific pursuit. Yet over the centuries many scientists have tried to turn a base metal such as lead into gold. Both lead and gold are dull, heavy, malleable metals. It does not seem unreasonable that one could be converted to the other. The great Newton himself spent hundreds of hours in his laboratory doing experiments in alchemy. Now we know that chemical reactions can join and separate elements, but cannot transform one element to a different one.

In the early 1800s, only about 30 elements were known. This was not enough to see any patterns in the chemical behavior of different elements. The invention of the battery gave a new way for compounds to be separated into their constituent elements. Several dozen new elements were soon recognized. Soon the Swedish chemist Jacob Berzelius introduced the modern notation for compounds, based on the numbers of atoms (such as H₂O). In 1869, the Russian chemist Mendeleyev arranged all the known elements into a table according to the weights of their atoms and similar chemical properties—the basis of the **periodic table** of the elements used in chemistry today (Figure 4–2). Mendeleyev's organization of the elements by atomic weight firmly established the atomic theory of matter.

Rutherford and the Atomic Nucleus

The ancient Greeks thought of atoms as indivisible particles of matter, like tiny beads. Even Dalton imagined that atoms were little hard spheres. However, around 1900, Joseph John Thomson discovered particles called electrons. An **electron** has a negative electric charge and is far lighter than the lightest atom known. Electric charge is a fundamental property of matter and anything with an electric charge feels the electrical force. Charge comes in two varieties—positive and negative—and particles with the same charges repel while particles with opposite charges attract. Thomson proved that the atom has structure and that atoms are not truly fundamental building blocks of matter. The stage was set for Ernest Rutherford.

We have seen in the opening story that Rutherford created an experiment to penetrate the atom. His results showed that atoms have structure and are not tiny hard spheres. Rutherford used tiny projectiles called *alpha particles* for his experiment. Alpha particles are emitted at high speed by some radioactive materials. They have a positive electric charge, and they are much more massive than the electron. In fact, an alpha particle is a helium atom with the

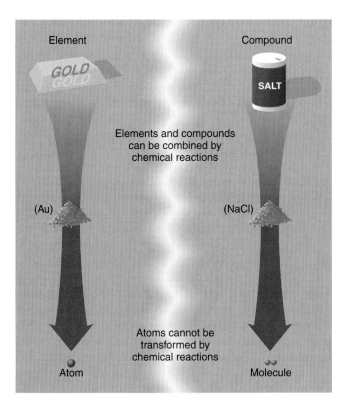

FIGURE 4–1
Dalton realized that an atom is the microscopic counterpart of a chemical element, and a molecule is the microscopic counterpart of a chemical compound. Chemical reactions dictate the ways that atoms and molecules combine to form different compounds. An atom is indivisible and cannot be converted to an atom of a different element by a chemical reaction.

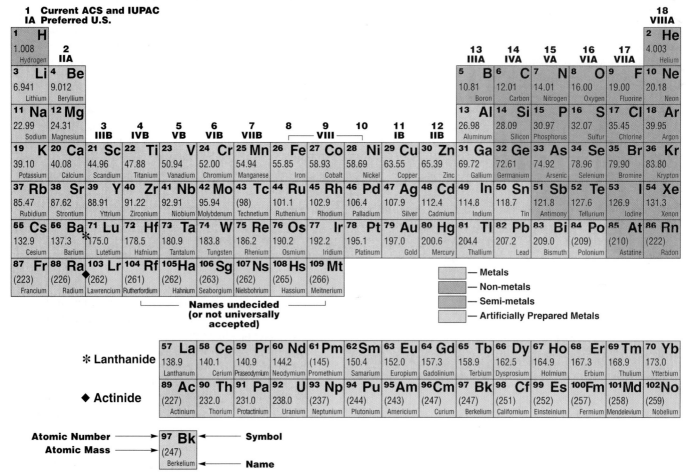

FIGURE 4-2
The periodic table of elements. Only 90 of these occur in nature; the rest are created fleetingly in the laboratory. Most of the universe is made of the two lightest and simplest elements—hydrogen and helium. Over 90 percent of the elements are rare, with a cosmic abundance of less than 1 part in 1 million. The atomic number is the number of protons in the nucleus, and the atomic mass is the total mass in units of proton mass.

electrons stripped off. Rutherford bombarded the gold foil with alpha particles and saw three things. (1) Most of the alpha particles passed straight through the foil as if it were not there. (2) A small fraction of the alpha particles passed through but were deflected by a small angle. (3) An even smaller fraction recoiled completely to reverse their direction of motion. How did Rutherford make sense of these results?

The first result told Rutherford that atoms are not hard spheres which fill space. Since the great majority of the alpha particles passed through the foil, the atom must be made of mostly empty space. The small deflections of some alpha particles told Rutherford that the center of the atom had a positive electric charge. This is the **atomic nucleus**. He figured this out because the deflections were just the right size to be explained by positively charged alpha particles flying in and being repelled by a stationary, positively charged nucleus. The deflections were caused by the well-known electric force. The third result—a few alpha particles bouncing back the way they came—told Rutherford that almost the entire mass of the atom was concentrated in a tiny nucleus at the center. We summarize the results of his experiment in Figure 4-3.

After Rutherford's work, a familiar picture of the atom emerged. The small, dense atomic nucleus is surrounded by orbiting electrons. This picture—the atom as a miniature solar system—has become a symbol of our modern scientific age. Yet the analogy with gravitational orbits is flawed. Rutherford had no explanation for why the electrons should stay in their orbits. In fact, a simple consideration of the electric force shows that atoms made of positively charged nuclei and negatively charged electrons should not be stable. Positive and negative charges attract each other, so the electrical force should cause all atoms to collapse! A full explanation of atomic structure requires quantum theory, which will be introduced in Chapter 10.

Structure of Atoms and Molecules

We do not need quantum theory to summarize the main points of atomic structure. All atoms consist of the following: Each atom has a positively charged nucleus at its center and a swarm of much less massive, negatively charged electrons circling the nucleus.

The scales of atomic structure are truly tiny. Hydrogen is the smallest atom, with a diameter of about 5.3×10^{-11} meter. About 10 million hydrogen atoms fit across the head of a pin! The nucleus of a hydrogen atom consists of a single positively charged particle, called a **proton**. Heavier atoms have a second type of nuclear particle, called a **neutron**. The neutron is similar in mass to the proton but is electrically neutral. The proton is about 10,000 times smaller that the atom itself. The mass of a hydrogen atom is 1.7×10^{-27} kilogram. Even the smallest mote of dust contains trillions and trillions of atoms. Since the electron is nearly 2000 times less massive than the proton, it is a good approximation to say that the nucleus carries most of the mass of every atom. However, the tiny electron has an equal but opposite electrical charge to the more massive proton. These charges cancel exactly, and normal matter is electrically neutral.

The number of protons in the nucleus defines the element. For example, as shown in Figure 4–4, an atom of hydrogen—the simplest element—has one proton in its nucleus and a single electron. An atom of helium has two protons and two neutrons in its nucleus, and two surrounding electrons. Each different element has a different number of protons in the nucleus. The periodic table is just a sequence of atoms with increasing numbers of protons. The number of protons and the number of electrons are always equal. There is no simple rule for the number of neutrons, but in any case neutrons affect the weight but not the chemical properties of an element. Figure 4–4 shows the five most abundant elements in the universe.

The electrons act as a shield. Under ordinary, everyday conditions, the electrons swarming around the nucleus prevent the nucleus of an atom from making contact with the nucleus of any other atom. If two atoms collide, only the electron swarms interact, as illustrated in Figure 4–5. If the nucleus of an atom were represented by a tennis ball on the 50-yard line of a football stadium, the electrons would be tiny particles whirling around the outskirts of the stadium. The neighboring atoms in a solid would be other tennis balls several hundred yards away, each with its own swarm of electrons. This analogy holds even for a dense material such as lead or gold—the solidity of everyday objects is an illusion due to the electric force within atoms. The emptiness of normal matter is one of the amazing consequences of atomic theory!

Recall that atomic theory also states that elements are fundamental and that one element cannot be changed to another by chemical means. What stops us from just adding or subtracting a proton from the nucleus to make a different element? It turns out a special force within the nucleus acts to keep it tightly bound. And the electrical repulsion between two positively charged nuclei stops them from ever merging. We will see later in the book that atomic nuclei can merge or be ripped apart under conditions of extreme temperature and pressure. But in the everyday world, an element never changes. You can think of the atomic nucleus as a little fortress—no particle can leave or enter.

What about more complicated forms of matter? Unlike atomic nuclei, electrons are gregarious. They travel readily

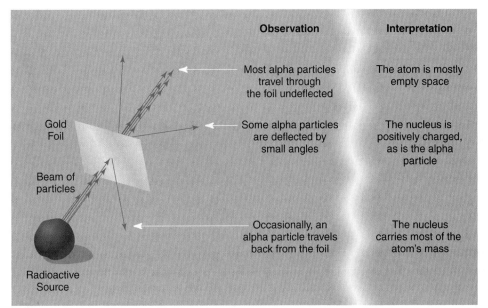

FIGURE 4–3
Ernest Rutherford's experiments probed the "hidden structure" of nature. A beam of positively charged alpha particles penetrates a thin sheet of gold. Most of the particles travel through the foil with their paths unaltered. A few are deflected by small angles, and the occasional alpha particle travels back in the direction from which it came. These observations lead to a model of the atom very different from the idea of atoms as tightly packed, hard spheres.

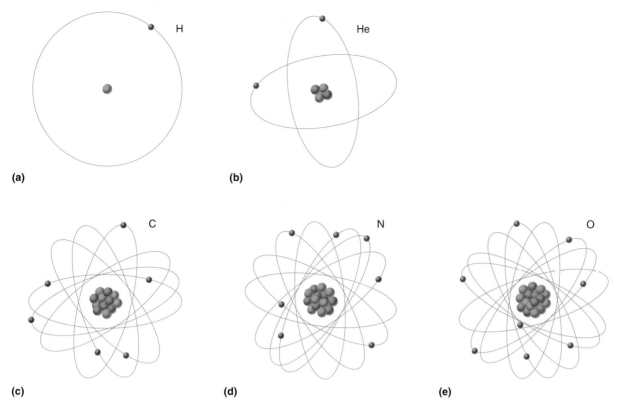

FIGURE 4-4
Schematic representations of the five most abundant elements in the universe: **(a)** hydrogen (H) with an atomic number of 1, **(b)** helium (He) with an atomic number of 2, **(c)** carbon (C) with an atomic number of 6, **(d)** nitrogen (N) with an atomic number of 7, and **(e)** oxygen (O) with an atomic number of 8. In each case, there are equal numbers of protons, neutrons, and orbiting electrons. Elements have rarer forms, or isotopes, with extra neutrons in the nucleus.

from one atom to another. Molecules form when atoms share their electron structures. If we consider elements to be building blocks of different size and mass, then molecules are groups of atoms joined by their electrons. Some molecules have as few as two or three atoms—water (H_2O), carbon dioxide (CO_2), nitrogen (N_2). Organic molecules can have hundreds of atoms. If we follow the building block analogy, there are about 110 different elements or different-sized building blocks. About 90 of these occur in nature; the rest have been made fleetingly in the laboratory (see Figure 4–2). On the other hand, the building blocks can be combined in an enormous number of ways. There are thousands of naturally occurring molecules and tens of thousands that have been created in the laboratory.

Chemical reactions occur when atoms or molecules give up or receive electrons. Remember that the central nuclei never meet or collide in a chemical reaction. Elements with quite different masses can have similar chemical behavior if

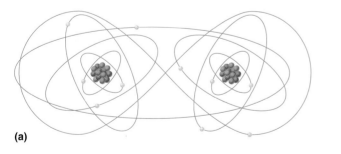
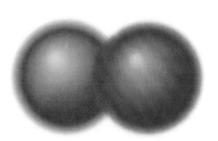

FIGURE 4-5
Schematic representations of a molecule. **(a)** The sharing of electrons between two atoms, which keeps them together in a chemical bond. In this diagram (not intended to represent any specific molecule), two electrons are being shared, out of a total of six associated with each atom. **(b)** A less accurate, but common, way to think of a molecule, as a clump of tiny, spherical atoms. The atoms are represented as somewhat fuzzy in outline because their swarms of electrons do not have fixed positions.

they have similar numbers of electrons in their outer regions. Look again at the periodic table in Figure 4–2; elements in the same (vertical) column often have very similar chemical properties. Chemical reactions are going on around us all the time.

The periodic table provides a wonderful example of how science works. Notice that humans have been conducting innovative experiments in chemistry for thousands of years—think of the invention of gunpowder or the many types of alcoholic beverages, for example. Our culture has been successfully building and manipulating chemical compounds without knowing anything about the "hidden structure" of atoms. For example, the Bronze Age was the first great age of technology when cultures moved beyond flints and fashioned a wide variety of tools and weapons from bronze. Bronze is a mixture or alloy of copper and tin, both of which are very malleable metals. How can the sum of two soft metals make a hard metal? Ancient cultures knew only that it worked. Now we know that soft metals have planes of atoms that easily slide over one another. It turns out that tin atoms provide the "grit" in the copper lattice that stops the planes from moving freely and gives bronze its stiffness.

Late in the 19th century, scientists began to see the patterns in the periodic table. Elements at the far right share the properties of being colorless and odorless gases that do not react with other elements—the noble gases. Elements at the far left are highly reactive, but they combine with elements in the seventh column to form stable compounds such as common salt (NaCl). Scientists knew that the periodic table "worked," but they had no idea why. As we discussed in Chapter 1, science often starts with this process of recognizing patterns in nature. Let's revisit our card analogy. Someone who did not know the rules might watch a poker game for quite a while before figuring out the value of the different hands. We are also trying to solve a puzzle and figure out the "rules of nature."

In the rest of this chapter we will review some simple principles about the behavior of matter. If matter is just made of atoms and molecules, how can there be so many forms of matter, such as solids, liquids, gases, and electrically charged plasmas? What makes a crystal different from a cloud? Are the rocks on Mars exactly like those on Earth? Are the elements on Earth the same as the elements seen out in space? Why does the Sun's material give off light, while Earth's material does not? These and other questions will be answered as we consider how matter and radiation interact.

This discussion of particles so small that they are invisible begs an obvious question. Are atoms real? Or are they just a theoretical idea that scientists find useful to describe nature? There is plenty of indirect evidence for tiny units of matter, and in the past few years we have direct evidence, too (Figure 4–6). We can use electrons to make extremely fine images of matter. We can also use electric fields to sense the positions of individual atoms. A new discipline called *nanotechnology* is emerging, which will enable engineers to fabricate devices from single atoms and molecules. We should have no doubt—atoms are real.

Nature of Energy

The next time you drive your car, consider the source of its energy. Millions of years ago, radiation flowing through space from the Sun was intercepted by Earth. That sunlight allowed plants to grow, and after the plants died and were buried deep underground, the stored energy was turned into petroleum. Humans pumped up the petroleum and refined it into the gasoline that fills your tank. The stored chemical energy is being transformed to forward motion by your engine. As you stop to ponder this, your hot engine cools. The heat energy that it releases rises through the atmosphere and eventually escapes and continues its journey through space.

Scientists define **energy** as the ability to do work (Figure 4–7). Perhaps a more intuitive way to express this is to say that energy is anything that can cause a change. This sounds vague, but in fact scientists have found concrete and quantitative ways to define energy. Let's look at several important aspects of energy. First we will see that energy can come in an amazing variety of forms. Then we will show how energy and heat are related. Last we will discuss the fact that energy can change forms but is never created or destroyed: This is one of the great unifying ideas that governs our universe.

Different Forms of Energy

Energy comes in various forms, many of which are familiar to us in the everyday world. Scientists in the 19th century had the great challenge of showing how these forms are related. Recall from Chapter 1 that science tries to look for patterns and relationships in nature and to describe them with simple physical laws. Energy is a simplifying concept that allows us to relate phenomena that are on the face of it very different. Let's look at some examples.

One broad type of energy is called **potential energy.** We can think of this as stored energy or energy waiting to be released. We have already discussed the basis of chemical energy. When paper or wood burns, oxygen in the atmosphere reacts with the carbon to release energy. In the case of gasoline, the combustion in your engine releases the energy stored in long molecules of hydrogen and carbon. Chemical energy is released when electrons in atoms change their configuration. This type of stored energy can be substantial. In addition to being the source of power in gasoline and dynamite, chemical energy powers all the rockets of the world's space program.

Another major type of potential energy is the ability of gravity to cause a change. Gravitational energy is a form of energy that depends on an object's response to gravity. If we pick up a brick from the floor and put it on the edge of a table, we have used our own energy by moving the brick upward in Earth's gravity. That energy is stored in the brick as gravitational energy. If the brick is knocked off the table, it will release its gravitational energy as motion. A falling brick clearly has the ability to cause a change! The higher a body is raised against gravity, the more gravitational energy it has, because it has greater potential to release energy as it crashes downward. Hydroelectric power plants, for

FIGURE 4-6
Indirect and direct evidence for the microscopic structure of matter. **(a)** A single large crystal of table salt (NaCl). In a crystal, the atoms pack in an orderly structure that repeats itself. The basic unit of atomic structure is a cube about 0.5 nm on a side; here it manifests on a scale 10 million times larger. (SOURCE: E. Hecht.) **(b)** The first photograph ever taken of a solitary atom. The blue-green dot at the center is a single barium atom slowed down or "cooled" by a laser beam. The red surrounding structure "traps" the atom with radio waves. (SOURCE: Warren Nagourney.)

(a)

(b)

example, convert the energy of water falling over a dam to electric energy.

Another familiar type of potential energy is the energy stored in a coiled spring or a stretched elastic band. There are two other types of potential energy that we will encounter later in the book. Changing electric and magnetic forces create radiation that can travel through space or a thin gas. Light is one example of electromagnetic energy. Humans have learned to create light in various ways, but the Sun is by far the most bountiful source of this type of electromagnetic energy, as we will learn in Chapter 11. We will also discover that energy is stored in the form of mass. When the "frozen" energy stored as mass is released, it is the most efficient energy source known.

The second broad type of energy is **kinetic energy**—the energy of motion. Think of a flying bird, a falling tree, a speeding car, and a running child. All have kinetic energy. The more massive or faster the object is, the more kinetic energy it has. For example, a bowling ball rolling across the floor has a greater kinetic energy than a billiard ball moving at the same speed. A car moving at 60 miles per hour (mi/h) has more kinetic energy than the same car moving at 20 mi/h. Examples of both kinetic energy and gravitational potential energy are given in the Science Toolbox on page 94.

Energy is a useful concept in science only if it can be quantified. In ordinary speech, we use the word *energy* in many ways. For example, the term *chi* from Chinese medicine is translated as "energy," but means something closer to bodily vitality and well-being. People speak of "focusing your energy" or "putting mental energy" into some effort, and a mystic might speak of "psychic energy." These concepts may or may not be meaningful, but they are definitely not quantitative. The scientific concept of energy is very specifically defined. Kinetic energy is given by a specific formula, and another formula describes gravitational energy. Energy is measured in a single set of units (mass × distance^2/time^2), called calories (abbreviated cal) in the English system of units or joules (abbreviated J) in the International System of units (see Appendix A-2).

Energy and Temperature

Two hundred years ago, scientists were puzzled by another type of energy: **heat.** In many ways, heat seems like a fluid. It flows from one place to another; some objects absorb and give off a lot of heat, and other objects absorb or give off very little. But if heat is a fluid, then every object must contain a fixed amount. A U.S. cannon maker called Count Rumford showed that this was not true. He observed that a lot of heat was generated in boring the barrel of a cannon. But blunt

FIGURE 4-7
Energy is required to raise a rocket against the force of gravity. Chemical energy is released to launch the Space Shuttle, just as chemical energy is used to power most vehicles (and us, via the food we eat). Chemical energy is stored in the way electrons are shared between atoms. (SOURCE: NASA.)

tools generated more heat than sharp tools. So heat is just the result of friction—the energy of motion when two surfaces rub against each other. For a quick example, rub your hands together briskly. Even two pieces of ice make heat when they are rubbed together.

Can heat or thermal energy be quantified? It certainly can. In 1840, James Prescott Joule did a famous experiment. The British physicist arranged a weight so that it turned paddles in water as it dropped. He suspected that the spinning paddles would heat the water, and he was right. Joule measured a rise in the temperature of the water each time he dropped the weight. "The work done by the weight of 1 pound dropping through 772 feet . . . will, if spent producing heat by friction in the water, raise the temperature of 1 pound of water by 1 degree Fahrenheit," he concluded. Joule's simple experiment showed that kinetic energy, or energy of motion, can be converted directly to heat.

Let's look at the implications of Joule's experiment. Moving particles have kinetic energy, even on the scale of molecules. Think of the molecules and atoms as tiny particles, darting around and hitting other particles. Each one has a mass and a velocity, hence a kinetic energy. The motion of the paddles in Joule's experiment caused all the water molecules to move faster. Heat is just a microscopic form of kinetic energy.

Now we can bring in the idea of **temperature.** We all have an idea of "hot" and "cold," but what aspect of matter does temperature actually measure? The answer surprises many people when they first learn it. Temperature is merely a measurement of the motion of molecules and atoms. (For convenience, let's refer only to molecules, since an atom is a special class of molecule with only one atom instead of several. When we say *molecule* in the following discussion, you can think of either molecules or atoms.) The higher the temperature, the faster the molecules are moving. The colder the temperature, the slower the molecules are moving. The English botanist Robert Brown made this visible in 1827 when he looked at pollen suspended in water through a microscope. The pollen particles were in constant random motion. They were kept in motion by the constant buffeting of the water molecules all around them.

Notice that thermal energy and temperature are not the same thing. A drop of boiling water and a cup of boiling water have the same temperature, but one clearly has more heat energy—as you would notice by the change each caused on your skin. Temperature does not measure the total amount of heat in an object. It is a measure of the microscopic motions of the molecules or atoms. The Science Toolbox on page 96 shows this numerically and gives some applications of this important concept.

Scientists use a different temperature scale than those you are used to. In the United States most people still use the archaic Fahrenheit temperature scale, introduced almost 300 years ago. Most of the world uses the Celsius (or centigrade) scale, where 0°C is the freezing point of water and 100°C is the boiling point of water at sea level. But matter can clearly be colder than the freezing point of water, so there must still be some heat below 0°C. Scientists use a physical scale of temperature, where zero corresponds to *minimum* motion of atoms or molecules whatsoever. This is the Kelvin scale, named after Lord Kelvin (William Thomson), a famous English physicist who made many contributions to our understanding of heat. You can see from the many examples in physics—force in newtons, power in watts, temperature in kelvins, energy in joules—that one way we commemorate scientists is by naming units of measurement after them (see Appendix A-3). Unfortunately for modern physicists, the most important units have already been named!

One kelvin (1 K) is the same size as one degree Celsius (1°C), but the zero point of the scale is far lower. On the Kelvin temperature scale, water boils at 373 K and freezes at 273 K and room temperature is at about 295 K. To go from a temperature in kelvins to a temperature in degrees Celsius, subtract 273. The different temperature scales are compared in Figure 4–8. When temperatures are very high, say in the millions of degrees, the difference between the Kelvin and

SCIENCE TOOLBOX

Potential and Kinetic Energy

Throughout this book we will encounter many examples of kinetic energy and gravitational potential energy. Every object moving in space has kinetic energy, and everything that is subject to the gravitational force of a star or planet has potential energy.

Examples of kinetic energy are all around us. A fast-moving object has more kinetic energy than a similar object moving slowly. In fact, kinetic energy is proportional to the square of the velocity. If we double the speed of a moving object, the kinetic energy increases by a factor of 4. At the same speed, a heavy or massive object has more kinetic energy than a light object. In fact, kinetic energy is proportional to mass. These two relationships get combined in the simple equation for kinetic energy

$$\text{Kinetic energy} = \frac{1}{2} m v^2$$

In this equation, m is mass and v is velocity. If we insert velocity in meters per second and mass in kilograms, the kinetic energy comes out in units of joules. Let's look at a couple of examples. A soccer ball with a mass of about 1 kilogram (kg) can be kicked at about 15 meters per second (m/s). The kinetic energy is therefore $½ \times 1 \times (15 \times 15) = 112.5$ joules (J). A baseball has a mass of about 250 grams (g), or 0.25 kg, and can be hurled at about 50 m/s, or 100 mi/h. A fastball has a kinetic energy of $½ \times 0.25 \times (50 \times 50) = 312.5$ J. Because of its speed, the small object has much more kinetic energy; you would be better off being hit by a soccer free kick than by a fastball.

How does a human compare to a speeding bullet? A fast sprinter with a mass of 70 kg can run 100 m in 10 s, or at a speed of 10 m/s. This is a kinetic energy of $½ \times 70 \times (10 \times 10) = 3500$ J. A bullet has a mass of about 50 g, or 0.05 kg, and travels with a typical velocity of about 400 m/s, or 900 mi/h. The bullet has a kinetic energy of $½ \times 0.05 \times (400 \times 400) = 4000$ J. A human really is nearly as "powerful" as a speeding bullet!

Any object that is subject to the force of gravity will have an amount of gravitational potential energy. Here are some examples of gravitational potential energy— a barbell held over your head, a boulder balancing precariously on top of a hill, a roller coaster at the highest point in its path, a lake held above a valley by a dam. In each case, something with mass has the potential to move under gravity. In each case, the motion will be toward the source of that gravity, which is effectively the center of Earth.

The gravitational potential energy is given by the weight or force of gravity exerted by an object multiplied by its height above the ground. Newton's second law of motion, which we encountered in Chapter 3, says that gravity force is equal to mass times acceleration. Combining these relationships, we see that

$$\text{Gravitational potential energy} = m g h$$

In this equation, m is mass, g is Earth's gravitational acceleration, and h is the height above the ground. We have already seen that acceleration at Earth's surface, denoted g, is 9.8 meters per second per second (m/s²). A top weight lifter can lift 300 kg over his head. The potential energy of this weight lifted 2 m off the ground is $300 \times 9.8 \times 2 = 5880$ J. This is the amount of energy a human must provide to lift such a weight. When the weight is released, the gravitational potential energy is converted to kinetic energy as the weight falls.

Swinging a bat or running quickly takes more power than doing those tasks slowly. Scientists define *power* as the rate at which energy is expended. Mathematically, power is the energy generated divided by the time it takes to generate it. If the energy is expressed in units of joules, the power is in joules per second, or watts. It is very appropriate that the unit of power is named after James Watt, the Scottish inventor of the steam engine. This invention was the force behind the industrial revolution.

Let's look at our weight-lifting example again. It takes only 2 s to raise the weight from the hanging position to the arms-raised position, a vertical distance of 1.5 m. So in that second $(1.5 / 2) \times 5400 = 4050$ J is generated, or a power of $4050 / 2 = 2025$ watts (W). This is enough to illuminate 20 lightbulbs!

Finally, consider a practical example of gravitational potential energy. Water that falls in the mountains can be harnessed to create energy. You could, for example, use a waterfall to turn a turbine and generate electricity. A modest-sized waterfall might have a drop of 50 m and a flow rate of 1000 gal/s (this converts to about 4 m³ of water per second, or 4000 kg/s). The available energy is therefore $4000 \times 9.8 \times 50 = 2.0 \times 10^6$ J/s, or 2 million W—enough to power a small town.

[handwritten: $g = 9.8 \frac{m}{s^2}$ → only valid near earth's surface]

Celsius scales is negligible. Absolute zero is 0 K or −273°C. Experimenters in the laboratory have chilled materials to within one-millionth of a degree of absolute zero.

To sum up, all atoms and molecules are in constant motion. This is true of any form of matter—solid, liquid, or gas. We call the description of materials in terms of microscopic motions the *kinetic theory of matter*. Heat is a measure of the amount of thermal energy contained in these microscopic motions. Temperature is a measure of the kinetic energy of the individual atoms or molecules. Scientists often call this description the kinetic theory of matter. The Science Toolbox on page 96 tells you how to calculate the speeds and ener-

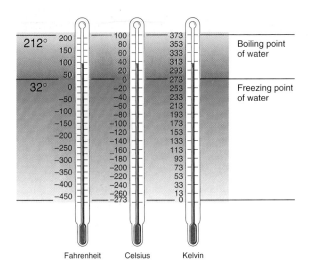

FIGURE 4-8
A comparison of the Fahrenheit, Celsius, and Kelvin temperature scales. The Kelvin scale is useful because it is related to the microscopic motions of atoms and molecules; zero on the scale corresponds to no motion.

be stored in the wind-up spring of a toy car or a clock, then released in the form of motion to move the car or the hands of the clock. A battery stores electric energy that can then be released to move the shutter of your camera. The head of a match stores chemical energy that can be released to heat. Energy can also change forms continually. A child on a swing is constantly changing gravitational potential energy to kinetic energy and back again. At the top of the swing's arc, the motion momentarily stops, and all the energy is potential energy. At the lowest point in the arc, all the potential energy has been converted to kinetic energy. An elliptical orbit is similar. Kepler's second law relates the speed of a planet's orbit to its distance from the Sun. A planet has the most gravitational potential energy and the least kinetic energy when it is farthest from the Sun. It has the least gravitational potential energy and the most kinetic energy when it is closest to the Sun. The orbit is a continuous conversion of the two types of energy.

Let's consider another cosmic example. Suppose an asteroid hurtles through space at 10 km/s and crashes into a planet. Before the asteroid hits the planet, it has a certain amount of kinetic energy. It also has a significant but much

gies of molecules corresponding to different temperatures. Such equations in turn are useful in predicting what kinds of phenomena will happen in planets, stars, and other materials at different temperatures.

Transformation and Conservation of Energy

There are many forms of energy in the universe. At first glance, the types of energy we have discussed seem very different. What could possibly relate the energy in a coiled spring to the energy in a cup of gasoline? What do the energy in a hot coal and the energy in a falling meteor have in common? First, all these types of energy are quantifiable and measurable. Second, scientists have shown that all the different forms of energy are interchangeable. Energy changes form all the time.

The idea of the **transformation of energy** relates to our theme of how science works. One of the basic goals of science is to provide simple explanations for the diversity of the natural world. Scientists can show that kinetic energy and gravitational potential energy are not fundamentally different. Both have the ability to cause a change. Both can be measured in units of joules. And one can be converted to the other. The transformation of energy is an important unifying principle throughout science.

Let's go back to Joule's experiment. He started by raising a weight in his apparatus. This then stored gravitational potential energy. The falling weight caused a paddle to rotate. The gravitational energy was transformed to the kinetic energy of the moving paddle. The paddle agitated the water molecules, giving them each a bit more energy. So the kinetic energy was transformed to heat energy in the water.

Energy is transformed to many forms and can even be stored (Figure 4-9). For example, mechanical energy might

FIGURE 4-9
Energy is stored and transformed in many ways. The roller coaster uses an engine to raise the cars high above the ground—the engine has released chemical energy from a fossil fuel, which originally stored the Sun's energy. At the top of the curve, the cars have maximum gravitational potential energy. This is then converted to the kinetic energy of rapid motion. At the end of the ride, brakes dissipate the kinetic energy as heat. (SOURCE: Busch Entertainment Corp.)

SCIENCE TOOLBOX
Velocities of Atoms and Molecules in a Gas

We have seen that all gas particles have kinetic energy due to their motions as they bounce around. Temperature is a measure of this microscopic kinetic energy. But how exactly is the temperature scale related to the velocities of the particles? The answer was worked out around 1860 by physicists James Clerk Maxwell from Scotland and Ludwig Boltzmann from Austria. Their simple law gives us insights that will be useful throughout our discussion of planets.

Now recall that the kinetic energy of any moving particle is $\frac{1}{2}mv^2$, where m is the particle mass and v is its velocity. Maxwell and Boltzmann deduced that the mean kinetic energy is proportional to T. This statement is usually written

$$\frac{1}{2}mv^2 = \frac{3}{2}kT$$

In this equation, k is a fundamental constant called the *Boltzmann constant*, which has the tiny value of 1.38×10^{-23} joule per kelvin (J/K). We can see the main features of this equation easily. The temperature is proportional to the square of the average velocity and to the mass of the particle.

As an example, consider the air that surrounds you. What is the typical velocity of a molecule? Let's assume we are dealing with nitrogen, since most air is composed of nitrogen. A single nitrogen molecule has an atomic weight of 28 and so a mass 28 times that of a hydrogen atom, or 4.68×10^{-26} kg. Assume the air is 20°C, or 293 K. We can rearrange the equation above to solve for velocity

$$v = \sqrt{\frac{3kT}{m}}$$

Now we plug in the numbers. If we use the correct units, the answer will come out in meters per second. The result is that

$$v = \sqrt{\frac{3 \times 1.3 \times 10^{-23} \times 293}{4.68 \times 10^{-26}}}$$
$$= 509 \text{ m/s}$$

It sounds like an amazingly high speed. But of course the kinetic energy of each molecule is only $(3/2)kT = 1.5 \times 1.38 \times 10^{-23} \times 293 = 6.1 \times 10^{-21}$ J, so that each of the hits on our skin is very puny. Nevertheless, the cumulative effect of all these tiny collisions is the pressure of ordinary air.

The velocity we have calculated is a typical, or close to an average, velocity for the particles in a gas. However, in practice, there is a broad *distribution* of velocities. Some gas particles travel much faster than the typical velocity, and others travel much more slowly. Figure 4–A shows the bell-curve distribution around the average for two different temperatures.

Now we can use the equation above to show how the velocity of a gas particle depends on the type of particle and the temperature. Particle velocity is proportional to the inverse square root of the mass. Gas of low atomic or molecular weight moves faster than gas of high atomic or molecular weight. How fast does hydrogen move at room temperature? In this example, the temperature is the same, and the mass changes. Since a nitrogen molecule is 28 times more massive than a hydrogen atom, the hydrogen atom moves $\sqrt{28} = 5.3$ times faster, or at a speed of $509 \times 5.3 = 2700$ m/s.

What about nitrogen at a lower temperature? Particle velocity is proportional to the square root of temperature. Cold gas particles move more slowly than hot gas particles. In this second example, the gas is the same, and the temperature changes. The coldest temperature ever recorded on Earth is around −90°C, or 183 K. The nitrogen molecules in that miserable place move $\sqrt{183/293} = 0.79$ times more slowly than the nitrogen molecules in the air you are breathing. Their velocity would be $0.79 \times 509 = 402$ m/s.

Now we can explore an interesting consequence of the spread of particle velocities in a gas. As a rule of thumb, there

(continued)

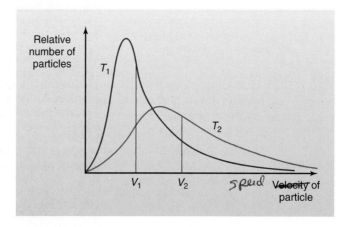

FIGURE 4–A
Velocity distribution of atoms or molecules in a gas. There is a broad distribution of thermal velocities for the particles in a gas. The average velocity increases with temperature (T_2 is higher than T_1), but there is substantial overlap between the distributions. Note the long tail of the distributions: There are particles moving several times faster than the average speed.

SCIENCE TOOLBOX

Velocities of Atoms and Molecules in a Gas
(continued)

are significant numbers of particles up to about 6 times the typical velocity. In Chapter 3, we saw that the escape velocity of Earth is 11.2 km/s. This applies to any moving object, from a rocket to a single atom. The fastest nitrogen molecules will travel at $509 \times 6 = 3050$ m/s, or about 3.1 km/s. This is well under the escape velocity.

However, hydrogen in Earth's atmosphere will move as fast as $2700 \times 6 = 16{,}200$ m/s, or 16.2 km/s. This is well above Earth's escape velocity. So the fastest hydrogen atoms, those in the tail of the distribution in Figure 4–A, are energetic enough to overcome the grip of gravity. Hydrogen will therefore seep into space. Earth can retain heavy gases but will lose light gases. This same idea applies to other planetary atmospheres, as we will see in upcoming chapters.

---

smaller amount of gravitational energy due to its position above the planet. As it falls toward the planet and speeds up, it gains additional kinetic energy. After it hits the planet, its material comes to rest and thus has no kinetic energy or gravitational energy with respect to the surface. Does the energy disappear? No, it is simply transformed. The kinetic energy of the asteroid heats the impact site, makes an explosion crater, and blasts out small, hot fragments that fall back on the planet. (If the asteroid is large enough, there is enough energy to liquefy or even vaporize significant amounts of rock.) Soon after the impact, the total system would have the same amount of energy, but the asteroid's initial energy has been converted to excavate material from the crater, and to heat material around the crater.

On Earth almost all energy originates from the Sun. Light energy from the Sun reaches Earth, where a fraction of it is stored as chemical energy in the cells of plants. Some of that energy is retained in fossil fuels, which we then release to drive our cars and heat our homes. We take in more of that chemical energy when we eat. Our bodies break down the chemicals and store the energy in our cells. Some of that stored chemical energy is eventually released as heat to maintain the temperature of our bodies at 37°C (equal to 98.6°F, or 310 K). In this way we can see on Earth a great chain of energy conversion that leads back to the Sun. It is an interesting exercise to take some activity in the natural world and see how many transformations separate it from the Sun's energy.

Scientists went from the idea of the convertibility of energy to an important physical law. In any closed system, the total amount of energy is constant. This is the **law of conservation of energy**. What do we mean by a *closed system*? We mean that if you draw a giant box around the objects you are considering, the energy contained within the box will not change. Energy can change between any of the forms we have described, but the total amount of energy is conserved. The conservation of energy is one of the most important principles in science; we will see it many times throughout this book.

This law operates in our previous examples. The total energy in a closed orbit is constant. An elliptical orbit has a continual exchange between kinetic and potential energy, but the sum does not change. In a circular orbit, neither kinetic nor potential energy changes. In the case of the asteroid hitting a planet, the energy of motion is converted to an equal amount of heat, which then converts to thermal energy that radiates away slowly into space. Energy does not come and go arbitrarily. Many systems appear to lose energy, but if you consider them carefully, you will see that the lost energy has turned into another form. A rolling ball or a swinging pendulum eventually come to a halt; in each case friction turns kinetic energy into heat. This is more obvious in the commonplace example of the brakes of your car. When you apply the brakes, the kinetic energy of your car is converted to heat in your brake linings through friction. The design of the linings must take into account the conservation of energy.

We can bring the discussion down to earth with a human example. People get their energy from the food they eat. Food energy is measured in calories, which is the amount of energy needed to heat 1 kg of water by 1°C. In our physical system of units, 1 calorie (cal) = 4186 J. The chemical energy stored in the food you eat is used in various ways by your body, but since it keeps your temperature high, most of it is released as heat. The average adult consumes 2500 calories per day. This is $2500 \times 4186 = 1.05 \times 10^7$ J. Now we note that there are $24 \times 60 \times 60 = 86{,}400$ s in a day. So the rate at which the human body generates heat is $(1.05 \times 10^7) / (8.64 \times 10^4) = 121$ J/s, or 121 W. Each of us radiates as much energy as a lightbulb! If you consume more than 2500 cal/day, the extra energy has to go somewhere. If you exercise, the extra calories can be converted to heat and radiated away. If you don't exercise, the extra calories are stored as chemical energy in fat. The issue of exercise and diet comes down to the conservation of energy.

States of Matter in the Universe

As we have shown above, temperature is just a way of measuring the average kinetic energy—or microscopic, random motion—of molecules in a material. The amount of energy or motion we put into a material has important effects on the structure of the material. What determines whether a substance is a solid, a liquid, or a gas?

It is instructive to arrange forms of matter in the universe by temperature, as is done in Table 4–1. Although this arrangement oversimplifies certain effects of pressure and other variables, it is very useful for understanding the forms of material we are dealing with in this book. The table is arranged from the bottom up, in order of increasing temperature.

Starting at the bottom, at the lowest temperatures, we find that matter is "frozen." It exists as a **solid.** The particles barely move, and they stick together in a rigid structure. This is true regardless of whether the substance is made of atoms (e.g., carbon) or molecules (such as glass), or a chemical compound (e.g., cement). The bonds are formed by the sharing of electrons (shown as tiny dots in the table) between nuclei. The nuclei consist of protons (blue) and neutrons (white). Solids have two general forms. They can be *amorphous,* like glass or plastic. The atoms in amorphous solids do not have any regular pattern but are jumbled up in a fixed structure. Solids can also be *crystalline,* like sand or salt. The atoms in a crystalline substance lie in patterns that repeat over and over within the solid. Sometimes the regular structures continue up to a scale much larger than the atom—the cubic crystals of salt can be seen with a magnifying glass (or see Figure 4–6a). Most of the mass of the Earth, for instance, is in this solid state. The crystals that form rocks are good examples of atoms bound together in lattice patterns.

Now recall that temperature is merely a way of measuring the rate of motions of the atoms. The faster the motions, the higher the temperature, and vice versa. At absolute zero temperature, or 0 K, atomic particles have virtually no motion (except for a tiny residual predicted by quantum theory, which we will discuss more in Chapter 10). Atoms in a solid are held in place, so they cannot move freely. But as we raise the temperature of a solid, its atoms will vibrate faster and faster.

At room temperature, around 300 K, atoms in a typical substance are vibrating at a speed of around ½ km/s. This energy allows many of the atoms to break loose from their lattice. When this happens, many substances melt into a **liquid.** In the liquid state, atoms are still in close contact, but chains or groups of atoms may move among one another. A liquid takes the shape of its container. Earth's oceans are in a liquid state. At even higher temperatures—400 to 600 K—the atoms are moving at several kilometers per second. The chains of atoms break apart, and atoms have enough energy to leave the surface of a liquid, creating a **gas.** The individual atoms or molecules in a gas are moving freely. There are large spaces between the particles, and they interact rarely by violent collisions. The densities of

TABLE 4–1
State of Matter

APPROXIMATE TEMPERATURE SCALE	VELOCITY OF TYPICAL ATOMS AND MOLECULES		STATE OF MATTER	TYPICAL LOCATION	TYPICAL RADIATION EMITTED
5000 K	10 km/s		IONIZED GAS (electrons knocked free)	Atmosphere of a star (the Sun)	Visible light
600 K	3.5 km/s		GAS (separate atoms)	Atmosphere of a hot planet	Infrared radiation
300 K	2.5 km/s		LIQUID (some atoms linked in chains)	Water (planet Earth)	Far infrared radiation
100 K	1.5 km/s		SOLID (atoms linked in lattice)	Rock on a cold planet	Far infrared radiation

NOTES: The temperature ranges when different substances are solid, liquid, and gas vary considerably. But in general, most substances are solid below 100 K and gaseous above 1000 K. The velocities are quoted for hydrogen atoms; other elements would have lower velocities at the same temperature. The wavelength of the typical radiation at any temperature is given by Wien's law.

most solids and liquids are similar; the densities of gases are many orders of magnitude lower. This is the state of matter of the air we breathe.

Notice that substances go from solid to liquid or from liquid to gas at very different temperatures. The transitions depend on a detailed consideration of the chemical bond. At room temperature, only the lightest atoms or molecules can be in the gaseous state. Heavy or complex molecules are always liquids or solids. But the state of a set of atoms or molecules depends only on the amount of thermal energy they contain. If you raise the temperature of a solid such as iron, you can melt it (at 1813 K) and then even boil it into a gas (at 3033 K). On the other hand, if you lower the temperature of a gas, you can turn it into a liquid and even a solid. The nitrogen in the air you are breathing liquefies at about 77 K (or −196°C) and solidifies at an extremely chilly 54 K (or −219°C). As we will see later, most of the universe is at temperatures that are either much higher or much lower than what we are used to.

As an analogy for solids, liquids, and gases, imagine people in a gymnasium. If many people are all standing close together in their places doing exercises, they are like atoms in a solid, which vibrate but do not move from their positions. If the people are lined up in rows, they are like atoms in rock crystals, arranged in a symmetric lattice pattern. If the people scatter at random to do their exercises, they are like atoms in an amorphous solid, such as glass. Now imagine a crowd of people spilling out of the stands onto the gym floor after a game, bumping into one another as they leave. This is like the behavior of a liquid. To imagine a gas, think of a few blindfolded people scattered on the gym floor. Most of the time they may move in straight lines, but occasionally people collide and careen off in different directions. They are like molecules in a gas, which travel relatively long distances (compared to the size of a single molecule), but eventually hit another molecule.

All these forms of matter—solid, liquid, and gas—are cooler than matter in the Sun or in most stars. If we keep heating a gas, the speeds of the atoms increase. They hit harder and harder. At a temperature of a few thousand kelvins, atoms hit one another with such force that the electrons break free from their orbits. The Sun's surface, for instance, has a temperature a little less than 6000 K. There the hydrogen nuclei, many of them having lost their electrons, move at speeds around 10 km/s or 20 times faster than the air molecules you are breathing. Gas that has had its electrons knocked off is called *ionized gas,* or *plasma.*

Gases are heated to even higher temperatures inside stars. Recall our discussion of Rutherford's experiment earlier in the chapter, where positively charged alpha particles recoiled off the positively charged nuclei of gold atoms. The electric force usually stops atomic nuclei from coming too close together. But subject to the enormous pressure and temperature inside a star, atomic nuclei can collide and stick. As we will see in Chapter 11, a gas at a temperature of millions of kelvins can transform elements and yield prodigious amounts of energy. This is the source of sunlight.

Heat Transfer and the Concept of Equilibrium

We have seen that heat is just one of many forms of energy. We have also seen that energy can change forms and that the total amount is always conserved. Let us see what the rules are for the behavior of heat.

Whenever materials of unequal temperature are in contact, heat will flow from one material to the other. Drop an ice cube in a glass of water, and the ice warms up and melts while the water cools down. Pour a kettle of boiling water into a tepid bath, and the boiling water cools while the bath warms up. This is true of any two materials. Touch a car that has been sitting in the hot sun, and the heat will flow from the metal to your finger. Or think of what happens when an auditorium fills with people. Human body temperature is warmer than room temperature, and as we saw earlier in the chapter, humans radiate heat. So heat flows from the people to the air in the room, and the room gets hotter.

Our everyday experience shows that heat tends to flow from a hotter material to a colder material. The study of how heat moves is called *thermodynamics.* Some of the first experiments in this subject were done by Isaac Newton, who found that the rate of heat transfer between two adjacent objects or regions is proportional to the difference in their temperatures. If there is no temperature difference between an object and its surroundings, there will be no heat flow. But the larger the temperature difference, the faster heat transfers from the warm region to the cool region. Using the mathematical form of this law, Newton calculated how much time is needed for various objects to cool from a given temperature.

Thermal Equilibrium

These everyday examples illustrate the concept of *equilibrium*. When different regions of a system have different temperatures, heat energy will flow and the system is said to be *out of equilibrium*. When all regions have the same temperature, no heat energy flows and the system is *in equilibrium*. So while the ice cube is melting, heat energy is flowing from the water in the glass to the ice. After the ice cube has melted, all the water has the same temperature, cooler than it was before. As you pour boiling water into a bath, heat energy is flowing into the bath water. After the boiling water is mixed in, all the water has the same temperature, warmer than it was before. The end result of both actions is a situation of equilibrium.

This principle also operates at the microscopic level. We have seen that temperature is just a measure of the violence of the motion of individual atoms or molecules. The tiny particles in a piece of very hot metal are moving faster than the particles in the skin of your hand. When you touch the hot metal, the fast-moving metal particles agitate the particles in the skin of your hand and make them move faster. Heat is transferred, and your finger gets hot. The molecules in an ice cube are vibrating more slowly than the molecules in water. As the ice melts, the molecules in the surrounding water slow down and the water cools.

We can even make this concept visible by imagining a set of balls on a pool table. Let's imagine that they are analogous to atoms. Suppose that they are not moving, which would correspond to zero temperature or no heat. Now roll a cue ball fast across the table so that it bounces hard off all the cushions. This incoming "atom" has a high velocity and so a high temperature. As the cue ball collides with the stationary balls, the stationary balls start to move and the cue ball slows down. The cue ball has transferred energy to the other balls (Figure 4–10). As a result, the cue ball gets "colder," and the other balls get "hotter." This is just like the example where a drop of boiling water falls into a cooler liquid.

Now we have a clearer idea of equilibrium. Hot atoms or molecules move faster than cold ones. If we have a hot body of material next to a colder body of material, then energy will flow from the hot one to the cold one. The average kinetic energy of molecules in the hot one will decrease, and the average kinetic energy of molecules in the cold one will increase. When the average kinetic energy per molecule is the same in both objects—that is, when they reach the same temperature—they are in equilibrium, and no more net heat flow will occur between them. Nature displays a universal tendency to move toward equilibrium.

So far we have only looked at everyday examples, but heat transfer and the idea of equilibrium will come up repeatedly as we consider planets and stars. Equilibrium is a quite general physical principle, so when dealing with a situation of heat flow, we will talk about **thermal equilibrium**. Put in simple words, this means that unequal temperatures will tend to even out and become equal.

Three Modes of Heat Transfer

The idea of thermal equilibrium is one of the most important and simple ideas in understanding the physical universe around us. Heat always flows so as to equalize temperature differences. Let's look at the three different ways that heat can be transferred from warm material to cold material: conduction, convection, and radiation. These three processes are happening around us all the time (Figure 4–11).

Conduction occurs when heat is transferred through atomic collisions. This can occur in a solid or a liquid; the atoms or molecules of a gas are too far apart for conduction to be effective. We have seen some examples of conduction —ice cooling down water, hot metal burning your finger.

Let's go back to the pool table analogy. Suppose the balls are all stationary and spread at random over the table. You stand at one end of the table and give the balls at that end high speed. Every time a ball comes within your reach, you give it an extra kick and speed it up. You are acting as a heat source, creating "hot" material at the your end of the table while leaving the other end "cold." Energy is transferred among all the balls as they hit one another. Eventually, the velocities all over the table are about the same, as "heat" transferred to the "cold" end, creating equal temperatures at both ends. You have given the balls at the far end energy without ever leaving your end of the table! This energy traveled by conduction. In the same way, if you (unwisely) held a metal poker with one end in a fire, then heat would travel up the poker by conduction and you would be rapidly forced to drop it.

Not all materials are equally good conductors. In general, metals are good conductors while plastics and porous materials such as wood are poor conductors. Suppose you have a metal and a plastic surface at room temperature. They are both at the same temperature, so why does the metal surface feel cooler? You know that your body is hotter than room temperature, so heat will try to flow from your finger to the material you touch. The long and complex molecules in wood are inefficient at conducting heat, so your finger is not affected much. Metal atoms are very efficient at conducting

FIGURE 4–10
Transfer of thermal energy, using pool balls as analogs for atoms. **(a)** The incoming ball has high velocity or "temperature" while the stationary balls have zero temperature. **(b)** After the collision, the temperature of the cue ball has decreased while the temperature of all the other balls has increased. In this way, atomic collisions can carry thermal energy from one place to another. (SOURCE: Copyright Tom Pantages.)

heat. The result is that heat is rapidly transferred from your finger to the metal. Your finger cools down, and to you the metal feels cold! When we want to trap heat, we use a material that is a poor conductor—Styrofoam for a coffee cup and wool for a sweater, for example.

<u>Convection</u> occurs when heat is transferred by the movement of masses of material. Since solids cannot move freely, convection is effective only in liquids and gases. When warm material mixes with cool material, energy is transferred to the cool material. Convection is a very efficient way to transfer energy. Think about it the next time you take a bath. As you pour hot water into the bath, you could wait for the heat to diffuse out toward you by the collisions of water molecules (conduction), or you could speed up the process by stirring the hot water into the cool water (convection). Now consider what happens when you boil a pan of water. At first, the temperature rises with no visible motion of the water. During this phase, heat is transferred from the heating element to the pan and from the pan to the water by conduction. As the water reaches 100°C, heat can no longer be transferred fast enough by conduction, so it begins to move by rolling convection motions. Hot pockets of water rise and gravity makes them sink again. We see the water boil.

Many familiar weather conditions are caused by convection. The Sun heats the ground, which heats the air just above it. Warm air is less dense than cold air, so it rises, cools, and then sinks again toward the ground. This movement of air in great circulating patterns is a good example of convection. Sit and watch cumulus clouds some afternoon—they are the puffy, fair-weather clouds. As warm air rises, it will reach an altitude where liquid water droplets can condense from the gaseous water vapor in the air (just as water beads form on the outside of a cold glass of water). The droplets, of course, form what we see as a cloud. The slow billowing of a cumulus cloud is evidence of convection.

Just as in other modes of heat transfer, convection is energy flow via moving fluid (gas or liquid) that tries to equalize the temperature difference between two places. The Sun heats up the ground, and convection transfers heat upward through the atmosphere. The clear air turbulence we sometimes experience while flying is a sign of convection. As Newton predicted, convection increases with an increasing temperature difference between the ground and the air. Turbulence is generally worse in summer than in winter. Not only is convection all around us, but it is beneath our feet, too. As we will see in the next chapter, geological activity is largely the result of churning convective motions of molten rock!

<u>Radiation</u> is another important mode of heat transfer. Radiation in general terms will be discussed in Chapter 10. For the moment we will confine our attention to thermal radiation, which is the energy emitted by any object that relates to its temperature. The word *thermal* refers to heat: The hotter an object is, the more thermal radiation it emits. Heat transfer by radiation is quite different from the types of heat transfer we have already discussed. Conduction takes place in a solid or a liquid. Convection takes place in a liquid or a gas. Both of these mechanisms require a material medium to

FIGURE 4–11
The three modes of heat transfer are illustrated as skiers gather around a fire. Heat is carried through the metal of the stove by conduction. Convection currents of warm air rise above the stove and heat the cabin. The skiers are warmed directly by radiation from the hot metal. (SOURCE: Copyright Tom Pantages.)

transmit the energetic motions of atoms and molecules. Radiation can travel through a gas, but it can also travel through a pure vacuum. Our most familiar experience of heat transfer by radiation is our daily exposure to the Sun. The Sun's radiation travels across space and strikes Earth, providing Earth's main source of heat. Without the Sun's thermal radiation, Earth would be in a deep freeze.

We will see many examples of these modes of heat transfer in the chapters ahead. Figure 4–12 summarizes the microscopic view of the processes. Astronomers study radiation, convection, and conduction in order to understand the universe better.

A Closer Look at Radiation

Radiation is the principal way that heat and energy travel through the universe. The energy of each and every star, including the Sun, is carried across space in the form of

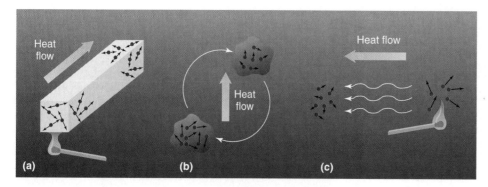

FIGURE 4-12
The three modes of heat transfer. In all situations, heat flows from hotter to cooler regions. **(a)** Conduction is the transfer of energy by atomic collisions and vibrations. The atoms themselves do not travel in the direction of heat flow. Conduction occurs most efficiently in a solid or a liquid. **(b)** Convection is the transfer of energy by large-scale motions of gas or fluid. Fast-moving atoms mix with slow-moving atoms. **(c)** Radiation is the transfer of energy by light waves or other kinds of waves; it occurs in a gas or in a vacuum.

radiation. With our telescopes on Earth, we capture and analyze that radiation. In Chapter 10, we will discuss how radiation arises and how it can be analyzed with telescopes. But, for now, we will focus on the role of radiation in the transfer of heat and energy. This will also help in our study of planets in the next few chapters.

What basic terms and concepts do we need to talk about radiation? Newton was the first to describe the components of radiation emitted by the Sun. He let a narrow beam of sunlight into a dark room and passed it through a prism (Figure 4–13). The light spread into the same array of colors that you can see in a rainbow. Newton proved that the visible radiation from the Sun is made up of a mixture of light of all colors. The array of colors that Newton saw—red, orange, yellow, green, blue, indigo, and violet—is called the **visible spectrum.** (Many people use the mnemonic Roy G. Biv to remember this sequence.) Newton was not the first person to disperse light into a spectrum, but he was the first to systematically deduce light's properties. Some scientists suspected that the colors were not part of white light but were introduced by the prism itself. So Newton passed the visible spectrum through a second prism and showed that it recombined to give white light. White light really is a superposition of colors. But are the colors fundamental? Newton selected one color from the spectrum and tried to disperse it further with a second prism. Blue light remained blue light, and red light remained red light. The colors therefore represent a fundamental property of light. These elegant experiments are shown in Figure 4–14.

Newton thought of light as a stream of tiny particles. Other scientists noticed that light had many of the properties of waves. As we will see in Chapter 10, it is equally valid to think of light as a wave or as a particle.

In 1800, astronomer and composer William Herschel did an interesting experiment. He passed sunlight through a prism, as Newton had done before. When he placed a thermometer in each color, the thermometer heated up, since sunlight of any color carries warming energy. Then he placed the thermometer beyond the red end of the spectrum, where no sunlight is visible. Would it heat up, Herschel wondered? Amazingly, it did. Herschel had discovered that there is radiation "beyond the rainbow" that cannot be detected by our eyes. It is called *infrared radiation*.

The easiest way to think about radiation is to consider its wavelike qualities. When light is spread out into a spectrum, each color corresponds to a different **wavelength.** Wavelength refers to the length of the wave—the distance between any two peaks or troughs in the wave (Figure 4–15). Whenever you see the word *wavelength* in reference to light, you could substitute the word *color* if it helps make the idea clearer. Notice, however, that it is just for convenience that we specify seven colors in the spectrum as listed above. There is actually a smooth and continuous change of color across the spectrum. Similarly, there is a smooth and continuous change of wavelength. Blue light has the shortest wavelength—about 0.0004 mm. Red light has longer wavelengths—about 0.0007 mm. Infrared radiation has wavelengths that are too long for the eye to see—longer than 0.001 mm.

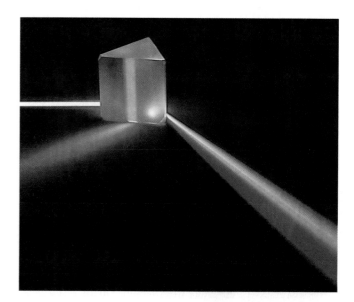

FIGURE 4-13
The familiar visible spectrum of light created by a prism. White light is split into its component colors. (SOURCE: David Parker/Photo Researchers Inc.)

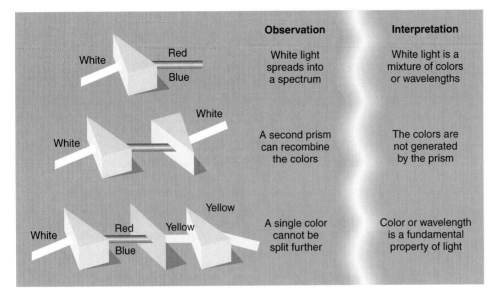

FIGURE 4-14
Newton's experiments to understand the nature of light. A prism disperses white light into colors arrayed from red to deep blue. A second prism can recombine the colors so they are not produced by the prism but are contained within white light. A second prism does not reveal any more structure within a single color, so color or wavelength is a fundamental property of light.

Figure 4–16 shows the spectrum of sunlight. Above the image of the spectrum we show a plot of the amount of radiation that arrives at each wavelength. Scientists usually represent a spectrum as a plot in this way. Spectra and other important ways of representing astronomical data are discussed in Appendix A-7. Figure 4–16 shows that the maximum amount of radiation from the Sun comes in the wavelengths we call yellow—the wavelengths to which our eye's receptor cells are most sensitive. In fact, this is an example of the way that humans adapt to their environment by evolution. The plot also shows that the intensity of radiation declines gradually toward longer and shorter wavelengths. From the combination of wavelengths, we see the Sun as yellowish white. The spectrum of radiation extends beyond the wavelength range to which our eyes are sensitive. Wavelengths too short for our eyes to detect are called *ultraviolet radiation*. The Sun emits invisible radiation at both ultraviolet and infrared wavelengths.

Wien's Law and Thermal Radiation

Now we are ready to relate the idea of **thermal radiation** to our earlier discussion of energy and temperature. The key principles were discovered in 1898 by the German physicist Wilhelm Wien (pronounced *Veen*). Wien discovered that all bodies constantly emit thermal radiation. Thermal radiation is always concentrated at certain wavelengths; you can see maximum intensity of the solar spectrum at the color yellow in Figure 4–16. Wein also found that the spectrum of thermal radiation does not depend on what the body is made of. He and several other German physicists experimented with bodies of different temperatures and deduced the following principles:

1. All bodies emit thermal radiation spanning a broad range of wavelengths.
2. The amount of radiation and the wavelength of maximum emission of radiation depend on the temperature of the body, but not on its composition. The wavelength of maximum emission is inversely proportional to the temperature.
3. The higher the temperature, the more radiation is emitted and the shorter (or bluer) the wavelength of the bulk of the radiation. The amount of radiation emitted is proportional to the fourth power of the temperature.

As shown in the Science Toolbox on page 105, the mathematical form of the second principle is unusually simple, considering that it expresses a very sophisticated truth about radiation. It is called **Wien's law**. The third principle is called the **Stefan-Boltzmann law**, and we will see its applications when we consider planets in Chapter 6 and stars in Chapter 12.

We saw earlier in the chapter that temperature is related to the microscopic motions of atoms and molecules. The

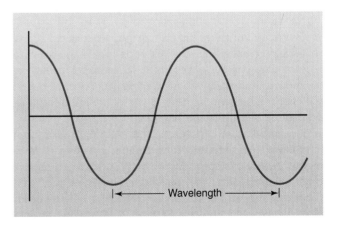

FIGURE 4-15
Light can be thought of as a wave carrying energy from one place to another. A wave is characterized by its wavelength, the distance between two successive peaks or troughs. The wavelength of visible light is less than 0.001mm.

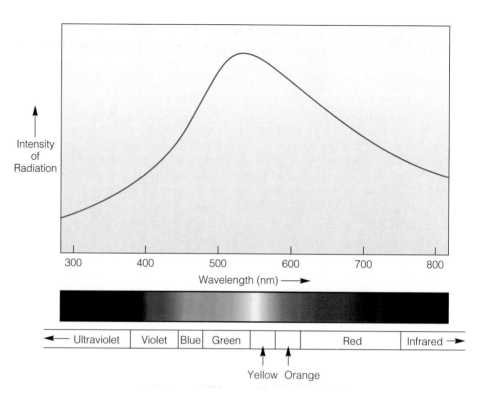

FIGURE 4–16
The spectrum of sunlight is the arrangement of colors according to their wavelength. Ultraviolet radiation has the shortest wavelength, visible light is in a restricted midrange, and infrared radiation has the longest wavelengths. Ultraviolet and infrared are examples of an enormously wide spectrum of radiation discussed further in Chapter 10. An informative way to show the spectrum is by a plot that shows the intensity or amount of the radiation occurring at each wavelength. The corresponding colors are shown at the bottom.

larger the kinetic energy of the particles, the higher the temperature of the material. Now we see that particles in motion emit a smooth spectrum of radiation. The larger the kinetic energy of the particles, the shorter the peak wavelength of the radiation. The thermal spectrum depends on temperature in a simple way, given by Wein's law. Since all atoms and molecules are in constant motion, all objects emit thermal radiation. We can also see why the radiation does not depend on composition. If we have a lump of iron and a lump of gold at the same temperature, the iron atoms and the gold atoms have the same kinetic energy. Therefore the iron atoms and the gold atoms emit the same thermal spectrum.

If everything is constantly emitting thermal radiation, why don't we see it? Objects at room temperature emit mainly infrared radiation that we cannot see. Not enough of the radiation comes out in the visible part of the spectrum to be detected by our eyes. We have the technology now to detect and make images with infrared radiation just as we do with visible radiation. Figure 4–B shows the infrared radiation that we all emit, made "visible."

As temperatures increase, the dominant radiation shifts along the spectrum toward bluer or shorter wavelengths. As you can see in Figure 4–17, only when objects reach high temperatures does the dominant radiation move into the visible region of the spectrum. In other words, we can see a radiant glow only from a very hot object. Figure 4–17 also shows the important consequences of the Stefan-Boltzmann law. The amount of energy radiated by an object—given by the area under the curve of the spectrum—increases in proportion to the fourth power of the temperature.

A good example of Wien's law in action comes when you turn on an electric stove. The coil on the stove starts out at room temperature (about 300 K) and is dull gray. This gray color is not emitted by the coil; it is merely the color of the metal as seen by the ambient light in the room. But then the coil heats up, and eventually we begin to see a dull-red glow. As the coil gets hotter, the glow becomes brighter and eventually becomes a slightly orange-red. The dull red glow of a heating element represents the short wavelength tail of the thermal spectrum; the maximum emission is at infrared wavelengths. (Molten lava has a similar red glow and has about the same temperature, about 1100 to 1500 K.) If the coil could get hotter, the radiation would get yellower and finally shift to a mix of colors similar to sunlight, which we perceive as white light. Because most objects in daily life are too cool to be "red-hot," their thermal radiation is in the infrared range, invisible to us.

It is easy to get confused when you are thinking about color and thermal radiation. We see most ordinary objects by *reflected light* from the Sun or from lightbulbs. A blue book is not hotter than a red book; it is just reflecting a different part of the spectrum of a light source. In a room with no light source, a book has no color because there is no light to reflect! We also see the Moon and the planets by reflected sunlight. The only objects that emit their own visible radiation have a temperature of a few thousand kelvins, like the Sun or the filament of a lightbulb. It is important to understand this difference. Now you may be wondering—what about a fluorescent lightbulb or tube, which feels cool to the touch? The gas inside this kind of light source has a very low density. So while the gas atoms have a high kinetic energy that corresponds to a high temperature (see the earlier

SCIENCE TOOLBOX

Using Wien's Law

Atoms and molecules are in ceaseless motion. This is true of all materials and for all states of matter. In a solid, the motions are purely vibrations. In a liquid or a gas, the motions also involve collisions of freely moving particles. We have seen that temperature is a measure of the kinetic energy of the particles in a material; motion only ceases at absolute zero on the Kelvin scale. Atoms and molecules emit radiation as a result of their motion. Wien discovered how this radiation depends on the temperature of the material.

The mathematical form of Wien's law identifies the dominant wavelength, or color, of light coming from a body at a given temperature. It is surprisingly simple. Suppose we designate the temperature of the body as T, given in Kelvins. The wavelength at which the maximum amount of radiation is emitted can be called W, given in meters. By using the variables T and W, Wien's law can be expressed as

$$W = \frac{0.0029}{T}$$

The number 0.0029 is a *constant of proportionality*, and it is the same in all applications of the law, as long as T is given in kelvins and W in meters. This is an inverse relationship between wavelength and temperature. So the higher the temperature, the shorter or smaller the wavelength of the thermal radiation. The lower the temperature, the longer or larger the wavelength of the thermal radiation. For visible radiation, hot objects emit bluer light than cool objects.

For example, if the Sun has a surface temperature of 5700 K, what is the wavelength of maximum intensity of solar radiation? If we substitute 5700 K for T, we have $W = 0.0029 / 5700$, or 5.1×10^{-7} m. Given that violet light has a wavelength of about 4.0×10^{-7} m, yellow about 5.6×10^{-7} m, and red about 6.6×10^{-7} m, what can we say about the color of the Sun's peak radiation? The peak wavelength of the Sun's radiation is at a slightly shorter wavelength than the color yellow, so it is a slightly greenish yellow. To see this greenish tinge to the Sun, you would have to look at it from space. It turns out that Earth's atmosphere scatters some of the shorter waves of sunlight, which shifts its peak wavelength to pure yellow.

Remember that thermal radiation always spans a wide range of wavelengths; the mathematical form given above specifies the single wavelength that is the peak of the spectrum. So although the Sun appears yellowish white, when you disperse sunlight with a prism, you see radiation with all the colors of the rainbow. Yellow just represents the wavelength of the maximum amount of emission.

Now let us consider the thermal radiation from a much cooler object. What are the wavelength and the dominant type of radiation that your body gives off? As a hint, the body's "normal" temperature is 98.6°F, or about 37°C. First convert T to kelvins. Since 1 K is 1°C plus 273, T is about 310 K. Your radiating skin might be a bit cooler, say, 305 K. Using the equation for Wien's law, we get $W = 0.0029 / 305 = 9.5 \times 10^{-6}$ m. The longest red wavelengths of visible light are around 7×10^{-7} m, so this is far beyond the red end of the spectrum. The human body emits thermal radiation at

(continued)

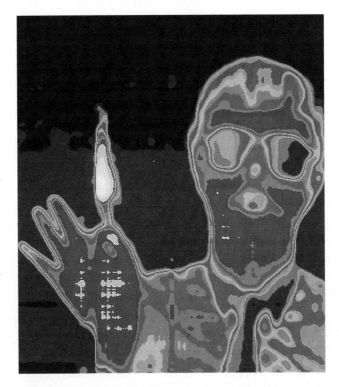

FIGURE 4-B
Technology originally developed for military applications allows us to detect and make images with infrared radiation. The colors in this image represent different temperatures. White is the hottest—the flame. Red is next-warmest—the palm and face. Blue is the coolest—the necktie and the glasses (the glasses are just reflecting the cooler regions of the room away from the man). (SOURCE: NASA/JPL).

SCIENCE TOOLBOX

Using Wien's Law (continued)

wavelengths that are too long for the eye to detect.

Any object at a temperature of a few thousand degrees emits visible thermal radiation and any object at a few hundred degrees emits infrared thermal radiation. Another example of Wien's law at work is the technological development called *night goggles* or *infrared vision*. These devices pick up infrared light and show it to us by converting it to visible light, such as a TV or photo image. In this way we can "see" the infrared thermal radiation of cool objects (see Figure 4–B).

Now suppose a certain star has a noticeably red color to the naked eye. Given that deep-red light has a wavelength around 700 nanometers (nm), what temperature is the star? (A nanometer, abbreviated 1 nm, is 10^{-9} m.) Compare the answer to the 5700 K temperature of the Sun, mentioned above. If the wavelength of dominant radiation W is about 700 nm, or 7×10^{-7} m, then Wien's law gives us $7 \times 10^{-7} = 0.0029 / T$. Solving for T, we have $T = 4140$ K. So the red star is around 1600 K cooler than the Sun.

Science Toolbox), the rate of collisions with the enclosing tube is low, so there is little heating effect.

Another type of confusion arises from popular culture. Artists talk about red as a "hot" color and blue as a "cool" color. Musicians use the same terminology—jazz is cool and associated with the color blue, and salsa is hot and associated with the color red. Blood is hot and red, but ice is cool and blue. Unfortunately, this subjective description of color is opposite to the scientific description of color based on thermal emission. According to Wein's law, objects that radiate blue light are hotter than objects that radiate red light.

Table 4–1 shows the type of radiation emitted by states of matter at different temperatures. We will encounter the full range of these types of thermal radiation in Chapter 10.

Radiation from Planets and Stars

Wien's law is a powerful physical principle that applies across the universe. Every object in space—every star, every planet—has a temperature that defines the peak wavelength of its thermal radiation. For astronomers, the most exciting application of Wien's law is that it gives us the ability to measure the temperatures of remote objects without visiting them.

Using a telescope, astronomers can measure the exact colors of the radiation emitted by a star to figure out its surface temperature. Wien's law describes the mathematical relationship precisely. Each factor-of-2 increase in wavelength is a factor-of-2 decrease in temperature. You can even make a rough estimate of a star's temperature without a telescope. Stars that appear reddish have temperatures around

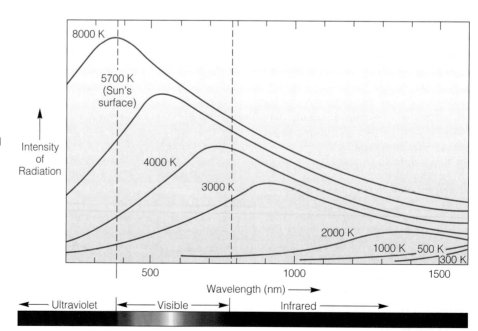

FIGURE 4–17
Wien's law shows that as temperature is increased, the dominant form of radiation shifts to shorter wavelengths, that is, toward bluer and bluer colors. At the far right, an object at 300 K or room temperature glows with wavelengths too long for the eye to see. At temperatures of 1000 to 2000 K, the peak of the thermal radiation is still at long wavelengths; however, the tail of the spectrum extends into the visible range, and objects glow a dull-red color. At temperatures from 4000 to 8000 K, the peak of the thermal spectrum shifts through the visible spectrum, and objects emit visible light. As an object is heated up, the total amount of radiation grows according to the Stefan-Boltzmann law (in proportion to the area under the different curves).

3000 to 5000 K, white stars are in the range 5000 to 10,000 K, and bluish stars are closer to 10,000 or 20,000 K. Not only are the colors of stars in the night sky pretty to look at, but they are also telling us about the temperature of enormous balls of gas trillions of miles away!

Similarly, astronomers use telescopes equipped with infrared detectors to measure the infrared thermal radiation emitted by planets. Cold planets radiate mostly at long infrared wavelengths (see the lowest temperatures in Figure 4–17). Warm planets have more thermal radiation at short infrared wavelengths. In a similar way, an instrument on a spacecraft can scan a planet for geothermal hot spots, such as an active volcano. Reconnaissance satellites use the same principle to detect the heat pulse from nuclear tests or rocket launches.

Internal Heat in Planets and Stars

With this introductory background to radiation and Wien's law, let us revisit briefly the question of heat transfer inside planets and stars.

Over geological time, each planet creates heat in its interior due to several processes, including the decay of radioactive minerals. The interiors of massive planets are also hotter than the surfaces due to the enormous pressure of gravity within the planet. If we have a hot interior and a cool exterior, our discussion of equilibrium tells us that heat will flow outward. The heat is transferred to the surface by heat-transfer processes such as we have discussed. For example, in Earth, internal heat travels through the rock by conduction and also by convection in molten rock masses, which rise and descend in sluggish convection currents. Energy then leaves the surface as thermal radiation. Thus, a planet (like any warm rock) slowly cools over geological time as it radiates its heat into space.

A star creates much more energy in its central region through nuclear reactions than a planet, processes we will discuss in Chapter 11. This intense energy release heats the central core of the star to a temperature of millions of degrees. The heat is transferred from the hot central core to the cool surface (a few thousand degrees) by a combination of radiation and convection. Finally, it is radiated into space as thermal radiation from the star's surface layers.

Summary

To study the nature of the universe, we must understand the nature of matter and its interaction with energy. Atoms are microscopic constituents of matter, composed of a tiny and massive nucleus and a circulating cloud of electrons. The positively charged protons in the nucleus are balanced by an equal number of negatively charged electrons—normal matter is electrically neutral. Atoms combine into molecules of different chemical compounds. We define an element by the number of protons in its nucleus. The processes in which atoms combine to form molecules, or those in which molecules break into their constituent atoms, are called chemical reactions. During chemical reactions, only the electrons in the outer part of each atom interact; the nuclei remain separate.

Energy is a general term to describe anything that can cause a change. Two important forms of energy in astronomy are kinetic energy—the energy of motion of an object, and gravitational potential energy—the ability of an object to move under the force of gravity. There are other forms of energy, too. Throughout the universe, energy changes forms, but the total amount of energy is conserved. Another form of energy is the thermal energy due to the constant motions of atoms and molecules. Temperature is just a measure of the kinetic energy or movement of the atoms and molecules of a substance. We define the zero point of the physical (Kelvin) temperature scale as the absence of all motion. The higher the temperature, the faster the movements of atoms and molecules. At any given temperature, lighter molecules will move faster than heavy molecules, on average.

Whenever two regions have different temperatures, heat will flow in a way to try to equalize the temperatures, a situation called thermal equilibrium. Equilibrium can be upset by the addition of new energy or heat that warms up one region, or by the removal of heat that cools a region. The three modes of heat transfer from one region to another are conduction, convection, and radiation. All bodies emit a broad spectrum of thermal radiation, which reflects energy of motion of the individual atoms and molecules. The higher the temperature, the more radiation is emitted and the shorter the wavelength of the peak radiation.

Important Concepts

You should be able to define these concepts and use them in a sentence.

element	energy
compound	transformation of energy
atom	law of conservation of energy
molecule	solid
chemical reaction	liquid
electron	gas
atomic nucleus	thermal equilibrium
proton	conduction
neutron	convection
periodic table	radiation
energy	visible spectrum
potential energy	wavelength
kinetic energy	thermal radiation
heat	Wien's law
temperature	Stefan-Boltzmann law

How Do We Know?

These questions and answers show how the scientific method is used to learn about the universe.

Q How do we know that matter is made of atoms and molecules?

A Around 1810, Dalton established that compounds are always composed of the same proportions of elements (such as 2 units of hydrogen and 1 unit of oxygen in water), indicating that the elements exist in discrete proportions. Seventeen years later, the botanist Brown observed pollen suspended in water through a microscope. He saw the continuous agitation of the grains of pollen by water molecules that were clearly much smaller than could be seen by a microscope. Within the last 20 years, the evidence that atoms are real has become much more direct. We can use electron and very short X-ray waves to make images of atoms and molecules. The most sensitive of these techniques can make out individual atoms.

Q How do we know that molecules in a material move and that they move faster in a high-temperature material than in a cool material?

A You can actually see direct evidence of motion in the molecules of a gas. Watch dust motes floating in still air in a beam of sunlight. They will follow random paths, moving in different directions. Sudden changes in direction are caused by dust motes being hit by unusually energetic air molecules. As mentioned above, Robert Brown observed similar types of collisions in a liquid in 1827. This motion is called *Brownian motion,* after its discoverer. Moreover, the kinetic theory of matter, based on the idea of moving molecules, gives a perfect description of the relation between pressure, temperature, and density in a gas. For instance, it allows us to predict the pressure that will be exerted by a gas (e.g., inside a tire or a balloon) if the gas has a specified temperature and density.

Q How do we know that energy is always conserved?

A There are many reactions and interactions in nature, but whenever scientists set up a careful experiment to measure the energy involved, it is found to be constant, as accurately as we can measure. A simple astronomical example is an elliptical orbit of a planet around a star. In this case the kinetic energy goes up and the gravitational potential energy goes down as the planet moves closer to the star; and then the kinetic energy goes down and the gravitational potential energy goes up as the planet moves farther from the star. The total energy—the sum of these two quantities—is constant. The hardest thing about such experiments is keeping track of all the forms of energy entering and leaving a system, since energy can be transformed to many different forms. When a rocking chair slowly comes to a halt or a ball stops rolling, it seems as if energy has been lost. But in each case the energy of motion is being converted to heat by the process of friction. With a careful accounting of the different forms of energy, you will find that the total amount is conserved. This, incidentally, is the problem with claims of perpetual-motion machines: Friction always transforms some of the output energy of the machine to heat, and machines that run indefinitely without additional input of fuel to make up that loss are therefore impossible.

Q How do we know hotter objects emit bluer (shorter-wavelength) thermal radiation?

A We can repeat Wien's experiments, by heating objects and measuring the radiation they give off at various temperatures, and get the same results he did. More directly, we can observe existing objects, from electric stove elements to stars, to show that objects at room temperature give off infrared radiation, and much hotter objects give off visible light. We can also see that as we heat a filament, the radiation it emits starts at the red end of the visible spectrum and shifts to bluer (shorter) wavelengths as the temperature rises.

Q How do we know that objects at room temperature emit infrared radiation?

A William Herschel made the first step 200 years ago when he placed a thermometer beyond the red end of a visible spectrum of sunlight. The thermometer registered heat from invisible waves that were longer than red light. Wien established the relationship that predicted that objects at room temperature emit thermal radiation with peak intensity at infrared wavelengths. In the past 30 years, we have developed the technology to make images with infrared waves. These images allow us to "see" that everyday objects are constantly emitting infrared radiation.

Problems

Use these problems to test your understanding of the information and concepts in this chapter. The * indicates a more advanced or mathematical problem.

1. Viruses and bacteria contain large, sometimes fragile chains of carbon atoms that attract other atoms. Using this fact and kinetic theory, explain why heating or boiling tends to destroy many harmful viruses and bacteria.

2. **(a)** If you triple your car's speed from 20 to 60 mi/h, how many times more kinetic energy is contained in the car's motion? **(b)** Which does more to increase a battering ram's kinetic energy and effectiveness, doubling its speed or tripling its mass?

3. The Sun is made mostly of hydrogen gas. Suppose your teacher gives you a flask of such gas at room temperature. Of course, it is not giving off light as the Sun does. What are the main differences between it and the gas in the Sun?

4. Suppose you hit a baseball or kick a soccer ball. What happened to the kinetic energy that you put into the ball? What was the source of the energy that let you move the ball? Can you trace it all the way back to the Sun? How many times is the energy transformed?

5. Describe the nature of the energy conversion that takes place in the following situations; note that there may be more than one transformation. **(a)** The motion of a child on a swing. **(b)** Hydroelectric power. **(c)** The motion of a wind-up toy. **(d)** A car driven by an electric motor and powered by a battery. **(e)** The changes that occur from the time you speak a word to the time a person hears your voice across a room. **(f)** An athlete pole-vaulting. **(g)** Boiling water over a campfire. **(h)** Applying the brakes in a car. **(i)** A forest fire started by the impact of a meteor. **(j)** A roller coaster ride. **(k)** Light from an electric lightbulb. **(l)** The Moon in its elliptical orbit around Earth.

6. Suppose someone gives you a sphere of iron in a special magnetic containment vessel that maintains the iron's spherical shape and size, no matter what the temperature. Suppose you increase its temperature from 100 to 1000 K, then to 5000 K, and finally to 20,000 K. What might you observe about the appearance and state of the iron at these four temperatures?

7. Any gas exerts a pressure against its surroundings. For example, if you put more gas in a balloon, it exerts more pressure and expands the balloon. Describe what causes the pressure in a gas, in terms of the fact that molecules of gas are moving and have kinetic energy. What happens when you leave the balloon in a refrigerator for a while, and why?

8. If you increase the temperature of the gas in a balloon, what happens to the speed of the molecules of the gas? What happens to the pressure? What happens to the balloon?

*9. (a) If you double the temperature of a gas, by what factor does the typical velocity of a molecule in the gas increase? (b) Think carefully about the phrase *double the temperature*. Is doubling the Fahrenheit or Celsius temperature reading the same thing as doubling the temperature on a Kelvin scale? Which corresponds to a more sensible physical concept of doubling the temperature? Why? (c) By what factor would you cool a gas from room temperature to make the molecules move 6 times more slowly?

*10. Recall that nitrogen (N_2) molecules are the most common molecules in air. Calculate the speed at which N_2 molecules hit your skin on a cold day when the temperature is $-1°C$ and on a hot day when the temperature is $35°C$. (*Hint:* Convert the temperatures to kelvins).

*11. Suppose an incandescent lightbulb has a temperature of about half that of the solar surface, or around 2850 K. (a) Qualitatively, compare the color of light from the bulb and from the Sun. (b) Quantitatively, calculate and compare the wavelength of the peak radiation of the two light sources. Look at the thermal spectra in this chapter. Will most of the lightbulb's energy come out as visible radiation? Is a lightbulb an efficient energy source?

*12. Suppose an explosion on the surface of the Sun doubles the temperature of the gas in that region. Describe the type of radiation that would come from the exposed hot gas in the explosion. What is the wavelength of the peak of the emission?

Projects

Activities to carry out either individually or in groups.

1. You can do a simple experiment to show that atoms and molecules are much smaller than the eye can see. Take a large pan or a pie plate, and fill it with 1 in. of water. Sprinkle pepper over the surface—this is to let you see the surface. Now let a small drop of dishwashing liquid fall onto the center of the water. What happens to the pepper? What has happened to the drop of dishwashing liquid? A small spherical drop has spread out into a wide but shallow cylinder on the surface of the water. You can equate the two volumes. A sphere's volume is $(4/3)\pi r^3$, where r is the radius of the drop, and a cylinder's volume is $\pi R^2 h$, where R is the radius of the circular surface of the water and h is the thickness of the layer that you want to measure. Make your measurements and estimate of the radius of the initial drop of liquid. Now deduce the thickness of the layer of dishwashing liquid $h = 4R^3/(3r^2)$. And remember to use the same units for r and R. How does this number compare to the size of a grain of sand or the thickness of a human hair? The results will be even more impressive if you repeat the experiment with pepper sprinkled over water in a bathtub.

2. With an outdoor version of the previous project, you can get quite close to measuring the size of a single molecule. Wait for a clear day with a slight breeze, and go to a small pond or other sheltered body of water. Take with you a small, known volume of a light oil such as olive oil (1 cm^3 or a small teaspoonful is sufficient). Carefully spill the oil onto the surface of the water 10 or 20 yd from shore. Wait for the breeze to spread the oil over the surface. Even though the oil is colorless, you should be able to see it when sunlight reflecting off it creates rainbow diffraction patterns. When the oil slick reaches its maximum extent, roughly measure its area and repeat the calculation above. What is the thickness of the oil layer? If the experiment is done carefully, you can spread the oil into a monolayer—a thin film that is a single molecule thick!

3. Keep a diary to discover the energy flow in your own body. For one week, write down a list of all the food you eat. Look up or buy a dieting book that lists the calorie equivalent of various foods—this is how you estimate the energy you take in from the food you eat. Also write down all your physical activities. Look up in or buy an exercise book that gives the number of calories burned in various physical activities. What happens to the energy in the food you eat? How much of it is used in exercise? How much of it is used just to keep your body functioning? What happens to this energy?

4. On a day when you can see a large, puffy cumulus cloud, or thundercloud, study the cloud structure for about 5 or 10 min. Observe the structure of the rounded bulges at the top of the cloud. Can you see movement and changes in structure? Can you see uplift? Explain the structure and motions in terms of heat transfer by convection. Where does the energy that drives the cloud come from? Why do thunderstorms tend to happen in the late afternoon?

Reading List

Atkins, P. W. 1984. *The Second Law of Thermodynamics.* New York: Scientific American Library.

Burke, J. 1978. *Connections.* Boston: Little, Brown, and Company.

Cardwell, D. S. L. 1971. *From Watt to Clausius: The Rise of Thermodynamics in the Early Industrial Age.* Ithaca, NY: Cornell University Press.

Englart, B.-G., Scully, M. O., and Walther, H. 1994. "The Duality in Matter and Light." *Scientific American,* vol. 271, p. 86.

Fowler, J. M. 1984. *Energy and the Environment.* New York: McGraw-Hill.

Hazen, R. M., and Trefil, J. 1990. *Science Matters.* New York: Doubleday.

Miller, G. T. 1997. *Environmental Science.* Belmont, CA: Wadsworth Publishing, Inc.

Schwarz, C. 1992. *A Tour of the Subatomic Zoo.* New York: American Institute of Physics.

Various authors. 1990. "Energy in the Global Economy." *Scientific American,* September issue.

von Baeyer, H. C. 1992. *Taming the Atom.* New York: Random House.

CHAPTER 5

The Earth-Moon System

CONTENTS

- Buffon and the Age of Earth
- How Old Is the Earth-Moon System?
- Internal Structures of Earth and the Moon
- Influences That Shape a Planet
- Geological Time Scale
- Earth's Atmosphere, Oceans, and Environment
- Explorations of the Earth-Moon System
- Summary
- Important Concepts
- How Do We Know?
- Problems
- Projects
- Reading List

The first images of Earth as a globe hanging in space led to a new awareness of our planet as a fragile, finite world, and catalyzed the modern environmental movement. This image shows both Earth and the Moon as seen from the *Galileo* space probe.
(SOURCE: Galileo Imaging Team/NASA.)

WHAT TO WATCH FOR IN CHAPTER 5

This chapter tells the story of the geological evolution of the twin Earth-Moon system. You will see the steady progress scientists have made in understanding the truly enormous age of Earth. Our discussion of the age of Earth stresses how scientists have used better and better techniques over several centuries to arrive at an accurate measurement of age. Understanding the geological activity of Earth requires the concepts of energy covered in the last chapter. You will see basic geological principles set in a broad planetary context. We will compare the interior structures and surface processes of Earth and the Moon. The chapter also discusses properties of Earth's atmosphere and oceans that make Earth unique in the solar system.

In our recent history, humans have inadvertently altered the chemistry of the atmosphere. This in turn raises issues of how scientific results apply to our personal lives and, at an even broader level, to decisions about global environmental policy. Last, we see an emerging awareness that Earth's evolution has been affected by cosmic influences such as debris that rains in from space. Examples of this include the mass extinction of species that occurred 65 million years ago and the catastrophic origin of the Moon early in Earth's history.

PREREADING QUESTIONS ON THE THEMES OF THIS BOOK

OUR ROLE IN THE UNIVERSE
How have cosmic influences affected the evolution of life on Earth?

HOW THE UNIVERSE WORKS
What physical processes drive Earth's geological activity?

HOW WE ACQUIRE KNOWLEDGE
How did scientists come to know the true age of Earth?

BUFFON AND THE AGE OF EARTH

One summer day in 1741, in a scene worthy of a Frankenstein movie, a French aristocrat entered the foundry on his estate to conduct a strange experiment. He ordered the casting of two dozen small, solid-iron globes. As the globes were removed from the furnaces at white-hot heat, the curious aristocrat held a watch and observed the glowing spheres carefully while they cooled to the temperature at which they could first be touched with bare fingers. According to one account, he sought out the people in his village with the most sensitive skin to determine when the spheres could be touched. The name of the aristocrat was Georges-Louis Leclerc, Comte de Buffon. The goal of the naturalist's odd experiment was to figure out the age of planet Earth.

It was a cleverly designed experiment. Buffon measured how long it took for the globes to cool and then extrapolated those numbers to Earth-sized globes. By seeing how the time to cool depended on the volume of the sphere, Buffon hoped he could reliably predict the cooling time of an immense sphere such as Earth. To the casual observer, this must have seemed like a very unusual idea. Real-life scenes like this probably led to the idea of the mad scientists "experimenting with the forces of nature." Perhaps such characters inspired writer Mary Shelley two generations later, in 1816, to create her own character, Victor Frankenstein, in his remote laboratory.

We think of science today in terms of complex instruments and precise measurements. But much of our research today can be understood by going back to the simple questions that the visionary scientists of the past asked. Buffon's experiment reflects two of the themes of the book. He approached the question "How old is our planet?" as a problem that someone might actually be able to solve! Buffon suspected that Earth was a sphere of rock and metal, and he wanted to determine its properties. Today we see Earth as a beautiful 4.5 billion-year-old world hanging in space (Figure 5–1). Buffon's experiment also shows how scientists answer simple questions about nature. Buffon started by imagining Earth as a solid metal sphere and by working with a scaled-down model of the planet. Let us look more closely at the scientific investigations of the Comte de Buffon.

The Comte de Buffon, whose title means "Count" in French, is better known in history simply as Buffon. He was a colorful character with eclectic interests.

Almost disinherited after a duel at a French university, he acquired lands at age 25 and became wealthy enough to pursue his experiments in science. He charmed Paris society and had Thomas Jefferson as a dinner guest. Jefferson reported that Buffon was "a man of extraordinary powers of conversation." One of Buffon's lady friends wrote that "Monsieur de Buffon has never spoken to me of the marvels of the Earth without inspiring in me the thought that he himself was one of them"—not a bad testimonial, especially to describe a scientist.

Buffon's wide interests led him to study all aspects of the natural world, and he produced a 36-volume encyclopedia, called *Theory of Nature,* which was published over the years from 1749 to 1785. He can best be described as the Carl Sagan of his day. His books included some of the first bestsellers of popular science, which fed an immense public interest in the new discoveries of the age.

A major question in Buffon's day was whether Earth was really only a few thousand years old, as most Biblical scholars claimed. Many scientists thought the question was unanswerable. Buffon addressed this question directly. He assumed that Earth had formed in conjunction with the much hotter Sun and that Earth had therefore started in a molten state. He had a bright idea, inspired by earlier remarks of Newton. It was clearly impossible to simulate the situation of cooling Earth. But if he could measure how long it took his solid globes to cool and show as well how the cooling time depended on the size of the globe, then he could figure the cooling time for a much larger globe the size of Earth. Following the ideas of the scientific method we have seen earlier, he repeated his experiment many times to get more reliable numbers. Using his estimates of the cooling time, Buffon correctly concluded that Earth has to be much older than the few-thousand-year estimate popular in those days. But the story was more complex than he supposed, as we will soon see.

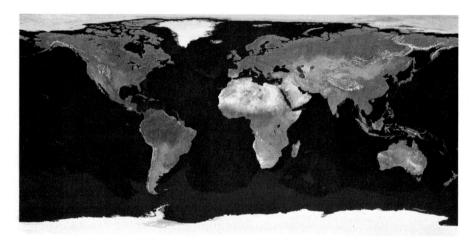

FIGURE 5-1
The surface of Earth, in a mosaic of images taken at different times to get a cloud-free view. The discovery of the age of this planet was a great intellectual adventure. (SOURCE: Tom van Sant/The Geosphere Project)

How Old Is the Earth-Moon System?

Buffon's experiment is important not for the specific answer he derived, but for his role as a scientific pioneer. If you look at the long march of human understanding, science begins with the idea that we can understand our role in the universe by using logic and observation. For example, the stars and planets were mysterious objects for most of human history. Yet we saw in Chapter 2 that Greek philosophers were able to use reasoned arguments and simple geometry to estimate the sizes and distances of our neighbors in space. Similarly, the most important scientific step in understanding Earth's history was not one particular set of measurements, but rather the mental leap of realizing that Earth had a history that could be unraveled by scientific observations and measurements. This happened mostly in the 16th and 17th centuries with the Renaissance and then with a period of time called the Enlightenment. Buffon showed that we can use direct experiments to learn about our place in time and space. As an example of how science works, let us look at humanity's step-by-step search to find the **age of Earth** from such a perspective.

First Estimates of Earth's Age

In the Middle Ages, scholars thought they could calculate Earth's age by finding out how long humans had lived on Earth. They assumed that humanity had been around almost since Earth itself formed. Therefore, scholars analyzed ancient records, especially Biblical scriptures that listed the generations since Adam and Eve. They concluded that humanity, Earth, and the whole cosmos were only a few thousand years old. The most famous calculation was made by Irish archbishop James Ussher in 1650. Ussher deduced that the cosmos formed on Sunday, October 23, in 4004 B.C., and that humanity was created on Friday, October 28, the same year. Even today, some people—especially fundamentalist religious groups—still believe that Earth is only a few thousand years old, based on this method of reasoning.

In the Renaissance, scientists began to question this method. They realized that a vast panorama of geological processes must have occurred, for which 6000 years was simply not enough time. In 1519, Leonardo da Vinci noticed fossil seashells embedded high in mountains, indicating that rocks on the peaks had once been under the sea. Such observations led to a new field of study, called *natural history*. It is certainly possible to imagine that Earth has always existed and has never changed. It is also possible to imagine that a creator put everything in place just as we see it today. Natural history implied that the world is changing all the time, and physical processes shape and alter our surroundings. The world we see today might not be the way the world was originally created. Scientists worked hard to figure out how long geological processes had been going on.

A breakthrough occurred in the 1670s, a time when Newton was at the height of his powers. Around 1671, the "father of English Natural History," John Ray, observed a bed of sand 100 ft thick deposited on top of cockleshells at Amsterdam. This discovery meant that 100 ft of sediment had been deposited after the cockles had been alive. Ray realized he could calculate how long this deposition process had been active. We can easily follow his logic. He had to make a separate determination of the rate at which sand is deposited. Then he divided the depth of the sediments by their measured rate of accumulation. By making a simple division he was also implicitly assuming that the average rate of deposition did not change much. According to his calculations, it would have taken at least 10,000 years for the sand layers to be deposited. This was "a strange thing," John Ray wrote, "considering [that the age of Earth] according to the usual account is not yet 5600 years."

The next step toward pinning down the age of our planet was to think about the cooling of Earth. Scientists believed Earth had once been molten—volcanoes were considered to be evidence of this. This hypothesis suggested a simple question: How long would a molten Earth take to cool to present-day temperatures? Newton approached the problem theoretically. In the 1680s he calculated that "a globe of red-hot iron equal to our Earth.... would scarcely cool... in 50,000 years." Newton's calculation inspired Buffon's odd experiments. Buffon made actual measurements of the cooling times of small solid globes. He then extrapolated to the much larger size of Earth and concluded that Earth's age must be at least 75,000 years—10 times as long as the Biblical chronology. Buffon even suggested that his result might be too short. He speculated that Earth might be as much as 3 million years old!

Evidence mounted that Earth was far older than previously believed. The new field of geology added to this evidence. The method of using layers and deposition rates to date rocks improved in the 1800s. Scottish scientist Charles Lyell, known as the father of modern geology, and others made two important discoveries. First, they studied sedimentary layers exposed in canyons in many parts of the world and realized that the total depth of sediments is immense. One estimate gave 72,000 ft of sediments as a typical figure in some regions. This meant the old estimates of the time needed to deposit the sediment of a single river were much too short for the age of the whole Earth. Second, geologists found evidence that mountains had gone through many cycles of erosion, subsidence, and uplift. Using deposition rates to calculate the age of one area might represent only one cycle, not the age of Earth itself. Lyell, in 1830, like Buffon before him, wrote that "millions of years" would have been required to form all the features seen on Earth. Thus, by the late 1800s, most scholars accepted that Earth must be at least millions of years old.

You can see why the idea of an old Earth is a conceptual leap. Except for violent events like earthquakes and volcanoes, the geological changes of Earth are imperceptible. Even over the span of a lifetime, erosion and deposition sculpt our surroundings very subtly. We see a snapshot in a long geological history, and we must use the patterns in Earth to unravel that history.

Measuring Rock Ages by Radioactivity

In the late 1800s, the crucial discovery of radioactivity made possible even more accurate estimates of Earth's age. The process is worth discussing in detail because it allows us to date not only Earth, but also rock samples from other worlds, such as lunar rocks and meteorites. Such dating sheds light on the age of our whole solar system and even of the universe itself.

The discovery of radioactivity happened by accident. In 1896, French physicist Antoine Becquerel left some photographic plates in a drawer with some uranium-bearing minerals. Later he opened the drawer and found the plates fogged. Being a good scientist, he did not dismiss the event, but investigated further. He found that the uranium emitted "rays," which, like X-rays (discovered the previous year), could pass through cardboard. The new "rays" turned out to be not electromagnetic radiation, like ultraviolet light or X-rays, but rather energetic particles emitted by unstable atoms. It might seem that serendipity plays no part in the scientific method, as discussed in Chapter 1. Yet a number of important discoveries—X-rays by Roentgen, penicillin by Fleming, and others—happened by accident. Perceptive scientists will pursue unexpected phenomena. Often, they will find a flaw in the equipment. Occasionally, they will make an important discovery. A lesser scientist than Bequerel might have seen the fogged photographs, shrugged, and carried on with his work.

This is how **radioactivity** works. A radioactive atom is an unstable atom that spontaneously changes (usually into a more stable form) by emitting one or more particles from its nucleus. The original atom thus becomes either a new element (changed in the number of protons in the nucleus) or a new form of the same element, called an **isotope** (changed in the number of neutrons in the nucleus). The original atom is called the *parent* isotope, and the new atom is called the *daughter* isotope. The decay is illustrated in Figure 5–2a.

The time required for one-half of the atoms of any original radioactive parent isotope to decay into daughter isotopes is called the **half-life** of the radioactive element. If 1

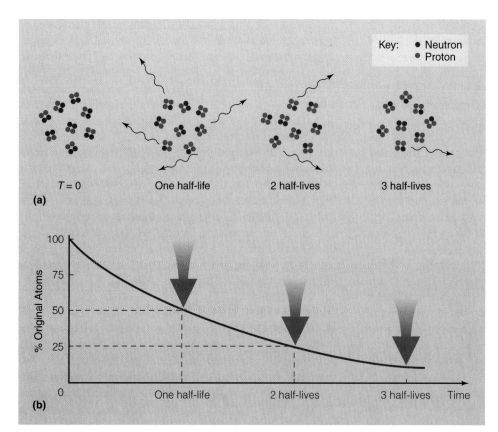

FIGURE 5-2
(a) A schematic diagram of radioactive decay. At the left, atoms of an imaginary radioactive element are shown at initial time zero. After 1 half-life, one-half of the nuclei have decayed, meaning that the nuclear structure has changed (note that neutrons have changed to protons in some nuclei). After 1 more half-life, another one-half of the nuclei have decayed, leaving only one-quarter of the original parent isotope. Wavy lines represent energy being given off by the nuclei that decay. (b) In any sample of radioactive atoms, one-half will decay within 1 half-life, three-fourths within 2 half-lives, and so on. The time of decay of any individual atom cannot be predicted. Energy like this powers much of Earth's geological activity.

billion atoms of a parent isotope were present in a certain mineral specimen, 0.5 billion would be left after one half-life, 0.25 billion after the second half-life, and so on (Figure 5–2b). Sometimes the result of the decay is another radioactive element, so the decay continues. But in every case the final product of the decay (or chain of decays) is a stable element. Here are some examples of the half-lives of unstable elements. The number after the name of the element is the atomic weight of the isotope—the sum of the number of protons and neutrons in the nucleus.

1. Rubidium-87 decays to strontium-87 with a half-life of 49 billion years.
2. Uranium-238 decays (in a series of steps) to lead-206 with a half-life of 4.5 billion years.
3. Potassium-40 decays to argon-40 with a half-life of 1.25 billion years.
4. Carbon-14 decays to nitrogen-14 with a half-life of only 5730 years.

Radioactivity is a random process. This means that the exact time when an individual atom decays is impossible to determine. Yet the average time for one-half of a very large number of atoms to decay is well determined. In other words, the concept of half-life is only meaningful in a statistical sense. Let's take an everyday example. If you had a pan of popcorn cooking, it would be possible to measure the time at which one-half of the kernels had popped. If you repeated the experiment many times, this "popcorn half-life" would be a well-measured number. Yet it would be impossible for you to stare at a particular kernel and predict when it would pop. This situation is similar to the random radioactive decay at the subatomic level.

Early in the 20th century, physicists realized that the process of radioactive decay could help determine the date when a given rock formed. Here's how the method works. Suppose we could determine the original number of parent and daughter isotope atoms in a rock, meaning at the time when the rock first formed. This can be done by counting the relative numbers of different isotopes in the minerals of the rock. Then if we simply count the present numbers of parent and daughter isotope atoms in the rock, we can tell how many parent atoms have decayed into daughter atoms and hence tell how old the rock is. For instance, if one-half the parent atoms have decayed, the age of the rock equals one half-life of the radioactive parent element being studied. This technique of dating rocks is called **radioactive dating**.

Radioactive dating is a powerful technique that is used throughout science. Since radioactive elements have half-lives that range from a fraction of a second to billions of years, we choose a parent isotope that is matched in half-life to the approximate age of the phenomenon we want to measure.

One important example is the use of carbon isotopes to measure the ages of organic materials. Every living thing takes in carbon—animals by the carbon in their food and plants by the carbon dioxide they absorb from the air. Most carbon is in the stable form of carbon 12. However, carbon in the natural environment has 1 in 1 million atoms of the

SCIENCE TOOLBOX

Working with Half-Lives

Using the general principles of radioactive decay, scientists have developed a simple equation to calculate a rock's age. Let us examine how a scientist might use these principles to develop such an equation. Our goal is to give the age of a rock sample in terms of the number of atoms that have decayed.

Suppose a radioactive isotope has a half-life of 1 million years. A certain number of these atoms are trapped in a crystal as molten lava cools to form a rock. After 1 million years, one-half of them would be left. After 2 million years, one-half of that amount would be left. This would be $1/2 \times 1/2 = 1/4$ of the original number. How many would be left after 3 half-lives, or 3 million years? It would be one-half that number again, or $1/2 \times 1/2 \times 1/2 = 1/8$.

From this progression, it is easy to see how to convert from this specific description to a formula for the general case. Let us call the fraction of the atoms that are left F. Suppose that we wait N half-lives, and ask how many atoms would be left. Based on the paragraph above, we see that the answer would be

$$F = \left(\frac{1}{2}\right)^N$$

To make the same equation even more useful if you have a calculator, we can take the logarithm of both sides (base-10 logarithm), which gives

$$\log F = N \log \frac{1}{2} = -0.301 N$$

Let us check this result. From the discussion above, we know that when $N = 3$ half-lives, the fraction of atoms left is $1/8$. Substituting $N = 3$, we get $\log F = -0.903$. A calculator confirms $F = 0.125$, or $1/8$.

Now suppose we want to determine the age of a rock crystal and its potassium atoms. A particular radioactive form of potassium decays with a half-life of 1.25 billion years (known to three significant digits), yielding a certain form of argon atoms. Suppose we measure the argon and potassium in the rock crystal, and we find that 58 percent of the radioactive potassium has already decayed into argon, while 42 percent of the original radioactive potassium atoms are left in the crystal. How old is the rock? Our measurement has told us that F is 0.42, and so our equation gives $-0.376 = -0.301 N$. Thus, $N = 1.25$ half-lives. So the rock is 1.62 billion years old.

We can also see why the half-life of the tracer used should be roughly the same as the expected age of our rock sample. Suppose we have a lava sample that is suspected to be about 100,000 years old. Potassium 40 has a much longer half-life. So N is given by $10^5 / (1.25 \times 10^9) = 0.000077$. Using the equation above to solve for F, we get $F = 0.99994$. The problem in this case is that very few potassium atoms will have decayed into argon atoms—only 6 out of every 100,000. Since potassium-40 is rare to start with, the measurement becomes very difficult. We should look for a more appropriate tracer with a shorter half-life.

In practice, there are more complexities in measuring the original numbers of radioactive atoms and the numbers that have decayed, but this discussion shows the way in which the general principles can be converted to a simple equation that covers all cases. This is often a goal in scientific work—to formulate a general equation that describes a phenomenon. Others can then make careful measurements, plug in the numbers, and get answers easily.

---

isotope carbon-14, which has a half-life of 5730 years. When living things die, no more carbon-14 is introduced and the existing amount diminishes steadily by radioactive decay. The remaining fraction of carbon-14 gives a good estimate of how long it has been since the living material died. This is how we know the age of wooden artifacts taken from an Egyptian tomb, for example. In another celebrated case, the Shroud of Turin—reputed to be the burial cloth of Jesus—was shown to be a fraud dating from the 12th century. By using this idea and a different isotope with a longer half-life, we can measure the age of dinosaur fossils.

Measurements of Earth's and the Moon's Age

How did scientists use the radioactive dating method to finally pin down the ages of Earth and the Moon? In the dating of rocks we measure the time since the rock became solid. The parent and daughter atoms must be fixed in location for the technique to work. Before the rock solidified, daughter isotope atoms might escape from the liquid lava in one way or another; but once the rock solidified, they were trapped in the solid mineral structure. Radioactive decay can only measure the time since the rock last solidified. See the Science Toolbox that follows.

Armed with this technique, geologists set out to find the oldest rocks on Earth. These oldest rocks are very rare because, after billions of years, most have been eroded or destroyed by Earth's active geology. The oldest intact rocks were found to be about 4.0 billion years old. We have to add to this an uncertain number representing the time during which Earth was molten and was being heavily bombarded by debris within the solar system. The best estimate is 4.6 billion years. Scientists also wanted Moon rocks to see if they were as old as Earth rocks.

The flights to the Moon constituted one of the greatest human adventures of all time. There were six landings during the Apollo program. The first landing, by Neil Armstrong and Buzz Aldrin, took place on July 20, 1969, and the last landing was in 1972 (Figure 5–3). Astronauts commented on the bright, harsh sunlight and the desolate beauty of the surface as they bounced along cheerfully in the Moon's low gravity. Each Apollo mission brought back a precious cargo of rocks, soil samples, and drill cores. Together with smaller samples brought back by earlier, unstaffed Soviet probes, scientists had many kilograms of rocks from nine different lunar sites of varied terrain.

They eagerly tested these samples and began to unravel the history of the Moon. Many of the Moon rocks were about 4.0 billion years old, with a few chips dating back to 4.4 or 4.5 billion years. Apparently, rocks formed during the first few hundred million years of the Moon's history were pulverized and destroyed by intense impact cratering that occurred at that time, which explains why the earliest rocks and chips are rare. Scientists add about 100 million years to this age to account for the time it took for the molten Moon to solidify for the first time.

The radioactive dating technique applied to all sorts of samples—terrestrial rocks, lunar rocks, and meteorites from deep space—yields an age of 4.6 billion years. The technique is based on the well-understood physics of radioactivity, which can be tested in the laboratory. Scientists consider this proof of the age of Earth. Consider this vast number for a moment. It shows that all the generations of humanity are only a blink of the cosmic eye; we've been around for less than 0.01 percent of the history of our planet.

Internal Structures of Earth and the Moon

The Engine That Drives the Planet

Radioactivity provided the crucial evidence for dating Earth and the Moon. Radioactivity has another critical connection to an understanding of Earth and other planets. It is one of the main sources of heat inside Earth, and heat is the ultimate source of energy that drives geological activity on any planet. How do radioactive minerals inside the planet produce heat? Imagine an atom of a radioactive element buried inside Earth. When the atom breaks down during the process of radioactive decay, it shoots subatomic particles and photons of energy outward into the neighboring material, as shown in Figure 5–2. Just as fragments of an exploding billiard ball might set neighboring balls into motion, the "exploding" atom's debris hits other atoms and increases their motions. Remember from the last chapter that increased motion results in a higher temperature. Radioactive material is therefore an energy source. When it is trapped inside any planet, it heats the interior of the planet.

The heating action of radioactive elements inside Earth explains why all the early models of Earth's age, based on cooling time, were far too short. As Newton and Buffon showed, Earth would cool within millions of years if it had no radioactivity. But because it has radioactivity inside, the insides are kept hot for a factor of thousands of times longer—several billion years.

The internal heat of Earth fueled by radioactivity is the "engine" that drives the geological activity of a planet. Volcanoes, for example, erupt because the heat inside Earth melts rock and continually creates supplies of molten magma; the magma is less dense than surrounding rock and thus tends to rise to the surface. Earthquakes are also caused by pressure associated with the movements of molten rock. Without radioactivity, Earth's interior would have cooled long ago. Volcanoes would be extinct and there would be no earthquakes. Earth would be geologically dead.

Layering of Earth and the Moon—Core, Mantle, and Crust

How did we uncover the secrets of Earth's interior? The deepest holes that geologists have drilled go only 10 or 15 km into the crust—a mere pinprick in the 6400-km radius of the planet. Information about deeper structure cannot come from drill holes; it comes from seismic waves, which are vibrations produced by earthquakes, volcanism, or artificial

FIGURE 5–3
Apollo 17 astronauts obtained samples from huge boulders, apparently dislodged in the past from bedrock farther up the slopes of the upland hills near their landing site. It is nearly 30 years since humans last set foot on the Moon. Dating of such rock samples by radioisotopes helped reconstruct the Moon's history. Apollo 17, in 1972, was the last human flight to the Moon. (SOURCE: NASA.)

explosions. Just as light deflects when it passes into water, seismic waves bend according to the density of the rock they pass through. Just as light slows down when it travels from air into glass, seismic waves slow down when they travel through dense rock. Nature provides the signals for these experiments. After a major earthquake, scientists around the world will compare the strength and arrival time of the seismic waves at many locations in different countries. They use the information to build a map of the internal structure. More limited experiments have been carried out on the Moon using seismometers brought by the Apollo astronauts.

Seismic studies on both Earth and the Moon have revealed that the density increases toward the centers of these bodies. The variation can be explained by internal heat, which accumulated in Earth and the Moon during their formation processes and melted both bodies very early in their history. How did this happen? Space was full of rocky debris, and when these chunks slammed into Earth, their kinetic energy was turned into heat. This heat spread through the young planet, making it liquid or at least very soft. Heavy materials, such as metals, tended to sink to the center, and light rock-forming minerals accumulated at the surface. This process of separating minerals is called **differentiation.**

Without differentiation, all rocks on Earth, the Moon, and the planets would be one uniform material. Instead, differentiation produced a characteristic triple layering inside Earth: a very dense **core** of iron and nickel at the center, which is surrounded by a **mantle** of dense rock and a thin surface **crust** of lighter rock. Gravity is the force that has separated these materials.

The smelting of metal ores produces a similar layering. When rock containing iron, for example, melts, the metal sinks to the bottom of the vat—like the core of Earth. Lower-density rock fills the upper part of the vat—like the mantle. On the surface floats a thin scum of slag, or low-density rock—like Earth's crust. You could picture planets as giant smelting vats. (There is one caveat: Certain metals are chemically attracted to low-density minerals in the crust and hence have tended to stay with those minerals as they formed. This explains why we still have some iron and other metal ores in the crust, even though most of Earth's iron is in the core.)

If you want an everyday example of differentiation, observe a well-mixed jar of oil and water. This experiment works best if you use syrupy oil with some color to it, such as olive oil. As the oil separates out from the denser water, it floats to the surface of the jar. This works with solids, too. Fill a jar with a mixture of marbles and ball bearings of the same size. Then stir the contents of the jar, and you will see the ball bearings congregate at the bottom and the less dense marbles "float" to the top.

Can we see examples of differentiation at work in Earth and the Moon? Yes. One of the lowest-density and most common minerals to solidify as magma cools is called *feldspar*. Feldspar solidifies in molten lava and then floats toward the surface. We see surface features of both Earth and the Moon that are extremely rich in feldspar, in the form of a rock called **basalt.** Basalt is a dark-colored rock and is most common in lava that erupts from volcanoes tapping the crust and upper mantle. We also find basalt on the Moon's surface. The dark patches that make up the features of the "man in the Moon" are actually huge plains of dark, basalt lava that erupted about $3\frac{1}{2}$ billion years ago. The brighter highlands of the Moon are rock made mostly of feldspar crystals.

When you take a hunk of rock in your hand, you do not usually think of it as a light material. Yet **granite** has even lower density than feldspar. Granite is a pale, quartz-rich rock commonly formed from molten materials in continents. Erosion concentrates these low-density, silica-rich quartz minerals near Earth's surface. These minerals then melt and resolidify to form granite. To put it crudely, continents are a low-density granite "scum" floating on the denser rocks of the basaltic lower crust and upper mantle, similar to icebergs floating in the ocean. The Moon has no granites, because it lacks the erosive processes that concentrate the silica-rich materials.

In the core, by contrast, differentiation has concentrated the heaviest and most abundant metals. Despite the famous Jules Verne fantasy, we have no prospect of ever traveling to the center of Earth. We only know about this remote region from the study of seismic waves. The core is made of an alloy of iron and some other metals, notably nickel. The outer part of the core is a cauldron of molten metal. The inner part is even hotter, but it is under such high pressure due to the force of gravity that it remains solid in spite of the heat. Earth's iron core has a radius of about 3500 km, just over one-half of the planet's radius.

Figure 5–4a shows the layering of Earth. Note that all the diverse features of our familiar geography—mountains and canyons and cliffs and valleys—are confined to the slender zone of Earth's crust. From the top of the highest mountain to the bottom of the deepest ocean trench is less than 0.1 percent of the radius of the planet. Earth is nearly as smooth as a billiard ball would be on the same scale.

The Apollo astronauts placed seismometers on the Moon's surface, which operated for several years. The instruments registered some meteorite impacts and very few moonquakes, showing the Moon to be less geologically active than Earth. Scientists studying these data also got a surprise. The interior of the Moon (Figure 5–4b) is mostly like Earth's mantle. The Moon has only a small iron core, if any. The mean density of the Moon is much less than that of Earth, matching that of Earth's mantle rocks. This was suspected even before the seismic experiments, based on a simple consideration of the Moon's orbit. If we know Earth's mass, the force that keeps the Moon in its orbit tells us the Moon's mass. Divide this mass by the volume of the Moon, and you have the mean density. This density is too low for the Moon to be structured as Earth is. At the end of the chapter, we will see that the lack of a metal core is an important clue to the way the Moon was formed. The lunar crust is even more concentrated in feldspars than Earth's crust is, which is consistent with the idea that the low-density feldspars floated to the surface during the formative days when the outer layers were molten.

face of the planet (Figure 5–5). Erosion breaks rock down into mineral grains, which are tiny pieces of rock compounds. These materials may be dissolved in water and transported and deposited as sediments in new locations. For example, floodwaters deposit soil in new areas, and desert winds blow sand into dunes. When such deposits solidify, they form **sedimentary rocks.** Sedimentary rocks can form by compression due to overlying rock or the formation of cement when watery solutions percolate through the sediments. Such rocks are extremely common on Earth's surface because of the continual effects of water and wind. Two common sedimentary rocks are sandstone, which is formed from compressed sand grains, and limestone, which is made from deposits of calcium carbonate precipitated out of solution in water. This latter process is similar to the calcium carbonate bathtub rings left by evaporating water in a tub. Sedimentary rocks are not found on the Moon because the Moon lacks the wind- and water-driven erosive processes that create and deposit sediments.

The chemical composition of Earth's crust reflects the fact that most of the rocks are silicates—there is 46 percent oxygen and 27 percent silicon by mass. The metals aluminum and iron are the next most abundant, at 8 and 6 percent respectively. The only other elements with more than 1 percent of the crust's mass are calcium, magnesium, sodium, and potassium, all of which exist in mineral compounds rather than as pure elements. The deep interior of Earth is mostly made of iron and nickel. These heavy elements sank to the center due to the process of differentiation.

Earth is a restless planet. Few rocks have remained unchanged over 4.5 billion years of geologic time. The third type of rocks, **metamorphic rocks,** is formed when igneous or sedimentary rocks are modified by heat, pressure, or hot water containing acidic solutions or dissolved minerals. These strong forces change the texture of the mineral formations and create new types of rock. Marble, for example, is limestone or a limestone-related rock modified by heat and pressure. Metamorphic rocks are usually formed at considerable depth and uncovered by erosion or movements within Earth. Again, they are common on Earth but essentially unknown on the Moon, because the Moon does not have the circulating fluids and large-scale rock movements that help create and expose them.

Physical Layering of Earth and the Moon

We have seen how differentiation leads to changes in the composition of rocks as we go deeper into Earth. A second type of layering is revealed by seismic studies. It is a layering not by composition, but by how rock moves. In general, the outer layers of Earth and the Moon have cooled enough to become almost completely solid. Therefore, regardless of composition, these layers are relatively rigid and brittle, while the hotter, interior layers are more plastic and deformable. The brittle surface rock layer is called the *lithosphere,* from the Greek word *lithos,* or rock. The lithosphere extends about 100 km down through the crust into the upper mantle.

The regions below the lithosphere, however, especially from about 100 to 350 km in depth, are warmer and may

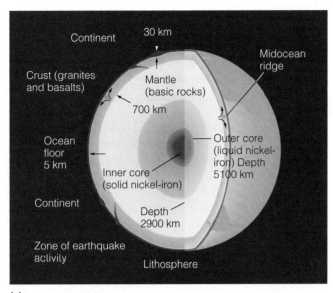

(a)

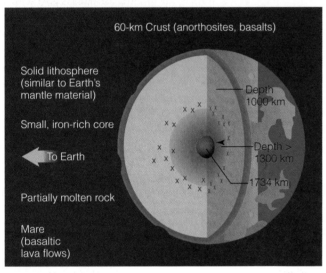

(b)

FIGURE 5–4
The interior structure of Earth and the Moon. **(a)** A simplified diagram of Earth's interior as revealed by modern seismic studies, showing the core-mantle-crust structure. Dots indicate typical earthquake positions. **(b)** The Moon's interior is simpler. The Moon's iron-rich core is very small, if present at all. The rigid layer, or lithosphere, is thicker than that on Earth. Crosses indicate typical earthquake positions.

Three Basic Rock Types in Earth's Crust

Geologists studying Earth's crust over the last few hundred years have identified three basic varieties of rock. **Igneous rocks** form as molten material cools and solidifies, either above or below ground. These include granite and all volcanic rocks, such as basalt.

Erosion is a relentless process. We might think that wind, sunlight, and rain playing against rock could have little effect. Yet over millions of years this attrition sculpts the sur-

FIGURE 5-5
The sharp peaks and canyons of Earth are not found on other worlds. Crumpling of the crust creates uplifted areas, which are eroded by water and wind. Water carves valleys (right center) and deposits the eroded materials at the valley mouth (light triangular deposit); these are the French Alps above Val d'Isére. (SOURCE: Photograph by WKH.)

include pockets of molten lava. This region can deform plastically. Putty is a material that behaves plastically. Pull on it sharply and it will break, but pull on it slowly and it will stretch and flow. You can think of most of Earth's mantle as something like a glacier: A piece of a glacier can be broken off with a hammer; it seems to be solid and brittle. Yet, over long periods, the glacier flows sluggishly. In the same way, the mantle seems rigid, but it can flow slowly in massive currents.

The presence of a cool, rigid lithosphere over a more plastic lower layer is typical of many planetary bodies. As we will see in the next section, the thickness of the lithosphere is an important factor in explaining many planetary surface features.

Influences That Shape a Planet

The Changing Earth and the Static Moon

When Charles Lyell was a young man, early in the 19th century, he traveled to a seashore in England that he had visited as a child. His keen eye noted that the shape of the shoreline had subtly changed. Erosion by wind and sea was steadily eating away at the landscape. As we have seen, Lyell played a major role in arguing that Earth had a long geological history. He believed that gradual change exerted over a very long time could dramatically alter Earth's surface.

Lyell was not a theorist, alone with his equations in a book-lined study. He was an active explorer. Until his seventies he roamed up volcanoes and scrambled into canyons all around the world. In Chile he decided that if one earthquake could elevate a terrain by a few feet, then thousands of earthquakes spread over millions of years could raise a chain of mountains. At Mount Etna in Sicily, he observed the freshly hardened lava, and he concluded that the great volcano had been built from many lava flows. Back in England, engineers were cutting through hillsides to lay tracks for the newly invented railways. Lyell saw the many layers of corrugated sediments that were revealed. He knew that he was looking at the record of a constantly changing Earth.

The idea of change extended to living creatures. A **fossil** is the remains of a living organism, after minerals have seeped in and hardened. Bone and shell are replaced by minerals which then turn to stone. Fossils were trapped in the layers of rock below Earth's surface, and they often seemed to correspond to creatures that were not alive today. Perhaps these creatures had migrated to distant lands? Thomas Jefferson told pioneers headed west to search for wooly mammoths, whose bones had been found on the east coast of the United States. Georges Cuvier, the French founder of the field of paleontology in the early 19th century, cataloged dozens of extinct species. Lyell also studied fossils and speculated that climate change had caused the extinction of many species. Like rock, life was subject to the changing Earth. What causes all this change?

Earth's interior is hot and active, and the thin lithosphere is easily broken or bent by the churning of the layers underneath. This churning causes the steady replacement of the surface rock, so Earth's large-scale surface features are relatively young, averaging a few hundred million years old.

The Moon, on the other hand, is so small that it has cooled off and has a relatively dead interior and thick lithosphere. It shows fewer signs of surface sculpting from below, and those date from its early history, when it was still warm inside. The dominant process that has sculpted the Moon's surface in the last 3 billion years comes from outside, not from inside. Eons of asteroids and comets have slammed into the Moon, creating its characteristic cratered surface. The granular soil that the Apollo astronauts walked on is the result of small meteorites pulverizing the surface. No internal processes exist to "recycle" the surface. Earth is subject to the same "rain" of asteroids and comets, but the atmosphere burns up the smaller projectiles before they land, and erosion by wind and rain has obscured the cratering record.

Mass explains most of the difference between Earth and the Moon. Earth is so massive that a lot of energy is released by radioactive decay within the interior rocks. This heats and liquefies the rock, which then drives the activity of the crust. The Moon is 80 times less massive, so it has proportionately less energy from radioactive decay. The heat generated within the Moon is insufficient to melt the rocks and drive geological activity. This simple difference illustrates the fundamental competition between internal and external forces in shaping the surface conditions on planets. In general, a massive planet is more likely to retain a hot interior, and internal geological forces win the contest to shape the surface. Smaller worlds have lost their heat and have little internal geological activity, and there the external impacts play the dominant role in shaping the surface features.

Plate Tectonics: The Hidden Sculptor of Earth's Surface

Every child who stares at a map may notice that the continents seem to fit together, as the pieces of a jigsaw puzzle do. Is this significant? Or is it random chance? For more than a century, geologists studied earthquakes, mountain building, and volcanic eruptions without understanding the underlying causes. The first glimmer of the truth came in the early 1900s, when German geophysicist Alfred Wegener puzzled over the fit of coastlines of some continents, especially South America and Africa, which seemed to fit together. Wegener boldly proposed that those two continents once were joined, and had split and drifted apart in the ancient past. He called this phenomenon *continental drift*.

Few scientists accepted the seemingly crazy notion that continent-sized blocks of the lithosphere could move around on Earth's surface. What forces could possibly drive such drift, they asked. Wegener could match up distinctive rock formations on opposite sides of the Atlantic Ocean, but he could suggest no reasonable mechanism for the motion of the continents. Finally, in the 1960s, scientists used new techniques to study the seafloor and to learn that large, continent-sized blocks of the lithosphere actually do move across the face of the globe. First, they found that the ocean floors are not featureless plains, but have mountains and canyons as mighty as those on the continents. Second, they used patterns of magnetized rock on the seafloor to show that the Atlantic seafloor was slowly spreading. Last, they used the technique of radioactive dating to show that the underwater mountains that bisect the Atlantic Ocean are young. All this evidence provided support for Wegener's hypothesis.

The blocks of the lithosphere that move are called *plates*, and they include both continental masses and large regions of the seafloor. The data showed that where two plates move apart, fresh basalt wells up from the mantle and forms new crustal rock to fill in the gaps. This happens, for example, along the mid-Atlantic ridge, a mostly underwater seam where young lava extrudes. Iceland, a highly volcanic island, is an exposed section of the mid-Atlantic ridge. As seen in the middle of Figure 5–6, the collision of two plates compresses the lithosphere laterally and crumples it to form mountain ranges. For example, the Himalayas are young

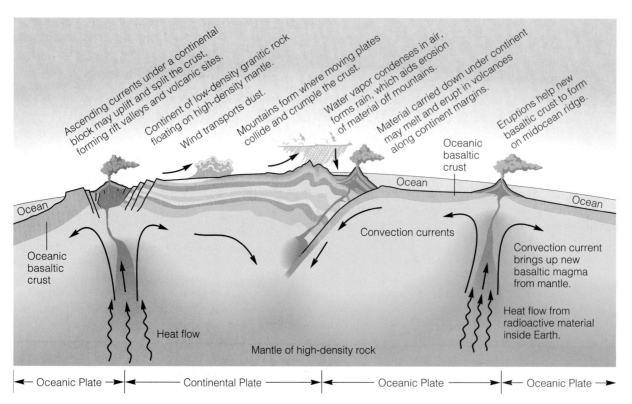

FIGURE 5–6
A schematic diagram of processes that make Earth more geologically active than most other worlds. Heat flow by convection from the interior drives currents of hot material flowing beneath the lithosphere. These currents can stretch, crack, and compress the surface rock layers, and cause movements of the plates. Volcanoes erupt where molten material flows through fractures to the surface. Plate movements cause earthquakes. Collisions of plates build up crumpled mountain ranges. Atmospheric processes and water flow cause rapid erosion. The vertical scale is exaggerated for clarity.

mountains formed where the Indian plate has pushed into the southern Asian plate.

The theory that Earth's surface features are molded primarily by moving plates, dragged along by currents in the underlying plastic mantle, is called the theory of **plate tectonics**. This theory is now universally accepted among geologists. Its acceptance was the major revolution in geological science in the 20th century.

We get some insights into how science works from the story of plate tectonics. At a personal level, the story was not a happy one for Alfred Wegener. He had only fragmentary evidence to support his radical hypothesis, and some of his ideas were wrong. He also had no idea what force moved the continents. Wegener was ridiculed by some of his colleagues and he died young without seeing his hypothesis accepted. Similarly, we might recall from Chapter 2 the story of Aristarchus, who had the correct idea of a heliocentric universe but very little evidence to support it. Over the longer term, scientists do accept new ideas. Theories that were once radical become conventional wisdom, but it requires evidence. The evidence for plate tectonics—especially the seafloor data collected in the 1960s—eventually became so overwhelming that geologists embraced the new idea of an active Earth.

Geological Processes in Earth's Evolution

Let us summarize the internal geological processes that shape Earth's surface. The driving agents for all this activity are gravity and energy released from the decay of radioactive elements. As we saw in the last chapter, heat flows from hot to cool regions. So the energy from Earth's core must move outward and eventually be radiated into space. Recall the three modes of heat transfer, also discussed in Chapter 4. Some heat is carried out by conduction within solid rock. However, there is too much heat to be transferred by conduction alone, so convection also carries heat from the hot interior toward the cool surface. That means that huge, sluggish masses of hot rock—in a molten or plastic state—flow slowly upward until they hit the bottom of the lithosphere. Then they spread laterally, cool, and eventually sink again, as shown by the cross section in Figure 5–6. It is just like a pot of water as it begins to boil and the energy reaches the surface by rolling motions of the water. In the case of the viscous mantle, however, the rolling cycle takes about 200 million years!

This churning of the semiliquid layers underneath the lithosphere creates tremendous stresses on the brittle rock layers of the lithosphere. Depending on the nature of the stresses, the lithosphere may compress or may stretch. If the stresses are compressive (middle of Figure 5–6), folding and buckling of the lithosphere may create mountain ranges, such as the Rockies. A traveler through such ranges can see the huge curved folds in the rock layers. If the stresses stretch the lithosphere, it may thin like stretched taffy or even pull apart in a series of fractures, called *faults* (left side of Figure 5–6). An example of this occurs along Baja California and the California coast, where stretching forces have pulled Baja California away from the mainland in the last few million years. Stretching also causes fractures such as the San Andreas fault.

If the compressive or stretching stresses become too great, the lithosphere will be deformed beyond its limit—like taffy or putty pulled too fast or too hard. As the lithosphere fractures, it sends off seismic waves that travel through the crust and mantle. This is the phenomenon we perceive as an earthquake. The faults are usually underground, but sometimes they can be seen at the surface. Once an earthquake occurs, it will relieve the stresses, but if the sluggish convection currents continue, the forces will build up again. This is why a given fault line can have continued sporadic earthquakes for hundreds of years.

Often the fractures allow molten lava to gain access to the surface from the interior (Figure 5–7). **Volcanism** is the result—the erupting of molten lava onto the surface. (Usually this material is called *magma* while it is underground and *lava* once it reaches the surface.) This explains why volcanoes are often concentrated in regions of fractures and earthquakes, where the lithosphere is being deformed most rapidly by underlying mantle currents. Alternatively, volcanoes can occur where the lithosphere happens to be thin and the magma can penetrate. The volcanoes of the Hawaiian Islands chain were created when a thin lithosphere plate moved over a hot spot in the mantle, and the magma punched a series of holes through the thin crust.

Geological processes occur all over the world. All the continents have mountain ranges, and mild earthquakes are

FIGURE 5–7
Molten rock from the interior reaches the surface through vents and fissures in the lithosphere, emerging as lava. Eruptions of lava create new landmasses, even as the sea erodes the coastlines of the old landmasses. This is a view of the 1988 Kalapana lava flow in Hawaii. (SOURCE: Photograph by WKH.)

occasionally felt even in unlikely places such as England and Kansas. However, the big picture of geological activity is given by the dozen or so lithospheric plates that fit across Earth's surface. Most of the world's earthquakes and volcanoes occur where these plates meet, as you can see in Figure 5–8. Mountain ranges are found where these plates collide (e.g., the Andes and the Himalayas) or separate (the mid-Atlantic range). Unfortunately, over one-half of a billion people live on the "ring of fire" that traces the plate boundaries at the rim of the Pacific Ocean. The motions of the plates might be imperceptible, but they are real. As we will see in a later chapter, the techniques of radio astronomy have been used to directly measure the motion of continents. Europe and North America are moving apart at a rate of about 5 cm/yr. About 150 to 200 million years ago, they were locked in a geological embrace. In fact, all the continents were joined in a primeval land mass called Pangaea (Figure 5–9).

Our discussion of Earth's geology brings into focus two of the book's themes. All the processes we have covered—gravity, internal heating, volcanism—will be met again when we consider other planets and their satellites. The laws that govern how rocky and icy worlds behave seem to apply everywhere in the universe. Also Earth is just one of many worlds, each with its own rich geology and fascinating landscapes.

Other geological processes play a role in Earth's evolution. Rain and flowing water erode the landscapes of Earth, continually altering the sculptures created by plate tectonics. **Erosion** is the term that includes all processes by which rock materials are broken down and transported across a planet's surface, generally from higher to lower regions. Deposition includes all processes by which they are laid down and accumulated. Erosion can wear down an entire mountain range in 100 million years or so. On Earth, almost as fast as mountains can be pushed up by plate tectonics, water carves valleys and canyons into them. Many of the flat plains of Earth are simply masses of sediments washed down from mountains and plateaus. Wind, too, moves sediments and deposits them elsewhere.

What limits the size of mountains on Earth? Gravity does. You could imagine a mountain 10 times higher than Everest. But if such a mountain existed, the pressure due to gravity at its base would melt the rock, and the mountain would gradually slump. Even Everest is slumping. Mountain topography is a constant war between dynamic mountain-building processes like plate tectonic motion and volcanism, and mountain-flattening processes like erosion and gravitational settling.

Earth's surface remains in a process of constant renewal. We have seen that all the major geological processes—the

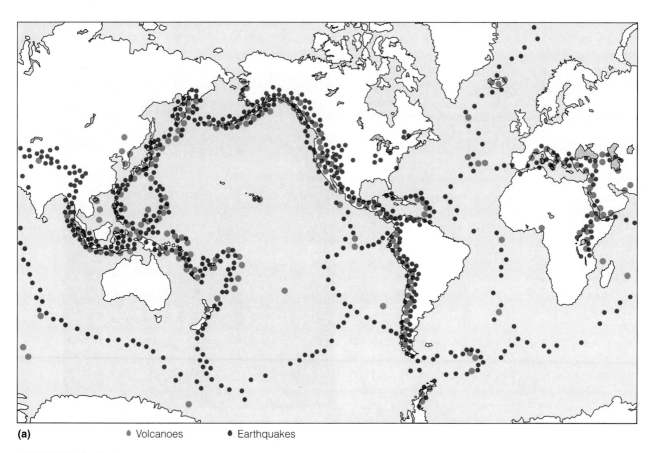

FIGURE 5–8
(a) Earthquake and volcano sites are distributed in curving lines across Earth's surface and under the oceans, too. These lines correspond exactly to the plates of the lithosphere. **(b)** The plate boundaries are shown as red lines. Geological activity is caused when plates merge, separate, and shear. (SOURCE: Courtesy of G. Miller.)

moving of continents, the churning of the mantle, the raising and erosion of mountains—take place on a time scale of 100 to 200 million years. This amounts to only a few percent of the age of Earth. It is hard to find surface features that are more than about 800 million years old. This is why the search for the oldest rocks on Earth, discussed earlier in the chapter, was a major piece of detective work. Even though Earth is 4.6 billion years old, like other planets, it has a very "young" surface, geologically speaking!

Impact Cratering: Sculptor of Geologically Dead Worlds

As can be seen in Figure 5–10, the Moon's surface features are very different from Earth's. Most of the Moon is covered with bright lunar uplands, pocked with deep, circular features called **impact craters.** Rocks brought back by Apollo astronauts showed that the lunar uplands are 4-billion-year-old masses of feldspar-rich rock, which solidified from a Moon-wide ocean of molten material after the Moon formed. Impact craters scar the landscape of the uplands. These craters formed when asteroids or comets hit the surface.

The dark patches that are visible with the naked eye from Earth, colloquially known as the "man in the Moon," are lunar lava plains, generally smoother than the bright uplands and at lower elevations. Analysis of lunar rock samples shows that the lava eruptions occurred about 3.5 billion years ago, marking the last gasps of lunar internal heating. After this period, the Moon had cooled too much to be geologically active. Since then the surface has been scarred by impacts. The large range in craters sizes is seen clearly in Figure 5–11, a photograph taken by Apollo astronauts in orbit over the Moon.

Why are the Moon's features so different from Earth's? Since the Moon cooled off 3.5 billion years ago, its surface has not been significantly modified by internal geological activity during that long period. The only agent acting to sculpt the surface has been the external rain of comets and asteroids, creating craters. Why does Earth not show craters, since it is obviously exposed to the same rain of interplanetary bodies? The answer is that the continual reforming of Earth's surface by plate tectonics, volcanism, and erosion has kept Earth's surface young. Earth has only a few hundred million years' worth of craters, while the Moon's craters have accumulated for billions of years. Craters are present on Earth, but most are hard to detect from aerial photographs because the craters have been filled in by sediments.

Figure 5–12 shows a typical lunar crater in detail. The figure shows some typical features, including the circular rim and depressed floor. Dust and debris blown out of impact craters fall back onto the Moon, forming the secondary craters scattered nearby. The impact also forms bright "streamers," streaks of ejected material that radiate from the bright crater, which you can see in Figure 5–10.

Impact craters have a wide range of shapes, depending on their sizes. Small craters, up to a few kilometers across, are generally bowl-shaped. Larger craters have central mountain

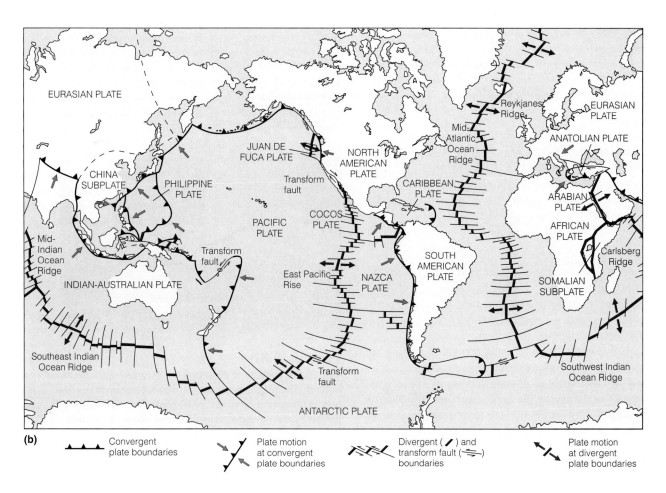

INFLUENCES THAT SHAPE A PLANET 123

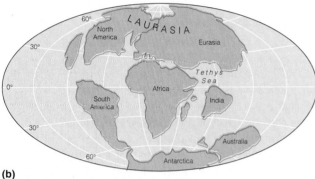

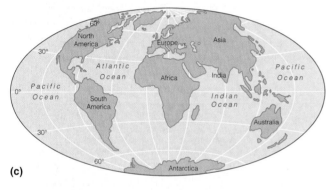

FIGURE 5-9
Maps showing how the continents of Earth have shifted over geological time. **(a)** The view about 240 million years ago. **(b)** The view about 70 million years ago. **(c)** Modern Earth.

A very important finding from the lunar rocks is that the rate of impacts was much higher soon after Earth and the Moon formed than it is today. From 4.5 billion years ago until 4 billion years ago, the average impact rate was thousands of times higher than it is today. After 4.0 billion years ago, it declined rapidly to today's much lower level. The early intense bombardment occurred because the planets swept up interplanetary debris that was left over after the planets formed. Only a small amount of debris is left today, but it is enough to provide occasional impacts onto Earth.

The number of craters can help determine the age of a planet's surface. Imagine interplanetary bodies such as comets and asteroids drifting through the solar system and occasionally crashing into planets and moons. If each impact makes a crater, then it follows that the longer a surface is exposed, the more craters will appear on it. In other words, the older the surface, the more craters are crowded onto it, per unit area. Since the bombardment rate was so high just after formation, most craters on an old surface must themselves be very old.

In general, if you see many impact craters on a planet's surface, that surface is old—more than 1 or 2 billion years old. If you have trouble finding any impact craters on a surface, that surface is young—less than 1 billion years old. Note that the larger number of craters on the Moon than on

FIGURE 5-10
The surface features of the Moon are very different from those of Earth. The bright, rough areas are called highlands and are heavily cratered, as seen at the left edge of the figure. Dark patches (upper and lower right) are lava plains. A relatively young crater has blasted streamers of bright dust across much of this hemisphere. (SOURCE: Lick Observatory photograph, rectified to south polar view at the Lunar and Planetary Lab, University of Arizona.)

peaks, and the largest impact features may have multiple concentric rings. The central peaks and rings are extraordinary evidence of the conservation of energy, a concept discussed in the last chapter. When a large asteroid or comet hits the lunar surface, the enormous kinetic energy of motion is converted instantly to heat. The heat vaporizes most of the incoming object. The rest of it is liquefied, along with some of the lunar surface. A wave of liquid rock travels out; and at the point where the rock cools enough to solidify, a crater wall forms. Other waves show up as concentric rings. The central peak is the result of a wave that reflects off the crater wall and converges at the center. Droplets of molten rock spray off from the central point to form the surrounding smaller craters. All this violent activity takes only a fraction of a second! It is somewhat similar to what happens when a water drop hits a pool of water (Figure 5–13).

FIGURE 5-11
In orbit over the lunar surface. The sculpting of the lunar highlands from large numbers of meteorites of all sizes can be seen in this nearly vertical view. The photograph was taken from the *Apollo 16* landing module, looking down on the command module, just prior to proceeding to the lunar surface. (SOURCE: NASA.)

Earth does not mean the Moon itself is older, only that the surface is older—in other words, geological processes are not very active. Imagine you are flying around the universe in your spaceship. Whenever you see a heavily cratered planet, you can say it is geologically dead. If you see an uncratered planet, you can suspect that it has some form of geologic activity that is continually modifying the surface.

Geological Time Scale

Scientists have been able to describe the history of not only Earth, but also the whole Earth-Moon system. This history is usually expressed in terms of rock layers or *strata* formed at different times; it is called the **geological time scale.** When geologists in the early 1800s began to inspect the layers in Europe, they realized that each layer had different fossils. Young strata near the surface had fossils and bones of contemporary animals and plants. Older strata at deeper depths had unfamiliar animals and planets, now extinct. The different fossils meant that there had been a succession of different species of animals and plants.

The turbulent history of Earth means that it is difficult to sort out the different layers. The strata below our feet are not always arranged neatly as the layers of a cake. In two different locations, a layer of a particular age might be made of two different types of rock. It might be near the surface in one location and far below the surface in another location. A violent movement of Earth might even invert the order of some layers! In addition, the fossil record is very incomplete.

Specific conditions are required for an intact living organism to die and be turned into stone. The extreme heat and pressure under Earth's surface can easily obliterate any traces of life. It is believed that only 1 in 10,000 of the earliest lifeforms is represented in the fossil record. Decoding this information is painstaking detective work.

Geologists and paleontologists gradually realized that the older layers had only simple animals and plants, such as fishes and primitive organisms that lived on the seafloor. They concluded that the unique fossils in different rock layers could be used to arrange the layers in order of age. Based on the dominant fossil organisms in different layers, geologists

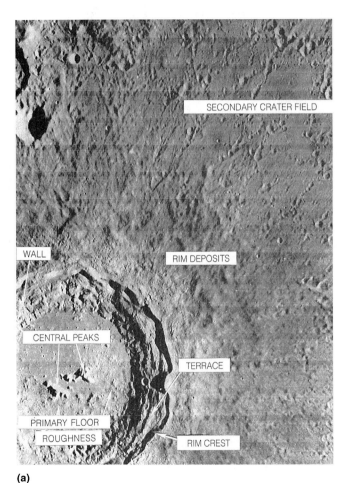

FIGURE 5-12
(a) Typical features of a large lunar crater can be seen in this view of the prominent crater, Copernicus, which is about 90 km (55 mi) across. The main bowl of the crater (lower left) was hollowed out by a meteorite a few kilometers in diameter. Rim deposits and secondary craters are caused by debris blasted out of the main crater. (SOURCE: Courtesy of James W. Head III, Brown University.) **(b)** An approximate cross section of such a crater.

FIGURE 5-13
Time-lapse photograph of a drop hitting a liquid surface. The analogy with the formation of a crater millions of times larger is appropriate because the high pressures and temperatures of a large asteroid impact cause the surface to deform plastically, like a fluid. At the point where the rock cools enough to solidify again, a crater rim and (in larger craters) a central peak form. (SOURCE: Dr. D. Gorham and Dr. I. Hutchings/Photo Researchers, Inc.)

divided the known history of Earth into several geological eras, and the eras into smaller subdivisions called *geological periods.* This is shown in Table 5-1. The periods are tens of millions of years long, and certain characteristic species dominate each period.

When radioactive dating became available, geologists discovered that the oldest layers with abundant, simple fossils did not date to 4.5 billion or even 4.0 billion years ago, but only to 600 million years. Prominent animals and plants on the land surfaces go back only about 400 million years. Before that, the land was essentially barren. In other words, prominent species have existed in the sea for only about 13 percent of the age of Earth and on land only 9 percent!

Because the record of well-developed fossils covered only about 13 percent of Earth's history, and because the earliest rocks were missing altogether, little was known of the Earth-Moon system's earliest history until 1969. In that year, Apollo astronauts and Soviet robotic probes began collecting rock and soil samples from the Moon. The rocks dated from about 3 to 4.4 billion years ago, which means that the Moon gives us information about the time periods that are poorly represented on Earth. The samples revealed the cosmic environment during those early eras. The counting of impacts offered some important clues about how planets might have formed from the accumulation of smaller bodies. The lunar samples also showed that the Moon was volcanically active for the first third of its history, then cooled off and became geologically dead.

The combination of terrestrial and lunar data has allowed us to piece together a fairly complete geological history of the Earth-Moon system, shown in Table 5-2. Following the various columns from bottom to top, we see the formation of both Earth and the Moon, the dates of the oldest rocks, the emergence of oxygen, sea life, land plants, and so on. Early Earth would have seemed like an alien world to us. Soil chemistry and other data indicate that the oxygen content in our atmosphere did not become substantial until plants emerged and began converting CO_2 to O_2, around 2.6 to 3 billion years ago. Tiny, single-celled organisms probably first emerged in the seas 3.5 to 4 billion years ago. If you had a time machine and could visit our planet in the first 85 percent of its history, you would find a barren and alien planet with an unfamiliar atmosphere and no creatures larger than your thumb! We will revisit the amazing emergence of life and place it in a cosmic context in the last chapter of the book.

Sudden Changes During Geological History

Until about 1980, geologists concentrated on defining the main strata and their corresponding geological eras and periods. They paid less attention to the transitions from one geological period to another, which showed marked changes in the relative numbers of different species. No one knew if these changes occurred gradually or abruptly.

As the evidence accumulated, geologists decided that some of the changes were very dramatic. Study Table 5-1 and notice that the boundaries between the five eras end in the suffix *-zoic,* referring to life. Modern studies show that the Paleozoic era ended in a huge catastrophe about 250 million years ago, when about 90 percent of the then-existing plant and animal species died out in no more than a few million years. After that, reptiles rose fairly rapidly and dominated the ensuing 175 million years. Then, 65 million years ago, another catastrophe wiped out about 75 percent of the then-existing species, including the giant reptiles, or dinosaurs. Such events are called **mass extinctions**—relatively brief intervals in which a large fraction of species become extinct. The ends of the Paleozoic era and the Mesozoic era are the two best examples of mass extinctions. They mark such dramatic breaks in the fossil record that geologists define the geological eras as beginning or ending with them. Smaller examples of mass extinctions also occur between certain geological periods and are used to define them.

For years, the two largest mass extinctions got surprisingly little attention in geology textbooks. Part of the reason was the fragmentary nature of the fossil record. The layering of rocks and fossils is not neat and orderly throughout geological time. There are times when volcanism or metamorphosis of rocks destroys the fossil evidence. Also the method of radioactive dating has limited precision. Let us say we

TABLE 5-1
Geological Time Scale

ERA	AGE (MILLIONS OF YEARS)	PERIOD	LIFE FORMS	EVENTS
Cenozoic	0	Quaternary	Humanity	Technological environmental modification; extraterrestrial travel; ice ages
	3	Tertiary	Mammals	Building of Rocky Mountains
	65	*Rapid extinction of ~75% of species, including dinosaurs*		
Mesozoic		Cretaceous		Large meteorite impact (65 million years ago); continents taking present shape
	130	Jurassic	Dinosaurs	
	180	Triassic	Reptiles	
	240	*Rapid extinction of ~90% of species*		
Paleozoic		Permian	Conifers.	Building of Appalachian Mountains
	280	Pennsylvanian	Ferns	Pangaea breaking apart
	310	Mississippian		
	340	Devonian	Early land plants	
	405	Silurian	Fishes	
	450	Ordovician		
	500	Cambrian	Trilobites	Earliest Abundant fossils (trilobites, etc.)
	570			
		Ediacarian	Small soft forms	
	640		First macroscopic life forms; sexually reproducing life forms	Growth of protocontinents; scattered fossils
Proterozoic	1000			
	2000		Oxygen-producing microbes	Oxygen increasing in atmosphere
	2600			
	3000		Microscopic life	Crustal and atmospheral evolution Earliest Fossils (algae)
Archeozoic	3600			Oldest rocks; crustal formation? Heavy meteoritic cratering
	4500			
Formative	4600			Formation of Sun and planets
Presolar	12,000?			Formation of the Milky Way Galaxy
	14,000?			Origin of universe

SOURCE: Data in part from Schopf and Morris.

could determine the age of 100-million-year-old rocks with an accuracy of 0.1 percent. The measurement error is still 100,000 years. This means that 100,000 years is the limit of our ability to resolve time. If we see the disappearance of many species over that interval, should we consider it gradual or catastrophic change?

The Catastrophe 65 Million Years Ago—A Giant Impact

The first mass extinction to be explained by direct scientific evidence was the one at the end of the Mesozoic era, 65 million years ago. As we have seen, approximately 75 percent of all species of plants and animals became extinct within less than a few million years! In the early 1980s, geochemists made an interesting discovery about the thin, 65-million-year-old layer that divides the Mesozoic era from the Cenozoic era. It contained an excess of elements that can be found in meteorites, or debris that falls on Earth from space. This discovery suggested that a giant meteorite impact might have been connected with the end of the Mesozoic. Further evidence supported this. The layer was found to have glassy spheres formed by melted rock, and quartz grains that had

TABLE 5-2
History of the Earth-Moon System

TIME (BILLIONS OF YEARS AGO) — (PRESENT)

EARTH
- Formation (~4.5)
- Intense cratering*
- Crustal formation
- Shorter day
- Early life
- Oldest surviving rocks
- Length of day increasing (due to tidal interaction with moon)
- Atmospheric changes (O₂ increase due to plants)
- soft-bodied life forms
- ?←Continental evolution→
- Earliest substantial fossils
- Mammals→| |←
- Humans→||←

MOON
- Formation*
- Magma ocean*
- Intense cratering*
- Early heating*
- Rapid motion away from Earth
- Magma ocean cools*
- Mare formation*
- Interior cooling*
- ←Continued slow movement away from Earth→
- Sporadic cratering*
- Occasional moonquakes and gas emissions*

*Information discovered or improved through Apollo expeditions to the Moon.

---

been suddenly exposed to extreme pressures reached only during impacts. Scientists also discovered concentrations of soot that seemed to reflect worldwide forest fires.

For years scientists argued about how to interpret this evidence. The father-and-son team of Walter Alvarez, a geologist, and Luis Alvarez, a Nobel Prize–winning physicist, had originally discovered the meteoric element excess. They used the total amount of meteoric elements to estimate that a 10-km asteroid hit Earth and caused devastation that killed off the dinosaurs and other species. This was a very controversial idea. As we learned in the first chapter, there are almost always rival explanations for any set of data. Because radioactive dating has limited time resolution, some paleontologists argued that the extinction was caused by intense volcanism 65 million years ago. Scientists could not prove that the extinction was catastrophic, occurring over a period of days or years rather than tens of thousands of years. The "smoking gun" in this detective story was missing.

If an impact was to blame, where was the impact crater? Around the Gulf of Mexico, scientists found that the 65-million-year-old layer contained deposits from large tsunamis, or tidal waves. The tsunami deposits suggested an impact in this area. Eventually scientists discovered a large crater completely buried under sediments, straddling the coastline of the Yucatan peninsula in Mexico. Its diameter was at least 160 to 180 km (100 to 110 mi) from rim to rim. But did it have the right age? After a delicate period of negotiations involving core samples taken by a Mexican oil company, scientists had the evidence they needed. The age of the crater was indeed 65 million years. It was created by impact of an asteroid about 10 km (6 mi) across—the size of a small town!

What happened after the asteroid hit, to cause the extinction of so many species? The details are exceedingly complex and are still poorly understood. The huge impact probably blasted a cloud of dust and other debris into the high atmosphere and then into space. Debris lofted into space fell back into the atmosphere all over the world in the following hours. An imagined but realistic view of this event is shown in Figure 5–14. Scientists calculate that if you were

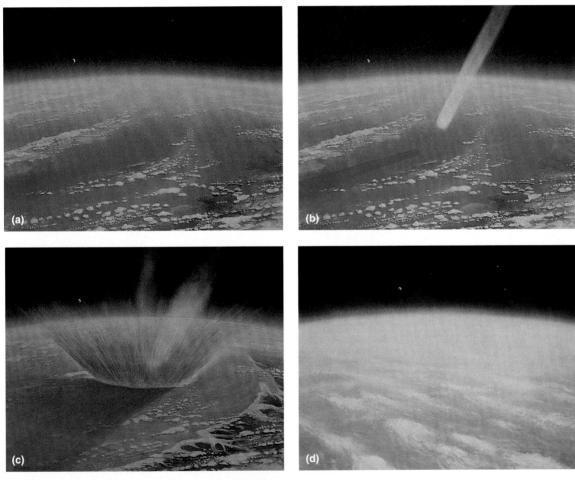

FIGURE 5-14
The impact that probably wiped out the dinosaurs and ended the Mesozoic era 65 million years ago. Similar events have occurred during other portions of geological time. **(a)** A view from about 100-km altitude 1 minute before impact. **(b)** Two seconds before impact, a 10-km asteroid streaks through the atmosphere. **(c)** One minute after impact, debris is being blasted up through the atmosphere. **(d)** A month after the impact, dusty debris and smoke have spread through the atmosphere, blanketing much of Earth and blocking sunlight at the surface. **(e)** Years after the impact, the dust and smoke have cleared, and a 180-km crater is revealed. On the ground, many old reptilian species have not survived the event, and previously obscure mammals are beginning to fill environmental niches. (SOURCE: Paintings by WKH.)

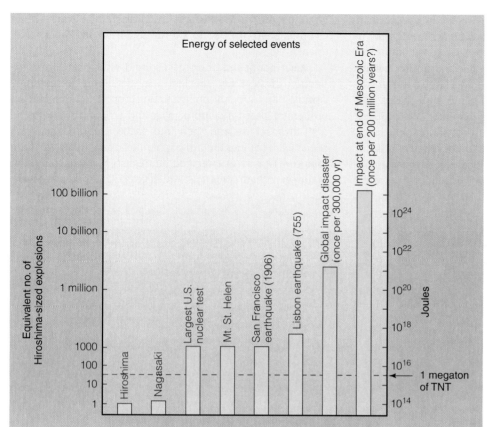

FIGURE 5-15
Comparison of the energy released in various human experiments and natural disasters. In general, the larger the energy, the rarer the event. See the text for discussion. (SOURCE: After a diagram by Kring with additional data from Hartmann and Morrison and others.)

standing on Earth some minutes after the impact, you would have observed the sky light up as the debris fell like brilliant fireballs and shooting stars. The radiant energy from the sky caused a strong heat pulse that may have killed many land animals and touched off forest fires in areas that were not protected by cloud cover, rainstorms, or snowstorms. More importantly, enormous amounts of fine dust were injected into the stratosphere, which probably drifted there for some time and shut out sunlight for weeks or months. All these short-term changes wreaked havoc among plant species and the food chain. Reptiles had ruled Earth for 150 million years. Now they died out due to a catastrophic change in their environment. Previously minor species, including certain small mammals, had less competition. They proliferated into this ecological vacuum, evolving into many new species, including a certain upright mammal that goes to college.

The Catastrophe 250 Million Years Ago—Volcanic Eruptions?

A still larger mass extinction occurred roughly 250 million years ago, ending the Paleozoic era. As many as 90 percent of then-existing species vanished in less than a few million years. In the oceans, the fraction has been estimated as high as 95 percent of all species! This event was so dramatic, it has long been called "The Great Dying" in geology textbooks. Its cause is still uncertain. No positive proof of an impact has been found. Most hypotheses center on internally caused geological events. One leading hypothesis is that large plumes of hot magma rise from the mantle sporadically, as blobs rise in a lava lamp. When they arrived under the crust, they may have caused major episodes of volcanism and plate tectonic shifts, altering Earth's climate. These would not be sudden disasters like the transient climatic effects of large impacts, but they could cause changes that might increase the rate of the extinction and emergence of new species in new environmental niches.

Another hypothesis involves a sudden turnover or disturbance in the ocean, which brought oxygen-poor water to the surface and increased CO_2 concentrations in the air. In 1996, Oregon paleontologist Gregory Retallack presented evidence of strongly fractured or shocked quartz grains from the layer at the end of the Paleozoic. Shocked quartz grains are usually accepted as evidence of an impact. Conceivably, an impact in the oceans or continental margins might have disturbed the oceans and made a crater subsequently destroyed by plate tectonic activity. Remember that the fossil record this far back in time is quite patchy. Also, since radioactive techniques cannot date rocks with very high precision, it is difficult to distinguish between the rival hypotheses of a sudden catastrophe and a somewhat slower geological change. The race to explain the most dramatic dying in Earth's history continues.

How much energy is released in a giant catastrophic impact? The answer is shown in Figure 5-15, a bar graph comparing the energy of various natural disasters. The energy scale on the right is in the international standard unit of joules; the energy scale on the left conveys this information in terms of the A-bomb dropped on Hiroshima as a standard. An asteroid impact that might disrupt agriculture and kill one-quarter of the human population is estimated at perhaps 20 million A-bombs, and probably occurs about once every 300,000 years, according to astronomers' estimates. The size of the Mesozoic-ending catastrophic impact was closer to 100 billion A-bombs. Such a catastrophe probably has occurred only about every 100 or 200 million years, on average.

The Cosmic Connection

The discovery that asteroid craters as well as volcanoes, plate tectonics, and erosion have seriously affected the history of our planet led to an important concept that became fully understood only within our own lifetimes. The geological sculpting of Earth, the Moon, and other planets is influenced by internal and external forces. These same forces influence the evolution of biological and environmental conditions on Earth. Our external cosmic environment has shaped the history of life on Earth.

Further studies have indicated that some smaller-impact craters coincide with the dates of more modest species' extinctions in the geological time scale. For instance, one study found evidence of an impact coinciding with the end of the Devonian period, 340 million years ago, when 70 percent of species become extinct in less than a few million years. The most recent extinction was only 11 million years ago, when 30 percent of species disappeared. Up to a dozen mass extinctions have been identified in the fossil record, although an impact crater of the correct age has been identified in only a couple of cases.

Evidence of mass extinctions has led to a radical adjustment of our theories of evolution. The work of Darwin and his successors, starting in the 1850s, showed that evolution was driven by competition among species, along with mutations in the genetic structure of organisms and the exploitation of new environmental niches. The result is the continual emergence of new species. In classic Darwinian theory, evolution is gradual, and it is driven by events within Earth's biosphere. The discovery of the event 65 million years ago indicates that major changes in the environment—and hence in the course of evolution—can also come from random external astronomical phenomena, such as the fall of an asteroid. **Random processes** can range from the decay of an atom to the impact of an asteroid, as we will see in the following Science Toolbox.

This realization dramatically transforms our conception of our role in the universe. Medieval scholars assumed that Earth was stable and unchanging. As natural science developed from the Renaissance to the time of Darwin, scientists thought that all changes on Earth were gradual and that all the important influences on Earth's environment and biology (not counting the heat and light from the Sun) came from within Earth's system itself. Now we realize that capricious events such as asteroid or comet impacts can radically alter the course of evolution. One reminder, which you can visit, is the famous impact crater in Arizona, nearly a mile wide, which formed only 20,000 to 50,000 years ago. The chance of a catastrophic impact within a human lifetime is slight. We rarely see the effect of the cosmic forces on our planet. Nevertheless, over the long term of geological history, these forces have shaped the world we live in.

Earth's Atmosphere, Oceans, and Environment

The last section shows that Earth's environment has had a long history of change. Studies of our atmosphere and oceans show that this story isn't over. One of the unique features of our planet is the dominance of liquid water on its surface, a situation illustrated by Figure 5–16. Even this circumstance relates to our cosmic environment: If Earth were a little closer or a little farther from the Sun, the water would be either vapor or frozen and all land surfaces would be dry desert.

Origin and History of Earth's Atmosphere and Oceans

The primitive atmosphere 4 billion years ago was different from today's atmosphere. It was dominated by carbon dioxide (CO_2) and water vapor (H_2O) in gaseous form. This primitive atmosphere would have been quite alien to us. Studies of ancient sediments showed that the primitive atmosphere lacked oxygen (O_2)—early Earth was not even suitable for human life! Our metabolism requires oxygen to breathe; carbon dioxide is toxic to us. In fact, the earliest substantial fossil lifeforms (clumps of algae), dating from about 2.5 to 3.5 billion years ago, are types of algae found today only in oxygen-poor environments, such as salt marshes along sea coasts.

FIGURE 5–16
The most common "landscape" on Earth is unique in the solar system. An open ocean of liquid water covers about three-fourths of our planet. In general, planets much closer to the Sun are too hot for liquid water, and planets much farther from the Sun, too cold. (SOURCE: Pacific Ocean; photograph by WKH.)

SCIENCE TOOLBOX

Random Processes

In Chapter 3 we discussed the wide range of physical phenomena that are repeated or cyclical. Our examples of periodic processes included waves, vibrations, oscillations, and orbits. In this chapter we have encountered several phenomena that are random. The examples are fantastically different in scale—the random decay of a radioactive atom and the random arrival time of space debris on Earth! Let's look at random processes in a bit greater detail.

A random process cannot be understood without dealing with the notion of probability. Random implies unpredictable. We are not sure of the outcome of a particular random event; we can only assign a probability to the outcome. However, while a *single* random event may be unpredictable, the net result of a *large number* of random events may be very well behaved and predictable.

Go back to our example of the deck of cards in Chapter 1. We reproduce in Figure 5-A*a* an experiment in which an attempt is made to place a card halfway through the deck. The result is not perfect, but it is rare for the card to be more than three or four places from the central position. Now suppose you are blindfolded and have to insert the card randomly. The likely outcome of 40 separate placements is the histogram in Figure 5-A*d*. This is a random distribution. In other words, the card has an equal probability of appearing at any position in the

(a) Estimating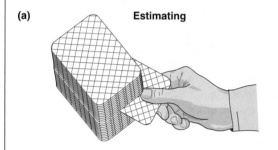

(b) Random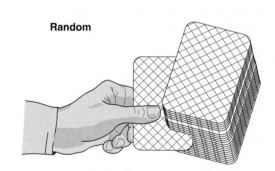

(c)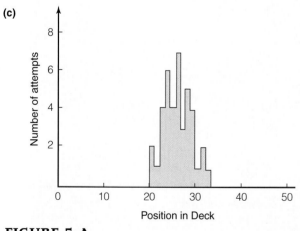

(d)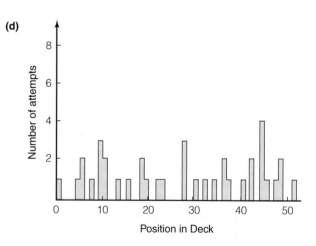

FIGURE 5-A
Forty cards are inserted in a deck by (a) estimation and (b) random placement. (c) Estimation yields a result in which the card position peaks near the middle of the deck. The scatter is a measure of the error in card placement. (d) Random placement yields a distribution with no preferred position in the deck. While any position has an equal probability of occurring, some positions do not occur at all in 40 placements and others occur several times—the distribution is not uniform.

SCIENCE TOOLBOX

Random Processes (continued)

deck from first to last. In mathematical shorthand we say

$$p_n = \frac{1}{n}$$

In this equation, p_n is the probability that the card will appear in any one of n different positions in the deck. Since a deck has 52 positions, $p_n = 1/52 = 0.019$, or 2 percent. That is, each of the 52 positions has about a 2 percent probability of being selected. The probability is the same each time you repeat the random placement. For example, suppose your first placement is position 32. The next placement has a probability of $1/52$ of being position 32 (or any other position)—the outcome of a random process does not depend on what came before. If you prefer, think of the random process of tossing a coin. Since there are two possible outcomes, the probability of either heads or tails is $p_n = 1/2$. The next time you toss the coin there is still an equal 50 percent probability of either outcome. In the case of throwing a die, the die has six sides, and $p_n = 1/6$. Once again, each successive throw has a $1/6$ chance of turning up any particular number.

Notice that in each of these examples—random card placement, coin tossing, dice rolling—the result of a single event cannot be predicted with certainty. The outcome each time is random. However, the pattern of results of a large number of random events is well behaved and predictable. We know that after tossing a coin many times, the percentage of heads or tails will be very close to 50% (Try it!) In other words, a large number of experiments allows us to accurately measure that $p_n = 1/2$. A large number of rolls of the dice will confirm that the fraction of times we roll a 2 or a 5 will be very close to $1/6$.

Look again at Figure 5-A*d*. The distribution is uneven or "lumpy." Even though we placed the card 40 times, the card did not go into 40 different positions with 12 left over. Some positions got selected twice or even three times, others not at all. If there are equal numbers of each outcome, we call it a *uniform* distribution. A random distribution is not the same as a uniform distribution.

How does this discussion relate to our physical examples of random processes? Radioactivity is a process that is random in time. If we could stare at a single radioactive atom, we would not know exactly when it would decay. However, after a time equal to the half-life, the atom would have a 50 percent chance of having decayed. We can write this

$$p_t(1 \text{ half-life}) = \frac{1}{2}$$

The subscript t shows that our event is random in time. This probability applies to a single atom, but it applies equally to a large number of atoms. After a half-life has

(continued)

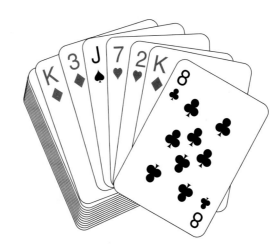

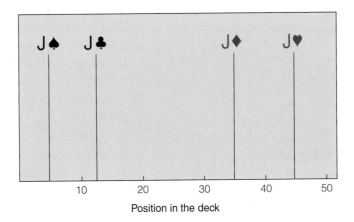

FIGURE 5-B
Random occurrence of jacks (or any other value of card) in a well-shuffled deck of cards. Although 4 out of 52 cards are jacks, the spacing of the cards is not uniform and there is no way of predicting where any particular card will appear.

> # SCIENCE TOOLBOX
>
> ## Random Processes (continued)
>
> passed, one-half of a very large number of radioactive atoms will have decayed. After twice the amount of time, p_t (2 half-lives) = $3/4$. We know from our discussion of radioactivity that 75 percent of the atoms will have decayed after 2 half-lives.
>
> In a similar sense, the sequence of arrivals of large interplanetary objects that impact Earth is scattered essentially randomly through time. We can make a clear distinction between a random and a periodic process. In a Science Toolbox in Chapter 3, we used an *ordered* deck of cards as an example of a periodic process.
>
> If you turn the cards over one by one, a jack (or a 4 or any value of card) will turn up every 13th card. This sequence is completely predictable if the cards are in order. Now suppose we turn over the cards of a well-shuffled deck. Where in the sequence will the jacks turn up? Figure 5–B shows the possible outcome. There is no fixed spacing or periodicity of the jacks; their position in the sequence is random.
>
> The impacts of large pieces of space debris that cause mass extinctions may well look like Figure 5–B (think of the occurrence of jacks as impacts and the sequence from 1 to 52 as a time sequence). Studying the fossil record might tell us how many large impacts occur over the past few billion years. Let's say this analysis implies that there is a 50 percent probability of an impact within any 100-million-year period. To use our math terminology from above, p_t(100 million years) = $1/2$. This probability gives the impact rate *on average*, but it gives us no prediction of exactly when the next impact will occur. The arrival rate of doom and gloom from the skies is random.

How did today's atmosphere reach its present state? Three processes were important. First, the water vapor condensed and fell as rain, forming oceans. Had Earth been closer to the Sun and hotter, the water would have stayed in gaseous form. Had Earth been farther from the Sun and colder, the water would have frozen. So the most common "landscape" on our planet is unique in the solar system. The oceans then initiated a second process. Large amounts of carbon dioxide dissolved in the ocean water (like the carbon dioxide dissolved in soft drinks), to make weak carbonic acid. That acid solution eventually reacted with seafloor sediments to create carbonate-rich sedimentary rocks. With the carbon dioxide and water vapor removed, the dominant remaining gas in the atmosphere was nitrogen (N_2). Third, and most important of all, plant life evolved. The plants consumed much of the remaining carbon dioxide and emitted oxygen. Oxygen content rose to the levels we enjoy today. Today's atmosphere is 76 percent nitrogen and 23 percent oxygen by weight, with only modest traces of carbon dioxide and water vapor. This atmosphere is also unique in the solar system.

The Ozone Problem

The mass extinctions at the end of the Paleozoic and Mesozoic eras show that biological evolution can be disrupted if changes are too large or too abrupt. That is why there is such concern today about environmental changes caused by worldwide industrial growth. The time scale of such changes (decades) is much faster than the time scale for noticeable evolutionary change (roughly 100,000 to 1 million years). Rapid change of the environment is a threat because evolution cannot keep up. Note an important point: The problem is not just change (which occurs all the time), but rather the rate of change. Current estimates suggest we are losing about 5 percent of the planet's plant and animal species per decade, implying that within a few centuries, Earth will have experienced a mass extinction comparable to the biggest ones in the history of the planet. To put it another way, if someone looks at the fossil record of Earth a million years from now, our period may stand out as another Great Dying. The same economic growth that supports a large human population is also killing off many species.

One example of a human-caused environmental problem is the damage to the ozone layer. At 20 to 30 km above the ground, ultraviolet rays from the Sun are absorbed by the ozone layer, as shown by Figure 5–17. The fact that most of these rays do not reach the surface is good for us, because ultraviolet rays damage organic molecules. Energetic ultraviolet photons that hit organic molecules can actually break them apart and cause skin cancer. Certain chemicals called chlorofluorocarbons (or CFCs), such as freon, are widely used in air conditioners and other devices. As is now well known, these chemicals slowly seep into the atmosphere and reach the upper layer, where they eventually break down the ozone molecules. This destroys the ozone layer, lets the ultraviolet light through, and increases the risk of skin cancer and random genetic mutation for everyone. Continued CFC use could create great dangers in the next century. Therefore, international agreements have been reached to phase out CFC production.

The discovery of the CFC threat to the ozone layer is a good example of spin-off from one area of science to another. The recognition of the damage from CFCs to the ozone layer came in part from geochemists and planetary astronomers studying chemical processes in the atmospheres of Venus and Mars. They predicted damage to the ozone layer from CFC chemicals, and this was first detected over Antarctica, where a natural winter "hole" in the ozone

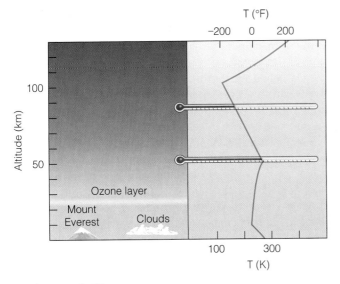

FIGURE 5-17
The left-hand panel shows the altitude of most of the atmosphere's water vapor (in the form of clouds) and the higher-altitude ozone layer. The right-hand panel shows the temperature profile of the atmosphere. The Sun's ultraviolet radiation, which is damaging to our cells, is almost completely absorbed by the ozone layer, high above the ground.

layer broadened dramatically through the 1980s. Satellite data in the early 1990s showed that the winter breakdown of ozone had begun at high northern latitudes as well. A record winter ozone depletion over Antarctica occurred in 1993; since 1995 the depletion seems to have stabilized (Figure 5-18). Most scientists attribute the widening of the ozone hole to CFCs and its stabilization to the phaseout of CFCs. If the phaseout continues, experts in this field predict that ozone damage will begin to decline by about 2010. The ozone issue shows that scientific data and international cooperation can be used to deal with an environmental problem in the nick of time.

The CO_2 Increase and Global Climate Change

Another example of rapid, human-caused change is the increase in carbon dioxide in our atmosphere. The burning of fossil-fuel, the clear-cutting of millions of acres of forests, and the release of pollutants from factories all lead to the buildup of carbon dioxide in the atmosphere. As shown in Figure 5-19, a carbon dioxide increase of more than 20 percent has been clearly observed in data taken since the beginning of the industrial revolution in the 1700s. Why does this increase in carbon dioxide matter?

The problem is that carbon dioxide, though a minor constituent of air, is one of the most effective of the so-called greenhouse gases. The effect of these gases is shown in Figure 5-20. In both a greenhouse and Earth's biosphere, incoming sunlight warms the ground, and the ground reradiates thermal infrared radiation. Thus, the **greenhouse effect** is a warming of the air that occurs because the thermal infrared radiation cannot get back out of the system easily. In a greenhouse, glass blocks it and seals in the warm air. In Earth's atmosphere, carbon dioxide molecules absorb the infrared radiation, heating the air itself. Certain other gases, including water vapor and methane (CH_4), also absorb some of the thermal infrared radiation, adding to the effect. The water vapor effect explains why a cloudy, humid night does not cool down as fast as a clear, dry night. Methane is also increasing; its concentration in the air has doubled since the early 1800s.

We are witnessing a lively, sometimes acrimonious, debate over global warming. Science, economics, and politics are all mixed up in this debate. It seems difficult to answer the simple question: Is Earth actually warming up? Even though current computer models of the global climate are very sophisticated, they are less complex than the real

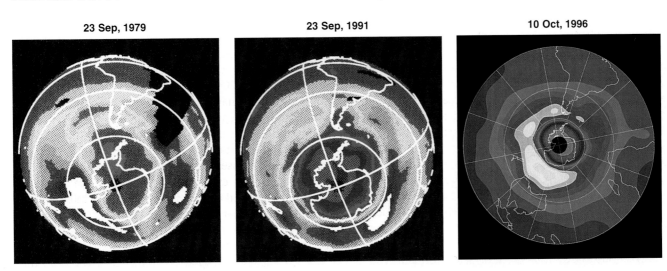

FIGURE 5-18
Seasonal thinning of the ozone layer (shades of pink) in the upper stratosphere over Antarctica, as measured by the *Nimbus*-7 satellite. The area where the ozone concentration has been reduced by more than 50 percent is larger than the continental United States. Since 1995, the ozone layer appears to have stabilized due to the phase-out of ozone-depleting chemicals. (SOURCE: NASA/GSFC and NOAA.)

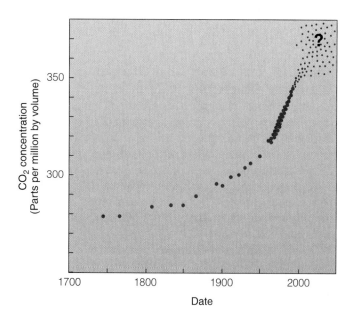

FIGURE 5-19

The measured CO_2 increase in Earth's atmosphere since the industrial revolution. A dot marks a measurement taken from gases frozen in Greenland and Antarctic ice; the closely-spaced dots are direct measurements of air since 1958. Currently, industrialized nations produce most of the CO_2. Projections into the future are controversial, but the amounts will increase as poorer countries rapidly industrialize to emulate living standards of wealthier countries. (SOURCE: Data adapted schematically from Raynaud and others, 1993; the measurement uncertainty is comparable to the size of the symbols.)

atmospheric data; computer models are only as reliable as the data fed into them. Perhaps the most worrysome issue is the influence of the oceans. The oceans store nearly 10 times as much of the Sun's energy as the atmosphere, and they largely drive Earth's climate. Yet temperature data for the oceans are even sparser than for the atmosphere. We are trying to measure small changes in a large and complex system and to extrapolate those changes into the future. Prediction from past data is always tricky.

This debate perfectly reflects one of our themes—how science works. We are dealing with a situation in which data are limited and models of Earth's atmosphere are incomplete. Moreover, while Figure 5-19 shows a clear and undeniable increase in carbon dioxide, the increase in global temperature is far less certain (Figure 5-21). The complexity of the climate system means we are unsure how carbon dioxide changes translate to global temperature changes. Our certainty will increase with more data and better models.

Most climate experts expect that the increasing carbon dioxide will cause an average warming of the climate by the next century, and many believe it has already started. The exact amount of climate change during the next few decades, and its economic effect, is hard to predict. Nonetheless, virtually all studies predict climate changes as a result of projected increases in carbon dioxide. Most models predict net warming by 1 to 3.5°C by the year 2100. Most models also predict increasing extremes in rainfall and temperature (see projections in Figure 5-21). One degree does not sound like much, but even changes in average temperature of a degree or less can have big effects on agriculture by changing growing seasons and increasing the spread of tropical diseases. Some idea of the effects of a 3°C change comes from finding out that the ice ages involved average temperature drops of only about 5°C!

world. For example, significant additional climatic variations arise from eruptions of volcanic dust, and subtle variations in solar radiation affect the ozone content and structure of the stratosphere. Another problem is the incomplete

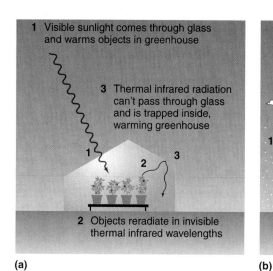
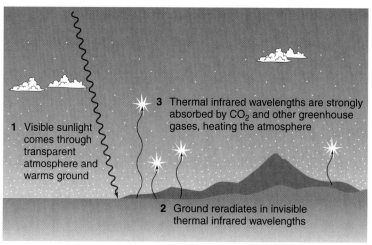

(a) (b)

FIGURE 5-20

Explanation of the greenhouse effect. The greenhouse effect warms both **(a)** the inside of a greenhouse and **(b)** Earth's atmosphere. Sunlight enters the system, but reradiated thermal infrared wavelengths cannot get out of the system easily.

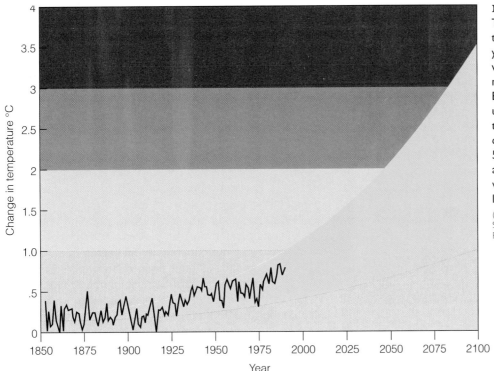

FIGURE 5-21
The measured rise in Earth's surface temperature since 1850. The curved yellow region shows the global warming predicted by computer models of Earth's climate systems. Even though the projections are uncertain, all models predict a temperature increase in the next century. Compare this with Figure 5-19. The rises in CO_2 concentration and temperature are correlated, but we cannot say with certainty that the latter is caused solely by the former.
(SOURCE: Data from the National Academy of Sciences and National Center for Atmospheric Research.)

Translating Scientific Research to Public Policy

We have encountered an area where science impacts public policy. Politicians would like a clear and unanimous statement from scientists about global warming, but the nature of science means that such certainty is elusive.

The problems with the ozone layer and possible global climate change illustrate the difficulties of using recent scientific discoveries to make good public policy in a democratic society. The ozone issue is a success story. Faced with clear evidence and a broad scientific consensus that CFCs were affecting the ozone hole, politicians decided to act. The international action to phase out CFCs, and the apparent reduction in the ozone loss rate, seems like a textbook case of applying new findings to benefit society. It perhaps helped that the economic costs of phasing out CFCs were not too high since replacement chemicals were available.

In the case of global warming, the evidence is less persuasive. (However, the consequences of not acting could be far worse!) Carbon dioxide is the culprit, and nations still argue about what steps to take to reduce carbon dioxide production. The United States and western Europe produce most of the world's carbon dioxide—23 and 14 percent, respectively. The technology exists to move away from the use of fossil fuels, but the economic impact would be enormous. On the other hand, the highest *per capita* production of carbon dioxide is in newly industrialized countries. These countries are projected to become the largest producers of carbon dioxide, and they claim that curbs on carbon dioxide would hamper their industrial growth. In the interests of short-term economic prosperity, and because the data on global climate are still poor, several countries (including the United States) have called for delay in action until more studies can be made. In this political environment, it is difficult to transform scientific observation to policy. Even if the predicted temperature changes are detected, researchers will continue testing whether the changes are due to the greenhouse gases or some other cause. Uniform observations of the whole Earth from space, over a period of years, will help clarify the issue.

What are the broad implications of the relationship between science and society? Science itself is value-neutral. In other words, the general conduct of scientific research is not considered to be good or evil, and the scientific enterprise has no political agenda. However, the consequences of scientific knowledge can be both good and bad. The energy stored inside every atom can be used to create clean power, as in a fusion reactor, or it can be used to make weapons of mass destruction. Understanding of genetic structure can be used to help fight disease, or it can be used for controversial forms of genetic engineering. It is up to societies to decide how they use the fruits of scientific knowledge.

Remember that scientists are people who have opinions, too. The trick is to distinguish scientific issues—which can be decided by the use of evidence and logic—from the matters of opinion. Scientists will often disagree, because of competing theories or incomplete evidence. But if you hear a scientist emphasizing a personal or political opinion, be suspicious. It is the duty of good scientists to keep their scientific knowledge and their opinions separate. The best answer is for citizens to stay well informed. If people are aware of the basic scientific issues, then they can participate in these important debates.

FIGURE 5-22
Spaceship Earth. We live on a small and delicate planet with finite resources. The modern environmental movement dates from the late 1960s, when the Apollo voyages showed us this view for the first time. (SOURCE: NASA.)

Let us return to the science of Earth. The biggest shift in public and political perception of our planet occurred in 1968, when the astronauts of *Apollo 8* were the first humans to ever see the Earth from lunar orbit (see Figure 5-22). It is a striking vision, an isolated world with delicate oceans and a thin sheath of atmosphere. It is not a coincidence that the environmental movement dates back to this time; it was shaped by the awareness created by our trips to the Moon. Now close to 6 billion people jostle on the planet, and we have the ability to alter our global environment in many ways. Our biggest challenge is to use this power wisely.

Explorations of the Earth-Moon System

We think of ourselves as living on a single planet, but in reality, we live in a system of two worlds. Our sister world, the Moon, is easily visible in our sky, and we see its effects on ocean tides. The reality of the double system was dramatized beginning in 1968, when human exploration of the other half of our system began.

In Chapter 2, we considered the monthly movement of the Moon around our planet. To summarize briefly here, the changing illumination of the Moon is a result of the changing relative positions of the Sun, Moon, and Earth. This phenomenon is called the *phases* of the Moon. When the Moon nearly passes between the Sun and Earth, it is lost in the Sun's glare. We have a new Moon. When the Moon is in the opposite direction to the Sun, the illuminated side of the Moon faces back at Earth. We see a full Moon. At times between a new and full Moon, we see a smooth progression from crescent toward full. Remember that the Moon is always half-lit by the Sun. However, the fraction of that lit surface that we see depends on the relative positions of the three objects. The Moon orbits Earth, so that the same features always point toward Earth.

Tides and Their Effects

Many people superstitiously believe that the Moon influences human behavior by some unknown force, causing people to act strangely during a full Moon. However, this has never been convincingly proved (although there may be some tendency for more people to be out on nights of full Moon, and hence for more interesting events to happen then).

On the other hand, there is a real gravitational force between Earth and the Moon that has many important effects. As Newton discovered, Earth's gravity attracts the Moon toward Earth and keeps it in orbit around Earth. We followed his reasoning in Chapter 3. But recall the expression of Newton's gravity law; it says that gravity is a mutually attractive force. So the Moon is attracting Earth, too. Since the force of gravity depends on the inverse square of the distance, the side of Earth facing the Moon has a stronger force pulling toward the Moon than the opposite side, because it is closer to the Moon. This is shown in Figure 5-23a. The two unequal forces cause a net stretching force along the Earth-Moon axis. This stretching force is called a **tidal force.**

Tidal forces occur anytime two celestial bodies are near each other. The actual effect is to stretch the whole planet into a slightly football shape, as shown in Figure 5-23b (exaggerated for clarity). At this point you may wonder about the role of the Sun. The gravitational force of the Sun on Earth is much larger than the gravitational force of the Moon on Earth. So the Sun keeps Earth in orbit around it. However, the *stretching* force caused by the Moon is larger than the stretching force caused by the Sun. So the Moon is the larger influence in causing the tides. We calculate this directly in the next Science Toolbox.

Of course, the liquid ocean can move much more freely in response to these forces than the solid rocks inside Earth. Water thus flows until it is "piled up" in tidal bulges on each side of Earth. You may wonder about the fact that tides occur on both sides of Earth. Why does the attraction of gravity not just cause the water to pile up on the side closest to Earth? Look at the analogy to a spring at the top of Figure 5-23. When the spring is stretched, the distance between all parts of the spring increases. In the same way, the tidal stretching force applies to the oceans on both sides of Earth. Since there are tides on opposite sides of Earth, and Earth rotates once per day, a given spot on rotating Earth passes through two high-tide zones in one day.

If this explanation were the full story, the high tide would occur whenever the Moon was most nearly overhead, and also 12 h later, as you can see from Figure 5-23b. In fact, the ocean tides are complicated by three effects. First, as shown in Figure 5-23c, Earth's rotation drags the tidal bulges out of line with the direction to the Moon. Second, coastlines complicate the flow of water so that the actual high tides occur in a com-

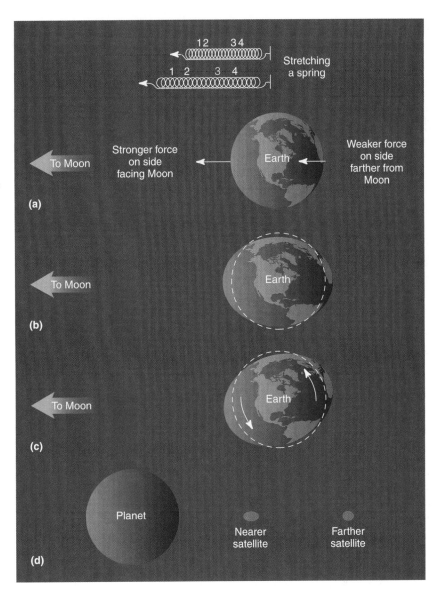

FIGURE 5-23
An explanation of tides. **(a)** The difference in forces between the near side and far side of a body orbiting around another body. In the analogy with a spring, positions 1 and 2 are the top and bottom of the ocean on one side of Earth, and positions 3 and 4 are the bottom and top of the ocean on the other side of Earth. As the spring stretches, the distance between all points increases. Thus there are land tides and ocean tides, and the ocean is piled up on *both* sides of the Earth. **(b)** The resulting stretching of the body; blue areas schematically show ocean tides raised on the Earth. **(c)** The ocean tides are dragged off the Earth-Moon line by Earth's rotation. **(d)** A satellite close to a planet has more tidal stretching than one far from a planet. (See text for further explanation.) Note that the distortions due to tides in this figure are highly exaggerated.

plex rhythmic pattern. Third, the Sun contributes its own tidal forces, which are smaller than the Moon's. Note that extra-large tides occur at new Moon and full Moon, because the Sun and Moon are aligned and both are raising tides along the same axis.

Although the liquid oceans can move easily to respond to the Moon's influence, the rocky mass of Earth feels the tidal force, too. Land tides put extra stresses on the brittle rocks of the lithosphere. This can lead to earthquakes. Geologists have found that earthquakes around the world do not occur randomly. There is a slightly higher chance of earthquakes near full Moon or new Moon, when the tidal force is largest.

So far we have discussed the familiar effects of tides on Earth, but the tidal force is a general consequence of Newton's gravity and we will see its effects throughout the universe. Anytime two large objects orbit each other closely, tides are important. This is true of planets, stars, and even entire galaxies! Let us look at some examples of the general effects of tides.

The same tidal force that stretches a satellite tends to slow a satellite's rotation until the longest axis of the satellite lines up with its planet, as shown in Figure 5–23d. This is the reason that most satellites face toward their planet, as the Moon does. If a satellite (or a passing body) comes very close to a planet, tides can stretch the object enough to break it. The closer two objects are in space, the stronger the gravity between them and the stronger the tidal force. As shown in Figure 5–23d, the closer the object comes to the planet, the more the object gets stretched. Within a certain distance, the stretching can break it.

Tidal forces can also heat the interior of a satellite in an elliptical orbit. In that case, the satellite comes close to the planet and undergoes strong stretching on one side of the orbit, then moves away from the planet and relaxes back toward a more spherical shape. This flexing of the satellite heats it through internal friction; in the same way, if you flex a tennis ball enough times, it becomes warm. This effect is called *tidal heating*. The more elliptical the orbit, the stronger the tidal heating.

SCIENCE TOOLBOX

Tidal Forces

If the Sun keeps Earth in its orbit, why is it the Moon that causes tides? To understand this, we need to compare the strength of the gravity of the Sun and the Moon on Earth.

In Chapter 3 we saw that the force of gravity is proportional to the mass of two bodies and inversely proportional to the square of the distance between them. In this equation there is also a numerical constant G. We will use the subscripts S, E, and M to represent the Sun, Earth, and Moon. The force of gravity caused by the Sun on Earth is

$$F_{SE} = \frac{GM_S M_E}{R_{SE}^2}$$

The force of gravity caused by the Moon on Earth is

$$F_{ME} = \frac{GM_M M_E}{R_{ME}^2}$$

We save ourselves some work by noticing that some of these quantities will cancel out when we take the ratio of the Sun's force on Earth to the Moon's force, or, F_{SE}/F_{ME}. (In general, when you are doing algebra problems, try to simplify the relations as much as you can before you plug in numbers and solve the equation.) The ratio of forces is

$$\frac{F_{SE}}{F_{ME}} = \frac{M_S}{M_M}\left(\frac{R_{ME}}{R_{SE}}\right)^2$$

Now we can insert the values to see which force is strongest, the Sun's or the Moon's. The masses of the Sun and Moon are $M_S = 2.0 \times 10^{30}$ kg and $M_M = 7.4 \times 10^{22}$ kg. The distances from Earth are $R_{SE} = 1.5 \times 10^8$ km (1 astronomical unit, or 1 AU, by definition) and $R_{ME} = 3.8 \times 10^5$ km. We get the result $F_{SE}/F_{ME} = 173$. So the Sun's attractive force on Earth is 173 times the Moon's attractive force. There is no question that the Sun controls the orbit of Earth.

So how can the Moon cause the tides on the earth? Gravity depends on the inverse square of distance. So the force of gravity on the side of Earth facing the Moon is larger than the force of gravity on the far side. The tidal force is a stretching force. Tides are caused by the *difference* between the gravity force on one side of an object and that on the other side.

We can make a good approximation for the strength of the tidal force by taking the gravity force we have just calculated and multiplying it by the ratio of the front-to-back distance of Earth divided by its distance from the Sun or Moon. (Calculus is needed to derive the result precisely.) Let us call Earth's diameter D_E. For the stretching due to the Sun on Earth, we get

$$\frac{D_E}{R_{SE}} = \frac{12,700}{1.5 \times 10^8} = 8.5 \times 10^{-5}$$

For the stretching due to the Moon on the Earth, we get

$$\frac{D_E}{R_{ME}} = \frac{12,700}{384,000} = 0.033$$

The ratio of these two numbers is 390. The size of Earth is a much larger fraction of the Earth-Moon distance than it is of the Earth-Sun distance.

In other words, while the Sun exerts a larger force on Earth than the Moon does, the Moon exerts a larger stretching force. By what factor? By the ratio of 390 to 173, or roughly a factor of 2. Even though the Moon controls Earth's tides, the Sun is a significant contributor. This is the reason that tides are more extreme near a full Moon or a new Moon, when the stretching forces due to the Moon and Sun line up in the same direction.

The tidal force is a universal consequence of gravity. The force that causes our oceans to move operates elsewhere in the solar system, and beyond. Even when there is no water to respond to the force, the solid mass of a planet feels the stress caused by this force. Large objects in close proximity exert the strongest tidal forces.

In general, we know that a force is something that can cause a change. Tidal forces cause water to move around on the surface of our planet. They can also cause heating of a satellite or planet.

Tidal forces can also change a body's orbital motion. Through a complex interplay of gravity, the tidal forces act in such a way as to slow down the rotation of Earth, and at the same time to make the Moon slowly spiral farther away from Earth. You can look at this as yet another example of the conservation of energy. As Earth loses rotational energy by spinning more slowly, this energy appears as the increased gravitational energy of the Moon, which moves to a larger distance in Earth's gravity. In the early history of the Earth-Moon system, these effects meant that Earth rotated faster and the Moon was closer to Earth so that it orbited Earth in fewer days.

What evidence do we have to support this idea? The Apollo astronauts placed laser reflectors on the Moon in order to measure lunar motions precisely. These measurements confirm that the Moon is moving very slowly away from Earth, as predicted by the theory of tidal forces. In addition, paleontologists have studied the daily and monthly growth rings in certain fossil organisms. The results show that 1 billion years ago, the Moon took only 23 days to go around Earth and Earth rotated in only 18 hours! More controversial fossil data from 2.8 billion years ago suggest that the month was only 17 days long!

Throughout this book, we will encounter many examples of these various tidal effects. This illustrates our theme that simple physical ideas have widespread applications. We can test these ideas on well-studied local phe-

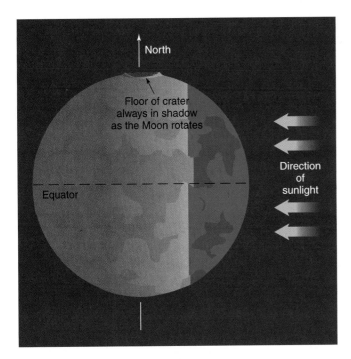

FIGURE 5-24
Because the Moon has no axial tilt to the Sun, craters at the north and south poles of the Moon can have floors that are permanently shadowed and very cold. Evidence for ice has been found in these cold crater floors.

nomena and then apply our knowledge throughout the universe.

Ice on the Moon?

It is conventional wisdom that the Moon has no water. There is no liquid water on the surface because liquid water cannot exist in a vacuum. Rocks and soil samples returned by the Apollo mission contain no water of hydration (water molecules trapped in the minerals), such as is found in many minerals on Earth. Basically, the Moon is made of very dry material. In the early 1960s, however, a few scientists pointed out that craters at the poles of the Moon could be in permanent shadow and extremely cold, as shown in Figure 5-24. These researchers speculated that if water molecules had been released on the Moon (e.g., during eons of impacts by ice-bearing comets), the molecules could be trapped in the cold pockets of these craters, forming accumulations of ice hidden in the permanent shadows.

There was much excitement in 1996, therefore, when researchers announced that radar equipment on a small spacecraft orbiting the moon had returned exactly the signals expected for ice deposits in the soil of crater floors at the Moon's south pole. The initial results were controversial, but concentrations of hydrogen atoms were confirmed in 1998 by an inexpensive NASA spacecraft, the *Lunar Prospector* orbiter. Many researchers believe these H atoms are within H_2O molecules. Ice crystals maybe mixed in with the soil near each of the lunar poles. Such ice could be a crucial resource in supporting a future research station on the Moon, because astronauts would not have to haul water from Earth. It could not only support human life, but also aid in the manufacture of rocket fuel. Future missions will keep "prospecting" for lunar polar ice.

The Origin of the Moon

Where did the Moon come from? People have asked themselves that question for thousands of years. Samples from the Apollo and Soviet expeditions, combined with Earth-based observations, provided a number of clues about the origin of the Moon:

1. As a whole, the Moon is deficient in iron, compared to Earth.
2. The Moon is made out of material similar to Earth's mantle.
3. The lunar surface is deficient in water and other volatile compounds, compared to Earth.
4. Very large bodies hit the Moon early in its history.
5. Although the evidence is masked by erosion and geological upheaval, very large bodies also hit Earth.
6. The proportions of different oxygen isotopes on the Moon (O_{16}, O_{18}, etc.) are exactly the same as in Earth's minerals, but different from those in other parts of the solar system.

These clues led to the current hypothesis about the origin of the Moon—a hypothesis that is motivated by the discovery of a high rate of cratering in the early solar system.

According to this hypothesis, planets formed from collisions and aggregations of many bodies. The young Sun was very hot, and the rock fragments flying through space were semimolten. This is why they could stick together to build large objects like planets. Scientists think that an interplanetary body as large as Mars crashed into Earth at a late stage of Earth's formation (Figure 5-25). This giant impact occurred after the iron in each body had drained into a central core. Figure 5-26 shows a simulation of this event, realized with a supercomputer. After the impact, most of the iron from the core of the impacting body joined the iron at Earth's core. The lighter debris from Earth's mantle (and from the impacting object's mantle) was ejected into a swarm of material around Earth, where it aggregated to form the Moon.

Notice how this hypothesis fits the available facts. It explains the Moon's deficiency of iron and its similarity to Earth's mantle, because the Moon actually formed from mantle debris. Most of the iron ended up at the center of the most massive object, leaving the Moon iron-poor. The hypothesis explains the Moon's relative lack of water, because the debris would have been heated to a high temperature and water vapor would have escaped into space. The oxygen isotopes would match Earth's since much of the debris was from Earth itself, and since the interplanetary body would have formed at the same distance from the Sun.

The hypothesis of a catastrophic origin of the Moon leads to an obvious question: Where is the scar of this enormous impact on Earth? We do not see a crater for several reasons. The impact happened so early in Earth's history that the rock was still hot and plastic. Such semimolten rock would alongside the larger Earth, from the same materials. You can see that this explanation does not explain the differences and similarities between Earth and the Moon as neatly as the hypothesis of a catastrophic origin does. For instance it doesn't explain why Earth has lots of iron, but the Moon has very little.

We should also recognize that science is not able to provide certain answers for questions from the distant past. The historical record is incomplete, and the evidence is fragmentary. Although we may have very good and plausible ideas about ancient events—the death of the dinosaurs, the formation of the Moon—we may never be absolutely sure. Most of the progress we make in science is based on evidence such as the list of facts given above. The correct interpretation of these facts is a puzzle that lends science a lot of its excitement. But science also moves forward by using new techniques. The use of supercomputers to simulate complicated situations is a good example. We will see more evidence of the power of computers in science later in the book.

FIGURE 5-25
The giant impact that created the Moon. The impact, about 4.5 billion years ago, blew mantle material into a cloud of debris around Earth. This view, based on a computer model, shows luminous matter, some as hot as the Sun's surface, spraying outward about ½ hr after the impact. (SOURCE: Painting by WKH.)

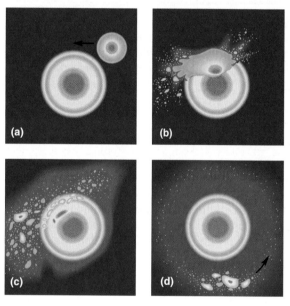

FIGURE 5-26
The giant impact hypothesis for the origin of the Moon. **(a)** A large interplanetary body approaches and collides with primordial Earth about 4.5 billion years ago. **(b)** The iron in both bodies is then concentrated in cores. **(c)** The debris blown into space consists of mantle material. **(d)** Part of the debris goes into an orbiting cloud around Earth, where the Moon begins to aggregate. (SOURCE: All views based on computer models by A. Cameron, W. Benz, W. Slattery, M. Kipp, and J. Melosh.)

have flowed under the influence of gravity to fill in any large crater. Furthermore, the impact was then followed by more than 4 billion years of geological activity that concealed any traces of the impact scars.

This discussion of the origin of the Moon reveals some aspects of how science works. The hypothesis that the Moon formed by a catastrophic impact on young Earth is not unique. There are always a number of ways to explain a set of data. Scientists have the job of finding the best explanation. For example, some scientists have proposed that the Moon formed

Summary

Scientists have been able to piece together the history of Earth and the Moon. Our planet and its satellite are a double system that formed 4.6 billion years ago. The Moon probably originated during a gigantic collision in the late stages of Earth's formation, after the iron core formed. The Moon formed close to Earth from the material that was ejected, and it has been slowly moving outward in its orbit ever since, due to tidal forces. We measure the age of the Earth-Moon system and the chronology of Earth's history by using the technique of radioactive decay. This well-understood physical process also provides the energy that drives most of Earth's geological evolution.

Both Earth and the Moon at one time had molten or partially molten interiors, which allowed differentiation—the separation of dense and less dense rocks by gravity within a planet. The differentiation explains the overall composition of Earth, with a dense metal core at the center and lighter rocks forming a crust at the surface. After the initial cooling, internal heat is caused by energy release due to radioactive decay. Unlike the Moon, Earth is large enough to have trapped its internal heat for 4.5 billion years. Only the outer lithosphere is rigid. Much of its mantle is hot and plastic, with a slow circulation of molten rock that create stresses in the lithosphere, causing earthquakes and plate tectonic activity. Plate tectonics describes the constant shifting and reformation of plates, including continents, on Earth's surface. This geological activity explains why most of Earth's surface is no more than a few hundred million years old. By contrast, the neighboring Moon's surface is 3 to 4 billion years old and heavily cratered. With much less gravity and much less internal heat than Earth, the Moon is geologically dead.

By combining studies of rock strata, the fossil record, and radioactive ages, we have constructed a chronology of Earth, known as the geological time scale. The layers of Earth reveal a succession of prehistoric species, generally from less complex to more complex, with distinct breaks in the fossil record. The vast majority of these fossil species are now extinct. The impact of an interplanetary body 65 million years ago caused one of these breaks, or mass extinctions. The largest mass extinction was about 250 million years ago, and its cause is uncertain. The evolution of life on Earth has not been slow and gradual, but has been punctuated by catastrophes caused by space debris. This gives us a new view of our place in the universe. Our history not only is defined by the closed environment of Earth, but also is subject to cosmic influences.

Most of Earth's environmental changes have occurred slowly, over many millions of years. This includes the buildup of oxygen in the atmosphere due to the respiration of tiny organisms several billion years ago. Environmental changes continue, and some are caused by human activity on a very short time scale (compared to the time scale that allows biological evolution to respond). Human activity has caused the ozone layer to be depleted, and it has caused the carbon dioxide content of the atmosphere to rise, which may lead to global climate changes in coming decades. We bear a great responsibility as caretakers of our planet.

Important Concepts

You should be able to define these concepts and use them in a sentence.

age of Earth	sedimentary rocks
radioactivity	metamorphic rocks
isotope	fossil
half-life	plate tectonics
radioactive dating	volcanism
differentiation	erosion
core	impact craters
mantle	random processes
crust	greenhouse effect
basalt	tidal force
granite	geological time scale
igneous rocks	mass extinctions

How Do We Know?

These questions and answers show how the scientific method is used to learn about the universe.

Q How can we know how old a rock is?

A Every rock contains tiny concentrations of radioactive elements. Radioactive elements or isotopes decay by a well-understood physical process that has been extensively studied in the laboratory. Although the decay of an individual atom is random, we can accurately predict when half of "parent" isotopes will have decayed into "daughter" isotopes. This is the half-life. We can measure amounts of radioactive parent and daughter isotopes in the rock and then apply the known half-life for the radioactive decay process of that parent isotope to prove how long it has been since the rock formed. The technique is very reliable, although it does presume that the parent and daughter isotopes have stayed in the same place, so it can only measure the time since the rock last solidified.

Q How do we know the age of the Earth-Moon system?

A We start with the technique of radioactive dating, described above. Geological processes renew Earth's surface every few hundred million years, so most rocks you might pick up off the ground are only a small fraction of the age of Earth. Geologists search in a few special places to find rocks that date back to early in Earth's history (e.g., the northern parts of Greenland); these establish that the Earth is at least 3.9 or 4.0 billion years old. In addition, scientists use special techniques to combine data from different rocks to get the age of the whole Earth. The final answer is 4.6 billion years. In the case of the Moon, the process is simpler because the Moon does not have geological activity and very old rocks sit near the lunar surface. The oldest lunar rocks confirm that the Moon also formed about 4.6 billion years ago.

Q How do we know global warming has started due to CO_2 and other greenhouse gases?

A We do not know this for sure because it is so hard to measure or predict global average temperatures and because we do not fully understand other natural processes that may also cause major climate changes. However, in terms of theory, it is very simple to see that the more CO_2 in the atmosphere, the more of Earth's thermal infrared radiation will be absorbed by the air. The precise effect on specific local climates is hard to predict. Observationally, several international studies in the mid-1990s concluded that a general global rise in average temperature has begun.

Q How do we know that continents drift due to plate tectonics?

A The fact that the continents seem to fit together as pieces of a jigsaw puzzle is not really direct evidence of continental drift. The actual motions of continents have been measured by precise laser-based measurements from satellites and by techniques of radio astronomy that we will discuss later in the book. Mapping of the seafloor has revealed areas such as the mid-Atlantic ridge where new lava wells up from the mantle and pushes older blocks outward, causing the Atlantic basin to become wider. Regions where plates, or continent-sized blocks, collide are marked by crumpled mountain chains, such as the Himalayas and Andes. These geological motions are imperceptible on a human time scale, but they are dramatic over tens of millions of years.

Q How do we know that a major impact from space wiped out the dinosaurs and many other species 65 million years ago?

A There are many independent lines of evidence. The soil layer that formed 65 million years ago contains excesses of elements found in asteroids. It also contains little beads of rock, quartz grains that were apparently melted in a huge explosion, and soot that appears to have come from global forest fires. A large 65-million-year-old impact crater has been found buried under sediments off the coast of the Yucatan peninsula. This evidence does not make conclusive proof of the impact hypothesis. In particular, it is difficult to show that the mass extinction was truly instantaneous or catastrophic due to the patchy fossil record and the limited precision of the radioactive dating technique.

Q How do we know how the Moon formed?

A We have a plausible hypothesis: The Moon formed from the debris of a catastrophic impact early in Earth's history. This hypothesis explains many of the basic facts of the Moon and Earth. In particular, it explains why the Moon has no iron core and is made of material like the Earth's mantle. Supercomputer simulations show that a body with the mass and orbit of the Moon could have formed from a collision with molten Earth. However, we are less certain of the Moon's origin than we are of the cause of the mass extinction 65 million years ago. We may never be sure of a sequence of events that occurred in our solar system so long ago.

Problems

Use these problems to test your understanding of the information and concepts in this chapter. The * indicates a more advanced or mathematical problem.

1. Suppose a mineral sample contains a radioactive isotope that has a half-life of one billion years. Suppose that chemical tests prove the mineral has one-fourth of its original content of this radioactive element. How long ago did the mineral form?
2. Fossils of apelike ancestors of the genus *Homo sapiens*, found in Africa, are believed to date back about 2 to 4 million years. What percentage of the age of Earth is this? Human civilizations date back about 10,000 years. What percentage of the age of Earth is this? For what percentage of Earth's age have humans been able to alter Earth's environment through industry and agriculture?
3. Imagine that Earth was so close to the Sun that the mean surface temperature was above 373 K (100°C). How would the surface and atmosphere be different?
4. When there is a new Moon, what phase would Earth appear to have to an observer on the Earth-facing side of the Moon? At what time of day or night does **(a)** the first-quarter Moon pass highest above the horizon, **(b)** the full Moon pass highest above the horizon, **(c)** the full Moon rise?
5. Explain why the Moon keeps one side toward Earth.
6. Why are earthquakes more likely to occur near a new Moon or a full Moon?
7. If Earth and the Moon are nearly identical in age, as is believed, then why are rocks on Earth so much younger than rocks on the Moon?
*8. Suppose you are using a radioactive isotope with a half-life of 320 million years to date rocks. Based on the fossils they contain, you believe one rock to be 100 million years old and a second rock to be 700 million years old. What would you expect to be the ratio of daughter atoms between the two rocks?
*9. Assume that the temperature of Earth is given by its distance from the Sun, and assume that the temperature scales with the inverse square root of the distance from the Sun. Assume that the mean temperature of Earth currently is 293 K (20°C). How much closer to the Sun would Earth have to be to have a mean temperature of 373 K (100°C)?
*10. Suppose that Earth is smaller and denser, but has the same mass and orbit with respect to the Sun and the Moon. Would the tidal forces on Earth be larger or smaller? Why?

Projects

Activities for students to carry out either individually or in groups.

1. Get some geological maps (the U.S. Geological Survey is a good source) or talk to members of the geology department of your school to learn the geological features of your region. Are the rocks primarily volcanic or nonvolcanic? Are earthquakes common or uncommon in your area?
2. You can easily simulate radioactive decay with popcorn. Take a frying pan and heat a thin, 1/4 in. layer of oil to a constant temperature just below boiling point (or use an air popper for less mess). Place 10 kernels in the pan. With a clock or stopwatch note the time it takes for 5 kernels to pop. This is the "half-life" of popcorn. Try it several times to see how much your measurement varies. If you take an average, you will get a more reliable number. Try to tape record the sounds; you can then review the tape or play it at a slower speed. This popping is a random process just as radioactivity is. Put a single kernel in the pan. Can you predict exactly when it will pop?
3. Study the Moon with a pair of binoculars. Many modern binoculars show a view comparable to what Galileo could see. Can you see the craters? Can you see how far the shadows project from the crater walls? Knowing the size of the Moon, try to use the length of the shadows and simple geometry to figure out the approximate size of the biggest features on the Moon.
4. Study the Moon with a telescope at about 100X magnification. Sketch a lunar crater. Find a large and bright crater such as Tycho or Copernicus, and compare its appearance at full Moon (high lighting) with its appearance at low lighting some days earlier or later. Count the largest craters visible on the Moon—those with rays of material extending out from them. Now roughly count craters that are 4 to 5 times smaller. Are there many more? What does this tell you about the size distribution of space debris?
5. Make an artificial lunar landscape as follows. Fill a shallow pizza pan with flour, and then sprinkle its surface with a thin layer of dark powder, such as a colored spice. Throw one or more 1-cm dirt clods into the surface as hard as you can. (If you are feeling adventurous and do not mind the mess, try it with raw eggs.) Compare the resultant craters with the appearance of lunar craters. Illuminate your "lunar surface" with a bright lightbulb from above (full Moon) and then from the side at a low angle (partial phase Moon), and compare the appearances. Comment on similarities to and differences from the real lunar situa-

tion. (For example, real lunar craters are made by projectiles hitting at speeds of 10 to 20 km/s and exploding, while your dirt clods hit at only a few meters per second—about 1/10,000 as fast.)

6. Arrange for a class debate on the issue of global warming. This might take the form of a debate on the scientific evidence. It might incorporate other aspects, such as the economic costs and consequences of controlling the emission of greenhouse gases. Alternatively, have some students act as a congressional panel. Other students can then make 3-min presentations to the panel to put the case for or against acting on the issue of global warming.

Reading List

Badash, L. 1989. "The Age of the Earth Debate." *Scientific American,* August, pp. 90–96.

Beatty, J. K. 1996. "Clementine's Lunar Gold." *Sky and Telescope,* vol. 93, pp. 24–5.

Gleason, J. F., and 13 others. 1993. "Record Low Ozone in 1992." *Science,* vol. 260, pp. 523–4.

Hartmann, W. K., and Miller, R. 1991. *The History of Earth.* NY: Workman Publishing.

Huber, B. T. 1998. "Tropical Paradise at the Cretaceous Poles?" *Science,* vol. 282, pp. 2199–2200.

Ida, S., Canup, R., and Stewart G. 1997. "Lunar Accretion from an Impact-Generated Disk." *Nature,* vol. 389, pp. 353–357.

Jeanloz, R., and Lay, T. 1993. "The Core-Mantle Boundary." *Scientific American,* vol. 268, pp. 48–55.

Kaiser, J. 1998. "Possibly Vast Greenhouse Gas Sponge Ignites Controversy." *Science,* vol. 282, pp. 386–387.

Kasting, J. F. 1993. "Earth's Early Atmosphere." *Science,* vol. 259, pp. 920–6.

Kellog, L. H., and 2 others. 1999. "Compositional Stratification in the Deep Mantle." *Science,* vol. 283, pp. 1881–1884 (and following articles).

Kerr, R. A. 1995a. "It's Official: First Glimmer of Greenhouse Warming Seen." *Science,* vol. 270, pp. 1565–7.

Kerr, R. A. 1995b. "Greenhouse Report Foresees Growing Global Stress." *Science,* vol. 270, p. 731.

Kerr, R. A. 1995c. "U.S. Climate Tilts Toward the Greenhouse." *Science,* vol. 268, pp. 363–4.

Kerr, R. A. 1996. "A Shocking View of the Permo-Triassic." *Science,* vol. 274, p. 1080.

Knoll, A.H., Bambach, R. K., Canfield, D. E., and Grotzinger, J. P. 1996. "Comparative Earth History and Late Permian Mass Extinction." *Science,* vol. 273, pp. 452–7.

Larson, R. L. 1995. "The Mid-Cretaceous Superplume Episode." *Scientific American,* vol. 272, p. 82.

Monostersky, R. 1996. "1995 Captures Record as Warmest Year Yet." *Science News,* vol. 149, p. 23.

Morrison, D., Chapman, C., and Slovic, P. 1994. "The Impact Hazard." In *Hazards due to Comets and Asteroids,* ed. T. Gehrels. Tucson: University of Arizona Press.

Newsom, H. E., and Sims, K. 1991. "Core Formation during Early Accretion of the Earth." *Science,* vol. 252, pp. 926–1003.

Raynaud, D., Jouzel, J., Barnola, J. M., Chappellaz, J., Delmas, R. J., and Lorius, C. 1993. "The Ice Record of Greenhouse Gases." *Science,* vol. 259, pp. 926–34.

Robock, A. 1996. "Stratospheric Control of Climate." *Science,* vol. 272, pp. 972–3.

Taylor, G. J. 1994. "The Scientific Legacy of Apollo." *Scientific American,* vol. 271, pp. 40–7.

Wilhelms, D. E. 1993. *To a Rocky Moon.* Tucson: University of Arizona Press.

Zebrowski, E., Jr., 1997. *Perils of a Restless Planet: Scientific Perspectives on Natural Disasters.* New York, NY: Cambridge University Press.

CHAPTER 6

The Terrestrial Planets

CONTENTS

- Landing on the Surface of Mars
- Terrestrial Planets
- Mercury
- Venus
- Mars
- Comparative Planetology: Rules of Planetary Evolution
- Summary
- Important Concepts
- How Do We Know?
- Problems
- Projects
- Reading List

This view of Mars' surface from the *Viking 1* lander shows basalt boulders and beautiful sand dunes in the background. After a long absence, we returned to Mars in 1997 with the *Pathfinder* mission. (SOURCE: NASA, courtesy of Sara Smith.)

WHAT TO WATCH FOR IN CHAPTER 6

This chapter emphasizes the physical nature of the four planets closest to the Sun: Mercury, Venus, Earth, and Mars. We will build on the previous chapter by applying our knowledge of terrestrial geology to other planets. Our study of Venus contains lessons in what can cause the composition of an atmosphere to change. Mars is the most Earthlike planet, in terms of climate and geological history. The evolution of our knowledge of Mars offers a strong example of how science works; it shows the progression of increasingly accurate knowledge of Mars over several centuries. Scientists have developed an ever more sophisticated picture of climate changes on Mars and have even speculated about the existence of life on the planet. We discuss five rules that let us explain similarities and differences among these planets. These rules represent physical principles that should apply to planetary systems throughout the universe.

PREREADING QUESTIONS ON THE THEMES OF THE BOOK

OUR ROLE IN THE UNIVERSE
Is Earth the only planet with life?

HOW THE UNIVERSE WORKS
What are the general rules that govern the evolution of planetary atmospheres and surfaces?

HOW WE ACQUIRE KNOWLEDGE
Is there any evidence that Mars has had life in the past?

LANDING ON THE SURFACE OF MARS

The morning of July 20, 1976, dawned cold and clear in the deserted, rocky plain called Chryse, on the planet Mars. This particular morning was 7 years to the day after the first humans had walked on the Moon. The dawn temperature on Mars was around −84°C (−120°F), but by afternoon the air had warmed to about −34°C (−27°F). At about 4:11 P.M., a very unusual thing happened.

High in the apricot-colored sky, a white parachute opened. Beneath the parachute dangled a sturdy, three-legged spacecraft about as tall as a person. As the parachute and its cargo neared the surface, the parachute cut away and the craft dropped. At the last moment, rocket engines in the underbelly of the craft flared to life, stirring up clouds of reddish dust and depositing the craft onto the surface of Mars.

The name of the spacecraft was *Viking 1*. Covers popped off *Viking 1*'s cameras as the dust settled. Back at the Jet Propulsion Laboratory in Pasadena, California, scientists controlling the craft waited for *Viking 1*'s cameras to come to life and send back the first surface photographs of Mars.

Viking 1's survey of Mars marked the first successful landing of instruments on the surface of the neighboring planet that is most like our own planet. From Earth, scientists had for decades charted moving clouds, polar ice caps, and seasonally changing dark markings on the surface of Mars. Some had hypothesized that life existed on the planet—vegetation that changed with the seasons. Others even speculated about the existence of intelligent civilizations. The only way to test the theories was to get instruments onto the surface of the red planet. Scientists waited with anticipation as the first images streaked across the million of miles back to Earth. These images showed a beautiful, but lifeless landscape.

Terrestrial Planets

Viking 1 marked the beginning of a new era in the direct exploration of our solar system. In the 1960s and 1970s, the first robotic probes reached the nearby planets and began sending back data. As mentioned in Chapter 3, the four planets between the Sun and the asteroid belt—Mercury, Venus, Earth, and Mars—are called **terrestrial planets**. Like Earth, the terrestrial planets are made out of a mixture of rocky and metallic material. By contrast, the planets in the outer solar system are made of a lower-density, icy material, as we will see in the next chapter. Mercury, Venus, and Mars all rank between Earth and the Moon in size. We get a good idea of scales in the solar system by seeing everything in one picture. Figure 6–1 shows the 27 largest solar system objects. Full data for

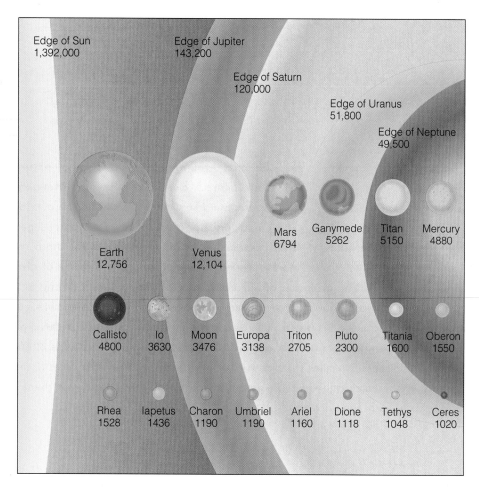

FIGURE 6-1
The 27 largest bodies in the solar system, shown to true relative scale. They include one star, nine planets, 16 satellites, and one asteroid. Diameters are in kilometers.

all the planets and the most important moons are listed in Appendix B-1.

At first glance, the four terrestrial planets seem to have few similarities. Mercury is very hot, airless, and Moonlike; it has no natural satellites. Venus has a very thick atmosphere of carbon dioxide and white clouds that hide its surface; it also has no satellites. Blue Earth has a modest atmosphere of nitrogen and oxygen and is mostly covered with oceans of liquid water; it has one relatively large Moon. The red planet, Mars, has only a thin atmosphere of carbon dioxide and is much colder than Earth; it has two very small satellites. Only Earth has any large and active life-forms. In fact, it is probably the only planet to have any life at all. One planet with no atmosphere, two with carbon dioxide, one with nitrogen and oxygen—these planets seem to have a random mix of properties. Are there any connections or links between them?

Scientists have learned the answer steadily with a sequence of space probes launched by the United States and the former Soviet Union since the 1960s. The observations collected by space probes have revealed many new facts about these worlds. The new facts suggest some unifying principles. As we describe the planets one by one, try to look for such connections. You can view it as a mystery story about our cosmic neighborhood. For example, what underlying principles control the geological features on the surfaces of these planets? What principles explain the different atmospheric state of each body?

We learned about the unifying principle of gravity in Chapter 3. Gravity is the force that alters the layering of a planet while it is still molten, as heavier materials sink toward the center. The pressure and heat caused by gravity help drive geological activity in a planet. The attractive force of gravity allows a planet to retain an atmosphere, and it also determines what type of gases can be retained.

Another unifying principle was mentioned in Chapter 4. The same basic elements and chemical processes exist throughout the universe, although in different proportions. Scholars once thought different planets might be made of totally unfamiliar materials. Recent discoveries have shown more or less familiar minerals and rock types on the various planets, although the chemistry is often influenced by unusual environments of temperature and pressure. For example, similar forms of basaltic lava exist on all four terrestrial planets.

A unifying principle that we will stress in this chapter emerged in the last chapter: Heat inside a planet influences its internal geology. This heat is caused by the gravity of the entire planet and by the radioactive decay of heavy atoms in the interior. The surface features are shaped by a contest

FIGURE 6–2
A view of Mercury, photographed from the U.S. space probe *Mariner 10*. This view reveals Mercury's craters and general resemblance to our own Moon. (SOURCE: NASA.)

between the internally driven geological activity and the external process of impact cratering by interplanetary debris. Let us see how these principles play out on each of the terrestrial planets.

Mercury

Mercury is the planet closest to the Sun. Not surprisingly, it is very hot, with surface soil temperatures well above 500 K (441°F) on the sunlit side. (Science fiction writers used to describe pools of molten metal on Mercury, but this is unlikely; the temperature is not quite high enough to melt common metals, especially in the subsurface soil.) There is essentially no atmosphere, and without a blanket of air to hold in the heat, the surface temperature drops to an extreme low of about 100 K (−279°F) at night.

Mercury is about one-third the size of Earth but about one-third bigger than the Moon. It is so small and far away that its angular size is too small for even the *Hubble Space Telescope* to show many surface features. However, it was photographed at close range in 1974 and 1975 by the U.S. robotic space probe *Mariner 10*. As shown in Figure 6–2, the photographs revealed that Mercury closely resembles the Moon: It is heavily cratered and has areas that appear to be lava plains. Further analysis of the *Mariner 10* data in 1997, as well as Earth-based observations, strengthens the view that the dark plains are basaltic plains superimposed on a more feldspar-rich crust—very similar to the features of the Moon. Only one hemisphere was photographed by *Mariner 10*. The other still remains to be explored. Figure 6–3 shows a close-up of one region of Mercury containing not only craters, but also a large, multiring impact basin, similar to examples found on the Moon. The similarity of craters on Mercury and on our Moon shows that impact cratering of planets has happened widely throughout the solar system.

A Day on Mercury

Mercury takes 88 Earth days to go around the Sun. In 59 days, it makes one complete turn on its axis with respect to the distant stars (in other words, the time it takes for any star to return to the same place in the Mercury night sky). Do you notice a relationship between the two numbers? The 59-day rotation (a Mercury "day") is exactly two-thirds of Mercury's 88-day "year." Is it only chance that the ratio is exactly two-thirds? No. Due to tidal forces and complex gravitational effects called resonances, Mercury's 59-day rotational period has stabilized at

FIGURE 6–3
A close-up of Mercury. The left side shows curved cliffs and fractures that surround one of the largest large impact features, known as Caloris Basin. The basin is about 1300 km (810 mi) across, and the cliffs rise to about 2 km (6000 ft). (SOURCE: NASA.)

two-thirds of the 88-day year. We will learn more about this phenomenon in the next chapter, when we will learn about other patterns in the orbits of moons and rings.

These numbers lead to an odd noontime-to-noontime day on Mercury. The combination of the 59-day spin and the 88-day year means that the time from one noontime to the next at any spot on Mercury's surface averages 176 days! The number 176 is not a coincidence either; it is the lowest number that is an even multiple of the spin rate and the rotational rate around the Sun. In other words, Mercury experiences about 88 days of burning daylight followed by 88 days of frigid night. In addition, the Sun's movement across Mercury's sky is not as regular and steady as that across Earth's sky because Mercury's orbit is not as circular as Earth's. This creates a slow "wobble" in the Sun's movement across the sky, relative to the horizon.

Mercury, Dr. Einstein, and Mr. Spock

Because Mercury's orbit is not circular, at a certain point in its orbit, called the *perihelion* (from the Greek roots *peri* = around, and *helios* = the sun), it is closest to the sun. In the 1800s, scientists were surprised to find that this point shifts in position slowly around the Sun, by a tiny angle, from year to year. The shift could not be explained by Newton's laws of gravity. In the late 1800s, scientists thought the shift must come from the gravitational force of an unknown planet between Mercury and the Sun. (This hypothetical planet was called *Vulcan*. This is the origin of the name for Mr. Spock's home planet on the TV show *Star Trek*.) But later observations showed that no such planet exists.

The solution to the mystery came in 1915, when Albert Einstein modified and improved Newton's laws with his new theory of relativity. Einstein's modifications describe the relationships between gravitation, space, and time differently than Newton did. Einstein's laws predicted exactly the rate of shift that was observed in Mercury's orbital orientation. Thus, the solution to the mystery of Mercury's orbit played a major role in worldwide acceptance of Einstein's theory of relativity early in this century. We will study Einstein's theories in greater detail in Chapter 13. This is a dramatic example of how science works. It is rare that a single observation leads to a new theory. Yet this example of a small shift in the orbit of a single planet led to a new conception of gravity.

A Surprising Polar Cap on Mercury

Mercury is generally known as a very hot planet, since the daytime side gets broiled by the nearby Sun. It was all the more amazing, then, what was discovered in 1991 by bouncing radar signals off Mercury. Radar images, shown in Figure 6–4, revealed strange deposits located in the shadows of craters. These deposits suggested the existence of ice caps at

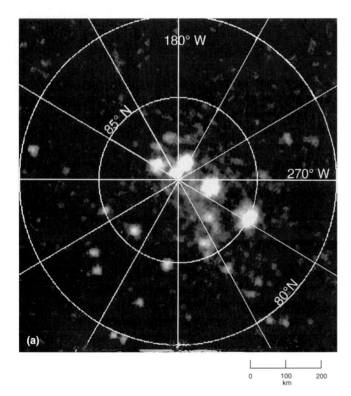
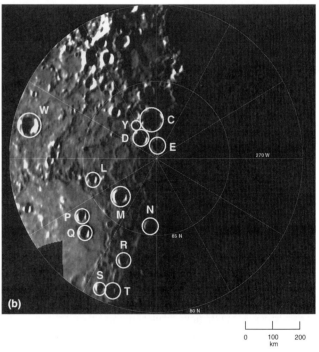

FIGURE 6–4
(a) Radar images of Mercury give evidence of deposits of ice, sulfur, or some other material trapped in permanently shadowed craters at the poles. This view shows radar images of the north pole (obtained from the Arecibo radio telescope in Puerto Rico), reprojected as if seen from directly above the pole. **(b)** Reprojected photographs of the south pole of Mercury (sunlit only on one hemisphere), with circles showing craters that match locations of radar-bright spot. This proves that the deposits lie in the floors of permanently shaded craters. (SOURCE: Images courtesy of Martin Slade, Jet Propulsion Laboratory, and Mert Davies, Rand Corporation.)

the poles, which are reminiscent of the ice deposits on polar craters on the Moon discussed in the last chapter. Like the Moon, Mercury has polar craters whose floors are in permanent shadow. These cold craters may trap water molecules released when comets hit the surface of the planet. We will see that familiar terrestrial substances such as water exist throughout the solar system—even on barren, lifeless worlds such as Mercury and the Moon.

Think about what it would be like to visit the north pole of Mercury. Unlike Earth, with its 23.5° axial tilt, Mercury has a 0° axial tilt. This means that if you stand at the north pole of Mercury, the Sun is exactly on the horizon all year long. (As seen from Earth's north pole, in contrast, the Sun stays above the horizon throughout spring and summer—rising as high as 23.5° above the horizon in June—and then drops below the horizon throughout autumn and winter.) If you stood on the depressed floor of a crater at Mercury's pole, the Sun would always be hidden by the crater walls, and the floor would be in perpetual shadow.

Researchers studying the deposits debate whether the deposits are frozen water or some other material. Some suspect that the deposits might be sulfur or a mixture of the two materials. Both might have been brought to Mercury by comets, which sometimes hit the planet. Ice and sulfur contained in the comet might have vaporized and created a cloud of gas molecules near Mercury's surface; the water vapor or sulfur gas molecules might then have condensed into solid frost on the cold rocks in the dark craters at the poles. If the deposits really contain H_2O ice, they could be a welcome resource if human explorers ever attempt to land on the otherwise forbidding planet.

Venus

Venus is often known as Earth's sister planet because it is the closest planet to us and the most similar in size. Its diameter is 95 percent of Earth's, and its mass is 82 percent of Earth's. Both planets have atmospheres and whitish clouds. But at that point, the resemblance ends.

Venus' Infernal Atmosphere

If you view Venus through a backyard telescope, you can see the disk and phases of Venus, but no surface details. The entire surface is hidden by nearly blank, whitish clouds (Figure 6–5). For many years, astronomers assumed that these clouds were water vapor clouds, like Earth's clouds. Science fiction writers pictured the Venusian surface as a swamp with perpetual rain or as a rainforest with giant tropical plants and dinosaurs.

In 1932, speculation was replaced by data. Astronomers from the Mt. Wilson observatory in California were able to identify features in the spectrum of Venus' atmosphere. Surprisingly, the gas was not water vapor, oxygen, or other gases important in Earth's atmosphere, but carbon dioxide. (In Chapter 10 we will describe in greater detail how gas properties are identified through spectra.) Venus' atmosphere is about 96 percent CO_2, as measured by volume percentage, with nitrogen the largest trace constituent.

FIGURE 6–5
A view of Venus, showing the white cloud formations. This view was made using ultraviolet light, which makes the cloud features more prominent than they would be to the naked eye. (SOURCE: NASA *Pioneer Orbiter* photograph.)

In the 1960s, scientists learned another surprising fact about Venus when they first detected its thermal infrared radiation. They could use this measurement and apply Wien's law to calculate the temperature of Venus. The temperature turned out not to be slightly warmer than Earth, as you might expect for a cloudy planet 72 percent as far from the sun as Earth, but a hellish 750 K (891°F)—even hotter than Mercury! Moreover, the atmospheric winds distribute this heat, keeping the temperature about the same on the night side as on the daytime side. Soon after, astronomers used Earth-based telescopes for spectroscopic studies of the clouds and found that the clouds are not composed of water droplets, like Earth's clouds. They are made primarily of sulfuric acid (H_2SO_4) droplets! With its acid rain and blazing temperatures, Venus differs markedly from the paradise envisioned by science fiction writers. It is unfortunate that Venus is named after the goddess of love.

The CO_2 Greenhouse Effect: Cause of Venus' High Temperature

As soon as scientists measured the high temperature of Venus, they tried to figure out why our neighbor planet is so hot. Astronomer Carl Sagan was among the first to deduce that the dense CO_2 of Venus' atmosphere might create a very strong greenhouse effect—much stronger than the one we worry about on Earth. This work was one of the earliest and most noteworthy of Sagan's scientific contributions. He became known to many people through his book *Cosmos* and TV programs, and through his work as a popularizer of science. It is a rare and happy occurrence when a scientist has the enthusiasm and ability to convey his or her subject to a wider audience.

The **greenhouse warming of Venus** should remind us of the studies of global warming on Earth. As we discussed in the last chapter, a debate raged in the 1980s over the greenhouse effect on Earth caused by CO_2. Really, there were two debates. One took place in scientific journals and at professional meetings, where all the strengths and limitations of the scientific method were on display, as we discussed in Chapter 1. The evidence for global warming was impressive but not overwhelming. Scientists could not draw firm conclusions due to incomplete data and to the limitations of computer models of the atmosphere. The second debate took place in public, and it contained aspects of politics as well as science. Some ideologues claimed that the possibility of greenhouse warming on Earth was a fraud invented by scientists and environmentalists to attract more research funding. But the greenhouse warming of Venus from a buildup of CO_2 is real and directly observable.

In the accompanying Science Toolbox, we show how to calculate what the temperature of Venus would be if there were no greenhouse effect. In that case, the only difference in heating between Earth and Venus would come from the fact that Venus is closer to the Sun. A planet in Venus' position, with no greenhouse effect, would have a temperature in the general range of 100 to 163°F, depending on the cloud cover—but not 900°F! Venus thus gives us a chance to study climate modification by carbon dioxide.

Of course, any foreseeable greenhouse warming on Earth would be only a few degrees because the total CO_2 amounts are much less than those on Venus. But the physical principles are well understood, and Venus proves that they work. The next time someone tries to tell you there is no such thing as climate modification by greenhouse gases, ask him or her why Venus has a temperature of 900°F!

This whole discussion is a good example of how the study of other planets clarifies our knowledge of our own planet. Venus is in effect a "natural laboratory experiment" that shows what happens to an Earth-sized planet when its atmosphere is altered and it is moved closer to the Sun. Venus allows us to test our theories. If we had never undertaken exploration of the universe around us, we would never have obtained such a clear picture of how planets' climates can evolve under strikingly different conditions.

The Volcanic Landscape of Venus

No one knew what lay beneath the clouds of Venus until the 1970s, when the former Soviet Union sent a series of robotic probes onto the surface of the planet. In 1967, the Soviet *Venera 4* ("Venus 4") probe became the first human artifact to reach another planet, but it crashed on the surface without returning data. Three years later, *Venera 7* became the first human artifact to land successfully on another planet, transmitting data for 23 minutes from Venus' surface. Notice that this initial Venus landing occurred 6 years before the initial Mars landing described at the beginning of this chapter—Venus was the first planet successfully reconnoitered by human devices. The Venus lander confirmed the high temperature and showed that the atmospheric pressure is an incredible 90 times as great as the air pressure at sea level on Earth. This is equivalent to the pressure endured by a diver nearly a kilometer (3000 ft) below the terrestrial ocean surface!

Surface conditions on Venus indicate that it would be very hard for humans to land there in the foreseeable future. Any spacesuit, for example, not only would have to withstand the 900°F temperature, but also would have to be rigid enough to withstand the crushing pressure of the atmosphere.

From the mid-1970s to the 1980s, Soviet scientists landed a number of additional Venus probes that took photographs and measured soil compositions. Figures 6–6 and

FIGURE 6–6
Six landscapes on Venus. The photographs were taken from three different Russian space probes. All the photographs, taken in different regions of Venus, show barren landscapes of boulders and gravel, with horizon and bright sky at the top. (SOURCE: Courtesy of C. Florensky and A. Basilevsky, Vernadsky Institute, Moscow.)

SCIENCE TOOLBOX

Venus and the Greenhouse Effect

By applying the principles of thermal equilibrium and conservation of energy that we discussed in Chapter 4, scientists can easily calculate what temperature Venus would have if it had no greenhouse effect.

First, we imagine that Earth is simply moved to the position of Venus, at 0.72 AU, and ask how much more radiant energy would be received from the Sun at that distance. The inverse square law says that if Earth were 0.72 times as far from the Sun, the radiation would be $1/0.72^2$ times as strong, or 1.93 times stronger.

How much would this raise the planet's temperature? If we assume that a typical spot on Earth is 68°F, then we can ask how much hotter it would be on Venus, with the incoming radiation boosted 1.93 times. This is a somewhat tricky calculation, but we can approach it by thinking about the principle of equilibrium: The temperature would rise until the total energy radiated by the surface equaled the energy coming in from the Sun. As we saw in Chapter 4, Wien's law states that hot objects emit bluer or shorter-wavelength radiation than cool objects do. Hot objects also emit more radiation per unit area of their surface—this property of radiation is called the *Stefan-Boltzmann law*. It states that for bodies emitting thermal radiation (discussed in Chapter 4), the total energy radiated per unit area is proportional to T^4, which is the material's temperature T raised to power 4.

For example, at Earth, we could say that the radiation emitted by 1 cm^2, which we designate as S_{Earth}, would be proportional to Earth's surface temperature, designated T_{Earth}, raised to the 4th power:

$$S_{Earth} \propto T_{Earth}^4$$

Similarly, at Venus

$$S_{Venus} \propto T_{Venus}^4$$

A useful trick in this kind of calculation is to divide the second equation by the first, so that we deal only with the ratios of the quantities that are changing, and the constants of proportionality cancel out. In other words,

$$S_{Venus}/S_{Earth} = (T_{Venus}/T_{Earth})^4$$

From the discussion above, we know that the quantity on the left side, that is, the ratio of radiation received by (and emitted by) the two planets, is 1.93. Plugging in 1.93 and solving for T_{Venus}, we learn that T_{Venus} will be $(1.93)^{1/4}$, or 1.18, times higher than T_{Earth}. Now we must remember that in all physical calculations about temperature, we must use the absolute temperature scale, measured in kelvins. The Kelvin scale is designed so that zero corresponds to zero energy and no thermal radiation. This is not true of the Celsius and Fahrenheit scales. A characteristic temperature on Earth is about 68°F, which can be expressed as 293 K. Multiplying 293 K by the factor 1.18, we get that Venus would be 346 K, or roughly 163°F, if the temperature difference between the planets were due only to solar distance. This calculation is actually confirmed by the heating of spacecraft when they fly toward the Sun to approach Venus.

If we think about the situation more carefully, we realize that the surface of Venus is shaded by dense white clouds that reflect most of the sunlight back into space. Thus, the radiation reaching the surface of Venus is even less than assumed above. Measuring the high reflectivity of Venus' global cloud layer, we find that Venus absorbs only about 41 percent of the sunlight hitting it, while Earth absorbs about 61 percent. (These figures are approximations.) Because Venus absorbs so little sunlight, our corrected answer comes out that cloudy Venus ought to be more like Earth than was calculated above. Depending on the uncertainty in the cloud cover, the temperature might be as low as 100°F—no more than a hot summer day on Earth!

Yet the measured temperature is nearly 900°F! Why the difference? Perhaps Venus has more internal heat than Earth, which leads to a hotter surface. This explanation fails because Venus has no more geological activity than Earth. The difference can only be due to the greenhouse effect of the massive CO_2 gas content in Venus' atmosphere. When the CO_2 greenhouse effect is factored into the calculation (a much more difficult set of equations), then the theoretical calculation finally gives the correct, observed answer.

6–7 show a barren, lifeless planet covered with rocks and gravel similar to barren volcanic terrain on Earth. The chemical measurements taken by the Venera probes suggested that most of these soils and rocks are made of basaltic lava, and are somewhat similar in composition to those that cover much of the seafloor crust of the Earth. The probes also found a few samples of granite similar to continental rocks on Earth. There is clear evidence for **volcanism on Venus**.

From the late 1970s to the early 1990s, U.S. and Soviet probes began a new kind of study: radar mapping of Venus. Space probes in orbit around Venus bounced radio signals down through the clouds and off the surface of Venus. Astronomers then used the returned radio waves to construct images of the entire surface. Figure 6–8 shows a mosaic of radar maps. The sphere is deprojected and turned into a relief map in Figure 6–9. The radar measured altitudes of all spots on the planet, revealing that most of the planet is

FIGURE 6-7
An analog of Venus' landscape on Earth. These weathered basaltic plains on the Mauna Kea volcano in Hawaii are very similar to the Venus landscapes shown in Figure 6-5. High and thick clouds provide a diffuse lighting that heightens the resemblance to cloud-covered Venus. (SOURCE: Photograph by WKH.)

FIGURE 6-8
A mosaic of the surface of Venus made using radar maps from the *Magellan* spacecraft. The maps reveal impact craters and large volcanoes like Beta Regio in the upper right center. Faults and lava flows are bright in this false color image—the true color of the rocks is a dull gray. (SOURCE: NASA.)

covered by low, rolling plains, with only about 1 km (3000 ft) of relief in these areas. These plains may be analogous to the low-elevation, basaltic seafloor crust of Earth. The other 40 percent of Venus is covered by uplands, including a few Australia-sized, continentlike plateaus standing a few kilometers above the surface. These uplands contain a few huge volcanic peaks, towering as high as 10.6 km (34,500 ft)—higher than Mount Everest rises above sea level.

Are these volcanoes active? Some evidence, including rapid changes in amounts of certain gaseous sulfur compounds in Venus' atmosphere, indicates that the volcanoes probably have been active in recent geological time (Figure 6-10). They may have produced fresh lava and sulfurous gases within the last few decades.

Surface Ages and Internal Activity of Venus

The most detailed radar mapping of Venus, by the U.S. *Magellan* probe from 1990 to 1992, revealed modest numbers of impact craters. As you learned in Chapter 5, the numbers of impact craters give clues to the age of the surface. There are somewhat more craters on Venus than on Earth, but Venus has no ancient, heavily cratered areas, like the high-

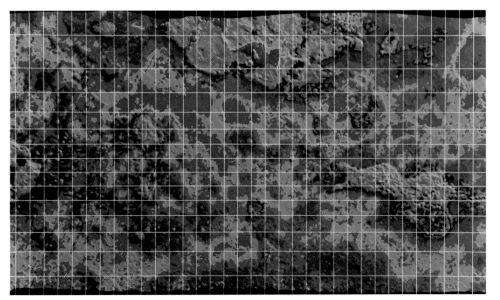

FIGURE 6-9
A map of the surface of Venus, showing rolling lowlands (blue) and several elevated areas that may be analogs of continents on Earth (green-yellow). Highest peaks (red) may be volcanoes. The grid of lines shows latitude and longitude in intervals of 10°. (SOURCE: NASA map from *Magellan* spacecraft radar data; Jet Propulsion Laboratory.)

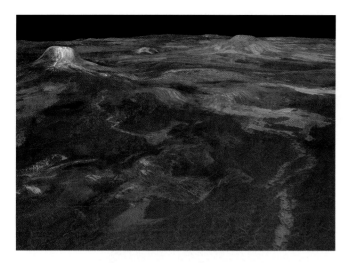

FIGURE 6-10
This image, based on the *Magellan* radar data, shows the volcanoes Gula Mons (left) and Sif Mons (right) with their associated lava flows. The vertical scale has been stretched by a factor of 10, so the features are not as prominent or steep as they appear here. (SOURCE: NASA.)

lands of the Moon or Mercury. These findings indicate that Venus' surface is geologically young. Estimates range from 500 to 800 million years old, older than most surface regions of Earth.

More strikingly, there are no areas that seem clearly much older or younger than the average. If the whole surface is only 500 to 800 million years old, it may mean that much of the planet was resurfaced by lava about 500 to 800 million years ago. No one knows why such a dramatic resurfacing occurred during such a relatively short interval. Some researchers believe that such a dramatic resurfacing episode might have been caused when convective "superplumes" of hot magma rose in Venus' mantle, leading to lava eruptions in many spots on the surface. Remember that although the surface features are geologically young, the planet as a whole, like Earth and the Moon, is 4.6 billion years old. It is possible that Venus has undergone many episodes of resurfacing. Earth's surface seems different, in the sense that it has more variety in ages. Young, active areas, such as California, contrast with 2-billion-year-old continental core areas of Earth, such as central Canada, where a number ancient impact craters can be found.

Perhaps the greatest geological difference between Earth and Venus is that Venus does not seem to have as much active plate-tectonic or continent-building activity as Earth. Many of Venus' plains show tectonic fractures, but there are no signs of the large-scale movement of entire plates. This is probably because Venus does not have the same internal structure of a thin, rigid lithosphere broken into mobile plates. Gravity data collected by the orbiting probes suggest that Venus may have a thicker lithosphere than Earth does, which may cause a different style of interaction with mantle plumes of magma. Figure 6-11 gives a good idea of the variety of surface features on Venus. In summary, studies of Venus' surface and interior may eventually shed more light on the structure and history of both Venus' and Earth's interiors. The exact cause of differences in tectonics and continent growth on the two planets, however, remains to be discovered.

Mars

In many of its surface properties, **Mars** is the most Earthlike planet. The planet is roughly one-half the size of ours, but it has many Earthlike features, such as a nearly 24-hour day, clouds, sediment layers, sand dunes, volcanoes, and polar ice caps that grow and recede with the changing of the seasons. Mars even has dry river channels—evidence that liquid water flowed on its surface in the past. For two centuries, scientists have wondered whether Mars harbors life, or perhaps had life in the past. Such questions have excited the imaginations of scientists, philosophers, and science fiction writers. They continue to excite anyone with an interest in the red planet, and they continue to be the focus of modern exploration of Mars.

The Mythic Mars—An Example of Evolution in Scientific Ideas

In its orbit around the Sun, Mars comes within 56 million km (35 million mi) of Earth, closer than any other planet but Venus. Because it is farther from the Sun than Earth, its sun-lit side faces Earth when it is closest to us. (You might want to draw a diagram of the two orbits to confirm this fact.) This illumination when Mars is relatively close offers Earth-based observers a good chance to study its surface features.

Such studies started soon after Galileo first used the astronomical telescope in 1610. Dutch physicist Christian Huygens first sketched the dark, patchy markings on the planet in 1659 (Figure 6–12a). A few years later, Italian astronomer Giovanni Cassini tracked the markings to determine that Mars turns once on its axis in 24 h and 37 min. In other words, the length of the Martian day is almost the same as that on Earth. The tilt of the axis of Mars is 25°, slightly more than the 23.5° of Earth, so that Mars also has seasons as Earth does.

By the end of the 1700s, astronomers had observed even more features that resembled Earth's. Mars had scattered clouds and bright, white polar caps. The **polar ice caps of Mars** shrink in summer and grow in winter, as do the polar ice caps of Earth. The surface markings observed by Cassini and others underwent seasonal changes, growing darker in the summer and changing shape from year to year; some observers even thought they might be patches of vegetation.

All these observations supported a Renaissance idea discussed in Chapter 3—the plurality of worlds. This hypothesis suggested that other planets were actually worlds like Earth. To see how important this idea was, think about the long march of ideas over the centuries. In Greek times, the planets seemed to be supernatural entities or mystical, crystalline lights that traveled in a celestial realm above mundane Earth. Then, in the 1600s, the Copernican revolution established that the planets and Earth were all bodies moving

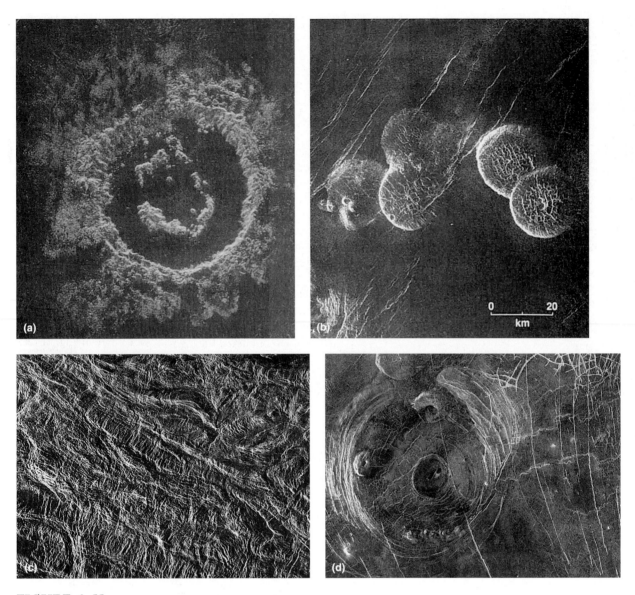

FIGURE 6-11
A variety of surface features on Venus. **(a)** This impact crater 50 km (30 mi) across was named after the founder of the Red Cross, Clara Barton. The double-ring structure is often seen in larger craters and is caused by mechanical processes during the impact explosion—see the drop-of-liquid analogy in Chapter 5. **(b)** These unusual circular domes are about 750 m (2500 ft) high. They are probably pancakelike eruptions of viscous lava. Diagonal bright lines are tectonic fractures, which are common on Venus. **(c)** This complex pattern of intersecting ridges and cracks is caused by repeated fracturing of the crust. **(d)** The large circular feature is about 200 km across, caused by upwelling magma from below. Smaller "pancake" domes are seen on top.
(SOURCE: NASA *Magellan* radar images; Jet Propulsion Laboratory.)

around the Sun. The idea of the plurality of worlds took the Copernican revolution one step further: Other planets might be worlds as Earth is, with familiar geological surfaces, climates, clouds, weather, and even life-forms!

In the case of Mars, some scientists took this idea a little too far in the late 1800s. Various observers saw streaky markings on Mars, and some drew them as straight, narrow lines, which they called *canals*. The term was not meant to imply that the streaks really were canals, but was chosen to maintain a tradition of naming Mars' dark markings after bodies of water (see Figure 6-12c). In the United States, the wealthy Bostonian Percival Lowell built his own private observatory in Flagstaff, Arizona, to study Mars under clear desert skies. Lowell mapped the supposed canals as a mysterious network of perfectly straight lines stretching across the red planet. By 1895, Lowell announced a shocking new theory. He noted correctly that smaller planets, with their weaker gravity, cannot hold their atmospheres as long as Earth has. Therefore, he said, Mars was a dying, drying planet, where water was scarce and locked in the polar ice caps most of the year. Lowell reasoned that the equator of Mars was the only place warm enough for life to flourish on the planet. He hypothesized that the "canals" were really artificial channels built by a Martian civilization to bring water

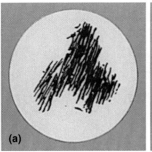 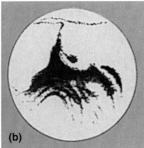 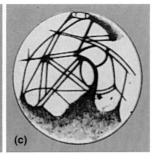

FIGURE 6–12
Early drawings of Mars, showing progress in our knowledge of surface features on the red planet. **(a)** One of earliest known sketches shows the largest dark area of Mars, called Syrtis Major, in 1659. **(b)** Sketch in 1865 showed a streaky extension north from Syrtis Major. **(c)** Italian observer Giovanni Schiaparelli popularized the conception of fine, straight lines, or "canali" on Mars. (SOURCE: After drawings by Huygens, Dawes, and Schiaparelli.)

from the polar ice fields to the warm, habitable equator! (More specifically, Lowell said that the canals themselves were too small to see, but the lines glimpsed through telescopes were the bands of vegetation and irrigated fields along the sides of the canals.)

Lowell's ideas won him worldwide fame—and controversy. Some astronomers perceived straight-line "canals" on Mars, but others saw only wispy or patchy streaks. Photographs did not resolve the issue, because photographic plates had limited sensitivity. Photographs of Mars required long exposure times, and atmospheric shimmering blurred the images. Visual observers can take advantage of instants of sharp seeing—moments when the shimmering of Earth's atmosphere is low and features come into sharp focus. Until the advent of late 20th century electronic imaging, only the naked-eye observers could see fine details on Mars. They could not agree on whether straight-line "canals" really existed.

We can use the small-angle equation from Chapter 2 to show how limited our view of Mars must be. From the Science Toolbox, we note that the diameter of a distant object is $d = \alpha D/206{,}265$, where α is the angular size in seconds of arc, and D is the object's distance. At Mars' closest approach, $D = 5.6 \times 10^7$ km, and the very best we can do from the ground is to make out an angle of about $\frac{1}{2}$ second of arc. Therefore, $d = (2.8 \times 10^7)/(2.1 \times 10^5) = 133$ km, or about 80 mi. In other words, an Earth-based observer could barely see 80-mile-wide streaks, and could not distinguish engineered structures or geological features such as canyons.

Lowell's theory of canal-building Martians had a profound effect on our culture. His ideas forced humans to expand their imaginations to consider the possibility of a civilization on another planet. English poet Alfred Lord Tennyson wrote a poem about whether the Martians could detect civilization on Earth. British writer H. G. Wells published an early science fiction novel in 1898, *The War of the Worlds*, in which Martians invade Earth, only to be defeated by terrestrial germs. Wells noted that the native populations of the Pacific and the Americas had been decimated in the 1700s and 1800s by European diseases (and syphilis went in the other direction, from America to Europe). He raised the possibility that contact between worlds could prove fatal to one species or the other, through either disease or conflict. Wells wrote: "The Tasmanians, in spite of their human likeness, were entirely swept out of existence in a war of extermination by European immigrants . . . Are we such apostles of mercy as to complain if the Martians warred in the same spirit?"

Wells was brilliant in exploring the consequences of scientific ideas, but there was also more fevered speculation. An unidentified flying object (UFO) scare in the 1890s, after the publication of *The War of the Worlds*, produced a wave of reports of Martian spaceships. They were described as looking like the first dirigibles, which were then flying. Notice how UFOs have been reported to have the form of the aerial technology of the day—dirigibles in the 1890s, and sleek and shiny metallic objects since the 1950s, when the first spy planes and jet aircraft were being produced. This suggests a social element in UFO reporting—that images of UFOs reflect images in our own culture. Figure 6–13 shows peaks in the number of reported UFO sightings at the time of

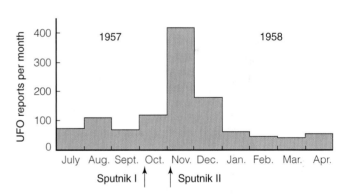

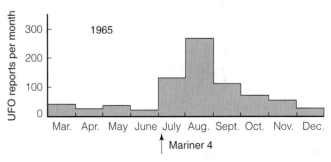

FIGURE 6–13
Examples of sudden increases in the number of UFO sightings correlated with social factors such as the first satellite launches (top) and the first close-up photographs of Mars (bottom). This correlation suggests that most UFO sightings have nothing to do with alien visitations.

two major events in the early space program: the first artificial satellite and the first Mars probe. The next major peak in UFO sightings occurred—did you guess it?—when humans reached the Moon in 1969. We can be confident that most of these sightings were a product not of a real alien visitation, but of interest in other worlds.

From the 1910s to the 1940s, Edgar Rice Burroughs (creator of Tarzan), Ray Bradbury, and others wrote more science fiction novels about Martian civilizations. Around Halloween in 1938, the brilliant young movie director Orson Welles broadcast a radio dramatization of *The War of the Worlds,* in which the Martians landed in New Jersey and advanced on New York. It caused a panic on the East Coast when listeners mistook it for reality. Mars had, in effect, captured the world's imagination and had become deeply embedded in popular culture.

Scientific information about Mars continued to evolve. Better astronomical techniques proved Mars to be a more forbidding environment than had been thought. It was colder than Lowell had thought and the air was thin. Nevertheless, some argued that evolution might have produced Martian vegetation adapted to the difficult conditions. Many scientists, perhaps a majority, believed that such vegetation caused the seasonally changing dark markings. But data from Earth-based telescopes in the 1960s made Mars seem still more hostile: Spectroscopic measurements revealed that the air pressure on Mars was comparable to that 100,000 ft above the Earth, which is higher than commercial jets can fly. In 1965, *Mariner 4* became the first space probe to reach the red planet. As it zipped past Mars, it snapped a few close-up photographs of local regions showing an impact-cratered surface that looked like the Moon. Most scientists then speculated that Mars must be a frozen, dead, Moonlike world. They were closer to the truth than Lowell was—but Mars has turned out to be more interesting than they thought.

The truth about Mars began to emerge more clearly in the 1970s with probes that mapped the planet from orbit and two Viking probes that landed on its surface in 1976, as described in the opening of this chapter. After a long hiatus, mapping and exploration on Mars continued in the 1990s. Mars has yielded up many of its secrets, but not all.

The Real Mars—A Modern View

The vegetation and Martian civilization pictured by Lowell turned out not to exist. Modern spacecraft mapping of Mars, as well as *Hubble Space Telescope* images, show that patchy, streaky markings do cross Mars, as shown in Figure 6–14. These markings account for the early reports of canals—Lowell and his supporters simply misconstrued the patchy streaks as thin, straight lines. In reality, the patterns of streaks and patches are deposits of wind-blown dust, drawn out into streaky markings by prevailing winds. Seasonal dust storms occasionally sweep the planet, altering the configurations of these deposits.

Modern space missions have cleared up other Martian "mysteries." One of the Viking orbiters in the 1970s photographed a mountain that looked like a human face under certain lighting conditions. This did not excite planetary re-

FIGURE 6–14
Modern image of Mars from the *Hubble Space Telescope* in orbit around Earth, showing the same region sketched in the historic drawings in Figure 6–12. At the right are some of the streaky markings once charted as canals. (SOURCE: NASA.)

searchers, since shadow features of natural mountains on the Moon and Earth often have a vague resemblance to human or animal forms. However, some writers and tabloid newspapers seized on the "face on Mars" as an artificial structure built by aliens. Extremists in this group even claimed that NASA had sabotaged its own *Mars Observer* mission, which failed in 1993, to prevent the "truth" from coming out! In 1998, the *Global Mars Surveyor* took a much more detailed picture of this region among many others, and showed that it is an entirely natural feature (Figure 6–15). Many Martian hills are capped with a layer of resistant rock, which erodes into broken ridges and valleys that cast intricate shadows.

There are three serious points to this story, involving media culpability and public gullibility. First, the human eye and brain tend to see patterns because that is what they evolved to do. A baby must learn to recognize its mother's face, so pattern recognition is a finely tuned skill. Second, the news media often sensationalize so-called scientific "mysteries" or "controversies" that reflect the views only of a few fringe writers, not the vast majority of active researchers. The "face on Mars" spawned a lot of TV programming, but was only a hill of dirt! Third, improved data are what allow us to choose between rival hypotheses. In this case, better imaging skewered the hypothesis that the face on Mars was anything other than a natural rock formation.

What did the landers and orbiters reveal about the real Mars? It is cold and dry—a frozen desert. The temperature of the thin air ranges from −123°F (187 K) at night to −20°F (244 K) on a typical afternoon. The red soil is warmer, because it absorbs sunlight, which mostly passes through the atmosphere without being absorbed by the thin

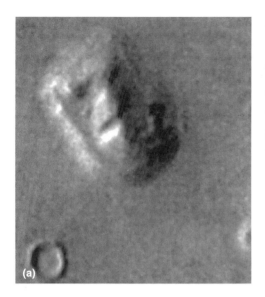
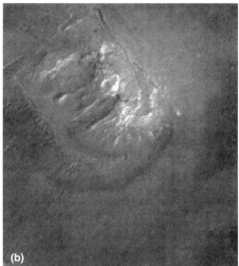

FIGURE 6-15
(a) One of the Viking orbiters imaged this topographical feature, illuminated from the left, which some people dubbed the "face on Mars." **(b)** *The Mars Global Surveyor* was capable of distinguishing features 100 times smaller than was possible with *Viking*. The "face," this time illuminated from the right, is revealed as a normal geological formation of eroded hills and gullies. (SOURCE: NASA.)

air. Soil temperatures can get well above freezing on a typical summer afternoon. The atmosphere is extremely thin and composed of mostly carbon dioxide (95 percent CO_2 by volume). The air is so thin that the air pressure on the surface is only about 0.7 percent of that on Earth at sea level.

These conditions seem inhospitable for supporting life on Mars. However, the probes that went into orbit around Mars in the 1970s discovered many intriguing features. Photographs from orbit above Mars showed not only old cratered regions, but also younger, sparsely cratered regions (Figure 6–16). There are canyons and landslides, vast fields of sand dunes, and enormous young (dormant or active?) volcanoes rising as high as 24 km (78,000 ft) above Mars. In comparison, Mount Everest on Earth rises only 9 km above sea level and 13 km above the deepest ocean floor. The polar ice caps (Figure 6–17) are more complicated than Earth's. At one or both poles are small, permanent deposits of H_2O ice. In the winter, the polar weather is so cold that CO_2 also freezes out of the atmosphere and makes much larger caps of dry ice deposits. In a similar way, our winter snow and ice deposits spread across Canada and the northern United States, only to shrink in spring. We should note that carbon dioxide is unlike water in that it does not exist as a liquid under conditions of low pressure. CO_2 goes directly from solid to gas as it heats and from gas to solid as it cools. Measurements of Mars also suggest permafrost ice deposits in the soil below the surface at moderate latitudes on Mars, analogous to Canadian tundra. Mars today has no water in the liquid state—water is frozen in the ice caps and permafrost, and it is incorporated into molecules that are trapped in mineral deposits.

Most exciting of all, the mapping of Mars in the 1970s revealed ancient, dry **riverbeds on Mars,** as shown in Figure 6–18. They are common in older regions, but not in the youngest regions. This evidence of erosion means that water flowed one or more times across the surface during some ancient era, carving the river channels. Some of the riverbeds may have been frozen over with ice, but virtually every planetary scientist agrees that liquid water did flow on ancient Mars.

Where did the water come from? Scientists are still trying to figure that out. Some of the riverbeds emerge from closed, collapsed regions, which suggests that the water came from local melting of underground permafrost, perhaps caused by local geothermal activity, similar to that in Yellowstone Park

FIGURE 6-16
Photograph from the *Viking* spacecraft in orbit around Mars. The vast canyon system in the lower center is long enough to stretch across most of the continental United States. Several volcanic mountains form dark spots in the desert at the left edge. Clouds can be seen at upper left, impact craters at right center. (SOURCE: NASA, courtesy of A. S. McEwen, U.S. Geological Survey.)

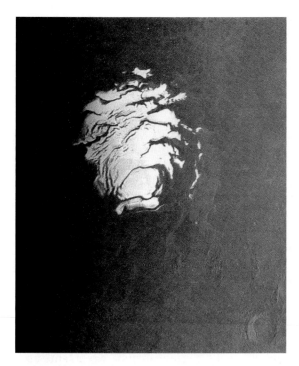

FIGURE 6-17
The south polar ice cap of Mars, seen from orbit. This summer view shows the small permanent cap that probably consists of a mixture of frozen H_2O and CO_2. It is roughly 360 km across. In winter, condensation of CO_2 frost expands the cap to 10 times this size. Night side of the planet is at the left. (SOURCE: NASA *Viking Orbiter* photograph, courtesy of Tammy Ruck and Larry Soderblom, U.S. Geological Survey.)

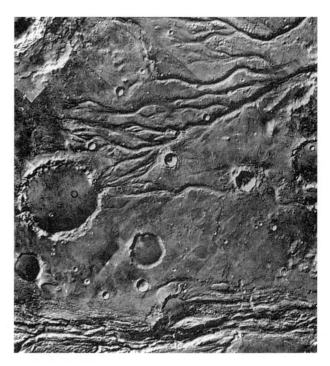

FIGURE 6-18
Ancient riverbeds on Mars. The area is about 180 km (110 mi) wide and drops about 3 km from left to right, in the direction of flow. Water apparently cut through some older craters, but predated other craters. (SOURCE: NASA *Viking Orbiter* photograph.)

in the western United States. Other riverbeds appear to have gathered their water from widely dispersed tributaries, suggesting rainfall on ancient Mars. A few researchers have mapped areas they believe contain glacial features and even shoreline deposits of long-vanished seas! In 1998, the *Mars Global Surveyor* orbiter showed that the topography dropping off from the Martian uplands onto the smooth lowland plains is remarkably similar to profiles dropping from Earth's continents onto smooth seafloor sediments. The issues of how much liquid water existed on Mars, and whether there were once lakes and seas, are still controversial. Ongoing research uses orbiting instruments to search for massive deposits of minerals such as carbonates and salts, which are often left on the surface after lakes evaporate.

In general, the images and spectral data from orbit show that Mars must have had a different climate or surface conditions in the past, involving running water and possibly warmer temperatures. Until 1976, we had only indirect evidence about the surface of Mars. There is a limit to what we can deduce, even from a spacecraft in a tight orbit around the planet. Direct evidence required the landing of instruments onto the surface of Mars. Although the former Soviet Union crash-landed the first human probes on Mars, the first successful landers were two U.S. spacecraft that touched down in 1976. *Viking 1,* described in the opening paragraphs of this chapter, was followed a few weeks later by *Viking 2*.

Mars remained unvisited by human machines for 21 years. Finally, on July 4, 1997, the *Pathfinder* lander and its small rover created a media sensation and introduced a new generation to Mars. All three probes sent back many images and measured the chemistry of the soil, and the *Pathfinder* probe measured the chemistry of six boulders. The pictures revealed a desolate beauty, with apricot-colored skies tinged by pink airborne dust and a surface of sand dunes and scattered boulders (Figure 6–19). *Pathfinder* photographed several small dust devils, or small twisters, on the horizon. The soil was sterile, with no signs of plant or animal life. *Pathfinder* was targeted to land on a plain at the mouth of a Martian river channel, and its data offered some support for the idea that rocks in the channel had washed down from the highlands onto the plains. There are close terrestrial analogs for this type of terrain (Figure 6–20).

A second type of direct evidence is sitting right under our noses. Scientists have catalogued many rocks from all over Earth as meteorites—rocks which fall from interplanetary space. Among these are a handful of unusual samples that did not seem to fit among other meteorites. Scientists made a surprising discovery during the 1980s: These rocks are from Mars! The best proof is the fact that the gases trapped in some of these rocks exactly match the atmospheric composition measured by the Viking landers on Mars, in terms of both the mixture of chemicals and the proportions of different isotopes of each element. We believe that these rocks were blasted off Mars when asteroids hit Mars and excavated craters (Figure 6–21). The fragments drifted through space

FIGURE 6-19
The desert landscape of Mars. At the Mars *Pathfinder* site, the twin peaks are hills on the horizon about 1 km away. Windblown dust has made a bright deposit on one side of the left peak. (SOURCE: NASA.)

believe they were blasted out of the large lava plains that cover much of Mars. This evidence of 1.4-billion-year-old lava confirms that Mars has had major volcanic eruptions in the last third of its history. We know that another Martian meteorite is 4.5 billion years old; it must date from the oldest parts of Mars' surface. This old rock is rich in carbonates that were probably deposited by evaporating moisture (in the same way that evaporating water in your tub leaves a carbonate "bathtub ring").

As tools for understanding Mars, these rocks have one big advantage and one big disadvantage. The advantage is that they are essentially free! Compared to the cost of a return mission to Mars, the cost of sending expeditions to the Antarctic icepack is tiny. If we look hard enough, we might find more of these rare samples. But the big disadvantage is that we do not know exactly what geological locations they came from on Mars. Thus, we cannot use them to date or describe specific geological units. For this reason, scientists would still like to land probes on Mars that could return rocks from specific known features, such as the channels, volcanoes, and polar regions.

The radical change in climate on Mars, from an ancient planet with flowing water to the modern planet of arid, frozen deserts, raises two fundamental questions about Mars that have implications for planets in general. With the evidence we have in hand, we can attempt to answer these questions: What caused the apparent ancient climate variations? and Could ancient Mars, with its liquid water, have supported life?

for hundreds of millions of years before landing on Earth. Martian rocks are rare. Of the thousands of meteorites that have been recorded, only about a dozen of these rocks are believed to come from Mars. Most of the few known Martian rocks were discovered on the Antarctic icepack, where extraterrestrial material is well preserved and easy to identify.

The Martian meteorites are lava rocks. Most of them formed about 1.4 billion years ago, as measured by the technique of radioactive dating described in Chapter 5. Scientists

Martian Mystery Number 1: Why Did the Climate Change?

The discovery of ancient riverbeds on Mars raises an exciting question about Mars and our own planet. For ancient rivers to exist, the climate must have been different in the past. Why did the climate change? What can we say about **climate change on Mars?** Liquid water cannot exist on present-day Mars. The air pressure today is so low in most areas that liquid water is not stable. A pan of liquid water would either evaporate immediately (or actually boil away at low temperature,

FIGURE 6-20
Terrestrial analogs to Martian landscapes occur in very arid regions with a history of weathering. **(a)** Glacier-topped boulders on a volcanic plain in Iceland. **(b)** Basaltic lava at 10,000-ft altitude on Mauna Loa volcano in Hawaii. (SOURCE: Photographs by WKH.)

FIGURE 6–21
Blowing rocks off Mars. An asteroid impact onto one of Mars' lava plains ejects boulders and dust from the planet. Some of the ejected rocks eventually reached Earth as meteorites and were identified as containing tiny amounts of Martian air. (SOURCE: Painting by WKH.)

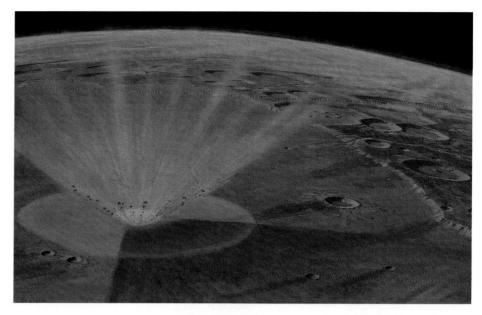

just as it would do 100,000 ft above Earth) or freeze. If substantial amounts of liquid water existed in the past on Mars, it probably means that the air pressure was greater at that time and the temperatures may have been warmer.

No one knows why the Martian climate changed. The number of impact craters superimposed on the riverbeds implies that many or most of them are probably 1 billion years old or more. Chemical measurements suggest that during the early history of Mars, volcanoes probably emitted much more CO_2 than is present today—enough to make an atmospheric pressure as much as 30 percent or more of ours, rather than the present-day value of only 0.7 percent. Thus, Mars may have had a denser atmosphere early in its history, which was slowly lost as gases leaked into space due to the low gravity.

Scientists have proposed several theories to account for climate change on Mars. Some theories suggest that the "wet period" may have been transient and minor. Localized geothermal activity might have melted ice on Mars and created short-lived rivers in those past ages, rather than global wet conditions. Other theories, which are based on analyses of Mars' orbital motion, note that the tilt of Mars' polar axis and ice caps toward the Sun is quite variable. Under some conditions, the tilt reaches as much as 46° or more. Perhaps during brief ancient periods, when the pole was tilted farthest toward the Sun, the polar ice melted completely during summer. This melt-off would have released large amounts of water and carbon dioxide into the air, possibly creating rivers and a thicker, warmer atmosphere. A moderate CO_2 greenhouse effect may have increased temperatures and perhaps even aided water flow. A solution to the mystery of the ancient Martian rivers and the ancient climate will probably require future missions to Mars, and it could tell us a great deal about planetary climate evolution.

Embedded in our language is an idea that a planet's environment is unchanging. For instance, we speak of "terra firma," and Shakespeare wrote that Earth's a stage and we are merely players on it—the metaphor is that Earth is a forever-fixed backdrop. Now we know from planetary science that a planet's climate is not fixed forever, and the continents move. A global change in Earth's climate occurred when primitive life evolved and put large amounts of oxygen into the atmosphere. We also know that Earth's climate has changed during ice ages and perhaps under the influence of human technology. Mars' climate has changed even more dramatically. Perhaps planets are not such stable backdrops to life as humans have assumed; the planet itself may be an active player in the story.

Martian Mystery Number 2: Life on Mars?

The second major mystery of Mars involves the possibility of **life on Mars.** Evidence from Mars allows us to address the profound question of whether we are alone in the universe. All research to date suggests that life can arise only if liquid water is available. Laboratory experiments show that when water is available, the large organic molecules and basic building blocks of life form quite easily from ordinary chemical reactions. Therefore, if ancient Mars had liquid water, did life arise there? If not, why not? If it did, how advanced did it get? Where is it today? Did it become extinct? Or is it somewhere out of sight?

By the time the Viking landers were built, scientists suspected that Martian life might exist not as large organisms, but as microscopic bacteria or other simple organisms in the soil. Therefore, both Viking landers contained equipment to make chemical tests of the soil (Figure 6–22). The first test was to look for organic molecules in the soil. If there is life on Mars, then living or dead organisms should create a residue of organic molecules in the soil, as on Earth. None were found, to an accuracy of a few parts per billion. In other words, the Martian soil is sterile.

Such an emphatic result might seem to rule out any life on the planet. However, Viking scientists had another trick

FIGURE 6-22
The Viking landers scooped up soil for chemical analysis inside the spacecraft, in a search for organic material. Here the *Viking 2* lander takes a sample of the soil. No organic material was found. This experiment occurred on October 9, 1976. (SOURCE: NASA.)

The possible discovery of past life on Mars electrified the scientific community, and it quickly became one of the biggest and most controversial news stories of the year. Carl Sagan once said on the subject of life beyond Earth, "Extraordinary claims require extraordinary evidence." We have seen in Chapter 1 that claims of UFOs as alien visitations are not supported by this level of evidence. How does the claim for ancient Martian life stack up?

Although the odyssey of ALH84001 from Mars to the Antarctic ice cap seems extraordinary, we know that the gas trapped within the rock matches the atmosphere of Mars, not Earth. There is almost no doubt that the rock is from Mars. But what about the organic compounds in the rock? Perhaps they reflect contamination of the rock by terrestrial chemicals, which seeped in after the meteorite arrived on Earth. The chemicals are not generally concentrated toward the surface of the rock, as usually results from outside contamination. After the 1996 pronouncements, detailed measurements showed that some of the organic material may represent terrestrial contamination, but some of it may still come from Mars.

At this point the evidence gets less convincing. The 1996 NASA team noted curious microscopic structures, and suggested that these may be fossils of actual Martian microbes. Examples of these possible alien fossils are seen in Figure 6-23. In the view of some researchers, microbial life may have evolved on Mars billions of years ago and died out later, or perhaps even persisted in primitive form in deep rock layers on Mars. Other teams have found organic concentrations in at least one other Martian meteorite. It sounds good, so why do many planetary scientists consider the evidence for ancient Martian life still controversial?

Part of the concern stems from the difficulty of interpreting the chemical evidence. There are many complex chemical reactions that do not involve living organisms. So it is difficult to *prove* that chemical traces were caused by a living organism. Another concern is the interpretation of the "fossil" structures. The forms in Figure 6-23 are much smaller than normal terrestrial bacteria, although they are similar in size to bacteria that exist in underground or nutrient-poor environments on Earth. Yet the evidence is ambiguous. The picture does not show the telltale signs of cell walls. Moreover, some scientists have argued that only traces like these would be left when a rock is subjected to extreme forces and sudden heating—which must have occurred when the rock was blasted off the surface of Mars. The argument rages on.

The issue of life on Mars illustrates how science works. While trying to unravel the mystery of ALH84001, we can reach some conclusions with near certainty (the rock is from Mars), and we cannot reach other conclusions with much certainty at all (did microbial life exist in the rock?). Most of the key evidence—the chemical traces, the fossil forms—has more than one possible explanation. Other future tests will involve better chemical and isotopic analysis of the composition of the organic molecules. These tests might show more conclusively whether the chemical traces were created by living matter. Many research groups are now working

up their sleeves. They included three more experiments on each lander that scooped up soil, "fed" it nutrients, and watched for signs of metabolic activity (such as plants growing and releasing carbon dioxide). All three experiments did detect some unexpected chemical activity! However, careful analysis indicated it was probably not from living organisms, but from unusual chemical conditions in the soil that relate to an interesting fact: There is no ozone layer on Mars. This means that strong ultraviolet light from the sun reaches the surface and destroys any exposed organic molecules. (Remember that Earth is protected by its ozone layer.) The same ultraviolet radiation may help create unusual oxides that caused the chemical reactions in the Viking soil experiments.

Other scientists suggested that the soil within a meter or so of the surface might not be the best place to look for life on Mars. Annual storms stir up the surface dust and blow it around on Mars, effectively sterilizing the whole surface layer of the planet. A similar process occurs in Antarctica, where icy winds sterilize the soils. There, tiny life-forms live not in the inhospitable soil, but in fractures inside rocks. The rocks offer an environment protected from the environmental extremes of the exposed surface. Thus, in the years after the Viking experiments, scientists speculated that life—or life's remnants—might be found hidden in subsurface layers or other protected environments of Mars as well.

As we saw in the story that opens Chapter 1, this story took an exciting turn in 1996. NASA scientists in Houston, Texas, were studying one of the oldest Martian meteorites, a Martian rock more than 4 billion years old. This rock—cataloged as ALH84001—had deposits of carbonates inside fractures in the rock, where liquid water had percolated. The carbonate contained concentrations of organic molecules as well as certain minerals that are produced on Earth by microbes that thrive in oxygen-poor environments. Were these features produced by nonbiological chemical processes on Mars, or could they have been caused by forms of life?

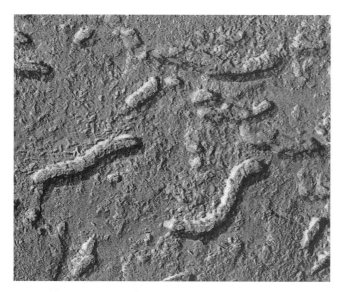

FIGURE 6-23
Alien fossils? The microscope images show tiny structures that some researchers suggest may be fossil remnants of Martian microbes. The tubelike forms are only 20 to 100 nanometers (nm) (20 to 100 billionths of a meter) long and are shaped like some types of terrestrial microbes, although on average they are much smaller than most terrestrial cells. They were found inside carbonate minerals rich in organic compounds inside a rock from Mars. Whether they are products of life is still controversial; they may represent stages of organization of organic matter prior to life's origin. (SOURCE: Everett Gibson, NASA.)

with fragments of the few Martian rocks we have. We may never be sure until we go back to Mars for more evidence.

So what has humanity learned about the question of life on Mars? First, the theories of a century ago have been ruled out. There are no Martian civilizations, nor are there any large animals or plants. Second, although biological evolution has produced incredible adaptations to a range of conditions on Earth—from the dark, cold seafloor to hot geothermal pools in Yellowstone Park—it apparently was not able to evolve advanced forms under the surface conditions on Mars. On the other hand, during the moister early history of Mars, biochemical evolution may have produced microbial Martian life, which may have died out since then. Most of the research on Mars over the next few decades will be aimed at confirming or refuting this hypothesis. If it is confirmed, we will have exciting proof that life can evolve on other worlds. On the other hand, if we find that life never existed on Mars, that may help us refine our estimates of the prevalence of life elsewhere in the vast universe.

Future Exploration of Mars

In the late 1980s, President George Bush announced an ambitious goal for NASA to send humans to Mars in the next few decades. The former Soviet Union, too, with its large space station, seemed to be inching toward human exploration of Mars. However, these plans were never well funded, and they languished with the collapse of the Soviet Union. In the 1990s, President Bill Clinton determined that no major effort to land humans on Mars would be pursued for the next decade, although human exploration is not ruled out as a long-term goal.

Therefore, the United States and various international partners are pursuing vigorous plans to send various unstaffed probes to Mars in the next decade. These missions will send cameras, spectrometers, and geochemical instruments to gain more data about the climatic history of Mars. They will follow on from the spectacular success of the relatively cheap Mars *Pathfinder,* which sent a small vehicle roaming over the surface of Mars in 1997. Other missions from the Japanese and the United States are scheduled for 1999, 2001, and 2003. Russia also planned major Mars missions, but the launch failure of its international Mars probe in 1996 devastated that nation's financially strapped planetary program. Several more missions are proposed by the United States, Europe, and Russia to set up a network of weather and seismic instruments to monitor Martian conditions. An attempt to return rock and soil samples is still in the discussion stage. Such a mission would allow us to get exact dates and geochemical histories for different geological features, and would allow a powerful new test for ancient life on Mars.

The Two Small Moons of Mars

The two small moons of Mars would make interesting places to visit by a future expedition on its way to the surface of the red planet. The two moons are named **Phobos** and **Deimos** (Fear and Trembling, the names of the horses who drew the chariot of Mars, god of war). The moons are shown in Figure 6–24. Both are small, dark, potato-shaped chunks of rock, heavily cratered by meteorite impacts. Phobos, the inner moon, is 27 by 19 km (17 by 12 mi). Deimos, the outer moon, is smaller, only 15 by 11 km (9 by 7 mi). As moons, they would be somewhat disappointing to an observer on Mars, because of their small size. From the surface of Mars, Phobos would look only a bit larger than one-third the size of our Moon, and Deimos would look even smaller.

The origin of Deimos and Phobos is uncertain, but many astronomers suspect that they started out as asteroids in the nearby asteroid belt and later got captured into orbit around Mars. The black color is believed to come from carbon-rich material, similar to that found on certain asteroids. This dark color is strikingly shown in Figure 6–25, a photograph from a Russian spacecraft that caught Phobos silhouetted in front of the red planet. A mission to gather data on these moons would help us figure out their chemical composition. This in turn might tell us whether mining asteroids would be a good idea.

The moons of Mars are the first two objects we have encountered in the solar system that are not perfectly round. Why? The answer lies with gravity and the idea of equilibrium. In Chapter 4, we discussed the idea of thermal equilibrium—the tendency of heat energy to flow so that temperature differences are equalized. Equilibrium applies to gravity, too. Any part of a planet's surface has a gravitational potential energy (for a review, see the Science Toolbox in Chapter 4). This potential energy increases with the mass of

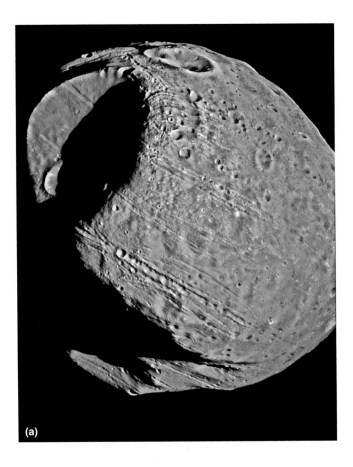

the planet and with the distance from the planet's center. Gravity pulls all the parts of a large object toward its center of mass. As an everyday example, this is why the cereal in a box settles when you shake it. The cereal is trying to get as close to the center of Earth as possible! Gravity always works to equalize the potential-energy differences in a planet. So the high places with larger gravitational potential energy are pulled downward more strongly than the low places. Take a large tray, and cover it with sand or gravel heaped high into "mountains." If you shake the tray, the sand will quickly settle to a flat surface where the gravitational potential energy is the same everywhere. This is equilibrium.

So what does this have to do with Phobos and Deimos? Large bodies in the solar system have sufficient mass that gravity can force their surfaces to have the same potential energy everywhere. The result is the most symmetric shape possible: a sphere. As we saw in the last chapter, Earth is as round and very smooth. The largest surface features are a small fraction of Earth's size. Small moons like Phobos and Deimos, however, do not have strong enough gravity to overcome the strength of the rock they are made of. So they can have irregular shapes.

Comparative Planetology: Rules of Planetary Evolution

In the 1970s, as space probes sent data to us about other planets, researchers developed the concept of **comparative planetology.** According to this concept, we no longer study each planet as a unique world, but rather compare one with another and look for unifying principles. The comparison of

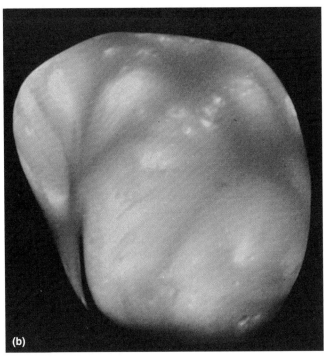

FIGURE 6-24
The moons of Mars. **(a)** Phobos is 28 km long. The impact that created the largest crater, shown at the top, almost shattered the satellite; fractures can be seen extending from it. **(b)** Deimos is 16 km long and has a smoother surface texture than Phobos, perhaps because of a longer duration of micro-cratering since the last major impacts. The older craters appear softened by such sandblasting. (SOURCE: NASA.)

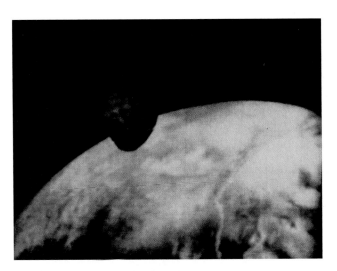

FIGURE 6-25
The black color of the moons of Mars is emphasized by this color view of Phobos, hanging in front of ocher-colored Mars. The black color probably indicates carbon-rich compounds. The photograph was made from the Russian probe, *Phobos-2,* in 1989. (SOURCE: Courtesy of B. Zhukov, Institute for Cosmic Investigations, Moscow.)

greenhouse warming on Venus, Earth, and possibly Mars is one example. While it is fun to think of the planets as having distinct "personalities" (and you may want to remember a few distinctive aspects of each one), it is scientifically more powerful to look for common features. Nature has performed various experiments for us, placing planets of different sizes at different distances from the Sun. We compare the results of the experiments to learn more about how planets work in general. Based on the material in the last two chapters, we can formulate some simple rules of thumb that help us understand planets.

1. Older surfaces have more impact craters.
2. Bigger planets have greater internal heat and more geological activity.
3. Bigger planets have stronger gravity and retain more atmosphere.
4. CO_2 is the "normal" composition for the atmosphere of a terrestrial planet.
5. Free oxygen means something strange is going on.

These rules are all based on simple physical principles related to gravity and chemistry and the kinetic theory of matter. Since we have evidence that these physical properties are universal, it is a good bet that the five rules apply to planets beyond the solar system. The exciting idea that we can predict the properties of yet undiscovered planets is an example of the long reach of the scientific method. Let us look at these five rules more closely.

Older Surfaces Have More Impact Craters

We have already met with this rule in the last chapter. The idea is simple. The longer a surface has existed in the solar system, the more often it has been hit by interplanetary debris. By looking at different worlds, or different parts of the same world, we learn whether the surface is ancient or young, and thus whether the world is geologically dead (ancient surfaces) or geologically active (young surfaces). For example, counts of impact craters show us that Earth has the youngest surface features and greatest degree of activity, Venus has the next-youngest surface, Mars is intermediate, and Mercury and our Moon have ancient, dormant surfaces. Remember that we are not talking about the age of the planet in general, but about the characteristic age of its visible surface features. All the planets formed 4.6 billion years ago.

Counting craters leads us to revisit the idea of uncertainty in measurement. We believe that cratering is caused by an essentially random process: impacts caused by interplanetary debris. Cratering is random in time and space—the craters on a planetary surface have no pattern. As craters accumulate, the variations depend on the square root of the number of impacts. (See the following Science Toolbox.) To get an idea of how counting statistics work, take a well-shuffled deck of cards and deal 13 cards. Consider a face card to represent a crater of a certain size. You know that *on average* the 13 cards should include 3 face cards. But what are the actual numbers in the hand you dealt? Reshuffle the deck, and deal 13 cards a number of times, each time noting the number of face cards. Figure 6–26 shows the experiment. The distribution of the numbers of face cards and numbered cards is an illustration of random counting statistics. In the same way, on several surfaces of the same age, there may be some random variation in crater numbers, even though the numbers generally increase as the surface grows older.

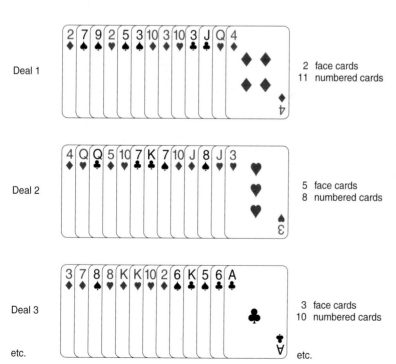

FIGURE 6–26
An illustration of counting statistics. Deal 13 cards from a well-shuffled deck, count the face cards (jack, queen, king) and the numbered cards; shuffle and repeat the experiment. There are 12 face cards and 40 numbered cards in the deck, so dealing out one-fourth of the deck should give you 3 face cards and 10 numbered cards on average. But what is the distribution of actual numbers when you do the experiment? How do the numbers vary if you count aces, for example, where you only expect one to turn up every deal?

SCIENCE TOOLBOX

Counting Craters on the Moon

In Chapter 1, we looked at some simple but important aspects of statistics. We saw that every measurement is made up of a number, a unit of measurement, and an uncertainty or observational error. The uncertainty is an integral part of the measurement—there is no such thing as a perfect measurement in science. We also saw that the random error or scatter in measurements goes down as the number of independent measurements goes up. In the Chapter 1 Science Toolbox on errors and statistics, the standard error is proportional to $1/\sqrt{N}$.

If we are just counting things, there are no units because a count is a pure number. Simply,

$$\text{Count} = N$$

If the items being counted are governed by a random process, the counting error is given by

$$\text{Random error} = \sqrt{N}$$

As noted in Chapter 1, the word *error* is a bit misleading since no mistake has been made. *Uncertainty* is a better term, but we will stick to the standard terminology in this book. The fractional error is therefore

$$\text{Fractional error} = \frac{\text{random error}}{\text{count}}$$
$$= \frac{\sqrt{N}}{N} = \frac{1}{\sqrt{N}}$$

The correct way to quote a counting measurement with the error attached is

$$\text{Count} = N \pm \sqrt{N}$$

Counting statistics can apply to microscopic phenomena. For example, if a sample of radioactive material yielded 285 decays within a fixed interval of time, the error in this count would be $\sqrt{285} = 17$. We quote this measurement as 285 ± 17. The fractional error is $1/17 = 0.059$, or about 6 percent. If we waited 10 times longer, the radioactive material might yield 2792 decays with a statistical error of $\sqrt{2792} = 53$. We quote this measurement as 2792 ± 53. The fractional error is now $1/53 = 0.019$, or about 2 percent. Notice that as we accumulate 10 times more events, the fractional error has reduced by a factor of about $\sqrt{10} = 3.2$ from 6 to 2 percent. This is directly analogous to the standard error of multiple observations reducing by a factor of $\sqrt{10}$ if we collected 10 times more observations.

We can see what the rules of thumb are for random counting errors. With 100 events, the statistical uncertainty is $\sqrt{100} = 10$, or a 10 percent accuracy. With 10,000 events, the statistical uncertainty is $\sqrt{10,000} = 100$, or a 1 percent accuracy. With 1,000,000 events, we have a 0.1 percent accurate measurement, and so on. For a very small number of events, the uncertainty becomes large. For a single event, $\sqrt{1} = 1$, and the mathematics confirms what we

(continued)

100 km

FIGURE 6-A
(a) A heavily cratered portion of the highlands of the Moon. Use a ruler to measure the diameter of craters in millimeters; then group your counts into size ranges and plot a histogram of the distribution. Are there equal numbers of craters in equal-size intervals? If you divide the image into four quadrants and count craters in each separately, how do the results compare?
(b) A similar-sized region of the Moon that has been resurfaced by volcanic activity. Notice the much smaller number of craters.

SCIENCE TOOLBOX

Counting Craters on the Moon (continued)

already know: We can learn very little from a single event!

Science often operates at the limit of counting errors. Astronomers collect "particles" of light called photons (which will be discussed in Chapter 10) with telescopes and try to accumulate enough counts to drive down the errors as just described. We get plenty of light from the planets, so this limitation of light detection rarely applies in the solar system. However, it becomes important as we explore the distant universe later in the book.

Remember that these counting errors only apply to events governed by a *random* process. There are nine planets in the solar system, but we would not quote this measurement as 9 ± 3, implying that sometimes we might count 11 or 6 planets. We have located all the planets in space, and there is no random uncertainty as to how many are out there.

Let us apply counting statistics to the material in this chapter. Suppose we try to establish the representative number of craters in a unit area, such as 1 square kilometer, on a particular geological region of the Moon, such as the lava plains.

Figure 6–A*a* shows a portion of the heavily cratered lunar highlands. Now try counting craters on this image to see if the statistics are random. Pick the size of the smallest crater that you can reliably measure from the photograph. Now divide the image into four quadrants, and count all the craters in each quadrant. We can call these numbers N_A, N_B, N_C, and N_D. You will find that they are not equal. However, the differences between them should not greatly exceed the counting errors, given by $\sqrt{N_A}$, $\sqrt{N_B}$, $\sqrt{N_C}$, and $\sqrt{N_D}$. Check that this is the case. By way of contrast, look at a similar-sized region of the lunar lava plains in Figure 6–A*b*. Geological processes can create a young surface where the smaller number of craters reflects a shorter cratering history.

You can also see in Figure 6–A*a* that there are many more small craters than big craters. This means that the fractional counting error for small craters is smaller than for large craters. The *distribution* of crater sizes tells about the nature of interplanetary debris. There will be more small interplanetary bodies hitting the Moon than big ones. We will revisit the subject in Chapter 8.

Bigger Planets Have Greater Internal Heat and More Geological Activity

This rule of thumb explains why Earth's surface is young and other terrestrial planet surfaces tend to be older. The rule follows from the principles of heat transfer discussed in Chapter 4. One type of heating applies to all planets, regardless of their size. Soon after it formed, each planet was partially heated by the kinetic energy of countless meteorites crashing into it. These meteorites were left over from the formation of the solar system.

In general, planets produce heat according to their size. Radioactive atoms decay in their interiors, and conduction and convection transport this heat to the surface. All this heat is transported from the interior region to the surface. Bigger planets have greater gravity, and the energy release due to radioactive decay creates a molten interior that can drive geological activity. Also, the bigger the planet, the longer it takes this heat to get out. There is no mystery here. If you pull a large rock and a small rock out of a campfire, the small one will cool off faster, while the big one will stay warm. Planets are the same.

This simple idea tells us a lot about planetary evolution, since the geological activity on planets is driven by heat from the interior. A small planetary body cools off very quickly, and its interior produces little heat because there is less radioactive material. The surface of a small planet will show countless crater scars from the impacts that happened throughout geological time. On the other hand, a large planetary body retains its heat, and its interior produces a lot of heat from the large mass of radioactive material. This heat will drive volcanism and even plate tectonics, which constantly reshape the surface. The surface of a large planet will be young and show fewer impact craters.

Note that this general rule has nothing to due with solar heat reaching the planet from outside. The Sun is the dominant heat source only for the meter-thick layer at the surface: This layer warms in the day and cools off at night. But the Sun's heat does not penetrate more than a few meters (or yards) into the planet. As you may know, the inside of a cave stays about the same temperature all day long and even all year long, because the cave is insulated by the surrounding rock layers. So is the inside of a planet. The heat inside a planet that melts rock and creates volcanic lava comes almost entirely from the heat sources in the planet itself, not from the Sun.

The rate and mode of heat transfer from the center of the planet to the surface control the rate of geological activity on the surface. If the planet is small and has cooled off, then the center is not much warmer than the surface, and the heat will be carried outward by the process of conduction in the mantle. However, if a planet is large and has a lot of energy produced by radioactive decay, the center is still much hotter than the surface. There will be a strong temperature difference between the center and surface which favors convection in the mantle, or sluggish mass movement of the hot, slightly plastic material. (If you have trouble with these ideas, review the discussion of heat transfer in Chapter 4.) As we saw in the last chapter, the hot, inner rock layers of a planet's mantle can actually flow slowly. Hot plumes of convecting material rising in the mantle can create "hot spots" under the lithosphere. Such hot spots produce volcanoes. The Hawaiian Islands are a terrestrial example, and the giant volcanoes on Venus and Mars may also be located over hot spots created by convection. A smaller planet with a colder interior has less internal heat to create convective movement in the mantle, and thus less internal energy to drive seismic activity, tectonics, or volcanoes.

Thus, simple ideas of heat generation and transfer inside planets allow us to understand the differences in geological features from one planet to another and the origin of volcanoes on the larger planets. The features caused by volcanism and plate tectonics can be seen in Figure 6–27. If we list the terrestrial planetary bodies in order of increasing size, we see exactly the kind of progression mentioned in these first two general rules:

Deimos	Cold throughout	Heavily cratered
Phobos	Cold throughout	Heavily cratered
Moon	Mostly cold	Heavily cratered, some lava flows (3.5 billion years old)
Mercury	Mostly cold	Heavily cratered, some old flows (3.5 billion years old?)
Mars	Warm interior?	One-half of the planet cratered; one-half is covered by young lava flows (1.4 billion years old?). Several very young volcanoes (0.5 billion years old, possibly still active?)
Venus	Still hot	Few impact craters. Interior with intense volcanism, lava-covered surface (mean age 0.7 billion years)
Earth	Still hot	Few impact craters. Interior with active volcanism, plate tectonics (mean age of crust 0.4 billion years)

Bigger Planets Have Stronger Gravity and Therefore Retain More Atmosphere

This rule is self-explanatory. As Newton discovered in the 1600s, all planets have gravity, and the gravity increases as the mass of the planet increases. Remember that according to the kinetic theory of matter, all gases in a planet's atmosphere are composed of atoms and molecules that are in motion.

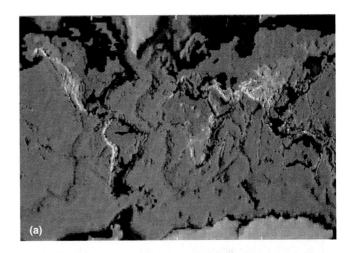

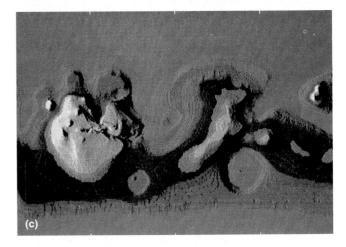

FIGURE 6–27
Comparison of geology—Earth, Venus, and Mars shown on maps with a similar topographic scale. Highest regions are yellow, lowest regions are blue. The altitude scale in kilometers (shown by the color bar) is the same in all three cases. Earth and Venus are similar in size, but Mars is one-fourth the size of Earth. Each of these planets is massive enough for interior heat to drive volcanoes and movements of the crust. **(a)** Earth has continental blocks with raised mountain chains where plates collide. **(b)** Venus has lesser development of continental masses. **(c)** Mars shows a more primitive surface, with a few impact basins and giant volcanic peaks.

(SOURCE: NASA; Jet Propulsion Laboratory.)

If the atoms or molecules near the top of the atmosphere find themselves moving in an upward direction faster than the escape velocity of the planet, they will escape into space, never to return. Thus, planet atmospheres are continually and slowly leaking away into space.

However, planets with stronger gravity have larger escape velocities, so gas molecules have to move in order to exceed the escape velocity. For example, a molecule moving upward at 4 km/s would easily escape from the Moon, but it would be retained by Earth's stronger gravity. Therefore, a large planet will hold onto its atmosphere longer than a small planet will, all other things being equal. We know also from Chapter 4 that the kinetic energy of a gas molecule is proportional to its mass and proportional to the square of its velocity. So in a mixture of gases at the same temperature or energy, light molecules will be moving faster than heavy molecules. Therefore, a light gas such as hydrogen or helium is more likely to escape into space than a heavy gas such as oxygen. The atmospheres of the three terrestrial planets massive enough to retain gases are shown in Figure 6–28.

Let us examine the terrestrial worlds by size to see if this rule really works. In the following list, the moons and planets

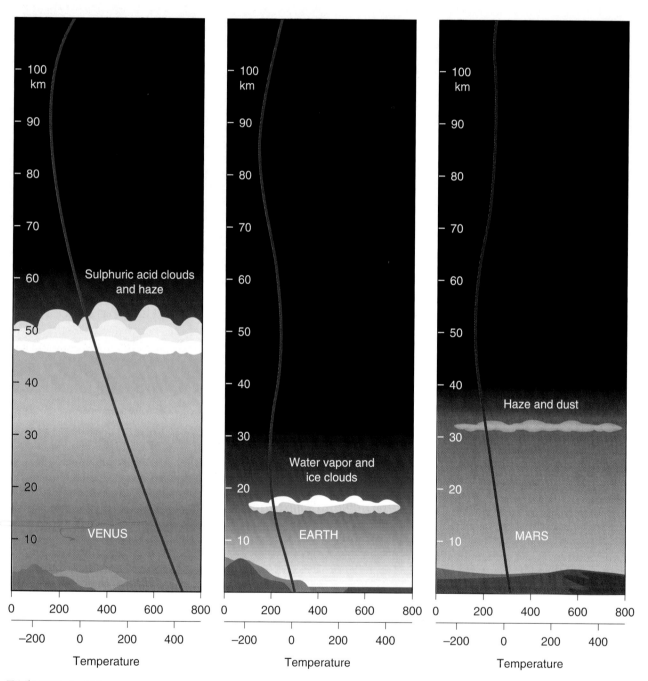

FIGURE 6-28
Comparison of atmospheres—the vertical profile through the atmospheres of Venus, Earth, and Mars. Mercury has almost no atmosphere, so it is not shown here. The chemical components of each atmosphere are listed in Table 6–1; Venus and Mars have atmospheres that are dominated by CO_2, and Earth's atmosphere is mainly N_2 and O_2.

TABLE 6-1
Atmospheric Compositions for Venus, Earth, and Mars

VENUS		EARTH		MARS	
GAS	PERCENT BY VOLUME	GAS	PERCENT BY VOLUME	GAS	PERCENT BY VOLUME
CO_2	96.5	N_2	78.1	CO_2	95
N_2	3.5	O_2	20.9	N_2	2.5
SO_2	0.015	H_2O	0.05–2	Ar	1.6
H_2O	0.01	Ar	0.9	CO	0.6

are listed in order of increasing size, and the atmospheric pressure at the planet's surface is indicated in units of bars, which are defined such that 1 bar equals the air pressure at sea level on Earth.

Deimos	No atmosphere
Phobos	No atmosphere
Moon	No atmosphere
Mercury	Essentially no atmosphere—slight concentration of some gases
Mars	Thin CO_2 atmosphere (0.006 bar)
Venus	Massive CO_2 atmosphere (90 bars)
Earth	Moderate N_2 plus O_2 atmosphere (1 bar)

Everything checks—massive planetary bodies have atmospheres, and less massive planetary bodies do not. The atmospheres are composed of heavy molecules. The only exception is Venus, which is slightly less massive than Earth but has a much thicker atmosphere. What accounts for this anomaly? We will see the solution in our next rule.

CO_2 Is the "Normal" Composition for the Atmosphere of a Terrestrial Planet

At first sight, this rule seems wrong. The Moon and Mercury have no atmospheres, Mars and Venus have CO_2 atmospheres, and Earth has a nitrogen and oxygen atmosphere with very little CO_2. Table 6–1 lists the atmospheric constituents. But let us look more closely at the physical conditions on each planet. All the terrestrial planets (including the Moon) started out warm enough to have volcanism, as we can see from the lava plains on the Moon, Mercury, Mars, Venus, and Earth. As volcanoes erupt, they emit carbon dioxide and water vapor gases. Thus, the natural by-products of molten terrestrial planet interiors are CO_2 and H_2O gases.

So why do all planets not have CO_2 and H_2O atmospheres? The Moon and Mercury are small and have weak gravity, so they have lost their original atmospheres. It is possible for energetic photons from the Sun to break down H_2O molecules into H and O atoms. Now look at the atomic weights of the atoms and molecules we are dealing with. Hydrogen has a weight of 1, oxygen has a weight of 16, water vapor has a weight of 18, and carbon dioxide has a weight of 44. From our discussion of the kinetic theory of gases in Chapter 4, we know that in a mixed atmosphere of gases, the light atoms or molecules are more likely to exceed the escape velocity than the heavy atoms or molecules. So we would expect H to escape easily and CO_2 to be retained. Let us combine this principle with the chemical properties of gases to explain why the terrestrial planets have such different atmospheres.

On Venus and Mars, the H_2O was lost and the CO_2 molecules were left behind. If water molecules are broken into H and O atoms, the hydrogen quickly leaks away, and the water molecules cannot re-form. The O atoms left behind do not accumulate in the atmosphere because they are so reactive. They oxidize rock minerals and end up trapped in the rocks and soil. This was one mechanism for oxidizing the iron minerals on Mars, producing Mars' red color. Any water on Mars that did not break up and escape was frozen and trapped in the polar ice caps and permafrost layers.

Then why did Earth not follow the same history? The answer is that Earth was at just the right temperature for the H_2O to condense into liquid water and form oceans instead of remaining gaseous. But CO_2 readily dissolves into liquid oceans, forming a weak carbonic acid solution; the carbonic acid reacts with rocks and forms carbonate mineral deposits. In this way, huge amounts of terrestrial CO_2 were removed from the atmosphere and deposited as rocks. There is an elegant test of this explanation for the distinctiveness of Earth. Earth and Venus are similar in size, so they should have produced similar amounts of CO_2. The CO_2 would be in the atmosphere on Venus but trapped in rocks on Earth. Scientists have made inventories of the total amounts of CO_2 trapped in terrestrial rocks in the form of carbonates and other minerals—and the answer is that Earth has almost exactly the same amount as Venus does.

In other words, CO_2 is the "normal" composition for the atmosphere of a terrestrial planet. The details depend on the size of the planet and its distance from the Sun (which controls the temperature of the atmosphere). Earth would have had a thick CO_2 atmosphere of perhaps 90 or 100 bars, as Venus did, except for the fact that the water formed liquid oceans. Mars had gravity too weak to retain more than a thin CO_2 atmosphere. So Earth was left with an unusual atmosphere of only 1 bar, mainly nitrogen (N_2), which would otherwise have been an obscure trace gas. Note that N_2 is a heavy gas with an atomic weight of 28, so it is easily retained by Earth's gravity.

Free Oxygen Means Something Strange Is Going On

While Earth's atmosphere is mostly nitrogen, about one-fifth of it is oxygen. This is very strange, if you think about it. Oxygen is one of the most reactive gases. As noted above, any oxygen left in an atmosphere is likely to react with other materials, for example, oxidizing iron minerals in rocks to form red, rust-colored minerals typical of Mars. This is likely to happen on a geologically short time scale of millions of years. Therefore, if you are flying around the universe in a spaceship and see a planet with lots of free oxygen, you can bet that something must be acting to replenish that oxygen supply.

That "something" is probably life. The living organisms of Earth consume CO_2 and release O_2. Plant life is the main reason that Earth has large amounts of free oxygen. Without plant life, we would have negligible oxygen in the atmosphere. Notice, too, that microbial and plant life evolved on early oxygen-poor Earth. After these organisms built up the oxygen supply, animals evolved who could use oxygen to produce the energy for their biological processes. We will discuss the evolution of life on Earth in the last chapter of this book. We will also see that we are close to the technology that would allow us to detect planetary atmospheres that have been modified by life processes. The search for life beyond the solar system is one of the most exciting prospects in science.

Summary

The inner solar system has four terrestrial planets of rocky and metallic composition. Mercury is small and Moonlike. Its surface is heavily cratered, and its gravity is too weak to retain an atmosphere. Venus is roughly Earth's size. Our sister planet has internal geological activity, which has resurfaced the planet and concealed most of the evidence of bombardments from space. Venus has an extremely dense atmosphere of CO_2 that causes an intense greenhouse effect, which heats the barren, volcanic surface to hundreds of degrees.

Mars has the most Earthlike environment, but it is a cold, frozen desert with only a thin atmosphere of CO_2. Mars has abundant water trapped in three forms: polar ice, underground permafrost, and water molecules inside hydrated minerals. In spite of the fact that there is no liquid water now, dry riverbeds and possible lakeshore features show that ancient Mars at least temporarily had a climate that allowed liquid water to flow across the surface. Two mysteries about Mars remain unsolved: the cause of the Martian climate change and the question of whether life evolved on Mars. Possible evidence of ancient microbes has been reported in certain rocks from Mars, but this evidence needs to be better understood. If it is confirmed, scientists want to know how far life evolved on Mars and whether it might still exist under the surface.

Differences among the terrestrial planets can be explained in terms of some general principles. These principles should apply to other planetary systems as well. Larger planets hold heat longer and have more energy produced by radioactive decay in their dense cores. Therefore, they have more active volcanism, younger surface features, and fewer impact craters. Larger planets also have stronger gravity that is able to retain atmospheric gases. Smaller planets are likely to be geologically dead and have heavily cratered surfaces. On smaller planets, the gravity is weak and the gases escape into space. Volcanic gases emitted from the larger planets are rich in CO_2 and H_2O. The H_2O evaporated and was lost on Venus, remained liquid on Earth, and froze on Mars. The CO_2 became the main atmospheric gas on all three planets, though on Earth the CO_2 dissolved in the oceans and oxygen was added by plants.

Important Concepts

You should be able to define these concepts and use them in a sentence.

terrestrial planets	greenhouse warming of Venus
Mercury	
Venus	volcanism on Venus
Mars	life on Mars
polar ice caps of Mars	Phobos
riverbeds on Mars	Deimos
climate change on Mars	comparative planetology

How Do We Know?

These questions and answers show how the scientific method is used to learn about the universe.

Q How do we know that the temperature of Venus is 900°F and that of Mars is mostly below freezing?

A Even without direct evidence, we know that the temperature of a planet with no atmosphere depends on its distance from the Sun. The intensity of the Sun's radiation diminishes with the inverse square of the distance. Before spacecraft landings, astronomers were able to measure the thermal infrared radiation from planets and estimate their surface temperatures. The Russian probes that landed on Venus and the U.S. probes that landed on Mars measured air temperature directly and confirmed these values.

Q How do we know that the greenhouse effect is the cause of the high temperature on Venus?

A Calculations and spacecraft measurements near Venus both confirm that the "normal" temperature of a body at that distance from the sun would be more like 100 to 163°F. Some additional effect must raise the temperature. The thick CO_2 atmosphere of Venus raised suspicions that the greenhouse effect was the culprit. Next, scientists calculated how much radiation would be trapped by an atmosphere as thick as that of Venus if it had a composition primarily of CO_2. These calculations confirm that the global greenhouse heating due to the thick CO_2 accounts for Venus' high temperature.

Q How do we know that Venus is geologically active?

A There are several lines of reasoning. First, we can make an analogy with Earth. Venus is similar in size and mass to Earth, so we expect that it will also have a heavy and molten core that can drive geological activity. We also see trace gases in the atmosphere of Venus which are typical of volcanic activity. The most direct evidence comes from radar mapping, which lets us "see" through the dense atmosphere. Radar mapping of the surface shows clear signs of volcanoes and lava flows and extensive resurfacing of Venus in the past.

Q How do we know that ancient Mars had running water and a more fertile climate?

A The evidence is not conclusive, but one plausible interpretation is that Mars was wet in the past. The widespread existence of dry riverbeds on this now-frozen planet suggests strongly that liquid water once ran on Mars. Also, soil samples tested by Viking landers, and the few meteorites from Mars, show evidence of carbonates and other deposits left when water evaporated. Such evidence shows that Mars had at least temporary periods of surface water. The amount of water involved is highly uncertain. The real question is whether these were only eruptions of local melted ground ice that lasted only days or months or whether they reflect long-term global climatic episodes lasting for millennia.

Q How do we know that the so-called Martian meteorites really came from Mars?

A The dozen or so rocks that are claimed to have come from Mars all have certain characteristics of meteorites—surface features, crystal formations, and mineral compositions that are not found in terrestrial rocks. One was seen to fall out of the sky, so we know they are from space. Most are basaltic lavas around 1.4 billion years old, which is much younger than common lunar rocks or other meteorites (see Chapter 8). By 1980, scientists had posed the question, Where in the solar system could there have been volcanism 1.4 billion years ago? Some researchers suggested Mars. The large numbers of craters on the Martian lava plains suggest that they are about 1.4 billion years old. Finally, the Viking landers measured the composition of the Martian atmosphere, which exactly matches the traces of gas trapped in the rocks.

Q How do we know there was once life on Mars?

A We do not. The issue is still controversial, and every piece of evidence has more than one possible interpretation. Most of the attention has centered on one of the Martian meteorites. This rock shows chemical traces that are consistent with a primitive metabolism at work. The rock also has tiny elongated forms that might be fossils of ancient bacteria. None of the evidence is conclusive. Only one rock has been studied in great detail, and we do not even know where on Mars it came from. To answer this question beyond doubt, we will probably have to retrieve more rocks from Mars.

Problems

Use these problems to test your understanding of the information and concepts in this chapter. The * indicates a more advanced or mathematical problem.

1. Which planet's orbit comes closest to Earth? (*Hint:* Look up the orbital characteristics in the table of planetary data in Appendix B–1 on p. 148.) Give two reasons why Mercury, Venus, and Mars are difficult to see when they are farthest from Earth.
2. In terms of a percentage of Earth's size, how big are Mercury, Venus, and Mars? How would you expect their gravities to compare? What has their size got to do with whether they are geologically active?
3. Which planet has a hotter surface, Mercury or Venus? Why?
4. Venus and Earth are nearly the same size, and early volcanoes on each probably produced primarily CO_2 and H_2 gas (plus some water vapor). **(a)** Why is CO_2 not a major constituent of Earth's atmosphere, as it is on Venus? **(b)** Why is H_2O not a major constituent of either atmosphere?
5. What do numbers of impact craters tell us about the relative ages of the surfaces of Mercury, Venus, Earth, and Mars? List these four in order of increasing average surface age.
6. What qualities of Mars might make operating a long-term base easier on Mars than on the Moon, assuming that the same amounts of initial materials could be supplied?
*7. Based on its distance from the Sun compared to Earth, what would you expect the surface temperature of Mars to be? Use the same reasoning as in the first Science Toolbox in this chapter. For a challenging exercise, calculate the range in distances, compared to the Earth-Sun distance, where the temperature would allow liquid water to exist on a planetary surface.
8. What scientific questions would you try to answer if you were the director of a planned scientific base that would remain on Mars for 10 years? Write a short report with your scientific plan, imagining that you must present it to a skeptical funding agency.
*9. Imagine you were an alien sending robotic landing probes to Earth in order to study it. **(a)** If two probes parachuted through the atmosphere to random spots on the surface (as was the case with the Viking landers on Mars), how well might they characterize the whole Earth? **(b)** How many probes might be needed to give a good idea of all the environments on our whole planet? **(c)** Do a simple statistical test of how many probes it would take to reach a landmass of the Earth. Take a die; consider the numbers 1, 2, and 3 to be water landings and 4 to be an earth landing; ignore 5 and 6. How many rolls of the die—separate space probes sent randomly to Earth—before you roll a 4?
10. Describe in what ways Mars is of special interest in terms of the Copernican revolution and for establishing whether Earth is unique in the universe, or only one planet among many that might have terrestrial climates, life, or even intelligent life. What does Mars show us about the existence of liquid water on other planets? What does it tell us about the probability of finding a climate habitable by humans? What can we learn about the stability of planetary climates? Is life on Earth unique?
*11. Imagine that we discover a star just like the Sun within easy view of our telescopes. **(a)** There is a hypothetical planet around this star that is twice the mass of Earth and is one-third of the Earth-Sun distance from its star. What would the surface temperature of this planet be? Is it likely to have an atmosphere? Geological activity? **(b)** Now consider a second hypothetical planet, at the Earth-Sun distance from its star but only 10 percent of Earth's mass. Answer the same questions asked of the first planet.

Projects

Activities to carry out either individually or in groups.

1. If Venus is visible, observe it with a modest telescope. **(a)** If the telescope shows Venus to look like a thin crescent, where is Venus relative to Earth and the Sun? What important information does the crescent shape give you about the layout of the solar system (you might want to review Chapter 2)? **(b)** If the telescope shows it to look fully or mostly illuminated, where is it relative to Earth and the Sun? **(c)** Based on the phases you observe, draw a map of the inner solar system as seen from the north side of the system, with the sun at the center and Earth at the right. Then show

the position of Venus in its orbit at the time you observed it. If Venus is lost in the glare of the Sun when you try this experiment, use the orbital information for Venus to estimate how long you would have to wait before it was visible again.

2. If Mars is visible in the night sky, compare its color to those of nearby stars. (*Hint:* You do not need a telescope for this observation, but it is best done away from city lights.) Can you see a difference in color? What color is Mars? Why is it that color?

3. If Mars is visible, observe it with a moderate to large telescope. **(a)** Can you see the disk of the planet? **(b)** Can you see a polar ice cap? **(c)** Can you see any markings on the surface? (d) Sketch the appearance of the planet. If Mars is lost in the glare of the Sun when you try this experiment, how long will you have to wait before it was visible again?

4. Observe and sketch the positions of two planets in the sky, relative to the background star pattern. Label the horizon with the points of the compass. Repeat your observations every clear night for several weeks. **(a)** Can you see movement over the course of a day or so? **(b)** Can you see movement over the course of a week or so? Estimate the rate of angular motion. Can you use this to estimate the period of the planet's orbit of the Sun? **(c)** Which planet moves faster, compared to the background stars? What is the direction of their motion compared to the motions of the Sun and Moon? Explain any difference in rate of movement, in terms of the planets' and Earth's orbital movements around the Sun.

Reading List

Beatty, J. K. 1996. "Life from Ancient Mars?" *Sky and Telescope,* vol. 92, October, pp. 18–9.

Bullock, M. A. and Grinspoon, D. H. 1999. "Global Climate Change on Venus." *Scientific American,* vol. 280, March, pp. 50–7.

David, L. 1999. "A Master Plan for Mars. (Future Missions)" *Sky and Telescope,* vol. 97, pp. 34–40.

Gibson, E. K., and 3 others. 1997. "The Case for Relic Life on Mars." *Scientific American,* vol. 277, December, pp. 58–65.

Golombek, M. P. 1998. "The Mars Pathfinder Mission." *Scientific American,* vol. 279, July, pp. 40–9.

Kargel, J., and Strom, R. G. 1996. "Global Climatic Change on Mars." *Scientific American,* vol. 275, November, pp. 80–8.

Kaula, W. M. 1995. "Venus Reconsidered." *Science,* vol. 270, pp. 1460–4.

Luhmann, J. G., Pollack, J., and Colin, L. 1994. "The Pioneer Mission to Venus." *Scientific American,* vol. 270, pp. 90–7.

Mulin, M. C. 1999. "Visions of Mars. (Mars Global Surveyor Mission)" *Sky and Telescope,* vol. 97, April, pp. 42–9.

Nelson, R. M. 1997. "Mercury: The Forgotten Planet." *Scientific American,* vol. 277, November, pp. 56–67.

Pepin, R. O. 1991. "On the Origin and Early Evolution of Terrestrial Planet Atmospheres and Meteoritic Volatiles." *Icarus,* vol. 92, pp. 2–79.

Phillips, R. J., Grimm, R., and Malin, M. 1991. "Hot-Spot Evolution and the Global Tectonics of Venus." *Science,* vol. 252, pp. 651–8.

Robinson, C. 1995. "Magellan Reveals Venus." *Astronomy,* vol. 23, pp. 32–41.

Robinson, M. S., and Lucey, P. 1997. "Recalibrated *Mariner 10* Color Mosaics: Implications for Mercurian Volcanism." *Science,* vol. 275, pp. 197–200.

Slade, M. A., and 2 others. 1992. "Mercury Radar Imaging: Evidence for Polar Ice." *Science,* vol. 258, pp. 635–40.

Trefil, J. 1995. "Ah, But There May Have Been Life on Mars," *Smithsonian,* vol. 26, pp. 70–7.

CHAPTER 7

The Giant Planets and Their Moons

CONTENTS

- Discovery of Volcanoes on Io
- Jupiter, Saturn, Uranus, and Neptune
- Comparative Planetology: Why Giant Planets Are Giant
- Ring Systems of the Giant Planets
- Satellite Systems of the Giant Planets
- Pluto: Ninth Planet or Inter-planetary Body?
- Summary
- Important Concepts
- How Do We Know?
- Problems
- Projects
- Reading List

Jupiter's strange volcanic moon Io is mottled by yellow, orange, and white sulfur flows. Dark spots are believed to be currently active volcanoes. The doughnut at the center is a view down through the debris that rises out of the central dark vent and falls in a ring around it. This volcano is named Prometheus after the Greek god of fire. (SOURCE: NASA *Voyager 1* photograph; courtesy of A. S. McEwen, U.S. Geological Survey.)

PREREADING QUESTIONS ON THE THEMES OF THE BOOK

OUR ROLE IN THE UNIVERSE
How do Earth and the terrestrial planets differ from the gas giant planet?

HOW THE UNIVERSE WORKS
How can gravity help us to understand ring systems and moon systems?

HOW WE ACQUIRE KNOWLEDGE
Can we predict the properties of unexplored worlds?

WHAT TO WATCH FOR IN CHAPTER 7

This chapter continues our description of the physical nature of the planets. Note the contrasts between the four gas giant planets in this chapter and the four rocky or metallic terrestrial planets in the last chapter. The gas giants are the largest four planets and have massive atmospheres of hydrogen and helium. The gas giants also have systems of satellites that are like miniature solar systems, with worlds of different sizes orbiting around larger central bodies, and mostly following nearly circular orbits. Exploration of these moons has revealed distinctive worlds with surprising characteristics, including water, atmospheres, and erupting volcanoes. This information broadens our sense of our place in the universe—there are moons in the solar system which have properties reminiscent of terrestrial planets. The opening story on the discovery of Io's volcanoes is a good example of how science progresses, having the power to make specific predictions that may later be confirmed by actual observation.

Much of this chapter tells the story of recent discoveries of exotic features in the outer solar system. The chapter is organized in three broad sections: the four gas giant planets themselves, their systems of rings, and their systems of satellites. The chapter ends with a look at the outermost and tiniest planet, Pluto, and its moon. In spite of the variety of these diverse worlds, many of their characteristics can be understood from only a few general principles of planetary science and chemistry.

DISCOVERY OF VOLCANOES ON IO

Early in 1979, a seemingly routine issue of a premier U.S. general scientific periodical had just hit the shelves. That issue of *Science* magazine included a research paper by Stanton Peale, of the University of California, and his two NASA colleagues Pat Cassen and Ray Reynolds. The team had spent several months studying Jupiter's four big satellites. These moons had been crucial in reinforcing the Copernican revolution when Galileo discovered them. After 3½ centuries, these moons had become the subject of modern physical studies. Peale and his colleagues had calculated the amount of tidal heating inside these moons—the phenomenon explained in Chapter 5 in the discussion of tides. On Io (pronounced *eye-oh*), the innermost of the four moons, they noticed something odd. The gravity of nearby moons forces Io into a slightly elliptical orbit. As a result, tidal heating is unusually strong inside Io. The researcher predicted that the heating in Io's interior might cause "widespread and recurrent surface volcanism...." This was a totally unexpected prediction, because astronomers had previously assumed that in the region of distant Jupiter, 5 times farther from the Sun than Earth, all satellites would be cold, icy, and geologically dead.

The research team's article was motivated by one of the great voyages of space exploration. As that issue of *Science* hit the stands on March 2, 1979, a spindly spacecraft called *Voyager 1* was approaching Jupiter and Io after 1½ years of flight from Earth. This would be humanity's first clear look at the moons Galileo discovered so long ago. On March 3, the spacecraft was close enough to Io to reveal its odd, splotchy surface with orange markings unlike those of any other moon. Photographs on March 4 showed more details. Strangely, there were no impact craters, which showed that the surface must be geologically young. Photographs taken the next day revealed Io as a world that looked like a pizza, in the words of one investigator. Patchy deposits of sulfur compounds stained the surface white, orange, and black. Still no impact craters were visible. *Voyager 1* sailed past Io as scientists struggled to explain these surface features, which were unlike those seen on any other world.

On March 8, *Voyager 1* photographed

Io in crescent phase, back-lit by the sun. Later that day, Linda Morabito, an engineer using the images to help navigate *Voyager 1*, discovered a faint bulge projecting from the edge of Io. Within a few days, excited scientists recognized this as a cloud of volcanic debris shooting up from the surface. They soon located several more such clouds. Thermal infrared data from other instruments onboard *Voyager 1* revealed hot spots at these erupting sites—200°C hotter than their surroundings. Instead of being a dormant, icy world, Io had the first erupting volcanoes discovered beyond Earth. In fact, Io is the most active volcanic world in the solar system! The theoretical prediction by Peale, Cassen, and Reynolds was dramatically proved to be correct.

This story is one of the most impressive successes of the scientific method. Using only mental tools—the laws of gravity and theoretical ideas about internal tidal heating—humans on Earth were able to predict the existence of active volcanoes on a distant world. The prediction was confirmed by direct observation a few weeks later. No other method of investigation about the nature of the universe has had such success.

Jupiter, Saturn, Uranus, and Neptune

The terrestrial planets, which we studied in the last chapter, are huddled relatively close to the Sun, in the inner solar system—the region between the Sun and the asteroid belt. Now we move beyond the asteroid belt to the outer solar system. This region is dominated by four giant planets, which range from about 4 to 10 times the diameter of Earth. Beyond those four giants is tiny Pluto, which is smaller than our Moon, and a number of other still smaller objects. The four largest planets—**Jupiter, Saturn, Uranus,** and **Neptune**—have massive gaseous atmospheres and are often called **gas giant planets**.

The name of the monarch of the Roman gods is fitting for Jupiter, because it is the biggest planet. Jupiter contains 71 percent of the total mass of all the planets—nearly 2½ times as much as all other planets combined! Jupiter's diameter is a little over 10 times Earth's diameter. Placed on the face of Jupiter, Earth would look like a dime on a dinner plate. Saturn's diameter also dwarfs Earth's by almost 10 times, while Uranus and Neptune have diameters about 4 times that of Earth. In general appearance, the four giants look radically different from one another (Figure 7–1).

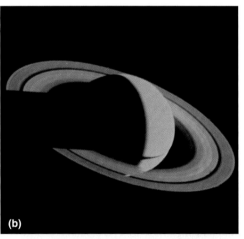
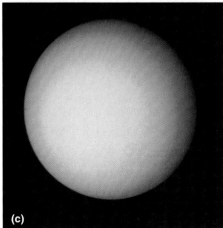
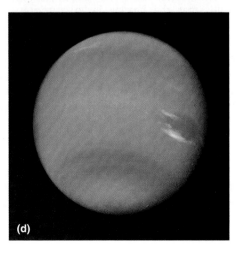

FIGURE 7–1
Photographs of the four gas giants taken from the *Voyager* space probes.
(a) Jupiter. The Great Red Spot storm system is visible at the lower left, along with many other atmospheric disturbances and bands. **(b)** Saturn. This backlit view and crescent phase can never be seen from Earth because Earth is always closer to the Sun than Saturn. Note that the globe of the planet can be seen through the lower part of the rings, proving that the rings are not solid.
(c) Uranus. The planet is pale blue, indicating a methane atmosphere, and almost featureless. **(d)** Neptune. The intense blue color is caused by methane, which absorbs red light more efficiently than blue light. (SOURCE: NASA photographs.)

Jupiter and Saturn are predominantly tan-colored, and Uranus and Neptune are blue. Saturn has a huge and prominent ring system. To appreciate the beautiful images in Figure 7–1, you have to recall that Uranus and Neptune are so far away that they barely show as disks in the largest telescopes. Little was known about Uranus and Neptune before these pictures were taken by the *Voyager 2* space probe in the 1980s.

The gas giants have many other characteristics that set them apart from the terrestrial planets. All four giant planets rotate faster than Earth, with rotation periods ranging from about 10 to 17 h, as contrasted with Earth's 24-h rotation. Each giant planet also has a whole family of moons or satellites. There are at least 59 moons in the outer solar system. (Additional smaller moons are only a kilometer or so across and barely merit the name *moon,* and additional tiny moonlets are discovered from time to time.) To summarize, the four giant planets together contain over 99 percent of the total planetary mass, and they play host to more than 90 percent of the known satellites.

A Variety of Icy Worlds

The gas giant planets are far from the Sun and very cold. They are composed primarily of hydrogen (H) and helium (He). As we will see in later chapters, these are the two most common elements in the Sun and in the universe as a whole. Hydrogen and helium remain gases down to extremely low temperatures. However, according to the laws of chemistry, other compounds such as water vapor, which would be gaseous in the warm inner solar system, condense into frozen solids—ices—in the cold outer solar system. This is the most important single thing to remember about the outer solar system: gaseous or **icy material** is a major building block of worlds located there, and rocky material is only a secondary component.

What do we mean by *ices?* The gases that formed the Sun and its planet-spawning surroundings were mostly hydrogen and helium, but helium is inert and does not form chemical compounds. So the ices that formed in the outer solar system are a variety of frozen compounds of hydrogen. One is very familiar: frozen water (H_2O), or ordinary ice. Others are less familiar, for example, frozen methane (CH_4), and ammonia (NH_3). Although hydrogen compounds dominate, there are still more ices in the outer solar system. One is frozen carbon dioxide (CO_2), which is a component of the polar regions of Mars, as we learned in the last chapter. In everyday life we know it as *dry ice.* Instead of forming rocky worlds as the terrestrial planets did, the outer solar system formed worlds of rock plus ice, often in roughly equal mixtures.

Ices even dominate the moons of the gas giant planets. As seen from Earth, each of the large moons of the giant planets subtends an angle of 1 second of arc or less—more than 2000 times smaller than the angular diameter of Earth's moon. Telescopes on Earth cannot see details smaller than a few tenths of a second of arc because of atmospheric shimmer. This means that Earth-bound astronomers were forever barred from seeing these satellites as anything more than fuzzy disks. However, within our lifetimes, these moons have gone from being tiny blurry images to exciting worlds with unique personalities.

We described one example, Jupiter's volcanic moon Io, in the opening of this chapter. Spacecraft have now given us sharp, close-up images of these worlds. Most people think of the solar system as containing "only" nine planets, but now we know that the satellites of the outer solar system include another dozen substantial worlds, many with spectacular properties. We will visit two moons bigger than either the planet Mercury or the planet Pluto. We will learn more about Io, the sulfur-dominated moon with the most active volcanoes in the solar system. We will hear about a moon covered with smooth, featureless ice, and another moon with a nitrogen atmosphere denser than Earth's, possibly with oceans on its surface. There is a moon with a thin atmosphere and columns of smoke rising from geyserlike vents. There are several moons that orbit their planets in a "backward" direction—opposite to the direction of all planets and most other moons. All this diversity makes the outer solar system an exciting place to explore.

General Properties of the Atmospheres

The atmospheric composition of the giant planets is mostly hydrogen, ranging from 63 to 93 percent hydrogen by mass. Most of the rest is helium, with only tiny traces of other compounds (as measured in terms of the total mass of each material). This is very different from the terrestrial planets, which are mostly made of rocks and metals, with only a thin sheath of light gases in their atmospheres.

How do we know? The density of a planet gives a good indication of its composition. Density is mass divided by volume. We measure the volume by knowing the distance to the planet and converting its angular size to a true size, using the small-angle equation (see the Science Toolbox in Chapter 2). We measure the mass by using the law of gravity to analyze the orbit. The giant planets are much less dense than the terrestrial planets. The mean density of the giant planets is 700 to 1600 kg/m^3 (kilograms per cubic meter), compared to 3900 to 5500 kg/m^3 for the terrestrial planets. Remember that the density of water, a useful benchmark, is 1000 kg/m^3. Saturn, at only 700 kg/m^3, would float as an ice cube if we could find a big enough ocean! This analogy is significant. The mean density is an important clue to the internal composition of these worlds. The density of all four worlds is much lower than that of rock. It proves that the giant planets are made not of rock, but largely of low-density ices, liquids, and gases.

We can make more direct measurements of composition by using spectrometers on Earth-based telescopes and on spacecraft. As we will study in greater detail in Chapter 10, each element or compound absorbs light with a specific set of wavelengths. The pattern of absorption is a direct guide to the chemical composition of the planetary atmospheres. We find a hydrogen-helium mix of roughly the same basic composition as found in the Sun and all other stars and gas in our region of the galaxy. This is no coincidence. This was the composition of the thin gaseous material that filled the solar system as the planets formed. As giant planets grew to massive size, their gravitational attraction became strong enough to capture and hold some of the surrounding gaseous material. This is how they acquired massive, hydrogen-rich atmospheres. None of the terrestrial planets grew big enough to do this, and so their atmospheres consist mostly of gases added by internal processes. We will look at this formation process in Chapter 9.

The best evidence comes from a direct probe. Imagine the awesome cloudscapes that might confront a spacecraft flying into the atmosphere of one of the giant planets. It might look like the view seen in Figure 7–2. In 1995, such a flight was made for the first time when the *Galileo* spacecraft parachuted a probe into the atmosphere of Jupiter. Unfortunately, the probe had no camera to send back images such as Figure 7–2, but it did measure cloud composition and other environmental conditions. Because the probe got hotter than expected when it slammed into the atmosphere, scientists required several months to interpret the data from the instruments, which were operating beyond the range of their usual temperatures. Eventually, the data revealed Jupiter's atmosphere had roughly the same percentage of hydrogen and helium as the Sun. Jupiter appears to have more carbon, nitrogen, sulfur, and other heavy elements than the Sun; these may come from interplanetary bodies that strike Jupiter. Few organic molecules were found.

The Nature of the Clouds

What can we say about the clouds in the atmospheres of the giant planets? The gaseous atmospheres are thick with clouds and haze, so thick that we cannot see the surfaces of these worlds. The clouds on all four giant planets are organized in systems of dark belts and bright zones that run parallel to their equators. These can be best seen on Jupiter, Saturn, and Neptune, as shown in Figure 7–1. The clouds of the giant planets are believed to be composed primarily of ice crystals of ammonia, ammonium hydrosulfide, and frozen water. These cloud crystals do not represent the overall composition of hydrogen and helium. They are minor constituents that have condensed out of the atmospheres, in the same way that the minor constituent, water vapor, condenses out of our atmosphere to form clouds.

The *Galileo* probe confirmed some earlier ideas, but it also raised further questions. For example, investigators were surprised that the probe did not detect clear signs of the three main cloud layers hypothesized from earlier studies. It detected no more than thin mists at the levels of the ammonia and ammonia hydrosulfide clouds and observed no thick layer at the lower level, where water clouds were predicted. It also measured less water vapor in the air than expected. Scientists are still puzzling over whether the probe happened to enter a particularly dry area of thin clouds, or whether all Jupiter is drier than had been thought. The strong winds of up to 400 mi/h measured at still deeper levels were also a surprise. As the probe descended farther, it measured higher temperatures. By the time of the last contact with the probe, it was bouncing on its parachute shroud lines, buffeted by the high winds. It had plunged into a region below the clouds where the air pressure was about 20 times that on Earth (equivalent to the pressure around 230 m below Earth's ocean surface) and the temperature was about 420 K (300°F).

Colors and patterns also reveal conditions in the atmospheres. Differences in the cloud patterns and colors of the giant planets are explained by differences in atmospheric temperature, chemistry, and structure. On Jupiter and Saturn, the bright *zones* are whitish or yellowish, and the darker *belts* show gray-brown and reddish tinges. The colored compounds in the clouds—including polysulfides, phosphorus, and organic compounds—stain the clouds these various colors. Remember that the word *organic* does not imply the presence of life, but merely indicates molecules with carbon-hydrogen bonds created by the mix of chemicals in the clouds of these planets. Also remember that many of the published images of planet atmospheres have had their colors somewhat exaggerated. The true colors of the giant planets are quite delicate and subtle.

The clouds on Uranus are indistinct, because they are hidden beneath a thick, hazy upper atmosphere. The blue colors of Uranus and Neptune are caused by two phenomena: the absorption of red light by methane, which is an important constituent of the haze above their clouds, and the same scattering of blue light that exists in Earth's atmosphere. The blue color is less prominent for Jupiter and Saturn because their uppermost clouds have less of a methane haze above them.

The clouds are dynamic, and we can show that they are in constant motion. Even a backyard telescope can reveal that Jupiter's cloud belts change. Figure 7–3 shows the long-term variations. The clouds change from month to month and year to year because of turbulent movements of the atmosphere, just as clouds on Earth change from day to day. For example, the *Voyager* probes discovered eastward jet stream winds on Jupiter blowing along the equatorial and

FIGURE 7–2
Spectacular cloudscapes could greet visitors in the atmospheres of the giant planets. This view shows a nighttime scene in the upper atmosphere of Saturn, at the top of its cloud decks. We are at midlatitude above the planet. Overhead, illuminating the scene like a glowing silver rainbow, arches the ring system of Saturn. Below the horizon, the Sun lights the ring system. (SOURCE: Painting by Ron Miller.)

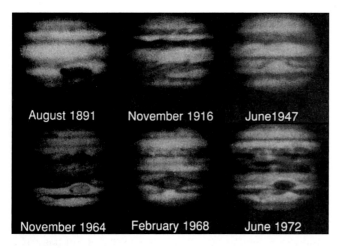

FIGURE 7–3
Photographs of Jupiter spanning 81 years, showing the changing array of Jupiter's belts and zones. The dark north temperate belt is relatively permanent, but the equatorial zone changes from bright (1891) to dark (1964). Many of these images show the Great Red Spot, which has been noted since the telescope was invented nearly 400 years ago. (SOURCE: Lowell Observatory.)

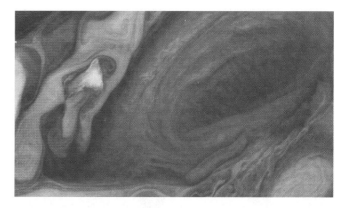

FIGURE 7–4
Nature's abstract painting. Jupiter's awesome Great Red Spot was photographed at close range by one of the *Voyager* space probes. It is a turbulent storm system bigger than Earth. It has been observed for several centuries. The smallest details in this view are about 100 km (60 mi) across. (SOURCE: NASA.)

temperate zones at speeds 150 m/s (338 mi/h) faster than the movements of the clouds in the dark equatorial belts. Saturn has even higher differential wind speeds; the eastward jet stream along the equator moves up to 450 m/s (1010 mi/h) faster than the neighboring belts. The colors, wind shear patterns, and clouds upwelling from below combine to produce fascinating, swirling patterns of color as in abstract paintings.

The most famous and dramatic feature on any of the giant planets is Jupiter's **Great Red Spot** (Figure 7–4), which is an oval storm system that has existed for at least 330 years, since it was first reported by G. D. Cassini in 1665 (Figure 7–3). The Great Red Spot became more strongly visible around 1887 when it was rediscovered and given its current name. It is 3 times the size of Earth! This and other smaller, transient spots are probably hurricanelike systems. Small clouds approaching the Red Spot get caught in a counterclockwise circulation as do leaves in a giant whirlpool. Figure 7–5 shows that Neptune has a similar feature, a giant elliptical storm, bluish in Neptune's case. The storm is capped by higher, white condensation clouds.

There is also electrical activity in these violent storm systems. *Voyager's* photographs of the night side of Jupiter revealed mighty blasts of lightning playing among the clouds. The parachuted probe also detected radio static from distant lightning. Typical lightning strokes were more energetic than those on Earth, but there were fewer per unit area than on Earth. Also, *Voyager* photographed enormous auroral displays like the northern lights and southern lights of Earth, flickering high above the clouds in Jupiter's polar regions. Saturn's atmosphere has less lightning and less auroral activity, and the atmospheres of Uranus and Neptune are still less active.

Modeling weather systems of giant planets is a good test of our understanding of climate and air circulation on our own planet. A particularly active area of current scientific research is the comparison between theoretical models of the giant planet's atmospheric circulation and theoretical models of Earth's atmospheric circulation. Why does Earth have circular patterns of weather rather than belt patterns? Why do storms last for hundreds of years on Jupiter but only weeks on Earth? How does energy flow through atmospheres of different compositions? The answers to these and similar questions may help us clarify our understanding of Earth.

Internal Structure of Giant Planets

As we might expect, the temperatures of the giant planets reflect their increasing distance from the Sun. Radiation decreases with the inverse square of distance from an energy source, so more distant planets receive less of the Sun's warming rays. If we compare at the depth in the atmosphere where the pressure is equal to sea-level pressure on Earth, the typical temperatures of Earth, Jupiter, Saturn, Uranus, and Neptune are 290, 170, 150, 78, and 69 K. (The sequence from Earth out to Neptune is 63, −153, −189, −319, and −335°F.) The trend toward colder conditions as we move away from the Sun is clear.

How do we know what lies below the clouds of the giant planets? Results from the Jupiter parachute probe and the *Voyager* probe flights by all four giant planets gave us a better understanding of the atmospheric conditions of all four giants. Their upper atmospheres are cold, but their lower atmospheres below the cloud layers are hot and have high air pressure. The vertical structure of any planet atmosphere follows a simple physical principle that relates temperature, density, and pressure. The more deeply you go in an atmosphere, the higher are the temperature, the density, and the pressure. The Science Toolbox on page 183 describes this gas law in greater detail. Major differences between the vertical structures of the atmospheres of different planets are caused by gravity. A more massive planet has stronger gravity. Stronger gravity creates conditions of crushing pressure and blistering heat at the base of the atmosphere of a giant planet.

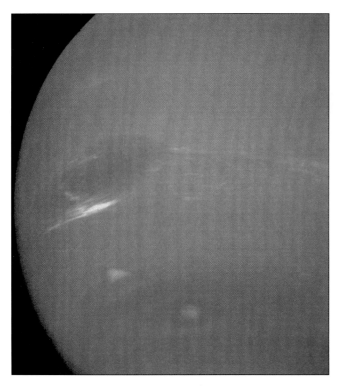

FIGURE 7-5
Color view of a portion of Neptune shows the bluish color of its atmosphere. Dark oval clouds, such as the prominent "Great Dark Spot," mark circulating storm systems, capped by white condensed clouds higher in the atmosphere. (SOURCE: NASA *Voyager 2* photo)

As the most massive planet, Jupiter has the hottest atmosphere at lower levels. This heat provides the energy to drive storm systems such as the Red Spot, and more upwelling of colorful clouds from the warmer regions below. On all the gas giants, the zones usually represent higher, brighter clouds, and the belts represent gaps in the high clouds where we see lower, darker, and more colorful clouds.

What lies below the deepest levels we have probed? Are there alien surfaces where explorers could land? We could hardly imagine a greater contrast with our experience of terrestrial planets. On all the terrestrial planets we would travel down through a gaseous atmosphere to a hard, rocky surface. If we descended far below the visible clouds of any of the gas giants, we would find the atmosphere growing thicker, hotter, and with higher and higher pressure. The state of the hydrogen and other matter inside the giants is unfamiliar to us in ordinary life. No place on Earth duplicates the very high pressures deep in these planets, caused by the extreme weight of the overlying layers. If the temperature were cool enough, an ocean surface of liquid hydrogen could exist perhaps 100 km below the clouds; but the temperature is too high, and so the gas simply gets denser at lower depths, turning into a mush resembling a thick, hot liquid, but with no well-defined surface. Sunlight cannot penetrate far into these atmospheres. Deep within the giant planets, we would enter a murky "twilight zone" between vapor and liquid. Deeper still, we would find liquid gradually turning into solid—nothing firm to stand on, but a slushy mixture of ice and rock.

By carefully applying simple physical principles, scientists have revealed the internal structures of the giant planets. They know the chemical composition and the overall density (mass divided by volume, as described above). They apply the idea of differentiation that we saw in Chapter 5—the tendency for heavier elements and compounds to sink to the center of a planet. They use the law of gravity and the gas law from the Science Toolbox to figure out how the temperature varies with depth in the atmosphere. To determine the state of the hydrogen, helium, and other materials inside these planets, scientists use laboratory data on the properties of these materials at high pressure. The results give a rough idea of the interior structure of the four giant planets. Figure 7–6 on page 182 shows the layers in a cutaway view (compare to Figure 7–1). We can use basic physical and chemical data to deduce the conditions in places that have never been visited!

Jupiter has a bulk internal composition of roughly two-thirds hydrogen, with the rest being helium mixed with small amounts of rocky silicates, metals, and other impurities. In Figure 7–6, you can see that Jupiter's cloudy atmosphere grades into a liquid form of hydrogen, called *liquid molecular hydrogen*. At these levels it is cold enough that hydrogen atoms stick together as H_2 molecules, but dense enough that the hydrogen is a liquid. At 20,000 km below the clouds of Jupiter, the pressure is about 4 million times Earth's atmospheric pressure. The high pressure and high temperature cause hydrogen atoms to collide frequently at high speed, so that their electrons are stripped away. Loose protons are surrounded by loose electrons. This exotic state of hydrogen is called *liquid metallic hydrogen*—metallic because the electrons flow freely and could carry an electric current, as in a metal. This mass of high-pressure liquid metallic hydrogen covers the deeper rocky and icy core.

Saturn has a composition and interior structure similar to those of Jupiter. In Uranus and Neptune, the pressures are much lower and not great enough to produce liquid metallic hydrogen. On these two planets, vast oceans of liquid molecular ammonia, methane, and hydrogen extend from the base of the atmosphere down to the icy and rocky core. These oceans are at very high pressure and reach temperatures of several thousand kelvins. Uranus and Neptune have a still higher proportion of silicates, metals, and impurities. All four of these giant planets are believed to have central cores that are rocky and metallic and about 1.5 to 3 times the diameter of Earth's core. In Saturn, Uranus, and Neptune, these Earthlike cores are surrounded by icy outer cores made of frozen water, methane, and ammonia. In all four giant planets, deep mantles of hydrogen in various forms surround these rocky and icy cores.

In essence, you can visualize the giant planets as cold super-Earths, roughly 2 or 3 times as big as our planet, buried in vast oceanlike mantles of high-pressure hydrogen in various forms. Surrounding each planet is a deep hydrogen atmosphere full of clouds. It is these massive envelopes of hydrogen-rich materials that make the total diameter of the giant planets extend to 4 to 10 times Earth's diameter.

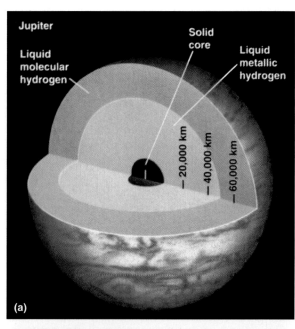
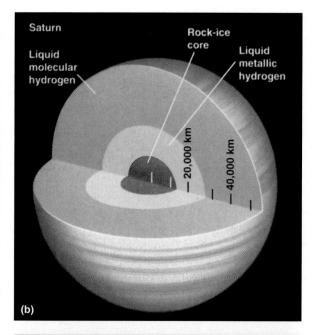
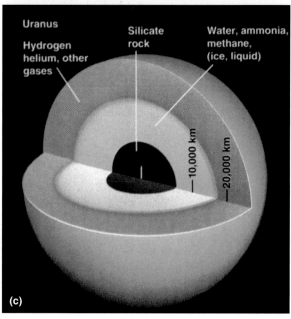
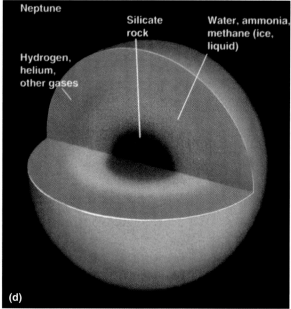

FIGURE 7-6
Internal structures of the four gas giant planets. (a) Jupiter and (b) Saturn have small rocky cores and layers of hydrogen under increasing pressure lower in the atmospheres. (c) Uranus and (d) Neptune have insufficient pressure to create hydrogen in the liquid metallic state, but as on Jupiter and Saturn, differentiation has caused the heavy elements to sink and form a rocky core. For scale, note that Earth is slightly smaller than the rocky cores of each of these giant planets.

Discoveries of Uranus and Neptune

Jupiter and Saturn have been known to humans for thousands of years. They lend their names to two days of the week, which we noted in Chapter 2 were chosen to commemorate the brightest moving objects in the sky. Since Uranus and Neptune are so far away, they are not visible to the naked eye. Uranus was discovered in 1781, when astronomer William Herschel accidentally detected the planet during an ambitious star-mapping project. Herschel thus became the first known human to recognize a new planet. Of course, his telescope could reveal no details on the planet. J. Bode (of Bode's rule) suggested that the planet be named for the Greek god Uranus. In Greek mythology, Uranus was the father of Saturn, who in turn was the father of Jupiter. Because of Uranus' long 84-year period of revolution around the Sun, it has made only two orbital circuits since its discovery. Compare this to Mercury, named after the fleet-footed messenger god, which zips around the Sun several times a year!

The discovery of Uranus led to the discovery of Neptune. Astronomers were surprised to find that Uranus did not move in a perfectly elliptical orbit, as would be predicted by the laws

SCIENCE TOOLBOX

Gas Laws

Try this trick. Fill a glass with water to its brim, and place a playing card over the top. Hold the card in place with your finger, and invert the glass. Now remove your finger. The card does not move, and the water stays in the glass!

The trick works because air exerts *pressure* with a force on the card that is greater than the weight of the water. It would not work on the Moon where there is no atmospheric pressure. Scientists realized in the 17th century that we live under an ocean of air. Pressure is one of the attributes of air or any other gas. Otto von Guericke, the mayor of the German town of Magdeburg and a talented amateur scientist, studied effects of air pressure. He invented a vacuum pump and showed that feathers and stones fall at the same rate in the absence of air resistance, confirming a prediction of Galileo. He also showed that air is essential to life by pumping the air out of a jar containing small animals. Von Guericke demonstrated the power of air pressure with a spectacular experiment. He built a copper sphere 0.5 m across composed of two hemispheres. After he removed the air from the sphere with a pump, two teams of eight horses could not pull the hemispheres apart. Yet when he opened the valve and let the air back in, the hemispheres fell apart.

We define pressure P as force F acting over an area A. Mathematically,

$$P = \frac{F}{A}.$$

So the same force acting over a smaller area makes for higher pressure. Distributing force (or weight) over a larger area is the reason why you sink into snow when wearing shoes but not when wearing snowshoes. It also explains why it is possible to lie safely on a bed of nails whereas sitting on any one nail would cause it to penetrate the skin. The air pressure exerted on a large window in your house is equal to about 10 times your weight! However, the air inside exerts the same pressure outward so the two forces are balanced.

The first systematic studies of the properties of gases were carried out by Robert Boyle in the middle of the 17th century. Boyle was the youngest of 14 children of the Earl of Cork, one of the richest men in Ireland. He compressed and expanded a fixed amount of air—the only known gas at the time—and measured its pressure. Under conditions of constant temperature, he observed a simple inverse relation between pressure and volume

$$P \propto \frac{1}{V}$$

If we increase the pressure on a gas, its volume will shrink. You can observe this easily with a bicycle pump. If you cover the valve and press on the plunger, the volume of the column of air in the pump decreases. Boyle concluded that "there is a spring or elastic power to the air we live in," and you can feel it, too, as you press up and down on the pump. As you apply larger force and therefore more pressure, the air compresses even more.

In the late 18th century, several French scientists made another important discovery about gases. If a gas is heated (and the pressure is fixed), it expands by 0.3 percent of its volume for every degree Celsius it is heated. If we use the Kelvin temperature scale, the simple proportionality is

$$V \propto T$$

Gas expands as it gets hotter. Double the temperature (on the Kelvin scale) and the volume will double. You can see this behavior with a bicycle pump also. Press as hard as you can on the plunger and hold it down. The gas inside heats up at once, and after a minute or so, the barrel of the pump will be noticeably warmer. The relationship holds for any gas. Combining these two relationships, we have

$$PV \propto T$$

This equation is a powerful tool for understanding and predicting the behavior of gases. It is often called the *ideal* gas law because it applies to gases under certain ideal conditions. These ideal conditions require that (1) the number of molecules is very large, (2) the molecules occupy a negligible fraction of the gas volume, (3) the molecules are in constant random motion, (4) there are no forces between molecules, and (5) the molecules bounce off one another without losing energy.

The ideal gas law describes the temperature and density variation of Earth's atmosphere with altitude. It also describes the variation of temperature and density of *any* planetary atmosphere. This in turn allows us to predict the conditions under which a gas might be compressed into a liquid or a solid. The exotic structure of giant planet atmospheres thus can be predicted by a well-tested law of physics.

We can also connect our discussion of gas properties to the microscopic properties of atoms and molecules. Daniel Bernoulli was the first to get this insight, in the mid-18th century. Bernoulli was the head of an extraordinary family—no fewer than eight of his sons and grandsons became noted scientists! Bernoulli realized that gas pressure was caused by the continual microscopic collisions of gas molecules with their container. Compressing a gas increases the number of collisions per second, hence the pressure. Compressing a gas also increases the energy of each collision, hence the temperature. Bernoulli's calculation showed that the product of pressure and volume is related to the number of molecules N and the average kinetic energy of each one:

$$PV = \frac{2}{3} N \left(\frac{1}{2} mv^2 \right)$$

But we have already seen in the Science
(continued)

SCIENCE TOOLBOX

Gas Laws (continued)

Toolbox on gas molecule velocities in Chapter 4 that the kinetic energy of a particle is related to temperature by $\frac{1}{2} mv^2 = (3/2)kT$. Substituting in the equation above, we get

$$PV = NkT$$

The constant of proportionality is given by the Boltzmann constant k and the number of molecules N. This elegant little law applies for any gas and for any of the physical conditions we have discussed in this book.

---

of Kepler and Newton. They recognized that the gravitational attractions of Jupiter, Saturn, and other planets would slightly disturb Uranus' orbit, but even these effects did not explain the problem. What could be wrong? Researchers in the 1800s hypothesized that the gravitational forces of still another planet, as yet unseen, disturbed the motion of Uranus. They used the disturbances of Uranus' motion to *predict* roughly where the unseen planet would be. After several years of complex hand calculations of the orbits by French and English mathematicians, the new planet was discovered by French astronomers in 1846. With its pale blue color, it was named Neptune after the god of the oceans. The fact that the position of an unseen planet could successfully be predicted simply from mathematical application of the laws of gravity confirmed the power of the scientific method and the application of physical laws throughout the universe.

Uranus is unique for its axis of rotation, which is highly tilted to the plane of the solar system, as shown in Figure 7–7. Most planets, as does Earth, have their north pole pointing more or less "upward," at a steep angle to the plane of the solar system. Uranus, on the other hand, has its pole lying nearly *in* the plane. This means that Uranus' north or south pole can point almost directly at the Sun at different times during its 84-year trip around the Sun, as shown in Figure 7–8. Remember that seasons on Earth or any other planet are caused by the tilt of its polar axis (review the material in Chapter 2). Thus, Uranus has a unique seasonal sequence. When the north pole points almost directly toward the Sun, the southern hemisphere is in a long, dark winter, lasting for about one-quarter of the planet's revolution. After this 21-year south polar winter and north polar summer, the Sun shines on the equatorial regions. Each point on the planet now goes from day to night during the planet's 17-h rotation. During this season, the situation is much like that on Earth or Mars, but much colder. After 21 more years, the south pole points approximately toward the Sun, and the southern hemisphere experiences a 21-year summer.

Another feature of Uranus that is shared only with Venus is its rotation. As we show in Figure 7–8, Uranus rotates from east to west, which is "backward" compared to Earth's rotation. This is called **retrograde rotation**.

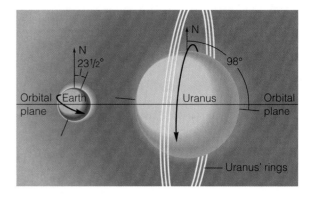

FIGURE 7–7
Comparison of the sizes and rotations of Earth and Uranus. Earth has an axial tilt of 23½° and a prograde (west-to-east) rotation. Uranus has a much steeper axial tilt and retrograde rotation.

What causes the extreme axial tilt and backward rotation of Uranus? Scientists speculate that just as Earth was probably hit by an unusually large planetary fragment that created its Moon (see Chapter 5), Uranus was hit by an unusually large planetary fragment that tilted its rotation axis. The difference in outcome was probably a result of the different geometry of the two collisions. Earth may have been hit more head-on in a way that blasted out mantle material. By chance, Uranus was apparently hit in a way (perhaps a glancing blow on the pole?) that affected its axial tilt. Notice that the scientific explanation we present is tentative, because the only evidence we have is the peculiarity of the orbit itself. Most planets show a high degree of regularity in their orbits; they have low-axial-tilt, **prograde rotation** (the planet spins in the same direction as its orbit of the Sun), and near-circular orbits. However, a few random large collisions may have given distinct "personalities" to some planets.

Infrared Thermal Radiation of Giant Planets

Scientists were greatly puzzled in the 1960s by one strange aspect of Jupiter. The biggest planet emits twice as much heat as it receives from the Sun!

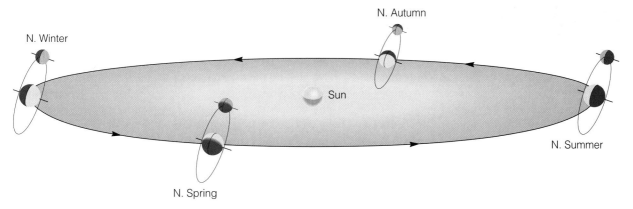

FIGURE 7-8
Schematic drawing of Uranus' unusual seasons (not to true scale). The extreme axial tilt of Uranus produces extreme seasons. During northern summer (right), Uranus' north pole points almost exactly toward the Sun and the equator remains in twilight all day. In spite of many years of direct Sun at the pole, the temperature remains frigid because of Uranus' great distance from the Sun. Note the seasonal effect on the satellites, one of which is shown schematically. Because each satellite orbits above the equator and keeps one face toward Uranus, the satellites undergo the same extreme seasons, as can be seen here.

Think about this for a moment. We see Jupiter by its reflected light from the Sun. Since no planet is a perfect reflector, like a mirror, Jupiter must send off less visible light than it receives from the Sun. The same is true of heat. We have seen that the giant planets have a sequence of decreasing temperature that is due to their increasing distances from the Sun and its energy. Each planet is heated by the Sun and emits thermal radiation according to its temperature—we saw the relation that describes this in the Science Toolbox in Chapter 4. With a temperature of a few hundred kelvins, the giant planets emit their thermal radiation at infrared wavelengths. Although the cloudtops are very cold, the interior of each planet is hot. This heat slowly radiates through the atmosphere and into space. Measuring this thermal infrared radiation gives us information about the internal heat of the planet. By the 1960s, space scientists had the technology to directly measure the infrared thermal radiation from Jupiter. They could then compare this to the thermal energy Jupiter received from the Sun.

In every case the internal heat energy was larger. Jupiter, Saturn, and Neptune generate thermal radiation that is 1.67 to 2 times the solar input. Although Uranus doesn't produce as much internal heat as the other three gas giants, it still produces more heat than it gets from the Sun. This is very different from Earth and other terrestrial planets, where the internal heat (generated by radioactivity) is only 0.005 percent of the heat received from the Sun.

Where does the extra heat of the giant planets come from? Theorists believe that Jupiter and Saturn are slowly contracting, which generates heat in their interiors and causes a flow of infrared radiation outward. The same is true of Uranus and Neptune to a lesser extent. This heating of a contracting body is well understood; we will encounter it again when we consider how stars form. It is also an example of the law of conservation of energy, which we encountered in Chapter 4. When the giant planets formed, gravity caused them to contract, and it is still causing them to contract to a much lesser degree. A shrinking planet has decreasing gravitational potential energy, and the process of contraction converts that to heat energy. Any gas that is compressed will heat up. Think of a bicycle pump if you cover the valve and press firmly on the plunger for a while. The air in the barrel will heat up, and the barrel will start to feel warm.

This extra thermal radiation due to gravitational contraction must have been most intense when the giant planets formed. It has declined ever since, to the modest amount observed today. Although Jupiter and Saturn radiate their own energy, they are not true stars. Even Jupiter's great mass is not enough to create the central pressure and heat necessary for nuclear reactions to occur in its interior. And the extra energy that comes from this slow contraction is far less than the prodigious energy that comes from the heart of the Sun.

The different amount of internal heat among the giant planets explains another puzzle—why does Uranus have a much more bland appearance than the other three giants? The answer may have to do with the amount of heat coming out of the planet's interior. On the giant planets, this internal heat flow is stronger and plays a more important role in stirring the atmosphere than it does on any terrestrial planet. The smaller rate of flow for Uranus may be one reason that its storm clouds do not convect up to high positions in the atmosphere, remaining hidden in the low atmosphere and giving the planet a uniform look.

Life on Giant Planets?

Physical conditions on the giant planets seem unlikely to have produced life. However, a few scientists have speculated that reactions among organic chemicals might have produced aerial, floating life-forms in Jupiter's lower atmosphere, something analogous to jellyfish floating in our oceans. Critics of this theory point out that updrafts and downdrafts might carry the materials to regions too hot or too cold to allow life. Early results from the Jupiter parachute probe suggested fewer organic molecules than expected, diminishing hopes for any biological processes in its atmosphere. As for Uranus and Neptune, there is still too little information even to speculate.

This pessimism is tempered by the realization that we know very little about the range of physical conditions that can support life processes. While we might imagine that life requires a terrestrial environment—liquid water, oxygen, and moderate temperatures—we cannot be sure. In the past few decades we have found life in some very inhospitable environments on Earth. So we should be open to the possibility of life in unusual places. We will revisit this fascinating topic in the last chapter of the book.

Comparative Planetology: Why Giant Planets Are Giant

Chapter 6 listed general rules that explain important properties of planetary atmospheres. We can apply these rules to the giant planets to explain their massive atmospheres. The third rule noted that bigger planets have stronger gravity and thus are better able to keep gas from drifting away into space.

Recall also the relation between temperature and particle kinetic energy discussed in Chapter 4. The higher the temperature of a gas, the higher the speed of the molecules that make up the gas. Therefore, the cold outer planets that are far from the Sun will generally have low gas particle speeds. Remember also that among all molecules at any given temperature, the lightest ones (such as hydrogen and helium) will be the fastest-moving ones. Hydrogen—the lightest gas—will have the fastest-moving molecules, on average. Helium will rank second-fastest. Combining these ideas, we can say that hydrogen and helium are the gases most likely to escape from a planet and can be retained by only the largest and coldest planets.

Applying these principles, we can understand why giant planets became so large, in contrast to the terrestrial planets, such as Earth. On small Earth, any hydrogen molecules that found themselves in our atmosphere moved fast enough to escape into space. Thus Earth has virtually no hydrogen in its atmosphere. Heavy gases, such as oxygen, nitrogen, and carbon dioxide, make up the atmospheres of the terrestrial planets. In contrast, on the four cold giant planets, hydrogen and helium moved more slowly due to the lower temperature. Hydrogen and helium were also more easily retained by the more massive cores. So hydrogen and helium and all other gases were retained. Since hydrogen was originally more abundant than all other gases put together, hydrogen dominates on all four of these planets, and the hydrogen-rich atmospheres today are very massive and very deep.

Figure 7–9 combines all this information on one plot. The vertical axis is the escape velocity of a planet, and the horizontal axis is the temperature. All the planets and three large moons are plotted. The sloping lines show the highest typical velocities of various gas atoms and molecules. If the line falls below a particular planet, then those gas molecules never reach the escape velocity and the gas is retained. Thus gravity and the thermal motions of atoms neatly explain the composition of planet atmospheres. The physical principles are completely general. In other words, they would apply to *any* planet in *any* solar system. All we need to know is the mass of the planet, which fixes the escape velocity, and the distance from the central star, which fixes the temperature.

We can continue with this line of reasoning and ask, How do you make a giant planet? It may surprise you to realize that we have covered all the basic physical ideas to let us predict roughly how big such a body should be. Let us be quantitative by using results from previous chapters. Every step is a simple scaling; we have seen them all before.

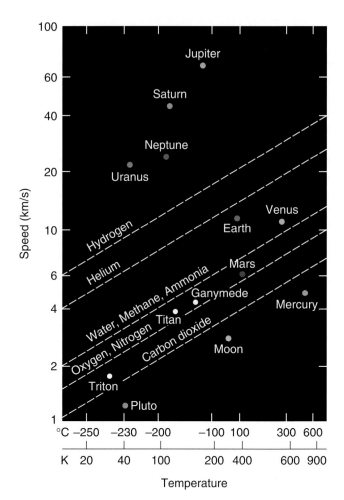

FIGURE 7–9
The retention of planetary gas depends on the mass of the gas atom or molecule and the temperature and escape velocity of the planet. Various planets and large solar system bodies are plotted. The lines show the highest typical velocities (10 times the average velocity) for different atoms and molecules. The giant planets have high escape velocities and can hold onto even the lightest gases such as hydrogen and helium. Terrestrial planets have lost hydrogen and helium but can retain heavier gases.

A Science Toolbox in the last chapter showed how to estimate how the temperature of a planet varies with distance from the Sun. Jupiter is 5.2 AU from the Sun, so a planet at that position receives $(1/5.2)^2 = 1/27$ of the radiation that Earth receives. We also saw that the temperature is related to the radiation received by $S \propto T^4$. Therefore, $T \propto S^{1/4}$, so the temperature at Jupiter's position is $(1/27)^{1/4} = 0.44$ times Earth's surface temperature of 293 K. This gives a temperature of $293 \times 0.44 = 128$ K.

How does this relate to the formation of a giant planet? Remember from our discussion of the kinetic theory of matter in Chapter 4 that the kinetic energy of an average atom is $1/2\ mv^2 = (3/2)kT$. So the atom velocity is proportional to the square root of temperature, $v \propto T^{1/2}$. Now we use the result, derived in the Science Toolbox in Chapter 4, that the velocity of hydrogen in Earth's atmosphere would be 16.2 km/s. Therefore, the velocity of hydrogen around an Earth-size object at the distance of Jupiter would be $16.2(0.44)^{1/2} = 10.7$ km/s. Since this is close to the escape velocity, little hydrogen would be retained on such a planet.

So if we had an Earth-size planet at the distance of Jupiter, it would stay an Earth-size planet. But a rocky planet several times the mass of Earth will retain a hydrogen atmosphere. Such a planet would be somewhat bigger in diameter than Earth. And, of course, there would be a feedback effect: As it trapped hydrogen, this would increase its mass, thus increasing its gravity and its escape velocity. So it would trap still more hydrogen. We can predict that if there is a rocky planet in the outer solar system that is bigger than Earth, its gravity will be so strong that the planet not only will retain hydrogen in its atmosphere, but also will trap the abundant hydrogen from the space around it. (Most of the thin gas in interplanetary space is hydrogen.) This is how giant planets are built—we will hear the full story in Chapter 9.

Ring Systems of the Giant Planets

The outer planets are much more massive than the inner planets, but they have another feature that distinguishes them from terrestrial planets: rings. People have been familiar with the rings around Saturn since the invention of the telescope. Today we know that all four giant planets have **ring systems.** There are many questions about the rings, and we understand only some of the answers. For example, what are they made of, and how did they form? Why is only Saturn's ring system prominent? What explains differences in ring systems from one planet to another?

The beautiful ring system of Saturn, well seen in Figure 7–1, has intrigued scientists since 1610, when Galileo first turned his telescope on Saturn. With his poor optics, he could make out only a fuzzy blob on either side of Saturn's disk, and he drew Saturn as a triple planet. Not until 1655 did Christian Huygens realize that the blobs were a ring system encircling the planet along its equator. Seen with a modern backyard telescope, Saturn's rings present a changing appearance from year to year in a 29-year cycle. Because the plane of the rings is tipped with respect to the plane of the solar system, we see the rings from different directions as Saturn orbits the Sun every 29 years.

FIGURE 7-10
Saturn's rings are not made of a solid sheet of material but are composed of millions of tiny moonlets. In some regions, the moonlets are scattered thinly enough that you can see through the ring, as visible in this photo. (SOURCE: NASA *Voyager* photograph.)

Contrary to appearances, Saturn's rings and all other ring systems are not solid flat sheets, but are composed of billions of separate pieces. These smaller particles are much too small to be seen individually from Earth. Each particle in the rings is a tiny moonlet on its own orbit around the planet, following Kepler's laws of orbital motion. The fact that Saturn's rings are not solid is demonstrated by Figure 7–10, where you can see through the thinner parts of the ring.

Four Distinctive Ring Systems

Each giant planet has its own distinctive set of rings, and only Saturn's are bright enough to be prominent, as seen from the distance of Earth. The other three ring systems are thin and faint and were not discovered until the 1970s and 1980s, primarily by observations from the two *Voyager* space probes. The rings are transparent, and we can see through the space between the particles they are composed of. In addition, the inner parts of a ring system orbit the planet faster than the outer parts do. Each particle in a ring system orbits with a speed that depends on its distance from the planet, in a miniature application of Kepler's laws.

Jupiter's ring is a single ring with a sharp outer edge and a diffuse inner edge. It is much fainter and narrower than Saturn's rings. It is composed of dark, microscopic dust particles, probably of rocky composition. The sparseness of the ring and darkness of its particles explain why we cannot see it from Earth.

The composition of Saturn's rings is very different from Jupiter's. The ring particles are not dark rocky material, but bright icy particles. Scientists had expected that the rings would be relatively smooth and uniform, but they were sur-

FIGURE 7-11
In this imaginary view, we are floating among the icy particles within Saturn's rings. Saturn is at the left, and the distant Sun (right) backlights the scene. The rings, probably less than 100 m thick, stretch off into the distance. (SOURCE: Painting by WKH.)

prised when *Voyager's* close-up photographs showed thousands of individual thin ringlets, often separated by gaps. The widest gap in the rings is easy to see with an amateur telescope from Earth; it is called *Cassini's division,* after its discoverer. The icy particles in Saturn's rings range from golf ball size to as large as a house. Larger sizes are probably rare. A visitor to Saturn's ring system would be surrounded by an amazing swarm of floating bodies (Figure 7-11). No probe has yet traveled into such an environment. Although this system is 274,000 km (171,000 mi) from tip to tip, dynamic forces keep the rings less than about 100 m thick! This is in-

credibly thin—imagine something the size of a pizza that is 1000 times thinner than a human hair!

Uranus' rings present yet another model, as seen in Figure 7-12a. Here, instead of a broad sheet, the ring particles are concentrated into several stringlike rings, separated by wide empty spaces. Eleven such rings have been identified on such photographs from the *Voyager 2* spacecraft. In Figure 7-12b, you can see that Neptune's rings are basically similar to Uranus' rings. However, the rings of Neptune have thicker clumps of particles along the "strings," which form distinct arcs.

Structure within Ring Systems

What causes ring particles to cluster into ringlets with well-defined edges? What keeps particles out of the almost-empty gap known as Cassini's division? Why does Neptune's ring system show concentrations into arcs along part of each ring?

Scientists are still debating these questions, but it now seems clear that most of the ring structure is caused by gravitational forces acting on the particles, either from known, nearby moons or from smaller, unseen moonlets within the rings themselves. The *Voyager* spacecraft provided support for this theory by discovering moonlets straddling several narrow rings in the Saturn and Uranus systems. Two such moons can be seen among the rings in Figure 7-12a. Moons of this type, which confine rings into narrow zones, are called **shepherd satellites.** Scientists have studied their gravitational effects on myriad nearby particles. It turns out that they confine the flock of neighboring ring particles into an orderly group, as pairs of sheepdogs do. The outer shepherd moon slows down any particle that strays outward and so drops it into a lower orbit. The inner shepherd moon speeds

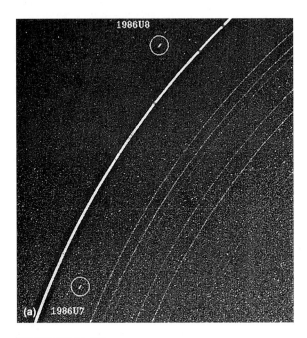
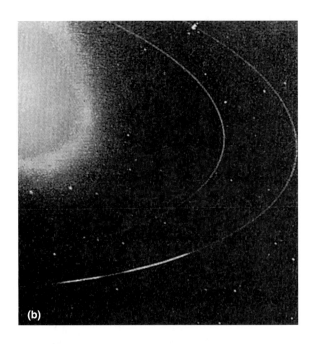

FIGURE 7-12
The narrow rings of Uranus and Neptune. **(a)** A portion of the ring system of Uranus, along with two shepherd moons (circled). The gravitational pull of the moons helps to keep ring particles confined to stringlike arcs. Uranus itself is out of the picture to the lower right. **(b)** A portion of the ring system of Neptune, showing concentrated arcs along the rings. Glare at upper left is from overexposed disk of Neptune. (SOURCE: NASA photographs.)

up any particle that strays inward and so boosts it into a higher orbit. Thus, the thin rings of Uranus and Neptune are probably confined by small moonlets.

Another piece of evidence that rings are controlled by moons comes from the biggest gap in Saturn's rings—Cassini's division. Particles in that region find themselves orbiting Saturn in about one-half the orbital period of the nearest large moon, Mimas. Therefore those particles would feel a repeated gravitational tug from Mimas, at the same position, every two trips around Saturn. This phenomenon, when one body has an orbital period equal to a simple fraction of the orbital period of another, such as one-half, one-third, or two-thirds, is called **resonance.** The regularly repeated resonant tug of the larger body disturbs the orbit of the smaller body, often kicking the smaller body clear out of its original orbit. You can see a general discussion of this idea in the Science Toolbox on resonance. A more familiar example of resonance is a child on a swing. If you push on a swing at random times, you do not "pump up" the swing's motion because the pushes tend to cancel one another. But if you push at exactly the resonant interval that matches the swing's natural period, the swing goes higher and higher. The swing is like the small particle in the ring, getting pushed resonantly at just the right intervals to "pump up" its velocity and change its orbit. This phenomenon helps explain how nearby moons could clear gaps in the rings.

Roche's Limit

We have seen that tidal forces are very strong on a body that is very close to a planet. For example, in the Science Toolbox in Chapter 5, we showed that the Moon's tidal force on Earth causes the ocean tides. Now study Figure 5–23d. Notice that a planet's tidal forces distort a close satellite more than one farther away. If a satellite spends much time inside a certain critical distance from a planet, the satellite can actually be torn apart by the tidal forces. The critical distance at which the satellite can be destroyed is called **Roche's limit.** The distance depends on the density and strength of the bodies, but generally it is about two to three planetary radii out from the center of a planet. Inside this limit, it is difficult for a large satellite to exist. See this simple calculation demonstrated in the Science Toolbox that follows. (Spaceships and small satellites less than a kilometer or so in size and consisting of metal, unfractured rock or ice *can* exist inside the limit because they have enough effective strength to resist being pulled apart. Larger bodies, given enough time, can stretch plastically until the stresses are so great that the bodies break apart.)

There is an important implication of this idea. If a massive swarm of particles were in orbit around a planet outside Roche's limit, they might aggregate into a moon. If they were inside Roche's limit, they could not aggregate and would remain as a dispersed ring. This lets us understand why rings persist once particles are established in orbit close to a planet: The particles are too close to planets to aggregate into moons. Well-understood forces bring such particles into a thin disk over the equator of the planet.

These seemingly abstract principles are supported by direct observation. All ring systems are inside Roche's limit for their planets. Some ring systems, such as Saturn's, have their outermost edges very near Roche's limit.

Where Did the Ring Particles Come From?

Most scientists now believe that the particles composing ring systems come from erosion the or breakup of small satellites on the outskirts of the rings. Throughout the history of the solar system, interplanetary debris of various sizes has rained onto all moons. The concentration of debris is especially great close to a giant planet, because the planet's gravity attracts the chunks of rock and ice. Smaller particles constantly "sandblast" the inner moons of each giant planet, dislodging dust grains. The smallest particles knocked off a moonlet close to the planet tend to spiral inward toward the planet and spread into a ring.

Occasional large impacts may break up a moon, throwing a swarm of fragments into the space near the planet and forming a thick ring like Saturn's. The ring systems we see today thus may date not from the beginning of the solar system, but from more recent collisions near the planet. An alien visitor to our solar system 3 billion years ago might have seen a different set of ring systems. When we look at Saturn's rings today, we may be seeing the debris of a moon-shattering catastrophe in the "recent" geological past.

Satellite Systems of the Giant Planets

We have seen that the ring systems of the gas giant planets are essentially "failed moons." But what about the satellite systems themselves? Most of the moons in the outer solar system are made of ice-rock mixtures. Of these moons, four are larger than our own Moon, and others are comparable to it in size. Most have white, icy surfaces or darker, tan-gray dirty-ice surfaces.

The *Voyager* probes traveled through all satellite systems of the four giant planets between 1979 and 1989, revolutionizing our knowledge of the satellites. (*Voyager 1* was intentionally deflected out of the solar system after visiting Jupiter and Saturn; *Voyager 2* visited all four giant planets.) The *Voyager*s discovered many new moons in each system. Sporadic additional discoveries using *Voyager* data and other techniques turned up more moons, bringing the total to at least 16 for Jupiter, 18 for Saturn, 17 for Uranus, and 8 for Neptune. Moonlets have been spotted on the edge of Saturn's rings. They may be nothing more than temporary aggregations of particles that congregate and later disband, making it hard to specify the total number of moons in the solar system. The closest moons to each planet orbit the fastest, in accordance with Kepler's laws.

While the smallest moons are merely overgrown examples of debris among the ring particles, the largest moons are bigger than the planets Mercury and Pluto. How big does a hunk of rock have to be to be called a moon or a satellite? There are a broad range of sizes, so there is no

SCIENCE TOOLBOX

Resonance and Harmonics

As we consider the ring systems of the giant planets, an obvious question arises. Why do they have such complicated structure? We might expect a disk of material orbiting a planet to be more or less smooth and uniform. Instead, we see a complex pattern of concentric rings with hundreds of gaps and occasional very narrow rings. Sometimes we see gaps in the pattern. What causes this complexity?

We have seen in our discussion of periodic processes that almost every object has a natural frequency of vibration or oscillation, whether it is a star, a planet, a bridge, an iron bar, a wine glass, or a guitar string. We will use a string as our example because the music that results from its vibration is so familiar—for example in a guitar. Figure 7–A shows two phenomena that occur with vibrations and waves. On the left side we see that one vibrating string may set another string vibrating at the same frequency (it works with bells, too). This is called *resonance*. Soldiers have to break step when crossing a bridge because their marching rhythm might match the natural frequency of the bridge, sending it into violent oscillation. Most mechanical objects have a natural resonant frequency—perhaps there is an engine rotation frequency at which your car vibrates or a wheel rotation frequency at which your steering shakes?

On the right side of the figure we see that a string has many modes of vibration. When the string is divided into 2, 3, or another integer number of parts, the frequency or number of vibrations per second increases. As the length of each vibrating segment gets shorter (a shorter wavelength), the frequency of vibration and pitch of the corresponding sound increase. In addition to the fundamental frequency f, the string can vibrate through more complete cycles each second, corresponding to higher frequencies.

As we mentioned in Chapter 1, these phenomena were known in ancient times. Pythagoras discerned that a string emits harmonious sounds when it is divided into a small number of equal lengths. But when it is divided at an arbitrary point, say at 57 percent of its length, the sound is discordant. This profound result led to the idea that there are certain harmonious relations in Nature. Kepler believed that he could find harmonious relationships between planet orbits and this led him to discover elliptical orbits. And all our music derives from harmonious and discordant relationships among notes. As the figure shows, these *harmonics* have integer multiples of the fundamental frequency—f, $2f$, $3f$, $4f$, and so on.

$$\text{Harmonic frequency} = nf$$

The period of a vibration (time in seconds) is just the inverse of the frequency (number of vibrations per second), so we have

$$\text{Harmonic period} = \frac{P}{n}$$

In the musical scale, if the fundamental frequency were middle C, then $2f$ would be an octave higher, $3f$ would be another fifth higher, and $4f$ would be two octaves above middle C. This idea works for orbits too! The resonance between orbits is caused by gravity.

Suppose we had a moon orbiting outside a ring system. Let us say the moon has an orbital frequency f. Remember from Kepler's laws that the ring particles will move faster than the moon since they

(continued)

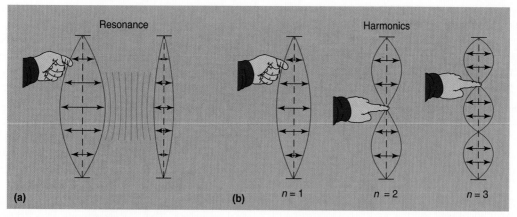

FIGURE 7–A
Vibrations of a string. **(a)** Like any simple object, a string has a simple mode of vibration that corresponds to the resonant frequency. One vibrating string will set in motion an identical string nearby in an example of resonance. **(b)** Strings and other objects have more than one mode of vibration. The simplest mode is called the *fundamental* ($n = 1$). Higher harmonics mean a larger number of shorter waves on the string and so higher frequencies.

SCIENCE TOOLBOX

Resonance and Harmonics (continued)

are closer to the planet. There is a place in the ring system where the orbital frequency is twice that of the moon, or $2f$. At this radius a ring particle will line up with the moon every other orbit and will be tugged out of position. Each tug is tiny, but the effect is cumulative, and over many orbits the particles will be driven away from that radius. A gap is created. The same thing occurs when the ring particle has an orbital frequency 3 times that of the moon, or certain other simple multiples. This orbital resonance may operate wherever the particles have 2 or 3 or 4 (or more) times the orbital frequency of the moon, creating a gap where the ring particles have periods of $1/2$ or $1/3$ or $1/4$ (or less) times the moon's period. Resonances can be extremely complicated and not all simple ratios of orbital periods act this way.

Let us test the idea. The first big moon beyond Saturn's rings is Mimas, at a radius of 185,000 km from Saturn. As we saw in Chapter 3, Kepler's third law relates period and semimajor axis or distance: $P^2 \propto a^3$. We can manipulate this to give $a \propto P^{2/3}$. For a higher harmonic, the period is reduced to P/n and the distance decreases by

$$a \propto \left(\frac{P}{n}\right)^{\frac{2}{3}}$$

So for the $n = 2$ harmonic, the distance where gravitational resonance will clear out the orbit is $(1/2)^{2/3} = 0.63$ times the moon distance, or 117,000 km. This is exactly the radius of the largest gap between the A and the B rings, discovered by Giovanni Cassini in 1675. The next harmonic at $n = 3$ is close to the gap between the B and the C rings. Resonances like this have been located in all the ring systems. The resonance when the ratio of ring period to moon period is $1/2$ or $1/3$ or $1/4$ is just one set of harmonics. Resonance occurs whenever the ratio of ring period to moon period is a ratio of whole numbers, such as $2/3$ or $3/7$ or $4/9$. Just as the large number of harmonics and overtones leads to the richness of music, the large number of resonances leads to the rich complexity of planet ring systems.

As noted above, Pythagoras and other early scientists believed that the universe was governed by simple relationships between numbers. They even speculated on the "harmony of the spheres"—a celestial music that mirrored the harmonics produced by a plucked string. Shakespeare gave a nod to Pythagoras in *The Merchant of Venice*:

> There's not the smallest orb which thou behold'st
> But in his motion like an angel sings.
> Such harmony is in immortal souls,
> But whilest this muddy vesture of decay
> Doth grossly close it in, we cannot hear it.

In reality, the planets, moons, and rings produce no sounds to fly through the vacuum of space. However, the harmonics that Pythagoras and Kepler sought are present in the dance of gravity.

---

simple answer to this question. It is a matter of definition rather than science. However, objects less than 10 km across are too small to call moons. (Deimos, the smaller of Mars' two moons, is just larger than this.) Most of the named moons in the solar system are between 10 and 100 km across. The moons larger than 1000 km across can have atmospheres and other distinctive features. In effect, the solar system contains not just nine planets, but two dozen world-class objects that are large enough to have individual geological and astronomical personalities.

Satellite systems of the giant planets are very different from satellites of the terrestrial planets in another regard. Satellite systems of the giant plants are like miniature solar systems. Each giant planet formed in some way similar to the solar system itself, with a massive central body surrounded by a family of orbiting smaller bodies of different sizes. In contrast, Earth has only one Moon that was probably spawned by a giant collision, and Mars has two tiny moons that are captured asteroids.

As a group, the moons have greater variety than anyone anticipated. To explain our approach to describing the satellite systems, we point out that each giant planet is surrounded by an imaginary spherical volume of space, called its **sphere of gravitational influence.** Within this imaginary sphere, the planet's gravity has a bigger effect on satellite motions than the Sun's gravity does. For example, if you tried to put a satellite in orbit around Jupiter, but *outside* Jupiter's sphere of gravitational influence, the satellite would eventually drift away from Jupiter into an orbit around the Sun, because the Sun would dominate its motion.

Thus, permanent satellites occur only inside the sphere of gravitational influence. Inside this volume, we find that as a rough rule of thumb, each giant planet's family of moons can be divided into four groups:

1. Countless mini-moonlets and dust particles involved in the ring systems. We have already discussed these in the section on rings.
2. Small moons close to the planet, on the outskirts of the ring systems (or in some cases inside the ring system). These are sources of ring material. Most are too small to be seen from Earth and were discovered by the *Voyager* probes.

SCIENCE TOOLBOX

Roche's Limit and the Outer Edge of Ring Systems

All ring systems have outer edges located at about 1.8 to 2.5 times the planetary radius from the center of the planet. Why are the edges of some ring systems not at larger or smaller multiples of the planetary radius?

Do you think you know enough science to predict the position of ring systems' outer edges (and amaze your friends)? It is easy to predict this basic fact of nature from what we have learned; it all goes back to Newton's universal law of gravitation. We will generalize the calculation from the Science Toolbox in Chapter 5 on the Moon and Earth's tides. Edouard Roche, a French mathematician, worked it out in 1850. Imagine that there are two tiny ring particles at a distance X from a planet. Each particle has mass m and radius r (the two particles have the same mass and size). The planet has mass M and radius R. Suppose the two particles are touching, as in Figure 7–B. Then the gravity force pulling them together is (from Newton's law of gravity)

$$F = \frac{m^2}{2r^2}$$

Now think about the gravitational force pulling either particle toward the planet. It is

$$F = \frac{Mm}{X^2}$$

Now, from what we have learned about tidal forces, the nearer particle will be pulled harder toward the planet than the farther particle, and differential calculus allows us to differentiate the gravity law. We can write the expression for this differential force (if you have not had calculus, you will have to take this statement on faith!) as

$$\Delta F = \frac{-2Mm}{X^3} \Delta r$$

The Greek capital letter Δ (delta) is often used in mathematics and physics for a small increment or difference. So ΔF is a differential force, and Δr is a differential distance. The minus sign means that the differential force gets bigger as the distance from the planet gets smaller. Since the difference in distance is just 2 particle radii, we have

$$\Delta F = \frac{-2Mmr}{X^3}$$

Remember that the differential force is the force pulling the two small particles apart, and the gravity between them F is the force pulling them together. From our study of rings, we can *predict* that the outer edge of the ring will be where these two forces are equal. Outside that distance, the particles will attract one another by gravity and clump into a single satellite. Inside that distance, the tidal forces will pull the particles apart, and so they will orbit separately as a ring system.

So, let us set the two forces equal and solve for the distance X. We have

$$\frac{m^2}{4r^2} = \frac{-2Mmr}{X^3}$$

or

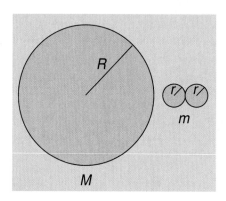

FIGURE 7–B
Two small ring particles (masses m) are just touching. Roche's limit is the distance X where the gravity holding the two particles together is equal to the tidal force from the planet, mass M, pulling them apart. Compare with Figure 5–23d.

$$X^3 = \frac{-8Mr^3}{m}$$

The minus sign is no longer important because we are only interested in the size of the distance. Now, remember that the mass of the planet is M = volume times density = $(4/3)\pi R^3$ (planet density), and the mass of a particle is m = volume times density = $(4/3)\pi r^3$ (particle density). If we substitute these two expressions for M and m, we notice that the factors of 4/3 and π cancel. If the density of the particle is essentially equal to the mean density of the planet, the densities cancel, too! The expression becomes much simpler

$$X^3 = 8R^3$$

or

$$X \approx 2.5R$$

This last simple relation says that the two forces are equal, and the particles will start to aggregate if the particles are located at about 2.5 times the planet radius. In other words, the ring can exist only inside 2.5 planet radii. This is essentially what is observed!

This distance is called Roche's limit. As Roche concluded in 1850, the rings are essentially defined by the tidal forces, which act to keep the ring particles apart. If the particle densities are somewhat different from the planet density, the outer edge of the ring will be found at different distances, which is also what is observed. The theory can be made more sophisticated by considering variations in particle density, strength, and shape. The important point is the astonishing fact that we can look through the telescope, observe the dimensions of Saturn's rings, or other rings, and then sit down with a pencil, paper, and Newton's law of gravity (dating from 1687) and explain why the rings look the way they do!

3. Large moons at intermediate distance from the planet. These are world-class bodies, and many have distinctive geology.
4. Small moons on the outskirts of the planet's sphere of gravitational influence. These moons seem not to be native to the planet, but rather passing interplanetary bodies that were captured by the planet's gravity.

There are too many moons to enumerate details of each. The larger moons, in the third group, are the ones we want to emphasize, since they tend to have their own geological personalities. To understand such moons, we first need to review some principles established earlier.

Surfaces with fewer impact craters are younger. Thus, whenever we see an uncratered region on a satellite, we know that some geological forces have been at work to modify and renew its surface. We know that worlds with greater internal heat have greater geological activity. That is roughly true among the moons we will describe. The smallest ones are just cratered ice balls with no internal activity, while the bigger ones have a variety of surface features with at least some young regions.

We also have seen that tidal forces are very strong near giant planets because of their strong gravity. These tidal forces can heat the interiors of satellites, as mentioned in the opening story of this chapter. As a result of this effect, some satellites close to planets show signs of geological activity even though they are quite small.

The last issue we will explore is the surface composition of each moon. The two basic materials of giant planets' moons—bright ice and black soil—are important in explaining their properties. The black soil is colored by carbon-rich compounds called **carbonaceous material.** Even a small percentage of the material can stain the ice black. Over many years, countless impacts from large to small have vaporized the ice component on the surface, leaving soils enriched by the black, carbonaceous material. But if modest heating occurs, the ice component can melt and erupt on the surface as a watery mixture. As it freezes, any dark soot sinks out, and a white surface of ice forms. The interacting roles of black carbonaceous soils and white ices help to explain why the moons of the outer solar system are diverse worlds with many unexpected properties. Satellites composed of these two materials may show either dark surfaces or bright, icy surfaces. Watch for these effects as we discuss some individual moons. During our discussion, refer, too, to the table in Appendix B-1 for specific physical properties of these bodies.

The Largest Moon in the Solar System: Jupiter's Ganymede

We show the orbits of most of Jupiter's moons in Figure 7–13, which serves as a road map to Jupiter's system. The four large satellites observed by Galileo can be seen moderately close to the planet. The third of these is Ganymede, the biggest moon in the solar system, with a diameter of 5262 km. It is about 8 percent bigger than the planet Mercury and more than twice as big as little Pluto. As seen in Figure 7–14, Ganymede has old regions that are darker and heavily cratered, with many grooves and fissures. Watery material has apparently erupted to form fresh, brighter icy deposits in swaths across the

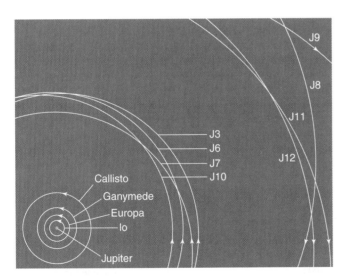

FIGURE 7–13
The miniature solar system of Jupiter and its satellites. Orbits of the four large Galilean moons are shown close to Jupiter (lower left). Still closer, smaller moons and the rings are too close to show on this scale. Two groups of outer moons lie in closely clustered orbits. Note that the outer group moves in retrograde direction—opposite to the usual sense of orbital motion in the solar system. See text for further discussion.

FIGURE 7–14
Jupiter's largest moon, Ganymede. Part of the surface consists of old, dark, cratered terrain. The dark terrain is broken, however, by swaths of brighter, younger, more ice-rich material. At the upper left corner we can see part of a white polar cap of frost. (SOURCE: NASA *Voyager 2* photograph.)

FIGURE 7-15
Imaginary panorama on the icy surface of Jupiter's moon Ganymede. Ganymede's surface is split by parallel fissures, where water may have erupted and refrozen. This view, from such terrain, shows a ridge of jagged ice blocks thrust up during movements of Ganymede's crust. In the sky are Jupiter itself and Jupiter's inner moons Europa (left) and Io (in front of Jupiter). Jupiter would look about 15 times as big in Ganymede's sky as our Moon does in our sky, and Europa would look slightly smaller than our Moon. (SOURCE: Painting by Ron Miller.)

planet. Many impacting bodies have blasted away the gray-brown surface soils to reveal brighter, fresher, whiter ice below.

What heat source could melt Ganymede's ice and allow water eruptions? Ganymede's orbit today is so circular that it does not change its distance from Jupiter by a large enough amount to cause tidal heating; but researchers believe it may have had minor orbit variations in the past that could have produced tidal heating. This may have caused subsurface melting and shifting of the ice "plates" on the surface, creating the fractures and causing water to well up and then freeze into fresh icy swaths. A view from the desolate, icy surface of Ganymede would be impressive, with Jupiter and other moons resplendent in the sky (Figure 7–15).

Jupiter's Icy Moon: Europa

Europa is Ganymede's next-door neighbor, located closer to Jupiter, as shown by Figure 7–13. It is about the size of our Moon. However, photographs from the *Voyager* probe, such as Figure 7–16, show that it is very different from our Moon or Ganymede: It has a nearly uncratered, bright surface of frozen water. No other satellite has so few impact craters. This means that the surface must be relatively young, geologically speaking. Still closer photographs, from the *Galileo* space probe in 1996 and 1997, indicate that the surface is a floating ice pack, constantly jostled, broken, and re-formed (Figure 7–17). Apparently the surface is a young ice layer, perhaps only kilometers thick, floating on a deep global ocean. Water erupts through the fractures and resurfaces Europa, perhaps on a timescale of tens of hundreds of millions of years. What keeps the ocean from freezing, and how could the young, icy surface have formed? Most scientists believe that tidal heating has kept Europa's interior warm enough to keep the interior water from freezing. The discovery of a liquid water ocean under a thin ice crust on Europa is one of the most exciting results of solar system exploration; it shows that liquid water oceans on Earth are not unique in the universe.

What exists in Europa's hidden ocean? Arthur C. Clarke, in his novel *2010*, speculated that life-forms might have evolved, as they did in our primeval oceans. There is no proof of this idea, but scientists remain excited about the possibility of biology under Europa's ice, especially after re-

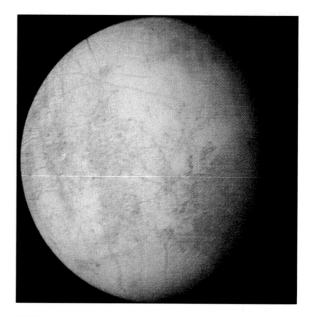

FIGURE 7-16
Jupiter's icy billiard-ball moon Europa. This world is about the size of our Moon but is covered by nearly smooth, whitish ice. Dark streaks appear to mark fractures. (SOURCE: NASA *Voyager* photograph.)

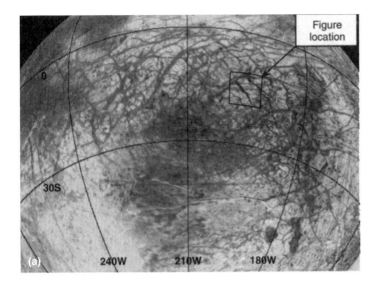
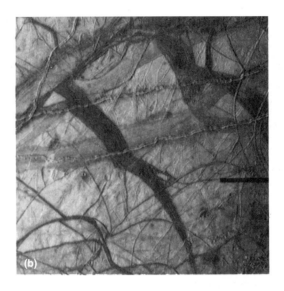

FIGURE 7-17
Solving the Europa jigsaw puzzle. (a) Global view of Europa showing location of detailed figure. (b) Close-up view of present-day fractured system. The darker bands mark fractures that have separated the brighter platelike masses of ice. Contrast is enhanced in Figures a and b. (c) Reconstruction of ice plates in Figure b. Fracture spaces have been filled in with black (left) and then the plates have been reconstructed into the original surface (right). This shows dramatically how Europa's ice crust is constantly fracturing and ice plates separating into new positions, with fresh ice filling in the fractures. (SOURCE: NASA Galileo images courtesy of Robert Sullivan, Cornell University, see Sullivan and others, in press)

cent work showing that microbes live in many seemingly hostile conditions on Earth and beneath its surface. For example, bacteria have been found in the Antarctic ice pack. Future missions to intriguing Europa might clarify the question of what lies beneath the ice.

Jupiter's Volcanic Moon: Io

Io is the closest to Jupiter of the four large moons. It is about the size of Europa and our Moon. It is often considered the strangest satellite in the solar system. As described in the opening of this chapter and shown in Figure 7-18, the *Voyager* photographs of Io revealed erupting volcanoes, which had been predicted from the theory of tidal heating. Subsequent observations show that volcanoes are always erupting on Io—it is, in fact, the most volcanically active world in the solar system. Material shot up by the many volcanoes rises high above the surface of the little moon and then falls back in a lazy arc due to Io's weak

FIGURE 7-18
The volcanic moon, Io. The arrow marks one of the largest of Io's volcanoes, named Pele. It is ejecting dark, ashy debris in the direction of the arrow. Debris forms a huge, domelike cloud, which is shown as a dark arch between the volcano vent and the edge of the planet. This fan of debris could cover most of the United States east of the Mississippi! A more irregular cloud of gas and debris, ejected by a volcano just over the horizon, can be seen at left edge as a bright, bluish glow. Special processing has been used to enhance the brightness and color of that cloud, which would be much fainter if seen by the naked eye from a spaceship over Io. (SOURCE: NASA *Voyager 1* photograph; courtesy of A. S. McEwen, U.S. Geological Survey.)

FIGURE 7-19
An ordinary day on Jupiter's volcanic moon, Io. On the horizon, an immense umbrella-shaped plume sprays out of a volcanic vent. Beyond, the sun is just emerging from behind Jupiter, backlighting its atmosphere and ring. (SOURCE: Joint painting by Ron Miller and WKH.)

gravity. About 10 cm of new material is added to the surface each year by the volcanic activity.

The heating and volcanism have apparently melted and blown away the water and other volatile materials of Io. As the more volatile compounds were lost, sulfurous, silicate, and metallic compounds remained. Sulfur compounds became the dominant surface materials in a process that happened on no other moon. These minerals cause the range of yellow, orange, and white colors that we see. Pure liquid sulfur is black or dark brown, and the darkest spots on Io are volcanic craters of molten or near-molten sulfur. Instruments on *Voyager* measured the temperatures of some dark spots as 600 to 700 K (621 to 800°F), consistent with near-molten sulfur deposits. Some local "warm spots" have soil temperatures around room temperature. A visit to Io would truly reveal a strange world, with erupting volcanoes and Jupiter dominating the sky at close range (Figure 7-19).

Io's core is also very different from that of other satellites. In 1996, scientists analyzed gravitational accelerations acting on the *Galileo* probe as it flew very close to Io. They found that Io has a large iron core, probably in the range of 36 to 52 percent of Io's radius. This piece of evidence agrees with the theory that the interior has been melted by volcanism and explains why Io has a larger density than any other known satellite (3530 kg/m^3, almost as high as Mars' mean density).

Saturn's "Missing Link" Moon: Enceladus

Enceladus is one of several modest-size inner satellites of Saturn. It is significant because of its unusual combination of old cratered terrain and smooth, icy plains, seen in Figure 7-20. In this sense it is a "missing link" between the cratered moons such as Jupiter's Ganymede and the smooth surface of Jupiter's Europa. Tidal heating apparently caused enough watery eruptions to resurface about one-half of the satellite, leaving the other one-half in its more primitive, cratered state. The ice of Enceladus is some of the whitest and brightest in the solar system, reflecting over 90 percent of the sunlight that strikes it. Enceladus is about the color of a field of snow.

Saturn's Smoggy Moon: Titan

Saturn's largest satellite, Titan, is the second-largest satellite in the solar system. Its unique feature is its dense atmosphere. Although this world has only 40 percent the diameter of Earth, the surface pressure of its atmosphere is 1.6 times greater than that on Earth! Intriguingly, the atmosphere is composed mostly of nitrogen, the same gas that makes up most of Earth's atmosphere. This does not mean that Titan's environment is like ours, however. Because of the large distance from the Sun, the temperature is very cold, around 93 K (−292°F).

As seen in Figure 7-21, Titan's atmosphere is a nearly featureless orange haze, which, like smog, is caused by sunlight's triggering reactions among molecules in the atmosphere. Minor constituents detected in Titan's atmosphere include organic molecules such as methane, ethane, acetylene, ethylene, and hydrogen cyanide.

FIGURE 7-20
Saturn's icy moon, Enceladus, 502 km across, offers a missing link between the smooth billiard-ball surface of Jupiter's moon Europa and the fissured surface of Jupiter's moon Ganymede. Portions of the ancient cratered surface seem to have been cracked as water erupted and froze into smooth ice plains, obliterating older craters (bottom central region). This suggests that Enceladus may have been heated enough to melt portions of its icy interior at some time in its history. (SOURCE: NASA *Voyager 2* photograph.)

FIGURE 7-21
Saturn's giant, hazy moon, Titan, is slightly bigger than the planet Mercury. Titan has a smoggy, orange atmosphere. In this crescent-lit view, the illuminated atmosphere can be seen all the way around the night side. (SOURCE: NASA *Voyager* photograph.)

FIGURE 7-22
An imaginary view on the dark, cloud-swept surface beneath the clouds of Titan. Measurements suggest snowstorms and rainstorms of methane-based compounds—with possible rivers and "waterfalls" of liquid methane and liquid ethane. (SOURCE: Painting by Ron Miller.)

Meteorological calculations suggest weird conditions on Titan. The temperature and pressure are near the point where methane can exist as gas, liquid, or solid. Methane may play a role like that of water on Earth, existing as vapor, "rain" drops, or ice. Rain and snowflakes of methane and complex, gasolinelike compounds may fall from the dark sky, as seen in Figure 7-22. Infrared views from the Hubble Space Telescope partially pierced the haze layer and showed a vague, patchy structure, suggesting a non-uniform surface, with more ice-rich deposits in some regions. Lakes of liquid methane and ethane possibly exist in other regions. The most exciting thing about Titan is that its rich organic chemistry might give us some clues to the origins of life on Earth, or possibly elsewhere in the universe. A joint U.S.-European project named Cassini is scheduled to parachute a probe through Titan's clouds to explore the surface environment in 2004. Figure 7-23 shows the expected structure of the atmosphere.

Uranus' Fractured Moon: Miranda

Of Uranus' several modest-size satellites, Miranda—the inner moon—is the most noteworthy. Some 470 km across, the moon's icy surface is heavily cratered, like that of other Uranian moons. However, it has unique swaths of younger, ridged, and fractured terrain, as seen in Figure 7-24. These features are reminiscent of the icy swaths on Jupiter's much larger moon, Ganymede. One cliff face is among the highest in the solar system, rising as high as 5 km (16,000 ft) in a smooth, 45° slope. The question that puzzles researchers is how Miranda became so heavily fractured. Like Jupiter's moon Io, Miranda may have been heated by tidal forces, leading to internal tectonic activity that caused the fractures.

Neptune's Smoking Moon: Triton

Neptune's largest moon, Triton, is the most exotic moon in its satellite system. Triton is strange in several ways. It moves

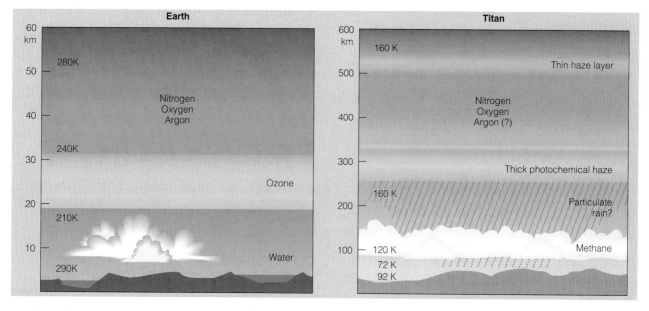

FIGURE 7-23
Earth's atmosphere is compared to the likely structure of Titan's atmosphere. Note that the vertical scales are different; Titan has a much deeper atmosphere than Earth does. The rich brew of simple organic molecules means Titan is an important laboratory for prebiological chemistry.

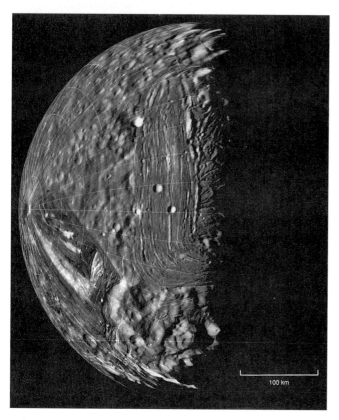

FIGURE 7-24
Uranus' 470-km moon, Miranda, surprised scientists with its strange swaths of fractures and linear ridges, cutting through older cratered terrain. This image has been reprojected by computer to place one of the fracture arrays at the center; faint latitude and longitude lines reveal that the pole (left, along with the pole of Uranus itself) was pointing nearly toward the Sun during Voyager 2's encounter. (SOURCE: NASA Voyager 2 photograph.)

around Neptune in a "backward," or retrograde, direction, opposite to that of most other satellites. This motion suggests that Triton may have originated as an interplanetary body and may have been captured into its present orbit around Neptune. At a diameter of 2705 km, it is about 18 percent bigger than Pluto, which orbits the Sun in the same region (its orbit crosses Neptune's). Triton and Pluto may be two surviving examples of a whole group of bodies that formed near Neptune's orbit, with Triton being captured and Pluto being left behind. Others may have crashed into Neptune long ago. Another unusual trait of Triton is its sparse atmosphere. The atmospheric composition resembles that on Saturn's moon, Titan—mostly nitrogen with some methane. However, the amount of gas and the surface pressure are very tiny, barely one-thousandth that of Mars; you could not survive without a spacesuit on Triton.

The greatest surprise in the *Voyager* pictures of Triton was the discovery of smoking volcanic vents. The photographs show columns of dark smoke rising vertically 8 km (26,000 ft) into the sky, then shearing off in high-altitude winds. The near absence of impact craters also indicates that the surface has been actively resurfaced in current geological time (Figure 7-25).

Where does the energy that drives the eruptions come from? Triton is the coldest body yet studied at close range, with a measured surface temperature of 38 K (−391°F) in the Sun! Theorists have pointed out that if Triton were originally captured into an elliptical orbit around Neptune, tidal forces would have altered its orbit to the present circular shape, heating the interior at the same time. This internal tidal heating was probably the source of the heat that resurfaced Triton with fresh ice, and left smoking vents on its surface (Figure 7-26).

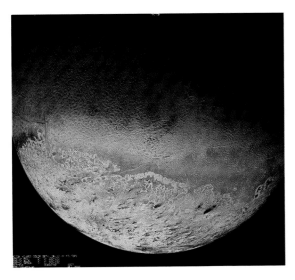

FIGURE 7-25
Neptune's largest moon, Triton, 2700 km across, shows a strangely mottled surface of bright ice and frost deposits. The brighter top area surrounds the south pole and is believed to mark a polar frost cap. Dark streaks are surface deposits of sootlike material blown downwind from geyser vents. The smoke columns (see text) are too delicate to be seen in this picture. (SOURCE: NASA *Voyager 2* photograph.)

FIGURE 7-26
On the surface of Triton. In the distance of this imaginary view is one of the smoking vents of Triton, with a column of dark fumes rising many kilometers and crossing the disk of Neptune, which illuminates the scene. The foreground is an icy chasm venting fumes. (SOURCE: Painting by Ron Miller and WKH.)

Pluto: Ninth Planet or Interplanetary Body?

When **Pluto** was discovered in 1930, it was hailed as the ninth planet, a great discovery of the 20th century. Its discoverer, Clyde Tombaugh, was for many decades the only living person to have discovered a planet. Today, some scientists feel that it would be more accurate to view Pluto merely as one of the largest interplanetary bodies. There are several reasons to downgrade Pluto: (1) Pluto is smaller than our own Moon and less than one-half as big as Mercury. (2) Its orbit is more elliptical than other planets' and crosses the orbit of Neptune. (3) Other small bodies, up to about one-tenth the size of Pluto, have been found in that region; Pluto is merely the biggest one known. (4) As mentioned above, Neptune's satellite, Triton, is about the size of Pluto and may be a captured, Pluto-class body—again indicating that Pluto is not a unique planet. (5) Although Pluto has a moon, general interplanetary bodies have also been found to have moons. Officially, Pluto remains a "planet."

Pluto's satellite, Charon (pronounced *kehr*-on), is one-half as big as Pluto itself—the smallest ratio known between two coorbiting bodies in the solar system. Pluto rotates in 6.4 days, and Charon moves around Pluto in the same period of 6.4 days, so that the two are locked in a dance always facing each other. Pluto has a density close to 2000 kg/m^3, indicating that it is composed of about 70 percent rock and 30 percent ice. Charon's density may be a bit lower; it may be more ice-rich. The daytime surface temperature is about the same as that of Triton, within a few degrees of 38 K ($-391°$F).

Pluto is the only "planet" not yet visited by a spacecraft, so we have no close-up photographs. Even as imaged by the Hubble Space Telescope, it is only a tiny, fuzzy disk (Figure 7-27). Such photographs—plus studies of the brightness changes as Pluto rotates and is eclipsed by its moon—reveal that Pluto has patchy, bright and dark markings that may mark deposits of different kinds of frosts.

Curiously, spectra show that Charon's composition of ice and frost is slightly different than that of Pluto. Pluto has mostly frozen nitrogen ice, with a few percent of frozen methane and carbon monoxide. It is a bright mixture, reflecting 60 percent of incoming sunlight. Charon's surface is a bit darker and is mostly frozen water. Why would the two intimately linked worlds have different surface materials? Researchers think that the lighter molecules, such as methane, escaped from Charon because of its lower gravity. Once the molecules escaped into space within the Pluto-Charon system, some of them collided with the surface of Pluto itself. Thus, over time, Charon has lost methane molecules and Pluto has gained some of them. This would explain the difference in surfaces.

Pluto and Charon are frontier outposts of the solar system. From that cold distance, the Sun would appear only as bright as a streetlight seen at night several hundred yards away.

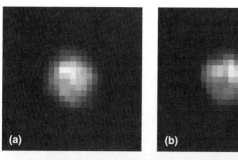

FIGURE 7-27
The sharpest views of distant Pluto are those made with the Hubble Space Telescope. **(a)** and **(b)** These two examples were made at different stages of Pluto's rotation. The nature of the dusky patches, barely resolved, remains uncertain. **(c)** This false-color view shows Pluto and its smaller satellite, Charon. The angular separation is 0.6 arc second; these two objects can barely be separated from the ground. (SOURCE: NASA, Space Telescope Institute.)

Are there any other substantial planetary bodies beyond Pluto? Several astronomers have sought dynamic or photographic evidence of a planet there, sometimes called *planet X*. Clyde Tombaugh, the discoverer of Pluto, conducted a long search and ruled out any planet as large as Neptune near the plane of the solar system, out to a distance of around 100 AU.

However, there may still be some surprises in store. Surveys in 1979, 1988, and following years found no more Pluto-size objects, but did turn up several smaller objects in the region near Pluto. The diameters of these objects are a few hundred kilometers, or about one-tenth the size of Pluto. These discoveries convinced many astronomers that the outer fringe of the solar system is full of icy bodies, of which Pluto may be only the largest (or one of the largest). We will probably not find a full-fledged planet X, Earth-size or larger. However, additional Pluto-size bodies may lurk in the outermost solar system, waiting to be discovered as astronomers look for fainter and fainter bodies on the outskirts of the solar system.

Summary

The four giant planets—Jupiter, Saturn, Uranus, and Neptune—and their satellite systems show some similarities as well as differences. All four of the giant planets have dense, cloudy atmospheres, averaging from 63 to 93 percent hydrogen by mass, with the rest being mostly helium. All have rings composed of billions of small particles ranging in scale from microscopic to house-sized. All have systems of satellites. Most of the satellites contain abundant ice as well as darker soils, because this part of the solar system is so cold that ices were a main constituent of the material available for building moons and planets.

There are also important differences. The colder, more remote planets have hazier, less colorful cloud patterns. Uranus and Neptune are smaller than Jupiter and Saturn, and have a blue color caused by a methane-rich haze, instead of the reddish-brown and tan cloud forms of the larger bodies. Uranus and its satellite system are tipped so that the equator and satellite orbits lie nearly perpendicular to the plane of the solar system. The planets also differ in the structure and composition of their ring systems. Rings are probably composed of tiny pieces of satellites that were fragmented during the history of the solar system by impacts of interplanetary bodies. The appearance of ring systems at any era of history may depend on the time since the latest satellite disruption and the composition of that satellite.

The dozens of satellites of the giant planets were assumed only a generation ago to be geologically dead balls of rock and ice. Direct exploration has shown much greater variety than expected. Tidal heating is important in driving geological activity and even erupting volcanoes on some of these satellites—generally those that are large and close to their planets. Some of the most distinctive satellites include Jupiter's moon Ganymede, the largest known moon; Io, with erupting volcanoes; Saturn's Titan, with its dense, smoggy, nitrogen-rich atmosphere; and Neptune's Triton, which has smoking vents and a thin atmosphere. Pluto, usually classified as the ninth planet, may be only the largest of many interplanetary bodies in that region. It is smaller than our Moon, and it is unique in having a moon about one-half its own size.

The diversity of worlds in the outer solar system is fascinating. It can also be bewildering. So remember that there are simple physical principles that govern their behavior. Gravity is the most important. Gravity causes the vertical structure of giant planet atmospheres, where gas compresses into liquid and eventually a slushy solid. The stretching action of gravity explains why some moons have fractured surfaces and hot interiors. This tidal force also explains why ring systems are only found close to a planet. Moons and ring particles all orbit their planets in miniature application of Kepler's laws. We expect that these ideas will apply to more distant solar systems that we might discover in the future.

Important Concepts

You should be able to define these concepts and use them in a sentence.

Jupiter	prograde rotation
Saturn	ring systems
Uranus	shepherd satellites
Neptune	resonance
gas giant planets	Roche's limit
icy material	sphere of gravitational influence
Great Red Spot	Pluto
retrograde rotation	carbonaceous material

How Do We Know?

These questions and answers show how the scientific method is used to learn about the universe.

Q How do we know the composition of the inside of a giant planet?

A The low mean density shows that the gas giants must be made of light elements and icy compounds. Density is mass divided by volume, so the giant planets are not massive enough for their size to be mostly made out of rock. The *Galileo* parachute probe directly measured the elements and compounds in Jupiter's atmosphere in 1995. Laboratory measurements and chemical studies have shown the various forms these materials would take under different pressures. Calculations using the law of gravity and the gas laws allow us to estimate the pressures and the forms of matter that would exist at different depths inside Jupiter.

Q How do we know that the rings of Saturn and other planets are composed of millions of small moonlets, rather than being a single flat plate?

A Observations from Earth (as early as the 1800s) and from space probes have shown that the rings do not revolve around the planet as a single plate. The rings are transparent with a lot of space between the many particles. Also, the inner part of each ring system moves faster than the outer part, in accordance with Kepler's and Newton's laws for movements of independent satellites around planets. Last, photographs from spacecraft have revealed the largest independent moonlets within the rings.

Q How do we know that Jupiter's satellite, Io, has active volcanoes?

A The *Voyager* space probes photographed the plumes of debris erupting from volcanic vents. Later, infrared detectors on large Earth-based telescopes in the 1980s and the *Galileo* space probe in the 1990s confirmed the eruptions by measuring the heat from the vents. The same observations revealed cycles of volcanic activity, with some vents shutting down and new eruptions occurring elsewhere on Io. We also have a theoretical explanation for the activity—the volcanoes are driven by tidal heating from nearby, massive Jupiter.

Q How do we know that many of the satellites of giant planets are covered with ice?

A We do not have any samples of the ice. However, the colors, reflectivity, and properties of the spectrum of the reflected light are different for ices than for rocky minerals. Measurements of all these properties show that the surface materials of many satellites of giant planets are dominated by frozen ices of various compounds such as water (H_2O), methane (CH_4), and other materials. Chapter 10 will give more details on the techniques of identifying compounds from the spectra.

Q How do we know that there are only nine planets?

A We can be very sure that we have not "missed" any planets closer to the Sun than Neptune. A planet inside Earth's orbit would appear close to the Sun in the sky (as Venus and Mercury do), and if sizeable, it would be bright because it received and reflected a lot of the Sun's light. In the outer solar system, a planet would be fainter, but we know we have not missed anything large because if we had, the orbits of the outer planets would show large irregularities. Astronomers have searched for remote bodies, especially along a swath of sky centered on the ecliptic. They take photographs every few days or weeks and look for faint images that move from one photograph to the next with respect to the fixed stars. This kind of painstaking research has ruled out anything much larger than Pluto in the frigid depths of the outer solar system. And remember, Pluto has been downgraded in the minds of many astronomers to the status of a mere interplanetary body. The case can be made that there are only eight true planets.

Problems

Use these problems to test your understanding of the information and concepts in this chapter. The * symbol indicates a more advanced or mathematical problem.

1. Scientists believe that no life exists in the outer solar system, but the possibility has not been ruled out. Suppose you were searching for such life. Which locations among the planets and satellites of the outer solar system might have conditions most like those on Earth and thus be possible locations for at least primitive microbes? What conditions would argue against life in each of these sites?
2. Why does Earth not have a hydrogen and helium atmosphere as the giant planets do?
3. **(a)** Suppose two giant planets have the same mass but are located at different distances from the Sun. Temperature measurements reveal that the upper atmosphere of the closer planet is much hotter than that of the other planet. Which one would you expect to have the most hydrogen in its atmosphere? Explain your reasoning. **(b)** Titan is smaller than Earth but has a denser atmosphere. Use the relative temperatures of Titan and Earth to make at least one hypothesis of why Titan might retain a denser atmosphere.
4. Describe the seasons and other effects that would occur if Earth had the same axial tilt as Uranus.
5. Jupiter's moon Europa is almost the same size as our Moon, but contains a large percentage of ice instead of rock. Would the gravity on Europa be greater than, the same as, or less than that of the Moon? Explain your reasoning.
6. Why might it be more dangerous to fly through the rings of Saturn than through the ring of Jupiter?
7. The techniques for detecting planets around other stars are advancing rapidly. A number of extra-solar planets are already known. Suppose that we find a star like the Sun with five planets in orbit around it. **(a)** A planet with the mass of Earth at one-half the distance of Mercury. **(b)** A planet of Jupiter's mass at the distance of Earth. **(c)** A planet with

the mass of Earth at the distance of Jupiter. **(d)** A planet 10 times the mass of Jupiter at the distance of Neptune. **(e)** A planet with the mass of Jupiter at twice the distance of Pluto. In each case, speculate on general physical state of each planet in terms of temperature and atmospheric composition.

*8. Take the hypothetical planets from the previous problem, and calculate in each case the likely surface temperature compared to that of Earth. What are the likely atmospheric constituents of each planet?

*9. Look up densities and make a quantitative problem out of them.

*10. Imagine a planet with a radius of 50,000 km and a surrounding ring system and two moons. **(a)** What would you predict the outer radius of the ring system to be, based on Roche's limit? **(b)** If one moon has a distance of 1 million km from the planet and an orbital time of 30 days, what is the orbital time of the second moon at a distance of 3 million km?

Projects

Activities to carry out either individually or in groups.

1. With large binoculars or a small telescope, observe Jupiter. Can you see the satellites? How many do you see? Why might you see fewer than the four large moons that Galileo discovered? Sketch the positions of the satellites relative to the disk of Jupiter, and record the positions from hour to hour (or have different class members observe at different hours). Can you see changes in the positions? Of the four, why would Io be the fastest-moving and Callisto be the slowest-moving? This is the experiment carried out by Galileo in the early 17th century that played a large role in the Copernican revolution. See if you can find a copy of Galileo's book *Siderius Nuncus*, in which he described his observations.

2. Observe Saturn with a modest-size telescope. Can you see the rings? Sketch the appearance of the rings. Can you estimate their inclination angle? Are any of Saturn's many moons visible?

3. Observe Jupiter with a telescope that has an aperture of at least 6 in. At first it will look like a blank, bright disk. Keep studying the disk, and you will see darker cloud bands crossing the disk. Can you see any further details in the cloud bands? Can you see the Great Red Spot? Why might the Red Spot not be visible? Sketch the appearance of the planet. How do the colors appear compared to the NASA photographs in this book? What do the colors indicate about the atmospheric composition of Jupiter?

Reading List

Anderson, J., Sjogren, W., and Schubert, G. 1996. "Galileo Gravity Results and the Internal Structure of Io." *Science,* vol. 272, pp. 709–712.

Araki, S. 1991. "Dynamics of Planetary Rings." *American Scientist,* vol. 79, pp. 44–59.

Binzel, R.P. 1990. "Pluto." *Scientific American,* vol. 262, p. 50.

Cruikshank, D. P., and seven others, 1995. "Ices on the Surface of Triton." *Science,* vol. 261, pp. 742–5.

Esposito, L. W. 1993. "Understanding Planetary Rings." *Annual Review of Earth and Planetary Science,* vol. 21, pp. 487–523.

Hubbard, W. B. 1997. "Neptune's Deep Chemistry." *Science,* vol. 275, pp. 1279–80.

Jewitt, D. C. 1994. "Heat from Pluto." *Astronomical Journal,* vol. 107, pp. 372–8.

Johnson, T. V. 1995. "The Galileo Mission." *Scientific American,* vol. 273, pp. 44–51.

Littmann, M. 1989. "Where Is Planet X?" *Sky and Telescope,* vol. 78, p. 596.

Lunine, J. 1993. "Triton, Pluto, and the Origin of the Solar System." *Science,* vol. 261, pp. 697–8.

Morrison, D. 1980. *Voyage to Jupiter.* Washington: NASA Special Publications, no. 439.

Muhleman, D. O., Grossman, A. W., Butler, B. J., and Slade, M. A. 1990. "Radar Reflectivity of Titan." *Science,* vol. 248, pp. 975–980.

Owen, T. C., and nine others. 1993. "Surface Ices and the Atmospheric Composition of Pluto." *Science,* vol. 261, pp. 745–8.

Peale, S., Cassen, P., and Reynolds, R. 1979. "The Melting of Io by Tidal Dissipation." *Science,* vol. 203, p. 892.

Stern, S. A. 1996. "The Pluto-Charon System." *Annual Reviews of Astronomy and Astrophysics,* vol. 30, pp. 185–233.

Tittemore, W. C. 1990. "Chaotic Motion of Europa and Ganymede and the Ganymede-Callisto Dichotomy." *Science,* vol. 250, pp. 263–7.

Interplanetary Bodies

CHAPTER 8

CONTENTS
- The Tunguska Event
- Interplanetary Bodies: Many Types
- Comets
- Meteors and Meteor Showers
- Asteroids
- Meteorites
- Interplanetary Threat or Opportunity?
- Summary
- Important Concepts
- How Do We Know?
- Problems
- Projects
- Reading List

Reconstruction of the scene during the meteor explosion over Siberia in 1908, based on eyewitness accounts collected by Russian scientists. This view is from a trading station about 60 km (40 mi) from the blast, where witnesses described the sky opening up with fire and felt a searing blast of heat. One witness was knocked 20 ft off a porch. Scientists believe the explosion was caused by the impact of a stony meteorite 50 to 60 m in diameter, which exploded above the ground because of the stress of hitting the atmosphere. (Painting by WKH.)

PREREADING QUESTIONS ON THE THEMES OF THE BOOK

OUR ROLE IN THE UNIVERSE
How have interplanetary bodies affected life on Earth?

HOW THE UNIVERSE WORKS
What governs the orbits and composition of interplanetary bodies?

HOW WE ACQUIRE KNOWLEDGE
With little physical evidence, how did we learn the true nature of comets, meteors, and asteroids?

WHAT TO WATCH FOR IN CHAPTER 8

Over the past three chapters we have explored the major bodies of the solar system, starting with the familiar Earth and Moon and continuing with the terrestrial planets and the gas giant planets. This chapter describes the physical nature of the millions of small bodies that move in orbits around the Sun, among the planets. The steady process of scientific discovery has shown that these bodies have a range of different orbits and vary in composition from rocky to icy. Scientists have tracked these small bodies and measured orbits for them through telescopes. Spacecraft have come close enough to take pictures in a few cases. Scientists have also studied the fragments left behind when the smallest ones crash into Earth.

All the major interplanetary bodies combined—meteors, asteroids, and comets—amount to a tiny fraction of the mass of all the planets. Their importance stems from two aspects that relate to the theme of "our place in the universe." Cosmic debris has rained down on Earth and other planets, with occasionally catastrophic consequences. We consider the probabilities of impact from the full range of interplanetary bodies. Also interplanetary bodies can be very old, and they have been well preserved in the deep freeze of the outer solar system. They provide clues about the origin of Earth and the other planets. These clues will be addressed further in the next chapter.

THE TUNGUSKA EVENT

On the morning of June 30, 1908, a mysterious explosion occurred in the skies over Siberia. Herdsmen 500 km (300 mi) away reported "deafening bangs" and a fiery cloud on the horizon. Three times closer, the explosion caused thunderlike noises, flung carpenters from a building, and knocked crockery off shelves. An eyewitness 60 km from the blast reported that "the whole northern part of the sky appeared to be covered with fire.... I felt great heat as if my shirt had caught fire [and] there was a... mighty crash.... I was thrown onto the ground about 7 m from the porch.... A hot wind, as from a cannon, blew past the huts from the north.... Many panes in the windows were blown out, and the door of the barn was broken." A reconstructed view of the explosion from this location, based on eyewitness reports, is shown in the painting that opens this chapter.

Probably the closest observers were a group of reindeer herders asleep in their tents in several camps about 30 km (20 mi) from the site. They were blown into the air and knocked unconscious; one man was blown into a tree and later died. "Everything around was shrouded in smoke and fog from the burning fallen trees." What could have caused this dramatic and frightening event?

The explosion was caused by an object from space that struck Earth's atmosphere and exploded. Seismic vibrations were recorded 1000 km (600 mi) away. About 170 km (110 mi) from the explosion, observers saw the object in the cloudless sky as a brilliant, fiery ball much larger than the full Moon in apparent size. The mysterious object was a modest bit of interplanetary debris, probably rocky in composition, with a diameter of about 50 to 60 m. Many such fragments circle the Sun; the Siberian object was merely the largest to hit Earth in the last century or so. It was sheer luck that it hit one of the most remote parts of Earth's surface and that few people were hurt or killed. Had it hit a populated area, the devastation could have been enormous.

Recent studies reveal that such explosions may happen every couple of centuries (and less often over populated land, which represents less than one-seventh of Earth's area). As discussed in Chapter 6, even larger objects have hit Earth, but these occurrences are rarer. For example, an iron asteroid fragment perhaps 100 m across hit Arizona about 20,000 years ago, leaving a well-preserved crater 1 km wide, and it is likely that a 10-km asteroid hit Earth 65 million years ago, ending the reign of the dinosaurs. Small impacts are much more common than large ones. Brick-size interplanetary stones fall from the sky in various locations every year. Several houses, one person, and a car have been hit in recent decades. Tiny dust grains are even more common; they can be seen every night if you watch long enough. These bright streaks of light are sometimes called *shooting stars.*

Interplanetary space contains many small bodies of different sizes. All move in elliptical orbits around the Sun, as prescribed by Kepler. Occasionally their orbits intersect those of planets, leading to a collision. Large enough bodies leave sizable craters that we can see on the surfaces of planets and moons throughout the solar system. Our study of interplanetary bodies will echo the themes of the book. We will see that science has been able to explain these "visitors in the sky," which had mystified people for thousands of years. We will see that the physical nature of comets, meteors, and asteroids gives us clues to the formation of the solar system. Finally, we will learn more about the large pieces of debris that have interrupted the progress of life on Earth.

Interplanetary Bodies: Many Types

Scientists considered interplanetary bodies only a minor curiosity a generation ago. Today, there is growing interest because we realize that they affect planet histories in general and the evolution of life on Earth in particular. As we saw in Chapter 5, our presence on Earth depends on the history of impacts by interplanetary debris. Furthermore, these bodies contain many clues about the origin of our solar system. We will use this information later in the book to estimate the likelihood that planets orbit other stars.

Interplanetary bodies range in composition from icy to rocky and metallic. Among the larger bodies, the icy ones are called **comets,** and the rocky and metallic ones are called **asteroids.** As with so many words, there are historical reasons for these names. When the ices of a comet are warmed by the Sun, they change into gas and evaporate away into space, or **sublimate.** This makes comets give off gas and dust particles. The gas and dust are what makes the fuzzy, luminous "tail" of a comet. To ancient people, the tail looked like long hair blowing in the wind. The name *comet,* therefore, comes from the Greek word *coma,* for hair. An asteroid, on the other hand, has no gas or tail and appears in the telescope as a faint star. Its name derives from the Greek root *aster-,* for star. We know now that the light we see from an asteroid is all reflected from the Sun—asteroids are cold chunks of rock that emit no visible light of their own.

Until the last century or so, comets and asteroids seemed to be totally distinct phenomena. But our study of how science works has shown that we can make progress by looking for connections between things. Today we realize that both comets and asteroids are examples of interplanetary debris left over from the period of planet formation; they differ mostly in composition. Smaller bits of debris (either icy or rocky) in interplanetary space typically range from microscopic grains to bodies up to a few meters across.

When a piece of space debris hits the atmosphere or surface of Earth or another planet, it is typically traveling at 10 to 40 km/s. This is a range of up to more than a hundred times the speed of sound! As a massive object, it therefore has tremendous kinetic energy. What happens next is an excellent example of the transformation of energy from one form to another (review Chapter 4). Hitting a planet's atmosphere at this speed causes friction with the air. The projectile slows and its surface heats due to the friction with the atmosphere. In this process, its energy is transformed from kinetic energy to heat energy. (The heating you feel when you rub one hand against the other is another example in which kinetic energy is turned into heat.) The incoming object heats and begins to glow. As a result of the shock of hitting the atmosphere and heating, it may break into many pieces. This is also why a spaceship reentering the atmosphere from orbit becomes hot, and why reentry is a critical and dangerous procedure.

We may be lucky enough to see these phenomena. Scientists distinguish between the pieces of interplanetary bodies that reach Earth's atmosphere and the fraction that actually hit the Earth. **Meteors,** or so-called shooting stars, are pea-size and smaller rocks that burn up in the atmosphere and do not hit the ground. **Meteorites** are larger rocky or metallic bodies, or pieces of them, that do hit the ground. These two words come from the same root as the word *meteorology,* (the study of weather). For hundreds of years, people thought that shooting stars were purely terrestrial phenomena that originated in Earth's atmosphere. Thousands of meteorites have been collected and studied. You can see them in museums and planetariums. They are free samples of the distant reaches of the solar system.

When scientists study a meteorite, they recognize that it is just a fragment of something larger, and they try to deduce the nature of the object that it came from. This larger object is called the **parent body.** Studies of meteorites strongly support the idea that their parent bodies are asteroids. Asteroids have collided with one another, as well as with planets, throughout geological time. The biggest collisions smash the asteroids and produce fragments that drift in space. Some of these fragments have paths that cross Earth's orbit, and a few fragments reach the ground as meteorites.

An additional challenge is to analyze each meteorite and find out the location and orbit of its parent body. One way of answering such questions is to study the orbits of the objects, illustrated in Figure 8–1. Table 8–1 compares some orbits for the four classes discussed in this chapter: comets, asteroids, meteors, and meteorites. This detailed information

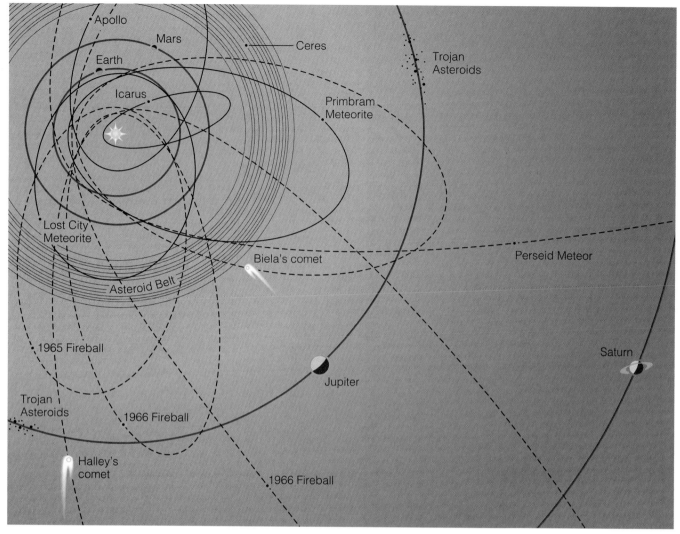

FIGURE 8–1
A portion of the solar system, illustrating a small fraction of the crowded population of interplanetary bodies. Planet orbits are in red; orbits of selected comets, main belt asteroids, Trojan asteroids, Earth-approaching asteroids, meteorites, and large meteors (fireballs) are in black.

is the result of years of scientific studies and it gives us clues to the origins of the interplanetary bodies.

One important thing to notice about Figure 8–1 is the concentration of asteroids in the region between the orbits of Mars and Jupiter. This group of asteroids is called the **main asteroid belt**. In Table 8–1, notice the column labeled "Maximum Distance from Sun"; this is the farthest each body can get from the Sun as the body travels in its elliptical orbit. As you can see, most meteorites have orbits whose farthest point lies in the main asteroid belt; this supports the idea that meteorites come from asteroids. Asteroids and meteorites are rocky bodies that formed closer to the Sun where it was too warm for ice to exist. On the other hand, the table also shows that many comets and meteors have traveled from much vaster distances, far beyond Pluto. This indicates that most meteors are bits of debris dislodged from comets. The greater distance of origin also gives a clue about the differences in composition between comets and asteroids.

Comets formed in the cold, outer reaches of the solar system, far from the Sun, and that is why they are icy.

The comet and asteroid parent bodies can trace their origins to the birth of the solar system 4.6 billion years ago. As the planets formed, the solar system was filled with innumerable small, preplanetary bodies, ranging up to 1000 km across. **Planetesimals** is a generic term used to refer to these ancient preplanetary bodies, whether icy or rocky. Thus, comets, asteroids, and their fragments are all descended from the original planetesimals from which the planets formed. As a context for the material in this chapter, we should recall that 99.85 percent of the mass of the solar system is locked up in the Sun. Jupiter accounts for 0.1 percent, and all the other planets together account for another 0.04 percent. The interplanetary bodies we will discuss in this chapter amount to no more than 0.01 percent, or 1 part in 10^4, of the solar system mass. Yet, as we will see, they can have spectacular effects on Earth.

TABLE 8-1
Properties of Selected Small Bodies in the Solar System

CLASS	EXAMPLE	DIAMETER (KM)	ORBIT SEMIMAJOR AXIS (AU)	MAX. DISTANCE FROM SUN (AU)	ECCENTRICITY	INCLINATION	REMARKS
Comets							
Short-period	Encke	1–8	2.2	4.1	0.85	12°	Probably dirty ice
	Halley	Few?	18	35	0.97	162°	Probably dirty ice
Long-period	Kohoutek	8 × 15	Very large	Very large	1.0	14°	Probably dirty ice
Meteors							
Shower	Perseid	10^{-6}	40	79	0.97	114°	Cometary debris
	Taurid	10^{-6}	2.2	4.0	0.80	2°	
Fireballs	July 31, 1966	?	32	63	0.98	42°	May be related to comets or asteroids
	May 31, 1966	?	3.0	5.4	0.80	9°	
Asteroids							
Belt	1 Ceres	1020	2.8	3.0	0.08	11°	Carbonaceous rock surface
	2 Pallas	538	2.8	3.4	0.23	35°	Rocky surface
	3 Juno	248	2.7	3.4	0.26	13°	Rocky Surface
	4 Vesta	549	2.4	2.6	0.09	7°	Lavalike surface
	14 Irene	170	2.6	3.0	0.16	9°	Rocky surface
Trojan	624 Hektor	100 × 300	5.1	5.2	0.02	18°	Unusual shape
Apollo	433 Eros	7 × 19 × 30	1.5	1.8	0.22	11°	Elongated
	1862 Apollo	1.4	1.5	2.3	0.56	6°	
Comet/asteroid	2060 Chiron	100–320?	13.7	19	0.38	7°	Erupted in 1988[a]
Meteorites (chondrites)							
	Pribram	10^{-4}	2.5	4.2	0.68	10°	
	Lost City	10^{-4}	1.7	2.4	0.42	12°	
	Leutkirch[b]	10^{-4}?	1.6	2.2	0.40	2.5°	

[a]Chiron was cataloged as an asteroid after its discovery in 1977. It brightened and developed a coma in 1988, proving it is really an icy body. It illustrates the "fuzziness" of comet/asteroid nomenclature.

[b]An object photographed over Europe in 1974. No fragments were recovered at the time, but it is believed to be a stone meteorite.

SOURCE: Data from Hartmann, Binzel; and collaborators; Beatty and Chaikin. Diameter of Comet Encke measured with radar by Kamoun and collaborators.

Comets

Comets are the most spectacular of the small bodies in the solar system. They can pass close to Earth and appear as bright objects visible from much of the world (Figure 8–2). A bright comet appears, on average, about once per decade. When a comet passes through the inner solar system near the Earth, it can be seen drifting slowly from night to night among the stars. (Writers sometimes incorrectly describe comets as "flashing across the sky" like shooting stars. They do not. They hang motionless and ghostly among the stars, and seem to change position only from hour to hour or from night to night.) We saw in Chapter 2 that young Tycho Brahe used his observations of a comet to show that it passed among the planets. This insight shattered the ancient Greek idea of crystalline spheres and set the stage for the Copernican revolution.

Comets have several parts, as seen in Figures 8–2 and 8–3. The brightest diffuse part is the **comet head,** sometimes called the *coma*. The **comet tail** is a fainter glow extending out of the head, usually pointing away from the Sun.

Although bright comets are widely discussed in the media, many people are disappointed when they see one because they observe from a city. As shown in Figure 8–3, the glow from city lights washes out the delicate glow of a comet. A drive to a dark site in the country, away from the city glow, produces a completely different impression; a bright comet's tail may extend across 20° or more of the sky, like a delicate, arcing searchlight beam.

A telescope reveals a brilliant, starlike point at the center of the comet head. At the center of this bright point is the **comet nucleus,** which is the only substantial, solid part of the comet. The nucleus is too small to be resolved by telescopes on Earth. A typical comet nucleus is a tiny world of dirty ice only about 1 to 20 km (a few miles) across—tiny compared to most planets and moons! The gas and dust that make up the rest of the comet's head and tail are material emitted from the nucleus. A comet's faintly glowing head is usually far larger than the largest planet, but it is only a thin cloud of gas and dust emitted from the nucleus.

As a comet nucleus moves from the outer solar system to the inner solar system, the sunlight warms it and causes the ice to sublimate into gas. Expansion of the gas carries it, to-

FIGURE 8-2
Halley's comet during its approach in 1986, when it was in the Southern hemisphere and not well placed for northern observers. The photograph shows the pinkish dust tail and the bluish gas tail. (Photograph by William Liller from Easter Island for NASA's International Halley Watch; 4-min exposure on Fujichrome 400 film with an 8-in Schmidt telescope at f1.5.)

FIGURE 8-3
Proof that city dwellers miss out on comet displays. These two photographs of Comet Hale-Bopp were made with the same camera, lens, and roll of film less than an hour apart. Photograph **(a)** is from a dark sky country location, while photograph **(b)** is from the center of Tucson, only about 5 mi away. Comet Hale-Bopp was a brilliant comet with a 5° to 10° tail, easily visible to the naked eye in 1997, but most city dwellers saw only the fuzzy head as a pale blob, as in photograph **(b)** because the sky is washed out by the glow of city lights. Notice how many more faint stars are visible from the dark sky site even though the photographic exposures are the same. (Kodak ISO 1600 film, 35-mm tripod-mounted camera, 100-mm telephoto lens, f2.8, 60-s exposures with no guiding; photographs by WKH.)

gether with dislodged dust grains from the dirt in the nucleus, away from the nucleus. As this cloud of gas and dust spreads away from the nucleus, it meets radiation and thin, ionized gas rushing outward from the Sun. The out-rushing solar gas is called the **solar wind**. Because solar wind and radiation are moving away from the Sun, they always carry the comet gas and dust in a direction away from the Sun. Only microscopic grains get caught in the solar wind; larger grains are too heavy. (Similarly, if you release a handful of gravel and dust, the dust is blown by the wind but the stones are not.) The gas and fine dust from the nucleus get swept into a long tail.

Comet tails often stretch more than an astronomical unit—longer than the distance from Earth to the Sun! Caught in the solar wind, the tail streams out behind the comet as the comet approaches the Sun but leads as the comet recedes from the Sun (Figure 8-4). It might be likened to a woman's long hair streaming out behind her as she walks into the wind, but in front of her if she reverses direction. In fact, we have simplified the description slightly—a comet actually has two components to its tail. The ions are swept directly away from the Sun by the solar wind. The dust particles have complex interactions with solar radiation and can form a gently curving structure.

Comet Names and Orbits

In ancient times, before science explained the phenomenon, people interpreted comets as evil omens. The mysterious appearance of a visitor in the night sky was as frightening as the sudden chill and darkness of a solar eclipse. For example, Halley's comet has been recorded throughout history. In the year 66, its arrival was said to have heralded the destruction of Jerusalem. Five circuits later it was said to mark the defeat of Attila the Hun in 451. In 1066 it presided over the Norman conquest of England. In 1456, the appearance of a comet coincided with a threatened invasion of Europe by the Turks, who had already taken Constantinople three years before. Pope Calixtus III ordered prayers for deliverance "from the devil, the Turk, and the comet."

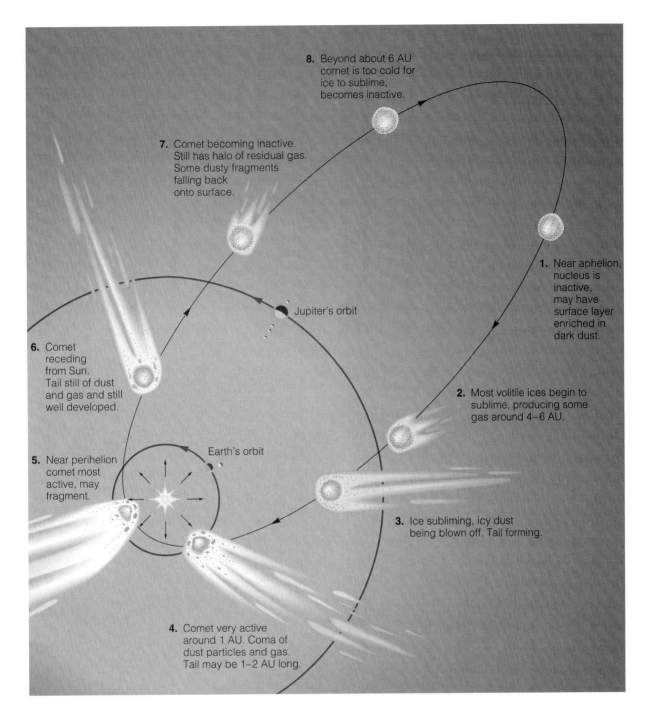

FIGURE 8-4
Stages in the development of a typical comet as it travels from the outer solar system into the inner solar system, where it loops rapidly around the Sun and moves outward again. The tail develops only within a few astronomical units of the Sun, where sunlight is warm enough to sublimate the ice in the nucleus. In order to show comet detail, the figure is not drawn to scale.

These ancient superstitions were debunked by English astronomer Edmond Halley. Halley had thought long and hard about planetary orbits, and he played a key role in the emergence of Newton's theory of gravity. One day in 1684, Halley was having lunch in a London pub with two other scientists. He felt sure that Kepler's elliptical orbits could be explained by a gravity force that diminished with the inverse square of the distance from the Sun. They all made a wager over who could prove this assertion. They each failed. Halley then visited Newton in Cambridge and posed the same question to him. He found that Newton had already done the proof, and he urged Newton to publish it. From then on, Halley was the irresistible force that helped Newton to finish the *Principia*—perhaps the greatest science book ever written. Halley paid the publishing costs himself and sent copies to many leading scientists and philosophers throughout Europe.

In 1704, Halley was working with Newton's theory of gravity and some newly developed methods of computing orbits. He discovered that comets travel in long, elliptical orbits around the Sun and that certain comets reappear many times. Calculating the orbits of 24 well-documented comets, Halley found that 4 comets (seen in 1456, 1531, 1607, and 1682) had the same orbit and appeared approximately 75 years apart. Halley correctly inferred that these appearances were by a single comet. He even showed that slight irregularities in the time of appearance of the comet were caused by the gravitational influence of other planets, especially Jupiter. Halley predicted that the comet would return late in 1758. It did on Christmas night of that year.

The discovery that comets are visitors on ordinary elliptical orbits, with predictable motions, helped dispel the superstition that comets are evil omens. In fact, the story of Halley's comet is an excellent example of several themes of this book. Halley showed that Newton's theory of gravity applied just as well to the elongated orbit of a comet as to the nearly circular orbit of a planet. His discovery extends the application of this simple physical law. His discovery is also a classic example of how science works. Before Halley, comets were viewed as random and mysterious visitors. Halley linked four separate events to a single object and revealed a pattern in nature. He was then able to provide a mathematical description of the pattern—in terms of elliptical orbits—and to make a successful prediction for the next appearance in the sequence. For many people in the 18th century, this cemented the idea that Nature was predictable. It seemed that we live in a "clockwork universe."

Since the days of Halley, the brightest comets have been named for their discoverers. Many comets are not visible to the naked eye, so every comet is given an anonymous, scientific name. Also telescopes and computational techniques have advanced greatly since Halley's time. We do not have to wait for a return visit to recognize a comet. The orbits of most comets are calculated within a few weeks of their initial discovery. Halley's comet renewed its fame in 1910, when Earth passed through its tail. It was dimmer in 1986 because it did not pass as close to Earth (Figure 8–2). A recent bright comet was comet Hale-Bopp, which was well placed for observation in 1997 (Figure 8–3). Comets are beautiful visitors to our nighttime skies (Figure 8–5).

FIGURE 8-5
Comet Hyukatake near the time of its closest approach to Earth, in late March 1996. Hyukatake passed near Polaris and was a circumpolar object for many nights; its trajectory was very far from the ecliptic. (Courtesy of R. W. Doty.)

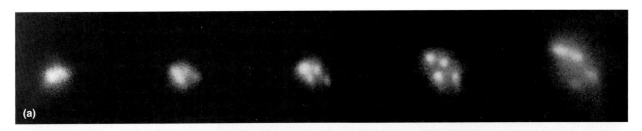
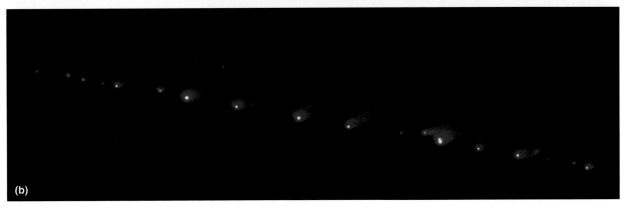

FIGURE 8-6
Several comets have been seen to break into pieces. **(a)** These five photographs show the breakup of Comet West (1975n) into four pieces in 1976. The four nuclei separated at relative speeds of about 1 to 5 m/s (2 to 11 mi/h) over a period of 3 days. One piece lasted only 2 weeks; others were followed for over 6 months. (New Mexico State University.) **(b)** The extraordinary comet Shoemaker-Levy 9 was pulled apart into a string of fragments by tidal forces when it passed close to Jupiter in 1992. The 21 icy fragments are stretched across 1.1 million km, or 3 times the distance from Earth to the Moon. (Hubble Space Telescope photograph; H. Weaver, T. Smith, and NASA.)

The Comet Nucleus: A Dirty Iceberg in Space

By the 1700s, after astronomers had demonstrated that comets were objects orbiting the Sun, the next mystery to solve was the physical nature of the comet itself. Progress on this question came in 1868 when English astronomer William Huggins first studied comets through a spectroscope. Spectroscopes spread visible light into an array of wavelengths, or a spectrum, revealing the amount of light reflected from an object in different colors. The spectrum in turn allows astronomers to measure properties of the material in a celestial body. This method will be discussed in greater detail in Chapter 10.

Huggins found that the comets he observed were made up in part of gaseous carbon. Modern astronomers have shown that comets contain the same atoms we see elsewhere in the solar system, including H (hydrogen), O (oxygen), N (nitrogen), and C (carbon). Comet spectra reveal many of the molecules that can be formed from these four elements, including CN, CH, OH, H_2O^+, CN^+, CH^+, OH^+, N_2^+, CO^+, and CO_2^+. (The superscript + indicates an ion, atom, or molecule that has lost an electron due to heat. When an atom loses an electron, it acquires a positive charge.) Recent studies have revealed complex organic molecules, such as CH_2CN. This discovery is important because it shows that organic molecules—the building blocks of life—form elsewhere in the universe besides Earth.

Around 1950, using all the available observations, Harvard University astronomer Fred Whipple was the first to deduce that a comet nucleus must be a block of dirty ice. He noted that the composition of gas streaming off the nucleus indicated sublimating ice, and that the dust particles must have been trapped in the ice of the nucleus. Whipple's theoretical picture of a comet nucleus came to be called the **dirty iceberg model of comets,** and it has been confirmed by close-up observations from spacecraft.

The ice in many comet nuclei is weak, so that some nuclei can spontaneously break into pieces, either from tidal gravitational forces that stress them when they pass too close to the Sun or a planet, or from pressure that builds up as the ice turns to gas. Figure 8-6 shows two examples of comets that broke apart in this way.

Do you see the connection between the atoms and molecules observed in comets and their largely icy composition? Note that all these atoms (H, O, C, N) and molecules are just the ones that would be expected if the gas were to form from the sublimation of common solar system ices such as H_2O (water), CH_4 (methane), NH_3 (ammonia), and CO_2 (carbon dioxide). In other words, the H, O, C, and N atoms and molecules streaming off comets are coming from the sublimating ice. In addition to revealing the composition of comet gases, the spectroscope shows that some particles in comet tails are much bigger than molecules. These particles, or dust grains, form slightly curved tails. Gas ion tails, by contrast, are straight (Figure 8-7). Magnetic forces straighten the stream of ions in the gas tails, while the uncharged dust grains curve.

Direct confirmation of Whipple's model came when an international fleet of five spacecraft (two Japanese, two

FIGURE 8-7
Comet Hale-Bopp photographed in March 1997. The bluish tail of ionized gas is straight and streams directly away from the Sun. The more creamy-white tail of dust particles is curved. (Photograph by Nick James.)

Russian, and one European) probed Halley's comet in 1986. The European probe came closest, flying only about 600 km from the nucleus. That probe was named *Giotto,* after the Italian artist whose 1304 painting of Halley's comet made him one of the first artists to paint a comet. The probe's close-up pictures revealed a dark-colored, floating iceberg-like object, about 15 km long and 8 km wide (Figure 8–8). For comparison, this is about the size and shape of Deimos, the smaller moon of Mars.

Giotto and the other probes showed that the region around the nucleus was a hazy, dusty environment. The haze made the pictures fuzzy, and *Giotto* was hit by a dust grain when it was about 960 km from the nucleus, knocking it partially out of commission for 32 min during the closest approach! However, the spacecraft recovered in time to take spectacular images of the nucleus. Halley's nucleus is as dark as black velvet. It reflects only 4 percent of the light that strikes it, compared with 60 to 80 percent for clean ice. Earth-based observers have found similar results for other comet nuclei. The black material is believed to be the carbon-rich dust, finely distributed through the ice. Jets of gas and dust shoot off "active" spots on the black nucleus. The gas in the head of the comet, around the nucleus, was found to be about 80 percent water vapor by volume.

Giotto also gave us excellent information on the chemical composition of a comet. The carbon-rich dust particles in the nucleus were also found to be rich in hydrogen, oxygen, and nitrogen, somewhat resembling meteoritic particles collected on Earth. These extraterrestrial dust particles are believed to be rich in organic molecules (large, carbon-based molecules) and are different from familiar terrestrial dust, which is richer in silicon, iron, other metals, and their oxides.

FIGURE 8-8
Giotto space probe's close-up view of the 15-km (9-mi) long nucleus of Halley's comet. In the weeks after the encounter, computer processing yielded this enhanced color view based on the closest images. Bright jets shoot out of localized active vents on the sunlit side. Smallest details are about the size of a football field. The nucleus is dark gray, and its dark side can be seen silhouetted against the softly glowing comet head (right side). (*Giotto* photograph courtesy of Harold Reitsema and Alan Delamere, Ball Aerospace; copyright 1986 MPAE.)

Interest in this organic-rich cometary dust is growing. As comets collided with the young planets, comets may have provided water and organic molecules and they may even have aided in the formation of life on Earth.

The Oort Cloud and the Kuiper Belt

In ancient times no one knew how far away comets were. Many people thought that they were phenomena in our own atmosphere. Seneca, the Roman contemporary of Jesus, wrote: "Someday there will arise a man who will demonstrate in what regions of the heavens the comets take their way." That man was Tycho Brahe. We saw in Chapter 3 that he arranged observations of a bright comet in 1577 from two different locations. The comet showed a parallax shift with respect to the stars as seen from the two places. Brahe proved by triangulation that the comet was more distant than the Moon. This ruled out the old theory that comets were terrestrial. He used direct observations and simple geometry to sweep away centuries of mystery and speculation. This is the way that science works as we try to make sense of the universe.

But where did comets go when they were not visible to the eye or the telescope? The solution to this puzzle began with the work of Dutch astronomer Jan Oort and the Dutch-American astronomer Gerard Kuiper in the 1950s. Oort was a remarkable astronomer who worked on everything from comets to cosmology; he was an active researcher well into his nineties. He also established a strong tradition of astronomy in Holland that continues to this day. Kuiper was a visionary. He spent a number of years searching for the best places to do observational astronomy—the ideal sites are high, dry, and dark. Kuiper was convinced of the merits of Mauna Kea in Hawaii, but at the time it was a barren, extinct volcano at the inhospitable altitude of 14,000 ft. Today it is host to the largest collection of telescopes in the world.

Oort and Kuiper studied a large collection of data on comet orbits with particular attention to how much time comets spent at large distances from the Sun. They also thought carefully about the way the solar system might have formed. They deduced that there must be two major sources for comets, both in the extreme outer edges of the solar system. You will recall from Chapter 3 that, according to Kepler's second law, an object travels more slowly when it is far from the Sun. Thus, even a comet that comes from the outermost solar system to the region of Earth will spend most of its lifetime moving slowly in the outermost area. Examples of the relationship between orbital periods and orbital sizes are given in the accompanying Science Toolbox.

After studying the statistics of comet orbits and the time they spend in different parts of the solar system, Oort proposed that there is a spherical swarm of comet nuclei far beyond Pluto that orbits the solar system as bees swarm around a hive. The comets are 50,000 to 150,000 AU from the Sun, much too far from Earth to be seen. They take 10 million to 60 million years to go around the Sun. What led Oort to this striking hypothesis? He first noted that most comets are one-time visitors. Unlike Halley's comet, they have not repeated their orbits within human history. He also knew that comets could arrive from any direction; this suggests that they occupy a spherical region rather than the flat plane defined by the planet orbits. From Kepler's second law, Oort knew that each comet must spend most of its orbit far from the Sun. So for every comet seen in the inner solar system, there must be many more lurking in the depths of space beyond Pluto. From statistics of comet orbits and their rate of appearance, Oort calculated that roughly 100 billion inactive comet nuclei, invisible from Earth, lie in this frigid, distant region!

Many comets that we see in the inner solar system are temporarily "dropping in" from the **Oort cloud.** They travel in huge elliptical orbits that bring them from distances of thousands of astronomical units to the inner solar system, where they loop around the Sun at a distance of only a few astronomical units or less and then return to the deep freeze of the Oort cloud.

Kuiper hypothesized a second group of comets just beyond Neptune and Pluto, arising from debris left over after planet formation and lying roughly in the plane of the solar system. Called the **Kuiper belt,** this group is concentrated at a distance of 30 to 100 AU from the Sun. Comet nuclei located in this group are mostly too faint to be seen even with large telescopes and are too cold to give off gas or form tails. Nonetheless, sensitive searches with electronic detectors began to turn up comets in this region in 1992. Astronomers David Jewitt and Jane Luu, working in Hawaii, discovered more than a dozen such objects in the mid-1990s. These are the largest and closest comets in the Kuiper belt; most are located at distances from 35 to 45 AU from the Sun. Jewitt and Luu estimated that the Kuiper belt contains 35,000 comet nuclei bigger than 100 km and many more small ones.

Observations of Sun-like stars sparked renewed interest in the Kuiper belt. By carefully blocking out the light from the star, astronomers revealed that a number of these stars have disks of dust—possibly comet debris—surrounding them. The disks reach out to several hundred astronomical units from the star. If stars like the Sun have Kuiper belts of cometary debris, it could be a sign that they have planets, too. Perhaps planets and comets are a natural consequence of star formation. Perhaps our rocky vantage point in space is not unique. This speculation spurs astronomers in their search for debris around stars.

How do comets get from the Oort cloud or the Kuiper belt into the region of Earth, where we can see them? We do not know for sure, but we have a plausible scenario. Comets are probably nudged from their initial, distant orbits by the gravity of passing stars. Some comets may move directly into orbits that pass inside Earth's orbit; other comets may get deflected only into the region of the giant planets. The comets that come close to the outer planets are called *centaurs,* after the half-man, half-horse creatures of Greek mythology. Centaurs pass among the giant planets, and some get deflected toward Earth, too. All these orbits are complex and difficult to predict.

SCIENCE TOOLBOX

Kepler's Laws and Comet Orbits

Kepler explained the regularity of planet motions in terms of elliptical orbits. In Chapter 3, we saw the relationship he discovered that relates the period (or orbital time) of a planet to the semimajor axis (or one-half of the long axis of the orbit). The semimajor axis is the average distance from the Sun. If the period P is in years and the semimajor axis a is in astronomical units, we have

$$P_{\text{Years}}^2 = a_{\text{AU}}^3$$

Obviously this is correct for Earth, since the equation gives $1^3 = 1^2$, which is true. (As an exercise you can verify it for the other planets as well.) Kepler's third law applies with an accuracy of about a percent; below this level planets interact with each other by gravity and deviate from the simple relationship.

Kepler made his discovery long before Halley recognized that comets were cyclical phenomena. But the relationship above applies for all motions around the Sun, including the orbits of comets. For example, Chiron (originally classified as an asteroid but now known to be a comet) moves on an elliptical orbit between Saturn and Uranus, with a semimajor axis of 13.7 AU. The cube of this number is 2571. Thus, if the period squared is 2571, the period must be $\sqrt{2571} = 51$ years. That is how long it takes Chiron to go around the Sun.

You can use the law the other way round. For example, we know from historical records that Halley's comet returns about every 75 years. What is its average distance from the Sun? The square of the period is 5625. Therefore, the cube of the semimajor axis is also 5625 AU. Taking the cube root on a calculator, we find that the semimajor axis is about 18 AU, which puts the most distant point near the orbit of Uranus.

Let us think a bit more carefully about what we mean by average distance. We have seen that the planets do not have extreme elliptical orbits; they look like slightly squashed circles. In this case, the foci of the ellipse are close together, the eccentricity is close to 0, and the semimajor axis is not much different from the radius of a circular orbit. Comet orbits are entirely different. Their ellipses are extremely elongated, and the eccentricity is close to 1. The foci of the ellipse are very far apart. In fact, the distance of closest approach to the Sun may be much smaller than the semimajor axis (see Figure 8–A). It is therefore a good approximation to say that the maximum distance from the Sun is nearly twice the semimajor axis. At its farthest point, Halley's comet reaches about 36 AU from the Sun, between the orbits of Neptune and Pluto.

As we have seen, comets in the Kuiper belt and Oort cloud are at distances of roughly 100 and 100,000 AU, respectively, from the Sun. What would be the period of such comets? Remember that the maximum distance is nearly twice the semimajor axis. If you insert the semimajor axis values of 50 and 50,000 in the equation above, you deduce periods of about 350 years and 11 million years, respectively! Oort cloud comets are very occasional visitors to the inner solar system.

The enormous period and elongation of orbits in the Oort cloud has an important implication. Kepler's second law states that an orbiting body will sweep out equal areas in equal amounts of time. With a highly elliptical orbit, this means a comet

(continued)

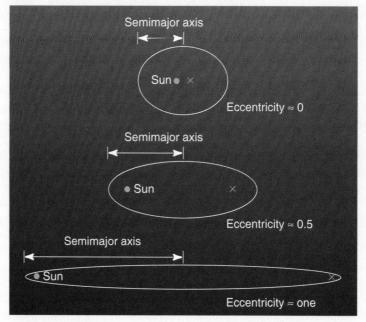

FIGURE 8–A

Orbits can be slightly or extremely elongated. The top view is like that of the planets in the solar system; the eccentricity is small, and the semimajor axis is close to the radius of a circle. (In fact, most planets' orbits are more circular than this top view.) The middle view shows a more eccentric orbit in which the two foci of the ellipse are well separated. The bottom view is more typical of comet orbits; the eccentricity is large, the Sun lies close to one end of the orbit, and the farthest distance from the Sun is nearly twice the length of the semimajor axis.

SCIENCE TOOLBOX

Kepler's Laws and Comet Orbits (continued)

must travel much more slowly when it is far from the Sun than when it is close to the Sun. How much more slowly? Newton's gravity law can be used to show that the velocity v in a circular orbit is

$$v = \sqrt{\frac{GM}{r}}$$

In this equation, G is the gravitational constant, and M is the mass of the Sun. We can approximate the answer just by scaling from this relation where $v \propto 1/\sqrt{r}$. Taking ratios so that G and M cancel, we can say that

$$v_{\text{Earth}} / v_{\text{comet}} = (r_{\text{comet}} / r_{\text{Earth}})^{1/2}$$

At 1 AU, Earth has an orbital velocity of about 30 km/s. In other words, we are whipping along at just over 67,000 mi/h! So at 50,000 AU, a comet will have an orbital velocity of 30 $(1/50,000)^{1/2}$ = 0.14 km/s or 300 mi/h. The result is that a comet spends the vast majority of its orbit crawling along at an enormous distance from the Sun. A comet with a period of a million years will only spend one or two of those years within the planet orbits!

The logical consequence of these extreme orbits is that for every comet that we see on a highly elongated orbit in the inner solar system, there must be a much larger number at the outer regions of their orbits. These distant frozen comets are far too faint to detect, but our understanding of gravity makes us confident that they exist.

The Life Story of Comets

Now we can turn these facts into a history that explains how comets formed. The story starts 4.6 billion years ago with a large cloud of gas and dust that is destined to form the Sun and the solar system. Comets formed by aggregations of ice crystals and carbon-rich dust grains in the cold, outer regions where the giant planets were also forming. We will present all the aspects of planet formation in the next chapter.

Most of the material surrounding the Sun occupied a disk-shaped system. This is why the planets, which formed from that material, have orbits that lie roughly in a plane. Comet nuclei that formed in the region beyond Pluto remained with a flattened distribution, forming the Kuiper belt. However, many of the comet nuclei approached close to a giant planet and were flung by its strong gravity into highly elliptical orbits that took them hundreds or thousands of astronomical units from the Sun. Since they could be flung in any direction, they took up a spherical distribution. This is the Oort cloud.

Comets remained "stored" in the Oort cloud and Kuiper belt during much of the solar system's history. However, as a result of random disturbances by the gravitation of passing stars or of giant planets, comets in the Oort cloud or Kuiper belt occasionally found themselves on trajectories back into the outer or inner solar system. Once on an elliptical orbit among the outer planets, a comet may have wandered for a million years or more before being kicked into the inner solar system by a close encounter with a giant planet. As comets passed through the inner solar system, they were warmed by the Sun, their ice sublimated, and they temporarily formed the familiar heads and tails that give comets their traditional appearance.

A fraction of the comets that made it into the inner solar system had additional close encounters with planets, with further changes to their orbits. In a few cases, the gravitational tug of a planet greatly reduced the size of the elliptical orbit. This process created comets that repeatedly pass among the terrestrial planets, such as Halley's comet, with its 75-year period, or Encke's comet, with its 3-year period. Note in Table 8–1 that Halley's comet makes it back to its "home" in the Kuiper belt, with a farthest distance from the Sun of 36 AU, but Encke's comet has its orbit modified so much that now it never gets farther than 4.1 AU from the Sun.

The life story of a comet is a complex gravitational ballet. For most comets the excitement comes early on when they are flung into the Oort cloud by a close encounter with a giant planet. They travel in lazy, looping orbits of the Sun, with periods of hundreds of thousands or millions of years. Most of their lives are spent in the deep freeze of space, so far beyond Pluto that the comet cannot be seen from Earth, even with big telescopes. Occasionally, comets plunge into the inner solar system. The Sun's heat brings the frozen rock to life, liberating gases into a luminous coma and tail.

This behavior is very different from the familiar motions of the planets. Normally, the orbit of any body in space can be predicted far into the future, by using the laws of Kepler and Newton. The motions are very orderly. We are accustomed to the idea that the orbits of Earth and other planets do not change much over time. However, as noted, when interplanetary bodies drift very close to planets, their orbits will be radically changed. For example, if a Kuiper belt body passed near the top of Neptune's atmosphere, it might be flung into an orbit that would lead to a close approach to Saturn. That encounter might fling it into the inner solar system, where it could hit Earth millions of years later. Yet if it approached Neptune on a path altered by only a few kilometers, it might miss Saturn and never come into the inner solar system at all. In other words, a tiny change in initial conditions could lead to two radically different histories. This interesting and dynamic behavior is very different from the concept of a "clockwork universe."

Orbits that can change radically with only a tiny change in initial conditions are called **chaotic orbits,** because they are almost impossible to predict. Chaotic orbits are examples of a wide range of physical phenomena that are not easily predictable—the weather is a familiar example. We look at the physical idea of chaos more closely in the next chapter. Most interplanetary bodies are not on chaotic orbits, but the existence of even a few bodies on chaotic orbits makes the future fascinatingly uncertain. Perhaps there are bodies today near Saturn or Uranus for which a change in path of only 100 m might determine whether Earth has a catastrophic impact a million years from now!

Meteors and Meteor Showers

Catastrophic impacts in the solar system are extremely rare, and most interplanetary bodies are much smaller than comets. So a far more common event occurs when a tiny piece of rock from space hits our atmosphere. The typical meteor, or "shooting star," that you might see flashing across the sky is likely to be a sand-grain-size bit of debris from a comet. Meteors do not occur with the same frequency on different nights of the year. On an average evening, you may see about 3 meteors per hour. But on certain dates each year you may see 60 meteors or more per hour, all radiating from one direction in the sky (Table 8–2).

These concentrated bursts of meteors are called **meteor showers.** The best-known example is the Perseid shower, which occurs every year around August 12, when bright meteors streak across the sky every few minutes from the direction of the constellation Perseus. (A shower is named for the constellation most prominent in the area of the sky from which the shower radiates.) Occasionally the showers are so intense that meteors fall too fast to count. During the Leonid shower of November 17, 1966 (in the constellation Leo), meteors fell like snowflakes in a blizzard for some minutes, at a rate estimated to be more than 2000 meteors per minute (Figure 8–9).

In 1866, G. V. Schiaparelli (of Martian canal fame) discovered that the Perseid meteor shower occurred whenever Earth crossed the orbit of a particular comet. The relation between meteor showers and comets is that the meteor particles in each individual shower are debris from *a single comet.* The shower occurs as Earth crosses the orbit of the comet, passing through the swarm of debris. Other relationships were soon found between specific meteor showers and specific comets, as Table 8–2 shows. In 1983, an infrared astronomical telescope in orbit imaged the meteor dust spread along the orbit of Comet Tempel 2—the first direct observation of the comet dust scattered in space along a comet's orbit. The debris is not uniformly spread along the comet's orbit. When Earth passes through a sparse patch of debris, we see slightly more than the average number of shooting stars. When Earth passes through denser debris, we get a spectacular light show.

TABLE 8-2
Dates of Prominent Meteor Showers

SHOWER NAME (AFTER SOURCE CONSTELLATION)	DATE OF MAXIMUM ACTIVITY[a]	ASSOCIATED COMET
Lyrid	April 21, morning	1861 I
Perseid	August 12, morning	1862 III
Draconid[b]	October 10, evening	Giacobini-Zinner
Orionid	October 21, morning	Halley
Taurid	November 7, midnight	Encke
Leonid	November 16, morning	1866 I
Geminid	December 12, morning	"Asteroid" 1983 TB[c]

[a]Showers can last several days before and after the peak activity on the listed date. Observations are best when the constellation in question is high above the horizon, usually just before dawn.

[b]The Draconids are now weak because their orbits have been disturbed by the gravity of planets, but further disturbances may again strengthen the shower in the future.

[c]This object was discovered in 1983 by the IRAS satellite and cataloged as an asteroid, but it is probably a "burnt-out" comet nucleus.

FIGURE 8-9
A rare meteor shower; the Leonids of November 17, 1966. The rate of meteors visible to the naked eye was estimated to exceed 2000 per minute. The brightest star (upper left) is Rigel, in the constellation Orion. This exposure of a few minutes' duration was made with a 35-mm camera. (D. R. McLean.)

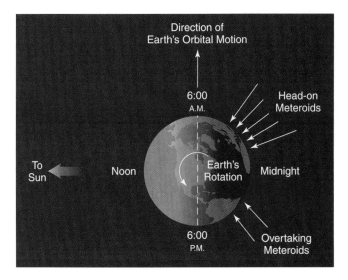

FIGURE 8-10
Meteors that approach Earth at a low rate from random directions are more likely to be seen after midnight, for the same reason that a car's windshield gets more hits than the rear window. Earth's leading side, in its orbital motion, is the dawn side.

In addition to meteor showers, there is a much lower rate of encounters with objects that are traveling in random directions through space. As a practical tip, you are most likely to see these sporadic meteors in the hours between midnight and sunrise. Figure 8-10 shows why. Since Earth travels in its orbit at 30 km/s, only rare high-velocity objects will be able to overtake Earth with sufficient residual speed to create a glowing trail. It is similar to driving down the highway in a rain shower, where more drops will hit the front windshield than the rear window. On any Moon-less night at a dark site, you can see about 5 meteors per hour before midnight and about 15 per hour just before dawn. Meteors approaching Earth at shallow angles are more likely to burn up without leaving a noticeable trail, so you are most likely to see the meteors in a shower that are coming straight in. This creates the perspective effect in which the meteor shower appears to diverge from a point in the sky (Figure 8-11).

The smallest interplanetary particles are microscopic dust grains and molecules spread out along the plane of the solar system and concentrated toward the Sun. If you look west in a very clear rural sky as the last glow of evening twilight disappears (or look east before sunrise), you can detect the diffuse glow of sunlight reflecting off the cloud of these particles. It appears as the **zodiacal light**—a faint, glowing band of light extending up from the horizon and along the ecliptic plane, shown in Figure 8-12. It is brightest at the horizon. The zodiacal light is sunlight reflecting off a swarm of countless dust grains—released by comets in trails along comets' orbits, but eventually spreading into a uniform disk-shaped cloud of dust around the Sun.

Interplanetary debris contains objects with an enormous range of sizes. The bodies that rain in on Earth's atmosphere every year range from dust size to house size. But there are many millions that are the size of a dust grain and only a couple as big as a house. Most pieces of interplanetary dust are far too small to reach the ground, burning up at altitudes around 75 to 100 km. This is the conservation of energy at work, turning kinetic energy to heat and light. Occasional large ones, called **fireballs,** are very bright and spectacular. They generally explode in the air instead of hitting the ground, again indicating that they are too fragile to survive atmospheric entry. Fireballs have sometimes been reported as UFOs.

The stability of Chinese civilization means that we have many centuries of records of celestial phenomena. In Figure 8-13 records of fireballs are shown for four selected centuries, divided by the month of the sighting. The distribution is neither uniform nor random! The two major peaks are centered on the times of major meteor showers, the Perseid shower in August and the Taurid and Leonid showers in October and November (look again at Table 8-2). Ancient evidence confirms that Earth has been sailing through comet debris for millennia. The small pieces of debris give a light show in the sky; the larger pieces may cause substantial damage.

Asteroids

Now we turn our attention to the interplanetary bodies located between Mars and Jupiter. Asteroids are rocky and metallic objects that lack enough ice to give off gas and turn into comets. The largest interplanetary bodies are asteroids.

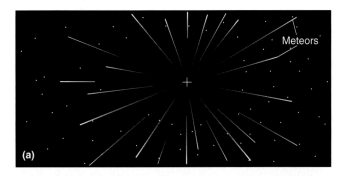

FIGURE 8-11
(a) Meteor showers often seem to come from a fixed point in the sky. **(b)** Perspective causes this illusion—most of the visible meteors are arriving from a distant point.

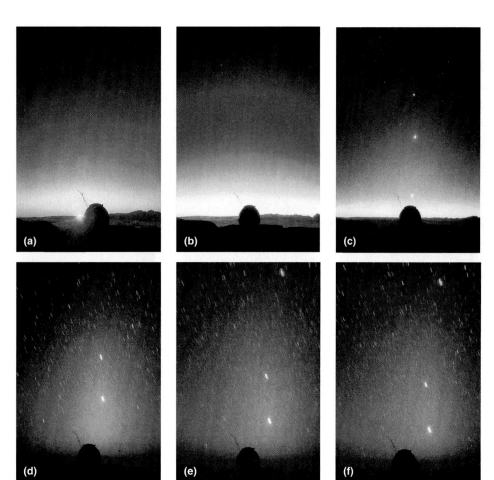

FIGURE 8-12
Sunset and the emergence of the zodiacal light. **(a)** Sunset. **(b)** Blue color still dominates the sky ½ h later and obliterates the faint zodiacal light. **(c)** The zodiacal light begins to be visible along the zodiac 1 h after sunset. The Moon, Venus, and Jupiter, in order up from the horizon, are emerging and are aligned along the zodiac. **(d)** The zodiacal light reaches its best visibility about 1½ h after sunset, appearing as a diffuse band of light rising nearly vertically from the horizon and running near Venus and Jupiter. **(e)** and **(f)** The zodiacal light slowly sets 2 and 2½ h after sunset, respectively. The Pleiades star cluster appears at top right. Features in these photographs can be seen only away from cities under dark, clear skies. [Exposures **(a)** and **(b)** on Kodachrome 64, **(c)** through **(f)** on 3M ASA 1000 film; 28-mm wide-angle lens with nearly 90° vertical field of view; exposures **(c)** to **(f)** 5 min at f2.8.] (Photographs by WKH from Mauna Kea Observatory.)

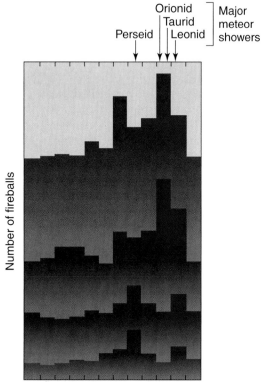

FIGURE 8-13
Chinese astronomers have 2000 years of records of astronomical events, including fireballs. The cumulative records for four selected centuries, divided by the month of each fireball sighting. There are peaks that correspond closely to the times of the major meteor showers. These historical data show that the arrival rate of interplanetary debris varies over the centuries. They also show that comet debris probably accounts for a range of phenomena, from meteor showers to fireballs to the occasional large impact. (Data gathered by S. V. M. Clube.)

The biggest, Ceres, is about 1000 km (600 mi) across, while the smallest asteroids are only a few meters across. Detailed images from passing spacecraft show that asteroids smaller than 100 km or so are irregularly shaped, cratered objects (Figure 8–14). They appear throughout the solar system but are most abundant in the main asteroid belt—a region between the orbits of Mars and Jupiter.

Discovery of Asteroids

Astronomers discovered asteroids while looking for a new planet. After Uranus was discovered in 1781, some astronomers felt that there might be a planet in the large empty zone between Mars and Jupiter. This was suggested by the roughly geometric spacing of the planets (refer to the discussion of Bode's rule in Chapter 3). Astronomers set out

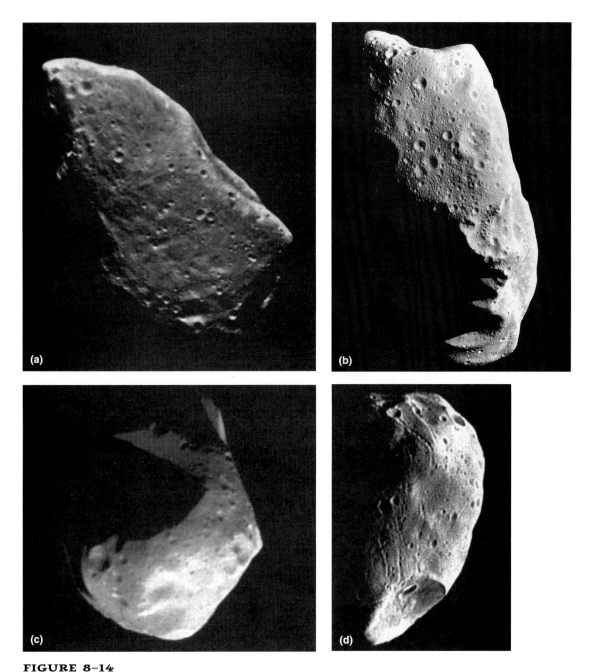

FIGURE 8-14
The first three closeup photos of asteroids by spacecraft, and a closeup photo of Mars' satellite, Phobos. **(a)** Asteroid 951 Gaspra, about 12 × 16 km in size. **(b)** Asteroid 243 Ida, about 52 km long. **(c)** Asteroid 253 Mathilda, about 59 × 47 km. Scientists were somewhat surprised by the large craters and on the asteroids, especially Mathilda; the impact-explosions needed to make these craters are almost big enough to have fragmented the asteroid. **(d)** Mars' satellite, Phobos, 19 × 27 km in size, is believed possibly to be an asteroid captured into orbit around Mars. The similarity in appearance gives some support to this idea. (Source: NASA missions, (a) and (b) are from the *Galileo* spacecraft; (c) from the *NEAR* spacecraft; and (d) from the *Viking* orbiter.)

to find the "missing planet" in 1800. Ceres was discovered on the first night of 1801 by Sicilian monk Giuseppe Piazzi, located 2.8 AU from the Sun between Mars and Jupiter.

Arthur C. Clarke, the famous science fiction writer, has noted an interesting fact about this discovery. Ceres was found just after the philosopher G. W. F. Hegel had "proved" philosophically that there could be no more than the seven then-known planetary bodies, Mercury through Uranus. As you can see, the scientific method of going out and looking tells us more about the physical nature of the universe than the philosophical method of sitting at home and speculating!

Between 1802 and 1807, three more small, planetlike bodies turned up between 2.3 and 2.8 AU from the Sun. Because of their small size, they came to be called minor

ASTEROIDS 219

planets, or asteroids. All are too dim to be seen with the naked eye. Photographic searches began in 1891, and more are discovered each year by using modern electronic detectors. More than 6000 of these bodies have been cataloged in the main asteroid belt. Astronomers estimate that more than 100,000 are big enough to be detected.

Asteroids are known by numbers (assigned in order of discovery, but only after the orbit has been accurately identified) and a name (chosen by the discoverer). Ceres is thus more properly called 1 Ceres, for example; the second-discovered asteroid is 2 Pallas. The names are sometimes Latinized, and they cover a wide range of human interests, including cities, mythology (Quetzalcoatl, Odysseus), politicians (Hooveria), celebrities (Zappafrank), family members, and lovers. (Some years ago we learned of a new astronomical amusement: making sentences using only asteroid names. Our favorite—not our own creation, we are sorry to say—is "Rockefellia Neva Edda McDonalda Hamburga.")

Several subgroups of asteroids have orbits outside the main asteroid belt. The **Trojan asteroids** are one such subgroup. They orbit the Sun in two swarms that lie in Jupiter's orbit, 60° ahead of and 60° behind the planet. These locations are called *Lagrangian points,* after the French astronomer Joseph Louis Lagrange. Lagrange discovered that particles can be held there by the combination of gravitational forces from the Sun and Jupiter. The name *Trojan* comes from the tradition of naming these particular asteroids after heroes in Homer's epic poem of the Trojan War. Many dozens of Trojan asteroids are known. Most are bigger than 50 km across. Astronomers estimate that the two Trojan swarms contain nearly as many asteroids as the main asteroid belt, but most Trojans are too far away to see easily. The largest Trojan is 624 Hektor, estimated to be 100 km wide and 300 km long, tumbling end over end in Jupiter's orbit.

Some asteroids come close enough to be of special interest to Earth dwellers. These are the objects that have been perturbed by gravitational encounters into the inner solar system. We use the word *perturb* to mean that the orbit is disturbed or deviated slightly, and so the object takes a new trajectory. We have now seen a number of situations in which a **gravitational perturbation** affects the orbit of a planetary body. This idea is explored further in the Science Toolbox that follows. Astronomers know of hundreds of asteroids that approach Earth's orbit. The two biggest are 1035 Ganymede at roughly 30-km diameter (it comes within about 0.2 AU of Earth's orbit) and 433 Eros at roughly 10 × 30 km (it comes within about 0.1 AU of Earth's orbit). Asteroids that actually cross Earth's orbit are the only ones that have a chance of hitting us in the foreseeable future. The biggest of these **Earth-crossing asteroids** is 1627 Ivar at about 8-km diameter. It would cause a global disaster if it actually hit Earth—but that is unlikely to happen within the next million years.

Nearly 200 Earth crossers were known by 1993, and only about 10 percent of those 1 km across or bigger have been discovered yet. Many small ones have been seen, including some that are only tens of meters across. Some of these may hit Earth within a million years. One such body passed by in 1991 at one-half the distance to the Moon! The best step we can take now to evaluate the risk is to continue the survey programs that discover and catalog Earth-approaching asteroids, until we have cataloged all those larger than 500 m across. By calculating their future orbital trajectories, we could determine the likelihood of a major threat in the coming decades or centuries. In the longer term, the ability to fly to these asteroids and slightly alter their orbits would safeguard civilization against an asteroid disaster.

Earth crossers generally survive only 10 million years or so before hitting Earth or being deflected by Earth's gravity to hit some other planet, making huge explosions and creating impact craters. The supply is constantly replenished as other asteroid fragments are thrown out of the belt by the gravitational force of massive Jupiter. Still other asteroids that approach Earth may be remnants of burned out comets that have lost their ices and are no longer active.

Perhaps the best sense of the welter of asteroids in the solar system comes from a map showing the actual positions of thousands of cataloged asteroids on a specific, randomly chosen date. This is shown in Figure 8–15. The asteroid belt stands out clearly. Additionally, the figure shows the two somewhat vaguely defined swarms of Trojans in

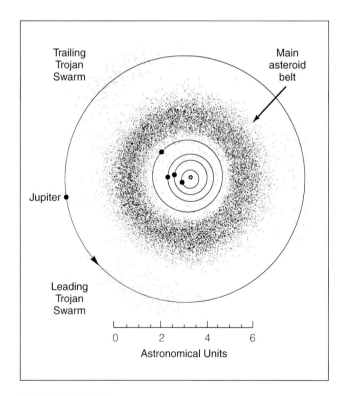

FIGURE 8–15
A map of the positions of thousands of cataloged asteroids on a specific day (March 19, 1993). Each small dot represents one asteroid. Large black dots show the positions of five planets that happened to be on the same side of the Sun on this date. (Adapted from a diagram courtesy of Mark Sykes, Steward Observatory.)

SCIENCE TOOLBOX

Gravitational Perturbations

Newton's law of gravity is a powerful tool we use to understand the universe. It describes the force between any two masses—planets, stars, or galaxies. We have seen, however, that Kepler's laws do not describe the orbits in the solar system exactly. It is an approximation to say that the planets orbit with the Sun stationary at one focus of an ellipse. As you read the discussion that follows, remember that we are talking about the minor effects of gravity. Even though Kepler's laws are not exact, they are very, very good at describing *most* orbits.

Newton's third law says that Jupiter exerts the same gravity force on the Sun that the Sun exerts on Jupiter. So Jupiter tugs the Sun into its own slight motion. It is also an approximation to say that the planet orbits are ellipses. We have seen in the last chapter that Neptune was discovered by scientists trying to find out what was tugging Uranus from its purely elliptical orbit. Neptune itself shows departures from an elliptical orbit that cannot be explained by Pluto.

These small extra forces are called gravitational perturbations. For gravity in the solar system

$$F_{total} = F_{Sun} + \Delta F$$

The gravity force in the solar system is given by the force due to the Sun plus a little bit extra. (Recall that scientists use the Greek letter Δ to indicate a small increment.) The extra bit of gravity, ΔF, is a gravitational perturbation. The perturbation is much smaller than the main force,
so we can write $\Delta F << F_{Sun}$, or equivalently $\Delta F / F_{Sun} << 1$. We can see that the equation above can be written

$$F_{total} = F_{Sun}(1 + \Delta F / F_{Sun}) \approx F_{Sun}$$

The main perturbing force in the solar system is caused by the gravity of the most massive planet. Jupiter pulls the Sun into a slight motion. Jupiter tugs the planets near it away from pure elliptical orbits. Jupiter deflects comets into slightly different orbits. Jupiter interacts with the asteroid belt to create gaps in the asteroid distribution with distance from the Sun. All these phenomena are gravitational perturbations. It is a good approximation to say that the Sun controls orbits in the solar system. But the extra influence of Jupiter causes some interesting and important phenomena.

The description we have just given does not apply within the sphere of gravitational influence of a planetary object—Earth controls the motion of telecommunication satellites, Saturn controls the motion of its rings, Pluto controls the motion of Charon. However, the spheres of planetary influence of the planets take up a tiny fraction of the volume of the solar system. Throughout most of the volume, the Sun is the main influence, and Jupiter is the perturbing influence.

The idea of a perturbation occurs throughout physics. In fact, we could generalize the first equation to write $X_{total} = X + \Delta X$, where X might be anything from a particle velocity to a magnetic field. We can use this expression anytime in nature
that a small influence is added to a large influence. In mathematics, we have powerful tools for manipulating situations in which a small quantity is added to a large quantity—the differential calculus is just one example.

A gravitational perturbation leads to a small deviation in the orbit; the position and velocity are slightly changed. What happens over time as the orbit repeats? Each perturbation is a little gravitational "kick," and as the number of kicks accumulates the effect on the object can become large. Let us say a chunk of rock in the asteroid belt receives a small gravitational kick from Jupiter every time the rock orbits the Sun, which increases the rock's velocity by a tiny 1 percent. We can relate the velocity before the kick to the velocity after the kick by $v_{after} = v_{before} + \Delta v$, or equivalently

$$v_{after} / v_{before} = 1 + \Delta v / v_{before}$$

The effect of successive kicks is cumulative. After n kicks, the velocity becomes

$$v_{after} / v_{before} \approx (1 + \Delta v / v_{before})^n$$

A 1 percent velocity boost means $\Delta v / v_{before} = 0.01$. If we take 100 orbits, $n = 100$, and we get $v_{after} / v_{before} = 2.7$. So in only a few hundred years the velocity of a perturbed rock might triple, which would be enough to send it on a completely different orbit. In this simplified discussion, you can see that gravitational perturbation over time can lead to radically altered orbits.

Jupiter's orbit and a scattering of Earth-approaching asteroids among the terrestrial planets. A plot of the numbers of asteroids at different distances from the Sun shows a lot of structure (Figure 8–16). The regions that are almost free of asteroids are called *Kirkwood's gaps,* after their discoverer. The gaps are caused by gravitational resonance with mighty Jupiter, in exactly the same way as Mimas clears out regions of Saturn's rings. You can see the gaps at the primary harmonics of 2:1, 3:1, and 4:1, but they also occur at other ratios such as 5:2 and 7:2. The same gravitational physics occurs in the asteroid belt, on a scale 3000 times larger than Saturn's rings!

Remember that in spite of the apparently dense crowding in the asteroid belt seen in Figure 8–15, distances are

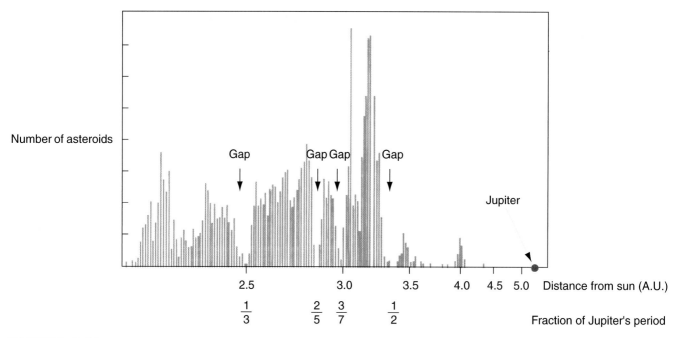

FIGURE 8-16
The map from Figure 8-15 is turned into a radial plot, showing that there are "gaps," or minima in the distribution of asteroids with distance from the Sun. The gaps (named after the astronomer Daniel Kirkwood, who discovered the pattern in 1866) are examples of resonance, where asteroids would have one-half or one-third or some other simple fraction of the orbital period of Jupiter. The resonant relations deplete asteroids from orbits with these periods and average solar distances.

vast and asteroids are small. Even if you were cruising in the asteroid belt, you would rarely see another asteroid passing close by, contrary to Hollywood versions of the belt as a forest of rocks. Several unstaffed spacecraft have already flown through the belt with no serious consequence, although they did pick up an increase in number of impacts by dust grains.

Asteroids' Moons and Compound Shapes

The second close-up photograph of an asteroid, that of the 52-km-long asteroid 243 Ida taken by the *Galileo* spacecraft, revealed a 1-km moon orbiting roughly 100 km from Ida's center. Another asteroid moon was confirmed by ground-based observers in 1999. Existence of asteroid moons explains an older discovery: Earth, the Moon, and Mars show numerous pairs of adjacent impact craters of the same age. These must have formed when an asteroid and its nearby satellite hit at the same time. Studies suggest that perhaps 10 to 20 percent of asteroids have sizable moonlets moving around them.

A related discovery came from a new radar technique that creates images of nearby asteroids by means of signals bounced off them from large radio telescopes as the asteroids pass close to Earth. Results of this technique have astonished astronomers. In 1989, when the 1.5-km asteroid 4769 Castalia passed by Earth, a fuzzy radar image showed a two-lobed, dumbbell shape rotating end over end (Figure 8–17). Such images, together with other techniques, show that some asteroids must be loosely bonded clumps of two or three large fragments. These compound asteroids are

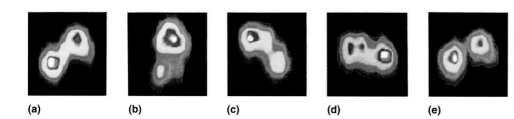

(a)   (b)   (c)   (d)   (e)   (f)   (g)

FIGURE 8-17
(a) to **(f)** Sequence of false-color radar images of asteroid 4769 Castalia taken about 27 minutes apart, showing the asteroid's rotation and its extraordinary dumbbell shape. White is the strongest radar reflection, followed by red, yellow, and blue as the weakest signal. (Courtesy of Steven Ostro and colleagues at Arecibo Observatory in Puerto Rico.) **(g)** Possible close-up appearance of Castalia as it passes the Earth-Moon system (upper left). The asteroid probably formed by the low-speed collision of two rounded bodies. (Painting by WHK.)

within reach of our spaceships; it would be a strange experience for an almost weightless astronaut to float among the rocks of an asteroid built like two rounded mountains, just touching each other!

The compound asteroids and asteroid satellites may be part of the same type of phenomenon. Imagine an asteroid hitting another body and blowing it apart in a titanic collision. Try to picture an asteroid passing too close to a planet and being pulled apart by the larger body's tidal forces. In both cases, a jumble of fragments race outward, and adjacent fragments may bump into one another. Some may fall together as pairs and make dumbbell-shaped compound objects, and others may go into orbit around one another. Thus, the seemingly disconnected discoveries of asteroid moons and compound shapes may both be clues to asteroids' violent histories.

Rocky, Metallic, and Carbonaceous Asteroids

We have also found out what asteroids are made of. Astronomers use telescopes to measure the colors, reflectivity, and spectra of asteroids. Then astronomers combine this with radar and spacecraft data to estimate the mineral properties. Using such techniques, astronomers in the 1970s and 1980s discovered that asteroids and interplanetary bodies can be grouped by composition. There are three broad classes of asteroid: stony, metal-rich, and black carbonaceous. Some of the outer belt asteroids, Trojans, and even centaur comet nuclei were later found to be more dark-brown than black, probably due to coloration from organic compounds. The three classes of asteroids are concentrated at different distances from the Sun (Figure 8–18).

Remember that the asteroid main belt runs from about 2.1 AU to about 3.4 AU. How does the composition of asteroids vary with distance from the Sun? The Sun's radiation varies with the inverse square of distance, so the inner edge of the belt receives $(3.4 / 2.1)^2 = 2.6$ times as much radiation as the outer edge. The inner one-half of the asteroid belt is dominated by stony and metal-rich asteroids. These asteroids are fairly light-colored, reflecting 10 to 20 percent of the light that strikes them, like many familiar rocks on Earth. A few additional minor classes of asteroids have been found in this region.

In the middle of the asteroid belt, at about 2.7 AU, lies a **soot line** beyond which black carbon-rich minerals dominate most asteroids and comets. In other words, most interplanetary bodies beyond 2.7 AU are black, reflecting only about 4 percent of the light that strikes them. This low reflectivity is caused by the sooty carbonaceous materials mixed into these bodies. The two groups of Trojan asteroids clustered in Jupiter's orbit also have this dark color, as do comet nuclei associated with the Kuiper belt and Oort clouds even farther from the Sun.

Somewhere between 3 and 4 AU is another important dividing line that does not show up as clearly in asteroid appearance. This is called the **frost line,** where frozen water is stable. Asteroids beyond this distance probably contain substantial amounts of ice as well as black soot—in other words, they are really comet nuclei, even though they might not be active. They might turn into active comets if they are deflected into new orbits that take them closer to the Sun, where the ice will warm up and release gas and dust. In fact, most interplanetary bodies beyond 5 AU, even if cataloged as asteroids, are probably comet nuclei—or potential comet nuclei—with large amounts of ice. (Indeed, several distant objects that were initially classified as asteroids, such as 2060 Chiron at 10 AU, were later seen to give off telltale puffs of gas, proving that they are really comets. But the asteroid designation stays in place until actual cometary emissions have been observed.)

The composition of asteroids depends on their distance from the Sun. This fact shows that the asteroids have not been strongly mixed up; they lie at their original distances. Those closest to the Sun got so hot that they melted and produced iron cores. On the other hand, asteroids farther from

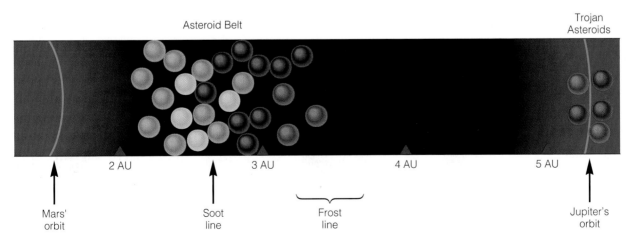

FIGURE 8–18
Schematic diagram of the distribution of asteroid classes in the solar system. The inner asteroid belt consists of light-colored stony objects (tan-colored dots) and metal-rich objects (darker shades of brown), while the outer belt and Trojan asteroids consist of black and dark-brown carbonaceous objects. See the text for further discussion.

the Sun contained sooty carbonaceous material, water molecules, and even ice. They did not get warm enough to melt the rock, but we know from mineral studies of carbonaceous meteorites that the ice in some of them melted and percolated through the rock fractures as liquid water. Still farther from the Sun, the ice solidified and produced comet nuclei. From this point of view, comets are nothing more than "icy asteroids" that formed so far from the Sun, in such cold regions, that they retained much ice mixed with the carbonaceous material.

Interplanetary bodies are not just dull pieces of rock and ice; they have had complicated geological stories that reveal the history of the solar system. Some asteroids melted and resolidified, forming nickel-iron cores, mantles of dense rock, and surface crusts with basaltic lava—just like miniature versions of Earth! Other asteroids are never-melted primitive rock, and some of these contain frozen water or other ices. Earth crossers are a mixture of rocky objects from the asteroid belt and burned out comets. Meteorites bring us samples of all these types of material except the ices, which vaporize when an object hits Earth. Astronomers look forward to each new interplanetary body as a detective story to unravel.

Meteorites

Much of what we know about asteroids comes from meteorites, the stony and metallic objects that fall from the sky. Modern evidence suggests that most meteorites are fragments of Earth-crossing asteroids, originally ejected from the asteroid main belt. Thus, by fortunate chance, we get "free samples" of the cosmic debris left over from the formation of the solar system. They allow us to reconstruct many of the processes that led to the formation of Earth and the other planets.

The Scientific Discovery of Meteorites

Long ago, stones from the sky were a source of awe. Meteorites have been found buried at prehistoric Native American sites in Arizona and elsewhere (Figure 8–19) and kept as a sacred possession of a tribe in Alaska. A stone venerated in the Temple of Diana at Ephesus (one of the seven wonders of the ancient world) reportedly fell from the sky and was probably a meteorite. The "Black Stone," enshrined around 600 or earlier in the sacred Muslim shrine at the Ka'aba in Mecca, is also believed to be a meteorite. When a meteorite fell at Ensisheim, France, in 1492, the emperor Maximilian, in residence nearby, decided he should go on a Crusade. The main mass of the meteorite was enshrined in the Ensisheim church.

It took a long time for a scientific explanation for meteorites to emerge. In the 1700s, many naturalists still felt that the idea of stones falling from the sky was superstition. However, a German physicist, E. F. F. Chladni, questioned this belief. In 1794, he reported that the Ensisheim stone and other supposed celestial stones seemed similar to one another and different from normal terrestrial stones. He concluded that these "meteorites did indeed fall from the sky."

This conclusion was controversial, and many people still refused to accept celestial stones. Upon hearing that a meteorite had fallen in Connecticut, Thomas Jefferson, himself an accomplished naturalist, is supposed to have joked, "It is easier to believe that Yankee professors would lie than that stones would fall from heaven." In France, the French Academy—that bastion of rational thinking—dismissed meteorites as superstition; but when a meteorite exploded over a French town in 1803, pelting the area with stones, the Academy sent the noted physicist J. B. Biot to investigate. His report is one of the historic documents of science. Biot constructed an irrefutable chain of evidence from eyewitness accounts, measurements of the 2 × 6 km area of impacts, and specimens of the meteorites themselves. This report established that stones can indeed fall from the sky.

Meteorite Impacts on Earth

Rocks in space may seem remote from human affairs. They are, except when they collide with Earth. Because of their relative orbital motions, interplanetary bodies collide with the Earth at very high speeds, usually 11 to 60 km/s (24,000 to 134,000 mi/h). At such speeds, material is heated by friction with the air. Dust grains and pea-size pieces burn up before striking the ground. Larger pieces are usually slowed by drag. Since meteorites pass through the atmosphere too fast for their interiors to be strongly heated, stories of meteorites remaining red-hot for hours after falling are untrue. The types of meteorite falls are listed in Table 8–3.

FIGURE 8–19
Burial of the Winona stony meteorite in a crypt constructed by prehistoric Indians in northern Arizona. The Indians probably saw it fall and thus attached special significance to it. (Museum of Northern Arizona, Anthropological Collections; photograph by WKH.)

TABLE 8-3
Types of Meteorites and Frequency of Observed Meteorite Falls

TYPE	PERCENTAGE OF METEORITE FALLS	PROBABLE SOURCE REGION
Unmelted stones	81	Asteroid main belt
Basaltlike stones (melted)	9	Asteroid main belt
Stony-iron meteorites	1	Asteroid main belt
Iron meteorites	4	Asteroid main belt
Unmelted carbonaceous stones	5	Outer main belt and outer solar system

Human encounters with meteorites are extremely rare. Nobody has been killed by a meteorite in recorded history, although a dog was killed in Nahkla, Egypt, in 1911. In this century, a few people in the United States have experienced close encounters. In 1938, an Illinois woman heard a crash in her garage and found a meteorite lying on her car seat. Some years later, an Alabama woman was seriously injured by a ricocheting meteorite that hit her hip. A house in Wethersfield, Connecticut, was hit in 1971; then another house just a mile away was hit 11 years later! In 1991, two boys in Noblesville, Indiana, heard a whistle and a thud, and they looked down to see a small 4-in meteorite lying on the sidewalk. One year later, a woman found a 30-lb meteorite which had smashed through the trunk of her 1980 Chevy Malibu (see Figure 8–23c). She was offered $69,000 for the wreck, far more than the car was worth! Do not let this summary worry you. Your chances of being hit by a meteorite are tiny—statistically speaking, your bathtub is far more dangerous.

Large meteorites are rare. A few boulder-sized meteorites are recovered each year. In 1972 an object weighing perhaps 1000 tons just missed Earth, skipping off the outer atmosphere like a stone off a pond; it was filmed from the ground and detected by Air Force reconnaissance satellites. Had it continued falling instead of skipping back into space, it would have caused a bomb-sized explosion in Canada. Objects weighing 10,000 tons, like the Siberian object of 1908, which are large enough to cause nuclear-scale blasts, fall every few centuries. Larger blasts, thousands of years apart, may form craters many kilometers across.

Types and Origin of Meteorites

After decades of study, scientists have begun to understand how meteorites formed and what they have to tell us about the ancient history of the solar system. Meteorites are among the most complex rocks studied by geologists. There are many different types, but generally the types correspond to the three main asteroid types: stony, metallic, and carbonaceous. Sometimes they appear as *brecciated meteorites*—meteorites formed from often radically different fragments of rock, jumbled together and welded by pressure into one rock! We now recognize that meteorites are fragments of asteroids blown apart at various times during the history of the solar system. Such fragmentation is caused when two aster-

FIGURE 8-20
Imaginary view of the collision of two asteroids. Fragments created by the collision may ultimately fall on Earth and the planets as meteorites. Since the two asteroids are of different mineralogical types, some of the meteorites produced will be mixtures of different rock types. The larger body is about 30 km across.
(Painting by WKH.)

oids collide, as shown in Figure 8–20. The brecciated meteorites contain different kinds of rock because pieces from two different types of asteroids may be jumbled together during the collision. Fragments are ejected in various directions during these collisions, a few being thrown into orbits that eventually intersect Earth's.

This violent history also explains the different types of meteorites. The stony types are samples of surface and mantle rock layers, and the metallic ones are samples of dense cores inside these bodies, analogous to Earth's iron core. The carbonaceous ones are samples of the black, carbonaceous asteroids (and perhaps also comets) of the outer solar system. Table 8–3 summarizes the major types and how abundant they are.

More than 80 percent of meteorites are stony bodies that never melted, which indicates that their parent asteroids (or portions of them) never got hot enough to melt. About 1 in 10 meteorites is a lavalike rock that once melted and then solidified. We have proof that some asteroids reached high enough temperatures to melt while others did not. The heating mechanisms are unknown; concentrations of radioactive minerals and certain magnetic and electrical effects have both been suggested. Whatever the heat source, the iron meteorites (4 percent) and stony-iron mixtures (1 percent) show that some asteroids differentiated completely, forming iron cores and a stony-iron interface between the metal core and the rocky mantle.

Scientists are particularly interested in the primitive, unmelted types, which include most stony meteorites and all carbonaceous meteorites. These rocks are primeval fragments of the solar system's original mineral grains. Dating by the techniques discussed in Chapter 5 shows that the most primitive meteorites formed during a "brief" interval of only 20 million years, some 4.6 billion years ago. In other words, if the history of the solar system were reduced to 1 year, the asteroid-size building blocks of planets had formed and melted by January 2. *Formed* means that they aggregated from dispersed dust grains to solid, rocklike objects.

This interval, then, marked the birth of planetary material. Other types of dating show that major collisions among these bodies happened throughout the interval from 4.6 to about 4.4 billion years ago. Additional collisions smashed other meteorites' parent bodies at other points in solar system history.

Interplanetary Threat or Opportunity?

Figure 8–21 summarizes modern data about interplanetary bodies of all sizes. The figure combines data based on statistics about meteorites, asteroids, comets, and lunar craters. The scale at the bottom gives the size of the crater caused by an impact; the scales at the top give the size of the original

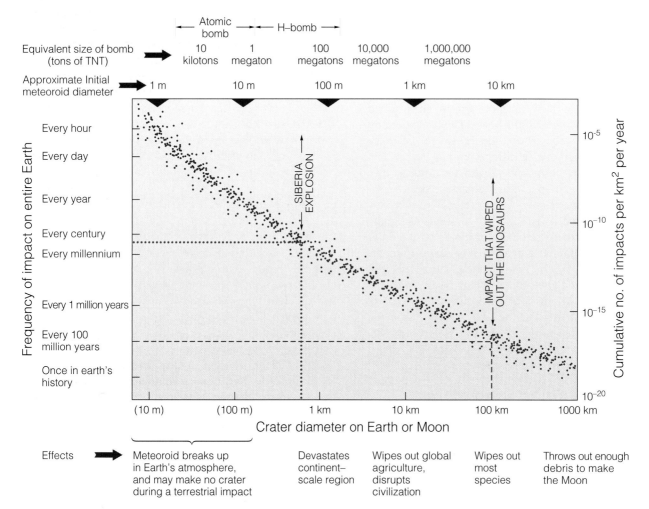

FIGURE 8–21
The asteroid threat, or the frequency of impacts of asteroids and comets of various sizes. The increasing frequency of impact with decreasing size shows that there are many more small objects in space than large objects. The bottom scale gives the size of the explosion crater that results; the top scales give the size of the original planetesimal and the equivalent size of a nuclear bomb. The stippled band shows the approximate rate of impact of different sizes, according to best available data. See the text for discussion. (From diagram by Hartmann, and from data by G. Shoemaker, C. Chapman, D. Morrison, G. Neukum, and others.)

planetesimal and the equivalent size of the explosion in terms of nuclear bombs. The scale at the left gives the average frequency of impact of various sizes. The graph shows the division between meteors and meteorites: Most bodies at the left end of the diagram are so small that they burn up or break up in the atmosphere and do not reach the surface. These are meteors and fireballs. Only the tiny particles that cause meteor showers and the zodiacal light are missing; they would be found off the left side of the plot.

The Interplanetary Threat

Asteroids and comets have hit Earth throughout geological time, with small ones hitting frequently and large ones hitting rarely. In the last decade, researchers have realized the important effects of these random cosmic impacts on the biology and history of our planet. This knowledge also penetrated popular awareness due to popular articles, TV shows, and movies on "killer asteroids" and "killer comets." So-called killer asteroids and killer comets are those larger than 1 km or so across—big enough to make explosions that would disrupt agriculture and civilization, killing perhaps one-quarter of Earth's population. Any asteroid or comet that passes close to Earth's orbit has a fair chance of eventually hitting Earth, usually within a few million years, because other planets can perturb its orbit until it hits Earth itself. New asteroids and comets are constantly being injected into the inner solar system from the asteroid belt and from the Oort cloud, so there is a constant supply of these objects. The estimated interval between impacts of these killer asteroids and comets is 300,000 years—an interval so long that there has never been one in recorded history.

The blue dotted line in Figure 8–21 gives one example of how to read this graph. It shows that about every century or two, somewhere on Earth, we can expect an explosion on the scale of an atomic bomb or a hydrogen bomb. The most recent example is the 1908 Siberian event you read about in the opening story. (Of course, most of these would happen over oceans and have minimal effect. Such an explosion over populated land may happen only once every 500 to 1000 years.) The blue dashed line gives another example: It shows that every 100 million years or so we may get an impact big enough to wipe out most species and reset the evolutionary clock. In fact, as discussed in Chapter 5, there is strong evidence that such an impact happened 65 million years ago and that it caused temporary climate changes big enough to wipe out the dinosaurs and about 75 percent of species that existed at that time. Note also that the giant impact of a body about 6000 km across—big enough to trigger the formation of the Moon—would lie off the lower right corner of the diagram and be likely only during the enhanced cratering rate that existed at the beginning.

This single diagram ties together a variety of phenomena whose connection most people fail to realize: nightly shooting stars, fireballs, the Siberia explosion, the extinction of dinosaurs, the origin of the Moon, and the threat of future impacts. These events are all part of our bombarded planet's interaction with interplanetary debris.

Large impacts are not predictable. We can only estimate their occurrence statistically, in averages over enormous spans of time. Let us return to our familiar example of the deck of cards. We looked at periodic (and predictable) sequences of cards in Chapter 3 and at random (and unpredictable) sequences of cards in Chapter 5. Say that each card represents a time span of 10 million years, and an ace represents a major impact that occurs in the span. There are four aces, so a well-shuffled deck has one ace every 4 / 52 = 13 cards on average. In our analogy, a major impact occurs every 130 million years. Turn cards over one at a time. How long do you have to wait for an ace (an impact)? How long do you have to wait for the next one? We presume that impacts are random because the orbits of interplanetary bodies are mostly chaotic. However, there is tentative evidence that large impacts may not be truly random. Figure 8–22 shows that a variety of indicators of large impacts—mass

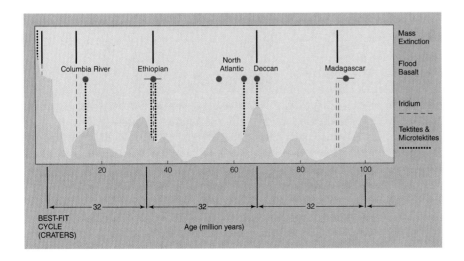

FIGURE 8–22

Suggestions that mass extinctions and large geological changes on Earth might be periodic are controversial. The bars at the bottom of the plot show a fixed 32-million-year spacing). The bars at the top of the plot are mass extinctions over the past 100 million years. The dots refer to major changes in the flood basalt or the geological layering of Earth. The dashed lines correspond to layers in the rock strata with high concentrations of iridium (of extraterrestrial origin). The dotted lines correspond to high concentrations of tektites and microtektites—small glassy spheres formed during large impacts. The shaded region at the bottom shows changes in sea level over this span of time. Do you see evidence for periodicity in any of the phenomena? (Data from Rampino and Caldeira.)

FIGURE 8-23
The threat from the skies. **(a)** A German woodcut commenting on the appearance of a comet in 1521 and foretelling of seismic events. (S. V. M. Clube.) **(b)** This illustration from a 17th-century book also connects the appearance of a comet with calamity. Due to the correlation between cometary debris and impacts, these historical records reflect more than just superstition. (The Granger Collection.) **(c)** Michelle Knapp studies the damage done to her car when it was hit by a stony meteorite in 1992. The probability of any impact's affecting any individual person is tiny. (Courtesy of John Bortle, W.R. Brooks Observatory.)

extinctions, drastic changes in sea level, deposition of meteoric material—may line up with the 32-million-year oscillation of the Sun in and out of the Milky Way. During these times, the solar system is more likely to pass close to stars than at other times; these close passages could disturb motions of comets and cause enhanced impact rates on Earth. This idea remains controversial.

The *interplanetary threat* refers to the chance of damage caused by a massive impact. The odds are high for a disaster on the scale of a state or county within a few centuries, for a regional disaster within a few thousand years, and for an impact big enough to end civilization within every 100,000 or 1 million years. There is perhaps a 50 percent chance that another explosion of the magnitude shown in the first image of this chapter will make the front pages within your lifetime. Figure 8–23 illustrates the fact that "space junk" can affect human affairs.

How should we respond to this threat? We have seen that one approach is to continue our survey programs as part of ordinary international research funding. Within the next decades, especially if new search telescopes can be brought on-line, we should have a complete inventory of the larger Earth-approaching asteroids and comets. Better yet, we will know if one is on a collision course with Earth with enough

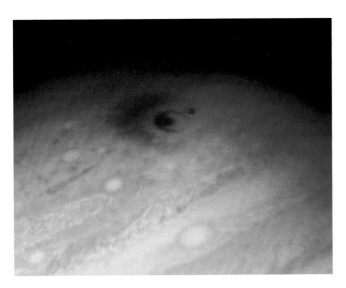

FIGURE 8-24
The Hubble Space Telescope obtained numerous images of explosions as fragments of comet Shoemaker-Levy 9 hit Jupiter. This color image shows a black cloud left by one of the explosions. (H. Hammel, MIT, and NASA.)

warning to plan some response. Unfortunately, science and politics have become mixed in this debate. Although the chances of a catastrophic hit are small, some Defense Department experts have suggested developing a system to deflect or blow up any approaching body before it can hit Earth. Edward Teller, known as the father of the H-bomb, and members of the Strategic Defense Initiative ("Star Wars") project have been active in suggesting an H-bomb-based asteroid defense system. However, unless it was very carefully done, blowing up a small asteroid might create a cloud of fragments would actually *increase* the threat from numerous small impacts! Critics also have suggested that such ideas are merely an attempt to maintain funding for military programs in a post-cold-war world; China in 1996 cited the asteroid threat as a reason to continue its nuclear testing.

The rational view says that we should not take the overheated rhetoric and the media flights of fancy too seriously. Existing and planned surveys should be able to detect all the large Earth-crossing asteroids. The vast majority of these will not be headed directly for Earth. For those asteroids that appear to be headed toward us, a survey with careful orbital observations combined with supercomputers to calculate the orbits would allow us to detect any sizeable objects that might hit Earth, years in advance. Within a decade or two, humanity may have the capability to fly out to such threatening asteroids and deflect them so they miss Earth. Look at Figure 8–21; the probability of such an event in our lifetimes is tiny. The issue is discussed in greater detail in a Science Toolbox. Do not lose sleep over the possibility of a catastrophic impact from space!

Observing a Major Impact on Jupiter

For many years, the asteroid and comet threat seemed purely theoretical, because impacts larger than the 1908 Siberia explosion had never been recorded in history. However, in 1993, U.S. astronomers Carolyn Shoemaker, Gene Shoemaker, and David Levy discovered an unusual comet on a collision course with Jupiter. It was already broken into fragments from tidal forces during a previous close encounter with Jupiter. Astronomers around the world watched breathlessly as the fragments plowed into the planet over a period of days in the summer of 1994. Unfortunately, the impact sites were just around the edge of Jupiter on the far side, but photographs from the Hubble Space Telescope revealed the huge explosion clouds rising above the edge of Jupiter's disk. Jupiter's rotation brought the impact sites into view hours later, and photographs revealed huge black clouds at the explosion sites, probably created by the black dust in each fragment (Figure 8–24). The

FIGURE 8-25
Reconstruction of the scene as fragments of comet Shoemaker-Levy 9 crashed into the cloud layers of Jupiter and exploded. Explosions are visible in the middle distance and on the horizon as fireballs streak across the sky. Additional fragments of the comet are approaching in the sky and will hit Jupiter in the coming days. (Painting by WKH.)

SCIENCE TOOLBOX

Probability and Impacts

Science allows us to predict some events with virtual certainty. For example, the Sun will certainly rise tomorrow. If you let go of a brick, it will certainly fall to the ground. The winter will certainly be colder than the summer. But there are other events in nature which do not occur with certainty. Thus, we need to discuss probability.

A probability is a pure number with no units. Its maximum value is 1—the event definitely occurs or always occurs. Its minimum value is 0—the event does not occur or never occurs. You might hear talk about odds of 10 to 1 against a team's winning a game, or a 10 percent chance of rain; these are both ways of describing a probability of 0.1. If there is more than one possible outcome, each outcome can be assigned a probability. A coin can land with heads or tails facing up. In terms of probability, we would say $p_{heads} = p_{tails} = 0.5$ and $p_{heads} + p_{tails} = 1$. (We are saying that we know all the possible outcomes and that the coin is exceedingly unlikely to land on its edge!) If we were rolling a die, we would describe the possible outcomes as $p_1 = p_2 = p_3 = p_4 = p_5 = p_6 = 1/6 = 0.167$. We also know that $p_1 + p_2 + p_3 + p_4 + p_5 + p_6 = 1$.

We have used examples of coin tossing and dice rolling in which selection of outcomes is random. Remember that the use of probability does not *imply* a random process. (A professional baseball player might have a batting average of 0.330, which means any particular time he comes to bat, he has a 1-in-3 chance of getting a hit. But if you stood at the plate and swung randomly at 90 mi/h fastballs, your probability of getting a hit would be much lower). In general, if there are n equally probable outcomes, the probability of each outcome is

$$p_n = 1/n$$

We have just used two examples of the rule of addition for probabilities, which says that

$$p_{A \text{ or } B} = p_A + p_B$$

This rule applies if the probabilities are mutually exclusive or disjoint (in other words, a coin can not land *both* heads and tails at the same time). For independent events or independent properties, we must use the multiplication rule for probabilities

$$p_{A \text{ and } B} = p_A \times p_B$$

For example, suppose your astronomy class has 200 students. Suppose also that one-half of the people in the class are women, one-half are men, and one-quarter of the class are seniors. If the class sits down randomly, the probability that the person next to you is a woman is 100 / 200 = 0.5. The probability that the person next to you is a senior is 50 / 200 = 0.25. Using the multiplication rule shows that the probability that the person next to you is a female senior is 0.5 × 0.25 = 0.125. To go back to an earlier example, the probability of tossing three heads in a row (or three tails) is 1/2 × 1/2 × 1/2 = 0.125. Similarly, the probability of rolling three 6s in a row is $(0.167)^3 = 0.0047$, or about 0.5 percent. You would not expect to see this happen until you had rolled the dice about 200 times.

Random events are independent. The probability of one event does not depend on the history of previous events. The probability can only be defined as an average over a large number of events. (We have seen this idea before—we cannot assign a probability to the decay time of a single radioactive atom, but we can determine the average probability of decay for a large number of atoms.)

Now we can relate this discussion to the probability of impacts from interplanetary debris. Figure 8–21 shows that a catastrophic impact occurs about once every 100 million years. The last one occurred 65 million years ago. Does that mean another catastrophe will not occur for 35 million years? Not necessarily! Large impacts appear to be random, not periodic. If catastrophes occur *on average* every 100 million years, the next one might occur in 35 million years or 150 million years or 1 million years.

This all sounds very abstract. What is the personal danger? If $p_{impact} = 1$ over 100 million years, then over a human lifetime (in round numbers), $p_{impact} = 100 / 10^8 = 10^{-6}$. If 1 in 10 on the planet die in such a disaster, our chance of dying is 10^{-7} or 1 in 10 million. Let us put this tiny number in perspective. It is somewhat less than your probability of dying by botulism. It is 1000 times less likely than your probability of dying in a plane crash, and it is 100,000 times less likely than your probability of dying in a car crash. Asteroid impacts may be good movie fodder, but they do not represent a credible hazard.

---

reality of the impact threat was dramatically demonstrated by this event: Large impacts really do happen (Figure 8–25).

The Interplanetary Opportunity

While asteroids and comets pose a potential threat, they also present a long-term interplanetary opportunity. Scientists project that we will have depleted many of Earth's natural resources by the middle of the 21st century. The solution to resource shortages may be the mining of asteroids. Based on meteorite samples, some Earth-approaching asteroids must contain not only pure nickel-iron alloys, but also ores of economically important platinum-group metals. These asteroids

FIGURE 8-26
Beyond the Moon, the closest planetary bodies are Earth-approaching asteroids. Small examples, a few hundred meters across, approach close enough to the Earth-Moon system (lower right) to be visited relatively easily by spaceships. Astronauts transfer from a "parked" ship to the asteroid surface. Some researchers believe their materials could be economically exploited for use on or near Earth. (Painting by WKH.)

could be mined in space by using "free" solar energy that streams through space 24 h/day. Masses of metal worth billions of dollars—enough to supply Earth's needs of certain metals for decades—should be obtainable from individual kilometer-scale and smaller asteroids. Studies are already underway to examine the feasibility of flying to Earth-approaching asteroids for economic and scientific exploration, as suggested by Figure 8–26. In terms of energy expenditure, some are actually easier to reach and return from than the Moon!

If asteroid resources could be "harvested" and processed in space, then mining, processing, and consequent industrial pollution could decline on Earth. Thus, instead of a purely defensive response to the asteroid threat, we could invest in a positive (if admittedly visionary) program to respond to our new knowledge about asteroids. The ability to reach asteroids and deflect them from one orbit to another would emerge as a by-product of such a program and would thus solve the problem of the asteroid threat. At the same time, such a program could help to solve Earth's environmental problems.

In fact, missions to expand our knowledge of asteroids and comets are on the drawing boards. The first mission to place a probe in orbit around an asteroid is expected to be NASA's *NEAR* (Near Earth Asteroid Rendezvous) mission, which plans to orbit asteroid 433 Eros in 1999; and similar missions to comets are under international study.

Summary

All of the solar system's small bodies—comets, asteroids and their fragments, meteors, and meteorites—began as planetesimals, the preplanetary bodies that formed in the solar system 4.6 billion years ago. These bodies formed during a relatively short period that lasted about 20 million years. The fact that meteorites of many different types, together with the Moon and Earth, share a common age of 4.6 billion years is one of the most important observational facts of solar system astronomy; it reveals that the solar system itself formed 4.6 billion years ago. To keep the interplanetary bodies in context, remember that added together they amount to a minute fraction of the mass in the planets. They are important because of the clues they provide to the history of the solar system.

Four main groupings of interplanetary bodies are the Oort cloud of comets, the Kuiper belt of comets, the main belt of asteroids, and the Trojan asteroids. Figure 8–27 diagrams the evolution of these various bodies, from icy and rocky planetesimals to today's comets, asteroids, meteors, and meteorites. This figure shows that icy bodies were perturbed or deviated from their regular orbits into the Oort cloud, and later they became today's comets. Dusty debris from comets creates meteors and meteor showers. The figure also shows how rocky and metallic debris from the asteroid belt is occasionally perturbed toward the terrestrial planets, making meteorites and meteorite craters.

Meteorites are extremely valuable because they represent free samples from inside different kinds of ancient parent bodies. They show that many parent asteroids—at least many of those larger than a few hundred kilometers across—melted soon after they formed, producing iron cores and rocky mantles. Other asteroids apparently lacked strong heating and thus preserve their original materials. The existence of water-

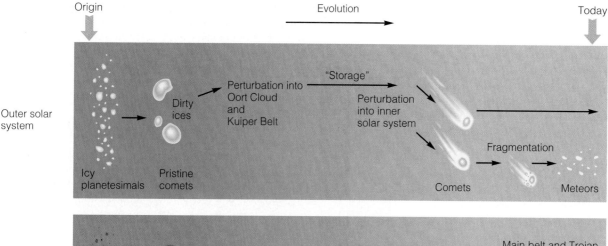

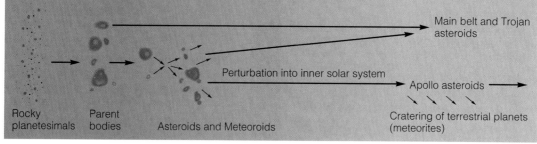

FIGURE 8-27
Schematic histories of comet and asteroid material. The view on the left shows how the material condenses into bodies that are many kilometers across. The view on the right shows how these bodies fragment and have their orbits altered, which explains the interplanetary phenomena that we see today. The diagram illustrates names used for the various types of objects.

bearing minerals on some of these asteroids also proves that some asteroids were never strongly heated, because the water would have been driven off by heating.

Recent discussions about the importance of asteroids and comets have tended to emphasize the threat created by their occasional collisions with Earth. It is certainly true that occasional, catastrophic impacts have altered the history of life on Earth. At the same time, however, asteroids may offer an opportunity for future resource gathering that may solve some of the environmental problems of Earth. In any case, the asteroids and comets offer some of the best clues to the ancient history and formation of the solar system, a point we will revisit in the next chapter.

Important Concepts

You should be able to define these concepts and use them in a sentence.

comets	dirty iceberg model of comets
asteroids	Oort cloud
sublimate	Kuiper belt
meteors	chaotic orbits
meteorites	meteor showers
parent body	zodiacal light
main asteroid belt	fireballs
planetesimals	Trojan asteroids
comet head	gravitational perturbation
comet tail	Earth-crossing asteroids
comet nucleus	soot line
solar wind	frost line

How Do We Know?

These questions and answers show how the scientific method is used to learn about the universe.

Q How do we know that comets contain ices such as frozen H_2O or frozen CO_2?

A When comets get close enough to the Sun for these ices to sublimate, the comets give off gas, and the gas has the right composition to come from frozen H_2O, frozen CO_2, and other ices. In particular, we can gather the light from the glowing tail with a telescope, spread the light out into a spectrum, and look for spectral features that are characteristic of these two molecules. Also, the *Giotto* probe measured water vapor and other products of sublimating ice as the probe flew through the head of Halley's comet. This direct measure of gas chemistry was done in the same way as it is on Earth.

Q How do we know that meteor showers are caused by debris from comets?

A Meteor showers do not occur randomly. They occur at specific times each year, which is a clue that they are related to the position of Earth in its orbit of the Sun. The major showers occur when Earth crosses the orbit of specific comets. Meteor showers vary in intensity from year to year, because the debris is not uniformly distributed along a comet's path and Earth moves through slightly different parts of the path each year. Scientists got strong evidence supporting this idea when an infrared satellite was able to directly image tiny particles strewn out along the path of a comet.

Q Does the Oort cloud actually exist?

A Evidence for the Oort cloud—the large spherical region where most comets live for most of the time—is indirect. When comets travel in the inner solar system, we can follow enough of their paths to predict their entire orbit. In many cases the orbit is a very elongated ellipse. By Kepler's laws, a comet in this type of orbit must travel very far from the Sun. Since the comet travels very slowly when it is far from the Sun, it will spend most of its orbital period in the deep freeze of the outer solar system, at a distance of thousands of astronomical units. An important consequence of this idea is the fact that the comets we see whipping through the inner solar system are representatives of a much larger population of slow-moving, distant comets. Since the comet orbits do not have any preferred orientation, the Oort cloud must be a sphere. Thus, our confidence that the Oort cloud actually exists is a reflection of our confidence in Newton's law of gravity and Kepler's laws. A small, rocky object at a distance of thousands of astronomical units would be incredibly faint, so we have little prospect of directly detecting the Oort cloud.

Q How do we know that most meteorites are fragments of asteroids?

A There is a combination of circumstantial evidence. To some extent, there is a match between the spectroscopic properties of meteorites and asteroids. For a number of meteorites whose orbits are known, the orbits have their farthest point from the Sun lying in or near the main asteroid belt. Also, properties of minerals in certain meteorites, such as the irons, show that they crystallized inside bodies about the size of asteroids.

Q How do we know that meteorites and asteroids formed 4.6 billion years ago?

A The same radioactive dating methods used to date Earth rocks can be used on meteorites. Many meteorite samples have been measured, and there is a strong concentration of formation dates at 4.6 billion years ago. Since we believe that many meteorites are fragments of asteroids, this argument extends to the age of asteroids.

Q How do we know when the next 1-km or 10-km meteorite will strike Earth?

A We do not know exactly when the next large meteorite will hit Earth. Large impacts occur randomly, due to the chaotic obits of many interplanetary bodies. Even though we know that such impacts have occurred in the past and we can calculate their approximate frequency, we cannot predict when any *particular* impact will occur. Although it is conceivable to discover an object that will hit Earth a certain number of years from now, no such object has yet been found. However, knowing the number of Earth-approaching asteroids and their typical orbits, we can calculate statistically the rate of large impacts, and this agrees generally with the characteristic ages of large craters on Earth. Roughly speaking, we can expect a 1-km asteroid to hit Earth within about the next 100,000 years and a 10-km asteroid within the next 100 million years.

Problems

You should use these problems to test your understanding of the information and concepts in this chapter. The * indicates a more advanced or mathematical problem.

1. If a comet should happen to pass through Saturn's satellite system but got no closer to the Sun than the vicinity of Saturn, why is it not likely to be detected from Earth?
2. Kepler's third law states that $a^3 = P^2$, where a is the semimajor axis of a body orbiting the Sun (expressed in astronomical units) and P is the period (expressed in years). Should this result apply to comets? Why or why not? If a typical comet in Oort's cloud has a semimajor axis of about 100,000 AU (10^5 AU), how frequently will it return to the inner solar system?
3. In terms of measuring and reporting useful scientific information, what actions would be appropriate if you observed an extraordinarily bright meteor or fireball? In two columns, list examples of useful and nonuseful descriptions of the fireball's speed, brightness, and apparent size. Could you make any quantitative measurements? How would you make sure you had not observed a known astronomical object? What actions would be appropriate if you saw a meteor strike the ground?
4. Summarize relations between the bodies and particles responsible for **(a)** meteors, **(b)** comets, **(c)** the zodiacal light, **(d)** asteroids, and **(e)** meteorites. What is the typical size or range in size of each type of object? Draw a diagram to show which part of the solar system each type of object inhabits.
5. Suppose future astronauts could match orbits with a comet and reach its nucleus. Describe the possible surface appearance of a comet. Consider gravity, surface materials, sky appearance, and so on. Which would be easier to match orbits with, a long-period or short-period comet?
*6. Answer the following questions, assuming that typical interplanetary material moves at about 15 km/s relative to the Earth-Moon system. **(a)** If a kilometer-scale asteroid were discovered on a collision course with Earth when it was 15 million km away, how much warning time would we have? **(b)** What would be the potential dangers of such an asteroid's collision with Earth? **(c)** If a much smaller asteroid were similarly discovered at the distance of the Moon (384,000 km), how much warning time would we have? **(d)** Roughly how long would the objects take to pass through the 100-km thickness of the atmosphere?
7. What is the danger of an event such as that in Problem 6 compared with the danger of other natural disasters, such as earthquakes? If there are so many pieces of interplanetary debris out in space, why is the probability of a large impact so low? Use a library or the Internet to look up the rate of occurrence of natural disasters. Is a large-scale meteorite disaster a plausible source of myths during the 10,000-year-long history of human civilization? Defend your answer.
8. In what ways are the orbits of comets and asteroids different from the orbits of the planets? Why is it so hard for us to predict the orbits of interplanetary bodies far into the future? Why does Jupiter have such a large influence on the orbits of comets and asteroids?
*9. Suppose a small asteroid of 50 percent nickel-iron, with radius equal to 100 m, could be located and exploited. If the value of the alloy were $0.90 per kilogram, what would be the potential economic value of the asteroid? Compare this with the $20 billion cost of the Apollo program. Assume that the density of the nickel-iron is 8000 kg/m^3.
*10. How many asteroids would be required to provide enough material to make one planet the size of Earth? Assume that

a typical asteroid is 120 km across and Earth is about 12,000 km across.

*11. If 100,000 meteorites capable of killing someone strike Earth's surface every year, what is the probability that anyone has been killed in recorded history? This problem is simple, but you need to make some assumptions to solve it. Assume that the meteorites hit random positions on Earth's surface. You will need to estimate an average population over the time period you consider. You will also need to estimate the cross-sectional area (as seen from above) of a person. *Hint:* What is the fraction of Earth's surface that is "covered" by people? Why is it more likely that someone would have been hit by a meteorite in this century than at any other time in history?

Projects

1. Use a large piece of cardboard to make a model of the inner solar system out to the orbit of Jupiter. Assume that the planets travel approximately in the plane of the cardboard. Use orbital properties listed in Table 13–1 to cut out scale models of the orbits of various interplanetary bodies. (A slit through the first cardboard could be used to show how comet or asteroid orbits penetrate the ecliptic plane. Note that the Sun must always occupy one focus of each orbit.) **(a)** Show how the geometry of passage of a comet (or other body) through the ecliptic plane, especially for highly eccentric orbits, depends on the angle between the point where the comet is closest to the Sun and the ecliptic plane, measured in the orbit. (This angle is fixed for each body but is omitted from Table 13–1 for simplicity.) **(b)** Show how the prominence of a given comet may depend strongly on where Earth is in its orbit as the comet passes through the inner solar system.

2. If you can manage it, visit the Barringer Meteor Crater in northern Arizona. Why is this feature misnamed? Observe the blocks of ejected material and deformation of rock strata as explained in museum signs and tapes. What would any prehistoric observers (the estimated impact date was 20,000 years ago) have witnessed at various distances from the blast that formed this crater nearly 1 mi across? (Other impact sites are known in various states, but they are eroded or undeveloped.)

3. If you cannot get to an impact crater, examine meteorite specimens in a local museum. Compare the appearance of stones and irons. In iron samples that have been cut, etched, and polished, look for the distinctive crystal patterns. The sizes and patterns of the crystal structures give information about the cooling rate and environment when the meteorite formed inside its parent body. In fact, the large iron crystals in a meteorite are unique products of slow cooling in a deep space environment—bear that in mind if you buy a chunk of a meteorite since there are many fake specimens for sale. Do any of the samples have both stone and iron? Offer a hypothesis on how they might have formed in terms of the core-mantle structure of the parent bodies.

Reading List

Beatty, J. K., and Levy, D. H. 1995. "Crashes to Ashes: A Comet's Demise." *Sky and Telescope,* vol. 90, pp. 18–26.

Belton, M. J. S., and 19 others. 1994. "First Images of Asteroid 243 Ida." *Science,* vol. 265, pp. 1543–7.

Chapman, C., and Morrison, D. 1994. "Impacts on the Earth by Asteroids and Comets: Assessing the Hazard." *Nature,* vol. 367, pp. 33–40.

Gallant, R. A. 1994. "Journey to Tunguska." *Sky and Telescope,* vol. 87, pp. 38–43.

Gehrels, T., ed. 1994. *Hazards due to Comets and Asteroids.* Tucson: University of Arizona Press.

Gehrels, T. 1996. "Collisions with Comets and Asteroids." *Scientific American,* vol. 274, pp. 54–9.

Hartmann, W. K., Miller, R., and Lee, P. 1984. *Out of the Cradle.* New York: Workman Publishing Co.

Jewitt, D. C., and Luu, J. 1995. "The Solar System beyond Neptune." *Astronomical Journal,* vol. 109, pp. 1867–76.

Levy, D. H., Shoemaker, E., and Shoemaker, C. 1995. "Comet Shoemaker-Levy 9 Meets Jupiter." *Scientific American,* vol. 273, pp. 85–91.

Luu, J. X., and Jewitt, D. 1996. "The Kuiper Belt." *Scientific American,* vol. 274, pp. 46–52.

Morrison, D., Chapman, C., and Slovic, P. 1994. "The Impact Hazard." In *Hazards due to Comets and Asteroids,* ed. T. Gehrels, Tucson: University of Arizona Press.

Ostro, S., and 13 others. 1995. "Radar images of Asteroid 4179 Toutatis." *Science,* vol. 270, pp. 80–6.

Pieters, C. M. 1994. "Meteorite and Asteroid Reflectance Spectroscopy." *Annual Reviews of Earth and Planetary Science,* vol. 22, pp. 457–97.

Schaefer, B. E. 1997. "Comets that Changed the World." *Sky and Telescope,* vol. 93, pp. 46–51.

Scotti, J., Rabinowitz, D., and Marsden, B. 1991. "Near Miss of the Earth by a Small Asteroid." *Nature,* vol. 354, pp. 287–9.

CHAPTER 9

How Planetary Systems Form

A scene in the early solar nebula. Planetesimals of rocky and icy debris orbit in the foreground. The young Sun is partially obscured and reddened by dust and gas in the inner nebula. (SOURCE: Painting by WKH.)

CONTENTS

- Safronov and the Theory of Planet Formation
- Archaeology of the Solar System
- The Protosun
- The Solar Nebula
- From Planetesimals to Planets
- Planets Orbiting Other Suns
- Summary
- Important Concepts
- How Do We Know?
- Problems
- Projects
- Reading List

PREREADING QUESTIONS ON THE THEMES OF THE BOOK

OUR ROLE IN THE UNIVERSE
How many plausible sites for life can we find?

HOW THE UNIVERSE WORKS
Are the processes that led to the formation of the solar system unique?

HOW WE ACQUIRE KNOWLEDGE
How well can we understand events that took place 4.6 billion years ago?

WHAT TO WATCH FOR IN CHAPTER 9

This chapter tells the story of the formation of the solar system. We synthesize information from the previous chapters in order to describe how our Earth as well as the Sun and other planets formed, 4.6 billion years ago. The chapter also exemplifies how scientists move from observational data to a theory that solves a complex mystery. Physical evidence reveals that the planets grew within a period of 50 million years—the first 1 percent of the solar system's age—by the accretion of dust grains that were left over after the Sun formed. Measurements of the age and formative processes come from meteorites and other sources. The physical processes that led to planets in the solar system may work during the formation of other stars. Discoveries in the past few years confirm that planets do exist near many other stars. In terms of the Copernican revolution, the evidence suggests that planetary systems, and even Earth-like planets, may be common in the universe; systems like ours may not be unique or "special."

VICTOR SAFRONOV AND THE THEORY OF PLANET FORMATION

Bold scientific theories can have complex origins. In the 1960s, the Soviet scientist Victor Safronov, little known in the West, wrote numerous papers and a book on planet formation. A translation into English was published in Jerusalem in 1972 and distributed through the U.S. Department of Commerce. The gray paperback book summarized several decades of work by Safronov and his colleagues. The Russian scientist proposed a theory of how the solar system formed. Safronov described how dust grains orbiting the early Sun would have collided with one another and aggregated, or joined, into asteroid-like bodies and eventually into planets. What seemed like an obscure book was destined to play an important role in our understanding of the origin of the solar system. It inspired much subsequent work by American scientists. The story of this theory illustrates both the international character of science and how science builds broad theories by combining the research of many individuals.

Two decades before Safronov's book was published, scientific study of the origin of planets was just beginning. Few scientists worked on this problem. Space exploration had not yet started, and there was not the wealth of data we now have about the solar system. In the United States, a handful of scientists were making telescopic observations of comets and asteroids and were studying the chemistry of meteorites to get clues about how the planets formed. Meanwhile, Russian scientists were working in almost complete isolation from scientists in the West because of the cold war. They also approached the problem differently, from a mostly theoretical perspective. The mathematician Safronov was the most important member of this Russian group. He concluded that the planets must have formed in a disk-shaped system of dust grains, ice grains, and gas particles, all orbiting the early Sun in the same direction.

Safronov's big breakthrough was to calculate what would happen to such a system as the particles collided with one another. He knew that the particles would follow elliptical orbits, but he also realized that gas in the system would cause drag effects that would slow down the particles. Developing the equations that represented such reasoning, he calculated the speeds at which neighboring particles would collide with one another. He concluded the particles would tend to stick together as snowflakes cluster in a snowstorm. As the particles grew bigger, they would stick together due to their own gravity. Using the rates and speeds of collisions, Safronov calculated that a few large, planet-size bodies would grow from the initial small bodies during this aggregation process, a process often called *accretion*. He realized that the nature of the collisions also explained the general rotation rates of planets and the tilts of their axes.

If not for the cold war, Safronov's results would have been published in international journals and would have had an immediate effect on scientists in the West. However, because of cold war competition, Soviets did not publish much in the West. In 1969, Safronov gathered his many papers and wrote his book in Russian, summarizing his theories. Finally, in 1972, the circuitous translation program produced the English-language edition of his life's work. At last, Safronov's science had overcome the sociopolitical hurdles and appeared on a world-wide stage.

Even then, other scientists were slow to accept the implications of Safronov's work. In 1975, the suggestion that the Moon had been formed by a giant impact on Earth was roundly criticized. To Western researchers, giant impacts seemed like a wild speculation, dreamed up solely to explain the Moon. Finally, in 1984, after Safronov's ideas had been read and accepted, researchers agreed that impacts probably played a key role in forming our Moon and shaping other features of the solar system. U.S. and European researchers have extended Safronov's ideas into complex computer models, which show how systems of particles orbiting the early Sun could have aggregated into a handful of planets. Eventually, Safronov traveled to the United States and other countries, and he was awarded scientific prizes for his work.

Safronov's story shows how international politics can create barriers to the spread of scientific ideas. And yet, in the end, the pen was once again mightier than the sword. The cold war did not prevent Safronov's theoretical work from being accepted. The key to acceptance was the usual hallmark of science: observational evidence and confirmation by independent workers. In this case the evidence was gathered from studies of meteorites, asteroids, lunar samples, laboratory experiments, and other sources. The questions we want to answer in this chapter include these: What does the combination of theory and observational evidence tell us about how our home system of planets formed? How long did it take? Why do planets, satellites, and interplanetary bodies have so many different properties? Do the same processes form planets around other stars?

Archaeology of the Solar System

Nobel laureate Hannes Alfven, who spent years researching the solar system's origin, once said: "To trace the origin of the solar system is archaeology, not physics." In other words, in this field, we need to search for clues from ancient materials to help us reconstruct the events that occurred long ago.

The most important clues are the properties of the solar system that cannot be explained by present-day conditions but must have arisen as the solar system formed. Table 9–1 lists some of these clues. These properties do not reflect ordinary geological evolution over 4.5 billion years; rather, they reflect the *primordial,* or original, conditions. In this chapter, we will show how these clues help us piece together the origin of the solar system. Sometimes our way of thinking about our problem affects our ability to come up with the answer. For a long time, scientists viewed the formation of the solar system as a catastrophic event. Such a view implies that the creation of planets is rare, random, and difficult to predict. Now we believe that the formation of planets is a "normal" process that is a by-product of how stars form.

Imagine that you are a great detective and you have just arrived at a crime scene. The body is cold; the house is empty. The murder took place hours before, but the room around you contains clues as to what happened. Some clues are obvious, others are subtle—all clues are important! You may find the body in a particular position. Objects in the room may be out of place. There may be chemical traces on the carpet or a distinctive odor in the air. In the same way, the inhabitants of planet Earth are faced with a mystery. The events that led to our existence took place long ago. Yet all around us in space are telltale signs of the formation process.

TABLE 9-1
Solar System Characteristics to Be Explained by a Theory of Origin

1. Most of the mass of the solar system is contained in the Sun; the planets contribute only 0.2 percent.
2. All the planets' orbits lie roughly in a single plane.
3. The Sun's rotational equator lies nearly in this plane.
4. The planets and the Sun all orbit in the same west-to-east direction, called *prograde* (or *direct*) revolution.
5. Planetary orbits are nearly circular.
6. Planets differ in composition.
7. The composition of planets varies roughly with distance from the Sun: Dense, metal-rich planets lie in the inner system whereas giant, hydrogen-rich planets lie in the outer system.
8. Meteorites have different geological properties from those of all known planetary and lunar rocks.
9. The Sun and all the planets except Venus and Uranus rotate on their axes in the same direction (prograde rotation). Obliquity (tilt between equatorial and orbital planes) is generally small.
10. Planets and most asteroids rotate with rather similar periods, about 5 to 10 h, unless obvious tidal forces slow them (as in Earth's case).
11. Distances between planets usually obey Bode's simple rule, which is a roughly geometric spacing with distance from the Sun.
12. Outer planets are more massive than inner planets, and most of the mass of outer planets is composed of hydrogen and helium.
13. Planet-satellite systems resemble the solar system in miniature.
14. Impact craters are common throughout the solar system, and the cratering rate was much higher in the first few hundred million years of solar system history.
15. As a group, most comets' orbits define a large, almost spherical swarm around the solar system (the Oort cloud). Other comets reside in the Kuiper belt, near Pluto and just beyond it.
16. The planets have much greater angular momentum (a measure combining orbital speed, size, and mass) than that of the Sun.

Let us look at the clues briefly now, and then in greater detail as we unravel the mystery of how the solar system formed.

First, we can consider clues from geometry. Most of the mass of the solar system is contained in a central spherical object—the Sun. The planets are not distributed randomly around the Sun. All the orbits lie roughly in a plane. The distances of the planets from the Sun obey a roughly geometric spacing. However, minor components of the solar system are distributed differently in space. In particular, the swarm of comets called the Oort cloud fills an enormous sphere centered on the Sun and the planets.

Next, there are clues from motions. We must first distinguish clearly between two important motions. A **revolution** is the motion of a planet around the Sun. A **rotation** is the spin of a planet on its axis. The planets all revolve around the Sun in the same direction in orbits that are nearly circular. In addition, the Sun and all but two of the planets rotate on their axes in this same direction. In other words, the planets spin in the same sense as they travel around the Sun. The spin axes of the planets are in most cases perpendicular to the plane of the orbits.

We must also look at clues from composition. Planets near the Sun are small, dense, and rich in metals. Planets far from the Sun are large, low-density, and rich in hydrogen and helium. Each planet differs slightly in composition. Meteorites give us our only direct evidence of the composition of interplanetary bodies—meteorites differ in geological properties from all known terrestrial and lunar rocks.

Last, there are clues from morphology. Planet-satellite systems resemble the solar system in miniature; satellites orbit in the equatorial plane of the planet and in the same sense as the planet's rotation. Ring systems behave in the same way. Rings seem to be evidence of a failed planet (in the case of the main asteroid belt) or moon (in the case of the ring systems of the four outer gas giants). Impact craters are seen throughout the solar system, but most were created in the first 10 percent of solar system history.

The Protosun

Let us consider the first four facts about the solar system listed in Table 9–1. We have known since the time of the Copernican revolution that the Sun is the dominant object in the solar system. Our tour through the solar system has revealed some impressive worlds, but they are all dwarfed by the Sun. The sum of the mass of all the planets combined is barely 0.2 percent of the mass of the Sun. People have known for thousands of years that the planets all move across a thin strip of sky called the ecliptic. In terms of our heliocentric picture of the solar system, the planets occupy a volume shaped like a disk, not like a sphere. The plane of the planet orbits is also the plane of the Sun's rotation. Also, the planets all orbit the Sun in the same direction, and it is the same direction as the Sun spins. Is there any simple idea that can explain these facts?

Our hypothesis for the formation of the solar system starts with a large cloud of gas and dust. What do we mean by *dust?* We refer to tiny solid particles about 10^{-4} to 10^{-5} mm across. Before the Sun existed, its material must have been distributed thinly in interstellar space. The Sun began to form a little more than 4.6 billion years ago when this cloud (or part of it) began a process of contraction and flattening. We will first describe that process and then show how it led to the formation of the Sun.

Early Contraction and Formation of the Disk

Newton's law of gravity is a powerful tool that we use in the archaeology of the solar system. Remember from Chapter 3 that each atom in the universe attracts every other atom. According to this law, if the atoms are far apart, the forces of attraction are very small, but if the atoms are closer together and numerous enough, their attraction for one another can become important.

Why is a cloud of gas likely to contract? Particles in the middle of the cloud have particles on all sides, so they feel gravity equally in all directions. However, particles on the edge of the cloud have few particles outside them and more particles toward the center of the cloud. So on average they feel a gravity pull toward the center of the cloud. As they move inward in response to the gravity force, the cloud gets a little smaller (Figure 9–1). Recall that gravity is an inverse square force—the attraction between particles increases as the distance between them gets smaller. So in the slightly smaller cloud, the inward tug on the particles near the edge gets larger. And so the cloud contracts even more. Which makes the gravity force stronger. You can see that the contraction increases the gravity, which increases the contraction, and so on—this is a runaway process! The cycle

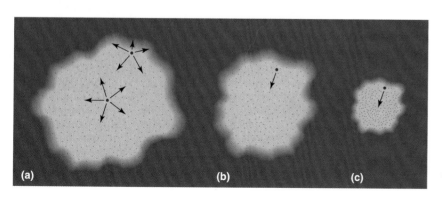

FIGURE 9–1
Simplified and idealized version of gravitational collapse. **(a)** A particle near the center of a diffuse cloud will feel a roughly equal gravitational pull in all directions. However, a particle near the edge feels more gravity toward the center of the cloud. **(b)** In response to the unequal gravity, particles near the edge of the cloud move toward the center and the cloud shrinks. **(c)** As the cloud shrinks, the density goes up and so the gravity force on any particle goes up. This acts to accelerate the contraction, leading to a gravitational collapse.

continues and the initial shrinkage of the cloud turns into a **gravitational collapse.**

What causes the initial contraction? It does not take much. The cloud only needs to feel a random disturbance (or perturbation) from outside to start the process of gravitational collapse. The disturbance could be the gravity from the passage of a nearby star. It could be the death of a nearby star in a great explosion. One little "push," and the collapse is underway.

Another important feature of a gas cloud in space is rotation. Step outside and think of the air around you. It is never perfectly still. Air has two types of motions. One is the *microscopic* motion of the individual molecules; we saw in Chapter 4 that every molecule has a random velocity that depends on the temperature of the gas. The other motion is large-scale, or *macroscopic*. The Sun's energy causes the air to move in large circulating patterns. These rotating patterns can occur in either direction and on any scale—from a little dust devil in your backyard to a hurricane hundreds of miles across! You can understand this better if you do a simple experiment using a cup of coffee or a pan of water. Stir the liquid vigorously but as randomly as you can. Then wait a moment and put a drop of cream in it. Notice that the cream rotates smoothly in one direction. This phenomenon occurs because the sum of the random motions creates a small overall rotation in one direction or the other.

A diffuse gas cloud in space has these same two types of motion. The gas atoms or molecules have random velocities corresponding to their temperature. But the cloud also has a small amount of overall rotation. What happens to the rotation as the cloud collapses?

Angular momentum is the quantity that measures the total rotary motion of a system—it increases with the amount of rotating material and with the rotation rate and size of the rotating system. Mathematically, angular momentum is the product of the mass, the rotation rate, and the size or radius of an object. A powerful principle of physics, called the **law of conservation of angular momentum,** states that angular momentum is constant as a system changes. The Science Toolbox on the next page explains this in greater detail. If an object shrinks, the rotation rate must increase to compensate. You are familiar with the example of a figure skater who spins faster when pulling in his or her arms.

Let us see what this means for our large, slightly rotating cloud in space. As the cloud begins to contract, the rotation rate must increase because angular momentum is conserved. In general, a big, slowly rotating system will turn into a small, rapidly rotating system. The collapse *amplifies* the rotation rate. But in the rotating cloud, the gas cannot collapse equally well in all directions. Gas can easily collapse along the rotation axis or perpendicular to the plane of rotation. However, along the equator of the rotating cloud, the gas meets a resistance due to the rotation. This is the same force that presses you against a car door when you go around a tight curve, or keeps the tension in a string when you whirl an object over your head. So the cloud collapses more along the poles of its rotation than along its equator, and the effect is to squash it into a disk shape.

The result of these two simple physical forces—contraction and rotation—is an amazing change in the size and shape of the cloud. A large, slowly rotating cloud that is more or less spherical is transformed into a small, rapidly rotating disk! This is shown schematically in Figure 9-2.

There is a final important feature of the gravitational collapse we have just described. At first, when the cloud was large, its atoms were far apart and fell more or less freely toward the center of gravity in the middle of the cloud. This state is called *free-fall contraction*. The runaway collapse process will naturally concentrate a lot of mass near the center of the cloud. Regions of high density will exert a strong gravity, which will pull nearby layers in faster. This in turn will increase the gravity, pulling in more gas, and so on. Most of the gas will therefore pile up rapidly in the center of the cloud, and a small fraction of the gas will be "left behind" on the outskirts where the gravity is weaker. If free fall per-

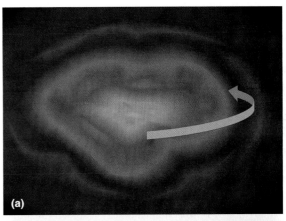

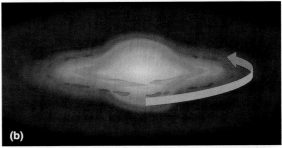

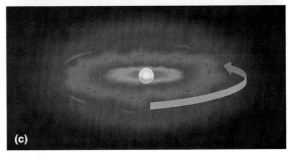

FIGURE 9-2
Three stages in the evolution of the protosun. **(a)** A slowly rotating interstellar gas cloud begins to contract because of its own gravity. **(b)** A central condensation forms, and the cloud rotates faster and flattens. **(c)** The Sun forms in the cloud center, surrounded by a rotating disk of gas.

SCIENCE TOOLBOX

Momentum and Angular Momentum

As we saw in Chapter 3, Newton masterfully summarized the science of motion in three laws. He had an additional insight: he realized that many aspects of motion are controlled by a concept called momentum. *Momentum* is simply defined as the product of mass M and velocity v:

$$\text{Momentum} = Mv$$

Normally we call this *linear* momentum because it refers to motion in a straight line. Momentum incorporates mass and motion. A stationary object has no momentum. The faster an object moves, the larger its momentum; and the more massive a moving object is, the larger its momentum. The idea is familiar to you: A car moving at 20 mi/h has much more momentum than a bicyclist moving at 20 mi/h, and can do a correspondingly larger amount of damage!

Newton knew that momentum of individual objects could change when objects collided or interacted. But he realized that the total amount of momentum was the same before and after the interaction. We call this the *law of conservation of momentum*.

If you hold a gun, it has no momentum. After it is fired, the bullet travels at very high speed, so its momentum is the product of a very small mass and a very large velocity. The recoil of the gun is exactly the same amount of momentum in the opposite direction—it has the opposite sign so it cancels the bullet's momentum. To keep the recoil small, the gun is made heavy so that its momentum is the product of a large mass and a small velocity. The key point is that the total amount of momentum has not changed. Zero momentum before has become the sum of two equal and opposite amounts of momentum after, which is also zero. You would see the same effect if you stood on the ice wearing skates and threw a heavy rock away from you. The momentum you gave the rock would be balanced by an equal momentum propelling you backward. But since your mass is much larger, your backward velocity would be smaller than the rock's forward velocity.

Many people wonder what makes a rocket fly. How does it stay up in the air? Does it push against something when it flies? The answer is clear if you think of momentum. To create upward motion and momentum of the rocket, fuel is burned and ejected at high speed backward. In fact, rocket and jet nozzles are specifically designed to make the exit speed of the hot gas as high as possible. At every point in its flight, the forward momentum of a rocket is exactly balanced by the backward momentum of the fuel vapor. The rocket has large mass and moderate velocity, and the gas has small mass and extremely high velocity. A rocket does not fly by pushing against anything. That is why, unlike an airplane, it can fly in space. It makes no difference whether the rocket is in air or in the vacuum of space!

Angular momentum is like linear momentum, but it applies to rotating or orbiting systems. It is defined as the product of mass M, velocity v, and radius r:

$$\text{Angular momentum} = Mvr$$

In this case the radius is the size of the rotating object or the distance of an orbiting body from the center of gravity. The law of conservation of angular momentum says that angular momentum will stay constant as a system changes (Figure 9–A). For example, in a spinning object, if the radius r decreases, the rotation speed v will increase. Let us look at a couple of examples.

If the solar system really collapsed from a gas cloud that originally extended at least to the orbits of Neptune and Pluto, then the rotation speed must have increased greatly. By how much? The mean distance of Neptune from the Sun is 30 AU or $30 \times 1.5 \times 10^8 = 4.5 \times 10^9$ km. If we assume that all the material within this orbital radius ended up in the Sun, then after the collapse, the size of the cloud is just the radius of the Sun, or 700,000 km. In the product Mvr, the mass has not changed. So if the radius r has decreased by a factor $4.5 \times 10^9 / 700,000 = 6500$, then the rotation velocity must have increased by the same factor. We will see later in the chapter that the Sun does not spin at the high rate we would expect because of other effects that slowed down the Sun's spin after it formed.

Now let us look at Earth in its orbit of the Sun. As we discussed in Chapter 3, in the Science Toolbox on Kepler's laws, Earth's orbit deviates from a circle by 3.4 percent. This means it varies from 1.017 times the mean Earth-Sun distance to 0.983 times the mean Earth-Sun distance. Since angular momentum is conserved, when the distance goes up, the velocity must go down; and when the distance goes down, the velocity must go up. The mean orbital velocity of Earth is

(continued)

FIGURE 9–A
The conservation of angular momentum is illustrated by a skater. She can increase her rate of spin by pulling in her arms and her free leg. (SOURCE: Bob Burch/Bruce Coleman, Inc.)

SCIENCE TOOLBOX

Momentum and Angular Momentum
(continued)

$2\pi r / (1 \text{ year}) = (2 \times 3.14 \times 1.5 \times 10^8) / (365 \times 24 \times 3600) = 14.94$ km/s. So when Earth is closest to the Sun, Earth's speed is $14.94 \times 1.017 = 15.18$ km/s. When it is farthest from the Sun, Earth's speed is $14.93 \times 0.983 = 14.69$ km/s. This slight variation in orbital speed affects timekeeping based on the stars.

Now we have a concrete and physical explanation for Kepler's second law of planetary motion. This law says that an imaginary line that connects a planet (or a comet or any other orbiting body) to the Sun sweeps out equal areas in equal intervals of time. Kepler's law is no more than a statement of the law of conservation of angular momentum. Orbital velocity depends on distance to the Sun, but both vary so as to keep the angular momentum constant.

---

sisted, the cloud could have completely contracted to form the Sun in only a few thousand years (see Figure 9–3 for an imagined view).

If you review the material we have just covered, you will see that we can explain each of the first four facts in Table 9–1. The runaway process of gravitational collapse naturally dumps most of the mass in a central object. So we can understand why the Sun contains most of the mass of the solar system. The collapse leads to a disk type of geometry, so the planets inhabit a single plane. Because the Sun itself was an integral part of this disk, we can explain the fact that the Sun's equator aligns with the disk. Furthermore, we account for the orbital motion of all the planets in the same direction, because the entire disk of material rotated in a single direction. We have accounted for some of the most important clues with a few simple physical ideas.

Helmholtz Contraction of the Protosun

As the gas cloud collapses, a new force comes into play. In the free-fall phase of the contraction, gas and dust particles stream freely toward the center of gravity. The density is so low that collisions are rare. But as the gas gets denser, the distance between particles gets smaller and the rate of collisions gets higher. Now the gas starts to behave as a gas, according to the law we covered in a Science Toolbox in Chapter 7. Compress a gas, and the pressure and temperature will rise. When you push on a bicycle pump with the valve covered, the gas will shrink, but it will also exert a pressure that can counter your entire weight. In the forming solar system, gas pressure competes against the inward force of gravity.

German astrophysicist Hermann von Helmholtz showed how these competing forces shape the contraction and evo-

FIGURE 9–3
The appearance of the disk-shaped solar nebula at the time when the Sun and planets were just forming. A cloud of dust and gas surrounds the Sun. The Sun's light is reddened by the dust, in the same way as it is reddened at sunset. In the background is a vast region of gas and dust in which other nearby stars are also forming at the same time. Observations of distant newly forming stars confirm many of these features, including the jets of high-speed material shooting "up" and "down" along the axis of the disk.
(SOURCE: Painting by WKH.)

lution of the cloud. He published his first calculations in 1871, which was even before the energy source that powers the Sun had been revealed. Helmholtz showed that pressure and heat would build up in the contracting cloud. This is an example of the conservation of energy, a familiar idea we first met in Chapter 4. Gravitational potential energy is converted to thermal or heat energy. He was able to use physical principles to calculate the rate at which temperature increased inside the contracting protosun. Gravitational contraction in which the shrinkage is slowed by outward pressure is called *Helmholtz contraction.*

Calculations based on the Helmholtz theory describe the evolution of the collapsing cloud. The shrinkage of the disk is eventually halted by the increasing pressure of the hot gas, at a temperature of a few thousand kelvins. The disk stabilized at a size corresponding to the size of the present-day solar system. Meanwhile the bulk of the gas continued to collapse under strong gravity. Eventually the cloud's central temperature rose to 10 million K or more, starting the nuclear reactions that made it a star and not just a ball of inert gas. You might wonder how sure we are about events that took place so long ago. The previous discussion describes a plausible scenario based on simple physics. Our confidence that it is correct is bolstered by actual observations of newly formed stars surrounded by clouds of gas and dust.

The Solar Nebula

So far we have described the collapse and rotation of a large, diffuse cloud. After it reaches a stable configuration with a young Sun and a surrounding disk of gas and dust, we are ready to account for the properties of the planets. Any cloud of gas and dust in space is called a **nebula** (plural: *nebulae*), from the Latin term for mist. The disk-shaped nebula that surrounded the contracting Sun is called the **solar nebula.** The molecules of gas and grains of dust in the solar nebula moved in circular orbits. This must have been true because noncircular orbits would have crossed the paths of other particles, leading to collisions that would have damped out the noncircular motions. As an analogy, imagine a large number of racing cars moving around a circular track. If any racing car tries to move on a noncircular path, it will move from one lane to another and collide with other cars. In a like manner, if the orbit of a particle started to get elliptical, the orbit would overlap other orbits. As with racing cars changing lanes, collisions would occur that would, on average, bring the particle back toward a more circular orbit. Thus, large-scale motions of the gas and solid particles in the solar nebula were generally in near-circular paths around the sun, accounting for the fifth fact in Table 9–1.

Condensation of Dust in the Solar Nebula

The next thing to happen in the solar nebula was the slow and steady formation of planets from microscopic particles. What clues do we get from the chemical composition of the solar system? We are seeking to explain the next two facts in Table 9–1: the fact that planets vary in composition and the fact that the composition depends on the distance from the Sun.

We have seen that Helmholtz contraction heated the outer, dusty regions of the solar nebula to at least 2000 K. Once the Sun formed and the contraction stopped, there was no new source of gravitational energy. As a result, the gas in the nebula began to cool. At such a high starting temperature, virtually all elements were in gaseous form. Like all cosmic gas, most of the solar nebula was hydrogen and helium, but a few percent of the atoms were heavier elements, such as silicon, iron, and other planet-forming material.

How did solid particles of planet-forming rock arise in this gas? The answer can be seen on Earth. When air masses cool, water molecules within them *condense* to form particles. In a cool cloud, water vapor condenses into snowflakes, raindrops, hailstones, or the ice crystals in cirrus clouds. Condensation occurs more easily when there is a grain of dust or solid material to start with. Similarly, as the solar nebula cooled, different molecules condensed into droplets. As it cooled even more, the droplets solidified into tiny solid particles.

Different mineral compounds condense at different temperatures—this is called a **condensation sequence.** The condensation sequence is important in understanding the differences among planets. Given that the gas in the solar nebula is cooling from about 2000 K, let us see the order of the particles that form. At temperatures of about 1600 K, heavy elements such as aluminum and titanium condensed. They formed microscopic particles of metallic oxides, as shown in Table 9–2. At about 1400 K, a more important constituent, iron, condensed. Microscopic bits of nickel-iron alloy formed as mineral grains or perhaps as coatings on the earlier aluminum and titanium grains. Still more important, at about 1300 K, abundant silicates began to appear in solid form. These silicate minerals are the common rock-forming materials, containing complex mixtures of magnesium, calcium, and iron—all bound to oxygen. Black carbonaceous minerals condensed at the much lower temperature of 300 K. Finally, at temperatures in the range of 100 to 200 K, hydrogen-rich molecules condensed into ices—water ice, frozen methane, and frozen ammonia.

You can see in the condensation sequence a clear relationship between temperature and the atomic or molecular weight of the materials that become solid. Above 1000 K, we see oxides of aluminum (molecular weight 102), pure flakes of iron and nickel (atomic weights, respectively, of 56 and 59), and silicate rocks such as enstatite and olivine (molecular weights, respectively, of 100 and 172). At 300 K and below, we see carbon in the form of pure soot (atomic weight 12), water ice (molecular weight 18), and ices of methane and ammonia (molecular weights 16 and 17, respectively).

If the solar nebula had cooled in the same way throughout, there would be no composition variations in the solar system. The same minerals would have condensed everywhere. But the nebula stayed hotter near the center, due to the energetic young Sun, and the nebula cooled more rapidly

TABLE 9-2
Condensation Sequence in the Solar Nebula

APPROXIMATE TEMPERATURE (K)	ELEMENT CONDENSING	FORM OF CONDENSATE (WITH EXAMPLES)	COMMENTS
2000	None		Gaseous nebula
1600	Al, Ti, Ca	Oxides (Al_2O_3, CaO)	
1400	Fe, Ni	Nickel-iron grains	Parent material of planetary cores, iron meteorites?
1300	Si	Silicate and ferrosilicate minerals [enstatite, $MgSiO_3$; pyroxene, $CaMgSi_2O_6$; olivine, $(Mg, Fe)_2SiO_4$] in form of microscopic grains	First stony material, combined to form meteorites; some still preserved in primitive meteorites
300	C	Carbonaceous grains	Forms black, carbonaceous soils and rocks
300 to 100	H, N	Ice particles (water, H_2O; ammonia, NH_3; methane, CH_4)	Large amounts of ice; still preserved in outer planets and comets

SOURCE: Data from Lewis, Grossman, and others.

on the edges. Figure 9–4 shows how the temperature at which condensation stopped depends on the distance from the Sun.

Now we can understand why the inner and outer solar system bodies are so different. The inner nebula formed dust grains that were complex mixtures of magnesium-, calcium-, and iron-rich silicate minerals—the basic materials of rocks. Lighter gases and molecules were driven out by the Sun's radiation. The outer parts of the nebula, far from the Sun, cooled to much lower temperatures and formed other important types of solid grains. At distances corresponding to the middle of the asteroid belt, black carbonaceous minerals condensed. Beyond the outer asteroid belt, snowflakes of water ice condensed. This explains the soot line and the frost line that we covered in the last chapter. At 100 to 200 K, in the outermost nebula, ammonia and methane ices condensed. In the outer solar system, these ices remain solid even in direct sunlight; they survive today in comets and in the atmospheres and on the satellites of the giant planets. The condensation sequence tells us why there are icy worlds in the outer solar system, but not in the inner solar system. We have neatly explained facts 6 and 7 in Table 9–1!

Meteorites as Direct Evidence

Remember from the last chapter that the small, solid bodies that formed in the primordial solar system are called planetesimals. Comets represent one type of survivor of the ancient population of planetesimals. The material in icy comets has probably never been heated significantly since the early days of planetesimal formation and preserves clues to that era. Many interplanetary dust grains may be such debris from comets. They filter into Earth's upper atmosphere and have been captured for study by U.S. Air Force sampling programs—imagine a high-altitude aircraft trawling something like an air filter to scoop up the tiny particles. Figure 9–5 shows a clump of grains collected in this way and also

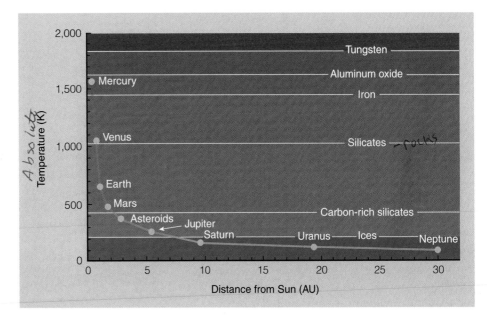

FIGURE 9-4
The condensation of different chemicals depends on the distance from the Sun. This plot shows the temperature at various distances from the Sun at the time that condensation stopped. At Earth's distance, only metals, oxides, and silicates condensed. At Saturn's distance and beyond, all materials including ices condensed. As a consequence, moons in the outer solar system have a significant ice content.

(a)

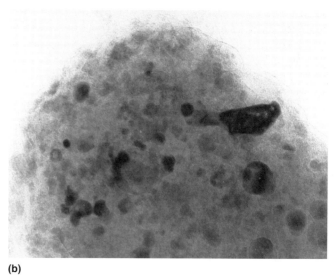
(b)

FIGURE 9-5
(a) Clumps of microscopic particles collected by high-flying aircraft. They may have originated in comets. The largest clump is 0.1 mm across or 60 micrometers (μm). (SOURCE: Courtesy of D. E. Brownlee, University of Seattle, Washington.) **(b)** A bit of interplanetary dust, shown at much higher magnification by a transmission electron microscope. The larger dark blobs are mineral grains, probably magnetite (Fe_3O_4) or iron-nickel sulfide. They are surrounded by extremely fine-grained, carbon-rich particles formed at lower temperatures than the iron grains. This microscopic structure supports the theory that tiny mineral grains condensed and then accreted into larger bodies as planetary material formed in the dusty nebula around the primordial Sun. (SOURCE: Courtesy of Roy Christoffersen and Peter Buseck, Arizona State University.)

shows how microscopic carbonaceous grains have aggregated around iron grains.

Asteroids are also leftovers of the planetesimals. In the asteroid belt, the planetesimals never completed the process of aggregation, because of the disturbing effects of gravitation from nearby Jupiter, as mentioned in the last chapter. Meteorites (which are fragments of asteroids) thus give us evidence about the planetesimals. Many big meteorites have been identified, but scientists are always looking for more evidence. A fine rain of meteorites—anything from the size of dust grains to pebbles—reaches Earth all the time. They would be impossible to notice on most terrain, but they stand out on the cool blue-white surface of the Antarctic ice pack. Scientists regularly travel there and sweep the surface with sticky rollers, to gather more of this valuable evidence from space.

What do they find when they analyze these samples? Primitive carbonaceous meteorites (Figure 9–6), which were

FIGURE 9-6
A piece of the Allende carbonaceous meteorite, showing white inclusions. The inclusions are examples of materials that formed in one environment and were later trapped in another material. The black matrix is composed of microscopic dust grains condensed at a few hundred kelvins. The white inclusions contain aluminum-rich minerals condensed at high temperatures. As recognized in the 1970s, the inclusions shed light on the earliest formative conditions in the solar system. (SOURCE: Courtesy R. S. Clarke, Smithsonian Institution.)

never melted, retain much of their original structure. Their composition supports the aggregation process described above. For instance, some carbonaceous meteorites, such as the one in Figure 9-6, contain inclusions of material believed to be among the earliest substantial solid objects in the solar system. These pebblelike inclusions are rich in elements that condensed first (at the highest temperatures), such as osmium and tungsten. Their minerals formed at temperatures of about 1450 to 1840 K, consistent with fact 8 in Table 9-1. Yet the surrounding matrix of these carbonaceous meteorites contains microscopic grains formed at lower temperatures. The high-temperature minerals must have condensed first and aggregated into pebblelike objects, only to be overwhelmed and surrounded by the sooty carbonaceous material, which condensed later after the nebula had cooled, as seen in Figure 9-7. Just as in archaeology, we can trace a complex history in a single rock from space!

A Nearby Presolar Explosion?

Scientists began to study meteorite compositions in the 1970s, and an interesting mystery soon emerged. They found that inclusions of high-temperature minerals in certain carbonaceous meteorites, such as pictured in Figure 9-6, had peculiar abundances of certain isotopes. Remember that most elements in nature occur in one main, stable isotopic form and several rarer forms, both unstable (radioactive) and stable. Researchers found certain isotopes in the meteorites that could only have been created very shortly *before* the solar system itself.

For instance, they found xenon 129, a form of xenon that arises from the decay of radioactive iodine 129. This decay process is very fast, geologically speaking, once the iodine is created—the half-life of iodine 129 is only 17 million years. Xenon is chemically inert and unlikely to form mineral grains, so the chemical must have been trapped before it decayed when it was still iodine. That leaves only a brief interval—in the range of 1 to 20 million years—between the creation of the iodine and its incorporation into meteorite material. Remember that the inclusions were among the first minerals condensed in the solar nebula, and they are the main minerals to get a dose of the mysterious iodine. We must conclude that the iodine was injected essentially at the time the solar system was born.

Where did the radioactive iodine come from, and how did it get trapped in the first planetary material? We believe the iodine and other isotopes involved were created by nuclear reactions inside certain massive stars, which burn their nuclear fuel quickly and then explode. The evidence indicates that such a star was located near the presolar nebula and exploded, spewing short-lived radioactive isotopes and other debris into the cloud that was becoming the solar nebula. Indeed, the blast from the explosion probably helped compress the presolar cloud, pushing the atoms closer together and thus helping to initiate the collapse that produced the Sun. We may have not only solved the mystery but also found out why the presolar cloud collapsed; the birth of our star was triggered by the death of another star. This same sequence may be common among stars. As we will see in greater detail in later chapters, stars form in clusters embedded in a galaxy's region of densest interstellar gas; the largest stars in the cluster tend to be unstable and explode, compressing nearby gas and promoting still additional star formation.

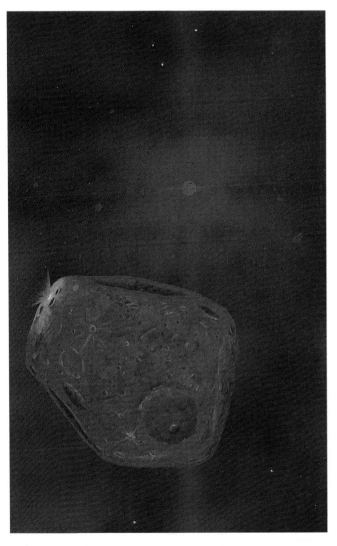

FIGURE 9-7
A scene in the early solar nebula showing a planetesimal with an impact. In the background we see the dust-reddened Sun and the disk that formed from the solar nebula. (SOURCE: Painting by WKH.)

From Planetesimals to Planets

We have followed the story of the solar system from the initial diffuse gas cloud to the collapse of a young star with its surrounding disk of gas and dust. Molecules condense into droplets and then solid grains as the nebula cools. The process of condensation takes us up from the molecules and clusters of molecules to grains about 1 mm across. Rocky and metal-rich grains survive throughout the nebula, but icy and hydrogen-rich grains can survive only in the outer regions. But this leaves a large question unanswered: How do we get from tiny particles to planets? After all, even the smallest planet is trillions of times larger than a dust grain!

We would also like to use the next four clues in Table 9–1. So let us see what facts 9 through 12 tell us. All the planets except two spin in the same direction as they revolve around the Sun, and they all have similar rotation periods. Fact 11 is especially intriguing. Planets do not form with equal spacing from the Sun, nor do they form at random distances from the Sun. Rather, each planet is a factor of 1½ to 2 times farther out than the previous one. This simple progression must be a clue to the process of planet formation. Last, the outer planets (omitting Pluto) are 15 to 300 times more massive than the inner planets, and most of their large mass is in the form of hydrogen and helium.

As the solid microscopic dust grains began to form into substantial bodies, the scene in the solar nebula must have resembled Figure 9–7. The emerging solid planetesimals drifted around the Sun in a dusty, gaseous, disk-shaped nebula. Most of the mass of the nebula was in the form of gas; only a few percent of the mass was solid material. Why? First, most of the mass in the contracting protosolar cloud concentrated in the middle of the cloud and ended up in the Sun. The nebula probably had no more than one-tenth the mass of the Sun. Furthermore, hydrogen and helium comprise about 98 percent of the mass of the original Sun-forming cloud. Therefore, only about 2 percent of the nebula was composed of the heavier elements that more readily formed solid matter. These were elements such as silicon, magnesium, carbon, and oxygen, which condense into solid mineral and ice particles. Figure 9–8 summarizes the main differences between the terrestrial planets and the gas giant planets that must be explained by any successful theory. In size, mass, and density, there is a major difference between the planets of the inner and outer solar system.

The Process of Accretion

Now we come to an amazing part of the story. Although the preplanetary particles may have formed as microscopic grains, they clearly aggregated into bigger bodies; otherwise there would be no planets. Making mountains out of molehills is not sufficient—we must make entire worlds out of motes of dust! The gradual accumulation of material to make large objects was the key process in producing the solar system as we know it. This process is called **accretion**.

As you read in our opening story, the Russian scientist Safronov was a key player in working out the process of collisional accretion. As the grains collided and aggregated, they formed intermediate-size planetesimals, which ranged in size from millimeters to many kilometers. These planetesimals were abundant throughout the solar system, as we can judge from the following three pieces of evidence. First, craters on planets and their satellites indicate impacts of planetesimals with diameters of at least 100 km

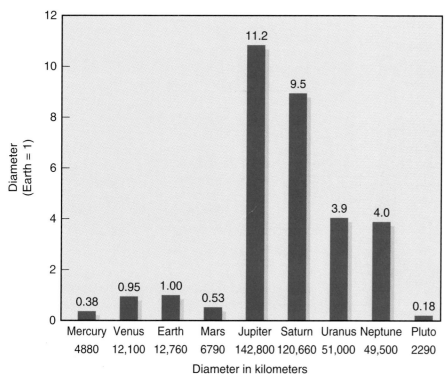

FIGURE 9-8
(a) Diameter, **(b)** mass, and **(c)** average density of the planets. The distinction between the inner terrestrial planets and the outer giant planets is clear. Also, see Table 9–1.

(a)

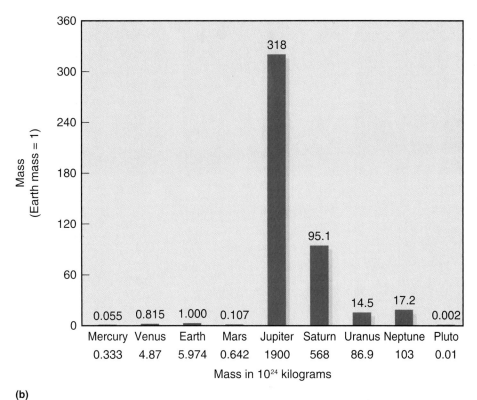

(b)

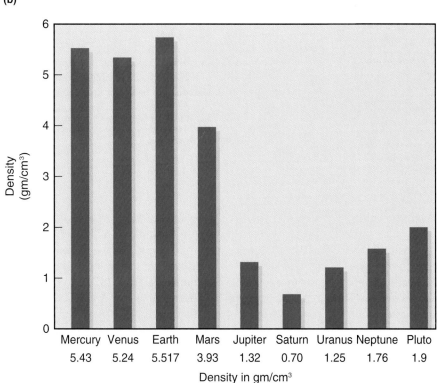

(c)

(Figure 9–9). Also, data from many types of meteorites show that meteorites were originally contained in asteroid-sized bodies up to a few hundred kilometers across. Last, asteroids and comet nuclei, reaching several hundred kilometers in diameter, are surviving examples of the remnants of planetesimals.

What did the newly forming planetesimals look like? As they grew in diameter from several meters to hundreds of kilometers, they probably resembled asteroids—perhaps with irregular shapes due to mergers of bodies and fractures due to collisions.

We can speculate as to how the accretion took place. The details, however, are very complex. In the beginning, planetesimals orbiting the Sun were like the racing cars mentioned earlier, moving on adjacent, parallel lanes of a circular racetrack. If they collided, they hit in low-speed sideswiping

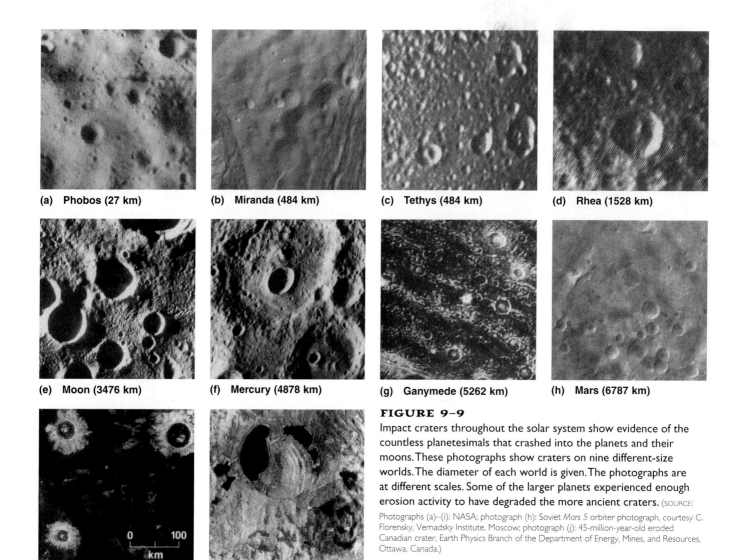

(a) Phobos (27 km)
(b) Miranda (484 km)
(c) Tethys (484 km)
(d) Rhea (1528 km)
(e) Moon (3476 km)
(f) Mercury (4878 km)
(g) Ganymede (5262 km)
(h) Mars (6787 km)
(i) Venus (12,104 km)
(j) Earth (12,756 km)

FIGURE 9-9

Impact craters throughout the solar system show evidence of the countless planetesimals that crashed into the planets and their moons. These photographs show craters on nine different-size worlds. The diameter of each world is given. The photographs are at different scales. Some of the larger planets experienced enough erosion activity to have degraded the more ancient craters. (SOURCE: Photographs (a)–(i): NASA; photograph (h): Soviet *Mars 5* orbiter photograph, courtesy C. Florensky, Vernadsky Institute, Moscow; photograph (j): 45-million-year-old eroded Canadian crater, Earth Physics Branch of the Department of Energy, Mines, and Resources, Ottawa, Canada.)

collisions. Initially, nothing disturbed their circular motions or forced them to "change lanes," so all particles moved together on nearly circular paths. As the largest planetesimals grew, their gravity got stronger and drastically perturbed the motion of any smaller planetesimal that passed nearby, sending it on a new course across the orbits of other bodies, similar to a car changing lanes. You can think of this in terms of the conservation of energy, as discussed in Chapter 4. If a massive body transfers kinetic energy to a much smaller body, the massive body slows down slightly but the small body speeds up greatly. This transfer occurs not through a collision but by a gravitational interaction. In this way, the planets themselves "pumped up" the collision speeds of the smaller planetesimals. Later, the speeds got so high that planetesimals sometimes smashed each other, bringing the growth process to an end.

How long did planet building take? You might think it would take a very long time to build a planet by the accretion of tiny pieces. Let us imagine it was a linear process—adding one piece at a time. Let us also suppose that we start with 1-m chunks of rock. The biggest planetesimals are about 100 km across. Volume increases with the cube of the size, so it would take $(100 / 0.001)^3 = 10^{15}$ small chunks to make a large planetesimal. A medium size planet is about 10,000 km across. This is another factor of $(10,000 / 100)^3 = 10^6$ in volume. So a planet is made of $10^{15} \times 10^6 = 10^{21}$, or a thousand billion billion, small chunks. And if you are building a planet one piece at a time, it will take a thousand billion billion times longer than it does for two small chunks to come together. It seems hopeless.

Yet, as worked out by Safronov and other scientists, accretion is actually a very efficient process. The largest planetesimals grew fastest, sweeping up the others. The process is nonlinear: The larger they grew, the more their gravity pulled in neighboring planetesimals, causing the growth to accelerate. The time needed for formation of solid bodies can be measured from the fact that certain radioactive materials with half-lives of a few million years were trapped inside meteorite parent bodies before the isotopes decayed. Studies in the 1990s show that some parent bodies of some meteorites had reached diameters on the order of 100 km, and partially melted, all within in a few million years of the

Sun's formation. This time is strikingly short compared to the history of the solar system. If you represent the 4.6-billion-year history of the solar system by 1 Earth year, asteroidlike solid bodies had formed from the dust by noon on January 1. Similar isotopic studies indicate that the largest of these bodies continued to grow and reached planet size in 50 to 100 million years—the first 1 or 2 percent of solar system history. In the analogy just given, Earth and the other planets would be fully grown by around January 4.

Dynamical studies indicate that this process tended to form bodies rotating in the same prograde direction as the disk rotates. This research also suggests rotation periods of typically 5 to 20 h, consistent with observations of planets and asteroids. The studies also suggest that if two large bodies started growing in orbits that were too close together, the bodies would eventually attract each other gravitationally, collide, and merge. The final surviving planets are therefore spaced apart. In this way, the solar nebula divided into doughnut-shaped zones around the Sun, each about 1.5 to 2 times as wide as the one closer to the Sun, with one planet dominating each zone. (As mentioned in Chapter 3, Bode noticed this spacing several hundred years ago.) We have successfully explained facts 9 through 11 in Table 9–1.

Using this scenario of gradual growth through collisions, we can account for the major classes of bodies in the solar system—the terrestrial planets, giant planets, asteroids, and comets. The explanation is as follows:

1. *Terrestrial Planets.* In the inner part of the solar system, the process of accretion continued until Mercury-size to Earth-size planets formed. Most of the planetesimals in this region were made of silicate rocky material, familiar to us from the rocks of Earth and the Moon. The terrestrial planets grew until most of this material was swept up; out-rushing gas and radiation from the newly formed Sun blew away the remaining gas and dust.

2. *Giant Planets.* The giant planets formed in the same way as the terrestrial planets, from accreting planetesimals. The giant planet zone contained ices as well as rocky material, augmenting local planetesimal masses. Thus the *embryo planets* that were to become Jupiter, Saturn, Uranus, and Neptune grew larger than Earth. When they reached about 10 to 15 times the mass of present-day Earth, they had such strong gravity that they began to pull in gas from the surrounding solar nebula. Thus they accreted not only planetesimals to make a solid and liquid planet, but also massive atmospheres of gas whose composition approximately equaled that of the nebular gas. Hence the giant planets can be thought of as two-phase planets, with their initial cores of icy and silicate materials and their massive hydrogen-rich atmospheres added on later. Notice that we are *predicting* that Jupiter and Saturn have solid cores, even though we do not have the evidence yet. We have also explained the large mass and light gas composition of the outer planets, fact 12 in Table 9–1.

3. *Asteroid Belt.* Asteroids are planetesimals that never made it all the way to planet status. Why were asteroids stranded in the zone between Mars and Jupiter? Likely it was the planet-forming zone closest to the largest planet in the solar system. According to the rough geometric spacing expressed by Bode's rule, a planet should have grown here. Ceres, the largest asteroid, did grow to 1000 km in diameter by the time the growth process stopped in that zone. It stopped growing because nearby Jupiter had grown so huge that Jupiter's gravity disturbed the motions of the asteroids and pumped up their collision speeds. This caused them to smash into innumerable fragments during collisions, instead of coalescing into an even larger body.

4. *Meteorites.* The largest objects in the asteroid belt reached sizes of a few hundred kilometers across. They were internally heated, melted, and differentiated into metal and rock portions—just as Earth was. The transfer of energy from nearby Jupiter increased their speeds, causing some of them to shatter in collisions with other large or fast neighbors. Fragments of the iron cores, rocky cores, and mantles sprayed out; much of this debris left the asteroid belt, and a few pieces ended up with Earth-crossing orbits. We can therefore explain the origin of iron and stony-iron meteorites. The violent origin of these rocks is particularly clear in the brecciated meteorites, with their fused jumble of rock types.

5. *Comets.* As noted in the last chapter, comets are icy planetesimals from the outer solar system. They avoided colliding with the giant planets but were flung by close encounters with these planets into the Oort cloud—fact 13 in Table 9–1. As the most massive planet, Jupiter is mainly responsible for creating these elongated orbits. The transfer of only a tiny bit of Jupiter's kinetic energy is enough to kick a comet into deep space. Remember that the "kick" is not an actual collision; it is the transfer of energy by the long arm of gravity. (NASA used this same technique, called the *gravitational slingshot,* to send the *Voyager* spacecraft close to Jupiter and boost them into the outer solar system.) Icy planetesimals in the region beyond Neptune had no giant planets to throw them into the Oort cloud; these formed the Kuiper belt.

Origins of Satellites

From what we have learned about the origin of the solar system, we can explain the origins of the satellite systems of the various planets. At first glance, satellites as a group appear to be a chaotic mix of different kinds of bodies, with no rhyme or reason: Giant planets have families of diverse moons that are small compared to the planet; some of the moons move in prograde directions, others retrograde. Other planets have no moons. Earth and Pluto are more like "double planets," with single moons that are one-quarter to one-half the size of the planet.

The most important process for forming satellites was similar to the accretion process that formed the planets themselves. Each of the four giant planets attracted enough nebular gas to form its own miniature solar nebula. In this disk-shaped cloud of gas and dust, the accretion growth process was repeated on a smaller scale. That is, each giant planet became an analog of the Sun, and moons grew

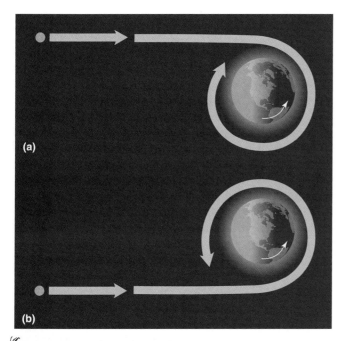

FIGURE 9-10
If a planetesimal approaches a planet and gets slowed in an extended atmosphere, it may end up in a **(a)** retrograde orbit or **(b)** prograde orbit. Such capture, however, requires "braking" by some mechanism (such as atmospheric drag, shown here) and is difficult to achieve. In spite of the thousands of planetesimals that may have been available, only a few captured satellites are known, such as the outermost moons of Jupiter and Saturn.

around it, analogous to planets. In these miniature planetary systems, the most abundant building materials were ice and the black carbonaceous dirt common to the outer solar system. Thus giant planets spawned systems of dirty-ice satellites in prograde, circular orbits, with the plane of the satellite orbits aligned with the equatorial plane of the planet. Thus we account for fact 14 in Table 9–1.

Planets also gained satellites by capturing them. As the last planetesimals altered their orbits in close encounters with giant planets, a few of the planetesimals were captured into orbits around the planets. When a planetesimal approached a planet, it had a 50 percent chance of passing on either the prograde side or the retrograde side, as shown in Figure 9–10. If it passed closely enough to be slowed by the extended primordial atmosphere of the planet, or if it suffered a collision that reduced its relative velocity to the planet, then it had a chance of being captured into an orbit around the planet. Orbits could be either prograde or retrograde. The change in velocity of the approaching planetesimal needed to effect capture is lowest at large distances from the planet, which may explain why several planets have outermost satellites that seem to have been captured. For instance, Jupiter's eight outermost moons appear to be black, carbonaceous objects similar to asteroids and comet nuclei in the region. One-half of the eight are in prograde orbits, and the other four are in retrograde orbits.

A third process of satellite formation was accretion of debris ejected by large impact, as we already explained in Chapter 5 for the case of the Earth-Moon system. If a large planetesimal hit a planet late in the planet-forming process, it could partially disrupt the planet and blow material into orbit, and a large satellite could form from the debris. Pluto's satellite, which is large relative to Pluto itself, may also have formed in this way, although many scientists think it is more likely to have been captured.

Gradual Evolution plus a Few Catastrophes

We now have a broad view of the formation and evolution of the solar system. Gas and dust condensed quickly into small particles throughout the collapsing solar nebula. Many planetesimals were accumulated by the planets, scattered into the Oort cloud, or left stranded in the asteroid belt within 50 to 100 million years—the first 1 to 2 percent of the solar system's age. Most of those remaining then crashed into the planets and their moons, forming craters. As lunar rocks taught us, this process continued throughout the solar system (see Figure 9–6) for first 500 million years—about 10 percent of the solar system's age. At this point, the mayhem stopped. Most planetesimals had disappeared from interplanetary space (explaining fact 15 in Table 9–1). The remaining gas and dust in the solar nebula had been blown away by the outward pressure of radiation and gases streaming out from the Sun. Today's comets and asteroids are representatives of the leftover planetesimals, which hit planets at a lower and roughly constant rate.

The theory we have described accounts for many features of the solar system. But there is a fly in the ointment—the last fact to be explained in Table 9–1. The Sun carries over 99 percent of the mass of the solar system but less than 1 percent of the angular momentum. Put simply, the Sun does not spin as fast as it should, given the angular momentum of the system as a whole. Why is this a problem? The law of conservation of angular momentum governs how rotation must change as an object changes size. As the presolar nebula collapsed, slowly moving material far from the center sped up as it got closer to the center. So either the planets somehow acquired "extra" angular momentum, or the Sun somehow "lost" angular momentum. Scientists currently favor the idea that the magnetic field of the Sun interacted with fast-moving particles to "brake" the Sun's rotation speed.

We have stressed that the growth of planets produced certain regularities in the solar system. The most important features are planetary orbits that lie in the plane of the Sun's equator, regular spacing of planet orbits, orbits that move in the same direction as the Sun's rotation, planets that mostly spin in the same sense as they move around the Sun, and small axial tilts of planets in most cases. These regularities of the solar system require smooth evolutionary growth from a system of many small planetesimals, not a catastrophic creation in some chaotic system. Many earlier theories of solar system formation tended to invoke rare catastrophes in order to explain the existence of planets. For example, French naturalist Georges Buffon, whom we met at the beginning of Chapter 5, suggested in the mid-1700s that the Sun had crashed into a passing star and the resulting debris formed planets. Recent calculations show that such collisions would be very rare. If planets required such an accident, fewer than 1 in 1 million stars might have planets. As we will see, however, the evidence suggests that planet formation may be a fairly "normal" process that is a by-product of how stars themselves form.

One of the beauties of the modern picture of planet formation is that it can explain the few irregularities of the solar system by a few catastrophic events. These catastrophic events involved collisions or near misses of the planets with the remaining planetesimals. Most planetesimals were very small compared to planets, but the largest one or two near any given planet were an appreciable fraction of the size of that planet. Thus, as each planet grew, it experienced collisions of different magnitude. If the largest collision was strong enough, it could have had important effects.

We have already seen several examples of this idea in previous chapters. A Mars-size planetesimal may have hit Earth late in its growth and blown off material from which the Moon formed. Uranus was probably also hit by a relatively large planetesimal, tipping its rotation axis to lie nearly *in* the plane of the solar system instead of nearly perpendicular to it, as is the case for most other planets. Other properties of the solar system, such as the different types of ring systems and the geological differences between hemispheres of some planets (e.g., Mars), may trace back to large impacts. The largest impacts were not big enough to randomize the properties of planets and their orbits, but they were large enough to give planets individuality of character!

Despite our confidence that we understand the solar system, there is one sense in which we cannot exactly predict orbits. Many small bodies in the solar system—including comets, Centaurs and even Pluto—are on **chaotic orbits.** Objects on chaotic orbits will eventually undergo a close approach with a much more massive body, leading to a radically different trajectory. These unpredictable orbits are discussed in the following Science Toolbox, along with the general issue of determinism in Science. Remember that the solar system as a whole is not chaotic. The planets and most of the asteroid belt are on regular orbits that can be predicted with a high degree of precision using Newton's law of gravity.

Let us look again at how science works in terms of a theory for the formation of the solar system. Our detective work has been very successful. We can account for the general features of the solar system and almost all the facts in Table 9–1. But this success leads to some general questions about the scientific method. Do we know that the theory just described is unique, or could there be other explanations for these facts? How many "peculiarities" do we have to explain with subsequent events before the formation scenario is compromised? Is the angular momentum issue a serious or a minor problem? Remember that we are engaged in a form of archaeology—all the interesting events took place several billion years ago. The legend of the cyclops started long ago when hunters found the skull of an elephant and misinterpreted the hole where the trunk connects for a large, central eye socket. Could we be misinterpreting evidence too?

Most scientists think not. The physical principles of gravitational collapse and accretion are well understood. Our theory may not get all the details of the solar system right, but it accounts for the main features and has great explanatory power. Better yet, as we will see, many of the features we have deduced for the ancient solar system have actually been observed around other embryo stars that are in the process of formation. We might even have enough confidence to predict that planets are a fairly frequent consequence of star formation. Our theory cannot tell us *how many* planets might form around a star or what their *characteristics* might be, but it indicates that they should occur. There is only one way to find out. We have to look.

Planets Orbiting Other Suns

The Sun is a star like other stars. So an inevitable question arises: Do other planets orbit these other suns? The search for **extrasolar planets** is one of the most exciting adventures in science. It addresses a central theme of this book: our place in the universe. Life evolved on a moist, rocky planet called Earth. It may also have existed in a primitive form on ancient Mars. Its presence in more extreme environments in the solar system has not yet been ruled out. Are there other potential sites for life, far beyond the solar system?

The idea that planetary material is a by-product of star formation gets a strong boost from observations of newly forming stars. Many of these have disk-shaped systems of gas and dust, including mineral grains of silicates and ices, just as the hypothetical solar nebula does. (We will learn more about these systems in later chapters about the stars.) But many questions are unanswered. Do these systems necessarily produce planets like Earth or Jupiter? What is a typical number of planets? How does planet formation depend on the amount of rotation of the initial nebula? How varied is the chemical composition of these disks?

It is unlikely that planets will form around *every* star. Some stars may be so hot that they drive out most of the gas and dust before accretion takes place, inhibiting planet formation. Planets will not form if planetesimals have their collision velocities pumped up too high, too soon. This could happen if a second star were nearby. More than one-half of all stars have companion stars orbiting them, as we will see in a later chapter—perhaps no planets exist in these environments. Still, in our galaxy alone, that leaves hundreds of millions of Sun-like, single stars that might have planetary companions. Do they harbor planets as the Sun does, or is our system unique? Definite proof of either situation would be awesome. In the first case there may be millions of planets like Earth, but in the second case we might be like the only system of islands in a global ocean.

Direct Detection

The most direct way to detect a planet around a nearby star is to image it directly. If we imagine looking at the solar system from the outside, we can demonstrate how difficult this

SCIENCE TOOLBOX

Chaos and Determinism

Newton's law of gravity gives the force between two masses separated in space. The equation we saw in Chapter 3 is exact. If we could measure the exact masses and the separation, the force and the subsequent motion would be exactly determined. Of course, we know that observations have uncertainties. But *in principle,* two objects in a gravitational embrace work as a perfectly predictable machine. The success of Newton's theory led to the metaphor of a *clockwork universe.*

What happens if you add just one more object? When three bodies orbit one another, we must use a separate equation to describe the force between each pair. There is no exact solution to this set of equations. If you had a powerful computer, you could calculate the orbits with a high degree of precision, but it would take an infinite amount of time to reach infinite precision. When we are dealing with the gravity of three or more objects, Newton's law does not allow us to precisely predict the outcome of a complex situation.

Newton would probably have been amused by the strong reaction of poets of the Romantic era to his theory; they considered his theory to be tyrannical because it seemed to rob humans of free will by suggesting a specific outcome to any set of initial conditions. But that is not how gravity works. In any realistic situation, the motions of objects cannot be predicted exactly from the starting conditions. We cannot prove that the future flows rigidly from the past.

In the past few decades, scientists have gained a second insight about gravity. In any complex system (which might only be three objects!) a *slight* change in the starting conditions can give a *large* change in the final outcome. Change an asteroid's position by a few meters and it may have a totally different future! This is the idea of chaos. In some situations, Nature can be wildly unpredictable!

Chaos is a new field of science, with many applications. Perhaps the most familiar example is the weather. You may wonder why weather forecasts are mostly useless more than four or five days ahead, when giant supercomputers and armadas of satellites are working on the problem. Weather involves motions on a wide range of scales—from dust devils to tornadoes to continent-size hurricanes. A small effect on a small scale can become a large effect on a large scale. Meteorologist Edward Lorentz called this the *butterfly effect*: A butterfly stirring the air in New York can eventually affect the weather in London. This cascade of small changes to large effects even has a place in folklore:

For want of a nail, the shoe was lost;
For want of a shoe, the horse was lost;
For want of a horse, the rider was lost;
For want of a rider, the battle was lost;
For want of a battle, the kingdom was lost!

You might think that chaos is synonymous with disorder. But order and disorder can coexist in the simplest physical system. Even the motion of a pendulum—a classic timekeeping device—is not strictly periodic. It displays some chaotic behavior. Or consider a dripping faucet. A careful experiment will show that the drops do not fall at exactly equal time intervals. It is too simplistic to think only in terms of peri-

(continued)

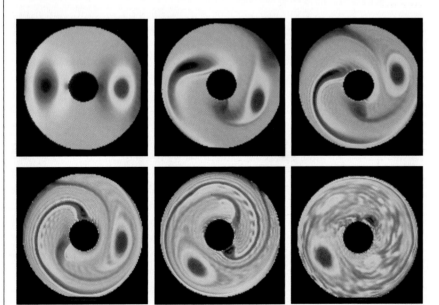

FIGURE 9–B
A computer simulation of Jupiter's atmosphere successfully recreates the Great Red Spot. A pair of vortices rotating in opposite directions (upper left) are modified by rotation and global winds. One vortex disappears, and the other persists—an island of structure surrounded by chaotic motion.
(SOURCE: NASA, Jet Propulsion Laboratories.)

SCIENCE TOOLBOX

Chaos and Determinism (continued)

odic or random processes. Scientists have been able to use computers to simulate Jupiter's Great Red Spot as a vortex in a rotating flow (Figure 9–B). After several cycles, most of the circulation pattern has broken up in turbulence, but a long-lived storm persists—an island of order in a sea of chaos. In the last chapter of this book, we will see that these ideas may even help us understand the emergence of life.

In considering chaotic orbits, it is important to remember that the solar system is not fundamentally chaotic. The orbits of the major planets can be accurately predicted over thousand of cycles and millions of years. However, the behavior of small solar system bodies can be very sensitive to initial conditions.

Let us take several examples. The orbit of Pluto may be chaotic due to successive small interactions with other planets and with solar system debris. Pluto's long-term behavior is therefore impossible to predict with precision.

Also, the orbits of asteroids near Kirkwood resonant gaps are difficult to predict with any certainty. Occasional collisions among asteroids may send them into resonances and then far from the asteroid belt. Some of those may then travel on paths that bring them close to a planet, resulting in a radically different orbit.

Last, the process of accretion that gave rise to the solar system is fundamentally chaotic. The process by which small rocky chunks collide and form planets can only be understood in a statistical sense. It is not possible to start with a proto-planetary disk and "predict" the formation of Earth or any other planet.

We conclude that small changes in the initial configuration of the solar nebula would have had effects on the final outcome. If we could "replay the tape" of the history of the solar system a hundred times with slightly different initial conditions, we would probably see planets each time but not the same planets or even the same number of planets.

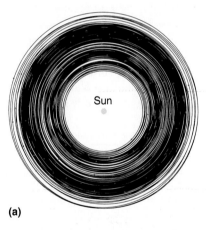
(a)

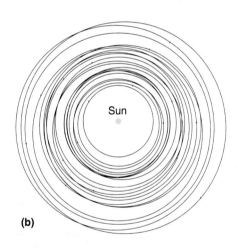
(b)

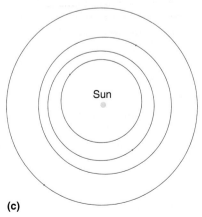
(c)

FIGURE 9–C
These drawings represent the results of a single computer simulation of the formation of the inner planets. Simulations with slightly different starting conditions give quite different results. **(a)** The simulation starts with 100 planetesimals orbiting the Sun. **(b)** After 30 million years, the planetesimals have accreted into 22 proto-planets of different sizes. **(c)** By 150 million years, four planets have formed.
(SOURCE: Adapted from G. Wetherill.)

is. The star's glare overwhelms the more feeble reflected light of the planet. Let us figure out what percentage of the Sun's light is intercepted by Jupiter. The surface area of a sphere around the Sun at the distance of Jupiter is $4\pi R^2$, where R is 5.2 AU, or 7.8×10^8 km; this equals 7.6×10^{18} km^2. The area of that sphere that is taken up by Jupiter is πr^2, where r is Jupiter's radius, or 71,400 km; this equals 1.6×10^{10} km^2. Jupiter therefore intercepts $(1.6 \times 10^{10}) / (7.6 \times 10^{18}) = 2 \times 10^{-9}$ or 2 billionths of the Sun's light.

But the situation is worse than that. We would only see one-half of Jupiter's lit surface when it was at its most favorable position of large apparent separation from the Sun (think of the phases of Venus). Also, Jupiter is not a perfect reflecting surface like a mirror. It reflects only about 30 percent of the sunlight that falls on it. So we lose another factor of 6, and the planet shows us only 3×10^{-10} of the Sun's light. What about looking for Earth-like planets? Since the Earth is 5.2 times closer to the Sun than Jupiter, the Sun's radiation at Earth's distance is about 5^2, or 25, times stronger by the inverse square law of radiation. However, Earth is also about 10 times smaller than Jupiter and so intercepts 10^2, or 100, times less light, giving 4 times less reflected light overall.

The bad news continues. As we look at the solar system from afar, the planets have a small angular separation from the Sun. We have not yet discussed the distances to the stars. However, we argued in Chapter 3 that even at the time of Copernicus we could put a lower limit on stellar distances from the fact that parallax was not observed. This lower limit was about 7000 AU. The small-angle equation in Chapter 2 tells us that a separation of 5.2 AU seen from a distance of 7000 AU subtends an angle of $206,265 \times 5.2 / 7000 = 153$ seconds of arc, or a bit less than one-tenth of the Moon's diameter. But as we will see in a couple of chapters, the nearest stars are actually 30 times farther away, reducing the angular separation of a Sun-Jupiter system to about 5 seconds of arc. A Sun-Earth system would, of course, appear a factor of 5 times closer together, or about 1 second of arc. Our task is like trying to detect a candle flame less than an inch away from a stadium floodlight!

We can do better by moving the experiment to longer wavelengths. The Sun's spectrum peaks at visible wavelengths, and the thermal spectrum falls off toward longer infrared wavelengths. However, a planet has a temperature of a few hundred kelvins. Wien's law from Chapter 4 tells us that the thermal spectrum from a planet will peak at long infrared wavelengths. So rather than look for the Sun's reflected light, a better strategy is to look at a wavelength at which a giant planet's *intrinsic* radiation will peak. The gain in contrast over the optical experiment is a factor of 1000. This brings Jupiter's radiation up to "only" 3×10^{-6}, or 3 millionths, of the Sun's radiation.

One ingenious idea is to catch a giant planet when it passes in front of a star. We saw in Chapter 3 that similar *transits* of Venus were used to measure the scale of our solar system over 200 years ago. Seen from afar, a Jupiter-like planet would block a tiny amount of the light from a Sun-like star. The percentage is given by the ratio of the areas, or about 1 percent. However, the reality is not as favorable. Giant planet atmospheres will diffuse much of the star's light around its circumference, so it blocks light very poorly—the actual reduction in the star's intensity as it suffers one of these partial eclipses might be as low as 0.01 percent.

Also, the distant planetary system would have to be lined up just right for us to see the event. Using the small-angle equation, we can deduce that we would only see the reduction in light when the planet and star lined up within an angle of $206,265 d / D$, where d is the diameter of the Sun and D is the Sun-Jupiter distance. This gives $206,265 (1.4 \times 10^6 / 7.8 \times 10^8) = 370$ seconds of arc. Such a small angle is only $370 / 206,265 = 0.0018$ of a full circle. So if planetary systems are flung at random orientations in space, only 1 in 500 will have a favorable alignment for us to see the eclipse. And the eclipse would only occupy a small fraction of time of each orbit. The circumference of a Jupiter orbit is $2\pi D = 4.9 \times 10^9$ km, so the planet would cross the star for $(1.4 \times 10^6) / (4.9 \times 10^9) = 0.0003$, or 0.03 percent, of the orbit time. This is a single day out of a 12-year orbit.

It all sounds impossible. We have used simple estimation to show that the direct detection of planets around other stars is enormously challenging. Yet you may be surprised to hear that there are telescopes and instruments being planned that should be able to carry out *each* of the experiments described above. The philosophic and scientific stakes are so high—the detection of Earth-like planets and a possible answer to the question "Are we alone?"—that a lot of research money is being devoted to this effort.

Indirect Detection

Direct imaging is not the only way to find a planet. We can use our knowledge of gravity to detect a planet by its influence on the star it orbits. This is a more subtle technique, but it is potentially very powerful.

When any body orbits a star, the star itself does not stay still. The two bodies move around their common **center of gravity.** The easiest way to think of the center of mass is in terms of the balance point between two objects (Figure 9–11). The situation with orbits is shown in Figure 9–12. If the orbiting body equals the mass of the star, the center of gravity will be halfway between them. But if it is much smaller than the star (as in the case of a planet), the center of gravity can be close to the star or even inside the star itself.

You can think of the center of gravity as a *balance point* between two objects. Two children holding hands and spinning around will move around a point midway between them. A grownup swinging a young child by the arms will have to lean backward slightly because the balance point is slightly moved in front of the adult toward the child. But you can swing a small object on a string over your head without having to move much at all. In the same way, two equal-mass stars orbit a point midway between them. If a giant planet orbits a star, the center of gravity of the two bodies will be much closer to the star than to the planet. If a small planet orbits a star, the center of gravity of the two will be very close to the center of the star itself. Astronomers

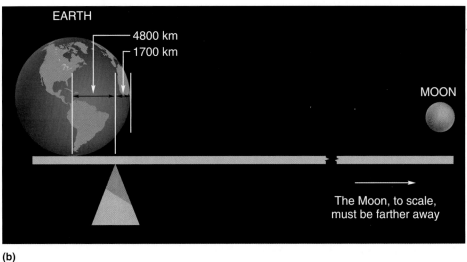

(a) (b)

FIGURE 9-11
(a) Two balls at the ends of a slender rod balance at the center of mass of the system, where the string meets the rod. **(b)** The center of mass of the Earth-Moon system is 4800 km from the center of Earth. The figure shows an imagined balance beam, not to scale. For two equal-mass objects, the center of mass is halfway between them.

calculate the position of the center of gravity in the same way as you would figure the balance point on a seesaw. With two equal-size people, the balance point is in the middle. If one of the people is very heavy, that person must sit close to the pivot to balance the weight of the light person (see Figure 9-11 again).

Let us see what might be observable for solar systems. Jupiter has 0.1 percent of the mass of the Sun—no other planet exerts nearly as much influence on the Sun. Therefore we can neglect the other planets and think of the Sun-Jupiter system. With a mass ratio of 1000:1, the center of gravity is one-thousandth of the distance from the Sun to Jupiter, or $5.2 / 1000$ AU $= 7.8 \times 10^5$ km. This is slightly larger than the Sun's radius, so Jupiter causes the Sun to pivot around a point just outside its surface. A star's wobble can betray the presence of an unseen planet. The size of the wobble indicates the mass of the planet. The time for one complete wobble is the orbital period of the planet.

Could we observe such a wobble? The angle of the wobble is given by the diameter of the wobble—1.6×10^6 km—at the distance of a nearby star. We can use the small-angle equation to calculate this angle as 0.01 second of arc, an incredibly small angle. It is the wobble of a hula hoop seen 10,000 mi away! The correct term for the wobble is a **reflex motion,** and no existing telescope has the ability to detect such a small motion. As you can imagine, the reflex motion is even tinier for an Earth-like planet. Earth is 5 times closer to the Sun than Jupiter and 300 times less massive, so Earth's leverage on the Sun is 1500 times smaller than Jupiter's.

However, reflex motion has another observational signature. When any source of waves approaches you, the waves are compressed, or made shorter. When the source travels away from you, the waves are stretched out, or made longer. This phenomenon is called the **Doppler effect.** You are familiar with the way a siren or car horn sounds more high-pitched when it is approaching you and less high-pitched when it is receding. A higher pitch corresponds to a shorter sound wave, and a lower pitch corresponds to a longer sound

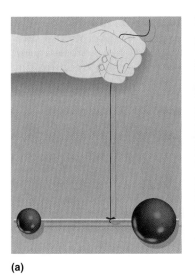

FIGURE 9-12
Newton's laws show that when a planet orbits a star, both bodies end up moving around the center of gravity of the system. Several consecutive positions of the two bodies are numbered. If the two bodies were of equal mass, the center of gravity would be halfway between them; in the case of the Sun and Jupiter, the center of gravity is just outside the Sun's surface.

PLANETS ORBITING OTHER SUNS 255

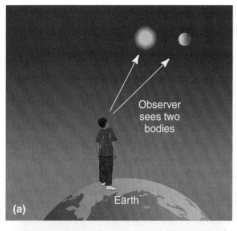
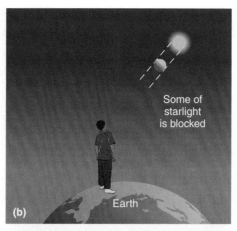
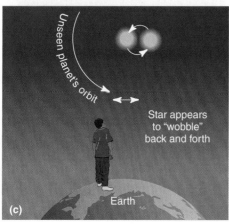
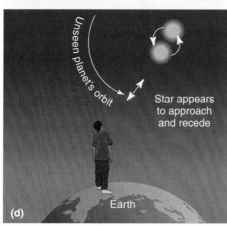

FIGURE 9-13
Four possible ways of detecting the presence of a faint or unseen planet moving around another star. **(a)** The planet might be bright enough and far enough from the star to be detected directly with an image through a telescope. **(b)** The planet might pass in front of the star, causing a drop in the star's brightness. **(c)** Detailed imagery of the star might detect its movement back and forth on the plane of the sky. **(d)** Another type of instrument can detect the star's movement toward and away from Earth as the planet pulls it around the center of gravity.

wave. Light waves work the same way. The Doppler effect shifts light to either shorter or longer wavelengths. A light source moving toward you is shifted to shorter or bluer wavelengths, and a light source moving away from you is shifted to longer or redder wavelengths. The mathematical form of the Doppler effect is presented in the following Science Toolbox. We will meet the Doppler effect a number of times throughout this book.

How big is the reflex motion of the Sun? We can easily work it out. The Sun pirouettes in a little circle of circumference $2\pi r$, where $r = 7.8 \times 10^5$ km, or 4.9 million km. This circuit takes 12 years, the same time as Jupiter's orbit. The average speed is therefore $4.9 \times 10^6 / (3600 \times 24 \times 365 \times 12) = 0.013$ km/s, or 13 m/s. By the same reasoning, the influence of other planets on the Sun is even smaller. Uranus moves the Sun by only 0.3 m/s over a period of 84 years, and Earth moves the Sun by a miniscule 0.09 m/s—about the walking speed of an ant!

Geometry is also an issue, as it was with detecting eclipses by planets. The Doppler effect only applies to the component of the motion that is directly toward or away from the observer. Real motions in space will not always be lined up so conveniently. For example, if we were looking directly down on our own solar system, all the orbits would appear perpendicular to the line of sight. We would see no Doppler effect at all. Looking at the solar system edge-on, we would see the full Doppler effect of the reflex motion. If extrasolar planets exist, their orbits will be randomly oriented in space, so we will generally see some fraction of the full Doppler effect.

To summarize, astronomers use four techniques to search for possible planetary companions of stars, as shown in Figure 9–13. The first is direct imaging, to measure the reflected light of a planet; the second is to catch the slight dimming of a star if a planet passed in front of it. The third way is to measure the precise position of the star and detect its wobble compared to other stars as it is tugged by the gravity of an unseen companion. The fourth technique is to detect the motion of a star toward and away from Earth, as a planet moves the star around its center of gravity. Each of these strategies works best for a large and massive planet. But they all challenge the limits of astronomical measurement.

The Payoff: Discoveries of Extrasolar Planets

In recent years, researchers have made great progress in the detection of extrasolar planets. The conclusion: Our solar system is not unique! The technique that finally worked was the indirect detection of planets by the Doppler effect. Over two dozen extrasolar planets are now known, with five or six more discovered each year. Several research groups are working hard on the search.

SCIENCE TOOLBOX

The Doppler Effect and Planet Detection

In 1842, Austrian physicist Christian Doppler realized that we measure waves to have a wavelength that depends on the motion of the source of waves. When waves leave a moving source, they get "bunched up" in the direction of motion and "stretched out" in the direction away from the motion. Imagine stones being dropped into a pond at regular intervals. The ripples will spread out in concentric circles, and the wavelength—the distance between successive crests or troughs—will be the same in every direction. Now see what happens if the source of waves is moving, as in Figure 9-D. In this case the stones drop in at a different position each time, and the waves are off-center as they travel outward.

If the source is approaching you, there is a smaller distance between successive ripples; the wavelength is shorter. If the source is moving away from you, the wavelength is longer. Viewed from any direction between these special cases, the wavelength varies smoothly between the maximum compression of waves and the maximum stretching out of waves. Another special situation occurs when the view is at 90° (or transverse) to the direction of source motion. The wavelength is unaltered, regardless of how fast the source of waves is moving. This wavelength shift is called the Doppler effect.

The Doppler effect applies to any source of waves. The waves can be ripples on the surface of a pond or sound waves traveling through air. Since light is also a wave, a moving light source will experience the Doppler effect. Light emitted from a source moving toward us experiences a *blueshift*. Light emitted from a source moving away from us experiences a *redshift*. As a mnemonic aid, remember that *recession* produces *redshifts*.

The size of the Doppler effect is given by the simple formula

$$\frac{\Delta\lambda}{\lambda} = \frac{v}{c}$$

Here, $\Delta\lambda$ is the small change or shift in wavelength, and λ is the normal wavelength when the source is not moving. The fractional (or percentage) wavelength shift is equal to the velocity of the source as a fraction of the velocity of light.

In Doppler's time, this equation was derived by using a sound experiment. Scientists hired a brass band and got the trumpeters to belt out a single sustained note from an open railroad car as it passed by at a known speed. The scientists then measured the change in pitch of the note as the train approached and receded, compared with the note of a stationary musician. By repeating the experiment with different train speeds, they deduced the relationship given above.

We can start with an example using sound waves. In this case, c is not the speed of light but the speed of sound—330 m/s. How fast would a police car have to be moving toward you for the pitch of its siren to increase by 50 percent? (This is called a *fifth* on the musical scale, or a change in the note from C to G.) Increasing the pitch by a factor of 1.5 means decreasing the wavelength by a factor of 1 / 1.5 = 0.67, so $\Delta\lambda / \lambda = 1/3$. To make this happen, a police car would have to travel at one-third the speed of sound, or 110 m/s (225 mi/h)!

The Doppler effect for light is generally very small. Imagine a supersonic jet fighter was heading toward us at night at 1000 mi/h. By how much would its wing lights be blue-shifted? Since 1000 mi/h is 0.46 km/s and the speed of light is 300,000 km/s, the blueshift would only be $0.46 / 300,000 = 1.5 \times 10^{-6}$, or about 1 part in 1 million. You would never see such a small effect.

Let us consider the example of planet detection. As seen from a distant vantage point, the reflex motion of the Sun caused

(continued)

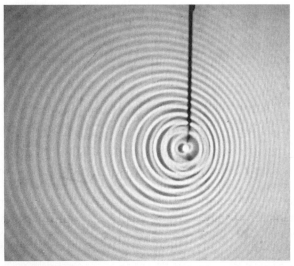

FIGURE 9-D
If a source of waves is moving, the waves are compressed in the direction of motion and stretched out in the direction opposite to the motion. (SOURCE: Education Development Center.)

SCIENCE TOOLBOX

The Doppler Effect and Planet Detection (continued)

by Jupiter is 13 m/s—slightly faster than a sprinter can run. The Doppler shift is $0.013 / 300,000 = 4.3 \times 10^{-8}$, or 4 parts in 100 million. Visible light has a wavelength of about 5×10^{-5} m, or 500 nanometers (nm). If a Sun-like star were being tugged by a Jupiter-like planet, the star's light would be shifted from 500 to 500.00002 nm when the star wobbled toward us, and shifted from 500 to 499.99998 nm when the star wobbled away from us. Planet detection is difficult because it requires a fantastically precise measurement.

Figure 9-E shows how the variation of this Doppler shift would appear over time. The nearly circular motion of the planet creates a nearly circular reflex motion of the star. This circular motion gives a sine wave variation of the Doppler effect. We would have to make measurements for 12 years to see a complete cycle of variation. And to be safe, we would want to observe for longer than 12 years to make sure that the phenomenon repeated and was cyclical.

The solid curve in Figure 9-E assumes that we are lucky enough that the plane of the planet orbit is parallel to our line of sight. In this case the planet is moving directly toward us and directly away from us as it orbits the star. The star's wobble is therefore directly toward us and directly away from us, and we see the full Doppler effect. In general, this will not be true. Suppose that the plane of the planet orbit was inclined at 45° to the line of sight. By simple geometry, the component of reflex motion in our direction is cos 45°, or 0.7, times smaller at every point in the orbit. The dashed line shows the reduced Doppler effect that we would observe. If a planet (of any size) orbits a star perpendicular to the line of sight, we observe no Doppler effect at all.

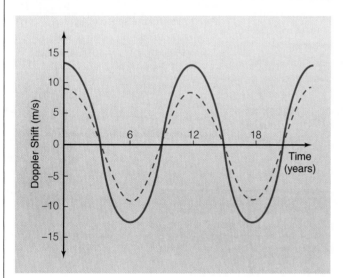

FIGURE 9-E
The reflex motion of a Jupiter-like planet orbiting a Sun-like star. The Doppler shift varies with a 12-year cycle, equal to the orbital period of the planet. The solid curve applies if the plane of the planet orbit is in the line of sight. The dashed curve applies if the plane of the planet orbit is inclined by 45° to the line of sight. If the plane of the planet orbit is perpendicular to the line of sight, no Doppler effect of the reflex motion can be detected.

How did they manage to detect such a tiny effect? You will recall from our discussion in Chapter 4 that a star such as the Sun emits a broad thermal spectrum. There is no way to measure the shift of such a broad spectrum by one-millionth of a percent. However, a stellar spectrum also has narrow spectral features that are associated with specific elements in the star—we will discuss this in greater detail in the next chapter. Wavelengths of these narrow spectral features can be measured with high accuracy. The researchers use a clever trick at the telescope to be sure of what was actually moving. They pass light from the star through a glass cell of iodine vapor. The vapor imprints a set of narrow spectral features on the spectrum, which act as a set of tick marks against which the spectral features of the star can be measured. With care, it is possible to reach an accuracy of 3 m/s for a single velocity measurement.

A second requirement of the search is patience. Giant planets are expected to be far from a star. By Kepler's third law, large orbits correspond to longer orbital periods. The searches that are currently bearing fruit were begun a decade ago.

The characteristics of the first 20 extrasolar planets to be discovered are shown schematically in Figure 9-14, with the inner part of our own solar system shown on top for comparison. Examples of typical orbital data are listed in Table 9-3. Each planet mass estimate is a minimum since we can

FIGURE 9-14

The first 20 extrasolar planets to be discovered, all within the past few years. Our own inner solar system is shown for comparison; note that Jupiter would be off the diagram to the right at 5.2 AU. The systems are shown in order of increasing distance of the planet from the star. Notice that the planets are all massive and relatively close to their stars, compared with our solar system. The star and planet sizes have been exaggerated so as to be visible in this plot. (SOURCE: Courtesy G. Marcy and R. P. Butler.)

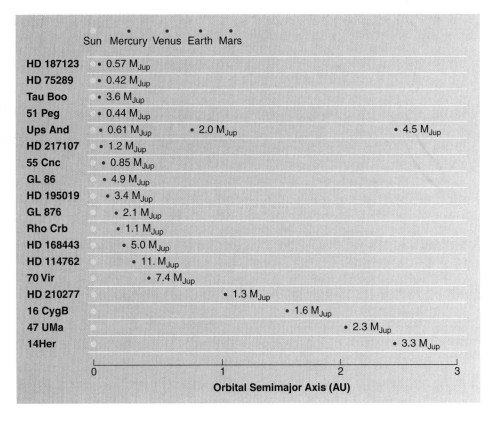

measure only the component of the reflex motion that is along the line of sight. The unknown inclination leads to an upward correction on the mass estimate that is no more than a factor of 2 on average. Bearing in mind the uncertainties in these difficult measurements, what have we learned about extrasolar planets?

1. Extrasolar planets are not extremely rare, but they are not ubiquitous either. About 10 percent of Sun-like stars have at least one giant planet.
2. Most of the planets detected so far have masses from about Jupiter's size to about 10 times that of Jupiter.
3. About one-half of the extrasolar planets have rapid orbital periods of well under a year and orbits within an astronomical unit of their host star.
4. Several of the planets have orbits that are highly elliptical. In three cases the orbital eccentricities are larger than for any planet in our solar system.
5. Multiple planets have been discovered around at least one extra-solar star. Our solar system is not unique.

Now that the science of extrasolar planets has been established, we have a new set of clues to work with. Scientists

TABLE 9-3
Extrasolar Planets

STAR	DISTANCE (LIGHT-YEARS)	STAR MASS (SUNS)	ORBITAL PERIOD (DAYS)	ORBITAL ECCENTRICITY	SIZE OF DOPPLER-SHIFT (M/S)	MINIMUM MASS (JUPITERS)	AVERAGE RADIUS OF ORBIT (AU)
47 UMa	46	1.1	1,098	0.03	46	2.4	2.1
16 Cyg B	72	1.0	802	0.57	44	1.7	1.7
70 Vir	59	0.95	116.6	0.40	308	6.8	0.47
HD 114762	90	1.15	83.9	0.34	613	11.6	0.36
Rho CrB	57	1.0	39.6	0.04	67	1.1	0.23
Rho¹ Cnc	44	0.85	14.64	0.03	77	0.93	0.11
Upsilon And	57	1.25	4.61	0.1	74	0.65	0.056
51 Peg	50	1.0	4.23	0.01	55	0.45	0.051
Tau Boo	49	1.25	3.31	0.006	469	3.7	0.045

(Notes: Typical data for the first 9 extrasolar planets to be detected. Several new examples are discovered each year. Upsilon And has 3 planets, the orbital data for the planet closest to the star is listed. Data from G. Marcy and R. P. Butler.)

are puzzled. These planets tend to have masses of 1 to 10 Jupiters, yet much larger planets of 10 to 100 Jupiters could have been detected. Since they were not, we conclude that they are rare. Apparently, there is some mechanism that stops planets from growing much bigger than 10 times Jupiter's mass. Several of these massive planets are in highly noncircular orbits. Can we be sure that the layout of our solar system is typical?

Many of the extrasolar planets are moving in fast, tight orbits close to the host stars. For example, the star 51 Pegasi has a planet of roughly one-half the mass of Jupiter at a distance of only one-sixth the distance of Mercury from the Sun. It whips around in its orbit in just over 4 days! The star 70 Virginis has a planet of 7 times Jupiter's mass, located at about Mercury's distance from the Sun. These examples go against our conventional idea of how the solar system formed. Theorists are struggling to explain how a giant planet can form or survive in the hot inner regions of a solar system. Unfortunately, the Doppler technique reveals the mass of a planet but not its size. Without a size estimate we cannot say whether these giant planets are gassy, icy, or rocky.

You should remember that the hunt for extrasolar planets has only been underway for about a decade. By Kepler's third law, the period for the Doppler variation of a planet at 3 AU (the right hand edge of Figure 9–14) is 5 years. We can anticipate the discovery of planets at distances corresponding to our outer solar system as the searches continue to accumulate data.

There are many questions that remain to be answered. Do planets form around any kind of star? So far, only stars like the Sun have been studied. Can planets form around binary stars? So far, we know of one example: 16 Cygni B has a planet of about twice Jupiter's mass in a highly elliptical orbit that goes from 0.6 to 2.8 AU. The star is nearly identical to the Sun, but has a companion star at roughly 900 AU. How common are multiple planets? The Doppler technique has now achieved the sensitivity and time coverage to detect true solar system. Can Earth-like planets be detected? Earth-like planets are beyond the sensitivity of the current generation of imaging and Doppler experiments, but there are always surprises. The first two extrasolar planets discovered were Earth-size objects in orbit around an exploded star remnant called PSR 1257+12. Most astronomers think the explosion should have disrupted any planetary system, and that the two Earth-size objects must have formed by a process different from the process that spawned our solar system. They may even have aggregated from debris left over after the explosion; in any case they do not seem very Earth-like in terms of history or environmental conditions.

What does the future hold? With telescopes in space, it may be possible to directly image planets. Figure 9–15 shows a computer simulation of the infrared detection of Earth-like planets using multiple 1m telescopes in space. The same instrument could measure spectra of extrasolar Earths, and in principle detect absorption due to atmospheric constituents like carbon dioxide, ozone, and water vapor. Science has just taken a major new step in the Copernican revolution. Earth is a rocky cinder sheltering in the glow of a nearby star. It is natural to wonder how often the universe has created such planets and how often the history of planets involves the creation of life. For the first time we have the evidence to move beyond pure speculation. It will be a few years before we know how common planets are. It will be even longer before we know the conditions on these planets and the likelihood of their harboring life. A new adventure is beginning.

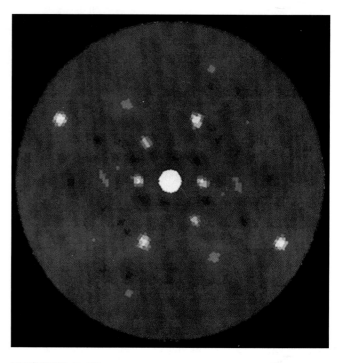

FIGURE 9–15
A computer simulation of Earth-like planets around another star, as imaged by a set of telescopes in space. The instrument would work at 10 microns in the infrared, where the contrast of a planet with respect to the nearby star is favorable. Light from the separate telescopes is combined so as to efficiently blank out the central star; this process leads to two image of each planet, on opposite sides of the star. (SOURCE: Courtesy J. R. P. Angel and N. Woolf.)

Summary

During the last third of the 20th century, scientists have unraveled how Earth and the rest of our planetary system came into being. The Sun formed about 4.6 billion years ago from a contracting interstellar cloud, composed mostly of hydrogen and helium gas with only a few percent of heavier elements. The Sun formed in the central part of this cloud, incorporating over 99 percent of the material. As outer parts of the cloud cooled, solid grains of various minerals and ices condensed from the heavier elements and accumulated into small rocky fragments called planetesimals. All this happened during a relatively brief interval, lasting roughly 50 million years. As the planetesimals collided during this interval, countless small bodies aggregated into a few larger bodies, and eventually planets.

Simple physical ideas describe the early evolution of the solar system. In its gravitational collapse, the solar nebula contracted by a large factor, and the gravitational potential energy was turned into the heat energy of the newly formed Sun. The conservation of angular momentum governed the transition from a large gas cloud with a small degree of rotation to a small and dense disk of gas that was rapidly rotating. The moon systems of the giant planets formed by a similar process in miniature. A few additional moons were added when interplanetary bodies were captured, and the moons of Earth and Pluto may have formed as the result of giant collisions.

The fate of the planetesimals explains many features of the solar system. Many of the last surviving planetesimals crashed into the planets and moons, forming craters. Other planetesimals were stranded in the asteroid belt and the Kuiper belt. As the planets grew, their increasing gravity disturbed the orbits of the neighboring planetesimals, causing the planetesimals to collide at speeds high enough to smash them. The debris generated in asteroid collisions is the set of fragments we know on Earth as meteorites. When icy planetesimals in the outer solar system made near-miss encounters with giant planets, the planets' gravity flung many of them almost out of the solar system, forming the Oort cloud reservoir of comets. The outer planets, with their high masses and strong gravity, retained the light, hydrogen-rich gases of the original cloud, explaining the nearly solar composition of their hydrogen-rich atmospheres. Lower-gravity terrestrial planets lost these gases, which were replaced with volcanic gases such as carbon dioxide.

Unraveling the mystery of how our planet formed has offered a good illustration of how science works. By combining knowledge on diverse subjects such as orbital motions and the chemistry of meteorites, scientists have learned how our planetary system formed. Since the origin processes do not require special circumstances, it seems likely that they operated around many other single stars. In an exciting recent development, astronomers have detected massive planets around other stars—proof that our solar system is not unique. All these results have led to another step in the Copernican revolution, reaffirming that Earth may not occupy a unique place in the scheme of things. If even 1 percent of stars have planetary systems like ours, then millions of Earth-like planets could exist among the stars of our galaxy.

Important Concepts

You should be able to define these concepts and use them in a sentence.

revolution	condensation sequence
rotation	accretion
gravitational collapse	chaotic orbits
angular momentum	extrasolar planets
law of conservation of angular momentum	center of gravity
	reflex motion
nebula	Doppler effect
solar nebula	

How Do We Know?

These questions and answers show how the scientific method is used to learn about the universe.

Q: How do we know that the whole solar system—the Sun and the planets—formed over a short interval of time 4.6 billion years ago?

A: The basic evidence for the age of the solar system is radioactive dating. Rock samples from Earth, the Moon, and many different types of meteorites from different parent bodies prove that all these bodies formed about 4.6 billion years ago. In addition, certain isotopes with half-lives of only a few million years were trapped in meteorites of various sizes before the isotopes had time to decay. This shows that the process of accretion took place in the first couple of percent of the solar system's life. More indirect evidence from theory and simulations indicates that the progression from dust-size bodies to planet-size bodies took 50 to 100 million years. We do not have such good evidence for the formation time of the Sun itself, but all theory and observation agree that the planets and Sun formed at the same time.

Q: How do we know that planets formed from accretion of small particles?

A: We do not have a lot of direct evidence, and of course no one was around to witness the spectacular creation of the solar system. Much of the confidence that we know how planets were built stems from the successful theories of condensation and accretion—these theories do a good job of explaining the composition of the planets. There is also some direct evidence for the accretion process in the microscopic granular structure of meteorites that never melted. Some meteorites contain clumps of microscopic or millimeter-scale grains; others contain centimeter-size inclusions that seem to have formed earlier and then to have been trapped in clumps of later material.

Q: Why do planets and satellites have so many different properties if they all formed by the same process?

A: Planets and satellites formed in quite distinct regions. Worlds far from the Sun are formed largely from ices; planetesimals close to the Sun were too warm to contain ices, so worlds in that zone formed from rock. Small planets cool off and end up geologically dead; large planets stay hot inside and maintain volcanism. Also, large planets are more likely to retain atmospheres. Furthermore, the last impacts by the largest planetesimals tended to introduce variations in axial tilt, spin rates, and surface features.

Q: Can we really explain the diverse properties of the solar system with a single theory?

A: The "theory" of the formation of the solar system is not a single idea like the law of gravity. Our explanation for the formation of the solar system involves several physical principles. These include the concept of gravitational collapse, the conservation of angular momentum, the way small objects build into large objects by accretion, and the chemistry that determines how different elements and compounds condense. However, the way these principles operate in a complex region of gas and dust is not deterministic. In other words, we cannot be confident that one set of initial ingredients would always give us a solar system just like the one we live in! Scientists aim only to explain the general characteristics of the Sun and the planets.

Q Why do we believe that planets are likely to form when stars form?

A All the planets and interplanetary bodies combined make up only about 0.2 percent of the mass of the solar system. The vast majority of the mass is locked up in the Sun. Planets therefore represent the debris left over from star formation. Scientists think it is unlikely that star formation anywhere is perfect—with no material left over to form planets. One reason is that a solar nebula is likely to have some small rotation, which will inevitably lead to material in a disk far from the central star. After debris forms into a disk, the process of accretion will naturally sweep it into planets. However, we cannot predict how many planets will form or what their sizes and spacings will be. We also do not know how much material will be left over for planet formation. It might be 0.2 percent or 0.01 percent or 10 percent; our theory is not good enough to predict this number.

Q How do we know that extrasolar planets exist?

A The evidence is indirect, but it is based on our well-understood theory of gravity. Planets cannot be seen directly—they would be too close in the sky to a vastly brighter star. Jupiter-mass objects can be detected by their gravitational tug on the host star. The star has a wobble with a period equal to the orbital period of the planet. The wobble or reflex motion is detected via the Doppler effect. We cannot be sure that they formed in the same way as our planets, or that they represent systems like ours with Earth-like planets as well. Doppler detection can only tell us the mass and orbit of an extrasolar planet, not the size, structure, or composition of the planet. However, we do have exciting proof that our solar system is not unique.

Problems

You should use these problems to test your understanding of the information and concepts in this chapter. The * indicates a more advanced or mathematical problem.

1. Since all planetary material derives from matter that condensed from the same nebula, including terrestrial rocks, why do meteorites have different chemical and geological properties from those of rocks you might find in your own yard?

2. Because of heating by the Sun and by the contraction process, gases in the inner solar system never got as cool as gases in the outer solar system. In terms of the condensation sequence, relate this to the estimated or observed composition of the planets.

3. Suppose we discover six planets orbiting around another star in randomly inclined orbits, both prograde and retrograde. Would you say that the planets of that system originated by the same processes as our solar system? Why or why not? What role would angular momentum play in the formation of such a strange solar system?

4. (a) How are theories of solar system origin different in principle from theories of the origin of the universe? (*Hint:* In principle, do we have prospects for testing origin models of either our planetary system or the universe by studying other examples of them?) (b) Can a theory of the solar system predict the exact number of planets that will form, or the detailed properties of each one? If not, why not? (c) Has the discovery of extrasolar planets increased our confidence that we know how our solar system formed? Can we explain all the properties of these new solar systems? (d) List some of the obvious peculiarities of the planets that cannot be easily explained in terms of our theory for how the solar system formed. How do we account for these features?

5. (a) How do we know that the solar system formed 4.6 billion years ago? (b) What reasons do we have for believing that the entire solar system formed during a single, relatively short interval of less than 100 million years? (c) Was the moon in existence 150 million years after the solar system formed? How do we know?

6. The formation of the solar system involved some complicated physics, but it is a useful skill to convey complex natural phenomena in simple language. If a 7-year-old member of your family asked where the world came from, how would you answer?

*7. The gas in the early solar system was roughly three-fourths hydrogen (Table 9–3). Use the kinetic theory of gases in planetary atmospheres (see Chapter 7) to explain the absence of abundant hydrogen in the terrestrial planets' atmospheres. If we define a hydrogen atom (H) as having a mass of 1 atomic mass unit (amu), then a hydrogen molecule (H_2) has a mass of 2 amu, and a carbon dioxide molecule (CO_2) has a mass of 44 amu. Use these facts plus the information in Chapters 4 and 7 to explain why a moderate-size planet (such as the terrestrial planets) can retain carbon dioxide in its atmospheres, while quickly losing hydrogen.

*8. Suppose you represent the history of the solar system by a calendar of 1 year (where the Sun began forming 4.6 billion years ago on January 1, and we are living now at the end of December 31). (a) The best evidence from meteorites and other sources indicates that the solar system had reached approximately its present appearance by 100 million years after the Sun formed. By approximately what date was the solar system in approximately its present appearance? (b) According to fossil evidence, the human species began to evolve about 1 to 4 million years ago. On the calendar above, when did humans appear? (c) If we say that recorded human history goes back 5000 years, on the calendar above, how long has human history lasted? (d) Repeat the previous three calculations, using scientific notation (expressed in terms of powers of 10). If you are unsure about this, get your instructor to do a problem this way. Note how much faster it is than working it out by long division.

*9. If the original solar nebula extended out to 100 AU, how fast was material at the edge of the nebula rotating if it eventually moved inward to become part of Earth? (*Hint:* Use the law of conservation of angular momentum.) How fast should this material rotate if it fell far enough to become part of the Sun? Why does this calculation imply a problem for our theory of the formation of the solar system?

*10. Imagine you are hunting for planets around stars like the Sun. (a) What fraction of the star's light would be reflected by a planet 10 times the size of Jupiter at the distance of Earth from the Sun? What fraction of the star's light would it eclipse if the orbit alignment permitted us to view an eclipse? What angular separation would the planet have, as seen from a distance of 10,000 AU? (b) Repeat these same three calculations for a planet the size of Earth at the distance of Mercury from the Sun.

*11. Take each of the hypothetical planets from the previous problem, and estimate the Doppler effect of its reflex motion. How long would you have to observe each system to be sure of measuring a full cycle of Doppler shift? Why might the Doppler signal be unobservable even if you had the sensitivity to detect it?

Projects

Activities to carry out either individually or in groups.

1. See if your geology department or a local museum or planetarium has meteorite samples. Examine different meteorite types such as iron meteorites, stony irons, basaltlike stony meteorites, or the chondrite stony types that have never been melted. Relate each type to the sequence of events in the history of planetesimals as they formed, heated, and differentiated 4.6 billion years ago. For example, how were pure iron meteorites created?

2. Try to construct an experiment to demonstrate the law of conservation of angular momentum. Find a well-oiled swivel chair that can rotate freely. Position yourself as closely above the rotation axis as you can, and hold two books at arm's length. This experiment works best if you are light and can hold up heavy books; if you cannot do it, find a small friend with good upper-body strength. Get someone to spin you fast. Now pull the books into your chest, and notice what happens to your rotation rate. Why does this experiment work better with a big hardcover book than with a small paperback? If you are in a group, go to a children's playground and have a number of people distribute themselves evenly around the edge of a merry-go-round. Get it moving, and have someone time the rotation period with a watch. Now get all the people to move as close as they can to the center of the merry-go-round. Watch the rotation speed increase. By what factor did they decrease their distance from the center of the merry-go-round, and by what factor did the rotation speed increase? (*Warning:* Science experiments can be fun!)

Reading List

Angel, J. R. P., and Woolf, N. J. 1998. "Searching for Life in Other Solar Systems." In *The Magnificent Cosmos.* New York: Scientific American, Inc., pp. 22–25.

Black, D. C. 1991. "Worlds Around Other Stars." *Scientific American,* January, pp. 76–82.

Black, D. C. 1996. "Other Suns, Other Planets?" *Sky and Telescope,* vol. 92, pp. 20–27.

Boss, A. P. 1996. "Extrasolar Planets." *Physics Today,* vol. 49, pp. 32–38.

Henry, T. J. 1996. "Brown Dwarfs—Revealed at Last!" *Sky and Telescope,* vol. 91, pp. 24–28.

Lee, D.-C., and Halliday, A. 1996. "Isotopic Evidence for Rapid Accretion and Differentiation in the Early Solar System." *Science,* vol. 274, pp. 1876–1879.

Levy, E. H., and Lunine, J. I. 1993. *Protostars and Planets III.* Tucson: University of Arizona Press.

MacRoberts, A. M., and Roth, J. 1996. "The Planet of 51 Pegasi." *Sky and Telescope,* vol. 91, p. 38.

Marcy, G., and Butler, R. P. 1998. "The New Diversity of Planetary Systems." *Sky and Telescope,* March, pp. 30–37.

McSween, H. Y. 1989. "Chondritic Meteorites and the Formation of Planets." *American Scientist,* vol. 77, pp. 146–153.

Safronov, V. S. 1972. *Evolution of the Protoplanetary Cloud and Formation of the Earth and Planets.* NASA publication TTF-677 (original in Russian, Nauka Press, Moscow, 1969).

CHAPTER 10

Detecting Radiation from Space

CONTENTS

- Beyond the Visible Spectrum
- Radiation and the Universe
- The Electromagnetic Spectrum
- Radiation and the Structure of the Atom
- Spectral Lines
- Astronomical Uses of Radiation
- Telescopes and Detectors
- Summary
- Important Concepts
- How Do We Know?
- Problems
- Projects
- Reading List

The largest telescope in the world. The Arecibo radio dish is 305 m across and is built into a natural depression in the dense forests near the center of Puerto Rico. The antenna that detects radio signals from space is suspended from three tall towers. This is the most sensitive device we have for detecting invisible radiation from space. (SOURCE: National Astronomy and Ionosphere Center.)

WHAT TO WATCH FOR IN CHAPTER 10

In previous chapters, we learned about Earth and its Moon, the planets, and how the solar system might have formed. In the second half of the book, we will learn about more distant regions of the universe. How do we know so much about the universe? Much of our astronomical knowledge depends on detecting signals from space. These signals include light, or the visible spectrum. In this chapter you will learn that light is just one type of electromagnetic radiation, which spans an enormous wavelength range from tiny gamma rays to long radio waves. Light and the other types of electromagnetic radiation sometimes behave as waves and sometimes as particles. You will learn how radiation emitted from astronomical objects can reveal the temperature and chemical composition of the distant bodies. The study of radiation from space is an important aspect of our theme of how we know.

The last part of the chapter deals with telescopes, which astronomers use to collect and focus radiation from space. The most important features of a telescope are its collecting area and its resolution. A telescope's collecting area allows it to gather light, while its resolution enables it to distinguish objects closely spaced in the sky. Astronomers use a variety of instruments to record radiation from space. For visible light, traditional photography has been mostly replaced by the use of electronic detectors. Since only visible light, radio waves, and some infrared radiation can penetrate Earth's atmosphere, we must use satellites in space to detect other regions of the electromagnetic spectrum. As you will see in this chapter, the study of light and invisible information from space has transformed our view of the universe.

PREREADING QUESTIONS ON THE THEMES OF THIS BOOK

OUR ROLE IN THE UNIVERSE
How does light represent a tiny fraction of the information that reaches us from space?

HOW THE UNIVERSE WORKS
What simple properties of radiation apply across the universe and in all astronomical objects?

HOW WE ACQUIRE KNOWLEDGE
What can we learn by studying light and the invisible universe?

BEYOND THE VISIBLE SPECTRUM

We take in the world through our eyes. Sight is arguably the most powerful and sophisticated sense, the one many of us feel we could not do without. Yet the visible spectrum that we see—the richness of the rainbow from red to blue—is just a tiny slice of an enormous array of types of radiation. There is an unseen universe waiting to be explored.

As you have read in Chapter 4, English astronomer William Herschel opened the first chapter in the story of human discovery of invisible radiation. In 1800, he dispersed the Sun's rays with a prism and placed a thermometer beyond the red end of the spectrum. The temperature rose, showing that the thermometer had absorbed invisible energy from the Sun with a wavelength longer than that of red light. The next year, German chemist Johann Ritter created a spectrum in the same way and placed paper soaked with silver chloride beyond the violet end of the visible rays. The paper darkened, indicating it had absorbed invisible energy from the Sun with a wavelength shorter than the shortest wavelength of blue light. Like explorers, these scientists had traveled beyond the rainbow, measuring waves that the eye cannot see.

Nearly one hundred years later, another pair of discoveries pried open the spectrum even further. In his darkened laboratory, Wilhelm Roentgen passed electricity through a tube filled with gas at much lower density than that of air. To his surprise, a chemical-coated screen on the other side of the room glowed whenever he passed electricity through the tube. The discovery was accidental, but as any good scientist would have done, Roentgen used logic and further experimentation to try to understand his observation. Light could not be responsible; the room was darkened, and the tube was encased in thick cardboard. When his hand passed between the tube and the screen, he was startled to see the bones in his hand, as if the flesh had been stripped away! Newspapers gave prominent coverage to this spectacular discovery. Roentgen had discovered a strange new form of high-energy waves—X rays. He was awarded the first Nobel Prize in physics. In the same year, 1895, young Guglielmo Marconi experimented with long-wavelength radio waves that traveled through space and walls and people unimpeded.

People considered the types of radiation studied by Roentgen and Marconi to be wonderful and mysterious. Today, we take them for granted. X rays are one of the essential elements of modern medicine, and radio waves are the basis for worldwide communication. How can we use these waves, which are much shorter and much longer than the waves of visible light, to explore the invisible universe?

Radiation and the Universe

By sending astronauts to the surface of the Moon or a robot past Neptune, we can learn directly about regions of space beyond Earth. However, most parts of the universe are too far away to reach in person or with space probes. In fact, very little of the information in this book is derived from travel or direct contact. To obtain information about remote regions, we often rely on messages from them in the form of light. When this light is dispersed, there are two aspects of the spectrum that are of importance to astronomers. First, radiation that is smoothly distributed in wavelength gives us evidence of the thermal properties of its source, as discussed in Chapter 4. The peak wavelength of this radiation measures the temperature of a remote body. Hotter objects have radiation with shorter peak wavelengths. Second, electrons moving from one energy state to another in atoms create a pattern of sharp lines arrayed in wavelength. Each chemical element has a unique pattern of sharp lines. The overall pattern of lines is like a "fingerprint" that reveals the chemical composition of a remote object.

In this chapter we will study the broad spectrum of radiation of which visible light is just a small slice. For all human history we have watched light from the Sun and the stars and the patterns of the night sky. Yet this light that the eye can see only spans a factor of 2 in wavelength. In the past 50 years, we have developed the technology to discover that the universe is filled with many kinds of invisible radiation. Now we can explore the universe with radiation ranging from radio waves to gamma rays—a factor of 10^{15}, or 1000 trillion in wavelength!

The study of the invisible universe illuminates the themes of this book. A few simple physical ideas govern a vast spectrum of radiation. We will see that all types of radiation travel at the same speed and that any type of radiation is specified by its wavelength. You can recall from Chapter 4 that any material that is not transparent—a solid, a liquid, or a dense gas—will radiate waves that are determined solely by its temperature. An object at a temperature of 5000 to 6000 K will glow with a yellowish color, whether it is molten metal in a furnace, the Sun itself, or a star similar to the Sun on the other side of the galaxy. In this chapter you will learn that any low-density gas produces the same set of sharp spectral lines. The spectrum of helium is the same whether it is in a sealed tube in the laboratory, or in an enormous gas cloud that is so far away in the universe that its light has taken billions of years to reach us.

We also see the scope of the scientific method. Physicists study the interactions between matter and radiation, using carefully controlled experiments in the laboratory. Astronomers apply those principles throughout the universe. You should look again at the material in Chapter 1. Using radiation to understand remote objects is induction on a large scale! We must remember that astronomers make a big assumption: The universe works the same everywhere and at all times. So far, this assumption has been justified.

Last, we see another perspective on the role of humans in the universe. We have learned that we are sensitive to only a tiny fraction of the range of waves that the universe contains. After several thousand years of visual astronomy, spacecraft and new detector technologies have given us a flood of information at other wavelengths. Imagine that you had to live with only one of your senses—the sense of sight. Think how many worlds would be closed to you: music, conversation, the taste and smell of food. You can probably understand, then, the excitement of astronomers who have new tools that extend their "senses" to explore the electromagnetic spectrum. These scientists are answering the intriguing question: How would the universe look if we had eyes that could see the invisible rays we discussed in the opening story? We will answer this question as we study the full spectrum of signals from space and the ways in which astronomers detect them.

The Nature of Light

What is the nature of light? Experiments in the 18th century showed that it is related to the forces of electricity and magnetism. This sounds unlikely, so it is worth recounting how physicists came to this realization.

Michael Faraday was a bookbinder's apprentice who rose to become the director of the most prestigious scientific organization in science, the Royal Institution in London. He was a masterful experimenter and a wonderful communicator. Every Christmas he would deliver a public lecture series crammed with demonstrations, a series that continues in his name today. Faraday discovered that a changing electric field could create a magnetic force. Once, the Prime Minister visited his laboratory and complained about abstract research and the lack of any use for all the gizmos in the laboratory. Faraday replied that the results were important, so important that one day Her Majesty's government would tax their many applications. He was right. The connection between electric and magnetic forces is the basis for electric motors and the generation of all electric power.

Faraday also knew that a changing magnetic field could generate an electric current. These two results suggest an intimate connection between electric and magnetic forces. Both are familiar to us in the everyday world; the current in an electric motor creates a magnetic force that drives a rotor, and water flowing though a turbine creates a changing magnetic force that generates electricity. But what do these forces have to do with light?

In the 19th century, Scottish physicist James Clerk Maxwell answered this question with his theory of electricity and magnetism. He derived an elegant set of equations linking the two forces. In the theory, a changing electric or magnetic force could generate a disturbance that behaved as a wave. Since the electric and magnetic changes are always linked, this disturbance is called an electromagnetic wave. Maxwell used the theory to predict that electromagnetic waves should travel through space at 300,000 km/s. This is exactly the speed of light! Maxwell therefore speculated that light was just one form of **electromagnetic radiation.**

We will use the word *radiation* to describe light, radio waves, X rays, and the other types of electromagnetic waves. You should be aware that physicists and engineers speak of streams of subatomic particles, such as protons or electrons, as radiation. For example, the dangerous particles escaping from radioactive material are sometimes called radiation. Electromagnetic radiation has a general property that is of great interest to us in this book—it carries energy and information from one place to another across the vast reaches of space.

The Electromagnetic Spectrum

Light and other forms of electromagnetic radiation are organized into a spectrum. You will recall from Chapter 4 that the visible spectrum (plural: *spectra*) is an arrangement of all the visible colors in order of wavelength. You can see the spectrum of visible light by passing a shaft of sunlight through a glass prism. Water droplets in a rainstorm also act as little prisms and allow us to see the visible spectrum—the same arrangement of colors from violet to red—in the rainbow. Newton discovered that "white" light from the Sun is really a mixture of all the different colors of the spectrum.

As we saw in the opening story, scientists discovered after Newton's time that there is radiation beyond the red and violet ends of the visible spectrum. The **electromagnetic spectrum** extends to far shorter and longer wavelengths than those to which the eye is sensitive. *Visible light* has a tiny wavelength, around 0.0000004 to 0.0000007 m (or 4×10^{-7} to 7×10^{-7} m in scientific notation). In related units this is 400 to 700 nm, where 1 nanometer (nm) is one-billionth of a meter, or 0.4 to 0.7 micrometer (μm), where 1 μm is one-millionth of a meter (see Appendix A–3). Electromagnetic radiation with still shorter wavelengths exists; it is called *ultraviolet radiation*. The shortest wavelengths of all belong to *X rays* and the highly energetic *gamma rays*. Beyond the red end of the visible spectrum lies electromagnetic radiation called *infrared radiation*. Still longer than infrared waves are *microwaves* and then *radio waves*.

The electromagnetic spectrum is vast, and the visible spectrum is just a sliver of it. Imagine displaying the electromagnetic spectrum with a highly compressed or logarithmic wavelength scale (see Appendix A–1). If the distance between radio waves and gamma rays were equal to the length of a football field, the visible spectrum would be just 2 yards (yd) wide. Figure 10–1 shows the names attached to different regions of the electromagnetic spectrum. As this figure indicates, the radio waves we receive on our radios and TVs are simply long-wavelength versions of the light waves we see with our eyes. Similarly, X rays are just short-wavelength versions of visible radiation. It is remarkable that phenomena as different as radio waves and X rays have the same physical basis.

Characteristics of Radiation

For convenience, we can think of light and other forms of electromagnetic radiation as waves that carry energy from one place to another. Electromagnetic radiation is a repeating or periodic phenomenon—review our discussion in Chapter 3. The wave property of light was also mentioned briefly in Chapter 4. Let us look at it in greater detail. We can describe any wave motion by four numbers, two of which are related. Look at Figure 10–2a. The distance between two successive crests (or troughs) is the **wavelength.** The number of crests (or troughs) that pass a fixed point every second is the **frequency.** The wavelength and frequency of any wave are two measures of the same thing; multiplying them gives you the velocity of the wave.

A third important number is the amplitude of the wave, shown as one-half the height of a wave from peak to trough in Figure 10–2a. A wave is a disturbance—the amplitude quantifies the amount of the disturbance. For example, the amplitude of a water wave is the amount by which the wave deviates above or below a smooth, undisturbed surface. For a sound wave, the amplitude is the amount by which the air density goes above or below the average density. In the case of light or other electromagnetic waves, the amplitude is given by the positive or negative amounts of electric or magnetic force. A more useful quantity is **intensity,** or the strength of the wave that we actually measure. The intensity of the radiation is proportional to the square of the amplitude of the wave.

Figure 10–2b shows a useful analogy for remembering an important point about wavelength and frequency: The two have an inverse relationship. Imagine trains traveling at the same speed but with different-length cars. The shorter the cars (smaller wavelength), the more cars that pass per second (higher frequency). Conversely, a smaller number of long cars (longer wavelength) pass each second (lower frequency).

All electromagnetic radiation travels at the same velocity in a vacuum, 300,000 km/s (186,000 mi/s). This universal

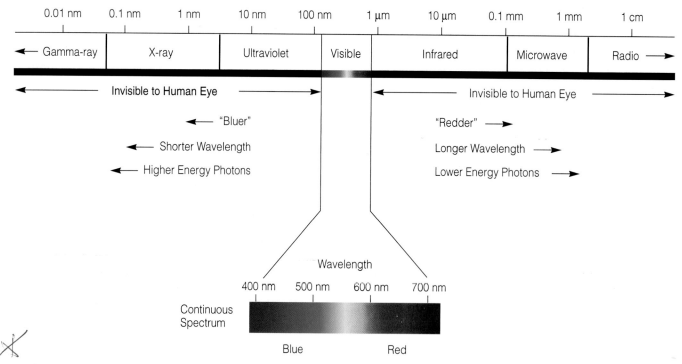

FIGURE 10-1
The electromagnetic spectrum. Top chart shows a wide range of wavelengths with names attached to different regions. Lower chart shows the order of colors in increasing wavelength from violet to red, representing the narrow range that is visible to the eye. Recall from the opening story how we first knew that there were invisible forms of radiation. The logarithm of wavelength is plotted so that a small move across the spectrum corresponds to a large change in wavelength.

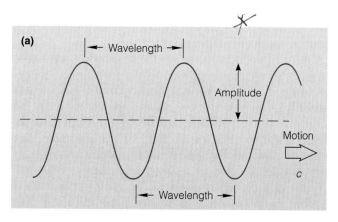

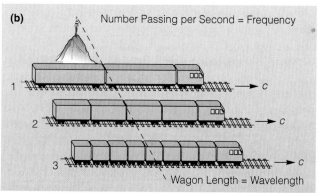

FIGURE 10-2
(a) The wavelength of a wave is the distance between successive crests or troughs. The frequency is the number of crests or troughs that pass per second, given a velocity c. One-half of the height of the wave from peak to trough is the amplitude, or square root of the intensity. (b) Railcar analogy for the inverse relation between frequency and wavelength. Given three trains traveling at the same speed c, the train with the longest cars has the longest wavelength and the lowest frequency, and the train with the shortest cars has the shortest wavelength and the highest frequency.

physical constant is the **speed of light,** and it is given the symbol c. Remember that the speed of light is also the speed of all other types of electromagnetic wave in a vacuum. (We should note, however, that light traveling through air or glass or any other transparent medium travels at different, slightly slower speeds.) It is a striking fact that all electromagnetic waves travel at the same speed in a vacuum. To use the analogy just given, imagine that we have trains with cars of an enormous variety of sizes traveling at the same speed. The number of cars passing per second (frequency) goes up at the same rate at which the size of the cars (wavelength) goes down.

How can we ever measure such an amazingly high speed? Galileo was a masterful experimenter, and he tried to measure the speed of light nearly 400 years ago. One night, he stationed a friend on a hilltop several miles away. Both Galileo and his colleague had lanterns with a shutter that could be lowered over the light. They planned that when Galileo sent a flash of light between the hilltops, his friend would immediately respond with a flash from his own lantern. Galileo tried to measure the time for a light signal to make the round trip between hilltops, but he soon realized that it was too short to measure. Light apparently traveled many miles in a second. Beyond that, Galileo could say no more.

The first real measurement of the speed of light used an ingenious observation by Danish astronomer Ole Roemer in 1675. Roemer began with Galileo's careful observations of the moon of Jupiter. As the moons move in their orbits of Jupiter, it is possible to carefully time the moment when they pass behind Jupiter and are eclipsed. Roemer discovered that the eclipses occur slightly earlier when the distance between Earth and Jupiter is shorter (see Figure 10–3 for the geometry of this experiment). The reason is that the light takes a shorter time to travel the shorter distance to our eyes. Roemer deduced that it takes 11 min for light to cross the distance between Earth and the Sun—the actual number is 8 min, but his logic was correct. You can see here an echo of the ancient Greeks who used logic and geometry to learn about the nature of the universe. Roemer's observations provided additional evidence that Earth was in motion around the Sun. Modern laboratory measurements give a very accurate value for the speed of light. With a prodigious speed of 299,792 km/s, it is no wonder we consider light travel to be instantaneous in the everyday world.

Waves carry energy from one place to another. The everyday world contains some remarkable examples of the ways that wave energy can be transformed from one kind to another. Recall from our discussion in Chapter 4 that energy is conserved—it can change forms but can never just appear or disappear.

Let us take a familiar example. When you listen to a voice on the radio, a remarkable series of transformations of energy occurs. The radio announcer talks, and the sound energy travels out as compression waves in the air. These waves move a small magnet inside a microphone backward and forward, creating a changing electric current (as Faraday showed long ago). The changing electric current is really a varying flow of electrons down a wire. Moving electrons can be used to make an electromagnetic radio wave that travels out from an antenna at the radio station. (The feeble power in the original electric signal is amplified many times—from thousandths of a watt to many millions of watts—before it can be broadcast.) After traveling through the air for miles, the electromagnetic radio wave is detected by the antenna on your radio. Inside your radio the electromagnetic wave is converted to a varying electric current. (It must be amplified again because the signal has been diluted by its travel.) The varying electric current moves a small magnet—this time in the loudspeaker of your radio. The moving magnet creates compression waves

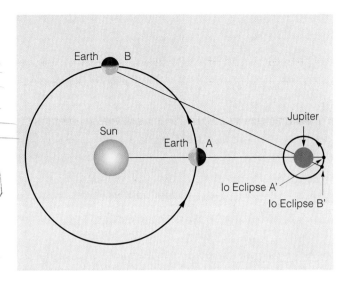

FIGURE 10–3
Roemer's method for measuring the speed of light. As Earth travels around the Sun, the eclipse of a moon of Jupiter occurs earlier as the line of sight moves from A to B. The difference is the smaller amount of time it takes light to travel between Earth and Jupiter when they are on the same side of the Sun. The actual timing of the eclipse uses Kepler's laws applied to the orbits of Earth, Jupiter, and Jupiter's moons. This diagram is not drawn to scale.

in the air, which you hear as sound! The voice on the radio sounds just like a person's voice across the room, but electromagnetic waves have been used to send the signal over large distances.

Waves and Particles

Scientists have puzzled over the nature of light for centuries. Some of the properties are simple—light carries energy from one place to another, and it travels in straight lines. But other properties are quite unexpected. The wavelength of light is a tiny fraction of a millimeter, so light's wavelike properties manifest themselves on a scale quite unfamiliar to us. After his pioneering experiments, Newton decided that light behaved more as a particle than as a wave. We will consider the ways in which light can act as either a wave or a particle with a series of analogies. Remember that the properties we will describe apply to *all* types of electromagnetic radiation—not only light but also radiation from radio waves to X rays.

Let us begin with the ways in which light acts as a wave. Suppose you stand by a perfectly calm swimming pool in which a cork is floating. You disturb the cork by jiggling it. You notice that this disturbance causes a set of waves to move out across the water. These waves provide a useful analogy, but not a perfect description, of some of the properties of light. For instance, just as a water wave expands from its source, light spreads out in all directions from its source. Both light waves and water waves can bend slightly around corners. If a swimming pool has an inward-protruding

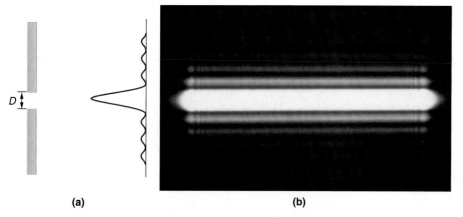

FIGURE 10-4
Diffraction of light. **(a)** When light of a single wavelength passes through a slit, the profile of the images shows a central peak and smaller and dimmer peaks on either side. **(b)** An image of the light through a slit shows the diffraction pattern clearly
(SOURCE: P. Urone.)

corner or wall, you can observe that a wave bends slightly as it goes past the corner. Light behaves similarly. The property whereby light bends its path slightly as it passes an edge is called *diffraction.* For astronomers, diffraction limits the sharpness that can be achieved in the image formed by a telescope. Figure 10–4 shows that light passing through a slit does not form a perfect image of the slit; some of the light is bent to either side.

Also, we know that a water wave carries energy from one point to another at a certain speed. But notice that the entire body of water does not travel in the direction of the wave's motion—the water at any place is only moving up and down. For example, surfers can ride the energy in a wave, but the same volume of water is not carrying them from the beginning to the end of the ride. In just the same way, you can transmit energy in a wave by flicking the end of a tightly held rope. Take another example. Sound is a wave of air compression that travels from one place to another. Yet the air molecules themselves do not travel; they just vibrate or oscillate in one place. This is analogous to holding one end of a stretched-out slinky and pushing quickly to send a pulse from one end to the other. These examples illustrate an important property of waves: Energy is carried by a disturbance traveling in a medium, but the medium itself does not travel. With electromagnetic waves we do not even need the material medium. Light waves travel through the vacuum of space even though nothing carries them!

Light has another property that is characteristic of waves. If we have two sources of waves, the waves can "add" together. Figure 10–5 shows two waves with the same wavelength. If the peaks and troughs of each wave line up (Figure 10–5a), the result is a wave with twice the amplitude (and 4 times the intensity). Alternatively, the peaks of one wave might line up with the troughs of the other wave (Figure 10–5b), and the final wave has zero amplitude or intensity everywhere. If the waves are shifted by a different amount, the result is an intermediate intensity. You can see interference in water waves when two sources of waves sit side by side (Figure 10–5c).

Thomas Young demonstrated this wave nature of light over 150 years ago with his discovery of light *interference.* In his experiment, light arriving at a screen from two small sources side by side produced a pattern of alternating bright and dark stripes (see Figure 10–6a). You can think of this as

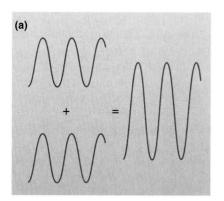 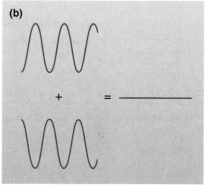 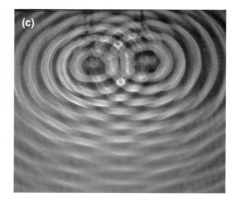

FIGURE 10-5
Waves interfering schematically. **(a)** If two waves arrive at one place so that the peaks and troughs line up, the sum of the two waves is a wave with twice the amplitude. **(b)** If two waves arrive so that the peaks of one wave line up with the troughs of the other wave, the resulting wave has zero amplitude. Note that if the two waves have different wavelengths, the interference pattern is more complex. **(c)** Interference of water waves from two sources. If you measured the wave amplitude along a vertical path between the two sources, you would see alternating regions of twice the wave height and zero wave height. (SOURCE: Richard Megna, Fundamental Photographs)

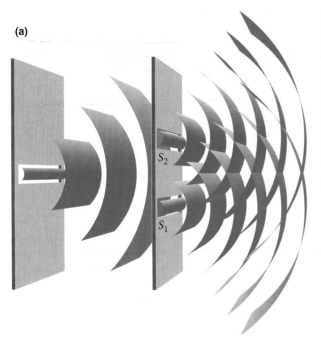
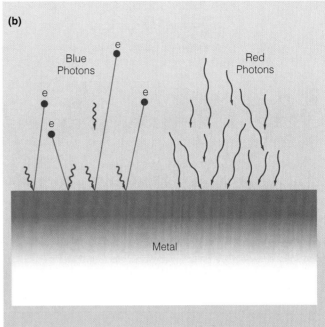

FIGURE 10-6
Complementary ways of looking at light. **(a)** In Thomas Young's experiment, a point source of light passes through two slits to a screen. Light emerges as spherical waves centered on each slit. Alternating bands of light and dark are seen on the screen, which can be explained as the enhancement and cancellation of waves added together. If we imagine the light as particles, they should travel in straight lines through the slit to form only two bands. **(b)** The photoelectric effect, which was interpreted by Einstein in terms of photons of light. The electron is kicked out as if the incoming light particle were a tiny bullet that carried energy. Blue photons have sufficient energy to liberate an electron. If the wavelength of the light is too long, the electron will not be kicked out. Red photons have too little energy to liberate an electron, regardless of the number that arrive. This observation cannot be easily explained in the wave interpretation of light.

wave arithmetic. The amplitude of a wave can be positive (a level above the mean) or negative (a level below the mean). In Young's experiment, there are directions where the positive amplitude from one source combines with the negative amplitude from the other source to give no signal—a dark stripe. There are also directions where the positive amplitudes of both sources combine to give extra signal—a bright stripe.

However, electromagnetic radiation has some properties that are not well described in terms of a wave. For example, when light strikes the surface of a metal, some of the light's energy is absorbed. This energy can release electrons (Figure 10-6b). Longer-wavelength light releases lower-energy electrons. However, when the light has a wavelength longer than a certain value, no electrons are released from the surface of the metal, no matter what the intensity of the radiation. Light acts as a stream of tiny bullets, and unless each of the bullets has a certain minimum energy, no electrons can be released. This is called the *photoelectric effect*. Albert Einstein won the Nobel Prize for this work before his later and better-known work on relativity.

The particle description is more helpful when we think of light traveling through the vacuum of space. If light is a wave, what is doing the "waving"? Yet you can easily think of light as a stream of tiny bullets streaking through the universe. Scientists use this description all the time; the particles of light are called **photons**. However, just as there are situations in which the wave description seems inappropriate, there are also situations in which the particle description seems inappropriate. Let us imagine Young's experiment if light behaves as a stream of particles. The photons would leave the two side-by-side sources and would travel in all directions equally. In this scenario, the illumination should be uniform with no bright and dark stripes. Look at the arithmetic of waves in Figure 10–5, which leads to a natural explanation of interference. Put in terms of the arithmetic of particles, there is no way that two particles can add to give no particles!

Which is light, a wave or a particle? Consider a platypus. It has some ducklike and some beaverlike properties, but it is neither. Similarly, light has some wavelike and some particle-like properties, but it is neither a pure wave nor a pure particle. Perhaps the problem is our need to classify the nature of light into categories that we can visualize easily in everyday life. We might have to accept that light has "dual" properties. Considered in microscopic detail, light is a phenomenon not wholly familiar in terms of analogs in our everyday world. We will see this a lot in the microscopic world of the atom where behavior may be well described by physical laws and mathematics, but not easily grasped by intuition.

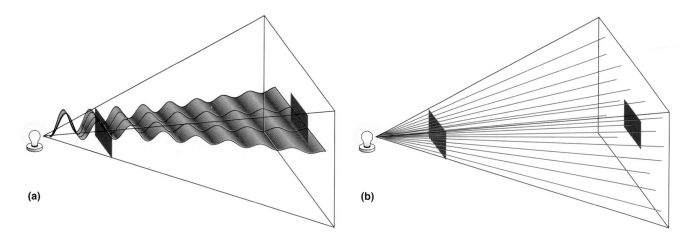

FIGURE 10-7
The inverse square law of light or other forms of electromagnetic radiation. As the distance from the source increases, the radiation passes through successively larger areas. However, the intensity of radiation measured through any fixed area (shaded square) reduces as the square of the distance. The inverse square law can be thought of either as **(a)** the dilution in intensity of a light wave or as **(b)** the thinning out of photons as they travel away from a source.

How Radiation Travels

How does light behave as it travels through space? If we could focus light into a narrow beam, as with a flashlight, it would travel with undiminished intensity through space. But most light sources in astronomy emit their radiation equally in all directions, as a lightbulb does. Figure 10–7 shows the example of light traveling away from a star in all directions. As electromagnetic radiation moves through space, it covers a continually expanding area. The intensity of light per unit area therefore follows an inverse square law. Increase your distance from a light source by a factor of 2, and the intensity drops by a factor of 4. Increase your distance by a factor of 3, and the intensity drops by a factor of 9.

This inverse square behavior fits both wave and particle descriptions of light. For a wave, the energy in the wave is spread out over a larger area as the waves travel from the source (Figure 10–7a). The area covered by the wave increases proportionally to R^2 and the intensity of the wave at any point decreases proportionally to $1/R^2$. Similarly, the best way to alleviate a loud noise is to move away from it! In the particle description of light, you can see that photons "thin out" as they move away from a light source (Figure 10–7b). Since the photons stream out through an ever-increasing area, the number of photons passing through any particular patch decreases as the square of the distance from the source. The two shaded squares in Figure 10–7 show the effect of a factor of three increase in distance from the source of light.

Remember that light is electromagnetic radiation with a certain range of wavelengths. The inverse square law applies to all forms of electromagnetic radiation. It describes how the signal falls off as you drive away from your favorite radio station or how the infrared radiation falls off as you move away from a warming fire. We met the inverse square law in Chapter 3 when we learned that the force due to gravity diminishes as the inverse square of the distance to the mass.

Other forces or sources of energy in nature also follow an inverse square law—electricity and magnetism are two more examples. There is nothing mysterious about this behavior. It is a geometric property of the space we inhabit. Any force or signal will dilute according to the inverse square of the distance it has traveled through space.

Thermal Spectra

We can now relate the temperature of an object to the type of electromagnetic radiation it emits. Recall from Chapter 4 that every atom in the universe is in motion. This is true whether the atom is in a star or a planet or in deep space. Temperature is just a measure of the motions of atoms. Every object emits a smooth thermal spectrum of radiation with a peak wavelength that is inversely proportional to temperature—this relationship is called *Wien's law*. Let us explore the type of thermal radiation from objects with a wide range of temperatures.

The surfaces of most stars have temperatures of several thousand kelvins, and so the surfaces emit thermal radiation that peaks in the visible spectrum. For example, the Sun has a surface temperature of about 5500 K, and the Sun's radiation peaks at 5.5×10^{-7} m, or 0.55 μm (review Figure 4–17). We can also use Wien's law to predict the thermal spectra that peak at invisible wavelengths. By contrast, consider a planet at a temperature of 300 K (like Earth). This is 300 / 5500, or about one-eighteenth, of the temperature of the Sun; so the peak radiation from the planet will be at a wavelength 18 times longer than the peak wavelength of sunlight. This is 18 × 0.55 = 10 μm, in the infrared part of the electromagnetic spectrum. Later in this book, we will see that the entire universe emits a thermal spectrum at the frigid temperature of 2.7 K—this peaks at microwave wavelengths. At the other extreme, hot stars such as white dwarfs can have surface temperatures of 50,000 K. This is 50,000/5500, or 9 times the temperature of the Sun; so the peak radiation will be at a

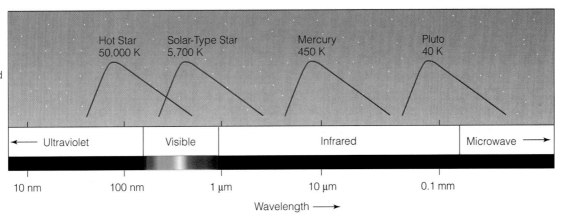

FIGURE 10-8
Wien's law across the electromagnetic spectrum. The wavelength of the peak radiation of a thermal spectrum is inversely proportional to the temperature (see discussion in Chapter 4). Thermal spectra can be produced by objects with a wide range of temperature, which leads to a wide range in the types of radiation emitted. Compare with Figure 10-1. Stars have surface temperatures of thousands of kelvins and so have the peak of their thermal emission at optical or ultraviolet wavelengths. Planets have temperatures of hundreds of kelvins and so have the peak of their thermal emission at infrared wavelengths.

wavelength 9 times shorter than the peak wavelength of sunlight; $0.55/9 = 0.06$ μm is in the far ultraviolet. Four examples are shown in Figure 10-8, which shows Wien's law applied over an enormous range of temperatures. Later in this book, we will encounter examples of thermal spectra that peak at wavelengths as short as X rays and as long as microwaves.

What can we learn about the universe from invisible forms of electromagnetic radiation? Plenty! Everything in space that emits visible light has a temperature of a few thousand degrees. Remember that when we see planets, we are just looking at reflected sunlight—like an imperfect mirror. Any object colder than a star will emit thermal radiation in the infrared part of the spectrum. This is true of planets, and it is true of us, too! We are at a temperature of several hundred kelvins, and we emit infrared waves. There is even colder gas and dust between stars that emits microwaves. At the other extreme, a gas at a temperature of millions of kelvins will emit a thermal spectrum that peaks at X ray wavelengths. The invisible forms of electromagnetic radiation are our signposts to the very cold and very hot regions of the universe.

Radiation and the Structure of the Atom

Astronomers study large objects such as planets and stars and galaxies, but their understanding depends on knowledge of the microscopic interactions of matter and radiation. For example, we have seen that temperature is a measure of the microscopic motions of atoms and molecules. The electromagnetic spectrum is our window on the universe, but to understand how this radiation is produced, we must revisit the tiny world of the atom. More than 100 years ago, scientists who studied radiation from atoms were faced with two mysteries—one was observational and the other was theoretical.

Astronomers studying the radiation from stars discovered a curious phenomenon. They noticed narrow features in visible spectra of the Sun and other stars. In addition to the smooth thermal spectrum that we have already discussed, there were narrow, bright lines at certain wavelengths and narrow, dark lines at other wavelengths. Similar features were seen in the spectra of hot gases in the laboratory. Recall that a spectrum is a map of wavelengths, but since wavelength is inversely related to energy, a spectrum is also a map of energy. These "spectral lines" were a mystery because classical physics held that atoms could have *any* energy. Since energy can take any value and varies smoothly and continuously, there is no way to explain sharp features at particular energies.

The second issue was even more puzzling. We learned in the opening story of Chapter 4 how Ernest Rutherford developed a model of atoms consisting of a small, dense nucleus (of protons and neutrons) surrounded by a cloud of much lighter electrons. Protons carry a positive electric charge, and electrons carry an equal negative electric charge. An atom of the simplest element—hydrogen—has a single proton in the nucleus and a single orbiting electron. It seems as if this orbit might continue forever. However, in the classical theory of radiation, any charged particle emits radiation if it is accelerated. (This is how we make radio waves—by sending electrons racing up and down a wire.) A simple calculation showed that the electron in a hydrogen atom should rapidly lose energy and spiral in toward the proton, pulled by the electrical attraction between the two particles. The same argument goes for any atom. Classical theory predicted that atoms should collapse in a fraction of a second, yet normal matter is obviously stable. Atoms seemed to mock physicists by their very existence!

About 100 years ago, German physicist Max Planck developed a radical new theory that solved both of these mysteries. Just as matter has a fundamental unit called an atom,

SCIENCE TOOLBOX

Properties of Electromagnetic Radiation

The nature of electromagnetic radiation can be confusing. On one hand, it acts as a stream of particles carrying energy from one place to another. On the other hand, it acts as a stream of waves by displaying the properties of diffraction and interference. If electromagnetic radiation acts as a wave, what is doing the waving? In physics, the word *field* describes something that extends through space, having a magnitude or value at every point. Fields can exert influence on a distant object. For example, because of their field properties, magnets can attract or repel each other without any apparent connection; static electricity, too, exerts an attraction across space. A gravitational field holds Earth in its orbit of the Sun across the vacuum of space.

Luckily for us, we can use simple mathematics to describe the properties of electromagnetic radiation without worrying about whether we should visualize it as a wave or as a particle. In terms of waves, electromagnetic radiation is described by its wavelength. The velocity, or speed, of all electromagnetic waves including light is 300,000 km/s, written as the symbol c. (*Note:* It looks as if we quoted the velocity with a small number of significant figures. We actually know the speed of light with a high level of accuracy. It is 299,792 km/s, which just happens to be within 0.1 percent of a round number. See Appendix A–4 to review the concept of precision.)

The velocity is the product of the wavelength and the frequency. Equivalently, there is an inverse relationship between the wavelength and the frequency of the wave. In equation form,

$$\text{Wavelength} \times \text{frequency} = c$$

We can look at some examples. Red light has a wavelength of 7×10^{-7} m, hundreds of times smaller than the thickness of a human hair. Expressed in meters per second, the speed of light is 3×10^8. So the frequency of a red light wave is $c\,/\,\text{wavelength} = 3 \times 10^8\,/\,7 \times 10^{-7} = 4 \times 10^{14}$ cycles per second, or hertz (Hz). This is an extremely rapid wave or oscillation! Your favorite radio station might have a frequency of 400 megahertz (MHz), which is 4×10^8 Hz. These radio waves have a wavelength of $c\,/\,\text{frequency} = (3 \times 10^8)\,/\,(4 \times 10^8) = 0.75$ m. This is a wave big enough for you to easily imagine (but not see, because your eyes cannot detect radio waves). Since there is a simple relationship between wavelength and frequency, you can specify an electromagnetic wave by either quantity. Radio engineers tend to characterize waves by their frequency (radio stations are listed in hertz). Optical astronomers, however, tend to refer to waves by wavelength—a practice we will follow in this book.

Now if we switch to the particle description, we can consider the properties of photons. The energy of a photon is proportional to the frequency of the radiation. Higher frequencies or shorter wavelengths carry more energy than lower frequencies or longer wavelengths. It makes sense that shorter waves carry more energy than long waves. We know that it is the ultraviolet radiation from the Sun that can damage our skin with sunburn, and the highest-energy gamma rays can cause even more tissue and cellular damage. However, the air is full of low-energy radio waves, which apparently do us no harm. In equation form,

$$\text{Energy} = h \times \text{frequency}$$

If the frequency is given in units of cycles per second, or hertz, the energy comes out in units of joules (abbreviated J). The number h is *Planck's constant,* named after a German physicist who was one of the first winners of the Nobel Prize in physics. Its value is 6.63×10^{-34} joule-seconds (J·s). This is a fantastically small number—the tiniest we have encountered in this book.

Light and all other forms of electromagnetic radiation come in tiny bundles of energy called photons. This minimum unit of radiation or energy is called a *quantum* (plural: *quanta*), and it refers to a fundamental "graininess" in the physical world. Just as all matter is made up of indivisible particles in the atom, such as electrons, so radiation is made up of indivisible units of energy called quanta. The idea that light is distributed in discrete, or *quantized,* units was one of the major discoveries in physics this century.

How small is the quantum of energy? Let us look at a single photon of white light. It has a wavelength of 5×10^{-7} m, so it has a frequency of $c\,/\,\text{wavelength} = (3 \times 10^8)\,/\,(5 \times 10^{-7}) = 6 \times 10^{14}$ Hz. The energy is therefore $6 \times 10^{14} \times 6.6 \times 10^{-34} = 4 \times 10^{-19}$ J. The energy of a single visible photon is tiny indeed, as indicated by the extremely small value of Planck's constant. For example, using the fact that 1 watt (W) of power is 1 J/s of energy, a 100-W lightbulb emits $100\,/\,(4 \times 10^{-19}) = 2.5 \times 10^{20}$ photons per second. This is a rate of 250 billion billion photons per second! In other words, the "graininess" of light and matter is so fine that we cannot see it in the everyday world. We will return later to the implications of this incredibly tiny quantum unit.

Planck proposed that energy has a fundamental unit called a *quantum* (plural: *quanta*). Energy is not smooth and continuous, but it comes in discrete packets called photons. See the Science Toolbox above for more details. This revolution in physics was called the **quantum theory of radiation**. Neils Bohr combined Rutherford's picture of the atom—a tiny nucleus with electrons around it—with Planck's quantum theory to show how matter emits and absorbs photons.

In some ways, the atom is like a tiny solar system, that is, a relatively massive central object and tiny orbiting particles. But there is an extraordinary difference. In the solar system we can imagine a planet in any orbit around the Sun; each orbit would have its own velocity and hence energy. However, in an atom, electrons can occupy only specific orbits, called **energy levels.** Physicists describe this by saying that the atom has quantized energy levels, because only certain quantities of orbital energy are possible.

How does the quantum theory explain the two mysteries we have just described? The quantum theory holds that there is a lowest energy that any electron in an atom can have. The electron cannot lose energy continuously, so it does not spiral in toward the nucleus, and the atom does not collapse. The answer to the second mystery is particularly interesting. If an atom could gain or lose any amount of energy, it could emit radiation with any wavelength. All these wavelengths would fill in to make a smooth and continuous spectrum. But electrons are held in specific energy levels. So atoms can only gain or lose the energy corresponding to an atom moving between two energy levels. This creates a set of rules for the emission and absorption of radiation in atoms. Let us see how it works.

Emission is the process in which an atom loses energy in the form of a photon. Absorption is the complementary process in which an atom gains energy in the form of a photon. Figure 10–9a is a schematic view of the energy levels of an atom. Levels or orbits that are farthest from the nucleus have the highest energy. Each transition of an electron between levels corresponds to a photon with energy equal to the energy difference between the two levels. The classical picture corresponds to the right side of Figure 10–9a, where the energy levels are so finely subdivided that any energy level is possible. In the quantum picture shown in Figure

FIGURE 10-9
Simplified schematic of the energy levels of electrons in an atom. **(a)** The upper panels show the energy levels as horizontal lines. Energy increases going upward, and above the dashed lines, the electron is freed from the atom. Going from left to right, the number of energy states increases (as it would, e.g., if we considered heavier and heavier elements), and so the number of possible transitions increases. The far right approximates the classical case, where the electron can have any energy. Only a few of the possible transitions are shown. **(b)** The lower panels show how the energy levels translate to spectral lines. Each possible jump of the electron is a specific energy difference, which corresponds to a specific wavelength. Going from left to right, the number of spectral lines increases. The far right is the classical case; when the electron can change its energy by any amount, all wavelengths are represented and the spectrum fills in, to become nearly smooth.

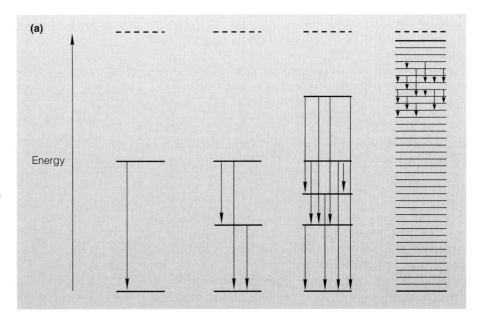

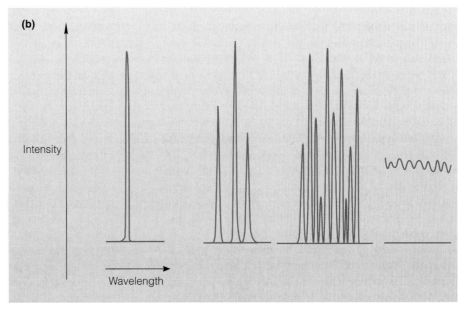

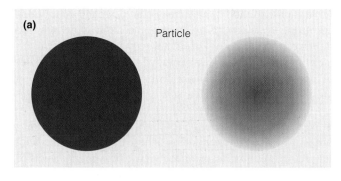
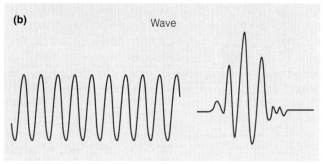

FIGURE 10-10
Classical and quantum views are contrasted in this schematic view of a wave and a particle. (a) The left side shows a classical hard-edged particle; the right side shows a fuzzy particle where Heisenberg's uncertainty principle states that its position in space cannot be determined exactly. (b) The left side shows a classical wave; the right side shows a wave where the energy is concentrated in space, giving it properties similar to those of a fuzzy particle.

10–9b, the energy levels are discrete, so an electron can only lose specific amounts of energy. This creates sharp lines in the spectrum since each specific energy loss corresponds to a particular wavelength of radiation. It is like the difference between sliding down a smooth slope and walking down steps. In the classical case (the far right of Figure 10–9b), an atom moves to a lower energy level by losing any amount of energy. Consequently, the spectral transitions have any wavelength, and so the spectrum fills in to be smooth and featureless.

As an analogy for the quantized world of the atom, think of two different types of arithmetic. In one system you can add or subtract any fraction or decimal number. Starting with 8, say, you can add $1/2$ or $1/9$ or 2.37, or you can subtract $1/18$ or 10.4 or 3.006. Do this many times, and you will fill in the number line—this is the smooth and continuous world of classical physics. Now think of integer arithmetic. Starting with 8 again, you can add 1 or 4 or 7, or you can subtract 2 or 5 or 6. Do this many times, and you will have a set of integer values—this is the quantum world. (An integer is analogous to a quantum, and the number line is analogous to a spectrum of energy or wavelength.)

In the everyday world, the quantized nature of energy and matter is not apparent (see the last Science Toolbox). Crouch down on a beach, and you can see the tiny particles of rock that make up sand. From a distance, the texture appears smooth and uniform, and the structure is not visible. Let us go back to the analogy of the number line. If we step far back from the number line so that we can only see billions or trillions marked off, the individual integers are no longer visible. Seen on this large scale, the integer number line appears smooth and continuous. Similarly, transactions in the everyday world involve many trillions of atoms or photons. The graininess of the quantum world is not visible to us.

The quantum theory of radiation may seem strange to you. It was strange and uncomfortable for many physicists, too! The theory became accepted because it describes very well how nature works. In the 1930s, physicists were still reeling from the idea of the quantum when German physicist Werner Heisenberg came up with an equally bizarre idea. He showed that quantities such as energy, momentum, and position cannot be defined with absolute precision. This idea is enshrined in physics as the **Heisenberg uncertainty principle.** Scientists believe that the solar system analogy of the atom, with particles like hard balls in specific orbits, is a weak one. In the quantum model of the atom, the electron is not a discrete object like a billiard ball. It has a certain probability of being at various positions, so the electron is not just at one position in its orbit, as a planet would be. The imprecision described by Heisenberg is only substantial on the scale of atoms or subatomic particles (see the Science Toolbox on page 278). It does not limit our knowledge of the everyday world. The "fuzziness" of the subatomic world is not visible to us.

We have seen that light can be thought of as either a wave or a particle. In the quantum view of the subatomic world, a particle can also be thought of as a wave! Let us look at the difference between the classical and quantum ways of looking at particles and waves. In Figure 10–10a, the left side shows the classical view of a particle as something hard-edged like a marble or a ball bearing. On the right side, the quantum view of the particle is somewhat fuzzy, because it has a probability of being found over a range of space. Figure 10–11b shows the same comparison for a wave. On the left side is a classical wave, extending off in space in either direction. On the right side, the quantum view considers the energy of the wave to be more concentrated in space. In the classical view, particles and waves appear to be totally distinct entities. In the quantum view, they are quite similar!

Now we can compare two alternative views of electrons in an atom. In Figure 10–11a, the classical model of a miniature solar system does not explain why the electrons are confined to certain orbits. In Figure 10–11b, we see that the quantum model is based on the wave properties of particles. A useful analogy for imagining the quantized energy levels of an atom is to think of the set of vibration modes of a plucked string. The vibrations of a guitar string are made of a whole number of waves. In the same way, each energy level corresponds to a whole number (as opposed to a fraction) of waves, and the orbits therefore have quantized wavelengths (or energies). The guitar string analogy is also appropriate because the wave's energy is distributed in space, not concentrated as a hard-edged particle. The quantum idea of particles as waves may seem counterintuitive, but it helps explains *why* the atom should have quantized energy levels.

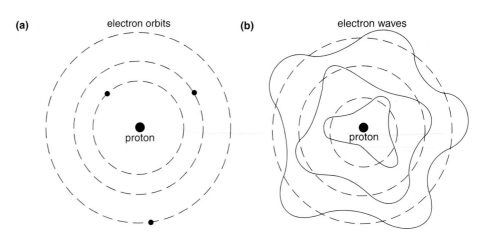

FIGURE 10-11 Complementary ways of looking at electrons in an atom. **(a)** The atom as a miniature solar system. This classical model does not explain why the electron has specific energies and is confined to specific orbits. **(b)** In the wave view of an electron, the best analogy is the vibration of a plucked string. The string can only vibrate with a pitch (frequency) or energy corresponding to integer numbers of complete oscillations. If we imagine these vibration states as wrapped around the nucleus of the atom, we have a quantum analogy for the behavior of the electron.

Spectral Lines

We are now in a position to describe the processes of *emission* and *absorption* of radiation. Emission occurs when an atom lowers its energy by ejecting a photon. The result is that an electron moves from a higher to a lower energy level. Absorption is a complementary process, in which a photon adds energy to an atom. The result is that an electron moves from a lower to a higher energy level. Atoms can also gain or lose energy by collisions with other atoms. The effect is the same—if the atom gains energy in the collision, an electron will move to a higher energy level; and if the atom loses energy in the collision, an electron will move to a lower energy level.

These processes occur as atoms try to reach equilibrium. Recall from Chapter 4 our discussion of heat flow—energy moves from hotter to cooler objects until they reach the same temperature. We saw a second example in our discussion of planets. Gravity tends to even out the surface of a planet so that all points on the surface have the same gravitational potential energy. If you take a tray with a pile of sand and jiggle it rapidly, you will see the sand come into equilibrium on a nearly flat surface. Another way to state the principle of equilibrium is that objects tend to move to their state of lowest energy. This happens every time your cup of coffee cools off or you slump down in a chair. What does this have to do with atoms? Atoms behave the same way. Whenever an atom has its energy raised—by an impact from a photon or another atom—it will try to recover its lowest-energy state by emitting a photon. We can now see how the emission and absorption of photons lead to spectral lines.

Emission Lines

As an atom drops to a lower energy state, it loses energy by emitting a photon. The photon's energy equals the difference in energy between two energy levels of an electron. As stated earlier, electrons in an atom can experience only certain, fixed changes in energy level, as shown in Figure 10–12a. For instance, if an electron starts in level 3, it can drop only to level 2 or level 1. Atoms can emit only photons of energy corresponding to these differences in energy level.

As a result, each element can emit only certain wavelengths of light. Revisit Figure 10–9 if you are unsure of this idea. Figure 10–12b shows, for example, the radiation produced by a sample of hydrogen gas containing neutral (or uncharged) atoms with their electrons at different energy levels. **Emission lines** result from the emitted photons and appear in a projected spectrum as *lines* or narrow bars of color.

Spectral emission lines are extraordinarily useful in astronomy. Each element has a unique number of protons in the nucleus and electrons orbiting the nucleus. This means in turn that each element has a unique set of electron energy levels that create a specific set of wavelengths represented by the emission lines. If you see a set of emission lines, you can match it with a single element and infer that atoms of that element are present and glowing in the distant object. If more than one element is present, the pattern is more complex because there are more lines, but the principle is the same. You can deduce the object's *composition,* even without having a sample! By contrast, the smooth spectrum of thermal radiation only gives information about the *temperature* of the object. At the same temperature, a lump of iron or a carbon rod or a cloud of hydrogen all emit the same thermal spectrum. Thus spectral lines are much more useful if we want to determine the chemical composition of a distant object.

Note that if an atom is in its lowest possible energy level, that atom cannot produce an emission line. This is because the electrons cannot drop to any lower energy level. An atom in which all electrons are in the lowest possible energy level is said to be in its *ground state.* An atom in which one or more electrons are in energy levels higher than the lowest available ones is said to be in an *excited state.* Excited states usually last only a fraction of a second before the electrons decay to the lowest available energy level—trying to reach equilibrium. Atoms generally need to be disturbed to produce and maintain excited states. This can happen in two ways. Radiation will add energy to a gas and so cause electrons to raise their energy state. In a hot gas, the same role can be served by collisions of the atoms or molecules themselves. Heating a gas enclosed within a certain volume increases the velocity of atoms and so increases the probability that they will collide with one another.

SCIENCE TOOLBOX

Uncertainty and the Quantum World

There is a ball somewhere in a completely darkened room. How would you find it? You could blunder around in the dark, but very likely you would bump into it and send it off in some unknown direction. A smarter way would be to search for it with a flashlight. However, we know that light in the form of photons carries energy. So when the photons bounce off the ball (which allows you see it), the ball gains a tiny amount of energy. In the macroscopic world full of big objects, this tiny amount of energy carried by photons is negligible. As far as we are concerned, the flashlight lets us see the position of the ball as accurately as we want. But what about when we are looking for a tiny particle, instead of a ball?

As we search for smaller objects, the energy and momentum of the light beam become significant. Light reflecting off the object gives it energy and so moves it slightly. Now if we are looking for a tiny subatomic particle, we might try to illuminate it with a single photon. But the photon hitting the particle moves it, and we no longer know where the particle is! So our knowledge of its position comes at the expense of our knowledge of its motion. We could be clever and shine a photon of lower and lower energy at the particle. But at some point the photon, which must have a long wavelength if it has a low energy, becomes too big to interact with the particle. Thus, there is a fundamental uncertainty in the subatomic world: *The act of measurement always alters the objects being measured.* In the everyday world, this effect is too small to be noticed, but it dominates our understanding of the microscopic world.

Werner Heisenberg expressed this indeterminacy in a simple and elegant form. The equation describes a tradeoff between measurements of time and motion. The product of the uncertainty in position times the uncertainty in velocity (or momentum) is related to Planck's constant. In equation form, it is stated

$$\Delta x \times \Delta v \times \text{Mass} \geq \frac{h}{4\pi}$$

where the Greek letter Δ is the commonly used symbol for the error or uncertainty in a quantity. In this equation, Δx is the uncertainty in the position in one direction, Δv is the uncertainty in the velocity in that same direction, and h is the universal constant named after Planck that we have already encountered. The equation is an inequality, saying that the product of the uncertainty in the position and the uncertainty in the momentum is about equal to or greater than a very tiny number.

What do these mathematical relationships mean? They mean that we can know the position of a tiny particle, but only at the cost of not knowing its motion at all. Or we can know its motion, but at the cost of not knowing where it is! The amount of the uncertainty is given by Planck's constant, a tiny number that has little or no effect in the everyday world.

Let us see how the equation works in practice. A pitcher throws a 90 mi/h fastball (about 40 m/s). The radar gun measures its speed as it passes through the 10-cm beam of radar. What is the quantum uncertainty in the speed of the baseball? Rearranging the equation above, we see that the uncertainty in speed $\Delta v = h / (4\pi \times \text{mass} \times \Delta x)$. In matching units, the position uncertainty is 10 cm, or 0.1 m; the mass is about 0.3 kg, and $h = 6.63 \times 10^{-34}$ J · s. So $\Delta v = 6.63 \times 10^{-34} / (12.6 \times 0.3 \times 0.1) = 1.7 \times 10^{-33}$ m/s. In other words, we could theoretically measure the speed of the fastball to 32 decimal places! The quantum uncertainty plays no role in this measurement.

What about a much smaller object? How accurately could we measure the speed of an electron in a hydrogen atom? An electron has a mass of 9×10^{-31} kg, and a hydrogen atom has a rough size of 10^{-10} m. Working as before, we get $\Delta v = 6.63 \times 10^{-34} / (4\pi \times 9 \times 10^{-31} \times 10^{-10}) = 5.8 \times 10^{5}$ m/s. Our uncertainty is larger than the velocity of light! This means that we have no knowledge of the electron's velocity once we have determined that it is in a hydrogen atom.

Heisenberg believed his result expressed a fundamental aspect of nature; it has nothing to do with poor observations or imperfect measuring equipment. *There is a limit to the precision with which we can know the physical world.* The uncertainty principle is not just an exotic theoretical idea. It is demonstrated in practice everyday in the physics laboratory. Measurements of subatomic particles are always limited by the equation we have just seen. In terms of measuring position, the uncertainty principle implies an unavoidable fuzziness to the subatomic world.

Needless to say, this view has larger philosophical implications. We have seen many examples in this book of how science works to gain knowledge. Scientists have explored space, delved into the world of the atom, and opened up the electromagnetic spectrum. We know far more about the natural world than we did 100 years ago. Surely this long march of progress will just continue. Heisenberg's uncertainty principle places a limit on knowledge. Some scientists do not like the fact that there is a limit to our knowledge, a veil beyond which we cannot see. Other people are reassured, because it means that *determinism* has no place in physical theories. In other words, since the behavior of atoms is not completely predictable, we can never predict the future behavior of *any* physical system with certainty. (However, we can predict the behavior of everyday objects with high precision.) Either way, we are faced with understanding the strange world of the quantum.

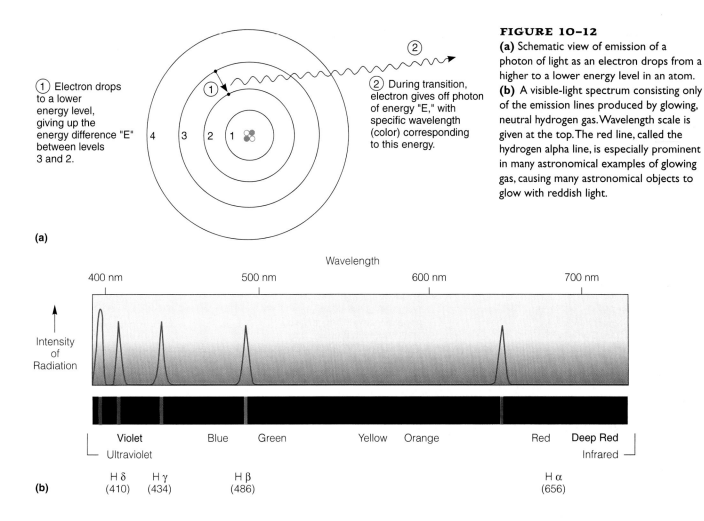

FIGURE 10-12
(a) Schematic view of emission of a photon of light as an electron drops from a higher to a lower energy level in an atom.
(b) A visible-light spectrum consisting only of the emission lines produced by glowing, neutral hydrogen gas. Wavelength scale is given at the top. The red line, called the hydrogen alpha line, is especially prominent in many astronomical examples of glowing gas, causing many astronomical objects to glow with reddish light.

Where in the electromagnetic spectrum would we look for emission lines from atoms? Let us take hydrogen, the simplest element. The energy required to raise the electron from the ground state to be free of the proton is the largest amount of energy that can result in a spectral line. Therefore, it corresponds to the shortest-wavelength feature we might see. This wavelength is about 90 nm, which is in the ultraviolet too blue for our eyes to see. Other electron transitions in a hydrogen atom have smaller energy differences, so they yield redder spectral lines. Heavier elements have more electrons and more energy levels. They therefore have more possible transitions and a denser thicket of emission lines. But for the most common elements such as carbon, nitrogen, oxygen, and silicon, the spectral lines fall in the same region of the electromagnetic spectrum. Roughly, you can consider that most common emission lines of the most abundant atoms fall in the decade of wavelength from 100 to 1000 nm (or 0.1 to 1 μm). This spans the visible spectral range but also extends to ultraviolet and infrared wavelengths.

Molecules and Emission Bands

Molecules are groupings of atoms that can share their electrons. Therefore, the electron structure in a molecule is more complex than that in an atom. The electron's path may take it around two or more nuclei. As a result, the emission line structure of a molecule can be complex. For example, a gas containing water molecules (H_2O) has many more emission lines than a gas containing single H and O atoms. The molecule has various ways of responding to a disturbance in addition to having its electrons change energy levels; for example, it may vibrate as two balls linked with a spring, or it may rotate. As a result, the energy levels from a molecule are vastly more numerous, and the resulting emission lines can blend together into a broader emission feature called an *emission band*. The rest of the story is the same. A given molecule (such as H_2O) can produce only certain emission bands, allowing us to identify the molecule in a remote source.

In order for atoms and molecules to produce clear emission lines and bands, they must be detached from one another, as in gases. If the atoms were linked together, they would form molecules; and if the molecules were linked, they would form a solid or liquid. In most solid or liquid substances, the electron structure is so complex that emissions are not confined to one wavelength, but are smeared out. Therefore, emission features of solids and liquids are barely discernible. Most emission lines and bands arise from gases.

Where in the electromagnetic spectrum would we look for emission features from molecules? Some spectral features from molecules are seen in the visible spectrum.

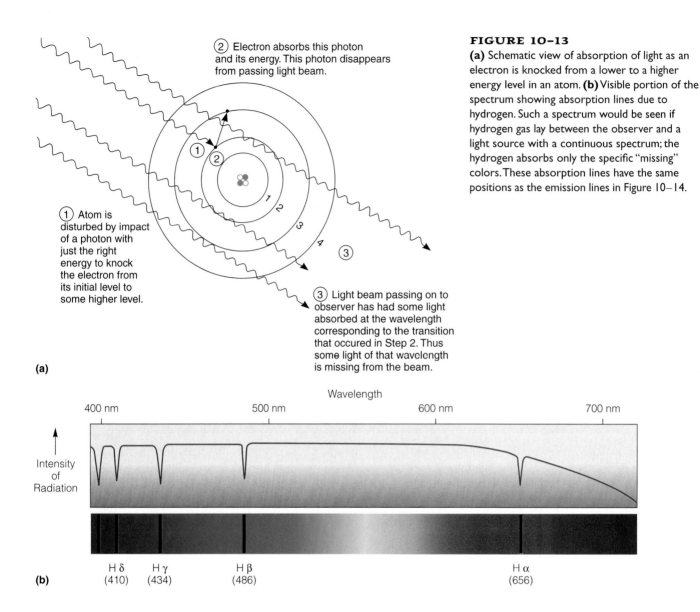

FIGURE 10-13
(a) Schematic view of absorption of light as an electron is knocked from a lower to a higher energy level in an atom. (b) Visible portion of the spectrum showing absorption lines due to hydrogen. Such a spectrum would be seen if hydrogen gas lay between the observer and a light source with a continuous spectrum; the hydrogen absorbs only the specific "missing" colors. These absorption lines have the same positions as the emission lines in Figure 10-14.

However, molecules have shapes that allow them to vibrate and oscillate in many different ways. Most of these modes involve less energy than a typical electron transition, so the spectral features are typically found at longer wavelengths. Many of the most important spectral features from molecules are found in the infrared or even the microwave part of the electromagnetic spectrum.

Absorption Lines

Let us return to an atom with an orbiting electron, as shown in Figure 10-13a. This time the atom is disturbed by a photon that bumps it up into a higher energy level. Energy was removed from the beam to do this. This process of energy removal from a light beam is called *absorption*. Only the specific energy corresponding to a specific electron transition (level 2 to 3 in the case of Figure 10-13) can be absorbed in an atom. Thus only a photon of specific energy or wavelength can be removed from the beam. A beam of light passing through a cloud of such atoms will have many of these photons removed, thus absorbing some light of that color. As a result, the light in a narrow interval of the spectrum is lost, and this missing interval is called an **absorption line**. Different various absorption lines result from the various possible upward transitions of electrons, for example, level 2 to 3, 2 to 4, 1 to 2, 1 to 3, and so on. The visible part of the spectrum, with absorption lines due to hydrogen, is shown in color in Figure 10-13b.

There is one important detail yet to explain. When an atom gains energy by absorbing a photon, the atom will then return to a lower energy by reradiating that photon. Every photon that is absorbed is eventually emitted again. So why does the emission not fill in the absorption line and leave no visible feature in the spectrum? The answer is that the original light beam is traveling through the gas and has light removed at specific absorption wavelengths. However, the atoms of the gas reradiate photons in *all* directions, so not enough photons are emitted in the original direction of the light beam to fill in the absorption.

Since the *intervals* between energy levels in a given atom are the same regardless of whether absorption or emission is occurring, the pattern of emission and absorption lines for a given element is the same. An element can be identified from either its emission lines or its absorption lines.

Molecules and Absorption Bands

Much of the preceding discussion of absorption also applies to molecules. As light penetrates a cloud of molecules in a gas, transitions of electrons among its numerous, closely spaced energy levels produce *absorption bands*. The molecules in the gas can be identified if the absorption bands themselves can be detected and identified. The absorption bands of a substance have the same wavelength intervals as its emission bands.

Astronomical Uses of Radiation

A major goal of astronomy is to detect and understand nature's messages from space. We have seen that spreading out the wavelengths of the radiation from an astronomical object yields two types of information. The smooth thermal spectrum indicates the temperature of the atoms of the star or planet. The sharp spectral lines have patterns that indicate the chemical composition of the star or planet. These are powerful tools in understanding parts of the universe that we can never travel to and touch.

We have also seen that the radiation we can see is just a small slice of the much larger spectrum of electromagnetic radiation. However, not all this radiation from space reaches us. Earth's atmosphere is opaque to most wavelengths of electromagnetic radiation (Figure 10–14). Only a thin band of optical and near-infrared wavelengths and a wider band of radio wavelengths penetrate to the ground. For most of the history of astronomy, our knowledge of the universe has been confined to the messages coming through the narrow window of visible light. In the past 30 years, this situation has been transformed by new technologies and the space programs of NASA and the European Space Agency. We are now free to explore the entire electromagnetic spectrum.

Telescopes and Detectors

For most of human history, we have learned about the universe with our eyes. As we saw in Chapter 3, around 400 years ago Dutch opticians learned to place two lenses together to make a **telescope.** Galileo used this new invention to magnify and sharpen our view of the heavens—and provide vital evidence in support of the Copernican hypothesis. Astronomers have been building larger and larger telescopes ever since. Astronomy has been transformed again in the past few decades. Advances in technology have changed the way we detect radiation and have enabled us to see the full range of the electromagnetic spectrum.

The important features of a telescope have not changed since the time of Galileo. The most important function is to collect light—this is controlled by the **collecting area.** Telescopes have the ability to gather light and reveal fainter

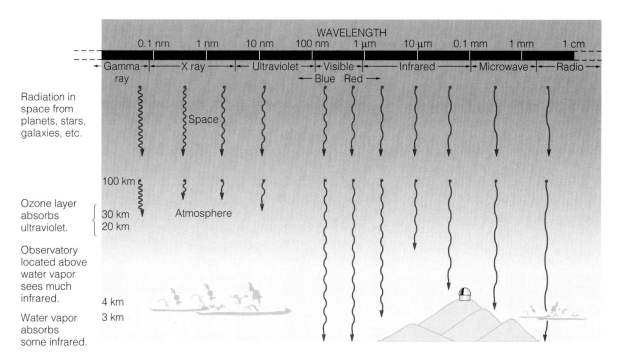

FIGURE 10–14
Electromagnetic spectrum is shown at the top. (The wavelengths are distributed across the top in this schematic view; but in fact all wavelengths arrive at all points at the top of the atmosphere.) All these types of radiation can be detected by a telescope in space (upper middle), but many are absorbed by various gases in our atmosphere (bottom).

objects than the naked eye can see. When light from a distant planet or star reaches Earth, a certain number of photons strike each square centimeter of Earth each second. The pupil of the eye has a diameter less than 0.5 cm and can receive only a limited number of photons per second. But a telescope collects the photons that strike a lens or mirror many centimeters across. When all this light is brought to a focus, it produces a much greater level of illumination. The light-gathering power depends on the collecting area. The collecting area increases in proportion to the square of the diameter of the light-gathering device. Increasing the light-gathering power helps us collect more *information*.

A second function of a telescope is its **resolution**—the ability to discriminate fine detail. Whereas the eye can resolve angular details only a few minutes of arc across, a telescope might show details 100 times smaller, or only 1 second of arc across. The resolution of a telescope is proportional to the ratio of the wavelength of the light being measured to the diameter of the telescope. In equation form, the resolution in arc seconds is 250,000 × (wavelength / diameter), where the wavelength and diameter can be expressed in any units as long as they are the same. Larger telescopes can resolve smaller angles and are said to have higher resolution. If you use this equation, you will see that a telescope with a diameter of 4 m should have a resolution of 0.02 second of arc. However, in practice, the blurring of incoming light due to turbulent motions in Earth's atmosphere at most sites reduces the resolution to about 1 second of arc. Many backyard telescopes are large enough to resolve to this 1 second of arc limit. To get a sense of this level of angular resolution, remember that 1 second of arc is the angle subtended by a dime at a distance of $1\frac{1}{2}$ mi.

Telescopes designed for visual observation have a third function: to *magnify* an object; magnification is the apparent angular size of a distant object seen through the eyepiece relative to its apparent size seen by the naked eye. If a telescope makes something look 10 times larger, we say it has a magnification of 10×. Note that magnification is not a fundamental property like the first two we described. When an image is magnified, the blurring is magnified, too. Also, a magnified image is dimmer since the same amount of collected light is spread over a larger area. Early telescopes were designed entirely for observers to look through. In modern professional telescopes, the human eye is replaced by instruments that can make more precise measurements. Thus modern astronomers rarely look through their giant telescopes!

Telescopes have a similar function regardless of the wavelength of radiation collected. Radio, infrared, optical, ultraviolet, and X-ray telescopes all must gather radiation and bring it to a focus to form an image. Detectors are placed at the focus of a telescope. Again, regardless of the wavelength of observation, a detector should register the photons at the focus as efficiently as possible.

Optical Telescopes

The first type of telescope to be built was the *refractor*. It uses a lens to bend, or refract, light rays to a focus, as in Figure 10–15a. Galileo first used this type astronomically in 1609. The second type, the *reflector,* uses a curved mirror to reflect light rays to a focus, as in Figure 10–15b. Isaac Newton built the first reflecting telescope in 1668.

The first large telescopes were refractors, but almost every major research telescope now uses the reflector design.

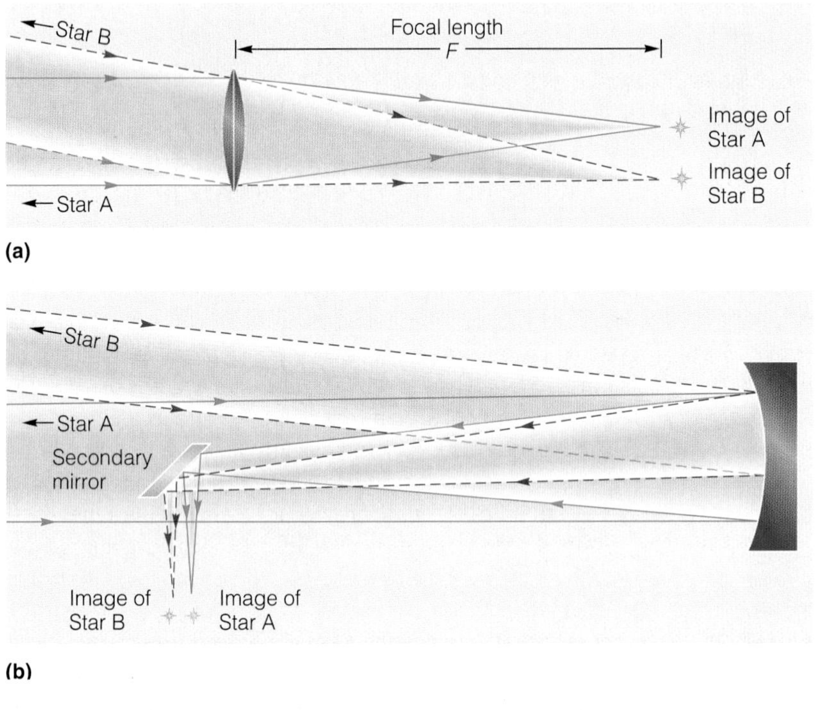

FIGURE 10–15
(a) Cross section through a lens, showing image formation in a refracting telescope (and most cameras). Light rays from two stars are focused into two images. **(b)** Cross section through a mirror system, showing image formation in a reflecting telescope. Light rays strike curved mirror (right) and are reflected back toward the focus. The focus would normally lie in an inconvenient position in front of the mirror, but a secondary mirror is used to bounce light rays back through a hole in the primary for easier access to the image.

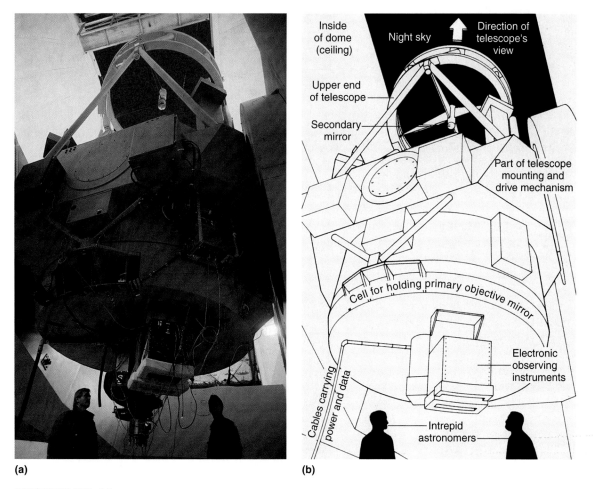

FIGURE 10-16
(a) A view of a modern telescope of 2.2-m aperture, at Mauna Kea Observatory, Hawaii. In this telescope design, light is focused through a hole in the center of the main mirror, where the astronomer mounts electronic instruments to record data. Data are fed through cables to computer equipment, which performs initial analysis and stores data for further analysis. The dome (top) is open to night air in order to give the clearest observing conditions; temperatures inside the dome often drop below freezing during observing sessions. (SOURCE: Photograph by WKH.) **(b)** Schematic diagram of *(a)*.

There are three reasons for this. First, each little segment of a lens acts as a prism, splitting white light into its component colors. Since red and blue wavelengths are bent by different amounts, there is no single place in a refracting telescope where light of all colors is in focus. Reflecting telescopes avoid this problem because a mirror reflects *all* wavelengths to a single focus. Second, large refractors have large and heavy lenses, which must be supported around the edge, and they may sag slightly in the middle, distorting the image. Reflectors use a curved mirror, which can be supported across its back surface. The largest refractor ever built had a diameter of 1 m, whereas astronomical mirrors as large as 8 m across have been constructed. Third, large refractors are long and thin, so they suffer from *flexure*—bending of the telescope structure that can distort the image. Reflectors are more compact, because the light is reflected back down through a hole in the primary to a focus point below the telescope (see light path in Figure 10–15b). A typical modern reflector is shown in Figure 10–16.

Driven by the need for greater light-gathering power and better resolution, astronomers have built larger and larger optical telescopes ever since Galileo first used his small refracting telescope. (You could duplicate his best effort with a dollar's worth of components from a drugstore.) This is an exciting time for optical astronomy. After many decades when the Palomar 5-m telescope was the largest available, a dozen telescopes of 6-m or larger diameter are being built or planned. Each of these new devices has a factor of 100,000 more collecting area than that of Galileo's telescope!

Today, there are two main approaches to building large telescopes. The first is to construct large, single mirrors, trying to make them as big, light, and accurate as possible. One of the most ingenious facilities sits under the football stadium at the University of Arizona. Glass chunks are placed in an enormous rotating oven. As the oven is heated, the glass flows freely over a set of molds that give the glass a honeycomb structure. The oven spins and the mirror takes a roughly parabolic shape. As the oven cools, the mirror solidifies in its final shape, and the

FIGURE 10-17
Photographs taken inside the rotating oven at the University of Arizona Mirror Laboratory during the casting of the 6.5-m mirror for the Multiple Mirror Telescope upgrade in 1992. **(a)** The oven was filled with a total of 10 tons of borosilicate, a type of glass with excellent thermal stability. **(b)** As the oven was heated up to a peak temperature of 1185°C, the glass melted and flowed into a nearly parabolic shape. **(c)** The honeycomb pattern is due to a set of molds that allow the mirror to be very strong and light for its size. (SOURCE: Courtesy of Roger Angel, University of Arizona.)

molds can be removed to give a lightweight structure. The mirror is then ready for polishing and aluminizing. Figure 10–17 shows views from a camera inside the rotating oven during the casting of a 6.5-m mirror in 1992. This was the largest mirror to be cast in the United States in more than 50 years.

Alternatively, a large collecting area can be built up out of a number of smaller mirrors. This approach was first used at the Multiple Mirror Telescope in 1980 in southern Arizona.

FIGURE 10-18
A view of the Keck telescope on Mauna Kea, Hawaii. The Keck telescope has 36 hexagonal mirrors that give it an effective diameter of 10 m, making it the largest optical telescope in the world. A second, similar telescope is on a nearby site. (SOURCE: Roger Ressmeyer, Starlight.)

The largest optical telescopes in the world are at the Keck Observatory on Mauna Kea in Hawaii, funded by a bequest from the world's largest privately owned oil company. Each of the two Keck telescopes has 36 hexagonal segments, each with a diameter of 1.8 m (Figure 10–18). The effective diameter of the telescope is 10 m. The individual mirrors are controlled by an active system that maintains the mirror's shape to a high degree of accuracy.

Amateur astronomers can share in this excitement by building their own telescopes. Many amateurs (sometimes working in local clubs) grind their own mirrors and other optical elements—15-cm or 20-cm or larger—by methods like that used by Newton. With such telescopes, they can see lunar craters in detail, Saturn's rings, the colored gas of various gas clouds, and many other wonders of the universe. With systematic study of the skies using a small telescope, amateur astronomers can discover comets and dying stars and can track the behavior of variable stars. You do not have to be a professional astronomer to contribute to the journey of discovery described in this book.

Optical Detectors

Astronomers need quantitative information about distant objects. Pictures of astronomical objects are interesting, of course, but much astronomical work requires measuring an object's brightness—the amount of light coming from it at all wavelengths or over a range of wavelengths. This is called **photometry**. Astronomers use photometry for a precise measurement of the amount of light at various wavelengths; this allows them to measure temperature, composition, and other properties of a remote object. Photometry must be calibrated with a target of known brightness, perhaps the Sun, or one of the planets, or a nearby bright star. Photometry is not confined to optical wavelengths; astronomers working across the electromagnetic spectrum from radio waves to X rays need to measure the precise amount of radiation received by their detectors. **Spectroscopy** is the technique in which radiation in a particular part of the electromagnetic spectrum is dispersed in wavelength and recorded behind a telescope.

For the first 200 years of telescope use, observations were made with the naked eye and subsequently recorded on paper. Naturally, the reliability of these observations could be questioned. When photography was developed in the middle of the 19th century, astronomers were quick to use properly exposed film to make a permanent record of their observations. Many small telescopes today come equipped with photographic attachments for this purpose. Faint objects such as nebulae and galaxies are usually better represented in photographs, because light can be accumulated in long exposures. By contrast the eye only "stores" light for a fraction of a second. Many photographs of star fields and nebulae in this book were taken with small telescopes or ordinary cameras. Consult the picture captions, most of which describe the equipment and exposure used. In many cases, you can duplicate or improve on the results with your own equipment.

In recent years, photographic techniques have been supplanted at most research telescopes by electronic detectors. The most important type of electronic detector is a *charge-coupled device* (CCD), commonly known as a **CCD detector**. CCDs are extremely light-sensitive detectors, made possible by microelectronic technology (CCDs are found, e.g., in most camcorders). The device works by converting incoming photons to electrons, which are then stored and accumulated until they are read out and converted to an electric signal or current. The electric signal is then converted to intensity units that can be displayed as an image. CCDs consist of arrays of typically $2000 \times 2000 = 4{,}000{,}000$ light detectors in a device the size of a postage stamp. Each tiny detector is called a picture element, or *pixel*. Each pixel has its own measure of brightness, and the many pixels combine to form an image much as dots of different sizes combine to form a photographic image in a newspaper. Figure 10–19 shows a mosaic of four CCDs combined to view a larger area of sky. This device has 16 million independent pixels for recording information.

CCDs have revolutionized optical astronomy. On a large telescope a CCD can produce images of light sources that are roughly 1 billion (10^9) times fainter than the eye can see! Where does this fantastic gain come from? CCDs are 10 times

FIGURE 10-19
An assembled mosaic of four 2048×2048 pixel CCDs. The actual size of the array is approximately 70 mm on an edge. The signals are brought out from the two sides of each CCD on gold wires and then out of the vacuum system in which the CCDs are kept cold with liquid nitrogen. (SOURCE: NOAO.)

as efficient as the eye in detecting incoming photons, and the diameter of a large reflector is 1000 times the diameter of the pupil of the eye. An additional factor of 100,000 in sensitivity comes from the fact that the brain "reads out" the eye's image every $1/30$ s (allowing us to see continuous motion), whereas a CCD can collect data on a single target for 1 h or more. Astronomers mount CCDs in sealed cameras cooled with liquid nitrogen, which reduces the background signal due to thermal noise. CCDs are nearly perfect detectors, converting almost every incoming photon to an electric signal. For this reason, the only way astronomers can see even fainter objects is to build larger telescopes and so collect more light.

Photographs are not obsolete; they are still the favored medium for wide-angle imaging. The heart of a CCD is a silicon wafer that converts light to electrons—it is difficult to make this device larger than a couple of centimeters across. By contrast, photographic emulsion can be made up to 1 m across. In this book, many of the images that cover more than 10 minutes of arc are photographic, and almost all the images that cover smaller bits of the sky in detail are electronic. Some of the photographs in this book were taken with ordinary cameras or amateur astronomer equipment; in the former case, we list information on the films and exposures so that you can try repeating the pleasures of backyard sky photography.

Image Processing

Extraordinary advances have occurred in the last decade in our ability to process images in order to gain maximum information from them. Most of this is due to the advent of cheap and powerful computers. The widespread use of computers has permeated every field of science. Astronomy makes extensive use of **digital information**. This profound change in our way of viewing the world is discussed in a Science Toolbox that follows. It is now easy to convert each

SCIENCE TOOLBOX

Digital Information

We take in information about the world all the time. Reading a book, looking at a picture, listening to a conversation or a piece of music—all these activities convey information. We have seen over the past few decades a revolution in technology that allows information to be transmitted around the world. Radios and telephones and televisions are ubiquitous appliances that send information from one place to another. This requires the conversion of light and sound to measurable amounts of information. In science, information that can be measured and counted is called *digital information*.

The most basic way to define information is in a two-way, or binary, sense: yes or no, on or off, 1 or 0, white or black, up or down. The fundamental unit of information is called a *binary digit,* or a *bit*. With many pieces of information, or bits, we can describe very complex aspects of the natural world. Let us see how it works with some simple examples.

Suppose you ask a friend to think of a number from 1 to 100. How many "yes or no" questions would you have to ask to determine the number? The smart way to play the game is to divide the number range in 2 each time. "Is it more than 50?" No. "Is it more than 25?" Yes. "Is it more than 37?" Yes. "Is it more than 43?" No. "Is it more than 40?" Yes. "Is it more than 41?" No. "Is it 41?" Yes! If you play this game by just guessing a number each time, it might take you close to 100 guesses (or you might get lucky). But you can *always* determine any number up to 100 with 7 guesses. Try it! If you think of it in terms of information, the number your friend knows—any number from 1 to 100—can be defined by 7 yes-no questions or 7 bits of information.

In mathematical terms, the information content depends on the number of binary choices or bits multiplied together. A single bit specifies 2 things, 2 bits specifies $2 \times 2 = 4$ things, 3 bits specifies $2 \times 2 \times 2 = 8$ things, and so on. In general, with N bits of information,

$$\text{Information content} = 2^N$$

In the example we just gave, $N = 7$ bits gives an information content of $2^7 = 128$, so 7 bits allows us to easily specify the numbers from 1 to 100. This way of describing information uses a counting system with a base of 2 rather than the familiar decimal system with a base of 10. But it is easy to convert any decimal number to a binary number. Here is how the counting goes:

Decimal number		Binary number	
	0		0
	1		1
	2		10
	3		11
	4		100
	5		101
	6		110
	7		111
	8		1000
	9		1001

And so on. It is just like decimal counting, only with 2 digits instead of 10. Each column is a higher power of the base—instead of ones (10^0) and tens (10^1) and hundreds (10^2) and thousands (10^3), we have ones (2^0) and twos (2^1) and fours (2^2) and eights (2^3) and so on. You can see that a bit in the binary system is like a significant digit in the decimal system (review Chapter 1). The number from our guessing game can be written in the binary system as 101001 ($32 + 8 + 1 = 41$).

Why is the binary system the basis of the information age? In 1947, three researchers at Bell Laboratories discovered how to switch an electric current on and off by using a tiny silicon sandwich called a *transistor*. A computer has no fingers, so it does not count in a base of 10 as we do. A computer contains many thousands of transistors; each one can switch a current on or off millions of times per second. The on-and-off status of the current is the 1- and-0 counting system of a computer. By doing binary operations at blinding speed, computers do most of the calculations of the modern world. Equally important is the fact that they can manipulate and transmit *any* information that can be converted to a binary form. Let us look at the everyday examples of words and pictures.

What is the information content of written language? There are 26 letters in the alphabet. With capitals and punctuation there are about 60 different items, so we can describe writing with 6 bits of information, since $2^6 = 64$. The average word has close to 5 letters or $6 \times 5 = 30$ bits of information. This book has about 350,000 words in it, so the total information content is $30 \times 350,000$, or about 10 million, bits. For historical reasons, computer switches are grouped in 8s, so that a computer "word" is 8 bits, called a *byte*. This book contains $10,000,000 / 8 = 1,300,000$ bytes or 1.3 megabytes. It could fit on a floppy disk (the term *floppy* is an anachronism from the earlier days of computers; today's storage media are rigid).

Now we can look at an example with astronomical relevance. What is the information content of an image? If each picture element, or pixel, can be black or white (on or off), then it contains 1 bit of information. As we increase the size of the array of pixels, the information content increases (see Figure 10–A). With a 2×2 array (4 bits), not much information can be conveyed. However, with a 10×10 array (100 bits), we could make a visual display of a letter or several numbers. And with a 15×15 array (225 bits), we could transmit slightly more complex information. As you can see from Figure 10–A, information with only 1 bit per pixel (black or white) is very crude. There are no gradations of intensity, and color cannot be represented. Figure 10–B shows an image with a range of 64 colors, so each pixel represents 8 bits of information ($2^8 = 64$). The array is 29×21 pixels, giving a total information content of $29 \times 21 \times 8 = 4872$ bits. You can see how the ability to

(continued)

SCIENCE TOOLBOX

Digital Information (continued)

convey complex information increases with the number of bits.

The CCDs used for imaging in astronomy do not measure colors directly. The intensity is measured in filtered light of a specific color, and the colors are combined afterward. A CCD can measure very fine gradations of intensity by counting the electrons created when photons hit each pixel. A typical CCD can count up to 250,000 electrons. From the equation above, we can see that this is 16 bits of information because $2^{16} = 262,144$. So each pixel contains 16 bits of information. (Not all these grades of intensity or bits are conveyed by the printing process of this book, but they were present in the original electronic images.) A CCD with an array of 8000 × 8000 pixels (like the device in Figure 10–19) therefore creates an image with 8000 × 8000 × 16 = 1 billion bits of information! Recall that a word has only 30 bits of information—a picture is worth a *lot* more than a 1000 words.

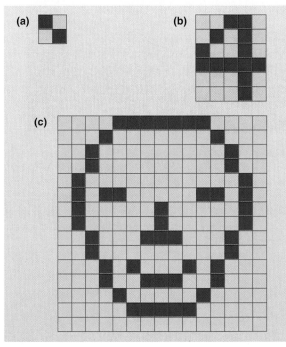

FIGURE 10-A
Digital information can be represented as the pixels of an image. In this case, each pixel can have two levels—black and white—so the information content of each pixel is 1 bit. The information content clearly increases in the progression from **(a)** 4 bits to **(b)** 30 bits to **(c)** 225 bits.

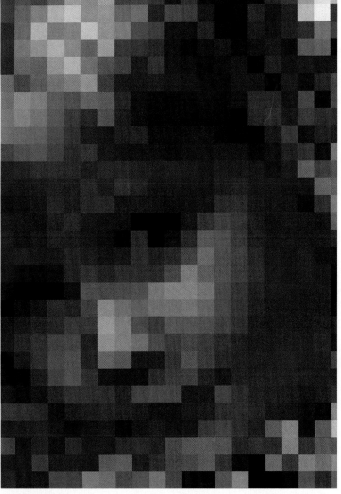

FIGURE 10-B
An image represented by a grid of 609 pixels, each of which has 64 possible colors or 8 bits of information. Increasing the information content allows subtler details to be conveyed. To see the young girl more clearly, hold the images at arm's length. (SOURCE: E. Hecht.)

picture element or pixel in an image to a number. Such a digital image may come not only from a CCD; an ordinary photograph can be scanned, measured, and converted to such pixels, which is called *digitizing the image.* The position of each pixel and its brightness are stored in a computer. Then the image can be processed in many ways. A technique that helps to emphasize subtle details of images is to assign different colors to each brightness level. In this way, we could produce an image in which the colors have nothing to do with the original colors but are used merely as a code to allow us to separate features of slightly different brightness. Such images are called *false-color images.*

False color can be misleading, because it does not necessarily contain any physical information. For this reason, "true color" images are used as much as possible in this book, although many texts and magazines are full of false-color imagery. The term *true color* is enclosed in quotation marks because at some level of detail it is difficult to reproduce all the nuances of natural color, especially in faint objects. But most images in this book do give some impression of the actual colors of the objects. A number of the CCD images in this book display true color. As with photography, this is done by combining two or three CCD exposures through different color filters into one final color image. Images made at wavelengths outside the visible spectrum are by definition false color, since the eye cannot detect these types of radiation! However, some attempt is made to adhere to the sense of the electromagnetic spectrum, so that longer infrared wavelengths are shown in red and shorter infrared wavelengths are shown in blue, for example. See Appendix A–7 for more information on the different ways that astronomers and other scientists represent data.

Radio Telescopes

A *radio telescope* is a device that gathers and concentrates radio waves; it is analogous to an optical telescope, which gathers and concentrates light waves. The U.S. engineer Karl Jansky built the first radio telescope at Bell Telephone Laboratories in 1930. Jansky's radio antenna was 30 m long, and it rotated on four wheels from a Model T Ford. He used it to detect radio "hiss" from the sky. In the process, he measured radio emissions from the Milky Way, toward the constellation Sagittarius. Grote Reber was an amateur U.S. astronomer who built a 10-m radio telescope in his backyard. For nearly a decade after Jansky's invention, Reber was the only radio astronomer in the world! As an optical astronomer does, he curved the metallic surface to better focus the radio waves. He observed almost every night from midnight to 6 A.M., then drove to his day job with a radio company. In 1940, Reber produced a beautiful map of radio waves coming from the Milky Way; this was the first map ever made beyond the optical spectrum.

Radio and optical telescopes work in basically the same way. Radio waves have much larger wavelengths than those of visible light waves, and radio waves carry much less energy per photon. For these reasons, radio telescopes must have larger surfaces to collect sufficient energy to give a strong signal. However, since the wavelength is larger, the

FIGURE 10–20
View of one of the radio telescopes in the Very Large Array in central New Mexico. This view shows the underside of the "dish" (top), or curved reflector, that collects the radio waves. This telescope (along with its neighbors) can be repositioned in various arrays along the railroad tracks, giving suitable effective resolution for different observing projects. (SOURCE: Photograph by WKH.)

surface need not be as accurately shaped as the mirror of an optical telescope. Radio telescopes have a curved surface (or dish) of metal or wire mesh, which reflects the radio waves to a focus, where they are converted to an electric signal. Figure 10–20 shows one of the 27 radio dishes of the Very Large Array (VLA) in central New Mexico. The largest telescope ever built is the 305-m Arecibo dish, built into a natural topographic depression in Puerto Rico. It is not steerable, but can still track objects across a narrow strip of the sky using Earth's rotation and movable "feed" antennas.

Telescopes in Space

Large portions of the electromagnetic spectrum are observable only from the cold vacuum of space. Even the visible light and the radio waves that do make it through the atmosphere can be masked by the terrestrial environment. Optical astronomy is hampered by light pollution from cities, and radio astronomy is hampered by interference from TVs, radios, and microwave ovens. Figure 10–14 showed schematically how far the different wavelengths of radiation penetrate Earth's atmosphere. Most infrared radiation is

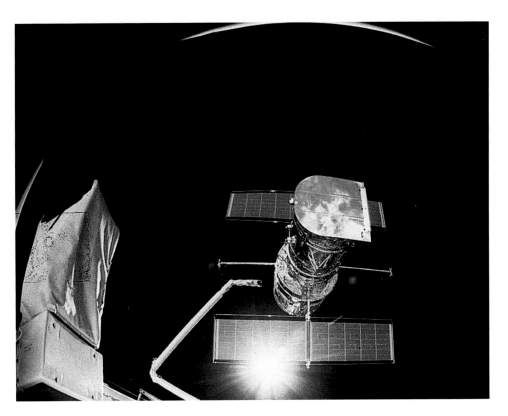

FIGURE 10-21
The Hubble Space Telescope, launched in 1990 by the Space Shuttle. The HST has a primary mirror 2.4 m across and a set of five instruments that can do imaging and spectroscopy at ultraviolet and visible wavelengths. The telescope is floating in the middle of the image with its solar panels above (behind front end) and below; details around the edges are part of the Space Shuttle's cargo bay. (SOURCE: NASA.)

absorbed a few kilometers above the ground by water vapor in the lower atmosphere. Ultraviolet radiation and all shorter wavelengths are extinguished higher up in the ozone layer. We must go into space to observe most of the infrared and microwave regions and all the ultraviolet radiation, X rays, and gamma rays.

After World War II, U.S. scientists used captured German V2 rockets to send instruments high into the atmosphere. Since then, a series of grander and more expensive satellites have pried open the short-wavelength sky and given us a view of a violent, high-energy universe. In the past three decades, scientists have made brief sorties into space with small telescopes mounted on balloons and rockets, but the best data come from satellites launched into low Earth orbit. Because **space astronomy** is expensive, many missions are joint ventures between two or more countries. Government agencies cooperate on the funding, and scientists from the countries collaborate on interpreting the data.

The dramatic gains possible from space astronomy are best illustrated in the mid-to-far infrared part of the electromagnetic spectrum (i.e., the longest infrared wavelengths). Working from the ground has two problems. First, water vapor in the atmosphere absorbs most infrared wavelengths so effectively that we cannot observe all the infrared spectrum, even from a high, mountaintop observatory like that on Mauna Kea (review Figure 10–14). Second, Earth and its atmosphere are at a temperature of about 300 K and emit thermal radiation in the infrared band (see Figure 10–8). All astronomical sources are therefore seen against a bright background of thermal infrared radiation. The background is *1 million times darker* in space. In infrared astronomy, the difference between working on the ground and working in space is like the difference in optical astronomy between observing in the day and at night!

Space astronomy offers three distinct advantages. It opens up spectral regions that cannot be observed from the ground. It offers an environment where the optical and infrared noise is very low. It allows telescopes to operate outside the blurring effects of Earth's atmosphere, where they can achieve their resolution limit. These advantages are set against the much greater cost and limited lifetime of an observatory in space. Many of the images in this book were taken with the premier astronomical satellite—the Hubble Space Telescope (HST); see Figure 10–21. The originally faulty optics were repaired by astronauts in 1993 in a daring series of space walks; since then, it has been taking spectacular pictures of planets, stars, nebulae, and galaxies.

Interferometry

In recent years, astronomers have applied a clever technique to improve telescope performance. This trick makes it appear as if we have a bigger telescope than we really do. The technique, called **interferometry,** uses widely separated telescopes in a special arrangement to increase the resolution of fine details in astronomical objects. In our earlier discussion of telescopes, we noted that the resolution of a telescope increases with the telescope's diameter (see also the following Science Toolbox). If we could build a good-quality optical telescope or radio telescope 1 mi wide, for example, we could achieve fantastic resolution (especially if the telescope were in space, where the full resolution could be realized without being ruined by atmospheric shimmering). We cannot build a 1-mi-wide telescope, but cunning astronomers have devised a partial solution to the problem.

SCIENCE TOOLBOX

Collecting Area and Resolution of Telescopes

Why do astronomers continue to build bigger and bigger telescopes? The electronic detectors in common use today are nearly perfect, so gains can only be made by collecting more light with a larger mirror. The number of photons gathered by a telescope is proportional to the collecting area. So we can compare the light-gathering power of two telescopes (A and B) by calculating the ratio of their collecting areas

$$\text{Gain in light gathering} \propto \frac{\pi R_A^2}{\pi R_B^2}$$

$$= \frac{D_A^2}{D_B^2}$$

The proportionality holds whether we use the radius R of the mirror or the diameter D, but astronomers generally quote the entire aperture or diameter. So we see that a 4-m telescope has $(4/1.5)^2 = 7$ times the collecting area of a 1.5 m telescope. Since light from an object in space diminishes with the inverse square of the distance, there is another way of looking at this comparison. If a 1.5-m telescope can collect a certain number of photons per second, say, 500, from a star, then a 4-m telescope can collect $500(4/1.5)^2 = 3560$ photons per second from that same star. The 4-m telescope would collect 500 photons per second from that star when it was $\sqrt{3560/500} = 2.7$ times farther away. This is, of course, the ratio of the apertures. A larger telescope allows an object to be detected out to a larger distance in space. So we get the important result that

$$\text{Gain in limiting distance} \propto \frac{D_A}{D_B}$$

For another example, the Hubble Space Telescope has a mirror 2.2 m across—this was the largest structure that could fit in the cargo bay of the Space Shuttle. Each Keck telescope on Mauna Kea has an aperture of 10 m, so each Keck telescope has $(10/2.2)^2 = 20.7$ times the collecting area of the Hubble Space Telescope.

However, light-collecting area is not the entire story. The other important attribute of a telescope is resolution, or the ability to discriminate light arriving at the telescope from slightly different angles on the sky. We might want to distinguish two stars with a small angular separation. Or we might want to discern details on the surface of a planet. In either case, good resolution is required. The fundamental limit to a telescope's ability to resolve small angles is due to diffraction—the slight bending of light as it passes through a telescope. As we have seen in the text, the angular resolution is

Angular resolution =

$$2.5 \times 10^5 \times \frac{\text{wavelength}}{\text{diameter}}$$

In this equation the angular resolution is in seconds of arc, and the wavelength of radiation and diameter of the telescope must be in the same units. Let us begin with the eye as an example. The pupil has a diameter of about 2 mm, or 0.002 m. Visible light has a mean wavelength of about 5×10^{-7} m, so the resolution of the eye is $(2.5 \times 10^5)(5 \times 10^{-7} / 0.002) = 63$ arc seconds, or about 1 arc minute. This accounts for the limited accuracy of the measurement of star and planet positions before the invention of the telescope.

Now consider a modest-size telescope that a hobbyist might use. A telescope of aperture 0.2 m (about 8 in.) gives a resolution of $(2.5 \times 10^5)(5 \times 10^{-7} / 0.2) = 0.6$ arc second. This is excellent—the angular diameter of a dime seen at a distance of 3 mi! Larger telescopes than this gain in light-gathering power but have effectively no gain in resolution because turbulence in the atmosphere jumbles all images to an angular size of about 1 arc second. To do better, we must go into space. The Hubble Space Telescope has an aperture of 2.2 m and an angular resolution of $(2.5 \times 10^5)(5 \times 10^{-7} / 2.2) = 0.05$ arc second, which is about 10 times better than can be done from the ground. Equivalently, the Hubble Space Telescope can distinguish a pair of objects that is 10 times farther away than the resolution limit of a ground-based telescope.

You can see that the angular resolution varies inversely with the telescope diameter—the larger the telescope, the smaller (or better) the resolution. You can also see that the angular diameter varies in proportion to the wavelength of radiation—images made with short waves have better resolution than images made with long waves. The same telescope will make sharper images in blue light than in red light. Interestingly, this principle also controls the miniaturization of electronics. Integrated circuits are etched onto silicon wafers by using chemicals and light, and the semiconductor industry strives to use shorter or bluer wavelengths for this process so that more components can be packed onto a single wafer. Also, information is stored and read from compact disks by using lasers, and engineers devote much effort to developing blue or ultraviolet lasers that can pack more data or music or video onto a compact disc.

What about other parts of the electromagnetic spectrum? No waves beyond the blue end of the visible spectrum—shorter than 3×10^{-7} m—can penetrate Earth's atmosphere. So all ultraviolet, X-ray, and gamma ray astronomy must be done from space. In fact, these very short waves are absorbed by glass, so other techniques must be used to focus the waves and the simple equation for angular resolution given above does not apply.

At the other end of the electromagnetic spectrum radio waves are gathered and focused in a similar way to light waves. A large radio telescope has a diameter of 100 m, so if it imaged radio waves with a wavelength of 1 cm, or 0.01 m, the resolution would be $(2.5 \times 10^5)(0.01 / 100) = 25$ arc seconds. So the much longer waves overcome the benefit of a larger aperture, and radio telescopes give

(continued)

SCIENCE TOOLBOX

Collecting Area and Resolution of Telescopes
(continued)

a blurry view of the universe. As described in the text, radio astronomers have invented a clever trick to bypass this limitation. Interferometry allows radio waves from different telescopes to be *combined as if they were coming from a single extremely large telescope.* (Of course, the collecting area does not increase by such a large factor; it is the sum of the areas of the individual telescopes.) The effective diameter can be the size of Earth (or even larger, using radio telescopes in orbit). Using a diameter of 10,000 km in the example above, we get a resolution of $(2.5 \times 10^5)(0.01/10^7) = 0.00025$ arc second. This corresponds to the angular size of the head of a pin at a distance of 100 mi! Radio astronomers have perfected an amazing tool for viewing the radio sky in fine detail.

Engineering obstacles and high construction costs make it impossible to build a 1-mi-wide telescope. However, it is much easier to build two big telescopes, locate them 1 mi apart, and have them work together to observe the same patch of sky. The trick is that their light (or radio waves in the case of a radio telescope) must be fed into a single detector, just as all the light from a single telescope goes into one detector. The two telescopes must act as one, even though they are separated. This blending is achieved by converting radio waves into electric signals and sending them from separated telescopes to one central facility, where they are processed together. You can see why optical interferometry is harder than radio interferometry by recalling the wave nature of light. To combine waves of light from separate telescopes, we must line up the waves to within a faction of a wavelength; otherwise the waves might cancel as they do with interference. This requires us to know the distance between two telescopes to better than one-millionth of a meter! With longer radio waves, the requirement is a lot less severe.

Using the technique of interferometry, astronomers have been able to measure incredibly small angular details—2 or 3 ten-thousandths of a second of arc—in distant stars and galaxies. The technique has been especially fruitful for radio telescopes, which have made beautiful images of the details of distant galaxies. One of the most advanced such observatories is the VLA (Very Large Array) of the National Radio Astronomy Observatory, with 27 movable radio dishes (see Figure 10–22) spread over an area up to 36 km across. Astronomers in 1986 achieved the first space-based interferometry, using a radio telescope in space linked to one on the ground, thus increasing the effective size of the telescope to a distance wider than the diameter of Earth! This technique holds much promise for the future.

FIGURE 10-22
The 27 antennas of the Very Large Array shown in their most compact configuration. Each antenna is 25 m in diameter, and the 27 antennas are arranged in a Y shape that can cover an area 36 km across. The radio interferometer is located in New Mexico, in a remote region ringed by mountains so as to be free from radio interference. (SOURCE: NRAO.)

FIGURE 10–23
Artist's impression of a lunar observatory. The environment of the Moon offers substantial advantages for astronomy at many wavelengths. A lunar observatory could be operated in robotic mode, with minimal human intervention. (SOURCE: NASA.)

New Frontiers

These are exciting times for astronomy. Large new telescopes are being built, and NASA is planning to complete the launch of a series of space telescopes that will cover the electromagnetic spectrum from the far infrared to gamma rays. The narrow visual window has been opened wide, giving access to electromagnetic information greater than a factor of 10^{15} in wavelength. Astronomers are often attracted to wilderness frontiers, far from mainstream civilization. Visionaries are already thinking of the next step forward. Observatories on the Moon offer great advantages for most types of observation (Figure 10–23). Optical and infrared telescopes on the Moon could collect data in conditions of excellent darkness, with low backgrounds and no atmosphere to distort the incoming images. Radio telescopes could work free from human-made interference on the lunar far side. Plans for robotic lunar observatories have been made, with possible deployment early in the 21st century.

Using these technological advances, astronomers are hoping to detect not only the full spectrum of electromagnetic waves, but also nature's messages from space that reach Earth in other forms. For example, *cosmic rays* are high-energy charged particles produced in distant parts of the galaxy. Traveling at very close to the speed of light, they have energies up to 1 million times larger than can be produced in the largest particle accelerators. As they reach the upper atmosphere, they hit atoms of our air and create showers of secondary particles. Optical radiation from these cascades can be collected by optical "concentrators" on high, dry mountaintops, thus allowing astronomers to monitor cosmic ray activity.

We have also begun trying to detect a tiny, ghostly particle called the *neutrino*. These tiny particles are produced in the cores of stars, in active galaxies, and as a relic from the early universe. Vast tanks of fluid are used as detectors, hoping to catch the light flash from the very occasional neutrino interaction with atomic particles in the fluid. The detectors have been placed deep under the sea, under mountains, and in mine shafts to shield them from signals due to other particle events. Neutrino astronomy began with an observation of neutrinos from the Sun (see Chapter 11), and it received an enormous boost with the detection of a burst of neutrinos from a distant supernova in the Large Magellanic Cloud in 1987 (see discussion in Chapter 13).

Perhaps the most exotic signal from space is a *gravity wave*. General relativity, Einstein's theory of gravity, predicts that rapid motions and changes in the state of matter should lead to the emission of gravity waves. Gravity waves are like ripples in the structure of space. They travel at the speed of light. The U.S. Congress has funded an ambitious project to detect gravity waves; they are important because they open a new window on the universe that is completely independent of the types of radiation we have discussed in this chapter. Gravity waves are created by the death of stars, by the interactions of black holes, and by the behavior of the entire universe when it was small and dense billions of years ago. The Laser Interferometer Gravitational Observatory (LIGO) will try to detect gravity waves by measuring the vibration or "ringing" of a large mass as the gravity wave passes through. The challenge is to rule out the inevitable vibrations associated with restless Earth and to amplify the tiny astronomical signal so that it can be detected. LIGO is being built in the California desert in a geologically quiet area. The universe is alive with the gravitational signals of astronomical catastrophies, from supernovae to collisions of neutron stars. Detecting these signals is one of the most exciting frontiers in astronomy.

Summary

We have seen that studying the spectrum of radiation—work begun by Newton and others centuries ago—can vastly expand our knowledge of the universe. Astronomy is undergoing a transformation comparable in scope to the Copernican revolution. For nearly 2500 years, we have had only one window on the universe: the light gathered by our eyes and then by our telescopes. Recently, our horizons have broadened vastly. We have glimpsed a universe full of exotic phenomena: radio waves from galaxies halfway across the universe, X rays from binary stars locked in a tight embrace, gamma rays from sources of unknown origin. The universe we see with our eyes is only a small part of the story.

Astronomers learn about the universe by deciphering messages carried by the radiation from extraterrestrial bodies. Electromagnetic radiation travels across the vacuum of space at 300,000 km/s. The light we see with our eyes is just one example of this radiation. Light spread out in order of wavelength is a visible spectrum of the object. However, there are also wavelengths far too short (too blue) to see, and wavelengths far too long (too red) to see. The entire electromagnetic spectrum covers a tremendous wavelength range, from gamma rays the size of an atomic nucleus to radio waves hundreds of meters long.

We can describe electromagnetic radiation as a wave or a particle, depending on the kind of interaction it has with matter. Waves have a wavelength, or typical distance between peaks. Their frequency is a measure of the number of peaks that pass per second. Wavelength and frequency are inversely related. We can equally well think of the radiation as a particle, or a photon, whose energy is proportional to its frequency. Atoms absorb and release energy in the form of photons—these discrete units of matter and energy mean that the microscopic world is "grainy." Most of what we know about the universe depends on the ways that matter and radiation interact with each other.

Objects in space reveal their nature by the radiation they emit. The smooth thermal spectrum has a peak wavelength that indicates the temperature. In addition to a smooth spectrum, any hot gas has sharp spectral features. These sharp lines appear because the electrons around an atomic nucleus can only inhabit certain fixed energy levels. When electrons change their energy level, they emit a spectral line of a specific wavelength. Even though scientists can make very accurate measurements, the microscopic world is fundamentally "fuzzy." There is a limit to the precision with which we can measure microscopic quantities, a concept expressed in the Heisenberg uncertainty principle. This means that subatomic particles cannot have their positions and motions measured with certainty—a situation that requires us to recognize that both our data and our knowledge of the physical universe may be limited.

Atoms and molecules each produce emission lines and absorption lines and bands in the spectrum as photons interact with them. These derive from the energy levels of their electronic orbits. Each element and compound has its own lines and bands. The spectral lines act as a fingerprint that helps us figure out what a hot object is made of. The technique works equally well for a discharge tube in the laboratory as for hot gas around a quasar billions of light-years away.

Telescopes are devices that enable us to improve on the light-gathering power of the eye and to resolve finer details of an astronomical target. Thus we can see objects much fainter than those visible to the unaided eye. We are currently witnessing a surge in the construction of large telescopes on mountaintop sites around the world. Since optical detectors are almost perfectly efficient, astronomers need a larger collecting area to see more deeply into the universe. The technique of interferometry gives astronomers far better resolution than can be achieved with a single telescope. Other telescopes have been placed in orbit to give ultra-sharp optical images and to detect long and short wavelengths that cannot penetrate Earth's atmosphere. Perhaps the most exciting revolution in our knowledge is the peeling back of the electromagnetic spectrum, revealing for the first time the invisible universe.

Important Concepts

You should be able to define these concepts and use them in a sentence.

electromagnetic radiation
electromagnetic spectrum
wavelength
frequency
intensity
speed of light
photons
quantum theory of radiation
energy levels
Heisenberg uncertainty principle
emission lines
absorption lines
telescope
collecting area
resolution
photometry
spectroscopy
CCD detector
digital information
space astronomy
interferometry

How Do We Know?

These questions and answers show how the scientific method is used to learn about the universe.

Q How do we know whether light is a wave or a particle?

A Newton was the first to study light in detail; he actually believed that it behaved as a stream of particles. In the 19th century, simple optics experiments showed that light deflected when it shone past a sharp edge (diffraction) and that two light sources would produce alternate bright and dark bands on a screen (interference). Both are typical behaviors of a wave. Light is an extremely short wave, about 0.5 millionth of a meter long. On the other hand, light carries energy from one place to another as a particle, and light can travel through the vacuum of space. There are some situations in which light displays wavelike properties and other situations in which it displays particlelike properties. Both descriptions are valid.

Q How do we know that radio signals are electromagnetic waves?

A The first electromagnetic waves to be studied experimentally had long wavelengths. The German physicist Heinrich Hertz (whose name is used for the unit of frequency) was able to produce radio waves from an oscillating electric field. Earlier, Faraday had shown that varying electric and magnetic fields were intimately related. Subsequent experiments have shown that there is no fundamental difference between radio waves and visible light. Both types of radiation are composed of linked, varying electric and magnetic fields. Both types of radiation travel at 300,000 km/s in a vacuum. The only difference is that visible light has a wavelength millions of times shorter than that of radio waves.

Q How do we know the speed of light?

A Light travels so fast that in the everyday world its speed seems infinite. Galileo tried to measure the speed by positioning a friend on a distant hilltop at night and signaling to him with a lantern. When the friend saw the lantern flash, he signaled back, and Galileo tried to measure the round-trip travel time of the light. It was so short he could not measure it. Today, fast electronics allows us to accurately measure the distance that light travels in a short time. For example, light reflecting off a mirror will travel a 1-m round trip in only 3×10^{-9} s [or 3 nanoseconds (ns)]. Fast electronics are used to both emit and detect the pulse of light. With a time measurement accuracy of less than one-trillionth of a second (also called a *picosecond*, or 10^{-12} s), these experiments can measure the speed of light with an accuracy of a fraction of a percent.

Q How do we detect invisible types of radiation?

A Over many millions of years of evolution, many animals have evolved eyes that can detect visible radiation. However, the eye can only detect electromagnetic waves of a factor of 2 in wavelength, which is a tiny fraction of the factor of billions between radio waves and gamma rays. In the past century, we have developed the technology to detect electromagnetic radiation beyond the visible spectrum. In most cases, the waves are focused on a solid material such as silicon where the energy they deposit is converted to an electric signal. The electric signal is then converted to digital information that can be displayed on a computer. We can then make an image of the digital information so that we can effectively "see" the invisible radiation.

Q How do we know what remote objects are made of?

A Our knowledge for the most part depends on the reach of the scientific method. The Moon is the only astronomical object from which we have retrieved physical samples (in another lucky example, we have a few chunks of Mars which were blown off the surface by impacts, traveled through space, and fell to Earth). In addition, space probes have sampled the atmospheres of several planets and a comet. In every other case, we use indirect information to measure compositions. The spectral features of each element and molecule make a unique fingerprint. By studying the pattern of spectral features of a remote object, we can deduce the chemical composition. This method has shown us what the atmospheres and surfaces of many of the planets are made of and has shown us that stars are composed of mostly hydrogen and helium, as is the Sun. It has also revealed that the elements found billions of light-years away in the universe are the same as those found on Earth.

Problems

You should use these problems to test your understanding of the information and concepts in this chapter. The * indicates a more advanced or mathematical problem.

1. Rank the following in order of increasing wavelength **(a)** the output of an ultraviolet tanning lamp, **(b)** a radio broadcast of Saturday's big football game, **(c)** leakage from a very cheap microwave oven, **(d)** a beam of red light, and **(e)** the X rays in your dentist's office.

2. **(a)** If the velocity of light is 300,000 km/s, what is the wavelength of the photon that you detect when you listen to a station at 1200 MHz (1.2×10^8 cycles per second) on your radio dial? **(b)** What is the frequency of an X-ray photon with a wavelength of 1 nm (10^{-12} km)?

3. Give one example of a property of light that is best understood if light is a particle. Give another example that is best understood if light is a wave.

4. **(a)** Why can an atom in the ground state not produce emission lines? **(b)** Use this fact to explain why the colorful emission-line glows of certain nebulae occur near hot stars, but not in the cold gas of interstellar space.

*5. Use the equation for the energy of a photon to calculate the number of photons per second emitted by **(a)** the arc lights at night at a football stadium, a power of 1 megawatt (MW), **(b)** a night light with a power of 2 W, and **(c)** the glow of a firefly at 0.01 W. One watt is equal to an energy rate of one joule per second.

6. Suppose that we lived in a crazy world where quantum effects were large and Planck's constant was equal to 1 J · s. Redo the calculation of the velocity uncertainty for a fastball traveling at 90 mi/h (from the Science Toolbox). How would this level of uncertainty affect the game of baseball? Given the same equation, if you knew the speed of your car accurate to 10 m/s, how uncertain would you be about its position?

7. A spacecraft passes close to a hitherto-unknown planet. A camera system on the spacecraft photographs surface details quite clearly in all wavelengths except the absorption bands of carbon dioxide. At these wavelengths, the image is featureless. Interpret these results.

8. Why has there been more emphasis on getting ultraviolet and infrared telescopes into orbit than on getting radio telescopes into orbit? Optical radiation reaches Earth's surface, so why have astronomers spent so much money building optical telescopes in space? Why is it just as easy to do infrared astronomy in the day as at night? Why was X-ray astronomy not possible until a few decades ago?

*9. The United States has a radio interferometer consisting of two 10-m-diameter telescopes separated by 500 m. Japan has a radio interferometer consisting of five 4-m-diameter telescopes separated by 800 m. France has a single 16-m-diameter radio telescope. Where would you go if you had a very faint target to observe? Where would you go if you needed the highest resolution?

*10. If you had a backyard telescope that could resolve 1 second of arc, how small a mountain or crater could you see on the Moon? How small a Sunspot could you resolve? How many resolution elements or pixels would you have if you made a map of the Moon? Now suppose you had access to a radio interferometer that could resolve 0.1 millisecond of arc (10^{-4} second of arc). How many resolution elements or pixels would you have within the area that your backyard telescope could resolve?

11. The opening picture of this chapter shows the largest telescope in the world, a radio dish in Puerto Rico with an aperture of 305 m. If it makes an image using radio waves of 21-cm wavelength, what is the angular resolution? Is this better or worse that the resolution of your eye using visible light?

Projects

Activities to carry out either individually or in groups.

1. Compare views through binoculars of different sizes. (Binoculars carry a designation such as 7 × 35, where the first number is the magnifying power and the second is the diameter of the lens that lets in the light.) Confirm that at fixed power, larger lens diameters give brighter images.
2. Observe and sketch a rainbow. Confirm that the colors appear in order of wavelength as seen in Figure 10–1. Use a prism to disperse a beam of white light. By seeing how the light deflects as it passes through the prism, try to figure out which wavelengths of light travel the most slowly in glass. You can demonstrate that light carries energy with a simple experiment. (This piece of equipment can be bought in a hobby store, but it is more fun to make it yourself!) Invert a thimble on a long nail or needle that is standing upright. Attach to the top of the thimble four vanes that are black on one side and white on the other, making sure that all the white surfaces face the same way. Use light materials such as toothpicks and thin pieces of card, and make sure you balance the vanes carefully so that the thimble can spin without wobbling. Shine a strong flashlight toward the white side of the vanes. Now explain why the thimble rotates. Remember that a white surface reflects photons of light while a black surface absorbs them.
3. Study a room light equipped with a dimmer switch. Turn the switch from bright to the faintest visible setting. Note the redder color of the lightbulb at the dimmest position. Relate this effect to Wien's law.
4. An optical telescope needs a very smooth surface to form an image. Take a roll of aluminum foil, and tear off a large square. Spread it out flat and look at your reflection in it. Notice that the quality of the image is close to that of a mirror. Now scrunch the foil up into a ball without tearing it, and carefully spread it out again on a flat surface. (*Warning:* Patience is required!) Notice the inferior image of yourself, and explain what is happening the to light as it bounces off the foil. Explain why this foil would still be an excellent reflector of longer-wavelength radio waves. Now try to gather light with the foil. Go to your kitchen and line the insides of bowls of different shapes with the foil; then see which shapes focus Sunlight the best. (*Warning: Do not put your eye anywhere near the devices you make—even very poorly focused Sunlight can damage your eyes!*) Think about how light reflects off a surface, and explain why the inside of a reflective sphere is not a good shape for making a telescope.

Reading List

Bowyer, S. 1994. "Extreme Ultraviolet Astronomy." *Scientific American,* vol. 271, p. 32.

Bunge, R. 1993. "Big Scopes: Dawn of a New Era." *Astronomy,* vol. 21, p. 48.

Chaisson, E. 1994. *The Hubble Wars.* New York: HarperCollins.

Davies, P. C. W., and Brown, J. R. 1986. *Ghost in the Atom: A Discussion of the Mysteries of Quantum Physics.* Cambridge: Cambridge University Press.

Englart, B. G., 1994. "The Duality in Matter and Light." *Scientific American,* vol. 271, p. 86.

Fugate, R., and Wild, W. 1994. "Untwinkling the Stars." *Sky and Telescope,* May, p. 24, and June, p. 20.

Gehrels, N. et al. 1993. "The Compton Gamma-Ray Observatory." *Scientific American,* vol. 270, p. 68.

Kellerman, K. I. 1997. "Radio Astronomy in the 21st Century." *Sky and Telescope,* February, p. 26.

Krisciunas, K. 1988. *Astronomical Centers of the World.* Cambridge: Cambridge University Press.

McLean, I. 1995. "Infrared Arrays: The Next Generation." *Sky and Telescope,* vol. 89, p. 19.

Robertson, L. J. 1992. "Spinning a Giant Success." *Sky and Telescope,* vol. 84, p. 26.

Tucker, W., and Tucker, K. 1986. *Cosmic Inquirers: Modern Telescopes and Their Makers.* Cambridge, MA: Harvard University Press.

CHAPTER 11

Our Sun: The Nearest Star

CONTENTS

- Kelvin and the Age of the Sun
- Understanding the Nearest Star
- The Sun's Properties
- Energy from the Atomic Nucleus
- Solar Energy from Nuclear Reactions
- The Sun's Interior
- The Sun's Atmosphere
- Effects of the Sun on Earth
- Summary
- Important Concepts
- How Do We Know?
- Problems
- Projects
- Reading List

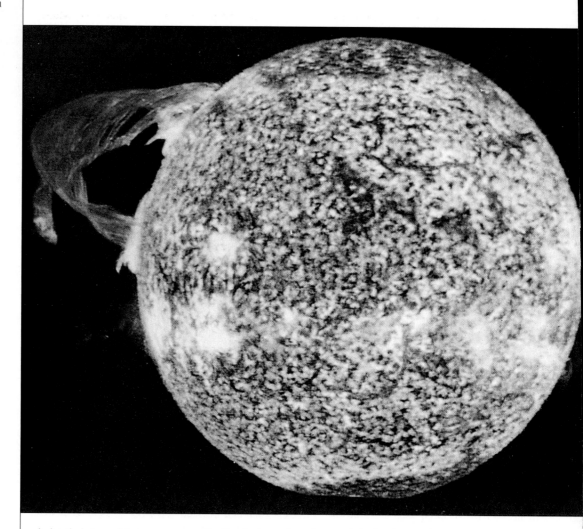

A detailed view of the nearest star. The Sun is shown here in a spacecraft image made in the light of ionized helium at 30 nm. The surface is seething with activity, and the loop prominence to the left is many times the size of Earth. (SOURCE: NASA Skylab.)

WHAT TO WATCH FOR IN CHAPTER 11

The Sun is the nearest star and the engine for all life on Earth. This chapter discusses the properties of the Sun. We start by summarizing the laws of radiation that lead to the different types of spectra of a hot gas—continuous, emission line, and absorption line. You should understand how these laws of radiation help us interpret the Sun's spectrum and learn properties of the Sun. The central scientific question in this chapter is the search to determine the origin of the Sun's enormous energy output. You will discover that energy source lies deep within the Sun, where hydrogen nuclei (or protons) undergo fusion and release nuclear energy. With knowledge of the cause of the Sun's energy output, it is possible to calculate the physical conditions throughout the sphere of hot gas. For example, the core where fusion occurs must be at a prodigious temperature of more than 10 million K, while the much cooler surface is at 5700 K. You will learn that the Sun is not a smooth and featureless ball of hot gas. The Sun's surface is marked by dark spots and arcs of gas. The gas below the surface is in constant, seething motion. This apparently chaotic activity on the surface actually follows a cycle of 22 years, and variations in the Sun have profound effects on Earth. Our discoveries about the physical properties of the Sun have helped us understand Earth's climate.

PREREADING QUESTIONS ON THE THEMES OF THE BOOK

OUR ROLE IN THE UNIVERSE
How does the Sun affect Earth's climate?

HOW THE UNIVERSE WORKS
What is the energy source that powers the Sun?

HOW WE ACQUIRE KNOWLEDGE
How do we know what goes on deep within the Sun?

KELVIN AND THE AGE OF THE SUN

Lord Kelvin was not amused. Having made important contributions to the study of energy and heat, he was one of the most famous scientists of his day. He was a towering intellect, and he held strong views on the age of the Sun and of Earth. His displeasure was aroused at a meeting of the Geological Society of London in 1871, as he debated the age of Earth with Thomas Huxley.

Huxley was standing in for Charles Darwin, a shy young man who had proposed a radical new theory of evolution. Species changed in response to the pressures of the environment, Darwin said. He believed that the process of natural selection caused species to pass on survival traits to their offspring and so continually to adapt to the world. Darwin suspected that the gradual evolution of life on Earth from single-celled organisms to the present array of elephants and eagles and whales must have taken a very long time. Huxley presented Darwin's hypothesis that life had been evolving on Earth for billions of years.

Kelvin strongly opposed the radical ideas of Darwin. The great physicist had a different way to estimate the age of Earth and of the Sun. He focused on the ideas of thermodynamics, the science of the measurement and transfer of heat. The Sun is the incubator that keeps all life on Earth alive. However old Earth is, the Sun must be at least that old. Kelvin knew that the answer depended on the energy source of the Sun. Earlier physicists had speculated that the Sun was burning fossil fuels, as a gigantic furnace would. However, it would have lasted only a few thousand years with a chemical energy source. Following a suggestion of Hermann von Helmholtz, Kelvin studied whether the Sun may in fact be fueled by an alternative energy source: gravitational contraction. If the material of the Sun were slowly settling toward its center, he reasoned, gravitational energy would be converted to heat and then radiated into space. Using these ideas, Kelvin calculated a maximum age for the Sun of 500 million years.

The Kelvin-Huxley debate ended inconclusively. Kelvin's age estimate was still less than the span of time believed to be needed for the diversity of species to be produced by natural selection. Darwin was bitterly disappointed. He died not knowing if the age of Earth would accommodate his theory. This halting progress—the clash of ideas and the difficulty of choosing between hypotheses—is typical of how science works.

Kelvin was a brilliant man. He had entered the University of Glasgow at age

10, published his first research paper at age 16, and was a full professor by the age of 22. The temperature scale used by scientists is named after him. However, this time he was wrong. It turned out that Kelvin was missing something vital: an understanding of the processes of nuclear energy. It would be another 40 years before Ernest Rutherford discovered the structure of the atom, and 40 years more before scientists would harness energy by splitting the atomic nucleus. The energy produced by the atomic nucleus is vastly more efficient than the chemical energy that heats our homes and powers our cars. It is the basis for weapons of massive destruction, but it is also the basis for all life on Earth. Our star the Sun is powered by the fusion of hydrogen nuclei. The Sun keeps shining as long as its fuel reserves are not depleted—about 10 billion years, and long enough to encompass Darwin's estimate of the time scale of biological evolution. In this chapter, we will answer an important question: How does our knowledge of nuclear fusion help us understand the properties, light, and life span of the Sun?

Understanding the Nearest Star

Much of modern astronomy deals with stars—how they generate their energy, what kinds of light they radiate, how they form, and how they evolve. How can we study such remote objects as the stars? We can study the closest one, which is nearer to Earth than five of the planets. Humans pondered the stars for many centuries before they realized that the Sun is just another star, and the stars are all suns. Light from its surface reaches us in only 8 min. Our eyes are dazzled by it. Earth is bathed in its flow of radiation, washed by the winds of its outer atmosphere, blasted by seething swarms of atoms blown out of it, bombarded by bursts of X rays and radio waves emitted by it. It is our Sun, a ball of hydrogen and helium whose diameter is roughly 100 times Earth's size.

In this chapter, we will study the Sun, our nearest star. Despite being separated by an almost perfect vacuum, the Sun and Earth are intimately linked because almost all life on Earth depends on the Sun's energy. The source of that energy lies deep within the Sun, far below the surface we see. Energy is produced from reactions among the nuclei of atoms in the Sun's core.

Although the Sun is 150 million km from Earth, it is 300,000 times closer than the next-nearest star. This chapter describes how we have learned to see it in great detail. We now understand the physical conditions in this great ball of gas and the ways in which energy travels from the core to the surface. We will see that the Sun and Earth are linked in more subtle ways and that small changes in the Sun produce profound changes in Earth's climate. In human terms, the Sun has a virtually inexhaustible supply of energy. This has implications for the management of our own energy resources.

Two of the major themes of this book are important in our discussion of the Sun. The search for answers about the Sun illustrates how science works to understand remote objects. Nobody has traveled to the Sun. No space probe has ever brought back a gas sample. No optical telescope can see past the surface layer. So how do we know so much? We use the technique of spectroscopy to identify the Sun's chemical composition. We also use the Sun's electromagnetic radiation beyond the visible spectrum to make images of the tenuous outer atmosphere. We even use tiny high-energy particles to "see" deeply into the Sun's core. Thus, astronomers can probe the interior structure with a wide range of observations and can make physical and mathematical models of many kinds to describe the data.

The Sun also exemplifies our theme of applying physical laws to understand the universe. Atomic nuclei fuse to make solar energy, a process that we have observed and tested in making hydrogen bombs. The processes of radiation and convection carry the energy away as heat, which is then radiated into space. We can use the inductive method of science to apply this knowledge to suns much more distant than our own. Turn back to our first view of the distant universe, Figure 1–2. Fusion powers all the billions of stars in each galaxy in that image.

This chapter begins with a description of the Sun's properties. The heart, or central core, of the Sun is the ultimate source of its energy—and ours. We then discuss how energy can be released from an atomic nucleus and how the Sun's core exploits and regulates the fusion process. Next, we consider the Sun's structure, from its core to its outer atmosphere. Finally, we look at the various ways that the Sun interacts with Earth.

The Sun's Properties

We have not yet sent space vehicles to probe the Sun's atmosphere, as we have begun to do for the planets. To study the Sun or stars, for the present we must rely on indirect evidence: interpreting their light gathered by telescopes. Imagine you had to deduce as much as you could about a person from the contents of his or her house or apartment. That person's clothes, books, and music and the food in her or his kitchen would give you valuable clues. You could learn a lot without ever meeting the person. Deduction from indirect evidence is one of the standard methods of astronomy. Although our evidence is indirect, powerful techniques of analysis help us create models that fill in gaps in our observations.

Rotation

In ancient times, astronomers in China and India noticed and recorded dark spots on the Sun, called **sunspots**. In the 1600s, when the first telescopes were pointed at the Sun, astronomers were able to closely track the sunspots. Galileo saw these blemishes as evidence that the Sun was not a smooth and perfect sphere. He used this evidence to argue against the ideas of Aristotle, who had thought that the celestial objects were perfect and unchanging. Although it seems like a simple insight, Galileo's observation marked a decisive break with the ancient Greek conception of the universe. Unfortunately, direct observation of the Sun is extremely dangerous. Galileo spent his last years almost totally blind from his years of observation of the Sun.

What do sunspots reveal about the Sun? For nearly 400 years, astronomers have counted sunspots to measure solar activity and have tracked them to measure solar rotation. Today we know that sunspots are magnetic disturbances on the Sun's surface. The Sun is close enough that we can watch storms develop on its surface. We know that the Sun rotates at its equator once every 25.4 days relative to the stars. It rotates every 27.3 days relative to Earth, since Earth's orbital motion is in the same direction as the solar rotation and must be added in. Like Jupiter, the Sun rotates differentially. Its equatorial region rotates faster (25 days) than the polar regions (33 days)—proof that the Sun has a gaseous, not solid, surface.

Studying the Sun's Spectrum

Ever since Newton showed that sunlight is a mixture of all colors (see Chapter 4), scientists have studied the Sun's spectrum. In 1817, German physicist Joseph Fraunhofer found that certain wavelengths were missing from the Sun's spectrum; the spectrum appeared to be crossed by narrow, dark absorption lines (see Figure 11–1). What were these lines? As we learned in the last chapter, when a gaseous element is heated, it emits radiation of certain wavelengths and no others. These very narrow wavelength intervals, unique to each element and as unmistakable as a set of fingerprints, are called *emission lines*. Researchers soon found that the emission lines of hydrogen exactly matched the position of some of Fraunhofer's solar absorption lines. The two sets of lines were caused by the same element!

In the 1850s, German physicist Gustav Kirchhoff discovered in the laboratory the conditions that produce the three different kinds of spectra described in Chapter 10: the continuum, absorption lines, and emission lines (illustrated in Figures 10–1, 10–12b, and 10–13b, respectively). When Kirchhoff looked through his spectroscope toward a sodium flame against a dark background, he saw a strong yellow *emission* line. But when he changed the background to a brilliant beam of sunlight passing through the same flame, he saw a strong *absorption* line at the same yellow wavelength. In each case, the lines came from the gaseous sodium atoms in the flame. From such observations, Kirchhoff derived three physical laws, called **Kirchhoff's laws of radiation:**

1. A gas, or a sufficiently heated solid, will glow with a continuous spectrum, or continuum.
2. A hot gas will produce only certain bright wavelengths, called emission lines. Each element emits a characteristic set of emission lines.
3. A cool gas, if placed between the observer and a hot continuous-spectrum source, absorbs certain wavelengths, causing absorption lines in the observed spectrum.

Now let us look at what is happening at the subatomic level in each of these situations. As a guide to understanding,

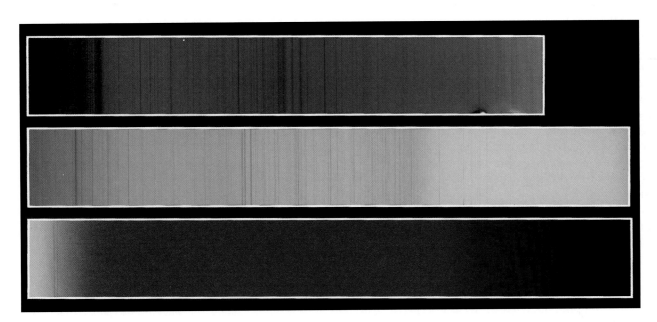

FIGURE 11-1
The spectrum of the Sun, showing the dark absorption features first noted by Joseph Fraunhofer. The three segments join together from top to bottom and left to right to give a visible spectrum from blue to red. The two lines at the far blue end are due to ionized calcium. Many of the features in the green and red portions of the spectrum are due to hydrogen. (SOURCE: K. Gleason, Sommers Bausch Observatory, University of Colorado).

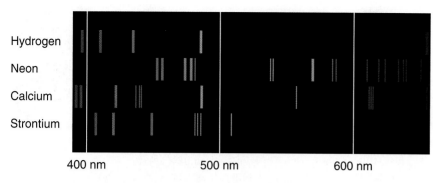

FIGURE 11-2
Spectral fingerprints. Each element has a unique set of emission lines, which help to identify the composition of any hot gas in space. The spectra show the emission lines of four elements; additional lines are found at ultraviolet and infrared wavelengths.

review the material in Chapter 10. You can think of electron energy levels as a kind of ladder, with energy increasing upward. The rungs of the ladder indicate the discrete, or quantized, energy levels. The lowest rung of the ladder is the *ground state;* all higher levels are called *excited states,* and the spacing of the rungs decreases as the energy increases. Below the highest rung (or energy level) the electrons can only have quantized energy levels—they are *bound* to the atom. Above the highest rung, the electron energy level can vary continuously—such electrons are called *free* electrons.

We have already seen in Chapter 4 that the microscopic motions of atoms and molecules lead to a smooth spectrum of thermal radiation. The peak wavelength of the radiation is an indicator of the temperature, according to Wien's law. If a material is hot enough that electrons are not bound to the atomic nucleus, the electrons can have a continuous range of energies. These free electrons will emit a smooth, featureless spectrum. The filament of a lightbulb is an example. The electrons in a gas can be liberated from the atomic nucleus either by energy from photons or by collisions with other energetic particles. The smooth continuum of the Sun's radiation tells us that the edge of the Sun is a layer of hot gas. Remember that Wien's law allows us to determine a body's temperature by its dominant radiation. The wavelength of the strongest solar radiation gives the temperature of the solar surface as about 5700 K.

Emission lines arise when a gas is hot enough that atoms are in excited states, but not hot enough that all the electrons are liberated. Using the ladder analogy, we say that each electron that drops from a higher rung to a lower rung emits a photon of a particular wavelength. Each different downward jump corresponds to a particular energy difference. The energy lost by the atom appears as a photon. The array of possible ways in which the atom can lose energy maps into a set of sharp emission lines. A neon sign is a familiar example. The neon gas inside the glass tube is excited (or has energy added to it) by an electric current. As the excited atoms drop back into lower energy states, they emit photons with particular energies. The purity of the color reflects the fact that the radiation is concentrated in a few red emission lines. Other gases have their own distinctive set of emission lines—think of the bluish color of mercury vapor street lights or the yellowish color of sodium vapor street lights (see the examples in Figure 11–2).

Absorption lines arise in a cooler gas, when atoms are in or near the ground state. If the cool gas is illuminated by a hot source of radiation, the gas will have photons with a smooth and continuous range of energies. However, photons can only be absorbed if their energy corresponds to the difference between two electron energy levels. To return to the ladder analogy, photons are absorbed if they can raise an electron to a higher rung. Each possible upward jump corresponds to a different wavelength of photon that can be absorbed. Since this energy is removed from the incoming radiation, it leaves a deficit of photons at a particular set of wavelengths. Photons of these wavelengths are emitted again as the excited electrons drop back down into lower-energy orbits, but in this case the emission takes place in *all* directions, including back toward the source. The result is that radiation at these specific wavelengths is subtracted from the background source, resulting in dark lines. The cool surface layer of the Sun is an example.

Figure 11–3 shows the different situations that lead to a continuum, emission lines, and absorption lines in a gas. Kirchhoff himself found that the absorption lines and emission lines of a given gas have identical wavelengths. This makes sense since the same atomic energy levels are involved in the two processes. What we see depends on the temperature and density of the gas relative to the radiation coming from behind it, as indicated in the third law. Later an important modification was made to Kirchhoff's laws: An absorption spectrum need not originate *in front of,* or in a cooler gas than, the continuous spectrum. It can arise within the same gas as the continuous spectrum. This is because within a single gas, electrons may be jumping upward in some atoms (making absorption lines) and downward from the free state into other atoms (forming the continuum spectrum). These light-emitting layers of gas form the well-defined visible surface of the Sun.

Astronomers usually make a spectrum by dispersing the light from the entire object. However, it is possible to get more detailed spectral information in a different way. We can form an image by filtering the light so that only a small range of wavelengths is detected. If the narrow wavelength range is centered on a particular spectral feature, we can make a map of the physical conditions that lead to the spectral transition. Figure 11–4a shows the Sun as viewed in the normal yellow-white light of the continuum, as well as an image made over a narrow range of wavelengths centered on the hydrogen alpha emission line (Hα emission). The Sun contains hydrogen over its whole surface, but the bright regions in Figure 11–4b show where the temperature and density are just right to cause electrons to change energy by just the right amount to give the Hα emission line.

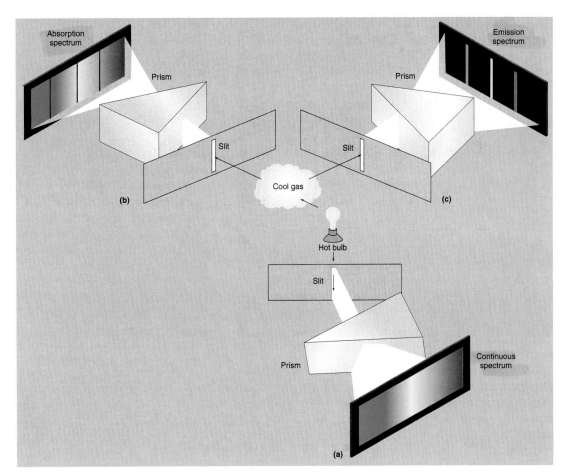

FIGURE 11-3
A diagram that illustrates Kirchhoff's laws of radiation. The continuum source is represented by the hot filament of a lightbulb. A prism is used to disperse the wavelengths of light in each case. When the lightbulb is viewed directly through the prism, a continuous spectrum is seen. When the gas cloud is viewed at 90° to the light beam, the spectrum is a series of sharp emission lines caused by electron transitions in the gas—an emission spectrum. When the lightbulb is viewed through the cool gas cloud, the continuous spectrum has radiation absorbed at wavelengths corresponding to the electron transitions in the gas—an absorption spectrum.

Composition

Kirchhoff's work was a ground-breaking achievement in the field of spectroscopy. Spectroscopy is the study of spectra. In particular, we can use this technique to determine the composition of the Sun and other stars. After more than a century of studying the Sun's spectrum, astronomers have identified the elements in the Sun, or the **solar composition**. The Sun is about 76 percent hydrogen (H) and 22 percent helium (He) by mass—roughly the same H/He proportions we found in the giant planets' atmospheres. Contrast this with Earth. The heavy elements that are common in Earth, such as iron, nickel, silicon, and carbon, make up only 2 percent of the Sun by mass. The most abundant elements, which probably reflect the overall material from which the solar system formed, are listed in Table 11-1.

In practice, identifying elements by using spectroscopy can be like solving a puzzle. Each element produces its own characteristic set of spectral lines. However, the sets of lines of many elements are present, and the resulting pattern can be very complex. We can collect so much radiation from the Sun that we can detect the presence of elements even in very low concentrations. For example, even though fewer than 1 in 1 million atoms in the Sun is a nickel atom, we can easily detect spectral lines of nickel in the solar spectrum. You can see that remote sensing by spectroscopy is a very sensitive technique.

Spectroscopy is a great example of the reach of the scientific method. Forensic chemists use spectroscopy to detect traces of particular elements in samples of human skin or hair. The same techniques are used to detect trace elements in giant gas balls trillions of miles away! In 1868, French astronomer Pierre Janssen and English astronomer Norman Lockyer independently found spectral lines in the Sun that

TABLE 11-1
Composition of the Sun

ELEMENT	PERCENTAGE OF MASS OF THE SUN	ATOMIC NUMBER
Hydrogen (H)	76.4	1
Helium (He)	21.8	2
Oxygen (O)	0.8	8
Carbon (C)	0.4	6
Neon (Ne)	0.2	10
Iron (Fe)	0.1	26
Nitrogen (N)	0.1	7
Silicon (Si)	0.08	14
Magnesium (Mg)	0.07	12
Sulfur (S)	0.05	16
Nickel (Ni)	0.01	28

Note: Based on spectroscopic measurements of the Sun and measurements of meteorites and other samples.

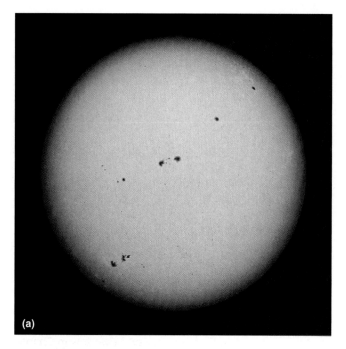 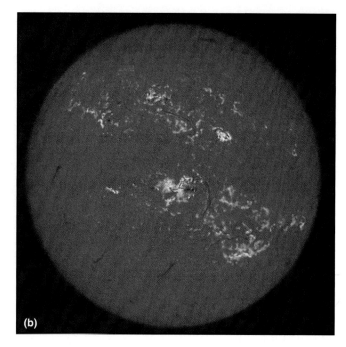

FIGURE 11-4
(a) The Sun in normal visible light, with a few sunspots visible. The yellow color is indicative of gas at about 6000 K. Shading around the edge, called *limb darkening*, is caused by the solar atmosphere's absorption of light. (b) The Sun photographed by a spectroheliograph in red light emitted by hydrogen. Bright areas involve intense hydrogen emission. (SOURCE: National Optical Astronomy Observatories.)

corresponded to an unknown element. This element, named *helium* (from the Greek *helios,* meaning sun), was the first to be discovered in space instead of on Earth. The discovery of helium raised an uncomfortable question about the scientific method as applied to astronomy. If the Sun contains strange materials that are unknown on Earth, how well can we ever understand the universe? Helium is exceedingly rare on Earth, but was eventually detected by Lockyer in 1895. As we saw in the discussion of planetary atmospheres, it escaped into space early in Earth's history. And the earlier concern has been put to rest—no element has been discovered in the Sun or any other star that cannot be measured in the laboratories of Earth.

Knowing what the Sun is made of does not tell us the source of its energy. To learn this, we return to the debate of the opening story in this chapter. The answer was provided by the pioneers who split the atom nearly 100 years ago.

Energy from the Atomic Nucleus

As we saw at the beginning of the chapter, the energy source of the Sun was for a long time a great scientific mystery. It was easy to show that chemical energy from burning fossil fuels could not be responsible. The Sun could be longer-lived if it got its energy from gravitational contraction, but no more than a fraction of a billion years. However, unknown to Lord Kelvin, there is an energy source of amazing efficiency: the atomic nucleus.

Early in the 20th century, Ernest Rutherford showed that the atomic nucleus was incredibly dense and that it occupied a tiny percentage of the volume of an atom (see Chapter 4). Imagine a football stadium where the nucleus is a pea on the 50-yd line and the electrons are whizzing around the outskirts of the stadium. Rutherford and Henri Becquerel demonstrated that the nucleus had structure by observing the phenomenon of radioactivity. They found that certain naturally occurring heavy elements could give off radiation and change their properties. In one elegant experiment, Rutherford sealed a small amount of radioactive material in a glass tube that otherwise contained a perfect vacuum. Several months later, he showed by a careful analysis that there was helium in the tube and that the radioactive material had changed its chemical properties. Here was the transformation of elements that alchemists had sought for centuries!

The early pioneers of radioactivity had little idea of the nature of the radiation they were studying. They dubbed the three types of radioactivity *alpha, beta,* and *gamma* after the first three letters of the Greek alphabet. Today, we know that these types of radiation are quite distinct. Alpha decay is the emission of helium nuclei, the kind trapped by Rutherford in his experiment. Beta decay occurs when a neutron turns into a proton, an electron, and a neutrino (this third particle will be discussed later in the chapter). Gamma radiation is energy released from the atom in the form of very high-energy electromagnetic waves. The young Polish physicist Marie Curie showed that radioactive material produced millions of times more energy per atom than any chemical process. She and her French husband Pierre were the first to isolate a radioactive element. Marie became the first female professor in the 600-year history of the Sorbonne University in Paris, and she was the first person to be awarded two Nobel Prizes. Because they knew little about the nature of radiation, the pioneers were also unaware of its damaging effect on human tissue. Many paid with their lives for their research, including Marie Curie.

The Conversion of Mass to Energy

In 1905, a young scientist named Albert Einstein startled the scientific world with a series of papers on the nature of mass, energy, and time. Extending the ideas of earlier scientists, he postulated that the speed of light is an absolute limit. From this he deduced that mass could be converted to energy, and energy to mass. In the everyday world, this remarkable process is not apparent. However, physicists observe it routinely when tiny particles are accelerated to very high speeds.

How can it be that the speed of light is never exceeded? As a particle is accelerated toward the speed of light, its mass steadily increases. The effect is only pronounced at extremely high speeds. Energy is being turned into mass. You will recall that a more massive object has greater inertia and so resists any change in its motion. But as a tiny particle is accelerated closer and closer to the speed of light, the energy given to it goes more and more into increasing its mass and less and less into increasing its speed. Kinetic energy turns into mass. The particle gets heavier and heavier and approaches—but never reaches—the speed of light. This process also works in reverse: Particles can convert mass to pure energy.

The next Science Toolbox demonstrates the fantastic amount of energy associated with even tiny amounts of matter. Normally, this energy is safely tucked away in the familiar mass of everyday objects. It can be liberated in two ways. In one process, naturally occurring heavy elements can split their nuclei and release energy. We have learned to purify and concentrate these rare elements and harness this energy source. The other process involves merging the nuclei of light elements. When atomic nuclei fuse, a small percentage of the mass is released in the form of radiant energy in a process called **mass-energy conversion.**

The nucleus of the atom contains a prodigious energy source. Mass and energy are related by Einstein's famous and simple equation $E = mc^2$ (see the Science Toolbox for applications of this equation). Since c is a very large number, a tiny amount of mass is equivalent to a fantastic amount of energy. The "frozen" energy in a mass about the size of an adult human is enough to power the country's entire energy needs for a year. Compare that to the energy you would get from the same mass of a fossil fuel, probably enough to run a single car for a month, or heat a single home for a week. Einstein doubted that there would ever be any practical use for this form of energy. Unfortunately for us in a world full of nuclear weapons, he was wrong.

The equivalence between mass and energy has some profound consequences for the way we look at the physical world. In Chapter 4 we saw that energy can be stored and can appear in many different forms. Mass is another of those forms. We can consider it *potential* energy since it is normally frozen in the form of stable particles. However, under certain extreme conditions this mass-energy can be released. The equation $E = mc^2$ works in both directions. Just as mass can be considered a form of energy, energy can be considered a form of mass. A rapidly moving car has slightly more mass than a stationary car. A spent battery has slightly less mass than a fully charged battery. These effects are tiny since mass equals energy divided by c^2, which is a very small number. Yet they are real. When we apply the law of conservation of energy, we must include mass-energy as well.

Energy from Nuclear Fission

To understand how energy can be released from the nucleus of an atom, we must become familiar with the concept of **binding energy.** Throughout this book we will see systems held together by forces—atoms, molecules, solar systems, and galaxies. The binding energy of a composite system is the energy required to take it apart. For example, objects are bound to Earth by gravity. As we saw in Chapter 3, to liberate anything from the gravity of Earth, we must give it a velocity of 11 km/s. This speed corresponds to a kinetic energy sufficient to counter the gravitational binding energy. Or consider an atom. The electron in a hydrogen atom is bound to the proton by the electric force. As we saw in the last chapter, the binding energy of the hydrogen atom is a tiny 2.2×10^{-18} J. So any photon with a frequency higher than 3.3×10^{15} Hz (or equivalently a wavelength smaller than about 100 nm) can liberate an electron. The next force we will consider is the force that binds an atomic nucleus.

Energy can be released from the atomic nucleus by the spontaneous decay of radioactive elements. This process is called **nuclear fission.** These heavy elements occur naturally in Earth's crust, but in very small concentrations. In nuclear fission, a nucleus splits into two or more pieces, and the fragments have a combined mass that is less than the mass of the original nucleus. The remaining mass is released as energy according to the equation $E = mc^2$. When a single uranium 235 nucleus decays, 3.2×10^{-11} J of mass-energy is released. This may not sound like much, but it is the nuclear binding energy released by a single atom—a fistful of uranium 235 could power a small city. We can compare this to chemical energy by considering the burning of coal. The electric binding energy released when a carbon atom combines with an oxygen atom to form a CO_2 molecule is 6.4×10^{-19} J. This is 50 million times less energy than we get from a single radioactive decay of uranium!

Energy release from radioactive elements within Earth contributes to the heating of the planet. This is why early estimates of the age of Earth were wrong. They assumed that Earth was a cooling ball with no internal energy source. When radioactivity is taken into account, the estimated age of Earth increases greatly. In fact, radioactive decay can be used as a clock, as described in Chapter 6. The age of Earth has been estimated by using this technique at 4.6 billion years.

SCIENCE TOOLBOX

Mass-Energy Conversion

Einstein proposed that each unit of matter has the same amount of energy associated with it. We can think of matter as "frozen energy." In the simple and famous equation

$$E = mc^2$$

E stands for energy and m stands for mass. Energy can be created from mass by using the conversion factor c. The conversion factor is a universal constant, the speed of light. Since the speed of light is an enormous number, 300,000 km/s, a tiny amount of matter is equivalent to an amazing amount of energy. Several examples will make this clear.

To calculate the conversion of mass to energy, we must use a consistent set of units. When units are in the international metric system, mass is in kilograms and energy is in joules. The speed of light is 3×10^8 m/s.

In the first example, let us consider the total energy consumption per year of all the people in the United States. It is about 8×10^{18} J (this is equivalent to each person having 10 lights turned on at the same time, since we are a very wasteful society). The equivalent mass is given by moving the factor c^2 to the other side of the equation, so that $m = E/c^2$. The result is $m = (8 \times 10^{18}$ J$) / (3 \times 10^8$ m/s$)^2 = 88$ J $\cdot$ s^2/m^2. Since 1 J $\cdot$ s^2/m^2 = 1 kg, the answer is 88 kg.

What does this number mean? It means that *if we could liberate the entire energy content* of the mass corresponding to a large person, we could power the country for a year. In a fusion process we only get about 0.1 percent of the energy out, so we would need 1000 times the mass to get the required amount of energy. That larger mass is still only about 88 × 1000 kg, or about 10 tons, compared to the millions of tons of chemical fuels that we actually burn to keep our homes lit and our factories working. Nuclear energy sources are about 1 million times more efficient than chemical energy sources.

Now let us recall the *Saturn V* rocket that carried astronauts to the Moon. A rocket the size of a 10-story building lifted a payload about the size of a minivan free of Earth's gravity. Over 90 percent of the mass of the rocket consisted of the highly volatile fuel needed to accelerate the payload. The energy source was chemical energy. If, however, the energy source had been pure mass-energy, only 10 grams (g) of fuel would have been needed to get to the Moon. NASA could have dispensed with a 10-story building full of chemical fuel in favor of a pocketful of nuclear fuel.

The contrast between mass-energy and chemical energy can be made clear with the simple example of a hamburger. A ¼-lb burger has a nutritional value of about 250 calories, which is 250 × 4186 = 10^6 J. This is the amount of chemical energy your body can extract from the burger. If we apply the equation above with a mass of 0.15 kg, the mass-energy of the meat is about 10^{16} J. Thus, we only extract 1 part in 10^{10} (or one-ten-billionth) of the possible energy in a hamburger by chemical means.

Finally, imagine something very small, such as the period at the end of this sentence. The ink in that period weighs roughly 1/100,000 g or 10 micrograms (μg). What is the equivalent amount of energy, using Einstein's equation? In the previous example, the annual U.S. energy consumption was about 8×10^{18} J, which is roughly 10^{19} J. We can take a shortcut by noticing that 1/100,000 g is about one-ten-billionth (10^{-10}) of the amount of mass in the first example. So the energy release from the dot of ink is about $10^{19} / 10^{10} = 10^9$ J. The unit of energy consumption is called a *watt*, after the famous Scottish scientist and engineer. One watt is one joule per second, so a lightbulb might consume 100 J/s. If the typical house consumes energy at a rate of 10 kilowatts (kW), or 10,000 J/s, then 10^9 J will last $10^9 / 10^4$ s. If only we could harness it, the mass-energy in the dot at the end of this sentence could run one family's home for a day.

Humans have learned to harness nuclear fission. Heavy nuclei can be split by a collision with a neutron. Since some radioactive elements also release neutrons as a decay product, the possibility exists that a nucleus will decay, releasing a neutron, which then triggers another decay, which releases another neutron, and so on. This is called a *chain reaction*. All that is required is a high enough concentration of the radioactive material that each decay triggers at least one more decay. We have learned to mine and process uranium to the required level of purity. (There is even evidence for a *natural fission reactor* in Gabon in Africa; a high underground concentration of uranium created a chain reaction there 1.7 billion years ago!) If a chain reaction happens in a controlled way, the energy can be harnessed, as in a nuclear reactor. If it happens in a catastrophic and uncontrolled way, the result is a bomb of terrible destructive power. In either case, we are extracting energy from the mass of nuclear particles as they undergo fission.

Energy from Nuclear Fusion

More relevant to our understanding of how the Sun shines is the process of **nuclear fusion.** In this process, energy is released by the fusing of light atomic nuclei into a single, heavier nucleus. For example, when hydrogen is converted to helium, some mass is lost: four hydrogen nuclei (or protons) have slightly more mass than a single helium nucleus (or alpha particle). The residual mass is released as energy according to Einstein's equation $E = mc^2$.

It is important to be clear on the difference between fission and fusion. Fission occurs when a massive atomic nucleus splits into smaller pieces. Since the pieces have less mass than the original nucleus, energy is released. Fission occurs spontaneously because massive atoms are unstable —we are familiar with the random nature of the decay process from our discussion of radioactivity in Chapter 5. A chain of fission reactions will continue until a stable element in the middle of the periodic table, such as iron, is reached. Fusion occurs when light nuclei are merged into a more massive nucleus. Since the merged nucleus has less mass than the starting pieces do, energy is released. Fusion does not occur spontaneously; it requires extreme physical conditions before the nuclear binding energy can be released.

Why is fusion so difficult to arrange? That is because protons have a positive electric charge and will strongly resist any attempts to make them collide. Enormous temperatures and pressures are required for protons to move fast enough that after a collision they will stick. It requires energies characterized by a temperature of at least 10 million K. Above that temperature, the electrical repulsion can be overcome, and the much stronger nuclear force acts as the "glue" to bind the nuclei together. Even then, the process cannot be understood without considering the quantum nature of matter. The positive charge on each proton is a strong electrical barrier against their fusing. But in the quantum view of matter, there is always a probability that some protons would fuse as if the barrier were not there. It is as if a cannonball had a significant chance of passing through a castle wall untouched!

On Earth, we have learned to produce, but not control, fusion reactions. If light elements are compressed by an explosion, it is possible to momentarily create a temperature and density sufficient for a fusion reaction to occur. The result is a hydrogen bomb, the most violent product of human creation. We have not yet succeeded in controlling this process, in a fusion reactor. It would involve the containment of hydrogen at a temperature of millions of degrees; since all known materials melt at a few thousands of degrees, you can see that this is a difficult problem! However, the incentives are great. Fusion power would be an efficient form of energy, using a cheap raw material, and with few of the radioactive by-products that are found in a fission reactor.

Solar Energy from Nuclear Reactions

Hermann von Helmholtz showed in 1871 that the energy output of the Sun would equal that released by the burning of 7000 kg of coal every hour on every *square meter* of the Sun's surface. Helmholtz realized that no ordinary chemical reactions can produce energy at this rate. Thus, the Sun is not "burning" in the normal sense of a chemical reaction.

What Helmholtz could not have known was that the Sun is a powerful fusion reactor. Nuclear reactions inside the Sun, as in all stars, do two important things: They generate energy, and they gradually change the Sun's composition because they build up increasingly heavy nuclei. We should remember that the temperature inside the Sun is so high that electrons have all been stripped from their atomic nuclei. The electrons are also far less massive than the nuclei and carry far less energy. We therefore need only consider the interactions between nuclei.

The principal nuclear reactions inside the Sun convert hydrogen to helium in three stages, as illustrated in Figure 11–5. Because this chain of reactions starts with two hydrogen nuclei, that is, two single protons, it is called the **proton-proton chain**. In step 1, two protons collide and fuse, forming deuterium, which is designated ^{2}H. Two additional particles are released: a positron and a neutrino. The positron is the antiparticle of the electron, identical except for having a positive charge. It is designated e^+. The positron quickly interacts with an electron, disappearing to produce radiation. The **neutrino** is a ghostly particle, with no charge and no mass. It is designated by the Greek letter ν. The neutrino interacts extremely weakly with matter; a neutrino from the Sun could pass through a mile of steel without being stopped. Therefore it acts as if other particles are not there, escaping the Sun's core traveling at nearly the speed of light. We will return to the neutrino later in the chapter.

In step 2, the hydrogen nucleus hits another proton and fuses into a form of helium known as helium 3, designated ^{3}He. More radiation is released. In step 3, two of the ^{3}He nuclei collide and fuse into the most common form of helium, helium 4, designated ^{4}He. This third step leaves two extra protons behind, which are available to participate in step 1 again. In each step, the reaction releases energy in the form of photons, and its end result is that the Sun creates helium out of the lighter element hydrogen.

The total amount of mass left at the end of the three-step chain is slightly less than the mass of the initial hydrogen atoms. During the fusion, a small amount of mass is converted to a large amount of energy. In the Sun's fusion sequence, about 0.007 kg of matter is converted to energy for each 1 kg of hydrogen processed. In other words, the efficiency of mass-energy conversion is 0.7 percent, and the amount of energy released is $0.007mc^2$. The synthesis of a single helium nucleus creates 4.5×10^{-12} J of radiant energy. The entire Sun radiates 4×10^{26} J each second. This corresponds to 400 trillion trillion W, which equals a lot of lightbulbs!

In fact, every second, the Sun converts 4 million tons of hydrogen to energy and radiates it into space! Yet, the reservoir of hydrogen in the Sun is so large that, even if its fuel is used at the rate of one tall skyscraper's worth per second, there has been no detectable change in the Sun's output in recorded history. According to a recent calculation, the Sun will not run out of hydrogen for about 4 billion years. The consumption of hydrogen inside the Sun tells us that stars are not permanent, but must evolve and run down.

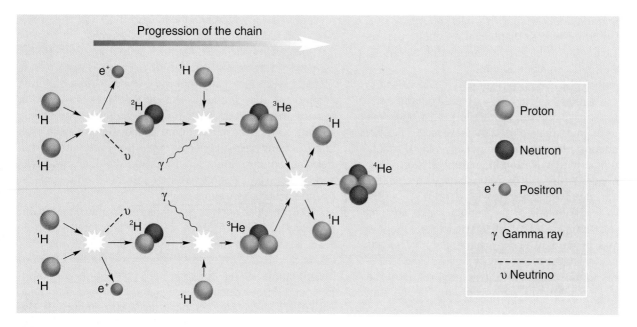

FIGURE 11-5
The proton-proton chain is the primary source of energy from the Sun. The chain begins with four protons and ends with a helium nucleus. Additional subatomic particles are produced by the initial reaction, and high-energy photons are released at every stage in the chain. The protons released in the final step in the chain are available to participate in a new set of reactions.

The discovery of fusion as the power source of the Sun is a good example of how science works. We can see how scientists piece together information into a coherent physical description of a remote object. The starting point is our knowledge of the distance to the Sun. This combined with a measure of how much radiation reaches Earth's surface gives the Sun's energy output. Since Earth only intercepts one-billionth of the Sun's light, the Sun must be a truly impressive energy source. It is easy to show that no chemical energy source can produce the required output for more than a few thousand years. The evidence points to a new energy source: the fusion of atomic nuclei. Sure enough, we find from spectroscopy that the Sun is made mostly of hydrogen and helium, the nuclear fuel and its by-product.

The Sun's Interior

Armed with knowledge of the Sun's size, composition, and energy source, astronomers can work out a model of the physical conditions at every point within the Sun. This is often called the *standard solar model*. As a result of gravitational contraction (see Chapter 9), the temperature at the Sun's center reached 15 million K as the proton-proton reactions became established. This remains the current temperature in the **solar core,** where nuclear energy is generated. Approximate conditions in the region between the core and the surface can be calculated by using equations that describe the interior pressure and density and the energy generation rates inside the Sun. The relationship between temperature and pressure and volume is the same gas law that we encountered in Chapter 7 in our discussion of the atmospheres of giant planets. The results of the model are plotted in Figure 11-6.

These calculations indicate that the gas pressure at the Sun's core is about 250 billion times the air pressure at Earth's surface. This high pressure compresses the gas in the core to a density about 158 times greater than that of water and about 20 times greater than that of iron. One cubic inch of this gas would weigh 2.5 kg, or nearly 6 lb! Although it may seem strange that material this dense can be a gas, it is inevitable, given the very high temperatures. The particles are moving too rapidly to stick together in the kind of lattice that we find in a solid. Such a high-temperature gas, where all the electrons have been stripped from the atomic nuclei, is called a *plasma*.

Despite the violent reactions going on in its core, the Sun itself is stable. Even though we associate fusion on Earth with nuclear weapons, the Sun is not a bomb and it is not exploding. At every point within the Sun, the inward pull of gravity is balanced by outward pressure due to energy released from nuclear reactions. The Sun is in equilibrium (see Chapter 4 for a discussion of this important concept). Very early in its history, the Sun must have collapsed from a giant gas cloud, a process we described in Chapter 9. However, the collapse phase took only 50 million years, and it finished over 4.5 billion years ago. Ever since then, the Sun has been converting hydrogen to helium in a steady nuclear reaction.

The fusion core of the Sun occupies about the inner one-quarter of the Sun's radius. This one-sixty-fourth of the Sun's volume contains about one-half of the solar mass and generates 99 percent of the solar energy. We cannot see the Sun's

FIGURE 11-6
Variation in the density and temperature of the Sun with distance from the center of the core. The physical conditions vary smoothly, with no sharp edge, and the material remains a hot gas throughout, even in regions of very high density. This plot is based on a model of the Sun's properties, since density and temperature in the interior are not observed directly.

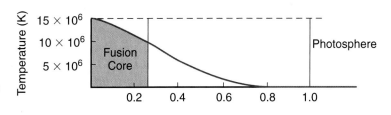
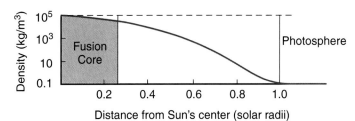

core directly, so these statements are based on models of the composition and energy generation.

How Energy Gets from the Sun's Core to Its Surface

Energy in the Sun travels outward, using two of the three modes of heat transfer discussed in Chapter 4. Since heat energy always flows from hot to cool regions, solar energy travels outward from the hot core and through a cooler zone of mixed hydrogen and helium toward the surface. Although the core is the place where helium is currently being produced, some helium exists in the cooler surroundings because the Sun formed with some helium already in the solar nebula, and because helium can slowly diffuse out from the core. Throughout most of the Sun's volume, this energy moves primarily by *radiation*. That is, the energy radiates through the gas in the form of electromagnetic radiation, just as light travels through our atmosphere.

In the outer part of the Sun, energy moves outward by *convection*. Convection occurs when the temperature difference between the hot and cold regions is so great that radiation cannot carry off the outward-bound energy fast enough. Large bubbles of gas, having become heated enough to expand, become less dense than their surroundings and rise toward the surface. They cool by radiating their energy into space (as sunlight) and then sinking. This is just what happens in a pan of water when you bring it to a strong boil. The only way the water can carry the heat from the burner to the air above is by rolling convection motions.

It takes several hundred thousand years for energy to diffuse to the surface, by which time the temperature has dropped from 15 million to a few thousand kelvins. The visible surface of the Sun, or **photosphere**, corresponds to a low enough density that photons no longer collide on their way out. At this point they travel freely through space. The result is that energy takes a huge number of years to diffuse out from the core of the Sun to the surface, but only 8 min then to travel the large distance to Earth! See the next Science Toolbox for more details. Figure 11-7 shows a schematic cross section of the Sun.

FIGURE 11-7
A cross section of the Sun and its atmosphere, showing (to approximate scale) the energy-producing core (the yellow center), the radiation zone (purple), the outer convective zone (red), and the tenuous corona (orange).

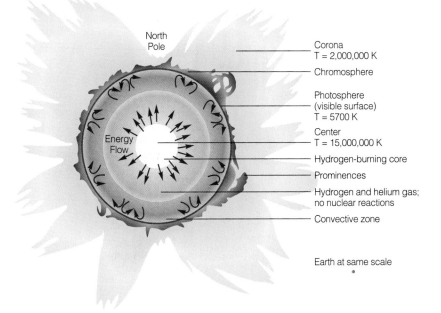

THE SUN'S INTERIOR

SCIENCE TOOLBOX

Collisions and Opacity

Why does the Sun appear to have a sharp edge if it is made of gas? Why does a cloud appear to have an edge? In each case, there is no sharp boundary. In the Sun, the density of hot gas increases smoothly right as you go down through the region we see as the edge. In a cloud, the density of air inside the cloud is not very much higher than it is outside the region we see as the edge. In both cases, the answer lies in the way in which light interacts with particles.

Opacity, or optical depth, is the degree to which a material transmits light. If a material transmits all light, it is transparent and the optical depth is zero. If it transmits no light, it is opaque and the optical depth is a large number. Everyday examples are obvious: glass is transparent, iron is opaque; water is transparent, ink is opaque. A gas can have opacity, too. We cannot see to the center of the Sun, so we know it is opaque.

In a very dense gas, like the hot core of the Sun, photons travel only a very short distance before colliding with a particle. The rate of progress of the photon is proportional to the square root of the number of collisions it suffers. Imagine you were standing blindfolded at the center of a large circular park. You could leave the park just by walking purposefully in any direction. Now imagine there are people standing randomly across the grass. If as you tried to leave, you would collide with one person, get disoriented, move off in another direction, collide with another person, move off in yet another direction, and so on.

This type of motion is called a *random walk,* analogous to the lurching movement of someone who is totally dizzy trying to find the way home. After 100 direct steps, you would of course be 100 steps away from the center of the grassy area. If you collided with 25 people, however, your rate of progress would be $\sqrt{25} = 5$ times slower, so 100 steps would only take you $100/5 = 20$ steps away from the center. If you had the misfortune to bump into 100 people, your rate of progress would be $\sqrt{100} = 10$ times slower. Your 100 steps would only take you 10 steps from the center. In probability theory, the distance traveled per second is inversely proportional to the square root of the number of collisions per second. And since the number of collisions depends on the density, the distance traveled per second is inversely proportional to the square root of the density.

If x is the distance a photon travels between collisions, and it suffers N collisions, then the distance d that the photon travels in any one direction is

$$d = x \sqrt{N}$$

So it takes 100 collisions to travel a distance $10x$, 10,000 collisions to travel a distance $100x$, and 1 million collisions to travel a distance $1000x$. Diffusion of radiation is a slow and inefficient way to transport energy, which is why stars like the Sun move energy by convection, too.

In the center of the Sun, the density is so high that photons diffuse much more slowly than they would while traveling through empty space, even though photons travel at the speed of light between collisions (Figure 11-A). As the photon works its way out of the Sun's core, the photon loses energy in its collisions, gradually changing from an X-ray photon to an ultraviolet photon. At every point in its slow journey, the photon is in equilibrium with its surroundings, so its wavelength corresponds by Wien's law to the temperature of the gas surrounding it. Let us see how much more slowly a photon moves at the Sun's core than at the surface. The ratio of speeds is given by the inverse square root of the ratio of densities. The Sun's density is 158,000 kg/m³ at the center and 0.001 kg/m³ at the edge. Photons thus travel $(158,000/0.001)^{1/2} = 12,600$ times more slowly in the core than they do when they leave the Sun. It takes about 100,000 years for radiation to get from the Sun's core to the surface!

Now let us see how this type of diffusion relates to the reduction in the intensity of light as it travels through a gas. The change in intensity of a beam of light as it passes through a gas is

$$I = I_0 e^{-\tau}$$

(continued)

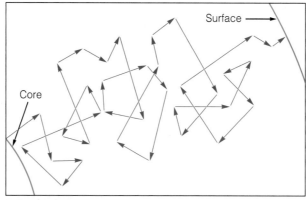

FIGURE 11-A
The random walk of a photon from the core of the Sun to the surface. The progress of the photon outward is much slower than the speed of light in this diffusion process.

SCIENCE TOOLBOX

Collisions and Opacity (continued)

In this equation I_0 is the starting intensity, I is the final intensity, and the Greek letter τ (called tau) is the opacity or optical depth. We can relate this to our previous discussion by noting that the opacity is the square root of the number of collisions, so

$$I = I_0 e^{-\sqrt{N}}$$

Let us plug in some numbers and see how light is diminished. Note that the reduction in intensity occurs for scattering of the photons *or* absorption of the photons. (Even if a photon is absorbed and raises an atom's energy level, the photon will eventually be reemitted in a random direction.) With a gas thin enough that no collisions occur, $I = I_0$ and the light intensity is not diminished. After 5 collisions, $N = 5$, and $e^{-\sqrt{5}} = 0.11$, so the light intensity is diminished by nearly 90 percent. After 10 collisions, $N = 10$, and $e^{-\sqrt{10}} = 0.04$. After 15 collisions, $N = 15$ and $e^{-\sqrt{15}} = 0.02$. You can see that the light is rapidly extinguished as the opacity increases.

At some point in its journey from the center of the Sun, radiation travels through rarefied gas. Photons will reach a region where the density is so low that they are unlikely to suffer any collisions at all. At this point, they travel freely in a straight line to Earth. Astronomers call this region the Sun's *photosphere*; it occurs at a temperature of 5770 K, which has thermal radiation in the form of visible light. We see it as the Sun's edge. Inside this region, light bounces around and the Sun is opaque. Beyond this region, the Sun is transparent, light travels directly, and we have the appearance of a surface.

The same principle applies to a cloud. There is water vapor throughout the atmosphere, and in some places the concentration is higher than in others. Inside a cloud, the concentration is high enough that light bounces around from one water droplet to another and cannot travel freely; this region is opaque. The edge of the cloud just defines the region where, on average, light will not collide with a water droplet and can travel directly to our eyes. To take another example, consider a glass of water. Water is transparent, but if you slowly add drops of milk, it will become opaque. Milk contains large fat molecules that scatter light. As you add more milk, the number of collisions between photons and large molecules increases, and the opacity increases.

Solar Neutrinos

The nuclear reactions that power the Sun not only release enormous amounts of radiation, but also release enormous numbers of elusive particles called neutrinos. Neutrinos interact so weakly with ordinary matter that they mostly pass through the whole Earth as if it were not there! After neutrinos are produced in the core of the Sun, they leave the Sun within a few seconds and streak toward Earth at the speed of light. Neutrinos therefore offer a great opportunity to "see" material coming directly from the thermonuclear heart of the Sun. The neutrino flux at Earth's surface is prodigious; about 10^{14} neutrinos pass through every square meter every second. About 10 trillion neutrinos pass through your body every second, and you do not feel a thing!

Detecting neutrinos is quite a challenge. They interact so weakly that vast detectors must be assembled to catch the rare interactions between a neutrino and an atomic nucleus. Huge vats of ultra-pure liquids are used; the occasional neutrino collisions create a distinctive flash of light. The detectors must be placed deep underground to shield them from confusing signals due to other particles from interstellar space. The U.S. physicist Ray Davis of Brookhaven National Laboratory was the pioneer of this type of experiment, and his original detector has been running for over nearly 30 years deep in a gold mine in South Dakota.

Astronomers have drawn several conclusions from the experiments of Ray Davis and others. Most important, neutrinos have been detected for the first time from beyond Earth, and there is no doubt that they come from the Sun. Their existence is extremely strong evidence that nuclear fusion occurs deep below the Sun's surface. We do indeed know what powers the Sun. The second conclusion is more unsettling. The experiments reveal only about two-thirds of the neutrinos expected from the standard solar model. Perhaps our model of the Sun needs to be modified, or perhaps our understanding of how neutrinos interact with matter is incomplete. Scientists are laboring in laboratories deep underground at several sites around the world, trying to detect the elusive neutrinos and refine our model of the Sun.

The detection of solar neutrinos is a good example of the way that science moves forward. We have a successful model of the Sun and its energy generation, but scientists continue to refine and test it. Neutrinos are an important tool in this process because they are produced in each nuclear fusion reaction and they emerge unscathed from the Sun's core. In the Science Toolbox on page 311, we present the data on neutrinos from the Sun, and we also discuss in more general terms how scientists test a model or a hypothesis.

Solar Oscillations

Another way to study the Sun's interior is to measure the way it vibrates. The Sun oscillates and vibrates at many frequencies, as an ocean surface or a bell. The wavelike solar oscillations can be used to infer the interior properties, just as geologists use seismic waves to study the structure of Earth. The oscillations are seen as volumes of gas near the Sun's surface that rise and fall with a particular frequency. The best-studied oscillation has a 5-min period, during which portions of the Sun's surface move up and down by about 10 km. This discovery has created a whole new field called **solar seismology**. A computer model of one of the many modes of oscillation of the Sun is shown in Figure 11–8.

Note that these rising and falling volumes of gas near the Sun's surface are another indication of the energy flowing outward through the Sun. The solar atmosphere is "boiling." It is like seeing the rolling motions of cells of water on the surface of a boiling pan of water. This can be detected because spectral lines emitted from gas moving upward will be slightly Doppler-shifted to the blue; spectral lines from gas moving downward will be slightly Doppler-shifted to the red (see the discussion of the Doppler effect in Chapter 9). In this way the rolling motions of convection near the Sun's surface can be mapped out.

FIGURE 11–8
A computer representation of one of the millions of modes of sound wave oscillations in the Sun. Blue regions are approaching, and red regions are receding; the gas motion is measured by using the Doppler effect. (SOURCE: NOAO.)

The Sun's Atmosphere

If the Sun is a giant ball of gas, why does it appear to have a sharply defined surface? The answer lies in the **opacity** of the gas—its ability to obscure light passing through it. The density and temperature of the Sun decrease smoothly and continuously, moving outward from the core (see Figure 11–6). Photons are continually colliding with particles, but as the photons migrate outward, there is a point at which the density is low enough that no more collisions occur. The photons travel unimpeded to Earth, and we see an "edge" at that point. A solid probe, if it could survive the high temperature, could drop directly through the photospheric "surface" and plunge into the Sun, like an airplane passing through the surface of a cloud. Inside a cloud, for example, light bounces off water droplets. The edge of the cloud corresponds to a region where the density of water droplets is low enough that light travels freely outward. As with a cloud, there is no sharp discontinuity in density or temperature at the photosphere of the Sun.

Energy ascending from inside the Sun heats the photosphere—the bright surface layer of gas that radiates the visible light of the Sun—to a temperature of about 5770 K. Other features of the surface layer of the Sun are shown in Figure 11–9. The convection of gas that mottles the solar surface (Figure 11–9a) is very similar to the convection of terrestrial clouds (Figure 11–9b). Sunspots indicate the presence of magnetic fields. The pattern of gas that links sunspot pairs (Figure 11–9c) is very similar to the pattern of iron filings that trace the lines of force in a magnet (Figure 11–9d).

Chromosphere and Corona

Just above the photosphere lies the **chromosphere** (which means color layer). The chromosphere is a pink-glowing region of gas at a temperature of 10,000 K. Its light is mainly the red Hα emission line described in Chapter 10. The chromosphere can be seen by the naked eye during a total solar eclipse. When the Moon covers the rest of the solar disk, this thin outer layer is visible as a ring of small, intense red emission, just visible in Figure 11–10 on page 314.

Above the chromosphere is the outermost, tenuous atmosphere of the Sun, the **corona**. Gas in the corona reaches the amazing temperature of 2 million K, due to heating by violent convective motions in the photosphere and chromosphere. At this very high temperature, even heavy elements such as oxygen and iron have all their electrons stripped off, and all atoms are highly ionized. The gas in the corona has very low density, and it seems surprising that it could possibly be at the very high temperature of 2 million K. One way for gas to reach such a high temperature is by compression due to gravity—the situation in the Sun's core. Another way is for a much less dense gas to have energy dumped into it by some other process. In the corona, energy comes from magnetic fields and convection in the Sun's atmosphere. A fluorescent gas tube is a familiar example of this phenomenon. As you know, a fluorescent gas tube is cool to the touch. Yet the gas inside has enough electric energy forced into it to emit visible light that indicates a thermal temperature of several thousand kelvins! The explanation is that a fluorescent tube is almost completely evacuated—the gas inside is at very low density. So the rate of collisions of gas atoms on the walls of the glass tube is very small, and the tube stays cool. However, the individual gas atoms have kinetic energy appropriate for a very high temperature.

SCIENCE TOOLBOX
Testing a Hypothesis

Very few scientists create new theories or discover laws of nature. The day-to-day business of science consists of gathering data and consolidating existing knowledge. Recall from Chapter 1 that a hypothesis is a proposed explanation for a set of measurements. The mathematical formulation of the hypothesis is called a *model*. Scientists test hypotheses by acquiring data of greater and greater scope and accuracy.

Let's take an important example from the history of astronomy. The Copernican hypothesis was that the planets travel around the Sun on circular orbits. The planet velocity is the same at every point in a circular orbit. Therefore the velocity of Mars, for example, as seen from the Sun should not change. (The actual measurement is an angular velocity on the sky, which can be converted to a space velocity, given the distance to Mars. An additional complication comes from the fact that the measurement is made not from the Sun but from the moving planet Earth.)

Kepler knew that the angular motion of Mars was not constant, and he hypothesized that the orbit was elliptical. We can use Kepler's laws to calculate what this implies about the velocity of Mars in its orbit. Mars has an orbital eccentricity of 0.093. So if r is the mean distance from the Sun, the orbit varies from $1.093r$ to $0.907r$. Elliptical orbits vary in speed with distance from the Sun according to $v \propto \sqrt{r}$ (see the Science Toolbox in Chapter 8), so the velocity varies smoothly from $\sqrt{1.093}\, v = 1.045v$ to $\sqrt{0.907}\, v = 0.952v$ throughout the orbit. In other words, the orbital velocity varies from 4.5 percent above the mean to 4.5 percent below the mean; a total variation of 9 percent. So to detect the elliptical motion—and rule out the hypothesis of a circular orbit—requires measurement with *at least* this accuracy. We can improve on the testing of a hypothesis with more measurements or more accurate measurements or both.

Now consider an example of great current interest—the detection of extrasolar planets. In Chapter 9 we learned that the wobble motion of a star provides an opportunity to detect an unseen planet by its gravitational influence. Suppose we accurately observe the velocity of a star just like the Sun and look for the undulation in velocity caused by an orbiting planet. Figure 11–B*a* shows observations of the velocity of this hypothetical star over 18 years. The data show a clear variation and the solid curve is a model with a planet mass equal to Jupiter's mass. In this case, we can confirm the hypothesis that a planet orbits the star and measure that it is a planet like Jupiter.

Now let us use this example to look at hypothesis testing in greater detail. Figure 11–B*b* shows a situation in which the data are not accurate enough to test the hypothesis of a Jupiter-size planet. You can see that the error bar on each observation is large, and so we do not have a sensitive test of the hypothesis, which is given by the dashed line. A similar situation arises when the effect being looked for is very small. The reflex motion of an Earth-size planet is only 0.09 m/s. If you imagined this motion plotted in Figure 11–B*a*, it would be a curve with 100 times less amplitude than the Jupiter curve shown, completing a cycle each year. As you can imagine, a large amount of extremely accurate data would be required to test the hypothesis of an Earth-size planet.

In Figure 11–A*c*, the problem is insufficient data. The observations span a
(continued)

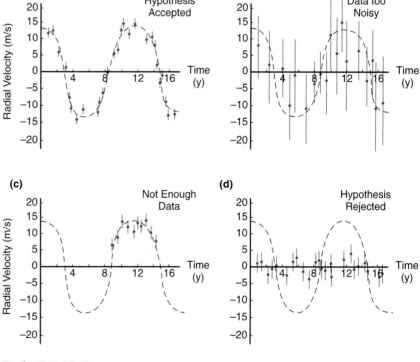

FIGURE 11–B
The technique of Doppler detection of extrasolar planets is used to illustrate the idea of hypothesis testing (see Science Toolbox text and Chapter 9 for details).

SCIENCE TOOLBOX

Testing a Hypothesis (continued)

5-year interval, but that is not long enough to see an entire cycle of reflex motion. Once again, the hypothesis of a Jupiter-size planet cannot be tested. The situation would be even worse if we were looking for a planet even farther from its star. Uranus has an orbital period of 84 years, so nearly a century's worth of data would be needed to look for such a planet. Figure 11–B*d* shows another possibility. The accurate data show no significant variation from a constant velocity, at a level smaller than the prediction for a Jupiter-size planet. These data allow us to *reject* the hypothesis of a planet with that mass.

The general description of hypothesis testing incorporates three features: the number of observations, the error in each observation, and the amount by which each observation deviates from the prediction of a hypothesis or model. Mathematically, we can say

$$\chi^2 = \sum \frac{(Q_{data} - Q_{model})^2}{\sigma(Q)^2}$$

In this equation, Q is any quantity we can measure. It might be an angle, a distance, a temperature, or any astronomical quantity. We take the square of the difference between the data and the model at every point, divide it by the square of the error in the observation, and sum this quantity over all data points. (A sum over quantities is represented in mathematics by the capital Greek letter sigma: Σ.) Essentially, we are calculating the amount by which the data deviate from the model. A good model closely represents the data and has a low value of χ^2 (χ is the Greek letter chi). A bad model poorly represents the data and has a high value of χ^2.

As a last example, we consider testing the hypothesis of solar energy generation by the detection of neutrinos. We are confident that we know what powers the Sun, but not all aspects of the solar model agree with observations. Figure 11–C shows nearly 20 years of solar neutrino measurements from the underground Homestake experiment. The actual quantity measured is the number of radioactive argon atoms produced by interactions with neutrinos. The error bars on each measurement are large because the number of particles detected is low (recall the limitations of counting statistics from the Science Toolbox on counting craters in Chapter 6) The upper colored band shows the range of predicted outcomes from the solar model. While it agrees with some individual measurements, it does not agree with the narrower band showing the mean and standard error of the combined measurements. As a result, we conclude that our model of solar physics is incomplete.

It now appears that neutrinos have a tiny mass that has not yet been incorporated into models for the Sun.

FIGURE 11-C
Twenty years of neutrino detections from the Sun, measured by the underground Homestake experiment. The large error bars on each data point reflect the very small number of radioactive atoms that are created due to a neutrino interaction. The narrow band shows the most probable range of all the observations combined. The upper wide band shows the range of predictions of the standard solar model. (SOURCE: National Optical Astronomy Observatories.)

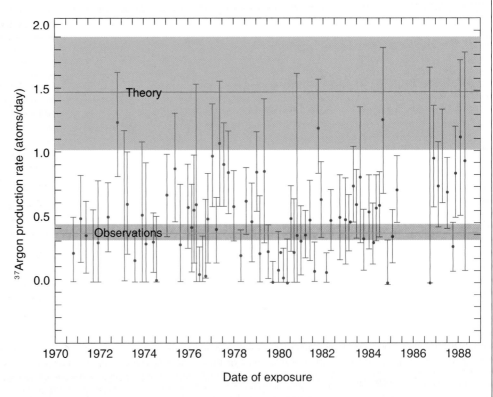

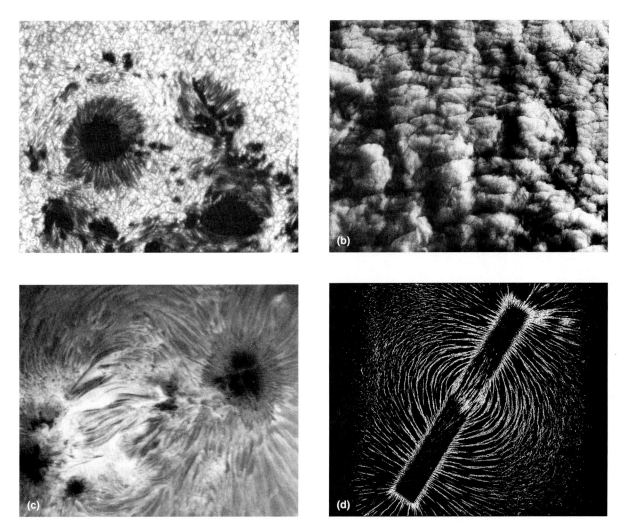

FIGURE 11–9
(a) A detailed photograph of the solar surface in the region of a sunspot. Outside the sunspot, the normal solar surface is mottled by granules believed to be convection cells in the solar gas. The main sunspots are comparable to Earth in size, and the large granules are comparable to continents. (SOURCE: Balloon-borne telescope photograph, Princeton University, Project Stratoscope, supported by NSF, ONR, and NASA.) **(b)** The solar convection is similar to terrestrial cumulus clouds, but on a much larger scale. This is a view of Earth from a 10-km altitude. (SOURCE: Photograph by WHK.) **(c)** This image, taken at the red wavelength of the Hα emission line, shows the chromosphere structures that link sunspots. (SOURCE: AURA, Sacramento Peak Observatory.) **(d)** The patterns connecting sunspots are caused by the same magnetic forces that align iron filings around a magnet. (SOURCE: Grundy Observatory.)

The corona is extremely hot, and by Wien's law the peak wavelength of thermal radiation scales with temperature. Since energetic particles tend to produce short-wavelength radiation, we might expect the corona to emit most of its energy as X rays. Imaging of the Sun at X-ray wavelengths from above the atmosphere has permitted us to visualize the remarkable appearance of the X-ray Sun, as shown in Figure 11–11, which vividly shows flares and active areas emitting X rays. The flares shoot material upward into the corona, disturbing the coronal structure.

Why are both the chromosphere and the corona hotter than the photosphere? After all, they are farther from the Sun's internal energy source. The answer is that energy can flow out from the Sun's surface in a way that can dump huge amounts into the tenuous outer atmosphere. The photosphere is laced with magnetic fields, and convective energy can travel quickly out through the thin gas, heating it up dramatically. The Sun's magnetic field controls motions of gas in the corona, creating delicate streamers. We see the effects from Earth as flares and arching prominences, often dwarfing Earth in size.

Sunspots

Sunspots sometimes can be seen with the naked eye when the Sun is dimmed by fog or a darkened piece of glass. (Warning: *Never point any telescope or binoculars at the Sun because the concentrated light will burn your retina. Even staring at the Sun for no more than a couple of seconds can damage your eyes.*) Their nature was not realized until 1613, when Galileo studied sunspots and concluded that they are located on the solar surface and are carried around the Sun

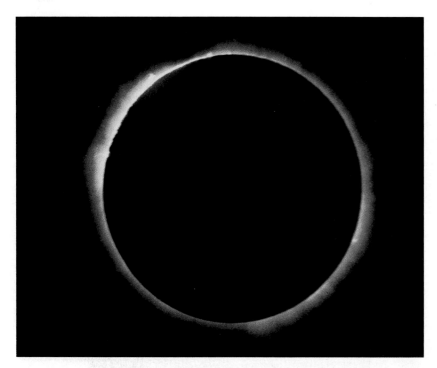

FIGURE 11-10
The Sun's inner atmosphere during an eclipse. Pink flames of hot gas, colored by the red Hα emission of excited hydrogen atoms, protrude from the Sun's surface at the top and several other points around the Moon's black silhouette. The pearly diffuse glow is the inner corona. (SOURCE: NASA photograph by astronauts using solar telescope in Skylab space station.)

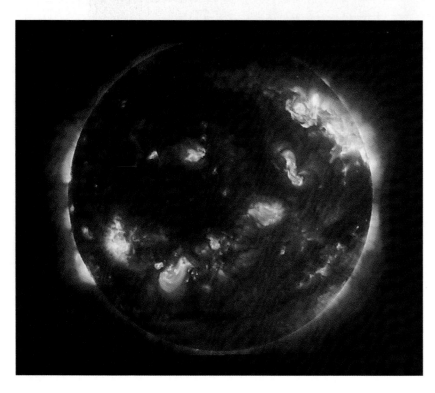

FIGURE 11-11
A view of the Sun in X-ray wavelengths showing flares as sites of intense high-energy radiation. X rays are invisible, and image processors here chose false-color orange tones usually associated with sunlight. This image was taken by a small X-ray telescope that flew on a rocket during the total solar eclipse of July 11, 1991. (SOURCE: L. Golub, Harvard-Smithsonian Center for Astrophysics.)

by solar rotation. Sunspots can be observed by magnifying the Sun's image with a telescope and projecting the image into a viewing chamber that is shielded from outside light.

A sunspot is a magnetically disturbed region that is cooler than its surroundings. A sunspot looks dark because its gases, at 4000 to 4500 K, radiate less than the surrounding gas at about 5770 K (see Figure 11–9). The motion of solar gas near sunspots is not controlled by atmospheric forces, as terrestrial storms are. Solar gas moves due to the influence of magnetic fields near the Sun's surface. Ions (charged atoms or molecules), which are common in the Sun, cannot move freely in a magnetic field but must spiral around the magnetic field lines and stream from one magnetic pole to the other.

As we have noted, a gas with many ions is called a *plasma*. Unlike a neutral gas, plasma motions are strongly

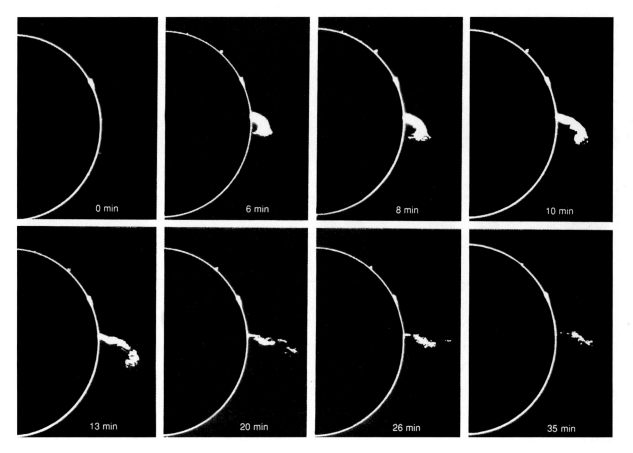

FIGURE 11-12
This sequence of photographs shows an eruptive prominence, or jet of gas, blasting off the Sun over a period of 35 min. The photographs were made with a coronograph, which obscures the bright solar disk and allows solar atmospheric activity to be monitored. (SOURCE: National Center for Atmospheric Research.)

influenced by magnetic fields. For this reason, plasmas in the sunspots and elsewhere in the solar atmosphere move in peculiar patterns that indicate the twisted patterns of the solar magnetic field. Huge clouds of gas, larger than the whole Earth, erupt from the disturbed regions of sunspots. These *prominences* can be seen when silhouetted above the solar limb, or edge, as in Figure 11-12. The largest blasts of material and their very active sunspot sites are called *flares*.

The Solar Cycle

Around 1830, an obscure German amateur astronomer named Heinrich Schwabe began observing sunspots as a hobby. For years he tabulated the number and position of sunspots. Schwabe noticed a regular pattern to the numbers and positions, and in 1851 he proposed the existence of a **solar cycle**, as shown in Figure 11-13. This discovery, which was followed a year later by the discovery that terrestrial magnetic

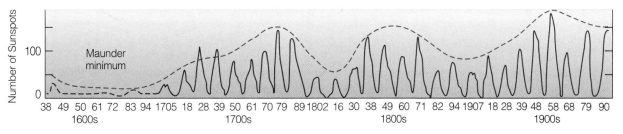

FIGURE 11-13
Sunspot counts since the 1600s show the cycle averaging 11 years for sunspot numbers (one-half the 22-year magnetic cycle), with evidence for a longer 80-year cycle (dashed line). The extensive period of low sunspot activity in the 1600s is believed to correlate with climate changes at that time and is called the *Maunder minimum*. (SOURCE: After data of M. Waldmeier, National Solar Observatory.)

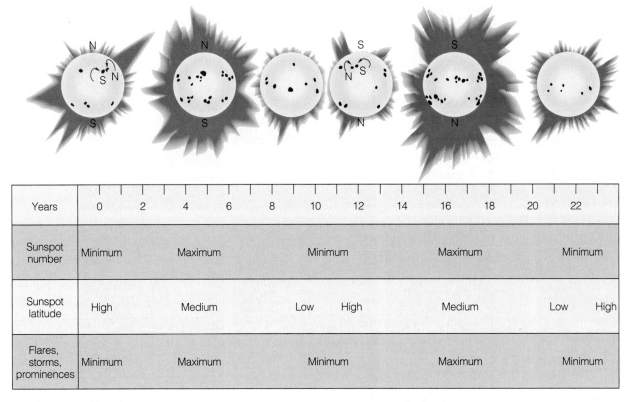

FIGURE 11-14
A schematic sequence of solar changes during the 22-year magnetic cycle. Note that the sunspot numbers go through two 11-year cycles during one full magnetic cycle. (See the text for a more detailed explanation.)

compass deviations exactly follow the same cycle, was a key step in understanding the Sun and its effects on Earth.

The magnetic cycle's duration averages 22 years and consists of two 11-year cycles of sunspot number, as shown in Figure 11–14. In a period of minimal sunspots, the few visible spots are grouped within about 10° of the solar equator. When a new cycle begins after a year or so, groups of new spots appear at high latitudes, about 30° from the solar equator. After a few years, the sunspot number reaches a maximum, and the spots are at intermediate latitudes, about 20° from the solar equator. After about 11 years, the spots appear mostly about 10° from the equator, and a sunspot minimum occurs again.

This pattern is intriguing, but even more remarkable is the fact that the magnetic field of *the entire Sun* reverses between each 11-year sunspot cycle! Imaginary observers on the Sun would find their compasses pointing north in one direction for 11 years (subject to disturbances by frequent magnetic storms) and in exactly the opposite direction for the next 11 years. This behavior is not entirely unknown: Earth's magnetic field reverses every few hundred thousand to few million years. Both patterns of reversal may be caused by cyclic flow patterns in the deep fluid cores of the two bodies (liquid metal for Earth, high-temperature gas with more than 150 times the density of water for the Sun). The sunspot cycle is important to us because during years of maximum sunspot activity, solar particles shooting off the Sun affect the magnetic field and upper atmosphere of Earth, disturbing radio communications and causing spectacular light displays in the upper atmosphere.

Solar Wind

Another aspect of the Sun that affects Earth is the solar wind. The solar wind occurs as particles that were blasted out of flares and spots rush outward through interplanetary space. The solar coronal plasma, having been heated to nearly 2 million K by the violence of photospheric convection, also expands rapidly into space (limited only by magnetic forces acting on charged particles). Together these effects cause the **solar wind,** which is an outrush of gas past Earth and beyond the outer planets. Near Earth, the solar wind travels at velocities near 400 km/s and sometimes reaches 1000 km/s. The gas has cooled only to 200,000 K, but it is so thin that it transmits no appreciable heat to Earth. According to spacecraft data, the solar wind extends significantly farther than Saturn's orbit.

We know that the Sun is made of mostly hydrogen and helium and that the atoms are hot enough to be highly ionized (they have their electrons stripped off). It follows that the solar wind is composed of protons, electrons, and helium nuclei. In addition to the solar wind, solar radiation itself exerts an outward force on small dust particles. This effect, which is greater on small particles, is called *radiation pressure*. Together these cause the forces that blow comet ion tails away from the Sun.

Effects of the Sun on Earth

There could be no life on Earth without the Sun. In this book we have explored the more inhospitable environments of the solar system and can appreciate Earth's beneficial relationship with the Sun. All human resources derive from the Sun's energy. This is true of our food, since the Sun is the bottom of the food chain, and it is true of our fuel, since most of it is fossil-fuel-derived from photosynthesis of the Sun's radiation. Interestingly, the Sun affects Earth in more subtle ways, as we learn next.

Auroras

Those who live at far northern and southern latitudes on the Earth's surface are occasionally treated to a spectacular light show in the sky. This is called an **aurora**. It is caused by high-energy particles crashing into the atmosphere near the magnetic poles, causing intermittent auroras in the Arctic and in Antarctica. These auroras are also called the northern and southern lights (or *aurora borealis* and *aurora australis,* respectively). See Figure 11–15 for two examples.

What causes this colorful display? Solar flares emit not only radiation, such as X rays, but also streams of atomic particles, such as protons and electrons. These join and enhance the solar wind. If a flare directs material toward Earth, the enhanced solar wind hits Earth after a few days' travel. During solar flares, the surge in the solar wind is often so strong that Earth's magnetic field is seriously distorted, affecting the motions of charged particles throughout Earth's vicinity. As the solar wind sweeps around the outer limits of Earth's outermost atmosphere, or ionosphere, it builds up voltages of 100,000 volts (V) or more between the outer regions of the magnetic field and the atmosphere. This voltage drives some charged particles along the magnetic field lines toward the poles (dashed lines in the middle of Figure 11–16). These particles crash into the upper atmosphere, excite the gas atoms there, and cause them to glow. This glow can be seen from the ground.

The Sun and Climate Change on Earth

In addition to giving Earth the haunting lights of the aurora, the Sun may profoundly affect Earth's climate. Researchers suspect that long-term changes in solar radiation cause substantial variations in global climate. For example, during the period from 1645 to 1715, sunspot numbers were unusually low (see Figure 11–13). Tree-ring patterns and other evidence suggest that Earth's climate altered during this period. In particular, records show that northern Europe was gripped by a mini-ice age during this period. Similarly, scientists have found a correlation between low sunspot activity and severe Ethiopian droughts that occurred over a period of more than four centuries (see Figure 11–17). Besides studying sunspots, astronomers are studying other changes in the Sun, using increasingly sophisticated methods.

It is straightforward to measure how much of the Sun's energy reaches Earth. We know that 1370 J of energy reaches every Sun-facing square meter of Earth every second. Although the overall rate shows variations of only 0.1 percent per year, the percentage changes in certain parts of the spectrum, such as X rays or ultraviolet radiation, can be much larger. Since ultraviolet radiation is absorbed by the ozone layer, a change in the ultraviolet output of the Sun can

FIGURE 11–15
Examples of the aurora, both viewed in Alaska. **(a)** A typical curtain display. (SOURCE: Photograph by Nancy Simmerman, Alaska Photo.)
(b) An unusual ring of vertical rays. (SOURCE: Photograph by Michio Hoshino, Alaska Photo.)

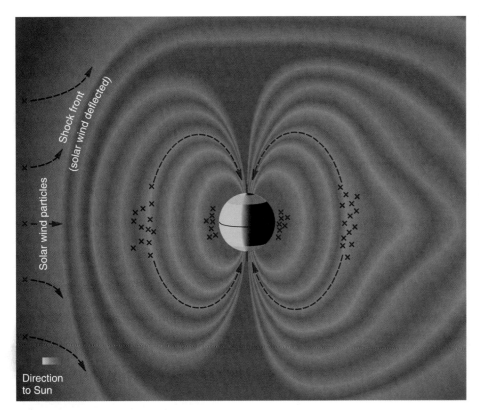

FIGURE 11-16
Interaction of solar wind ions (from left) with Earth's magnetic field. The shock front is analogous to the bow wave cut by a moving boat. Dashed lines show the typical paths of solar ions (×'s). Those ions that penetrate Earth's field accumulate in doughnut-shaped van Allen radiation belts around Earth. Concentrations of ×'s mark their positions. Ions in the belts eventually empty into Earth's polar atmosphere, colliding with air molecules and forming auroral zones near the north and south magnetic poles.

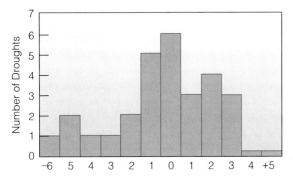

FIGURE 11-17
Correlations between sunspot activity and Earth's weather. A graph of suspected correlation between terrestrial weather and the solar sunspot cycle. Major droughts in Ethiopia, recorded from 1540 to 1985, appear to correlate with sunspot minima. Such studies may benefit agriculture by helping to predict weather cycles. (SOURCE: Data from Wood and Lovett, with updates from *Brittanica Yearbook*.)

change atmospheric heating and structure. Small changes in the heat budget of the atmosphere can have serious effects. Changes in mean annual temperature by only a degree can cause dramatic changes in climate and food production. This is why signs of global warming are causing such concern. The whole question of solar influence on climate and agriculture is attracting new research. Solar studies have growing importance for astronomy, meteorology, agriculture, and world economics. The relationship between variations in the Sun and climate on Earth is a scientific mystery that is still being unraveled.

Solar Energy and Other Cosmic Fuels

Solar energy has accumulated in the ground as coal and petroleum, the remains of living organisms. Industrial civilization has been built with this fossil solar energy, and a region's standard of living is correlated with its rate of consumption of such energy. But while it took about 300 million years to accumulate the fossil fuels, we have burned through a significant fraction of them in only 100 years.

With the Sun providing every day-lit square meter in Earth's vicinity with 1.37 kilowatts (kW), solar energy seems a likely alternative to fossil fuel. A giant test of such technology is the power station Solar One, near Barstow, California. It uses 1818 mirrors to focus sunlight on a central collector, generating 10,000 kW of electricity. More down to Earth is the fact that the roof area of a typical home intercepts roughly 50 to 100 kW of solar energy, equal to the average daily energy consumption of a U.S. household. Rooftop solar tiles, still under research, convert photons directly to electric current that can be stored in batteries. Widespread use of this technology would drive its cost down. This in turn would make homes more self-sufficient and relieve energy expenses, which average $1000 per year or more for many households in the United States.

The two energy sources that humans depend on are fossil fuels and nuclear fission energy. Unfortunately, the store of fossil fuels is finite, and their consumption leaves unwanted by-products. Nuclear fission has the overwhelming disadvantage of involving and creating deadly radioactive wastes that could wreak health havoc if they escaped into the environment, by either accident or terrorism. Nuclear fusion is attractive because it uses a solar-type reaction acting on hydrogen, which is in abundant supply in the form of water.

We have given other "cosmic" fuels less attention, but they are extremely promising. Geothermal energy taps heat from the interior of Earth, whose origin is the decay of naturally occurring radioactive elements. Two other clean and abundant fuels are tides (drawing from the gravitational energy of the Earth-Moon orbit) and winds (drawing from the solar energy that drives Ear… search, technologies are n… these renewable energy sour… lar to the cost of electricity g… only things standing in the wa… are inertia and, often, govern… tive fuels have an additional s… can be collected almost anywh… make countries energy-indepenc… planetary culture.

Summary

The Sun warms us and illuminates us, and we know more about it than any other star. All we can see directly is the yellowish surface layer, the photosphere. However, astronomers have pieced together information from a variety of observations to make a detailed model of the structure, composition, and energy source of the Sun. This process is typical of the application of the scientific method to study remote objects.

We have used spectroscopy to reveal the solar composition. Spectral emission lines are created in hot regions of the solar photosphere, and spectral absorption lines are created in the solar spectrum as radiation from the Sun's interior passes outward through the layers of cooler gas. The distinctive wavelength pattern of these lines shows that the mass of the Sun is about three-quarters hydrogen; most of the rest is helium, and less than 2 percent is heavier elements.

As the Sun formed by contraction of an interstellar gas cloud, it got so hot that atoms at the center collided at high speeds. Now it is a stable sphere, with a temperature that falls from 15 million K at the center to 5700 K at the edge. At a temperature above 10 million K, collisions between particles are violent enough to cause nuclear reactions in which hydrogen nuclei are fused into helium nuclei. This energy is released with remarkable efficiency by mass-energy conversion. We distinguish this process, called nuclear fusion, from nuclear fission, in which energy is released from the splitting of heavy nuclei. Astronomers believe that the most important reactions in the present-day Sun are a series called the proton-proton chain. By understanding the phenomenal energy that comes from nuclear reactions, we have solved an old problem concerning the age of the Sun. Astronomers predict that the Sun is 4.6 billion years old, a middle-aged star, about halfway through its conversion of hydrogen to helium.

Nuclear fusion is a universal energy source. Every star in every galaxy in the universe is powered by it. One of the signatures of the proton-proton chain of fusion reactions is the release of a tiny particle called a neutrino. Neutrinos travel freely from the Sun's fiery core to Earth and rarely interact with matter. The interior of the Sun is opaque to light, but neutrinos allow us to "see" the solar core. We have detected neutrinos on Earth that clearly come from the Sun. This is persuasive evidence that the Sun is indeed powered by nuclear fusion.

Energy travels very slowly from the Sun's core to the outer region. As photons diffuse outward, they degrade in energy from X rays to visible radiation. Energy is carried from the Sun's center to the outer layers by radiation (the direct travel of electromagnetic radiation) and convection (the rolling motions of large pockets of gas). We see the Sun vibrate and oscillate in many patterns, the focus of the subject called solar seismology. We see a sharp edge to the Sun, due to the opacity of the solar gas to radiation. In fact, the density and temperature decrease smoothly with increasing distance from the center, with no abrupt changes. Beyond the edge of the photosphere lies the chromosphere and the hot and tenuous corona. Energy flow through the Sun violently disturbs the surface, producing phenomena such as prominences, flares, and sunspots. Sunspots vary in a 22-year solar cycle, responding to widespread changes in the Sun's magnetic field. We see particles shot off the Sun in an outward-moving gas called the solar wind, which interacts with Earth, causing auroras and other phenomena.

The Sun supports all life on Earth. The Sun's energy also supports our technological lifestyle. Solar energy offers a limitless resource to replace our dependence on fossil fuels. In addition, subtle variations in the Sun's properties profoundly affect Earth's climate in ways that we are only just beginning to understand.

Important Concepts

You should be able to define these concepts and use them in a sentence.

sunspots	solar core
Kirchhoff's laws of radiation	photosphere
solar composition	solar seismology
mass-energy conversion	opacity
binding energy	chromosphere
nuclear fission	corona
nuclear fusion	solar cycle
proton-proton chain	solar wind
neutrino	aurora

How Do We Know?

These questions and answers show how the scientific method is used to learn about the universe.

Q How do we know what the Sun is made of?

A If the solar wind comes from the Sun, then its particle composition provides direct evidence of the surface composition of the Sun. More indirectly, we can carefully analyze sunlight by spreading it out in wavelength in the form of a spectrum. Superimposed on the smooth continuum are narrow features at specific wavelengths. The pattern of wavelengths exactly matches the spectrum we get when

...gen and helium in a glass tube and observed [with a] spectrograph in the laboratory. The pattern [of li]nes from each element is like a fingerprint, [so we c]an recognize the same patterns in the Sun's radia[tion.] We are therefore confident that the Sun is primarily made of hydrogen and helium, with small amounts of heavier elements.

Q How do we know what powers the Sun?

A Again, we have no direct evidence. However, we know the distance to the Sun, and that plus the apparent brightness of the Sun gives us the true or intrinsic brightness of the Sun, by using the inverse square law. Then, knowing the size of the Sun, from its distance and angular size (see the discussion in Chapter 2), we can calculate how much energy radiates from each square meter of the Sun's surface. It is easy to show that this is far more energy than any chemical source can produce. We know of a much more efficient energy source that takes hydrogen nuclei and turns them into helium nuclei—nuclear fusion. If the Sun is a gas ball held together by gravity, we can calculate that the temperature in the core is higher than 10 million K. This is hot enough for the fusion process to operate.

Q How can we measure the interior structure of the Sun?

A In one case we have direct evidence: The neutrinos that fly directly out of the Sun's core are evidence of the nuclear reactions taking place there. Otherwise, we must use inference, because light only shows us the cool outer layer of the Sun, its photosphere. The Sun does not collapse because the energy produced by nuclear fusion in the core produces an outward pressure that balances the inward force of gravity. The Sun's mass and composition are directly measured. We know the temperature of the edge (from the visible spectrum of the photosphere) and of the core (from the fact that fusion is the energy source). So we can construct a model that predicts the temperature, density, and pressure at every point between the edge and the core.

Q How do we know that the Sun is stable?

A Despite the violence of the collisions that take place within its hot gas, the Sun lives a dull life. Its phase of gravitational contraction took a short time and was completed nearly 4.6 billion years ago. Even though it is powered by a fusion reaction, the Sun is not exploding as a bomb. Once scientists know the mass and energy production mechanism of the Sun, they can make a model of its structure. This model can be used to calculate the temperature, density, and pressure at every point within the Sun. Even though the Sun contains rapidly moving particles and churning volumes of gas, it is in overall equilibrium—it is neither expanding nor contracting. As long as it continues to fuse hydrogen into helium, the Sun will be a stable star.

Q How does the Sun affect Earth's weather?

A All motions in Earth's atmosphere represent energy, and the source of that energy is the Sun. The Sun is therefore the engine that drives Earth's climate in general. Any pattern in the way the Sun's radiation arrives at Earth's surface has the potential to affect the weather. This is obviously true for the 24-h variation of day and night—daily temperature extremes are driven by the Sun, late afternoon thunderstorms are caused as the ground releases heat stored from the Sun, etc. It is also true of the 365-day variation in the angle that the Sun's radiation reaches Earth—the cause of the seasons. However, there is a correlation between climate change on Earth and much more subtle changes in the Sun. Two examples are the 22-year solar cycle and the much slower variations in the Sun's output that occur on the time scale of ice ages. At the moment we do not know how these longer-term changes affect Earth's climate.

Problems

You should use these problems to test your understanding of the information and concepts in this chapter. The * indicates a more advanced or mathematical problem.

1. Why do astronomers infer that the Sun's energy comes from nuclear fusion reactions of the proton-proton cycle? How do we know it does not come from chemical burning? Why is the Sun's energy generated mostly at its center and not near its surface? What is the temperature required for this fusion reaction, and why is it so difficult to re-create this reaction on Earth?

*2. Calculate the kinetic energy of a 1-kg meteor traveling at the speed of sound (330 m/s). The kinetic energy is given by $\frac{1}{2}mv^2$ (m is mass and v is velocity). If the meteor were to hit the ground, the meteor would have the destructive power of 0.1 kg of TNT. Now calculate the mass-energy associated with the same meteor. How much larger is it than the kinetic energy of motion?

*3. Use the equation for mass-energy conversion to calculate how many hydrogen atoms would have to be fused per second to equal the energy output of a 100-W lightbulb.

*4. In Chapter 4, we considered the kinetic energy of everyday objects in motion. For each of the following examples, calculate the mass-energy of the projectile and compare it to the kinetic energy. How much larger is the mass-energy in each case? **(a)** A fastball with a mass of 0.25 kg and a kinetic energy of 300 J. **(b)** A bullet with a mass of 0.05 kg and a kinetic energy of 4000 J. **(c)** A 300-kg barbell raised so as to have a potential energy of 5400 J.

*5. If you see a cluster of sunspots in the center of the Sun's disk, how long would the spots take to reach the limb (apparent edge of the disk), carried by the Sun's rotation? How long would it take for the cluster to appear again at the center of the disk?

6. Why is the solar cycle said to be 22 years long, even though the number of sunspots rises and falls every 11 years? Why is a radio disturbance on Earth likely to occur within minutes of a solar flare near the center of the Sun's disk, whereas an aurora occurs a day or two later, if at all?

7. Suppose you could make detailed comparisons of the appearance of the Sun at different times. What variations in appearance would you see at intervals of **(a)** 10 min, **(b)** 1 week, **(c)** 5 years, **(d)** 10 years, and **(e)** 100 million years? Why will the Sun change drastically in several billion years? Do changes in the Sun's surface features—sunspots, flares, and prominences—correspond to large changes in the total energy output?

8. How much more massive is the Sun than the total of all planetary mass (see Appendix B-1 for a table of planetary data)? If Earth formed in the same gas cloud as the Sun, why is Earth made from different material than the Sun? What is the most abundant element in the solar system? The second-most-abundant element?

9. It takes a photon many years to diffuse out from the place where it is produced by the fusion reaction to the place

where it leaves the Sun's surface. Why? What does this diffusion tell you about the likelihood that the Sun will vary significantly from one year to the next, or even over a human lifetime?

*10. In the text, it states that 1370 J reaches every sunlit square meter of Earth's surface every second. Also, it states that the U.S. energy consumption is 8×10^{18} J/yr. **(a)** Use these two pieces of information to calculate the collecting area of solar panels that could supply all the country's energy needs (for simplicity, assume that the solar panels are perfectly efficient and that the Sun shines for 8 h/day). **(b)** Suppose we could cover part of the surface of Mercury with perfectly efficient solar panels and beam the radiation back to Earth. What area of Mercury would we have to cover to meet Earth's energy needs?

11. Describe the effects of the Sun on Earth that go beyond the amount of visible radiation that reaches Earth. What evidence do we have that some terrestrial phenomena are linked to the sunspot cycle?

Projects

1. In a pinhole camera, light passing through a small hole casts an image if projected onto a screen at a distance that is much larger than the diameter of the hole. Confirm this principle by cutting a 1-mm hole of any shape in a large cardboard sheet and allowing sunlight to pass through the hole onto a white sheet in a dark room or enclosure several feet away. Confirm that the projected image is round, an actual image of the Sun's disk. Are any sunspots visible? How would you increase the contrast of your image to make it more likely to capture and view sunspots?

2. In the previous experiment, measure the diameter of the image and the distance between the pinhole and the screen. The size of the image divided by the distance between the pinhole and screen gives the same ratio as the diameter of the Sun divided by the distance between the Sun and Earth (the astronomical unit is 1.496×10^8 km). Do you know why? Multiply the ratio you measure by the value of the astronomical unit to get the diameter of the Sun. Measure several times, and take an average value to improve the accuracy. How does your answer compare to the answer in this book? Think of what might be limiting your accuracy, modify your apparatus, and repeat the experiment.

3. In the Science Toolbox on mass-energy, it was stated that the mass of the dot at the end of this sentence is about 10 μg. How could you check this claim with a simple experiment? Imagine weighing the dot with a postal scale, and think of how many dots you would need to get onto the page or pages to do the experiment.

Reading List

Bahcall, J. N. 1990. "Where Are the Solar Neutrinos?" *Astronomy,* March, p. 40.

Edberg, S. J. 1995. "Discovering the Daytime Star." *Astronomy,* vol. 23, p. 66.

Foukal, P. 1990. "The Variable Sun." *Scientific American,* vol. 262, p. 34.

Hathaway, D. H. 1995. "Journey to the Heart of the Sun." *Astronomy,* vol. 23, p. 38.

Kennedy, J. R. 1996. "GONG: Probing the Sun's Hidden Heart." *Sky and Telescope,* vol. 92, p. 20.

Kerr, R. A. 1986. "The Sun Is Fading." *Science,* vol. 231, p. 339.

Lang, K. R. 1996. "Unsolved Mysteries of the Sun: Parts 1 and 2." *Sky and Telescope,* vol. 92, p. 38, and vol. 92, p. 24.

Marsden, R., and Smith, E. L. 1996. "Ulysses: Solar Sojourner." *Sky and Telescope,* vol. 91, p. 24.

Meadows, J. 1984. "The Origin of Astrophysics." *American Scientist,* vol. 72, p. 269.

Mechler, G. 1995. *The Sun and the Moon.* New York: Knopf.

Mims, F. M. 1990. "Sunspots and How to Observe Them Safely." *Scientific American,* vol. 262, p. 130.

Nesme-Ribes, E., Baliunas, S. L., and Sokoloff, D. 1996. "The Stellar Dynamo." *Scientific American,* vol. 275, p. 45.

Noyes, R. W. 1982. *The Sun, Our Star.* Cambridge, MA: Harvard University Press.

Wentzel, D. G. 1991. "Solar Chimes: Searching for Oscillations inside the Sun." *Mercury,* vol. 20, p. 77.

CHAPTER 12

Properties of Stars

CONTENTS

- Fingerprinting Stars
- The Nature of Stars
- Distances to Stars
- Observed Properties of Stars
- Fundamental Properties of Stars
- Classifying Stars
- The Evolution of Stars
- Summary
- Important Concepts
- How Do We Know?
- Problems
- Projects
- Reading List

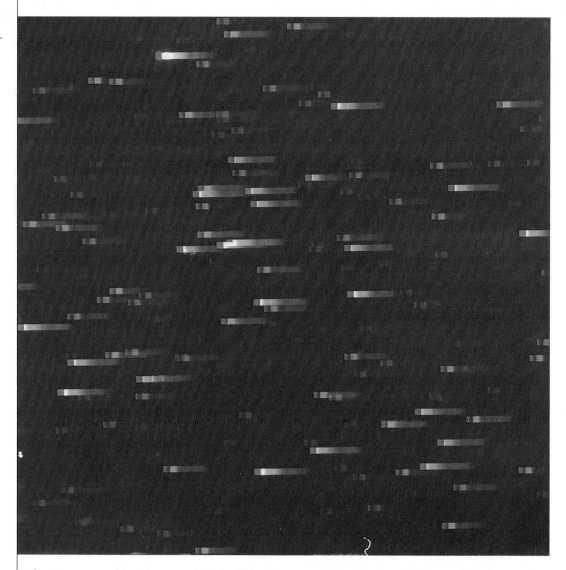

Spectra were used by pioneers early in the 20th century to unlock the secrets of the stars. This photograph was made using a thin prism in the light beam, so that the light from each star was spread into a little spectrum. By studying thousands of spectra like these, astronomers determined the chemical composition and physical properties of the stars in the night sky and showed that many of them were like the Sun. (SOURCE: University of Michigan Observatories.)

WHAT TO WATCH FOR IN CHAPTER 12

In Chapter 10, we described the characteristics of the Sun, a typical and well-studied example of a star. In this chapter, we discuss the properties of stars in general. You will learn that a star's distance from Earth is a key property. Stars have a very wide range of intrinsic brightness or luminosity, so a star that is far away and luminous may appear to have the same brightness as one that is nearby and has low luminosity. Distances distinguish between these two possibilities. You will learn how astronomers measure the distances of stars near the Sun by the parallax technique—a simple application of trigonometry. Spectroscopy yields further information, such as the star's temperature, chemical composition, and motion toward or away from the observer. We will see that other important properties, such as mass and size, cannot be measured directly; they must be inferred from a model of the structure of the star. From these specific properties, we can then proceed to classify stars generally according to their surface temperature and luminosity. Stars like the Sun have a clear relationship between temperature and luminosity; these are called main-sequence stars. Some stars have sizes and luminosities that differ greatly from the Sun. Finally, you will learn that stars are not unchanging, as was once believed. All stars must evolve as they use up their cosmic fuel supply. Models show that the diverse life stories of stars are dictated primarily by their mass.

PREREADING QUESTIONS ON THE THEMES OF THE BOOK

OUR ROLE IN THE UNIVERSE
How does the Sun compare to other stars?

HOW THE UNIVERSE WORKS
What are the physical rules that govern the different types of stars?

HOW WE ACQUIRE KNOWLEDGE
How can we understand the properties of remote objects like stars?

FINGERPRINTING STARS

One hundred years ago, dozens of women labored in the basement of the Harvard College Observatory, paving the way for the classification and understanding of stars. The women examined photographic plates on which many stellar images had been smeared out into tiny spectra. They painstakingly measured hundreds of thousands of stellar spectra, noting the wavelength of each of the prominent lines. Known as "computers," these women were paid 25 cents an hour (less than half of a man's rate of pay for similar work). Not only did they analyze data, they also carried out many tedious calculations manually, in the days before electronic calculators. At the time, women could not be staff members at the Observatory, nor could they take classes or earn a degree at Harvard University.

The photographic plates being studied were part of a large survey of the properties of stars. Spectroscopic pioneers of the 19th century noticed that few stars had spectra exactly matching the spectrum of the Sun. Some had more powerful lines of hydrogen; in others, the hydrogen lines were extremely weak. The chemical composition revealed by the "fingerprint" of spectral lines varied from star to star. Astronomers wondered whether the chemical composition related to any other property of a star.

Annie Jump Cannon was one of the most important members of the classification project. Born in 1863, Cannon was one of the first women from her state to attend a university. After graduating from college, she was hired by Harvard College Observatory as a computer. Working with incredible speed and precision, she steadily began to make bigger and bigger contributions to the project. Cannon proposed a spectral classification system that is still used in astronomy today. She arranged stars by temperature in alphabetical categories. Her sequence runs from white-hot A stars, through yellowish G stars, such as the Sun, to cool red M stars. Cannon personally classified 225,300 stars, which was a heroic contribution to her subject. Later in her career, Cannon received many honors and awards. She was the first woman to receive an honorary doctorate from Oxford University. She used the cash portion of one award to establish a prize that is still given to honor the achievement of young women in astronomy.

In the 1920s, Cecilia Payne-Gaposhkin took the data from the classification project and extended it. Payne-Gaposhkin was born in England and educated at Cambridge University. The first person to receive a Ph.D. in astronomy at Harvard University, she eventually became a professor there. Payne-Gaposhkin realized that the spectral variations among stars were mostly a reflection of temperature. Her unifying idea allowed the apparent diversity of stars to be understood in terms of a temperature sequence.

Science is an incremental process. Fundamental advances in science often begin with the apparently tedious act of classification. Observations are compiled and sorted in the hope that patterns will emerge in the data. The patterns suggest a hypothesis, which in turn leads to a theory based on well-tested physical ideas. Much the same process was followed in the subject of paleontology, where patterns in the fossil record were eventually understood in terms of the theory of evolution and the mechanism of natural selection. Likewise, the classification of spectral smudges on photographs was just as essential as the insights of theorists to the understanding of how stars work.

Recognition for women in science came slowly. Some 100 years after the work of the female "computers," the first woman became a full professor of physics at Harvard University. Thanks to pioneers like Cannon, Payne-Gaposhkin, and others, we now know that the stars are hot balls of gas like the Sun. They are made of hydrogen and helium with small traces of heavier elements, just like the Sun. They get their energy from the fusion process. Our ideas of how stars work can be tested against observations of the full range of stellar types. We have made the night sky our laboratory.

The Nature of Stars

Now we are ready to take the great leap out of the solar system and into the realm of the stars. **Stars** are objects held together and powered by gravity, with so much central heat and pressure that energy can be generated in their interiors by nuclear reactions. It is quite a leap to the stars—the nearest star is 260,000 times as far from us as we are from the Sun. The nearest star is 6800 times as far away from us as we are from Pluto. If we made a model solar system the size of a half-dollar, the neighboring stars would be dots smaller than the period at the end of this sentence, scattered about half a block apart.

How do we know the properties of such distant stars? Optical limitations and turbulence in Earth's atmosphere stop even the world's largest telescope from distinguishing details smaller than a few hundredths of a second of arc (about 10 millionths of a degree). But the stars with the largest apparent angular size are no larger than about 0.06 second of arc. Therefore, the disks and surface details of nearly all distant stars are hidden from us, in contrast to the great detail we can study on the Sun. In spite of these limitations, we know that there are giant stars bigger than the whole orbit of Mars, stars the size of Earth, and stars the size of an asteroid. There are red stars and blue stars. There are stars of gas so thin you can see through parts of them, and stars with rocklike crusts and properties similar to diamonds. There are stars that are isolated spheres and stars that are locked in a binary embrace. Some stars are blowing rings of gas and dust, and other stars have exploded, leaving us with a view of the glowing debris.

All three themes of this book are echoed in the discussion of stars. Scientists use the full extent of the scientific method to understand stars. Radiation must be gathered from remote regions of space, focused with a telescope, and spread into a spectrum. Some stellar properties are measured directly, like temperature and chemical composition. Other properties can only be inferred using a model for the structure of the star, like size and mass. Stars also illustrate the widespread application of physical laws in the universe. We will see that the distant points of light in the night sky, and our fiery neighbor the Sun, share the same physical laws—the same law of gravity, the same type of energy source caused by the fusion of atomic nuclei, and similar chemical compositions. In addition, our knowledge of stars bolsters the Copernican worldview. The Sun is a typical star, of middle age and medium weight.

Stars are so diverse that they are best understood by assembling large samples to study. The study and classification of populations is a vital aspect of science. Although classification itself does not guarantee understanding, it can point the way forward. For example, paleontology starts with the collection and classification of fossils. The patterns revealed in the fossils of various geological strata give a sense of how species evolve on Earth according to the principles of natural selection. In a like manner, as soon as astronomers could begin to measure different properties of stars, they began to categorize the stars and speculate about their evolution.

In this chapter we will outline several interesting discoveries that have resulted from the study of large samples of stars. First, stars come in a wide variety of forms: massive and not so massive, large and small, bright and faint. Second, stars are using their nuclear fuel and must change with time, even though this change is not visible in a human lifetime. All stars pass through a variety of evolutionary stages. Third, stars of different initial mass evolve at different rates and into different final forms. The nature of star birth and star death will be covered in the next chapter.

Perhaps you can understand the difficulties of studying stars by considering a terrestrial analogy. Imagine yourself an intelligent squirrel living in a forest. All around you lies the bewildering diversity of nature. You notice tall trees, short trees, and tiny saplings. You also notice fallen logs; in some cases, those logs have decayed—helped by weather and insects, they have started to turn back into earth. A squirrel does not live long enough to witness the growth from a seed to a sapling to a tall tree, and the subsequent death and decay. Consider this question: Could an intelligent squirrel deduce the life cycle of a tree without living to witness it? Astronomers face such a situation when they study stars. Compared to the long life of a star, a human lifetime is the blink of an eye. However, by gaining an understanding of a star's structure and power source, astronomers can deduce the past and future of stars of different types. In this small way, at least, we get to escape the confines of human existence.

Names of Stars

Let us examine the first step in classifying stars—naming. Stars are named and cataloged according to several systems, some dating to ancient Greece. Ptolemy's *Almagest,* which contained a catalog of star names, was the main source of astronomical knowledge from antiquity. About a thousand years ago, this knowledge was passed on through Europe by Arabian astronomers. As a result, many of the brightest stars ended up with Arabic names. Since *al-* is the common Arabic article, many star names start with this prefix: Algol, Aldebaran, Altair, Alcor. (Other scientific words also have Arab origins: *algebra, alchemy, alkali,* and *almanac.*) Another reason why so many bright stars have Arabic names is connected with the nature of travel in desert countries. With blistering heat during the day, people in the Middle East used to travel at night, so they became familiar with the stars of the night sky as navigation aids. With trade and commerce, their names were passed on and used by other cultures. You can see the names of the most prominent stars shown in the sky maps in Appendix A–6.

Only the very brightest stars in the night sky have names, and the names are derived from the mythology of a variety of cultures. Under optimal viewing conditions, about 6000 stars are visible to the naked eye. Most of these stars are grouped in constellations and cataloged by astronomers in approximate order of brightness, using Greek letters. Thus, the brightest star in the constellation of the Centaur is called Alpha Centauri (α Centauri). A well-known variable star, Delta Scuti (δ Scuti) is the fourth brightest star in the constellation of the Shield. Fainter stars, or stars with unusual properties, are often known by letters followed by constellation names or by catalog numbers, such as T Tauri (the twentieth variable star cataloged in the constellation of the Bull). The many stars fainter than the eye can see are usually anonymous; astronomers give them labels based on the coordinates that describe their position in the sky.

Stars are named after their discoverers only in very special circumstances. The brightest stars in the sky were all named hundreds of years ago. You may see advertisements encouraging you to pay to "name a star" after yourself or a friend, but these names have no scientific legitimacy. Modern star catalogs do not use proper names; they typically use abbreviations of the catalog name followed by a number. The conventions for naming stars are established by the International Astronomical Union, the professional governing body of astronomy. No one can "own" a star—the night sky belongs to all of us.

Distances to Stars

The vast distances that separate stars and hard to observe are awkward to express in Astronomers use units appropriate to thes easiest to understand is the **light-year** (ab distance light travels in 1 year, which is about 6 million million miles, or 10^{16} meters. You can calculate this easily by recalling that the speed of light is 300,000 kilometers per second (km/s). A light-year is just the speed of light times the number of seconds in a year, or $300{,}000 \times (3600 \times 24 \times 365) = 9.5 \times 10^{12}$ km. Notice that a light-year is not a metric unit (see Appendix A–3); metric units were designed for terrestrial use, and astronomers often use units that are better suited to the scale of the cosmos.

Remember that a light-year is a unit of *distance,* not time. The common mistake of using the light-year as if it were a unit of time is like saying that the ball game lasted for 2 miles. On Earth, light travels so fast that it appears to arrive instantaneously. We saw in Chapter 10 how difficult it was for early scientists to measure the speed of light. Over the vast reaches of space, however, light takes a substantial time to travel from one star to another. The nearest star beyond the Sun, Proxima Centauri (which is in orbit around Alpha Centauri), is about 4.3 ly away. The Sun is 8 light-minutes away from Earth. The North Star, Polaris, is about 650 ly away. Polaris' light takes about 650 years to reach us, so we are seeing it now as it was in the 1300s! If Polaris had suddenly exploded in 1950, we would not know it until about A.D. 2600.

Astronomers more commonly use a still larger unit of distance called the **parsec** (abbreviated pc). Recall that a parsec is just over 3 light-years; the exact conversion is 1 pc = 3.26 ly. In metric units, 1 pc = 3×10^{16} m. The term *parsec* refers to the fundamental way that astronomers measure distance by geometry. A parsec is the distance that produces a *par*allax shift of 1 arc *sec*ond. (The concept of parallax was introduced in Chapter 3 and will be discussed later in this chapter.) We will use the parsec and its multiples—kiloparsecs (abbreviated kpc) = 10^3 pc; megaparsecs (abbreviated Mpc) = 10^6 pc—to express cosmic distances as we move to more remote parts of the universe. You can convert parsecs to light-years roughly by multiplying by 3. Another convenient fact to remember is that near the Sun, stars are roughly 1 pc apart. For instance, Alpha and Proxima Centauri, the closest stars to the solar system, are about 1.3 pc away.

We now have a better sense of the scale of space. A parsec is 40 million times the diameter of the Sun. Even though the stars can appear close to each other in the night sky, this is an illusion of projection. In three-dimensional space, stars are separated by distances millions of times their size. (In a similar way, if you viewed a region of sparsely planted trees from the side, they would appear to be closely spaced. Only when viewed from above would the sparseness be obvious.) A parsec is also about 200,000 AU, or 2500 times the diameter of the solar system. If we assume that other stars have planets too, then the full extent of the planetary orbits fills only about $1/2500^3$, or less than a ten-billionth, of the volume of space. Space is fantastically empty.

FIGURE 12-1
Apparent brightness can be a poor guide to distance. On a dark night, and with no further clues, a nearby bicycle light might appear to be as bright as a more distant truck light. The intrinsic brightness of the two lights is, of course, very different. The apparent brightness of each light varies according to the inverse square law. (SOURCE: Courtesy M. Seeds.)

Apparent Brightness

A nearby flashlight may appear to be brighter than a distant streetlight, but in absolute terms, the flashlight is much dimmer. This statement contains the essence of the problem of stellar brightness. A casual glance at a star does not reveal whether it is a nearby glowing ember or a distant great beacon. (Figure 12–1 illustrates this fundamental ambiguity.) Astronomers must distinguish between how bright a star *appears to be* and how bright the star *really is*. *Apparent brightness* is the brightness perceived by an observer on Earth, and *absolute brightness* is the brightness that would be perceived if all stars were magically placed at a standard distance. There can be a great difference between the total amount of radiation a star emits and the amount of radiation we measure at Earth's surface.

Apparent brightness can be defined as the number of photons per second collected at Earth from an astronomical source. Apparent brightness depends on the light-collecting aperture of the viewing object and on the distance to the source. A star appears much brighter when viewed through a telescope than when viewed with the naked eye; the larger aperture of the telescope can collect many more photons per second. Similarly, the closer a star is, the brighter it appears; the number of photons intercepted by our light-gathering device decreases by the inverse square of the distance. Figure 12–2 shows the effect of distance on the apparent brightness of a light source. (Also review the discussion of how radiation travels in Chapter 10, and look at Figure 10–7.) Now you can understand why astronomers build increasingly larger telescopes. A larger collecting area offsets the diminishing amount of light that reaches us from more and more distant sources. Larger telescopes can detect more distant sources, and they can capture more photons from nearby sources.

The most direct way to quote apparent brightness is in units of photons per second. However, the most convenient way to measure apparent brightness is to express it as a ratio to the apparent brightness of the Sun or some other prominent star. This ratio allows us to compare how much brighter or dimmer a star is than our Sun or other familiar star. Apparent brightness defined in this way requires no units, because the ratio of two measurements that have the same units is a pure number.

FIGURE 12-2
The difference between the apparent and intrinsic brightness of a star. The star emits light in all directions. The total amount of light radiated in all directions gives the absolute brightness, or luminosity, of the star. The apparent brightness of the star is given by the amount of light that crosses a particular area (shown here as area A) at a certain distance from the star (*d*). By the inverse square law, the light that passes through area A at distance *d* must pass through an area 4A at a distance 2*d* and an area 9A at a distance 3*d*. The apparent brightness measured through a fixed area A therefore diminishes with the square of the distance. At a fixed distance, the apparent brightness of a star varies with absolute brightness. If the distance is known, and the apparent brightness is measured, the intrinsic brightness of the star can be determined using the inverse square law.

Table 12-1 lists the apparent brightness of some well-known astronomical objects relative to the bright star Vega (astronomers traditionally use Vega to calibrate their brightness measurements). The table shows that the Sun is by far the brightest object in the sky. You can use the table to create ratios that show the relative brightness of different objects. For example, the Sun is 11 billion times brighter than Sirius, the brightest star. We can see this because $4 \times 10^{10} / 3.6 = 11 \times 10^9$. The best telescopes in space can detect objects about 800 million times fainter than the eye can see! We can see this because $0.0025 / 3 \times 10^{-12} = 8 \times 10^8$.

Let's see what else we can learn by simple estimation from the numbers in Table 12-1. We can see that three of the planets appear brighter than any star when they are closest to Earth. However, each of them intercepts only a tiny fraction of the Sun's rays and imperfectly reflects them back to Earth. Jupiter, for example, is about nine-billionths ($3.6 / 4 \times 10^{10}$) as bright as the Sun. You can also see the effect of the inverse square law in the relative brightness of the different planets. Planets farther from the Sun appear fainter (although the different size of the planets plays a role too). The three outer planets are all too faint to be seen with the naked eye unless you are far from city lights, and Pluto is over 50 million times fainter ($10^{-6} / 58$) than Venus at its brightest.

Can we use the information in Table 12-1 to estimate the distance to the nearest stars? Yes, if we make the bold *assumption* that the brightest stars are just like the Sun. We are therefore assuming that stars emit the same number of photons per second as the Sun, and the difference in apparent brightness is a measure of how the photons thin out with increasing distance. This assumption also implies that stars with the highest apparent brightness are the nearest. By the inverse square law, the brightest few stars must be about $\sqrt{10^{10}} = 100,000$ times farther than the Sun. This is a distance of $1.5 \times 10^8 \times 10^5 \approx 10^{13}$ kilometers, or about ⅓ parsec. The 6000 stars visible with the naked eye span a range of $3.6 / 0.0025$, or a factor of 1440. If we *assume* that they are also like the Sun, we infer a distance range spanning a factor of $\sqrt{1440} \approx 40$ for the stars you can see in the night sky. We can also relate star brightness to a more familiar terrestrial object—a lightbulb. From Table 12-1, we can calculate that the Sun is like a 100-watt lightbulb seen at a distance of $\sqrt{27,700 / 4 \times 10^{10}} \times 100 = 0.08$ m, or about 3 inches. (Don't try this; it will hurt your eyes, just as staring at the Sun would!) On the other hand, the brightest star is like a 100-watt lightbulb seen at a distance of $\sqrt{27,700 / 3.6} \times 100 = 8770$ m. This is like looking at a reading light in a house over 5 miles away. Thus, we can get a sense of the enormous range in brightness between the Sun and all the other stars.

In this book, we will calculate relative brightness using a linear scale, as in Table 12-1. However, astronomers use a relative brightness scale based on logarithms. Astronomers use a different system because they are victims of history. When Hipparchus cataloged 1000 stars in about 130 B.C., he ranked their apparent brightness on a *magnitude* scale of 1 to 6, with first-magnitude stars the brightest and sixth-magnitude stars the faintest, in terms of visibility to the naked eye. Viewed

TABLE 12-1
The Apparent Brightness of Selected Objects

OBJECT	APPARENT BRIGHTNESS (RELATIVE TO VEGA)
Sun	4×10^{10}
Full moon	100,000
100-watt lightbulb at 100 m	27,700
Venus (at brightest)	58
Mars (at brightest)	12
Jupiter (at brightest)	3.6
Sirius (brightest star)	3.6
Canopus (second-brightest star)[a]	1.9
Vega	1.0
Spica	0.4
Naked-eye limit in urban areas	0.025
Uranus	0.0063
Naked-eye limit in rural areas	0.0025
Bright asteroid	0.0040
Neptune	0.0008
Limit for typical binoculars	0.0001
3C 273 (brightest quasar)	8×10^{-6}
Limit for 15-cm (6-in.) telescope	6×10^{-6}
Pluto	1×10^{-6}
Limit by eye with largest telescopes	2×10^{-8}
Limit for CCDs with largest telescopes	6×10^{-12}
Limit for the Hubble Space Telescope	3×10^{-12}

[a]This lesser-known Southern Hemisphere star is used as a prime orientation point for spacecraft. A small light detector on spacecraft is called the Canopus sensor. Sirius is not used for this purpose because it is a binary star.
SOURCE: Data from Allen (1973 Astrophysical Quantities, Athlone Press, London).

with the naked eye, stars could only be classified with six gradations of brightness. A difference of one magnitude corresponds to a factor of roughly 2.5 in apparent brightness; five magnitudes represent a factor of 100 in brightness (see Appendix A-6). Vega defines the zero-point of the magnitude scale. Hipparchus invented this nonlinear scale to match the response of the eye. (A more familiar example is the loudness of a sound, which is represented by units of decibels. The decibel unit must be tied to a known sound intensity measured at a fixed distance, since loudness depends on how far you are from a sound.) The 2100-year-old magnitude system is so ingrained that astronomers continue to use it. We will use a linear brightness scale, because it is more intuitive and much easier to work with.

To be meaningful, apparent brightness must be specified at a particular wavelength. Stars have different colors, which means that the apparent brightness depends on the wavelength. Also, light detectors (the eye, photographic films, and electronic CCDs) have different sensitivities to particular colors or wavelengths. For this reason, astronomers specify the exact wavelength to which any set of measurements refers. Standards have been derived to express apparent brightness measured in blue light, red light, infrared light, radio waves, X rays, and so on. Here we will usually be referring to a system having the same color sensitivity as the human eye, which is sometimes called *visual apparent brightness* (corresponding to a range of wavelengths centered on the green part of the visible spectrum).

Absolute Brightness

We have just discussed the apparent brightness of stars and other objects as seen from Earth. Apparent brightness depends on the star's distance and thus does not express the star's true energy output. The true, or intrinsic, energy output is called the star's **absolute brightness,** or **luminosity.** By convention, astronomers have set up a brightness scale that relates the apparent brightness of a star to the brightness of the Sun as seen from a distance of 10 parsecs (Figure 12-2, with the distance set to 10 pc). The exact reference distance is not important, just as we could have chosen to refer the values of brightness in Table 12–1 to an object other than Vega. The relationship between apparent brightness, absolute brightness, and distance is fundamental to astronomy.

Note that the distinction between apparent and absolute brightness would not be important if all stars had the same energy output. Imagine a large darkened room scattered with 100-watt lightbulbs. Assuming you had measured the distance and absolute brightness (the wattage) of just one reference lightbulb, you could deduce the distances of all other lightbulbs by observing their apparent brightness and applying the inverse square law. For instance, a lightbulb that *appeared* to be four times brighter than the reference bulb must be two times closer, and a lightbulb that *appeared* nine times fainter than the reference bulb must be three times farther away. In this situation, apparent brightness is an exact indicator of distance.

Figure 12–3 is a graphical presentation of the inverse square law. In the top graph (Figure 12-3a), three light sources are shown, which have values of apparent brightness of 50, 100, and 200 watts at a distance of 1 meter. (The exact units of the apparent brightness measurement don't matter; they depend on whether the light is collected by eye, with a telescope, or so on.) Each curve diminishes with the inverse square of the distance. The vertical dashed line shows the apparent brightness of the three sources at a distance of 2 meters: $50/2^2 = 12.5$ units, $100/2^2 = 25$ units, and $200/2^2 = 50$ units for the 50, 100, and 200-watt sources, respectively. Notice that at any distance, the values of apparent brightness of the three sources are in the same ratio as their values of absolute brightness.

The lower graph shows how very different objects can appear to have the same brightness. Follow the horizontal dashed line in Figure 12–3b. The three light sources appear to have a brightness of 25 units if the 50-watt source is at a distance of $\sqrt{50/25} = 1.41$ m, the 100-watt source is at a distance of $\sqrt{100/25} = 2$ m, and the 200-watt source is at a distance of $\sqrt{200/25} = 2.82$ m. There are many other distances at which the sources can be placed and have the same apparent brightness, but they will always be in this ratio. Apparent brightness is not a measure of the true output of the source. With a lightbulb, it is easy; we can just look at the top and check the wattage. However, stars do not have their wattage written on them!

In the example just given, the lightbulbs only ranged over a factor of 4 in luminosity. In astronomy, stars differ enormously in luminosity, or the amount of energy they emit each second. Imagine now that a large room is scattered with bulbs of widely different wattage, ranging from 1-watt night lights to 10,000-watt arc lamps. A watt is a measure of energy output per second, or luminosity. In this case, a 1-watt bulb would have the same apparent brightness as a 100-watt bulb 10 times farther away, or a 10,000-watt bulb 100 times farther away. These calculations are simple applications of the inverse square law. Apparent brightness is therefore a very poor indicator of distance. Conversely, the intrinsic energy output of a star cannot be calculated without knowing the distance. Astronomers must deal with this challenging problem of studying stars with very different light output widely distributed through space.

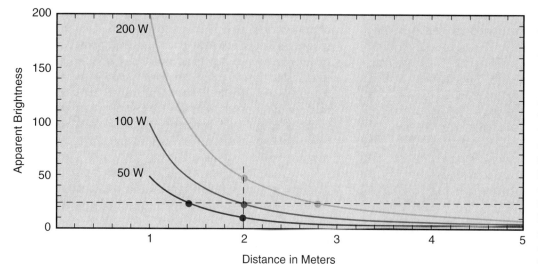

FIGURE 12–3
A graph of the inverse square law of radiation, showing how the apparent brightness of a light source decreases with distance. Three light sources are shown, of 50, 100, and 200 watts. The three places where the vertical dashed line intersects the curves (shown as dots) show the apparent brightness of each source at a distance of 2 meters. Notice that the ratio of the apparent brightnesses of the three sources is the same at any distance. The three places where the horizontal dashed line intersects the curves (also shown as dots) show the distances at which the three sources appear equally bright.

Measuring Distances Directly

The use of the light-year and parsec to measure stellar distances gives you a sense of the vast distances to the stars. To ancient cultures, however, the stars were relatively close to Earth. The Egyptians imagined them as points of light on a tented canopy, held up by mountain ranges at the corners of the kingdom. The Greeks thought of them as fiery embers carried overhead on crystalline spheres. Ancient cultures could not conceive of distances much larger than the size of Earth.

Early distance estimates were no more than educated guesses. In the late 17th century, Dutch scientist Christian Huygens made an image of the Sun through a pinhole in a darkened room. He varied the size of the pinhole until the image seemed equal in brightness to an image of Sirius, the brightest star. Since the pinhole admitted $1/27{,}000$ of the light of the Sun, Huygens concluded that Sirius was 27,000 times farther away than the Sun. Around the same time, Isaac Newton tried to use Saturn as a sort of reflecting mirror to measure the intensity of sunlight. He guessed the percentage of the Sun's light that Saturn reflects and assumed that bright stars have similar absolute brightness to the Sun. Newton concluded that the bright stars are about 18,000 times farther away than the Sun.

As we have seen, another crude method may be applied to the information in Table 12–1. The brightest stars have values of apparent brightness that are about 10 billion (10^{10}) times fainter than the Sun. In 1829, English scientist William Wollaston used this simple fact to estimate that most typical stars must be at least 100,000 (10^5) times more distant than the Sun, since the inverse square law indicates that dimming 10 billion times corresponds to increasing distance 100,000 times.

However, all of these estimates are flawed. The brightest stars in the sky are much more luminous than the Sun, so they are much more distant than we could calculate by assuming that they resemble the Sun. Also, these estimates give only typical distances, rather than accurate distances, for individual stars. Even so, this is an example of how a simple physical idea—the way light dims as it spreads through space—can be used to deduce remarkable information about the distance to the stars. At the time, this reasoning implied a universe thousands of times larger than previous estimates!

As we have seen earlier in the book, ancient astronomers were perplexed by the lack of any observable stellar parallax. It meant that the stars must be fantastically distant. The first successful measurement of stellar parallax did not come until 1838, when German astronomer Friedrich Bessel detected the slight seasonal shift of the star 61 Cygni—only 0.6 second of arc. The measurement of a **parallax distance** is a direct trigonometric technique that is independent of any assumption about the nature of the star being observed. Recall that *parallax* is the angular shift in the position of an object caused by a shift in the observer's position. The calculation of a parallax distance is another application of the small angle equation from Chapter 2. More details are given in the accompanying Science Toolbox.

Hold your finger in front of your face, and look past it toward a bookcase on the other side of the room. Your finger represents a nearby star, and the books on the far wall represent distant stars. Your right eye represents the view on one side of the Sun. Your left eye represents the view 6 months later, after Earth has traveled to a point on the opposite side of the Sun (a shift in Earth's position by 2 AU as seen from the star). First wink one eye and then the other. Your finger (the nearby star) seems to shift back and forth. Hold your finger only a few centimeters from your eyes; the shift is large. Hold your finger at arm's length; the shift is smaller. Likewise, the farther away the nearby star, the smaller the parallax angle. The parallax in this experiment can be measured in degrees, but the parallaxes of actual stars are all less than a second of arc. Parallaxes as small as 0.01 second of arc can be reliably measured, and the distance of such a star is 100 parsecs.

As the following Science Toolbox shows, the distance of a star in parsecs is simply the inverse of its parallax angle in seconds of arc. This explains the origin of the term parsec: it is the distance corresponding to a parallax of one second of arc. If a star is too distant, its parallax is too small to be measured. (Imagine a skinny triangle whose base is thousands of times shorter than its other two sides.) Parallaxes smaller than about 0.01 second of arc are difficult to measure accurately, because of the blurring of Earth's atmosphere. Therefore, stars farther away than about 100 pc are beyond the distance limit for reliable parallaxes. Scientists at the European Space Agency attempted to overcome this limitation when they launched the *Hipparcos* satellite in 1989. Its mission was to measure parallax in the precise viewing conditions of space. The unfortunate failure of a rocket motor placed the satellite in a highly elliptical orbit, which caused the detectors to degrade as they passed repeatedly through the radiation belts around Earth. Despite this handicap, *Hipparcos* successfully measured the parallax of over 100,000 stars.

Knowing accurate distances is the prime requirement for measuring most other properties of stars. Thus, the estimated 20,000 to 25,000 stars that lie within 100 pc are our main statistical sample for measuring stellar properties. Note that even this large sample is not sufficient for us to be able to measure the distance to certain rare stellar types. Other techniques have been devised for estimating distances to more remote stars, but they all depend ultimately on the accuracy of the parallax measures of nearby stars.

Observed Properties of Stars

In the last chapter we saw that the Sun's spectrum, or distribution of light into different colors (wavelengths), provides information about the Sun's atmosphere and surface layers. The same is true of other stars. Spectroscopy—the study of spectra—is a vital tool for understanding the physical properties of astronomical objects, and many astronomers devote their entire careers to it. Chapter 10 discussed the prin-

SCIENCE TOOLBOX

Parallax and Stellar Distances

As Earth moves in its orbit of the Sun, our perspective on the stars changes slightly. Nearby stars show a parallax shift compared to more distant stars. In other words, the apparent position of a nearby star will appear to oscillate slightly with respect to the backdrop of more distant stars. This oscillation repeats every year as Earth orbits the Sun; the full extent of the parallax shift is seen in observations made 6 months apart.

The small angle equation explained in Chapter 2 shows us how to relate angular size to the apparent size of an object as seen from a certain distance. In this application of the small angle equation, the angular size (α) becomes the parallax angle (ρ) and the linear size (d) is the Earth-Sun distance. If we use astronomical units (AU) for the distance, then the small angle equation can be rearranged as

$$D = \frac{206{,}265}{\rho}$$

The distance to a star in AU equals 206,265 divided by the parallax angle in seconds of arc. Figure 12-A shows the geometry of this triangle. Note that in practice we measure an angle of 2ρ because we make observations 6 months apart. The base of the triangle is then twice the Earth-Sun distance ($d = 2$ AU). You should also remember that the angle ρ is greatly exaggerated in Figure 12-A to make it visible; in the astronomical application, the triangle is long and extremely skinny. A typical stellar parallax is 1 second of arc, which is the thickness of a piece of paper as seen from across a large room. The detection of such a tiny angle required the resolution of a large telescope, which is why parallax was not detected until over 200 years after the invention of the telescope.

Now we can see where the distance unit of a parsec comes from. If we *define* one parsec to equal 206,265 AU, then the equation above simplifies to

$$D = \frac{1}{\rho}$$

The distance to a star in parsecs is 1 divided by the parallax angle in seconds of arc. A parsec is the distance to a star whose *par*allax is 1 *sec*ond of arc. This equation is useful because it relates a unit of distance measurement to the most common unit of angular measurement in astronomy. The parsec is a useful distance unit because it corresponds to the typical distance between stars (which is equivalent to saying that the parallax angles of the nearest stars are around 1 second of arc). Rounded to one significant figure 1 pc = 2×10^5 AV.

Note that parallax angle has an inverse relationship with distance. More distant stars have smaller parallax angles. For example, Alpha Centauri has a parallax angle of 0.77 second of arc, which implies a distance of $1/0.77 = 1.3$ pc, or 4.2 light-years. The bright star Altair has a parallax angle of 0.20 second of arc, so it is more distant at $1/0.20 = 5$ pc, or 16.3 light-years.

Below about 0.01 second of arc, the parallax angle becomes too small to measure reliably. As a consequence, stars beyond about $1/0.01 = 100$ parsecs cannot have their distances measured by this direct trigonometric technique. However, there are tens of thousands of stars that are close enough for a parallax measurement.

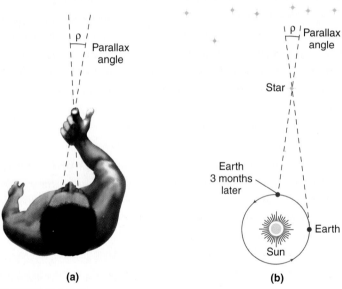

FIGURE 12-A
The principle of parallax determination applied **(a)** to one's finger and **(b)** to stars. Nearby objects observed from two positions appear to shift by a parallax angle ρ (the Greek letter *rho*) compared with background objects; measurement of ρ gives the distance of the object once the separation of the two positions is known. (By astronomical convention, the angle cataloged as a star's parallax corresponds to the angle diagrammed here; but astronomers actually observe from two positions 6 months apart. This yields twice as large a parallax shift, which is easier to measure.)

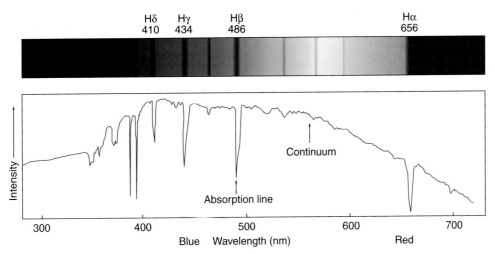

FIGURE 12-4
The visual portion of a typical stellar spectrum shown here as it would appear in color intensity representation (top) and in a graphical version of intensity plotted against wavelength (bottom). The spectral lines appear as dark or bright vertical lines in the color plot and as sharp valleys or peaks in the graph.

ciples of spectra in detail, and you may wish to review that material as you study this chapter. Let us see how spectra can be used to reveal the nature of stars.

Figure 12–4 shows the type of data that astronomers get when they disperse starlight with a spectrograph. The top bar is an idealized example of a spectrum where the wavelength of the light is shown by the appropriate color. By convention, short wavelengths are at the left and long wavelengths are at the right. Absorption lines appear as dark vertical lines; a series of hydrogen lines is marked by their wavelength in nanometers (nm). Emission lines are not present, but they would appear as bright vertical lines. The lower part of the figure shows the spectrum in graphical form, which is the way astronomers usually represent this information. Absorption lines appear as sharp valleys or notches in the spectrum, and emission lines appear as sharp peaks. The graph shows intensity as a function of wavelength. The intensities of the image represent intensities of light, or radiant energy. Usually the blue end is to the left, as in the figure. The general level of brightness between absorption or emission lines is called the *continuum*. You can see that it has the broad distribution of wavelengths typical of thermal emission from gas at a temperature of a few thousand degrees (K).

In the earliest days of spectroscopy, spectra could be viewed only with the naked eye. By the late 1870s, astronomers began to use photographic plates to record spectra observed through a telescope. For 100 years, astronomical data could be recorded only by photography. Photographic spectra have now been superceded by spectra recorded with electronic detectors, or CCDs. The wavelength range of the spectrum that can be recorded depends on the wavelength range of sensitivity of the material of the CCD—typically, silicon.

Spectra of Stars

In 1872, Henry Draper was the first to photograph stellar spectra. This represented a tremendous advance. Instead of sketching or verbally describing spectra, astronomers could directly record, compare, and measure spectral features. Draper began an ambitious project to photograph and catalog all the bright stars in the sky, but he died long before it was complete. His widow then donated enough money to Harvard to continue the work. Harvard Observatory astronomer Edward C. Pickering headed the project to create the Henry Draper Catalog. He hired a large group of women to do the painstaking work of measuring the spectra of thousands of stars.

As described in the opening story, Annie Cannon and a group of young women assistants invented a system of **spectral classes** based on the appearance of spectral lines. Spectra were classified alphabetically according to the strength of the hydrogen absorption lines. Stars with the deepest hydrogen lines were A stars, stars with the next-deepest hydrogen lines were B stars, and so on through the alphabet (the original sequence stopped at P). The most numerous spectral features were lines of hydrogen and lines of helium. Absorption lines of other elements are also seen, and astronomers traditionally lump together all elements heavier than helium under the term "metals." This is clearly a misnomer, because only a few of the elements are actually metals!

When Annie Cannon did her survey, modern atomic theory had not been developed, and astronomers had a poor understanding of the causes of spectral lines. In Chapter 10, we gave a simplified description of how emission and absorption lines depend on the orbital structure of electrons in atoms. Now we look at the process in more detail by considering the simplest and most abundant type of atom in the universe: hydrogen. As shown in Figure 12–5, the orbits of hydrogen atoms can be numbered. Since a neutral hydrogen atom has just one electron, an atom in the ground state would have one electron in the orbit $n = 1$. If the atom had been bumped by other atoms or if it had absorbed radiation, the electron might be in the orbit $n = 2, 3$, and so on. Further absorption of energy might cause it to jump from $n = 3$ to $n = 4$, creating an absorption line, or it might spontaneously revert from $n = 3$ to $n = 2$, creating an emission line. Each possible transition (1 to 2, 2 to 3, 4 to 2, and so on) creates a different line. As Figure 12–5 shows, transitions between the $n = 2$ level and higher levels create the lines prominent in the visible part of the electromagnetic spectrum. The famous line *hydrogen-alpha* (abbreviated Hα) lends its brilliant red color to many astronomical gases, including the solar chromosphere and many nebulae.

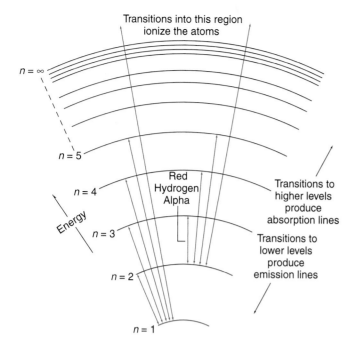

FIGURE 12-5
A hydrogen atom, showing different possible electron orbits or energy levels ($n = 1, 2, 3$, and so on). Each change of orbit by an electron produces a spectral line, because energy is removed or added to the light beam. The series of lines starting from or ending on $n = 1$ occur in the ultraviolet part of the electromagnetic spectrum and are called the Lyman series. Lines starting from or ending on $n = 2$ occur in the visible part of the electromagnetic spectrum and are called the Balmer series. The well-known red Hα line of the Balmer series involves transitions between $n = 2$ and $n = 3$.

Temperature

Spectroscopy is vital to understanding the temperatures of stars. We can study stellar spectra and measure which color is the most strongly radiated. We can then use Wien's law to calculate the temperature. There are clear patterns in the prominence of the different spectral features—this was the puzzle that Cannon and Payne-Gaposhkin had to solve. They rearranged the spectral classes to bring them into a true physical sequence based on temperature, as shown in Figure 12-6. The sequence that was finally adopted begins with the

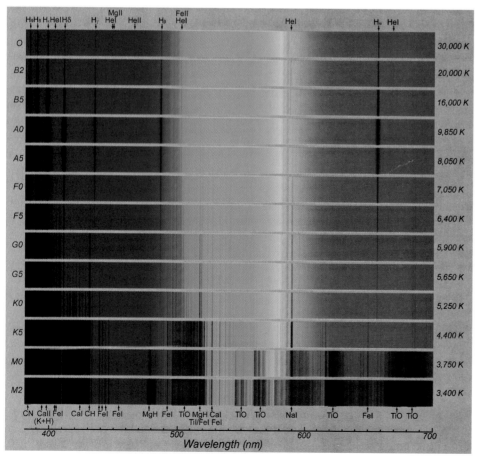

FIGURE 12-6
This computer simulation of the spectra of stars shows the absorption lines that are commonly seen in a wide range of spectral types. The spectral type is shown along the left edge, and the wavelength scale is shown along the bottom. The spectral lines of different atoms are marked along the top, and spectral lines for molecules are marked along the bottom. By convention, He I refers to neutral helium, He II refers to singly ionized helium (one electron removed), and so on. Also by convention, the set of transitions that go to the same lower energy level are assigned names in a series defined by the Greek alphabet, yielding the sequence Hα, Hβ, Hγ, Hδ, and so on. The coolest stars at the bottom of the diagram show spectral bands due to different molecules.
(SOURCE: Roger Bell, University of Maryland, and Michael Briley, University of Wisconsin at Oshkosh.)

TABLE 12-2
Principal Spectral Classes of Stars

TYPE	SPECTRAL CLASS	TYPICAL TEMPERATURE (K)	SOURCE OF PROMINENT SPECTRAL LINES	REPRESENTATIVE STARS
Hottest, bluest	O	30,000	Ionized helium atoms	Alnitak (ζ Orionis)
Bluish	B	18,000	Neutral helium atoms	Spica (α Virginis)
Bluish-white	A	10,000	Neutral hydrogen atoms	Sirius (α Canis Majoris)
White	F	7000	Neutral hydrogen atoms	Procyon (α Canis Minoris)
Yellowish-white	G	5500	Neutral hydrogen, ionized calcium	Sun
Orangish	K	4000	Neutral metal atoms	Arcturus (α Bootes)
Coolest, reddest	M	3000	Molecules and neutral metals	Antares (α Scorpii)

hottest stars, class O, which show ionized helium lines in their spectra. The sequence of classes is O, B, A, F, G, K, and M, from the hottest to the coolest. (The rearrangement of the letters and the omission of many of the original types make the sequence harder to remember, another example of the historical baggage that astronomers carry around. Generations of astronomy students have created phrases as an aid to remembering the sequence, including *Oven-Baked Ants: Fry Gently, Keep Moist; Overseas Broadcast: A Flash! Godzilla Kills Mothra;* and *Oh Boy! Astronomy Final's Gonna Kill Me!* You can make up your own.) About 99 percent of all stars can be classified into these groups. For finer discrimination, the classes are subdivided from 0 to 9. The Sun, for example, is classified as a G2 star.

We can summarize the features of the spectral classes (see Table 12-2). Ionized helium is seen only in the hottest stars, above 30,000 K. Hydrogen absorption lines appear most strongly in stars of classes A and B. At even lower temperatures, lines of ionized calcium, sodium, and iron appear. By class M, at 3000 K, the energies are so low that atoms can stick together into molecules, such as titanium oxide (TiO) and magnesium hydride (MgH). Even water molecules have been identified in the spectra of such cool stars, where of course they take the form of steam! A schematic view of the information from Figure 12-6 is shown in Figure 12-7.

How can we understand the appearance of different spectral features in terms of microscopic physics? Each spectral class corresponds to a different temperature of the photosphere gas of the star. Recall from Chapter 4 that as a gas gets hotter, it emits bluer and more copious amounts of radiation. At a certain temperature, the electrons get stripped from the atomic nuclei. This process is called *ionization* and the positively charged nuclei are called *ions*. Heavier elements require higher temperatures in order to be ionized. You can see from Figure 12-7 that the different spectral features get stronger and weaker as the temperature varies. The physical conditions have to be just right to produce a particular type of electron transition. For example, molecules are seen only in the coolest M stars, because at higher temperatures most molecules are broken into their constituent atoms. Spectral lines of metals are seen in cool G and K stars, but hydrogen lines are not seen because the temperature is not high enough to excite electron transitions in hydrogen atoms. Stars as hot as A stars show strong hydrogen lines, but metal lines are not seen because the heavier atoms are mostly ionized. It takes a high temperature to ionize helium, so helium lines are seen only in O and B stars.

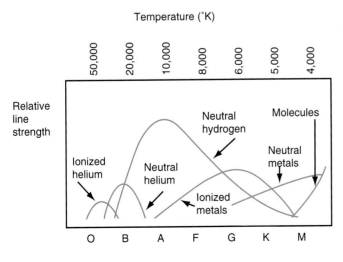

FIGURE 12-7
This diagram shows which absorption lines appear prominently in the spectra of different stellar classes (and temperatures). It takes more energy to ionize helium than hydrogen, so only the hottest stars show helium absorption. It takes very little energy to ionize a molecule, so the coolest stars show molecular absorption bands. In the hottest stars, molecules are fully ionized and therefore do not give absorption lines. Note that the absence of a certain type of spectral line only means that the physical conditions are not right to excite the line; it does not necessarily mean that the corresponding element is absent.
(SOURCE: Courtesy T. Snow and K. Brownsberger.)

Chemical Composition

Astronomers deduce what a star is made of using spectroscopy. There are two separate problems: detecting the *presence* of an element and measuring its *amount*. The presence of an element is detected by identifying at least one— or preferably several—of its absorption lines or emission lines in the spectrum of the star. Astronomers deduce the amount of the element from the appearance of the spectral lines. Generally, wider and darker (or deeper) absorption lines indicate that more atoms of the element are present. Likewise, wider and brighter emission lines indicate more atoms of a particular element. The relative proportions of all the different elements give the chemical composition of the star. The classification of enormous numbers of spectra at the Harvard College Observatory was the first step in discovering the elemental composition of stars. Payne-Gaposhkin was the first to demonstrate that nearby stars are primarily made up of hydrogen and helium and therefore are approximately like the Sun in composition.

Note the power of the spectroscopic techniques we have discussed so far. Identification of the spectral lines tells us what elements are present in a star, because elements display patterns of lines as unique as fingerprints. Measurement of the relative line strengths of a particular element tells us about the temperature, density, and pressure in a star's photosphere, or surface layer. This method supplements the temperature information gained from other techniques, such as Wien's law applied to the peak of the radiation from the star.

We have also learned something very important about the universe we live in. Recall from Chapter 11 that helium was a mysterious element when it was first discovered, because it is so rare on Earth. Now we know that hydrogen and helium are the dominant elements of the solar system. These two simple elements comprise the bulk of the Sun and the bulk of the giant planets. It is Earth that is unusual— Earth is a large rock rich in carbon and silicon and metals and too low in mass to retain the light gases. When astronomers ventured to understand the stars, they did not know what to expect. Would there be strange elements or unusual states of matter at those vast distances? No. Most stars are put together just like the Sun. Hydrogen and helium are the most common elements everywhere in the universe that we have studied so far. This reassures us that the laws of physics and chemistry really are universal.

Stellar Motion

Spectroscopy also reveals another important stellar property: motion. As we saw in Chapter 9, the Doppler shift gives a very simple way to detect part of a star's motion—its **radial velocity**, or motion along the line of sight. The Doppler shift occurs because waves from a source get bunched up if the source is moving toward us and stretched out if the source is moving away from us. The Doppler shift of a star is proportional to its velocity, expressed as a fraction of the velocity of light (see the Science Toolbox in Chapter 9 to review this important idea). Be careful to clearly distinguish the Doppler shift from the shift in the thermal spectrum of an object that occurs when it gets hotter or colder, described by Wien's law. A Doppler shift is an indication of motion, not an indication of a change in the temperature of an object. Since the Doppler shift for stars is very small—usually well under 0.1 percent—we use the sharp spectral features as markers to measure the shift.

A consistent blueshift or redshift of all of a star's spectral lines proves that the star is moving either toward or away from us. For example, if the star is receding at 0.1 percent of the speed of light (300 km/s), the light will be redshifted by 0.1 percent of its normal wavelength. A line normally found at wavelength 500.0 nm would appear at 500.5 nm. Or, if the star is approaching at 0.02 percent of the speed of light (60 km/s), the light will be blueshifted by 0.02 percent of its normal wavelength. In this case, a line normally found at 500.00 nm would appear at 499.98 nm. Stars near the Sun have radial velocities that are 10 to 20 km; roughly half are redshifts and roughly half are blueshifts.

The other component of a star's motion is its *tangential velocity*, or motion perpendicular to the line of sight. It cannot be measured as simply as radial velocity. In order to measure tangential velocity, we must measure the distance of the star and its rate of angular motion across the sky, called **proper motion**. From the numbers given above, we can calculate the size of the proper motion. If 20 km/s is a typical velocity for a star, it will move $20 \times 3600 \times 24 \times 365 = 6.3 \times 10^8$ km in a year. The angle moved is given by the small angle equation from Chapter 2. In seconds of arc, it is 206,265 (d/D), where $d = 6.3 \times 10^8$ km. We will take the distance to be a parsec, or $D = 3 \times 10^{13}$ km. So the angle is $1.3 \times 10^{13} / 3 \times 10^{13} \approx 0.4$ second of arc per year. This is small but measurable. Of course, the angular motion will be smaller for slower or more distant stars, or if the star happens to be moving almost entirely in the radial direction. On the other hand, we can com-

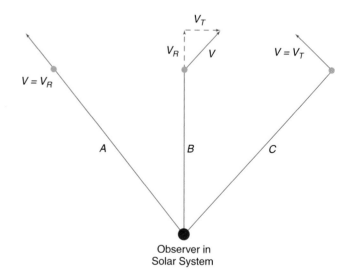

FIGURE 12–8
Three examples of a star's space velocity *(V)* compared with its radial velocity *(V_R)* and tangential velocity *(V_T)*. In A, the star's space velocity is aligned with the line of sight; no proper motion would be seen. B is the most common case, in which both V_R and V_T are appreciable. In C, the space velocity is perpendicular to the line of sight; no radial velocity or Doppler shift would be seen.

pensate for a small motion with patience by making observations over a period longer than a year.

These measurements of stellar motion are often lumped together with parallax in a branch of astronomy called **astrometry,** the precise study of the positions and motions of stars. If both radial and tangential velocities are known, they can be combined to give the star's space velocity (its true speed) and direction of motion in three-dimensional space relative to the Sun (Figure 12–8). The two directions are separated by 90° since one is along the line of sight and the other is across the line of sight. The famous theorem of Pythagoras allows us to combine the lengths of the sides of a right-angled triangle ($A^2 + B^2 = C^2$). It also allows us to combine the two perpendicular components of velocity ($V_A^2 + V_B^2 = V_C^2$, where V_A is the radial velocity, V_B is the tangential velocity, and V_C is the space velocity). Space velocities of most stars near the Sun are a few tens of kilometers per second and are directed nearly randomly.

Fundamental Properties of Stars

Thus far, we have discussed properties of stars that can be measured directly from the spectrum. Given the wavelength of the peak intensity of the spectrum, temperature can be calculated. The chemical composition comes from observing the pattern of the spectral lines and figuring out which elements are responsible for the lines. The star's motion along the line of sight is given by the Doppler shift. To determine other fundamental properties of a star, however, we require additional information, such as the star's distance or a theoretical model of the star.

Luminosity

One fundamental property of a star is the total amount of energy it radiates each second. As we saw earlier in the chapter, this energy output is called luminosity, or absolute brightness. Unlike apparent brightness, luminosity is intrinsic to a star. The most useful concept of luminosity is **bolometric luminosity,** the total amount of energy radiated each second in all forms at all wavelengths. Since many stars radiate mostly visible light, visual luminosity and bolometric luminosity are often roughly the same. Note, however, that objects much hotter than the Sun radiate primarily at ultraviolet wavelengths, and objects much cooler than the Sun radiate primarily at infrared wavelengths. In these cases, visual luminosity represents only a small part of the bolometric luminosity. In this book, luminosity will generally mean bolometric luminosity, abbreviated L.

The most basic method of estimating luminosity derives from the measurement of distance. A faint light in the night may be a candle 100 meters away, a streetlight a few kilometers away, or a brilliant lighthouse beacon 100 km away. Once we know the distance to an object, we can determine its absolute brightness. This method becomes less accurate for very distant stars. For one thing, as mentioned earlier, parallax distance measures become unreliable beyond about 100 parsecs. More importantly, the space between stars contains a thin haze of dust that dims starlight over larger distances. Just as smog might make it difficult to estimate the distance to a distant mountain range, an unknown amount of interstellar dust complicates estimates of stellar luminosity. Since interstellar dust reduces the apparent brightness of a star, it results in an *underestimate* of the luminosity.

As discussed previously, the luminosity, the distance, and the apparent brightness of an object are all interrelated. If we know any two of these quantities, we can estimate the third because they are related by the inverse square law. These calculations are basic to stellar astronomy.

For example, if two stars have the same apparent brightness but one is three times more distant than the other (as determined by a parallax measurement of the two stars), then the more distant star has nine times the luminosity of the nearer star. Alternatively, if two stars have equal luminosity and one appears to be four times fainter than the other (in this case, we might have evidence that both stars are like the Sun), then the fainter star is twice as distant as the brighter one. Or, if two stars have equal distance (this might be a case of binary stars or stars in a cluster), then the star that appears five times brighter is five times more luminous.

Diameter

Another fundamental property of a star is its size. It is extremely difficult to measure the size of a star by observation. We can use the small angle equation to see why. Imagine a star like the Sun at a distance of a parsec. The angle subtended by a solar-type star at that distance is 206,265 (d/D), where the diameter of the Sun, d, is 1.4×10^6 km, and $D = 3 \times 10^{13}$ km. We see that the star is only 0.01 second of arc across. More distant stars will appear even smaller. Since Earth's atmosphere blurs the images of all stars by about 0.5 to 1 second of arc, we have no hope of resolving such a distant star from the ground. Even from space, a large telescope would be needed to make an image of a star with high enough resolution to see surface features.

However, we can also calculate a star's diameter by using another method. Once we know a star's temperature and luminosity, we can apply the **Stefan-Boltzmann law** to measure its diameter. We have seen in this book that all objects emit thermal radiation with a peak wavelength that relates to the temperature. Hotter objects emit shorter wavelength radiation (Wien's law). The Stefan-Boltzmann law says that, in addition, *hotter objects emit more photons per second.* In other words, as an object heats up, both the frequency and the amount of thermal radiation increase. The following Science Toolbox explores this important relationship, from which we can determine a star's diameter. Such measures indicate a vast range of stellar diameters, from less than the size of Earth to huge stars bigger than the diameter of the orbit of Mars!

Mass

A third fundamental property of a star is its mass. Mass is one of the most difficult stellar properties to measure

SCIENCE TOOLBOX

The Stefan-Boltzmann Law

We have seen that the smooth spectrum of the Sun is due to thermal emission from the photosphere, at a temperature of 5700 K. Other stars in the sky can be either bluer or redder than the Sun; they have photospheres that are respectively hotter or cooler than the Sun's. But how large are these stars compared to the Sun? The Stefan-Boltzmann law relates a star's size to its temperature and luminosity; it applies not just to stars but to *any* object emitting a thermal spectrum. The mathematical form of the law states that the luminosity (L) is proportional to the star's surface area and the fourth power of its surface temperature:

$$L = 4\pi R^2 \sigma T^4$$

If the radius (R) of the star is given in meters and the temperature (T) is in Kelvin, the numerical constant $\sigma = 5.67 \times 10^{-8}$ and the luminosity will come out in watts. (The Greek letter *sigma*, σ, is called the Stefan-Boltzmann constant.) We can see that a star's luminosity is related to its surface area ($4\pi R^2$) and the amount of energy emitted by each square meter of the surface (σT^4). If the temperature of a star doubles, the amount of energy radiated increases by 2^4, or a factor of 16. Thus, while doubling the area of a star would increase the output by a factor of two, doubling its temperature would increase its output 16 times. Hotter stars not only radiate bluer light than cooler stars (a result that was predicted by Wien's law) but also *more light per unit area*.

Let us check out this equation for the Sun first. The Sun's luminosity is 3.8×10^{26} watts, and the surface (or photosphere) temperature is 5700 K. Rearranging the equation above,

$$R = \sqrt{\frac{L}{4\pi\sigma T^4}} = \sqrt{\frac{3.8 \times 10^{26}}{4\pi \, 5.67 \times 10^{-8} \times 5700^4}} = 7 \times 10^8 \text{ m}$$

This works for *any* star. Just plug in the luminosity and the surface temperature, and you can calculate the radius.

Now that we have checked this equation using the Sun as an example, let us see what the Stefan-Boltzmann law implies for stars with other values of luminosity and temperature. If we rearrange the equation above to put radius on the left-hand side, we find that

$$R \propto \frac{\sqrt{L}}{T^2}.$$

In this equation, we use the symbol $\propto$ meaning "proportional to." This allows us to form ratios in comparing the Sun to other stars, in which case the numerical constants in the Stefan-Boltzmann law cancel out. If we let the subscript $*$ refer to any star and the subscript $\odot$ refer to the Sun, we can write

$$\frac{R_*}{R_\odot} = \frac{\sqrt{(L_*/L_\odot)}}{(T_*/T_\odot)^2}$$

As we will see later in the chapter, there is a whole family of stars that produces energy by the fusion of hydrogen into helium. The most massive of these stars is about a million times as luminous as the Sun and has a surface temperature of about 40,000 K. Therefore, $L_*/L_\odot = 10^6$ and $T_*/T_\odot = 40,000/5700 = 7$. Substituting in the last equation, we get $R_*/R_\odot = \sqrt{10^6/49} \approx 20$. Thus, there are stars 20 times larger than the Sun that fuse hydrogen into helium. The least massive of these stars is about one-thousandth the luminosity of the Sun and has a surface temperature of only 2300 K. We see that $L_*/L_\odot = 10^{-3}$ and $T_*/T_\odot = 2300/5700 = 0.4$. By the same reasoning, we get $R_*/R_\odot = \sqrt{10^{-3}/0.16} \approx 0.2$. In other words, there are stars one-fifth the size of the Sun that fuse hydrogen into helium.

We obtain even more remarkable results if we consider stars that have the same spectral type as the Sun but very different values of luminosity (the physical conditions inside such stars will be discussed in the next chapter). Stars with the same spectral type have the same surface temperature. In this case, the Stefan-Boltzmann law simplifies even further to $R \propto \sqrt{L}$. There are stars with the same color as the Sun but 100,000 times the luminosity. These giant stars must be $\sqrt{10^5} \approx 300$ times the size of the Sun. There are stars with the same color as the Sun with one ten-thousandth the luminosity. These dwarf stars must be $\sqrt{10^{-4}} = 0.01$ times the size of the Sun.

Although the Sun is a typical star, the range of stellar types is enormous! In every case, the Stefan-Boltzmann law enables us to estimate the size without a direct measurement.

---

because of the problem of weighing a star in empty space. Astronomers estimate the mass of most stars based on a *model* of how stars evolve and radiate their energy, as we will see later in this chapter. The starting point is our knowledge of the fusion process that powers the star. We also know that hydrogen and helium are the two principal ingredients of the fusion process. A model of the power source enables us to predict how the temperature of the star varies in radius from the fusion core to the cooler photosphere. This prediction, in turn, allows us to hypothesize how density varies with radius. The final step is to mathematically convert the change of density with radius into a total mass. This sounds indirect—and it is! However, every step in the chain of logic is based on well-understood laws of physics.

It is much simpler to calculate the mass of stars orbiting each other. Then we can calculate stellar masses by an application of Kepler's laws. Binary stars are quite common, and this important way of measuring mass will be discussed in Chapter 14. As we will see in the next chapter, stars range widely in mass from about one-tenth to about 100 times the mass of the Sun.

Hydrostatic Equilibrium

Stars vary widely in their fundamental properties of luminosity, size, and mass. We might imagine that the release of fusion energy would blow a star apart. Or we might imagine that the relentless pull of gravity would cause a star to collapse. Yet we know that the Sun is a stable star that has been shining steadily for billions of years. It will continue to shine steadily for billions of years in the future. We can look for a physical principle that unites stars with very different properties. Then we might be able to answer this question: What makes a star a star?

Recall our discussion of thermal equilibrium from Chapter 4. Heat always tends to flow from hotter to cooler regions in order to equalize the temperature. A star like the Sun has a constant source of energy from fusion reactions in its core. Heat is therefore constantly flowing outward. A star cannot reach thermal equilibrium as long as there is an energy source at the core—the inner regions will always be hotter than the outer regions. Stars do not "cool off."

Stars that are in the process of converting hydrogen to helium are also stable. This stability is described by the principle of **hydrostatic equilibrium.** The term *hydrostatic* is a combination of *hydro-*, meaning that the gas in a star acts like a fluid, and *-static,* meaning that the star is not expanding or contracting. Hydrostatic equilibrium is a balance between two forces at every point within a star—the inward force of gravity, and the outward pressure of the gas caused by its temperature. (Recall from the Science Toolbox in Chapter 7 that the pressure in an ideal gas is proportional to its temperature.) The temperature within a star is governed by the heat flow from the nuclear reactions in the core. A star therefore stays "puffed up."

Figure 12–9 shows a schematic view of hydrostatic equilibrium. If we imagine the star as having a series of layers of gas, then the pressure in each layer must balance the weight on that layer. Deeper layers of the star have more weight pressing down on them, so the pressure must increase as we move toward the center. Increasing pressure means increasing temperature. The idea of a star as a series of layers makes the idea of a balance between pressure and gravity clearer, but a star does not actually have distinct layers—the pressure, density, and temperature all increase smoothly toward the center. Hydrostatic equilibrium governs the physical conditions at any position inside a star.

Classifying Stars

Now that we understand how to measure the properties of individual stars, how can we use this information to classify them? Classification can be a key to physical understanding. Classification consists of gathering data, looking for common traits and features, comparing like with like, and trying

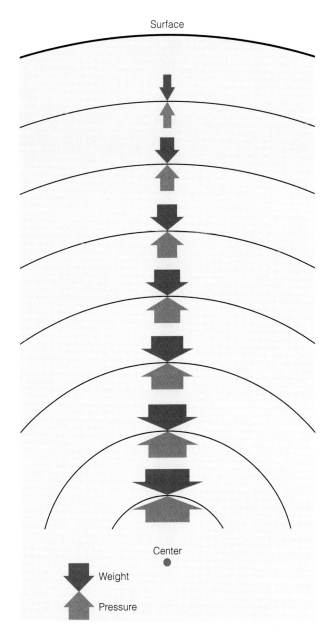

FIGURE 12–9
The principle of hydrostatic equilibrium. Gravity gives each layer of the star a weight, which is an inward force. The inward force is countered by pressure in the gas caused by the energy release from fusion reactions in the core. Layers are shown only for convenience; the gravity and pressure increase continuously toward the center. (SOURCE: Courtesy M. Seeds.)

to make sense of the patterns. It may sound more exciting to invent a new theory than to classify phenomena in the natural world. Classification can be mundane, but in the hands of an expert, it is an art. As we saw in Chapter 5, this kind of work led to the realization that Earth and the species that inhabit it evolve. Classification was also the key to demonstrating that stars evolve. In this section, we build on the opening story of the chapter and discuss what astronomers have learned by classifying stars.

To review a simple example of classification, examine Figure 12–10a, which shows motorized vehicles plotted in

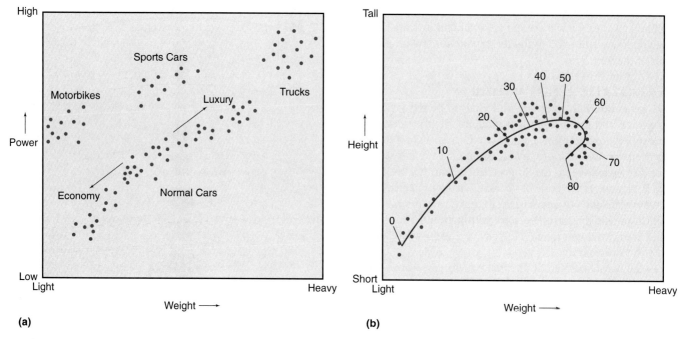

FIGURE 12-10
Measurement and classification can be an aid to understanding. **(a)** Motor vehicle weight plotted against engine power. Not all combinations are useful and most cars have a nearly fixed ratio of power to weight, represented by the diagonal track in the diagram for normal cars. **(b)** Height plotted against weight for a human population. Most adults fall within a relatively small range of height and weight. However, people change their attributes as they age, and some rare groups of people are quite different from the bulk of the population. The variable that relates these two measurements is time, or age. Age in years is marked along the track that follows the measurements.

terms of their weight and engine power. The sequence of cars increases in weight and power, running from compact to mid-size to luxury cars. Special categories, such as trucks, have a lot of power for their large weight, and sports cars have a lot of power for their modest weight. Not every combination of weight and power exists; any vehicle in the lower-right region of the graph would be too underpowered to be useful! Classification and measurement teaches us something about the power demands of vehicles.

Another example is more analogous to classifying stars. Imagine that you live in a small town. You gather information on each of the inhabitants—age, height, and weight. A graph of everyone's height and weight would show a clear and obvious trend. Figure 12-10*b* represents this kind of data, where the degree of shading indicates the fraction of people with a particular combination of height and weight. People spend most of their lives as adults, so most measurements show adults of average height and average weight. A tail of measurements off to the lower left represents children. The line shows the average characteristics of people as they age, with ages marked. This line reverses itself; old people actually shrink. Extreme groups are also plotted—those who are very tall (members of the high school basketball or volleyball team, perhaps), and people with congenital conditions that retard growth. Also included are people with eating disorders.

There are two important points to note in this example. First, humans do not exist with all conceivable combinations of height and weight; you will never see someone who is 3 feet tall and weighs 300 pounds, or 6 feet tall and 60 pounds! The pattern tells us something about the shape of the human organism. Second, the location of an individual on this plot is primarily fixed by age. In other words, although we have plotted height against weight, the hidden variable is time. As humans age, they grow so that their measurements will shift on this diagram.

These conclusions are obvious to us, but what if intelligent aliens visited this same small town on a fact-finding mission? The aliens would notice small and large people, and they might notice a trend in the plot of height against weight. But could they deduce from a quick visit that the small (young) turned into the large (old)? This is analogous to the problem of the intelligent squirrel mentioned earlier in the chapter. Astronomers are faced with the same situation in unraveling the life stories of stars. We get a snapshot of stellar properties and must deduce how and why they change over time. In the following section, you will learn about a powerful method of classifying stars and what it reveals about nearby stars, prominent stars, and stars of very different sizes.

The Hertzsprung-Russell Diagram

When astronomers realized the variety of stars that exist, they needed to devise a sensible way to arrange and study the data to find relationships among the various stellar forms. This could have been done in many ways, but one method has become traditional. Building on the work of Annie Cannon, Danish astronomer Ejnar Hertzsprung and American astronomer Henry Norris Russell independently conceived the idea of plotting the spectral classes of stars according to their luminosity. They did this work from 1905 to 1915. The plot they produced is usually called the **H-R diagram** in honor of Hertzsprung and Russell, and an example is shown in Figure 12–11. We use schematic color images to show stars of different types.

As we saw earlier in the chapter, temperature is the principal factor that governs the differences among spectra of O, B, A, F, G, K, and M stars. For this reason, most astrophysicists make H-R diagrams by using a temperature scale instead of a scale showing spectral class. In this book, we identify temperature, spectral class, and luminosity along the edges of the diagram as an aid to interpretation. (Following Russell's original version, and the tradition of astronomers ever since, the spectral classes are arranged with *cooler* stars to the right.)

Different locations on the H-R diagram correspond to different types of stars. It is important to understand why this is true. First, remember from Wien's law that stars of different temperature have different color. The right part of Figure 12-11 corresponds to redder, cooler stars and the left part to bluer, hotter stars. Similarly, the upper part corresponds to more luminous stars, and the lower part to less luminous stars.

Notice that the Sun is near the middle of the diagram (luminosity 1 $L_\odot$, spectral type G). If we move upward from the Sun's position, we encounter stars more luminous than the Sun, even though they have the same temperature. These stars must be bigger than the Sun, since the Stefan-Boltzmann law tells us that each square meter on a star at the same temperature as the Sun will radiate as much energy as a square meter on the Sun. Thus, to get more total luminosity, we must have more square meters of surface area and hence a larger size. Similarly, a star below the Sun on the diagram must be smaller than the Sun. To summarize, luminous stars are hot and big, while less luminous stars are cool and small.

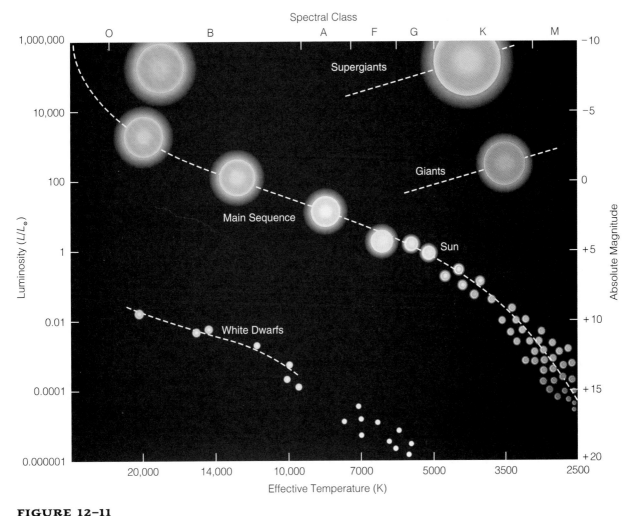

FIGURE 12-11
An H-R diagram for a selection of different types of stars. Stars are plotted schematically to show colors and relative sizes, but they are not to true scale. (This would require much larger red giants and much smaller white dwarfs.) Main-sequence stars, which release energy by fusing hydrogen into helium, run diagonally across the diagram.

Do stars appear *throughout* the diagram? Or are there only certain combinations of temperature and luminosity? The answer is that only a few regions of the H-R diagram are crowded with stars. Stars that would fall in other parts of the diagram are rare or nonexistent. Even the earliest H-R diagram revealed an important discovery about stars: Among most stars, there is a smooth relationship between surface temperature and luminosity. Hertzsprung called these stars **main-sequence stars.** They lie on the diagonal band from upper left to lower right in Figure 12–11. The main sequence runs from hot, luminous, blue stars (upper left) to cool, faint, red stars (lower right). It turns out that main-sequence stars all have something in common—they all get their energy from the fusion of hydrogen into helium, just like the Sun.

Nearby Stars

It is important to remember that the H-R diagram is not a pictorial representation of where stars lie in space. It is formed by graphing two measured properties of a star, with each point in the diagram representing the properties of a different star in the sky. The H-R diagram is a tool or method that can be applied to any sample of stars. Figure 12–12 shows a modern H-R diagram for the 100 stars closest to the Sun. Stars at this range are near enough for us to measure accurate distances and to detect examples of very low luminosity. Stars in such a sample are believed to be representative of all stars in the neighborhood of the Sun. Over 90 percent of this representative sample of stars fall on the main sequence. The Sun is plotted near the center, at spectral class G, luminosity 1 $L_\odot$. Figure 12–12 also reveals the important fact that *most stars are fainter and cooler than the Sun.*

Appendix B–3 lists more detailed properties of the 27 stellar systems within 4 parsecs of the Sun. One column lists fainter stars that have been found very close to the primary stars. The table reveals another interesting fact about representative stars: Most of them are inhabitants of systems in which two or three stars orbit one another. Stars without stellar companions, such as the Sun, are in the minority. This important aspect of stars will be discussed further in Chapter 14.

Prominent Stars

If we try to expand our statistical sample of stars by tabulating more distant ones, we run into a problem. At great distances the less luminous stars, like those in Figure 12–12, are too faint to see. The only distant stars we see are unusual, luminous ones. In fact, the prominent stars in our night sky are mostly distant, unusually luminous stars. These stars are quite rare in any volume of space. The Science Toolbox following shows the very different visibility of giant and dwarf stars.

Why is the visibility of stars an important issue? The simple answer is that we are trapped in space. We can only view the universe from our fixed location on Earth. Some stars are so dim that they elude detection with our telescopes. Other stars are so bright that we can see them from very far away. When astronomers carry out surveys of stars, they are limited to the light they can capture with their telescopes. This type of census gives a very different answer from a count of stars throughout a volume. Imagine, for example, that you could jet through space and count every one. In general, we will overcount luminous or giant stars compared to their true space density and undercount dim or dwarf stars compared to their true space density. There may be a big difference between what we easily see and what is truly out there in space.

Look at Appendix B–4. Each of the 17 brightest stars in the sky is more luminous than the Sun. All but three of them are farther away than all of the 27 nearest stellar systems listed in Appendix B–3. Sirius is the only star that is common to both lists, which means that the brightest stars and the nearest stars are quite different groups of objects. The brightest stars are luminous and distant, while the nearest stars are mostly less luminous than the Sun. Figure 12–13 shows a histogram with the numbers of stars of every type and luminosity in a volume of 1 million pc^3. The enormous predominance of cool and low-luminosity stars is clearly seen. Hertzsprung aptly called the luminous stars "whales among the fishes." Thus, the statistics of prominent stars do not give us a sample of representative stars. Rather, they are biased toward the "whales." Nonetheless, a tabulation of such stars is important because it reveals that some of the "whales" do not occupy the main sequence of the H-R diagram.

Non-Main-Sequence Stars

As astronomers studied the properties of larger samples of stars, they found stars with many different combinations of luminosity and temperature. Most of these stars had properties that placed them at different positions on the main

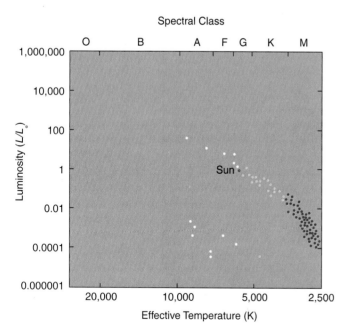

FIGURE 12–12
An H-R diagram for the 100 nearest stars, a representative sample from a volume of space around the Sun. This diagram shows that most stars are on the main sequence. Notice also that most are fainter and cooler than the Sun. Therefore, they fall in the lower-right corner of the diagram.

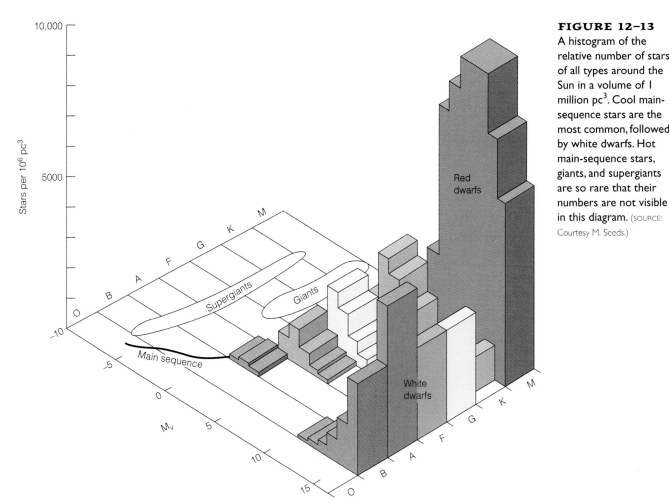

FIGURE 12-13 A histogram of the relative number of stars of all types around the Sun in a volume of 1 million pc^3. Cool main-sequence stars are the most common, followed by white dwarfs. Hot main-sequence stars, giants, and supergiants are so rare that their numbers are not visible in this diagram. (SOURCE: Courtesy M. Seeds.)

sequence of the H-R diagram, slanting from high luminosity and high temperature to low luminosity and low temperature. However, some stars had properties dissimilar from any main-sequence star. Let us revisit Figure 12-11 to examine more closely the types of non-main-sequence stars.

Some red stars with spectral types K and M have much higher luminosity than the faint, main-sequence K and M stars. They radiate more light than the fainter stars of the same temperature. According to the Stefan-Boltzmann law, they must have larger areas. Hertzsprung named them **giant stars.** Most are called *red giants,* because their outer regions are cool. Other stars were found to be brighter than the giants or the main-sequence O stars. Application of the Stefan-Boltzmann law showed that they were larger than giant; they came to be called **supergiant stars.** The enormous size of the red giant Antares, for example, can be visualized in the imaginary close-up view shown in Figure 12-14. Still other stars had combinations of high temperature and low luminosity that placed them below the main sequence of the H-R diagram. By using the same radiation law to derive their sizes, astronomers recognized that these dim stars were unusually small. They came to be called **white dwarf stars.**

Figure 12-11 schematically dramatizes the relative brightness and size of these stars. The work of Hertzsprung, Russell, and their followers showed the systematic properties of different types of stars: One type was the main sequence, and the giants, supergiants, and dwarfs were distinctly different types of stars. Appendix B-4 shows that nearly half of the most prominent stars in the sky are giants or supergiants. Although the whales among the fishes are rare, they account for many of the familiar stars. Further work showed that these groups occur all over the sky and at many distances from the Sun. This meant that universal physical processes could explain the different non-main-sequence stars.

Occasionally we find groups of stars at about the same distance from Earth and formed at about the same time. These *star clusters,* such as the one in Figure 12-15, offer us, so to speak, living H-R diagrams in the sky. Just by studying the colors and brightness of stars in Figure 12-15, we can see the pattern of the H-R diagram. Most of the stars are on the main sequence; the brightest ones are blue-white, and the faintest ones are red. This particular scene, known as the Jewel Box cluster, is all the more attractive because the fourth brightest star is an orange-red giant or a supergiant. The result is a startling color contrast, hot stars scattered like gems on a velvet backdrop.

Stellar Radius and the H-R Diagram

The H-R diagram only shows temperature and luminosity. However, we can use the Stefan-Boltzmann law to calculate

FIGURE 12-14
An imaginary view of the red M-type supergiant star Antares, as seen from a hypothetical planet 9.5 AU away, the same distance Saturn is from our Sun. Although the Sun would resemble only a brilliant dot from such a planet, Antares is so big that it would cover an angle of almost 40°. It would look 80 times bigger than our Sun in Earth's sky and have a reddish-orange color. The fainter star in the upper right is a bluish B-type star that orbits Antares at a distance of several hundred AU. (SOURCE: Painting by WKH.)

FIGURE 12-15
The Jewel Box cluster, cataloged as NGC 4755, displays the different star types charted on the H-R diagram. The cluster is named from Sir John Herschel's remark that it resembles a superb piece of jewelry. It is about 8 pc across and 2400 pc away. The brightest stars are mostly hot, bluish-white, young B stars; the faintest stars are red, low-mass stars. (SOURCE: Copyright Association of Universities for Research in Astronomy, Cerro Tololo Interamerican Observatory.)

SCIENCE TOOLBOX

Comparing Giants and Dwarfs

The dramatic difference between what we can easily see and what really lies out in space is an important issue in astronomy. To use Hertzsprung's analogy, the whales are much easier to see than the fishes. We can see whales out to large distances, but in any volume of water, there are far fewer whales than fishes. We can use the inverse square law and simple geometry to help us understand what the most common stars are in any region of space.

We have seen that the apparent brightness of a star decreases with the square of the distance. This means that more luminous stars can be seen from farther away than fainter stars. A simple example will show the difference between counting stars to a certain apparent brightness, and counting all the stars in a volume. Figure 12-B*a* shows two stars, one like the Sun and one five times the luminosity of the Sun. The horizontal line shows the level at which any star becomes invisible to the eye (or telescope). We can see all stars like the Sun out to a distance of 1 parsec. However, by the inverse square law, we can see all stars of 5 $L_\odot$ out to $\sqrt{5/1} = 2.24$ pc. In all directions from Earth, this distance defines a sphere.

Let us call D the visibility distance, or the maximum distance we can see a star of a particular luminosity. The ratio of the spherical volume in which we can see the 1 $L_\odot$ (Sun-like) star to the spherical volume in which we can see the 5 $L_\odot$ stars is

$$\frac{4\pi/3\ (D_5)^3}{4\pi/3\ (D_1)^3} = (2.23/1)^3 \approx 11$$

In this equation, D_1 is shorthand for the visibility distance of a star like the Sun, and D_5 is shorthand for the maximum visible distance of a star with five times the luminosity of the Sun.

Therefore, if there were equal numbers of stars like the Sun and stars five times more luminous, we could see and count about 10 times more of the more luminous stars. Or, if the luminous stars were 10 times rarer in any chunk of space, we would count about equal numbers of each type. In other words, surveys to a certain brightness limit always lead to an *overestimate* of the true numbers of luminous stars. Intrinsically dim stars are counted over a much smaller volume than intrinsically luminous stars.

You will notice that when we take a ratio, the numerical factors cancel out. We can also make this equation more useful by remembering that the inverse square law says that $L \propto D^2$, and so the limiting distance $D \propto \sqrt{L}$. If we substitute this in the equation above and keep everything in solar units, we find that the number of stars of luminosity L that we count compared to the Sun goes up proportionally to $L^{3/2}$. If we call N_V the true number of stars of luminosity L (compared to stars

(continued)

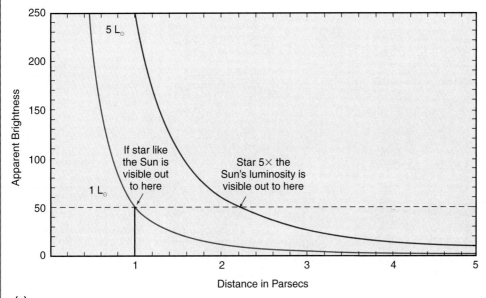

FIGURE 12-B
The visibility of stars of different luminosity. **(a)** Here a star like the Sun and one five times the luminosity of the Sun are compared. Each star is visible out to the distance at which its apparent brightness falls below the observable limit (the horizontal dashed line). The ratio of the volumes defined by these two distances (or radii) gives the relative visibility of the two types of star.

SCIENCE TOOLBOX

Comparing Giants and Dwarfs (continued)

like the Sun) and N_B the number we count to some apparent brightness limit, then

$$N_V = N_B L^{-3/2}$$

Figure 12-B*b* shows an even more extreme example. One star is like the Sun and the other is 20 times more luminous. We can see the more luminous star out to a distance of $\sqrt{20/1} = 4.46$ pc. By the equation above, $N_V = N_B (20)^{-3/2} = 0.01 N_B$. In this case, even if the stars were equally sprinkled through space, we would count nearly 100 times more of the more luminous kind! Or, if stars with 20 $L_\odot$ were 100 times less common than stars like the Sun, we would count very luminous stars and stars like the Sun with roughly equal frequency.

Here is an example that involves a low-luminosity, or dwarf, star. A star one-tenth the luminosity of the Sun can only be seen out to $\sqrt{0.1/1} = 0.32$ pc. The smaller visibility distance corresponds to a smaller volume for seeing these dim stars. We would expect surveys to a certain brightness limit to *underestimate* the true numbers of low-luminosity stars. Using the equation above, we find that $N_V = N_B (0.1)^{-3/2} = 32 N_B$. If dwarf stars were equally sprinkled through space, we would undercount them by a factor of over 30 in a survey limited by apparent brightness. Or, if space was peppered with 30 times more 0.1 $L_\odot$ stars than Sun-like stars, we would count them with equal numbers.

These sample calculations show why the lists of nearby stars and prominent stars are so different. However, once we have a true census of the true numbers of high- and low-luminosity stars in a representative volume of space, we can correct *any* survey for the bias in favor of high-luminosity stars. The result is a true census of whales and fishes.

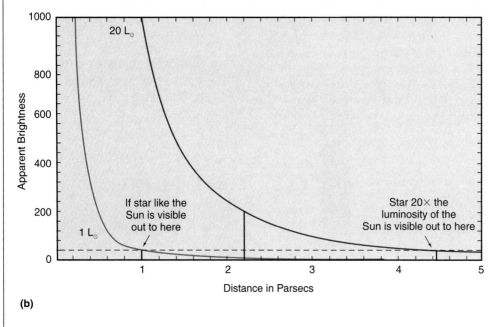

FIGURE 12-B (continued) **(b)** As in part (a), except that in this case a star like the Sun and one 20 times more luminous are compared.

stellar radius and show it on the H-R diagram as well. Figure 12-16, shows the positions of the major stellar types; the slanting lines have been added to show positions of constant radius. The important concept is this: Any point on the H-R diagram corresponds to a star of a certain radius. The reasoning should be familiar. Any point on the diagram corresponds to a certain temperature T. According to the Stefan-Boltzmann law, any surface at temperature T must radiate a certain amount of energy per square meter each second. But any point on the diagram also corresponds to a particular luminosity L, which is the total amount of energy radiated each second. This luminosity fixes the number of square meters involved, giving us the total area and thus the radius.

Now you can see the enormous difference in the physical properties of stars in the H-R diagram. Astronomers have the practical problem that stars with the same color might be quite different in their other properties. To see this, consider stars of different luminosity that have the same surface tem-

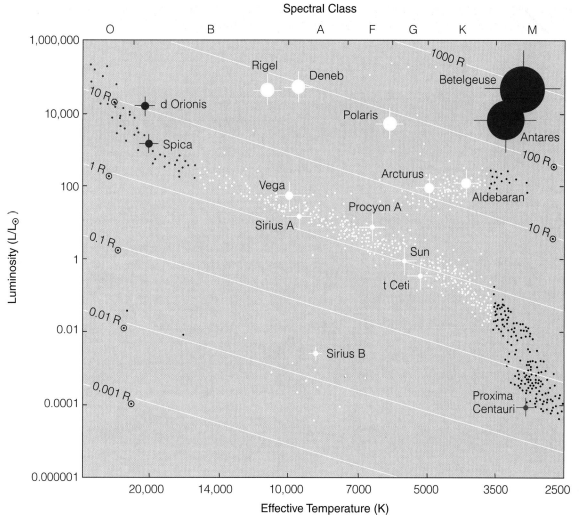

FIGURE 12-16
An H-R diagram with lines of constant radius showing the dimensions of stars in different parts of the diagram. Selected well-known stars are marked. Their relative sizes are schematically indicated, but there is not enough room to show them to true scale.

perature (stars aligned vertically in the H-R diagram). Arcturus is over 10 times larger than the Sun but has the same color. Betelgeuse is more than 10,000 times larger than Proxima Centauri, but they both have a reddish color. Without knowing distances, is there any way to distinguish main-sequence stars, giants, and supergiants of the same color? Yes! Figure 12-17 shows the spectra of a main-sequence star, a giant, and a supergiant of the same spectral class. The more diffuse atmosphere in the larger stars yields fewer atomic collisions and a narrower absorption line. Spectroscopy can be sufficient to

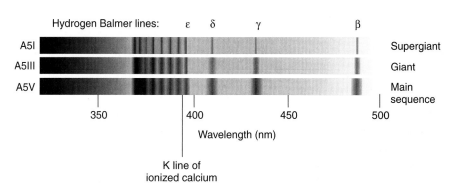

FIGURE 12-17
A comparison of the spectra of a supergiant, giant, and main-sequence star of the same spectral class. The lower pressure in the atmospheres of the larger stars gives narrower absorption lines.

establish a *luminosity class* of a star. This allows a distance to be estimated, even if no parallax measurement is possible.

The Evolution of Stars

Up to this point, we have considered the static properties of stars. We see each star at a moment in its life, like a snapshot. We have information that can tell us the mass, size, and luminosity of a star. We know that main-sequence stars get their energy by the fusion of hydrogen into helium. Now we are ready to see if we can use this information to deduce how stars live their lives (recall the analogy of the squirrel in the forest). It turns out that the key property that controls the evolution of a star is its mass.

Stellar Mass and the H-R Diagram

What causes the distinctive types of stars found on the H-R diagram? Stars have different masses, as shown in Figure 12–18. High-mass main-sequence stars are hotter, brighter, and bigger than low-mass main-sequence stars. You can see that massive stars are more luminous than less-massive stars. Is there any simple relationship between mass and luminosity? How do stars in different parts of the H-R diagram evolve? Can one type of star turn into another type of star? We will begin to answer these questions using the H-R diagram as a diagnostic tool, and stellar evolution will be discussed in more detail in the next chapter.

How did we discover the principles that control the structure and evolution of stars? Much of this work was begun by English astrophysicist Arthur Eddington in the 1920s and was carried on by other scientists through the 1950s. Eddington showed that the same two opposing influences at work in the Sun are competing in any star: Gravity pulls *inward* on stellar gas, while gas pressure pushes *outward*. This is the principle of hydrostatic equilibrium, discussed earlier in the chapter. In any stable star, these forces are balanced at every point within the star. The internal gravity of a star is determined by its mass; therefore, mass is a fundamental property of a star, even though it may be hard to measure directly.

As we have already seen, heat is produced during the gravitational contraction of a star. A stable main-sequence star is one that has contracted until the inside is hot enough to start nuclear reactions among hydrogen atoms. At this point the interior becomes a stable heat source, radiating light and creating enough outward pressure to counterbalance the inward force of gravity. What had been a contracting ball of gas now becomes a star with a constant size, governed by the release of energy in the interior.

Explaining the Main Sequence

Main-sequence stars convert hydrogen to helium by the fusion process. The main source of energy in a main-sequence star's interior is nuclear reactions in which hydrogen is consumed. In this sense, all main-sequence stars are like the Sun. But you can see from Figure 12–18 that main-sequence stars can be 1 million times more luminous than the Sun or 10,000 times less luminous than the Sun. We need to explain the amazing range of energy output on the main sequence. You can also see from Figure 12–19 that the masses of main-sequence stars fall in a particular range. The histogram shows that there are few main-sequence stars less than one-tenth the mass of the Sun and few with mass more than 50 times the mass of the Sun.

Because stars form with a huge supply of hydrogen, stars remain stable on the main sequence for a relatively long time at a fixed size. If a stable star were magically expanded, its gas would cool and the nuclear reactions would decline, reducing the outward pressure. The outer layers would fall back to their original state. If it were magically compressed, the inside would get denser and the reactions would increase, raising the outward pressure and expanding the star. A main-sequence star tends to stay in a stable state because of the hydrostatic equilibrium that exists at each point within the star. The star remains stable as long as its internal composition and energy production rate stay the same.

Calculations based on these ideas revealed a **mass-luminosity relation** among main-sequence stars. A hydrogen-fusing star more massive than the Sun has higher luminosity and surface temperature than the Sun. Hence, on the H-R diagram it would lie to the upper left of the Sun. Similarly, a star of lower mass would lie to the lower right of the Sun. Accordingly, these stars fall along a line on the H-R diagram—the main sequence. The *main sequence* is, therefore, the

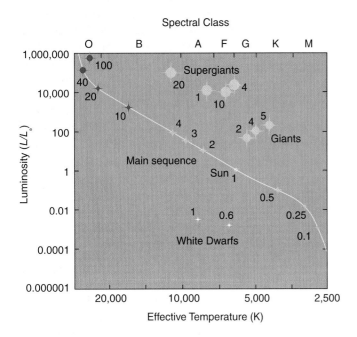

FIGURE 12–18
An H-R diagram with measured masses of stars (in solar masses) marked in different parts of the diagram. A smooth progression of masses is found along the main sequence, but masses in other parts of the diagram are mixed because stars of different mass evolve into the same regions. In other words, evolved stars of different masses can have the same temperature and luminosity. Schematic sizes are shown (not to true scale).

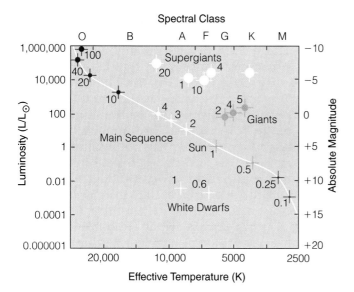

FIGURE 12-19
A histogram showing the frequency of occurrence of different stellar masses in the solar neighborhood. The most common stars have about $\tfrac{1}{4}\,M_\odot$, and there are few stars more than fifty times the Sun's mass. The abundance of the lowest-mass objects is uncertain.

TABLE 12-3
Properties of Main-Sequence Stars

SPECTRAL TYPE	MASS ($M_\odot$)	RADIUS ($R_\odot$)	EFFECTIVE TEMPERATURE (KELVIN)	LUMINOSITY ($L_\odot$)
O5	40	18	40,000	5 × 10⁵
B0	18	7.4	28,000	2 × 10⁴
B5	6.5	3.8	15,500	800
A0	3.2	2.5	9900	80
A5	2.1	1.7	8500	20
F0	1.7	1.4	7400	6
F5	1.3	1.2	6580	2.5
G0	1.1	1.1	6030	1.3
G5	0.9	0.9	5520	0.8
K0	0.8	0.8	4900	0.4
K5	0.7	0.7	4130	0.2
M0	0.5	0.6	3480	0.03
M5	0.2	0.3	2800	0.008

Mass is in units of solar mass (2×10^{30} kg), radius is in units of solar radius (7×10^8 m), effective temperature is a measure of surface temperature, and luminosity is in units of solar luminosity (3.8×10^{26} watts).

group of stars of different masses that have reached stable configurations and are generating energy by consuming hydrogen in nuclear reactions.

Mathematically, a star's luminosity is proportional to the mass to the 3.5 power. In solar units, $L = M^{3.5}$. This is a very steep relationship, as a few examples will show. A star with five times the mass of the Sun will have a luminosity of $5^{3.5} = 5 \times 5 \times 5 \times \sqrt{5} = 280$ times that of the Sun. Conversely, a star with only one-tenth the mass of the Sun will have a luminosity of $0.1^{3.5} = 0.1 \times 0.1 \times 0.1 \times \sqrt{0.1} = 1/3160$ times that of the Sun. Massive stars are like powerful beacons. Low-mass stars are like puny flashlights. Why does luminosity depend so strongly on mass? A modest increase in mass corresponds to a star with a substantially higher pressure and temperature in the core. Since the rate of nuclear reactions depends sensitively on temperature, more massive stars have much larger rates of nuclear fusion. This in turn leads to a much higher luminosity.

The detailed properties of main-sequence stars are summarized in Table 12-3. The main attributes of main-sequence stars can be summarized as follows:

1. Main-sequence stars are converting hydrogen to helium by nuclear fusion.
2. Main-sequence stars are stable, with an internal structure that is governed by the principle of hydrostatic equilibrium.
3. Massive main-sequence stars are larger, more luminous, and have hotter photospheres than low-mass main-sequence stars.
4. Most main-sequence stars lie between one-tenth and fifty times the mass of the Sun.
5. The main sequence on the H-R diagram defines a sequence of mass, with luminosity and mass related by $L = M^{3.5}$.

Figure 12-18 shows this last trend more clearly by charting masses of some stars on the H-R diagram. Notice the smoothly increasing mass and luminosity as we ascend the main sequence. We can now follow the chain of reasoning that explains the main sequence of stars. More massive stars have greater gravity, which creates higher pressure in the stellar interior. The higher pressure results in higher temperature, which causes higher energy output by the fusion process, giving both higher luminosity and higher surface temperature. (The same reasoning does not apply to stars that lie off the main sequence; a star's position on other parts of the H-R diagram is not a simple indicator of mass. In other words, giants, supergiants, and dwarfs are stars that do *not* have exactly the same energy generation, composition, or energy transport as main-sequence stars.)

The Rate of Stellar Evolution

Stars must evolve, because they are gradually using up their fuel supply. In 1926, astrophysicists Henry Norris Russell and Heinrich Vogt showed that the structure of an ordinary (main-sequence) star is determined uniquely by its mass and chemical composition. Therefore, a certain mass of material with fixed composition—for example, one solar mass consisting of $\tfrac{3}{4}$ hydrogen and $\tfrac{1}{4}$ helium—can reach only one stable configuration. The temperature and luminosity of such would place it on the H-R diagram at the position occupied by the Sun. But if the composition were altered to $\tfrac{1}{2}$ hydrogen and $\tfrac{1}{2}$ helium, the rate of nuclear fusion would be different, as would the configuration of the star and its location

ble, plus the physics of models of stars. As-res, temperatures, and surfaces, and atmos-never directly explored quite accurately what together.

and Vogt explains the form to another: As a changes its composition um structure. All nuclear ition, and all changes in new structure. We can bring this discussion down to Earth by using the example of our own Sun. The Sun has brightened by about 60 percent since it formed about 4.6 billion years ago. It is expected to brighten by another factor of 2 by the time it is finished as a main-sequence star 5 billion years from now. This will raise Earth's temperature by about 20°C. While the Sun brightens, the polar caps will melt, the oceans will slowly evaporate, and the atmosphere will leak away into space!

Stars evolve at different rates. The more massive a star, the higher its interior temperatures and the faster it will consume its nuclear fuel. A Sun-like star of 1 $M_\odot$ stays on the main sequence about 9 billion years (9×10^9 years), but a star of 10 $M_\odot$ stays there only about 20 million years (2×10^7 years). The entire history of a star of 1 $M_\odot$ (from protostar to white dwarf) takes about 11 billion years, whereas a 10 $M_\odot$ star lasts only about 24 million years. In spite of the differences in total time, the largest fraction of each star's life is still spent on or near the main sequence.

Since a star's hydrogen-burning lifetime depends mostly on its mass and luminosity, a simple formula gives the time the star will spend on the main sequence—which is most of the star's lifetime. The hydrogen-burning lifetime equals M/L times 9 billion years, where M and L are the mass and luminosity of the star in solar units. We will investigate this relationship more closely in the next chapter. Notice on Figure 12–18 that main-sequence stars span a large range in mass—a factor of 1000—but an even larger range in luminosity—a factor of 10 billion!

We can revisit the analogy shown in Figure 12–10a, the "main sequence" of car properties. Imagine that a mid-size sedan is a Sun-like star, with a fuel tank of average size and an engine of average fuel economy. Compared to this, a luxury car has a bigger fuel tank but far worse fuel economy, so it will not go as far on a tank of gas (analogous to a driving "lifetime"). A compact car, on the other hand, will have a smaller fuel tank but far better fuel economy, so it will go farther on a tank of gas. In these terms, low-mass main-sequence stars are misers, stretching their meager fuel supply for a very long time. Massive main-sequence stars are profligate, burning through their large fuel supply in a brief blaze of glory.

The most massive stars stay on the main sequence for only the twinkling of a cosmic eye. Some of them evolve into the supergiant region, and some less-massive ones become ordinary giants. All of them quickly evolve to unstable configurations; many may explode, and all disappear from visual prominence.

Determining the Ages of Stars

The ages of single stars are difficult to determine. The Sun's age was measured at 4.6 billion years by dating planetary matter that is unavailable in the case of other stars. Certain indicators, such as the amount of "unburned" light elements (lithium, for example) in a star's atmosphere, can also be used to estimate a star's age. On the other hand, we can use an H-R diagram and our model of the structure of a main-sequence star to date populations of stars.

Consider a group of stars that formed at the same time. After about 10 million years, stars larger than 20 $M_\odot$ will have disappeared from the main sequence; that is, the H-R diagram of the group will contain no main-sequence O stars. After about 100 million years, stars more massive than 4 $M_\odot$ will have evolved off the main sequence, and the H-R diagram will contain scarcely any main sequence B-type stars. The older the cluster, the more of the main-sequence stars will be gone. The missing stars will have been transformed into giants, white dwarfs, or even fainter terminal objects.

Thus, we reach an important conclusion: The H-R diagram can serve as a tool for dating groups of stars that formed together. This principle will be applied in later chapters, as we probe the **ages of stars** in our galaxy, and in other galaxies.

Summary

Our view outward from Earth has taken us far beyond the places that we can investigate in the near future by manned or robotic space missions. Even at the speed of light (which is 10,000 times greater than speeds our spacecraft have achieved), it would take years to reach the nearest stars. Instead of sending instruments to the stars, we must rely on the messages in the starlight that comes to us. Our abilities to interpret starlight are rapidly increasing with the construction of new large telescopes and the launching of telescopes into orbit. It is a measure of the power of the scientific method that we can gather and interpret starlight in order to understand the nature and histories of remote stars.

We have learned that stars are luminous balls of gas like the Sun, but hundreds of thousands of times more distant than the Sun. Stars are so distant we must use new units of measurement: the light-year and the parsec. We cannot see the size of a star directly; the images of all stars are blurred to about the same angular size by Earth's atmosphere. The apparent brightness of a star does not provide a good idea of its distance, since stars differ greatly in their absolute brightness or luminosity. Trigonometry is used to measure the distance to nearby stars out to about 100 parsecs, by the method of trigonometric parallax.

The study of stars began with the classification of stellar spectra. Spectroscopy reveals the temperature and chemical composition of the photosphere; the Doppler shift of the spectral lines reveals the star's motion in space. Stars have a wide range in temperature, from two times cooler to ten times hotter than the Sun. The spectral features reveal that all stars contain mostly hydrogen and helium, with small quantities of heavier elements. In this respect, all stars are like the Sun.

Other stellar properties are deduced indirectly. Luminosity can be calculated from apparent brightness and distance. Size is derived from the Stefan-Boltzmann law, which specifies how much energy flows through the surface area of a star. Starting from knowledge of the mass, the power source, and the chemical composition, theorists can predict the physical conditions throughout a star. The balance between gravity pulling in and gas pressure pushing out is called hydrostatic equilibrium. Every star remains a stable sphere as long as it is steadily converting lighter elements into heavier elements by the fusion process.

Stars display a range of mass, luminosity, temperature, and chemical composition. Temperature and luminosity do not occur in any possible combination; stars are grouped into distinct types on a plot called an H-R diagram. Stars like the Sun, which shine by energy released from the fusion of hydrogen into helium, are on a track called the main sequence. The most massive stars on the main sequence are large, hot, and blue; the least massive main-sequence stars are small, cool, and red. There are stars much larger than the Sun, the giants and supergiants, and stars much smaller than the Sun, the dwarfs. Differences among star types are conveniently displayed on the H-R diagram, which plots stellar luminosity against temperature. Stars are born, stars evolve, and stars die. The H-R diagram presents a "frozen" record of stellar evolution.

As stars form and begin to fuse hydrogen in nuclear reactions, they settle onto the main sequence: The more massive the star, the more luminous its main-sequence position on the H-R diagram. Stars evolve from one form to another. They begin with pre-main-sequence configurations, settle onto the main sequence for a relatively long time to convert hydrogen to helium, and then evolve off the main sequence. Mass is the fundamental quantity that controls the evolution of any star. The more massive a star, the faster it goes through its sequence of life stages. Massive stars evolve quickly because they have strong gravity, which leads to higher pressures and higher central temperatures; they therefore consume their hydrogen faster.

Most stars, including most of those in the Sun's neighborhood, have masses like that of the Sun or even smaller. Relatively faint and cool, they lie in the lower right part of the H-R diagram. The most massive stars are much brighter and can be seen from much farther away. These luminous stars will dominate any survey that is limited by apparent brightness. They lie in the upper part of the H-R diagram. Thus, many of the stars prominent in the night sky are prominent not because they are close but because they are the very luminous "whales among the fishes," far away among legions of distant fainter stars.

Philosophically, our achievement in understanding the stars is profound. The great English astrophysicist Arthur Eddington pointed out that humans (or other intelligent creatures) could deduce the existence of stars from purely physical principles, even without telescopic observation! His reasoning was that gravity would cause gas to clump into star-size masses. Even though humans are confined to Earth, we have reached out with our minds and our telescopes and understood the stars.

Important Concepts

You should be able to define these concepts and use them in a sentence.

star	bolometric luminosity
light-year	Stefan-Boltzmann law
parsec	hydrostatic equilibrium
apparent brightness	H-R diagram
absolute brightness	main-sequence stars
luminosity	giant stars
parallax distance	supergiant stars
spectral class	white dwarf stars
radial velocity	mass-luminosity relation
proper motion	stellar evolution
astrometry	ages of stars

How Do We Know?

These questions and answers show how the scientific method is used to learn about the universe.

Q How do we know how far away the stars are?

A The apparent size of stars in the sky gives no clue to their true size, since the images we see blur and twinkle in their passage through Earth's atmosphere. Direct measures of stellar distance use trigonometry. As Earth moves in its orbit of the Sun, the position of a nearby star shifts slightly on the sky—a parallax shift. Measuring the angles of this triangle gives the distance. Most stars are too far away to produce an observable parallax. However, if we know the intrinsic type of a star, in particular its luminosity, we can use the inverse square law to deduce its distance using its true and apparent brightness. Since stars have such a huge variation in intrinsic properties, this is a far less reliable procedure than a parallax measurement.

Q How do we know the mass of a star?

A In general, we do not. Only when a star is in a binary orbit do we get a direct measure of the mass from an application of Kepler's laws. We rely on our knowledge of the energy source—nuclear fusion of hydrogen into helium for a main-sequence star—to construct a model of the star. The model gives the mass required to match the energy output of the star. Main-sequence stars have a smooth and predictable relationship between mass and luminosity, which we can check with observations of binary systems. Mass is the fundamental variable that relates temperature and luminosity on the H-R diagram.

Q How do we know that stars evolve?

A We do not know by direct observation, although we have been lucky enough to see a few stars die as supernovae. Instead, we rely again on our knowledge of the energy source. We can predict that stars must eventually exhaust their nuclear fuel and evolve to a different configuration; the evolution is driven by mass. We can show that there are stars older than the Sun, so that other stars existed before the Sun was born. Some stars are so massive that our models predict they can only live for tens of millions of years as main-sequence stars. Others are emitting energy at such a feeble rate that they could continue relatively unchanged for many billions of years.

Q Do all stars produce energy by converting hydrogen to helium?

A No. We have seen that all stars that produce energy by the fusion of hydrogen nuclei (protons) into helium nuclei lie on the main sequence of an H-R diagram. The relationship between the luminosity and the temperature of a main-sequence star is governed by its mass. However, every point on the H-R diagram also corresponds to a star of a certain radius, since the rate of energy generation per square meter of surface area is given by the Stefan-Boltzmann law. In addition to stars like the Sun, we find main-sequence stars that are hotter and more luminous than the Sun and main-sequence stars that are cooler and less luminous than the Sun. However, some stars have combinations of temperature and luminosity than cannot be explained by the fusion of hydrogen into helium. Some giant stars need a more prodigious energy source to maintain their large size, and some dwarf stars have insufficient hydrogen to be powered by hydrogen fusion.

Q What is the true population of stars in space?

A It is surprisingly difficult to measure the number of stars of all types in each volume of space (a cubic parsec, for example). We do not have the ability to fly through space and search for dim and unrecognized stars. Instead, we must rely on light gathered by telescopes. We are limited by the inverse square law of light travel: Powerful stars can be seen out to large distances, but faint stars can only be seen when they are nearby. The result is that we will tend to overestimate the abundance of luminous, or giant, stars and underestimate the abundance of low-luminosity, or dwarf, stars. One strategy is to survey a small enough volume around the Sun that even the dimmest star is visible. However, stars that are very luminous and rare might not be contained within that volume. The other approach is to count stars down to an apparent brightness limit and try to correct for the effect of limited visibility.

Q How do we know the age of a star?

A Our measurement of age is indirect and requires a physical model for the star. In the case of the Sun, we have a detailed model that predicts the temperature, density, and pressure at every point, and consequently the rate of energy production and conversion of hydrogen into helium. Combined with our measurement of the chemical composition, this gives an estimate for the time that the Sun has been on the main sequence. (It is good news that this agrees with the age of the oldest rocks on Earth.) For other stars, there is less information, but the idea is the same. Mass and chemical composition are the two vital ingredients in the calculation of age. Massive stars evolve rapidly, so they must be much younger than the Sun. Stars of low mass evolve very slowly and will still be on the main sequence long after the Sun has died.

Problems

Use these problems to test your understanding of the information and concepts in this chapter. The * indicates a more advanced or mathematical problem.

1. If a telescope of 25-meter (1000-inch) aperture could be put in orbit or on the Moon, it could resolve an angle of only about 0.005 second of arc. Compare this resolution with the maximum angular size known for stars (0.03 second). Why would this telescope perform better on the Moon than on Earth? Why do astronomers never use the highest magnifications theoretically possible with earth-bound telescopes?

2. A star is 20 pc away. How many years has its light taken to reach us? Roughly how many years does it take light to reach us from the nearest stars? If you look in the night sky at a star that is 80 pc away, what major events were occurring on Earth around the time light left that star?

3. What is the chemical composition of most stars? How does the helium get produced in the Sun and other main-sequence stars, and why is it such a rare element on Earth?

4. (a) A reddish star and a bluish star have the same radius. Which is hotter? Which has higher luminosity? Explain your reasoning. (b) A reddish star and a bluish star have the same luminosity. Which is bigger? Explain your reasoning.

*5. (a) A certain star has exactly the same spectral features as the Sun, but is a factor of 10^{12} fainter in apparent brightness. List some conclusions you would draw about it and describe your reasoning. (b) Two stars appear to have equal brightness, but one star is 10^3 times farther away. Which star is larger, the nearer or the more distant one, and why? (c) Now suppose two main-sequence stars are in a binary pair, and one is like the Sun while the other is four times hotter. How much more luminous would the hotter star be? (d) Most main-sequence stars are cool and red. Explain the reason why hardly any of these common stars are seen in the night sky.

*6. (a) A star has a parallax of 0.05 second of arc. How far away is it? (b) The nearest stars subtend far too small an angle to be resolved by a ground-based telescope with a resolution of 1 second of arc. At what distance would a solar-type star have to be to be resolved by such a telescope?

*7. Suppose you could fly around interstellar space encountering stars at random. (a) Describe the stars you would encounter most often. Mention masses and H-R diagram positions. (b) About what percentage of stars would be as massive as, or more massive, than the Sun? (Hint: Use statistics from the tables in this chapter.) (c) Would the stars encountered be similar to bright stars picked at random in our night sky? If not, why not? (d) Suppose we can see the coolest main-sequence star out to a distance of 10 parsecs. Over a how much larger *volume* could we see the hottest main-sequence star?

8. Four stars occupy the corners of an H-R diagram. (a) Which one is the largest? (b) Which one is the smallest? (c) Which one is small and cool? (d) Which one is large and hot? (e) Which two are on the main sequence? (f) Which two are visible to the largest distance?

9. Two stars lie on the main sequence in different parts of the H-R diagram. Which is: (a) Larger? (b) More luminous? (c) More massive? (d) Hotter? (e) Which part of this problem could not be answered just from location in the H-R diagram if both stars were not on the main sequence?

10. Show that astrophysical estimates of the Sun's lifetime are consistent with meteoritic and lunar data on the age of the solar system. Why would the data be inconsistent if astrophysicists calculated that solar-type stars stay on the main sequence only 1 billion years?

11. Why do normal stars not all collapse at once to the size of asteroids or tennis balls under their own weight, since gravity always pulls their material toward their center?

12. Three billion years ago, how was the chemical composition of the Sun different from the way it is today? How will the chemical composition differ when the Sun ends its life as a main-sequence star in 5 billion years?

Projects

Activities to carry out either individually or in groups.

1. Using a telescope of fairly high magnification (such as 300× or 400×), examine the image of a bright star on several different nights, and sketch it. Are there differences from night to night? Can you identify the "seeing" disk, diffraction rings, or diffraction spikes? If so, label them on your sketches. Note any shimmering caused by atmospheric turbulence or air currents in and around the telescope. Run the eyepiece inside or outside the focal point, making the image a round blob. Shimmering and other turbulent effects are often more evident in this way.
2. Take several lamps with no shades to a long, darkened room. Gather lightbulbs of as many different power ratings as you can; bulbs of the same size with ratings of 25 W, 40 W, 60 W, 75 W, 100 W, 150 W, and 200 W should be easily available. Darken the room, and pretend that the lightbulbs are stars. Standing across the room, see if a classmate can distinguish between all the different levels of brightness. Can you find two values of apparent brightness that are too close to distinguish? See how the results vary for different members of the group. Do the results suggest why early astronomers invented a system in which neighboring brightness categories differed by a factor of 2.5 in intensity? Now take pairs of lighted bulbs—one of high wattage and one of low wattage—and hold them side by side. Have a classmate viewing from afar direct you to move away with the luminous bulb until it appears to be the same brightness as the less-luminous bulb. Measure the two distances. Do they follow the expected inverse square law? Repeat the experiment with different lightbulb combinations and different classmates to improve the statistics.
3. Locate or construct spherical objects that have the same proportional sizes as selected different types of stars, such as a supergiant like Antares, the Sun, and a white dwarf.
4. From a camping store, buy several small blocks of paraffin wax. (Paraffin wax is designed to have a very low melting point.) With the Sun nearly overhead, see how long it takes the block to melt under direct sunlight, and calculate the rate of melting of the block. Now try and melt a second block of wax with a 100-watt lamp held close to the block. To be a fair experiment, the inside and outside temperatures should be the same, or you can shield the second block from direct sunlight. What is the distance of the lamp that gives just the same melting time or rate of melting as the Sun? Can you use this information and the inverse square law to calculate the luminosity of the Sun in watts?

Reading List

Croswell, K. 1987. "Visit the Nearest Stars." *Astronomy,* vol. 15, p. 16.

Croswell, K. 1992. "The Grand Illusion: What We See Is Not Necessarily Representative of the Universe." *Astronomy,* vol. 20, p. 44.

Davis, J. 1991. "Measuring the Stars." *Sky and Telescope,* vol. 88, p. 361.

Dobson, A. K., and Bracher, K. 1992. "Urania's Heritage: A Historical Introduction to Women in Astronomy." *Mercury,* vol. 21, p. 4.

Furth, H. P. 1995. "Fusion." *Scientific American,* vol. 273, p. 174.

Kaler, J. B. 1991. "The Brightest Stars in the Galaxy." *Astronomy,* vol. 19, p. 30.

Kaler, J. B. 1991. "The Faintest Stars." *Astronomy,* vol. 19, p. 26.

Kaler, J. B. 1992. *Stars.* Scientific American Library. New York: W. H. Freeman.

Kaler, J. B. 1994. "Giants in the Sky: The Fate of the Sun." *Mercury,* vol. 22, p. 34.

MacRobert, A. M. 1992. "The Spectral Types of Stars." *Sky and Telescope,* vol. 92, p. 48.

Steffey, P. C. 1992. "The Truth About Star Colors." *Sky and Telescope,* vol. 84, p. 266.

Welther, B. 1984. "Annie Jump Cannon: Classifier of Stars." *Mercury,* vol. 13, p. 28.

CHAPTER 13

Star Birth and Death

CONTENTS

- Supernova 1987A
- Understanding Star Birth and Death
- Star Formation
- The Main-Sequence Stage
- Evolved Stars
- The Death of Stars
- Summary
- Important Concepts
- How Do We Know?
- Problems
- Projects
- Reading List

Pillars of gas and dust conceal the sites of star formation in the Eagle Nebula. In this spectacular image from the Hubble Space Telescope, we can see new stars forming in the densest regions at the top of each pillar. The new stars are revealed as the radiation from hot stars at the top of each tower erodes the concealing material. (SOURCE: Jeff Hester and Paul Scowen, Arizona State University/NASA.)

WHAT TO WATCH FOR IN CHAPTER 13

In the last chapter, we discussed the properties of stars. We learned that stars have an enormous range of luminosities, sizes, and masses. Most stars are on the main sequence, fusing hydrogen into helium like the Sun. However, there are also cool stars hundreds of times larger than the Sun and hot stars hundreds of times smaller than the Sun. In this chapter, we unravel the life stories of stars of all masses. Star birth and death occur quickly on a cosmic time scale, so most of the stars we see in the night sky are in the middle of their lives. We need to know how stars evolve to understand the cosmic abundance of the elements, since heavy elements are produced in massive stars and recycled into space as stars age and die. You will learn how the cycles of star birth and death follow the laws of thermodynamics—energy is conserved, but there is a universal tendency for organized forms of energy to turn into heat and disorder.

First, we discuss star birth, which occurs in regions of space shrouded by gas and dust. Young stars are energetic, emitting copious amounts of infrared radiation and often spewing jets of material from the poles. After the birth of a star, every aspect of its life is controlled by its mass. We will learn that massive stars can use nuclear fusion to build heavier elements than less-massive stars. All stars get their energy from the fusion process; starting with hydrogen, they create helium and then heavier elements. The cycle of star birth and death has distributed many of the chemical elements into space. Late in their evolution, stars once again begin to shed mass, either in a steady wind or by blowing large bubbles of gas. The final state of a star is determined by the mass of its core. Most stars have low mass, and they become white dwarfs that slowly cool over billions of years. The rarer, high-mass stars die suddenly in a supernova explosion. After the violence of a supernova, the core may be turned into dense and extraordinary states of matter: neutron stars and black holes. We will use Einstein's general theory of relativity to understand the intense gravity that results from the final collapse of a star.

PREREADING QUESTIONS ON THE THEMES OF THE BOOK

OUR ROLE IN THE UNIVERSE
How do stars produce the chemical elements that are needed for life?

HOW THE UNIVERSE WORKS
How does the battle between gravity and energy production govern the fate of a star?

HOW WE ACQUIRE KNOWLEDGE
How can we understand the life stories of stars?

SUPERNOVA 1987A

Oscar Duhalde tilted his head back and looked at the night sky. Duhalde was a telescope operator at the Las Campanas Observatory in Chile, and he had stepped outside around midnight on February 23, 1987, to check the sky conditions. The observatory was in the foothills of the Andes, far from civilization, and the stars burned brightly on a backdrop as black as velvet. He noticed a new star near the 30 Doradus Nebula in the Large Magellanic Cloud, which is a nearby galaxy named by the explorer Magellan on his first southern voyage. Since the star had not been there on previous nights, it must have brightened sharply. It was a delicious moment of discovery. This was the first supernova bright enough to be seen by the naked eye for nearly 400 years! Duhalde was probably the first human witness to the death of this particular star, in an extraordinary and violent event called a supernova.

The astronomical community spun into a whirlwind of activity. At Las Campanas, Ian Shelton developed a photograph of 30 Doradus that he had already taken and therefore was credited with recording the discovery—the first of the year, thereby designated SN 1987A. The supernova was independently spotted by an amateur astronomer working in New Zealand; several other amateurs missed making the first observation because of clouds at their observing sites. Around the world, telescopes took spectra of the dying star, and satellites recorded its high-energy emission.

The first trace of the supernova was not the dazzling light, however, but an invisible wave of neutrinos that passed through Earth. In Japan, researchers at a particle detector 3000 feet underground sifted through their data and found 11 events caused by interactions between neutrinos and the 2100 tons of highly purified water that filled their detector. The detection of these ghostly particles from the death throes of an object beyond our own galaxy ironically heralded the birth of a new field of astronomy.

Early *Homo sapiens* were just developing the first societies on the plains of Africa when the blue supergiant in the Large Magellanic Cloud exhausted its nuclear fuel. The subsequent core collapse triggered a prodigious explosion. A flood of radiation and neutrinos poured into space, in a sphere expanding at the speed of light. Nearly 170,000 years later, just after we developed the tools of modern astronomy, that sphere swept across Earth's path. SN 1987A provides a wonderful opportunity to check our theories of how a star ends its life. We also have ringside seats for the formation of the exotic object that is left behind when a massive star dies.

Understanding Star Birth and Death

We are trapped in both time and space as we try to understand the nature of stars. We cannot travel and probe the physical conditions inside stars of different types. For the most part, we must study the clues that are contained in starlight. Nor can we observe the evolution of stars. A human life is like a blink of the eye in the life span of any star; the night sky presents us with a frozen moment in long cycles of star birth and death.

Understanding the birth and death of stars is a particular challenge. Star birth is shrouded behind dense clouds of gas and dust; it is difficult to penetrate this material with telescopes. The collapse and violent death of a massive star occupies no more than a millionth of the main-sequence lifetime. Therefore, we would have to observe a million stars at various stages of their evolution to expect to find one case of star death. Sometimes we get lucky, and that is why SN 1987A is such a great opportunity.

Despite the lack of direct experimentation possible for stars, our study of stellar birth and death is still firmly grounded in the scientific method, which is a major theme of this book. Spectroscopy enables us to measure the surface temperature and luminosity of a star, which in turn allows us to estimate the star's size. We can then place the star on an H-R diagram, which tells us its evolutionary state. (Remember from Chapter 12 that an H-R diagram is a plot of surface temperature against luminosity. When we talk about stars "moving across" the H-R diagram or "leaving" the main sequence, it is shorthand for describing how the physical properties of a star change with time.) We can compare these data to stellar models and deduce the quantity that specifies the past and future of any star—mass. We then deduce more information by studying *populations* of stars and check our theories against the few direct observations of birth and death.

This chapter continues another theme of the book: how the universe works. The laws of gravity and radiation that we observe on Earth also apply across space and time. The life story of any star is a battle between two competing physical forces: gravity and the pressure caused by nuclear reactions in the core. Sometimes pressure wins and material is shot off into space—gently in the case of a giant star, and violently in the case of a supernova. More often gravity entombs material forever in a collapsed core: white dwarfs, neutron stars, and black holes. The entire life story of a star is determined by its mass. Thus, we can observe a nearby star and confidently predict its fate, even though we will not be here to witness it! We will also see that stars are fusion factories for the production of chemical elements. It is a remarkable fact that cosmic chemistry is the same wherever we look.

In this chapter, we continue to address another fundamental question: Where did we come from? We have seen that the Sun and other stars are mostly made of hydrogen and helium, but we know that life depends on a number of heavier elements, including carbon, nitrogen, and oxygen. Life-giving elements are produced deep inside stars and are then spat out into space during the late stages of a star's life. We cannot understand our cosmic origins without understanding the cycle of star life and death.

Figure 13–1 shows the **cosmic abundance** of all the stable elements. Notice that the scale is logarithmic. The abundance of hydrogen is almost a trillion (10^{12}) times the abundance of gold! These values were derived from the Sun, but the relative proportions of different atoms are similar in all stars astronomers have studied. Let us return to our analogy of the deck of cards from earlier chapters. Imagine each card is an atom. In a typical sample of star matter, most particles are hydrogen atoms and only 1 in 12 is a helium atom. This would be like a deck where the aces represent helium and all the other cards represent hydrogen. The next most abundant element is oxygen, found at a level of 1 in 1500 atoms. We would need to search through 28 decks of cards to find a single oxygen "atom." All other cards are either hydrogen or helium. Gold is much rarer still. We would need to search through 10 billion decks of cards to find a single gold "atom"! Spectroscopy is such a powerful technique that astronomers can routinely identify elements at a concentration of less than one part in a billion.

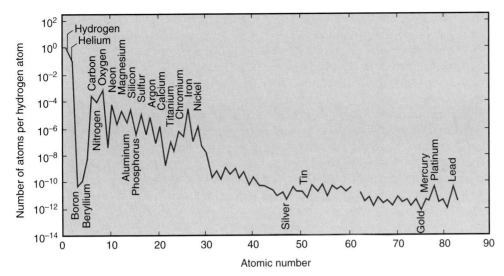

FIGURE 13-1
The cosmic abundance of the chemical elements. There is an enormous range in abundance; the heaviest elements are trillions of times rarer than hydrogen and helium, which comprise 99.9 percent of all material. Heavy elements are built up from lighter ones in the cores of stars. The jagged pattern is caused by the addition of a helium nucleus by the fusion process, which adds two protons to the original nucleus. The peak at iron is due to the stability of the iron nucleus. The heavy elements beyond iron are produced during the late evolutionary stages and the violent deaths of massive stars. (SOURCE: Data from A. Cameron.)

One of our important goals is to explain the features of Figure 13-1. Why are hydrogen and helium the most abundant elements? What explains the relatively high abundance of carbon and oxygen? Why is there a peak in the graph at iron? What explains the "sawtooth" pattern in the abundance distribution? Why are elements much heavier than nickel so fantastically rare? We will see that the creation of the materials of the everyday world takes place deep within stars. We cannot understand the cosmic abundance of elements without understanding the evolution of stars.

Star Formation

As we saw in Chapter 3, astronomers once believed that the stars were fixed and unchanging as they revolved around Earth on the outermost crystalline sphere. Today, we know that stars continue to be born, evolve, and die. How do we know that stars change and evolve? The universe offers us several lines of evidence that stars are constantly being born. First, we know that while the solar system is only about 4.6 billion years old, the entire Milky Way galaxy—our system of 40 billion stars—is at least 10 billion years old. Thus, our Sun formed much more recently than the oldest stars we can see in the sky. From this, we can infer that the stars around us did not all form at the same time.

Second, many young, massive stars are grouped into clusters. We know that the stars in these clusters formed at about the same time. Because of tidal forces and the tendency of each star in the cluster to follow its own orbit around the center of our galaxy, most clusters disperse into isolated stars in no more than a few hundred million years. However, the famous Pleiades cluster, for example, is only about 50 million years old. H-R diagrams show that a few of these clusters are much older. The existence of such clusters shows that star clusters did not all form at the beginning; some continue to form today.

Third, we saw in the last chapter that massive stars evolve fastest. Calculations show that stars of 20 to 100 times the mass of the Sun can last only a few million years in a visible state. Since we still see these massive stars shining, we can conclude that they must have formed less than a few million years ago. Thus, star formation has been a continuing process during the whole history of the galaxy, including the last million years. There are stars in the sky younger than the human species.

Molecular Clouds

Stars form from sparse material between other stars. In most cases, the region between stars is a near vacuum containing very thin, cold gas. Atoms and molecules number only about 10^6 per cubic meter. The temperature is typically around 10 to 20 K. However, certain regions of space have a density of gas and dust 10,000 times greater. With some 10^{10} atoms per cubic meter, the atoms are close enough to collide frequently. Moreover, the temperature is low enough that atoms will stick together to form molecules. The most important molecular gas is molecular hydrogen (H_2), and it forms on dust grains. These regions, called **molecular clouds,** are rich in molecular hydrogen, carbon monoxide (CO), water (H_2O), and more complex forms such as formaldehyde (H_2CO) and ethyl alcohol (C_2H_5OH). Such complex molecules are virtually absent in most parts of interstellar space. Although a molecular cloud is denser than most interstellar gas, it is still a near-perfect vacuum compared with ordinary room air, where there are about 20 trillion trillion (2×10^{25}) molecules per cubic meter.

Molecular clouds are important because they are the places where most stars form. In these regions, atoms, molecules, and dust grains are crowded closely enough to begin to attract each other gravitationally—a key step in forming stars from a gas. Let us look at what is happening in a molecular cloud in the Orion Nebula. Observations taken with a radio telescope show the dynamic activity of molecules in the cloud. Figure 13–2 shows a spectrum made at millimeter wavelengths, with many spectral features caused by different states of molecular vibration and rotation. The low energy of the transitions explains why the spectral lines are seen at long wavelengths, in the millimeter or submillimeter part of the electromagnetic spectrum. Astronomers have found that the chemistry within a molecular cloud can be very complex.

Certain dense regions in molecular clouds display particularly spectral lines that are as intense and pure as the light from a laser! Astronomers have observed OH and H_2O molecules that generate *maser* emission. Maser is an acronym for microwave amplification by the stimulated emission of radiation, analogous to the familiar phenomenon in optical light called a laser. Normally, gas in space is at such a low density that most of the atoms or molecules are in their lowest energy states. However, molecules in the dense cores of molecular clouds can be forced to overpopulate the excited energy levels. When this happens, a single photon can trigger a surge of radiation as the molecules drop into their lowest energy states.

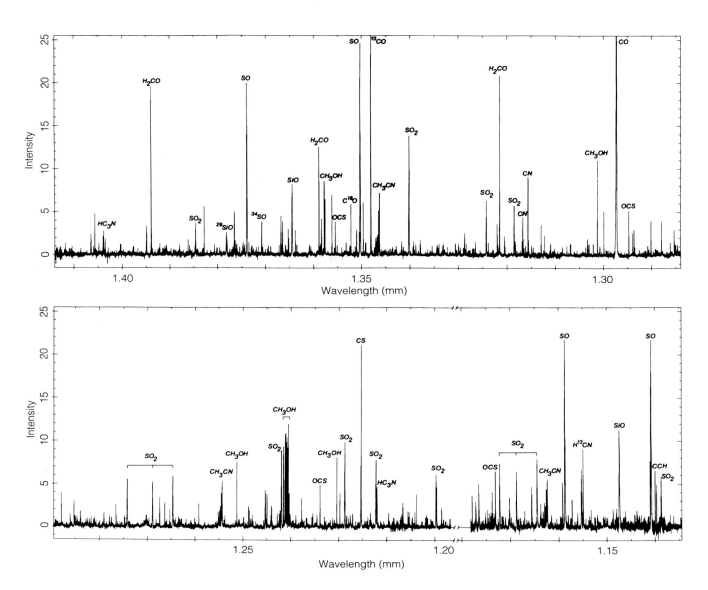

FIGURE 13-2
A radio-wavelength spectrum of the core of the OMC-1 giant molecular cloud, made at the Owens Valley Radio Observatory. Earth's atmosphere is usually opaque to millimeter waves; this spectrum covers a small wavelength range around 1.3 mm, where there is an atmospheric "window." Over 800 spectral features are visible, corresponding to 29 different types of molecules. Each spectral feature is caused by photons with a wavelength corresponding to the difference between two energy states. This type of data is needed to understand the chemistry of cold and dense interstellar regions. (SOURCE: Courtesy of T. G. Phillips, California Institute of Technology.)

Many remarkable molecules have been discovered in interstellar space, including the amino acid glycine (NH_2CH_2COOH), one of the building blocks of life. In the 1950s, British astronomer Sir Fred Hoyle wrote a science fiction novel called *The Black Cloud*, in which he speculated that life might form in the dark depths of a molecular cloud. Scientists do not think it likely that replicating molecules could form in the near vacuum of space, but the discovery of complex molecules such as glycine is certainly intriguing. The Sagittarius B2 molecular cloud contains a substantial amount of ethyl alcohol. American astronomer Ben Zuckerman noted that, if purged of impurities, this cloud would yield approximately 10^{28} fifths at 200 proof! This exceeds the sum of all human fermentation efforts in recorded history. A few years after this discovery, NASA received a proposal from a Texas businessman to "harvest" this alcohol. Molecular clouds are too diffuse to make this idea practical. It turns out that a spaceship pushing a funnel with a diameter of 1 km at 10 percent of the speed of light would take 1000 years to collect enough ethyl alcohol to fill a shot glass.

Toward a Theory of Star Formation

In Chapter 11, we learned how the Sun formed when a diffuse interstellar cloud contracted and produced a central star surrounded by a dusty nebula. Our solar system has left us with a set of clues that indicates a rapid collapse from a gas cloud to a star, taking only a few million years. Now let us examine a more general theory of this process that can explain other stars with different masses.

When scientists use the word *theory*, they usually mean a well-tested body of related ideas, often with a mathematical formulation that can be applied to a variety of cases. The theory must be backed up by observations. It should make predictions that can be verified. Astronomers have made some progress toward a **theory of star formation**. We might want this theory to answer several questions. What causes some clouds to contract and not others? How much of the material in a molecular cloud goes into stars and how much is left over? Why does this process only give stars with a mass range of a factor of a thousand—from one-tenth to 100 times the mass of the Sun? The theory involves the same two opposing forces we have considered before: gravity versus thermal pressure.

Gravity pulls all the atoms in a cloud inward. But even at only 10 K, the atoms are striking each other at a speed of 0.4 km/s (nearly 1000 mph) and creating an outward pressure that opposes the tendency to collapse. As a cloud contracts, gravity increases because a denser cloud packs more mass into the same amount of space. On the other hand, outward pressure also increases if the gas heats up, since this makes the atoms and molecules move faster. So how does a cloud actually collapse? Some of the heat that keeps a cloud puffed up can escape in the form of photons associated with particular energy transitions in atoms and molecules—in other words, emission lines can help cool the gas. Another trigger to cause the collapse of a cloud can be compression resulting from a nearby stellar explosion.

In 1902, English astrophysicist James Jeans contributed an important insight to this outline of star formation. He calculated the mass and temperature at which a cloud would begin its gravitational contraction. The calculation is idealized, because it does not consider the effects of magnetic fields or the possibility that the cloud might be rotating. The simple theory predicts that in the denser interstellar medium, masses contracting will be several hundred to a thousand solar masses. This mass is several times larger than the most massive stars and thousands of times larger than the least massive stars. In other words, the simple process of gravitational contraction cannot produce the full range of star masses that we see.

How, then, do individual stars form? Conditions in the interstellar medium are such that the gas subdivides into enormous concentrations that contain enough mass for hundreds of stars. We know of star clusters with this many stars, as noted at the beginning of the chapter. The new fragments become stars, and the whole mass turns into a star cluster. It is not hard to see why the interstellar gas might begin to contract. The material is not uniform; clots of dust and gas are constantly agitated by the galaxy's rotation, material is ejected from the later stages of stellar evolution, and so on. Naturally, this movement allows some clouds to accumulate enough material for contraction to begin. It has even been suggested that stars might form like an infection spreading through the body, where one region of star formation causes an adjacent region of gas to form stars, and so on. In this way, star formation might spread though an entire galaxy.

The preceding discussion does not tell the whole story of star formation, however. Star formation in molecular clouds occurs more slowly and less efficiently than the simple theory of gravitational collapse would lead us to expect. The presence of magnetic fields also complicates the process of star formation. Current thinking is that magnetic fields prevent gravitational collapse in molecular clouds. Magnetic fields create a "pressure" that opposes the inward force of gravity. Rotation of the entire cloud can also slow its collapse into stars. Scientists have taken many effects into account to build a complex theory of star formation.

Modern theory envisions the following four stages in the process of star formation in molecular clouds:

1. A slowly rotating core forms in a molecular cloud.
2. The core becomes unstable and collapses into a protostar and a surrounding disk, both of which are embedded in an infalling envelope of dust and gas. The collapse phase is inside-out, in the sense that the material nearest the core collapses first.
3. The first fusion reactions begin—producing deuterium. The energy released from nuclear reactions produces a stellar wind of outflowing material, which opposes the infalling material from farther out. The stellar wind rushes through the paths of least resistance at the rotational poles, leading to jets of material that flows out along the poles.

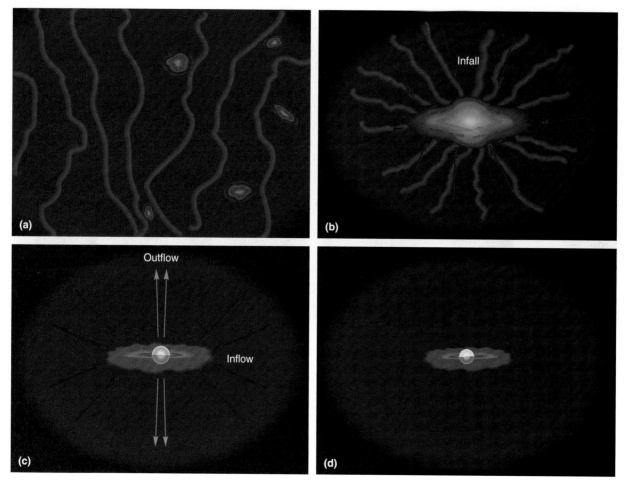

FIGURE 13-3
The main stages of star formation. **(a)** The dense cores of molecular clouds can collapse as the magnetic field slowly diffuses away and the cloud loses magnetic support. **(b)** The material around the protostar collapses from the inside outward. **(c)** A stellar wind blows along the poles of the rotating system, leading to bipolar outflows or jets. **(d)** The infall finally terminates, revealing the newly formed star. (SOURCE: Adapted from a diagram by F. Shu.)

4. The wind fans out until it flows in all directions. At this point, a young stellar object has formed, still surrounded by its nebular disk. Figure 13-3 illustrates these early stages of star formation.

Protostars and Pre-Main-Sequence Stars

In the second stage of star formation, a **protostar** forms. A protostar is a cloud of interstellar dust and gas that is dense and cool enough to begin contracting gravitationally into a star. An interstellar cloud may hover on the verge of this state for millions of years. A nearby exploding star or some other disturbance might compress a gas cloud and thus trigger the collapse, which then proceeds very rapidly. For instance, typical protostars contract to stellar dimensions in 100,000 years—a wink of the cosmic eye. This explains why astronomers use the term *collapse*: the initial contraction is very rapid in terms of astronomical time. The collapse may not be smooth; the inner core of the cloud may collapse first, later absorbing the surrounding material. Also, under certain conditions a single rotating cloud may break up into two or more stars. Of course, once the cloud begins to reach stellar dimensions (a few astronomical units, or 10^{-5} pc), the atoms of gas collide frequently enough to produce substantial outward pressure and slow the collapse. At this point, the object becomes a **pre-main-sequence star**.

What do pre-main-sequence stars look like? Where do they lie on the H-R diagram? Can they actually be found among the many stars in space? Can we actually see stars being born? In fact, sites of active star formation are often shrouded in gas and dust. The clearest view is at infrared wavelengths. Figure 13-4 shows M17, the Omega Nebula, which is obscured by a factor of 1 million in visible light, but only by a factor of 4 at the infrared wavelength of 2 microns. The pre-main-sequence star stage covers the evolution from the end of the protostar stage to the main sequence.

Japanese theorist Chushiro Hayashi performed pioneering calculations on the evolutionary tracks and appearance of stars contracting toward their main-sequence configurations.

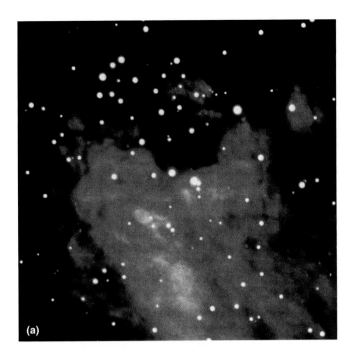 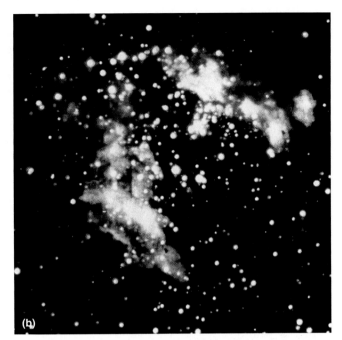

FIGURE 13-4
M17, the Omega Nebula. **(a)** An optical color photograph made with the Kitt Peak 4-m telescope, showing a region of vigorous star formation. **(b)** A composite infrared image, with 1.2-micron radiation colored blue, 1.65-micron radiation colored green, and 2.2-micron radiation colored red. This shows how the nebula would appear if the eye responded to infrared radiation. Most of the crucial star-forming activity is obscured by dust and shows up only in the infrared image, which was taken with the Kitt Peak 2.1-m telescope. (SOURCE: I. Gatley/NOAO and C. Lada/Harvard-Smithsonian Center for Astrophysics.)

Scientists have shown that the energy during most of the pre-main-sequence period does *not* yet come from nuclear reactions. Instead, gravitational contraction causes a release of energy. After only a few thousand years of collapse, surface temperatures reach a few thousand Kelvin, causing visible radiation. Hayashi's work showed that convection would transport large amounts of energy from the interiors of most newly forming stars, making them very bright for short periods of time. A star like the Sun, for example, contracts in less than 1000 years from a huge cloud to a size about 20 times bigger than the Sun, with a luminosity about 100 times greater than the Sun's present luminosity.

These calculations have three important implications. First, they show that stars have complicated, if short-lived, evolutionary histories even before nuclear reactions start. Second, they show that newly forming stars have properties that place them above and to the right of the main sequence on an H-R diagram. The position of a star on the H-R diagram helps us identify it as a new star by direct observation. Third, the calculations indicate that stars spend only a small fraction of their lifetime in the pre-main-sequence stage.

Massive stars evolve faster to the main sequence than less-massive stars. For instance, a 15 $M_\odot$ star reaches the main sequence in only 100,000 years. A 5 $M_\odot$ star takes about 1 million years, while a Sun-like star takes tens of millions of years. The lowest-mass stars, around 0.1 $M_\odot$, take over 100 million years. Figure 13-5 shows how the properties of stars of different mass change as they approach the main sequence. You can see that all stars all begin life as very cool objects, but with a wide range in luminosity. As in the early solar nebula, dust forms in the cooling cloud and blocks the outgoing starlight. After a few hundred thousand years, the nebula will eventually dissipate, revealing the star. Before the nebula is lost, like the cocoon cast off by a butterfly, the star will be invisible in optical light. It will only be seen as an infrared star.

Mass Limits, Large and Small

What factors determine the range of possible star masses? When astronomers conduct surveys, they find stars that span a factor of 1000 in mass. The most massive stars are about 100 times the Sun's mass, and the least massive stars are about one-tenth of the Sun's mass. Yet gravity can cause the collapse of gas clouds with an enormous range of masses. Why are there not enormous stars 1000 or more times the mass of the Sun? And why are there not pint-sized stars one-hundredth the mass of the Sun or less?

There are physical reasons for the mass limits on stars. If a cloud is dense and cool enough to contract but has less than about 8 percent of a solar mass, it will contract but never develop a high enough central pressure and temperature to reach a main-sequence state (i.e., no extensive fusion of hydrogen). Protostars from about 0.01 to 0.08 $M_\odot$ may heat up temporarily from their gravitational contraction and may even develop a few feeble nuclear reactions, but

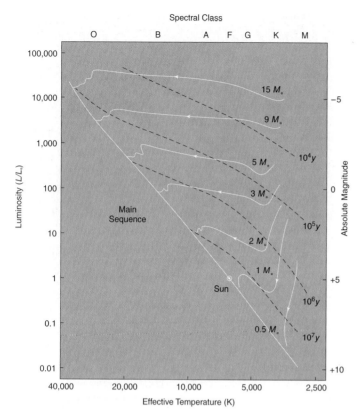

FIGURE 13-5
An H-R diagram showing simplified pre-main-sequence tracks for seven stars of different mass. Notice that high-mass stars settle on the upper end of the main sequence, while low-mass stars are on the lower end. Dashed lines show the states reached after the indicated number of years. More detailed calculations, taking into account the nebula around the star and other details, show more complex "squiggles" in the evolutionary tracks, but these details are not firmly established. (SOURCE: Data from I. Iben.)

eventually they fade without settling on the main sequence for any appreciable time. Smaller objects, including even Jupiter-sized planets, also warm up temporarily from their gravitational contraction; they may glow for a while in the infrared, but they fade before any nuclear reactions can start.

If the cloud is dense enough to contract but is more than about 100 $M_\odot$, the contraction is violent and produces an extremely high central temperature and pressure. Under these conditions, so much energy is generated inside the new star that the star is very luminous and may blow itself apart almost immediately without spending much time on the main sequence. This rapid destruction will be taken up again later in the chapter. We will see that explosions of massive stars explain many features of our starry surroundings.

The Transition from Planet to Star: Brown Dwarfs

Gravity tends to form more low-mass stars than high-mass stars. Therefore, collapsing gas clouds are expected to produce many objects near the bottom of the mass range of stars. We discuss them here because they are astronomically

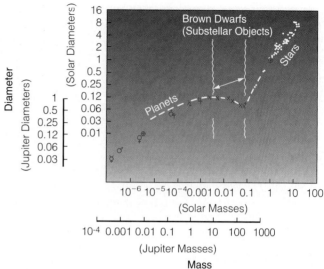

FIGURE 13-6
The transition from planets through brown dwarfs to stars. A combination of theoretical and observed data shows the changing size of objects as we add more mass. If we start with planetary masses (lower left) and add material of solar composition, the size levels off at about the size of Jupiter and then declines slightly from compression by gravity. As additional mass pushes the total mass over 0.08 $M_\odot$, where nuclear reactions begin, the size again expands.
(SOURCE: After data by T. D'Antona, J. Lunine, and W. Hubbard, private communication.)

important; there are many more of these objects than there are stars like the Sun, for example. Objects with less than 0.08 $M_\odot$ (about 80 times Jupiter's mass) but bigger than planets are called **brown dwarfs.** They are known as *substellar objects* because, lacking nuclear reactions, they are not true stars. The name *brown dwarf* comes from an early theory that they glow with a dull red light. In fact, they emit nearly all of their radiation at invisible infrared wavelengths.

Let us see how brown dwarfs relate to low-mass planets and high-mass stars. Consider a hypothetical experiment. Start with a planet with the mass of Jupiter, and steadily add gas with the same chemical composition as the Sun, ¾ hydrogen and ¼ helium. As we add mass, we progress from planets through brown dwarfs to true stars. The result is shown in Figure 13-6. As we add mass to a Jupiter-like planet, it gets only a little bigger until it reaches roughly 2 $M_{Jupiter}$, where it levels off. As we add even more mass, the object actually gets a little smaller, because gravity compresses the gas to a denser state. Then, as we approach 80 $M_{Jupiter}$, where nuclear reactions begin, it "turns on" as a star and again gets much bigger as we add more hydrogen fuel. The size increases because energy from nuclear reactions creates gas pressure which "puffs up" the star. Thus, conceptually, there is natural physical division between planets and brown dwarfs and between brown dwarfs and stars. We will use the word *planet* for anything smaller than 2 $M_{Jupiter}$ and the term *brown dwarf*, or substellar object, for objects from 2 to 80 $M_{Jupiter}$. What would a brown dwarf re-

FIGURE 13-7
An imaginary view of a brown dwarf star from a hypothetical nearby planet. The brown dwarf emits a dull glow from its barely red-hot surface. In the distance is a true star, the "sun" of this imaginary system. (SOURCE: Painting by Ron Miller.)

ally be like, if we could see it up close? An imaginary view is shown in Figure 13-7.

Astronomers have been searching for good examples of brown dwarfs, which would help us understand the relationship between planets and stars. Brown dwarfs, of course, are hard to detect because of their faintness. A few brown dwarf candidates have been reported in the past 5 years, including one with luminosity only 0.0004 that of the Sun—the lowest-luminosity object yet found outside the solar system. But there have been many false alarms, and it is hard to prove these objects are in the 2 to 80 $M_{Jupiter}$ mass range needed to call them true brown dwarfs.

Young Clusters and Associated Young Stars

Where can we look in the sky to find objects in stages before they become stars like the Sun? Because many stars form in a single large gas cloud, young stars are often found grouped together in star-forming regions. At the beginning of this chapter, we mentioned that clusters of stars have proven to be very young because of their dynamic properties—the stars are moving so fast that they will disperse in tens of millions of years.

Some of the best examples of young stars are found in the Orion star-forming region, a large area of the sky around the constellation Orion. This region, which is about 400 to 700 parsecs away, is a dense, cloudy hotbed of star-forming activity. When we look in this direction on a starry night, we can see the Orion Nebula, which contains the results of recent star formation (Figure 13-8). Many newly forming objects lie hidden from our eyes inside dust clouds in interstellar space. But these are now being mapped by detectors sensitive to infrared light that passes through the clouds, even though the clouds block visible light (see Figure 13-4).

The most important pre-main-sequence stars are the *T Tauri stars*, named after the 20th cataloged variable star in the constellation Taurus. Although a great many of them are found in the region of Taurus and the neighboring constellation Orion, they can be found in many other parts of the sky. T Tauri stars may represent a transitional stage between infrared stars surrounded by opaque nebulae and stable stars that have lost their "cocoons" and settled on the main sequence. The number of T Tauri stars alone in a star-forming cubic parsec may exceed the number of *all* stars per cubic parsec near the Sun by more than a factor of 10. T Tauri stars vary irregularly in brightness and are typically 20,000 to 1 million years old—younger than the human species!

After T Tauri stars form, they emerge slowly from their cocoon of dust and debris. For years, astronomers speculated that the pre-main-sequence stage of stellar evolution might be related to the disk-shaped solar nebula in which the planets formed. Starting in 1984, astronomers began discovering disk-shaped nebulae of cool dust near young stars. This infrared-emitting dust extends hundreds of astronomical

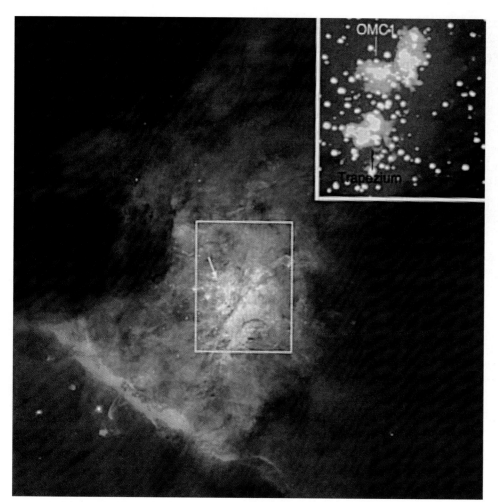

FIGURE 13-8
The Orion Nebula is a cloud of glowing gas and dust nearly 10 pc in diameter. This image shows the central 0.75 pc of the nebula. The gas is ionized by radiation from the Trapezium, a cluster of hot young stars near the center of the nebula. The inset shows the Trapezium and Orion Molecular Cloud 1 (OMC-1), where the view using infrared radiation penetrates to the heart of the star-forming region. (SOURCE: C. R. O'Dell, Rice University/NASA; inset courtesy of I. McLean, Joint Astronomy Center, Hawaii.)

units from the star. These discoveries seem to provide a "missing link" between an initial collapsing gas cloud and the later dusty disk that is required to form planets. The edge-on disk around Beta Pictoris can be seen in Figure 13–9, in a view where the central young star has been carefully blotted out. Beta Pictoris is an older star somewhat more massive than the Sun, so we have evidence that the debris left over from star formation can form disks in stars with a wide range of ages. In addition to their dusty disks, many T Tauri stars have bipolar outflows. During bipolar outflows, the gas shoots out in opposite directions, perpendicular to the disk, as seen in Figure 13–10. Stellar winds may also blow some material away from some T Tauri stars in all directions. In many T Tauri stars, however, some gas and dust apparently spiral inward toward the star, get caught in magnetic fields around the star, and then get accelerated; they then squirt "upward" and "downward" in two diffuse jets away from the disk.

Another nearby example of a young star cluster is NGC 2264. Most of the low-mass stars (spectral classes A to K) in this cluster lie distinctly to the right of the main sequence on an H-R diagram, as shown in Figure 13–11, and many of these are identified as T Tauri stars. The calculated lines of constant age (dashed lines in Figure 13–11) show that the T Tauri stars match the positions predicted for ages in the range of 3 to 30 million years. This demonstrates how the H-R diagram can be used, together with physical models of stars, to determine the ages of open clusters.

Recall our rule of thumb from Chapter 12 that massive stars evolve fastest. Brilliant, massive, bluish, O-type giants or supergiants last only a few million years. For this reason, they are found only in star-forming regions a few million years old. By the time stars have dispersed from such regions, the O giants or supergiants have already burnt out. Thus, they are rare in any typical volume of space, but they dominate star-forming regions. The Orion nebula and other star-forming regions like it will not last forever. The tight groups of stars will disperse. The brightest stars will exhaust their nuclear fuel and die. The surrounding gas will no longer be energized by young stars; it will cool and dim. Millions of years from now, the curtain will fall on this spectacular light show.

The Main-Sequence Stage

A star joins the main sequence when it begins to generate energy by consuming hydrogen in nuclear reactions deep in its central regions. Prior to that time, the star generates energy primarily by gravitational contraction that raises the temperature and pressure in the central regions. As the pressure and

FIGURE 13-9
The best example of planetary dust around a nearby star. Beta Pictoris is 16 pc away, and far infrared measurements from space had detected thermal radiation from dust in 1983. Optical astronomers blocked the glare from the central star using a disk and cross-hairs, revealing bright material that extended on either side of the star. This is an edge-on view of a dust disk that extends about as far as the Oort cloud in our own solar system. (SOURCE: S. Larson and S. Tapia, 1987 at Cerro Tololo Interamerican Observatory, Chile.)

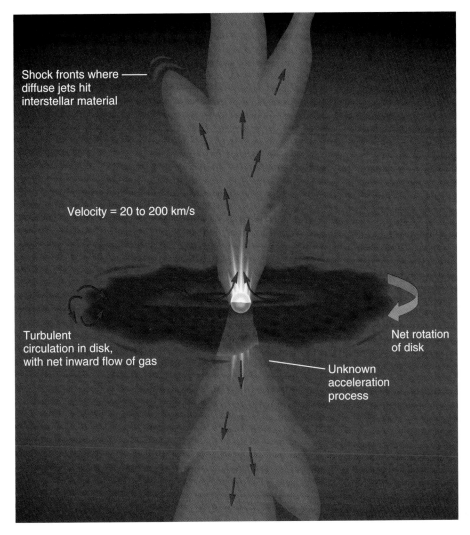

FIGURE 13-10
Recent evidence suggests that as a new star "turns on" in the center of its turbulent disk-shaped cocoon nebula, gas is blown outward along the disk's axes1 by an unknown mechanism. Outflowing clouds move "upward" and "downward" at speeds of up to 200 km/s, ramming into surrounding interstellar gas.

SCIENCE TOOLBOX

Lifetimes of Stars

In the last chapter, we saw that the enormous range of stellar luminosities has several important consequences. High-luminosity stars can be seen out to larger distances than low-luminosity stars. If you did a complete survey of all stars in a volume of space, you might find that luminous stars are rare. However, because luminous stars can be seen to larger distances and therefore over larger volumes, they get overrepresented in surveys to a fixed-brightness limit. Thus, the apparently brightest stars in the night sky are giants and supergiants, while the closest stars to the Sun are main-sequence stars or dwarfs.

We have seen that the main-sequence lifetime is given by the amount of fuel divided by the rate at which fuel is consumed. Let us do this calculation for the Sun. In Chapter 11, we learned that the Sun consumes 2×10^{19} kg of hydrogen per year. If a star has luminosity L in solar units, the rate of fuel consumption is $2 \times 10^{19} L$. What about the amount of available fuel? The mass of the Sun is 2×10^{30} kilograms, but not all of that mass is available as nuclear fuel. Fusion can occur only in the hot core that contains 10 percent of the Sun's mass, so the available fuel is $0.1 \times 2 \times 10^{30} = 2 \times 10^{29}$ kg. If a star has mass M in solar units, the available fuel is $2 \times 10^{29} M$. Now we can divide to get the lifetime:

$$t = \frac{2 \times 10^{29} M}{2 \times 10^{19} L} \approx 10^{10} \, (M/L) \text{ years}$$

For the Sun, since $M = 1$ and $L = 1$, we get 10 billion years. For a star like Algol, with a mass of 4 $M_\odot$ and a luminosity of 100 $L_\odot$, the main-sequence lifetime is $10^{10} (4/100) = 4 \times 10^8$ years. A star with a mass of 10 M and a luminosity of $10^5 L$ has a main-sequence lifetime of $10^{10} (10/10^5) = 10^6$ years. Massive stars have *much* shorter lifetimes than the Sun. We have made some approximations in this calculation, so the answers will only be accurate to about a factor of 2.

We also saw in the last chapter that the main sequence is in fact a sequence of mass, and that high-mass stars use their nuclear fuel at a much faster rate than low-mass stars. The relationship between mass and luminosity on the main sequence (in solar units) is $L = M^{3.5}$. Therefore, a main-sequence star with mass 20 times the Sun is $20^{3.5} = 36,000$ times more luminous than the Sun, and a star 0.5 times the mass of the Sun is $0.5^{3.5} = 0.088$ times the luminosity of the Sun. Since stellar evolution is driven by the mass of a star, it makes sense to substitute for luminosity in the equation above:

$$t = 10^{10} \, (M/L) = 10^{10} \, (M/M^{3.5})$$
$$\approx 10^{10} / M^{2.5} \text{ years}$$

Using the same two examples from above, the main-sequence lifetime of a 20 $M_\odot$ star is $10^{10} / 20^{2.5} = 6 \times 10^6$, or 6 million years, and the main-sequence lifetime of a 0.5 $M_\odot$ star is $10^{10} / 0.5^{2.5} = 5.7 \times 10^{10}$, or 57 billion years. Since the universe is less than 57 billion years old, no star with half the mass of the Sun has ever left the main sequence.

We can develop these ideas further. This chapter discusses the different stages of stellar evolution. The time spent in the different evolutionary stages varies greatly. For example, a star like the Sun took about 30 million years to reach the main sequence, and it will spend a total of about 9 billion years on the main sequence. Subsequently, the Sun will spend perhaps 1 billion years as a red giant and then a very long time cooling as a white dwarf. A star 10 times the mass of the Sun will spend 10 million years on the main sequence and only a few weeks in the extremely bright state that follows a supernova explosion.

Suppose stars were being born all around us continuously. You could look around and find stars in middle age and near death, too. But the fraction of stars you saw in each evolutionary phase would equal the fraction of its lifetime spent in that stage.

For example, if each star like the Sun spent 1 percent of its main-sequence lifetime in a planetary nebula stage, you would need to survey $1/0.01 = 100$ such stars to expect to find just one in the rare planetary nebula stage. By a similar reasoning, you would need to survey a volume with millions of high-mass stars to expect to find one that had gone supernova. Of course, if you could just wait long enough, you could watch a 10 $M_\odot$ or 20 $M_\odot$ star until it evolved off the main sequence and exploded. But astronomers must make do with a frozen moment in the history of the universe. Therefore, the only way to see transient phases of evolution is to observe large enough samples of stars that *statistically* you expect to detect the rare parts of the life cycle.

In any group of stars that are at the same distance, the visibility of stars in a particular evolutionary stage depends on the product of the luminosity of that stage and the time spent in that stage. The light from a large population of stars can have a large contribution from stars at fleeting evolutionary stages because the star puts out so much energy, even though for a short time. And even though we anticipate that the sky is full of long-lived white dwarfs, they are intrinsically so dim that they do not contribute much to the summed light of a large group of stars.

For stars that are *not* at the same distance, the visibility of short-lived giants and supergiants is further enhanced by the fact that they can be seen out to such large distances. (This is true for stars selected over a wide area of sky down to a particular limit in apparent brightness.) This idea was discussed in a Science Toolbox in Chapter 12.

fuels to react, and no new energy release to create a pressure that can oppose gravity. The star must therefore contract to a very dense state. We now discuss four main stages of stellar old age: the giants, variables, explosions of several types, and terminal entombment in a state of very high density.

Note that the most common stars—those about 1 $M_\odot$ or less—go through the giant phase and then contract rather smoothly to a small configuration of high density. But the rare, high-mass stars above about 8 $M_\odot$ go through explosive instabilities. Eventually they reach even higher density stages than Sun-like stars because they have more mass and stronger gravity. They produce some of the most exotic objects yet discovered by astronomers, such as pulsars and black holes.

Red Giants

As we have seen in this chapter and the last, stars of different mass evolve at different rates. But whether they have high mass and evolve quickly or low mass and evolve slowly, all main-sequence stars eventually run out of hydrogen. When this happens, their properties change in such a way that they leave the main sequence. The core collapses because it is no longer sustained by pressure from fusion reactions. At the same time, there is a thin shell surrounding the collapsing core in which hydrogen is still fusing. Outside the shell, temperatures have never risen high enough to "ignite" the hydrogen nuclei (i.e., fuse them into helium), so there is still plenty of hydrogen fuel. The collapsing core releases energy from gravitational contraction, creating pressure and driving the hydrogen-burning shell to layers further and further out. The increased temperatures cause a great expansion of the outer layers, making the star evolve rapidly toward the giant state. The outermost atmosphere becomes huge, thin, and cool, even though the inner core is smaller, hotter, and denser than ever. From the outside, the outer atmospheric layers are seen to glow with a dull red color, and the star is perceived as an enormous **red giant.**

The largest giants are truly immense—approaching 1000 times the size of the Sun. If the Sun were replaced by one of these stars, its thin, red-glowing outer atmosphere would reach nearly to the orbit of Jupiter! As charted on the H-R diagram in Figure 13–12, stars from all points along the main sequence display a funneling effect: As they evolve off the main sequence, their properties change so as to place them in the red giant region. They resemble patients from all walks of life crowding into the same hospital because they are victims of the same malady—hydrogen exhaustion.

Meanwhile, the cores of these evolved stars contract until they reach temperatures near 200 million K. This is hot enough to begin to fuse helium nuclei in the central regions of most stars, primarily by the **triple-alpha process** (named after the alpha particle, another name for the helium-4 nucleus):

$$^4He + {}^4He \rightarrow {}^8Be + photon$$

$$^8Be + {}^4He \rightarrow {}^{12}C + photon$$

In this process, three helium-4 nuclei combine to produce a carbon-12 nucleus. Because beryllium-8 is unstable, some

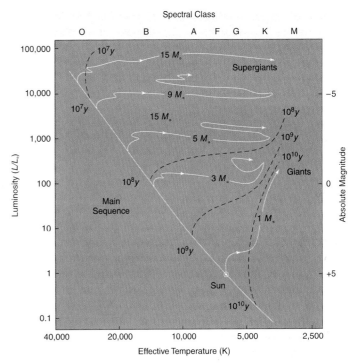

FIGURE 13–12
Evolutionary tracks off the main sequence toward the giant region, plotted on the H-R diagram for stars of different mass. Dashed lines represent the length of time since star formation; massive stars evolve fastest. All the stars evolve toward the upper-right corner of the H-R diagram, the region of giant stars. (SOURCE: After calculations by I. Iben.)

beryllium atoms may break up before completing the process, but this merely reduces the efficiency of the process. However, it does result in a deficit of beryllium and its neighboring elements in the periodic table, accounting for the steep valley in Figure 13–1. Notice that we are now fusing helium to make an even heavier element, carbon. Carbon is the basis of life on Earth, so this is a momentous step.

The triple-alpha process produces prodigious energy. In low-mass stars, this energy release rapidly heats the core, creating a burst of helium fusion called a *helium flash*. In some stars, the helium flash may consume the central core's helium in only a few seconds. Since the energy from the flash diffuses through the star slowly, the heating effects seen at the surface may last thousands of years. The flash takes place as the star puffs up to become a giant. In a Sun-like star, it occurs about 300 million years after evolution off the main sequence.

Note that the core's evolution begins to be independent of the evolution of the outer atmosphere. The characteristics of the star as perceived by an astronomer on Earth are those of the outer atmosphere. These characteristics determine the star's position on the H-R diagram. When you go outside at night and look at Betelgeuse or Antares, you are seeing the cool, red, outer atmosphere of a star. Hidden inside is a very hot, dense core. Thus, although we may say "the star" is cooling and getting redder, it is really its atmosphere that is cooling. All this time, the core is shrinking and growing hotter.

The Creation of Heavy Elements

While this chapter is primarily about the life stories of stars, its main subplot is the creation of elements. We have seen that main-sequence stars can create helium from hydrogen. More massive stars can also make carbon by the triple-alpha process. However, there are many elements heavier than carbon in Figure 13–1. Are these created in stars, or were they present at the birth of the universe? How can we explain the vastly varying cosmic abundance of different chemical elements?

The challenge to answer these questions was taken up in the 1950s by four young astrophysicists working at the California Institute of Technology. Three of them were from England: the husband-and-wife team of Geoffrey and Margaret Burbidge, and the iconoclastic Fred Hoyle. The fourth was William Fowler, a genial expert in nuclear physics. In a long paper published in 1957, they succeeded in explaining most of the features of Figure 13–1. They noted that the problem of element synthesis was closely related to the problem of stellar evolution. All the members of the team had stellar careers. The Burbidges both became directors of major observatories, Hoyle was knighted, and Fowler received the Nobel Prize. Their paper told the story of how stars had created the materials of the everyday world—the calcium in our bones, the nitrogen and oxygen in the air we breathe, the metals in the cars we drive, and the silicon in the computers we use.

Extreme temperatures are required in order to fuse heavy elements. Recall that the core of a giant star is contracting. As the core contracts and gets hotter, it initiates reactions involving even heavier elements. The elements synthesized depend on the mass of the star. Imagine a set of red giants with different masses: 4, 6, 8, 10 $M_\odot$, and so on. Those of about 4 or 6 $M_\odot$ will have helium-rich cores hot enough to ignite the helium nuclei and fuse them into carbon in the triple-alpha process just described. Those of around 8 $M_\odot$ will have hot enough cores to ignite the carbon and fuse it into heavier elements such as oxygen, neon, and magnesium. The reactions where helium nuclei are fused to heavier nuclei include:

$$^{12}C + {}^4He \rightarrow {}^{16}O + photon$$

$$^{16}O + {}^4He \rightarrow {}^{20}Ne + photon$$

These reactions require temperatures above 500 million K. Heavy nuclei can also fuse with each other, although in this case the electrical repulsion between protons in the nuclei is stronger, and a temperature above 1 billion K is needed for the reactions to proceed:

$$^{12}C + {}^{12}C \rightarrow {}^{24}Mg + photon$$

$$^{16}O + {}^{16}O \rightarrow {}^{32}S + photon$$

$$^{28}Si + {}^{28}Si \rightarrow {}^{56}Ni + photon$$

The last of these reactions produces nickel-56, which is radioactive. It decays rapidly, first into cobalt-56, then into a normal iron-56 nucleus. Note that the continued addition of helium nuclei leads to the preferential formation of heavy elements with nuclear masses that are multiples of four. A helium nucleus has two protons and two neutrons, so each time an alpha particle is absorbed in a fusion reaction, the atomic number increases by two. We can therefore explain the jagged sawtooth pattern of the cosmic abundance curve in Figure 13–1. In stars of around 10 to 12 $M_\odot$, a long series of reactions will fuse nuclei into elements as heavy as iron.

Therefore, if the star is massive enough, the core-building process leads to a core consisting of shells of different elements, surrounding an inner core of iron, as illustrated in Figure 13–13. We are used to thinking of iron as a solid metal, but the state of iron at the center of a massive evolved star is quite different. Stellar iron is actually denser than terrestrial iron, but the electrons have been stripped from all the atoms, and the nuclei are not held together in a lattice. Fantastic pressure keeps the iron as a high-temperature gas. The shell structure can be so complex that different fusion reactions can occur at the surfaces of different shells all at the same time.

With iron, the fusion process reaches an insurmountable obstacle. Iron has the most stable nuclear configuration of any element. This means that energy is *consumed*, not produced, as iron nuclei fuse into heavier elements. We can therefore account for the steep fall in the abundance of elements heavier than iron in Figure 13–1. Thus, the iron nuclei in stars do not continue to ignite and fuse as the core contracts and gets hotter. The heart of a star is like an iron tomb that traps matter and releases no energy to counter the continuing collapse. However, this is not yet the end of element creation. We will pick up the story when we consider the explosive deaths of the most massive stars.

Stars are like factories whose main business is the creation of heavy elements from lighter ones. Remember that main-sequence stars release less than 1 percent of the mass of a hydrogen atom as pure energy when they make helium. The same is true of the nuclear reactions at later stages of a star's life. Therefore, the heat and light we see from a star is just a minor by-product of the fusion process, just as the heat and light we see from a factory does not reflect its primary purpose of making large objects from smaller components. Stars are the "factories" of the matter we see around us.

Variable Stars

In 1595, amateur astronomer and Lutheran pastor David Fabricus noticed that a bright star in the constellation Cetus (the Sea Monster) was fading. Within 2 months, it had vanished from the sky. Fabricus was amazed as the star recovered its former brilliance several months later. The star (called Mira, meaning "wonderful") has a cycle with a period of 11 months. Nearly two centuries later, 17-year-old English astronomer John Goodricke discovered brightness variations in the star called Algol. While still a teenager, Goodricke found that Delta Cephei varies with a regular period of 5 days and 8 hours. The Royal Society of London awarded him a medal for his work. Goodricke overcame considerable odds in his short but brilliant career. He was born deaf and unable to speak, in an age when most deaf-mutes were hidden away

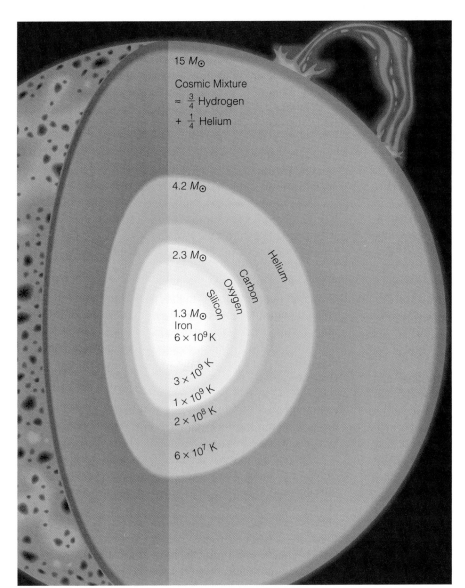

FIGURE 13–13
The onion-skin layering of different elements as calculated for the precollapse core of a 15 $M_\odot$ star. The dominant element in each layer is given at the top; each layer's mean temperature is given at the bottom. Figures also represent the mass of material out to a designated layer. Nuclear reactions in successively deeper layers fuse nuclei into successively heavier elements. The core and inner layers are greatly enlarged to show their structure.
(SOURCE: After diagram by A. Burrows.)

in mental asylums. Tragically, he caught pneumonia while observing Delta Cephei and died at the age of 21.

Both Fabricus and Goodricke had observed a **variable star**, a star that varies in brightness on a time scale of hours to years. Most variable stars apparently represent post-giant stages of evolution (although as we saw earlier in this chapter, some variables such as T Tauri stars are pre-main-sequence stars). Variable stars may pulse with a constant period, or they may flare up sporadically, often brightening by hundreds of percent and then fading again. Over 25,000 variable stars have been cataloged and divided into as many as 28 types.

Why do some stars vary? We have seen that main-sequence stars like the Sun are stable and do not vary. Main-sequence stars are in a state of hydrostatic equilibrium; pressure and gravity are in balance at every point. Think of stable stars as having a *thermostat*. If you compressed a main-sequence star, the density and pressure would rise, resulting in an increased nuclear reaction rate that would counter the compression. As long as energy flows smoothly and easily out of the star, this equilibrium is maintained. As an analogy, think of a boiling pan of water. Energy comes into the water from the heating element (core), travels through the water by a mixture of convection and radiation, and then radiates smoothly into the air (space).

Some giant stars have atmospheres that trap a portion of the radiated energy. When energy "dams up" in this way, the outer layers heat up and expand. The expansion cools the gas, which drops the pressure and allows gravity to pull the outer layers in again. The newly compressed gas once again dams up energy, which starts the cycle over again. Pulsation is *not* caused by any variation in the rate of energy generation in the core of a star, but by a variation in the rate that energy can escape from the star. Now think about a boiling pan of water with a lid on it. The lid traps steam and raises the pressure under it. At some point the lid will tip and release the steam. The pressure then builds up and the cycle continues. Meanwhile, the energy flowing into the water from the stove has not varied.

What determines the pulsation time of a star? Just like a bell (or any other object), a star has a certain frequency or time period in which it tends to vibrate in response to a disturbance. If the time required to "dam up" the radiation is close to the star's natural oscillation period, the star will begin to oscillate repeatedly. As the radiation is blocked, pressure builds up and the outer layers of the star expand. But once the expansion gets going, momentum and the star's natural tendency to oscillate at this speed carry the expansion too far. The atmosphere expands and becomes thinner, allowing radiation to escape easily. Then the star subsides, and the radiation once again begins to be dammed, restarting the cycle. In other types of stars, where the time scales of radiation damming and oscillation are not synchronized, the variation occurs irregularly.

There is a whole "zoo" of variable stars in astronomy. The long-period variables, of which Mira was the first known example, last several hundred days. John Goodricke discovered the Cepheid variable class. Cepheids have regular variations in brightness with periods from 1 to 50 days. (Polaris is, in fact, a Cepheid variable. Shakespeare was taking artistic license when he declared a love to be "as constant as the Pole star.") RR Lyrae stars vary with periods in the range of 1 to 24 hours. Some very small stars even have variations with periods as short as 1 or 2 minutes.

Cepheid variables are important for two reasons. First, because their variations are regular, they are somewhat better understood than stars whose brightness changes are unpredictable—called *irregular variables*. Second, and more important, the period of variation of each Cepheid directly correlates with its average luminosity. This relationship was discovered in 1912 by the famous astronomer Henrietta S. Leavitt, who measured hundreds of photographs of Cepheid variables in the first years of this century at Harvard College Observatory. (Leavitt was a colleague of the women whose work was described in the story at the beginning of Chapter 12.) Both Cepheid and RR Lyrae variables were eventually found to have distinct period-luminosity relationships. These relationships enable astronomers to determine the luminosity of any Cepheid at any distance simply by measuring its period. This, in turn, leads to a new way to measure the distances of stars (direct methods were discussed in Chapter 12): Find a Cepheid, measure its period and hence its luminosity, then measure its apparent brightness and derive its distance by the inverse square law. We will return to this subject in the next chapter.

Mass Loss in Evolved Stars

The more massive a star, the more rapid and violent its evolution. Stars of more than a few solar masses are very hot and evolve rapidly. When their atmospheres expand, these stars leave the main sequence with more luminosity than that of most giants. They are thus called *supergiants;* at a given temperature, they must have a larger surface area to emit a larger amount of energy. Supergiants are the largest and most luminous stars on the H-R diagram. As described at the beginning of the chapter, the collapsing core generates energy that drives the outer atmosphere of the star into space, resulting in mass loss. Because massive stars evolve quickly, mass is lost only a few million or tens of millions of years after the star's formation.

The most massive stars return a lot of material to the space between stars. A hot supergiant emits many ultraviolet photons that can drive gas away from the star. For example, a blue supergiant might lose a solar mass of gas about every 100,000 years. The most massive stars can therefore lose one-third to half of their total material into interstellar space. Red supergiants have cooler atmospheres in which the gas is not ionized and molecules can form. Tiny solid particles called dust grains can form, taking the visible light from the star and reradiating it as infrared emission. Massive stars can seed space with dust as well as gas.

More modest stars can also lose mass late in their lives. The gas being blown out of mass-shedding stars expands into space; it is often called a *stellar wind*. The gas may cool enough for grains of dust, such as carbon grains, to condense in it. It may collide with other nearby gas clouds at high speed, creating glowing shock waves. Often, however, it is shed in nearly spherical bubbles that surround the central star. Ultraviolet light from the star excites and ionizes the gas atoms, causing them to glow. Clouds of gas in space are called nebulae, and decades ago these particular nebulae came to be called **planetary nebulae** because the palely glowing bubbles looked like disks of planets in small telescopes. This term is a misnomer, however, because they have nothing to do with planets. As shown in Figure 13–14, they are among the most beautiful celestial features, with wispy symmetry and delicate colors.

The glowing gas that surrounds evolved stars is not just a pretty light show. When stars ejects gas into space, they are participating in an ongoing cycle of life and death. Massive stars create elements as heavy as iron, and convection churns those elements through the star. The ejected gas is therefore rich in heavy elements; the glowing gas of a planetary nebula shows emission lines from nitrogen, oxygen, carbon, neon, sulfur, chlorine, and iron. Chemically enriched interstellar material then becomes the raw material for a new generation of stars. This is what the Burbidges, Hoyle, and Fowler meant when they demonstrated that the history of matter is hidden in the abundance distribution of the elements. Notice that this includes the carbon, nitrogen, and oxygen atoms that are essential for life. Each of the heavy atoms in your body was once a part of another star. Without the creation and cosmic recycling of elements in stars, we would not exist!

The Death of Stars

All good things must pass. Every star in the night sky will one day exhaust its nuclear fuel and die. In the battle between gravity and the pressure caused by nuclear reactions, gravity is the victor. As a result, every star dies as a compact object. Low-mass stars die a quiet death, fading away slowly to become dark embers. High-mass stars die with a pyrotechnic flourish, lighting up the sky one final time. We will consider these different fates, but will first pause to consider the broader implications of stellar evolution.

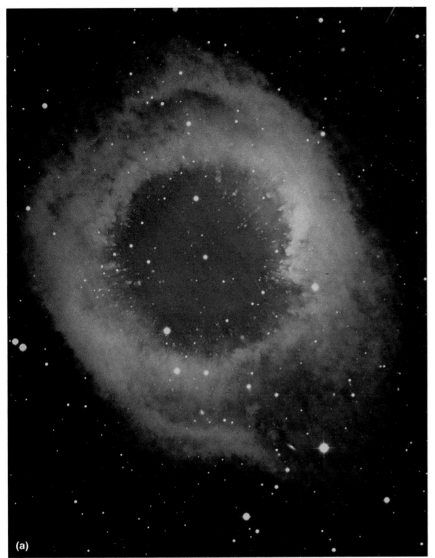

FIGURE 13-14
(a) The Helix Nebula (NGC 7293) in Aquarius is a glowing spheroidal shell of gas blown off the surface layers of an evolved star by strong stellar winds. The remainder of the star is visible at the center—a very hot, blue stellar core. Strong ultraviolet radiation from this star core excites atoms of different elements at different distances in the gas around the star; the dominant color is the red glow of hydrogen alpha emission from ionized hydrogen atoms. (SOURCE: Copyright Anglo-Australian Telescope Board, courtesy D. Malin.) (b) The Cat's Eye Planetary Nebula shows a complex structure when observed with the Hubble Space Telescope. The geometry is clearly more complex than a simple expanding shell. The blue-green color comes from doubly ionized oxygen atoms. (SOURCE: J. P. Harrington and K. J. Borkowski, University of Maryland/NASA.)

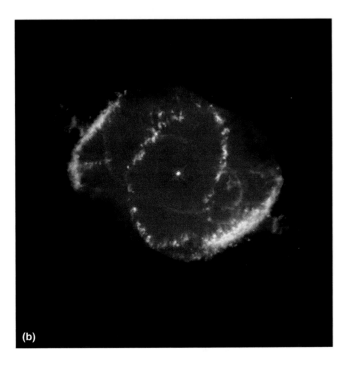

We have seen that stars are factories for creating heavy elements. Along the way, they release a tiny fraction of the frozen mass-energy of matter in the form of radiation. This radiation travels through space, never to return. In the case of our star the Sun, the radiation provides the heat and light that supports life on Earth. Let us consider the role of stars in the transformation and transmission of energy.

The Laws of Thermodynamics

Energy behaves in regular and predictable ways in the universe. In Chapter 4, we introduced the subject of thermodynamics—the study of the way heat flows. Now we will generalize that discussion to include heat and all other forms of energy. The rules scientists have devised are called the **laws of thermodynamics.**

We already encountered the first law of thermodynamics in Chapter 4. It states that the total amount of energy in a closed system is always conserved. You can recognize this as a statement of the law of conservation of energy. For example, an elliptical orbit has a constant interplay between kinetic energy and gravitational potential energy. When it is close to its star, a planet moves faster in its orbit—the kinetic energy is high, and the gravitational energy is low. When it is far from its star, a planet moves slower in its orbit—the kinetic energy is low, and the gravitational energy is high. The total energy of the planet remains constant. When energy *appears* not to be conserved, the answer is usually that some energy has leaked away in the form of heat. When a swing slows down or a rolling ball comes to a stop, kinetic energy has steadily been converted to heat.

When we look at the Sun blazing in the sky and consider that it will keep doing so for billions of years, it seems as if we get this energy for free. In what way do stars conserve energy? Stars create heavier elements in a fusion chain that moves from hydrogen to helium and, if the star is massive enough, to carbon and then to iron. Atomic nuclei are bound together by an attractive force that acts like glue. It takes energy to undo this glue. Since Einstein showed that the energy that binds an atomic nucleus must have an equivalent energy ($E = mc^2$), it follows that things stuck together have less mass than the same things pulled apart. Therefore, whenever a nucleus becomes more tightly bound, energy that has been frozen in the form of mass gets released as radiant energy. Going from hydrogen to iron, atomic nuclei are held together more and more tightly. Thus, each step up the fusion chain releases mass-energy. You can think of mass as a form of potential energy that stars convert to light.

The second law of thermodynamics is also familiar from Chapter 4. We described the principle of thermal equilibrium, where heat always flows from a hot object to a cold object. Equilibrium is established when both objects have the same temperature. The second law makes this strong proposition: Heat *cannot* flow spontaneously from a cold to a hot object. A cup of coffee cools down as it loses heat to the cooler atmosphere. An ice cube melts as heat flows into it from the warmer atmosphere. Notice that the first law—conservation of energy—does not require this behavior. Energy would be conserved if coffee actually got hotter at the expense of cooling the air, or if an ice cube got colder at the expense of warming the air. But we never see this happen. In nature, heat always flows in one direction, from hot to cold.

Once again, when the second law *appears* to be violated, it is because we are not considering the whole system. Heat always flows so as to warm things up, but that does not mean we cannot make ice cubes. A refrigerator can extract enough heat from water to make ice cubes, but at the expense of releasing a larger amount of heat into the air. If you are skeptical, feel along the top and bottom of your refrigerator. We can easily see that stars follow the second law. Energy is created in the core of a star by nuclear fusion, and then it flows out from hotter to cooler regions.

Another way to express the second law of thermodynamics uses our microscopic description of heat energy. We saw in Chapter 4 that heat is a measure of the kinetic energy (or amount of motion) of atoms and molecules. Fast-moving particles tend to share their motion with slower-moving particles, with the result that they all approach the same motion or temperature. This is the idea of thermal equilibrium. We should notice two important features of this description.

First, it is *possible* for a slow-moving particle to give up speed to a fast-moving particle, but this is *unlikely* to happen. When we consider a large number of particles, the energy will always flow from the quick to the slow. Thus, the second law of thermodynamics describes nature with probability rather than with certainty. A few water molecules might happen to lose enough energy to stick together as an ice crystal, but the chances of an ice cube spontaneously forming in your glass of water are incredibly small!

The second consequence of the microscopic description of heat flow is even more interesting. Heat is a measure of the random motions of particles, so it is really a measure of *disorder*. The second law of thermodynamics also states that the disorder of a physical system will always increase when it undergoes change. Scientists use the term **entropy** as a measure of disorder. Entropy is a profound scientific concept; you can learn more about it in the following Science Toolbox.

An increase in entropy is a fact of life as well as a law of nature. The papers on your desk tend to get disordered, odd socks turn up in your sock drawer, and the kitchen always seems to get messy. Why is it so much easier to break an egg than to put it back together? Why can you easily stir a sugar cube into your coffee, while no amount of stirring will make the sugar cube come back together? Let us use our familiar example of a deck of cards. Start with a deck in which all the cards are ranked by number and suit (Figure 13–15a). The deck is perfectly ordered and has low entropy. Anything you do to the deck will make it less ordered. After a few shuffles, there will be little groups of cards in sequence, but overall the deck will have more disorder and more entropy (Figure 13–15b). Shuffle many times, and the deck will begin to look truly random; the disorder and entropy are now very high (Figure 13–15c). Experience tells you that no matter how often you shuffle the deck, you will never return the cards to their initial, ordered state. Entropy always increases.

We can, therefore, make a clear connection between disorder or entropy and the probability of a system being in a certain state. Situations of disorder are more probable than situations of order. The Science Toolbox explores this idea further. When an ice cube (or a sugar cube) melts, the order in the crystal lattice turns into the disorder of randomly moving water molecules. Heat is a random or high-entropy form of energy. Change is all around us, and energy is always changing from one form to another. The first law tells us that energy is conserved. The second law tells us that energy tends to transform into disordered energy or heat.

Ordered energy changes systematically into disordered energy, as we can see in many everyday examples. The motion of a swing or a pendulum decays, and the lost kinetic energy turns into heat energy in the slight heating of air molecules. Our world runs on fossil fuel, which taps the energy stored in the structure of the chemical bond. No matter how efficiently we use that energy, much of it is turned into heat. The electricity in your house heats the wires it travels through, and later that heat leaks into the atmosphere. A magnet loses its strength with time and use. In this case, the aligned iron grains gradually become less aligned from heat and from the magnet striking other objects. (This also creates

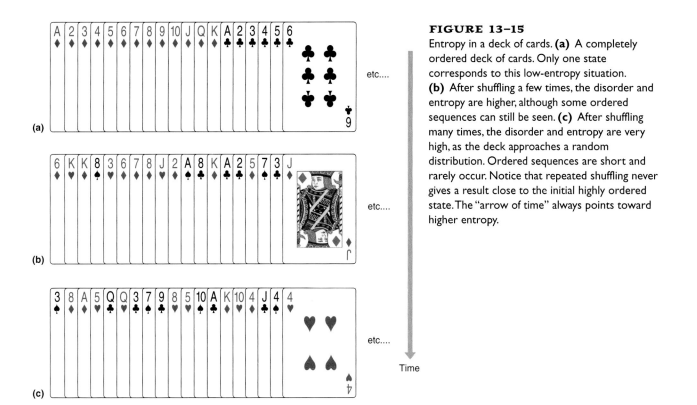

FIGURE 13-15
Entropy in a deck of cards. **(a)** A completely ordered deck of cards. Only one state corresponds to this low-entropy situation. **(b)** After shuffling a few times, the disorder and entropy are higher, although some ordered sequences can still be seen. **(c)** After shuffling many times, the disorder and entropy are very high, as the deck approaches a random distribution. Ordered sequences are short and rarely occur. Notice that repeated shuffling never gives a result close to the initial highly ordered state. The "arrow of time" always points toward higher entropy.

heat.) Magnetic energy is converted to heat energy. What does this have to do with stars? A star converts mass to energy. This is a very direct translation of order—a localized particle—into disorder—radiation, or heat energy that extends through space.

As before, when the second law *appears* to be violated, we are not seeing the whole picture. A refrigerator can make ice cubes, creating the order of a crystal lattice, but only at the expense of releasing a far larger amount of disordered (heat) energy. A star can build heavy elements, creating the structure of a massive nucleus, but only at the expense of disordered (radiation) energy sent into space. How does life comply with the second law of thermodynamics? Surely the creation of the structure of a brain, or a cell, or DNA itself, violates the trend toward disorder? It is true that living organisms have lower entropy than their surroundings. However, in the microscopic view, many scientists think that aging is caused by the gradual accumulation of damage to the DNA molecule. This increasing disorder in the genetic code compromises the operation of cells and limits the ability of the organism to repair itself. And in the larger view, Earth intercepts only a tiny fraction of the Sun's heat energy. Life eventually dies and returns that energy to the atmosphere, and then into deep space as a waste product. The second law is obeyed.

Physicists have identified another principle that is often called the third law of thermodynamics. It states that no physical system can be cooled to a temperature of absolute zero. Think of it: At absolute zero, there would be no atomic motion so there would be no heat and no friction. Machines could be perfectly efficient. Perpetual motion would be possible. But it is impossible to remove all the heat from a system. Take the coldest gas you can imagine. If you expand its container, the gas will cool. But you would have to expand it by an infinite amount to completely remove its heat content. Absolute zero cannot be reached.

The three laws of thermodynamics can be summarized as follows:

1. Energy can change forms, but the total amount of energy in a system, including heat energy, is always conserved.
2. The entropy, or disorder, of a system always increases, so as energy changes form, the amount of heat energy tends to increase.
3. It is impossible to remove all the heat from a physical system.

Here is a handy aid to memorizing these important laws. Think of the exchange of heat and other forms of energy as a game. The first law says you can't win. The second law says you can't break even. Worst of all, the third law says you can't get out of the game! We will revisit all the laws of thermodynamics later in the book when we consider cosmology.

White Dwarfs

Let us now consider the further evolution of Sun-like stars, by which we mean stars with initial mass in the range from 0.1 times the mass of the Sun to a few solar masses. The outer atmosphere of such a star is large and diffuse and may be shedding its outer layers, but most of the mass is still in a dense and hot core. As always, the fate of the star is determined by the battle between gravity and the pressure generated by nuclear reactions. Low-mass stars cannot generate nuclei heavier than carbon. Eventually, the core runs out of fuel, and without pressure support, the core collapses. Since the core cannot get hot enough to ignite any new nuclear

SCIENCE TOOLBOX

Understanding Entropy and Time

Imagine your favorite movie played backward. Could the events possibly ever occur that way? The people would seem to walk backward, which is silly but not impossible. But what about the smashed wine glass that leaps off the ground and assembles itself whole into a woman's hand? Or the man who flies out of the swimming pool and lands dry on the diving board? We would never see these things happen in real life. Yet none of the laws of physics that we have seen in this book mention *time* at all. Newton's laws are reversible and so are the conservation laws of energy, momentum, and angular momentum.

The second law of thermodynamics cannot explain why we sense time as an arrow that moves in only one direction. However, we can use it to explain why nature has a directionality that tends to increase entropy. Let us begin by defining entropy in terms of a microscopic description of order and disorder.

It is easy to see that the atoms in a crystal have more order than the atoms in a gas, but how do we actually quantify entropy? Consider a simple analogy using coins. An atom has many properties—energy, momentum, position, electric charge, spin—but a coin has only one: It can either be heads or tails. Suppose we are tossing four coins at the same time. Each coin can land two ways and the coins are all independent, so the number of possible results, or "states," is $2 \times 2 \times 2 \times 2 = 2^4 = 16$. (Review the way independent probabilities are combined in the Science Toolbox in Chapter 8.) The most orderly result would be four heads or four tails, which occurs 2 out of 16 times, or 13 percent of the time. The most disorderly or mixed up result is two heads and two tails, which occurs six different ways (check it!) or 38 percent of the time. Disorder is 3 times more likely than order.

The effect becomes more dramatic as the number of particles and possible results increase. With 100 coins, there are $2^{100} \approx 10^{30}$ possible results or states. So all heads can only occur by chance 1 in 10^{30} times. On the other hand, the most disordered result is 50 heads and 50 tails, and this occurs 1 in 12 times. What do these numbers mean? If you tossed 100 coins at a time every second, you would need less than a minute to see the most disordered state but 10^{22} years (far longer than the age of the universe) to see the most ordered state! We have related entropy to probability. In the real world, disorder is vastly more likely than order.

An ordered system will always tend to become more disordered, because disorder is a far more likely outcome. Let us go back to that well-shuffled deck of cards. What is the probability that you could shuffle it back into its perfect ordered sequence? About one chance in 10^{68}! It will never happen. For the same reason, you can never unscramble an egg or unstir the milk from the coffee. The great German physicist Ludwig Boltzmann derived the mathematical definition of entropy:

$$S = k \ln W$$

In this equation, k is the universal Boltzmann constant from Chapter 4, and $\ln W$ is the natural logarithm (to the base of $e = 2.718$ rather than to the base of 10) of the number of microscopic states of a system that gives a certain result. Entropy is higher for a result that has many possible microscopic states (like equal heads and tails) than it is for a result that has few microscopic states (like all heads or all tails). Boltzmann was so proud of this equation and the universal truth it represents that he had it inscribed on his tombstone.

Not everyone was as pleased as Boltzmann with the second law of thermodynamics. The idea that the universe was "running down" was badly received by English poets and philosophers in the mid-19th century. Algernon Swinburne wrote:

> We thank with weak thanksgiving
> Whatever gods there be
> That no man lives forever
> That dead men rise up never
> That even the weariest river
> Flows somewhere safe to sea.

The concept of entropy continues to fascinate writers and thinkers. The novelist John Updike wrote an ode to entropy that contains these words:

> Entropy! Thou seal on extinction,
> Thou curse on Creation
> All change distributes energy,
> Spills what cannot be gathered again.
> A ramp has been built into probability
> The universe cannot re-ascend.

---

fuel, this final collapse creates a very dense star the size of Earth or other planets. Such a star is known as a white dwarf.

In 1844, German astronomer Friedrich Bessel studied the motions of the brightest star in the sky, Sirius, and found that it was being perturbed back and forth by a faint, unseen star orbiting it. This star was not glimpsed until 1862, when American telescope maker Alvan Clark detected it, almost lost in the glare of Sirius. In 1915, Mt. Wilson observer W. S. Adams discovered that the companion was hot and bluish-white, with properties that place it below the main sequence on the H-R diagram. It has about the mass of the Sun (determined from the binary orbit), but it has such a low luminosity that the total radiating surface cannot be much more than that of Earth (determined by the Stefan-Boltzmann law). Astronomers have since discovered several thousand white dwarfs.

The theory that explains white dwarfs and other collapsed stars was developed by Subrahmanyan Chandrasekhar. Born in India, Chandrasekhar first began thinking about white dwarfs on the long boat voyage that took him to college in England. Although he had experienced prejudice many times in his life, more hurtful to him was the scorn that many senior astronomers poured on his ideas. But he persevered, and his ideas were shown to be correct. Chandrasekhar was not only a brilliant researcher but also a dedicated teacher and adviser, acting as the intellectual father to a whole generation of theoretical astrophysicists. Chandrasekhar received the Nobel Prize in physics in 1983. What was the strange state of matter that his calculations revealed?

Once no more energy is available to generate outward pressure, a star collapses until all its atoms are jammed together to make a very dense material. But what does "jammed together" really mean? At these temperatures and densities, electrons move at almost the speed of light, and matter loses its familiar properties. Stellar matter stops behaving like a perfect gas. In fact, it is no longer either gas, liquid, or solid, but a new form of matter composed of atomic nuclei held apart by a sea of electrons at densities of about 10^8 to 10^{11} kg/m^3. Such a density is quite extraordinary; a teaspoonful of white dwarf matter brought to Earth would weigh as much as an elephant.

At the microscopic level, positively charged atomic nuclei align themselves in orderly patterns, governed by the electrical forces that act between them. The electrons move freely through this crystalline lattice in the same way that electrons move through the copper lattice of a wire when an electric current flows. The quantum theory of matter says that no two particles can have exactly the same set of properties. This limitation means that electrons in close proximity experience a pressure that prevents the white dwarf from shrinking even further. The core of the Sun is destined to reach this strange state of matter in a little more than 5 billion years.

Most white dwarfs have temperatures of 10,000 to 20,000 K, and the hottest exceed 100,000 K. These hot stars have low luminosity because there is no new source of nuclear reactions. With such a large amount of stored heat and such a small surface area, white dwarfs take a long time to radiate enough energy to cool significantly. The oldest white dwarfs have cooled to temperatures under 4000 K and luminosities less than 0.0001 $L_\odot$, and they may have cooled for at least 10 billion years to reach these temperatures. Indeed, some white dwarfs may be among the oldest stars we can observe. Trillions of years from now, these cold stellar corpses will be dark objects with the mass of a star and the size of a planet.

White dwarfs cannot have more mass than 1.4 $M_\odot$ because the white dwarf structure becomes unstable at this point. If you tried to dump more mass on the surface of a 1.4 $M_\odot$ white dwarf, its gravity would become so strong that it would overcome the resistance of the electrons to denser packing. A still denser state of matter would arise. This critical mass is called the **Chandrasekhar limit,** and it applies only to the *core* mass of the star and not to the *initial* mass. Current data suggest that stars with initial masses of about 0.08 to about 8 $M_\odot$ evolve into white dwarfs. The more massive examples do so by developing strong stellar winds (like the solar wind) or eruptive explosions that blow off mass until they are below the Chandrasekhar limit. The Sun will probably blow off about 40 percent of its mass when it goes through its red giant stage about 5 billion years from now, collapsing into a white dwarf of about 0.6 $M_\odot$.

Recall from our discussion of stellar statistics that most main-sequence stars are less massive than the Sun. Therefore, *most* stars will end their lives as white dwarfs. In the classic Pink Floyd song "Shine On, You Crazy Diamond," a dying star is a metaphor for the brilliant but troubled life of one of the founding members of the group. With its crystalline carbon core, a white dwarf does resemble a massive diamond in some ways. However, each of these celestial gems is destined to lose its luster and fade from view.

Supernovae

At the end of its life, a massive red giant has consumed as much fuel as possible, and the fusion has worked its way into the outer parts of the star. At no point in the core is there both potential fuel and the possibility of creating enough heat to ignite it. The star has an "onion skin" structure, with more heavy elements concentrated toward the center. Since iron is the most stable element, it represents an insurmountable obstacle to further fusion. In other words, the iron "slag-heap" core will not ignite, even though it gets extremely hot and dense. In a star with an initial mass of 6 $M_\odot$, the core may include the central 1.1 $M_\odot$. In a star with an initial mass of 8 $M_\odot$, the core may encompass 1.4 $M_\odot$. In a larger star, the eventual core mass can exceed the Chandrasekhar limit. What is the fate of such a massive core?

The advanced evolutionary stages of a massive star represent a crescendo of activity. After millions of years of fusing hydrogen and helium, each of the subsequent fusion stages (creating carbon, neon, oxygen, and silicon) takes less than 1000 years. The conversion of silicon and sulfur to iron takes only a few days. As pointed out earlier, iron has the most stable nuclear configuration, so no more energy can be released by rearranging the nucleus. The core collapses at about one-quarter the speed of light, and the gravitational energy of the collapse is released as a prodigious burst of energy. The outburst of energy blows off all the outer layers of the star in a titanic explosion called a **supernova** (plural: *supernovae*).

A supernova represents a rare astronomical event of unspeakable violence. Astrophysicists have gained insight from supercomputer models, but the details are uncertain. Calculations suggest that the core collapse only takes a few seconds! The density rises by a factor of 1 million as a volume the size of Earth shrinks to a radius of about 50 km. As the core gets compressed to the density of an atomic nucleus, forces between particles cause it to rebound, and on its way out, it meets material that is still falling in. When matter meets matter traveling at supersonic speeds, the result is a *shock wave*, yielding compression and a rapid temperature rise up to billions of degrees. The energy of the explosion is easily sufficient to eject the outer envelope of the star, and to momentarily take the star to 10 billion times the luminosity of the Sun!

Up until now, we have presented iron as a dead-end in the fusion chain. Although helium capture in massive stars does not make elements heavier than iron, another mechanism can. Many of the nuclear reactions described so far release floods of neutrons that strike nearby nuclei. Since neutrons have no electrical charge, heavy nuclei can be built up by *neutron capture*. In massive, evolved stars, neutrons are slowly or gradually added to nuclei to build still heavier elements. These are the *s-process reactions* (s for slow), and they can make elements as massive as bismuth. However, a more spectacular form of neutron capture occurs in the white heat of a supernova explosion. These are the *r-process reactions* (r for rapid), and they make the heaviest elements such as radium, uranium, and plutonium.

Supernovae are wonderful creators and recyclers of heavy elements. The death of a massive star takes only seconds, but it has long-term consequences. The shock wave of the explosion allows the fusion of nuclei more massive than iron by **explosive nucleosynthesis.** The high flux of free neutrons leads to a rapid form of neutron capture, and the extra energy provided by the explosion causes the iron barrier to be overcome. Moreover, the explosion flings all the heavy elements formed in the star's brief lifetime deep into interstellar space. Supernovae are the source of most of the rare elements and precious metals in the world—the silver in the coins in our pockets and the gold in the jewelry on our bodies. We can now explain the remaining features of Figure 13–1. The rarity of supernovae explains the rarity of the heaviest elements. Atoms of precious metals like gold and platinum are 1 million times less abundant than iron and 10 million times less abundant than carbon, nitrogen, and oxygen.

The next time you look at a gold ring or earring, consider the exotic origin of those atoms. Many of these atoms were forged from lighter elements in the cauldron of a distant supernova and then surfed a blast wave into interstellar space. The atoms floated in deep, frigid space for millions of years before they became part of a collapsing gas cloud. Far from the center of the cloud, the atoms became part of a rocky mass, which churned for billions of years in the sluggish magma. Eventually, they were lifted in a concentrated nugget to the surface, where they came within the grasp of inquisitive human fingers. Also think of the way that the lighter elements on which life depends, such as carbon, nitrogen, and oxygen, get shot into space to be incorporated into a new generation of stars.

Another outcome of a supernova is a neutrino burst. Neutrinos are the weakly interacting particles that are produced in nuclear reactions of many kinds, including those in the Sun's core (see Chapter 11). During the core collapse, protons and electrons are forced to merge, creating a sea of neutrons and a huge number of neutrinos. The neutrinos flood into space at the speed of light, carrying away a huge amount of energy, about 10^{47} Watts, equivalent to the mass-energy of 50 Earths. For a brief few seconds, a supernova exceeds the luminous intensity of the entire rest of the universe! The final result comes months after the star itself fades. During the explosion, an expanding cloud of gas is launched at about 10,000 km/s, or 22 million mi/h! Initially it is too close to the star to be seen from Earth, but years later the site will be marked by a colossal, expanding nebula called a *supernova remnant*.

Many supernovae have been close enough to the solar system to produce temporarily prominent new stars that were recorded by ancient people. The ancient Chinese called them "guest stars." Some astronomers have suggested such an explanation for the Star of Bethlehem. The most famous supernova was the explosion that produced the Crab Nebula (Figure 13–16). It was visible in broad daylight for 23 days in July 1054 and at night for the subsequent 6 months. It was recorded in

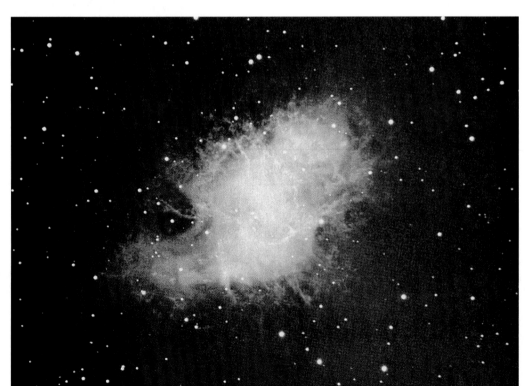

FIGURE 13–16
The Crab Nebula is the remnant of a supernova explosion recorded in A.D. 1054. Red outer filaments are splatters of excited hydrogen glowing with Hα emission. Other emission comes from synchrotron radiation from high-speed electrons moving in the nebula's magnetic field. The nebula is roughly 2000 pc away, 3 pc across, and expanding at a speed of about 1000 to 2000 km/s. (SOURCE: Copyright California Institute of Technology and Carnegie Institution of Washington; by permission of Hale Observatories.)

FIGURE 13-17
Native Americans may have recorded the appearance of the Crab Nebula supernova on the morning of July 5, 1054. **(a)** A vertical view up the cliff near the Chaco Canyon cliff-dwelling ruins in New Mexico, showing a pictograph painted on an overhead ledge. **(b)** A close-up showing the orientation perceived by the artist facing across the canyon toward the eastern horizon and leaning over backward to paint on a ledge. **(c)** A sketch based on a computer reconstruction of the appearance of the rising crescent moon and supernova, as seen to the east across the canyon shortly before dawn on July 5, 1054. (SOURCE: Courtesy WKH.)

Chinese, Japanese, and Islamic documents, and perhaps in Native American rock art, as seen in Figure 13–17. The nebula is the expanding, colorful gas shot out of this supernova. Other supernova remnants are scattered throughout our galaxy.

What are the chances of seeing a supernova in your lifetime? In recorded history, there have been 14 supernovae in our galaxy, and a careful examination of the statistics leads us to expect an average of one every 40 to 50 years (although many of these will not be visible from Earth because of obscuration in the plane of the Milky Way). By this reckoning, we are long overdue, as the last supernova in the Milky Way was nearly 400 years ago. This might be bad news, because a supernova within about 15 pc would produce enough high-energy radiation to destroy all life on Earth! The nearest massive star to us that might one day explode is Spica in the constellation of Virgo. Luckily, it is 80 pc away, which is probably a safe distance. Considering stars at larger distances, any one of us might live to see a supernova bright enough to be visible in daylight.

SN 1987A As we saw in the chapter's opening story, in 1987, Earth witnessed the collapse of the core of a 20 $M_\odot$ star in the neighboring galaxy called the Large Magellanic Cloud. Within a second, the central iron plasma core (see Figure 13–13), about the size of Mars, collapsed at one-quarter of the speed of light down to a size of about 100 km. As temperatures reached 30 billion K, iron nuclei were fragmented, and the star exploded in a blast of neutrinos and a flash of ultraviolet light brighter than any other stars in the whole galaxy. Although this event was some 52,000 pc away (further than any stars we have discussed so far), it was easily detected from Earth.

The first trace of the supernova came from the neutrino burst. The unseen wave contained so many neutrinos that about 10 billion passed through the body of every person on Earth (in the United States these neutrinos passed up through our feet, since the Large Magellanic Cloud is in the Southern Hemisphere). At the distance of the Large Magellanic Cloud, only 1 in 10,000 people suffered even a single (harmless) neutrino interaction. Sensors in Japan and Ohio detected a few of the elusive neutrinos, marking the first experimental confirmation of a 50-year-old theoretical prediction—and the first application of neutrino astronomy outside the solar system.

Light from the explosion began to arrive a few hours later. The extra time was required for the shock wave to climb from the center of the dying star to its surface. As brilliant as the sight is, the light from a dying star is less than 0.01 percent of the total energy of the event, much of which is carried off in the form of neutrinos. For many days it was bright enough to see with the naked eye, but only at equatorial and southern latitudes (Figure 13–18). Telescopes on Earth, on the Russian *Mir* space station, and on robotic satellites were pointed at the supernova in a coordinated observational effort.

Detective work has enabled researchers to recreate the history of the doomed star. The star was a B3 supergiant,

FIGURE 13-18
Supernova 1987A. **(a)** The region of the supernova before the explosion. Red wisps are gas excited by ultraviolet light of massive young stars in the region and glowing by the light of hydrogen alpha emission. **(b)** A few days after the explosion, the region is dominated by the brilliance of the supernova. (SOURCE: Copyright Anglo-Australian Telescope Board.)

which had a mass of about 20 $M_\odot$ when it left the main sequence. Calculations by Stan Woosley and collaborators at Lick Observatory suggest that the star had an iron core of 1.5 $M_\odot$ at the time of the detonation, and a temperature of 10^{10} K. Supercomputer calculations of the density of carbon and oxygen hours after the explosion are illustrated in Figure 13-19.

As the first nearby supernova in nearly four centuries, Supernova 1987A has given astronomers a ringside seat at a stellar explosion that can be studied with the full array of modern astronomical techniques. It is all but certain that the collapsed core of the supernova is an example of one of the most bizarre objects in the universe, a neutron star.

Neutron Stars and Pulsars

What is left after a supernova explosion? As early as 1934, American astronomers Walter Baade and Fritz Zwicky speculated that one result might be a **neutron star**. The concept of a neutron star is an extension of the concept of a white dwarf. We have discussed states of matter of increasing density. In an ordinary star, electrons are stripped from atomic nuclei and collide violently in the high-temperature and high-pressure gas. The collisions are violent enough to overcome the electrical repulsion between nuclei, and thus heavier elements are created by fusion. In a white dwarf, the energy supply from fusion reactions is exhausted, so gravitational forces cause the star to collapse. The new stable configuration is supported by the pressure of electrons, which cannot share exactly the same quantum properties.

But if a burnt-out star core is still more massive than a white dwarf, the gravity is so strong that it can overcome even the pressure of the electrons. Theory indicates that this could happen in objects between 1.4 $M_\odot$ and about 2 to 2.5 $M_\odot$ (the lower boundary may be as low as 1.2 $M_\odot$, since a neutron star formed by a supernova loses up to 0.2 $M_\odot$ in the form of neutrinos). The force of gravity causes electrons and protons to coalesce and form neutrons. In the absence of electrical forces and electron pressure, the core collapses to a state of dense neutron matter. The new stable configuration is supported by the fact that the neutrons are so tightly packed that they nearly "touch" each other.

Neutron stars are truly remarkable. Their matter is the densest in the observable universe, a phenomenal density of 10^{17} kg/m^3. A thimbleful brought to Earth would weigh 100 million tons! A neutron star could contain all the mass of the Sun but be no larger than a small asteroid—perhaps 20 km

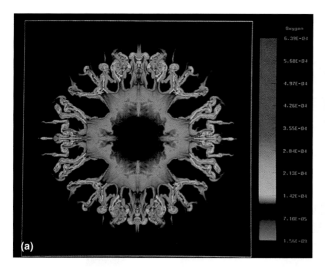 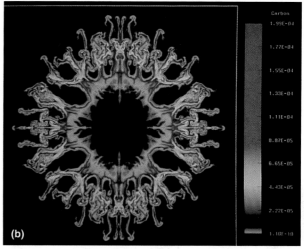

FIGURE 13-19
A supercomputer simulation of the physical state of a supernova 3.5 h after detonation. One quadrant has been calculated, and the image has been reflected to give a sense of the entire object. Red and orange colors correspond to the high densities of shock fronts in the expanding shell; blue colors correspond to low densities. **(a)** The density of oxygen. **(b)** The density of carbon. (SOURCE: D. Arnett, University of Arizona.)

across. Neutron stars rotate at up to 10 percent of the speed of light and have surface magnetic fields of 10^{12} Gauss, a million times stronger than the strongest magnetic fields that have been produced on Earth. These extreme properties are a natural result of the enormous shrinkage of the star. The rapid rotation reflects the conservation of angular momentum, and the modest magnetic field of a normal star is amplified when the magnetic lines of force are "squeezed" by the collapse. Because the density is comparable to the density of the nucleus of an atom, some astronomers have pictured a neutron star as a giant atomic nucleus with atomic mass around 10^{57}.

For decades, astrophysicists talked about neutron stars, but nobody did anything about them, because nobody *could* do anything. No known observational technique could detect them, and no one could prove they existed. But in November 1967, a large array of radio telescopes in England detected a strange new type of radio source in the sky. Analyzing the surveys (each equaling a 120-m-long roll of a paper chart), a sharp-eyed graduate student, Jocelyn Bell, was astonished to find that one celestial radio source (about 1 cm of data on the chart) emitted radio "beeps" every 1.33733 seconds! By careful analysis, the research team was able to rule out terrestrial sources for the unusual signals. Analysis showed that a rapidly repeating pulse could be produced only by a very compact source (as we will show in the next Science Toolbox). The estimated size was less than 4800 km across, much smaller than ordinary stars.

These pulsing radio sources came to be called **pulsars**. In February, Anthony Hewish and his colleagues published an analysis suggesting that pulsars might be super-dense vibrating stars. The mysterious pulsars turned out to be the long-sought neutron stars. In an exciting burst of research, the number of scientific papers on pulsars jumped from zero in 1967 to 140 in 1968. Over 600 pulsars have now been discovered. The codirectors of the original discovery project, Anthony Hewish and Martin Ryle, shared the 1974 Nobel Prize in physics.

Pulsars are just the subset of neutron stars that exhibit strong, pulsed radio emission. In other words, every pulsar is a neutron star, but there may also be neutron stars not yet detected because they are not pulsars. Why do neutron stars pulse? Following the core collapse, neutron stars spin very fast and have very strong magnetic fields. Recall from Chapter 9 that any spinning object, even a figure skater, spins faster as it contracts. Collapse by a factor of 1 million produces a corresponding increase in the spin rate. This is a consequence of the conservation of angular momentum. Similarly, the magnetic lines of force that thread a normal star will be greatly concentrated if it collapses to a neutron star. In 1968, Cornell researcher Thomas Gold showed how charged particles trapped in magnetic fields of spinning neutron stars produce strongly focused radio radiation. The pulsar acts like a lighthouse with a beam sweeping around rapidly. We see a pulsar only if the beam happens to periodically sweep across the direction toward Earth. This model of pulsar emission is illustrated in Figure 13-20.

The discovery of a pulsar in the center of the Crab Nebula (see Figure 13-16), and in some other supernova remnants, demonstrates that pulsars are related to supernovae. Each of the hundreds of known pulsars may be the neutron star corpse of an extinct massive star. A few of the youngest pulsars have been shown to pulse not only in radio waves, but also in X rays and visible light. In these cases, the visible and X-ray emission is caused by intense heating of material near the pulsar and does not represent the temperature of the pulsar itself.

Pulsars make excellent clocks. However, the rotation rate is not absolutely constant. The typical pulsar is slowing

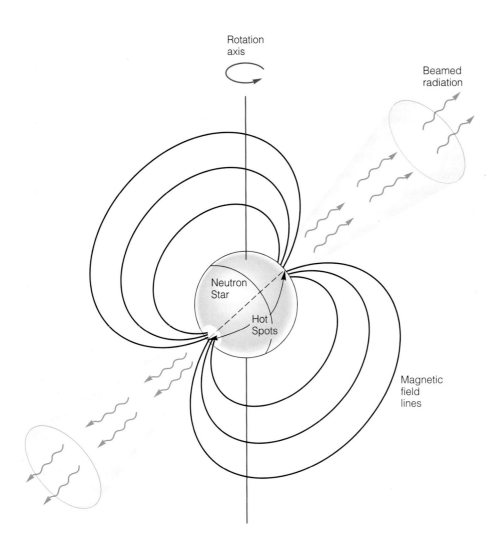

FIGURE 13-20
This model of neutron star emission accounts for many of the observed properties of pulsars. The neutron star is spinning rapidly and has a strong magnetic field. Charged particles are accelerated near the hot spots and produce a beam of radiation that travels out from each magnetic pole. The rotation axis is offset from the magnetic axis, so the beams sweep the sky as the pulsar rotates. If the beam sweeps across Earth, we see a pulsar.

down, with its rotation rate diminishing by a 30-millionth of a second per year. (No other timekeeping device can keep such good time!) Even this gradual slowing down corresponds to an enormous rate of energy release. Occasionally, radio astronomers have detected tiny abrupt changes in the spin rates of pulsars. These *glitches* are attributed to changes in the intense magnetic field, which alter the mass distribution and thereby the rotation rate. The solid crust can also undergo sudden shifts; it is strange to think of earthquake-like events happening in the solid crusts of star corpses! Most pulsars have periods in the range of 0.2 to 2 seconds. The pulsar in the Crab Nebula has a period of one-thirtieth of a second (Figure 13–21). However, in the next chapter we will discuss a small but interesting class of pulsars that completes a rotation in a tiny fraction of a second.

The final evolutionary state of a star depends critically on its mass. Stellar remnants less than 1.2 to 1.4 $M_\odot$ will evolve into white dwarfs, cooling embers supported by the pressure of electrons forced into close proximity. Neutron stars resulting from supernovae have masses below 2 to 2.5 $M_\odot$ and are supported by the pressure of neutrons forced into close proximity. These boundaries are uncertain because the physics of high-density matter is very complex, and the models depend not only on mass but also on rotation and magnetic fields. A stellar core of more than 2.5 to 3 $M_\odot$ has gravity strong enough to overwhelm neutron degeneracy pressure. Since no known force can resist the force of gravity, the collapse continues. The result is one of the strangest objects in astronomy: a **black hole.** We cannot understand black holes without talking about ideas of space and time that supercede those of Isaac Newton.

Einstein's Theories of Relativity

The man who changed the face of physics and revolutionized our ideas of space and time had an unlikely start as a scientist. Albert Einstein was a quiet and moody child, always wrapped up in his own thoughts. In one of the most infamous judgments in history, one of his schoolteachers pronounced, "Einstein will not amount to much." Einstein dropped out of high school at age 16, failed his college entrance exams, and failed to get into teacher training college. He took a clerical job as a third-class technician in the Swiss patent office in Bern. At age 26, when he wrote four papers that would rock the physics community, he was not a professional scientist and knew no other scientists. Each of his two famous theories started with an intellectual puzzle: one was a paradox and the other was a coincidence.

As a 5-year-old boy, Einstein marveled at the invisible power that moved a compass needle. Scottish physicist James Clerk Maxwell had described magnetic and electrical phenomena with an elegant set of equations in the 1860s. As we saw in Chapter 10, light and all other forms of elec-

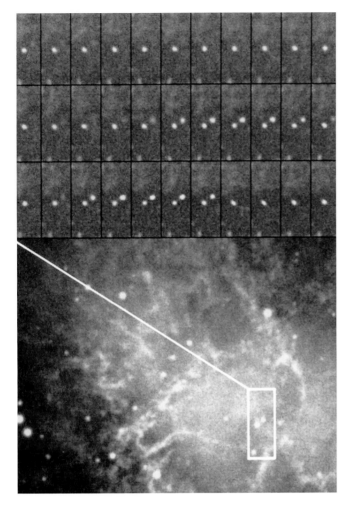

FIGURE 13-21
The Crab Nebula, with a close-up showing the region of the central pulsar. The entire cycle, including two flashes, lasts about one-thirtieth of a second, equaling one complete rotation of the pulsar. The Crab pulsar, at a distance of about 2000 pc, is one of the very few in which optical pulses have been observed. (SOURCE: N. Sharp/NOAO.)

tromagnetic radiation are caused by the oscillation of linked electrical and magnetic forces. In Maxwell's formulation, the speed of the radiation emerges as a universal constant, *independent of the motion of the observer.* Einstein's **special theory of relativity** started with his whimsical teenage conjecture: What would the world look like if I rode on a beam of light? Let us see why the constancy of the speed of light has bizarre consequences.

Suppose you are traveling in a car at 100 km/h and throw a ball forward at 50 km/h. You would expect that the ball would reach someone standing on the ground at 100 + 50 = 150 km/h. Similarly, if you are travelling at 80 km/h and throw a ball backward at 60 km/h, it will reach the stationary observer at 80 − 60 = 20 km/h. These are the familiar rules of motion defined by Galileo and Newton. What if you were traveling in a spaceship at one-third the speed of light, 100,000 km/s, and you shone a beam of light forward at the speed of light, 300,000 km/s? Intuition tells you that a stationary observer ahead of you would measure the light arriving at 300,000 + 100,000 = 400,000 km/s and an observer behind you would measure 300,000 − 100,000 = 200,000 km/s. There is the paradox: Either the speed of light is always constant, as Maxwell said, or motions add and subtract, the way Newton described. Can we ever catch up with a beam of light?

Observations and special relativity say no. In the 1880s, American physicist Albert Michelson and chemist Edward Morley showed that the light from a star always arrives at 300,000 km/s. It never shows the annual addition and subtraction of the motion of Earth as Earth orbits the Sun. Einstein elevated the constancy of the speed of light to a law of physics. He also argued that laws of physics do not depend on the state of motion of the observer. Newtonian concepts of absolute space and time must be abandoned. The universal constancy of the speed of light has some fascinating consequences.

Objects moving at close to the speed of light, or *relativistic* speeds, suffer a physical contraction and actually get shortened in the direction of motion. This shrinkage occurs in such a way as to preserve the constancy of the velocity of light. A baseball moving at exceptionally high speed would be flattened to a pancake. The second result stems from a thought Einstein had while riding home in a streetcar in Bern. He saw the clock tower passing behind him and wondered how the clock would appear as the streetcar moved faster and faster. At 300,000 km/s, the streetcar would be moving away as fast as the light wave that showed the time as 6 P.M., for example—time would be frozen! Yet the watch of the passenger would continue to tick at its normal rate. Clocks moving at close to the speed of light slow down relative to a clock held by a stationary observer. Astronauts travelling at 90 percent of the speed of light age only half as quickly as people left back on Earth. The third effect of relativity is the increasing mass of an object as it approaches the speed of light. We have seen that mass has an equivalent energy by the equation $E = mc^2$. As any object accelerates toward the speed of light, its energy goes into increasing the *mass,* not the *velocity.* Fast-moving objects acquire more inertia; they resist any change in their motion, and they can never travel at the speed of light.

None of these effects are tricks or illusions; they are real physical phenomena. Physicists have shown that particles moving at relativistic speeds "bunch up" in the direction of motion and have a larger mass-energy. They have confirmed that accurate atomic clocks traveling on a jet plane keep slightly slower time than an identical clock on Earth. The predictions of special relativity are confirmed in physics laboratories thousands of times a day. Although special relativity supersedes Newton's laws of motion, the unusual effects are significant only close to the speed of light. Thus, relativity is barely detectable when the velocity v is much less than c, in other words $v/c \ll 1$. This is the case for spacecraft and fast cars and well-struck baseballs. At everyday low speeds, Newtonian equations work extremely well.

After outlining the special theory of relativity, Einstein set about including gravity in its framework. His special theory

(a) Gravity of 9.6 m/s/s

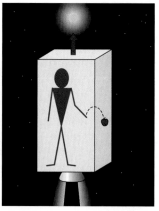
(b) Acceleration of 9.6 m/s/s

(c) Freefall to Earth

(d) Deep Space

FIGURE 13-22
Einstein's thought experiments that give insights into the nature of gravity. **(a)** In an elevator at rest on Earth's surface, an apple (or any other mass) will fall with an acceleration of 9.8 m/s^2. **(b)** Another elevator is in deep space far from any gravitational influence. If it is being accelerated at 9.8 m/s^2, the apple will fall exactly the same way. No experiment can distinguish between the two situations, so acceleration due to gravity and acceleration due to any other force are equivalent. **(c)** In an elevator falling toward Earth's surface at 9.8 m/s^2, local gravity is eliminated, and the person and the apple float freely. **(d)** Another elevator is in deep space, far from any gravitational influence. If this elevator is at rest or has a constant velocity, the apple and the person will float freely. No experiment can distinguish between the two situations, so inertial mass is equivalent to gravitational mass.

deals with objects moving at a constant relative speed. His **general theory of relativity** is a Herculean achievement that extends the discussion to nonuniform motion of any kind. Einstein began by thinking about an amazing coincidence. *Inertial mass* is the resistance of an object to any change in its motion. It is a permanent property of matter and does not depend on location; an object in deep space still has mass, and it still resists any change in its motion. *Gravitational mass,* or weight, appears to vary according to the environment. It depends on the local force of gravity; an object in orbit or in deep space has no weight, and its gravitational mass is zero. Whenever we measure gravitational and inertial mass on Earth, the numbers are identical with a very high degree of precision. Newton and Galileo were silent on this coincidence. Einstein saw that it must have a deeper meaning.

Einstein was sitting in his patent office in Bern when he had what he later called "the happiest thought of my life": A person who falls freely will not feel his or her own weight. He realized that motion or acceleration caused by gravity and acceleration caused by any other force were equivalent. Let us see why gravity is no different from any other force.

Imagine you are in a sealed elevator feeling your normal weight. Could you devise an experiment that would tell you whether you were accelerating through deep space at 9.8 m/s^2—Earth's gravitational acceleration—or just sitting stationary on Earth's surface? No! In either case, you would feel your normal weight. If you dropped an apple, it would fall to the floor normally. (In one situation gravity accelerates the apple toward the floor, and in the other case an outside force causes the floor to accelerate upward to meet the apple.) Figure 13-22a and b show this comparison. Now let us look at Einstein's "happy thought." Imagine you are in a sealed elevator with a sense of weightlessness. Could you devise an experiment that would tell you whether you were floating in deep space or plunging in free-fall toward Earth's surface? No again! In either case, an apple would float freely in the space around you. Figure 13-22c and d show the second comparison.

Einstein built his theory around the observation that acceleration due to gravity cannot be distinguished from acceleration due to any other force. General relativity has profound consequences for the way we think about the universe. The effects of special relativity that apply when objects are accelerated to near the speed of light—space being compressed, clocks slowing down, and masses increasing—must also apply to objects moving in intense gravitational fields. According to the general theory of relativity, gravity *distorts* space and time, causing space to become curved and time to slow down. Space and time are

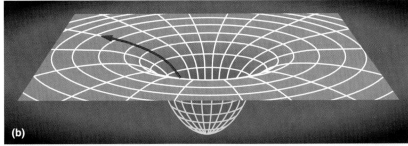
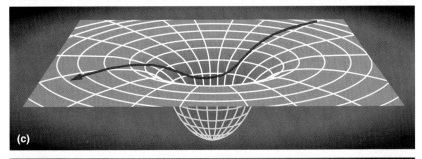
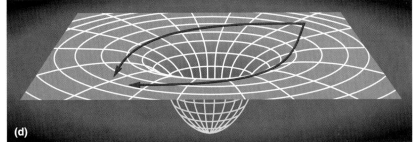

FIGURE 13–23
Four tests of general relativity that have been carried out in the relatively weak gravity near the Sun. The degree of space curvature and the size of the effects have been greatly exaggerated in order to be visible in this diagram. **(a)** The orientation of the orbit of Mercury advances by 0.5 seconds of arc per year. **(b)** Light leaving the Sun's gravity loses energy and therefore is very slightly redshifted. **(c)** Light traveling near the Sun suffers a slight time delay from the longer path through curved space. **(d)** Light can be deflected by gravity.

linked. Einstein's conception of gravity is totally different from the traditional Newtonian picture. As with special relativity, the consequences of general relativity often seem counterintuitive. Whereas Newton postulated absolute and uniform time and space, general relativity predicts that any mass alters the properties of time and space around it. As we will see in Chapter 17, this result extends to the discussion of the entire universe!

The general theory of relativity has passed all its observational tests with flying colors. Most of the tests so far have been carried out in the relatively weak gravity of the solar system, where the effects of curved space are subtle. The phenomenon that first led to a conflict with Newton's theory is the slight advance or precession in the point of nearest approach of Mercury to the Sun (Figure 13–23a). The effect is very small; only 0.5 second of arc per year. Since Einstein showed that radiation has an equivalent mass, it follows that light must respond to gravity, just as particles do. Photons leaving a massive object suffer a *gravitational redshift,* caused by the photons losing energy as they climb out of the gravitational "well" (Figure 13–23b). The gravitational redshift of the Sun is only 0.0002 percent, but it has been detected. A third prediction of general relativity is that clocks slow down in strong gravity. Atomic clocks *do* run more slowly on Earth's surface than at the top of a tall building, where gravity is slightly weaker! In space, we have found that the gravitational distortion of space slows a light signal that passes near the Sun by 250 millionths of a second (Figure 13–23c).

Finally, starlight is deflected slightly when it passes close to a massive object like the Sun (Figure 13–23d). Einstein's theory first gained wide acceptance after the solar eclipse expedition of 1919 showed that starlight was deflected by the predicted amount. (Most physicists were on the edge of

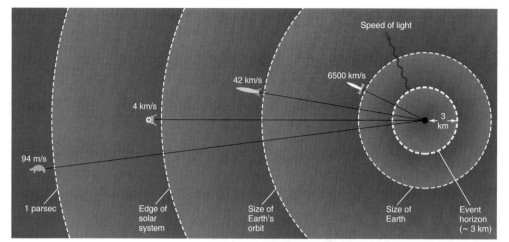

FIGURE 13-24
Exploring the gravitational field of a black hole of 1 solar mass (right). The schematic diagram (not to scale) shows the velocity required to escape from the black hole at different distances from it. At 1 parsec away (left), a mere 94 m/s would suffice, but this speed increases as we get closer. At a distance of about 3 km, a body would have to be moving at the speed of light to escape; this distance is called the event horizon. No object on a ballistic trajectory could escape from inside this distance.

their seats waiting for the announcement of the results of the experiment. Einstein went to bed; he was supremely confident that general relativity was the correct theory of gravity.) More recently, pulsars have been used as accurate clocks to confirm predictions of general relativity, resulting in the award of the 1993 Nobel Prize in physics to Joseph Taylor and Russell Hulse.

Later in his life, Einstein transcended the boundaries of his discipline to become the most famous scientist in the world. He was quoted in newspapers and appeared on the cover of magazines. He was offered the presidency of the new state of Israel (and declined it). He lobbied forcefully against the development of nuclear weapons. However, he often said that his happiest times were as a patent officer or as a young student daydreaming in the cafes of Zurich. Most importantly, he never lost the playful sense of curiosity that had led him to such masterful insights into space and time.

Black Holes

Black holes can only be understood using Einstein's theory of relativity, but observers hypothesized about their existence over 200 years ago. The Reverend John Mitchell, an English amateur astronomer, knew that Newton's law of gravity predicted that massive, dense objects would have high escape velocities. In 1784, he pointed out that a sufficiently dense object might have an escape velocity faster than light. Since all electromagnetic radiation travels at the speed of light, such an object would be completely dark.

We can understand the nature of a black hole in terms of the idea of escape velocity. Imagine that the Sun has somehow been compressed into a black hole of 1 $M_\odot$ (Figure 13-24). A rocket passing at a great distance would experience the same gravity field as a rocket at a great distance from the Sun. At 1 AU from the black hole, for example, the velocity needed to escape into interstellar space would be 42 km/s, the same as the speed required to leave Earth's orbit. You can see that the gravity far from a black hole is not very severe. It is not true that a black hole acts like a cosmic "vacuum cleaner," sucking up everything around it. But as we get much closer to the black hole, the escape velocity increases. Greater speed is required to escape the stronger and stronger gravity. At a distance of 3 km, the velocity required for escaping would be the speed of light. Since we know of nothing that can travel faster than light, nothing can escape this region.

In other words, crush a star like the Sun down to a radius of 3 km, and you have a black hole. The imaginary sphere with a radius of 3 km is called the **event horizon.** Inside this surface, no object, no particle, no information, not even light can escape. Any star that collapses within its event horizon disappears from the universe, betraying its presence only by its gravity. The radius corresponding to the event horizon is called the **Schwarzschild radius,** after the first astronomer to solve Einstein's equations of general relativity for a collapsed object. The properties of black holes are discussed in the following Science Toolbox. How is a black hole produced? An object of any mass can become a black hole if it is sufficiently compressed. However, black holes were *predicted* to exist as a consequence of stellar evolution. Any star that ends its life with a core mass of 3 $M_\odot$ or more will become a black hole, because no known force in nature can prevent its collapse within its event horizon.

Figure 13-25 illustrates the environment of a black hole and the effect of the gravity on light rays at different distances from the event horizon. At a large distance from a black hole, light travels away from a light source uniformly in all directions. As the black hole is approached, light passing near the hole will be slightly deflected. Closer to the event horizon, some light rays are deflected by the strong gravity and are captured by the black hole. At a distance of 1.5 times the Schwarzschild radius, half the light escapes. Photons emitted at right angles to the black hole are trapped in circular orbits. These orbits define the *photon sphere*. At the Schwarzschild radius, the deflection of light is so severe that no light can escape. The Schwarzschild radius defines the event horizon.

Another analogy can be used to convey the extreme space-time curvature caused by black holes. The general theory of relativity predicts that *any* mass will distort the space and time around it. Figure 13-26 presents the space curvature in two dimensions, as the distortion in a thin rubber sheet. In the absence of any matter, space will be flat and have no curvature (Figure 13-26a). For a concentrated

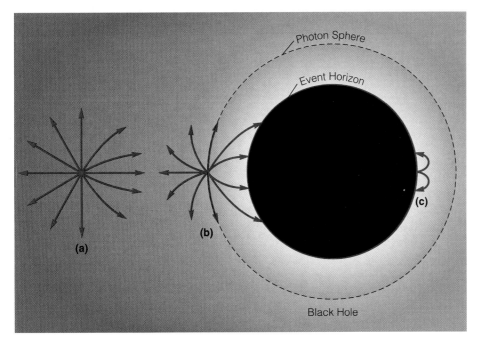

FIGURE 13-25
The environment of a black hole. **(a)** Far from a black hole, light travels mostly straight away from a source, except for light traveling near the direction to the black hole, which is deflected. **(b)** At a distance of 1.5 times the Schwarzschild radius, half the light escapes. The surface at this radius is called the photon sphere. **(c)** At the Schwarzschild radius, all light is trapped by the black hole. The surface at this radius is called the event horizon.

mass, the distortion is large enough to clearly deflect matter and radiation that pass near it (Figure 13–26b). In the extreme case of a black hole, the curvature is complete. We can imagine a piece of space and time being "pinched off" and permanently removed from communication with the rest of the universe (Figure 13–26c).

If you were unfortunate enough to fall into a 1 $M_\odot$ black hole, you would be killed by tidal forces long before you reached the event horizon. (Essentially, the difference between the gravity force on your head and that on your feet would rip you apart!) Assuming that somehow you could survive the descent, you would see clocks far from the black hole keeping slower and slower time, until, as you neared the event horizon, they appeared to stop altogether. Seen from the outside, your clock would appear to slow down as you took an infinite amount of time to reach the event horizon! If you carried a light source with you as you fell into the black hole, a distant observer would see the photons suffer a

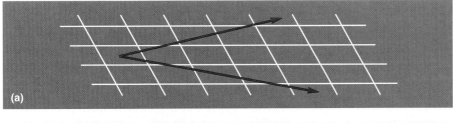
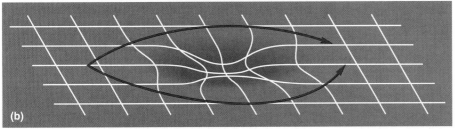
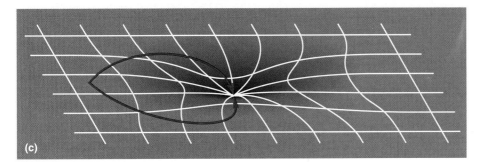

FIGURE 13-26
A rubber-sheet analogy for the curvature of space. **(a)** Empty space is flat and does not cause particles or radiation to deviate as they pass through it. **(b)** A concentrated mass distribution will distort space in its vicinity and cause the deflection of particles and radiation. **(c)** A black hole causes such extreme curvature of space that particles and radiation can be completely trapped.

SCIENCE TOOLBOX

Properties of Black Holes

A black hole is a very unusual state of matter. To see how strange it is, let us begin with the idea of escape velocity. A full treatment requires general relativity, but we can get an approximate answer using Newtonian gravity. The escape velocity of an object of mass M and radius R is

$$v = \sqrt{2GM/R}$$

In this equation, G is the universal gravitational constant, or 6.67×10^{-11} Newton $m^2 kg^{-2}$. If M is in kilograms and R is in meters, then v will be in meters per second. The escape velocity is higher for more massive objects of the same size or for a smaller object of the same mass; both correspond to a higher density. If you insert numbers for Earth, you will see that the escape velocity of Earth is about 11 km/s. For the Sun, the escape velocity is about 600 km/s.

To calculate the size of a black hole, we must set the escape velocity to be equal to the speed of light. Thus, $v = c = \sqrt{2GM/R}$. Squaring both sides and rearranging the equation, we get

$$R_S = 2GM/c^2$$

This is the Schwarzschild radius of a black hole. Let us see what we get for an object with the mass of the Sun. In this case, $M = 2 \times 10^{30}$ kg and $c = 3 \times 10^8$ m/s, so $R_S = 2 \times 6.67 \times 10^{-11} \times 2 \times 10^{30} / (3 \times 10^8)^2 = 2960$ m, or about 3 km. Crush the Sun down to the size of a small town, and it would be a black hole. The Schwarzschild radius is proportional to mass, so in the realistic example of a 3 $M_\odot$ stellar core, the black hole radius is 9 km.

Any object can become a black hole if it is sufficiently compressed. For example, Earth would become a black hole if it were compressed in an enormous vise down to a radius of 1 cm! Jupiter would be a black hole if its mass were contained in a region 6 m across. A human would be a black hole if concentrated into a region 10^{-25} m across—far smaller than a proton. None of these interesting possibilities has been found in the universe. As far as we know, nature can only make black holes out of objects that are star-sized or larger.

Mass is a fundamental property of a black hole, and it is a property we can measure. What other properties does a black hole have? All stars rotate, so when a star collapses by a large factor, the conservation of angular momentum dictates that the spin rate will speed up. Angular momentum is proportional to orbital velocity times radius (see the Science Toolbox in Chapter 9 for a review). Therefore, if the angular momentum is constant, orbital velocity must be inversely proportional to radius. Since a faster orbital velocity means a shorter orbital time, we deduce that the rotation period reduces proportionally to the radius as an object shrinks. The Sun rotates every 25 days and has a radius of 700,000 km. If it were to shrink to black-hole size of 3 km, it would rotate every $(3/700,000) \times (25 \times 24 \times 3600) = 9$ seconds! We can see that very rapid rotation must be a property of any star that becomes a black hole.

A black hole has no other properties that we can measure. For example, we cannot determine what a black hole is made of, because all information about chemical abundance is lost when the material enters the event horizon. The fact that information is lost in a black hole has an intriguing consequence. Jakob Beckenstein and Stephen Hawking showed in the 1970s that black holes carry enormous amounts of entropy. Black holes consuming matter are just another example of the universal tendency toward disorder.

Could someone survive a trip into a black hole? Do objects that disappear into black holes appear elsewhere in the universe? What can we say about wormholes in space? These ideas are the subject of books, movies, and TV shows, and they are fun for speculation. But in reality, we have no theory to describe what happens beyond the event horizon of a black hole. Since the event horizon is an information barrier, and because we do not have any black holes on which to experiment, we cannot apply the principles of the scientific method. For now, we must be content with speculation.

---

larger and larger gravitational redshift (to you, the light would stay the same color). The redshift occurs because light loses energy escaping from the intense gravity. Seen from the outside, the photons would be infinitely redshifted to zero energy as you reached the event horizon.

What lies within the event horizon of a black hole? Nobody really knows. The event horizon is not a physical barrier, just an information barrier. Einstein's theory predicts that matter will keep collapsing gravitationally until it has shrunk to a point of zero volume and infinite density! This endpoint is called a *singularity*, and it cannot be adequately described using general relativity. Black holes are not entirely black. In the 1970s, English physicist Stephen Hawking calculated that black holes could create subatomic particles near their event horizons and slowly radiate away their energy, or "evaporate." This so-called *Hawking radiation* is expected to be dramatic for microscopic black holes, but barely noticeable for solar-sized black holes. Far more important is the fact that any material falling toward the event horizon will be subject to enormous gravitational forces. The friction and heating of material that falls in will be released in the form of X rays. Therefore, a black hole

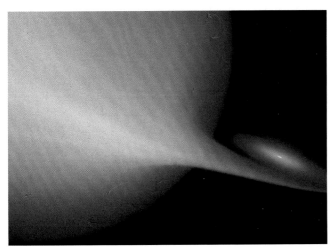

FIGURE 13-27
An imaginary view of a hypothetical star system to explain the X rays and gamma rays from sources such as Cygnus X-1. Gas shed from the surface of a hot supergiant star (left) flows toward a very dense star, possibly a black hole, orbiting it. It spirals around the dense object and accumulates into a disk-shaped nebula. New gas is dumped onto the nebula at a very high speed, causing extremely high temperatures and emission of blue and ultraviolet light, X rays, and gamma rays. (SOURCE: Painting by Adolf Schaller.)

may be a *source* of energy from the death spasms of matter falling into it.

Can we ever hope to detect a black hole? Yes. Outside their event horizons, black holes have gravity fields indistinguishable from those of ordinary stars of the same mass. Thus, they can orbit around stars just like planets or binary star companions. If we observed such a star from a distance, we would not see the black hole, but we could see the star's orbital motion and calculate the mass of the unseen companion, just as astronomers routinely do in the case of ordinary faint companions. The result would indicate an unusually high-mass companion for an X-ray source—maybe 5 or 10 $M_\odot$—which is a sign that we are dealing with a black hole candidate. Suppose a black hole is orbiting an evolved star that has expanded into the giant state and is shedding mass. Some of the expanding gas would fall toward the black hole at relativistic speeds. Because this gas would, on average, have some angular momentum around the star rather than falling directly toward it, it would form a disk of gas spiraling inward toward the black hole (Figure 13-27). This disk is called an **accretion disk.** Its gas would be extremely hot, because it would be constantly hit by new gas streaming in from the other star. Because of the high temperature, the disk would radiate at short ultraviolet or X-ray wavelengths.

The best evidence for a black hole would be a massive, high-temperature X-ray source orbiting another normal star. There are currently about a dozen good candidates. The argument for a black hole in these systems has three steps:

1. The X-ray emission must be consistent with accretion onto a compact object.
2. The calculation of the orbit of the binary system must lead to a mass for the compact object that exceeds 3 $M_\odot$.
3. It is assumed that general relativity is the correct theory of gravitation, and that neutron stars with masses above 3 $M_\odot$ cannot exist.

Is the evidence convincing? All the observations contain uncertainties, and the interpretation is indirect and circumstantial. Whereas the evidence for the existence of neutron stars is convincing, the evidence for stellar black holes is strong but not overwhelming. Black holes are the most exotic member of the stellar "zoo," and astronomers continue to work to prove their existence beyond doubt.

Summary

Stars are forming, even today, within a few hundred parsecs of the Sun and in more distant regions of space. The starry sky is not a static scene but the site of continual births of new stars out of interstellar dust and gas. Many stars and star systems are less than a few million years old—much less than 1 percent of the age of our galaxy. Some have become visible since humans evolved.

Star formation begins with protostars, which are clouds of dust and gas that begin to contract from their own gravity. They collapse fairly rapidly to stellar dimensions and become pre-main-sequence starlike objects. The process of star formation depends strongly on magnetic fields and rotation in the collapsing gas cloud. Substantial theoretical progress has been made in the last two decades toward understanding the basic features of star formation. Many of these features have been confirmed by observation, especially by infrared instruments, which detect the radiation from low-temperature dust in nebulae around the newly formed stars. Once the star reaches the main sequence, energy is generated in the star as hydrogen converts to helium by means of the proton-proton cycle for smaller stars and the carbon cycle for larger stars. All stars in the universe shine by energy released from the fusion of light elements into heavier elements.

The fate of all stars is governed by the irresistible force of gravity. Figure 13-28 summarizes the life histories of stars of different masses. At the low-mass end, stars are dim, red, and slow to evolve. The coolest-sequence stars have not yet evolved off the main sequence over the course of the entire age of the universe. Every star more massive than the Sun goes through a phase of mass loss, involving either a stellar wind or a more

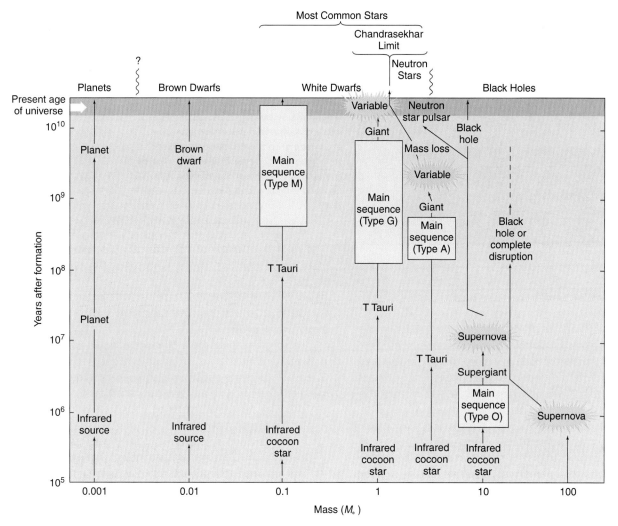

FIGURE 13-28
A summary of stellar evolution, showing the inexorable march toward high density. Evolutionary histories are shown for objects with different initial masses, with the terms described in this chapter. Notice the logarithmic scale on the vertical time axis; the short-lived evolutionary stages are at the bottom of the diagram. Some states are postulated on theoretical grounds, but they have not yet been firmly identified by observers.

violent explosion. Stars below 1 to 2 $M_\odot$ end life quietly, as cooling white dwarf embers. Massive stars are rare, and they evolve quickly toward a spectacular demise. Supernovae are not only responsible for the production of neutron stars and probably black holes, but they recycle rare and important heavy elements into the universe.

Stellar old age leads to two basic phenomena: high-energy nuclear reactions, as ever more massive elements interact and fuse, and inexorable contraction, as energy sources are eventually exhausted. During the first stages of old age, as stars evolve off the main sequence, the large release of energy causes an expansion of the stellar outer atmospheres, producing giants and supergiants. As heavier elements go through quick reaction sequences, various kinds of instability may produce variable stars and slow mass loss. After energy generation declines to a rate too low to resist contraction, low-mass stars contract to a dense state known as a white dwarf, with final mass less than 1.4 $M_\odot$. The less-common massive stars, initially with as much as 8 $M_\odot$ or more, undergo supernova explosions and blow off much of their original material. The entire course of stellar evolution involves the release of a small fraction of the mass-energy of a star, in the form of radiant energy.

If the remnant cores of stars end up between 1.4 $M_\odot$ and about 2.5 to 3 $M_\odot$, they form dense, rapidly rotating neutron stars known as pulsars. If the remnant cores have more than about 3 $M_\odot$, they may form black holes. Although virtually no radiation escapes from these strange objects, they may be detectable by orbital motions of their companion stars and by high-energy radiation from material falling into them. The search for the forms of dead and dying stars has yielded some of the most fascinating objects now being studied by both physicists and astronomers.

Take a minute to consider our cosmic origins. The carbon atoms on which life depends were produced in the cores of previous generations of stars. If you wear any jewelry, look at the gold or silver and consider the incredible journey those atoms have taken. All the iron and aluminum used to build our modern civilization, and all the precious metals we treasure and use for adornment, were forged in distant stars long ago.

Important Concepts

You should be able to define these concepts and use them in a sentence.

cosmic abundance	entropy
molecular cloud	Chandrasekhar limit
theory of star formation	supernova
protostar	explosive nucleosynthesis
pre-main-sequence star	neutron star
brown dwarf	pulsar
carbon cycle	black hole
red giant	special theory of relativity
triple-alpha process	general theory of relativity
variable star	event horizon
planetary nebula	Schwarzschild radius
laws of thermodynamics	accretion disk

How Do We Know?

These questions and answers show how the scientific method is used to learn about the universe.

Q How do we know that stars are forming all the time?

A It is extremely difficult to observe a star forming directly, because it occurs over a period of millions of years and takes place deep in clouds of dense gas and dust. But there are other pieces of reliable but indirect evidence. The universe is over 10 billion years old; because the Sun is younger, we know that stars did not all form in a big burst at the beginning of the universe. We also see massive stars that can only last about 1 million years on the main sequence; they must have formed within the history of the human species. Also, we see massive, hot stars with telltale evidence of recent formation—disks of dusty material and winds of hot gas. Moreover, we have theoretical reasons for expecting that gas clouds will collapse into stars.

Q How do we know the energy source in stars of various types?

A All stars shine by nuclear fusion; that is the only energy source efficient enough to produce the radiation we see. In general, we need two pieces of information to understand the energy source: chemical composition and luminosity. Spectroscopy can reveal the chemical composition, which indicates the raw material for the fusion and its end products. Because we can only observe the photosphere directly, we rely on churning gas motions within the star to bring the fusion products to the surface. The star's luminosity combined with the laws of radiation enables us to calculate a model for the star that can be used to predict the density and temperature at any point within it. The temperature determines which nuclear reactions can proceed—hydrogen fusing to helium above 10 million K, helium fusing to carbon above 200 million K, and so on.

Q How do we know that the heaviest elements are made inside a supernova?

A We know that the heavy elements were not around at the beginning of the universe, because the oldest stars show very low abundances of heavy elements. We also know that very heavy elements cannot be made by adding building blocks like helium nuclei. Iron is so stable that energy is required for this fusion rather than being released. Elements heavier than iron can be built up by the slow addition of neutrons within red giants. The nuclear physics for this reaction is understood from laboratory experiments. Many of the heaviest elements, such as gold, silver, and platinum, are made in the rapid process of neutron capture that occurs in the violent shock wave of a supernova explosion. The observed cosmic abundance of these elements agrees with the predictions of explosive nucleosynthesis. Despite the dreams of alchemists throughout history, these precious elements cannot be made in the laboratory!

Q Do black holes exist?

A Perhaps. An isolated black hole emits no detectable radiation, so we can detect them only in binary systems. If a binary star system has an unseen companion with a mass above 3 $M_\odot$, then it is very likely to be a black hole. However, the reliability of this conclusion always depends on how well we know the orbit and the mass of the visible star. This type of modeling always has uncertainties. At present, a small number of binaries probably harbor black holes, but we cannot be certain. If more such systems are found, our confidence will increase.

Problems

Use these problems to test your understanding of the information and concepts in this chapter. The * indicates a more advanced or mathematical problem.

1. Suppose you magically smoothed out all the density variations in the interstellar gas so that it was uniform. **(a)** Would this help or hinder star formation? **(b)** What processes would keep the gas from staying uniform indefinitely? **(c)** If stars began to form in this gas, would it ever return to a uniform state?

2. How do theories of solar system formation and theories of star formation support each other? Contrast the sources of information on these two subjects. (You may need to review the material in Chapter 9.)

3. Compare the time scale for the significant evolution of massive pre-main-sequence stars with the time during which astronomers have recorded observations of such stars. Is it reasonable that some young, pre-main-sequence stars might show evolution-related fluctuations in their properties within the time they have been observed?

4. Why do stars form in groups instead of alone? Suppose a cluster of stars formed 3 million years ago. Why would you expect the H-R diagram of the cluster to show no stars on either the very high-mass end of the main sequence or its very low-mass end?

5. Suppose a new 1 $M_\odot$ star began forming about 10 parsecs from the Sun. What would Earth-based observers see during the next few million years? Consider infrared observers as well as naked-eye observers.

6. **(a)** Why doesn't gravity immediately cause the collapse of all interstellar clouds? **(b)** Why does the structure of a star stabilize when it reaches the main sequence?

*7. **(a)** Why do stars just moving off the main sequence expand to become giants instead of starting to contract at once? Why does the force of gravity ultimately win?
  (b) Many red giants are visible in the sky, even though the red giant phase of stellar evolution is relatively short-lived. Why are so many red giants visible in the sky?

8. List examples of evidence that certain stars can lose mass.
9. What type of stars will eventually become **(a)** white dwarfs? **(b)** neutron stars? **(c)** black holes? What will be the ultimate fate of the Sun?
*10. A main-sequence B3 star has about 10 times the mass of the Sun and therefore has about 10 times as much potential nuclear fuel. Why, then, does it have a main-sequence lifetime only $1/200$ as long as that of the Sun?
*11. According to the law of conservation of angular momentum, a figure skater spins faster as she pulls in her arms. How does this principle help explain why neutron stars spin much faster than main-sequence stars? If the Sun collapsed to the size of a typical neutron star, how fast would it spin? Why will this never happen to the Sun?
*12. Comment on the roles of theorists and observers in the three decades of work on white dwarfs, pulsars, and black holes. Why are most stars destined to become white dwarfs? How good is the evidence that pulsars consist of pure neutron material? Are black holes fully understood today?

Projects

Activities to carry out either individually or in groups.

1. On a clear night (an early evening in February or a late evening in December is ideal), scan the region of Orion with your naked eyes and compare it to other regions of the sky. Notice the concentration of bright blue O-, B-, and A-type stars (such as Sirius and Rigel) and star clusters (such as Pleiades, Hyades, and the Orion Belt region) in this broad area. How do these features indicate that star formation has been going on in this general direction from the Sun in the last few percent of cosmic time?
2. Locate the star Mira (R.A. = $2^h 14^m$; Dec. = $-3.4°$) with a small telescope and determine whether it is in its faint or bright stage. If it is bright enough to see with the naked eye, record its brightness nightly by comparing it with other nearby stars of similar brightness. By checking the brightness of these stars with star maps, plot a curve of Mira's brightness over time.
3. With binoculars or a small telescope, locate the star Delta Cephei (R.A. = $22^h 26^m$; Dec. = $+15.1°$) and compare it from night to night with other nearby stars of similar brightness. Can you detect its variations by a factor of 2 in brightness in a period of 5.4 days?

Reading List

Cohen, M. 1988. *In Darkness Born: The Story of Star Formation.* Cambridge, Eng.: Cambridge University Press.

Eichler, D. J. 1994. "Ashes to Ashes and Dust to Dust." *Astronomy,* vol. 22, p. 40.

Frank, A. 1996. "Starmaker: the New Story of Stellar Birth." *Astronomy,* vol. 24, p. 52.

Jastrow, R. 1990. *Red Giants and White Dwarfs.* New York: Norton.

Kaler, J. 1991. "The Smallest Stars in the Universe." *Astronomy,* vol. 19, p. 50.

Kawaler, S. D., and Winget, D. E. 1987. "White Dwarfs: Fossil Stars." *Sky and Telescope,* vol. 74, p. 132.

Lada, C. 1993. "Deciphering the Mysteries of Stellar Origins." *Sky and Telescope,* vol. 85, p. 18.

Marschall, L. 1994. *The Supernova Story.* Princeton: Princeton University Press.

Soaker, N. 1992. "Planetary Nebulae." *Scientific American,* vol. 266, p. 78.

Stahler, S. 1991. "The Early Life of Stars." *Scientific American,* vol. 265, p. 48.

Thorne, K. P. 1994. *Black Holes and Time Warps: Einstein's Outrageous Legacy.* New York: Norton.

Whitmire, D., and Reynolds, R. 1990. "The Fiery Fate of the Solar System." *Astronomy,* vol. 18, p. 20.

Will, C. 1986. *Was Einstein Right? Putting General Relativity to the Test.* New York: Basic Books.

Woosley, S., and Weaver, T. 1989. "The Great Supernova of 1987." *Scientific American,* vol. 261, p. 32.

The Milky Way

CHAPTER 14

CONTENTS

- Herschel Scans the Skies
- The Distribution of Stars in Space
- Stellar Companions
- The Environment of Stars
- Groups and Clusters of Stars
- The Layout of the Milky Way
- Summary
- Important Concepts
- How Do We Know?
- Problems
- Projects
- Reading List

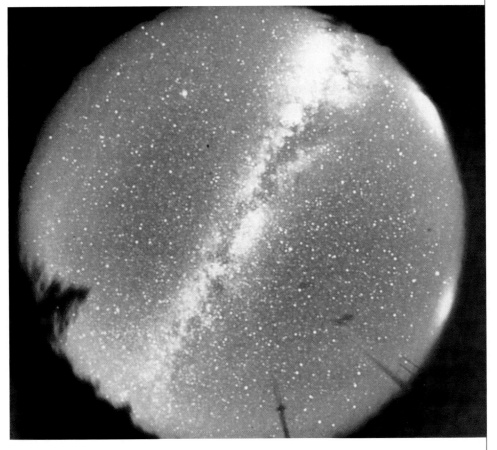

A wide-angle view of the night sky, taken with a fish-eye camera from Mt. Graham, Arizona. Around the horizon are the lights of Wilcox and Tucson; the bright object to the upper left is Jupiter. The bright band of the Milky Way has dark regions of obscuring dust. In directions away from the Milky Way, the stars are more thinly and uniformly distributed. (SOURCE: R. Angel, Steward Observatory/NOAO.)

PREREADING QUESTIONS ON THE THEMES OF THE BOOK

OUR ROLE IN THE UNIVERSE
Where are we in the large system of stars called the Milky Way?

HOW THE UNIVERSE WORKS
How does the material between stars affect starlight?

HOW WE ACQUIRE KNOWLEDGE
How do we know the size and shape of the Milky Way?

WHAT TO WATCH FOR IN CHAPTER 14

The last chapter presented the life stories of individual stars. In this chapter, we will learn how stars are distributed within the large system of stars called the Milky Way. Most stars are not alone; they are locked in a gravitational embrace with one or more companions. In close binary systems, each star can affect the evolution of the other. In fact, mass transfer between stars in a binary system can lead to some of the most spectacular phenomena in any star's life—eruption as a nova, detonation as a supernova, and intense X-ray emission from an accretion disk. You will see the different ways binary and multiple stars can be created. You will also learn that although the space between stars may seem empty, it is not a perfect vacuum. Space between the stars contains dust and gas that reddens and extinguishes starlight as it travels across vast distances.

In the last section of the chapter, we consider the layout of the Milky Way galaxy. Building upon William Herschel's pioneering observations in the 19th century, astronomers have mapped out various groupings of stars. Close to the Sun, we see open clusters and associations along with the hot gas and debris of active star-formation regions. These mostly hot and blue young stars tend to dwell in the flattened plane of the Milky Way. The most impressive groups of stars are the globular clusters, which are mighty swarms of old, red stars that orbit far from the plane of the Milky Way. By a combination of techniques, we can measure the distances to faint stars and map out the structure of the Milky Way. You will see that the Sun is offset from the center of an enormous disk of stars thousands of light years across. A spherical halo of stars and globular clusters surrounds this disk.

HERSCHEL SCANS THE SKIES

The man who established the place of the Sun in the realm of the stars was William Herschel. Born in 1738 into a musical family in Hanover, Germany, Herschel was an accomplished musician himself. He deserted from the German army during the Seven Years' War and got a job in England as the organist in a chapel in Bath. In his study of musical harmony, Herschel read a mathematics book by Robert Smith, a Cambridge professor. This led him to another book by Smith, on astronomical optics. Spurred by his long-time fascination with the stars, Herschel began to build his own telescopes.

Astronomy soon became a passion for Herschel. Throughout the year, he would spend each clear night scanning the skies. Caroline Herschel, William's sister, was his constant companion in this work, helping him document his observations. Herschel worked in weather so cold he had to break through ice on his inkwell to make notes. He rushed home during the intermissions of his concerts to squeeze in a few more observations. He developed an unrivaled familiarity with the night sky.

This dedication paid off in 1781. As he was scanning pairs of stars to look for the effect of parallax, he saw an object noticeably larger than a star that "did not belong" in the surrounding pattern of stars. His excellent telescope and sharp eye clearly resolved the disk of what was in fact a planet, even though it has a tiny angular diameter no larger than 3.7 seconds of arc as seen from Earth. In fact, this planet had been observed more than a dozen times previously, but each time the observer had marked it down as a star. Other astronomers soon determined that the planet's orbit was far beyond the distance of Saturn. The new planet was named Uranus after the oldest of the Greek gods. Herschel was the first person since antiquity to discover a new planet. At a stroke, he doubled the size of the solar system by finding a planet twice as far away as Saturn.

Herschel became famous after his discovery of Uranus. Backed by what we would call federal support—an annual stipend from King George III beginning in 1782—Herschel built a series of larger and larger telescopes. He ground his own mirrors and fashioned the fittings from fine wood, adapting the same techniques he had used to make cellos and oboes years before. Using the reflector telescope design of Isaac Newton, he built a telescope with a 1.2-m mirror in a few years. Forty times the size of Newton's first reflector, this size would not be surpassed until the 1840s. The construction was sometimes hazardous. Once, Herschel cast a mirror from molten metal using a mold made of dried horse dung after no foundry would take on the project. The mold cracked under the intense heat, and Herschel and his colleagues had to run to escape the pool of liquid metal.

Herschel wanted to figure out what he called "the construction of the heavens" —the number of stars in the galaxy and how they are distributed. To do so, Herschel allowed the telescope to sweep the skies, counting stars in parallel strips. (These were the days before telescopes had motors that could compensate for Earth's rotation and keep an object fixed in view.) Turning his attention to the Milky Way, Herschel was struck by the vast number of stars that were revealed by his telescope. "We find that the stars are crowded beyond imagination along the extent of the Milky Way,... so that, in fact, its whole light is composed of nothing but stars of every magnitude from such as are visible to the naked eye down to the smallest points of light perceptible with the best telescope."

Herschel developed a method of comparing the brightness of different stars. With two telescopes of the same size pointed at different stars, he would partially cover the aperture of one telescope until the stars appeared to have the same brightness. He discovered that the ratio of the uncovered areas of the two telescopes was equal to one divided by the relative brightness of the two stars. Herschel then assumed that all stars have the same true brightness, so the fainter a star appears, the farther away it is. He found far more stars and fainter stars in the Milky Way than in other directions (see the chapter-opening photograph). Herschel deduced from this observation that the stars in the direction of the Milky Way are distributed over larger distances than stars in other directions. He correctly mapped the Milky Way system as a disk of stars, with us inside it, but he did not know how big the system of stars was.

As he scanned the skies, Herschel was like a cartographer. Hundreds of years earlier, brave adventurers had mapped out the size and shape of the globe. Herschel continued the exploration by detecting stars across vast regions of space and by mapping the distribution of "this magnificent collection of stars."

The Distribution of Stars in Space

The night sky blazes with starlight. If you go to a site far from any cities, you can see over 6000 stars scattered across the sky. Long before the invention of the telescope, people could plainly see a band of light arching across midnight skies during certain seasons. Myths and legends have been molded around this prominent feature of the night sky. Nearly 2500 years ago, the Greek philosopher Democritus correctly attributed this glow to a mass of unresolved stars, which came to be called the *Via Lactea,* or **Milky Way**. In 1610, Galileo turned his telescope on the Milky Way and confirmed Democritus' idea. An entire set of stars held together by gravity is called a **galaxy**. The Milky Way galaxy contains the Sun and all the other stars in the night sky, and has its greatest concentration of stars in one strip of the sky.

How far away are all these stars? Does the distribution of stars have an end? What is the role of the Sun in this great assemblage? These questions get to the heart of one of the themes of this book: our place in the universe. In this chapter, we will build up a picture of the Milky Way galaxy, working up in scale. We will start by describing the immediate environment of stars, showing that most stars have companions. Then we will describe the thin medium that exists between stars. Next, we will look at the properties of groups and clusters of stars. At the end of the chapter, we will put all this information together to define the architecture of the Milky Way galaxy.

The stars seem to be arrayed above our heads on a two-dimensional backdrop. As we learned in Chapter 12, measuring the third dimension—distance—is a great challenge. Stars differ in absolute brightness by factors of 1 million or more, so a dwarf star might be 1000 times nearer than a supergiant of the same apparent brightness. Absolute brightness, or luminosity, is given by $L = d^2 F$, where d is distance and F is flux or apparent brightness. Remember that L is the true brightness of an object, or the number of photons it emits each second, while F is the brightness we measure on Earth, or the number of photons per second we collect with our telescope. Figure 14–1a shows how apparent brightness *is* a good measure of distance if we can identify stars of the same luminosity. (See also the example in Chapter 12 of 100-watt lightbulbs scattered through space.) Luminosity is a constant, so $F \propto d^{-2}$. This is a statement of the familiar *inverse square law.* The excellent correlation between F and d means that the flux can be used to calculate distance. In Figure 14–1b, the stars have a larger range in luminosity. The scatter is therefore larger, and the apparent brightness is a much poorer indicator of distance. Finally, Figure 14–1c shows the true situation an astronomer might face when stars of every type are measured. The range of luminosity on the H-R diagram is so large that any correlation has been washed out. Apparent brightness gives no useful measure of distance.

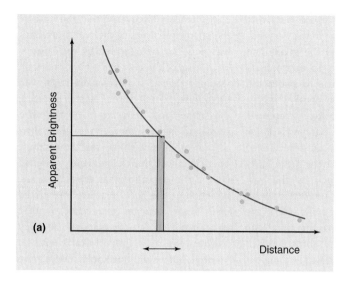
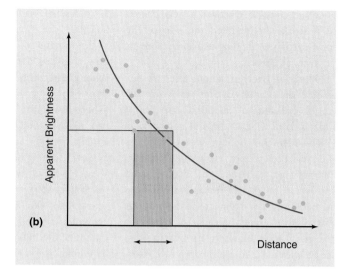
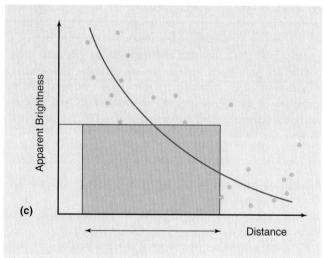

FIGURE 14-1
The relationship between flux, or apparent brightness, and distance. **(a)** If every star has the same luminosity, then apparent brightness and distance are related by the inverse square law. Apparent brightness is a perfect indicator of distance, with only a small scatter due to measurement error. **(b)** If stars have somewhat different values of luminosity, apparent brightness and distance are correlated, but with increased scatter. **(c)** If stars have widely different values of luminosity, there is almost no correlation between apparent brightness and distance. Apparent brightness is therefore an unreliable indicator of distance.

We can measure distances to some stars directly, using the method of trigonometric parallax described in Chapter 12. Unfortunately, most of the stars that are visible through a small telescope are too remote for a parallax to be measured. However, we can learn something about distance simply by observing the way stars are distributed on the plane of the sky. On a dark night, you can observe that some stars appear very close to each other. Are these pairs just caused by the chance alignment of two stars at different distances, or are they connected in some way? As early as 1767, John Mitchell, whom we have already met as the father of the idea of black holes, decided that there were too many alignments to be caused by chance. He believed that the stars in each pair were close enough in space to orbit each other by gravity. How did he reach this conclusion?

To understand Mitchell's reasoning, we need to start by understanding two types of distribution: uniform and random. For now, let us imagine that stars are distributed in only two dimensions, on a plane. A *uniform distribution* characterizes objects that are separated by equal distances. Figure 14–2a shows a uniform distribution of stars; they are spread out in a regular grid, with equal distances between each one. The distance from any star to its nearest neighbor is always the same; this is shown in the histogram below the star map. A uniform distribution may be a realistic way to describe the way atoms are laid out in a crystal, but not the way stars are laid out in space.

For a more realistic situation, let us consider a random distribution. A *random distribution* characterizes objects that are separated by random distances. Figure 14–2b shows the same stars randomly distributed in two dimensions. In this case, the distance between a star and its nearest neighbor star can vary quite widely, shown in the histogram below the star map. However, the *average* distance between stars is the same as in Figure 14–2a. To see that this must be true, remember that the number of stars and the total area are unchanged, so the average spacing is unchanged too. For any particular star, we can ask what the probability is that it will have a neighbor within a certain distance. The average spacing is the distance where the probability is 0.5—half the stars will have a nearest neighbor closer than the average spacing, and half will have a nearest neighbor farther away than the average spacing. The probability of a star having a

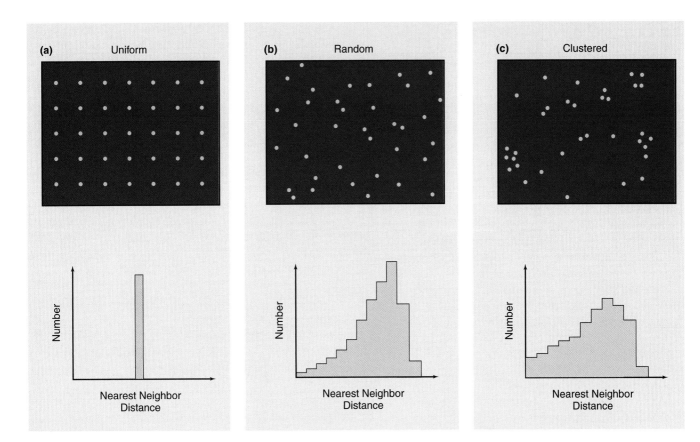

FIGURE 14-2
The separations of stars with different distributions. Two dimensions are shown, but the distributions also apply in three dimensions. The histogram below each star map is the distribution of nearest-neighbor distances for each star. **(a)** Stars with equal spacing would all have the same distance to their nearest neighbor. This grid—unrealistic in the case of stars—is a uniform distribution. **(b)** Stars that are randomly distributed would have distances to their nearest neighbors that peak at the same average value as a uniform distribution, but small separations will sometimes occur by chance. **(c)** When stars are found in binary pairs or groups, there is an increased probability of a small distance to the nearest neighbor when compared with a random distribution. The excess of small separations is a signature of clustering.

neighbor within any distance is proportional to the area considered, or the distance squared. The histogram below the star map shows this distribution of nearest-neighbor distances. In a random distribution, it is certainly possible for stars to be very close to each other, but it does not happen very often.

Now look at Figure 14–2c. Stars in this field show **clustering**. A *clustered distribution* characterizes objects that are separated by distances that tend to be smaller than for a random distribution. When stars are clustered, the nearest neighbors tend to be separated by smaller distances, shown in the histogram below the star map. Clustering is revealed by the higher probability of any star having a neighbor at a small distance, compared to a random distribution. Our example uses distributions in two dimensions, but the same argument works for a three-dimensional distribution as well. We have described a statistical measure of clustering based on a distribution of nearest-neighbor distances. In other words, we cannot claim that a *particular* pair of stars with small separation is clustered, because small separations will

occur by chance in a random distribution as well. However, the statistical approach is very powerful because it enables astronomers to detect departures from a random distribution that are quite subtle. We do not need a statistical test to tell us that a large clump of stars is clustered; it is obvious to the eye.

Now we can follow Mitchell's reasoning more easily. He measured the angles that separated each star in the sky from its nearest neighbor. He then calculated what angles he would measure if the same number of stars were distributed randomly in the sky. He compared the two distributions. Mitchell detected a clear excess of stars with close companions compared with what he would have expected from a random distribution. He concluded that some stars were physically associated—held together in one region of space by gravity. In particular, he identified many pairs of stars, which often differed greatly in apparent brightness, which must be at the same distance. Through his detective work, Mitchell provided the first clues to how stars are distributed in space.

Stellar Companions

As a first step in understanding the layout of the Milky Way galaxy, we consider the environment of stars. Most stars are not solitary wanderers in space. Four of the first six stars we encounter beyond the solar system have known companion stars. Surveys of this kind suggest that more than *two-thirds* of all stars are in a system of two or more stars orbiting each other. This is quite a surprise! Because the Sun is a single star, our immediate environment does not prepare us for the fact that most stars are not alone. Each pair of co-orbiting stars is called a **binary star.** Each set of more than two stars is a **multiple-star system.**

Why are these systems important for our understanding of the universe? First, binary stars allow us to measure stellar mass by the application of Kepler's and Newton's laws. This is the only way to measure a star's mass directly. Second, we can learn about stellar evolution by studying stars of different mass that are believed to have formed at the same time. We must be sure that our theories of star formation and evolution account for binaries and multiple systems. Third, in rare cases, two stars orbit close enough to transfer material between them. These systems act as fascinating laboratories for high-energy phenomena, as gas moves at high speeds and attains very high temperatures. Binary stars and multiple-star systems also let us find out if our own environment—the Sun and its planets—is typical. Do planets form just as readily around binary and multiple stars?

Types of Binary Systems

Some of the stars that are close together in the sky are at different distances and only *seem* to be aligned when viewed from Earth. Mitchell showed that there were too many pairs for all of them to be accounted for by such perceived alignments. However, his argument applies only in a general, or statistical, sense; it cannot be used to conclude that any *particular* system is a true binary star. We define a *binary system* as a pair of stars that we can prove to be bound by gravity by following their orbits. The first evidence for a physical association came nearly 40 years after Mitchell's work. In 1804, William Herschel noticed that Castor (the brightest star in the constellation Gemini) has a faint companion near it. By taking a series of photographs over several months, he demonstrated the two stars moved around each other in a binary orbit. Herschel's discovery was the first observation of gravitational orbits beyond the solar system—an important confirmation that the law of gravity is universal.

In a binary system, the brighter star is usually designated A and the fainter star B (for instance, Castor A and Castor B). The more massive star is usually called the *primary,* and the less massive one is called the *secondary.* Normally, the primary is also star A, the brighter star. Astronomers usually classify binaries according to how they are detected. We will discuss four types of detection.

A *visual binary* is a physical binary in which the orbiting pair can be resolved (seen separately) with a telescope. This type is the easiest to identify, and it is the type first discovered by Herschel. Some 65,000 have been studied. Images taken over a period of months or years can show clearly that two stars orbit each other. The motion of the pair on the plane of the sky is combined with the Doppler motion along the line of sight to give the stars' true three-dimensional orbit. However, in many other cases only one of the two binary stars is visible. For these binaries, detection of the second star is indirect. Imagine a giant and a dwarf star in orbit; these two stars may be so different in brightness that only one can be detected with a telescope. Or the two stars may appear so close together in the sky that a telescope cannot tell them apart. The separation of most visual binaries is 10 to 20 AU. At a separation of 20 AU, we can use the small angle equation to calculate the distance at which the binary would subtend an angle of 1 second of arc:

$$D = \frac{206{,}265 \times 20}{1} = 4 \times 10^6 \text{ AU} = \frac{4 \times 10^6}{2 \times 10^5} = 20 \text{ pc}$$

In other words, beyond about 20 pc, the typical binary would have too small a separation to be resolved with a ground-based telescope. Indirect methods of binary detection must be used.

In a *spectroscopic binary,* the individual stars are too close together in the sky to be resolved. Periodic Doppler shifts of the spectral lines reveal the orbital motion. In some cases, only the spectrum of one star can be detected. Figure 14–3a shows spectra taken on two dates, in which the absorption lines that mark the spectrum shift from red to blue and back again, indicating a periodic Doppler shift due to the orbit. We get more information when two sets of spectral lines are seen. Figure 14–3b shows a spectrum with a double set of absorption lines, with one star receding and one star approaching. As the orbit progresses and the stars are moving perpendicular to the line of sight, the spectral lines merge. About 1000 pairs have had their spectra measured in this way.

An *eclipsing binary* is a binary pair (generally unresolved) whose orbit is seen nearly edgewise. Because our line of sight lies in, or nearly in, the orbital plane, the stars alternately eclipse each other. Eclipses are detected by plotting **light curves,** or plots of brightness versus time, as shown in Figure 14–4. Depending on the relative brightness and size of the stars, the eclipse of the primary may produce a marked, short-term decrease in brightness. This may be followed by a decrease in brightness due to the eclipse of the secondary. The star then returns to its normal brightness, and the cycle repeats. As you can see from the figure, we can use the shape and timing of the eclipses to learn about the sizes of the two stars. Of course, we have to be lucky enough to see a system whose orbital plane is nearly parallel to the line of sight. Binary orbits occur at random angles in space, and the orientation is only favorable for eclipses in a small percentage of cases.

An *astrometric binary* likewise occurs when a bright star is moving around an unseen companion. Extremely careful measurements of its position, relative to background stars,

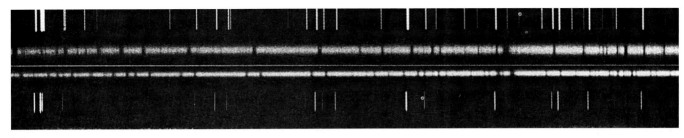

(a)

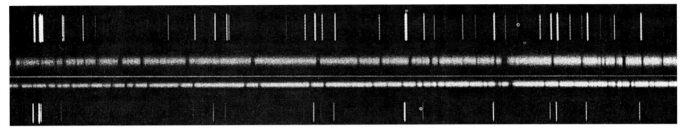

(b)

FIGURE 14-3
(a) A portion of the spectrum of a single-line spectroscopic binary, Alpha Geminorum. The bright vertical lines at top and bottom are reference emission lines produced in the spectrograph; the middle two bright spectra crossed by dark, vertical absorption lines are spectra of the star on two dates. Offsets of these lines to the red and then the blue show that the star is receding and then approaching because it is orbiting another star. (b) A portion of the spectrum of a double-line spectroscopic binary, Kappa Arietis. Two sets of absorption lines in the top stellar spectrum reveal two stars, one receding (or redshifted) and one approaching (or blueshifted). The lower stellar spectrum, in which orbital motions are perpendicular to the line of sight, shows lines merged, with no Doppler shift. (SOURCE: Lick Observatory.)

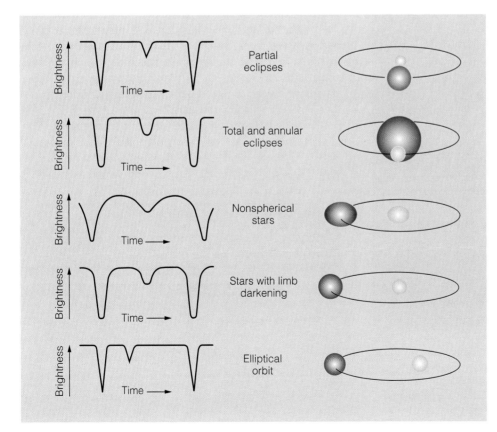

FIGURE 14-4
Various types of light curves (left) from eclipsing binaries reveal different geometric properties of the eclipsing stars and their orbits (right), even though the stars themselves cannot be resolved with a telescope. The stars are shown much larger than true size compared to the orbits in this diagram. Note that eclipses can only occur in a small percentage of cases in which the binary orbital plane is aligned nearly parallel to the line of sight.

STELLAR COMPANIONS 397

can reveal its motion, in turn revealing that it is a binary and has an unseen companion. We learn information about the unseen companion, as well as the visible star. One interesting facet of binary stars is that very different types of stars can be paired: massive and not so massive, giant and dwarf, red and blue. An imaginary example is shown in Figure 14–5. The variety offers challenges for both measurement and the imagination.

What can we learn from binary stars? The most important application of binary star studies is in determining star masses. We can determine the masses of the two stars from an analysis of the orbit (see the Science Toolbox on page 402). By contrast, the mass of an isolated single star can be determined only in terms of a model of stellar structure. Most eclipsing binaries also show Doppler shifts, which allow very detailed analysis of the motions, masses, and sizes of the stars. To take a simple example, suppose the Doppler shifts reveal that one star moves in a circular orbit at 100 km/s, and timing of the eclipse shows that it takes 10,000 s (about 3 h) to pass in front of the other star. Then the diameter of the primary star must be about 1 million km (distance = speed × time = $100 \times 10{,}000 = 10^6$). This type of observation is one of the most accurate checks on theories of stellar structure.

How Many Stars Are Binary or Multiple?

The question of pinpointing the number of binary and multiple star systems is a challenge to observers. The nearest stars are easiest to observe but represent too small a statistical sample to be reliable. At greater distances there are more stars, but faint companions might not be detected. Spectroscopic binary statistics are biased toward pairs with small separation distances, because according to Kepler's laws, these have the fastest velocities and greatest Doppler shifts, thus being the most likely to be discovered. Visual binary statistics are biased toward wide separation distances, which make the two stars easier to resolve. All these biases, which tend to make the data unrepresentative of the whole population, are called *selection effects*.

Table 14–1 lists two estimates of the percentage of systems ranging from single to sextuple. By the time we reach six-member systems, definitions of multiple systems become hazy. It is unclear whether close groupings like the Trapezium in the Orion Nebula should be counted as multiple systems. The 1982 Yale Catalog of 9096 prominent stars lists multiple systems ranging as high as one system with 17 members. Such systems would present an interesting spectacle (Figure 14–6). The variation in the percentages in each line of Table 14–1 illustrates our uncertainty about the frequency of binaries. In fact, the percentages are really lower limits; it is likely that many of the seemingly single stars have companions too small and dim to detect. Astronomers estimate that most stars have at least one companion. If true, this would imply that our Sun is slightly unusual!

This conclusion changes our view of stars entirely. The idea of the night sky full of single, separate stars is wrong.

FIGURE 14–5
Binary stars exhibit a wide variety of pairings. In this imaginary view, a white dwarf orbits a red giant that is shedding mass into a disk of gas and dust around itself. The white dwarf is a pinpoint of light relative to the swollen giant in the background. (SOURCE: Painting by WKH.)

TABLE 14-1
Incidence of Multiple Stars (Percentage of Systems with n Members)

n	25 SYSTEMS WITHIN 4 PC	AVERAGE OF PUBLISHED ESTIMATES*
1	48%	41%
2	36%	41%
3	12%	14%
4	4%	3%
5	—	1%
6	—	—

*Note: Data from various sources compiled by Batten and Abt.

FIGURE 14-6
A hypothetical view within the sextuple star system Castor, the brightest star in the constellation Gemini. Castor appears to the naked eye as a single star, but it is actually three pairs of close binaries all orbiting each other. Here we are on an imaginary planet orbiting the faintest pair, two low-mass red stars of about 0.6 $M_\odot$ each. The two other pairs are more than 1000 AU away (left). All four of the distant stars are whitish A stars about 50 percent more massive than the Sun and averaging 10 times more luminous. Although astronomers are searching near other stars for planets like that depicted in the foreground, none has been positively identified. (SOURCE: Painting by Ron Miller.)

We must understand the origin and evolution of systems of two, three, four, and more stars in order to claim any understanding of stars in general. As we will learn in the next section, binaries go through some interesting detours along the road of stellar evolution.

Mass Transfer in Binaries

Mass transfer is the movement of material from one star to another. Many binaries orbit slowly, with a large distance between the two stars. There is no interaction between stars that are at "arm's length." However, gravity holds some stars in a tighter embrace. As we would expect from Kepler's laws, close binaries orbit more rapidly. This tight configuration allows material to be transferred from one star to another, with interesting consequences.

How do astronomers describe mass transfer? We can picture a system of two co-orbiting stars as containing an imaginary **Roche surface**, or *lobe*—the boundary in three dimensions of the gas in a binary system. The cross section of the system is a figure eight, with one lobe around each star, as shown in Figure 14-7a. Slow-moving material inside either lobe orbits around the star in that lobe. Material that moves beyond either lobe is not gravitationally bound to either star. Mass can move from one star to the other through the point of contact between the Roche surfaces. When either star evolves into a red giant state, it expands until it fills its lobe, as in Figure 14-7b. Its outer layers then assume the teardrop shape of the lobe, and the Roche surface becomes the real surface of the star.

Now we can use the information from the previous two chapters to predict what will happen when two stars of different masses start life together as a binary system. Evolution can take a single binary through three stages. In the first stage, neither star fills its Roche surface. In the second stage, the more massive star leaves the main sequence and swells into a giant as it evolves. The giant may fill its Roche surface. Any further tendency to expand causes matter to be shed, mostly through the point common to the two lobes, as

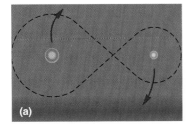 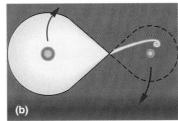 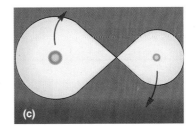

FIGURE 14-7
The evolution of a binary pair is related to the configuration of the Roche surfaces, or lobes, and the way that matter fills these regions of space. **(a)** In the first class of systems, neither star fills its lobe (dashed line). **(b)** After the larger star expands to become a giant, it may fill its lobe, taking on a teardrop shape and perhaps ejecting some mass through the tip, which interacts with the second star. **(c)** In the third class of systems, both stars fill their lobes. Mass usually travels from the more massive to the less massive star.

SCIENCE TOOLBOX

Binaries and Stellar Mass

With the typical separations of visual binaries, it takes many years of observation to measure an accurate orbit. It is worth the effort, however, because binaries give us the most direct way to measure stellar mass. Physical binaries also provide good laboratories for studying stellar evolution. Recall that mass is the single most important property of a star. The mass determines the structure, the rate of evolution, and the destiny of a star.

We learned about Kepler's laws of planetary motion in Chapter 3. Kepler's third law explains the relationship between the period of an orbit (P) and the semimajor axis of the orbit, or mean distance of the planet from the Sun (a). In mathematical terms, we write this as $P^2 \propto a^3$. From this equation, it follows that planets farther away from the Sun travel on slower orbits.

Newton came up with a powerful and generalized form of Kepler's law that applies to *any* two bodies in orbit around each other. When any two objects in space orbit each other, the period of revolution (the time it takes to complete an orbit) increases as the distance between them increases and as the sum of the masses of the two objects decreases. If M_A and M_B are the masses of the two stars in solar masses, and if we express the period P in years and the semimajor axis a in astronomical units, the equation is

$$P^2 = \frac{a^3}{M_A + M_B}$$

You can see that if M_A is one solar mass like the Sun and M_B is much smaller than M_A (as would be the case for any planet in our solar system), this equation reduces to $P^2 = a^3$, which is the familiar form of Kepler's law in the solar system.

We can rearrange this equation to get

$$M_A + M_B = \frac{a^3}{P^2}$$

This is a form we can use to calculate star masses. As we have seen, binary stars are quite common. For many of these pairs, we can measure both the period of revolution and the distance between the two stars. Thus, we can calculate the sum of the masses, represented by $M_A + M_B$, where A and B designate the two stars. But we want to know each individual mass, not the sum of the two. Newton showed that in a system of orbiting bodies, each body orbits an imaginary point called the *center of mass*, and that by measuring the distance of each star from the center of mass, we can measure the ratio of the masses. If we have both the sum of the masses and the ratio of the masses, we can get each individual mass.

Let us consider a couple of examples. The double star Sirius consists of the bright star Sirius A (which has the greatest apparent brightness of any star in the night sky) and the faint star Sirius B. The separation of the two stars is 20 AU, and the orbit takes 51 years. We can use the equation above to calculate that $M_A + M_B = 20^3/51^2 = 3.1$ times the mass of the Sun. The ratio of the distances of each star from the center of mass indicates that A is twice as massive as B, so $M_A/M_B = 2$. From these facts, we can deduce that Sirius A has a mass of 2 $M_\odot$ and Sirius B has a mass of 1 $M_\odot$.

The bright star Alpha Centauri orbits a fainter companion with a period of 80 years and a semimajor axis of 23 AU. We can calculate that $M_A + M_B = 23^3/80^2 = 1.9$ times the mass of the Sun. In this pair, the stars are at nearly equal distances from the center of mass, and $M_A/M_B = 1.2$. We deduce that $M_A = 1.1$ $M_\odot$ and $M_B = 0.9$ $M_\odot$. The case of Alpha Centauri is complicated by the presence of a third star—a very low-mass star of 0.1 $M_\odot$ at a distance of about 10,000 AU from the more massive pair. But we can easily use Newton's law of gravity to show that the influence of this third star is negligible. The gravity force is proportional to the masses and inversely proportional to the square of the distance between the masses. We know that the third star is 10 times less massive than A or B, and $10,000/23 = 430$ times farther away from them than they are from each other. Therefore, the force of the third star on A or B is $430 \times 10 = 4300$ times smaller than the force that keeps A and B orbiting each other. We can safely ignore the complication of the third star.

---

shown in Figure 14–7b. If the giant were a single star, it would be spherical, and what little mass it did lose would stream off in all directions. But pressure within the giant and the companion's gravity force the bloated giant to lose gas through its distorted, pointed tip. Like sand in an hourglass, this gas enters the lobe of the second star. Most of the gas spirals around the small star and crashes onto it, making it gain mass as the large star loses mass.

Mass drives stellar evolution, so the evolution of *both* stars in a binary system will be altered, compared to their destinies had they been single. Later, the second star evolves to the giant state, partly through its normal evolution, but partly because of the added mass. In the third stage, both stars fill their lobes, producing what is called a *contact binary*. Notice the joined teardrop regions of space in Figure 14–7c. The two stars actually touch as they orbit each other as a single unit! If two stars start very far apart, however, the expansion as a giant may not fill either Roche surface, and the binary may never evolve beyond the first stage.

Nova and Supernova

Mass transfer between the members of a binary pair can trigger even more spectacular effects: a nova or a supernova. Let us consider the case of a binary in which the more massive star has filled its Roche surface and the smaller star is a white dwarf of nearly 1.4 $M_\odot$. If the larger star dumps hydrogen onto the surface of a white dwarf, the hydrogen will be compressed by the intense gravity of the white dwarf. The gas is heated as it is compressed, and the hydrogen may ignite in nuclear reactions that blow excess gas outward. This type of explosion is called a **nova** (plural: *novae*).

The nova phenomenon can occur in cycles. Hydrogen may accumulate on a dwarf until nova explosions occur anywhere from 100 to 10,000 years apart. Since the individual explosions blow off only a small fraction of the star's mass, leaving the rest intact, the process can start over. Among novae that are closest to Earth, the cloud of expanding debris can sometimes be seen in telescopes a few years after the explosion.

The word *nova,* from the Latin root for *new,* was the term used for all "new stars" that appeared in the sky in past centuries. In ancient Chinese records, the novae were called "guest stars." In this century, astronomers discovered from their rate of light variation and other properties that there were two types of "new stars"—the novae and a much more energetic type called a supernova. These exploding stars were important in our discovery that the heavens are not immutable, as once thought, but are actually evolving, even as we watch.

What are the differences between a nova and a supernova? A nova involves the transfer of mass onto a white dwarf, resulting in a rapid ignition of nuclear fusion that brightens the star for a while. A nova can flare up repeatedly, although there is not a fixed interval between outbursts. Of the 200 billion stars in our galaxy, 30 to 50 explode as novae each year. The type of supernova that results from the death of a single massive star was discussed in the last chapter. However, a supernova can also occur in a binary system, when the mass transfer from a giant or supergiant pushes a white dwarf over the Chandrasekhar limit (also discussed in Chapter 13). The white dwarf collapses to form a neutron star or a black hole, blowing off a fraction of the excess mass in a titanic explosion. A supernova marks the brilliant death of a star. For a short time, a supernova releases enough light to rival the light of the entire Milky Way. These spectacular explosions are rare, with a supernova in the Milky Way occurring only once every 40 or 50 years. A supernova explosion is therefore about 1000 times rarer than a nova event.

Exotic Binary Systems

As you have seen, close binary systems are stellar "laboratories" that provide great opportunities for studying stellar evolution. Two stars in tandem may end their lives in quite a different way than if they had evolved in isolation. Each system

FIGURE 14-8
A hypothetical view of a contact binary system, as seen from a distant planet. The planet is imagined to circle a third star in the system, another red giant off to the right. (SOURCE: Painting by WKH.)

has a different story to tell. In addition, mass transfer can lead to some of the most spectacular high-energy phenomena that we know of in astronomy. Let us look at some exotic types of binary systems.

A contact binary would make a strange "sun" in the sky of some imaginary world (Figure 14-8). The nearest and most famous example is W Ursae Majoris in the constellation Ursa Major (better known as the Big Dipper). Contact binaries consist of stars of rather similar mass, both filling their Roche surfaces. They have total masses ranging from 0.8 to 5 $M_\odot$, and because they are so close together, their common revolution periods are very short, less than 1.5 days. Many of the shortest-period binaries are discovered by their X-ray emission. Material that falls from one star onto a compact companion accelerates and heats to millions of Kelvin, at which point it will give off radiation in the form of X rays. A binary called 4U 1820-30 with a period of only 11 minutes was discovered in 1986 by the European orbiting X-ray observatory EXOSAT. The binary is apparently a neutron star orbiting a white dwarf, about 6000 pc away. Theoretically, mass loss could produce contact binaries with periods as short as 2 minutes! Somewhere in our galaxy might be a planet in whose sky is a glowing figure eight doing cartwheels like some bizarre advertising gimmick.

Recall from Chapter 13 that pulsars represent high-density neutron material formed after the death of a massive star. In the 1980s, radio astronomers discovered a class of pulsars with a phenomenal spin rate. The fastest-spinning pulsar found so far was discovered in 1982, and it flashes with a 0.0016-s period. Among human technologies, not even the fastest turbine engine can spin at this rate. Yet stellar collapse has produced a mountain-sized ball of nuclear matter that spins 642 times every second! Nearly 100 *millisecond pulsars* are now known. Many millisecond pulsars have been found in clusters of stars that are 5 to 10 billion years old. This discovery is puzzling, because pulsars should spin down in only 10 million years. In other words, the rapid rotation of millisecond pulsars must have been caused long after their birth by some more recent event, such as a binary encounter.

Astronomers used to think that the evolution of neutron stars was very simple: They form as a result of a supernova explosion, they cool, and then they spin down. Now it is clear that pulsars can evolve in binary systems. A neutron star in orbit around a normal star will draw material from the normal star onto it. As the material spirals in, conservation of angular momentum dictates that the neutron star will "spin up" into a millisecond pulsar. In some cases, the binary system is disrupted by the close passage of another star, leaving an isolated millisecond pulsar.

The existence of millisecond pulsars is a confirmation that some of the most dramatic phases of stellar evolution result from binary systems. Enormous amounts of high energy can occur when a binary system contains two very massive stars; one possible outcome is twin supernova explosions leaving a binary neutron star. In the mid-1970s, astronomers discovered just such a class of objects called *X-ray bursters,* which they now believe to be binary neutron stars. Here is how we think they produce their X rays. As matter is torn from a normal main-sequence star onto a neutron star, the gravitational compression heats up the gas enormously. The pressure is so high that hydrogen is converted to helium by the fusion process and emits a steady level of X rays. The neutron star is then surrounded by a blanket of helium. When the helium layer reaches a certain thickness, its temperature is sufficient for the fusion of helium. However, in this case the thermonuclear reaction is explosive, and a strong pulse of X rays is produced. X-ray bursters can generate thousands of times more energy than the Sun in a pulse that lasts only a few seconds. An analogous process of explosive nuclear reactions on a white dwarf produces a nova.

The *gamma-ray bursters* are even more spectacular. Explaining these objects is one of the greatest challenges in high-energy astrophysics. In the 1970s, satellites designed to monitor Russian compliance with the Nuclear Test Ban treaty discovered gamma-ray flashes coming from the sky! When these sources are quiet, they are difficult to detect in any region of the electromagnetic spectrum, but several of them are at enormous distances from Earth—hundreds of millions of parsecs. Astronomers speculate that these bursts are the death spasms of two neutron stars spiraling in toward each other and merging. The fantastic release of gravitational energy creates a fireball that glows in gamma rays and rapidly fades to longer-wavelength radiation.

How Do Binaries and Multiple Systems Form?

Now that we have seen that our own single star may be atypical, we are left with a series of questions. What determines whether a single star, or one with companions, forms? What determines the distribution into systems of two, three, or more stars? What determines whether a companion to a star is a planet, a brown dwarf, or another star? There may be several processes at work to form binary and multiple systems: fission, capture, and subfragmentation.

Astronomers speculate that fission may create binary systems. *Fission* is a process in which a rapidly spinning protostar can split in two, leaving a very close pair or a contact binary. As evidence in favor of fission, close binaries have roughly equal masses, while widely separated binaries have the mass ratios expected from random pairings of stars. On the other hand, certain mechanisms might cause widely spaced pairs to evolve into contact binaries, and once they are in close proximity, stars might equalize their mass by the transfer of material. Most astronomers think that fission is not a plausible cause for most binaries.

Another possible explanation for binary systems is the *capture theory:* Binaries can form when one star enters the sphere of gravitational influence of another star. Unfortunately, the chance of encounters among random stars in the sky is far too slight to explain the observed numbers of binaries. Furthermore, randomly paired stars would be of widely different ages, but this is not observed among binaries. Where could stars of similar ages interact in a closely packed group? A newly formed star cluster is a very suitable environment; as many as half of the protostars in a cluster undergo collisions or close encounters. Many wide pairs probably formed in this way. Once two stars are bound together in an orbit, their combined gravity may attract other stars. Capture theory can probably explain multiple systems of three or more stars.

In a third category of theoretical process known as *subfragmentation,* a contracting cloud shrinks because of its own gravity, but instead of forming a flat disk with a central star, its mass distribution or angular momentum distribution may make it split into two or more clouds that orbit each other. These smaller clouds then collapse into separate stars. Since the smaller clouds were part of a large cloud that was gravitationally bound, the stars that form out of the smaller clouds are bound to each other also.

Adding to the puzzle of how binaries and multiples form is the question of how likely it is that they will harbor planets. We cannot yet answer this important question because of the difficulty of solving the equations of gravity for three or more objects. Recall from Chapter 9 that the planets of our solar system formed by a condensation of dust and ice particles, followed by accumulation during their collisions; the surrounding gas medium eventually blew away. We can think of planets as debris left over from star formation. There is no reason why this debris should not exist around binary or multiple stars. However, dynamical studies show that planet orbits may be unstable, and planets may be ejected from such systems. At the moment, astronomers are playing it safe by only looking for planets around single Sun-like stars. However, since most stars are not single, that leaves plenty of unexplored sites where planets might exist.

Each of the three theoretical processes mentioned may produce binaries of a certain type. A complication in sorting out binaries of different types is that orbits of binary and multiple stars evolve through gravitational influences. Mass transfer in close pairs can alter orbits. Widely spaced pairs formed inside larger star clusters can evolve into closely spaced pairs as the clusters break apart. In one theoretical study of some 800 imaginary triple-star systems, about 97 percent were found to be gravitationally unstable, eventually kicking out one star and becoming binary systems. Thus, the observed statistics of binaries and multiples may not reflect their original characteristics.

The Environment of Stars

We have seen that stars are not isolated. Binary and multiple systems share gas and interact in ways that can produce spectacular effects. Now we continue our inventory of the Milky Way galaxy by considering the material found in the environment of stars. Some of this material is close to a star and is lit up by a star's energy. Some of it is thinly distributed through the gulfs of space between stars.

Stars are surrounded by a thin but chaotic medium of gas, dust, and radiation. They form from this thin material, interact with it, recycle it, and expel it to form new interstellar material. Each individual cloud of gas and dust is called a **nebula**. A nebula is material that is concentrated by gravity and lit up by a nearby star. French astronomer Charles Messier published a catalog of the brightest and most vivid examples in 1781. They are thus known by their Messier numbers, or M numbers. Messier was a comet hunter, and he wanted a list of the nebulous objects in the sky that did *not* move with respect to the stars so he would know which objects to avoid in his searches. The familiar Orion Nebula, for example, is known as M 42 (Figure 14–9). Others have

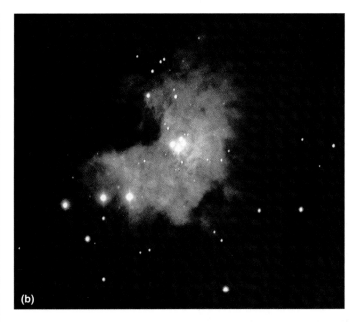

FIGURE 14-9
The Orion Nebula is the core of a star-forming complex about 460 pc away and about 5 pc across. **(a)** The larger view is a specially processed optical image showing the faint outer structure and the red colors due to Hα emission. This larger region is visible as a fuzzy object by the naked eye, near the belt of the hunter in the constellation Orion. **(b)** A close-up view shows a false-color composite of the central core, made from an image taken at near infrared wavelengths. The embedded young star cluster near the center of the nebula can be viewed only with long wavelengths that penetrate the dust. (SOURCE: Copyright Anglo-Australian Telescope Board, courtesy D. Malin.)

TABLE 14-2
Gas Densities in Different Environments

LOCATION	TYPICAL DENSITY (kg/m³)*	NUMBER OF PARTICLES PER m³	TYPICAL DISTANCE BETWEEN PARTICLES
Air at sea level	1.2	10^{25}	1 nm
Circumstellar cocoon nebula	10^{-5}	10^{22}	50 nm
"Hard vacuum" in terrestrial laboratory	10^{-9}	10^{18}	1 µm
Orion Nebula	10^{-18}	10^{9}	0.1 cm
Typical interplanetary space	10^{-20}	10^{7}	0.5 cm
Typical interstellar space	10^{-21}	10^{6}	1 cm
Interstellar space near edge of galaxy	10^{-25}	10^{2}	20 cm
Typical intergalactic space*	10^{-28}	10^{-1}	2 m

*Average density of observable matter in the whole universe is estimated to be about 3×10^{-28} kg/m³ but many astronomers believe the average density may be somewhat higher (3×10^{-27} kg/m³?) due to nonluminous unseen matter.

numbers from the more recent New General Catalog (NGC) of William Herschel and from the Index Catalog (IC). By tradition, astronomers have named most bright nebulae according to their appearance in small telescopes; examples are the Crab Nebula (M 1), the Dumbbell Nebula, and the Ring Nebula.

We sometimes casually say that "space is a vacuum," but this is not quite true. While space is a better vacuum than can be achieved in a laboratory (Table 14-2), the space between stars is filled with diffuse gas and dust. The material between stars is called the **interstellar medium**. What significance can this thin material have for us? For one thing, these vast clouds of dust and gas are landmarks in the night sky; some are twisted into beautiful wispy forms, some are dark, and some glow with different colors. For another thing, our solar system, planet Earth, and human beings are formed from atoms that were once part of the interstellar gas and dust. More provocatively, recent studies have shown that the interstellar medium contains complex organic molecules. These discoveries demonstrate the universal aspects of chemistry that make organic molecules common both in living things on Earth and in inanimate objects such as comets and interstellar clouds. Lastly, the interstellar medium dims and reddens starlight.

Composition of the Interstellar Medium

The interstellar medium includes gas (i.e., atoms and molecules), microscopic dust grains, and possibly larger objects. Atoms of interstellar gas were discovered in 1904, when German astronomer Johannes Hartmann detected their absorption lines. While studying the spectra of a binary star, he accidentally discovered absorption lines caused by ionized interstellar calcium atoms. Absorption lines caused by other elements, like sodium, were soon discovered. Since these lines had different Doppler shifts than the stars in whose spectra they appear, the atoms that cause the absorption must lie in the space between stars.

Further studies have convinced astronomers that, although the most prominent absorption comes from atoms such as calcium and sodium, the most common *interstellar gas* is the ubiquitous hydrogen. Like the Sun and stars, interstellar gas is about ¾ hydrogen and nearly ¼ helium by mass.

The most important tracer of interstellar gas is a spectral line that is emitted at radio wavelengths. In 1944, Dutch astronomer H. C. van de Hulst predicted that cold hydrogen gas would have a spectral feature at a wavelength of 21 cm, caused by a change in the spin of hydrogen atoms' electrons. A low-energy radio photon is emitted when the atom "flips" between two nearly equal quantum states. Harvard astronomers detected the predicted emission in 1951 using radio equipment. The **21-cm emission line** of atomic hydrogen is extremely important in astronomy for several reasons. First, the widespread detection of this line confirms that hydrogen is the main constituent of interstellar gas. Second, the long wavelength and low energy of the spectral line is a sign that the hydrogen is at a very low temperature. (Chapter 10 discussed why cold material emits long wavelength spectral features.) The 21-cm emission comes from gas as cold as 10 or 20 K. Third, the long wavelength of this radiation means that it can penetrate much greater distances through the interstellar gas and dust than ordinary light.

By 1940, astronomers at Mt. Wilson Observatory in California had detected another component of interstellar material: *interstellar molecules.* In the cold and low-density environment of an interstellar cloud, molecules have many low-energy transitions that create spectral features in the infrared or radio parts of the spectrum. Astronomers inherited the infrared and radio technology that had been developed during World War II, and after the war they used it to search for molecules in space. New detectors on large telescopes sparked an explosion of interstellar discovery during the late 1960s and 1970s. (Figure 13-2 showed the variety of molecular lines that are observed in the star-forming heart of the Orion Nebula.)

More than 100 different interstellar molecules have been cataloged. The most important in astrophysical processes are molecular hydrogen (H_2) and carbon monoxide (CO). Four atoms recur in these large molecules: carbon, hydrogen, oxygen, and nitrogen—the "building blocks" of life! It is intriguing that two of the detected molecules, methylamine (CH_3NH_2) and formic acid (HCOOH), can react to form the amino acid glycine (NH_2CH_2COOH). These large molecules can join to form the huge protein molecules that occur in living cells. Two exciting questions have been generated by research on interstellar molecules. First, does the existence of complex, carbon-rich molecules in space suggest that life could have originated elsewhere in space? The answer may be yes, and we will discuss this possibility in more detail in the last chapter of the book. Second, how do these molecules form? They are found both in ordinary interstellar gas, where collisions between atoms are extremely rare, and in denser regions, such as the clouds in which stars form.

Interstellar grains, or dust grains, are even bigger than interstellar molecules. Typical grains are about the size of smoke particles in our air. Grains account for about 1 percent of the total interstellar mass, although there is only 1 for every 10^{12} hydrogen atoms or molecules. These tiny particles range from $1/100$ to twice the wavelength of visible light, and they are often elongated. Grain composition has long been debated. In 1974, Sri Lankan astrophysicist Chandra Wickramasinghe proposed that grains are made of a form of carbon, like soot, condensed in cooling gas blown off by carbon-rich giant stars. Many different carbon compounds, silicates, and ices have in fact been identified by spectroscopy, especially for grains in star-forming regions. The grains may also contain iron, since elongated grains align parallel to each other, even though they are widely separated in space. Astronomers believe they are aligned by the weak magnetic field that permeates interstellar space, just as iron filings are aligned by a bar magnet.

With a composition of silicates, metals, and ices, interstellar grains resemble the dust in our primordial solar system. Intriguingly, reactions initiated by ultraviolet light in these materials can create complex organic molecules on grain surfaces. These microscopic particles are tiny laboratories that play an important, but still poorly understood, role in interstellar chemistry. Even though the tiny dust grains are rare compared to hydrogen atoms, we will see in the next section that they have a major effect on starlight.

Interstellar Medium and Starlight

The components of the interstellar medium—atoms, molecules, and dust grains—directly affect the light emitted by stars. When light from a distant star passes through the material between stars, the interaction changes the light's properties, such as its intensity and color. The physical laws that describe these changes are complex, but we can present a simplified version of the changes using several basic principles discussed in Chapters 4 and 10. Recall that when electromagnetic radiation of any kind (ultraviolet light, visible light, infrared, radio waves, and so on) interacts with individual atoms or molecules, radiation can only be absorbed or emitted according to the difference between fixed energy levels in the atom or molecule. The result is a set of sharp spectral features. When electromagnetic radiation interacts with much larger particles such as dust grains, the type of interaction depends on the chemical composition of the particles and their sizes relative to the wavelength of the light. Also, the appearance of the light from a star, and the appearance of the particles it illuminates, may depend on the direction from which the observer looks.

Let us recall the sizes of the different particles involved. A hydrogen atom has a radius of 5×10^{-11} m; typical interstellar molecules have sizes 10 to 20 times larger, ranging up to 10^{-9} m or 1 nm. Interstellar grains are tiny, but they are made of very large numbers of atoms or molecules. Grains range in size from 5×10^{-9} m to 10^{-6} m, or a wide range of 5 to 1000 nm. For comparison, the average wavelength of visible light is 500 nm.

Radiation interacts differently with each of the three types of interstellar particles: atoms, molecules, and dust grains. In reality, the interstellar medium is always a mixture of gas and dust, but it is easier to understand the effects if we imagine separate interactions of light with atoms, molecules, and dust grains. Figure 14–10 shows the typical situation when the interstellar medium is concentrated in a cloud. We show a cloud, or nebula, being illuminated by the light from a *hot* star because the Stefan-Boltzmann law tells us that hot stars emit far more energetic (short wavelength) photons than cool stars. A nebula is much more likely to be lit up by a hot star than a cool star. Luminous, hot stars are also young, and they are therefore less likely to have drifted away from the gaseous and dust-filled region of their birth.

Let us look at how starlight interacts with gas atoms in the interstellar medium. As light with a broad range of wavelengths enters the nebula, photons with certain wavelengths will have just the right energy to *excite* the gas, or knock electrons from lower to higher energy levels. Photons with a short wavelength may even have enough energy to *ionize* the gas, or knock electrons clear out of the atoms. (Review the material in Chapter 10 if you are not clear about this terminology.) Each time a photon excites or ionizes an atom, that photon is absorbed. When the atom re-emits the photon, it may do so in any direction. Thus, photons with wavelengths corresponding to the electron energy transitions in atoms are redirected, while the much greater number of photons that do not correspond to an electron transition pass through the gas undisturbed. The effect is to subtract energy from the light beam at specific wavelengths. Looking at the star through the nebula, we would see absorption lines created by the interstellar material. Study the perspective of observer A in Figure 14–10.

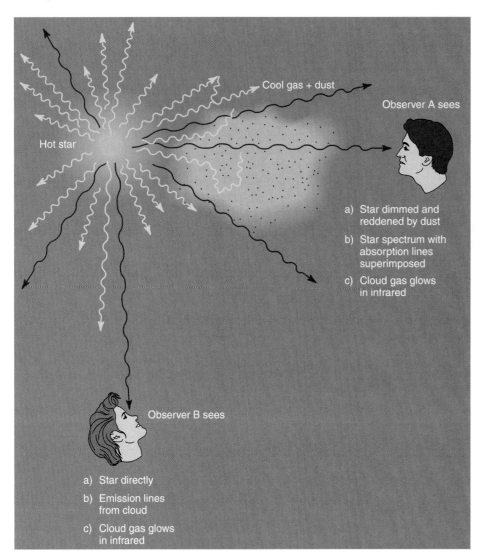

FIGURE 14-10
The effects of gas and dust in the interstellar medium on starlight. A hot star emits optical/ultraviolet thermal radiation in all directions. The starlight heats up the gas and dust particles in a cloud, which emits much cooler infrared thermal radiation. Observer A views the star through the cloud and sees the star spectrum with narrow absorption lines superimposed, caused by atoms in the cloud gaining specific amounts of energy. He also sees the star to be dimmed and reddened when blue light is scattered as it passes through the cloud. Observer B is at the same distance, but views the star and the cloud from the side. When this second observer looks at the cloud, he sees narrow emission lines caused by atoms in the cloud losing specific amounts of energy. Observer B sees the star directly without dimming and reddening by the dusty cloud.

As the electrons cascade back down through the energy levels of the atoms, they create emission lines. The photons in these emission lines leave the nebula in all directions, so that observer B off to one side would see the nebula glowing in the various colors corresponding to the strongest spectral lines. We are seeing an important astronomical example of Kirchoff's laws of radiation; the laboratory setup that was shown in Figure 11–3 also applies to stars and gas clouds in space! The colors of a nebula are hard to see, even with a large telescope, because the light's intensity is low and the eye's color sensitivity is poor at low light levels. (This is why a moonlit scene looks less colorful than in daylight.) However, sensitive films and electronic detectors can record the colors accurately. Some spectral lines are much more effective at removing the energy from a nebula than others. Since hydrogen is the most abundant gas, and since the red Hα emission line is one of its strongest transitions, many clouds of excited gas glow with a beautiful deep-red color (see the lower part of Figure 14–11). Another important transition of oxygen can give a nebula a blue-green tinge (for a striking example, see Figure 14–12).

Starlight can also interact with molecules in a nebula. Molecules are two or more atoms bound together by weak electrical forces. Like atoms, molecules have characteristic spectra that are related to the structure of their internal energy levels. Molecular spectra are typically more complicated than those of single atoms because the rotations and vibrations of molecules add together many closely spaced energy states. When observed with low spectral resolution, this can give the appearance of an absorption or emission *band* rather than a single narrow line (see Chapter 10 for a review). Tiny amounts of energy can cause vigorous rotation in molecules, which explains why many molecular absorption and emission features appear at infrared or submillimeter wavelengths. We therefore expect molecular spectral features to be observable even in very cold, low-density interstellar environments.

Starlight also causes a nebula to glow with thermal radiation. When a gas absorbs photons, it is gaining energy. As electrons are freed from atoms, they collide with other electrons and atoms. The effect of these collisions is to increase the speed of the average particle. As we saw in Chapter 4, this means that the temperature of the gas increases. Starlight can therefore heat up a remote gas cloud. The balance between heating by absorption and cooling by re-emission governs the temperatures of the gas and the dust. A cloud very close to a star is heated to a temperature of several thousand Kelvin, with thermal emission that peaks at

FIGURE 14-11
The Trifid Nebula, about 1600 pc away in the constellation Sagittarius, beautifully combines red and blue. The hot gas within the cloud glows with the red light of Hα emission. It is roughly 5 pc across. Dark lanes of colder dust superimposed in front create a flowerlike pattern. Associated cool clouds have a soft blue coloration from the scattering of starlight. The bright star in its center is a very luminous, hot O star with a blue-white light; it may be the star that excites the nebula to glow. (SOURCE: Copyright Association of Universities for Research in Astronomy/NOAO.)

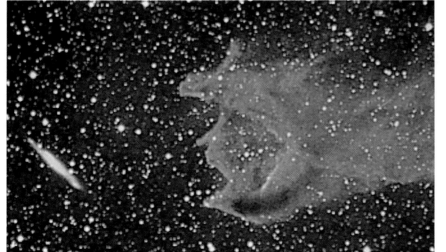

FIGURE 14-12
The dense molecular cloud (also called a cometary globule) CG4 is about 400 pc away and 2.5 pc across. The subtle colors are caused by emission from hot gas and by the scattering and reddening of starlight. The green tinge is due to a pair of spectral lines of ionized oxygen. The edge-on galaxy, which CG4 appears ready to devour, is actually millions of light years away, far beyond the Milky Way.
(SOURCE: Copyright Anglo-Australian Telescope Board, courtesy D. Malin.)

visible wavelengths or in the near infrared. A cloud far from a star is much colder—a temperature of tens or hundreds Kelvin—and the thermal emission peaks at far infrared or radio wavelengths. Thermal radiation travels in all directions and is seen by both observer A and observer B.

Finally, starlight can interact with interstellar dust grains. Dust grains are much larger than atoms and molecules, and their interactions with photons are quite different. The larger particles affect colors over a much broader range of wavelengths than individual spectral lines or bands. There are two important observational effects. Grains absorb incoming optical and ultraviolet radiation and re-emit it in the far infrared, with a thermal spectrum that indicates the cold temperature of the grains. The result is a general dimming of starlight at all wavelengths, called **interstellar extinction**. Since the dust grains are constantly colliding with the individual atoms and molecules, all these different-sized particles are in equilibrium, and they all have the same temperature.

In addition, grains can *scatter* radiation—absorb an incoming photon and re-emit it in a different direction. Scattering increases in efficiency for shorter wavelengths of visible light. As a result, red light (longer wavelengths) passes through clouds of dust, whereas blue light (shorter wavelengths) is scattered out to the side of the beam (see Figure 14–10). Thus, observer A looks through the dust cloud at a distant star and sees most of its red light but not much of its blue light. Observer B, who is at the same distance from the star, sees it directly with no loss of blue light. In this way, interstellar dust makes distant stars look redder than they really are—an effect called **interstellar reddening**. You can learn about the quantitative effects of dust grains in the Science Toolbox on page 413.

Many dust grains, thinly distributed throughout the interstellar gas, produce a general haze, or "interstellar smog." By contrast, when dust grains are concentrated in distinct clouds, their effects can be much more dramatic. Figure 14–13 shows two examples where dust almost extinguishes the background light. An observer who looks at the dust cloud from the side will see the blue light scattered out of the beam, however, so that a nebula illuminated in this way will have a bluish color. The light from a reflecting nebula arises from scattering by dust rather than by atoms or molecules. The blue color of the nebulosity is caused partly by dust scattering and partly by the fact that the light being reflected comes from a hot, blue star (see the example in Figure 14–14, or the dim nebulosity in the upper part of Figure 14–11).

You do not have to visualize remote regions of space to understand the phenomena of extinction and reddening. Just step outside! Let us ask these basic questions: Why is the sky blue? And why is the setting Sun red? Earth's atmosphere is full of gas molecules and tiny dust particles. The molecules and many of the dust particles are much smaller than the wavelength of light, so scattering is more efficient for short wavelengths of light. Light from the Sun must pass through these particles before it can reach our eyes. When the Sun is high in the sky, we look through the minimum amount of gas and dust, as shown in Figure 14–15a. Thus, the reddening is minimal, and the Sun is perceived as yellow. At sunset, as shown in Figure 14–15b, the sunlight passes through much more gas and dust. When the Sun is low in the sky, much of the blue light is scattered, thereby strongly reddens the Sun. Where does the blue light go? Air molecules and dust particles scatter the blue light many times until it reaches our eyes with nearly equal intensity from every direction. We therefore see the sky as blue; see Figure 14–15c. We know that the blue color is not a property of the air itself, since air in a jar or a room is transparent and colorless. In a city with bad smog, you can see that the Sun is dimmed and

FIGURE 14–13
(a) An example of interstellar reddening. A black dust cloud is silhouetted against a densely crowded field of stars in the constellation Sagittarius. Close examination of the dust cloud reveals that dimmed stars seen through it are strongly reddened. Near the cloud is the star cluster NGC 6520. **(b)** The effect of dust obscuration is dramatic in this picture of IC 4628. The clumps of dense gas and dust almost completely extinguish light from background stars, making those regions appear black.

(SOURCE: Copyright Anglo-Australian Telescope Board, courtesy D. Malin.)

FIGURE 14-14
The beautiful nebula NGC 1977 in the constellation Orion shines mostly by reflected light of nearby stars. Scattering of light in the nebula's dust makes it appear blue, just as scattering by atmospheric particles makes our sky blue. (SOURCE: Copyright Anglo-Australian Telescope Board, courtesy D. Malin.)

reddened. In a similar way, the interstellar medium dims and reddens the light of distant stars.

Structure of the Interstellar Medium

Interstellar gas and dust are far from uniform. In general, astronomers recognize four different types of regions defined by the condition of the gas within them. In dark clouds of dust and gas at about 10 K, most gas exists as molecules. We learned in the last chapter that these molecular clouds are the sites of most star formation. There are also slightly warmer regions ranging from a few hundred to a few thousand Kelvin that contain neutral hydrogen atoms. Hot regions at around 10,000 K surround hot O and B stars, whose ultraviolet photons excite and ionize the surrounding hydrogen atoms. These nearly spherical zones around hot stars are called *H II regions* after the symbol for ionized hydrogen. Last, there are extremely hot regions at about 10^6 K where gas has been superheated by a blast from a supernova. This is the hot interstellar medium.

Astronomers have learned a lot about the interstellar medium in the last few years. Radio telescopes work with the long waves that can penetrate to the heart of a cold molecular cloud. X-ray telescopes in space detect the

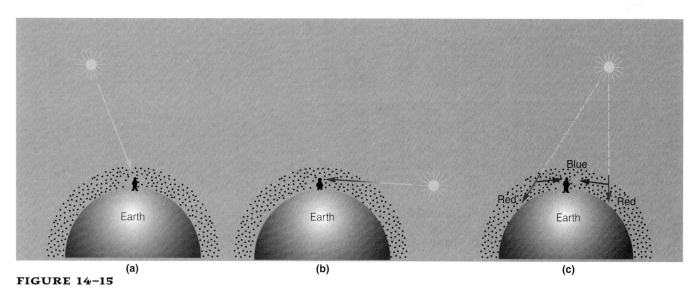

FIGURE 14-15
Explaining colors in the sky. **(a)** Light from the Sun high in the sky passes through minimal gas and dust and is minimally reddened. **(b)** Light from the Sun at sunset passes through the maximum amount of gas and dust and is strongly reddened. **(c)** Light from the sky at any time of day is the blue light scattered out of the sunlight beam.

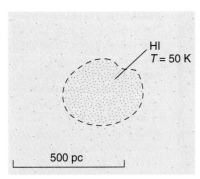
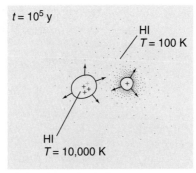
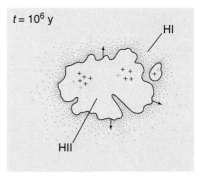

(a) Dense, cold molecular cloud, contracting

(b) Star formation, H II regions, molecular clouds

(c) More star formation, H II expansion by heating

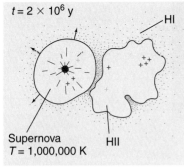
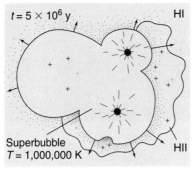
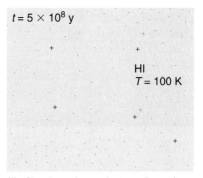

(d) Supernova expands, new star formation

(e) More supernovae, superbubbles, and star formation

(f) Clouds and star clusters dispersing, star formation and supernovae over

FIGURE 14–16
The evolution of interstellar material. The first panel shows a concentration of gas and dust that collapses and forms stars (shown as crosses). In the second and third panels, these young stars heat and ionize the gas that surrounds them. As the fourth panel shows, the most massive of the new stars then explode as supernovae, thereby triggering more star formation. After several million years, this star-forming activity reaches a peak, and the expanding shells around supernovae join to form superbubbles. Thin gas within the superbubbles is extremely hot, as shown in the fifth panel. The surrounding gas is eventually blown away, star formation ends, and the new stars that were formed gradually disperse.

short-wavelength radiation that a very hot gas emits. These instruments have shown that the interstellar medium is complex and chaotic. Thus, it is not yet clear exactly what fractions of space are occupied by gas of different densities and temperatures. Over the largest regions astronomers have studied, as much as 50 percent of the gas may be a cold, molecular form, and as much as 90 percent of the volume may be in the form of the hot interstellar medium.

The space between stars is a good example of an **inhomogeneous** medium, whose properties vary greatly from one place to another. By contrast, a **homogeneous** medium is smooth and uniform. (Most people see this word in reference to milk. Almost all milk is homogenized; you would have to go to a farm to experience the pleasure of inhomogeneous milk.) The interstellar medium is inhomogeneous in density and temperature. Astronomers have learned that the Sun is in the middle of a vast, low-density cavity, extending some 100 pc and containing over 100,000 stars. This large volume is called the *Local Bubble*. Astronomers believe that the cavity was caused by an ancient supernova explosion, which may have been witnessed by our human ancestors.

Finally, we can relate the interstellar medium to the continuing process of star formation. Figure 14–16 shows a time sequence of schematic pictures of a region of star formation. Astronomers cannot, of course, follow a single region over this length of time. However, they can piece together an evolutionary history by observing star-forming regions at different stages of development. The most massive stars die quickly as supernovae, depositing enormous energy into the interstellar medium. The expanding gas in turn triggers more star formation. Expanding supernova shells can merge and overlap to clear out large regions that contain thin and extremely hot gas, called *superbubbles*. After the most massive and luminous stars have died, the remaining gas cools down and the gas clouds and newly formed stars gradually disperse. We see from this picture that the pyrotechnics of star formation and glowing gas clouds is relatively short-lived. A typical bright nebula has only been lit up for a few tens of millions of years. We also see that the Milky Way must be evolving chemically. Massive stars eject heavy elements into the interstellar medium, and these atoms become part of succeeding generations of stars.

SCIENCE TOOLBOX

Dust Extinction and Reddening

Even though space is a remarkably good vacuum by terrestrial standards, it is not perfectly transparent. Tiny dust particles dim and redden the light from stars. As you may recall, we first encountered this characteristic *opacity*, the ability to obscure light, in Chapter 11 in the discussion of the Sun's atmosphere. Opacity is a measure of the fraction of photons that penetrate a gas when they suffer collisions or can be absorbed. The Sun's opacity prevents us from seeing any deeper than the thin outer layer of the photosphere. Dust clouds in interstellar space have opacity too.

As we saw in Chapter 11, the light intensity after passing through any material that scatters light is given by

$$I = I_0 \, e^{-\tau}$$

In this equation, I_0 is the initial intensity, I is the intensity after passing through the material, and τ (the Greek letter *tau*) is the optical depth. The optical depth is just the amount of interstellar dust the light must pass through, as you would expect. Notice that it does not matter whether the dust is thinly distributed in interstellar space or concentrated in a dense cloud. Starlight is affected either way.

Figure 14–A shows the dependence of light intensity on optical depth. It shows that the fraction of light transmitted by a cloud rapidly declines as the optical depth increases. For $\tau = 0.5$, $I = 0.61 I_0$. For $\tau = 1$, $I = 0.37 I_0$. For $\tau = 2$, $I = 0.14 I_0$. For $\tau = 4$, $I = 0.02 I_0$, and so on. The fraction of transmitted light drops more rapidly than the optical depth increases; dust is very effective at quenching light.

We can also relate optical depth to measurable properties of the particles in interstellar space. The optical depth is given by the simple equation

$$\tau = \sigma N = (\pi r^2) N$$

The optical depth is the product of the cross-sectional area of a single dust particle (the Greek letter sigma, where $\sigma = \pi r^2$ if we assume spherical particles) times the number of particles between us and the star. If we take the mean density of dust grains in a typical nebula like Orion (it is much lower than the number in Table 14–2, which is the number of *total* particles per cubic meter), we find that $\tau = 1$ over a distance of 0.5 pc. For a typical nebula size of 5 pc, the number of dust grains encountered by starlight traveling through the nebula is 10 times higher, so $\tau = 10$. This means that $I = I_0 \, e^{-\tau} = 5 \times 10^{-5} \, I_0$. A star behind this nebula will be reduced to 0.005% of its original intensity. This causes the kind of severe extinction that you can see in Figure 14–13.

Notice that the optical depth depends on the number of dust grains encountered along a line of sight, $\tau \propto N$. Optical depth does not depend on how the dust grains are distributed. In a nebula, the particles are relatively close together. (However, the environment is still much more rarified than the best terrestrial vacuum!) In interstellar space, the particles are more thinly distributed, but the same extinction can be seen over a longer path. If interstellar space is 1000 times less dense than a typical nebula, it will take a 1000-times longer journey for light to be extinguished by the same amount. Thus, $0.5 \times 1000 = 5000$ pc through interstellar space will also give an optical depth of $\tau = 10$. We cannot see more than about 16,000 light years through the Milky Way!

So far we have examined dust extinction, but what about the reddening effect? It turns out that dust grains are not very effective at scattering light when the wavelength is much bigger than the grain size. We can understand this by analogy to waves on a pond. If a wave is much smaller than an obstructing object like an island, the wave is blocked or reflected. However, if a wave is much larger than a small obstructing object like a pebble, the wave passes by

(continued)

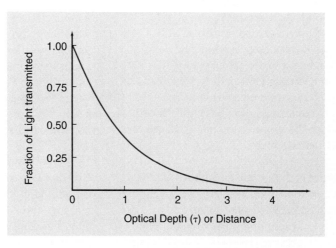

FIGURE 14–A
The relationship between the intensity of starlight and optical depth. The fraction of starlight transmitted depends on the number of dust grains in the interstellar medium between Earth and the star. If the obscuring material is smoothly distributed, optical depth is proportional to distance. Looking toward the Milky Way, the optical depth is 1 after a distance of about 500 pc.

SCIENCE TOOLBOX

Dust Extinction and Reddening (continued)

with no effect. Long waves do not "see" small particles.

Long-wavelength (or red) light is not scattered as strongly as short-wavelength (or blue) light. As a result, the farther starlight travels through a dust cloud or the interstellar medium, the more blue light is removed and the redder the star appears. Sometimes this scattered blue light reaches us indirectly, such as the reflected light from a nebula or the blue photons of sunlight that scatter in our atmosphere and make the sky look blue.

If we look at the scattering process in detail, we find that the efficiency of scattering is inversely proportional to wavelength. In terms of optical depth, $\tau \propto \lambda^{-1}$, where λ is wavelength. Dust extinction is much smaller at infrared wavelengths than at optical wavelengths. Equivalently, infrared radiation penetrates a dust cloud much more effectively than optical radiation. This is why astronomers use infrared techniques to study star-formation regions.

Groups and Clusters of Stars

If you could roam through space to ever-greater distances, you would eventually lose track of individual stars and see the Milky Way defined primarily by clusters of stars. Writing about clusters in 1930, Harvard astronomer Harlow Shapley pointed out that they are intimately interwoven with issues of stellar organization and galactic evolution. Yet at the same time, Shapley noted that a scientific study of them had hardly begun. Nobody had even known how to measure their distances or plot their distribution in space until the 1920s. Although some clusters, such as the Pleiades (or Seven Sisters), are easy to recognize with the naked eye, others are so far away that they require large telescopes to detect. Still others are so close that they cover much of our sky and were not even recognized until recent years.

In other words, our cosmic journey has brought us to clusters so far-flung that they were recognized as a class only in this century. Clusters have been of major importance because they have revealed the shape, size, and age of the vast assembly of stars in which we live.

Open Clusters and Associations

Open star clusters are moderately close-knit, irregularly shaped groupings of stars. They usually contain 100 to 1000 members and are about 4 to 20 pc in diameter. The Sun is possibly inside or on the edge of an open cluster centered only about 22 pc away toward the constellation Ursa Major, many of whose stars belong to this cluster. The best-known clusters, the Hyades and the Pleiades, lie 12° apart in our winter evening sky, about 42 and 127 pc away, respectively. Figure 14–17 shows the Pleiades in more detail. About 900 open clusters are concentrated along the Milky Way band, indicating that they lie in a flattened structure or plane.

As described in the last chapter, stars form in open clusters, and most open clusters have prominent young stars or associated clouds of star-spawning gas. Then why aren't all stars in clusters? The reason is that most open clusters break apart into individual stars within only a few hundred million years because of dynamic forces acting on them. Individual stars have space velocities that take them far from their birthplace. As they disperse, they no longer light up the residue of gas left over from their birth. In comparison with most cosmic lifetimes, open clusters are short-lived.

Stellar associations are cousins of open clusters. They often have fewer stars but are larger in size and have a looser structure. Some large associations include an open star cluster within them. They may have 10 to a few hundred members and diameters of about 10 to 100 pc. They are rich in very young stars, such as O and B stars (which burn their fuel too fast to last long), or T Tauri stars (which evolve toward the main sequence too fast to last long). Several hundred associations have been cataloged. The smallest associations grade into small, multiple-star-like groups, such as the Trapezium in the Orion Nebula, which might be a link between multiple stars and small clusters. Like open clusters, associations are involved with regions of recent star formation and are short-lived.

Globular Star Clusters

One of the most spectacular sights through a small telescope is a **globular star cluster.** Globular clusters are much more massive, more tightly packed, and more symmetrical than open star clusters or associations. They are also very old. Figure 14–18, on page 416, emphasizes the remarkable symmetry and compactness of globular clusters. They typically contain 20,000 to several million stars, although many of the stars in the central regions have images that are too close together and blurred to be resolved by Earth-based telescopes. Figure 14–19, on page 416, shows the high density of luminous stars in the core of 47 Tucanae, as observed with the Hubble Space Telescope.

Typical diameters of the central concentrations range from only 5 to 25 pc. To imagine conditions inside a globu-

FIGURE 14-17
The Pleiades, also known from mythology as the Seven Sisters, is an open cluster of young stars approximately 80 to 150 million years old. It is about 127 pc away and about 2 pc in diameter. The brightest stars are massive, hot, blue stars. Dust and gas clouds around them scatter blue light, creating wispy blue reflection nebulae. (SOURCE: Copyright Anglo-Australian Telescope Board, courtesy D. Malin.)

lar cluster, picture 10,000 stars placed around the Sun at distances no farther than Alpha Centauri, our nearest star. If we lived in the core of a globular cluster, our night sky would blaze with starlight ten times brighter than the light of the full Moon! Even the nearest globular clusters are thousands of parsecs away from us. Only because they have so many and such very bright stars can we see them at all. Yet a modest backyard telescope can reveal many prominent examples. Using larger telescopes, astronomers have found over 200 globular clusters.

Why are globular clusters globular in shape, as seen in Figure 14–18? In this book, we have encountered some cosmic systems that are flattened disks and others that are spherical. Systems that are disklike have a large amount of rotation—for example, the rings of Saturn and the planets of the solar system. The systems that are spherical have little or no rotation—for example, the Oort cloud of comets and globular clusters. The physical quantity that describes rotation is angular momentum. Astronomers have found that globular clusters are not exactly spherical, however, but slightly flattened. They therefore rotate slowly and have a small amount of angular momentum. Since angular momentum is a conserved quantity, the gas clouds from which globular clusters formed must have had small amounts of rotation.

A globular cluster is not a static collection of stars in space. In general, the stars are on elliptical orbits around the cluster center. However, the orbits of individual stars deep inside a globular cluster may be very complex. The cluster's overall gravity, the spatial distribution of its stars, their relative speeds, and the effects of near encounters among stars are all important in determining how an individual star orbits, in complex loops, the cluster's central regions. Remember that even in these crowded conditions, actual collisions between stars are rare or nonexistent. Furthermore, as the entire cluster orbits the galaxy, it passes through the galactic disk every 100 million years or so. We can imagine that such passages might allow spectacular close-up views of globular clusters from planets in the galactic disk, as shown in Figure 14–20 on page 420.

In the 1970s, globular clusters were found to be sites of strong X-ray radiation. Globular cluster X rays do not have the smooth periodic variations caused by orbital motions in binary pairs. Instead, they have irregular variations, sometimes over weeks or months, but sometimes doubling in intensity within a few minutes. These surprising findings led to new thinking about conditions inside globular clusters. Generally, the X-ray sources are not exactly at the centers of clusters. They seem to come from binary systems. Studies in the

FIGURE 14-18
The spectacular globular cluster 47 Tucanae is 4400 pc from Earth and contains over 1 million stars. The cluster's strong gravity force retains its spherical shape. (SOURCE: Copyright Anglo-Australian Telescope Board, courtesy D. Malin.)

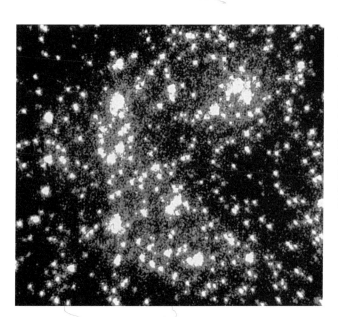

FIGURE 14-19
This ultraviolet image of 47 Tuc (Tucanae) was made with the Faint Object Camera on the Hubble Space Telescope. The view of the central regions covers only 11 seconds of arc, or only 0.2 pc—about 1 percent of the field of view of Figure 14-18. A high concentration of luminous blue stars in the cluster core indicates that the formation of binary systems has influenced the motion and evolution of many stars in the dense core. Binary capture is more likely in a dense star cluster, and mass transfer in a binary system can lead to one star becoming blue and luminous. (SOURCE: NASA.)

FIGURE 14-20
An imaginary view from a planet about 60 pc from a globular cluster. In such a case, the globular cluster would dominate the sky in a blaze of pinkish stars. Such a planet is unlikely to circle a star within the cluster, because globular clusters are deficient in the heavy elements that form solid planets. However, such scenes may occur as globular clusters pass through the galactic disk near stars that may (hypothetically) have planets. (SOURCE: Painting by Chesley Bonestell.)

1980s showed that inner-core regions of globular clusters are so crowded with stars that binary pairs may form more often than usual under these conditions. One idea is that globular clusters may contain many neutron stars and other stellar remnants, and, because of overcrowding, these may often capture passing stars into binary orbits. As the captured star evolves and blows off mass, gas is transferred onto the dense stars, and the resulting high-energy impact of gas onto the neutron star or its accretion disk may produce X rays. (Recall the discussion of millisecond pulsars earlier in the chapter.) There is little doubt that the dense packing of stars in the central few cubic parsecs of a globular cluster makes these regions extraordinary stellar environments.

Distances to Stars and Groups of Stars

As we saw in the chapter's opening story, Herschel's ambition was to map out the size and shape of the Milky Way, but he was hampered because he had no way of measuring stellar distance. Recall that apparent brightness alone is a very poor indicator of distance because stars have such a wide range of intrinsic properties (see Figure 14-1). The most direct measure of stellar distance is parallax. In a parallax measurement, we use the diameter of Earth's orbit of the Sun as the base of a triangle to make a geometric measurement of distance (see Chapter 12 to review this idea). By using data gathered by the European Space Agency's *Hipparcos* satellite, we can reliably calculate distances to just over 100 pc. The Hyades cluster, at about 42 pc, is pivotal in the galactic distance scale. The cluster can be measured not only by the parallax technique, but also by a variety of methods, yielding a more reliable determination of distance.

Astronomers have devised varied and sometimes ingenious methods to calculate the distances to individual stars and groups of stars. Here we describe some of the most important methods. First, we can use the collective properties of stars to determine the distances to open clusters. For example, we might use the angular size of a cluster to infer its distance. If all clusters have the same physical size, then simple geometry tells us that $D \propto 1/\theta$, where D is the distance to the cluster and θ is the angle subtended by the cluster. We can test the method with the Hyades, where parallax provides a direct measure of distance. If the Hyades subtends an angle of 7.5° at a distance of 42 pc, then a cluster that subtends only $\frac{1}{2}$° must be $(7.5/0.5) \times 42 = 630$ pc away. The problem with this method is that open clusters have a wide range in physical size—from 4 pc for the Pleiades to 16 pc for h Persei. The same is true of globular clusters. Therefore, the error in using angular size to predict distance can be as large as a factor of 4.

A better approach is using stellar spectra to determine properties that allow us to place the stars on an H-R diagram. We can then compare this H-R diagram of a more distant cluster to the H-R diagram for the Hyades. This technique is called **main-sequence fitting**. Each cluster has a concentration of stars that defines its main sequence on an H-R diagram. The main sequence of each cluster has the same shape. The amount the two main sequences have to be shifted vertically to "fit" on top of each other gives the apparent brightness difference between the two clusters. The apparent brightness difference is related to the relative distance by the inverse square law. Using the inverse square law and the known distance to the Hyades, the distance of the fainter cluster can be calculated easily. For example, if main-sequence stars in M 67 are 2000 times fainter than main-sequence stars of the same spectral type in the Hyades, then M 67 is $\sqrt{2000} \times 42 \approx 1900$ pc away. Main-sequence fitting gives relative distances with an accuracy of about 20 to 30 percent.

Another important technique uses the properties of variable stars to measure their distance. As you learned in Chapter 13, variable stars vary in brightness over periods ranging from less than an hour to several years. One type of variable star—Cepheid variables—has certain properties that allow distance to be easily calculated. Cepheid variables have periods of 1 to 50 days. There is also a tight relationship between the Cepheid's period and its luminosity, called the **period-luminosity relation**. Once you measure the period of one of them—say, 5 days—that tells you the luminosity of the star (Figure 14–21). Astronomers use a series of images taken nightly over a period of several months to measure the periods of Cepheids.

Variable stars, such as Cepheid variables and RR Lyrae variables, are very luminous, so they can be used to measure the distances to very remote regions. In Figure 14–21, you can see that RR Lyraes are 100 times as luminous as the Sun, so they can be seen out to $\sqrt{100} = 10$ times the distance we could see a Sun-like star. Cepheids range up to 10,000 times the Sun's luminosity, so they are visible out to $\sqrt{10,000} = 100$ times the distance we could detect a star like the Sun.

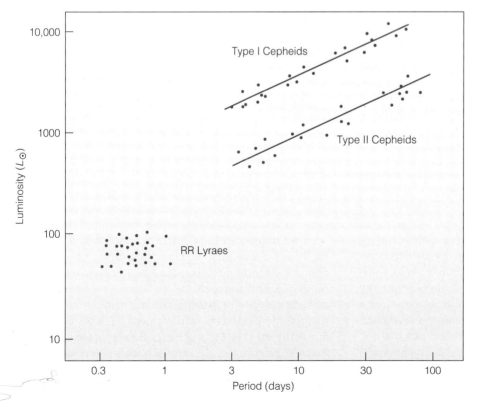

FIGURE 14–21
The relationship between period and luminosity for different types of variable stars. Type II Cepheids and RR Lyraes are found in older stellar populations and have few heavy elements. Type I Cepheids are found in young stellar populations and have more heavy elements. The unique relationship between period and luminosity allows Cepheids to be used as distance indicators.

Cepheids are also very rare, so a large collection of stars will contain very few. Luckily, they reveal themselves easily by their variations, since the vast majority of stars do not vary in brightness. In using variable stars to measure distance, we assume that stars in one part of the Milky Way function the same as stars in another part of the Milky Way. In other words, we assume the physics that leads to stellar pulsation is universal. With no way to test this assertion, it becomes an assumption in our use of the scientific method.

Astronomers can use Cepheid variables to measure the distance of a cluster in a process requiring several steps. First, a Cepheid in the cluster must be located and its period and type measured. Then its luminosity must be read from the appropriate period-luminosity diagram. Finally, the Cepheid's luminosity is combined with the apparent brightness to calculate the distance. Few astronomers are equipped to carry out all the necessary observations, such as measuring light variations and periods, determining spectral properties, and measuring interstellar reddening and extinction. Thus, astronomers have specialized. Some study periods of variable stars, some make measures of absolute brightness, some study interstellar reddening. The simple statement that a cluster is "this many" parsecs from Earth may represent years of work by many researchers.

Before 1930, astronomers calculated distances under the assumption that all intervening interstellar space is transparent. However, early estimates of the distance, brightness, and size of clusters yielded inconsistent results. In 1930, Robert Trumpler demonstrated that it was wrong to assume that space is clear. He showed that diffuse interstellar dust dims stars and clusters that are more than a few dozen parsecs away. The intensity of starlight falls off *more rapidly* than we would predict by the inverse square law. Since we see a star as dimmer than it truly is, we will overestimate its distance. This very important systematic error must be understood if we are to correctly map out the Milky Way. The obscuration is not the same in every direction because interstellar material is inhomogeneous. Fortunately, the total amount of dimming can be estimated by measuring the amount of interstellar reddening, or color change caused by the dust. The more reddening is observed, the greater the degree of dimming. Once the extinction is measured and taken into account, distances can be accurately measured if the luminosity of any star or class of stars in the cluster is known.

The Ages of Groups of Stars

In this chapter, we are exploring the different components of the Milky Way. In addition to studying how groups and clusters of stars are distributed in space, astronomers can learn about their ages. In this way, we get important clues to the history of the Milky Way galaxy. It is difficult to measure the age of an individual star. Astronomers have a direct age estimate for only one star: the Sun. The 4.5-billion-year age of the Sun is based on measurements of radioactive decay rates in Moon and Earth rocks, and on the knowledge that the Sun and the planets formed out of the same collapsing gas cloud. This age agrees with the age calculated from the rate of conversion of hydrogen into helium in the Sun's core.

Ages for other individual stars are far less certain because they depend entirely on stellar models.

Astronomers can measure the **ages of groups of stars** by constructing an H-R diagram. It is relatively straightforward but time-consuming work; the apparent brightness and surface temperatures of hundreds of stars must be measured. We assume that the stars in a cluster are at the same distance and that they formed at the same time. If the distance to the cluster is known, the apparent brightness of each star can be converted to luminosity. Astronomers have plotted the H-R diagrams of many clusters. As discussed in the last chapter, the most massive (and luminous) main-sequence stars exhaust their hydrogen first. They then change their properties in a way that moves them across the H-R diagram to a position far from the main sequence. As time goes by, less massive stars exhaust their hydrogen. Thus, the luminosity at which a star "turns off" the main sequence changes with time. This turn-off point is a chronometer that lets us measure the age of a group of stars.

Figure 14–22 brings together data on several clusters. The H-R diagrams are superimposed schematically; the sequence for each cluster really represents the combined measurements of hundreds of faint stars. Marked along the main sequence is the age at which various stars evolve off the main sequence. (Compare with Figure 13–12, where the dashed lines show predicted positions of stars of various ages.) Notice in Figure 14–22 that some clusters, such as NGC 2362, have an age of only 1 or 2 million years. These clusters are so young that few stars have evolved off the main sequence. It is remarkable to realize that a cluster like this is less than one-thousandth the age of the Sun, and its birth was witnessed by the first humans.

Open clusters are young in cosmic terms. Among published ages for 27 open clusters, about half are less than 100 million years old. The solar system is nearly 50 times as old as this. These results confirm that star formation is continuing to this day. Many open clusters are associated with dense nebulae where stars are currently forming out of collapsing gas clouds. The brightest stars are hot blue O and B stars, which formed recently, burn their fuel fast, and cannot last long. In Chapter 13, we mentioned two groupings in which the least massive stars have not yet even evolved onto the main sequence, the Trapezium association and NGC 2264. They are only about 1 and 3 million years old, respectively. In slightly older clusters, some of the more massive stars may have evolved into red giants, adding a colorful contrast. The cluster in Figure 14–23, for example, has two bright orange giants among a field of young blue stars.

Stellar motions can also indicate youth. Velocities of cluster stars, measured by Doppler shifts, show that some clusters are expanding or losing members, or both. The fastest-moving stars have velocities that exceed the escape velocity of their parent clusters. Thus, these clusters are breaking apart as we watch and must disperse after a few hundred million years. This disruption of open clusters occurs by several simultaneous mechanisms. Stellar winds and supernovae disperse the gas in the cluster and thereby reduce its hold on the member stars. The Hyades, for example, are

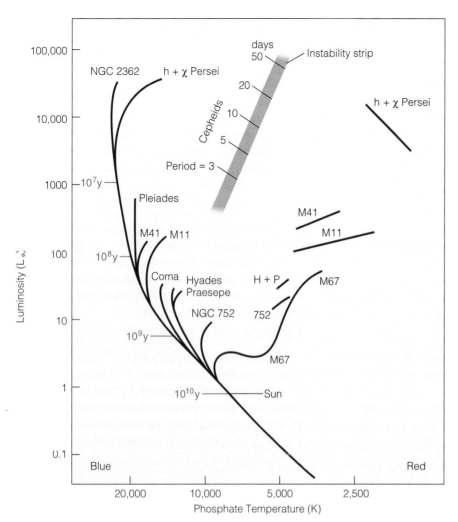

FIGURE 14-22
An H-R diagram for several clusters. Numbers along the main sequence represent the ages of stars just exhausting their hydrogen fuel and turning off the main sequence. The age of a cluster can be measured by the position of the main sequence turn-off point, the point at which stars change their properties and move upward and to the right in the H-R diagram. The part of the H-R diagram where stars become unstable and vary periodically is also shown. Cepheid variables have properties that place them in this region, although it is relatively short evolutionary stage. (SOURCE: Adapted from A. Sandage.)

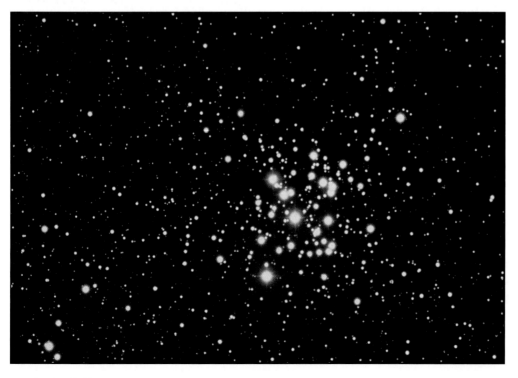

FIGURE 14-23
The open cluster NGC 3293 in the constellation Carina. The cluster is about 8 min of arc across and contains about 50 stars visible through a small telescope. At least two of the brightest are red giants, indicating that enough time has passed since the cluster's formation for a few of the more massive stars to evolve into giant status.

(SOURCE: Copyright Anglo-Australian Telescope Board.)

barely stable now, and the outer parts of the Pleiades are already dissipating (although the central, tighter grouping may be stable). Like open clusters, stellar associations are young. Because of their loose structure, they may break up even faster than ordinary open clusters. Most associations cannot last more than a few tens of millions of years, because of the disruptive tidal forces and the tendency for their stars to follow individual orbits around the center of the galaxy. Many associations have large masses of neutral hydrogen gas, some of which exceed the mass of the stars. The hydrogen may be debris left over from the formation of incorporated stars, or it may be material ejected from the fastest-evolving stars. Often this gas is carried out of the association by stellar winds and supernovae.

Open clusters and associations are mostly young, but astronomers have found globular clusters that are extraordinarily old. Star formation has ceased in them; all of the gas and dust have dispersed long ago. In all the globular clusters of our galaxy, the O, B, and A stars have evolved off the main sequence and have become red giants. The only main-sequence stars left are dim and red. For this reason, most bright stars in almost all globular clusters have a reddish color.

Determining the ages of globular clusters can be quite complicated. Spectra show that their stars contain fewer heavy elements than the more familiar stars of the solar neighborhood. Therefore, stars in these clusters display processes of energy generation and transport that are different than those of stars closer to Earth. Astronomers must incorporate these unique properties into a complex model of stellar evolution to date the globular cluster. The models are fed with details such as the heavy-element abundance, the amount of helium diffusion in the stellar cores (i.e., the amount of helium that migrates from the cores to layers farther out), and the amount of reddening in the cluster.

Errors in any of the numbers that these calculations depend on could lead to an error in the age estimate. For example, suppose we neglected to account for diffuse interstellar dust between Earth and the globular cluster. The dust dims the light of stars in a globular cluster, so we would underestimate the luminosity of the stars (the clusters contain RR Lyrae stars that are used to measure distance). Since luminosity is related to energy-generation rate in stellar models, we would underestimate the energy-generation rate. This, in turn, would lead to an error in the estimated age of the stars. Distance errors also affect age estimates. As we saw at the beginning of the chapter, a star's luminosity is proportional to its distance squared, $L \propto d^2$. Mathematically, this means that a 10 percent error in distance translates into a 20 percent error in luminosity. Once again, the age derived from the stellar models depends on the luminosity that is fed in to the models. Thus, the difficulties of distance determination affect our knowledge of the age of the oldest stars.

The best techniques set the age of globular clusters at about 9 to 13 billion years. This is a true range and not a measurement uncertainty; globular clusters did not all form at the same time. For example, Figure 14–24 shows an H-R diagram of the globular cluster M 92. The superimposed lines are stellar models and show how the main-sequence

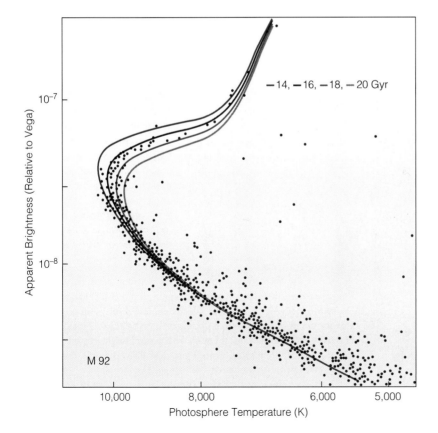

FIGURE 14–24
H-R diagram of the globular cluster M 92. The main-sequence turn-off is clearly seen, and the solid lines represent models fitted to the main sequence. The age estimate depends on details of metal abundance and energy transport in the stars, but in this model the age is estimated to be about 13 billion years.
(SOURCE: Adapted from A. Demarque.)

turn-off point evolves toward redder and fainter stars with time. The oldest stars in the Milky Way galaxy are therefore three times older than the Sun.

The Layout of the Milky Way

Our exploration of space has taken us out to distances of a few thousand parsecs. By looking at the distribution of stars and clusters throughout volumes of this size, we begin to perceive the Milky Way galaxy. As you can see from the photograph that opens this chapter, the sky is not the same in all directions. The Milky Way is a band of stars (and gas and dust) that circles the entire sky. The number and concentration of stars in the Milky Way are much higher than in other directions. Also, the Milky Way has variations, its highest density of stars occurs in the southern sky (see Figure 14–25). How can we understand our place in this vast collection of stars?

In 1750, English theologian Thomas Wright correctly hypothesized that our galaxy must be a slablike arrangement of stars. German philosopher Immanuel Kant also published this idea around the same time. In the 1780s, Herschel put enormous effort into counting stars in different areas of sky. He believed that the Sun was at the center of the Milky Way because he counted roughly equal numbers of stars all along that band of light. Unfortunately, Herschel was unaware of the obscuring effects of dust. It was not until the 1930s that Trumpler corrected for the dust and showed that the stars of the Milky Way were especially concentrated in the direction of the constellation Sagittarius. As discussed in the following Science Toolbox, when something is the same in all directions, we say it is **isotropic;** when it varies in different directions, we say it is **anisotropic.** The Science Toolbox shows how we can use the anisotropy of the stars in the night sky to deduce the shape of the Milky Way galaxy and our position within it.

The shape of the Milky Way is neatly revealed by the distribution of the two types of clusters we discussed earlier in the chapter. The clusters are not randomly distributed; the open clusters lie in the plane of the Milky Way and define the **galactic disk,** and the globular clusters form a spherical cloud around the disk called the **galactic halo** (Figure 14–26). Harvard astronomer Harlow Shapley, who pioneered measurements of the shape and size of the Milky Way, said that the clusters reveal "the bony frame of our galaxy." Dutch astronomer Jan Oort also did important work on galactic structure in the 1920s and 1930s. Shapley showed that globular clusters are distributed in a spherical swarm extending above and below the disk. He also showed that the halo is centered not on the Sun but on a distant point in the disk in the direction of the constellation Sagittarius. He correctly hypothesized that this point is the center of the galaxy. Astronomers also realized that there was a concentration of stars toward the galactic center, roughly spherical in shape but much smaller than the halo. These stars appeared to be mostly old and red. This third component of the Milky Way is called the **galactic bulge.**

Figure 14–26 shows the coordinate system that astronomers use to describe locations within our galaxy. The galactic equator runs along the center of the Milky Way band. Galactic longitude, designated l, measures the angular distance around the Milky Way, starting at a zero point defined to lie at the galactic center in the constellation Sagittarius. The direction $l = 90°$ lies toward the constellation Cygnus, near the top of the Northern Cross. Opposite the

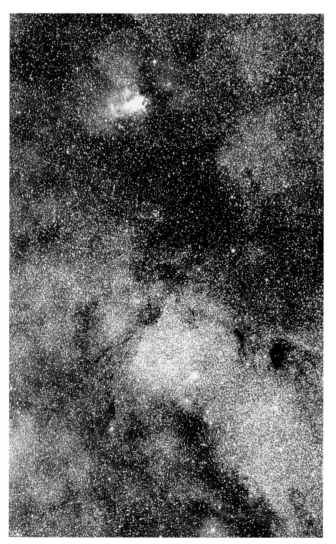

FIGURE 14–25
Proof that the Milky Way is composed of individual stars. To the naked eye, the Milky Way in this region of the constellation Sagittarius (looking toward the galactic center) displays fuzzy glowing clouds. This telescope photograph reveals that the clouds comprise innumerable distant stars. The central bright region (M 24) is about 5000 pc away, located just over halfway to the center of the Milky Way. At the top is the red-glowing nebula M 17, about 1800 pc away. The vertical height of the picture is 5°, so M 17 has about the angular diameter of the Moon. (SOURCE: Copyright Anglo-Australian Observatory.)

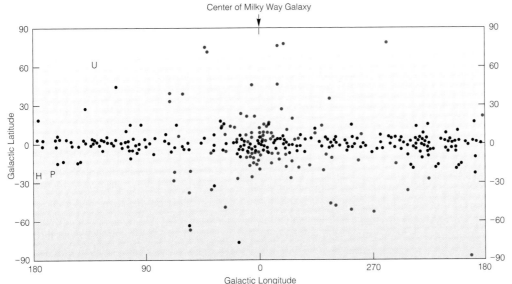

FIGURE 14-26
A map of the sky in galactic coordinates, showing the positions of selected open clusters (black dots) and the 93 most prominent globular clusters (red dots). The Ursa Major, Hyades, and Pleiades clusters are indicated by U, H, and P, respectively. The galactic center is in the middle of the plot. Open clusters concentrate along the plane of the Milky Way galaxy (the line at zero latitude), and globular clusters form a spherical swarm around the galaxy's center.

galactic center, at longitude $l = 180°$, is the direction of Taurus, near the Pleiades and Hyades clusters. The direction $l = 270°$ is just south of Canis Major. Galactic latitude, designated b, is defined to be zero along the galactic equator. The direction $b = +90°$ points "straight up" out of the disk in the northern sky. You can learn more about the position of prominent stars with respect to the galactic equator using the star maps at the end of this book.

We can relate the map of galactic coordinates in Figure 14-26 to our location in the Milky Way galaxy. Figure 14-27 is a schematic cross section through the galaxy. Looking toward the center of the disk, at $l = 0°$, we see the largest concentration of stars. Looking away from the center, at $l = 180°$ (sometimes called the anticenter), we see a lower concentration of stars in the galactic plane. At high galactic latitudes toward the center, we see a large number of globular clusters. At low galactic latitudes toward the center, we see a smaller number of globular clusters. Notice that if our view is confined to a few hundred parsecs (the small circle in the figure), we see roughly equal numbers of stars in every direction. Most of the bright stars in the night sky are relatively close. From our position within the galactic disk, we see nearby disk stars more or less equally in every direction.

We view the Milky Way from within the enormous disk of stars. The disk is about 30,000 pc across, 400 pc thick, and packed with open clusters, individual stars, dust, and gas, mostly arranged in ragged spiral arms. Globular clusters surround the disk in a spherical swarm, concentrated toward

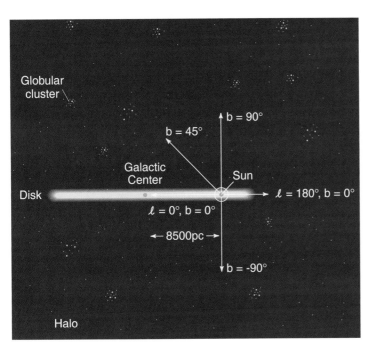

FIGURE 14-27
A cross section through the Milky Way galaxy, which can be related to Figure 14-26. The disk of the galaxy is oriented perpendicularly to the plane of the page, the galactic center is to the left, and the galactic anticenter is to the right. Looking at high galactic latitudes toward the galactic center, globular clusters are seen. Looking at high galactic latitudes away from the galactic center, few globular clusters are seen. The small circle shows that nearby stars are all in the disk; these are the bright stars scattered around the night sky.

SCIENCE TOOLBOX

Isotropy and Anisotropy

One of the limitations of astronomy is the fact that we are trapped on planet Earth. Even our best space probes have only covered a tiny fraction of the distance to the nearest star. Although we cannot roam through space, we can understand the size and shape of the Milky Way, and our place in it, by counting stars.

Isotropic means the same in every direction; *anisotropic* means different in different directions. You can use *anisotropy* to learn about your surroundings. If you are somewhere in a forest, you might notice fewer trees (or the trees thinning out) in one direction. This direction is probably the shortest distance to the edge of the forest. If you see equal numbers of trees in every direction, you have no clue where the edge is. Or imagine that you are lost in a fog bank. As you wander, you might find a place where sunlight is beginning to penetrate. This anisotropy—the fact that the level of light was not the same in every direction—is your clue to the location of the edge of the fog.

We are "lost" in the Milky Way galaxy. If there were equal numbers of stars in every direction, we would have no clue about the edge of the Milky Way. We might be tempted to believe that we were at the center of the galaxy. In fact, the distribution of stars in the sky is anisotropic, and this anisotropy helps us measure our galactic environment. (For simplicity, we will ignore the obscuring effects of dust.)

Figure 14–B shows how you might view the sky if the stars around you were arranged in a disk. If you are at the center of the distribution, you see few stars when you look directly out of the disk on either side, but you see larger and equal numbers of stars when you look in any direction through the disk. In this simple "toy" model, the band of the Milky Way would look the same in every direction (Figure 14–B*a*). You can tell you live in a slablike arrangement of stars, but you cannot tell whether it is a disk or an enormous sheet, and you cannot locate the edge. In Figure 14–B*b*, you are offset from the center of the disk. Now you see more stars when you look through the disk toward its center and fewer when you look through the disk toward its edge.

Now suppose you live in a spherical distribution of stars (Figure 14–C). If you are at the center of the distribution, you see roughly equal numbers of stars in every direction (Figure 14–C*a*). You would conclude either that you lived in the center of the universe, or that the universe extended so far that you could not see its edge. In Figure 14–C*b*, you are offset from the center of the sphere. Now you see more stars in all directions toward the center of the sphere and fewer stars in all directions away from the center. Each of these examples shows how we can use anisotropy in the counts of stars to deduce the size and shape of our galactic environment. In the real sky, we use hot young stars or open clusters to trace the shape of the galactic disk and globular clusters to trace the shape of the spherical galactic halo.

We can quantify the change in the number of stars with the change in the viewing angle as we look through a disk, a slab, or a sheet. Figure 14–D shows the viewer located in a slablike arrangement of stars that extend very far in every direction. The shortest distance to the edge of the slab is X. Using simple geometry, the distance through the stars at any angle θ from the vertical is

$$Y = \frac{X}{\cos \theta}$$

This relationship defines the way that the number of stars will increase if we look in directions closer and closer to the Milky

(continued)

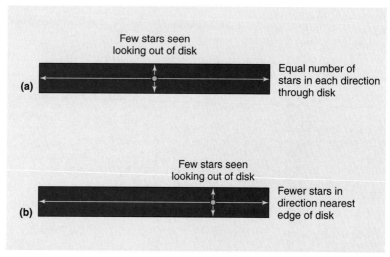

FIGURE 14–B
Anisotropy in a disk. **(a)** If you are located near the center of a disklike distribution of stars, there will be a higher number of stars in opposite directions looking through the disk. In all other directions, far fewer stars will be seen. **(b)** If you are nearer the edge of a disk and can see out past one edge, the number of stars will be higher looking toward the center than looking toward the edge.

SCIENCE TOOLBOX

Isotropy and Anisotropy (continued)

Way. In terms of our galactic coordinates, $b = 90 - \theta$. We can consider X the minimum number of stars we see at the galactic pole, where $\theta = 0°$ and $Y = X$. At a galactic latitude of $b = 45°$ (so $\theta = 45°$), $Y = X / \cos 45 = 1.4\, X$, so we see 40 percent more stars in that direction. At a galactic latitude of $b = 10°$ (so $\theta = 80°$), $Y = X / \cos 80 = 5.8\, X$, so we see 5.8 times more stars at such a low galactic latitude. The number of stars increases sharply as we approach the galactic plane.

There are two major simplifications in this argument. First, the Milky Way is a not a perfectly uniform disk of stars. Stars are grouped and clustered, which will cause variations in the counts of stars in adjacent directions. A more important complication is the effect of dust. Dust is found in the regions between stars. Therefore, when we look through areas with more stars, we will also be looking through more dust. Since dust dims starlight, we will tend to underestimate the numbers of stars at low galactic latitude. This explains why it took so long to correctly map out the Milky Way.

The same geometry shown in Figure 14-D can be used to explain why stars dim when they set in the night sky. Suppose that Figure 14-D now shows Earth's atmosphere, and the dots are dust particles in the atmosphere (we can ignore Earth's curvature and treat the atmosphere as a slab). The dust will dim and redden the light from all stars; recall that the opacity and amount of extinction just depend on the number of particles encountered or the path length through the particles. The minimum amount of extinction occurs looking directly overhead. This is the distance X, and an elevation angle of 90°. Stars overhead typically have their red light dimmed by 10 percent and their blue light dimmed by 30 percent; the extra loss of blue light shows the effect of reddening. A star at an elevation angle of 45° is dimmed by an extra 40 percent since $Y = X / \cos 45 = 1.4\, X$. A star at an elevation angle of 70° (only 20° above the horizon) is seen through an even larger amount of atmosphere: $Y = X / \cos 70 = 2.9\, X$. It is therefore dimmed by 65 percent compared to a star that is overhead. Astronomers prefer to observe targets that are high in the sky—now you can understand why. As a star sets toward the horizon, its light is rapidly dimmed and reddened by dust particles in Earth's atmosphere.

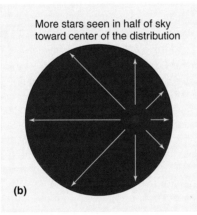

FIGURE 14-C
Anisotropy in a sphere. **(a)** If you are located at the center of a spherical distribution of stars, the counts of stars will be isotropic—equal numbers in every direction. **(b)** If you are offset from the center of the sphere, you will count more stars in all directions toward the center than in all directions away from the center.

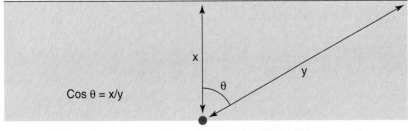

FIGURE 14-D
The view from within a disk or a slablike distribution of stars. The shortest distance to the edge is X. In any other direction, at an angle θ from a direction straight out, the path length is $Y = X / \cos \theta$. The number of stars encountered increases rapidly for shallow viewing angles through the distribution of stars.

the center of the disk. This swarm of clusters, which also includes some sparsely scattered individual stars and gas, defines the galactic halo. According to current estimates, the Sun is about 8500 pc from the galactic center. Although the International Astronomical Union has adopted 8500 pc as the distance to the center of the galaxy, recent studies of the space distribution of globular clusters and X-ray burster stars suggest that the number may be somewhat lower. We will continue to use 8500 pc in this book.

When we consider objects in distant parts of our galaxy, we are dealing with distances of thousands of parsecs. For this reason, astronomers often use a unit of distance even larger than other astronomical units, light-years, and parsecs. This unit is a **kiloparsec (kpc)** (notice the use of the same prefix, *kilo-*, that is used throughout the metric system to indicate 1000). Therefore, 1 kpc = 1000 pc, and astronomers say, for example, that we are 8.5 kpc from the galactic center and that our galaxy is roughly 30 kpc in diameter. These distances are difficult to comprehend. In a model of the Milky Way galaxy the size of North America, stars like the Sun would be microscopic specks less than one-thousandth of a centimeter across and scattered a block apart. The solar system would fit in a saucer. The quest to map out the size and shape of the Milky Way galaxy is another important step in the Copernican revolution. We are truly dwarfed by the vastness of the galaxy that surrounds us.

Summary

At least half of all the seemingly single stars in the sky are binary stars or multiple-star systems. Many of these may have formed by the interactions of stars in crowded new clusters, but some may have formed by other means, such as fission or capture. Binary and multiple-star systems can be detected in different ways. In visual binaries, both stars are visible and their orbit can be followed, given a sufficiently long span of observations. In other cases, one star is too faint to be detected, or the two stars lie too closely together to be resolved. Spectroscopy is a powerful tool for measuring the Doppler shifts of the orbiting stars. Binary star pairs provide the best way of determining certain stellar properties, such as mass and diameter.

The transfer of mass in close binary systems can alter the evolution of one or both stars. The classification of mass transfer is based on the way stellar material fills the Roche surface, an imaginary surface that defines gas that is gravitationally bound to either star. Once one lobe has been filled by expansion of a star to the red giant stage, gas may flow from one star to the other, causing flare-ups, X-ray emission, nova explosions, and supernova explosions. Mass transfer in binary systems produces some of the most spectacular phenomena in astronomy.

Space is not empty; it is thinly filled with atoms and molecules of gas, grains of dust, and possibly larger pieces of debris, which partly obscure distant stars and redden their light. These particles are not uniformly distributed through interstellar space. A nebula is a concentration of gas and dust that is lit up by nearby stars. Some nebulae glow with light emitted by excited atoms, some merely reflect the light of nearby stars, and some are dark silhouettes against distant backgrounds. The processes of emission, scattering, and absorption give nebulae different colors. Nebulae are made of the material from which stars are born and into which larger stars blow some of their material when their fuel runs out. Their existence shows that matter in our galaxy has not dispersed uniformly and smoothly. Rather, the interstellar medium is continually stirred, formed into clouds, dispersed, and disturbed by influences such as the formation of new stars and the explosions of old stars. We are witnesses to a vast cosmic recycling as interstellar matter forms clouds, clouds contract to form stars, and young and old stars blow out their material, replenishing and enriching the interstellar medium.

Open clusters and associations are young groups of about 10 to 1000 newly formed stars located in the galactic disk. These disk stars are forming all the time. The youngest groups are fractions of a percentage of the age of the Sun. Globular clusters, which are old groups of up to a few million stars, are located within a spherical volume above and below the galactic plane called the galactic halo. Stars in globular clusters formed as much as 12 to 13 billion years ago, so they are three times as old as the Sun.

Star clusters have played an extremely important role in mapping our galaxy and clarifying its history. They present us with groups of stars formed at about the same time, so that their H-R diagrams can be used to measure ages. The distribution of stars in the sky is not isotropic, and counts of stars (corrected for interstellar obscuration) can be used to map out the Milky Way galaxy. Open clusters and associations trace out the flattened structure of the galactic disk. Globular clusters define the spherical galactic halo. A set of old stars concentrated near the center of the Milky Way forms the galactic bulge. The Sun is located in the disk but offset from the center. Knowledge that the Sun is one of billions of stars in the Milky Way galaxy gives us a new sense of our place in the universe.

Important Concepts

You should be able to define these concepts and use them in a sentence.

Milky Way	homogeneous
galaxy	open star cluster
clustering	stellar association
binary star	globular star cluster
multiple-star system	main-sequence fitting
light curve	period-luminosity relation
Roche surface	age of groups of stars
nova	isotropic
nebula	anisotropic
interstellar medium	galactic disk
21-cm emission line	galactic halo
interstellar extinction	galactic bulge
interstellar reddening	kiloparsec (kpc)
inhomogeneous	

How Do We Know?

These questions and answers show how the scientific method is used to learn about the universe.

Q How do we detect binary star systems?

A Some of the evidence is statistical in nature. When we look at the way stars are scattered in the sky, there are too many with small angular separations to be due to chance. Many of these cases, in which two stars appear close to each other, are in fact binary systems, in which the two stars are bound by gravity. This hypothesis can be confirmed by observations of the orbits of the stars over a period of months or years. In other cases, we find that two closely separated stars have Doppler shifts that show a periodic variation. The cycle of velocity shifts can be used to determine the orbits. Sometimes the two stars are too close together to be resolved, but the Doppler shifts of the lines from both stars can be tracked in a single spectrum. When five or more stars are bound by gravity, it is usually obvious by the grouping on the sky. The Sun may be slightly unusual in not having any stellar companions.

Q How do we know that most stars are in binary or multiple systems?

A Careful measurements have shown that about two-thirds of the prominent stars in the night sky have faint companions. When both stars in a pair are bright or the angular separation is large, it is easy to measure a binary orbit. However, to get a true census of binaries, astronomers need a variety of approaches. Since stars vary so widely in luminosity, one star in a binary pair may be very much fainter than the other star. Alternatively, the stars may be too close together on the sky to be resolved. Spectroscopic surveys reveal binaries by the presence of two sets of spectral features at similar velocities, or by the periodic variation of the Doppler shift of a star due to a faint companion.

Q How do we know that the space between stars is not empty?

A The space between stars is far emptier than the best vacuum we can create on Earth, but it is not truly empty. Interstellar space contains diffuse gas and microscopic dust grains. Some of this material is concentrated in clouds or nebulae. These clouds are revealed when they glow from the illumination of hot stars, when they reflect starlight, or when they absorb or block out starlight. Dust grains dim and redden starlight. Both effects occur because these tiny particles absorb photons of visible light and reradiate them in the far-infrared. This process is more efficient for short waves, so blue light is extinguished more than red light. We know that this occurs because stars at a known distance are dimmer and redder than stars of the same spectral type nearby. Moreover, the dust grains are directly revealed by their far-infrared emission. Some dust is concentrated in clouds; dust is also thinly distributed throughout interstellar space. If the effects of dust are neglected, we will overestimate the distance to stars.

Q How do we know the distances of stars?

A Most stars are too far away to show a detectable parallax shift, which is the direct, geometric way to measure distance. Also, apparent brightness is a poor guide to distance because stars have such a wide range in luminosity. If we study a group of stars at the same (unknown) distance, their properties should reveal a well-defined main sequence on an H-R diagram. The brightness offset of the main sequence, compared with the main sequence of a set of stars at known distance, gives the unknown distance via the inverse square law. Certain variable stars exhibit a well-defined relationship between period of variations and luminosity. The period of the star, therefore, predicts the luminosity, which can be combined with the apparent brightness to give the distance. Since these variable stars are very luminous, they can be seen out to large distances.

Q How do we know the age of groups of stars?

A As stars on the main sequence exhaust their nuclear fuel, they change their properties in a way that places them in a different part of the H-R diagram. High-mass stars in a cluster or group use up their fuel the most quickly; they are on the upper part of the main sequence. Lower-mass stars use their fuel more slowly; they are on the lower part of the main sequence. Therefore, the point on an H-R diagram at which stars "turn off" the main sequence evolves to lower luminosity and redder colors with time. On a physical model of the stars, the position of the most luminous stars remaining on the main sequence can be converted to an age for the group or cluster. The oldest main-sequence stars are less massive than the Sun. This technique does not work unless we know distances, because luminosity is a key factor in creating the stellar models.

Q How do we know the size and shape of the Milky Way galaxy?

A The first clue to the shape of the galaxy is the Milky Way, a dense band of stars, gas, and dust that entirely circles the sky. William Herschel made the first quantitative measurements of the size of the Milky Way over 200 years ago, when he compared the counts of stars in different directions. We interpret the Milky Way as our view from within a flattened distribution or disk of stars. That is why we see a sparse scattering of stars when we look away from the Milky Way and numerous stars when we look into the plane of the Milky Way. The globular clusters define a different shape and are located in all directions in the sky, though they tend to be located in the half of the sky that has the densest stars in the Milky Way. It is not possible to get a complete view of the Milky Way galaxy, or a correct measure of its size, without taking into account the obscuring effects of dust.

Problems

Use these problems to test your understanding of the information and concepts in this chapter. The * indicates a more advanced or mathematical problem.

1. Describe how Kepler's laws can be verified from analyzing the motion of planets around the Sun. What was the first verification outside the solar system? How do binary and multiple-star systems generally differ from planetary systems? Give evidence of a type of binary or multiple-star system that might be generically related to planetary systems.

2. How are novae related to binaries? How are supernovae related to binaries?

3. How will the evolution of a 1 $M_\odot$ star in orbit close to a 3 $M_\odot$ star differ from the evolution of a 1 $M_\odot$ star by itself? Why are binaries more likely to have formed in star clusters than as isolated field stars? If you saw a single star in the sky and suspected it was part of a binary system, how would you verify your idea?
4. Since sunlight is white (a mixture of all colors), why does the Sun look red at sunset? What happens to the blue light? Why does the part of the sky away from the Sun look blue? Why does the Moon look different when it is low on the horizon than when it is high in the sky? Why was the sky red as photographed on Mars by *Viking* cameras?
*5. If a nebula contains dust with an optical depth of $\tau = 1$, what percentage of the light from a background star is lost when passing through the nebula? What percentage gets through? What happens to the missing light? What percentage of light gets through if $\tau = 5$? Why would the optical depth be lower if we looked through the nebula using infrared radiation?
6. How do massive stars help keep interstellar gas stirred up? Why are O-type supergiant stars likely to be associated with large-emission nebulae, whereas solar-type stars are not? Why do planetary nebulae often have simple, nearly spherical forms, whereas typical emission nebulae, such as the Orion Nebula, are ragged and irregular?
7. Why are O and B stars the brightest in open clusters? Why are red giants the brightest stars in globular clusters? If you saw the galaxy from a great distance, which would be brighter, open or globular clusters? Which would be redder? Which would be farther from the galactic disk?
8. Sketch the H-R diagrams of an open cluster and a globular cluster. Pay particular attention to how the main sequences differ.
*9. (a) Suppose we measure an H-R diagram for a cluster with an unknown distance, and find that the main sequence is offset by a factor of 35 times fainter from the main sequence of a cluster at a distance of 120 pc. What is the distance of the cluster? (b) Now suppose that the cluster is a factor of 76 times fainter—what is the distance?
*10. If a star overhead is dimmed by 20 percent, how much dimmer would it be when it is 60° above the horizon? How much dimmer would it be when it is only 30° above the horizon? At what angle above the horizon would it be dimmed by a factor of 4?
11. (a) Describe the night sky as seen from near the center of a globular cluster. (b) Describe the night sky as seen from the center of the Milky Way galaxy. (c) Describe the night sky as seen from a star just above the disk of the galaxy. (d) Describe the night sky as seen from the middle of a nebula that is forming stars.

Projects

Activities to carry out either individually or in groups.

1. Observe Mizar and Alcor with the naked eye. They are actually a close pair, seen as the middle "star" in the handle of the Big Dipper. Can you see the faint star Alcor? Sketch its position. (If you cannot see Alcor, it may be because of insufficiently keen eyesight, a hazy sky, or a sky illuminated by city lights.) Try to measure the angular separation of the two stars.
2. Observe the eclipsing binary Algol with a telescope or binoculars each evening for 10 to 20 nights in a row. (This can be done as a class project with rotating observers.) Using neighboring stars as brightness reference standards, estimate the brightness of Algol. Can you detect the eclipses, which occur at intervals of just under 3 days?
3. The star Epsilon Lyrae is famous as the "double double." It consists of a binary pair 208 seconds of arc apart, easily visible with a small telescope. But each star is a binary pair only 2 to 3 seconds apart. These pairs are a test of good optics and good atmospheric observing conditions. Does your telescope reveal the two close pairs? Sketch them. Assuming the stars are at the same distance, what do the different colors and levels of brightness tell you about the stars?
4. In a dark area away from city lights, with the naked eye, observe and sketch the Milky Way in the region of Cygnus. Can you observe the dark "rift" that divides the Milky Way into two bright lanes in this region? The rift is caused by clouds of obscuring interstellar dust close to the galactic plane between Earth and the more distant parts of the Milky Way galaxy. Can you see any stars in the region of the rift? Are they fainter or redder than stars outside this zone of obscuration?
5. Observe the Orion Nebula with a telescope. Sketch its appearance. Locate the Trapezium (four stars near the center). The dark wedge radiating from the Trapezium is a dense mass of opaque dust. If a large telescope (50 to 100 cm) is available, look carefully for color characteristics. Generally, the eye is unresponsive to colors of very faint light, but large telescopes gather enough light so that colors can sometimes be perceived, especially with fairly low magnifications, giving a compact, bright image.
6. (a) Observe the Pleiades with your naked eye, and make a sketch. How many stars can you count in the group? Can you see all Seven Sisters? (The number of stars seen depends on keenness of vision, darkness of the observing site, and the clarity of the atmosphere.) (b) Observe the Pleiades and Hyades, or open clusters h and χ Persei, in a telescope. Move the telescope and compare star fields in and out of the cluster. Estimate how many more stars there are in the cluster than in the background region. Are all the stars you see in the region of the cluster at the same distance as the cluster?
7. Locate a globular cluster with the telescope. Make a sketch. Can you resolve individual stars? What are the colors of the stars? Compare the view in the telescope with photographs, where the central region is often overexposed and "burned out."

Reading List

Binney, J. 1995. "The Evolution of Our Galaxy." *Sky and Telescope,* vol. 89, p. 20.

Blitz, L. 1982. "Giant Molecular Cloud Complexes in the Galaxy." *Scientific American,* vol. 246, p. 84.

Crosswell, K. 1995. *Alchemy of the Heavens.* New York: Doubleday/Anchor.

Hartquist, T., and Williams, D. 1995. *The Chemically Controlled Cosmos: Astronomical Molecules from the Big Bang to Exploding Stars.* New York: Cambridge University Press.

Henbest, N., and Couper, H. 1994. *The Guide to the Galaxy.* Cambridge, Eng.: Cambridge University Press.

Hodge, P. 1988. "How Far Away Are the Hyades?" *Sky and Telescope,* vol. 75, p. 138.

Hoskin, M. 1986. "William Herschel and the Making of Modern Astronomy." *Scientific American,* vol. 254, p. 106.

Knapp, G. 1995. "The Stuff Between the Stars." *Sky and Telescope,* vol. 89, p. 20.

Reddy, F. 1983. "How Far the Stars?" *Astronomy,* vol. 11, p. 6.

Scoville, N., and Young, J. 1984. "Molecular Clouds, Star Formation, and Galactic Structure." *Scientific American,* vol. 250, p. 42.

Trimble, V., and Parker, S. 1995. "Meet the Milky Way." *Sky and Telescope,* vol. 89, p. 26.

Verschuur, G. 1989. *Interstellar Matters.* New York: Springer-Verlag.

Wynn-Williams, G. 1992. *The Fullness of Space: Nebulae, Stardust, and the Interstellar Medium.* Cambridge, Eng.: Cambridge University Press.

CHAPTER 15

Galaxies

CONTENTS

- Hubble and the Nature of Galaxies
- The Milky Way Galaxy
- Discovering the Distances to Galaxies
- Properties of Galaxies
- How Galaxies Evolve and Interact
- The Puzzle of Galaxy Formation
- Summary
- Important Concepts
- How Do We Know?
- Problems
- Projects
- Reading List

A spiral galaxy like our own Milky Way galaxy, as seen from a distant region of space. NGC 2997 is a spiral galaxy in the southern constellation of Antlia. Each external galaxy is composed of billions of stars. Billions of galaxies like this make up the basic building blocks of the universe. (SOURCE: Copyright Anglo-Australian Telescope Board.)

WHAT TO WATCH FOR IN CHAPTER 15

All the stars we can see in the night sky are part of a vast collection of stars called the Milky Way galaxy. In the previous chapter, we saw how the stars are distributed in space. In this chapter, we will first consider the overall properties of our own galaxy. Later we will discuss the properties of other galaxies. The disk in which the Sun is located is rotating, and we can map out the rotation. The galactic center contains a high concentration of stars, and there is growing evidence for a massive black hole there—a collapsed object millions of times more massive than a stellar remnant. The halo of the galaxy seems inconsequential, but it contains far more mass than the disk. Yet the biggest surprise in our study of the Milky Way is the indirect detection of dark matter. This unseen material drives the rotation of the disk and gives the Milky Way most of its mass.

You will learn about Edwin Hubble's discovery, in the 1920s, that many of the faint and fuzzy "nebulae" in the night sky are also galaxies. They are so far away that their light takes millions of years to reach Earth. Hubble provided a scheme for classifying galaxies according to morphology (shape). We will survey our neighbor galaxies in the Local Group and learn about the main types of galaxies. Galaxies have an enormous range of masses and sizes, providing further evidence of dark matter; the mass of virtually every galaxy we can study in detail is dominated by this material, which only makes its presence felt by the force of gravity. Galaxies are separated by vast gulfs of space, but they interact by means of gravity in ways that can alter their evolution. Finally, while we know much about galaxies, the process of galaxy formation is still poorly understood.

PREREADING QUESTIONS ON THE THEMES OF THE BOOK

OUR ROLE IN THE UNIVERSE
Is the Milky Way a typical galaxy?

HOW THE UNIVERSE WORKS
Why do galaxies differ so much in their properties?

HOW WE ACQUIRE KNOWLEDGE
What is the evidence for dark matter?

HUBBLE AND THE NATURE OF GALAXIES

About 100 years ago, astronomers were becoming increasingly divided over the nature of "nebulae." Nebulae were fuzzy patches of light—some smooth and featureless, and some spiral in form—that were scattered around the sky. Many astronomers thought that nebulae were clouds of gas or star-forming regions distributed within the Milky Way, similar to the Orion Nebula. Others thought that they might be "island universes," or vast collections of stars located far beyond the boundaries of the Milky Way.

In April 1920, the National Academy of Sciences held a debate in Washington, D.C. Harvard astronomer Harlow Shapley argued that nebulae were nearby objects located in the Milky Way. The spiral forms were presumed to represent the swirling material of a star-forming region. By identifying Cepheid variable stars within the globular clusters, and using the period-luminosity relation to estimate their intrinsic brightness, Shapley was able to measure distances to many globular clusters. He found that the globular clusters traced an enormous sphere in space, with the Sun offset from the center of the sphere. Shapley's Milky Way was so huge that he thought it inconceivable that other objects in space as large as our own galaxy might exist. For Shapley, the Milky Way *was* the universe.

The minority view was presented by Heber Curtis from Lick Observatory. Curtis believed that spiral nebulae could be understood as enormous systems of stars. He pointed out that many nebulae resolved into numerous points of light when viewed through a large telescope. The result of the debate was inconclusive. There was no convincing evidence in favor of either hypothesis. One young man in the audience would soon provide the evidence, a brilliant astronomer named Edwin Hubble.

Hubble was highly accomplished—he had been a boxer, a Rhodes scholar, and a lawyer before turning his attention to astronomy. He joined the Mount Wilson Observatory in Southern California just after the First World War. The observatory had just completed construction of a 100-inch telescope—at that time, the world's largest. Hubble quickly focused this big "eye" on a number of nebulae and took photographic plates of unprecedented depth and resolution. Like others before him, he saw that they broke up into the combined light of individual stars. Hubble went further, taking photographs of the Andromeda nebula repeatedly over a period of several months. His aim was to discover variable stars embedded within the nebula.

Hubble's observations were immediately successful. He identified a number of Cepheid variables in the Andromeda Nebula and measured their periods. With the period-luminosity relation, each period could be used to predict luminosity for the Cepheid. Knowing that light dims with the inverse square of increasing distance, Hubble could calculate the distance to the Cepheids, and then to the Andromeda Nebula that contained them. He deduced that the nebula was about 1 million light-years away. Hubble published his paper in 1925, and astronomy was changed forever. Here was the evidence to prove that many nebulae are enormous systems of stars, as Curtis had argued in the 1920 debate.

The discovery that nebulae are galaxies of stars represents a spectacular increase in the size of the known universe. The Milky Way is about 30 kpc across, or 100,000 light-years. Hubble showed that Andromeda is 10 times farther away. Soon he found other galaxies that were millions of light-years away. His photographs showed galaxies that were even fainter, and possibly at even larger distances. Hubble's work conjured up the possibility of space without end, an unimaginable void filled with countless systems of stars. Welcome to the immense realm of the galaxies.

The Milky Way Galaxy

For several hundred years, astronomers mapped out the stars in the Milky Way in the belief that they were tracing the shape of the entire universe. But Hubble showed that many of the nebulae in the night sky are galaxies of stars separated from us by vast gulfs of space. We now realize that the Milky Way is just one of the galaxies of the universe; it is not unique. Our study of the Milky Way is important in helping us understand the other galaxies, just as our study of the Sun is important in helping us understand other stars. We will therefore look more closely at the properties of our own system of stars to infer the properties of other galaxies. The discovery of galaxies beyond our own is a continuation of the Copernican theme—the idea that our location in the Milky Way is not a special place in the universe. It also sets a challenge to astronomers to show that the same physical laws that govern the Milky Way also govern galaxies remote from our own.

Components of the Milky Way

As we saw in Chapter 14, different groupings of stars in the sky have particular properties. Astronomers refer to these as the stellar components of the Milky Way galaxy. The Sun is located in the disk, a flattened distribution of stars that appears as a band of light in the night sky. Many of the stars in the disk are young and blue, and assembled into groups, or associations. The disk is surrounded by a spherical swarm of stars and groups of stars called the halo. Halo stars are generally old and red. Toward the center of our galaxy is a concentration of generally old and red stars called the galactic bulge. We can learn about the shape and size of the Milky Way galaxy by measuring the distances to stars and groups of stars and by plotting the distribution of their positions on the sky. We can do this by using observations at different wavelengths.

We see different aspects of the Milky Way when we observe at different wavelengths. The most familiar view is the one you can see with the naked eye when you are outside at night, far from city lights. Figure 15–1a shows the panorama of the Milky Way in a mosaic of photographs taken with an ordinary 35-mm camera. Most people live in suburbs or urban areas and are unfamiliar with this view. In the lower right-hand corner of the figure, the southern horizon blocks our view of the Milky Way. In the Southern Hemisphere, you would see the Milky Way pass gloriously overhead. The best view of the galactic center occurs at southern latitudes because Earth's spin axis is tilted with respect to the plane of the Milky Way.

Using visible wavelengths, the photograph clearly shows the concentration of stars toward the galactic center. It also shows an uneven distribution of stars and a band of obscuring dust that marks the plane of the Milky Way. As we learned in the previous chapter, dust dims and reddens the light from stars in all directions. If we look directly up out of the disk, we can see past all the stars and through all the interstellar gas and dust to the top edge of the disk. The galaxy is transparent in this direction. However, the dimming is particularly severe in the disk of the galaxy, because that is where the gas and dust associated with young stars are concentrated. After a distance of 500 pc through the plane of the disk, the galaxy is becoming opaque. At 500 pc, the optical depth $\tau = 1$, indicating that a fraction of $1/e$, or about one-third, of the light reaches us (see Chapter 14). Therefore, we cannot see much farther than 500 pc. Infrared waves are less affected by dust, so the way to see deeper into the Milky Way is to use long-wavelength radiation. Visible light has a wavelength of about 500 nm. If we increase the wavelength by a factor of 4 to 2000 nm (or 2 μm), we can see 10 times farther. In other words, the optical depth does not reach a value of 1 until a distance of 5000 pc.

Images taken by radio telescopes, by contrast, can penetrate all the way to the galactic center. Figure 15–1b shows the cold atomic gas, measured with radio telescopes using the 21-cm line of neutral hydrogen. Notice that the thin band of cool

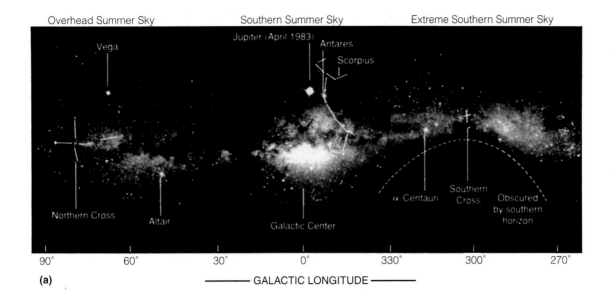

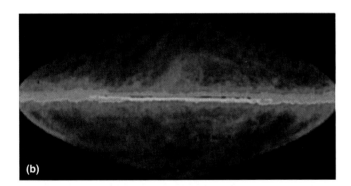

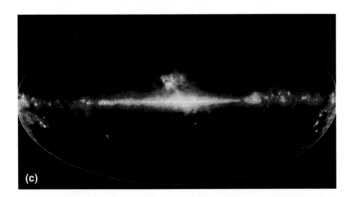

FIGURE 15-1
(a) A panorama of the inner Milky Way galaxy, from our position near the outer edge. This 180° view clearly shows the bright central bulge and the ragged dust clouds that lie along the plane of the galaxy. The galaxy's nucleus is hidden behind dark dust clouds at longitude 0°. (SOURCE: Mosaic of photographs from Hawaii by WKH.) **(b)** Atomic hydrogen in the Milky Way. Red indicates high hydrogen density; blue indicates regions of low hydrogen density. (SOURCE: NRAO/AUI.) **(c)** This map, at far-infrared wavelengths, was made with the Diffuse Infrared Background Experiment (DIRBE) on the Cosmic Background Explorer (COBE) satellite. Emission at 25, 60, and 100 μm is displayed as blue, green, and red, respectively. The Rho Ophiuchus cloud is at the center above the Milky Way, and Orion is at the right edge below the Milky Way. (SOURCE: Courtesy COBE Science Working Group, NASA/GSFC.)
(d) This DIRBE map at near-infrared wavelengths has emission at 1.2, 2.2, and 3.4 μm represented as blue, green, and red, respectively. The distributions of cool stars in the disk and bulge are clearly visible. Redder areas indicate larger amounts of obscuring dust; the central part of the Milky Way cannot be seen clearly at near-infrared wavelengths. (SOURCE: Courtesy COBE Science Working Group, NASA/GSFC.)

gas associated with disk stars is roughly uniform in all directions; the view is not obscured and chopped up by dust, as in Figure 15–1a. A wavelength of 21 cm is long enough to penetrate all the way to the galactic center. The thinness of the disk is clearly seen, but there are also complex structures of loops and filaments rising out of the galactic plane. These structures outline the ejected gas of supernova remnants. Notice that the filaments rise thousands of parsecs out of the galactic plane; a supernova can affect material over thousands of light-years!

Figure 15–1c shows the far-infrared emission of cool dust mixed in with the gas and stars that form the plane of the galaxy. Interplanetary dust particles in the solar system cause the S-shaped blue band. The intersection of the blue band and the red light shows the angle between the ecliptic and the galactic plane. Several nearby regions of star formation can be seen above and below the plane of the disk. Figure 15–1d shows the near-infrared emission from stars in the thin disk and the central bulge. You can see that near-infrared waves are not long enough to penetrate all the way to the galactic center. The peanut-shaped bulge is crossed by a band of obscuring dust.

Together, these images reveal structures traced out by gas, dust, and stars. Notice that the halo is extremely diffuse and does not show up in any of these pictures. Likewise, globular clusters are too small to show up clearly in the wide-angle views, and individual halo stars are too faint and too thinly distributed to be visible. Astronomers use images made at optical and radio wavelengths to estimate the galaxy's size. They deduce that we live in a truly enormous system of stars—30,000 pc (or 30 kpc) across, which translates into $30{,}000 \times 3.26 \approx 100{,}000$ ly, or $100{,}000 \times 9 \times 10^{12} \approx 10^{18}$ km. A size of 1 billion billion km is very hard to comprehend; no wonder it took astronomers more than a century to map out the size of the Milky Way galaxy!

Mapping the Rotating Disk

Astronomers have developed a method for converting observations of the radial motions of stars in different directions into a map of their positions in space. Let us see how this works. We have noted that cosmic systems of particles tend to become flattened if they are rotating. This is due to the principle of conservation of angular momentum: As any gas cloud collapses, the rotation speed will increase as the cloud gets smaller, while the collapse will be unimpeded along the axis of rotation. The result is a disk. The galaxy's flattened shape suggests that it, too, is rotating. All stars, including the Sun, are orbiting the massive bulge and the center of the disk.

How fast are we moving? It is difficult to tell by looking at the motions of nearby stars because they are all moving along with us in the rotating disk. The average Doppler velocities of nearby stars are 10 to 20 km/s. Recall that the Doppler effect shifts the wavelength of light and spectral features—to redder or longer wavelengths if the star is moving away from us, and to bluer or shorter wavelengths if the star is moving toward us. Roughly equal numbers of nearby stars are moving toward us and away from us. Remember that the Doppler shift only shows us the radial part of a star's motion. Equal numbers of redshifts and blueshifts mean that the motions are randomly distributed in three dimensions. However, if we look toward much larger distances, we can map the rotation of the entire galactic disk. The Sun is moving around the center of the galaxy at a speed of 225 km/s, or just over half a million miles per hour!

If the Sun travels at about 225 km/s, how long does it take the Sun to travel all the way around our circular orbit of the galaxy? We can calculate this easily because the length of the orbit is $2\pi r$, where the radius r is 8500 pc, and the orbital period $P = 2\pi r/v$, where v is the orbital velocity. Therefore, $P = 2\pi \times 8500 \times 3 \times 10^{13} / 225 = 7 \times 10^{15}$ s for one orbit. This is $7 \times 10^{15} / (3600 \times 24 \times 365) \approx 240$ million years. This orbit truly illustrates the vastness of our galaxy. Even traveling at half a million miles per hour, it takes 240 million years to complete an orbit. In its 4.6-billion-year history, Earth has repeated this journey 4.6/0.24, or nearly 19 times.

How fast do other stars orbit the galaxy? Nearby stars are at nearly the same distance from the center as we are, so they are moving around the center at almost the same speed as our Sun. Orbital speeds around the center vary at different distances from the center. The relation of orbital speed to distance departs from Kepler's third law, because the galaxy's mass is spread throughout many stars in the large central bulge, rather than being concentrated in one central object, as in the case of the solar system. Both linear and angular speeds around the center vary at different distances, which means the galaxy does not rotate as a solid disk; rather, the inner parts turn faster than the outer parts. This difference in speed at different distances is called *differential rotation* of the galaxy. Differential rotation, together with the random motions of stars, means that the stars of the galaxy do not move smoothly together but are ceaselessly changing their positions relative to each other as they move around the center.

We can view the radial part of the orbital speed at different distances if we use long wavelengths of electromagnetic radiation. The 21-cm radio emission produced by neutral hydrogen is especially useful for galactic mapping because it reveals the cold gas clouds that are concentrated where stars are forming. Similarly, it is possible to map out the galaxy in the millimeter wavelength emission of carbon monoxide (CO). Suppose we start scanning along the Milky Way with a radio telescope tuned to the 21-cm wavelength. Figure 15–2

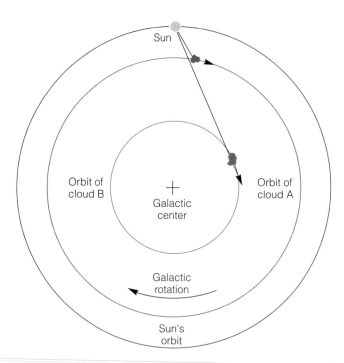

FIGURE 15–2
A view of the galactic plane from the north shows two hydrogen clouds, A and B, lying in nearly the same direction as seen from the solar system. Different Doppler redshifts, caused by different radial velocities, enable radio astronomers to distinguish the positions of the clouds. Clouds approaching the Sun on the other side of the disk would be blueshifted.

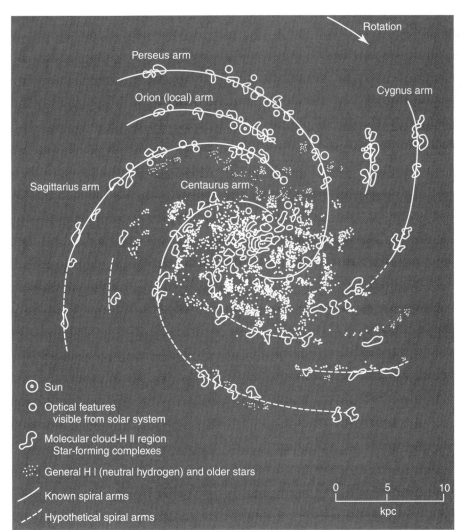

FIGURE 15-3
Currently known and estimated features of the Milky Way galaxy viewed from the north. Features nearest the Sun are the most well mapped; features on the far side of the galactic center are unmapped. Spiral arms, sometimes broken into pieces, can be traced out to a distance of 15 kpc, or about 50,000 light-years.

shows a situation in which we detect two clouds in virtually the same direction. However, cloud B should be moving faster than cloud A because it is closer to the center of the galaxy. This effect is enhanced because we view cloud B along a tangent to its orbit and see a Doppler shift that measures its full orbital speed. Cloud A is viewed at a larger angle to its orbital path, and so only a fraction of its speed is revealed as a Doppler shift. The mapping technique fails only if we look directly toward (or away from) the galactic center, when we see clouds moving on concentric orbits that show no Doppler shift.

Radio astronomers have used the 21-cm line to make a map *through* the disk of the Milky Way. As Figure 15–2 shows, the different velocities of clouds along a line of sight allow them to be placed on different orbits. Even though the Doppler shift usually gives only a fraction of the true space velocity of a cloud, astronomers can make a model of the rotating disk by assuming that all the clouds are on circular orbits. With this model, the measured Doppler shift along a particular line of sight translates into a distance for the cloud within the disk. The cold hydrogen clouds are located in a series of **spiral arms** and offer our first view of our galaxy's spiral shape. Additional evidence for spiral arms comes from young star clusters and associations. Many of these bright markers of star formation have distances measured by the techniques described in the last chapter. When astronomers plotted their positions in three dimensions, they saw that young stars and nebulae are not distributed at random, but lie in arms, coiling out at an angle from our galaxy's center.

Figure 15–3 shows the view of the Milky Way disk revealed by these different studies. The only part of the Milky Way that we cannot see clearly is the part of the disk on the opposite side of the galactic center. The spiral arms are named for the constellations in the directions of prominent features in each arm. The next arm beyond us is called the Perseus arm. Our arm is the Orion arm, sometimes called the Cygnus arm. The next arm in toward the center is the Sagittarius arm. The map shows us to be located on the inner edge of a spiral arm.

We can use this information to imagine what the Milky Way would look like if we could travel thousands of light-years above the disk. Astronomy can fuel our imaginations for a journey far beyond any we could take in a spaceship. If our galaxy could be seen from "above," it would probably look something like the view in Figure 15–4. We would see a delicate pinwheel of stars floating in black space, with the spiral arms trailing the rotation like hair swept back by the wind.

FIGURE 15-4
Our home, the Milky Way galaxy, as it might be seen from "above" the plane of the disk. The orientation is the same as in Figure 15-3. The solar system would be only one microscopic dot among the stars of the Orion arm in the upper-left center. The spiral arms are mostly made of bluish young stars. The central bulge is made of old red stars. A foreground globular cluster, one that happens to be high above the plane of the galaxy in its orbit, can be seen to the upper right. The brightest object is the nucleus. (SOURCE: WKH.)

Why does the galaxy have spiral arms? This question puzzled astronomers for many years. Some plausible "commonsense" analogies do not fit well with observations. For example, a rotating garden sprinkler sprays water out in jets that make a spiral pattern. The trouble with this model is that the material in galactic spiral arms is moving in a circle around the center, whereas the water droplets move on radial trajectories outward from the center. Another analogy is a cup of coffee with a few drops of cream stirred in. Because the coffee surface rotates faster at the center than at the rim, which slows it, the cream is sheared into long spiral streamers that look like galactic arms. The trouble with this model is that if arms are primordial features of the galaxy, twisted by rotation, they should be very old. Since the age of the galaxy is about 14 billion years and the rotation time is nearly 240 million years, there has been time for spiral arms to be twisted into about 60 complete windings. In contrast, the spiral arms in the Milky Way do not show more than one complete winding. Figure 15-5 shows that a spiral pattern would be rapidly scrambled by differential rotation.

Though the final answer is still uncertain, two modern theories help explain some features of galaxies. The *density-wave theory* emphasizes that spiral arms may not be fixed features of specific star groups, but rather waves in the galactic material. The crest of an ocean wave consists of certain molecules at one moment and certain other molecules the next, yet an outside observer sees a single wave that seems to have a history of its own. Spiral arms may be persistent concentrations of material, with individual stars entering an arm, passing through, and emerging on the other side. The galactic gas tends to pile up in the spiral arms, reaching high densities in the giant molecular clouds. Higher densities trigger more gravitational collapse, thus explaining why star formation occurs mainly in the arms. The concentration of star formation in turn explains the prominence of the arms, since newly formed massive stars are a galaxy's brightest stars. According to this view, the pattern of spiral arms does not rotate at the orbital speed of its constituent stars, but more slowly, with different stars defining the arms at different times. Perhaps the best analogy is the traffic "pile-up" that occurs on the highway when a slow-moving road crew is painting lines on the road (see Figure 15-6). Cars approach the obstruction at normal speed, bunch up as they pass the bottleneck, and then recover their normal spacing as they leave the obstruction. As seen from above, this would look like a moving traffic jam, where a region of high density involves a constantly changing set of cars.

A second theory about spiral structure is called the *stochastic star-formation theory*. It is based on the fact that star formation has not occurred smoothly and continuously since the galaxy formed, but rather in chain-reaction bursts. As discussed in the previous chapter, star formation occurs in open clusters, and the expanding gas from massive supernovae in one cluster compresses neighboring clouds of gas and dust, initiating the formation of new, adjacent clusters. Therefore, during a period of up to 100 million years, a large region of new clusters containing brilliant, hot, massive, short-lived stars may be produced in one part of a galaxy. During the galaxy's rotation in 200 million years or so, the inner edge of this region pulls ahead of the outer edge because of differential rotation (compare the rates of clouds A and B in Figure 15-2), and the region of bright, new stars is

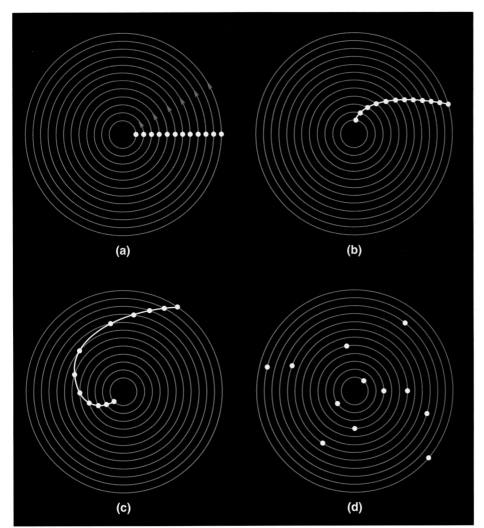

FIGURE 15-5
By Kepler's second law, stars near the center of a galaxy disk move faster than stars farther out in the disk. **(a)** Bright stars that start out in a straight line from the center of the galaxy rotate to **(b)** and **(c)** and show spiral structure after only a fraction of an orbit. **(d)** However, after a few revolutions, any sense of a pattern would be lost. (SOURCE: K. Kuhn.)

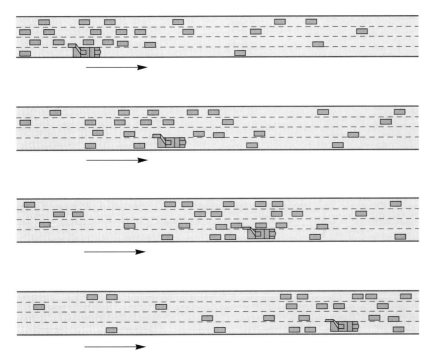

Direction of Traffic flow

FIGURE 15-6
An analogy for the density waves that contribute to the prominence of the arms in spiral galaxies. A slowly moving truck painting the lane marker on a highway represents a partial obstruction to the flow of traffic. Faster-moving vehicles slow down as they pass the obstruction. As seen from above, the bunched-up region of vehicles moves much more slowly than the vehicles themselves. Similarly, the prominent features in a spiral disk can be made of stars that move in and out of the feature. Traffic slows to the right; time increases going from top to bottom.

sheared into a spiral segment. Following a few hundred million years, this arm segment runs out of young stars and fades, which explains why the spiral pattern rarely achieves more than one winding. In summary, both theories suggest that spiral arms are shifting features associated with star formation; their individual member stars come and go, but the pattern persists.

The Mass of the Galaxy

We can use the motions of gas clouds and stars to measure the mass of the Milky Way galaxy. This measurement is another application of Kepler's third law, which we have previously used to study the motions of planets in the solar system and binary star orbits. As you can imagine, the motions in the galaxy are very complicated; instead of two or a handful of objects, there are many millions of stars moving over a very large region of space.

For example, the motion of the Sun is not only affected by all the stars between us and the galactic center, it must also be affected by stars that lie beyond the orbit of the Sun. How can we possibly take into account all those different gravitational forces? Isaac Newton had important insights into this problem. He showed that the motion of an orbiting object is controlled only by the mass of the system that lies *within* the object's orbit. In the case of the solar system, this is obvious because the Sun lies within the orbits of all the planets and therefore controls the motions. In the case of the galaxy, the Sun's orbital period is controlled only by the portion of the galaxy that lies within the orbit of the Sun. The stars exterior to the Sun's orbit do not affect its orbit; thus, the Sun, in effect, does not "feel" the outer regions of the galaxy. Newton proved this assertion for any symmetrical mass distribution, such as a disk or a sphere. If you could travel toward the center of Earth, the amount of material between you and the center would decrease. As a result, the gravity force on you would get smaller and you would weigh less. At Earth's center, with no material between you and the center, you would be weightless!

We can thus use the Sun's orbital motion to calculate the mass of the galaxy within a radius of 8.5 kpc from the center. The accompanying Science Toolbox shows the calculation. The result is about 10^{11} $M_\odot$, or 100 billion times the mass of the Sun. If we put together the information from radio and optical surveys, we can plot the average circular velocity of material moving away from the galactic center. Figure 15–7 shows the result, called a **rotation curve.** As anticipated by Newton, there is no net rotation at the center. The fast rotation speed within the inner kiloparsec is due to the gravity of the high density of stars in the galactic bulge. Then it levels off at a value just over 200 km/s out to the distance of the Sun from the center. What happens beyond the position of the Sun is unexpected and extremely interesting. The rotation curve continues to be flat, with no decline in orbital speed out to the most distant regions we can measure.

Why is this result so surprising? Because we expect the rotation of the galaxy to diminish once we are beyond the bulk of the material in the disk. As you can see from the

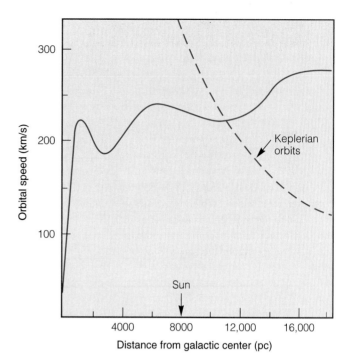

FIGURE 15–7
A rotation curve for the Milky Way galaxy. The circular velocity of the disk of the Milky Way remains roughly constant from the galactic center to a radius well beyond the position of the Sun. Motions associated with the spiral arms cause slight variations in velocity. The dashed line shows how the circular velocity would decline if the visible light from the disk traced the mass. The high disk velocities at large radii indicate dark matter in an extensive halo.

opening picture of this chapter, a spiral galaxy does not have a sharp edge. The clouds of gas and stars get sparser and sparser until they fade off into the darkness of space. We define 15 kpc as the radius of the galactic disk because that distance encompasses about 90 percent of the mass of the disk. Consequently, we expect the orbital speed of the disk to decline beyond that radius, as defined by Kepler's law (the dashed curve in Figure 15–7). We observe such a decline in the solar system; beyond the edge of the Sun, which includes almost all the mass in the solar system, the orbital speeds of the planets get smaller. In the Milky Way galaxy, we can find a few wisps of gas and groups of stars all the way out to 20 kpc. No decrease in orbital speed is seen; in fact, the speeds increase slightly! You can see from Figure 15–7 that the outermost regions of the disk are moving twice as fast as they should if the visible material of the galaxy represents the entire mass.

What do these motions tell us about the distribution of mass in our galaxy? The Science Toolbox shows that a simple application of Kepler's law indicates a total mass of 10^{11} $M_\odot$ within the orbit of the Sun. This area has a radius of 8.5 kpc from the galactic center. The continued rapid motions at the visible edge of the disk yield a sum of 2×10^{11} $M_\odot$ out to a distance of 15 kpc. Beyond this radius, clouds of gas and stars are so dim that it is difficult to measure their motions.

SCIENCE TOOLBOX

Weighing a Galaxy

How is it possible to weigh a galaxy? It is possible to measure the mass of a galaxy using the motion of one star within it! In the last chapter, we saw how Kepler's law could be used to measure the masses of stars in binary orbits. With mass in solar units, a in AU, and P in years, the equation is

$$P^2 = \frac{a^3}{(M_A + M_B)}$$

If we take M_A to be the mass of the galaxy within the Sun's orbit and M_B to be the mass of the Sun, then $M_A \gg M_B$. Therefore, we can replace the Sun by the single number M_G. We can also consider a to be r, the radius of the Sun's orbit around the galactic center. Newton showed that the Sun only responds to mass inside its orbit, so this calculation cannot give us the mass outside the Sun's orbit. Finally, we insert numerical constants to the equation:

$$P^2 = \frac{4\pi^2 r^3}{G M_G}$$

We also know that in a circular orbit, the period is given by the circumference of the orbit divided by the orbital velocity, or $P = 2\pi r/v$. Substituting this expression in the left-hand side of the equation above, and rearranging, gives the new result,

$$M_G = \frac{rv^2}{G}$$

G is the gravitational constant, with a value of 6.67×10^{-11} in units of Newton $\times$ m^2/kg^2. Now we can insert the known values of r and v, being careful to use meters for r and km/s for v. The orbital velocity of the Sun is $v = 225$ km/s, or 2.25×10^5 m/s. The distance to the galactic center is $r = 8500$ pc, or $8500 \times 3 \times 10^{16}$ m. The result is

$$M_G = 2.55 \times 10^{20} \times \frac{(2.25 \times 10^5)^2}{6.67 \times 10^{-11}}$$
$$= 1.9 \times 10^{41} \text{ kg}.$$

One solar mass is 2×10^{30} kg, so the total mass interior to the Sun's orbit is $1.9 \times 10^{41} / 2 \times 10^{30} \approx 10^{11} M_\odot$. This enormous number reflects only the material between our orbit and the galactic center, so it is obviously a lower bound on the total mass of our galaxy.

Now we can also see why the flat rotation curve of the Milky Way implies a steadily increasing mass in the outer regions of the galaxy. The equation above can be used to calculate the mass within any circular orbit in the disk. A flat rotation curve means that the velocity does not increase or decrease with radius. In effect, v is a constant. G is also a constant, so we deduce that

$$M_G \propto r$$

This result is remarkable. In the case of the solar system, the Keplerian orbits of the planets are responding only to the Sun. When the mass is all contained in the center of the system, the equation above shows that $v \propto 1/\sqrt{r}$. The orbital speeds of the planets reduce with distance from the Sun—a declining rotation curve. In our galaxy, the rotation curve is flat out to a large radius. Thus, the mass of the galaxy continues to increase out to the largest radii we can measure.

However, there are a few stars in the halo with distances measured out to 40 kpc. The speeds of these stars give a measure of the mass of the galaxy on the largest scales. The best measure of the total mass of the galaxy out to this distance is about $5 \times 10^{11} M_\odot$.

We have just encountered one of the deepest mysteries in astrophysics. Most of the mass of the Milky Way galaxy lies in an enormous halo beyond the visible edge of the disk. Since the 1970s, astronomers have known that stars and gas are just the tip of an iceberg of mass. The extra material that drives the motions of stars and gas clouds in the Milky Way does not shine. It does not emit radiation in any part of the electromagnetic spectrum. It reveals itself only by its gravitational influence. This fascinating material is **dark matter.**

How can we be sure that dark matter is not just normal material that is very difficult to see? One by one, astronomers have eliminated the possibilities. Perhaps the dark matter is composed of very cold gas atoms far from the galactic center. We can rule this hypothesis out, because cold gas would emit thermal radiation at radio wavelengths, which we do not see. Perhaps the dark halo is full of tiny dust particles. This hypothesis is ruled out because dust would re-emit starlight at longer wavelengths, and the galactic halo is not bathed in infrared radiation. Scientists have even suggested an exotic solution: Perhaps a large number of black holes comprise the dark matter. This suggestion is unlikely. Even though isolated black holes would be difficult to detect, they must be produced by the violent death of massive stars, and we would see other traces of the lives of these stars. Perhaps the dark matter is made of objects that are too small and cool for hydrogen fusion to take place in their central regions—from brown dwarfs down to planet-sized objects. This is the most difficult possibility to rule out, but recent observations by the Hubble Space Telescope suggest that there are too few low-mass stars near the limit of hydrogen fusion for this explanation to work.

So far, we can only say what dark matter is *not*. We can rule out normal material in forms ranging from atoms and

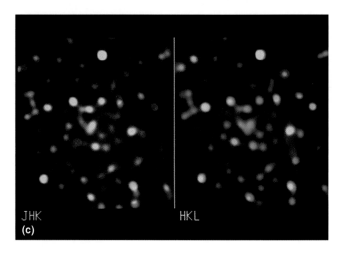

FIGURE 15-8
Homing in on the center of our galaxy. Each successive view zooms in by a factor of 100. **(a)** An optical picture of the central few kiloparsecs of the Milky Way. Dust obscures our view of the center of the galaxy, attenuating the optical light by a factor of 10^{12}. (SOURCE: National Optical Astronomy Observatories.) **(b)** An infrared image of the central half-degree toward the galactic center, showing the dense star cluster. A compact radio source lies close to the center of this cluster. The color-coding has 1.2, 2.2, and 3.4 μm as blue, green, and red, respectively. (SOURCE: Courtesy N. Sharp, NOAO.) **(c)** Composite infrared images of the central parsec of the galaxy, with effective image size of only 0.4 second of arc. The star density in this region is 10^7 times higher than in the solar neighborhood. Measurements at 1.2, 1.6, and 2.2 μm are blue, green, and red, respectively. (SOURCE: D. Depoy and N. Sharp, NOAO.)

tiny dust particles to substellar objects and massive stellar remnants. Astronomers are left with two unpalatable options. Newton's law of gravity might be wrong, in the sense that it does not correctly predict the motions of matter on large scales. Alternatively, we can postulate a new microscopic form of matter—a subatomic particle that has not yet been observed in the laboratory. As we will see later in the book, our very understanding of the universe depends on resolving this issue.

The Galactic Center

Our view of the night sky reveals a profusion of stars and dusty clouds in the direction of the galactic center. For a long time, astronomers have been trying to understand this dusty region, 30,000 light-years from us. In the 1930s, Bell Telephone Laboratories put the young researcher Karl Jansky to work on sources of radio static interfering with long-distance radio signals. Jansky built the first radio telescope and discovered that one major source was a steady hiss from the Milky Way galaxy. In the 1940s, an amateur astronomer and radio enthusiast, Grote Reber, built a radio telescope in his backyard and made the first radio maps of the Milky Way. He established that the strongest radio emission comes from the galactic center. This type of mapping reveals the nucleus because radio waves (along with infrared, X-ray, and gamma-ray wavelengths) can penetrate all the way from the galactic center. In fact, the extinction (the factor by which visual radiation is extinguished) due to dust toward the galactic center is a factor of 10^{12}. We might as well be trying to look through a closed door!

From the time of Jansky's first observations in 1932, it has been clear that something extraordinary is happening in the center of the Milky Way. At the long wavelengths he used, the galactic center is the brightest object in the sky, even brighter than the Sun. To understand this complex region, we will describe the different structures revealed by radio and infrared observations as we "zoom in" on the galactic nucleus (see Figure 15–8).

Satellite observations have clearly revealed the bulge of the Milky Way (see Figure 15–1d). Imaging shows that the bulge is elongated in the plane of the Milky Way; such an elongated distribution is called a *stellar bar*. The effect of this structure is to drive gas clouds in the inner few kiloparsecs away from their circular orbits, causing gas to fall toward the center of the galaxy. (Figure 15–8a shows the central few kiloparsecs of the Milky Way.) As a result, the inner 300 pc of the galaxy has 10^8 $M_\odot$ of molecular gas, 100 times the density of molecular gas of the galaxy as a whole. This same region is permeated with a high-temperature, high-pressure gas at 100 million K, which emits strong X rays. Star formation is very likely to occur in a region with dense and highly pressured molecules.

Now we zoom in by a factor of 100. There is an intense region of radio emission in the central few parsecs of the Milky Way called Sagittarius A. Radio astronomers have detected gas that appears to be swirling into these central regions. Figure 15–8b shows that the concentration of stars rises sharply toward the center of the galaxy. Radio astronomers have discovered a very compact radio source at the center of this distribution of stars—one of the densest star clusters ever discovered. Let us zoom in by another factor of 100 to see this star cluster in more detail. Figure 15–8c is a composite of images made at three near-infrared wavelengths. Infrared cameras resolve the center of the star cluster into discrete sources. Many of these individual stars are young, blue, and massive supergiants. The central parsec contains over 1 million stars; its density of stars is 10 million times that of the solar neighborhood. If we lived on a planet around a star near the galactic center, our night sky would be ablaze with stars as bright as the full Moon!

Something extraordinary is going on in the galactic center. Not only is the density of stars extremely high, but those stars are also moving very fast. Recent measurements of stellar velocities near the galactic center allow astronomers to make models of the mass distribution in the dense stellar core. Just as the speed of planet orbits tells us about the mass of the Sun, the speed of stellar orbits tells us about the mass of the central parsec of our galaxy. The result of the models is the prediction that a **supermassive black hole** lurks in the central regions. In other words, we cannot see nearly enough stars in the galactic center to account for the motions of those stars; they appear to be responding to a massive, unseen source of gravity. Notice that this "dark mass" in the galactic center is very different from the dark matter that dominates the outer regions of the galaxy. The dark matter in the galactic halo is widely distributed and cannot be identified with any known form of matter. The dark matter in the galactic center is the result of an extremely high density of stars. Theorists have calculated that stars can collide and merge and create a black hole with a mass far greater than any single star.

However, the evidence for a supermassive black hole is still circumstantial. Remember that stellar mass black holes are predicted as a consequence of stellar evolution. For small compact objects, the evidence is strong, but their existence is not proven beyond doubt. In the case of the galactic center, the models indicate a black hole mass 2 or 3 million times the mass of the Sun! We should be duly skeptical until the evidence is more conclusive. Meanwhile, we can imagine the extraordinary scene near the galaxy's nucleus (Figure 15–9).

FIGURE 15–9
A hypothetical panorama from a position only a few hundred parsecs away from the Milky Way's nucleus. Here we are surrounded by thousands of brilliant red giants and other stars of the halo population and by dense clouds of dark dust and glowing nebulosity lying along the Milky Way's plane. In the right distance is the actual nucleus, possibly a large black hole surrounded by a brilliantly glowing accretion disk and partly obscured by clouds. (SOURCE: Painting by WKH.)

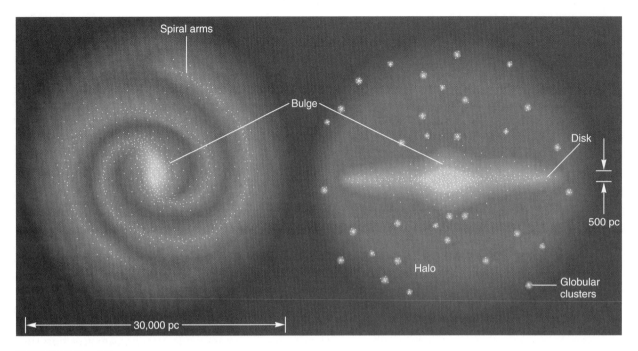

FIGURE 15-10
A cartoon view of the Milky Way looking face-on to the disk and edge-on to the disk, showing the dimensions and main components.

Stellar Populations

One of the most interesting discoveries about our galaxy is that it contains a range of star types of different composition, age, distribution, and orbital geometry. We have talked about collections of stars that have formed together and are held together by gravity—stellar groups and associations, and globular clusters. However, there are also major components of the Milky Way galaxy that share properties such as age, chemical abundance, or motions called **stellar populations**. Figure 15-10 shows a schematic view of the Milky Way, as seen from above and from the side, with the different stellar components marked.

Around us in the disk of the Milky Way we see stars that are relatively young. The spiral arms of the disk are traced by star-forming regions that are full of hot, young stars. These *disk population stars,* or *disk stars,* travel in circular orbits around the galactic center, like the Sun. (In fact, they have a slight vertical motion that makes them undulate as they orbit the galaxy.) Only a small percentage of the mass of disk population stars consists of elements heavier than helium, including carbon, oxygen, silicon, and iron. Astronomers often make a simple distinction between the primary elements of a main-sequence star—hydrogen and helium—and all the elements with higher atomic numbers in the periodic table. Elements beyond helium can only be found in a star after a star from a previous generation has left the main sequence and ejected those heavier elements into space. Therefore, the proportion of elements beyond helium is a measure of the total amount of previous star formation in the region of space that contains the star. Figure 15-11 shows the difference in the spectra between the Sun and a star that has very few lines from elements heavier than helium.

Stars in the halo—called *halo population stars,* or *halo stars*—live in a spherical swarm that extends far above and below the disk of the galaxy. The halo contains many indi-

FIGURE 15-11
A comparison of the spectra of a halo star (upper, HD 140283) and a disk star (lower, the Sun) of similar spectral type. (Bright lines at top and bottom are matching comparison spectra produced in the laboratory.) Both stars have prominent absorption lines of hydrogen (Hδ and Hγ), but the Sun has many additional absorption lines caused by various heavy elements. In the halo star these lines are very weak or absent, indicating that the heavy elements are virtually absent from its gases.

vidual halo stars and a few hundred majestic globular clusters. All these stars are moving on elliptical orbits that loop in and out of the galactic plane from any direction. Figure 15–12 shows the very different orbits of disk and halo stars. Halo stars were first identified by their rapid motions. Recall that the Sun is traveling around the disk at a speed of 225 km/s. Stars near us in the disk show small Doppler shifts of 10 to 20 km/s with respect to the Sun. Imagine that we are whipping around a circular racetrack. All the cars around us are going at roughly the same speed, but there is a smaller relative motion due to the cars that pass us and the cars that are being passed by us. Early in the 1900s, astronomers found stars at high galactic latitude with large Doppler shifts of 100 to 200 km/s. Such high speeds represent the motion of halo stars plunging toward or away from the disk. We can even find a few near the Sun that are speeding through the solar neighborhood. Stars of the halo population are nearly pure hydrogen and helium, with only a small percentage of heavier elements.

Astronomers have slightly confusing terminology for these stellar populations. *Population I* is the name for the disk population, not because these stars formed first, but because they were the first type of stars with which astronomers became familiar, located near the Sun. *Population II* is the name for the halo population, the type of star discovered second. In this book, we will use the more descriptive terms *disk population* and *halo population*. Also, astronomers refer to the elements other than hydrogen and helium as "heavy elements," even though they are not all very heavy, and they also refer to them as "metals," even though they are obviously not all metallic!

Astronomers now know that the division of all stars into two populations is an oversimplification. The orbits and volume of space occupied by disk stars and halo stars are different, but in other respects each population shows a range of properties. Halo stars generally have a tiny percentage of elements heavier than helium, whereas disk stars have a range of 1 to 5 percent in their abundance. Since halo stars formed with 10 to 50 times less of the important planet-forming materials such as silicon, oxygen, nickel, and iron, we might speculate that planets (and perhaps living organisms, too) are rare around halo stars. Halo stars are old, ranging in age from about 9 to 13 billion years, as indicated by the ages of globular clusters discussed in Chapter 14. Disk stars have an enormous range in age. The Sun is 4.5 billion years old, and a few disk stars are even older. Most disk stars are much younger, and the age range goes all the way down to zero—we see stars that are just forming now.

The bulge of our galaxy represents an intermediate population. Within a few kiloparsecs of the galactic center, we see not only stars that are old and red but also stars that are relatively young and blue. The *bulge population* is most similar to the halo population, but it includes younger stars with a larger proportion of heavy elements.

Table 15–1 lists the main properties of the three stellar populations in our galaxy. Notice that the halo has ten times more stellar mass than the disk. The many stars in the halo are not readily apparent for two reasons. First, they are often at large distances from the Sun, so their light is dimmed by the inverse square law of radiation. Second, the halo population is old enough that the most massive and luminous stars left the main sequence long ago. Halo stars are mostly

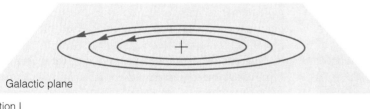

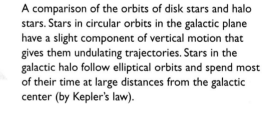

FIGURE 15-12
A comparison of the orbits of disk stars and halo stars. Stars in circular orbits in the galactic plane have a slight component of vertical motion that gives them undulating trajectories. Stars in the galactic halo follow elliptical orbits and spend most of their time at large distances from the galactic center (by Kepler's law).

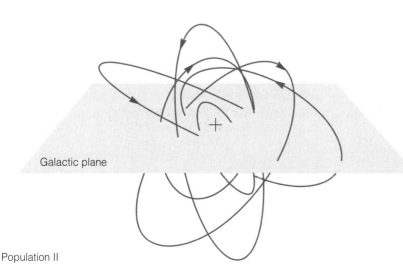

TABLE 15-1
Stellar Populations and Their Properties

PROPERTY	DISK POPULATION	BULGE POPULATION	HALO POPULATION
Stellar orbits	Circular	Elliptical	Elliptical
Distribution of stars	Patchy, spiral arms	Smooth	Smooth
Gas content	High	Low	None
Concentration toward galactic center	None	Slight	Strong
Typical extent (kpc)	15 (radius) × 0.4 (thick)	4 × 2	40 (spherical)
Heavy elements (%)	1–5	0.4–2	0.1
Total mass ($M_\odot$)	2×10^9	5×10^{10}	2×10^{10}
Typical age (years)	10^8	10^9	10^{10}
Typical star velocity relative to Sun (km/s)	10–20	20–100	120–200
Typical objects	Open clusters, stellar associations, gas and dust, young and hot stars	Sun, RR Lyrae stars, giant stars, novae, variable stars	Globular clusters, RR Lyrae stars, Cepheids

low-mass main-sequence stars and white dwarfs, stars that are intrinsically dim. Remember also that the stellar census in Table 15–1 does not include dark matter. We saw earlier that the total mass of the galaxy is about $5 \times 10^{11}\ M_\odot$, most of it lying far beyond the visible halo and disk. Adding the stellar mass of the three populations in Table 15-1, we see that they only account for about 15 percent of the mass of the galaxy. The bulk of the galaxy is composed of enigmatic dark matter.

The Formation of the Milky Way

The Milky Way has a very distinctive geometry: a rotating disk embedded in a spherical halo. The geometry of the galaxy is one of several clues that tell us how the Milky Way formed. The disk and halo populations also differ in heavy element abundance, in ages of the stars, and in the orbits of the stars. Is there any formation scenario that can account for all these differences?

The story our galaxy began 13 or 14 billion years ago. A diffuse gas cloud spread over an enormous region, at least as large as the present extent of the halo. A first generation of stars formed out of this material, which was composed of 75 percent hydrogen (by mass), 25 percent helium, and virtually no heavy elements. In fact, the halo may have assembled from smaller clouds even earlier, but it is difficult to probe this early epoch. When stars form out of a gigantic gas cloud, it is like a scaled-up version of our description of star formation. The remaining gas collapsed toward the center of the cloud. Since the initial gas cloud was rotating slightly, this rotation was amplified as the gas collapsed, due to the conservation of angular momentum. The disk probably formed slowly, and the high density and compression of the gas provided a continuing reservoir for star formation.

This picture accounts for the different shapes of the disk and the halo and the different stellar orbits in those two populations. As stars in the nearly spherical halo formed, they were traveling on elliptical orbits that spanned the extent of the halo. Since the remaining gas collapsed to a disk, the halo stars almost never meet or collide in their orbits. The motions of halo stars represent a "frozen" view of the earliest days of the galaxy. With no reservoir of gas to draw on, there is no continuing star formation, and the halo stars age and die. After more than 10 billion years, all that remain are low-mass, reddish main-sequence stars and dim white dwarfs. Meanwhile, the disk has a large reservoir of gas, and stars continue to form in that dense stellar environment. The churning of material from star birth and death causes the heavy element content of the disk to grow with time.

The key idea is that the chemical composition of stellar populations tells us something about the history of the Milky Way. Heavy element abundance is like the "fossil record" of astronomers as they study galaxies. As described in Chapter 13, nuclear reactions inside stars fuse light elements into heavier ones through successive stages of stellar evolution. Heavy elements are created inside stars and ejected by disruptive phases of stellar evolution, such as red giants, planetary nebulae, and supernovae into interstellar space. Later generations of stars form from this interstellar gas that has been enriched with heavy elements, and more heavy elements are steadily added to the gas.

This simple picture must be modified for several reasons. First, the globular clusters did not all form quickly and at the same time; there is a spread of about 3 billion years between the youngest and oldest globular clusters. Also, our picture has not yet explained the galactic bulge, which has a relatively large percentage of heavy elements, and therefore must have formed later than most of the halo stars. Astronomers now believe that the galaxy may have been formed from smaller units, which merged and formed stars over a period of several billion years. Another complication stems from the current star-formation rate in the disk. We can observe the disk in the 21-cm spectral line of hydrogen and accurately measure the gas content. This gas is the reservoir from which stars must form. It turns out that the disk is making stars so quickly that it could not possibly have sustained this activity for 12 to 13 billion years with the available gas supply. Either the disk formed stars more slowly in the past, or it has received fresh infusions of gas from beyond the Milky Way galaxy. Our galaxy may not have evolved as an isolated unit; its evolution has probably been affected by the external environment.

Our confidence in this formation scenario is tempered by the unknown nature of the dark matter. All this activity—the formation of disk stars, bulge stars, and halo stars—takes place in the presence of a much larger amount of dark matter. The gravity of the dark matter dictates the motions of these stars, just like soldiers marching to the beat of an unseen drummer. Until we discover the physical nature of the dark matter, our knowledge of the Milky Way will be incomplete.

Discovering the Distances to Galaxies

Speculation about the nature of galaxies began long before Edwin Hubble's observations with the giant 100-inch telescope at Mount Wilson Observatory. In the middle of the 18th century, German philosopher Immanuel Kant and German mathematician Johann Lambert each considered the possibility that certain fuzzy patches in the night sky might be enormous systems of stars remote from the Milky Way. Kant wrote:

> If such a world of stars is beheld at such an immense distance from the eye of the spectator situated outside of it, then this world will appear under a small angle as a patch of space whose figure will be circular if its plane is presented directly to the eye, and elliptical if it is seen from the side or obliquely.

In other words, Kant realized that a flattened system of stars like the Milky Way might appear to have different shapes depending on its orientation in space. He also realized that a galaxy seen from afar would subtend a small angle in the sky. Finally, he knew that the dimness of the starlight from a galaxy could be explained by the inverse square law of light traveling across a vast distance of space.

Speculation is not the same as evidence, however. For more than 150 years, astronomers struggled to understand the nature of nebulae. One hypothesis held that they were clouds or whirlpools of gas condensing to form stars. Spectroscopy provided some support for the local hypothesis. British amateur astronomer William Huggins showed that the spectra of a number of planetary nebulae and unresolved nebulae like that in Orion were caused by hot gas and not unresolved stars (recall the emission lines from a hot, glowing gas discussed in Chapter 5). The alternative, and much more radical, hypothesis held that nebulae were distant systems of stars. Spectroscopy of the Andromeda Nebula showed the features expected of a composite of stars, demonstrating that not all nebulae were purely gaseous. In addition, nebulae were distributed all over the sky, showing no preference for the plane of the Milky Way. Nor did they appear to be centered on the center of the galaxy, like globular clusters. As we saw in the Science Toolbox in the previous chapter, the way objects are scattered in the sky gives clues to their distribution in space. Objects that are very far from Earth will appear isotropic (the same in every direction) in their distribution in the sky.

Part of the confusion about nebulae stemmed from the fact that the catalogs of nebulae created by Herschel and Messier did not consist of a single class of object. Messier's catalog contains a mixture of star-forming regions like the Orion Nebula (M 42), stellar remnants like the Crab Nebula (M 1), and galaxies like the Milky Way (M 31 and M 51, for example). Most astronomers were reluctant to subscribe to the "island universe" hypothesis—the idea of nebulae as giant star systems remote from our own—for the simple reason that they could not conceive of a universe much larger than the Milky Way. We have seen this idea repeatedly in the history of astronomy since Copernicus. Each major discovery has made us minor players in a larger universe. The awareness of insignificance is not a comfortable feeling!

Resolving the controversy over nebulae required a new and reliable indicator of distance. After resolving the Andromeda Nebula into myriad individual stars, Hubble's crucial observation was to identify and measure the periods of Cepheids in Andromeda. Using the period-luminosity relation, he concluded that the Andromeda Nebula was a galaxy of stars about 300 kpc away, far beyond the periphery of the Milky Way. Since then, Hubble's estimate has been revised upward by a factor of 2, due to an error in the Cepheid distance calibration. Figure 15–13 shows the reasoning involved in his momentous discovery.

We have seen that a chain of techniques is used to measure distances in astronomy. Any property of a star or galaxy that can be used to estimate distance is called a **distance indicator.** Distances within the solar system are measured in the most direct way possible, by the timing of radar signals bounced off nearby planets. The distances to stars near the Sun are measured by their parallax. Very nearby stars can also show observable proper motions. Some of the stars with distances measured directly by geometry are in clusters, allowing the calibration of the technique of main-sequence fitting on an H-R diagram. Now we can plot H-R diagrams for other open clusters, and the amount by which the main sequence has to be shifted in brightness to overlay the cluster of known distance is a measure of the relative distance of the two clusters (see Chapter 14). This technique takes us with some reliability out to the edge of the Milky Way.

To calibrate distances to nearby galaxies, a more luminous distance indicator is needed than a star like the Sun. Populous clusters include RR Lyrae and Cepheid variable stars. The well-established period-luminosity relation for these rare stars enables astronomers to calculate distance, given the apparent brightness. Cepheids, in particular, are up to 20,000 times more luminous than the Sun and therefore are visible with large telescopes out to a distance of 10,000 to 20,000 kpc. Finally, it is possible to use supernovae (see Chapter 13), which can outshine the normal stars of an entire galaxy and can be seen out to distances of hundreds of thousands of kiloparsecs. Unfortunately, a supernova is such a fleeting stage of stellar evolution that one is observed only every 40 to 50 years in a galaxy. Therefore, there is no guarantee that we can use this measure in a specific galaxy.

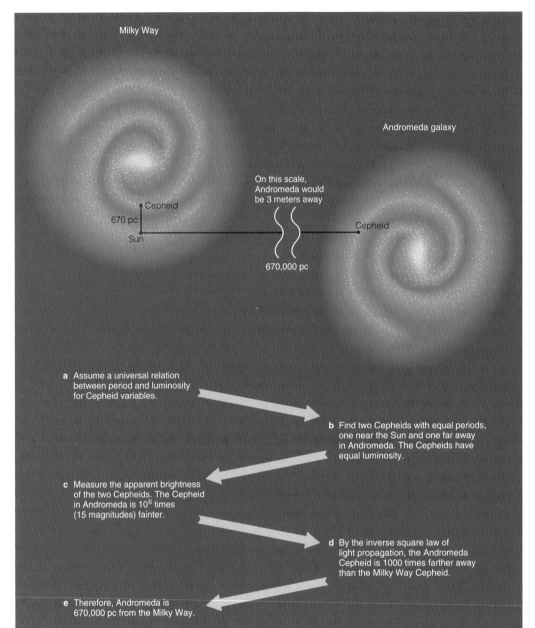

FIGURE 15–13
The chain of reasoning followed by Edwin Hubble in determining the distance to the Andromeda Nebula. A key assumption is that the period-luminosity relation for Cepheids is the same in the Milky Way and Andromeda. In general, this means assuming that the laws of physics apply uniformly across large distances in the cosmos.

The **distance scale** is the set of measurements that define distances from the solar system out to the most remote galaxies. Conceptually, the methods of the distance scale form a pyramid. Nearby methods are more direct and accurate. Moving farther from the Sun, we must use different techniques. Each technique depends on the reliability of those that work at smaller distances. A range of overlap in distance is required to calibrate each new method. Errors continue to accumulate and grow as we reach toward the galaxies. Figure 15–14 illustrates the overlapping ranges and increasing errors of distance estimators to the nearby galaxies.

Why must we use so many different methods to measure all distances in the universe? Part of the reason is the sheer immensity of empty space. The type of distance indicator found near the Sun must also be found 10,000 times farther away to map out the Milky Way. By the inverse square law, this increase in distance corresponds to a dimming by a factor of 100 million. The type of distance indicator found across the Milky Way must also be discovered 1000 times farther away to be useful in measuring the distances to other galaxies, a dimming by a factor of 1 million. Nearby distance indicators lose their usefulness as objects become too faint to observe. Distant ones are ineffective in the local universe because luminous stars are very rare. Cepheids can be seen in galaxies up to 20 million pc away, but the nearest one to the Sun is about 200 pc, too far for a reliable parallax measurement. Also, it is sheer bad luck that our galaxy has not hosted a supernova since 1604, since we might expect a supernova every 50 years or so.

The measurement of distance is absolutely fundamental to astronomy. Without knowing an object's distance, astronomers cannot derive basic parameters such as size, mass,

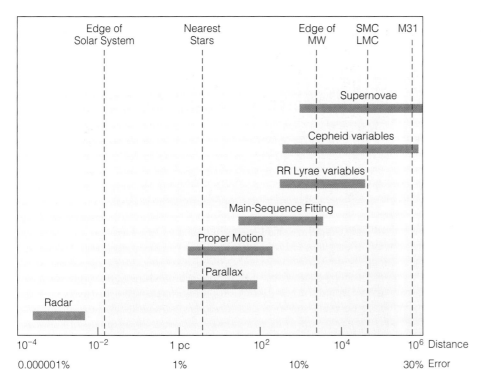

FIGURE 15-14
The chain of overlapping distance indicators plotted on a logarithmic scale, with a horizontal bar showing the range of applicability for each method. Important thresholds in distance are marked as vertical lines. Supernovae can be used to very large distances, but there are no calibrators in our own galaxy (see Chapter 16). SMC means Small Magellanic Cloud; LMC means Large Magellanic Cloud.

and luminosity. The goal in distance determination is to find objects whose intrinsic brightness or size is well known—distance indicators. Knowing the absolute brightness (from the well-understood physics of the source) and the apparent brightness (measured through a telescope), we can easily calculate the distance. It sounds simple, but the long journey to understanding the nature of nebulae shows that accurate distances are very hard to measure.

When astronomers measure distances, they must assume that the laws of physics are constant across the universe. Recall that the laws of physics are confirmed only in terrestrial laboratories, and to a more limited degree from our study of Moon rocks and particles in interstellar space. Beyond our galaxy, their application requires a powerful and unifying assumption. Hubble was relieved to find Cepheids and other familiar stars in distant galaxies, because it indicated to him that "the principle of the uniformity of nature thus seems to rule undisturbed in this remote region of space."

Properties of Galaxies

Having established the extragalactic nature of nebulae, Hubble took photographs of many galaxies and tried to classify them according to their appearance, or *galactic morphology*. The standard scientific approach in a new area of research is to classify objects by type and then arrange them in a system that shows smooth transitions from one type to another. The act of classification does not lead directly to a physical theory, but the hope is that the relationship between galaxy types will have a physical underpinning. For example, the galaxy types might be related by age, by gas content, by stellar populations, by mass differences, or by rotational differences. Hubble found three major types of galaxies in his study of the regions of space beyond the Milky Way: spiral, elliptical, and irregular galaxies.

A Classification System

The Milky Way is a **spiral galaxy.** The components of a spiral galaxy are the disk, the bulge, and the halo (some spirals also may have a nucleus or a bar). The disk has spiral arms with bright emission nebulae; large amounts of dust are often mixed in. The dust forms an obscuring band when the disk is viewed edge-on. When the disk is viewed face-on, the spiral arms appear clearly outlined by luminous, young, blue stars. Roughly one-third of spiral galaxies have bright barlike features in their central regions. In such cases, the spiral arms originate at the ends of the bar, rather than in the nucleus itself.

In both normal and barred spirals, Hubble noticed a gradual transition of morphological types. Among normal spirals, the sequence from *Sa* to *Sb* to *Sc* goes toward less tightly wound spiral arms and less prominent central bulges. The corresponding barred spirals are classified from *SBa* to *SBb* to *SBc*. Galaxies are flung at random orientations in space, so some disks will be seen face-on and some edge-on. Examples are given in Figure 15–15. The size of the central bulge and the degree of winding of the spiral arms go hand in hand, so we can classify spirals according to Hubble's scheme even if they are viewed edge-on. Spirals with more loosely wound arms have much more prominent luminous stars, star clusters, and emission nebulae that outline the

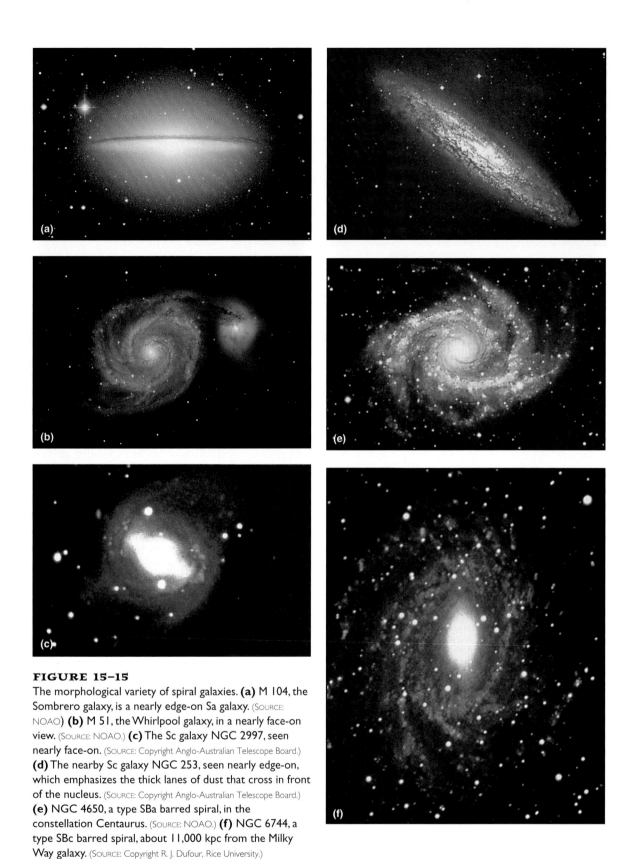

FIGURE 15-15
The morphological variety of spiral galaxies. **(a)** M 104, the Sombrero galaxy, is a nearly edge-on Sa galaxy. (SOURCE: NOAO.) **(b)** M 51, the Whirlpool galaxy, in a nearly face-on view. (SOURCE: NOAO.) **(c)** The Sc galaxy NGC 2997, seen nearly face-on. (SOURCE: Copyright Anglo-Australian Telescope Board.) **(d)** The nearby Sc galaxy NGC 253, seen nearly edge-on, which emphasizes the thick lanes of dust that cross in front of the nucleus. (SOURCE: Copyright Anglo-Australian Telescope Board.) **(e)** NGC 4650, a type SBa barred spiral, in the constellation Centaurus. (SOURCE: NOAO.) **(f)** NGC 6744, a type SBc barred spiral, about 11,000 kpc from the Milky Way galaxy. (SOURCE: Copyright R. J. Dufour, Rice University.)

arms. All spirals rotate in the sense that the arms trail, as observed for the Milky Way. The rotation of spirals can be mapped by Doppler shifts measured three different ways: using stellar absorption lines, using emission lines from H II regions, or using the 21-cm line of neutral hydrogen. Note that this rotation cannot be observed when a disk is face-on, because in this case the motion is transverse to the line of sight and there is no Doppler shift.

Another type of galaxy with prominent bulges has disks that lack spiral arms. These are classified as *S0 galaxies* and

FIGURE 15-16
(a) The star-bursting irregular galaxy NGC 1313. (b) Another irregular galaxy is IC 5152, roughly 610 kpc away. It is barely resolved into stars. The bright images with spikes and other scattered stars are foreground stars in our galaxy.
(SOURCE: Copyright Anglo-Australian Telescope Board, courtesy D. Malin.)

are sometimes called *lenticular galaxies* because of their lenslike shape. What about the third component, the halo? The halo is too diffuse to be visible, even on deep images of galaxies. The difficulty in seeing halos is important since, as we shall see, the halo contains most of the mass of any spiral galaxy! Hubble did not recognize S0 galaxies as a distinct class; modern astronomers consider them intermediate between spirals and ellipticals.

Spiral galaxies can vary in size, but the size increase from the smallest to the largest is only by a factor of 10. By contrast, one type of galaxy has an enormous range in size—an **elliptical galaxy.** Elliptical galaxies (also called *spheroidal galaxies*) range from tiny dwarfs to giants that are three to four times larger than the Milky Way. In general, elliptical galaxies have smooth shapes like spheres or squashed spheres and no spiral arms. They are classified according to how round or flat they look and according to their size. The numerical scheme from *E0* to *E7* runs from circular to highly elongated galaxies. Since galaxies are three-dimensional objects distributed in space, our perspective from Earth may not give us a true indication of the shape. An E7 elliptical is highly elongated, so it *must* be a relatively flat galaxy seen edge-on. However, an E0 elliptical is round, but it *need not* be a spherical galaxy. Both a flattened distribution of stars viewed face-on and a cigar-shaped distribution of stars seen end-on would be classified E0. Ellipticals have a single population of reddish stars, with little indication of gas, dust, or young, luminous stars. It was once suspected that rotation caused some elliptical galaxies to become flattened. We now know that though ellipticals can rotate, the amount is too slight to cause the observed flattening.

The third category of galaxies is the **irregular galaxy.** Irregular galaxies exist in many shapes, and they are usually small (Figure 15-16). Some irregulars have a degree of spiral structure but without the high symmetry of spirals. Others have chaotic morphologies without any obvious symmetry. Many of these chaotic irregulars appear to have undergone collisions or to be in the process of merging. Irregular galaxies have regions of intense star formation with conspicuous young stars. The Magellanic clouds are the only two examples visible to the naked eye (if you happen to live in the Southern Hemisphere). Finally, we note that a number of galaxies have unusual morphologies and resist being shoehorned into Hubble's classification scheme. *Peculiar galaxies* may have loops or tails or other extended structures not seen in irregulars.

Although classification is an important first step in organizing the richness of the extragalactic universe, it does not automatically lead to physical understanding. Hubble organized galaxies into a "tuning fork" diagram, in which the normal and barred spirals form parallel sequences (Figure 15-17), and the S0 galaxies are a transition type between the spirals and the ellipticals. This system correlated well with certain galaxy properties. For example, compared to spirals, elliptical galaxies have older stars and smaller amounts of gas and dust. For some time it was believed that the Hubble classification implied an evolutionary sequence, in which spirals gradually used up their gas, the stars aged and faded, and the final result was an elliptical. We now know that this cannot be true, since spirals contain old halo populations, and many must be as old as ellipticals. Morphology alone does not explain the differences in galaxy type.

The Local Group

To understand the neighborhood around the Milky Way, we will take a reconnaissance journey, describing the nearby galaxies in order of distance. Our journey will span a distance of 1000 kpc, or 3.25 million light-years. The galaxies in this volume of space make up the **Local Group.** A three-dimensional sketch of the Local Group is shown in Figure 15-18. Our cosmic neighborhood teaches us about a very important feature of the universe: Galaxies are not randomly distributed through space; they tend to cluster. Most of the galaxies in the Local Group are clumped into two subgroups, those around the Milky Way and those around the Andromeda galaxy. To visualize the distances, imagine the volume containing the Local

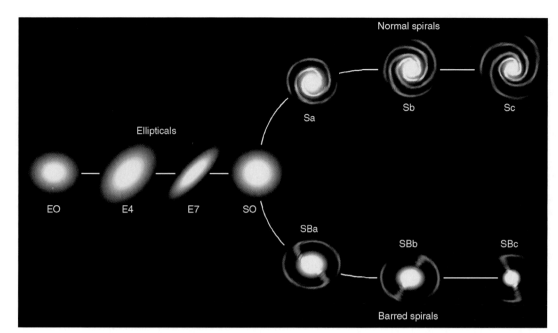

FIGURE 15-17
The simplified classification scheme for galaxies known as Hubble's tuning fork diagram, after the originator of the system. It is now clear that this sequence is not a simple evolutionary progression.

Group to be a medium-sized room. On this scale, the Milky Way would be the size of a dinner plate. The Magellanic clouds would be like crumpled balls of cotton 8 or 10 cm across within 1 m of the plate. A dozen or so galaxies of various shapes, from 1 to 15 cm across, would be scattered across the room. Andromeda, the nearest galaxy resembling the Milky Way, would be another dinner plate 7 m away.

Since prehistoric times, humans must have been aware of two glowing patches in the southern sky. Since there is no bright south polar star, the clouds helped navigators mark the pole. Europeans heard them described during Magellan's around-the-world expedition in the early 16th century, so they came to be called the **Magellanic clouds.** The clouds are in fact small galaxies moving in orbits around the Milky Way. The Large Magellanic Cloud is 50 kpc away, and the Small Magellanic Cloud is 63 kpc away, less than three times the distance to the far edge of our own galaxy. What kinds of galaxies are these companions of ours? They are several

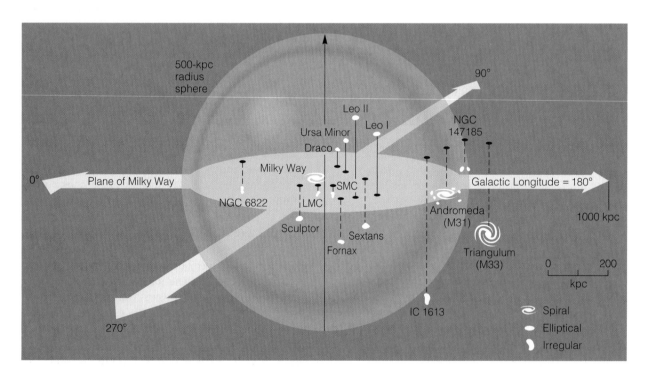

FIGURE 15-18
A three-dimensional map of most of the Local Group of galaxies. Solid lines extend north of the galactic plane; dashed lines extend south. The clustering of dwarf satellite galaxies around the great Milky Way and Andromeda spiral galaxies can be seen. It would take light just over 3 million years to cross the sphere.

FIGURE 15-19
(a) The Small Magellanic Cloud, about 63 kpc away and 8 kpc across, is an irregular galaxy exhibiting a barlike configuration of bluish stars. A long exposure reveals the extent of the faint outer regions and slightly overexposes the bright inner regions. **(b)** The Large Magellanic Cloud, about 50 kpc away and 19 kpc across, is the closest bright galaxy to us. A color view shows a barlike distribution of bluish stars and a scattering of red-glowing H II regions, including the massive Tarantula Nebula (center left).
(SOURCE: Copyright Anglo-Australian Telescope Board, courtesy D. Malin.)

times smaller than the Milky Way, judging by star counts and measures of neutral hydrogen that show that they are only a few percent as massive as the Milky Way. The clouds do not have the Milky Way's beautiful spiral structure; they are irregular galaxies. Each contains a softly glowing, barlike structure composed of stars, as shown in Figure 15–19. Somewhat off the end of the bar in the Large Magellanic Cloud is the spectacular Tarantula Nebula, also known as 30 Doradus. This luminous nebula can be seen with the naked eye. In fact, if it were moved to the distance of the Orion Nebula, it would fill the whole constellation of Orion and be bright enough to cast shadows on Earth! In its center is a cluster 60 pc in diameter containing thousands of massive, bluish supergiant stars.

We can best understand other galaxies and their evolution by considering their stellar populations. The Large Magellanic Cloud has a disk population similar to that found in our own spiral arm of the Milky Way. Faint-star photometry has also identified red giants and main-sequence stars. The bulk of the star formation in both clouds occurred 1 to 3 billion years ago, but it continues to the present day. There is little dust in the two clouds, except in the prominent young nebulae. Overall, the stellar populations are younger and more deficient in heavy elements than the solar neighborhood. The two clouds are connected by a bridge of diffuse hydrogen gas called the *Magellanic stream*. Australian radio astronomers showed that this long filament of H I also extends from the small cloud in an arc beyond the south galactic pole, and in the other direction it reaches into the plane of the Milky Way. This filament resembles the bridge between the two clouds themselves. The Magellanic clouds are satellites of our own galaxy, gravitationally bound to the Milky Way. Their orbits are likely to take them through the Milky Way disk, and astronomers speculate that the Magellanic stream is a tail of gas drawn out during such an encounter about half a billion years ago.

The Magellanic clouds are important to astronomers because they provide a cornucopia of stellar types at essentially identical distances. Use of the Magellanic clouds as a stellar laboratory was given a boost by the explosion of Supernova 1987A in the Large Magellanic Cloud. The Hubble Space Telescope has been used to study the ring of hot gas that was thrown off by the death of this star, deriving an accurate distance of 50 kpc using simple geometric arguments. The door is now open to calibrate many distance indicators, using the rich stellar nursery of the Large Magellanic Cloud.

The Local Group contains a number of small stellar systems (see Appendix B–5). Most of these are dwarf elliptical galaxies, and a few are irregulars. Elliptical galaxies resemble giant globular clusters. Two dwarf companions to the Milky Way are shown in Figure 15–20. Dwarf ellipticals are dominated by old halo stars and have little gas or dust. In most respects they are less impressive than our own giant spiral disk, with its chaotic clouds of gas and dust and regions of continuing star formation. However, they appear to be more active than globular clusters, whose stars are all around 9 to 13 billion years old.

FIGURE 15-20
(a) One of the closest galaxies is this loose dwarf elliptical called Leo I, about 1 kpc in diameter. At a distance of 180 kpc, it is close enough to be easily resolved as separate stars, and it looks somewhat like a very large globular cluster. (b) The nearby galaxy NGC 6822 is a small irregular galaxy about 3.5 kpc across and 500 kpc away. At this relatively close distance, as galaxies go, it is easily resolved into individual stars, pinkish H II regions, and a few ring-shaped supernova remnants. As in other photographs of external galaxies, uniformly scattered brighter stars are foreground stars in our own Milky Way. (SOURCE: Copyright Anglo-Australian Telescope Board, courtesy D. Malin.)

Analysis of H-R diagrams of the individual stars in dwarf ellipticals indicates that some of their stars are relatively young, only 3 to 9 billion years old. Dwarf ellipticals are the most common type of galaxy, but their diffuse light makes them difficult to detect. The satellite companions of the Milky Way are 5 to 30 times smaller and 1000 to 100,000 times less luminous than large spirals like M 31 and the Milky Way.

At 670 kpc, we encounter the first spiral galaxy truly comparable to the Milky Way, along with several of its smaller satellite galaxies. This galaxy must have been known since prehistoric times, because it is visible to the naked eye as a hazy patch on a clear, dark night. It was first recorded in a star catalog by Arab astronomer al-Sufe in A.D. 964. Hubble used Cepheid variables in the **Andromeda galaxy** (M 31) to finally settle the debate over the nature of the so-called spiral nebulae. The Andromeda galaxy is slightly larger than the Milky Way and similar in stellar content. The naked eye sees it as a faint patch of light (Figure 15-21a), but this is really only the brightest, innermost region, a few kiloparsecs across. Images made with large telescopes show that the spi-

FIGURE 15-21
(a) This color photograph of the central bulge of the Andromeda galaxy emphasizes the great brightness of the central region relative to the faint spiral arms. Only the inner edges of the spiral dust lanes are visible. Old stars of yellowish and reddish color dominate the central bulge. (SOURCE: Copyright Association of Universities for Research in Astronomy, Cerro Tololo Interamerican Observatory.) (b) A wide-field Schmidt telescope photograph of M 31, the Andromeda galaxy. M 31 is a type Sb spiral with a diameter of 30 kpc. The small galaxy below M 31 is a dwarf elliptical companion, M 32. (SOURCE: University of Oregon.)

SCIENCE TOOLBOX

Light Travel Time

In the everyday world, light seems to travel instantaneously from one place to another. In fact, the speed of light is not infinite, but it is an extremely large number: 300,000 km/s, denoted c. We can divide any distance by this number to figure out the time it would take light to cross that distance. In this way, we can see that light takes $1.5 \times 10^8 / 3 \times 10^5 = 500$ seconds to reach Earth from the Sun, or just over 8 minutes. It takes light about 40 times longer (Pluto at a distance of 39.4 AU) to leave the solar system, or about 5 hours.

These are large distances to us, but they are dwarfed by the distances between stars. As we consider larger regions of the Milky Way, a natural distance unit is the distance light travels in one year. This is called a light-year. We can easily calculate the size of this unit by remembering that distance has the dimensions of velocity times time. Thus:

$$D_{ly} = vt = c \times 1 \text{ year} = 3 \times 10^5 \times (3600 \times 24 \times 365) = 9.5 \times 10^{12} \text{ km}$$

A light-year is the typical distance between stars in the neighborhood of the Sun. It is nearly 10 trillion kilometers! The fundamental unit of distance defined by geometry is the parsec, equal to 3.1×10^{13} km. The units are related by a small numerical constant:

$$D_{ly} = 3.26 \, D_{pc}$$

Therefore, to roughly convert from parsecs to light-years, multiply by 3.3. Table 15–A gives the distance to various points within the Milky Way and beyond, in terms of both parsecs and light-years. Notice that distances are usually quoted with no more than two significant figures, because the precision of a distance measurement is rarely better than 1%. To fully appreciate how isolated we are in space, remember that light is the fastest thing we know. Our best spacecraft could not reach 1 percent of the speed of light. You would have to multiply the numbers on the right-hand side of the table by at least 100 to estimate how long it would take to send a probe through the Milky Way with current technology.

What do these values of look-back time mean? They mean that light from the nearest star has been traveling just over 4 years to reach us. When we look at the Orion Nebula in the night sky, the light we see is 1500 years old and left Orion when Europe was in the Dark Ages. Light from the Large Magellanic Cloud has been traveling for just over 160,000 years, since before human civilization began. If you observe the Andromeda galaxy with binoculars or a small telescope, that light is nearly 3 million years old. In other words, we see Andromeda as it was before humans evolved from apes on Earth!

What does Andromeda look like today? We cannot say! Since nothing travels faster than light (or other forms of electromagnetic radiation), there is no quicker way to send information from one place to another. We are stuck with collecting and measuring "old" light. While this seems like a limitation, we will see in the next chapter that light travel time is a wonderful tool. By looking farther out in space, we look farther back in time. In this way, astronomers get to explore the earlier stages of the universe.

TABLE 15–A
Distances Within the Milky Way and Beyond

LOCATION	DISTANCE (pc)	LIGHT TRAVEL TIME (y)
Nearest star to the Sun (Alpha Centauri)	1.3	4.2
Sirius	2.7	8.8
Vega	8.1	26
Hyades cluster	42	134
Pleiades cluster	125	411
Orion Nebula	460	1500
Nearest spiral arm (Sagittarius arm)	1200	3900
Center of the galaxy	8500	29,000
Far edge of the galaxy	24,000	78,000
Large Magellanic Cloud	50,000	163,000
Andromeda galaxy (M 31)	670,000	2.18×10^6

---

ral arms form a disk at least 30 kpc across (Figure 15–21b). As with the Milky Way, there are globular clusters and a halo of H I gas reaching perhaps 100 kpc in diameter. The Andromeda galaxy played an important role in the discovery of the two main stellar populations.

In the Local Group, galaxies are clearly made up of the light of many individual stars. Recall that stars in the neighborhood of the Sun are typically separated by about 1 parsec. That is a separation in three dimensions; when a galaxy disk is viewed face-on, projection effects will make the typical separation in the plane of the sky several times smaller, around 0.2 to 0.3 pc. We can use the small-angle equation to predict the distance out to which individual stars in a galaxy could be resolved. The best telescope on the ground can resolve stars with a separation of about 0.5 second of arc. The small-angle equation gives $D = 206,265 \, d/\alpha = 206,265 \times 0.3/0.5 = 1.2 \times 10^5$ pc, or 123 kpc. Hubble was able to resolve Cepheids at the larger distance of M 31 by working on the periphery of the galaxy and by using the fact that Cepheids are much brighter than the surrounding stars. The Hubble Space Telescope has angular resolution 10 times better than ground-based telescopes. This enables individual stars to be resolved in galaxies out to 10 times the distance, which encompasses the entire Local Group.

As we conclude our imaginary voyage through the galactic neighborhood, we should recall the vast distances that separate galaxies. When astronomers gather information about the universe, they use electromagnetic radiation that travels at 300,000 km/s. This enormous number—the speed of light—is nature's ultimate speed limit. Nothing can move faster. So far we have used parsecs and kiloparsecs to measure distances beyond the solar system. The parsec is a unit based on the geometry of stellar parallax. We could equally well use the time it takes light to travel a particular distance. The natural unit is a light-year, which is the distance that light travels in a year. Remember that 1 pc = 3.26 ly (remember also that a light-year is a unit of distance, *not* time). The distance to Andromeda is therefore $670,000 \times 3.26 = 2.8 \times 10^6$ ly. When you go out on a dark night and look at Andromeda through a small telescope, the light you are seeing has taken over 2 million years to reach Earth!

The time that light takes to traverse the vast distances of the universe is called **light travel time.** The preceding Science Toolbox shows how to calculate light travel time. As we explore regions more distant from the Milky Way in this book, the light travel time will increase. The farther out in space we look, the farther back in time we look. We will use large telescopes as "time machines" to view parts of the universe that are remote in both space and time.

Size and Luminosity

Although it sounds simple, it is not trivial to measure the size of a galaxy, since galaxies do not have sharp edges! As we can see from the photographs in this chapter, the light from a galaxy smoothly fades away until it becomes indistinguishable from the background sky level. The conventional way to measure a diameter involves summing up the total light from the galaxy through a very large aperture, so that the total light measured does not depend very much on the aperture's size. The diameter is then defined as the size enclosing half of the total light (the exact percentage does not matter, as long as it is large and applied consistently from galaxy to galaxy).

Measured in this way, galaxies vary widely in size. Dwarf ellipticals and irregulars can be as small as a few kiloparsecs; several companions to the Milky Way are this size. The disks of spiral galaxies range in size from 10 to 50 kpc. The largest galaxies in the universe are giant ellipticals with diameters up to 200 kpc, much larger than the distance from the Milky Way to the Magellanic clouds.

The luminosity of a galaxy is derived directly from its distance and apparent brightness. As with the size, measuring the apparent brightness of an extended object is not simple, because the brightness of the galaxy falls off toward the edge. Two strategies are used. The first is to measure the flux through a large enough aperture that essentially no light is missed. The other is to use the fact that the light from most galaxies falls off with distance from the center in a simple and predictable way. With this strategy, we can make a mathematical model of the light distribution and use it to calculate the total brightness and hence the luminosity. Galaxies have a wide range in luminosity, from $10^{11} \, L_\odot$ for giant ellipticals to $10^5 \, L_\odot$ for the puniest satellites of the Milky Way.

Colors, Stellar Populations, and Mass

The colors of galaxies provide useful information on the types of stars that dominate the light output. Stars that are young, blue, and rich in heavy elements are found in spiral disks. Stars that are old, red, and dominated by only hydrogen and helium are found in ellipticals, the bulges of spirals, and the globular clusters that orbit the halos of many galaxies. All this information can be seen in color photographs of the giant elliptical M 87 (Figure 15–22) and the spiral M 83 (Figure 15–23, on page 456). M 87 has a uniform halo population of old, red stars. On the periphery of this giant galaxy you can see a swarm of faint globular clusters. The nucleus of M 83 has the pale yellowish-orange color of giants. But the spiral arms are made out of bluish clusters of young stars, along with red-glowing regions of hot gas scattered like rubies in a jeweled brooch.

Each of the three galaxy types identified by Hubble has a particular combination of stellar populations and a characteristic color. Irregular galaxies have predominantly young disk stars (even though they may not have a well-organized disk or spiral arms) and are correspondingly blue and gas-rich. Spiral galaxies contain both old and young populations, and they follow a sequence from type Sc to type Sa of increasing amounts of light in the older bulge population. Elliptical galaxies contain mostly old halo stars and are correspondingly red and gas-poor. Therefore, the colors of a galaxy provide some indication of its age and history.

One of the most difficult properties of a galaxy to measure is its mass. We encountered this problem before in our discussion of stars. It is difficult to estimate the mass of a single star in space unless we have a lot of information about the evolutionary state of the star. With binary stars, we can apply the law of gravity to the orbits and learn about the masses. Similarly, when two galaxies are in a binary orbit, we can measure their masses. But what about the mass of a single, isolated galaxy? It turns out we can use our knowledge of stellar populations to get a useful estimate.

Let us characterize a galaxy by the ratio of its mass (in units of the mass of the Sun) to its luminosity (in units of the luminosity of the Sun). This quantity is called the **mass-to-light ratio.** Clearly, if a galaxy consisted entirely of stars like

FIGURE 15-22
The giant elliptical galaxy M 87 (NGC 4486) is about 13,000 kpc away and 20 kpc across (including its halo of globular clusters). It is some 5 to 10 times bigger than typical dwarf elliptical galaxies. The faint, fuzzy starlike images around M 87 are thousands of globular clusters swarming like bees around a hive. M 87 is about 40 times as massive as the Milky Way. Two background spiral galaxies appear in the image. Notice the reddish colors of the halo population stars. (SOURCE: Copyright Anglo-Australian Telescope Board.)

the Sun, it would have a mass-to-light ratio of 1. However, a galaxy is a *composite* of many millions of stars of differing ages and masses. The global mass-to-light ratio of a galaxy depends on the relative numbers of stars of different types. The visible light from galaxies of all types has a mass-to-light ratio in the range 1 to 30. Irregular galaxies, which have the largest percentage of young stars, are at the bottom of that range. Spiral galaxies, with a relatively large percentage of young stars, are in the middle of the range. Elliptical galaxies, with a majority of old stars, tend to be at the top of

FIGURE 15-23
The great barred spiral galaxy M 83 (NGC 5236) beautifully displays the consequences of stellar and galactic evolution. Here the yellowish central regions are surrounded by spiral arms dominated by bluish disk stars, massive and short-lived, forming in clusters. Scattered through the arms are red-glowing H II regions associated with local areas of star formation. M 83 is a type SBc barred spiral about 3000 kpc away and about 12 kpc in diameter. (SOURCE: Copyright Anglo-Australian Telescope Board.)

the range. The properties of galaxies are summarized in Table 15-2.

Why do galaxies of all types have a mass-to-light ratio larger than 1? You can see why if you review the material on H-R diagrams in Chapter 12 (particularly Figure 12-20). A young main-sequence star that is more massive than the Sun is enormously more luminous than the Sun. In our calculation of stellar lifetimes in Chapter 13, we showed that a star of mass 100 $M_\odot$ has $M/L = 100/10^6 = 10^{-4}$. By contrast, a main-sequence star less massive than the Sun has a faint luminosity. Again following our Chapter 13 calculation, a star of mass 0.1 $M_\odot$ has $M/L = 0.1/10^{-3} = 100$. Since we also know that high-mass main-sequence stars are relatively rare (see Figure 12-21), we anticipate an overall mass-to-light ratio above 1. Giants and supergiants are massive, but they are highly luminous, so they have low mass-to-light ratios like young main-sequence stars. White dwarfs have mass similar to the Sun but are low-luminosity stars, so they have large mass-to-light ratios. The calculation of a mass-to-light ratio for an entire galaxy is complex, but the general result is easy to state. The mass-to-light ratio is dictated by lower main-sequence stars and white dwarfs. As a stellar population ages and stars evolve off the main sequence, the mass-to-light ratio will evolve to higher values. Therefore, the sequence of in-

TABLE 15-2
General Characteristics of Galaxies

CHARACTERISTIC	SPIRALS	ELLIPTICALS	IRREGULARS
Mass ($M_\odot$)	10^9–10^{12}	10^6–10^{13}	10^8–10^{11}
Luminosity ($L_\odot$)	10^8–10^{11}	10^5–10^{11}	10^8–10^{11}
Mass-to-light ratio	2–10	3–30	1–3
Diameter (kpc)	5–50	1–200	1–10
Stellar populations	Old halo and intermediate age bulge, young disk	Old halo	Young and intermediate age
Composite stellar spectral type	A (Sc) to K (Sa)	G to K	A to F
Interstellar material	Gas and dust in disk	Small amounts of gas and dust	Copious amounts of gas, some dust
Large-scale galaxy environment	Small groups, low-density regions	Rich clusters	Low-density regions

creasing mean age going from irregular to spiral to elliptical galaxies is also a sequence of increasing mass-to-light ratios.

These calculations refer only to the visible stellar populations in galaxies. All mass estimates must be reconsidered in light of one of the most profound discoveries in astronomy: the detection of large amounts of dark matter. Our consideration of stellar populations gives us an important way to infer the presence of dark matter. We have seen that old stellar populations have larger mass-to-light ratios, as massive stars leave the main sequence and many white dwarfs are created. No stellar population is older than about 13 billion years, the age of the oldest globular clusters. Therefore, the oldest galaxies give us the *maximum* value of mass-to-light ratio that can be expected from any group of stars. A mass-to-light ratio much above 30 is evidence for dark matter that cannot be any form of normal stellar population.

The Evidence for Dark Matter

Earlier in the chapter, we learned that the Milky Way is mostly composed of a surprising form of matter that reveals itself only by its gravitational attraction. Dark matter is spread out over the entire galactic halo. As we have learned more about the universe, we have found nothing that is special or unique about our situation. Our Sun is an average star. Earth is part of a planetary system, and more and more planetary systems are being discovered. Our galaxy is one of many spiral galaxies. It would violate the Copernican idea that we have no privileged position in the universe if dark matter only existed in the Milky Way galaxy. Do we find dark matter in other galaxies?

The answer is an emphatic yes! Regardless of the technique used or the Hubble type of the galaxy, about 90 percent of the mass of each galaxy we study is dark matter. Dark matter is mass that makes its presence felt by gravitational forces but does not emit light or any other form of radiation. This is startling, because it means that all the visible light from galaxies represents a small fraction of the universe! Imagine an analogy to an accounting problem. On one side of the ledger, we derive the mass of a galaxy from the motions of stars in the galaxy. On the other side of the ledger, we use the mass-to-light ratios discussed above to estimate the mass of the galaxy in visible populations of stars.

The mass based on motions is always about 10 times larger than the mass associated with the observed starlight. We have seen that the mass-to-light ratios for stars in galaxies are in the range 1 to 30. The mass-to-light ratios of galaxies based on motions and gravity are in the range 50 to 200. This accounting discrepancy cannot be explained away; we must face the fact that the universe is made of material that we do not fully understand.

The mass distribution of spiral galaxies can be mapped out using a rotation curve, just as we described for the Milky Way. Figure 15-24 shows the rotation curves for three spiral galaxies. A rotation curve is a map of the velocity of different parts of a galaxy with respect to the nucleus. The velocity is measured from the Doppler shift of emission lines produced by gas, or absorption lines produced by stars. The striking feature of all these rotation curves is that they are flat; the velocities stay high out to the limits of the visible material in the spiral disk. We expect from Kepler's laws that the circular velocity of stars in a galaxy will begin to fall at a point where the orbit enclosed most of the mass of the galaxy. Yet we do not observe any decline in circular velocity. In many spirals, the gas disk extends farther than the stellar disk, in which case we can use the 21-cm line of neutral hydrogen to map the rotation curve out to very large distances. Since the velocity stays high and roughly constant out to the largest radii studied, we must conclude that even these outer orbits do not enclose most of the mass of the galaxy. The bulk of the mass must be invisible material in the halo.

Another way to probe the mass of a large galaxy is to measure the motions of any dwarf galaxy companions that are in orbit around it. These orbits cannot, of course, be followed in time because they typically take hundreds of millions of years. We have to make do with a snapshot of the instantaneous velocity of each companion. Since the companions are very small galaxies, their motions are dictated by the mass of the large primary galaxy—just as the motions of the planets in the solar system are dictated by the Sun. This technique has given us striking confirmation of the mass and extent of the dark halo of the Milky Way galaxy. Dwarf ellipticals in orbit around the Milky Way give a total mass for the Milky Way of about 10^{12} $M_\odot$, distributed up to 100 kpc from the center of the galaxy. This estimate doubles

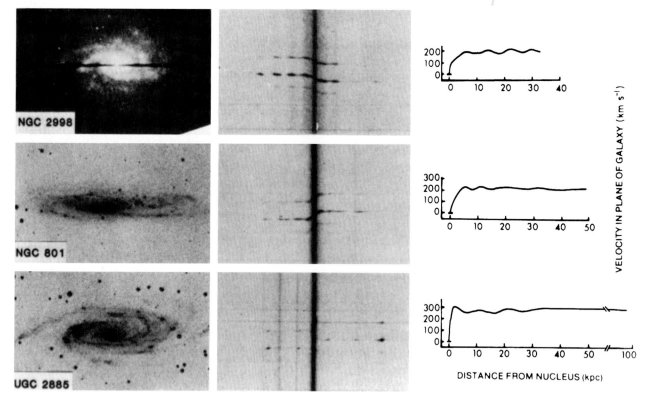

FIGURE 15-24
Rotation curves for three spiral galaxies: NGC 2998, NGC 801, and UGC 2885. The slit of a spectrograph is placed along the major axis of the galaxy (as shown on the left). In the spectrum for each galaxy (center), the wavelength runs vertically, and the position across the galaxy runs horizontally. Each of the resulting rotation curves (right) shows a constant amplitude of rotation out to the edge of the galaxy. The jump in wavelength near the center of each galaxy corresponds to the steeply rising part of the rotation curve. (SOURCE: Vera Rubin, Carnegie Institution of Washington.)

the size and extent of the Milky Way that was measured by the motions of individual halo stars. Since the starlight in the galaxy only corresponds to about 10^{11} $M_\odot$, more than 90 percent of the total mass of our galaxy is dark matter!

What can we say about other types of galaxies? Elliptical galaxies have no gas disk, so a rotation curve cannot be measured. However, large ellipticals are usually surrounded by a swarm of tiny dwarf elliptical galaxies. The motions of the tiny companions give a measure of the mass of the large galaxy. A larger spread in the Doppler shifts of the companions indicates a larger mass for the primary galaxy. The spread or scatter in motions in any gravitational system is called the **velocity dispersion.** Elliptical galaxies also have vast amounts of dark matter; their total masses of 10^{12} up to 3×10^{12} $M_\odot$ are distributed on scales of 100 to 200 kpc. Tiny galaxies also have dark matter. With the dwarf companions of the Milky Way, we can play this same game with the motions of individual stars. Spectra of individual stars in the galaxy yield Doppler shifts, and the spread in the Doppler shifts indicates the mass. If the mass is larger than expected from the summed light of the individual stars, then dark matter must be present. The velocity dispersion indicates dark matter in every dwarf galaxy that has been studied so far.

Everywhere we look in the universe, we see dark matter. Every type of galaxy we have studied—spirals, ellipticals, irregulars, dwarfs, and our own Milky Way—has a mass that is dominated by this strange material. Dark matter is a ubiquitous feature of galaxies. Various attempts are being made to detect dark matter directly, and the stakes are very high. The nature of dark matter is probably the largest unanswered question in astronomy.

Recent advances in CCD detectors have permitted a search for dark matter using the technique of **gravitational microlensing.** The idea is based on a simple prediction of Einstein's general theory of relativity. If a dark, compact object passes between Earth and a more distant star, the dark object can act as a gravitational "lens," amplifying the light of the star. In this way, the dark object reveals its presence. The problem is that stars are very small and space is very empty, so the probability of a crossing is tiny. The microlensing phenomenon is not like a conventional eclipse; the amplification is caused by the bending of light by an intervening mass (we will discuss lensing more extensively in Chapter 16). If the dark halo of the Milky Way is composed entirely of dark compact objects, only one in a million background stars will be lensed at any given time. To opti-

SCIENCE TOOLBOX

The Gravity of Many Bodies

Newton's law of gravity expresses the attractive force between two objects separated by some distance. The equation is exact: If we know the two masses accurately, and the distance between them accurately, we can calculate the force accurately. The orbit of a binary star is a situation with an exact solution, as we saw in the last chapter. However, if there are more than two objects close together in space, the gravity is no longer described by a single equation. A set of equations is required, and they no longer have an exact mathematical solution. Astronomers say that the definition of gravity in a set of three or more objects is a *computational* problem. In other words, the gravity force can be calculated using a powerful computer. Higher accuracy requires more computer power.

Up to this point in the book, we have been able to make an important simplifying assumption about gravity. In the solar system, the Sun is far more massive than any planet. Therefore, we can calculate the orbit of each planet as if it were affected only by the Sun. In principle, when we calculate Earth's orbit, we should consider the gravitational effect of Mars and Venus and all the other planets. When we calculate Jupiter's orbit, we should consider the gravitational effect of Mars and Saturn and all the other planets. In practice, the force between *any* two planets is tiny compared to the force between the Sun and any planet. (Try it! Use the gravity law, and Appendix B-1, and calculate the relative force in particular cases.)

What about the situation in which the objects in space have similar masses? Examples range from a small star cluster to an entire galaxy. No single star dominates the gravity of the system, so the gravity force between every pair of stars must be calculated. We require a very large number of calculations to model the system.

Let us demonstrate the complexity of the gravity of many bodies in steps. Figure 15–A illustrates the number of gravity calculations required to model the gravity of a system of stars. There is a progression in the figure from two to six stars. (The gravity of a single, isolated star can be calculated, but it is not very interesting!) We can make a little table for each situation to show how we would calculate the force F between every pair of stars. Of course, we do not need to calculate the gravity force of a star on itself, so we can ignore the entries along one diagonal. Also, we can notice that F_{13}, the gravity force of star 1 on star 3, is the same as F_{31}, the gravity force of star 3 on star 1. (Remember that you pull on Earth with the same force that Earth pulls on you.) This is true for any pair of stars. So we only need to calculate half the forces in each table, which we show as the forces below the diagonal lines.

How does the number of calculated forces increase with the number of stars? In the case of two objects, there is only one force, given by Newton's law as we

(continued)

(a) 2 Stars

	1	2
1	–	F_{12}
2	F_{21}	–

(b) 3 Stars

	1	2	3
1	–	F_{12}	F_{13}
2	F_{21}	–	F_{23}
3	F_{31}	F_{32}	–

(c) 4 Stars

	1	2	3	4
1	–	F_{12}	F_{13}	F_{14}
2	F_{21}	–	F_{23}	F_{24}
3	F_{31}	F_{32}	–	F_{34}
4	F_{41}	F_{42}	F_{43}	–

(d) 5 Stars

	1	2	3	4	5
1	–	F_{12}	F_{13}	F_{14}	F_{15}
2	F_{21}	–	F_{23}	F_{24}	F_{25}
3	F_{31}	F_{32}	–	F_{34}	F_{35}
4	F_{41}	F_{42}	F_{43}	–	F_{45}
5	F_{51}	F_{52}	F_{53}	F_{54}	–

(e) 6 Stars

	1	2	3	4	5	6
1	–	F_{12}	F_{13}	F_{14}	F_{15}	F_{16}
2	F_{21}	–	F_{23}	F_{24}	F_{25}	F_{26}
3	F_{31}	F_{32}	–	F_{34}	F_{35}	F_{36}
4	F_{41}	F_{42}	F_{43}	–	F_{45}	F_{46}
5	F_{51}	F_{52}	F_{53}	F_{54}	–	F_{56}
6	F_{61}	F_{62}	F_{63}	F_{64}	F_{65}	–

FIGURE 15–A
An illustration of the way the number of calculations increases when there are more objects whose gravity must be calculated. Each value of *F* represents the force between the object numbers given in the subscripts. The sequence goes from **(a)** two stars, to **(b)** three stars, to **(c)** four stars, to **(d)** five stars, to **(e)** six stars. Only half of these forces need to be calculated (either the red or the blue portions of each table) because the gravity force of object A on object B equals the gravity force of object B on object A. With *n* stars in a system, the number of forces that must be calculated is $\frac{1}{2} \times (n - 1)$.

SCIENCE TOOLBOX

The Gravity of Many Bodies (continued)

first encountered it, with $F_{12} = F_{21} = GM_1M_2/R_{12}^2$, where M_1 and M_2 are the masses of stars 1 and 2 and R_{12} is the distance between them. With three stars, there are three forces: F_{12}, F_{13}, and F_{23}. For shorthand, we use a notation in which the subscript joins the numbers of the two stars being considered. As you can see in the figure, the number of forces increases from 6 to 10 to 15 as the number of stars increases from 4 to 5 to 6.

The general form of this progression can be easily shown. We will call n the number of stars in the system, the number of either rows or columns in each table. By omitting the gravity of a star on itself, we reduce the number of rows (or columns) by one. Thus, the number of elements in the table is $n(n - 1)$. But we must divide by 2 because the force between each pair of stars only needs to be calculated once. In other words, $F_{14} = F_{41}$, $F_{35} = F_{53}$, $F_{24} = F_{42}$, and so on. The result is

$$\text{Number of forces} = \tfrac{1}{2} n (n - 1)$$

We can see that this works if we plug in a number from 2 to 6. The number of forces is 1, 3, 6, 10, or 15. But six stars constitute a puny star cluster. How does the number of forces increase as n becomes very large? Multiplying out, the number of forces is $(n^2/2 - n/2)$. If n becomes very large, then $n^2 \gg n$, and we can neglect the second term. For large numbers of stars, we find that

$$\text{Number of forces} \propto n^2$$

Now we can understand why the gravity of many bodies is such a difficult computational problem. The number of calculations increases with the square of the number of stars. Calculating the gravity of 50 stars is 25 times harder than calculating the gravity of 10 stars. Calculating the gravity of 1000 stars is 10,000 times harder than calculating the gravity of 10 stars. Can you imagine the computation involved in modeling the gravity of a galaxy with many millions of stars?

Here is how the astronomers create computer models of stellar systems. Stars are given initial positions in three dimensions—not in real space, but in the computational space of a computer. The external gravity on each star is calculated from the sum of the force exerted by all the other stars. This force changes each star's motion and its position. With the position of each star adjusted, a new set of forces is calculated. The result is once again a change in the position and motion of each star. This process is repeated over and over. Each set of calculations occurs at a particular time interval after the previous set of calculations. In this way, the motions of the stars can be predicted over a period of time.

Astronomers who do this kind of work are experts at computer programming. They use all kinds of computational tricks and shortcuts to make the calculations more efficient. For example, since gravity is an inverse square law, each star is more affected by its neighbors than by distant stars. Therefore, the accuracy of the force calculation can be lower for pairs of stars that are widely separated. Also, the force does not need to be calculated as often when a particular star is moving slowly. This cleverness saves a lot of computing power. Instead of the computing requirement going up in proportion to the square of the number of stars, it goes up only in direct proportion to the number of stars.

Nevertheless, to make a computer model of even a modest-sized galaxy requires the gravity of several million stars to be calculated at many time intervals. Ten years ago, such computations were impossible. The fantastic increase in the speed and power of modern computers has opened up the field of computational astronomy. Even a modest workstation can perform many millions of calculations per second. Our ability to simulate complex astronomical objects on a computer has given us a whole new way of looking at the universe.

---

mize the chances of detecting such an event, astronomers are searching fields toward the galactic bulge or the Magellanic clouds, which have the largest number of background stars.

The search for gravitational microlensing is the ultimate "needle in a haystack" experiment. Millions of stellar images per night must be examined in the hope of finding a few events. Variable stars are a source of potential confusion, but the lensing signature is very specific: The signal should rise and fall symmetrically, with the same amplitude at all wavelengths, and it should not repeat. This exciting technique is sensitive to everything from planets smaller than Earth to solar-mass black holes. Since 1993, nearly a hundred events have been reported by three international groups of astronomers. The lensing masses are in the range 0.05 to 0.30 $M_\odot$, from low-mass stars to substellar objects. Results thus far indicate that as much as half the mass of the halo may be made of dim white dwarfs or brown dwarfs. That still leaves a large amount of mass to explain.

We have seen that no single type of astronomical object can account for the dark matter. Some of it is likely to be made of stars that are not massive enough to liberate energy by hydrogen fusion, the so-called brown dwarfs. Cold, free-floating planets are also a possibility. More exotic suggestions include massive neutrinos and a variety of exotic particles. Around the world, physicists are conducting experiments in deep underground laboratories to look for subatomic particles that might solve this mystery.

The only alternative to the dark matter hypothesis is to jettison Newton's gravity law. It is important to remember that the key feature of Newtonian gravity—the fall-off of the gravity force with the inverse square of distance—is confirmed with high accuracy only in the solar system. We have no direct tests of Newton's theory beyond the neighborhood of the Sun. Perhaps the answer to the dark matter problem is the modification of our gravity theory on large scales. Few astronomers favor this option, but to be properly scientific, we should keep an open mind.

How Galaxies Evolve and Interact

Galaxies are so widely separated in space that it is natural to think of them as isolated entities. Yet astronomers have learned over the past two decades that galaxy morphology and evolution are strongly influenced by the larger environment. Our snapshot of galaxies frozen in time is misleading. Galaxies can interact and merge, they can accrete and eject gas, and they can collide to trigger bouts of star formation. It is quite a challenge to understand the behavior of galaxies when each one contains billions of stars, and each one moves on a slightly different orbit. Astronomers use the power of supercomputers to calculate the gravity and predict the stellar motions, as discussed in the preceding Science Toolbox.

We have seen that a simple way to think about **galaxy evolution** is in terms of stellar populations. Disk stars are young, mostly blue, luminous, and rich in heavy elements. Halo stars are old, red, less luminous, and poor in heavy elements. Disks rotate rapidly, and halos rotate very slowly. Irregular galaxies contain almost a pure disk population. Spiral galaxies contain a mixture of populations, with star-forming regions concentrated in the spiral arms, intermediate-age stars in the central bulges, and old stars in the halo. Elliptical galaxies contain almost purely a halo population. The galaxies that we see in the local universe reflect the consequences of stellar evolution. Imagine a population of stars that formed billions of years ago. At the beginning, the overall light would be dominated by young, hot stars. As the galaxy aged, the most massive stars would evolve off the main sequence, and the galaxy would become dimmer and redder. Astronomers construct theoretical models of stellar populations to match their observations. In principle, this approach can be used to measure the age of a galaxy.

Our observations of nearby galaxies have some interesting implications for their evolution. Elliptical galaxies have little gas or dust, and they have a low rate of current star-forming activity. However, the old stars out of which ellipticals are made will lose mass by stellar winds, replenishing the interstellar medium. Thus, gas and dust are somehow being *removed* from elliptical galaxies. The puzzle for spirals like the Milky Way is the converse. They are actively forming stars from gas and dust in the disk. However, the star-formation rate is so rapid that it could not have been going on for the entire lifetime of the galaxy (as measured by the oldest stars in the halo). One possibility is that star formation in the disk started relatively recently and therefore has not had time to exhaust the gas supply; however, there is no good evidence for this. The other possibility is that gas has been *added* to spirals over their lifetimes. A spiral could be rejuvenated either by gas gradually falling in from intergalactic space or by swallowing small gas-rich companions. Either way, environment plays an important role in the evolution of galaxies.

Can one type of galaxy turn into another? In general, astronomers believe the answer is no. A spiral galaxy cannot turn into an elliptical galaxy. Even if the star formation in a spiral ceased, and the stars all became old and red, the stellar orbits and shape of the disk would not change. The fundamental physical quantity that accounts for the shape of galaxies is angular momentum, which is related to the amount of mass and the velocity at which it is rotating. Spiral disks have high angular momentum, bulges have somewhat less, and halos have the least of all. Spiral and elliptical galaxies, therefore, have very different amounts of angular momentum. Since angular momentum is always conserved, it follows that a single galaxy cannot evolve from one type to another.

Interesting things can happen, however, when galaxies interact, collide, or merge—a set of possibilities known as **galaxy interactions.** The Milky Way and Andromeda galaxies are separated by only 15 to 20 of their own diameters. As galaxies drift through space, they can meet. Colliding galaxies are an extraordinary event, very different from the colliding of stars. Stars in galaxies are separated by millions of their own diameters. Therefore, despite their motions, few, if any, stars in a galaxy have collided with each other during the whole history of the universe.

If the random motions of two galaxies cause them to penetrate each other, individual stars are unlikely to collide. If two people stand ten yards apart and throw a handful of sand at each other, the "clouds" of sand will meet in the middle, but most grains will continue on to reach their target. The dusty gas in each galaxy, on the other hand, is more like a continuous medium, so that when galaxies collide at speeds of hundreds of kilometers per second, the gas interacts dramatically. You might think of the water in two buckets being sloshed toward each other—the two masses of water hit and make a splash. When gas atoms in a galaxy collide, the gas and dust in the interstellar medium gets heated. The gas atoms are excited or even ionized. The directed motion of galaxies (kinetic energy) is converted to the microscopic motion of particles (heat energy)—another example of the conservation of energy. Hydrogen is the most abundant element, so colliding galaxies are often intense sources of radio radiation from excited hydrogen and infrared radiation from hot dust.

Galaxies do not actually have to collide in order for their properties to be changed. Gravity has a long reach, so a close encounter will produce a *tidal interaction*. Just as the Moon

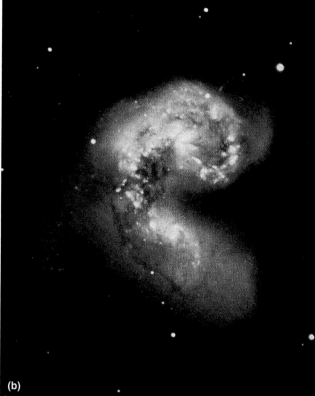
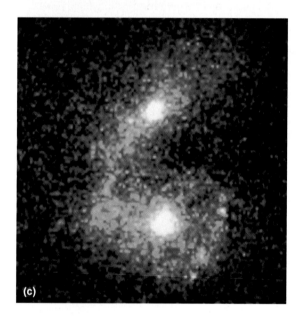

FIGURE 15-25
The interacting galaxies NGC 4038 and NGC 4039. **(a)** A deep photograph shows curved streams of material flung off during the interaction. (SOURCE: Hale Observatories.) **(b)** A color optical view shows the chaotic morphology of the two galaxies, where the interaction has triggered star formation. (SOURCE: Copyright Anglo-Australian Telescope Board, courtesy D. Malin.) **(c)** A near-infrared view is sensitive to the older, redder stars, and the effects of the interaction are less pronounced. (SOURCE: NOAO.)

raises a tidal bulge in Earth's oceans, a nearby companion may distort the shape of a galaxy. Stars and gas will be pulled out on either side, and rotation will sling the material out into graceful, curving arcs (Figure 15-25). When galaxies travel through space, a near miss is more likely than a head-on collision. We have already discussed the Magellanic stream, which is an arc of gas caused by an encounter between the Milky Way and the Magellanic clouds. The evidence of a tidal interaction does not last forever; within a few hundred million years (10 percent of the age of the universe), the flung-out stars will fade or lapse back into the bulk of the galaxy they came from. About one-third to one-half of all galaxies have had tidal interactions since they formed.

What happens in the rare case of a direct collision? In a merger of two nearly equal-sized galaxies, gas clouds collide and are compressed. They become hotter and denser, making star formation much more likely. In turn, star formation heats the surrounding dust, causing intense far-infrared emission. Roughly 10 percent of all galaxies have distorted shapes from past collisions or close encounters. Figure 15-26 shows a supercomputer simulation of the collision between two disk galaxies of similar size. The transient streamers that can be thrown off in such a merger are clearly visible. If two galaxies of very different size meet, the larger galaxy will gobble up the smaller one; this is called *galactic cannibalism*. As a small galaxy approaches a much larger one, two things will happen. First, the gravity of the small galaxy will be too weak to retain its own outer stars, so they will be ripped away onto the larger galaxy. Second, as the core of the smaller galaxy ploughs through the outer regions of the larger galaxy, it will lose energy and decelerate, while the stars in the large galaxy heat up and accelerate. As a result, the core of the small galaxy spirals into the center of the larger galaxy. Figure

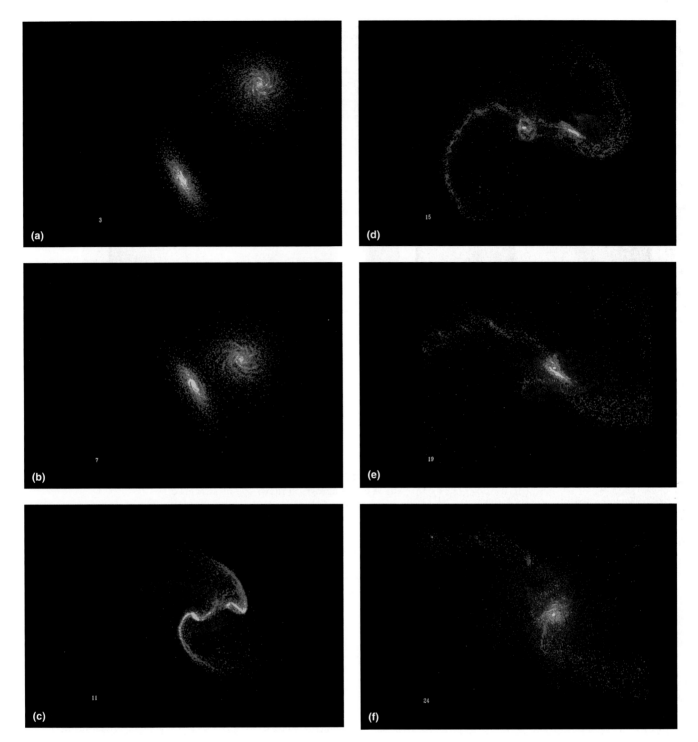

FIGURE 15-26
A supercomputer simulation of the collision and subsequent merger of two disk galaxies. Dark halos are shown in red, disks in blue, and central bulges in yellow. Elapsed time in millions of years is given in the lower left region of each frame. (SOURCE: Joshua Barnes, Institute for Astronomy, University of Hawaii; simulation created at the Pittsburgh Supercomputer Center.)

15-27 shows a supercomputer simulation of a small galaxy being devoured by a much larger disk galaxy.

The interaction and merger of galaxies is not just a game played on computers. In the past few years, astronomers have discovered several new dwarf companions of the Milky Way. One of these is almost certainly falling into the disk of our galaxy, nearly hidden from view through obscuring dust. We thus have solid evidence that the gas content of the disk is replenished by new material that falls in, which explains how the current star-formation rate could be sustained for so long. Astronomers have also discovered that the halo is not rotating as a single entity.

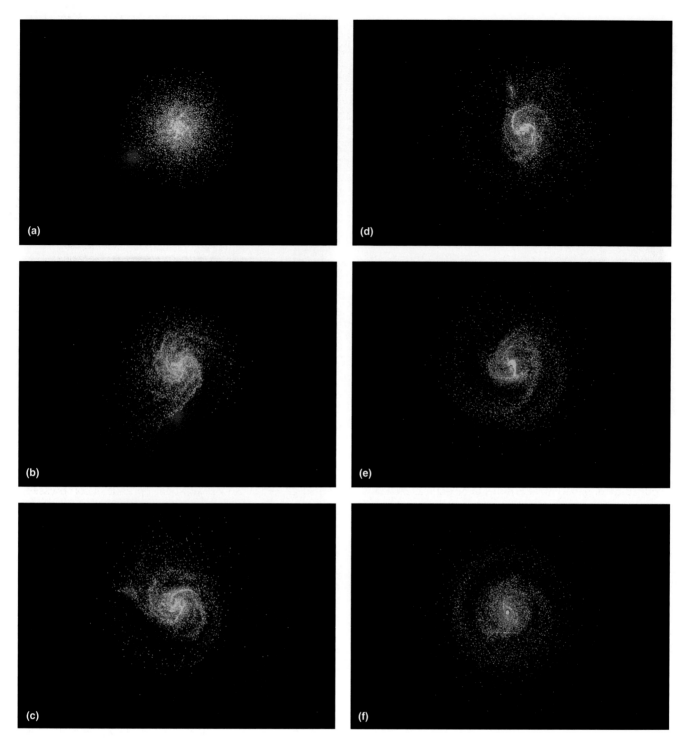

FIGURE 15-27
A supercomputer simulation of a small galaxy (stars in orange) being devoured by a larger disk galaxy (stars in blue, gas in white). The interval between pictures is 400 million years. As the small galaxy plunges into the core, a large amount of new star formation is induced in the larger galaxy. (SOURCE: L. Hernquist, University of California at Santa Cruz; simulation created at the Pittsburgh Supercomputer Center.)

Different regions of halo stars have different overall motions. Each set of halo stars that share a common motion may be a "fossilized" relic of a smaller galaxy that fell into the halo long ago (perhaps also explaining the age range of halo stars). Our galaxy may have been put together in bits and pieces.

The Puzzle of Galaxy Formation

The galaxies we see around us in space are well formed. A few galaxies appear to be merging or colliding, but none is

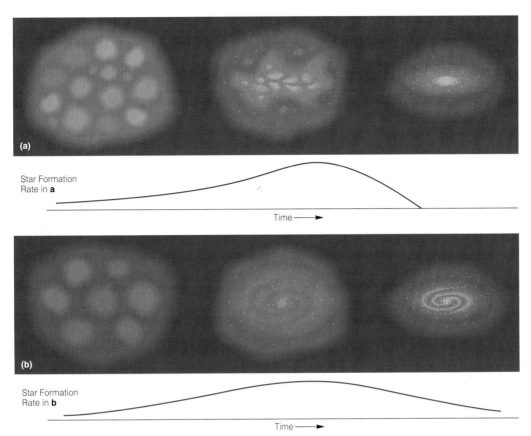

FIGURE 15-28
Possible formation processes leading to giant elliptical and spiral galaxies. **(a)** In a high-density region, collisions and mergers of many subclumps lead to efficient and early star formation, resulting in a smooth, gas-poor elliptical galaxy. **(b)** In an average-density region, star formation is slow and inefficient, and the continued infall of gas leads to the formation of a galactic disk, resulting in a spiral galaxy.
(SOURCE: Adapted from J. Silk.)

collapsing out of a gigantic cloud of gas for the first time. Thus, our study of **galaxy formation** is analogous to archaeology. We must look at the evidence of stellar populations, motions of stars in galaxies, and the shapes of galaxies to try to deduce the process of their formation. Most of the action took place billions of years ago.

Any theory about the formation and evolution of galaxies must explain the presence of two very different stellar components: (1) the thin, rotating disks of gas and young stars; and (2) the bulges and halos with older stars and much less rotation. Disks appear to form only in a relatively narrow range of masses, from 10^9 to 10^{11} $M_\odot$. By contrast, galaxies can have spherical or nearly spherical components—elliptical galaxies and the bulges and halos of spirals—covering a wide range of masses, from 10^5 to 10^{12} $M_\odot$. In addition, galaxies of all types have dark matter. Another important distinction is that elliptical galaxies are found primarily in regions that are dense with galaxies, whereas spirals are often quite isolated in space. Galaxy formation is analogous to star formation, only on a much larger scale, in that it involves the conversion of gas into stars.

We do not yet have a complete theory of galaxy formation. Current evidence suggests that galaxies formed billions of years ago by gravitational collapse from primordial gas clouds called *protogalaxies*. The causes and details of these events are still poorly understood. Theorists speculate that if star formation occurs early in the collapse (Figure 15-28a), the galaxy will retain its nearly spherical shape because the stars are too widely separated in space to collide. This scenario describes the formation of an elliptical galaxy. If star formation proceeds slowly (Figure 15-28b), the galaxy will remain gaseous. The gas will heat up as it is compressed. Then it will radiate, lose energy, and collapse into a disk. The conservation of angular momentum ensures that the slow rotation of the initial configuration speeds up in the eventual gas disk. This scenario describes the formation of a spiral galaxy. As mentioned previously, since both major galaxy types contain old stellar components (halos), we know that ellipticals do not evolve into spirals.

Why does star formation progress quickly in some cases and slowly in others? The initial rate of star formation probably depends on the average density of matter in a large region of space where the galaxy forms. In a dense region, the stars form early, and the shape of the galaxy remains "frozen" in a spherical form. This idea is consistent with the fact that ellipticals are found in high-density regions of the universe. We will survey the different galaxy environments of the universe in the next chapter. In a sparse region, the protogalactic cloud collapses into a disk before widespread star formation occurs. Therefore, we might expect spiral galaxies to be found in low-density regions of the universe—the Local Group is one such region.

Our picture of galaxy formation raises as many questions as it answers. Why did gas in the early universe form clumps of just the right size and mass to turn into galaxies? Why do disks from in a (relatively) narrow range of mass, while halos can be extremely large or extremely small? Where did the rotation of galaxies originally come from? How much might a galaxy change after 10 billion years or more of encounters and mergers with other galaxies? Speculative answers are deferred until we consider the large-scale structure of the universe in Chapter 16, and the formation of the universe in Chapter 17. The largest uncertainty of all is the role of the ubiquitous dark matter. The classification of galaxies is based on the roughly 10 percent of the mass of the galaxy that emits light. We are only beginning to map the large dark halos, and we still have no idea what they are made of! No theory of galaxy formation and evolution will be persuasive without a better understanding of dark matter.

Summary

The Sun is located in the disk of the Milky Way galaxy. The luminous stars and gas clouds we see in the solar neighborhood are signs of continuing star formation. The spiral arm pattern probably emerged after a number of galactic rotations, perhaps within 1 billion years. In the spiral arms, the densest clouds contracted and spawned associations and open star clusters. Each group broke apart into scattered stars a few hundred million years after its formation, but new star groups continued to form, so that the galaxy kept its present general appearance. Supernovae blew out gas laced with heavy elements created inside stars, so that later generations of stars had more heavy elements than the earlier stars. Perhaps 7 or 8 billion years after the galaxy's formation, in one of the spiral arms, a medium-sized star formed—our Sun—and in its surrounding dusty nebula, planet Earth was born.

Near the galactic center we see older, redder stars in a stellar component called the bulge. In the heart of the galaxy, behind veils of obscuring dust, there is evidence for a dark mass concentration in the form of a supermassive black hole. The outer parts of the galaxy are home to an even more mysterious form of dark matter. The rotation speed of stars in the disk stays high out to the largest distances measured, evidence that most of the mass of the galaxy is in an extended halo. Dark matter does not emit radiation at any wavelength, revealing itself only by its gravity. The nature of the dark matter is an outstanding puzzle in astronomy.

The stars in our night sky reveal our particular position in the Milky Way. But we can use our understanding of the structure of the Milky Way to imagine our viewpoint changed to elsewhere in the galaxy. In a star-forming region, the night sky would be lit up by young stars and delicate filaments of glowing gas. Deep within the bulge or in a globular cluster, our sky would be filled with old, red stars. Close to the galactic center, our sky would be dazzlingly bright with the light of millions of stars. Located next to a lone halo star, we would look down from a nearly starless sky onto the beautiful spiral of the disk.

The Milky Way is adrift in a vast volume of space that is loosely sprinkled with other star systems, or galaxies. Large new telescopes built early in this century demonstrated that many of the so-called nebulae were in fact distant galaxies. Measuring the distances to galaxies is still one of the greatest challenges in astronomy. The Local Group, out to about 1000 kpc, contains galaxies with a variety of shapes and sizes. Most galaxies can be classified as elliptical, spiral, or irregular. A more elaborate classification scheme accounts for the many morphological details. The largest galaxies are giant ellipticals with halo stars and little gas and dust. Spirals are intermediate in size, with young stars in a rotating disk mixed with gas and dust, and older stars in a central bulge and an extended halo traced out by the globular cluster system. Other galaxies of intermediate and small size have elliptical and irregular shapes. All considerations of galaxy properties are overshadowed by the presence of dark matter, which forms about 90 percent of the mass of all galaxies, large and small.

Astronomy has its own version of the nature-versus-nurture debate concerning the explanation of human behavior and personality. There is clear evidence that the history of a galaxy is affected both by how it formed and by its environment. Some galaxies do not fit into the simple classification scheme; they have peculiar shapes or appear to be interacting with a companion. The shape and the star-formation history of a galaxy can be influenced by tidal interactions, collisions, and mergers. This stately violence has played out over billions of years of the history of the universe.

The discovery of galaxies is yet another step in the Copernican revolution. Hubble made audacious use of the scientific method, using the properties of known types of stars to measure distances and extend the size of the known universe by a factor of 100. We have learned that the Milky Way was just one in a universe of galaxies, each containing billions of stars and scattered over millions of light-years of space.

Important Concepts

You should be able to define these concepts and use them in a sentence.

spiral arms	Local Group
rotation curve	Magellanic clouds
dark matter	Andromeda galaxy
supermassive black hole	light travel time
stellar population	mass-to-light ratio
distance indicator	velocity dispersion
distance scale	gravitational microlensing
spiral galaxy	galaxy evolution
elliptical galaxy	galaxy interactions
irregular galaxy	galaxy formation

How Do We Know?

These questions and answers show how the scientific method is used to learn about the universe.

Q How do know the mass of the Milky Way galaxy?

A The most direct way to estimate the mass of the Milky Way is to understand the types of stars it is made of and sum the mass of those stars. This method is hampered by the fact that dust obscures our view of much of the disk, and by the fact that many halo stars are too dim to detect. We must add to this stellar mass the gas mass of the disk, which can be measured using the 21-cm spectral line of neutral hydrogen. An independent way to measure the mass of the Milky Way disk interior to the Sun's orbit uses Kepler's law applied to the motion of the Sun around the galactic center. All of these estimates have to be reconsidered in light of the evidence for large amounts of matter beyond the visible edge of the galaxy.

Q How do we know that our galaxy is mostly composed of dark matter?

A The simple answer is that our knowledge of gravity leads us to this conclusion. We expect that motions in a gravitational system will decrease beyond the region where more of the mass resides. This is what we observe in the solar system—the Sun contains most of the mass, and the orbital speeds of the planets decrease moving away from the Sun. The disk of the Milky Way is also a rotating system. However, in this case the orbits of stars maintain a constant speed out to the edge of the disk. Astronomers call this a flat rotation curve. The stars seem to have their motion driven by an unseen form of matter in the galactic halo. The evidence for dark matter is confirmed by observations of the motions of individual halo stars. In addition, satellite galaxies to the Milky Way are observed to have rapid motions. We conclude that over 90 percent of the mass of the Milky Way galaxy is composed of dark matter, a total of 10^{12} $M_\odot$ distributed over a halo of radius 100 kpc. The dark matter does not emit detectable radiation in any part of the electromagnetic spectrum.

Q What is the nature of dark matter?

A We do not know. This is perhaps the major unanswered question in astronomy. However, astronomers have used careful observations to rule out some conventional possibilities. Dark matter cannot be made of stars, because we would see stars with our telescopes, even at a distance of 100 kpc. Dark matter cannot be made of stellar remnants like black holes, because we would see other evidence of the violent deaths of such massive stars. Dark matter cannot be made of brown dwarfs or other substellar objects, because they would reveal themselves by their amplification of the light of background stars. Dark matter cannot be made of small dust particles, because these particles would emit infrared radiation as they were illuminated by ambient starlight. This leaves open the possibility of a tiny, microscopic particle, quite different from the protons, neutrons, and electrons of normal matter. Since dark matter makes its presence felt only by the force of gravity, these tiny particles would have to interact very weakly with normal matter. Another possibility—not favored by most researchers—is that Newton's form of the gravity law is wrong on large scales, and we have been misled into supposing that dark matter exists.

Q How do we know the distances to galaxies?

A All the evidence that galaxies were distant stellar systems was indirect until the work of Edwin Hubble. Spiral nebulae show no preference for the plane of the Milky Way, and their isotropy on the sky is consistent with their being at large distances from our galaxy. Some spiral nebulae break up into the light of many individual stars when viewed through a large telescope. The brightness of the individual stars is consistent with their being beyond the Milky Way, dimmed by the inverse square law over a large distance. Hubble located Cepheid variable stars in the Andromeda Nebula (M 31) and measured their periods from a sequence of images. By assuming that Cepheids in Andromeda obeyed the same period-luminosity relation as Cepheids in the Milky Way, he was able to show that the Andromeda Nebula was a galaxy of stars far beyond the Milky Way. Cepheids have also been used to measure the distances to many other galaxies.

Q How do we know that dark matter is found in all galaxies?

A The evidence for dark matter in spirals is similar to the case of the Milky Way: The rotation curves are often flat out to the largest radii that can be measured. This indicates that over 90 percent of the mass is an unseen halo. Elliptical galaxies do not have rotating disks, but the velocity dispersions of stars or globular clusters are too high to be caused by the visible stars. This evidence is supported by studies of large galaxies in which the motions of small companion galaxies can be measured. The companions move too fast to be responding to only the visible matter. Small, irregular galaxies seem to have dark matter as well. We do not know that dark matter exists in *all* galaxies, but astronomers have found evidence for dark matter in every galaxy studied carefully thus far.

Q Can one type of galaxy change into a different type?

A In general, no. Galaxies do evolve as their stellar populations age, and the properties of galaxies can change by interactions and mergers with other galaxies. However, a spiral cannot turn into an elliptical. Even if star formation in the disk ceased, the aging stars would be visible for many billions of years. Similarly, an elliptical cannot turn into a spiral. Elliptical galaxies have no stellar component with as much angular momentum as the disk of a typical spiral galaxy. No galaxy is truly isolated, and the larger environment can affect the way a galaxy forms and evolves. Halo populations show a range of ages and may have been built up from a number of smaller units. Some giant elliptical galaxies may have been a merger of smaller galaxies. Disk populations may have had their gas supply replenished by material that fell in from outside.

Problems

Use these problems to test your understanding of the information and concepts in this chapter. The * indicates a more advanced or mathematical problem.

1. During what percentage of recorded human history (define as you think appropriate) have people *not* known that we live in an isolated galaxy of stars similar to other remote galaxies? How large did people think the universe was during most of recorded history?

2. **(a)** From the Milky Way's appearance, how do we know the solar system is in the galactic disk and not far above or below it? How might the central region's appearance differ in the latter case? **(b)** Compare the shapes of the volumes occupied by the swarm of open clusters and by globular clusters. Relate the difference to differences in stellar populations.

*3. **(a)** How do we know the size of the galaxy? **(b)** How do we know the location and distance of the galactic center? **(c)** What is the effect of dust on distance determinations in the galaxy? **(d)** If a star cluster is viewed through dust with an optical depth of $\tau = 1$, by what factor will we underestimate or overestimate the distance to the cluster, compared with a case in which there is no dust?

4. Describe evidence for spiral structure in the Milky Way. Which of the following types of objects reveal spiral structure when their positions are mapped on the Milky Way plane? **(a)** hot and luminous O stars; **(b)** cool and low-luminosity M stars; **(c)** H II regions or ionized gas clouds; **(d)** open clusters; **(e)** globular clusters; **(f)** star-forming regions; **(g)** supernovae. Which of these objects might be found in the halo of our galaxy?

5. Will the present constellations be recognizable in Earth's sky 100 million years from now? Why or why not? Imagine our solar system is moved to a different galactic environment. Write a paragraph describing the night sky in each of the following locations: **(a)** very close to the galactic center; **(b)** far above the disk in the sparse stellar environment of the halo; **(c)** at the center of a globular cluster; **(d)** near the center of the Orion star-forming region; **(e)** at the edge of the Milky Way disk. Which two of these locations are unlikely places to find a Sun-like star?

6. Summarize the evidence for violent, energetic activity in our galaxy's central region. Why do astronomers believe that it contains a supermassive black hole?

*7. Before Hubble demonstrated the true distance to the Andromeda galaxy, some astronomers claimed to have detected proper motions in the M 31 disk—in other words, they claimed to see stars in the disk participate in disk rotation over a period of several years. Assume M 31 is 670 kpc distant. Assume also that its disk rotates with the same 225-million-year period as the Milky Way and that it is 30 kpc in radius. Show that the motion of the M 31 disk is far too small to detect with a ground-based telescope over a 10-year period. What is the angular shift of a bright star at the outer edge of the disk over that time interval?

8. Which type of galaxy tends to be biggest? Which type can be the most luminous? Which type contains the fewest young stars? Why do almost all galaxies have a stellar mass-to-light ratio higher than 1?

*9. Suppose that we are using a telescope to find Cepheid variables in nearby galaxies. We can detect a star like the Sun out to a distance of 1 kpc. If Cepheids are 10,000 times more luminous than the Sun, could we detect them in the Magellanic clouds? Could we detect them in the Andromeda galaxy? By what factor increase in telescope aperture would it take to detect Cepheids in Andromeda?

10. **(a)** Two very faint and distant galaxies were detected on photographs but were too distant to allow identification of spiral arms and other typological features. If spectra showed that one was reddish with a spectrum of K-type stars, while the other was bluish with a spectrum of B- and A-type stars, what types would you expect the two galaxies to be? **(b)** Irregular galaxies are dominated by stellar associations, open clusters, and gas and dust clouds. All of these features indicate stellar youthfulness. Does this prove that the galaxies themselves formed recently? Why does the prominent light from star-forming regions in galaxies comes from massive, hot, blue stars? Why are these stars not seen in regions with no formation?

11. If you were to represent the Milky Way and Andromeda galaxies in a model by two cardboard disks, how many disk diameters apart should they be to represent the true spacing of the two galaxies? Where would the Magellanic clouds be, and what size would they be in this model?

*12. In general terms, how would Newton's law of gravity have to depart from an inverse square law to account for the dark matter in galaxies?

Projects

Activities to carry out either individually or in groups.

1. Compare views of the Milky Way with the naked eye, binoculars, and a small telescope. Scan along the Milky Way with each instrument, and record the number of stars per square degree in different constellations (or at different galactic longitudes). Relate these densities to the actual structure of the galaxy. Note that if observations of the summer evening Milky Way can be obtained in the regions of Scorpio and Sagittarius, the direction toward the center can be studied. Compare star counts with each instrument in the Milky Way plane with counts near a point 90° from the plane, where we are looking directly "up" out of the disk. Can you account for the differences?

2. Compare the Milky Way as seen with the naked eye or binoculars: **(a)** in the heart of your city; **(b)** on the edge of town; and **(c)** as far from lights as you can possibly get. City lights have little effect on the telescopic views of individual bright stars. But when it comes to broad areas of faint nebulosity or unresolved star clouds, even a single streetlight can illuminate local smog or fog and cause the iris of the eye to contract, thus destroying faint contrast and the ability to see faint glows.

3. Locate the Andromeda galaxy with the naked eye. Compare its visual appearance with its appearance in binoculars and telescopes of different sizes. Across what diameter can you detect the galaxy? (1° = 12 kpc at the Andromeda galaxy's distance.) Why do most photographs show a large central region of constant brightness, while visual inspection reveals a sharp concentration of light in the center? Can dust lanes or spiral arms be observed? (Check especially with low magnification on telescopes with apertures of 0.5 to 1 m or about 20 to 40 in., if available.)

4. If photographic equipment is available, take a series of pictures of the Andromeda galaxy with different exposure times, such as 1 min, 10 min, and 100 min. Describe some of the differences in appearance. What physical relationships are revealed among different parts of the galaxy? Make similar observations of other nearby galactic neighbors of the Milky Way.

Reading List

Barnes, J., Hernquist, L., and Schweitzer, F. 1991. "What Happens When Galaxies Collide?" *Scientific American,* vol. 265, p. 40.

Bartusiak, M. 1997. "Giving Birth to Galaxies." *Discover,* vol. 18, p. 58.

Binney, J. 1995. "The Evolution of Our Galaxy." *Sky and Telescope,* vol. 89, p. 20.

Crosswell, K. 1996. "The Dark Side of the Galaxy." *Astronomy,* vol. 24, p. 40.

Ferris, T. 1985. *Galaxies.* San Francisco: Sierra Club Books.

Hodge, P. 1993. "The Andromeda Galaxy." *Mercury,* vol. 22, p. 98.

Hodge, P. 1994. "Our New! Improved! Cluster of Galaxies." *Astronomy,* vol. 22, p. 26.

Lake, G. 1992. "Understanding the Hubble Sequence." *Sky and Telescope,* vol. 83, p. 515.

Miller, R. H. 1992. "Experimenting with Galaxies." *American Scientist,* vol. 80, p. 152.

Townes, C., and Genzel, R. 1990. "What Is Happening in the Center of Our Galaxy?" *Scientific American,* vol. 262, p. 46.

Trefil, J. 1988. *The Dark Side of the Universe.* New York: Scribner.

CHAPTER 16

The Expanding Universe

CONTENTS

- The Discovery of Quasars
- The Redshifts of Galaxies
- Expansion of the Universe
- Large-Scale Structure
- Active Galaxies and Quasars
- Summary
- Important Concepts
- How Do We Know?
- Problems
- Projects
- Reading List

A cluster of galaxies. Spiral and elliptical galaxies are clustered in a region tens of millions of light years across by the force of gravity. The entire cluster is receding from the Milky Way at a speed of hundreds of millions of miles per hour, carried away from us by the expansion of the universe. (SOURCE: N. Sharp, National Optical Astronomy Observatories.)

PREREADING QUESTIONS ON THE THEMES OF THE BOOK

OUR ROLE IN THE UNIVERSE
What is the position of the Milky Way in the larger universe?

HOW THE UNIVERSE WORKS
What physical processes cause the energetic events that take place in the heart of some galaxies?

HOW WE ACQUIRE KNOWLEDGE
How do we learn about the structure of the universe on the largest scales?

WHAT TO WATCH FOR IN CHAPTER 16

As you learned in the previous chapter, the Milky Way is a single galaxy in a vast universe of galaxies. Astronomer Edwin Hubble launched the modern study of the universe by showing that galaxies are systems of stars remote from the Milky Way. In this chapter, we will learn about Hubble's second major discovery: the expansion of the universe. Hubble observed that the light from galaxies is redshifted by an amount that increases with increasing distance from the Milky Way. Astronomers interpret the redshift as a sign that the universe is expanding. If we project the expansion back in time, we conclude that galaxies used to be much closer together and that the universe was much hotter and denser. You will see that galaxies are not randomly distributed in space. Astronomers have traced enormous structures that involve thousands of galaxies and span millions of light-years. Gravity has sculpted these structures over billions of years.

Certain galaxies have extraordinary events occurring in their nuclei. We will see the range of properties that characterize active galactic nuclei—rapid motions of stars and gas, strong emission all across the electromagnetic spectrum from radio waves to gamma rays, and evidence for a central compact object. The most distant and luminous examples of active galaxies are quasars. Quasars can be found out to the limits of the observable universe. Astronomers believe that the extreme emission from active galaxies involves a supermassive black hole that is devouring gas and stars near the center of the galaxy. You will see that there is growing evidence for supermassive black holes in nearby galaxies. Finally, you will learn that quasars can be used as beacons to probe remote regions of the universe.

THE DISCOVERY OF QUASARS

The universe always has the capability of surprising us. Rarely have astronomers predicted the extraordinary phenomena discussed in this book; more often, they have been left to scratch their heads at the puzzles provided by nature. The discovery of quasars is one such example.

Throughout the 1950s, radio astronomers used increasingly sensitive telescopes to compile lists of radio sources. Many were identified with distant galaxies, and some had no visible optical counterpart. Part of the difficulty stemmed from the inaccurate positions for radio sources, which meant that the origin of the radio waves was not from a well-determined angle on the sky. As a consequence, many optical objects were contained within the inaccurate radio position, and there was no way to decide which was the right one. Radio astronomers devised clever methods to improve the radio positions. One involved occultations of the Moon. As the Moon passes in front of a radio source, it dims the radio emission, and the well-determined position of the Moon puts a lock on the position of the radio source. By 1960, astronomers had pinpointed the location of several strong radio sources listed in the *Third Cambridge Catalog* (or 3C catalog) by this method.

At this point, the scene shifts to the California Institute of Technology, where astronomers began to focus their attention on two strong radio sources, 3C 48 and 3C 273, that appeared to coincide on the sky with bluish stars. They were called quasars, a contraction of the phrase "quasi-stellar radio source." At that time, no normal stars had been found to emit radio waves, with the exception of the Sun, whose radio emission is very weak and can be detected only because the Sun is so close. Senior astronomers then took spectra of the stars with the 200-inch telescope at Mount Palomar. To their surprise, the spectra showed broad emission lines that could not be identified with the lines of any known element. Were these lines evidence of a new element? Were these objects a new type of radio star? Astronomers were mystified.

It took a young researcher from Holland to solve the puzzle. In 1963, Maarten Schmidt noticed that four prominent lines in the spectrum of 3C 273 had the same relative spacing as the well-known series of hydrogen lines. However, the lines were at unusual wavelengths in the spectrum because they were shifted to the red by an unprecedented amount, about 16 percent. If the effect was a Doppler shift, it meant that the object is moving away from us at a speed of 45,000 km/s, or nearly 100 million miles per hour! The same lines in the spectrum of 3C 48 were redshifted even more, showing that object was traveling away from us at twice the speed of 3C 273. Astronomers were astonished. How could stars be moving away from us at speeds of hundreds of millions of miles per hour?

With further painstaking work, Schmidt and his colleagues showed that 3C 48 and 3C 273 were not truly points of light. Each stellar object was surrounded by a halo of fuzz, or nebulosity. According to the Hubble relation between distance and recession velocity, these quasars must be at distances of thousands of megaparsecs, and their light must have taken billions of years to reach us.

Schmidt showed that quasars are not radio stars but distant outposts of the universe. To appear as bright as they do at distances of billions of light years, quasars must be extraordinarily luminous. If the fuzz represents the light from a distant galaxy, then the bright quasar core must represent an even larger amount of light. What was this new phenomenon at the galaxy's center that could produce hundreds of times the entire light of a galaxy and emit huge amounts of radio waves? Astronomers had solved one puzzle, only to be presented with a deeper mystery.

The Redshifts of Galaxies

In 1912, American astronomer Vesto Slipher began a project to measure the spectra of several spiral nebulae. This was 10 years before Hubble would show that the spiral nebulae were actually distant systems of stars. Slipher worked at Lowell Observatory under the direction of Percival Lowell, who speculated about the "canals" on Mars. Slipher was looking for rotation of the nebulae with a new and efficient spectrograph. He hoped to see evidence of rotation using the Doppler effect—blueshifted light from material moving toward us and redshifted light from material moving away from us. He detected these wavelength shifts. Just as in the Milky Way, light from spiral nebulae is spread out over several hundred kilometers per second from the rotation of the disk. However, Slipher also encountered a big surprise: Most of the spiral nebulae were redshifted and moving away at incredible speeds.

Within a few years, Slipher had collected spectra of many spiral nebulae. In addition to the evidence for rotation, he noted that 21 out of the 25 spectra were shifted to red wavelengths by amounts of up to 1000 km/s. The high speeds were surprising. Several of the nebulae were moving at speeds of over 1 million mi/h! The predominance of redshifts was even more surprising. If the redshifts were due to the Doppler effect, then almost all of the nebulae were flying away from each other and from the Milky Way.

Were the nebulae gas clouds being ejected from the Milky Way? Or were they systems of stars receding from our own galaxy? Hubble took the crucial next step by measuring distances to spiral nebulae using Cepheid variable stars. In 1929, Hubble was able to combine Slipher's velocity measurements with his own distance measurements for several dozen galaxies. You can see the results in Figure 16–1. Each dot in the figure represents an entire galaxy!

As we consider the universe beyond the Milky Way, we will switch to a new unit that is more appropriate for the vast

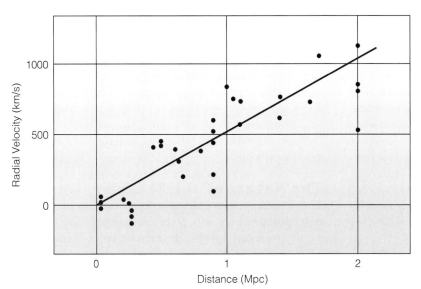

FIGURE 16–1
The linear relationship between radial velocity and distance of galaxies, discovered by Edwin Hubble in 1929. The slope of Hubble's original plot is steeper than the current value because of an error in the Cepheid calibration of the distance scale. Most galaxies have redshifts; only the nearest galaxies can have blueshifts. (SOURCE: *Astrophysical Journal*, courtesy University of Chicago Press.)

distances between galaxies. A **megaparsec (Mpc)** is a million parsecs, or a thousand kiloparsecs. (Note the use of the same prefix, *mega-*, that is used throughout the metric system to indicate 10^6.) A megaparsec is 3×10^{19} km! You should also remember that a megaparsec corresponds to a look-back time of 3.25 million years. Figure 16–1 shows that galaxies within 1 Mpc are moving at speeds of up to 1000 km/s away from us. The only four exceptions—the blueshifted galaxies with velocities less than zero—are galaxies bound by the cumulative gravity of the Local Group that happen to be moving toward us. Most importantly, we can see that, in general, more distant galaxies are receding faster from us. The correlation between radial velocity, or redshift, and distance is one of the foundation stones of modern cosmology.

Hubble's second discovery cemented his reputation as one of the giants of science, and it is the reason astronomy's premier observing facility, the Hubble Space Telescope, was named after him. We can see the three themes of this book illustrated in his work on the nature of galaxies. The awareness of the Milky Way as just one among many galaxies gives us a new sense of our place in the universe. The unexceptional nature of our star and our entire galaxy takes the Copernican revolution another step. We also see that physical laws can be applied over large distances. Hubble found that variable stars millions of light-years away behave similarly to variable stars near the Sun. And, we can see how scientists work with limited information to increase knowledge. Faint, fuzzy patches of light, mostly too faint to see with the naked eye, have been shown to be vast new worlds to explore.

The story of the discovery of galaxies also illustrates the interconnectedness of science. Some of the luster on Hubble's reputation is due to the force of his personality. He had a supreme self-confidence that often bordered on arrogance. In truth, Hubble's success depended on synthesizing the work of others. Slipher was the first to discover that most galaxies had redshifts, but he had no way of measuring their distances. Henrietta Leavitt first demonstrated the period-luminosity relation for Cepheids, and Harlow Shapley pioneered the use of Cepheids as distance indicators. Hubble benefited from the entrepreneurial energy of George Ellery Hale, who built the large telescopes that he needed to do his work. Hubble was not particularly adept as an observer. He depended heavily on the skills of Milton Humason at the Mount Wilson Observatory. Humason began as a mule driver during the construction of the observatory, and he worked his way up from janitor to telescope operator. By the 1940s, he had become an expert observational astronomer, and he worked with Hubble to take photographs that were littered with distant galaxies.

Expansion of the Universe

How can we interpret the observation of galaxy redshifts? We should concentrate on two basic facts: Almost all galaxies are redshifted, and the size of the redshift increases with increasing distance. In other words, galaxies are moving away from us, and the more distant galaxies are moving away faster. We are assuming that redshift is a measure of the Doppler velocity of a galaxy with respect to the Milky Way.

To help explain the redshift of galaxies, let us consider two different possibilities. First imagine a *static universe*, a universe with a fixed size that can contain galaxies in motion (if galaxies are not moving at all, no redshifts are measured). Suppose that galaxies are milling around randomly in space, separated by large distances (Figure 16–2a). They change position over time, but their typical separations are constant. In this type of universe, we would observe half the galaxies with a component of motion away from us and half the galaxies with a component of motion toward us. This scenario translates into equal numbers of redshifts and blueshifts. Also, galaxies in every region of space have the same random motions, so the results of the measurement do not depend on distance. At every distance, the average size of the redshifts and blueshifts is the same, and the numbers of redshifts and blueshifts are about equal. A plot of velocity against distance would have points (each representing an entire galaxy) scattered around the line of zero velocity, as shown in Figure 16–2b. This result is not what Hubble observed.

Now imagine an **expanding universe**, in which the distance between galaxies increases with time (Figure 16–2c). If every galaxy is moving away from the Milky Way, then we will measure a redshift for every galaxy. More distant galaxies move away more quickly. To see why this is true, let us consider the analogy of an expanding Earth. Suppose that Earth is expanding at a rate such that it will double in size in an hour. The distance between *any two cities* must also double in an hour. For example, Tucson is about 100 miles from Phoenix. After an hour, Tucson will be 200 miles from Phoenix, so the two cities are moving apart at a rate of 100 mi/h. Tucson is about 500 miles from Los Angeles. After an hour, Tucson will be 1000 miles from Los Angeles, so the two cities are moving apart at a rate of 500 mi/h. We can see there is a linear relationship between distance and recession velocity. A plot of the two quantities will show a correlation (Figure 16–2d), as observed by Hubble for galaxies.

Remember that all our illustrations and analogies are two-dimensional, whereas galaxies are moving apart in the three dimensions of space. Remember also that Doppler shifts can only measure the component of a galaxy's motion along the line of sight. Galaxies will also have transverse, or sideways, components of their motion that we cannot measure.

The Nature of the Redshift

The concept of an expanding universe leads to some important questions. Are we are at the center of the universe? Figure 16–3, on page 472, shows a section of the expanding universe at three successive times. You can see that the distance between a galaxy and *every other galaxy* increases with time. In other words, it does not matter on which galaxy an

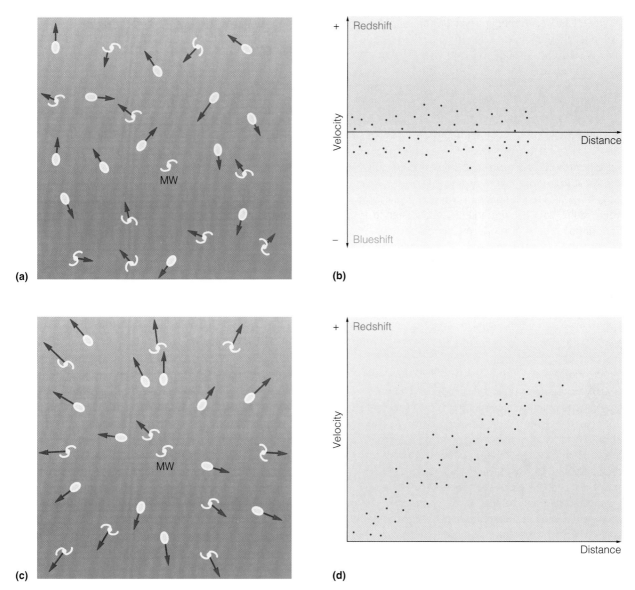

FIGURE 16-2
Static and expanding universes compared. **(a)** In a static universe, galaxies move randomly with respect to the Milky Way; the length of each arrow indicates the size of the velocity. **(b)** There is no correlation between distance and velocity, and the average velocity is zero. **(c)** In an expanding universe, galaxies move away from each other. **(d)** The recession velocity increases with increasing distance; this relationship is known as the Hubble relation.

observer is located; he or she will see the same expansion. To use our analogy of the expanding Earth, someone in Phoenix will see Tucson moving away at 100 mi/h, exactly the same recession velocity that someone in Tucson sees. Therefore, we would not be correct in assuming that Tucson is the center of the expansion. An observer in a distant part of this "universe" will also see the same expansion. On the far side of Earth, London is about 100 miles from Birmingham and 500 miles from Aberdeen. After an hour, someone in London will conclude that Birmingham is moving away at 100 mi/h and Aberdeen is moving away at 500 mi/h. Yet Earth's center is not London any more than it is Tucson. We conclude that there is no central position in this universe because all galaxies are moving away from each other. An observer on a planet around a star in a distant galaxy would measure exactly the same relationship between distance and velocity that we do.

Is everything in the universe expanding? Every galaxy is moving away from every other galaxy, but galaxies themselves are held together by gravity. Thus, galaxies themselves are not expanding. On a scale smaller than galaxies, normal stars are not expanding either. On familiar terrestrial scales, objects are held together by the electrical forces of matter. Earth is not expanding, nor is your hometown or your car or your head. Also, galaxies can be bound together by gravity so that they do not take part in the general expansion. The Local Group is one example. The expansion of the universe applies only on the largest scales. Notice that the galaxies plotted in Figure 16–1 do not form a perfectly narrow line. The expansion is not completely smooth and uniform. The scatter in the relationship between distance and velocity is caused by the gravitational interactions among galaxies.

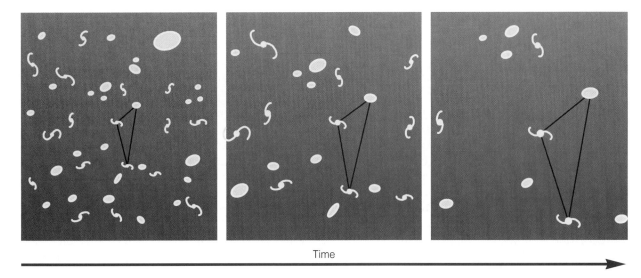

FIGURE 16-3
An expanding universe (in two dimensions) viewed at three successive times. The distance between any two galaxies increases with time. An observer on any galaxy would measure the same Hubble relation, and the expansion has no detectable center. Note that each galaxy remains the same size because it is held together by its own internal gravity.

What was the universe like in the past? If galaxies are moving away from each other, then they must have all been closer together in the past. Think of Figure 16-3 as the frames of a video. If we rewind the tape in order to see how our universe might have looked in the distant past, we would notice that the galaxies must have been on top of each other at an early point. Just as a gas will heat up when it is compressed, a universe of galaxies must heat up when it is compressed; galaxies are, of course, made of an enormous number of particles. The expansion that we now observe implies a time in the distant past when all the mass in the universe was concentrated in a state of extremely high temperature and density. We appear to be riding out the aftermath of an ancient and vast explosion of this dense concentration of matter.

Is the expansion of the universe like an explosion? Not really. The analogy of an explosion is flawed for two reasons. First, any explosion has a center and an edge. We have argued that every galaxy is moving away from every other galaxy, and an observer on any galaxy will measure the same expansion. The number of galaxies seen in any direction of the sky is similar, so we cannot use anisotropy to place ourselves within the distribution (as we did with globular clusters to show that we are not at the center of the Milky Way). Put simply, astronomers looking deeper and deeper into the universe do not "run out" of galaxies in any direction. We cannot detect a center or an edge to the expansion of the universe.

The second reason that an explosion is a poor analogy for the expanding universe has to do with the nature of the redshift. It is natural to interpret redshifts as Doppler shifts. The Doppler effect was first discussed in the context of the detection of extrasolar planets in Chapter 9. A large planet can cause a periodic Doppler shift in the light of a nearby star. Later, in Chapter 14, we saw that two stars in a binary orbit show a periodic Doppler shift as seen from Earth. In Chapter 15, we learned that we can map our position in the Milky Way by measuring the Doppler shift of stars in the rotating disk. In all of these situations, astronomers measure motions with reference to a particular object or a fixed point in space. In a more familiar example, the Doppler shift of sound waves is referenced to the air through which sound travels. By contrast, the modern theory of the universe holds that the redshift is caused by the expansion of space. In an explosion, the fragments fly *through* space, and their motion can be defined with reference to the center of the explosion and the medium through which the fragments travel. In the expanding universe, galaxies are carried apart by the expansion of space itself! We will explore the consequence of this intriguing idea in the next chapter.

The Doppler effect has certainly been confirmed with planets and stars. But the idea that the expansion of space explains *all* galaxy redshifts is called the theory of **cosmological redshifts**. There are other possibilities. Light escaping an intense gravitational field, such as the environment of a black hole, will suffer a gravitational redshift (see Chapter 13). Given what we know about galaxy masses, this effect is far too small to cause galaxy redshifts. Some researchers have proposed *tired light* theories, in which photons traveling toward us lose energy and get redshifted in traversing the vast distances of interstellar space. Such theories make few specific predictions and therefore are difficult to rule out.

Other researchers have argued that some galaxies have noncosmological redshifts, based on certain curious situations in which a galaxy with a large redshift appears to be close on the sky to a galaxy at a much lower redshift. Under the cosmological assumption, they should be at very different distances. Apparent associations are intriguing, but there are no reliable statistics about their occurrence. Some objects with very different redshifts will be projected close together on the sky purely by chance (just as stars close together in the sky are often at very different distances). Redshift is correlated with the apparent brightness and angular size of galaxies, both of which are indicators of distance. The great bulk of evidence thus favors the cosmological interpretation.

The Hubble Relation

When Hubble compared the radial velocities for galaxies with his distance estimates, he found a clear correlation between the radial velocity, measured from the redshifted spectral lines, and the distance. The relationship is linear, as shown in Figure 16–1. The slope of the correlation defines the expansion rate of the universe. This fundamental result of extragalactic astronomy is called the **Hubble relation.** Although it is sometimes referred to as Hubble's law, it is not a law of nature, like Newton's law of gravity or the laws of thermodynamics. There is no law of physics that requires galaxies to move apart or to have motions that increase with increasing distance. Hubble's relation is a purely observational result. However, it is a vital clue in helping us understand the universe we live in.

From now on in this book, we will use the term *redshift* as shorthand for a recession velocity thought to be caused by the expansion of the universe. Although the conceptual basis of Doppler redshifts and cosmological redshifts are quite distinct, the mathematical descriptions are identical, as long as the recession is much slower than the speed of light. We saw in Chapter 9 that the Doppler shift is given by $\Delta\lambda/\lambda = v/c$, where $\Delta\lambda$ is the difference between the wavelength of the observed light and the wavelength that light would have if the object were not moving. Astronomers define redshift by the symbol z, where $z = \Delta\lambda/\lambda = v/c$. Redshift is therefore a pure number with no units; it is equal to the fractional wavelength shift of galaxy light or the recession velocity (v) as a fraction of the speed of light (c). The Science Toolbox on page 475 presents the definition of redshift in more detail and looks at the relationship between redshift and distance.

Figure 16–4 shows the optical appearance and spectra of galaxies at five different redshifts. The spectra are plotted with wavelengths increasing (getting redder) from left to

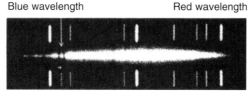
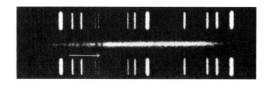
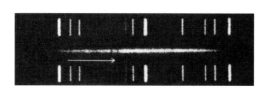
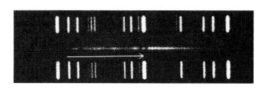
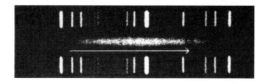

FIGURE 16–4
Photographic evidence of a relationship between redshift and distance for remote galaxies. The left column shows galaxies. The right column shows spectra; white lines at the top and bottom of each spectrum are emission lines produced in the instrument for comparison, being similar in each spectrum. A pair of dark absorption lines (the H and K lines of gaseous calcium) can be detected in each galaxy's spectrum, above the head of the white arrow. This pair is farther right (red) in each succeeding galaxy. The center column shows the inferred distance. (SOURCE: Hale Observatories.)

right. A pair of absorption lines from ionized calcium can be seen as dark notches in the galaxy spectrum (which is straddled by a set of emission lines that define zero redshift). For more distant galaxies, the spectral features are shifted to longer wavelengths. You can also see that the angular size of the galaxies decreases with increasing redshift, consistent with a larger estimated distance.

A more modern version of Hubble's relation can be seen in Figure 16–5. The data extend the linear relation between redshift and distance hundreds of times farther into space than Hubble's original work. The highest recession velocity in Figure 16–5 is 30,000 km/s, or 10 percent of the speed of light. The largest distance measure is 500 Mpc. In other words, we can detect galaxies that are over 1.5 billion light-years distant, moving away from us at 18 million mi/h! The two quantities plotted in a Hubble relation are not equally easy to measure. It is relatively simple to measure a redshift for a galaxy; only a spectrum is required. However, as we saw in the previous chapter, establishing a reliable distance can be very difficult. Therefore, we will revisit the issue of how astronomers measure the distances to galaxies.

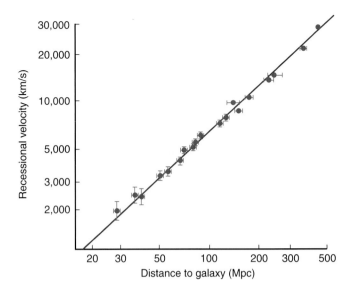

FIGURE 16–5
The Hubble relation for 20 spiral galaxies. Supernovae have been used as distance indicators, which allows the Hubble relation to be extended to a distance of 500 Mpc, or over 1.5 billion light-years. The slope of the correlation gives the expansion rate of the universe. (SOURCE: A. Reiss, W. Press, and R. Kirshner, Harvard University.)

Distances to Galaxies Revisited

Individual stellar types are used as distance indicators within the Local Group and out to about 10 Mpc, but they cannot be used at the enormous distances of the most remote galaxies, for two reasons. First, variable stars, such as Cepheids and RR Lyraes, are not luminous enough to be detected at such large distances. Second, the individual stars of a distant galaxy cannot be spatially resolved. The light from an individual star is hopelessly blurred into the overall light of the galaxy. Astronomers therefore look for properties of entire galaxies that can serve as distance indicators.

Remember the requirements of a good distance indicator: It should be based on well-understood astrophysics, it should involve a simple and direct observation, it must be possible to calibrate the technique against local distance indicators and thereby establish a chain of measurement that goes all the way back to parallax measurement in the neighborhood of the Sun. Which properties of galaxies should we use? Unfortunately, the most obvious ones are not good distance indicators. Galaxies span an enormous range in luminosity (see Table 15–2), so the apparent brightness of a galaxy gives very little indication of its distance. Galaxies also range in size from dwarfs 1 kpc across to giants 100 kpc across, so the apparent diameter of a galaxy gives very little indication of its distance. As it turns out, astronomers have developed different techniques for measuring the distances to spiral and elliptical galaxies.

The *Tully-Fisher relation* is based on the discovery that the luminosity of a spiral galaxy is correlated with the rotational speed of the gas disk. This fact be can established using galaxies with distance measurements that specify the luminosity. However, the Tully-Fisher relation can be used out to enormous distances when individual Cepheids cannot be resolved. The observation of rotation speed in a disk is relatively simple; it can be done with a single spectroscopic observation of the 21-cm line—the velocity width of the line equals twice the rotation speed of the disk. This distance indicator can be calibrated in the local universe and involves a simple measurement, but does it have a physical basis? As we saw in our discussion of the Milky Way, disk rotation speed is an indicator of mass. If we assume that the mass-to-light ratio of a spiral galaxy is fixed—that is, the stellar populations are constant—the mass then leads to an estimate of the galaxy's luminosity. The Tully-Fisher relation is thus a good distance indicator. When it is used on galaxies whose distance is known from another technique, it yields distances with a scatter of about 15 percent. The technique works best when used on galaxies with intermediate inclinations to the line of sight. A face-on spiral shows no Doppler motions due to rotation, because the disk motion is on the plane of the sky, and the total brightness of an edge-on spiral is difficult to measure because of the obscuring effects of dust in the disk.

The *Faber-Jackson relation* for elliptical galaxies is analogous to the Tully-Fisher relation for spiral galaxies. It is based on the fact that the range of stellar velocities of an elliptical galaxy, or the velocity dispersion, is correlated with the size of the galaxy. Once again, the observation is relatively simple. Astronomers use an image to measure the galaxy size (being careful to agree on their definition for objects without sharp edges!), and they use the width of stellar absorption features in an optical spectrum to measure the velocity dispersion. Unfortunately, Cepheids are not found in old stellar populations, so the Faber-Jackson relation cannot be calibrated with Cepheid variables. The physical basis for the Faber-Jackson relation is the fact that old stellar populations in elliptical galaxies are relatively simple. Both the size and

SCIENCE TOOLBOX

Relating Redshift and Distance

Hubble showed that the redshift of a galaxy is correlated with its distance from the Milky Way. Let us look at the implications of the Hubble relation in more detail, starting with the way redshift is defined.

Astronomers observe the light from almost every galaxy to be redshifted. The only exceptions are a few very nearby galaxies that are bound by gravity to the Local Group. Suppose we observe a galaxy and label the wavelength of any spectral feature as λ (lambda). The spectral feature will typically be an absorption line of hydrogen, calcium, or magnesium. The wavelength of that same spectral feature observed in a gas in the laboratory is λ_0. The difference in the two wavelengths $(\lambda - \lambda_0)$, which is a positive number, represents how much the wavelengths of the galaxy's light have been shifted to longer wavelengths $(\Delta\lambda)$. The redshift of a galaxy is defined as

$$z = \frac{\Delta\lambda}{\lambda_0}$$

Since $\Delta\lambda = \lambda - \lambda_0$, we get $z = (\lambda - \lambda_0) / \lambda_0$. Now we can use the equation for the Doppler effect from Chapter 9 ($\Delta\lambda/\lambda = v/c$) to define the redshift in terms of the recession velocity of the galaxy (v) and the speed of light (c)

$$z = \frac{v}{c}$$

This equation is actually an approximation that is only valid for describing the redshift of galaxies when the recession velocity is much smaller than the speed of light. Redshifts can be measured very accurately. A typical velocity measurement for a galaxy might be 12,540 ± 120 km/s, which corresponds to a redshift of $z = v/c = 12,540/(3 \times 10^5) = 0.0418$. The uncertainty in velocity corresponds to a redshift uncertainty of $120/(3 \times 10^5) = 0.0004$, so we would quote the complete measurement as $z = 0.0418 \pm 0.0004$.

Recall that the *conceptual* basis for galaxy redshifts is quite distinct from the Doppler effect. The Doppler effect is due to the relative motions of objects traveling through space or another material medium. The cosmological redshift is due to the expansion of space itself. Measuring the galaxy's redshift is one step toward determining its distance. The next step is using the Hubble relation.

As we have learned, the Hubble relation is a linear correlation between the redshift of a galaxy and its distance from the Milky Way. On a graph, the slope of the line is the Hubble constant, or a measure of the expansion rate of the universe. Mathematically, the Hubble relation can be expressed as

$$v = H_0 \, d$$

In this equation, v is the velocity of the galaxy in km/s, d is the distance in Mpc, and H_0 is the Hubble constant in km/s/Mpc. Refer to Figure 16-5 to see a recent measurement of the Hubble relation.

Real galaxies do not follow a perfect Hubble relation. There is always a scatter around the line caused by the fact that galaxies interact with each other by gravity, which gives them a component of velocity that is not due to the expansion of the universe, called a *peculiar velocity*. The amount of this extra velocity (which may be a redshift or a blueshift) is 100 to 200 km/s. Since galaxies tug each other around in space, there is no Hubble relation for the very nearest galaxies; velocity and distance are not correlated. For a galaxy with a recession velocity of 1000 km/s, the peculiar velocity is a significant fraction of the recession velocity, 10 to 20 percent. For a much more distant galaxy with a recession velocity of 10,000 km/s, the peculiar velocity is only 1 or 2 percent of the recession velocity. In other words, the peculiar velocity can be ignored for sufficiently distant galaxies, and the Hubble expansion is smooth and well determined.

Note that peculiar velocities have no systematic effect on the slope of the Hubble relation. The largest uncertainty in the slope of the Hubble relation, and therefore in the value of the Hubble constant, is caused by systematic errors that alter the distance scale.

Now we can combine the two equations above to relate the distance of a galaxy to its redshift. First, rearrange the terms of the Hubble relation to calculate distance as $d = v / H_0$. Then rearrange the terms of the redshift equation to get $v = z\,c$. Combining the two results, we get

$$d = \frac{z\,c}{H_0}$$

Again, this formula is only appropriate if the recession velocity is much less than the speed of light, or if $z \ll 1$. If we know the value of the Hubble constant, we can use redshift as a distance indicator. This technique is a very powerful tool, since we can measure redshifts of galaxies so faint that it is impossible to get more detailed information about them by other means.

While it is not unusual to measure a redshift to a precision of four significant figures or an accuracy of a few percentages, we cannot calculate distance with that level of accuracy. To use our previous example, suppose we measure a galaxy redshift of $z = 0.0418 \pm 0.0004$. For a Hubble constant of 65 km/s/Mpc, the galaxy is at a distance of $d = (0.0418 \times 3 \times 10^5) / 65 = 192$ Mpc. As we have seen, the Hubble constant is probably not known more accurately than 10 or 15 percent; thus, we would quote the distance only to two significant figures, or $d = 190 \pm 20$ Mpc. Remember that the use of redshift as a distance indicator relies completely on the assumption that redshift is caused by cosmological expansion.

the velocity dispersion are good measures of mass. The Faber-Jackson and Tully-Fisher relations are equally good distance indicators, with similar scatter on the measurement for a single galaxy. When galaxies are in a group or a cluster, they are at the *same* distance. As we saw in Chapter 1, with N measurements of the same quantity, the error on the combined measurement improves by a factor of $\sqrt{N}$. Astronomers therefore combine measurements of individual galaxies to get a more accurate result.

Astronomers have had great success with the use of supernovae as distance indicators. Recall from Chapter 13 that a supernova represents the violent death of a massive star. Enormous energy is released, and for a few days a supernova can rival an entire galaxy in brightness. Single stars with a range in masses do not yield supernovae with a tightly defined luminosity. However, supernovae that result from mass transfer in a binary system (as discussed in Chapter 14) are excellent standard "bombs." Here is how it works. If a white dwarf is in a binary system, it can slowly gather gas from its companion. As the white dwarf exceeds the Chandrasekhar limit of about 1.4 $M_\odot$, it must collapse. Carbon and oxygen in the collapsing star are compressed and heat up enough to start fusion. In these reactions, carbon and oxygen fuse to form silicon (^{12}C + ^{16}O → ^{28}Si), and pairs of silicon nuclei fuse into nickel (^{28}Si + ^{28}Si → ^{56}Ni). The rapid energy release blows the star apart; no remnant is left over. It is as if you had a jar of a highly volatile chemical and slowly poured more of the chemical into the jar. As you pass a particular threshold, the chemical becomes unstable and explodes. The result is the same every time—a predictable and standard explosion.

Figure 16–6 shows the very characteristic light curve of this type of supernova. The brightness is driven first by the decay of nickel-56 into cobalt-56, then by the decay of cobalt-56 into stable iron-56. This well-understood and regulated principle of atomic physics explains why the luminosity of supernovae has a small scatter and can be used as a distance indicator with an error of only 10 to 15 percent. Figure 16–5 shows a Hubble relation using supernovae out to a distance of 500 Mpc. As with other distance indicators, astronomers must be careful to understand the effects of dust both within the supernova and within the galaxy that contains the supernova. The disadvantage of this distance indicator is the fact that we cannot say with any certainty when or where a supernova will go off! However, if we monitor enough galaxies, they can be reliably found before they reach their maximum light. The advantage of this distance indicator is its enormous luminosity. At more than 10^8 $L_\odot$, or 10,000 times more luminous than a Cepheid, they can be used out to 1000 Mpc or beyond. They can even be detected at such a distance that the galaxy that contains them is invisible!

Finally, we point out that redshift can be used as a distance indicator. The Hubble relation lets us measure the redshift for a galaxy and assign it a distance without measuring the distance directly. Measuring a redshift only requires a spectrum of modest quality, so redshifts have been measured for extremely faint and distant galaxies. However, this distance indicator depends on the *assumption* that the observed redshift is caused by the expansion of space. In other words, redshift is only a distance indicator in the context of a cosmological model.

The Hubble Constant

As we have seen, the Hubble relation expresses how recession velocity increases with distance from the observer. The slope of the plot, the ratio of velocity to distance, is known as the **Hubble constant**. The Hubble constant, usually denoted by the symbol H_0, is a measure of the rate of the expansion, and it also indicates the size and age of the universe. Figure 16–7 shows the distinction between a high and a low value of the Hubble constant. It is one of the most im-

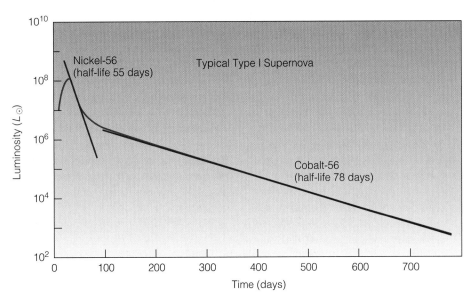

FIGURE 16–6
The light curve of a supernova that results from the addition of mass to a white dwarf in a binary system. The peak luminosity varies little from supernova to supernova, and the decline of the light curve is governed by radioactive decay first of nickel-56, then of cobalt-56. At its peak luminosity, a supernova can outshine the entire galaxy that contains it.

SCIENCE TOOLBOX

The Size and Age of the Universe

The Hubble relation gives the distance of a galaxy in terms of its recession velocity. If we accept that the recession velocity is caused by the cosmological expansion of space, we can show how the expansion rate, as measured by the Hubble constant, is related to both the size and the age of the universe.

If we assume that the expansion is uniform, we can derive an age for the universe. Essentially we are estimating the time in the past when all galaxies were together in space. We assume that space has been expanding at a constant rate, carrying galaxies apart smoothly and continuously. At a constant velocity, a distance traveled is equal to the velocity multiplied by the time ($v = dT$). Equivalently, the expansion time T is the distance divided by the velocity ($T = d/v$). Now we can use the Hubble relation ($v = H_0 d$) to substitute in the equation of T. We get $T = d/v = d/H_0 d$, or the simple result

$$T = \frac{1}{H_0}$$

Let us look at a couple of examples. Suppose a galaxy has a recession velocity of 7000 km/s, which we infer from the measured redshift and the use of the redshift equation $v = zc$. An independent measure of the distance using Cepheid variables gives a value of 115 Mpc. The Hubble constant based on these numbers is $v/d = 60$ km/s/Mpc. In other words, galaxies show 60 km/s of cosmological redshift for every Mpc increase in distance. The estimated age of the universe is $1/H_0$. Since the units of H_0 are km/s/Mpc, the units of $1/H_0$ are Mpc s/km. To remove the mixed distance units, we multiply by 3×10^{19}, the number of kilometers in a megaparsec. To convert to years, we divide by 3×10^7, the number of seconds in a year. The result is $T = (1/60) \times (3 \times 10^{19} / 3 \times 10^7) = 1.7 \times 10^{10}$, or 17 billion years.

Now suppose a group of galaxies has an average recession velocity of 12,500 km/s. The Tully-Fisher relation for spirals in the group gives a distance of 140 Mpc. This time we see that the measured Hubble constant is $v/d = 90$ km/s/Mpc. Following the same logic as before, we calculate that $T = (1/90) \times (3 \times 10^{19} / 3 \times 10^7) = 1.1 \times 10^{10}$, or 11 billion years.

We can now understand the implications of low and high estimates of the Hubble constant. In the first example above, the Hubble constant was a low value of $H_0 = 60$ km/s/Mpc. Thus, for each megaparsec of distance from the observer, the recession velocity increases by 60 km/s. A low Hubble constant represents a slow expansion and a long distance scale (a particular measured redshift corresponds to a large distance). Tracing the recession backward in time using the equation above, we calculate that the inferred age of the universe is 17 billion years. Therefore, a low Hubble constant means that we live in a large and old universe.

At the other extreme, assume that $H_0 = 90$ km/s/Mpc, in which case the recession velocity increases by 90 km/s for each megaparsec of distance. A high Hubble constant like this represents a rapid expansion, a short distance scale, and a smaller universe with an inferred age of 11 billion years. Therefore, a high Hubble constant means that we live in a smaller and younger universe. As we will see in the next chapter, the expansion is actually slowing down, so these estimates are actually upper boundaries for the age of the universe.

---

portant quantities in astronomy, and you can learn more about it in the Science Toolbox above.

How do astronomers estimate the Hubble constant? It is an enormously difficult research undertaking. As we have pointed out, the measurement of a redshift for a galaxy is the easy part. The hard part is the measurement of distance. Astronomers must avoid galaxies that are too close to the Milky Way because they are bound to each other by gravity and do not take part in the general expansion. They also must avoid galaxies elsewhere in space that are bound in a group or cluster, because they are all at roughly the *same* distance. Otherwise, it does not matter how the galaxies are selected. Since the expansion is isotropic, the same Hubble relation should be measured in every direction in the sky. Once the redshift and distance of a galaxy have been measured, each galaxy can be plotted as a point on a Hubble relation graph. The slope of the line that gives the best fit to the data is the Hubble constant.

All methods of distance determination need an accurate calibration at small distances. This calibration sets the luminosity of each new distance indicator; all of the uncertainty in the Hubble constant comes from this procedure. Once the calibration is known, relative distances are easy to calculate using the inverse square law. You can understand calibration by analogy from everyday experience. Suppose you were measuring distances around your house using a tape measure. The accuracy of all your measurements would depend on how accurately an inch had been defined, or calibrated, when the tape measure was made at the factory. Or suppose you were trying to measure the distance between two towns using the odometer in your car. The accuracy of your measurement would depend on how accurately the circumference of your

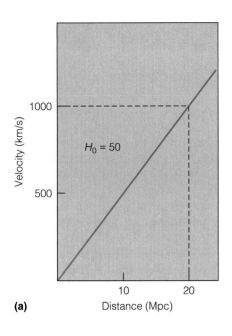
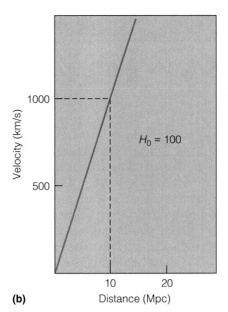

FIGURE 16-7
The distinction between low and high values of the Hubble constant. **(a)** With a low value such as $H_0 = 50$ km/s/Mpc, the distance corresponding to a redshift of 1000 km/s is 20 Mpc. This value represents a relatively large universe and a slow expansion. **(b)** With a high value such as $H_0 = 100$ km/s/Mpc, the distance corresponding to a redshift of 1000 km/s is 10 Mpc. This value represents a relatively small universe and a rapid expansion. Based on current research, the Hubble constant is probably within 10 percent of 65 km/s/Mpc.

wheel had been calibrated at the factory, since an odometer essentially counts wheel rotations. Likewise, in astronomy, the bedrock of all astronomical distance measurement is parallax. Trigonometry has been used to measure the distances to many stars in the neighborhood of the Sun; the calibration of *all* other distance indicators depends on this information.

Figure 16–8 shows the overlapping ladders of the distance scale, which extends the indicators beyond the Local Group (compare with Figure 15–14). There are several crucial benchmarks in the extragalactic distance scale. One is the Large Magellanic Cloud, which has an accurate geometric distance from measurements of the dust shell of Supernova 1987A. The LMC also contains a large enough stellar

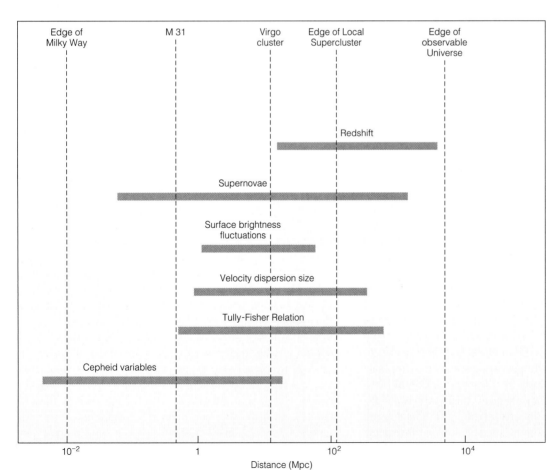

FIGURE 16-8
The ladder of distance indicators used to measure extragalactic distances, with the horizontal bar showing the range of applicability. Many of these indicators use the properties of an entire galaxy as a standard candle. Redshift is a useful distance indicator only if we assume a particular model for the expanding universe. Moreover, at distances less than 20 Mpc, redshift is a poor indicator of distance because of the gravitational interactions of galaxies in the Local Supercluster.

478 CHAPTER 16: THE EXPANDING UNIVERSE

"zoo" for the calibration of rare distance indicators, such as Cepheid and RR Lyrae variables. Another benchmark is the Virgo cluster, about 15 Mpc distant. The Virgo cluster is close enough that Cepheid variables can be detected using the Hubble Space Telescope. Since the Virgo cluster contains both spiral and elliptical galaxies, astronomers can calibrate the Tully-Fisher and Faber-Jackson relations for use out to even greater distances. We have seen that the light curve of a supernova can be used to measure distances out to hundreds of megaparsecs. Every galaxy that has both a supernova and measurable Cepheid variables can be used to calibrate the use of supernovas as distance indicators.

Astronomers have finally begun to reach a consensus as to the value of the Hubble constant. The subject of the distance scale has a long history of controversy and unrecognized systematic errors. Hubble's first measurements (see Figure 16-1) yielded a very steep slope and a high value for H_0: 540 km/s/Mpc. This high value was alarming because it implied such a rapid expansion that the universe had to be younger than the well-measured age of Earth! It turned out that Hubble had used an erroneous calibration of the Cepheid luminosity. During a vast observational effort over the past 30 years, published values of the Hubble constant have ranged from 50 to 100 km/s/Mpc.

Why this uncertainty of a factor of 2? One reason is dust. When we look out to the galaxies, we look through a veil of stars and dust in the disk of our own Milky Way. This relatively nearby dust dims and reddens the light from distant galaxies. Thus, dust makes a galaxy of a particular luminosity appear farther away than if it were viewed through no dust. The radial velocity is unaffected. The result is a lower value of the Hubble constant. Moreover, Cepheids and supernovae in distant galaxies are dimmed by dust in the galaxy that contains them as well as by patchy dust in our own galaxy. While all researchers agree that this effect is important, they disagree on the size of the correction for dust. Another cause of uncertainty is our imperfect understanding of many distance indicators. Our knowledge of the Hubble constant is only as good as our understanding of the physics that sets the luminosity of a distance indicator.

Despite this discouraging history, recent measurements have converged on a relatively narrow range for the Hubble constant. Most observations are consistent with a H_0 value of 65 km/s/Mpc. What does this mean in everyday units? It means that galaxies are being carried away from us by the expansion of space at a rate of 20 km/s, or $20 \times (3/5) \times 3600 = 43,000$ mi/h, for every million light-year increase in distance! This number has an uncertainty of about 15 percent, so we can quote it as 65 ± 10 km/s/Mpc. In other words, it is unlikely that the Hubble constant is much lower than 55 km/s/Mpc or much higher than 75 km/s/Mpc. The fact that many different measurement techniques give a similar expansion rate is a wonderful validation of our basic understanding of how the universe works.

Large-Scale Structure

At such vast distances from us, galaxies are so dim and distant that we cannot see their detailed properties. Instead, we can use them to learn more about the **large-scale structure** of the universe—in other words, the spatial distribution of galaxies on the largest scales we can measure. The distances between most galaxies are far greater than their sizes. As a result, we can treat the entire effect of the gravity of billions of stars within a galaxy as being equivalent to a single large mass in space. Galaxies are tracers of expanding space, just as corks thrown on a river would be tracers of the motions of the water. To help explain the largest structures in the universe, astronomers imagine galaxies as "particles" in a grand physics experiment that has been going on for billions of years.

Galaxies are not scattered randomly throughout the universe. They are clustered in space, and the clustering provides important information about how the universe evolved. The Science Toolbox on page 485 describes how astronomers measure clustering. Figure 16-9 shows the two hemispheres of the sky, where each point is a different galaxy from a large photographic survey. The blank bands mark the plane of the Milky Way; galaxies exist in these directions, but they are very difficult to identify because of the crowding of stars and the obscuration in the disk of our galaxy. It is obvious from these maps that galaxies are not uniformly or randomly distributed. There are regions with relatively few galaxies and regions with a high concentration of galaxies. We will discuss many of these features later in the chapter.

The clustering of galaxies is analogous to the clustering of stars described earlier, but it occurs on a much larger scale. Just as there are binary stars in the Milky Way, there are binary galaxies in the vast depths of space beyond the Milky Way. Binary galaxies are two stellar systems that are bound by gravity in a majestic waltz. It may take many millions of years for them to orbit each other. Like there are groups of stars in the Milky Way, there are **galaxy groups** consisting of anywhere from a few to a few dozen galaxies. As described in Chapter 15, we live in the Local Group, which includes three moderate-sized spiral galaxies, each with a few small companions, plus a few other dwarf galaxies. These galaxies are loosely sprinkled over a region about 1 Mpc across. Beyond the Local Group are the Sculptor and M 81 groups at 2 to 3 Mpc. Each is like the Local Group, with two or three large galaxies and a number of smaller ones. At a distance of 5 to 6 Mpc, there is a loose cluster centered on the giant spiral galaxy M 101, the Pinwheel galaxy. At the edge of our cosmic neighborhood, about 15 Mpc away in the direction of the Virgo constellation, we encounter the first truly impressive concentration of galaxies.

Clusters of Galaxies

The concentration of galaxies in the direction of Virgo is the nearest example of the many **galaxy clusters** found in the

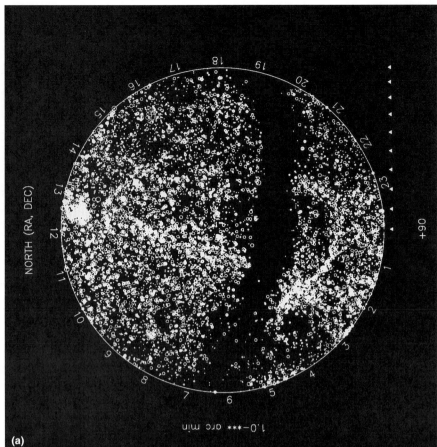

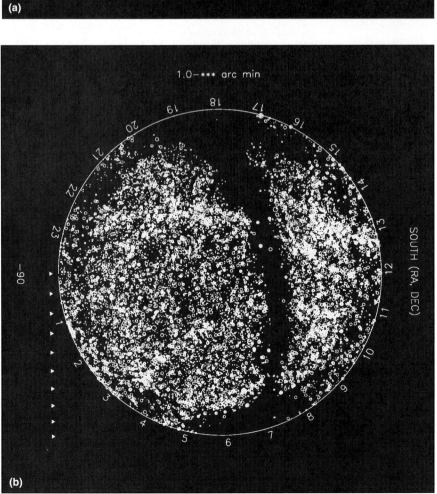

FIGURE 16-9
Equal-area projections of **(a)** the northern sky and **(b)** the southern sky, in which each point is a galaxy from large photographic surveys. Right ascension (R.A.) is marked around the circumference. Declination (Dec.) is the radial coordinate, with the pole at the center and 10-degree spacings represented by the triangular tickmarks. The blank bands mark the zone of obscuration by the Milky Way. (For a discussion of astronomical coordinates, see Appendix A-6.) (SOURCE: O. Lahav and D. Lynden-Bell, Institute for Astronomy, Cambridge.)

universe. Galaxy clusters contain hundreds or thousands of galaxies. There is no fixed demarcation between a galaxy group and a galaxy cluster. However, galaxy clusters not only have many more members than galaxy groups, they also have a much higher concentration of galaxies in space. This means that galaxy clusters can be recognized out to extremely large distances from the Milky Way. Thirty years ago, American astronomer George Abell cataloged over 2700 of the richest clusters of galaxies in the northern sky. Over 10,000 clusters have now been detected.

At a distance of 15 Mpc, or about 50 million light-years, we encounter the Virgo cluster (Figure 16-10a). Light from these galaxies began its journey well before humans evolved, not long after the catastrophic event that led to the death of the dinosaurs. Our modest Local Group is like a small suburb on the outskirts of a vast city of galaxies. The Virgo cluster is a sprawling mass of hundreds of galaxies, with three giant ellipticals near its center: M 84, M 86, and M 87. It is about 1.5 Mpc across (50 percent larger than the Local Group), and the dense part of the cluster covers 6° on the sky, or 12 times the diameter of the Moon. Despite these impressive numbers, Virgo is considered a *poor cluster* because it is only 2 or 3 times denser than the Local Group. Poor clusters usually have irregular shapes and a number of subgroups and concentrations within them. The bright galaxies in Virgo are more or less equally divided into spirals and ellipticals. The cluster also has a swarm of hundreds of dwarf galaxies, mostly dwarf ellipticals like those in the Local Group. The Virgo cluster shows up prominently in the map of the brightest galaxies in the sky (see Figure 16-9), along with the Centaurus cluster, which is a cluster of galaxies at a similar distance in the southern sky.

The nearest *rich cluster* is in the constellation Coma Berenices (Berenice's Hair). The Coma cluster, about 100 Mpc away and about 8 Mpc across, contains thousands of galaxies (Figure 16-10b). At its center are two giant elliptical galaxies. Like most rich clusters, the Coma cluster is spherically symmetrical, with a strong central concentration. Most of its galaxies are ellipticals, but it has a moderate number of S0 galaxies, which are like spirals but without the spiral arms and interstellar matter. Only the most luminous galaxies in a cluster appear in the photographs and CCD images in this book. We are seeing the "tip of the iceberg" of a much larger population of (mostly dwarf) galaxies concentrated in space.

Astronomers have discovered that the relative number of galaxies of different Hubble types in a region depends on the overall density of the environment. This **morphology-density relation** is a key to understanding how galaxies interact and evolve. Figure 16-11a shows the relative proportion of spiral, S0, and elliptical galaxies in regions that cover a range of 1 million in terms of galaxy space density. Below a density of about five galaxies per cubic megaparsec, the proportion of spirals is 60 to 70 percent and independent of density. Above this density, the spiral percentage drops steadily, until in the dense cores of the richest clusters (with thousands of galaxies per cubic megaparsec), virtually all the galaxies are S0s or ellipticals. We have learned that spiral galaxies inhabit low-density regions of the universe and elliptical galaxies inhabit high-density regions.

Do the very different properties of spiral and elliptical galaxies have something to do with their different cosmic environments? Astronomers think there is a connection. Figure 16–11b is a plot of the time it takes for galaxies to interact in different environments. We define an interaction as an encounter or close passage where the force of gravity leaves measurable effects on one or both of the galaxies. Below a density of a few galaxies per cubic megaparsec, this time is longer than the age of the universe. Spiral galaxies in low-density regions of the universe are relatively isolated. Thus,

 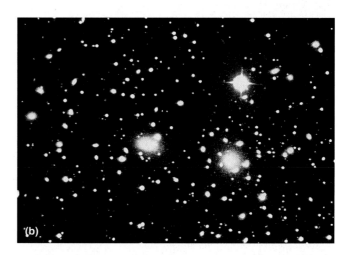

FIGURE 16-10
(a) The central portion of the Virgo cluster, which is 15 Mpc away and exerts a strong gravitational influence on the Milky Way. Giant ellipticals and dusty spirals can be seen. (SOURCE: Copyright Anglo-Australian Telescope Board.) **(b)** The Coma Berenices cluster in a photograph that has been computer-enhanced to simulate true color. The more than 1000 cluster members are mostly gas-poor galaxies. (SOURCE: NOAO.)

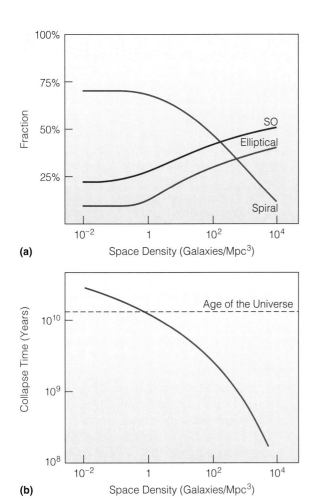

FIGURE 16-11
The morphology-density relation for galaxies. **(a)** The relationship between space density and the proportion of galaxies that are spiral, elliptical, and S0 types. The spiral proportion is constant at low density but falls off rapidly in rich clusters. **(b)** The time it takes galaxies to interact as a function of space density. At low densities, this time is longer than the Hubble time (the inverse of the Hubble constant), so galaxies are unlikely to have merged or interacted during the age of the universe. (SOURCE: Adapted from data of M. Postman and M. Geller/Harvard-Smithsonian Center for Astrophysics.)

they have weak gravitational interactions with other galaxies and rarely experience collisions. Encounters that do occur are gentle, with relative velocities of about 100 to 200 km/s —the typical amount of scatter in the Hubble relation. However, at the density of a rich cluster, the time scale for interaction is only a few hundred million years. Over the lifetime of the universe, there has been time for dozens of interactions. The strong gravity of a cluster causes the galaxies to have fast motions of 1000 to 1500 km/s. Collisions are relatively frequent and violent.

Now we can explain some of the puzzles described in Chapter 15. We saw that spirals like the Milky Way have too much star formation to be explained by their available reservoir of gas. In fact, spirals get injections of gas from intergalactic space, which is sparsely filled with cold gas. In low-density regions of the universe, this material can fall like a gentle rain onto the disks of spiral galaxies, replenishing the gas supply. Because of this material, spirals can be kept young and active. Elliptical galaxies have the opposite problem, with too little star formation. Their old stars lose mass, providing an ample supply of gas. Astronomers believe that as elliptical galaxies pass through the dense core of a cluster on their high-velocity orbits, they are continually swept free of gas and dust.

Presumably, the low fraction of spiral galaxies in a cluster results from collisions and mergers that strip away the gas from gas-rich galaxies. At the center of a rich cluster of galaxies, galactic cannibalism has a fascinating effect. Galaxies will gradually merge over time in the cluster core. The result of a large galaxy devouring a number of smaller galaxies could be a giant elliptical. This scenario can explain why the largest galaxies are found at the centers of rich clusters. Interactions, mergers, and collisions cannot explain *all* the differences between galaxies. Only a small percentage of galaxies are the remnants of mergers between massive galaxies.

We can understand how rare a galactic merger is if we consider how long it will take for the Milky Way to collide with the Virgo cluster. The Milky Way is about 15 Mpc from the mighty Virgo cluster. We can easily work out how long it would take to fall into Virgo, traveling at 200 km/s. A distance of 15 Mpc is $1.5 \times 10^7 \times 3 \times 10^{13}$ km, or 4.5×10^{20} km. Therefore, it would take $4.5 \times 10^{20} / (200 \times 3 \times 10^7) = 7.5 \times 10^{10}$, or 75 billion years—far longer than the age of the universe—for the Milky Way and the Virgo cluster to merge. We must look to the early universe to explain many galaxy properties; astronomers refer to these explanations as *initial conditions*. In the next chapter, we will see how galaxies might have evolved out of the very unfamiliar conditions of the universe when it was hot and dense.

Superclusters and Voids

Galaxy clusters can form into even larger structures, called **galaxy superclusters.** Gravity's long reach can cause galaxies to aggregate on an enormous scale. Astronomers have naturally speculated that even larger structures than galaxy clusters might exist. In the 1940s, Clyde Tombaugh, the discoverer of Pluto, plotted the positions of galaxies in the sky and made the first map of a structure of galaxies that spanned much of the northern sky. He showed it to Edwin Hubble, who refused to believe the observation, no doubt because no one at that time expected to see such large structures! Twenty years later, French astronomer Gerard de Vaucouleurs showed that the Milky Way is in fact part of an enormous flattened structure called the **Local Supercluster.** You can see it in Figure 16–9, an equal-area projection of the brightest galaxies in the northern and southern sky out to a distance of about 50 Mpc. The dashed line traces a high concentration of galaxies running in a strip across the sky, nearly at right angles to the Milky Way. This flattened structure of galaxies is analogous to the flattened disk of stars in the Milky Way.

SCIENCE TOOLBOX

Galaxy Clustering

As we saw in Chapter 14, the distribution of stars on the sky can be used to detect binary stars. When too many stars are seen at small angular separations compared to a random distribution, it is an indication that some stars are physically associated. We can use the same reasoning to measure the clustering of galaxies.

To illustrate the argument, we will return to the idea of random and clustered distributions that we used in our discussion of stars. (Objects in space are never uniformly distributed, so we will not consider that possibility.) For simplicity, we will view the distribution of the galaxies in two dimensions, without redshifts. Suppose that galaxies are randomly distributed in space, as in Figure 16–Aa. In a random distribution, the number of galaxies found in any patch of sky is proportional to the area of the patch. The probability of a galaxy having a neighbor within a certain distance is proportional to the area considered, or the distance squared. In a random distribution, it is certainly possible for galaxies to be very close to each other, but it does not happen very often.

Now look at Figure 16–Ab. Galaxies in this field are clustered, meaning that the average distance between galaxies is less than for a random distribution. When galaxies are clustered, the nearest neighbors tend to be at small separations. This increased likelihood of having a neighbor at small separations (compared to a random distribution) is a key signature of clustering. Astronomers define the clustering of galaxies in a simple way:

Clustering amplitude =
($N_{observed} - N_{random}$) / N_{random}

N is the number of galaxies observed in a particular area of the sky—or the number of galaxies in a particular volume, if we can measure galaxy distances as well. How does this measurement actually work? Astronomers begin with a galaxy catalog like the one in Figure 16–9. Take any galaxy and count its companions out to a certain angular radius. That number is $N_{observed}$. Now calculate how many galaxies would be found within a circle of the same radius in a random distribution; it is found easily from the average surface density of galaxies and the area of the circle. That number is N_{random}. The difference between the two is the excess number of galaxies within a certain angle. The *clustering amplitude* is the percentage of excess galaxies caused by clustering. We can calculate the clustering amplitude around each galaxy in the catalog and combine the numbers. We can also calculate the clustering amplitude on all angular scales. The result is called an *angular correlation function*.

However, we know that galaxies are distributed in three dimensions. In a two-dimensional map, there will be close alignments of nearby and distant galaxies that occur by chance. These situations where galaxies that are very far apart appear to be nearly coincident on the sky will act to dilute or wash out the correlation signal. If a sample of galaxies has measured redshifts, we can use the redshift as a distance indicator and calculate the clustering amplitude in three dimensions. The procedure and the calculation are the same as given above. This time, $N_{observed}$ is the number of companions a galaxy has within a sphere of a certain size; N_{random} is the average number of galaxies in a sphere of the same size if the galaxies are distributed randomly in space. The result is called a *spatial correlation function*.

If $N_{observed} = N_{random}$ on average, the clustering amplitude is zero. This would mean galaxies are randomly distributed. If $N_{observed} > N_{random}$, then the clustering amplitude is positive. Astronomers have found that the clustering amplitude is largest on small scales of about 1 Mpc. Galaxies are gregarious—the most likely place to find a galaxy is near another galaxy! The clustering amplitude is somewhat smaller on scales around 10 Mpc. By scales of 100 Mpc or larger, the clustering amplitude is very low. In principle, it is even possible to measure negative clustering amplitude. Figure 16–Ac shows galaxies arranged in a regular grid. Galaxies avoid each other, and there is a deficit of galaxies on small scales; this is *anticlustering*. However, the universe does not work this way. Gravity is an attractive

(continued)

(a)

(b)

(c)

FIGURE 16–A
Galaxy clustering in two dimensions.
(a) Randomly distributed galaxies.
(b) Clustered galaxies, where the average distance between galaxies is less than that for a random distribution. **(c)** Galaxies on a grid avoid each other (anticlustering), but this does not exist in nature.

SCIENCE TOOLBOX

Galaxy Clustering (continued)

force, and all surveys show positive clustering amplitude for galaxies.

You can also see that it takes a lot of data to measure clustering. Suppose you are interested in measuring clustering on a scale of 1 Mpc, but your galaxies are so thinly distributed that on average there are only a couple of galaxies in a 1-Mpc sphere. If $N_{observed}$ and N_{random} are both small numbers, the clustering amplitude will have large fluctuations because of the small number of statistics, and the measurement of clustering will be noisy.

In general, as we saw in our discussion of the statistics of counting in a Science Toolbox in Chapter 6, the random number of expected galaxies will have a statistical error of $\sqrt{N_{random}}$. To detect clustering, we require the excess number of galaxies to exceed the random fluctuations from counting statistics, or $N_{observed} - N_{random} < \sqrt{N_{random}}$. For example, if $N_{observed} = 115$ on scales of 10 Mpc and $N_{random} = 100$, the clustering amplitude is $(115 - 100)/100 = 0.15$, or a 15 percent excess of galaxies. However, the excess of $115 - 100 = 15$ is not much larger than the statistical fluctuation of $\sqrt{N_{random}} = 10$, so this is not a very convincing measurement. If $N_{observed} = 1150$ and $N_{random} = 1000$, the clustering amplitude is the same, $(1150 - 1000)/1000 = 0.15$. But now the excess of $1150 - 1000 = 150$ is five times larger than the statistical fluctuation of $\sqrt{N_{random}} = \sqrt{1000} = 33$. The ability to detect a clustering signal increases as the size of the galaxy sample increases. Astronomers need measurements of thousands of galaxies in order to measure clustering.

The Local Supercluster contains the Local Group, the Virgo and Coma clusters, and about 100 other clusters. It measures about 20 Mpc across by 2 Mpc thick and contains a total of 10^{16} $M_\odot$. In other words, it would take light 65 million years to cross the Local Supercluster, and it contains 10 thousand trillion stars! The center of the Local Supercluster is near the Virgo cluster; the Milky Way is near the outskirts. The gravity of the Virgo cluster is so strong that we—the Milky Way and the entire Local Group—are being pulled into it at a speed of 250 km/s. Do not lose sleep over this, however; it will take billions of years to get there!

Superclusters range in size from 50 to 100 Mpc across—imagine a physical structure more than 300 million light-years across! By comparison, galaxy clusters are usually in the size range of 1 to 5 Mpc. Figure 16–12 shows a map of 10 percent of the northern sky, with counts of galaxies made from deep photographs. The brightness limit of the galaxies is 10 times fainter than the brightness limit in Figure 16–9, so the typical distance of the galaxies is three times farther, or about 150 Mpc. We can immediately see that the distribution of galaxies is neither uniform nor random. There are knots or concentrations of galaxies, linear or filamentary features, and regions where the galaxy density appears to be very low. Much modern research is devoted to understanding this complex web of large-scale structure.

Because they represent only two dimensions, maps of galaxies are not ideal for studying large-scale structure. Nearby and distant galaxies are very far apart, yet they can appear close together in projection. The most detailed information on large-scale structure has therefore come from painstaking redshift surveys of galaxies in narrow "slices" of the universe. Rather than try the almost impossible task of measuring distances for thousands of galaxies, astronomers use recession velocity (or, equivalently, redshift) as a distance indicator. Position on the sky plus redshift allows astronomers to locate galaxies to be placed in three-dimensional space. Figure 16–13 shows the first of these surveys, carried out by Margaret Geller, John Huchra, and their collaborators at the Harvard-Smithsonian Center for Astrophysics. A series of separated slices can be used to reconstruct the distribution of galaxies over a large volume. Redshift surveys have been carried out for tens of thousands of galaxies, and telescopes dedicated to this type of science will soon increase this number to over a million. Astronomers are mapping out larger and larger regions of the universe, heirs to the pioneering cartographers who mapped our planet hundreds of years ago.

The bright concentration near the center of the wedge in Figure 16–13 is the Coma cluster (see also Figure 16–10b). Although the Coma cluster is spherical in shape, it appears elongated in redshift. This elongation occurs because the large mass of a cluster gives galaxies high velocities, and the range in velocity spreads the redshift distribution of galaxies toward and away from the observer; astronomers call this the "finger of god" effect. In other words, redshift is a poor distance indicator of individual galaxies in a cluster. Another striking feature in the slices is the presence of **voids** in the galaxy distribution. The voids are nearly circular in cross section and are

FIGURE 16–12
The distribution of 2 million galaxies covering 10 percent of the northern sky. The galaxies are not shown individually; the brightness of each dot represents the number of galaxies in each small patch of sky. Black is empty, gray is a number between 1 and 19, and white is 20 or more. A complex pattern of clusters, filaments, and voids is visible. (SOURCE: S. Maddox, W. Sutherland, G. Efstathiou, and J. Loveday, Oxford Astrophysics.)

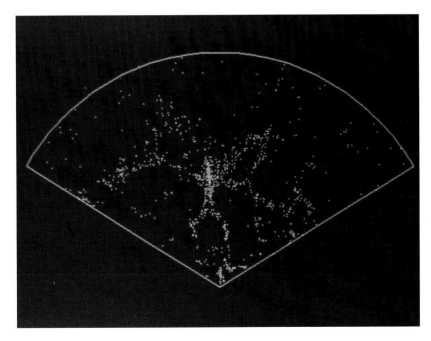

FIGURE 16–13
A three-dimensional "slice" of the universe. Redshifts were measured for about 1100 galaxies in a thin wedge stretching 120° across the sky. Earth is at the apex of the wedge, and its far edge is 200 Mpc away. The galaxies form strikingly thin sheets; regions between the sheets are nearly devoid of luminous matter. Surveys of adjacent wedges have shown that these structures are genuine "sheets" rather than "strings." The pattern is reminiscent of a cross section through a foam of soap bubbles.
(SOURCE: M. Geller, J. Huchra, M. Kurtz, and V. de Lapparent, Harvard-Smithsonian Center for Astrophysics.)

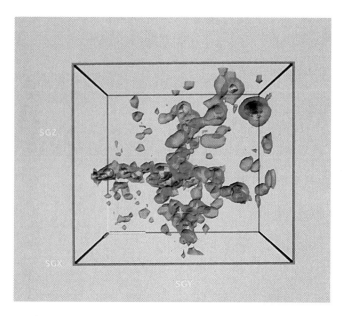

FIGURE 16-14
A volume 100 million light-years across in the local universe. This three-dimensional map is centered on the Virgo cluster, with the Milky Way at the left edge. The box is 40 Mpc wide by 35 Mpc tall by 50 Mpc deep. Individual galaxies are not shown, but the galaxy density is color-coded. Green surfaces are regions with about 0.5 galaxy per cubic Mpc; inner orange/brown surfaces are regions with about 1.5 galaxies per cubic Mpc. The horizontal concentration is the plane of the Local Supercluster. (Source: R. B. Tully, Institute for Astronomy, Hawaii.)

truly empty of bright galaxies. The Local Supercluster has been mapped out in three dimensions by Brent Tully at the University of Hawaii. Figure 16–14 shows the distribution of galaxies in a volume 40 Mpc across. Most of the volume is completely empty of luminous matter. As many as 98 percent of the galaxies occupy only 5 percent of the volume.

Surveys of galaxies reveal the architecture of the cosmos, patterns on the largest scales that have been sculpted by gravity over billions of years. Astronomers have used colorful language to describe the structures they see. Historically, it was thought that the rich clusters were set in a uniform sea of galaxies, like meatballs. The discovery of strings and filaments of galaxies added a component of spaghetti. The first slices of the universe showed that the galaxies form large connecting structures around empty voids 20 to 100 Mpc across, like Swiss cheese. The most recent research indicates a single analogy that cleans up these various terms. The general emptiness of space—where galaxies are distributed in sheets and filaments and clusters are found at the junctions of the sheets—is like a network of soap bubbles.

Astronomers have also used mathematics to describe large-scale structure. What is the typical shape, or *topology,* of the way galaxies cluster in space? A point has no dimensions, and in three-dimensional space, this would correspond to galaxies that were scattered with no clustering at all. A line has one dimension, so this would be the appropriate description of galaxies in stringy structures. A plane has two dimensions, so this would describe galaxies that are all contained in sheetlike structures. Recent analysis indicates that we live in a universe where the large-scale structure has a dimension of about 1.7, which is partway between 1 and 2. In geometry terms, a dimension of 1.7 corresponds to structures that are somewhat stringlike and somewhat sheetlike.

We have encountered an amazing array of structures in the universe. Some galaxy groups are about 0.5 Mpc across and contain dozens of galaxies. There are galaxy clusters up to 5 Mpc across that contain hundreds or thousands of galaxies. These are all part of superclusters up to 50 Mpc that may contain tens of thousands of galaxies. Does this progression ever stop? Or do we see ever-increasing structures as we consider larger and larger scales? Figure 16–15 provides analogies to terrain on Earth to guide our thinking. Perhaps the universe has an equal amount of structure on all scales, like the bends and wiggles of a coastline, which have the same appearance whether you look at an entire peninsula or a single beach. Mathematicians call a shape with structure on all scales a *fractal.* Perhaps the universe has a single preferred scale, like a flat terrain of lakes of roughly similar sizes. The true situation is closer to the third example, many small foothills and a set of mountains with only a few mighty peaks. Astronomers have learned that the universe is smooth and featureless on the largest scales they can study. Figure 16–16 shows counts of the galaxies in Figure 16–12 on four different scales. On scales of 10 or 30 Mpc, the observed variation in the number of galaxies in each box exceeds the random variation expected from counting statistics ($\sqrt{N}$)—an indication of gravitational clustering. (See the Science Toolbox in Chapter 6 on counting craters for a review of this concept.) However, by a scale of 300 Mpc, the number of galaxies shows little variation.

The study of large-scale structure reveals two important aspects of the universe on the largest scales we can measure. First, it is *isotropic.* We see a similar Hubble relation in every direction we look. We also see similar types of structures—clusters, superclusters, and voids—in every direction we look. Second, it is *homogeneous.* Imagine you are looking down on a beach from a mile above. The grains of sand and pebbles would not be visible and the beach would appear smooth and featureless. Similarly, a mountain range would appear smooth when viewed from a high Earth orbit. Gravity has obviously formed lumps that range in size from planets and stars to clusters and superclusters of galaxies. But the universe is smooth on scales larger than about a billion light-years.

Dark Matter Revisited

Now we turn to the role of dark matter in the largest structures in the universe. In Chapter 15, we saw that only about 5 to 10 percent of a galaxy's mass emits radiation at any wavelength. The part of a galaxy that emits visible light is embedded in a dark matter halo. Is dark matter only found

(a)

(b)

(c)

FIGURE 16–15
Terrestrial analogies for the types of structures that might be seen in the universe. **(a)** The bends and wiggles of a coastline are the same on all scales, which is a fractal structure. **(b)** A set of lakes might have a single characteristic size. **(c)** In a mountain range, there may be many small peaks but only a few high peaks.

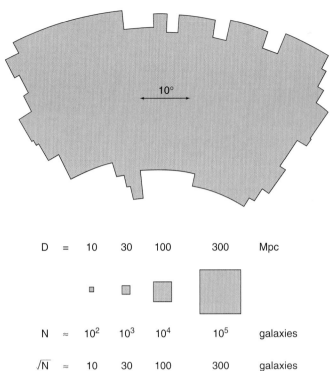

FIGURE 16–16
A schematic version of Figure 16-12, where 2 million galaxies are distributed over about 10 percent of the sky. At the typical distance of these galaxies, one degree corresponds to a transverse separation of about 15 Mpc. The boxes below the outline show four different scales D, with N as the average number of galaxies expected in each box. The scatter expected in the counts in each box is given by $\sqrt{N}$. On scales of 10 to 30 Mpc, the variation in galaxy counts from one part of the sky to another is greater than $\sqrt{N}$, indicating clustering due to gravity. On scales larger than 100 Mpc, the variation in galaxy counts is no more than $\sqrt{N}$, indicating that the universe is homogeneous at very large scales.

in association with galaxies, or is it also distributed on even larger scales?

Much evidence points to the existence of dark matter in galaxy clusters. Back in the 1930s, California Institute of Technology astronomer Fritz Zwicky observed that the velocity dispersion (the spread, or scatter, in velocities) of galaxies in the Coma cluster was too high to be accounted for by the visible matter. Galaxies are moving too fast to be held in one region of space by their mutual gravity. If the only mass in the cluster was that represented by the galaxies, the galaxies would fly apart—the cluster could not survive! The same argument has since been applied to many galaxy clusters. In each case, the velocity dispersion is 1000 to 1500 km/s. In each case, the mass of the visible galaxies is insufficient to hold clusters together. The implied mass-to-light ratio of clusters is in the range of 200 to 400, four to six times larger than the mass-to-light ratio of individual galaxies (see Chapter 15 to review mass-to-light ratios). Large mass-to-light ratios are direct evidence for dark matter in clusters, *in addition to* the dark halos surrounding the galaxies themselves.

A second probe of dark matter on a large scale involves **gravitational lensing.** One of the basic predictions of general relativity is that mass can bend light. When light from a distant galaxy passes through an intervening galaxy or cluster of galaxies, it is deflected by curved space in the vicinity of the cluster. Just like an optical lens, a gravitational lens can magnify or demagnify an image (corresponding to the

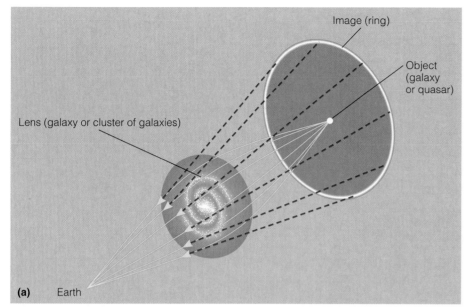
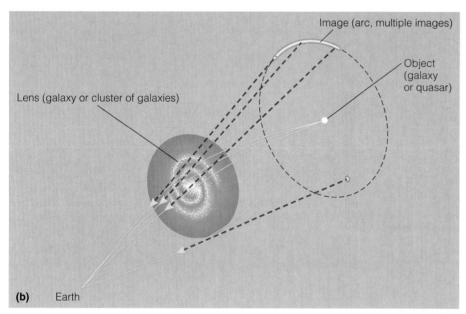

FIGURE 16–17
General relativity predicts that mass can bend light rays, thereby producing multiple images and image amplification, a phenomenon called gravitational lensing. If a massive galaxy or cluster of galaxies lies between Earth and a distant galaxy or quasar, the light can be deflected. **(a)** If the alignment between the background source and the lensing object is perfect, an image in the form of a ring is produced. **(b)** If the alignment is not perfect or the lens is not symmetrical, multiple images or small arcs are produced. The image splitting is small, rarely more than a couple of seconds of arc, and is greatly exaggerated in this picture.

amplification and deamplification of the light signal). Figure 16–17 shows the geometry that leads to this effect. Gravitational lensing creates a mirage: A single image can be distorted into a variety of shapes, including rings and arcs and multiple images. A computer simulation of gravitational lensing is shown in Figure 16–18, using an actual CCD image of the galaxy NGC 3992. When a nearby galaxy is perfectly aligned with a more distant galaxy and the lensing galaxy is symmetrical, the light is bent by the same angle at every orientation. The result is that the galaxy's image is stretched into a perfect ring, called an *Einstein ring* after the father of general relativity. When the alignment is not perfect, the ring breaks up into arcs of light.

Perfect alignments are expected to be rare. Indeed, only a dozen Einstein rings have ever been observed. However, there are so many faint galaxies in the sky that we might expect to see near alignment that leads to lensing arcs. A rich cluster of galaxies is expected to magnify and distort the light from more distant galaxies. Figure 16–19, on page 490, is a vivid example of the lensing phenomenon. Each little arc is the distorted image of a different background galaxy. The cluster galaxies are mostly red ellipticals, and the background galaxies are mostly blue spirals. Images like this one are spectacular confirmations of Einstein's theory of gravity—mass really does bend light! Taken together, all the lensing distortions can be used to construct a map of the mass distribution of the cluster. Since lensing is controlled by mass, the resulting map represents *all* the mass of the cluster, not just the visible component. Cluster mass measured by gravitational lensing agrees well with cluster mass measured by galaxy motions.

What do galaxy motions in clusters and gravitational lensing by clusters tell us about dark matter? When viewed on the

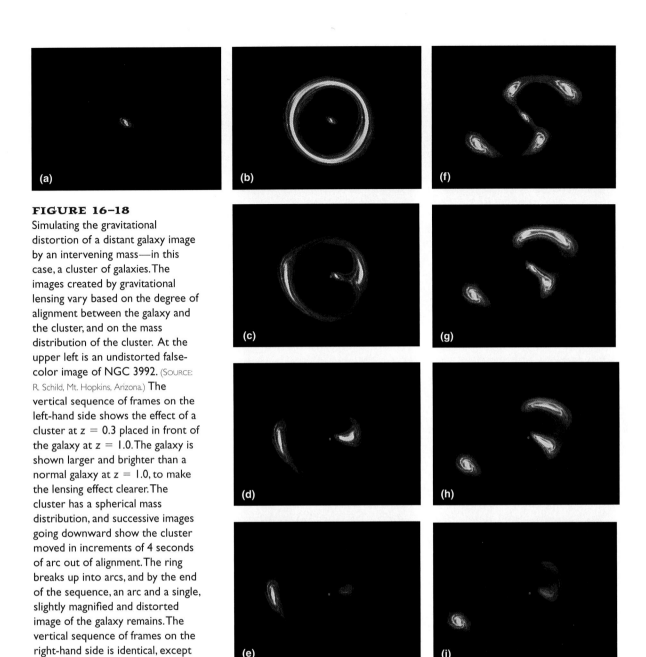

FIGURE 16-18
Simulating the gravitational distortion of a distant galaxy image by an intervening mass—in this case, a cluster of galaxies. The images created by gravitational lensing vary based on the degree of alignment between the galaxy and the cluster, and on the mass distribution of the cluster. At the upper left is an undistorted false-color image of NGC 3992. (SOURCE: R. Schild, Mt. Hopkins, Arizona.) The vertical sequence of frames on the left-hand side shows the effect of a cluster at $z = 0.3$ placed in front of the galaxy at $z = 1.0$. The galaxy is shown larger and brighter than a normal galaxy at $z = 1.0$, to make the lensing effect clearer. The cluster has a spherical mass distribution, and successive images going downward show the cluster moved in increments of 4 seconds of arc out of alignment. The ring breaks up into arcs, and by the end of the sequence, an arc and a single, slightly magnified and distorted image of the galaxy remains. The vertical sequence of frames on the right-hand side is identical, except for the fact that the cluster has an elliptical mass distribution. (SOURCE: Images courtesy of E. Falco, Harvard-Smithsonian Center for Astrophysics.)

largest scales, the universe is overwhelmingly composed of dark matter. On a galactic scale, about 90 to 95 percent of the mass detected by various means is dark. On the scale of clusters of galaxies, the proportion reaches 95 to 99 percent. Recall from the last chapter that astronomers do not know what material makes up dark matter. Many conventional astronomical possibilities have been ruled out, and most of the remaining options involve exotic subatomic particles. Even if this is true, the study of large-scale structure is one of the few ways of understanding the properties of dark matter.

The Most Distant Galaxies

The Hubble relation is well tested out to distances of about 1000 Mpc. At that distance, if we assume a Hubble constant of 65 km/s/Mpc, the recession velocity is $65 \times 100 = 65{,}000$ km/s, or about two-tenths the speed of light ($0.2c$). What does a galaxy this far away—at a prodigious distance of 3.3 billion light-years—look like? It looks small. Using the small-angle equation, we can deduce that with images 0.5 second of arc across, the smallest feature resolved by a ground-based telescope is about $0.5 \times 1000 / 206{,}265 = 0.0024$

FIGURE 16-19
Distant blue galaxies, which are mostly spirals, are gravitationally lensed by more nearby yellow galaxies in the cluster. Each little arc is the distorted and magnified image of a different background galaxy. Notice how the arcs are all centered on the massive center of the cluster. Taken together, these distortions can be used to map the cluster, including the dark matter component. (SOURCE: NASA/STScI.)

Mpc, or 2.4 kpc. Therefore, a galaxy like the Milky Way would only be 12 seconds of arc across. Ten times farther away, it would be unresolved and indistinguishable from a star. This galaxy also appears dim. If we make a deep image with a telescope in a direction away from the plane of the Milky Way, the faintest stars visible would be main-sequence stars such as the Sun, about 10 kpc away. By the inverse square law of light, they have the same apparent brightness as a 10^{10} $L_\odot$ galaxy that was $\sqrt{10^{10}} \times 10 = 10^6$ kpc, or 1000 Mpc distant. In other words, beyond about 1000 Mpc, a galaxy like the Milky Way is fainter than *any* star in our own galaxy. We "run out" of stars, and virtually every faint image is a distant galaxy!

How far away are the most distant galaxies? Beyond about 1000 Mpc, the assumption that we have used to characterize the redshift breaks down. We have used the Doppler equation to define the recession velocity of a galaxy ($v = zc$) and the distance via the Hubble relation ($d = zc/H_0$). In fact, the only true observable quantity is the cosmological redshift of spectral features in a galaxy, defined as $z = \Delta\lambda/\lambda = (\lambda - \lambda_0)/\lambda_0$ (see the earlier Science Toolbox). Rearranging this equation, the observed wavelength is 1 plus the redshift times the rest wavelength, $\lambda = (1 + z)\lambda_0$.

Galaxies are routinely observed with redshifts as high as $z = 1$. In other words, a spectral feature with a laboratory wavelength of $\lambda_0 = 327$ nm (singly ionized oxygen) is observed at 654 nm, and a spectral feature with a laboratory wavelength of 656 nm (the Hα line of hydrogen) is observed at 1.3 microns. The highest redshift galaxies have $z = 5$. At this redshift, the oxygen line is observed at 2.0 microns and the hydrogen line at 3.9 microns. Notice that very-high-redshift galaxies have most of their spectral features redshifted out of the optical window. Astronomers expect that the most distant galaxies will be detectable only at near-infrared wavelengths.

The distance to very-high-redshift galaxies is uncertain because it depends on the cosmological model, as we will see in the next chapter. However, for the currently preferred values of cosmological parameters, the age of the universe is about 13 billion years. In this model, $z = 1$ corresponds to a distance of about 10 billion light-years, or a look-back time of 60 percent of the age of the universe. The larger value of $z = 5$ corresponds to a distance of about 12 billion light-years, or a look-back time of 90 percent of the age of the universe.

Look back at Figure 1-2 and notice the many galaxies that litter this CCD frame. The Hubble Deep Field is the

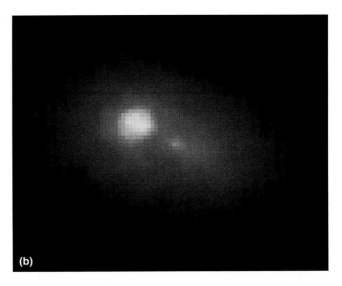

FIGURE 16-20
(a) An image of the nearby elliptical galaxy M 32, a companion of M 31, taken with the Hubble Space Telescope. The sharp central concentration of light is consistent with the presence of a massive black hole. The image is 50 pc on a side. **(b)** A high-resolution image of the nucleus of the Andromeda galaxy, taken with the Wide Field/Planetary Camera of the HST. The image is 12 pc, or 40 light-years, across. The fainter nucleus corresponds to the true center of the galaxy. (Source: T. Lauer/NOAO, NASA.)

deepest image of the sky ever made; a second field in the southern sky was recently imaged to a similar depth. About 2000 galaxies are visible to a level 4 billion times fainter than the visibility of the naked eye. If we presume that the census of galaxies in this small patch of sky is representative of the entire sky, we can calculate the total number of galaxies visible to this depth. The answer is about 40 billion galaxies! Multiplying by the average number of stars in each galaxy gives roughly 10^{20}. Even though we have not observed them all individually, we infer that there are about 100 billion billion stars in the observable universe. We can see the Copernican idea taken to a dizzying level. Earth, the Sun, and the Milky Way are all lost in the amazing vastness of the universe.

Active Galaxies and Quasars

It may reassure you to learn that almost everything we know about galaxies can be understood in terms of stellar processes, if we ignore for a moment the vexing issue of dark matter. Galaxies are made of stars and the gas and dust from which stars form. Galaxy morphology can be understood in terms of stellar orbits and the general features of stellar evolution. Galaxy spectra are simply the sums of the spectra of billions of individual stars.

However, the placid exteriors of some galaxies conceal events of great violence. A fraction of all galaxies harbor nonstellar processes in their centers. These processes can take several forms: torrents of radio or X-ray emission, high gas velocities, the implied presence of a massive black hole. A galaxy with any of these characteristics is referred to as an active galaxy. Our own galaxy has an active nucleus. In Chapter 14, we saw that the central parsec contains a compact radio source, a region of intense star formation, and a large quantity of fast-moving ionized gas. There is good evidence for a black hole at the galactic center, several million times the mass of the Sun.

Black Holes in Nearby Galaxies

Is the black hole at the center of the Milky Way unique? According to everything we have learned about the universe, we should be wary of thinking that there is anything special about our cosmic environment. The Copernican idea has proved valid on the largest scales of the universe, and astronomers have searched hard for compact objects at the center of nearby galaxies. In recent years, they have had great success. Careful study of the nuclei of some nearby galaxies reveals extreme concentrations of matter within the central few parsecs. Two types of evidence combine to point to supermassive black holes.

The first involves high-resolution imaging of the cores of bright galaxies. This type of imaging can be done at a level of 0.1 second of arc using the Hubble Space Telescope (HST), or at a level of 0.3 to 0.4 second of arc using ground-based telescopes on excellent sites. Figure 16–20a is an HST image of the core of M 32, in the Local Group. M 32 has a sharp spike of light within the central parsec. Alongside you can see the nucleus of M 31 (Figure 16–20b), which also has a central concentration of light. The density of stars in the center of M 32 is over 100 million times the density of stars in the Sun's neighborhood. Models of this light distribution indicate a black hole of 3 million $M_\odot$. M 31 is believed to harbor a black hole of 10 million $M_\odot$. However, a central light peak does not point uniquely to a black hole; it might also give evidence of a very dense star cluster.

The second type of evidence uses spectroscopy to measure the mean velocities of stars near the center of the galaxy. If the motions are too rapid to be accounted for by the stellar populations known to inhabit the nuclear regions, a case can be made for a compact object like a black hole. A steadily increasing number of galaxies have been found to show sharp increases in stellar velocities near the nucleus. We have seen how stellar velocities can be used to "weigh" a galaxy within a certain radius. High stellar velocities near a galaxy nucleus indicate a large mass concentrated in a small region.

Let us see how these lines of evidence combine to point to supermassive black holes. The velocity dispersion shows how the enclosed mass near the center of a galaxy varies with radius. An image shows how the light varies with radius. The ratio of the two quantities gives the variation in mass-to-light ratio with radius. Certain galaxies show a sharp rise in the mass-to-light ratio within the central parsec. We saw in the last chapter that normal stellar populations have mass-to-light ratios of 2 to 30, so any values higher than this range must indicate a dark mass concentration. This mass is extremely compact, so it is quite different from the dark matter that is distributed over large volumes in the halos of galaxies. Astronomers have seen evidence of black holes that range in mass from a few million $M_\odot$ in our galaxy and in M 32, to a few billion $M_\odot$ in M 87, a giant elliptical galaxy in the Virgo cluster.

What are the properties of such enormous black holes? We saw in a Science Toolbox in Chapter 13 that the size of a black hole is given by the Schwarzschild radius, $R_S = 2GM/c^2$. For the Sun, the Schwarzschild radius is 3 km. Therefore, a 10^6 $M_\odot$ black hole has a radius of 3×10^6 km and a 10^9 $M_\odot$ black hole has a radius of 3×10^9 km. The diameter of the more massive black hole is therefore 40 AU, or only 0.0002 pc. Imagine the mass of a billion suns packed into a volume the size of the solar system! We can use the small-angle equation to show how difficult it would be to resolve a supermassive black hole in a nearby galaxy. The smallest angle that can be resolved by the Hubble Space Telescope is about 0.05 second of arc. At a distance of $D = 10^6$ pc, in the Local Group, the size of the smallest feature that can be resolved is $d = D\alpha/206,265 = 10^6 \times 0.05 / 206,265 = 0.2$ pc. At the distance of the Virgo cluster, the minimum resolvable feature is 15 times larger, or 3 pc. Even for the nearest galaxies, we are restricted to looking on scales that are thousands of times the Schwarzschild radius. Our evidence for supermassive black holes is therefore indirect. We must infer the existence of a compact object from its effect on the stars and gas that surround it.

It seems extraordinary to hypothesize supermassive black holes, when the evidence for stellar-mass black holes is strong but not overwhelming. Yet the state of matter in such a compact object is not that unusual. Let us start with a black hole the mass of the Sun. If 2×10^{30} kg are squashed into a region 3 km in radius, the density is a phenomenal 10^{19} km/m^3. Density is proportional to mass over radius cubed, $\rho \propto M/R^3$, and we have seen that the Schwarzschild radius is proportional to mass. Combining these relationships gives the result that the density of a black hole is inversely proportional to the square of the mass: $\rho \propto M^{-2}$. In other words, supermassive black holes are far less dense than puny black holes. Thus, if a 1 $M_\odot$ black hole has a density of 10^{19} kg/m^3, then a 10^6 $M_\odot$ black hole has a density of $10^{19}/(10^6)^2 = 10^7$ kg/m^3, and a 10^9 $M_\odot$ black hole has a density of $10^{19}/(10^9)^2 = 10$ kg/m^3. This last number is only 10 times the density of the air you are breathing. Supermassive black holes are 100 times less dense than water! Theorists such as Martin Rees, at the University of Cambridge in England, have argued that a supermassive black hole is the inevitable result of the evolution of a massive star cluster.

Does every galaxy contain a supermassive black hole? A recent survey by John Kormendy and others in Hawaii finds evidence for black holes in about 25 percent of nearby galaxies. It also appears that the mass of a supermassive black hole depends on the mass of the spherical component of a galaxy. In spiral galaxies, the bulge mass dictates the mass of the black hole. In elliptical galaxies, the entire stellar mass dictates the mass of the black hole. Dark forces are at work in the hearts of many galaxies.

Active Galactic Nuclei

What was the first evidence of peculiarity in the centers of some galaxies? In 1908, Edward Fath discovered intense emission lines coming from the central regions of the bright galaxy NGC 1068. Vesto Slipher and Edwin Hubble discovered other galaxies with similar lines. By the 1940s, Carl Seyfert had studied active galaxies in detail and noted their common features. Active galaxies show one or more of the following features: a bright compact nucleus, strong and very broad emission lines, intense radio emission, and a peculiar morphology (structure).

Galaxies where the emission lines are strong and broad are called **Seyfert galaxies**, after their discoverer. Why are broad emission lines surprising? In a normal spiral galaxy, hot gas in the disk reveals itself through emission lines. These lines get broader by an amount corresponding to the rotation of the disk, thus revealing the velocity of the gas. We saw from our discussion of galaxy rotation curves in Chapter 15 that the total range in velocity in the gas is several hundred km/s. Gas moving toward us is blueshifted and gas moving away from us is redshifted, and the result is a smeared-out emission line. In a Seyfert galaxy, by contrast, the emission lines are much broader, indicating a gas velocity of *thousands* of km/s. Gas at such a high velocity could not be gravitationally bound; it would fly away from the center. There are two possible explanations for the high velocity of this gas. Either the gas is actually being ejected from the nucleus of the galaxy, or a dark massive object in the nucleus is keeping the gas bound at that high velocity. In either case, something unusual is going on in the nuclear regions. Spectroscopy is often required to reveal the activity; although many Seyfert galaxies have bright starlike nuclei, the images of others appear quite normal.

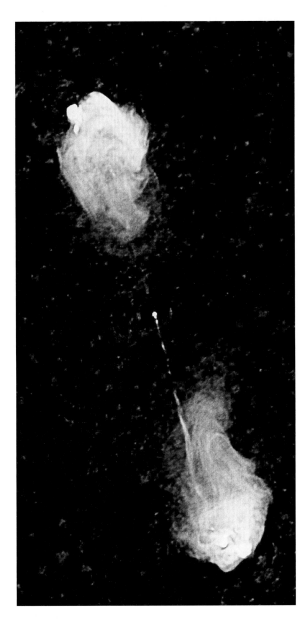

FIGURE 16-21
High-resolution radio image of the giant elliptical galaxy 3C 405, also known as Cygnus A. The jets of hot, fast-moving particles shooting out from the bright central nucleus expand into large, wispy radio lobes extending tens of kiloparsecs from the galaxy. (SOURCE: National Radio Astronomy Observatory, courtesy R. Perley, J. Dreher, J. Cohan.)

About 1 percent of all galaxies (and 10 percent of all active galaxies) have extraordinary levels of radio emission. By 1944, the amateur astronomer Grote Reber had detected strong radio sources in the constellations of Sagittarius, Cassiopeia, and Cygnus. The Sagittarius radio source corresponded to the Galactic Center, and the Cassiopeia source to a supernova remnant. However, the position of the Cygnus source could not be specified accurately until 1951, when Walter Baade and Rudolf Minkowski located a faint, distorted-looking galaxy at the position in the sky where the radio emission was centered. Cygnus A, the brightest radio source in the constellation of Cygnus, was the first known **radio galaxy**, with a radio luminosity 10 million times that of the Milky Way. Most radio galaxies are elliptical. Figure 16-21, a recent radio map of Cygnus A, shows the main features of a typical radio galaxy. The bright radio *nucleus* at the center corresponds exactly to the position of the optical galaxy. A pair of *radio jets*, one of which is only dimly visible, join the nucleus to two enormous *radio lobes*. The lobes have a complex structure, with large regions of wispy emission and intense emission at the outer edges. Cygnus A is 230 Mpc, or 750 million light-years away. Yet it emits radio waves strong enough to be detected by amateur astronomers with backyard equipment!

The intense radio waves that come from some galaxies are examples of **nonthermal radiation.** Most astronomical objects emit thermal radiation. Recall from Chapter 4 that all material emits thermal radiation. All atoms and molecules are in constant motion or vibration, and the average amount of energy of a particle is simply related to the temperature of the material. Thermal radiation covers a range of wavelengths, but there is a single wavelength where the intensity is highest—given by Wien's law. By contrast, nonthermal radiation has no simple relationship to temperature. The spectrum of nonthermal radiation covers an enormous range of wavelengths, often from radio waves all the way to gamma rays. Figure 16-22a and b show the two types of spectra.

The most important type of nonthermal radiation is called *synchrotron radiation,* which is emitted by particles moving in a magnetic field. All galaxies are threaded by weak magnetic fields. All galaxies have energetic particles that get their motions from the collisions of gas clouds or the energy of a supernova. We therefore have the ingredients for synchrotron radiation. A magnetic field makes charged particles move in curved trajectories, which accelerates them and causes them to emit electromagnetic radiation. This is how a radio transmitter works; electrons racing up and down a wire generate a radio wave. If energy is delivered to a hot gas threaded by a magnetic field, the electrons spiral around the magnetic lines of force at nearly the speed of light, and lose energy by emitting synchrotron radiation. (Protons are massive and not nimble enough to spiral very quickly, so they emit little synchrotron radiation.) Synchrotron radiation is variable; when the energy source for accelerating electrons varies, the amount of synchrotron radiation released can vary by a large factor.

Now we can understand the difference between thermal and nonthermal radiation at a microscopic level. Thermal radiation is the hallmark of a system that is in equilibrium, where the energy gained and lost by particles is always in balance (see Figure 16-22c). Examples include Earth, which is in equilibrium with the amount of radiation it receives from the Sun, or any point inside a star, which is in equilibrium with the amount of energy received from the star's central source of fusion. Photons are absorbed

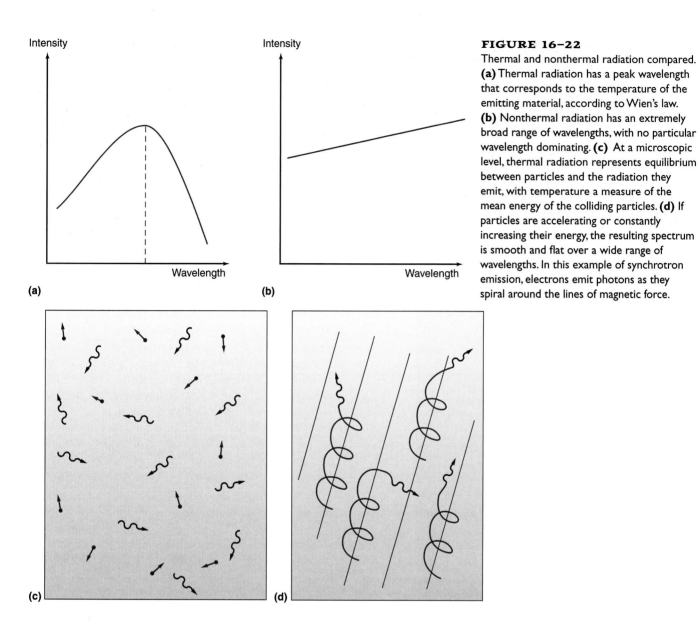

FIGURE 16-22
Thermal and nonthermal radiation compared. **(a)** Thermal radiation has a peak wavelength that corresponds to the temperature of the emitting material, according to Wien's law. **(b)** Nonthermal radiation has an extremely broad range of wavelengths, with no particular wavelength dominating. **(c)** At a microscopic level, thermal radiation represents equilibrium between particles and the radiation they emit, with temperature a measure of the mean energy of the colliding particles. **(d)** If particles are accelerating or constantly increasing their energy, the resulting spectrum is smooth and flat over a wide range of wavelengths. In this example of synchrotron emission, electrons emit photons as they spiral around the lines of magnetic force.

and emitted at equal rates. With nonthermal radiation, the system is not in equilibrium because the particles are constantly being accelerated (see Figure 16-22d). When energy is added to particles this quickly, they emit a broad spectrum of radiation, with no emission peak at any particular wavelength. The shortest wavelength of the nonthermal spectrum depends on the highest energy of the electrons. For highly relativistic electrons, with speeds close to c, the spectrum can extend to X rays or even gamma rays.

Besides broad emission lines and strong radio emission, active galaxies often have a *peculiar morphology*. The classic example is the irregular galaxy M 82, which harbors large amounts of nuclear gas and dust and has chaotic filaments of excited gas streaming out from the nucleus. The high-resolution imaging of the Hubble Space Telescope shows that the Seyfert galaxy NGC 1275, a peculiarly shaped elliptical in the Perseus cluster, has young and blue globular clusters (Figure 16-23a). Their formation is apparently connected with the violent events occurring in the nucleus. A close-up view of the peculiar galaxy Arp 220 reveals gigantic young star clusters near the nucleus (Figure 16-23b). It is important not to associate peculiar morphology with nuclear activity too strongly, however. Peculiar morphology can result from normal stellar processes (see Figure 15-25), and some apparently normal galaxies have active nuclei. But astronomers have discovered a fairly strong statistical connection between galaxy interactions (as revealed by peculiar morphology) and nuclear activity. It appears that interactions can fuel or "trigger" the formation of a supermassive black hole.

Quasars and the Nature of the Redshift

Now we return to the discovery of the most distant active galactic nuclei. In the chapter's opening story, we saw how Maarten Schmidt made the amazing discovery of objects with large redshifts. The recession velocities can be so high that they are a significant fraction of the velocity of light. These objects were originally called *quasi-stellar radio sources,* denoting the fact that they appeared almost starlike on photographic plates.

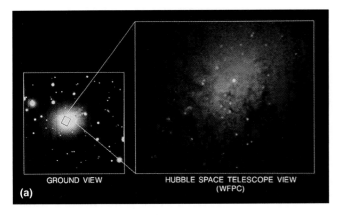
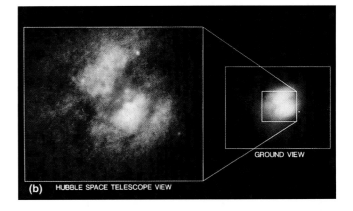

FIGURE 16-23
Peculiar galaxy morphology. **(a)** On the left, a ground-based image of the giant elliptical NGC 1275, which lies in the Perseus cluster of galaxies. On the right, a Wide Field/Planetary Camera image from the Hubble Space Telescope. The bright blue dots represent globular clusters full of young stars, which is unusual because most globular clusters contain old stars. (SOURCE: J. Holtzman, NASA.) **(b)** On the right, a ground-based image of the peculiar galaxy Arp 220, which appears double because of a dust lane running through the center. On the left, the Wide Field/Planetary Camera reveals complex structure less than 1 second of arc from the nucleus. (SOURCE: E. Shaya, D. Dowling, NASA.)

(You may also see the term *QSO*, for quasi-stellar object, which recognizes the fact that these distant galactic nuclei can have weak radio emission.) We will use the general term **quasar** for a distant galaxy in which the light from an active nucleus completely swamps the light from its surroundings.

The highest redshift quasar has a value of roughly $z = 5$, and most lie in the range $z = 1$ to $z = 3$. What are the implications of such large redshifts? If we assume that the redshift is caused by the expansion of the universe, then quasars must be very distant. The light-travel time is very large, and we must be observing light emitted when the universe was much younger than it is now. Remember that a journey out into space is a journey back in time. Large telescopes can be used as time machines. We see distant galaxies and quasars as they existed soon after their formation. Therefore, the study of the most distant objects brings us face to face with questions about the origin and large-scale structure of the universe itself. Also, if quasars are very distant, their luminosity, or intrinsic brightness, must be enormous. Assuming a cosmological redshift, 3C 273 is 620 Mpc, or 2 billion light-years, distant, and it shines with the light of 10^{14} $L_\odot$, or 100 trillion Suns. This is an almost unimaginable amount of energy. No wonder the interpretation of quasar properties has sometimes been controversial!

There is a crucial distinction between galaxy redshifts and quasar redshifts. Galaxies follow the Hubble relation, in which estimated distance is correlated with radial velocity. This correlation has a natural interpretation in terms of the expansion of the universe. In the nearby universe, measuring a galaxy's redshift and assigning it a distance is a relatively reliable procedure. But quasars do not have any property that reliably indicates distance. For example, the absolute luminosity of quasars ranges over a factor of more than 1000. This wide range means that the apparent brightness of any two quasars can differ by a factor of several thousand at the same redshift. Quasar brightness is a very poor distance indicator. Also, quasar redshifts are far higher than the range over which the Hubble relation has been tested. The distance to a quasar is meaningful only in terms of a cosmological model. The enormous distance estimates are the result of the *assumption* of cosmological redshifts.

Is there evidence to support the assumption of cosmological redshifts for quasars? Some astronomers have pointed out cases in which quasars with high redshifts appear to be associated on the plane of the sky with nearby galaxies that have low redshifts. If objects with different redshifts were shown to be physically connected or at the same distance, it would cast doubt on the assumption of cosmological redshifts. It all comes down to statistics. The sky is filled with thousands of quasars and hundreds of thousands of galaxies, so some alignments are bound to occur by chance.

In fact, a large body of evidence supports the idea that quasars *are* at the distances indicated by their cosmological redshifts. As their name implies, quasars are often not completely stellar in appearance. Images of quasars at redshifts from a range of $z = 0.3$ to $z = 1$ reveal fuzz, or nebulosity, surrounding the bright core. The nebulosity has the size and brightness expected of a luminous galaxy at that redshift. In addition, quasars often lie in clusters of galaxies at the same redshift; this association is strong circumstantial evidence that the quasars are at cosmological distances.

Perhaps the most convincing evidence that quasars are extremely distant comes from observations of gravitational lensing. As we have seen, general relativity predicts that mass can bend light, which leads to the distorted image of a background object. We saw earlier in the chapter that mass in galaxy clusters can distort the light of a background galaxy. About 1 in 300 quasars happens to lie directly behind an intervening galaxy. The lensing causes the magnification of the quasar light and the formation of multiple images. Gravity can deflect all forms of electromagnetic radiation, so we can observe the lensing of radio waves as well as visible light. Figure

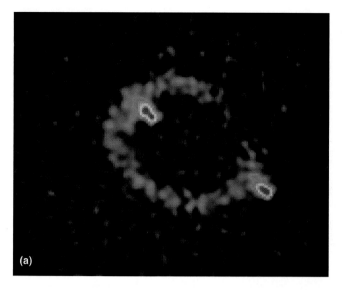 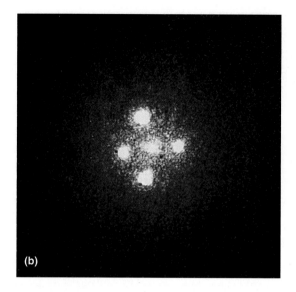

FIGURE 16-24
The gravitational lensing of distant quasars. **(a)** A radio map of the source MG 1131+0456 made with the Very Large Array. One component is split into two, and another component is perfectly aligned with the intervening lens, producing an Einstein ring. For a simulation that shows how this geometry can occur, see the top image of the left-hand sequence in Figure 16–18. (Source: J. N. Hewitt, MIT.) **(b)** An image of the lensed system G2237+0305, taken with the Faint Object Camera of the Hubble Space Telescope. The redshift of the quasar is $z = 1.70$, and the lensing galaxy is 20 times closer. Compare with the top image of the right-hand sequence in Figure 16–18. (Source: NASA.)

16–24 shows two examples of lensed quasars. In Figure 16–24a, the alignment is nearly perfect, and an Einstein ring is visible. In Figure 16–24b, the quasar light has been split into four images in a cross, with the lensing galaxy in between them. The vertical sequence on the right-hand side of Figure 16–18 is a simulation of the slight misalignment that can explain this image configuration. The general aspects of gravitational lensing are discussed in the Science Toolbox on page 500. The probability of finding four unrelated quasars as close together as those in Figure 16–24b, with identical redshifts and spectra, is infinitesimally small. This type of configuration can only be produced by the effect of gravitational lensing. The quasar must therefore be more distant than the lensing galaxy.

How are quasars related to other active galaxies? There is continuity in observed structure between normal galaxies, active galaxies at low redshift, and quasars at much higher redshift. Active nuclei are rare in normal galaxies, but a few nearby examples can be found. Searching larger volumes of space turns up even rarer but more luminous active nuclei, but at such large distances that the nebulosity from the surrounding galaxy is difficult to discern. At the greatest distances, only highly luminous objects can be seen, and the light from the surrounding galaxy is correspondingly weak. This progression is indicated in Figure 16–25. As redshift and distance increase, the brightness and angular size of a normal galaxy decrease. Also, the light from the active nucleus drowns out the starlight from the surrounding galaxy. By a redshift of $z = 0.3$ to 0.5, the host galaxy is barely visible. At this point, astronomers refer to the object as a quasar.

Properties of Quasars

The first quasars were discovered using radio surveys, and by the end of the 1960s, a few hundred were known. Radio searches for quasars are very efficient, because strong radio emission is a good indication of nonthermal activity. However, it turns out that relatively few quasars are strong radio sources; most must be found by their optical emission. Optical searches are more difficult because the sky is crowded with faint stars and galaxies. Luckily, the extremely blue colors and broad, and highly redshifted emission lines of quasars distinguish them from stars. Optical techniques are now the most efficient way of finding quasars. There are over 20,000 known quasars, and ongoing surveys will increase that number to over 100,000!

Quasars are extraordinarily distant. Their redshifts range from $z = 0.2$ up to $z = 5$. Using the approximation $d = z c / H_0$ (see the earlier Science Toolbox), the nearest quasar is therefore $0.2 \times 3 \times 10^5 / 65 \approx 1000$ Mpc, or over 3 billion light-years away! The distance to the highest redshift quasars depends on the cosmological model assumed, but it is in the range of 7000 to 8000 Mpc, or 20 to 25 billion ly. The range in look-back time is 20 to 90 percent of the age of the universe. Quasars span a large range in luminosity, from 10^{12} to over 10^{15} times the luminosity of the Sun—equivalent to up to 1000 times the luminosity of the Milky Way galaxy.

Astronomers have a clever technique for figuring out the size of a quasar's emission region. This technique relies on the fact that nonthermal radiation usually varies in intensity. Most quasars vary in brightness over a span of several weeks or a month. Because of the finite velocity of light, these data translate into an upper limit on an object's size. Imagine that the object is 1 light-year in diameter. If the light output from the entire object varies, it will take 1 year longer for the signal from the far side of the object to vary than the signal from the near side. An object's brightness cannot vary any faster than the time it takes light to travel from one side of the object to the other. If a quasar varies on

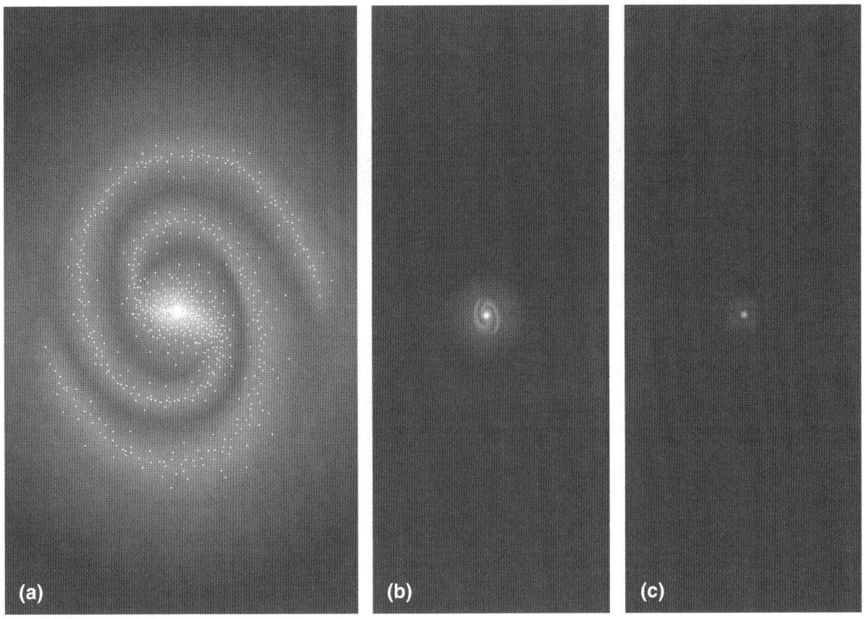

FIGURE 16-25
The appearance of an active galaxy as a function of distance. **(a)** At a distance of 20 Mpc, the active nucleus is a bright core in a well-resolved galaxy. **(b)** At 200 Mpc, only the basic features of the host galaxy are visible. **(c)** At 2000 Mpc, a redshift close to 0.5, the galaxy—now classified as a quasar—is so faint and small that it can barely be resolved. If we reversed this reasoning and placed a quasar in the nucleus of our own galaxy (about 10 kpc away), it would rival the Sun's brightness in the daytime sky!

Distance (Mpc)	20	200	2000
Angular Size (seconds of arc)	100	10	1
Apparent Brightness of Nucleus (Relative to Vega)	2×10^{-5}	2×10^{-7}	2×10^{-9}

a time scale as short as 1 week, its diameter must be less than 1 light-week, or $3 \times 10^5 \times 3600 \times 24 \times 7 = 1.8 \times 10^{11}$ km. This is roughly 1000 AU, which is only about 10 times the diameter of the solar system.

The fantastic energy output of quasars comes from a tiny volume at the center of the host galaxy. As an analogy for this remarkable concentration of energy, imagine flying above a large city at night with the lights of the city spread out below you. The city probably contains about 100 million house lights, streetlights, and car lights spread over a region 50 km across. Think of each light as a star. A quasar is then a light source only 1 cm across, with an intensity equal to 1000 times the sum of all the lights in the city! Packing so much energy into such a small volume is a real challenge for theories of the quasar power source.

The spectra of quasars show strong, broad emission lines. Figure 16–26, a composite spectrum of 740 different quasars, illustrates the typical spectral features of a quasar. If the width of the lines is due to Doppler shifts, then the hot gas emitting the lines must be moving between 10,000 and

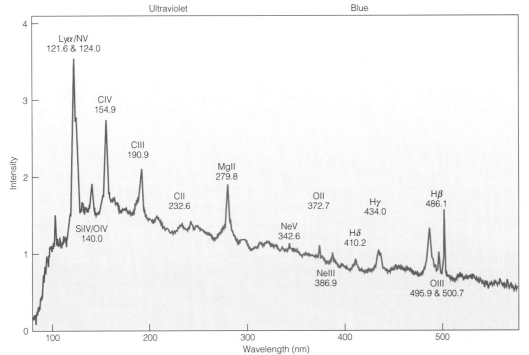

FIGURE 16-26
A composite quasar spectrum compiled from one of the largest published surveys of bright quasars. The composite includes data from 740 quasars, all with different redshifts. Each spectrum was blueshifted to zero velocity, and then the spectra were added together to give the spectrum of an "average" quasar. The numbers above each emission line give the wavelength of each line in nanometers, and the Roman numerals following the element designation give the level of ionization: II means one electron removed, III means two electrons removed, and so on. (SOURCE: P. J. Francis and others, reproduced from The Astrophysical Journal by permission of the authors.)

SCIENCE TOOLBOX
Gravitational Lensing

Einstein's theory of general relativity accurately describes the way light is affected by mass. In Einstein's theory, mass distorts space; light then bends to follow this distortion of space. General relativity received its first experimental confirmation in the observed deflection of starlight as it passed close to the edge of the Sun.

The equation that describes the deflection of light by mass is a fundamental result of general relativity. The square of the deflection angle in radians is given by

$$\theta_E^2 = 4GM / c^2 D$$

In this equation, G is the gravitational constant, c is the speed of light, M is the mass of the deflector, and D is the distance to the deflector. A quasar at a typical redshift of $z = 1$ has a distance of about 2000 Mpc. Let us calculate the deflection caused by a typical massive galaxy—roughly $10^{12} M_\odot$ if we include the dark matter. Converting to metric units of meters and kilograms, we get $\theta_E^2 = (4 \times 6.7 \times 10^{-11} \times 10^{12} \times 2 \times 10^{30}) / (3 \times 10^8)^2 \times 2 \times 10^9 \times 3 \times 10^{16}) = 10^{-10}$ radians. Thus, $\theta_E = \sqrt{10^{-10}} = 10^{-5}$. Multiplying by 206,265 to get the answer in arc seconds: $\theta_E = 2.0$ seconds of arc.

We have deduced that a single galaxy can deflect distant quasar light by 2 seconds of arc. If the quasar is perfectly aligned behind the galaxy and the galaxy is symmetrical (a good approximation for an elliptical galaxy), the deflection will occur equally in every orientation, and the deflected rays will trace back to form the image of a ring on the sky. Thus, a point source of light—a quasar—is transformed into an Einstein ring. The angular scale θ_E is known as the *Einstein radius*.

Lensing is rare. Only 1 in 300 quasars is gravitationally lensed. However, it is not as rare as you might expect, because a second consequence of lensing is the fact that the quasar light is *amplified*. A gravitational lens will magnify light just as an optical lens will. Lensing therefore makes a quasar brighter than it would have been in the absence of a foreground galaxy. This means that a lensed quasar can be seen to a greater distance than a non-lensed quasar. The greater volume out to that larger distance means a larger potential population to be lensed. Thus, there is a selection effect in favor of detecting lensed quasars.

A perfect alignment is, of course, rare, so we would not expect to see Einstein rings very often. Only a few have ever been observed. A more likely outcome is a slight misalignment, which leads to the ring's splitting up into four distinct images. If the misalignment is more severe, the result is a single magnified arc and a demagnified image on the other side of the lens. The same deflection equation applies to distant galaxies that are lensed by nearby galaxies. We have seen that a cluster can produce numerous lensing arcs of background galaxies. Remember that lensing is an excellent method for "weighing" a galaxy, since the deflection of light is caused by *all* the mass, including the invisible dark matter.

Now we can more easily understand the phenomenon of gravitational microlensing, discussed in the previous chapter. Imagine we are looking for dark matter in the halo of the Milky Way and suspect it might be made of substellar objects. What is the lensing signature of a 0.01 $M_\odot$ object observed at a distance of 10 kpc against a backdrop of more distant stars? We can scale from our previous result, if we note that $\theta_E^2 \propto M/D$, so $\theta_E \propto (M/D)^{1/2}$. The mass is 10^{14} times smaller and the distance is 2×10^5 times smaller, so the deflection angle is $(10^{14}/2 \times 10^5)^{1/2} = 22,000$ times smaller than the deflection angle for lensing by a galaxy. We get $2/22,000 = 10^{-4}$ seconds of arc, or 100 microarc seconds—hence the term *microlensing*. Since the deflection angle is much smaller than the smallest angle that can be resolved by any telescope (including the Hubble Space Telescope), the Einstein ring is not visible, and the image is not noticeably distorted.

If microlensing produces such a small deflection angle, how can the effect be measured? The answer is that astronomers look for the amplification of the light source, rather than image splitting. As a dark lens passes in front of a more distant star, the star will brighten considerably for the time it takes the lens to cross the area defined by the Einstein radius. Given the typical velocity of objects in the halo of the Milky Way, the background star will be brightened for about a week. Lower-mass dark objects have smaller Einstein radii, so the duration of the light amplification is shorter.

So what is the recipe for detecting microlensing? Look for a one-time brightening and fading of a background star with a light curve that does not depend on wavelength (the deflection angle predicted by general relativity is the same for any wavelength of light). This distinctive signature is unlike any variation of the known variable stars. The problem is the low probability of an alignment. Only about one in a million stars is microlensed at any particular time, so at least a million stars per night must be observed to have a chance of catching these rare events. Modern CCD detectors allow this kind of a wide-angle survey, and microlensing events have been detected in the halo of our galaxy. However, the rate of microlensing is too low for the dark halo of the Milky Way to be composed mostly of substellar objects like white dwarfs or brown dwarfs.

20,000 km/s, or roughly 5 percent of the velocity of light. The strongest emission lines are due to ionized hydrogen, carbon, magnesium, neon, oxygen, and nitrogen. Heavy elements have been observed in the spectra of even the most distant quasars, with look-back times of about 10 billion years. We conclude that the laws of physics seem to be unaltered over large distances; the elements we see in the far reaches of the cosmos are the same elements we find in the neighborhood of the Sun. Also, since we see these heavy elements at a look-back time of 95 percent of the age of the universe, a generation of stars must have lived and died in the first 5 percent of the lifetime of the universe. These elements were then dispersed into the interstellar medium to become ionized atoms in the gas swirling close to the quasar core.

To move beyond a simple description of quasar properties, we need to know more about their demographics. How many quasars are there in a volume of space? How old are they, and how do they relate to normal galaxies? The most important finding is that quasars *evolve*. There is a steep increase in the number of quasars at fainter brightness levels, indicating that quasars were both more luminous and more numerous in the past. Studying large and complete samples enables us to recount the history of the quasar population. Quasars were first born about 10 billion years ago. They increased in number for 2 or 3 billion years, reaching a maximum space density at $z = 2$. Since then, they have been gradually fading, like brilliant embers. The most luminous quasars are extremely rare now; you would need to search a volume of space about 500 Mpc across just to find one!

If we consider the relative numbers of quasars and galaxies in a fixed volume of space, about 1 galaxy in 1000 has quasar activity. This information can be interpreted two ways. Perhaps 1 galaxy in 1000 is special in some way that makes it develop a quasar nucleus. Or perhaps *every* galaxy goes through a phase of quasar activity lasting $1/1000$ of the age of the universe, or about 10 million years. We are limited to a snapshot of the universe, an inevitable consequence of our own short lives and the universe's great age. At any given time, only 1 in 1000 galaxies will be switched on to show quasar activity. The two results look the same, and we cannot tell the difference with a simple census.

Understanding the Power Source

Understanding the quasar power source is one of the most challenging tasks in astronomy. In the most luminous quasars, 1000 times the light from an entire galaxy is contained in a volume not much bigger than the solar system. Normal stellar processes are not efficient enough to produce this fantastic energy density. Can we show this? Yes! Suppose that we use the velocity of the hot gas near the quasar core to get an estimate of the mass in the central parsec. Then assume that normal stellar fusion is causing the energy release (without worrying for a moment about how a sufficient amount of stars could get into such a tiny volume). As we saw in Chapter 11, the Sun releases 0.7 percent of its mass in the form of radiant energy. Therefore, if M is the total mass of main-sequence stars within the central parsec, $E = 0.007 Mc^2$ is the maximum amount of energy that could come from stars within that small region. The observed amount of energy is 15 to 40 times larger. In other words, the energy source within a quasar is so efficient that it releases 10 to 30 percent of the mass-energy locked in matter. Stellar fusion cannot do this.

The basic ingredient for the model of the quasar power source is a supermassive black hole. This compact object is the gravitational "engine" that powers many of the observed properties of quasars. We have seen that the evidence for black holes in nearby galaxies is quite strong. Quasar black holes would have to be in the range of mass from 10^6 to 10^9 $M_\odot$. Figure 16-27 is a simplified cartoon of the central region of a quasar, showing the main components. This geometry is not observed directly, and there is no observation yet that probes scales close to the Schwarzschild radius. You should recognize that the black hole model for quasars is speculative and does not have the secure status of, say, the model of stellar evolution.

According to the model, the supermassive black hole at the center of a quasar forms and grows by accreting (gathering) gas and stars from the central parts of a galaxy. Since most galaxies rotate, the black hole is also likely to rotate. The law of conservation of angular momentum indicates that it will be spinning rapidly. No radiation can escape from the event horizon of a black hole. However, material that is falling in will be accelerated and will emit huge amounts of radiation. The region around a supermassive black hole is an enormous particle accelerator, converting the potential energy of material that falls in to radiation and kinetic energy of relativistic particles that flow out. This process is very efficient. In principle, a rotating black hole can convert 20 to 30 percent of the mass that falls in to pure energy, sufficient to explain the enormous luminosity of quasars. Most of this energy emerges in the form of nonthermal radiation, which can span the entire electromagnetic spectrum. Some quasars have been detected at every wavelength—from radio waves the size of a person to ultra-high-energy gamma rays the size of an atomic nucleus!

The second major ingredient of the quasar model is an **accretion disk,** the reservoir of hot gas and dust that feeds the black hole. Rotation of the material that falls to the black hole forms a disk rather than a spherical cloud. The proximity of an intense radiation source means that the disk will be hot. Thermal spectra with temperatures of 10,000 to 20,000 K have been observed in a number of quasars, consistent with the expected radiation from an accretion disk. Such a hot disk would "puff up" into a *torus,* or doughnut, rather than flatten like a pancake. This geometry would obscure our view of the black hole in the equatorial plane, making it visible only along the poles of the spin axis.

Outside the accretion disk are small dense clouds of gas orbiting the supermassive black hole. Clouds at distances of under 1 pc are partially ionized by strong ultraviolet radiation from the central engine, and they move at speeds of tens of thousands of kilometers per second in the strong gravity created by the black hole. This region gives rise to the broad lines seen in quasar spectra (Figure 16-26). All quasars show broad emission lines, but many lower-luminosity active galaxies also show narrower emission lines, with widths of

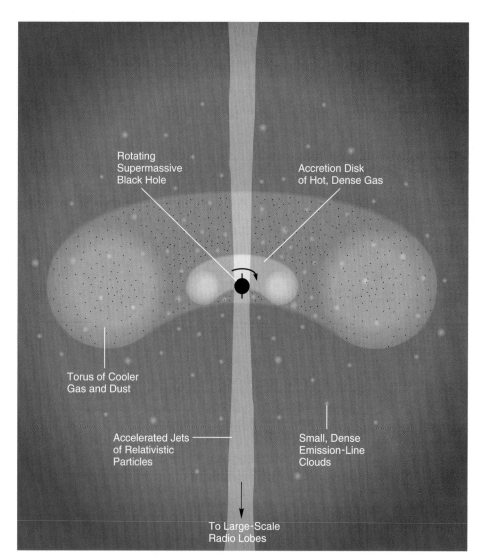

FIGURE 16-27
A cross section of the central parsec of a galaxy with an active nucleus and jets. The rotating black hole is fueled by gas falling in from the hot accretion disk. Along the poles of rotation, particles can be accelerated into two oppositely directed relativistic jets. Outside the accretion disk is a large dusty torus of obscuring material. Small, dense emission line clouds are distributed with a density that increases toward the center.

hundreds of kilometers per second. The region that gives rise to the narrow lines is much farther from the central engine; the clouds are distributed over kiloparsec scales.

Some quasars have the most spectacular signature of nuclear activity: radio jets. Most quasars are weak radio sources; jets are observed only in quasars that are strong radio sources. The region around a supermassive black hole can act as a powerful particle accelerator. The accretion disk blocks most of the outflow, but material can be accelerated in the two directions perpendicular to the disk, along the spin axis. Quasar jets are relativistic. The material that flows out can be moving at 90 to 95 percent of the speed of light, and the individual high-energy particles can be moving at over 99 percent of the speed of light. The region in which the jet forms is extraordinarily complex, with high-energy particles, radiation, shock waves, and magnetic fields. Supercomputers must be used to simulate the structure and evolution of a jet (Figure 16-28). Radio observations reveal a rich variety of astrophysical jets; examples are shown in Figures 16-29 and 16-30.

Can we relate the life story of quasars to our observations of nearby galaxies? Ten billion years ago, black holes grew quickly in the centers of massive galaxies, fueled by the frequent interactions between galaxies in the early, dense universe. Since their peak about 5 to 8 billion years ago, quasars have been gradually fading. Supermassive black holes cannot disappear! However, they can lose their ability to shine brightly if they have no fuel. As the universe expands, there are fewer interactions among galaxies, and as stellar populations evolve, the amount of mass loss available to fuel nuclear activity diminishes. By now, most quasars have been starved into silence!

If we add up all the quasar luminosity at the peak epoch of quasar activity and assume the supermassive black hole model, we can work out the mass of quasar remnants that must persist to the present day. The prediction is that most galaxies will have 10^6 $M_\odot$ black holes, 10 percent will have 10^7 $M_\odot$ black holes, 1 percent will have 10^8 $M_\odot$ black holes, and so on. Amazingly enough, searches for dark matter in nearby galaxies have found black holes at roughly the predicted rate, as we saw earlier in the chapter. The glory of quasars occurred billions of years ago, but we still see their dark shadows, lurking in the hearts of nearby galaxies.

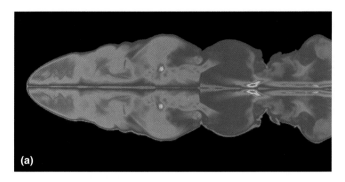 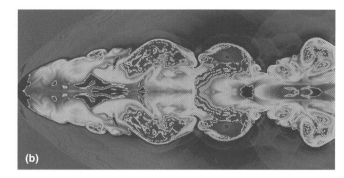

FIGURE 16-28
A supercomputer simulation of one of the jets in an active galactic nucleus. The black hole is on the left, and material flows to the right. The simulation deals with the case of a strong magnetic field. **(a)** Energy density in the jet, with blue the highest, followed by red, yellow, and green. **(b)** Matter density with the same color-coding. The complex details are caused by shocks and turbulence because the flow in the jet is not smooth. Emission from the jet shows knots or blobs. (SOURCE: D. Payne, California Institute of Technology; simulation performed at San Diego Supercomputer Center.)

Quasars as Probes of the Universe

Nearly 40 years after their discovery, quasars continue to perplex astronomers. However, even though astronomers do not completely understand the power source, they can use quasars as cosmological *probes* of intervening material. If we accept that quasars are at the vast distances indicated by their redshifts, we can think of them as powerful flashlights for illuminating the universe.

As we have seen, gravitational lensing can affect the light of distant quasars. Among thousands of known quasars, only a few dozen are multiply imaged. The lenses that cause the image splitting are elliptical galaxies in about 90 percent of the cases, and spiral galaxies in the other 10 percent. This result was anticipated, because only a very dense galaxy can bend light significantly, and ellipticals are the densest galaxies. The study of gravitational lenses has revealed the properties of elliptical galaxies at redshifts of $z = 0.5$ to $z = 1$. Lensing shows that distant ellipticals are embedded in halos made of dark matter, just as they are in the nearby universe.

Gravitational lenses are fascinating tools for studying the distant universe. Each system is like an optics experiment on a cosmic scale. We will mention one particularly interesting application here. The light output from most quasars varies. If the quasar is multiply imaged, the light will take slightly different times to reach Earth from each different image. A variation will be seen first in the image whose light has been deflected the least, corresponding to the shortest path between the quasar and Earth. The variation will be seen last in the image whose light has been deflected the most. With a good model for the gravitational lens geometry, astronomers can use the time delay (Δt) to measure the difference in light paths between any two images ($c\Delta t$). This is a *direct* measurement of the distance scale of the universe, which is independent of the normal chain of reasoning that goes into the determination of the Hubble constant. The Hubble constant measured by gravitational lensing agrees well with the Hubble constant measured using Cepheids as distance indicators. This is a striking confirmation of the basic picture of an expanding universe.

A second technique that uses quasars as probes is the measurement of quasar absorption lines. Quasar light travels for billions of years before it reaches Earth. If that light intercepts a galaxy along the way, the wavelength at which it will be absorbed indicates the redshift of the intervening

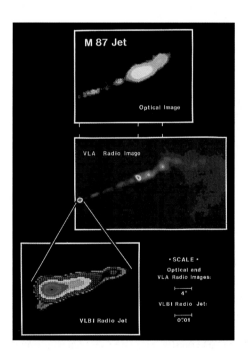

FIGURE 16-29
The core and jet of the elliptical galaxy M 87. The top view is an optical CCD image of the jet showing blobs of emission extending away from the nucleus to the left. The middle view is a false-color radio image of the jet made with the Very Large Array at a wavelength of 6 cm and a resolution of 0.35 second of arc. The lower view is zoomed in on the region close to the core. The false-color radio image, made with an intercontinental Very Long Baseline Interferometer network at a wavelength of 18 cm and a resolution of 0.008 second of arc, shows that the jet continues into regions close to the nucleus of the galaxy. (SOURCE: M. Reid, Smithsonian Astrophysical Observatory; National Radio Astronomy Observatory.)

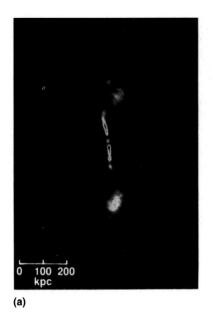
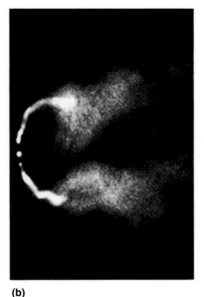

FIGURE 16-30
High-resolution radio images of jets being emitted from radio galaxies (bright central objects). **(a)** Linear jets moving out from elliptical galaxy 3C 449, at an estimated distance of 66 Mpc. Approximate scale is indicated. (Computer processing yielded dark cores in images of the galaxy and jets; they are actually bright.) **(b)** The jets from elliptical galaxy NGC 1265 are believed to be bent to the right from leftward motion of the galaxy at about 2000 km/s through intergalactic gas in the Perseus cluster of galaxies.

(SOURCE: National Radio Astronomy Observatory, operated by Associated Universities, Inc. under contract with National Science Foundation; image b, courtesy C. O'Dea, F. Owen, and M. Inoue.)

galaxy (Figure 16–31). There are two types of absorption lines. One type is caused by heavy elements like carbon, magnesium, and silicon, presumed to originate in the disks and halos of normal galaxies. The other type consists of hydrogen lines and is believed to represent unprocessed gas clouds of primordial abundance. The positions of absorption lines in a quasar spectrum represent a *map* of the absorbers in redshift. Assuming cosmological redshifts, we can convert redshift to distance and produce a map reaching out billions of light years from Earth. By measuring the properties of absorbers using lines of several different elements, it is possible to trace the chemical evolution of galaxy halos back to $z = 5$. (Since an average galaxy at that redshift is extremely faint, direct study would be impossible.)

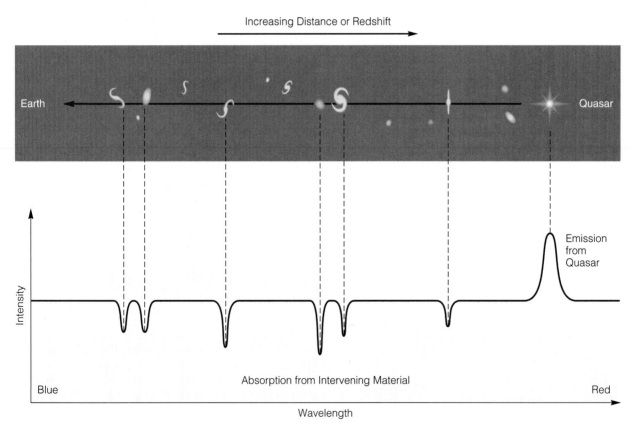

FIGURE 16-31
Occasionally, the line of sight from a quasar to Earth will penetrate the body or halo of an intervening galaxy. Since redshift is a measure of increasing distance, the most distant the absorbers are, the more the absorption lines in the quasar spectrum will be redshifted. The spectrum therefore provides a map in space of the absorbers, most of which are not visible directly.

Another application of quasar absorbers is to study the clustering properties of the intervening material. Notice that a single line of sight can only yield clustering in one dimension. Astronomers ask the question: Are the absorption lines distributed randomly along the spectrum, or are they grouped with small separations in wavelength or redshift? By combining lines of sight, astronomers can reconstruct the size and shape of the absorbing material in three dimensions. To use a somewhat frivolous analogy, imagine a magician plunging swords into a closed box as a way of determining what (or who) is inside. There is evidence that the heavy element absorbers associated with galaxy halos are clustered on enormous scales of 10 to 20 Mpc, providing unique data on the large-scale structure of the universe at high redshift. By contrast, the hydrogen absorbers are not clustered in space and do not appear to be associated with normal galaxies. Astronomers speculate that this gas might represent debris left over from galaxy formation. The absorption line technique is extraordinarily sensitive. As little as 100 solar masses of cold hydrogen can be detected at a distance of 10 billion light-years! Large new telescopes are enabling astronomers to use absorption lines to learn much about the distant universe.

Summary

The underpinning of modern cosmology is Hubble's discovery of a linear relationship between redshift and distance. All distant galaxies are receding from the Milky Way; in fact, according to the expanding universe model, every galaxy is rushing away from every other galaxy, carried by the expansion of space itself. If galaxy redshifts are caused by the expansion of space, we can use redshift as a distance indicator. The Milky Way is not at the center of this expansion; observers on any other galaxy would measure the same Hubble relation. Moreover, the enormous size of the universe, coupled with the finite speed of light, means that we do not see the distant universe as it is now. When we look far out in space, we observe ancient light from distant galaxies.

The slope of the relationship between redshift and distance is called the Hubble constant—a measure of the rate of expansion of the universe. The universe is billions of years old and billions of light-years across; the size of the Hubble constant tells us how old and how large. A high value of the Hubble constant indicates a rapid expansion and a smaller and younger universe. A low value of the Hubble constant indicates a slow expansion and a larger and older universe. After major observational efforts with the world's large telescopes, astronomers are converging on a value of the Hubble parameter around 65 km/s/Mpc, although significant uncertainties remain. One complication is the difficulty of finding reliable distance indicators and extending the accurate calibration of distances within the Milky Way to nearby galaxies. The best methods use luminous stars such as Cepheid variables and supernovae. At larger distances, the properties of entire galaxies are used as distance indicators.

Galaxy surveys have mapped out the large-scale structure of the universe in rich detail. Galaxies are gregarious. Most live in small groups or large clusters, and the clusters are often bound by gravity into enormous superclusters. These sheetlike structures are separated by voids up to 100 Mpc across. Yet, these impressive structures describe only 10 percent of the matter in the universe! The other 90 percent—the dark matter—does not emit radiation, and we know of its existence only through gravity. On scales larger than about 300 Mpc, the universe appears smooth or homogeneous.

A small percentage of galaxies called active galaxies have violent events occurring in their nuclei. This activity can take the form of strong radio emission or emission from high-velocity gas near the nucleus. There is a strong relationship between nuclear activity and galaxies with peculiar morphologies or strong bursts of star formation. A quasar is an active galaxy in which the active nucleus outshines the light from the entire galaxy. The most luminous quasars cram 1000 times the luminosity of an entire galaxy into a region not much larger than the solar system. Quasars are distant beacons in the universe; light from the most distant examples was emitted when the universe was only 5 percent of its current age. The best model for the power source involves a supermassive black hole feeding on gas and stars from the surrounding galaxy. The central regions of quasars and other active galaxies can act as enormous particle accelerators, spitting out relativistic jets of glowing material. Some quasars emit nonthermal radiation that spans the entire electromagnetic spectrum. Examples of supermassive black holes have been discovered in some nearby galaxies.

The simple fact of the Hubble expansion does not prepare us for the exotic phenomena of the distant universe. Gravity has sculpted structures of great complexity and delicacy, a pattern of sheets and knots and voids covering hundreds of millions of light-years. The sparse network of galaxies is like a froth of whitecaps on an ocean of dark matter. Another kind of darkness lurks in the hearts of some galaxies, in the form of supermassive black holes. Although the peak of the quasar era is billions of years past, the dead remnants of quasar activity can still be detected in nearby galaxies. This inventory prepares us for cosmology, the study of the universe as a whole.

Important Concepts

You should be able to define these concepts and use them in a sentence.

megaparsec (Mpc)	Local Supercluster
expanding universe	void
cosmological redshift	gravitational lensing
Hubble relation	active galaxy
Hubble constant	Seyfert galaxy
large-scale structure	radio galaxy
galaxy groups	nonthermal radiation
galaxy cluster	quasar
morphology-density relation	accretion disk
galaxy superclusters	

How Do We Know?

These questions and answers show how the scientific method is used to learn about the universe.

Q How do we measure the expansion rate of the universe?

A Astronomers observe almost every galaxy to have its light redshifted. The only exceptions are a few nearby galaxies that are bound gravitationally to the Local Group. However, not every galaxy is moving away from us at equal speed. More distant galaxies are moving away faster than nearby galaxies. Astronomers measure the recession velocity, or redshift, of a galaxy from its spectrum, and they measure its distance using a distance indicator like a Cepheid variable or a supernova. Redshift and distance are correlated, and the slope of the correlation gives the expansion rate. The slope is called the Hubble constant, a measure of how fast in kilometers per second a galaxy is receding for every megaparsec increase in distance from the Milky Way.

Q How do we know we are not at the center of the expansion?

A When people first encounter the Hubble relation, they naturally assume that we must be at the center of the universe. After all, every galaxy is moving away from the Milky Way. Yet if you imagine a smoothly expanding sheet with galaxies fixed to it, or a smoothly expanding loaf with galaxies like raisins in it, you can see that *every* galaxy sees the other galaxies receding. An observer on a distant galaxy would measure the same Hubble relation. Since no galaxy has a preferred position, no single galaxy is at the center of the expansion.

Q How do we know that supermassive black holes exist?

A We are not absolutely sure. The evidence is not as good as it is for black holes that are the remnants of massive stars. The center of the Milky Way has a large mass concentration and evidence of rapid gas motions, both of which point to a dark object several million times the mass of the Sun. Similar evidence in other nearby galaxies indicates mass concentrations that are too compact to be massive clusters of stars. Notice that these observations probe only the central parsec of a galaxy, which is a scale thousands of times larger than the size of the black hole itself. The indirect evidence for supermassive black holes in quasars is strong—we know of nothing else that could generate the fantastic energy output—but the direct evidence is weak. Quasars are so far away that we cannot directly observe their central regions.

Q How do we know that the universe contains mostly dark matter?

A We have already seen that galaxies contain large amounts of dark matter. When galaxies are bound into groups or clusters, we can use the speed of the galaxy motions to "weigh" the group or cluster. In every case, the motions are too fast to be accounted for by the visible galaxy material. We infer that there is a large amount of dark matter on scales larger than individual galaxies. Gravitational lensing in clusters provides direct confirmation of the hypothesis of dark matter. Distant galaxies can have their light distorted into arcs by a foreground mass concentration. The lensing geometry gives a measure of the total mass of the cluster, including the dark matter. Astronomers conclude that the universe contains a lot of dark matter beyond individual galaxies.

Q How do we know that quasars are extremely distant?

A Quasars are compact objects that live within normal galaxies. Astronomers use the redshift of a quasar as a distance indicator—in other words, the interpretation of quasars as distant object relies on the assumption that the redshifts are cosmological. At low redshifts, this assumption can be checked because it is possible to detect the galaxy that surrounds the quasar core, or to identify a cluster of galaxies at the same redshift as the quasar. However, at high redshift a quasar is an unresolved point of light with no other indication of its distance. When a quasar is gravitationally lensed, we have proof that the quasar is more distant than the galaxy that causes the quasar light to be deflected. A large body of evidence supports the hypothesis of cosmological quasar redshifts.

Problems

Use these problems to test your understanding of the information and concepts in this chapter. The * indicates a more advanced or mathematical problem.

1. Explain the distinction between a redshift caused by the expansion of the universe and a redshift caused by the Doppler effect.
*2. Suppose observers located in the Coma cluster of galaxies observe wavelength shifts in the spectra of our Local Group of galaxies, including the Milky Way. **(a)** Would they see redshifts or blueshifts? **(b)** What sizes of shift (in percent) will they observe if the recession velocity of the Coma cluster is 7000 km/s? **(c)** What would they conclude about our Local Group's velocity if they believed the theory of cosmological redshifts? **(d)** What Hubble constant would they deduce if they measured the distance to the Local Group to be 135 Mpc? **(e)** Could they use *both* the Tully-Fisher relation and the Faber-Jackson relation to measure the distance to the Local Group?
*3. If the Hubble constant is measured to be 65 km/s/Mpc, how fast a recession velocity do we expect for a galaxy at a distance of 100 Mpc? Why does the Hubble relation not apply at a distance of only 1 Mpc? What are the limitations in the use of the Hubble relation at a distance of 1000 Mpc? (*Hint:* Think of the idea of look-back time.)
4. Under the assumption of cosmological redshifts, galaxy velocities are telling us something about the state of the universe. **(a)** What would be implied if galaxies showed equal numbers of redshifts and blueshifts? **(b)** What would be implied if *all* galaxies showed blueshifts, with a size of blueshift that increased in proportion to the distance? **(c)** What could you deduce if all galaxies on one half of the sky showed redshifts, while all galaxies on the other half of the sky showed blueshifts?
5. Why is the study of the most distant galaxies we can see related to the study of conditions around the time our galaxy (and perhaps others) was forming?
6. Describe the large-scale structure of the universe as revealed by redshift surveys of galaxies. On what scale does the universe begin to appear smooth and featureless?
7. List the observational pieces of evidence that might be found to distinguish an active galaxy from a normal galaxy. In what ways do Seyfert galaxies bridge the gap between ordinary galaxies and quasars? Since our galaxy's nucleus

is a radio source, why is the Milky Way not considered to be a typical radio galaxy?

8. How does gravitational lensing give us information about the properties of galaxies and clusters of galaxies? Why do you think cases of lensing are so rare?

*9. Suppose we find a quasar that has a redshift of $z = 3.56$. **(a)** At what wavelength would we be able to observe the hydrogen Hα line? (*Hint:* It has a rest wavelength of 656 nm.) **(b)** If the quasar has light variations with a characteristic time scale of 3 days, what is the size of the quasar core? **(c)** Suppose we have measured the redshift accurately to be $z = 3.56 \pm 0.01$. Why is our estimate of the distance to the quasar not similarly accurate? **(d)** Why does a redshift greater than $z = 1$ not necessarily imply that the quasar is moving away from us faster than the speed of light?

10. What is our current theory of the quasar power source? Compare the evidence for supermassive black holes in quasars and in certain nearby galaxies.

Projects

Activities to carry out either individually or in groups.

1. Use a small telescope (with an aperture of at least 30 cm) to locate the Seyfert galaxy NGC 1068 (R.A. = 2 h, 42.7 min, Dec. = $-0°$ 01′) in the constellation of Cetus. What is distinctive about the nucleus of the galaxy? How does it compare with the nuclei of other, normal galaxies, such as Andromeda (M 31)?

2. Gather some graph paper with squares marked at millimeter intervals. Place 20 or 30 points more or less at random on a single sheet of graph paper, and label them all with numbers. Assume that the points are galaxies in two-dimensional space, and number 1 is the Milky Way. Now join together nine pieces of graph paper to make a much larger sheet. Mark out a grid of 1-cm squares on top of this larger sheet (use a color that is different from the color of the 1-mm squares). Carefully redraw the galaxies from your original single sheet on the centimeter grid, keeping their exact relative spacing. You have now made a "universe" 10 times larger than your original one. With a ruler, measure the distance in millimeters to the 10 nearest points to the Milky Way (point number 1) on both your original sheet of graph paper and your much larger sheet. Does the increase in distance follow a Hubble relation? Do you get the same result if you make measurements from a different point (or galaxy)?

Reading List

Blandford, R., Begelman, M., and Rees, M. 1982. "Cosmic Jets." *Scientific American,* vol. 246, p. 124.

Dressler, A. 1993. "Galaxies Far Away and Long Ago." *Sky and Telescope,* vol. 85, p. 22.

Eichler, D. 1995. "Galaxy Time Machine." *Astronomy,* Vol. 23, p. 44.

Freedman, W. 1992. "The Expansion Rate and Size of the Universe." *Scientific American,* vol. 267, p. 54.

Geller, M., and Huchra, J. 1991. "Mapping the Universe." *Sky and Telescope,* vol. 82, p. 134.

Kinney, A. 1996. "Fourteen Billion Years Young." *Mercury,* vol. 25, p. 29.

Macchetto, D., and Dickinson, M. 1997. "Galaxies in the Young Universe." *Scientific American,* vol. 276, p. 92.

Miller, R. 1992. "Experimenting with Galaxies." *American Scientist,* vol. 80, p. 152.

Phillipps, S. 1993. "Counting to the Edge of the Universe." *Astronomy,* vol. 21, p. 38.

Preston, R. 1988. "Beacons in Time: Maarten Schmidt and the Discovery of Quasars." *Mercury,* vol. 17, p. 2.

Veilleux, S., Cecil, G., and Bland-Hawthorne, J. 1996. "Colossal Galactic Explosions." *Scientific American,* vol. 274, p. 98.

West, M. 1997. "Galaxy Clusters: Urbanization of the Cosmos." *Sky and Telescope,* vol. 93, p. 30.

Cosmology

CONTENTS

- The Radiation of the Big Bang
- Early Cosmologies
- Relativity and Curved Space
- The Big Bang Model
- Measuring Cosmological Parameters
- The Early Universe
- Summary
- Important Concepts
- How Do We Know?
- Problems
- Projects
- Reading List

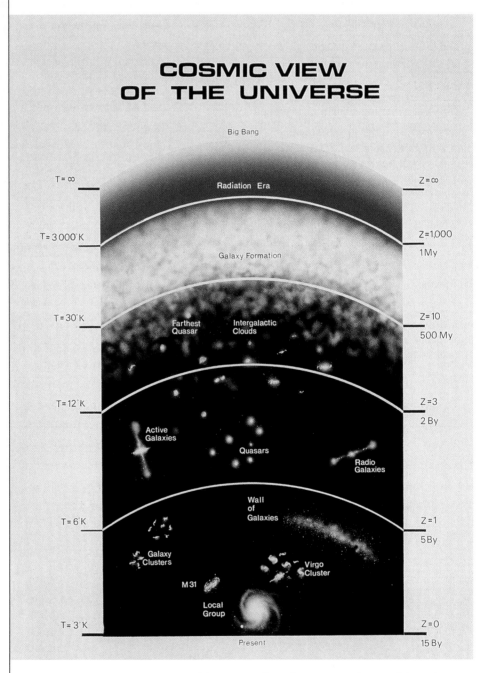

A schematic view of the evolution of the cosmos, from the intense radiation of the early universe, through the formation of the earliest structures, to the present day. (SOURCE: Space Telescope Science Institute.)

WHAT TO WATCH FOR IN CHAPTER 17

Perhaps the greatest intellectual adventure humans have undertaken is the quest to understand the universe. In this chapter, you will learn about cosmology—the study of the size, shape, and evolution of the universe. Modern cosmology began with the discovery that galaxies are distant systems of stars moving away from the Milky Way. The modern mathematical basis of cosmology—Einstein's general theory of relativity—governs the physical properties of the universe. This theory states that space can be curved as the result of the sum of all the matter in the universe. You will also learn about the scientific story of creation—the big bang model. According to the big bang model, the universe was once in a state of extremely high temperature and density. Since then, galaxies have been carried apart by the expansion of space for billions of years. Our confidence in the big bang model rests on three pieces of evidence. The first is the cosmological redshift of galaxies: The more distant a galaxy, the faster it is moving away from us. The second is the abundance of helium, created a few minutes after the big bang, when the entire universe had the temperature of the interior of a star. Finally, we observe the cold, featureless radiation left over from the big bang, redshifted by the expansion to microwaves.

Astronomers use two observable quantities to define the big bang model: the current expansion rate and the mean density of matter in the universe. You will see that these are difficult measurements to make with precision. The mean density appears to be smaller, by a factor of 3, than the critical density needed to stop cosmic expansion, indicating that the universe is open and will expand forever. General relativity tells us that the matter density of the universe is related to its space curvature. Space in our universe is nearly flat; the curvature is small enough that it has not yet been directly measured. Astronomers have studied the big bang model from the first second after creation. You will see that they have come up with tentative explanations for some basic aspects of the universe, including the four fundamental forces of nature, the flatness of space-time, and the excess of matter over antimatter. Most importantly, you will see that we can actually understand the universe of which we are such a small part.

PREREADING QUESTIONS ON THE THEMES OF THE BOOK

OUR ROLE IN THE UNIVERSE
What is the place of humanity in an enormous and ancient universe?

HOW THE UNIVERSE WORKS
What are the physical principles that govern the conditions in the early universe?

HOW WE ACQUIRE KNOWLEDGE
How do our limitations in time and space affect studying the universe?

THE RADIATION OF THE BIG BANG

In the 1940s, Russian physicist George Gamow was exploring the implications of the expansion of the universe. He realized that if the expansion were traced backward, it would point to a time when the universe was hot and dense. Perhaps it had been even hotter and denser than a star. Gamow knew that the hot early universe must have been filled with high-energy radiation. This alien world of blinding radiation and careening particles must have evolved into the large and empty universe of today.

Gamow was a brilliant scientist with an iconoclast's touch. Like Einstein, he never lost his sense of youthful curiosity and wonder at the workings of the universe. He wrote a series of books, illustrated by his own cartoons, that made abstract physics real by showing what the world would be like if you were shrunk to the size of a subatomic particle. In thinking about the evolution of the universe, Gamow realized that the early hot universe would expand and cool but would never become completely cold. Short-wavelength, high-energy photons of the early universe would be stretched out and lowered in frequency by the cosmic expansion. In effect, they would be redshifted to a much longer wavelength or a lower energy. He predicted that the universe should be bathed in radiation that was a relic of the big bang, redshifted to microwaves at a temperature only a few degrees above absolute zero. Unfortunately, Gamow's prediction could not be verified. At that time, radio astronomy was in its infancy, and no telescopes sensitive to microwave radiation existed. Gamow's prediction thus fell into obscurity for a decade.

Now fast-forward to the early 1960s. Arno Penzias and Robert Wilson, two engineers at Bell Telephone Laboratories, were testing a sensitive, horn-shaped microwave antenna designed to relay telephone calls to communication satellites orbiting Earth. As part of their tests, they were mapping the faint microwave radiation emitted by the Milky Way. An unknown source of noise affected their measurements even after they had carefully checked their equipment. The excess noise did not appear to change intensity with direction in the sky, time of day, or season, and it was not associated with any known astronomical source. At one point, they noticed that pigeons were roosting inside the antenna. Concerned that a residue left by the birds might be affecting their measurements (Penzias and Wilson euphemistically called it a "thin, white, dielectric film"), they cleaned the horn and started over. The noise persisted.

Through a colleague, Penzias and Wilson learned that a group of physicists at nearby Princeton University, led by Robert Dicke, was building a receiver to look for Gamow's proposed cosmic background radiation. Penzias and Wilson had been unaware of Gamow's work, but through luck and technical skill, they found the microwave signal first. The extra noise they detected had exactly the temperature predicted by Gamow. The two Bell Lab engineers were eventually awarded the Nobel Prize. Their discovery was dramatic confirmation that the universe had once been hot and dense—they had seen the "echo" of the big bang.

Early Cosmologies

In the opening pages of this book, we compared humanity to explorers on a strange island in an unknown sea. To extend the analogy, at this point in our explorations we know that the island is made of rocks and grains of sand. We know how big it is, we know that the sea is large, and we know that there are many other islands out there in the distance. Now we ask a series of deeper questions about our larger environment. Do the islands go on forever? Does the sea go on forever? Does our world have an edge, or is it infinite?

Cosmology is the study of the size and structure of the universe—in other words, the "geography" of the universe as a single system. This idea takes us to a new level of discussion, because our subject is the **universe,** defined as all matter and energy in existence anywhere, observable or not. The scientific method is stretched thin when we speculate about the universe. As far as we know, our universe is unique. We cannot learn about it by direct comparisons with other universes, as we can with planets and stars and galaxies. Instead, cosmologists approach their subject by making simple and testable assumptions and by developing theories that describe the present state, origin, and fate of the universe. This speculation is informed by observations, and theories must agree with known physical laws.

Cosmology begins with some fundamental questions about the nature of time and space, mass and energy. By answering these questions, cosmologists have encountered some startling ideas about the universe as a whole—ideas that lie far outside everyday experience. Cosmology is not the domain of scientists alone. For thousands of years, poets, priests, and philosophers have pondered the universe and tried to understand its nature. Innate in the character of curious, restless humanity is the desire to understand our surroundings and to know where we came from.

Cosmology is as old as the first ancients who looked at the stars set against the velvet backdrop of night. The universe is described in the earliest surviving writings of the Babylonian, Egyptian, Greek, Chinese, and Indian civilizations. According to Indian legend, the universe is a giant egg containing land, water, animals, gods, and so on, all brought forth from primordial waters by the creator god Prajapati. A Tahitian tradition says that a creator, Taaroa, existed in the immensity of space before any universe existed, and that he later constructed the heavens and the rocky foundations of Earth. A collection of Norse myths supposes that in the beginning there was nothing at all, with regions of frost to the north and fire to the south. The heat melted some of the frost, and from the drops of liquid grew a giant, Ymer, who created all the inhabitants of the world. This early phase of *mythological* cosmology linked celestial phenomena to the spiritual life of humans. We have found creation myths in almost all cultures; these cosmologies provide an enduring link with our early ancestors.

The next phase of cosmology dates back to the birth of scientific inquiry in the sixth century B.C. on the shores of Asia Minor (as described in Chapter 2). With the ancient Greeks, the beginnings of *philosophical* cosmology marked the first application of the power of reason to the universe. Observations were not central to Greek cosmology; the telescope would not be invented for another 2000 years. Instead, Greek thinkers proposed bold hypotheses, and used logic and abstract reasoning, to explain the universe. Anaximander believed the universe evolved from a state of primordial chaos to the order and structure of today. Others, like Aristotle, thought the universe was perfectly ordered and unchanging. The twin themes of chaos and order find a strong echo in modern cosmology.

 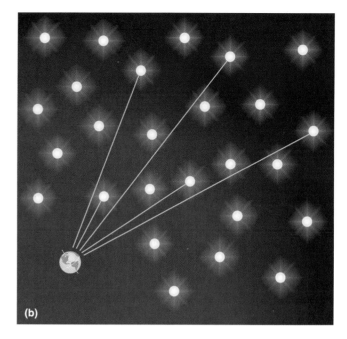

FIGURE 17-1
An analogy for Olbers' paradox. **(a)** In a large forest, every direction eventually ends with a tree, and we cannot see through the edge of the forest. (SOURCE: Janet Seeds.) **(b)** If the universe is infinite and filled with stars, every direction must eventually intercept a star at some distance, causing the night sky to be bright with starlight.

By the third century B.C., mathematical reasoning had become part of astronomy. Until then, most people believed the stars lay on a two-dimensional backdrop, nestled snugly around Earth. Euclid pulled together the theorems of geometry and laid the foundation for the idea of infinite space. Imaginary Euclidean triangles could be extended into space; distances to celestial bodies could be measured. Aristarchus used geometry to anticipate the Sun-centered cosmology of Copernicus by nearly 1800 years. The Greeks also wrestled with the uncomfortable implications of an infinite universe. Plato's colleague Archytus summed it up this way: "If I am at the extremity of the heaven of the stars, can I not stretch outward my hand, or my staff? It is absurd to suppose that I could not; and if I can, what is outside must be either body or space."

The 17th century ushered in the stage of *physical* cosmology, with Isaac Newton. Newton described gravity as a force by which every particle in the universe attracts every other particle. He realized that this principle might allow a simple description of the structure of the whole universe. Newton's theory of gravity was brilliant and audacious. For the first time in human history, the mundane motions of objects on Earth had been unified with the stately orbits of heavenly bodies. Newton viewed the cosmos as infinite in extent and filled with randomly moving stars. He argued that no other assumption would make sense. If the universe were not infinite, or if the stars were all in one part of the universe, then gravity would eventually cause all matter to clump together in one place. Only in an infinite universe does each particle feel balanced gravitational forces from other particles in all directions in the sky. There is motion in Newton's cosmology, but the universe is static—its appearance is unchanging in time.

However, Newton's framework had serious conceptual problems. The force of gravity declines with distance, but it has an infinite range. When gravitational forces acting on an infinite number of bodies spread over infinite space are added up, the amount of gravity is infinite too! Another objection to Newton's cosmology results from asking this simple question: Why is the sky dark at night? This concern was first raised by Thomas Digges in 1576, but it came to be associated with the German astronomer Wilhelm Olbers over 250 years later. **Olbers' paradox** can be described as follows. In an infinite universe filled with stars, every line of sight must eventually intercept a star (see Figure 17-1 for an analogy). Moving out from Earth, the brightness of a star *decreases* as the inverse square of the distance ($F \propto R^{-2}$). However, the number of stars in any spherical shell (which is $4\pi R^2 \Delta R$, where ΔR is the thickness of the shell) *increases* with the square of the distance. The result is that the amount of light in each shell is the same, regardless of distance. The contribution of light from distant stars continues to pile up. In an infinite universe, the sum of all light from distant stars is infinite: The night sky should be ablaze with light!

The modern response to Olbers' paradox is subtle, invoking the age of the universe and the recession of distant galaxies. First, there is a distance beyond which we can see no galaxies or stars. This distance does not represent an edge to the universe, but a distance corresponding to a light travel time of 12 to 15 billion years, beyond which we see no galaxies because none had formed that long ago. In other words, the total number of photons emitted by the galaxies in their finite lifetimes is too low to create the kind of pervasive bright glow described by Olbers. A secondary effect that helps explain Olbers' paradox is the expansion of the universe.

The redshifts of receding galaxies cause the apparent energies of the photons we receive from them to be reduced from high energies (short wavelengths) to low energies (long wavelengths), because the photons are "stretched" in the expanding space. Photons received from galaxies with redshifts close to the speed of light are strongly reduced in energy.

Newton's theory of gravity left another very basic question unanswered. What, exactly, was this force that operated across vast distances through the vacuum of space? Newton was acutely aware of this issue and went so far as to call the idea of gravity acting instantly at a large distance "an absurdity." He absolved himself of his ignorance about the cause of gravity by saying, "I have not been able to discover the cause of these properties of gravity from phenomena, so I frame no hypothesis." We did not gain a more profound understanding of the force of gravity until early in the 20th century.

Relativity and Curved Space

Modern cosmology began with Albert Einstein. First, his special theory of relativity showed that time and space are supple and not adequately described by the rigid, linear measures proposed by Newton (review the material near the end of Chapter 13). Experiment shows us that the speed of light is a constant number, regardless of the motion of the observer. There are three concrete but bizarre consequences of the fact that we cannot measure our motion with respect to a beam of light: (1) Time slows down for a fast-moving object; (2) a fast-moving object shrinks in the direction of motion; (3) the mass of a fast-moving object increases, with energy of motion converting to mass by the famous relationship $E = mc^2$. However, we cannot see these effects in the everyday world, where objects move slowly compared to the speed of light.

Special relativity deals with objects in constant relative motion. By contrast, general relativity deals with objects in nonuniform motion. Einstein was led to the theory of general relativity by pondering what seemed to be a coincidence in physics. The gravitational mass of an object, its response to a gravitational force, is identical to its inertial mass, the resistance it presents to a change in its motion. At first glance, these appear to be very different forms of mass. Imagine a large, smooth piece of iron at rest on a slick, icy surface. Gravitational mass dictates the force with which the iron presses down on the ice. Inertial mass is the resistance of the iron to a change in speed, trying to either speed it up or slow it down. Inertial mass has nothing to do with gravity, since the motion on the ice is horizontal and gravity acts vertically. Nevertheless, inertial and gravitational masses are measured to be equal to an exquisite degree of precision; modern experiments find the difference to be less than 1 part in 10^{15}. Einstein believed that this coincidence held the key to understanding gravity.

An analogy will clarify how difficult it is to distinguish between motion caused by gravity and motion caused by any other force. Suppose you are trapped in a windowless elevator. As we saw in Chapter 13, Einstein showed that there is no way to distinguish between the motion of objects in an elevator at rest on Earth's surface and in an elevator accelerating through distant space at a rate of 9.8 m/s². In each case, you would have your normal weight. He also realized there is no way to distinguish between the motion of objects in an elevator floating freely in space and an elevator in freefall toward Earth's surface at a rate of 9.8 m/s². In each case, you would be weightless. In short, there is no measurable difference between acceleration caused by gravity and acceleration from any other force. Gravity is just a convenient way to describe how the presence of mass causes an object to change its motion. Einstein "generalized" his special theory by showing how gravity could distort space and time. General relativity replaces Newton's force of gravity with the geometry of space itself. The familiar Newtonian idea of masses placed in smooth and uniform space is replaced with the counterintuitive idea of space that is distorted by the masses it contains. Matter curves space, and light and particles follow the undulating paths dictated by the curvature.

In 1917, Einstein applied the equations of general relativity to the universe as a whole. He assumed that the universe was static because astronomers at the time believed the universe to contain only the enormous Milky Way, with stars milling around in it. (Ironically, the astronomer Vesto Slipher was already gathering spectra that would reveal the recession of the galaxies and disprove the static model.) No matter how Einstein solved the equations, they stubbornly indicated a dynamic universe, one that was either expanding or contracting. To force a static solution to the equations, he added an arbitrary term. Einstein later admitted that this adjustment was "the greatest blunder of my life." Because of it, he missed the chance to predict the expansion of the universe 10 years before Hubble observed it.

In the 1920s, Russian mathematician Alexander Friedmann and Belgian mathematician Georges Lemaître independently solved the equations of general relativity and showed mathematically that the universe was expanding. Then, with Hubble's discovery of the redshift-distance relation, what had originally seemed to be a purely theoretical model was supported by observations. Galaxies are not moving apart *through* space in a large-scale version of the Doppler effect; they are being carried apart by the expansion of space *itself*. Galaxy recession velocities are indicators of a cosmological redshift. Once again we can use an analogy: the surface of a balloon with small beads glued on it to represent galaxies (Figure 17–2). (Keep in mind that this is a two-dimensional representation of a positively curved space that exists in three dimensions.) As the balloon is being blown up, its expansion reveals several relevant features of the expanding universe.

We can now answer the puzzling questions raised at the beginning of Chapter 16. The analogy in Figure 17–2 is accurate in the sense that the beads are carried apart by the stretching rubber of the balloon. In general relativity, the fabric of space is expanding, carrying the galaxies with it. The beads follow the Hubble relation, with recession velocity proportional to distance. No bead is at the center of the balloon,

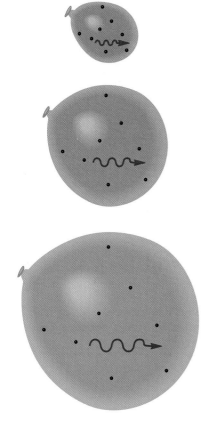

FIGURE 17-2
An expanding balloon as an analogy for the expanding universe. Beads glued to the surface of the balloon are galaxies. Radiation is represented as a wave drawn on the surface of the balloon. As the balloon expands, each galaxy is carried farther and farther away from every other galaxy. Radiation in the universe redshifts to longer wavelengths and lower energies as the universe expands.

nonexpanding space are well described by Newton's laws: The solar system is not expanding, nor is your house. Note also that as the balloon expands, the curvature of the space decreases; think of the difference in curvature between a balloon the size of your fist and one the size of a house. We can even add to the balloon an analogy for the cosmological redshift. Imagine a wave of light (or any electromagnetic wave) drawn on the balloon while it is small. As the balloon expands, the wavelength is stretched or reddened. In the real universe, light travels through expanding space, and we see a redshift that increases with the distance light has traveled.

We can best understand the shape and expansion of the universe by measuring **space curvature** through the deflection of light. If light and all other forms of electromagnetic energy have an equivalent mass, given by $E = mc^2$, then light should respond to space curvature just as particles do. A light beam sent across an elevator that is accelerating through space is deflected by a tiny amount (Figure 17-3a), because during the time it takes to cross the elevator, the elevator has moved. However, Einstein showed that this situation is indistinguishable from an elevator at rest on Earth's surface (Figure 17-3b). The same amount of deflection is predicted because of the gravity of Earth! Mass curves space, and both radiation and particles follow the trajectories dictated by the curvature.

We saw in Chapter 13 that space can be *locally* curved. The observation that starlight is deflected around the edge of the Sun by 1.8 seconds of arc (only 0.1 percent of the Sun's angular diameter!) was a dramatic confirmation of the general theory of relativity. Gravitational lensing occurs because a galaxy or a cluster warps space and causes the distortion of light from a background quasar or galaxy (see Chapter 16). In the extreme example of a black hole, space is so highly curved that it is "pinched off," and matter and radiation are trapped within an event horizon. However, general relativity also allows for the possibility that space is *globally* curved by all the matter and energy in the universe. To understand this, we must explore the difference between Euclidean and non-Euclidean geometry.

Newton's gravity relied on the familiar three-dimensional geometry of Euclid. In Euclidean geometry, space is flat, and

and no bead is at the edge. Although the space is expanding, the beads remain the same size. In our universe, although galaxies and clusters are moving farther apart, their internal gravity keeps them from expanding in size. These regions of

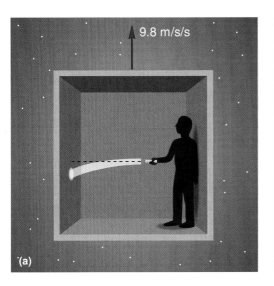
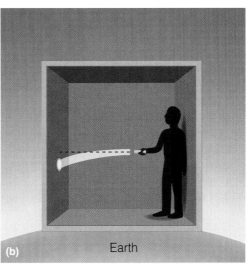

FIGURE 17-3
The gravitational deflection of light. **(a)** In an elevator accelerating through space at 9.8 m/s², a light beam will be deflected because of the motion of the elevator. **(b)** The same amount of deflection will occur in an elevator at rest on Earth's surface because energy has an equivalent amount of mass that must respond to gravity. The amount of deflection is greatly exaggerated here; in practice, it is a tiny fraction of a millimeter.

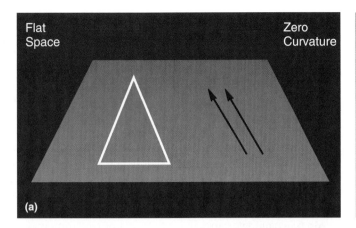

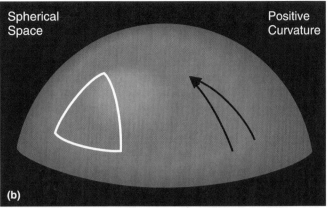

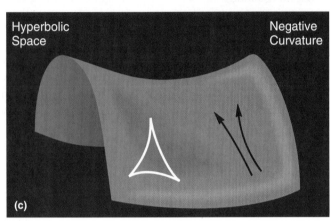

FIGURE 17-4
Two-dimensional analogies for different curvatures of space. **(a)** Euclidean geometry describes flat space, in which the angles of a triangle add up to 180°, and parallel lines never meet. **(b)** Spherical space has positive curvature. Angles in a triangle add up to more than 180°, and parallel lines converge. **(c)** Hyperbolic space has negative curvature; a saddle is an example of this shape. Angles in a triangle add up to less than 180°, and parallel lines diverge.

two-dimensional surfaces have no curvature (Figure 17–4a). The sum of the angles in a triangle is 180°, and parallel lines or beams of light will never meet. Euclidean space is flat.

In the late 1800s, mathematicians in Germany, Italy, and Russia became fascinated with types of geometry quite different from Euclidean geometry and that of everyday experience. None of those mathematicians dreamed that their esoteric work would be applied to the field of cosmology. Two classes of non-Euclidean space exist: spherical and hyperbolic. A positively curved geometry in our analogy is like the surface of a sphere (Figure 17-4b). The sum of the angles in a triangle is greater than 180°, and parallel lines or beams of light converge, so this spherical space is also called a "closed" space. Slightly less familiar is a negatively curved geometry, which in our analogy is shaped like a saddle or a hyperbola in two dimensions (Figure 17–4c). The sum of the angles in a triangle is less than 180°, and parallel lines or beams of light diverge, so this hyperbolic space is also called an "open" space. Table 17–1 summarizes the three types of geometry: Euclidean (or flat), spherical, and hyperbolic.

We will use a two-dimensional analogy for curved space, because experience and intuition help in understanding the two-dimensional situation, whereas the three-dimensional situation is difficult to grasp without mathematics. The surface of a flat or open space is infinite. Just imagine a sheet that continues forever. (In three dimensions, this corresponds to an infinite volume.) By contrast, the surface of a closed space is finite, and so is its volume. Earth's two-dimensional surface, for example, is just such a finite, closed space. However, Earth is also unbounded: It has a definite area, but you can travel in one direction forever without coming to an edge. By analogy, we can imagine the universe as a finite, closed space in which the galaxies stretch into space in every direction, but where there is nevertheless no edge. This analogy answers the age-old paradox of Archytus and other Greek thinkers: The universe can be finite *and* unbounded.

How can we prove that this analogy truly represents the universe? Everyday experience gives us no clue as to whether or not space is curved. Similarly, out on the desert or on the ocean, the planet we live on appears to be flat (Figure 17–5). No local surveying technique would show any departure from Euclidean geometry. However, observations over a large distance can indeed measure the curvature. If you were traveling along Earth's equator from east to west, you could make a right turn (a 90° angle) and you would be traveling directly toward the North Pole. At the North Pole, if you made another

TABLE 17-1
Euclidean and Non-Euclidean Geometries

TYPE OF SPACE	CURVATURE	VOLUME OR AREA	SUM OF ANGLES IN A TRIANGLE	PARALLEL LINES OR BEAMS OF LIGHT
Euclidean	Zero	Infinite	180°	Stay parallel
Spherical	Positive	Finite	Greater than 180°	Converge
Hyperbolic	Negative	Infinite	Less than 180°	Diverge

FIGURE 17-5
Earth is curved, but the curvature is not apparent when we view a small fraction of the size of the planet. The ocean horizon appears flat, not curved. Similarly, the geometry of the universe might appear to be flat on small scales even though it might be curved on large scales. (SOURCE: WKH.)

right turn from your direction of arrival, you would be traveling back down toward the equator. At the equator, if you made a third right turn, you would be traveling along the equator again and would arrive back where you started. The sum of the angles in your triangular journey is 270°—proof that Earth does not have a flat surface! Also, if you travel straight in any direction on Earth's surface and travel far enough, you will eventually return to your starting point. We cannot duplicate these experiments in the three-dimensional universe, but you will see later in the chapter that astronomers have invented clever ways to measure space curvature.

The Big Bang Model

How did the universe begin? Many cultures have avoided the notion of an origin of the universe by placing their creation myths in a cycle. Buddhist and Hindu legends measure the birth, death, and rebirth of the universe in units of 4 trillion years, which is a day in the life of Brahma. At night, all matter is absorbed into the spirit of the sleeping Brahma. At dawn, when Brahma awakes, matter reappears and the cycle continues. The Greek Stoics saw the universe as being created from fire, only to be destroyed by fire, and so on. Cosmological cycles are also found in the cultures of the Maya and the Aztec.

The modern theory of the origin of the universe starts with the idea of expanding space. Georges Lemaître was a priest and a mathematician, an unassuming man who had beaten the giants of physics to the punch in deducing that the universe could be expanding. In 1929, Lemaître was the first to hypothesize a revolutionary idea: At one time, the universe might have been as small as an atomic nucleus. He proposed that the universe derived from a *cosmic singularity*, "a day without a yesterday" when the universe was infinitely small and infinitely curved, and all matter and energy were concentrated in a single point. Many astrophysicists found the idea bizarre and distasteful, and English theorist Fred Hoyle disparaged the idea with the name "big bang." The label stuck, and scientists continue to call the description of the creation of the universe the **big bang model**.

The big bang model forces us to consider the idea of cosmic evolution. The universal recession of galaxies implies that the universe is evolving, and that it has not always been in the same state. Our observations of galaxies represent a single frame in a movie that has been playing for billions of years. Play the movie backward, and what would we see? All the galaxies move closer together. As the universe gets smaller, its volume contracts until galaxies are crushed together in a tiny universe. Galaxies and stars break down into a seething hot gas. The universe tends toward a state of infinite temperature and density. Lemaître described the big bang model lyrically in one popular account: "The evolution of the world could be compared to a display of fireworks just ended—some few red wisps, ashes, and smoke. Standing on a well-cooled cinder we see the slow fading of the suns and we try to recall the vanished brilliance of the origin of the worlds." Thinking of the big bang as an explosion is tempting, but it is also misleading. In an explosion on Earth, debris flies through space. In the big bang, the initial singularity contains *all* space and matter. Time itself begins with the big bang. The evolution of the universe is the unfolding of time and space from a condition of incredible heat and density to a cold and enormous state billions of years later.

The scientific story of creation says that everything—you and Earth and the Sun, the Milky Way and all the billions of galaxies—emerged from a tiny dense dot of energy and matter that unfolded into the universe we see now. It sounds as fantastic as any of the creation myths of older cultures. How do we know that the big bang actually occurred? There are three primary pieces of evidence:

1. Galaxies are taking part in a universal expansion, as indicated by the linear relationship between distance and redshift (the Hubble relation).
2. The abundance of the lightest elements can be explained by fusion in the universe when it was young, dense, and hot (cosmic nucleosynthesis).
3. Space is filled with the radiation from the early hot phase, now diluted and reduced in energy to the level of microwaves (cosmic microwave background).

We will consider each of these pieces of evidence in turn. But first, let us examine the important assumptions that astronomers must make in order to pursue cosmology.

The Cosmological Principle

Astronomers make certain assumptions when they study the universe as a whole. These assumptions may be difficult to prove or verify in practice, but they form an essential starting point for cosmology. The first is the idea that the laws of physics can be applied across the universe. It is a very bold assumption, because our laws of physics can only be precisely determined in laboratories on Earth, and they may not apply exactly over all time and space. As we noted in Chapter 15, Hubble had to assume that Cepheid variables always worked the same way in order to demonstrate that many of the nebulae were distant galaxies. Astronomers are quite confident that physics is not wildly different elsewhere in the universe. We see the same types of stars and galaxies everywhere we look. We see spectral lines from the same elements billions of light-years away that we do in nearby stars. These observations lend support to the ancient Greek idea that a rational order governs the universe.

Astronomers also assume something called the **cosmological principle**—the idea that the universe is everywhere homogeneous and isotropic (recall that homogeneous means uniform or evenly distributed). From our discussion of galaxy clustering in the last chapter, it is clear that this is not strictly true. If gravity had not formed clumps on the scales of planets and stars and galaxies, we would not be here! In cosmology, however, "homogenous" does not mean that all regions of space should appear identical or be smoothly filled with particles; it means that the same types of structures—stars, galaxies, clusters, and superclusters—are seen everywhere.

Figure 17–6 illustrates a homogeneous universe and a hypothetical example of an inhomogeneous universe. We saw in the last chapter that the universe *does* appear smooth or homogeneous on scales larger than about 300 Mpc. Viewed up close, a beach consists of grains of sand and shells and pebbles of many different sizes. From afar, all we see is a beach. The universe is isotropic if it looks the same in all directions. In other words, no observation can be made that will identify an edge or a center. Figure 17–7 shows an isotropic universe and a hypothetical example of an anisotropic universe. The concept of isotropy is supported by the fact that galaxies do not bunch up in any direction in the sky, and by the fact that we observe the same Hubble relation in different directions in the sky.

It is difficult to test the cosmological principle. The isotropy of the universe is reasonably well confirmed, because observers looking in different directions from Earth see essentially the same motions and structures. However, we cannot *test* homogeneity because we cannot travel to distant locations to see if things look any different. When we do look to great distances, we are also looking back to a time when the universe was hotter and denser, so the situations may not be comparable. All available evidence supports this principle, but our degree of certainty is not very high.

Cosmologists are like craftsmen designing a toy model of the universe. If the model is too realistic and tries to contain everything, it will be hopelessly complicated. If it is too simplified, it will not represent essential features of the universe. The assumption of homogeneity is particularly optimistic, because it relegates all the rich structures of matter, the stars, galaxies, and clusters to details of the model! In effect, we are

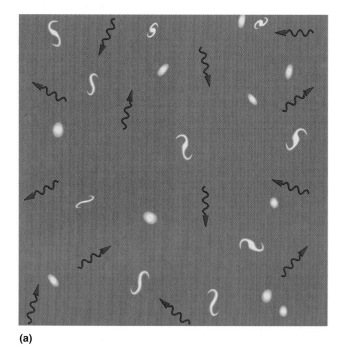
(a)

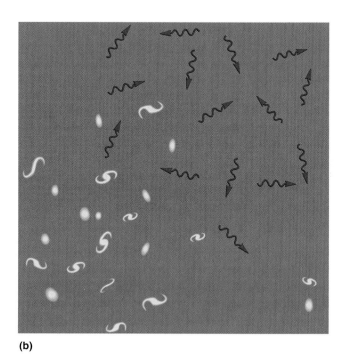
(b)

FIGURE 17–6
Homogeneous and inhomogeneous universes compared. **(a)** In a homogeneous universe, matter and radiation are distributed uniformly over large scales. The contents of any two nearby volumes of the universe are the same. **(b)** In an inhomogeneous universe, matter and radiation are not distributed uniformly. Different volumes can have quite different contents.

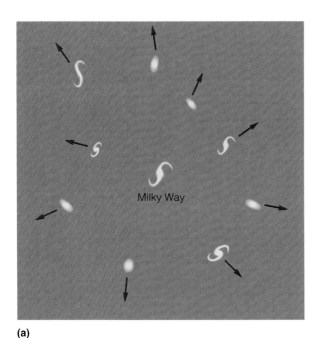 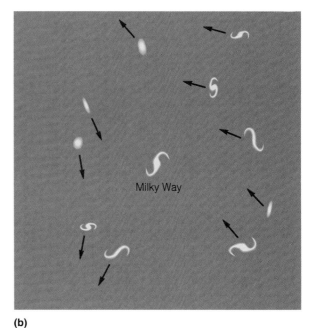

FIGURE 17-7
Isotropic and anisotropic universes compared. **(a)** In an isotropic universe, space has the same properties in any direction. The expansion is smooth and uniform, and the same Hubble relation is measured in any direction. **(b)** In an anisotropic universe, motions of galaxies may be systematic, but the motions are different in different regions of the universe. A Hubble relation is not observed.

using galaxies as markers to show us how space is behaving. We must also distinguish between the *observable* universe and the *physical* universe. Our view of the universe is limited to the region from which light has time to reach us in the age of the universe. Our view of the universe is not limited by space—we do not run out of galaxies or see an edge; it is limited by time. In other words, there are distant regions whose light has not yet reached us. This limitation means that the physical universe (all that there is) may be much larger than the observable universe (all we can see). Despite these limitations, modern cosmology is successful in explaining the basic features of the universe. Our bold assumptions have been rewarded with increased understanding.

The Expansion of the Universe

The Hubble relation is our primary evidence for the expansion of the universe. When Einstein, Lemaître, and others solved the equations of general relativity, they were able to describe how the size of the universe has changed with time. Astronomers use the symbol R to represent the scale or size of the universe at any time. You can think of R as the size of the universe, but more accurately it represents the distance between any two well-separated places. The cosmological principle states that any two points are moving apart at the same rate. Thus, the entire history of the universe is described by the way R varies with time. Since the universe is expanding, R has been continuously increasing for billions of years. Remember that R describes the expansion of space that carries galaxies apart; the galaxies themselves are not expanding. Galaxies are just markers of expanding space.

We now have a better way to define the cosmological redshift. If R is the scale of the universe at any time and R_0 is the scale of the universe now, then the redshift is the ratio of the present scale to the previous scale, minus 1. In equation form, $z = (R_0/R) - 1$. We see regions near us in space as they are now, so $R = R_0$ and $z = 0$. Remember that looking out in space corresponds to looking back in time. But the universe was smaller in the past, so a distant region of space has $R < R_0$ and $z > 0$. Conceptually, light left distant objects when the universe was smaller, and the waves have since been stretched by the expansion of space. We can now relate the expansion of space to the redshifts of distant galaxies and quasars discussed in Chapter 16. The light from a distant galaxy might have been emitted when the universe was half its present size. Using the definition of redshift given above, $R_0 = 2R$ and $z = 1$. The light from the most distant quasar was emitted when the universe was one-sixth of its present size. In this case, $R_0 = 6R$ and $z = 5$. Ancient light is light that has been reddened by the expansion of space.

We can also see the connection with the definition of the Doppler effect from Chapter 9. A redshift caused by the Doppler effect is defined as $z = \Delta\lambda/\lambda$, where $\Delta\lambda$ is the difference between the wavelength observed and the wavelength emitted. A cosmological redshift is $z = (R_0 - R)/R = \Delta R/R$, where ΔR is the difference between the scale of the universe observed now and the scale of the universe when the light was emitted. However, there are crucial distinctions between Doppler redshifts and cosmological redshifts. The Doppler effect applies to waves of any kind traveling through a medium. A cosmological redshift is caused by the

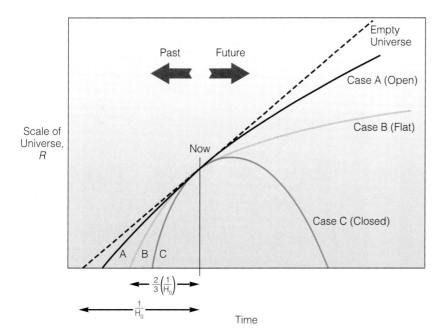

FIGURE 17-8
The change in the scale of the universe with time. The scale R represents the distance between any two points in an expanding universe. Case A is an open universe with less than the critical density, which will expand forever although it will continue to decelerate. Case B is a critical density universe, which will continue to expand at a decreasing rate toward a maximum size far in the future. Case C shows a closed universe with a density greater than the critical density. This universe will eventually recollapse. The dashed line represents an empty universe in which the expansion is not slowed by gravity. The age of an open universe is between $\frac{2}{3}(1/H_0)$ and $1/H_0$, and the age of a closed universe is less than $\frac{2}{3}(1/H_0)$.

expansion of the medium itself. The Doppler effect relates the redshift to the speed of the waves; at low redshifts, we saw that $z = v/c$. By contrast, the cosmological redshift is not related to the speed of light at all. The limitation imposed by special relativity—that nothing can travel faster than the speed of light—does not apply on the global scale of the universe, which is governed by general relativity. As a result, there can be remote regions of the universe that are moving apart faster than the speed of light! We will return to the consequences of this remarkable idea later in the chapter.

Figure 17-8 shows how the scale of the universe changes with time in the big bang model. The point at which the curves all meet represents the present. It takes only two numbers to describe all the possible models. The first is the current expansion rate, given by the Hubble constant (see Chapter 16). The second is the mean density of the local universe. In all possible expanding universes, the rate of expansion *decelerates* (slows down) with time, because galaxies are pulling on all other galaxies. The strength of the deceleration depends on the mean density of matter. The deceleration is also related to the curvature of space, which determines the fate of the universe. It is a fundamental consequence of general relativity that the structure, or curvature, of space is related to the amount of matter in the universe.

You can see in Figure 17-8 that the expansion rate changes with time, as measured by the slope of the curves at any epoch. The expansion rate we measure right now is the Hubble constant, H_0. When the universe was denser and smaller, it had a higher expansion rate, as shown by the steeper slope. If the universe had nothing in it, there would be no gravity to slow down the expansion. The result would be expansion at a constant rate—the straight dashed line in Figure 17-8. In an almost empty (i.e., low-density) universe, the deceleration is small, and R increases almost linearly with time. A low-density universe has negative space curvature in general relativity; it is called an *open universe,* and it expands forever. At a certain critical density, the universe continues to expand to some maximum size at an ever-decelerating rate, taking an infinite amount of time to come to a halt. This special case has zero space curvature; it is called a *flat universe*. A universe in which the mean density is above the critical density has a positive curvature and is called a *closed universe.* The mutual attraction of matter in such a universe is eventually enough to overcome the Hubble expansion. After R reaches a maximum value, the universe will collapse. As space begins to contract, all the galaxies will begin to rush toward each other and show blueshifts.

The critical density divides universes that will expand forever from universes that will eventually collapse. Thus, astronomers characterize the big bang model by a **density parameter** (Ω_0), which is the ratio of the observed density to the critical density. The density parameter is a dimensionless number. If $\Omega_0 < 1$, the universe is open and will expand forever. If $\Omega_0 > 1$, the universe is closed and will collapse. Remember that there are many possible values of the mean density that correspond to an open universe (in Case A in Figure 17-8, only one is plotted). Likewise, there are many possible values of the mean density that correspond to a closed universe (Case C in Figure 17-8; again, only one is plotted). However, the special case of a flat universe occurs only if the mean density equals the critical density, $\Omega_0 = 1$.

Consider the analogy of a rocket launched from the surface of Earth. We know that the escape velocity is about 11 km/s. A rocket launched with an initial velocity of less than 11 km/s will decelerate as it rises. Earth's gravitational attraction will eventually overcome the upward velocity and force the rocket back to the planet's surface. On the other hand, a rocket with an initial velocity above 11 km/s will escape from Earth forever. Of course, the rocket will continue to slow down since the planet's gravity has a long reach, but it will never reverse its direction and fall back to Earth. In our universe, the Hubble constant is analogous to the launch velocity of the rocket, and the mean density to the mass of the planet. For each possible launch velocity of a rocket, there is

TABLE 17-2
The Geometry of the Universe

TYPE OF UNIVERSE	CURVATURE	DENSITY PARAMETER (Ω_0)	FATE OF UNIVERSE
Flat	Zero	1	Expands toward a maximum size
Closed	Positive	Greater than 1	Reaches a maximum size and collapses
Open	Negative	Less than 1	Endless, decelerating expansion
Empty	Negative	0	Endless and constant expansion

a mass of planet where that velocity will just equal the escape velocity. Conversely, for every planet there is a single velocity that will allow the rocket to just escape gravity's pull. Larger planets have larger escape velocities. By analogy, for every possible value of the Hubble constant, there is a mean density that will be just enough to stop the universal expansion and close the universe. A larger Hubble constant requires a larger mean density to close the universe. Table 17-2 shows how the density parameter relates to the curvature of space and the future of the universe.

We can get an estimate of the age of the universe by tracing the expansion back to the time when all galaxies were on top of each other ($R = 0$). Space had zero size when the universe formed. In a Science Toolbox in Chapter 16, we showed that a linear Hubble expansion leads to an age estimate of $1/H_0$. For the current best estimate of $H_0 = 65$ km/s/Mpc, this age is roughly 15 billion years. We can now see that any expansion will have a deceleration, so the age estimate of $1/H_0$ is correct only in the artificial case of an empty universe (represented by the dashed line in Figure 17-8); that is, 15 billion years is an *upper limit* to the age of the universe in the big bang model. Mathematically, it turns out that a flat universe has an age of $2/3(1/H_0)$, or about 10 billion years. Therefore, the big bang model predicts that the universe will have an age between 10 and 15 billion years if the geometry is open, and less than 10 billion years if the geometry is closed.

There is one obvious and crucial test of the big bang model. We can measure the age of the oldest objects in the universe and compare that value with the age predicted by the model. Clearly, no objects in the universe should be older than the age given by the model. We will discuss this test later in the chapter.

Cosmic Nucleosynthesis

The abundance of light elements is one of the primary pieces of evidence for the big bang. In Chapter 13, we learned how stars turn light elements into heavy elements. Stars are the fusion factories that explain the cosmic abundance of elements (see Figure 13-1). However, a careful calculation of the rate at which stars turn hydrogen into helium reveals a mystery. About 25 percent of the mass of gas in the universe is made of helium. Yet fusion in main-sequence stars such as the Sun cannot come close to making this amount of helium over the lifetime of the universe. Helium shows a uniform distribution throughout the Milky Way and in other galaxies. By contrast, the abundance of heavier elements decreases with distance from the center of the Milky Way and is correlated with the number of supernovae. In other words, there is too much helium in the universe to be explained by stellar fusion. Where did the extra helium come from?

As we have seen in the opening story, Russian physicist George Gamow began speculating about the consequences of the big bang model in the 1940s. He realized that the density and temperature predicted for the early phases of the big bang would provide just the right conditions to produce helium nuclei by the fusion process. The creation of light elements such as helium in the big bang is called **cosmic nucleosynthesis**. Let us see how it worked.

In the first seconds of the universe, the temperature was billions of degrees. This was too hot for nuclei to stick together, so the universe was a dense hot broth of radiation and particles. The entire universe was contained in a volume about the size of the Sun! After about a minute, when the temperature had fallen to a billion degrees, nuclear reactions began to take place. Neutrons and protons combined to form deuterium nuclei, symbolized as ^{2}H and sometimes called heavy hydrogen. Deuterium can capture another neutron to make tritium (^{3}H, one proton and two neutrons) or another proton to make helium-3 (^{3}He, two protons and one neutron). After one more stage, helium-4 nuclei (^{4}He) were created, using up almost all the available neutrons. This process was complete 4 minutes after the big bang, by which time about 25 percent of the mass of the universe had turned into helium. During the next half hour, tiny amounts of lithium-7 (^{7}Li) and beryllium-7 (^{7}Be) were created. After this, the reactions stopped. The universe became too cool and diffuse to synthesize heavier elements. In the simple big bang model, all heavier elements are produced much later, in the interiors of stars.

Figure 17-9 is a comparison between theory and observation for four light elements: helium-4, deuterium, helium-3, and lithium-7. These calculations were carried out for a range of densities in the early universe; in the plot, these are converted to densities at the present time. There is a narrow range of density for which *all four* calculated curves agree with the corresponding observations. Since the four types of elements cover an enormous range in cosmic abundance and are found in a wide range of environments in the universe, this result is a strong confirmation of the big bang model. We thus have direct evidence of processes that occurred in the first few minutes after the creation of the universe!

Cosmic Background Radiation

Radio observations provide the third piece of evidence for the big bang. As we saw in the chapter's opening story, the **cosmic background radiation** was discovered accidentally by the radio engineers Arno Penzias and Robert Wilson. The microwave "hiss" that they measured did not come from any particular source in the sky. It was smooth and completely

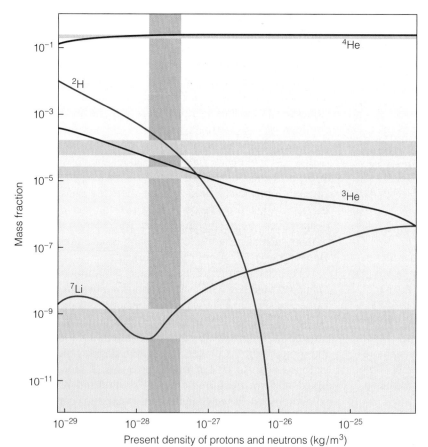

FIGURE 17-9
The predicted abundance of light elements from nucleosynthesis calculations (solid curves) compared with observational bounds on the abundances (light green horizontal bars). There is a relatively narrow range of density (dark green vertical bar) for which the observations of *all four* elements agree with the model. (SOURCE: Adapted from T. Walker et al.)

uniform across the sky. Observations of the cosmic background radiation have continued, with the most spectacular results coming from the Cosmic Background Explorer (COBE) satellite, launched in 1989. Observations of this radiation are so central to our understanding of the big bang model that a dozen experimental groups are making microwave measurements, using balloon-borne observatories or telescopes at dry sites like the high Antarctica plateau. Two ambitious new satellites are planned.

This radiation has two striking features. First, the spectrum is thermal, falling with great accuracy along a curve that defines a temperature of 2.726 K (Figure 17-10). Note that we can quote the temperature to four significant figures and an accuracy of a few thousandths of a degree—this is

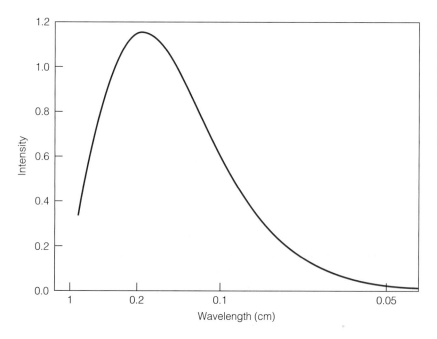

FIGURE 17-10
A spectrum of the microwave background radiation, as measured by the Cosmic Background Explorer (COBE) satellite. The radiation almost perfectly matches a thermal spectrum with a temperature of 2.726 K. Deviations from a thermal spectrum are less than 0.1 percent. This radiation dates back to 300,000 years after the big bang. (SOURCE: Courtesy COBE Science Working Group/NASA).

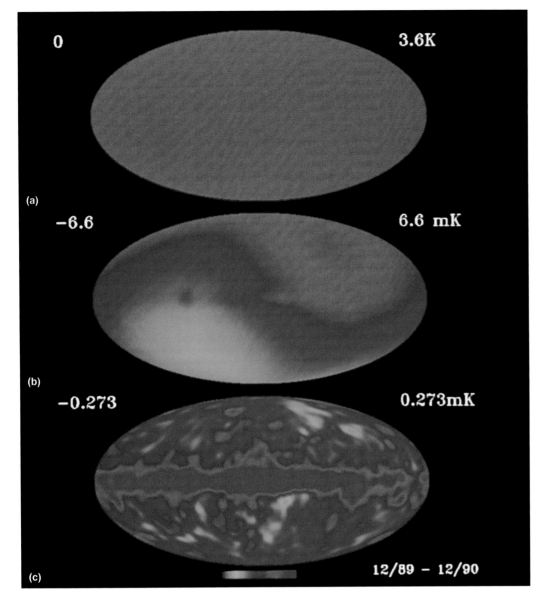

FIGURE 17–11
Measurements of the microwave background radiation made by the COBE Differential Microwave Radiometer (DMR). **(a)** The microwave background radiation is almost perfectly smooth. In this representation, where blue is 0 K and red is 3.6 K, no structure is seen. The mean temperature is 2.726 K. **(b)** In this map, the temperature range between blue and red is 250 times finer; therefore, blue is only 0.0066 K cooler than the mean temperature, and red is only 0.0066 K hotter than the mean temperature. Variations of a few thousandths of a degree can be seen. The microwave sky is slightly hotter in one direction and slightly cooler in the other direction, due to the overall motion of the Local Group of galaxies. **(c)** After subtracting the signal caused by the motion of the Local Group, an even finer temperature scale can be used. In this map, blue is 0.000273 K cooler than the mean temperature, and red is 0.000273 K hotter than the mean temperature. The plane of the Milky Way can be seen running across the center of the map. Away from the plane of the Milky Way is a noisy signal of less than a ten-thousandth of a Kelvin, due to the fluctuations from which galaxies eventually formed. (Source: NASA.)

the most accurately determined number in cosmology. The cosmic background radiation has the most perfect thermal spectrum observed anywhere in nature. Since astronomers believe that the radiation dates from about 300,000 years after the big bang, this means that the universe was in thermal equilibrium at that time. Second, the radiation is almost totally isotropic, meaning that it has the same intensity in all directions (Figure 17–11a). If we represent the intensity of the microwave radiation by the surface of a pond 100 m across, then the biggest ripples are only a few millimeters high.

There are two slight modifications to this isotropy. The radiation does reflect the Sun's motion around the center of the

THE BIG BANG MODEL

galaxy, the galaxy's motion in the Local Group, and the Local Group's motion toward the Virgo cluster. These combined motions cause a slight Doppler shift with respect to the cosmic background radiation (Figure 17–11b). The background is slightly warmer, or blueshifted, in the direction we are moving (toward the constellation Leo) and slightly cooler, or redshifted, in the opposite direction (toward the constellation Aquarius). The other departure from isotropy is the presence of tiny ripples, or fluctuations, in the intensity, which were revealed by COBE (Figure 17–11c). These fluctuations of less than 0.001 percent represent the seeds of galaxy formation. We will discuss the formation of structure later in the chapter.

We now have an answer to the question we asked earlier: Where was the big bang? The big bang was and is everywhere! Every particle in your body was once a part of this cauldron, along with every particle in Earth, the Milky Way, and all the other galaxies astronomers have discovered. Space is filled with the redshifted radiation from the big bang—a hundred million microwave photons in every cubic meter. In every breath you take, you inhale 100,000 or so of these ancient photons. The photons are so feeble that this radiation does not present a health hazard; the power is only 10^{-5} watts, or a ten millionth of the luminous intensity of a lightbulb. To see this radiation for yourself, tune your TV between channels. About 1 percent of the noise on the screen is caused by interactions with the cosmic background radiation. The big bang is all around us.

The discovery of the cosmic background radiation is a striking confirmation of the big bang model. The temperature matches that expected for redshifted radiation emitted by a hot gas soon after the universe began. We can see directly back to 300,000 years after the big bang, when the universe was a small fraction of its current age. The uniformity is a direct verification of the cosmological principle, an indication that at least the early universe was homogeneous and isotropic. Most importantly of all, the cosmic background radiation is evidence that the universe has evolved. It is a fossil that tells of a hot, dense, and featureless universe out of which galaxies, stars, planets, and life were forged.

Measuring Cosmological Parameters

With the basic features of the big bang confirmed by observations, cosmology becomes a quest to describe the type of universe we live in. Will the universe expand forever, or will it eventually collapse? Can we detect the curvature of space? What is the age of the universe? Astronomers have struggled to measure the parameters that describe the big bang model.

The Current Expansion Rate

As we saw in the last chapter, astronomers have worked diligently to measure the current expansion rate of the universe, characterized by the Hubble constant. Hundreds of hours of time with the Hubble Space Telescope have been invested in this research. Current measurements give $H_0 = 65 \pm 10$ km/s/Mpc. Remember that a variety of techniques have to be used to step out from the solar neighborhood to the realm of the galaxies. At small scales of about 100 pc, distances are calculated by the direct technique of stellar parallax. At large scales, the problem is that galaxies interact with each other by gravity in clusters, which tugs them away from revealing the pure expansion of the universe. Astronomers must measure distances out to at least 100 Mpc to get a measure of the smooth expansion, knowing that the universe is homogeneous on scales larger than this. Measuring distances over a range of a factor of 1 million is quite a challenge!

We also saw in the last chapter that gravitational lensing provides a direct measure of the Hubble constant on large scales. This technique is independent of the chain of reasoning that goes into conventional distance measurement. Astronomers have devised other clever ways to measure the expansion rate. It is encouraging that all these techniques give roughly the same answer. After 70 years, astronomers are finally homing in on the Hubble constant.

The Curvature of Space

Astronomers have also devised techniques to measure the curvature of space. The theory of general relativity predicts that the universe will have a global curvature according to the total density of matter in it. This curvature is subtle enough that it can be detected only with observations over a significant fraction of the observable universe. To test the curvature of our universe, we cannot measure the angles of gigantic triangles in space. However, we look for distant objects and compare them with nearby objects. As we will see, these measurements are extremely difficult.

One test of the geometry of the universe involves the way that the density of objects varies with distance from the Milky Way. We can use galaxies as markers of expanding space. Using the two-dimensional analogies in Figure 17–12, consider the surfaces with galaxies randomly scattered on them. Now imagine flattening the curved surfaces onto a plane so that we can measure the linear distance between any two points. First look at the map of the flat surface. You are familiar with this type of Euclidean geometry—the area out to any distance R is πR^2. Therefore, the number of galaxies out to a distance R increases in proportion to R^2, or $N \propto R^2$. To force a positively curved surface onto a plane, the edge must be stretched, thereby thinning out the density of galaxies far from the center. Thus, the number of galaxies out to a distance R increases *more slowly* than $N \propto R^2$. To force a negatively curved surface onto a flat plane, the edge must be compressed, which increases the density of galaxies far from the center. Thus, the number of galaxies out to a distance R increases *more quickly* than $N \propto R^2$. In three dimensions, the analogy holds. In other words, the curvature of the three-dimensional universe is revealed by whether the number of galaxies per cubic megaparsec increases more slowly or more quickly than $N \propto R^3$.

To measure the curvature of space, we can actually count the number of galaxies per cubic megaparsec at different

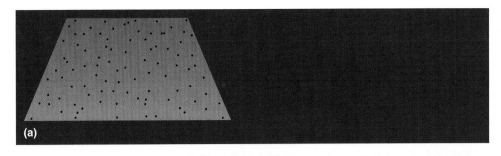
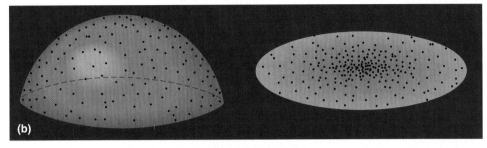
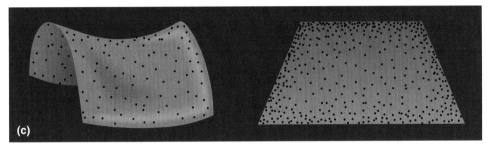

FIGURE 17-12
Two-dimensional analogies demonstrating that the distribution of objects depends on the curvature of space. **(a)** A flat, or zero-curvature, universe is represented by a flat sheet marked with a random distribution of points. The number of points increases as the square of the distance from the center of the sheet. **(b)** A closed, or positive-curvature, universe is represented by a spherical shape randomly marked with points. When the surface is stretched flat, the number of points increases *more slowly* than the square of the distance from the center. **(c)** An open, or negatively curved, universe is represented by a saddle shape randomly marked with points. When the surface is flattened out, the number of points increases *more quickly* than the square of the distance from the center.

distances (or redshifts) from Earth. Figure 17–13*a* shows that one such count revealed a density parameter of 1 or less. This density parameter is equivalent to a flat or open universe and a geometry that may be negatively curved. The results are promising, but there is a serious complication in applying this test that can lead to large systematic errors.

As we saw in Chapter 12, any survey limited by apparent brightness will be far more sensitive to luminous objects than intrinsically dim objects. For example, though the brightest stars in the sky are giants and supergiants, the most common stars in any volume of space are dwarfs (see one of the Science Toolboxes in Chapter 12 for this calculation). Exactly the same situation arises with galaxies. The most common type of galaxy is a dwarf galaxy, but giant galaxies can be seen at much larger distances, so they will be overrepresented in any census. In fact, the bias in favor of luminous galaxies is *worse* than the analogous bias in favor of luminous stars. Stars dim by the inverse square of the distance. Galaxies also dim according to the geometry of space, proportional to $(R/R_0)^2$. By manipulating the expression $z = (R_0/R) - 1$, we can see that $(R/R_0)^2 = (1 + z)^{-2}$. However, galaxies are dimmed by an additional factor of $1 + z$, because of the slowing of the arrival rate of photons. They are dimmed by yet another factor of $1 + z$ from the stretching of the spectral range; the range of galaxy wavelengths observed corresponds to a smaller wavelength range at the time the light was emitted.

The result of all these effects of cosmology is that light from a distant galaxy is dimmed by a factor of $(1 + z)^{-4}$. Therefore, a galaxy at a redshift of $z = 1$ appears $2^4 = 16$ times fainter than an identical galaxy in the nearby universe. A distant galaxy at a redshift of $z = 3$ appears $4^4 = 256$ times fainter than an identical local galaxy. When astronomers look for high-redshift galaxies, they tend to select more luminous galaxies than when they look locally. The bottom line is that it is very difficult to compare identical samples of galaxies at different redshifts, which is the basis of the test shown in Figure 17–13*a*.

A second test of the universe's geometry relies on the way that angles change in curved space. In flat, Euclidean space, more distant objects have smaller angular sizes on the sky, with the angular diameter inversely proportional to distance—the familiar small-angle equation. The curvature of space can distort the images of distant galaxies, and since it is mass that bends light, the distortion increases with the density of the universe. The curves in Figure 17–13*b* show the change in the angular size of an object with increasing distance, or redshift. You can see that the angular size of a distant object in a closed universe actually *increases* with redshift! How can an object appear larger as it gets farther away? If you think of the universe as a gigantic lens, then the gravity in a high-density universe can actually magnify the image of a distant object. We can also understand this bizarre effect by recalling that the universe was smaller at a high redshift.

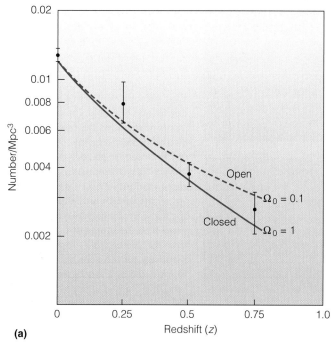

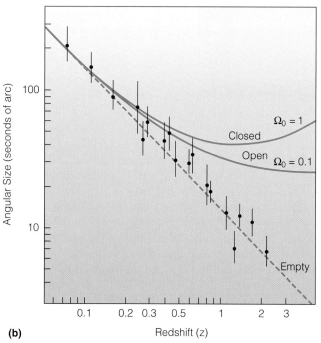

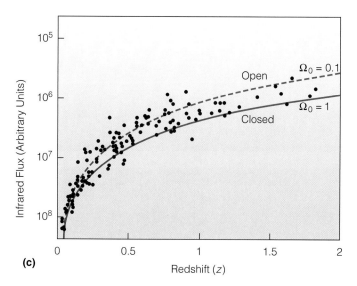

FIGURE 17–13
Three observational tests of the cosmological model. **(a)** The number of galaxies per cubic megaparsec depends on the mass density, or space curvature, of the universe. Here the density of galaxies (shown as dots with error bars) is plotted against redshift. At high redshift, a closed universe ($\Omega_0 > 1$) has a much smaller volume than an open universe, which is indicated by the much lower predicted density of galaxies at high redshift. The curves are derived from the big bang model. (SOURCE: Adapted from E. Loh and J. Spillar.) **(b)** The apparent angular size of an object of fixed physical size depends on the mass density, or space curvature, of the universe. The solid curves show the relationship between angular size and redshift for open- and closed-universe models. The data points show the size of the radio structures of active galaxies at different redshifts. The data do not allow a test of the cosmological model because the radio structures are *evolving* with cosmic time; see the discussion in the text. (SOURCE: Adapted from V. Kapahi.) **(c)** The apparent brightness of giant elliptical galaxies depends on the mass density, or space curvature, of the universe. The curves show the difference between the expectations for open- and closed-universe models. The cosmological test requires a good model for the evolution of galaxies. (SOURCE: Adapted from S. Lilly.)

Galaxies, however, have remained the same size during the expansion. At high redshift, a galaxy of a particular size would subtend a larger angle in what was then a smaller universe.

So far, astronomers have been unable to use the way angular diameter varies with redshift to measure the curvature of space. Part of the reason is the fact that galaxies do not have sharp edges. It is very difficult to accurately measure the size of a faint, fuzzy object in the distant universe. However, a more fundamental problem is **cosmic evolution.** Galaxies at high redshift are seen when they were young. A high redshift corresponds to a large look-back time, and cosmic evolution requires us to always view younger objects at larger distances. If galaxies are built up from smaller pieces, they might well have been intrinsically smaller when they were young. Once again, our comparison of nearby and distant objects is a comparison of apples and oranges.

A third test involves the way that apparent brightness changes with redshift. Our starting point is the fact that the distance to an object depends on the cosmological model. In the previous chapter, we saw that distance relates to redshift by the approximation $d \approx cz/H_0$. However, the effects of space curvature begin to be seen at redshifts of a few tenths, from $z = 0.2$ to $z = 0.3$. By redshifts of $z = 1$ to $z = 2$, the differences between cosmological models are substantial. A flat universe with zero curvature is *smaller* than an open universe with negative curvature. Therefore, any particular redshift in a flat uni-

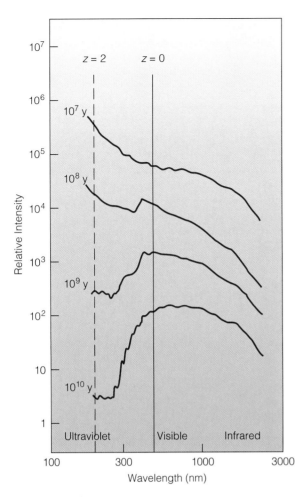

FIGURE 17–14
The optical and infrared spectrum of a burst of star formation at four different times after the star formation ceases. The energy distribution fades and reddens with time. Ten billion years after the burst, the ultraviolet radiation has faded by a factor of 1 million, 3000 times more than the fading of the infrared radiation. The solid vertical line corresponds to an observation of green light for a nearby galaxy. The same observation of a high-redshift galaxy (see the dashed vertical line) is measuring the ultraviolet light that was emitted long ago, because the light has been redshifted since the time it was emitted. (SOURCE: Adapted from S. Charlot and G. Bruzual.)

verse corresponds to a smaller distance than the same redshift in an open universe. Now imagine that we find an object of known luminosity far off in the distant universe. If the universe is flat, that object will be closer and appear *brighter* than if the universe is open. The application of the test requires that we find a particular type of object that can be found over a wide range of redshifts. By seeing how the apparent brightness changes with redshift, we can distinguish between different values of the density parameter, or space curvature.

These observations sound simple, but a serious complication arises because of cosmic evolution. Figure 17–13c shows the apparent brightness test applied to elliptical galaxies. The model curves show that at a redshift of 1, an elliptical galaxy would be brighter by a factor of 2 in a flat universe than in an open, low-density universe. The problem is that a high-redshift galaxy is a very distant galaxy. A very distant galaxy is a young galaxy since its light has taken billions of years to reach us. Look-back time has the unavoidable consequence of forcing us to compare nearby objects with distant ones that are younger. In other words, we cannot make a test of the cosmological model without first making a model of the evolution of galaxy light! Our test is not as direct as we would like.

Astronomers make the best of this situation by choosing the simplest stellar systems for the test. Elliptical galaxies are luminous—visible out to large redshifts—and they usually have old stellar populations. Assuming that most of its stars formed at the same time, an elliptical galaxy will fade and become redder. The galaxy fades quickly at first because the first light is dominated by luminous and short-lived blue stars. The galaxy then fades more slowly, as increasingly lower-mass stars move off the main sequence. Figure 17–14 shows an evolutionary model of a galaxy with a single burst of star formation. The change in energy distribution, particularly the loss of blue and ultraviolet light, is dramatic in the first billion years. Notice that the change in the amount of radiation strongly depends on wavelength. The red and infrared radiation declines by a factor of 100 after the initial burst of star formation, while the ultraviolet radiation declines by a factor of 100,000.

The evolution of stellar populations complicates cosmological tests for two reasons. First, more distant galaxies are likely to be brighter and bluer, simply because they are younger. Second, galaxies are measured over a fixed range of optical wavelengths, usually defined by a filter. Distant galaxies are redshifted, so we observe successively bluer parts of the energy distributions of more distant galaxies. For example, suppose we measure galaxies at a wavelength of 500 nm. A nearby galaxy has the intensity shown at that wavelength in Figure 17–14. However, a galaxy at $z = 2$ has had its ultraviolet light stretched out into visible light over time and space. The wavelength of the radiation we observe in this distant galaxy is $500/(1 + z) = 170$ nm. You can see that a youthful 10^7-year-old galaxy is putting out far more energy at this ultraviolet wavelength. By contrast, an old 10^9-year-old galaxy is putting out far less energy at this ultraviolet wavelength. (We cannot apply this reasoning to a 10^{10}-year-old galaxy because the universe at $z = 2$ is not old enough to contain a 10-billion-year-old stellar population!) We simply do not understand galaxy evolution well enough to apply the cosmological test with confidence.

Astronomers have not succeeded in accurately measuring the curvature of space. You can see in Figure 17–13 that the cosmological models only begin to diverge at $z = 0.2$ to 0.3. This corresponds to a distance of $d \approx cz/H_0$, or 1000 to 1500 Mpc. Multiplying by 3.3 to convert from parsecs to light-years, we see that these are look-back times of 3.3 to 5 billion years. Therefore, the *only* way we can detect space curvature is by comparing old and young objects. The evolution issue cannot be dodged; it is an inevitable consequence of the size of the universe and the finite speed of light. Currently, the most promising cosmological tools are supernovae. These "standard bombs" can be seen out to $z = 1$ and beyond, well past the distance where space curvature should show itself.

SCIENCE TOOLBOX

The Critical Density of the Universe

In the big bang model, the expansion of the universe is slowed down by gravity. If there is enough matter in the universe, the expansion can be overcome and the universe will collapse in the future. The density of matter that is just sufficient to eventually halt the expansion is called the *critical density*. The equation for critical density is

$$\rho_{crit} = \frac{3H_0^2}{8\pi G}$$

You can see that the critical density is proportional to the square of the Hubble constant; a faster expansion requires a higher density to overcome the expansion. We can calculate ρ_{crit} by inserting the gravitational constant, $G = 6.67 \times 10^{-11}$ $N\,m^2/kg^2$, and adopting $H_0 = 65$ km/s/Mpc. We first convert the Hubble constant to metric units: $H_0 = (65 \times 1000) / 3.1 \times 10^{22} = 2.1 \times 10^{-18}$ s^{-1}. Now we can solve to get $\rho_{crit} = 3 \times (2.1 \times 10^{-18})^2 / 8 \times 3.14 \times 6.7 \times 10^{-11} = 7.9 \times 10^{-27}$ kg/m^3 ≈ 10^{-26} kg/m^3. This amazingly low density is equal to about five hydrogen atoms per cubic meter, or a single hydrogen atom in a volume the size of an average TV set.

How well does this density match up to the density of large objects in the universe? The typical distance between large galaxies is about 2 Mpc, so we can calculate what the average density would be if a galaxy like the Milky Way were the only object in a box 2 Mpc across. The mass of the Milky Way, including dark matter, is about 10^{12} M$_\odot$, or $10^{12} \times 2 \times 10^{30} = 2 \times 10^{42}$ kg. The volume of a 2-Mpc cube is $(2 \times 3.1 \times 10^{22})^3 = 2.4 \times 10^{68}$ m^3. The typical density is therefore mass divided by volume, or $2 \times 10^{42} / 2.4 \times 10^{68} \approx 10^{-26}$ kg/m^3. Thus, we can see that the typical spacing of galaxies corresponds to a density roughly equal to the critical density.

The density parameter is the ratio of the observed density to the critical density:

$$\Omega_0 = \frac{\rho}{\rho_{crit}}$$

The density parameter is therefore a number with no units. If $\rho > \rho_{crit}$, giving a value of $\Omega_0 > 1$, the universe is closed and will eventually collapse. If $\rho < \rho_{crit}$, giving a value of $\Omega_0 < 1$, the universe is open, and will expand forever. The special case where $\rho = \rho_{crit}$ and $\Omega_0 = 1$ represents a flat universe, which will expand forever at an ever decreasing rate.

How do the actual measurements of matter density compare to the critical density? We have seen that mass measurements on the largest scales give $\Omega_0 = 0.1$ to 0.3, or a few tenths of the critical density. However, most of this mass is in the form of dark matter! The density of normal matter is actually 100 times lower, corresponding to $\Omega_0 = 0.002$ to 0.003. Therefore, if critical density is five hydrogen atoms per cubic meter, our universe only contains $5 \times 0.002 = 0.01$ hydrogen atom per cubic meter. This density is equivalent to a single hydrogen atom in a volume the size of your living room. The rest of the mass is made up of the ubiquitous dark matter; we do not yet know what kind of particles they are.

How does cosmic background radiation in the universe fit into calculations of critical density? We have seen that the microwave background radiation consists of 10^8 photons per cubic meter. There are 10 billion times more photons than hydrogen atoms in the universe. We can calculate the equivalent mass of all this radiation using the formula $E = mc^2$. The result is that all the microwave photons in space are equivalent to 5×10^{-31} kg/m^3, a density 20,000 times lower than the critical density. In other words, the density parameter from big bang radiation is only $1/20,000 = 0.00005$, far lower than the density parameter from matter. Despite the huge number of microwave photons, the universe is truly dominated by matter.

Despite all the problems, we can draw two conclusions from existing measurements. First, models with a large amount of positive curvature ($\Omega_0 > 1$) are ruled out. This kind of universe is too small, too young, and too curved to fit the observations. Second, the curvature is small enough that it may be consistent with a flat model and a universe of critical density.

The Mean Density

We have seen that observations of distant objects have not thus far yielded an accurate measure of the curvature of space. However, our gravity theory suggests an alternative way of measuring the geometry of the universe. General relativity predicts that the geometry or curvature of space is determined by the mass content of the universe. This suggests that if we measure the mean density of matter in the local universe, we can determine whether it is above or below the critical density needed to overcome the universal expansion. Matter in the universe is lumpy; atoms coalesce into stars, and stars are bound by gravity into galaxies. The key to the concept of mean density is to imagine all the matter in the universe broken up into atoms and *smoothly* distributed through space.

The value of the critical density depends on the Hubble constant, since a faster expansion requires a larger density to overcome it. For our assumed value of $H_0 = 65$ km/s/Mpc, the critical density is roughly 10^{-26} kg/m^3 (see the Science Toolbox above). This tiny number is equivalent to four or five hydrogen atoms in a cubic meter of space, or, analogously, to the density of a grain of sand distributed over the volume of Earth. The universe is fantastically empty! So all we have to do is measure the average density in a large volume and divide it by the critical density. This number is the density parameter, Ω_0. If the mean density is larger than the critical density, $\Omega_0 > 1$, and the universe will eventually collapse. If it is less than the critical density, $\Omega_0 < 1$, and the universe will expand forever. It sounds easy. Just count galaxies, add up their masses, and divide by the volume containing them to get the mean density. Then see if the mean density is larger or smaller than the critical density that we calculate using the best estimate of the Hubble constant. By now, you might suspect that nothing in cosmology is as simple as it seems—and you would be right!

First, we need to know how big a volume to survey. What region of space represents a "fair sample" of the universe? Suppose you were conducting a census of a large country. If you counted the people in a 100-square-mile area, you might get a very low density of people if the area was in the remote countryside. On the other hand, if your 100 square miles included a large city, you would calculate a very high density of people. Neither result would give a good estimate of the average density of people. You would need to take a number of samples covering the full range of living environments. It is the same with galaxies. Suppose we survey a volume of only a few cubic megaparsecs. We might happen to choose a region of the universe with lower-than-average density and get a biased result. In fact, since the average distance between large galaxies is about 1 Mpc, we might be unlucky and not find any galaxy in our volume!

We have seen that the universe becomes homogeneous on scales of 300 Mpc and higher; that is, the number of galaxies does not fluctuate very much. You will notice that 300 Mpc corresponds to a look-back time of about 1 billion years, which is less than 10 percent of the age of the universe. Therefore, a survey to this distance is still considered a "local" measurement, and the issue of galaxy evolution is not important. Do we have to count galaxies across the entire sky to this distance? No, because the universe is isotropic, and we will get the same census in any direction we look. Astronomers use surveys of slices, or "pencil beams," to measure the mean density of galaxies, as described in the previous chapter (see Figure 16–13). A typical survey might have a depth of 200 Mpc and dimensions on the plane of the sky of 5 Mpc in each direction. The total volume is therefore $200 \times 5 \times 5$, or 5000 Mpc^3. The resulting sample will contain several thousand galaxies.

Astronomers have added up the luminosity of galaxies in a large region of space. We saw in Chapter 15 that the mass-to-light ratios of stellar populations are well determined, so the summed luminosity can be converted into a summed mass. Next, the total mass is divided by the volume to derive the average density. Finally, the average density is then divided by the critical density to get the density parameter. The result is a very low number, $\Omega_0 \approx 0.0002$ to 0.003. In other words, stars make up only a few tenths of a percent of the critical density. We can use the cosmological principle to argue that our local volume is typical of every region of space. As remarkable as it sounds, the totality of stars in the universe—around 10^{20} of them contained in some 40 billion galaxies—does not come close to matching the critical density.

We can gain some insights by plotting all the estimates of matter density on scales ranging from the nuclei of individual galaxies to galaxy superclusters. The vertical axis of Figure 17–15 is shown as both the density parameter and the equivalent mass-to-light ratio. As we learned in the discussion on cosmic nucleosynthesis, there is a narrow range of density for which the big bang predicts the correct abundance of light elements (see Figure 17–9). The big bang model predicts this range of density from calculations that include all the conventional subatomic particles: protons, neutrons, electrons, and neutrinos. The density range predicted by nucleosynthe-

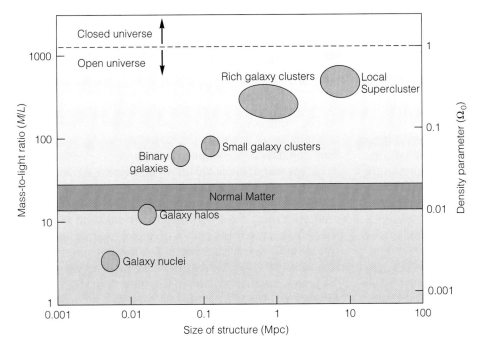

FIGURE 17–15
The amount of dark matter found in the universe on various scales. The mass is shown in two ways: as the mass-to-light ratio in units where the Sun has $M/L = 1$, and as the contribution to the density parameter Ω_0, or the fraction of the mass density needed to close the universe. Measured on the largest scales of superclusters of galaxies, the density parameter gets close to, but does not exceed, 1. The horizontal band shows the density of normal matter, not including dark matter, expected from the big bang model.

sis calculations—the low value of $\Omega_0 = 0.01$ to 0.02 shown as the horizontal band on Figure 17–15—leads us to two conclusions. First, the visible parts of galaxies account for only 15 to 25 percent of the matter predicted by nucleosynthesis calculations. In other words, most of the normal matter in the universe has not yet been detected! Astronomers speculate that this mass will be found as a thin gas lying between the galaxies. They predict that it will be heated to 100,000 K by ultraviolet radiation from young stars and rare but powerful quasars. Second, the density parameter from normal matter is far below the critical density. Normal matter is insufficient to close the universe and cause it to collapse in the future.

However, we cannot therefore conclude that the universe is open, because we have not yet taken account of dark matter. As we saw in Chapter 15, at least 90 percent of the mass of most galaxies is dark. (Remember that dark matter particles are known to interact very weakly with normal particles, so they do not enter into the nucleosynthesis calculations.) Including this mass raises the mean density derived from stars by a factor of 10, implying that $\Omega_0 = 0.02$ to 0.03. This is still only 2 to 3 percent of the mass needed to close the universe. In addition, a lot of dark matter is found on scales larger than individual galaxies, as we saw in the previous chapter. Astronomers weigh dark matter on scales up to 100 Mpc by observing how galaxies are tugged around in response to the way dark matter is distributed. It is as if the galaxies are shiny marbles rolling on an undulating surface of black velvet. The undulations represent the high- and low-density regions of dark matter. The marbles will roll toward concentrations of dark matter and away from regions where dark matter is sparse. A larger excess of density will generate a larger motion. The map of these motions is used to weigh the dark matter. Including this mass raises the mean density by another factor of 5 to 10, giving $\Omega_0 = 0.1$ to 0.3. Figure 17–15 summarizes all the measurements of the density parameter.

Measuring the density parameter is one of the most exciting and difficult tasks in observational cosmology. Astronomers do not consider the density parameter to be nearly as well determined as the expansion rate. Current evidence indicates that the universe contains only 10 to 30 percent of the critical density, indicating a negatively curved geometry. The vast majority of the matter content is dark. Based on this evidence, the universe is open and will expand forever.

The Age of the Universe

We continue our discussion of cosmological parameters by asking the question: How long has it been since the creation of the universe? The big bang model gives a prediction for the **age of the universe.** If we imagine the evolving universe as a movie, the birth of the universe is the time in the distant past when all matter and radiation were crushed in a state of infinite temperature and density. The scale factor, R, was zero. Space had not yet begun to unfold.

Earlier in this chapter, we saw that the age of the universe depends on the current expansion rate and the mean density. A fast expansion rate implies a young universe. Also, a large density implies a young universe. In other words, large values of H_0 and Ω_0 indicate a young universe. Review Figure 17–8 if you are not sure of this reasoning—basically, a large density means that galaxies have caused a larger deceleration of the expansion, which points to a more recent origin. For the currently favored value of $H_0 = 65$ km/s/Mpc, the universe must be less than 10 billion years old if it is closed, and between 10 and 15 billion years old if it is open.

Such reasoning points us to a critical test of the big bang model. Clearly, the universe cannot be younger than the oldest objects it contains. How does the age estimate from the big bang model compare with direct age measurements? Earth's age is well determined, but the Sun contains heavy elements produced in prior generations of stars. A number that will get us closer to the age of the universe is the age of the Milky Way. We saw in Chapter 14 that the oldest components of the Milky Way are globular clusters. Models of stellar evolution yield ages of 9 to 13 billion years. The relevant number is the age of the *oldest* globular clusters, about 13 billion years. We have to add to this the time between the big bang and the formation of the Milky Way. We can only crudely estimate this time scale, since our knowledge of galaxy formation is so rudimentary. However, it is likely to be in the range 0.5 to 1 billion years. We conclude that the age of the universe—independent of the big bang model—is 13 to 14 billion years.

Now we can combine these measurements, recognizing that each one has an error or uncertainty attached to it. Figure 17–16 plots the two observed quantities of the big bang model. The horizontal band shows the likely range of the Hubble constant, $H_0 = 55$ to 75 km/s/Mpc. The vertical band shows the likely range of the density parameter, $\Omega_0 = 0.1$ to 0.3. The dark shaded box is the area that defines the big bang models allowed by observations. On this plot, we can superimpose curved tracks that show the *independent* measurement of the age of the universe. The curved tracks of age neatly intersect the box of allowed big bang models. You can see that if the Hubble constant or the density parameter were much higher, the oldest objects in the universe would be older than the predicted age. As it is, the big bang model passes this important test with flying colors.

The Fate of the Universe

What does the future hold for our universe? Even though the brief flicker of human existence seems inconsequential in the eons of the universe, it is irresistible to ask the question. The outcome depends on whether the universe is open or closed. At the moment, the favored evidence points to an open universe, but a flat or closed universe cannot be ruled out. The fates could not be more different.

If the universe is closed, its rate of expansion will continue to decrease for another 5 to 10 billion years. The universe will pause momentarily at its maximum size and then begin to collapse, like a cosmic sigh. As space shrinks and all galaxies begin to approach each other, universal redshifts will be replaced by universal blueshifts. At first this will happen only to the nearest galaxies because those are the ones we see most recently. Eventually, however, distant galaxies will also be blueshifted. The photons that fill space will be squeezed to higher energies. As the universe accelerates toward its second appointment with a singularity, galaxies will crowd cheek by

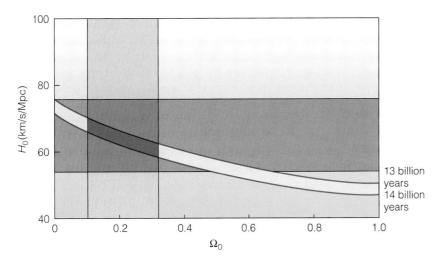

FIGURE 17-16
An important test of the big bang model comes from the requirement that the oldest objects in the universe match the age of the universe determined from the model. This plot shows the two important numbers that define a cosmological model: H_0 (the expansion rate) and Ω_0 (the mean density). The horizontal and vertical bands show the range of these parameters indicated by the best current measurements. The curved lines show the estimated age range of the oldest globular clusters, assuming that they formed at $z = 10$. The region between the curved lines intersects both the horizontal and the vertical bands (darkest shaded region), indicating agreement with the big bang model.

jowl. Humans or their descendants will not be around to care, because the Sun will have bloated into a red giant billions of years previously. The background radiation will be energetic enough to strip atoms of their electrons. Stars and planets will be obliterated. The final state of the contracting universe is often called the "big crunch"—a prospect of crushing finality.

If the universe is open or flat, galaxies will continue to retreat farther and farther from each other. Over many billions of years, stars will use up the store of hydrogen and helium for thermonuclear fusion. More and more mass will become locked up in stellar remnants; less and less will go into massive stars that can return gas into the interstellar medium. The cycle of star birth and death will be broken. After about 10^{13} years, all the nuclear fuel will be exhausted. All matter will be in the form of cold embers, such as white dwarfs, or collapsed objects, such as black holes. Over even longer periods of time, the dead stars within galaxies will interact. Some of the stars will gain energy from the encounters and leave the galaxy, "evaporating" into deep space. Others will lose energy and spiral into the center of the galaxy, becoming fuel for the massive black hole typically found there.

What happens next is highly speculative. Particle physics theories predict that protons are unstable, with a lifetime in excess of 10^{35} years. As the protons gradually decay, they release electrons, positrons, and neutrinos, and a small amount of energy as photons. Since the rules of quantum mechanics allow black holes to emit a small amount of radiation, they will eventually evaporate. The large black holes in the centers of galaxies might last as long as 10^{100} years, but the final state of an open universe would be a thin sea of electrons, positrons, neutrinos, and photons. As the temperature approaches absolute zero, it seems appropriate to call it the "big chill." It is a bleak, if distant, prospect.

The two possible fates of the universe could not sound more different, yet they share an expression of the second law of thermodynamics (see Chapter 13). Each result is a victory for entropy—the increase of disorder. In the "big crunch," the increase of entropy is obvious. Heat is the most disordered form of energy. As the universe approaches its second singularity, all structure is erased, and heat energy increases enormously. By the time of the "big crunch," the entropy in the universe has increased by a factor of 10^{35}. Although it is less obvious, entropy also increases in the "big chill." British physicist Stephen Hawking has shown that black holes not only lock up matter and radiation, they also have enormous entropy. A massive star that dies as a black hole increases its entropy by a factor of 100 million. The fate of an open universe is a disordered sea of particles interspersed with large and small black holes. An open universe increases its entropy by a factor of 10^{23} over its lifetime.

We can even speculate about the connection between the entropy of our universe and our sense that time flows forward. The perception of time as a river that flows in only one direction is called the **arrow of time.** We saw in Chapter 13 that all physical systems tend toward their most probable state, and that the most probable state is a disordered one. Let us use the shuffling of a deck of cards as an analogy for the evolution of the universe. In the first example, the perfectly ordered deck of cards represents a low-entropy universe. As we shuffle the deck, representing the continual interactions of matter and radiation, the deck becomes disordered. This progression can only go one way, since no amount of shuffling will reorder the deck. This scenario shows a clear arrow of time. The second example begins with a moderate degree of entropy—runs of cards in sequence, but also many that are randomly placed. After much shuffling, the ordered sequences will tend to get destroyed. This universe also has an arrow of time. In the third case, the deck is randomized at the start, representing a high-entropy universe. As the deck is shuffled, cards move around but the order remains random; there is no way to look at the different versions of the shuffled deck and say which came earlier and which came later. Perhaps it is the relatively low entropy of our present-day universe that gives us the sense of time's arrow.

The Status of the Big Bang Model

To many people, the story of creation told by astronomers is more fantastic than any legend or myth. It is important to ask these questions: How secure is the big bang model? Are there any other ways of explaining the universe we live in?

In the late 1940s, a group of theorists proposed an alternative cosmology called the *steady state model*. The steady state model proposed that the universe was homogeneous and isotropic over all time. There was no origin, no singularity. To account for the Hubble relation, the theory proposed that atoms were created steadily and spontaneously to fill the voids created by the recession of galaxies. Many scientists saw this hypothesis as an outrageous violation of the principle of conservation of mass, but defenders of the theory pointed out that it is hardly more outrageous than imagining that the whole mass of the universe appeared instantly, as in the big bang model! As mentioned earlier, Fred Hoyle, one of the steady state theorists, coined the term "big bang." Although he intended it as a put-down, the catchy name caught on. The steady state model died after the discovery of the cosmic background radiation, which had no natural explanation in a nonevolving universe.

Other ideas have been proposed and have come up short. The *tired light model* holds that light loses energy as it crosses the vast reaches of space. Redshifts are caused by this effect, rather than by the expansion of space. Observations rule out the tired light model. British physicist Paul Dirac, one of the founders of the quantum theory, proposed the idea of variable gravitation. If the force of gravity had decreased over billions of years, it would simulate some of the effects of an expanding universe. This theory has also been ruled out. In the 1930s, California Institute of Technology physicist Richard Tolman proposed an *oscillating universe model*, in which expansion and contraction follow each other in an endless cycle. However, Roger Penrose and Stephen Hawking later showed that the increase in entropy applies to an oscillating universe too. After many cycles, the universe would be much more chaotic than the one we observe. There is no physical mechanism to account for the oscillation.

We mention these failed models to show that scientists have continued to be ingenious in thinking up ways to explain the universe. The big bang model has survived because it has explanatory and predictive power. A web of evidence supports the idea of a dense and hot early phase of the universe. The big bang follows the precepts of a good theory that we laid out in Chapter 1. However, it would be arrogant of scientists to believe they have the final verdict on the creation event. The history of cosmology indicates that there are likely to be surprises in store for us.

One of the most interesting embellishments of the "vanilla" big bang model is the addition of a **cosmological constant** (denoted by Λ, the Greek capital letter lambda). As we saw earlier in the chapter, the cosmological constant was first introduced by Einstein in a misguided attempt to make a model of a static universe. It behaves like a constant energy density associated with the vacuum of space, independent of the matter and radiation content of the universe. Einstein used the cosmological constant as a repulsive force to balance the attraction of gravity. The addition of a cosmological constant creates a bewildering variety of models of the universe. The universe might have spent a long time at virtually the same size, meaning that the age might be much larger than the current expansion indicates. The universe might be accelerating at the present epoch rather than decelerating. Observations of distant supernovae actually imply that the universe is accelerating; these difficult observations need to be confirmed by other techniques. Most astronomers are uncomfortable with the idea of a cosmological constant, for two reasons. First, they consider it an arbitrary parameter and therefore a blemish on the beautiful theory of general relativity. Second, no known law of physics explains why the vacuum of space should contain enough energy to affect the expansion of our entire universe.

The Early Universe

The big bang is a successful model of the universe. As a result of this success, astronomers wonder how far they can push the model. Can it describe the first fractions of a second after the universe came into being? If so, the early universe would be an unmatched laboratory for studying the structure of matter. The first instants of the big bang reached extraordinarily high temperatures and densities, while all the matter in the universe was squeezed in a gravitational vise. The energies exceed any that can be produced in laboratories on Earth. These conditions may yield clues to the fundamental forces of nature. In this part of the chapter, we will attempt to answer a series of basic questions about the universe. Why is the universe nearly flat? Why is the universe made of matter and not antimatter? Why are there so many more photons than particles in the universe? What drives the expansion? Why are there four forces of nature?

Limitations of Space and Time

As we try to understand the universe, we face limitations of space and time. One important consequence of relativity is that space and time are linked; Einstein talked of a *space-time continuum*. We have already seen evidence of this in cosmology. Since light takes a long time to reach us from distant objects, we see them as they were at a younger age. Although our examples use light, the discussion applies to *all* electromagnetic waves, including radio waves, X rays, and so on. It makes no sense to ask what a quasar looks like "now." Information cannot be transmitted instantaneously in the universe, so we are limited in our knowledge by the finite speed of light. Looking out in space means looking back in time.

We will now introduce the idea of a **space-time diagram**, a graphic way of representing evolution in time and space. For simplicity, the three dimensions of space are collapsed into one; this should be familiar to you, because we have already used the scale factor, R, to describe the size of the universe. By convention, space is plotted horizontally and time is plotted vertically, increasing upward. If you think space-time diagrams sound esoteric, some everyday examples will show that every meeting or interaction must be specified in both space and time. Suppose you plan to meet a friend for lunch.

The two ways you might miss your appointment are by going to the right place but getting the day wrong, and by going to the wrong restaurant at the right day and time. In other words, you must consider *both* space and time. Suppose you are trying to catch a fly ball in baseball. You might miss the catch by either getting to the place the ball lands too late or misjudging the flight of the ball and running to the wrong place. Once again, a successful catch requires considering space and time together. Figure 17–17 shows a space-time diagram describing the travels of two businesswomen.

We can see that the finite speed of light imposes limitations in a space-time diagram. Look at Figure 17–18a. Imagine that a supernova has just exploded at position A. It will take time T_A for the light from the supernova to reach us. If A is 3 pc away, it will take just under a year to see the first light of the supernova. The shaded triangle shows the region that can see the new supernova as its light travels through space. Other events at larger distances, B and C, will take longer times, T_B and T_C, to be seen by us (Figure 17–18b and c). The slope of the two lower lines of the triangle is the speed of light. Since nothing can travel faster than light, all information about each object is contained within the shaded triangle. Thus, we cannot know about the supernova as it is *now*—that would require the instantaneous transfer of information, or a horizontal line in the space-time diagram. All we can do is wait a year for the information from position A to reach us, and then wait even longer times for information from positions B and C to reach us.

Now imagine you are located at A in Figure 17–18a. You start observing the universe at time zero, the apex of the triangle. At that same time, supernovae explode at positions B and C, remote from you in space. As time goes by, the light from those events travels through space. But it is not until time T_A that the supernova at point B becomes visible to you. The supernova at point C takes even longer to reach you (time $2T_A$, if the distance from A to C is twice the distance from A to B). The region of space within which you can observe events that occurred since time zero grows steadily. Astronomers call this region of space the **horizon**; its size is given by the length of the line that defines the top

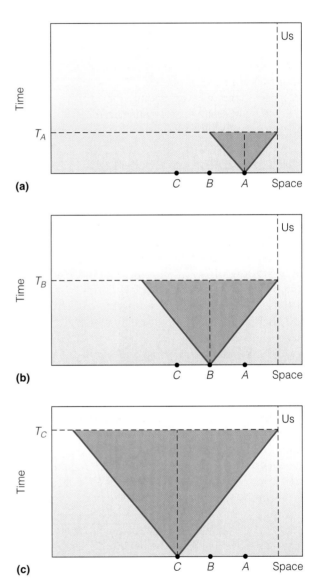

(a)

(b)

(c)

FIGURE 17–18
The limitation of time and space. As time passes, the region of visible space grows, and light reaches us from greater and greater distances. The slope of the line that defines the shaded triangle is the speed of light. **(a)** After a time T_A, light from position A reaches us, and we say that A becomes visible or "enters our horizon." **(b)** At a later time T_B, light from position B reaches us for the first time. **(c)** At an even later time T_C, light from position C reaches us for the first time. In general, the space within our horizon may only be a part of a much larger universe.

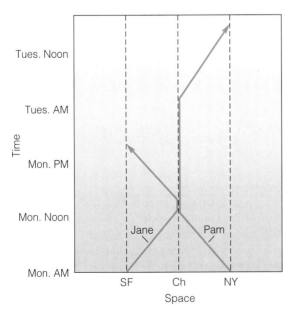

FIGURE 17–17
A simple space-time diagram illustrating the movements of objects (or people) in time and space. Jane lives in San Francisco; Pam lives in New York. They meet in Chicago for lunch on Monday. Jane flies home in time for dinner on Monday. Pam stays in Chicago overnight and arrives home by mid-afternoon on Tuesday.

THE EARLY UNIVERSE

of the triangle. We can know nothing about regions of space that lie outside our horizon.

We are ready to look at some of the bizarre consequences of the big bang model. Figure 17–19 is a space-time diagram of the expanding universe. Since the expansion has been decelerating, the boundary of the observable universe traces a curve in this plot. The most important point is that space was expanding more rapidly in the past. We measure the current expansion by the Hubble constant. The present-day value of the Hubble constant is the same everywhere in space—in other words, the linear Hubble relation applies at all distances—just as we would expect from the expanding balloon analogy. However, there was a time in the past when space was expanding *faster* than the speed of light! The

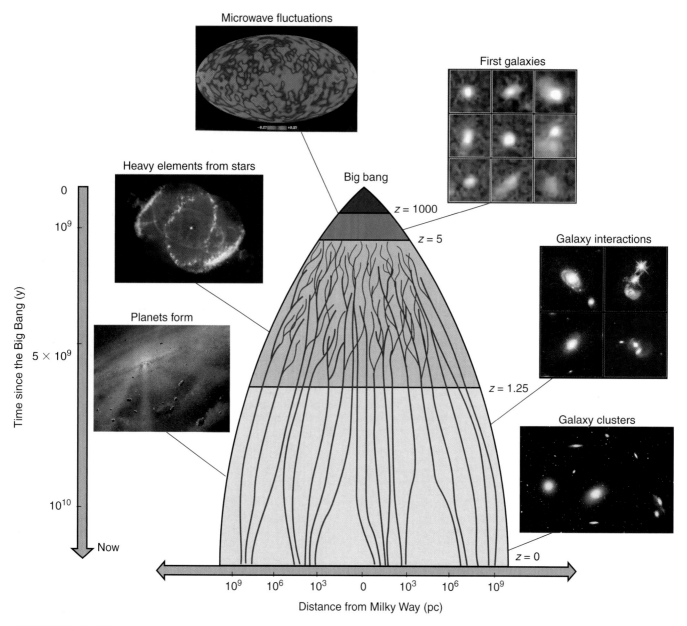

FIGURE 17–19

A space-time diagram showing the evolution of the observable universe. The curving envelope is the horizon; no information from outside the shaded regions can have reached us in the history of the universe. The dashed horizontal line at $z = 5$ represents the approximate epoch when the first galaxies formed. The horizontal line at $z = 1.25$ represents the epoch at which the expansion became slower than the speed of light. Major events in the history of the universe are shown. The formation of structures is represented schematically by the initial collapse of gas into galaxies, followed by the merging of galaxy lines into clusters separated by voids. Notice that the space axis is a compressed logarithmic scale. (SOURCE: "Microwave fluctuations" courtesy of NASA, Goddard Space Flight Center; "Heavy elements from stars" courtesy of NASA, STScI; "Planets form" painting by WKH; "First galaxies" courtesy of Rogier Windhorst and Sam Pascarelle; Arizona State University, NASA, STScI; "Galaxy interactions" courtesy of NASA, STScI; "Galaxy clusters" courtesy of W. Harris and N. Sharp, NOAO.)

speed limit of c imposed by the theory of special relativity only applies on local scales. (For the same reason, we cannot use the Doppler effect to describe cosmological redshifts.) On the global scales described by the theory of general relativity, space can be expanding at any speed. The expanding universe has no speed limit!

The solid horizontal line in Figure 17–19 represents the time in the past when the universe was expanding at the speed of light. It corresponds to a redshift of $z = 1.25$ and a time about 5 billion years ago when the universe was 40 percent of its current size. (These numbers depend on the cosmological model; we have chosen flat space-time for this illustration.) In other words, when the light from a galaxy or quasar at $z = 1.25$ was emitted, that object was moving away from us at the speed of light. Since then, the expansion has slowed, and photons have been able to traverse the large distance between the object and us. The most distant galaxies and quasars have $z = 5$. This redshift corresponds to a time about 13 billion years ago when the universe was 15 percent of its current size. When the light we see from them was emitted, they were moving away from us at nearly three times the speed of light! Thus, space was being created so rapidly that not only the quasar but also the photons it emitted were being "dragged" away from us by the expansion. Our intuitive notions of light travel time also come unstuck. In the local universe, light from a star 1 light-year away takes 1 year to reach us. Light from a galaxy 1 million light-years away takes 1 million years to reach us. But light in an expanding universe has to struggle to overcome the expansion. That is why the most distant objects in a universe 14 billion years old can be as much as 25 billion light-years away.

Extreme situations call for extreme analogies. Imagine that your top running speed is the speed of light. Now imagine that you are talking to a friend at the end of a long moving sidewalk at an airport. Suddenly the sidewalk starts moving rapidly, carrying you away from your friend. That situation represents the big bang. Space is being "created" between you and your friend. You put your bag down and run—your bag is the quasar, and you are the photon it emits. Initially, the sidewalk is moving so fast that, even when you run as quickly as you can, you are carried away from your friend. That scenario is analogous to the early universe. After a while, the moving sidewalk slows down and you begin to approach your friend—a situation representing the universe at half its present age. By the time you reach your friend, the sidewalk has slowed to a crawl, and you reach your friend at full tilt (the speed of light), a much faster speed than the speed at which the moving sidewalk is traveling. Meanwhile, your bag (the quasar) is far behind you, carried off by the expansion of space.

As time passes, larger and larger regions of the universe become visible. The steadily growing boundary of the observable universe is our horizon. Since there is no center to the expansion, we see out in space and back in time in every direction we look. As time goes by, light from more and more distant objects reaches our telescopes; we say the objects "enter our horizon." The observable universe grows a little larger each day. We can now see the distinction between the *observable* universe and the *physical universe.* Since the early expansion was faster than the speed of light, there are regions of space we have never seen. An open universe is infinite in size, but the observable universe is limited to the regions from which light has had time to reach us.

The Evolution of Structure

The early universe must have contained the "seeds" for today's large-scale structure. After all, if the universe were perfectly smooth and homogeneous, we would not exist! The contrast between the early universe and the present day is striking. The cosmic background radiation dates from just 300,000 years after the big bang, yet it is almost perfectly smooth. By contrast, matter today is clumped into stars and galaxies, with virtually nothing in between. Astronomers expected that the cosmic background radiation would show tiny departures from isotropy, representing the first irregularities out of which galaxies and stars and planets and people would eventually form. So it was with great excitement, and more than a little relief, that scientists associated with the COBE satellite announced the discovery in 1992 of **microwave background fluctuations.** These tiny variations amount to only 70 millionths of a degree between different parts of the sky. In general, the microwave background is almost perfectly smooth (see Figure 17–11a), but with a highly magnified temperature scale, departures from isotropy can be seen (see Figure 17–11c).

These tiny ripples were the starting points for galaxy formation. How long did it take for galaxies to form? The microwave background dates from a redshift of $z = 1000$, when the universe was one-thousandth of its current size. Galaxies have been found at redshifts as high as $z = 5$. The time between $z = 1000$ and $z = 5$ depends on the cosmological model, but is typically 1 billion years. This figure is the amount of time available for the transition from shimmers in the primeval fireball to the stark beauty of spiral and elliptical galaxies.

Galaxy formation is one of the most hotly debated topics in modern cosmology. Why is a galaxy the basic mass unit of the universe? Why is space not filled with planets, or grains of dust, or objects much larger than galaxies? Physics gives us some guidance. The temperature fluctuations of the microwave background are signs of matter beginning to clump. The slightly hotter and cooler regions on the sky represent radiation that is slightly blueshifted and slightly redshifted. Regions of slightly higher and lower density were interspersed. Denser regions had stronger gravity, thereby attracting nearby material. As the collapse began, it was resisted by higher pressure within the compressed region—the same tussle between gravity and pressure that governs the life history of a star. An object can form only if gravity wins the battle. At a time 300,000 years after the big bang, the size of the region where gravity and pressure were just balanced contained about 100,000 times the mass of the Sun. Lumps smaller than this could not form. There were no lumps larger than about a trillion times the mass of the Sun, because a gas cloud this size could not cool enough to collapse, and radiation kept it puffed up. This mass range—10^5 $M_\odot$ to 10^{12} $M_\odot$—is just the mass range of present-day galaxies!

However, we know that there are larger structures in the universe. Clusters and superclusters of galaxies range all the way up to 10^{17} $M_\odot$. Which formed first, galaxies or these larger structures? Not surprisingly, the answer depends on the nature of the dark matter. We do not know what type of particle constitutes dark matter, but if dark matter particles move quickly, close to the speed of light, they would have "washed out" anything smaller than a supercluster when structures started forming. Galaxies would have formed later, by fragmentation within the supercluster. This scenario is known as *top-down structure formation,* and it is based on *hot dark matter.* On the other hand, if dark matter particles move slowly, then galaxies would have formed first, with larger and larger structures forming later. This scenario is called *bottom-up structure formation,* and it is based on *cold dark matter.*

Which scenario is correct? Astronomers believe that clusters are considerably younger than the galaxies they

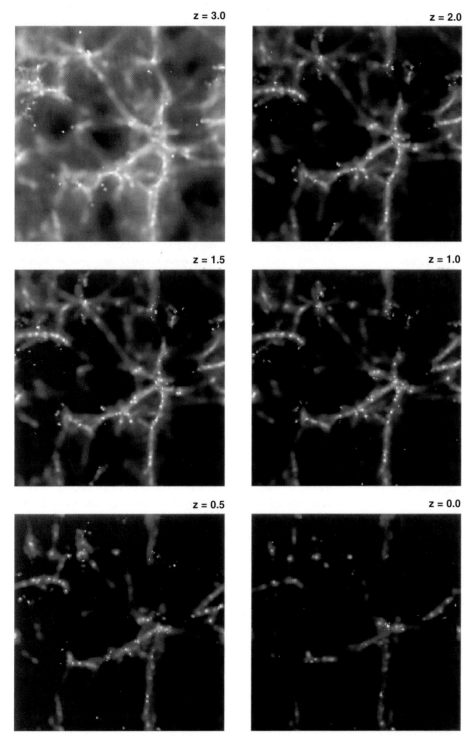

FIGURE 17-20
A supercomputer simulation of the evolving structure in the universe. Cold matter is assumed. Each box shows the density of hydrogen gas in a region 17 Mpc, or 55 million light-years, on a side. This visible gas is tugged by the gravity of a much larger amount of dark matter. The size of the box increases with decreasing redshift to compensate for the expansion of the universe. Gravity forms a "cosmic web" of structure, and as time goes by, more of the gas collapses into filaments, sheets, and individual galaxies. (SOURCE: Courtesy R. Davé, N. Katz, D. Weinberg, and L. Hernquist.)

contain. As a cluster evolves, it becomes increasingly symmetrical; the irregular shapes of some clusters suggest they are very young. On the largest scales, superclusters may only just now be forming. No set of equations can describe the rich structure of the universe. The most powerful way to test the models is to create a "universe in a computer." As described in a Science Toolbox in Chapter 15, it is possible to simulate a galaxy by calculating the gravity of many stars. It is also possible to simulate the universe by calculating the gravity of many galaxies! In the simulations, a galaxy is treated as a point, and the gravity among many galaxies is calculated according to Newton's law of gravity. The volume described by the calculation is steadily increased to represent the expanding universe. The enormous power of supercomputers allows tens of millions of calculations to be carried out each second. Figure 17-20 is an example of the beautiful structures that emerge within a computer model based on cold dark matter. These simulations give us an important clue about dark matter: Models with cold dark matter particles more closely match the clustering of the observed universe than models with hot dark matter particles. The simulations also reproduce the rich texture of clusters, voids, and filaments that we see in the real universe.

Particles and Radiation

Now let us look at the relationship between particles and (the much more numerous) photons in the early universe. The early universe behaved like a very hot gas. Photons outnumbered particles by about a billion to one, and most of the interactions were driven by the intense radiation; therefore, at this time we say the universe was *radiation-dominated*. As the universe expanded, the wavelength of the radiation was stretched, and the photons therefore lost energy because of the cosmological redshift. Figure 17-21a shows a portion of the volume of the early universe. Particles and high-energy photons were constantly colliding and interacting. Figure 17-21b shows the universe at a later time, when the expansion of space had caused the photons to lose energy and have a longer wavelength. Since the particles and the photons were in equilibrium, the particles had less energy too. Early on, gravity was too feeble to cause objects to collapse. Radiation was the boss.

When did radiation lose its eminent position in the early universe? As the temperature cooled, matter became dominant. We can relate the decreasing temperature of the universe to the increasing wavelength of the radiation by Wien's law (see the next Science Toolbox for examples). This is a good example of the widespread application of physical laws—the same relation that describes stars also describes the entire universe! One year after the big bang, the temperature was several million Kelvin and the thermal radiation of the universe consisted of X rays. After several thousand years, at a temperature of about 100,000 K, the radiation had stretched to become ultraviolet waves. As space expanded, the number of photons and particles in any volume stayed the same, but the photons lost *extra* energy from the effect of redshift. Therefore, the energy density of radiation went down *faster* than the energy density of matter as the universe expanded. About 10,000 years after the big bang, the universe became *matter-dominated*. This transition is illustrated in Figure 17-22. From this time on, matter was the boss.

After matter became the dominant component of the universe, matter and radiation followed separate paths. After 300,000 years, the temperature fell to about 3000 K, and thermal radiation from the big bang was redshifted to visible light. Figure 17-23 shows how the temperature has changed since the big bang. However, around 300,000 years after the big bang, radiation no longer had enough energy to keep negatively charged electrons and positively charged protons from combining to form stable hydrogen atoms. Free particles are very effective at interacting with photons. However, as we learned in Chapter 10, hydrogen atoms can absorb and emit light only at a few specific wavelengths, corresponding to the transitions between energy levels of the single electron. With such limited interactions, matter and radiation begin to ignore each other. This separation between matter and radiation is called **decoupling**.

The most dramatic effect of decoupling was that the universe became *transparent*. Consider a cloud. A cloud is opaque because light within it is bouncing around off many tiny water droplets. The cloud appears to have a sharp edge, but in fact its temperature and density change very gradually from inside to outside. At a certain point, the density of water droplets falls off to the point where light no longer interacts with them. From this point the light streams to us directly, and we see an edge. As we saw in a Science Toolbox in Chapter 11, the apparent edge of a star like the Sun is also produced by this effect. Here is another example of the

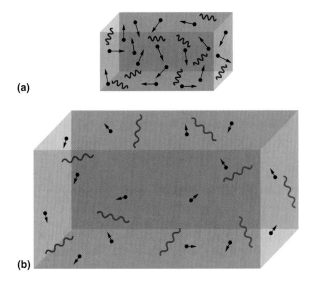

FIGURE 17-21
The loss of photon energy caused by the expansion of space. **(a)** A small volume of space in the early universe containing particles and photons (shown as waves). **(b)** Sometime later, the particles have less energy, and the photons have also lost energy by being redshifted from cosmological expansion.

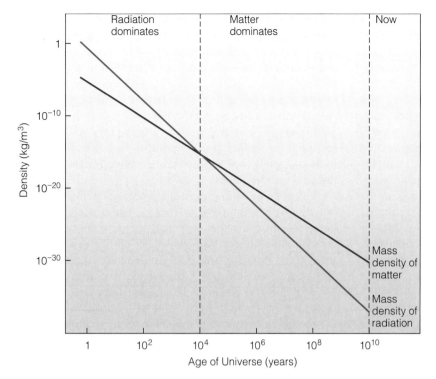

FIGURE 17-22
The time evolution of the density of the universe. Until 10,000 years after the big bang, the mass density of radiation exceeded the mass density of particles, making the universe radiation-dominated. The effects of cosmological expansion caused radiation to lose energy faster than matter. From 10,000 years onward, the universe was matter-dominated, and gravity could cause structures to form. Now the mass density of matter is thousands of times higher than the mass density of radiation.

widespread application of physics: The concept of opacity describes not only the edge of a star but also the transition to transparency of the early universe! After decoupling, big bang photons do not interact with matter, so they travel freely and the universe is transparent. These photons have not been altered, except by the redshift of the expansion, since 300,000 years after the big bang. Therefore, cosmic background radiation is a true relic of conditions in the universe when it was only 2 percent of its present age.

Cosmic background radiation permeates all space. It therefore provides the ultimate reference for all motions in the universe. As you can see in Figure 17–11b, the microwave radiation is 0.007 K hotter in one direction than in the opposite direction. This asymmetry is quite different

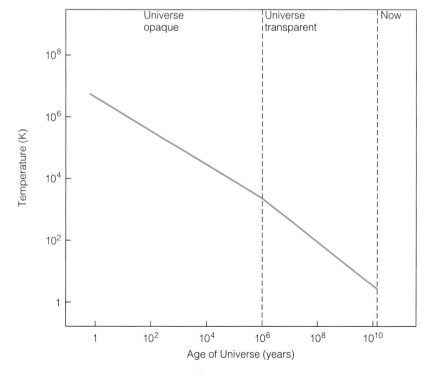

FIGURE 17-23
The time evolution of the temperature of the universe. Until about 300,000 years after the big bang, the radiation was intense enough to keep atoms ionized, and since photons interact frequently with electrons, the universe was opaque. After about 1 million years, the protons and electrons had all combined into hydrogen atoms, leaving photons to travel freely, and the universe became transparent.

SCIENCE TOOLBOX

Particles and Radiation in the Early Universe

We have seen that the cosmological redshift is related to the scale of the universe by $R_0/R = 1 + z$. Since the wavelength of the cosmic radiation stretches with the general expansion, $\lambda \propto R$. Combining these two relationships, we can see that $\lambda \propto (1 + z)^{-1}$. Wien's law gives us $\lambda_{max} \propto T^{-1}$, where λ_{max} is the peak wavelength of thermal radiation and T is the temperature of the radiation. We can combine these expressions to get the simple result

$$T \propto 1 + z$$

The temperature of the radiation in the universe is proportional to the redshift. Scaling this relationship to the observed temperature of the cosmic background radiation gives $T = 2.7 (1 + z)$. If you insert any redshift into this last equation, you get the temperature of the universe at that redshift.

Let's explore this relationship. At the present day, $z = 0$ and $T = 2.7$ K. We observe the peak wavelength of the cosmic background radiation at around 1 mm (see Figure 17–10). At the redshift of the earliest galaxies and quasars, $z = 5$, then $T = 2.7 \times 6 = 16$ K, which is still bitter cold. The wavelength of the peak of the radiation was a factor of $(1 + z)$ smaller, or $1/(1 + z)$ mm, or 0.17 mm.

Our view of the cosmic background radiation dates back to about 300,000 years after the big bang, or a redshift of $z = 1100$. At this redshift, the radiation temperature was $T = 2.7 \times 1100 \approx 3000$ K (once z is very large, the difference between z and $1 + z$ is negligible). Just as a star has a photosphere, you can think of this radiation as the photosphere of the entire universe! The wavelength of the peak of the radiation was $1/1100 = 0.001$ mm, or 1 micron, which is an infrared wavelength.

We can now take this scaling back farther still, to the time 10,000 years after the big bang when radiation was equal in importance to matter in governing the behavior of the universe. The redshift was $z \approx 20,000$. At this redshift, the temperature of the radiation was $T = 2.7 \times 20,000 \approx 55,000$ K, and the wavelength of the peak of the radiation was $1/20,000 = 5 \times 10^{-5}$ mm, or 50 nm. At this early time, the universe was awash with energetic ultraviolet photons.

If we follow this progression of increasing temperature back even earlier into the universe, we arrive at an important consequence of Einstein's equivalence between mass and energy, $E = mc^2$. Particles and antiparticles can be created out of pure radiation. The mechanism is called *pair creation,* and it dictates how physicists (and nature) create antimatter on Earth. Antimatter is created momentarily in the upper atmosphere by cosmic rays, or expensively and in minute quantities by enormous particle accelerators. The opposite process, annihilation, is defined as the disappearance of a particle-antiparticle pair, releasing high-energy radiation or gamma rays. The pairing up of particles and antiparticles, called *pair annihilation,* is a perfectly efficient process; 100 percent of the mass-energy is liberated. The processes of pair creation and pair annihilation are shown in Figure 17–A.

When did pair creation occur in the universe? About 1 minute after the big bang, when the temperature was 5×10^9 K, radiation had enough energy to create the lightest particle-antiparticle pairs. Electrons and their antiparticles—positrons—could be created in profusion out of the high-energy gamma rays. Even earlier, at an amazing 10^{-4} seconds after the big bang, the temperature was 10^{13} K, or 10 trillion degrees! At this time, protons and antiprotons, and neutrons and antineutrons, could be created in pairs. Thus, the early universe was an amazing factory for producing matter (and antimatter) out of radiation.

Finally, we look at an extraordinary aspect of the quantum theory of nature. As we saw in Chapter 10, there is a "fuzziness" in the physical world expressed by Heisenberg's uncertainty principle. We discussed this principle in terms of the properties of a tiny subatomic particle. We cannot simultaneously measure the position and motion of a subatomic particle with perfect accuracy. If we know one quantity, we must be uncertain about the other quantity.

A second version of Heisenberg's uncertainty principle is entirely equivalent to the version you saw in Chapter 10. It takes the form

$$\Delta E \times \Delta t \geq h / 2\pi$$

(continued)

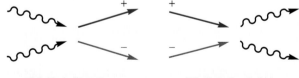

(a) Pair Creation **(b)** Pair Annihilation

FIGURE 17–A
The creation and annihilation of particle/antiparticle pairs. **(a)** High-energy photons can create particle-antiparticle pairs if their combined energy is more than the mass-energy of the particles they create. **(b)** If a particle and an antiparticle come into contact, they annihilate, and the energy is carried off in the form of high-energy photons.

SCIENCE TOOLBOX

Particles and Radiation in the Early Universe (continued)

In this expression, ΔE is the uncertainty in energy, Δt is the time interval over which the energy is measured, and h is the tiny Planck's constant. Over a small enough time interval, the energy of a particle or a system is uncertain by an amount related to the Planck constant. But Einstein's principle says that energy is related to mass, so uncertainty in energy corresponds to uncertainty in mass. Following the logic of the previous discussion, particle-antiparticle pairs can be created out of quantum fluctuations. The pairs are created (and disappear again) incredibly quickly. Electrons and positrons can be spontaneously created for no more than 10^{-22} seconds, and protons and antiprotons can be spontaneously created for no more than 10^{-25} seconds! Their ephemeral existence leads them to be called *virtual pairs,* as shown in Figure 17-B.

It is as if you could borrow an enormous amount of money from your bank as long as you pay it back extremely quickly. The faster you pay it back, the more you can borrow. Even if your balance is low (or zero), you can borrow the money. The quantum fluctuations we have just described can violate the law of conservation of energy, as long as it is done for a very short time.

Virtual pairs cannot be observed directly, but their effects have been detected on normal particles. Physicists have strong evidence that "empty space" is not really empty; it is filled with a seething froth of virtual pairs. All this goes on at such tiny levels of energy, and for such fleeting intervals, that it is unseen by us. However, in the early phases of the universe, these quantum fluctuations were very significant.

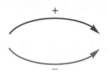

Virtual Pair

FIGURE 17-B
Virtual particle-antiparticle pairs. Heinsenberg's uncertainty principle allows for the spontaneous creation of matter and antimatter for tiny amounts of time. These virtual pairs can be observed indirectly. Even a pure vacuum is seething with this kind of activity.

---

from the much smaller fluctuations shown in Figure 17–11c. It is a true Doppler effect, occurring because Earth is not stationary. We are moving at 390 km/s with respect to the cosmic background radiation—our true motion in the universe.

In principle, if we search out in space far enough from Earth, we should find a set of galaxies that do not deviate from the smooth expansion of the universe and therefore are at rest with respect to the microwave background. Let us see if it works. Figure 17–24 shows all the motions we can detect. Earth spins, it orbits the Sun, and the Sun orbits the Milky Way. The Milky Way is being tugged toward M 31 at 100 km/s, and the entire Local Group is falling into the Virgo cluster at a speed of 220 km/s. Amazingly, the entire Local Supercluster is moving toward another set of more distant galaxies. It is enough to make you dizzy. However, on scales beyond 100 Mpc, it is possible to find a set of galaxies that has no net motion with respect to the microwave background. This discovery is very reassuring, since it allows us to verify the cosmological principle. Galaxies in the universe have isotropic motions on very large scales.

Inflation and the Very Early Universe

Despite its great success, the big bang model does not explain several important features of the universe. The explanations may lie in what happened in the first fractions of a second after the big bang. We will examine each question in turn.

Why does the universe have a geometry that is nearly flat? Evidence shows that the density parameter, Ω_0, is in the range 0.1 to 0.3. The observations are sufficiently uncertain that a flat universe ($\Omega_0 = 1$) cannot be ruled out. In other words, the universe is within a small factor of being flat. To see why this is remarkable, let us return to the analogy of the rocket launch from earlier in the chapter, when we introduced the density parameter. A flat universe is analogous to a rocket that is launched with a velocity of exactly 11 km/s; the rocket leaves Earth but slows smoothly to an eventual halt in deep space. Give the rocket a factor of 2 or 3 more starting velocity, and it will leave Earth easily and speed through space. Give it a factor of 2 or 3 less starting velocity, and it will quickly fall back to Earth. In other words, even a slight change in starting velocity from the special number 11 km/s produces a very different result.

Now we can reverse this argument. Imagine that we find the rocket long after it has left Earth, crawling along at a constantly decelerating rate. We can deduce that it must have had a launch velocity of *exactly* 11 km/s. Why? Because with a launch velocity even slightly different from 11 km/s, the rocket would be somewhere entirely different—either much farther out in space, or fallen back to Earth long ago. Return-

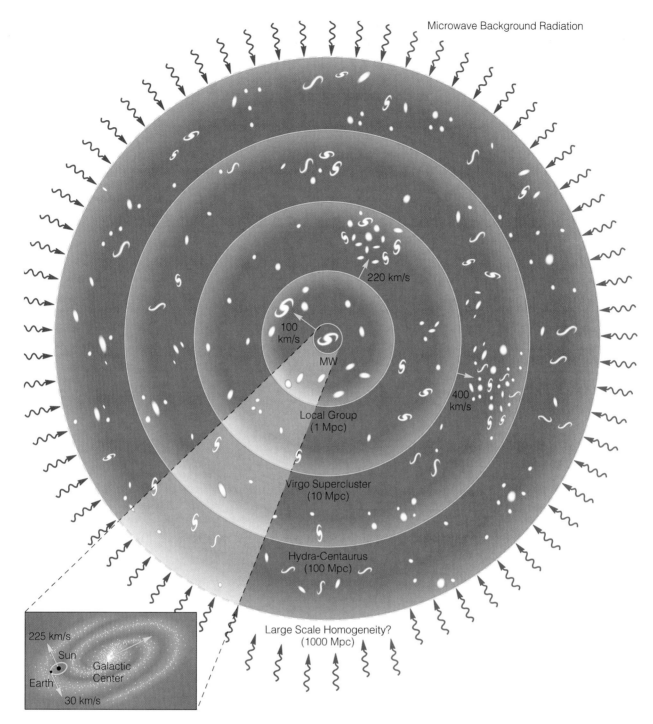

FIGURE 17-24
The cosmological frame of reference for measuring motions. Earth moves around the Sun at 30 km/s, and the Sun moves around the center of the Milky Way galaxy at 225 km/s (the inset shows an expanded view of the motions within the Milky Way). The Milky Way is falling toward the Andromeda galaxy (M 31) at 100 km/s. The entire Local Group, including the Milky Way, M 31, and a few dozen dwarf galaxies, is falling toward the Virgo cluster at about 220 km/s. The entire Local Supercluster is being tugged toward the region of the Hydra and Centaurus clusters. However, on a scale larger than 100 Mpc, the average motions of galaxies are at rest with respect to the expansion of space and the cosmic background radiation. (SOURCE: Adapted from J. Silk.)

ing to cosmology, we live in a very old and large universe in which the density parameter is close to 1. If the density parameter is close to 1 now, it must have been exactly 1 in the very early universe. Calculations show that the universe must have started with a density parameter equal to 1 with a precision of one part in 10^{59}! In terms of the expansion of space, space that is nearly flat now implies space that was exactly flat in the early universe. We must account for the

THE EARLY UNIVERSE 537

flatness of space; Ω_0 could have any value in the standard big bang model.

Why is the universe so smooth? Now, of course, the universe is lumpy with planets and stars and galaxies. However, we know from the uniformity of cosmic background radiation that the universe was very smooth 300,000 years after the big bang. The temperature is constant to within 0.007 percent everywhere in space. If we think of the universe as a hot gas, then the gas must have been well mixed. The only way to have a gas in perfect thermal equilibrium—all parts at exactly the same temperature—is for heat to be able to move from one part of the gas to another.

Yet space was expanding incredibly quickly soon after the big bang. At the time we see with the microwave background radiation, patches of space were separating at $61c$, over 60 times the speed of light! One year after the big bang, the separation speed was $1000c$! Look at Figure 17–25a. This space-time diagram shows regions of space that were close together at the time of the big bang. After 300,000 years, they have separated by a large distance equal to many times the light travel time between them. The horizons of these two regions do not intersect, which means that no signal can pass between them. No light or radiation can have traveled between regions of space that are outside each other's horizons, so there is no way they could have equalized their temperatures. Yet widely separated directions in the sky have virtually identical microwave background temperatures. This uniformity thus has no explanation in the standard big bang model.

In 1981, MIT physicist Alan Guth was pondering these problems—the smoothness and flatness of the universe—when he came up with the idea of the **inflationary universe.** In this adjustment to the big bang model, he proposed that the infant universe went through a period of extremely rapid expansion. During the inflationary epoch—the incredibly small iota of time from 10^{-35} to 10^{-33} seconds after the big bang—the universe expanded in size by 40 orders of magnitude (Figure 17–26). Inflation moved matter that originally was near us (within a few meters) to a position far outside today's observable universe (much more than 20 billion light-years). The inflationary model presents us with a cosmos that has been stretched far beyond our horizon. Therefore, the observable universe is probably a tiny fraction of the physical universe. Whatever the power of our largest telescopes, we are consigned to a humble corner of an immense cosmos. How does this truly audacious idea account for the smoothness and flatness of the universe?

Inflation increased the size of the universe suddenly and dramatically. Figure 17–25b shows the effect of inflation on a space-time diagram. Regions of space whose horizons are

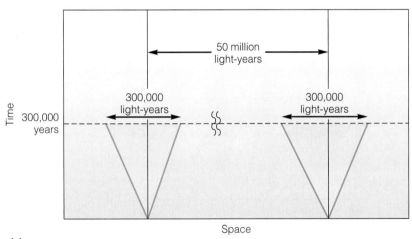

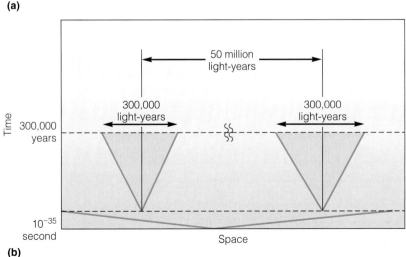

FIGURE 17–25
A space-time diagram showing two regions of space on opposite sides of the sky. By the time the universe became transparent to the microwave background radiation, 300,000 years after the big bang, these regions were separated by 50 million light-years. Light could not travel between them because their horizons do not overlap. **(a)** Two regions this far removed cannot establish thermal equilibrium, so there is no reason to expect their temperatures to be as uniform. They are separated by about 2 billion light-years now. **(b)** The inflationary model solves the horizon problem by a very rapid expansion of the universe in the first fractions of a second. Regions of space widely separated 300,000 years after the big bang were originally much closer together. Inflation stretches a small region of space that has a single temperature to the size of the observable universe.

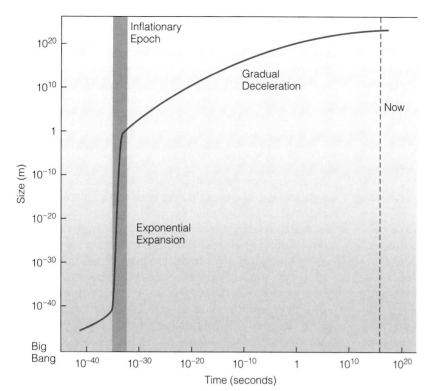

FIGURE 17-26
The enormous and rapid expansion of the universe predicted by the inflationary model. After the inflationary epoch, the expansion decelerated slowly because of the gravitational attraction of matter in the universe.

widely separated now were within the same horizon at the end of inflation. In other words, regions of space that are now out of contact with each other were in close contact in the very early universe, so it is no mystery that they share the same temperature. Whatever curvature the universe might have had before the inflationary epoch was stretched out by the prodigious expansion. Imagine a tiny balloon that is rapidly inflated to many times its original size (Figure 17-27). Any section we choose to explore is almost perfectly flat. The flatness of the universe is no longer a coincidence; it is an inevitable consequence of inflation.

Inflation seems incredible. If this idea is to be anything more than esoteric speculation, inflation must have a physical basis, and the inflationary big bang model must make predictions that can be tested by observation. Inflation predicts that space should be extremely close to flat—in terms of cosmological parameters, that $\Omega_0 = 1$. As we saw earlier in the chapter, the jury is still out on the difficult measurement

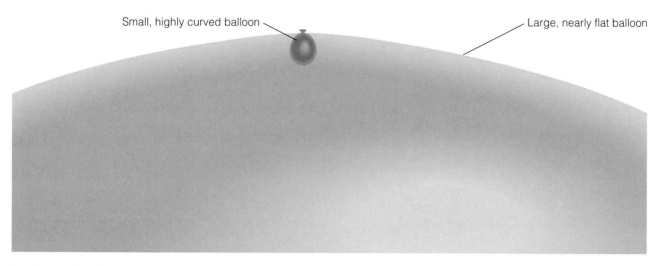

FIGURE 17-27
The inflationary model's solution to the flatness problem. By the end of the inflationary era, the universe had expanded by such a large factor that, whatever the original curvature, it was indistinguishable from being flat. In a similar way, measurements made over a small region of Earth cannot detect the curvature of the surface.

of the density parameter. Most research indicates that the density parameter is close to, but less than, 1. Cosmologists have tried to make up the deficit with a cosmological constant, but direct evidence for a cosmological constant is scant. Inflation also smooths the universe out very effectively. The inflationary model predicts that the remaining ripples, which are the seeds from which galaxies grow, must date back to the first tiny fractions of a second after the big bang. To understand inflation better, we must explore the subatomic world and discuss the fundamental forces of nature.

Symmetry and the Mass of the Universe

The universe contains roughly 10^{80} particles and 10^{89} photons. Why are there so many more photons than particles? To answer this fundamental question, we must explore **symmetry** in nature, the idea that phenomena that appear distinct share a similar basis. (In geometry, a sphere is the most symmetrical shape because it looks the same from every direction. By contrast, a lumpy rock looks different from every direction.) Scientists look for symmetry in order to unify the diversity of nature. Symmetry also has an aesthetic aspect—it expresses balance and harmony.

We already encountered several important examples of symmetry in Chapter 10. Electricity and magnetism appear to be quite different phenomena. Yet Michael Faraday showed over 200 years ago that a moving magnet could generate an electric current, and that a changing electric current could generate a magnetic field. The interplay between electric and magnetic forces finds its most beautiful expression in electromagnetic waves. Light is our everyday example of symmetry between electricity and magnetism. Particles and waves appear to be quite different phenomena. Yet, although light has obvious wave properties, it also behaves like a tiny particle of energy called a photon. Likewise, subatomic particles are "fuzzy" in the quantum view of nature and often have the properties of waves. The last example is the famous result of Einstein: $E = mc^2$. This equation expresses the fact that mass and energy are not separate and distinct. Mass is a "frozen" form of energy.

Although symmetry may be present in our theories or the equations that express the theories, it is not always obvious in the universe we live in. The equation $E = mc^2$ implies that mass and energy are interchangeable. Yet it is only in the extreme conditions inside stars that we see even a small percentage of mass turned into energy. In the everyday world, we never see the reverse process; light does not spontaneously turn into particles. In our universe, there is an enormous asymmetry between radiation and matter. Photons outnumber particles by about a billion to one. Of course, we have seen that these photons are so feeble in energy now that we can barely detect their effects.

Another way in which symmetry is concealed from us is the nature of matter. **Antimatter** is material whose particles have an opposite set of quantum properties. You can think of antimatter as the mirror image (or ghost) of matter. In the microscopic world of the quantum, every particle has an equivalent antiparticle. Charged particles have antiparticles with the opposite charge. For example, the antiparticle of the electron is the positively charged positron, and the antiparticle of the proton is the negatively charged antiproton. The counterpart of the neutron is the neutral antineutron, and the counterpart of the neutrino is the antineutrino. The photon is its own antiparticle. The laws of physics place matter and antimatter on an equal footing, so why is the ordinary world composed overwhelmingly of matter? When an antiparticle is created in the laboratory, it cannot survive long before encountering a particle. The result is that both particles disappear in a flash of gamma rays.

Astronomers have wondered whether large concentrations of antimatter exist somewhere in the universe. There are no antimatter stars anywhere near the Sun; they would react with the interstellar medium to produce far more gamma rays than are observed. Astronomers have even ruled out distant antimatter galaxies made of antimatter stars. Galaxies and antigalaxies would currently be separated by the depths of space, but in the past they must have been closer together, and we do not see gamma rays from their interactions redshifted by the expansion to lower energies. We conclude that the universe is made of matter.

The symmetries of nature were more readily apparent in the early universe. At a high enough temperature, mass and energy were freely interchangeable. Equality was established between the creation and annihilation of particles and between matter and antimatter. The last Science Toolbox described this situation. The present-day universe is almost entirely made of radiation and matter. Antimatter can be produced only in tiny, ephemeral amounts. In round numbers, the universe contains roughly 100 million photons for every particle, and no antiparticles. However, particles and antiparticles were created and destroyed in the early universe in equal numbers. One-millionth of a second (10^{-6} s) after the big bang, the temperature was an enormous 10^{14} K, and gamma-ray photons had enough energy to create particle-antiparticle pairs of many kinds. By about 1 second after the big bang, cosmological expansion redshifted the radiation to a low enough energy that particles and antiparticles could not be created.

At this point, if there had been exactly equal numbers of particles and antiparticles, they would have all disappeared in pairs, leaving a universe filled with only radiation! Thus, there must have been a tiny excess of matter over antimatter in the very early universe. For every 100 million antiparticles, there must have been 100 million plus 1 particles. After particles and antiparticles paired up to create a flood of gamma rays, a slight residue of matter remained. Now, billions of years later, the original gamma rays of the big bang have been redshifted to feeble microwaves. Although these photons have low energy, each cubic meter of space contains about 10^8 of them. Most of the particles in the universe were annihilated long ago with their corresponding antiparticles, and the residue of neutral matter has formed all the gravitational structure in the universe. In other words, our existence depends on a tiny asymmetry between matter and antimatter in the very early universe!

The Forces of Nature

For a deeper knowledge of the very early universe, we must consider the fundamental forces of nature. Like the ancient Greek thinkers before them, modern scientists hope to discover the simplest possible description of the physical world. Recall from Chapter 1 that the scientific method is based on the belief that diverse physical phenomena have a simple underlying basis. Think of the rich diversity of structures in the cosmos, from the swirl of spiral galaxies and the mystery of black holes to the filigree of arcs and voids in the distribution of galaxies. The single force of gravity has shaped all of these structures. Think also of the variety of the material world, from lustrous precious metals and inert and colorless gases to the radioactive fizzing of heavy elements. Atomic forces yield this variety.

The quest for symmetry continues with the **fundamental forces** of nature. There are four forces in the physical world. The two with which we are most familiar, *gravity* and *electromagnetism,* operate over an infinite range and contribute to the structure of the macroscopic world. The electromagnetic force is 10^{28} times stronger than gravity. If you doubt this, take a balloon and notice how the slight static charge of the balloon rubbed against your head will hold it on the ceiling against the full attractive force of the entire planet Earth. Similarly, a modest magnet holds a nail against Earth's gravity. In normal matter, electromagnetic forces are literally neutralized by the equal numbers of negatively charged electrons and positively charged protons. The other two forces of nature operate over very short distances within an atom. The *weak nuclear force* is actually far stronger than gravity, but its range is 100 times smaller than an atomic nucleus. It is responsible for the radioactive decay of massive nuclei and for the decay of neutrons when they are not locked in atomic nuclei. The *strong nuclear force* has a range about the size of an atomic nucleus, and it acts to bind neutrons and protons inside a nucleus. To do so, it must overpower the electromagnetic force, which would otherwise cause the positively charged protons in a nucleus to repel each other. Within an atomic nucleus, physicists have shown that protons and neutrons are made of even smaller particles called *quarks.* Table 17-3 summarizes the properties of the four forces.

As described so far, the four forces of nature could not appear more different. Two forces have infinite range, and two have short range. The forces are transmitted by particles with different masses, and they differ by a factor of nearly 10^{40} in strength! The ultimate expression of physicists' belief in symmetry is the search for **unification,** a theory demonstrating that the four forces of nature are manifestations of a single unified force.

Our low-energy world may represent a situation of "broken" symmetry, in which the underlying similarities between forces and particles are concealed. Imagine spinning a roulette wheel. At first, the ball has high energy and moves in circles. At the end, when the wheel has stopped, the ball rests in one of 37 numbered slots. We might watch the results of many spins and conclude that the ball has 37 different states. However, at high energy the ball always has the same motion; the symmetry is broken only at low energy. For another example, imagine a large number of pencils balanced on their flat ends. Viewed from above or the side, they present a picture of uniformity or symmetry. However, if they all topple (a state of lower potential energy), the result would be a tangle of skewed and overlapping pencils with no symmetry at all. We find ourselves in the low-energy worlds, staring at the jumble of forces and subatomic particles, trying to imagine the simplicity of the high-energy universe.

Physicists took the first step toward unification in the 1970s with a theory that united the electromagnetic and weak nuclear forces. This *electroweak theory* was convincingly confirmed by experiments using particle accelerators in 1983. In the early universe, the electromagnetic and weak nuclear forces were unified at temperatures above 10^{15} K, or a time within 10^{-12} seconds after the big bang. In the present-day universe, the symmetry is broken and the two forces appear quite distinct. Emboldened by this success, theorists have attempted to bring the strong nuclear force under the umbrella of unification in *grand unified theories.* (We use the plural "theories" because physicists

TABLE 17-3
The Fundamental Forces of Nature

FORCE	RELATIVE STRENGTH	INTERACTING PARTICLES	ROLE IN NATURE
Strong nuclear force	1	Quarks	Holds atomic nuclei together
Electromagnetism	10^{-12}	All charged particles	Holds atoms together; controls electromagnetic waves
Weak nuclear force	10^{-14}	Quarks, electrons, neutrinos	Involved in radioactive decay and the decay of neutrons
Gravity	10^{-40}	All particles	Holds planets, stars, and galaxies together and governs their motions

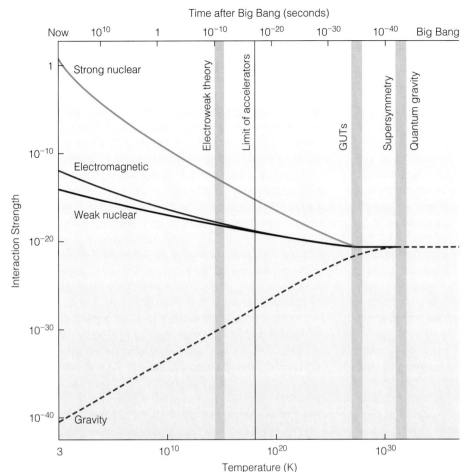

FIGURE 17-28
A plot showing the theoretical expectation that the interaction strengths of the four fundamental forces will become equal at a high enough energy or temperature. The top scale shows the time after the big bang corresponding to each temperature. The vertical bands show the unification temperatures of the forces, and the vertical line shows the maximum temperature that will be realized by planned accelerators. Gravity is shown as a dotted line, because unification theories that incorporate gravity are very speculative.

have come up with several versions of unification.) Figure 17-28 shows the theoretical expectation of the energy or temperature at which the forces of nature become unified. Along the top of the figure is the time after the big bang that corresponds to each temperature. The weak, strong, and electromagnetic forces are predicted to be equal at a phenomenal temperature of 10^{28} K, corresponding to a time only 10^{-35} seconds after the big bang! You can see from the figure that this temperature is far beyond the range of any existing or planned particle accelerator. The only possible laboratory for grand unified theories is the very early universe.

Grand unified theories might clear up several of the mysteries of the big bang. In 1964, experimental physicists discovered a slight asymmetry between the behavior of matter and antimatter. Grand unified theories predict an excess of matter over antimatter at temperatures near 10^{28} K, but those theories give no concrete prediction of the amount of the excess. Nevertheless, it is exciting when theories invented to explain a laboratory phenomenon might explain why the universe contains far more matter than antimatter.

Grand unified theories also contain an insight into why inflation might have happened. As the universe passed through the temperature at which the strong force separated out from the weak and electromagnetic forces, it changed its state—as in the way that water changes its state when it freezes and becomes ice. When water becomes ice, energy is released. (The analogy even extends to the loss of symmetry at lower temperatures. Water molecules move freely and point in all directions, while ice crystals are aligned in particular directions.) The very early universe cooled so quickly that it borrowed energy from the vacuum of space, in a manner described in the previous Science Toolbox. We have seen that the law of conservation of energy can be broken for tiny periods of time; these momentary violations of the physics of the everyday world are called **quantum fluctuations.** Borrowed energy drove the sudden inflation of the universe 10^{-35} seconds after the big bang. When the strong force became distinct, at 10^{-33} seconds, the energy from that transition flooded into the universe as 10^{89} photons, and the expansion continued more leisurely. (See Figure 17-26 to review the chronology.) This scenario sounds truly fantastic, but one of its predictions has already been confirmed. Inflation takes quantum fluctuations and boosts them into the seeds of large-scale structure. Theory predicts that there is no preferred size of a quantum fluctuation; they should occur equally on all scales. Observations with the COBE satellite showed that the tiny fluctuations in the microwave background have exactly this property.

The Limits of Knowledge

The story of the universe in the first second of its existence is highly speculative. However, we have paid attention to the scientific method in our discussion of the early universe, looking for theories that make unique and observable predictions. How far can we push into the very early universe? Is there is a limit to our knowledge?

Scientists believe the limit to our knowledge (and even to useful speculation) comes at the amazing temperature of 10^{32} K. This temperature occurred 10^{-43} seconds after the big bang, when the entire universe was 10^{-35} meters across. Imagine it: billions of galaxies in a space far smaller than the head of a pin! This time is called the *Planck era*, after one of the founders of quantum mechanics. You can see from Figure 17-28 that all the forces of nature become equal at the temperature corresponding to the Planck era. The melting of gravity into the other three forces is the ultimate symmetry. However, we currently have no gravity theory that works under such extreme conditions. General relativity treats space and time as smooth and continuous. However, the very early universe was dominated by quantum fluctuations. Space was so curved that there was no distinction between a particle and the space it occupied. The universe was a seething cauldron of matter and energy appearing and disappearing out of a vacuum, and space and time twisted and fractured like foam.

In the inflationary big bang model, quantum fluctuations drive the rapid expansion that smooths and flattens the universe. The entire universe emerges from almost nothing. According to Alan Guth, the architect of the model, "the universe may be the ultimate free lunch." Andrei Linde, a Russian physicist working at Stanford University, has proposed a variation of the model called *chaotic inflation* (Figure 17–29). In chaotic inflation, there are many "bubbles" in space-time during the Planck era. The bubbles are spawned by quantum fluctuations, and each bubble may contain different forces and particles. In our bubble, the conditions were right for space to inflate and become the observable universe. The idea of many possible universes, most of which are stillborn, is one of the most bizarre in cosmology.

What is the ultimate theory of nature? Nobody knows, but it must involve unification between gravity and the world of the quantum. We are made of atoms, and atoms are made mostly of electrons and atomic nuclei. The nuclei are made of protons and neutrons that are made of quarks. Current physics theories are incomplete because they treat

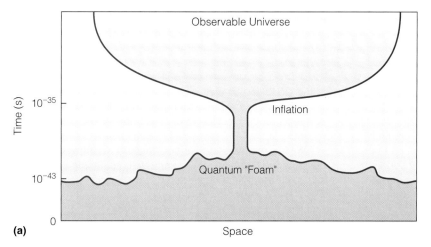

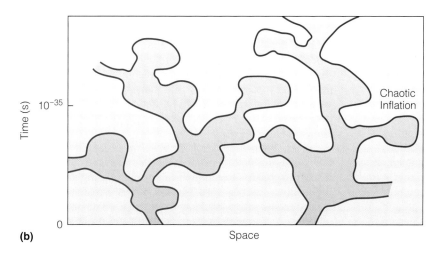

FIGURE 17–29

Space-time diagrams for conventional and chaotic inflation models. **(a)** In conventional inflation, the universe undergoes a phase of rapid expansion so that the entire observable universe comes from a region that was once microscopically small. **(b)** In the chaotic inflation model, quantum fluctuations at very early times create a chaotic space-time, where many universes can potentially be spawned, with very different properties from ours. The large and old universe we inhabit results from just one of these possible events. (SOURCE: Adapted from J. Silk, A. Lightman, and A. Linde.)

electrons and quarks as particles that have no size. If a charged particle is a point, it must have infinite density of mass and electrical charge. Physicists have been exploring theories that avoid this problem by treating particles as *strings* or *membranes*. These tiny entities can be open, or they can be closed in a loop; they can interact, merge, join, and split. Moreover, they exist in more than three dimensions, most of which are unobserved by us. These speculative theories have great mathematical beauty, but no one knows if they apply to the real world. We have come full circle in our story in this book. As modern scientists struggle toward an ultimate theory, we see an echo of the ancient Greek scientists like Democritus and Pythagoras. The quest to find the harmonies in nature continues.

Summary

Cosmology is as old as the first civilizations. The human mind has always been drawn beyond the pressures of simple survival to wonder about the size and nature of the cosmos. Modern cosmology began with Albert Einstein and Edwin Hubble. Einstein developed a new theory of gravity that replaced the idea of force with the geometry of space itself. Mass and energy both create and respond to the curvature of space. Hubble first demonstrated that galaxies are distant systems of stars, and he later discovered that galaxies are all being carried away from one another by the expansion of the universe. Cosmology rests on the cosmological principle—the assumption that the universe is homogeneous and isotropic, or appears the same from all points in space.

The scientific story of creation is the big bang model. The earliest parts of the story are speculative and uncertain. The beginning was a time of such high energy and density that mass and energy were freely interchangeable. Space was as curved as the particles that occupy it. The four forces of nature were unified into a single force. The universe emerged from seething quantum foam and began rapidly inflating, driven by quantum fluctuations. As a result of inflation, the universe became flat, smooth, and virtually empty. When the inflation was over, radiation flooded back into the universe. A slight asymmetry between the forces of nature left the universe with slightly more particles than antiparticles. As the universe cooled and matter could no longer be freely created from energy, the particles and antiparticles disappeared in a flood of radiation. The result was an expanding universe filled primarily with radiation, along with a residue of matter. (This residue was nevertheless sufficient to eventually form all the billions of galaxies in the universe.) This entire tumult of activity took little more than 1 second.

The next part of the history is supported by considerable observational evidence. Over the next 3 minutes, while the entire universe was still hotter than the center of a star, one-quarter of the mass of the universe was fused from hydrogen into helium. After 1 hour, all nuclear reactions stopped. As the universe expanded and cooled, radiation thinned out more rapidly than matter, until at 10,000 years, the universe became matter-dominated forever. All the events to this point occurred in a cauldron of careening particles and photons; our deductions are those of someone groping through a dense fog. After 300,000 years, the universe finally became transparent. Atoms became whole, as electrons were able to join atomic nuclei. Light could travel freely, deviating slightly from the laws of Euclid to follow the undulations that matter creates in curved space. Billions of years later, we see that radiation from this era diluted and redshifted to microwaves. The big bang model rests on a solid base of three pillars of evidence: the expansion traced by galaxies, the cosmic background radiation, and the abundance of helium created by cosmic nucleosynthesis.

Gradually, gravity sculpted the ripples in the early universe into structures on all scales. This process was influenced by the large amount of unseen dark matter. About 1 billion years after the big bang, enormous gas clouds collapsed into the first galaxies. Within them, legions of stars switched on for the first time. Stars followed a pattern of birth and death for 6 or 7 billion years. Most stars died a quiet death, as slowly fading embers. The massive ones exploded, seeding the space between stars with heavy elements. Many stars ended up surrounded by the debris of their own creation. On one such rocky fragment, life evolved out of a watery, organic broth. Some 4.5 billion years later, warmed by a yellow star and bathed in the microwave afterglow of the creation event, humans point their telescopes at the skies and ponder the meaning of it all.

We live in a vast and very old universe. The finite speed of light and the finite age of the universe limit our view to an observable portion of a larger and possibly infinite physical universe. Two numbers govern the past, present, and future of a universe described by the big bang model. One is the present rate of expansion, or the Hubble constant, H_0. The Hubble constant is determined to an accuracy of about 15 percent. The other is the mean density of matter; the mean density divided by the density required to just overcome the expansion is the density parameter, Ω_0. The best measurements of the density parameter show that the universe has only one-third of the amount of matter needed to overcome the expansion. By this reckoning, space is nearly flat and the universe will expand forever. Direct measurements of the subtle curvature of space are very difficult, primarily because observations must be made over large distances or look-back times, which forces us to compare old and young objects.

At the dawn of the scientific age, Plato was well aware of the limitations of our view of reality. He likened us to cave dwellers trying to discern the truth by interpreting the flickering shadows on cave walls. Our search for cosmic origins has acquainted us with a shadowy history that we do not completely understand. Perhaps the greatest surprise in cosmology is the fact that we can understand the universe at all, despite our microscopic part in it.

Important Concepts

You should be able to define these concepts and use them in a sentence.

- cosmology
- universe
- Olbers' paradox
- space curvature
- big bang model
- cosmological principle
- critical density
- density parameter
- cosmic nucleosynthesis
- cosmic background radiation
- cosmic evolution
- age of the universe
- arrow of time
- cosmological constant
- space-time diagram
- horizon
- microwave background fluctuations
- decoupling
- inflationary universe
- symmetry
- antimatter
- fundamental forces
- unification
- quantum fluctuations

How Do We Know?

These questions and answers show how the scientific method is used to learn about the universe.

Q How do we know that the big bang actually happened?

A Although astronomers cannot be absolutely certain, strong evidence indicates that the universe was much hotter and denser in the past. Galaxies are all moving away from one another, carried apart by the expansion of space. If we trace this expansion back in time, galaxies must have all been in the same place billions of years ago. Also, there is far more helium in the universe than can be accounted for by the output of all stars over cosmic time. The big bang model predicts that about one-quarter of the mass of the universe was turned into helium when the entire universe had the temperature of the center of a star. The big bang predictions of the abundance of helium and other rare light elements correlate with observations. Finally, the big bang model predicts that space is filled with relic radiation from the hot early phase of the universe. Astronomers have detected this redshifted radiation at exactly the predicted temperature, 3 degrees above absolute zero. The big bang also passes another important test: The age derived from the model agrees with the ages of the oldest objects in the universe.

Q How do we know that space is curved?

A Einstein's theory of general relativity is a theory of gravity in which mass distorts space. The curvature of space has been measured directly in the environment of compact stars. However, general relativity also predicts that space will have a *global* curvature, caused by the sum of all the mass in the universe. It is extremely difficult to measure the curvature of the universe, and no convincing result has been published so far. Instead, astronomers use the locally measured density parameter to infer space curvature. Space seems to be very close to flat.

Q What is the fate of the universe?

A The universe is likely to expand forever. It has been expanding at a smoothly decreasing rate since the big bang. Whether or not the universe will eventually collapse depends on the mean density of matter. Astronomical measurements indicate that the universe does not have enough mass to overcome the expansion, in which case it will expand forever at a decreasing rate. Measurements of the density are tricky because most of the mass is dark, and it may be quite smoothly distributed on large scales.

Q How do we know that inflation actually occurred?

A We do not know for sure. The evidence for inflation is far less convincing than the evidence for the big bang itself. Theoretical physicists believe that the separation of the forces of nature in the very early universe could have triggered a sudden and extreme expansion, which would later settle into the more leisurely expansion that we observe today. Inflation ensures that the universe contains radiation that is smoothly distributed, as observed. Inflation also causes quantum fluctuations to be expanded into the seeds for galaxy formation, with equal fluctuations on different scales, as observed. As a result of inflation, the universe should be exactly flat; it is not yet clear whether this prediction is consistent with observations. Inflation clears up some problems associated with the big bang model. However, as yet we have no direct observations that probe back to the epoch of inflation, and we have no complete theory of the physics that would cause inflation.

Q What caused the big bang?

A We do not know. If we project the big bang back to zero time, we reach a singularity—a point of infinite temperature and density. Our theories are inadequate to explain the universe at the time of the creation. Space and time did not exist as we understand them today. Asking what came before the big bang is like asking what is north of the North Pole on Earth's surface. Some theorists suppose that our universe was one of many that emerged from a kind of quantum "foam" of space and time; however, this is no more than speculation. In the very early universe, we reach the limits of knowledge. Science is silent on the question of why the big bang occurred.

Problems

Use these problems to test your understanding of the information and concepts in this chapter. The * indicates a more advanced or mathematical problem.

1. Progress in many scientific fields, such as the study of stars, plants, and animals, has included the classification of different types of specimens, followed by comparisons of the different classes. How does a cosmologist suffer a disadvantage in this regard?

2. Telescopes much larger than present-day designs, perhaps located in space, would have much more resolving power and light-gathering ability and could reveal much fainter objects. Give examples of how this development would clarify current cosmological problems.

3. **(a)** What is the cosmological principle, and how justified is it? **(b)** Why is the sky dark at night? **(c)** How does the cosmological redshift differ from a redshift caused by the Doppler effect? **(d)** How do we estimate the distances to objects at high redshift? **(e)** Why is the expansion slowing down in big bang models?

4. Many cosmologies, such as the big bang model, assume the cosmological principle that the universe is homogeneous at any given time, given measurements over a large enough scale. Yet quasars seem to be more common per unit volume at very great distances than near our galaxy. Does this refute the big bang theory? Why or why not?

5. The measurement of cosmological parameters requires distant astronomical targets whose properties are well understood. **(a)** Why is it difficult in practice to use galaxies for cosmological tests? **(b)** Quasars are the most distant known objects. Why do astronomers not use them to measure the size and shape of the universe?

*6. Suppose the Hubble constant is shown to be 65 km/s/Mpc, with a very small error, and the density parameter is shown to be equal to 1. Why is this bad news for the big bang model if we also know that the oldest globular clusters are 14 billion years old?

*7. How does the existence of dark matter affect the measurement of the density parameter, Ω_0? What type of observations have been used to map out dark matter on the scales of clusters and superclusters of galaxies? How do we know that there are not enough normal particles—protons and neutrons—to close the universe?

8. Why was the discovery of cosmic background radiation from all over the sky heralded as strong evidence in favor of the big bang theory? Does comparing the radiation from different parts of the sky give any evidence of asymmetry or inhomogeneity in the universe? How do astronomers interpret the departures from constant temperature?

*9. By what factor smaller was the universe at a redshift of 2.5 than it is today? If the hydrogen emission line with a rest wavelength of 122 nm is observed in the spectrum of a quasar at $z = 6$, at what wavelength would we observe the emission line? What was the temperature of the relic radiation of the big bang when the universe was smaller than it is today by a factor of 200?

10. Do you find any fundamental disagreement between the description of the universe's origin and structure, as described in this and the previous chapter, and any philosophical or religious beliefs you hold? If so, do you believe such a disagreement might be clarified by further observations, or do you believe further observations are superfluous?

11. The inflationary big bang model proposes that the universe underwent a brief period of rapid expansion only 10^{-35} seconds after the big bang. **(a)** How does this model solve the horizon problem? **(b)** How does this model solve the flatness problem? **(c)** Is there any observational evidence in favor of this model?

*12. Big bang cosmology involves some mind-bending concepts. Imagine that you are talking to a curious and intelligent 13-year-old cousin. Try to convey the following ideas as clearly as you can. Explain how we can look back in time by observing distant objects. Explain how the universe might have no center or edge, even though all galaxies are moving away from us. Explain how a flat universe contains objects whose angular size actually starts to increase with increasing redshift. Explain how distant quasars and galaxies can be moving away from us faster than the speed of light. Explain how the physical universe could be much larger than the observable universe. Explain how we can possibly have information on the universe in the first few minutes of its existence.

Projects

Activities to carry out either individually or in groups.

1. Construct two-dimensional analogs for the three types of universe that we might inhabit. The analog for flat space is just a sheet of white cardboard. The analog for positively curved space is a globe; find an old globe that you can glue over with strips of paper to make a white surface. The analog for negatively curved space is a saddle shape, which you can form out of papier-mâché or modeling compound and cover with white paper. Draw a grid of lines on each surface, and confirm that parallel lines remain parallel only on the flat surface and that the angles of a triangle add up to 180° only on a flat surface.

2. Go outside and find a way to suspend a piece of paper horizontally about 10 feet above the ground (the rim of a basketball hoop is a convenient starting point). Try to throw a small stone straight up so that it *just* touches the paper at the top of its trajectory. You may need to have someone else close to the paper to confirm this. A successful throw is one where the stone touches the paper but does not visibly move it. What percentage of throws can you perform this successfully? Can you do it at all? This is an analogy for the fine-tuning that is required to live in a universe that is close to flat long after the big bang.

Reading List

Brush, S. 1992. "How Cosmology Became a Science." *Scientific American,* vol. 267, p. 62.

Cowan, D. 1994. "The Debut of Galaxies." *Astronomy,* vol. 22, p. 44.

Davies, P. 1992. "The First One Second of the Universe." *Mercury,* vol. 21, p. 82.

Hogan, C. 1998. *The Little Book of the Big Bang.* New York: Springer-Verlag.

Kinney, A. 1996. "Fourteen Billion Years Young." *Mercury,* vol. 25, p. 29.

Odenwald, S. 1996. "Space-Time: The Final Frontier." *Sky and Telescope,* vol. 91, p. 24.

Parker, B. 1988. "The Cosmic Cookbook: The Discovery of How Elements Came to Be." *Mercury,* vol. 17, p. 171.

Peebles, J., Schramm, D., Turner, E., and Kron, R. 1994. "The Evolution of the Universe." *Scientific American,* vol. 271, p. 52.

Schramm, D. 1994. "Dark Matter and Cosmic Structure." *Sky and Telescope,* vol. 88, p. 28.

Talcott, R. 1992. "COBE's Big Bang." *Astronomy,* vol. 10, p. 42.

CHAPTER 18

Life in the Universe

A hypothetical depiction of the early oceans on a special place called Earth. Life formed on Earth billions of years ago in a watery environment as simple chemicals combined to form complex molecules, using energy derived from the Sun. We do not know if these conditions are common or rare beyond the solar system. (SOURCE: Painting by Ron Miller.)

CONTENTS

- Canals on Mars
- Are We Alone?
- The Nature of Life
- Sites for Life
- The Origin of Life on Earth
- Evolution, Intelligence, and Technology
- The Search for Extraterrestrial Intelligence
- The Anthropic Principle
- Summary
- Important Concepts
- How Do We Know?
- Problems
- Projects
- Reading List

PREREADING QUESTIONS ON THE THEMES OF THE BOOK

OUR ROLE IN THE UNIVERSE
Are we alone in the universe?

HOW THE UNIVERSE WORKS
What is the full range of life processes?

HOW WE ACQUIRE KNOWLEDGE
How do we search for life beyond the solar system?

WHAT TO WATCH FOR IN CHAPTER 18

Finally, we consider one of the most profound issues in science—the possibility of life in the universe beyond planet Earth. To study this interdisciplinary topic, we will synthesize what we have already covered about planets and stars with new material on aspects of chemistry and biology. We will also recapitulate the three themes of the book. The topic is fundamental to understanding our role in the universe. We consider life in the universe in several steps. Our progression will follow an order in which the initial issues are astronomical or chemical. In principle, we can use experimentation or direct observation to answer questions concerning astronomy and chemistry. Later issues involve sociology or psychology, and we have very little scientific basis for drawing conclusions about topics in these disciplines.

First, we discuss what we mean by "life," and what conditions are necessary for life to exist on planets. This initial speculation leads to the question of whether or not there are planets that could support a similar evolution of life elsewhere in the universe. We will use our theme of how the universe works to decide whether life is likely according to the laws of physics and chemistry. Next, we review the long process that led to the formation and development of life on Earth, as best we understand it. With considerably less certainty, we consider the factors involved in the evolution of intelligence and the eventual appearance of species that have the technology to explore their universe. We will examine these issues critically in terms of our theme of "how we know." All the arguments to this point can be used to estimate a probability about whether intelligent life actually exists on planets beyond the solar system. We discuss the optimal search strategy and summarize previous and planned attempts at communication. You will see that the idea of life in the universe has been the focus of human dreams and speculations for hundreds of years. Science is now beginning to provide the answer to the question: Are we alone?

CANALS ON MARS

Percival Lowell drove his construction team hard. It was 1894, and up on a high plateau in the northern part of the Arizona Territory near Flagstaff, the wealthy Bostonian was racing to complete an observatory with a 24-inch telescope. He wanted the observatory completed to observe Mars, which would be making its closest approach to Earth in 15 years, allowing its surface features to be resolved with a large telescope. Mars would loom large in the night sky only for a few months.

Finally, the building was complete. The thin air was cold and crisp, and the stars burned brightly in the dark night sky. Lowell trained his telescope on the red planet. The features shimmered and became sharp during moments of still air. He saw the polar caps and the irregular coloring of the surface. Best of all, he saw a set of linear markings—what he thought to be over 100 "canals" crisscrossing the planet.

Were these markings evidence of life on Mars? Lowell had no doubts. The linear features were obviously not natural formations. Lowell speculated that an intelligent civilization had constructed the canals to transport water from the frozen polar caps to the arid lower latitudes. He knew his telescope did not actually have the resolution to see the canals themselves; the dark markings must be strips of vegetation that were cultivated by irrigation from the canals.

Born in 1855 into an aristocratic Boston family, Lowell made money in trade. After graduating from Harvard, he traveled the world and became a scholar of Asian history and culture. However, his true love was astronomy. (We first met Lowell in our discussion of Mars in Chapter 6.) In 1895, Lowell wrote a popular book titled *Mars*, in which he presented his case for Martian life forcefully and eloquently. He wrote:

> That beings constituted physically as we are would find Mars a most uncomfortable place is pretty certain. But there is nothing in the world or beyond it to prevent, so far as we know, a fish with gills, for example, from being a most superior person. A fish doubtless imagines life outside water to be impossible; and similarly to argue that life of an order as high as our own, or higher, is impossible because of less air to breathe . . . is to argue, not as a philosopher but as a fish.

Lowell was not the first to see canals on Mars, and he would not be the last. In two previous close approaches by Mars, two Italian astronomers had observed scraggly markings they called *canali*, or channels. When American newspapers picked up the story, they translated the word as "canals," which carried the implication of intelligent engineering. Lowell had heard about the Italian observations and wanted to repeat them with a more powerful telescope. News of Lowell's discovery spread quickly. Even the sober *Wall Street Journal* picked up the Mars fever, writing in a year-end summary that "the most extraordinary event of the year is the proof afforded by astronomical observations that conscious, intelligent life exists on Mars."

Even as the debate raged, doubts were spreading. Astronomical photography was in its infancy, so images of the planet had to be laboriously drawn by hand. A few professional astronomers claimed to see the canals, but many more did not. The director of the Lick Observatory described Lowell's writings as misleading and unfortunate half-truths. The head of the Mars observing section of the British Astronomical Society wrote: "Had it not been for the foreknowledge that 'the canals are there,' I would have missed at least three-quarters of them." At the next close approach of Mars in 1910, astronomers had access to improved photography and larger telescopes, and they showed that the canals were irregular, disconnected features. Moreover, newer observations showed Mars to be far too arid to support large life forms. The polar caps are made of carbon dioxide, not water. Alfred Russel Wallace, co-creator of the theory of evolution with Charles Darwin, wrote emphatically: "Not only is Mars not inhabited by intelligent beings as Mr. Lowell postulates, it is absolutely uninhabitable." Lowell had been burned by the power of suggestion and wishful thinking.

However, the genie was out of the bottle. Within 3 years of Lowell's book, H. G. Wells published *The War of the Worlds*. In this influential work of science fiction, intelligent creatures live on Mars, but the canals are not enough to sustain life on the dying planet. In desperation, the Martians invade Earth with superior military technology, only to be beaten back as they succumb to terrestrial microbes. And so begins over a century of speculation about life in the universe.

Are We Alone?

One of the most intriguing questions in astronomy is whether or not Earth is alone in the universe in harboring life. Either extraterrestrial life exists, or it does not. Both possibilities have striking consequences. As American architect and designer Buckminster Fuller said, "Sometimes I think we're alone. Sometimes I think we're not. In either case, the thought is staggering." The universe is a large and bountiful place. There are an enormous number of potential sites for life, and the chemical constituents of life are distributed widely across space. At first, it seems unlikely that we are alone. If intelligent life exists beyond the solar system, our world might well be influenced by it, for better or worse. At the very least, as anthropologist D. K. Stern has pointed out, the discovery of alien life "would irreversibly destroy our self-image as the pinnacle of creation." The alternative is equally profound. If we are the only intelligent creatures in the universe, it would imply a universe of incredible grandeur with us as the only spectators.

The search for alien life is a new scientific adventure in the making. Astronomers study the possibilities of life beyond Earth as a direct way of questioning our role in the universe. How do we know the answer to the question, Are we alone? Admittedly, discussions of the existence, intelligence, psychology, or appearance of higher alien life forms are almost entirely speculative. But an interdisciplinary group of researchers has begun to take these issues very seriously. Astronomers have detected planets orbiting nearby stars. They have also found complex, carbon-based molecules in interstellar space and in meteorites, telling us that complex chemistry is active beyond our own solar system. Chemists have studied the pathways by which replicating molecules and simple life forms can be synthesized from simple constituents. Evolutionary biologists have considered how life might evolve in complexity from single-celled organisms. Physicists have even calculated the rates at which interstellar cultures might populate the galaxy. Philosophers have tried to gauge the likely impact on human culture of contact with intelligent aliens.

These are exciting issues. However, much of the published work on life in the universe has been tinged with undue optimism concerning the possibility of intelligent life beyond Earth. We are awash in a popular culture that encourages us to believe aliens exist. It is important to counter this natural tendency with realism and intellectual rigor. It is important to be clear about which questions can be answered by the scientific method, and which ones cannot. At the moment, we know of only one planet that harbors life: Earth. At the moment, we know of only one species with the intelligence to understand and explore the universe: humans. To apply the rigorous standards of the scientific method to speculation about extraterrestrial life, let us first begin by trying to define what we mean by life.

The Nature of Life

What do we mean by life? **Life** is essentially a process—a series of chemical reactions involving carbon-based molecules. In this process, matter and energy are taken into a system, used for growth and reproduction, and then expelled as waste products. Most biologists would agree on the following set of requirements for something to be considered alive:
1. Living things grow, evolve, and reproduce.
2. Living things are highly organized chemical factories.
3. Living things require energy, and they respond to their environment.

The most important characteristics of life are *growth* and *reproduction*. The smallest unit in which life processes occur is the **cell.** All known living things are composed of one or more cells, which contain an intricate array of molecules. Most biologists think that the simplest organisms are the simplest living cells: **bacteria.** How far can we stretch our definition of life? Is a virus considered a life-form? Although viruses are simpler than cells, they can reproduce themselves using materials from host cells. However, viruses cannot function independently of cells, and they appear to have evolved from living cells rather than the other way around. Similarly, can certain machines or computers be considered living entities? Technology has advanced to the point where machines and computers have taken on many of the attributes of living things. We will consider the provocative possibility of artificial life later in the chapter.

A key aspect of living things is their dynamic nature. We often make the mistake of thinking of ourselves as static beings, yet virtually every cell in the body is replaced over a 7-year period. Even our seemingly inert skeletons are living and changing, always replacing their cells. We *must* keep changing; our cells process new material in order to stay alive. When the processing stops, we call it death. The dynamic nature of living beings is illustrated by an analogy from Russian biochemist A. I. Oparin. Consider a bucket that has water pouring in at the top from a tap and flowing out at the same rate through a tap at the bottom. The water level in the bucket stays constant, and a casual observer would call it "a bucket full of water." But it is not simply a bucket standing full of water. The water at any instant is not the same water at any other instant, yet the outward appearance is constant. Humans are like buckets with water, nutrients, and air flowing through us. We also have more complex attributes, such as the ability to reproduce and be affected by structural changes that let us evolve from generation to generation.

What is life made of? The elements most common in living human tissue are listed in Table 18–1, along with the most common elements in the Sun, Earth's crust, and Earth's atmosphere. This table has several interesting aspects. First, over 99 percent of the atoms in your body are from just four elements: hydrogen, oxygen, carbon, and nitrogen. The large amount of hydrogen and oxygen in living organisms (and the ratio of two hydrogen atoms to every oxygen atom) indicates the high percentage of water that all life contains. Nitrogen is the most common element in Earth's atmosphere, as well as being important in all living things. Carbon is the critical element for life. **Organic chemistry** is a branch of chemistry relating to compounds containing carbon, regardless of whether or not a living organism is involved. As we saw at the beginning of Chapter 13, the life elements (except hydrogen) are created inside stars and are widely distributed through the cosmos. It is striking that the composition of life resembles a star more than it does the Earth below our feet. For example, carbon and nitrogen are essential to life, but they are more common in the Sun than they are in the Earth. Apart from oxygen, the most common elements in the Earth—iron, silicon, and aluminum—play only small roles in organic chemistry.

Why does life use so few of the 85 stable elements that exist in nature? The answer lies in carbon's unique ability to build complex molecules. Helium is the second most abundant element in the universe, yet it cannot form chemical bonds with other elements to build complex molecules. This trait makes it useless for creating life-forms. Among the most common elements in living organisms, hydrogen can combine with oxygen to produce only two molecules: water

TABLE 18–1
Most Common Elements in Life, Earth, and the Sun

HUMAN BODY		EARTH'S CRUST		EARTH'S ATMOSPHERE		THE SUN	
O	65%	O	46%	N	76%	H	75%
C	18%	Si	28%	O	23%	He	23%
H	10%	Al	8.3%	A	1.3%	O	0.009%
N	3%	Fe	5.6%	C	0.00013%	C	0.004%
Ca	2%	Ca	4.2%	Ne	0.000013%	Fe	0.0014%
P	1.1%	Na	2.4%	Kr	0.000003%	Si	0.0010%
K	0.35%	Mg	2.3%	He	0.0000007%	N	0.0009%
S	0.25%	K	2.1%	Xe	0.0000004%	Mg	0.0008%
Na	0.15%	Ti	0.006%	H	0.00000004%	Ne	0.0006%
Cl	0.15%	H	0.001%	S	0.000000001%	S	0.0004%

(H_2O) and hydrogen peroxide (H_2O_2). Similarly, hydrogen can combine with nitrogen to form only two molecules: ammonia (NH_3) and hydrazine (N_2H_2). On the other hand, the number of ways that hydrogen can combine with carbon is so large that it is unknown! The largest molecule listed in the *Handbook of Chemistry and Physics* has a chemical formula of $C_{90}H_{154}$. Carbon is thus the perfect building block for complex structures.

We can think of the chemical processes of life as a type of information storage. Organic chemistry uses only a small number of ingredients, yet these ingredients can combine with great complexity and therefore have the potential to store large amounts of information. Is carbon-based chemistry the only way to store information? Chemists (and science fiction writers) have speculated on life chemistry based on silicon, or some other element. Even more speculative is the idea of life based on some other organizing principle, such as electric or magnetic fields. We have never observed such life-forms, so we can say nothing substantive about them. However, the chemistry of the universal elements is well understood. In terms of a basis for life, carbon is superior to any other element in its ability to form complex chains and thereby store information.

Earth is a watery planet. We might wonder, therefore, whether water is essential to life. First, we note that water is potentially the most abundant liquid in the universe. Table 18–1 shows that oxygen is far more abundant than silicon, the main rock-forming element. Therefore, a rocky planet that uses up all its silicon by combining it with oxygen to make rocks will still have plenty of oxygen to combine with the most abundant element, hydrogen, to make water, or ice. Because it is liquid over a wide range of temperatures, water acts as a *solvent:* It dissolves other materials to make a solution. In this role, water maintains many vital cell functions—it dissolves and transports nutrients and waste products within a cell, regulates an organism's temperature, and even plays a role in shielding living things from harmful ultraviolet radiation (Figure 18–1). It is thus not surprising that

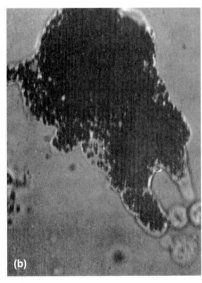

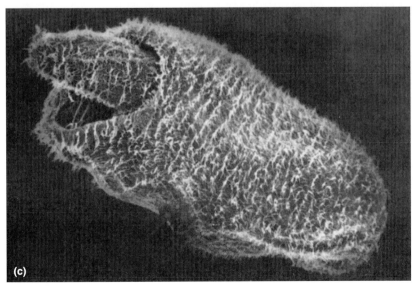

FIGURE 18–1
The importance of water to life on Earth.
(a) Tidewater pools (foreground) may have been a factor in the origin of life. Once organic molecules formed in seawater, the evaporation of water in the tidal pools would have concentrated these molecules, promoting reactions that would lead to more complex organic substances. The result was a broth rich in amino acids and complex organic molecules, which would be dumped back into the ocean at high tide, "fertilizing" the seas. (SOURCE: Photograph by WKH.) **(b)** Once primitive life evolved, water was an important medium for life processes. An amoeba flows to surround a nearby food particle. (SOURCE: Optical micrograph by S. Wolfe.)
(c) A liquid medium allows simple organisms to reach each other and interact. Here a protozoan *Woodruffia* ingests a paramecium. (SOURCE: Electron micrograph by T. Golder.)

complex life forms originated in the oceans of Earth. Even on land, a large percentage of the weight of plants (40 percent) and animals (70 percent) consists of water. Other solvents exist, such as ammonia and ethyl alcohol. However, water is the most abundant liquid, and it has special advantages for facilitating life processes.

We can also place the chemical basis for life in a cosmological setting. As we saw in the last chapter, the early universe contained only the lightest elements: hydrogen, helium, and trace amounts of lithium and beryllium. Such a universe is chemically *inert,* having no basis for complex chemistry. The carbon, nitrogen, and oxygen required to make organic molecules do not exist until a generation of massive stars has created these elements by fusion. Moreover, several generations of star birth and death are required before these elements return to the interstellar medium (by mass loss from giant stars, novae, and supernovae). In other words, the ingredients for life increase in abundance and become more widespread as the universe gets older. Also, we have good evidence that the CNO cycle (the means by which stars make elements heavier than helium; see Chapter 13) operates in stars across our galaxy and in stars in other galaxies as well. It follows that the chemical basis for life exists across the universe and is not due to a particular set of circumstances in our stellar neighborhood. This has important implications for the probability that life formed elsewhere.

Sites for Life

The next issue we address is whether or not *sites for life* exist outside the solar system. In essence, we are returning to our theme of how the universe works to understand whether or not our solar system is unique. Our discussion assumes that a planet is required for life—that life cannot develop on a small piece of rocky debris, in a nebula, or in the depths of interstellar space. Planets have a high concentration of the elements heavier than hydrogen or helium that are needed to set up a complex chemistry. Recall from our discussion of solar nebula in Chapter 9 that the hot zone closer than about 5 AU from the Sun allowed only metals, such as iron and nickel, and heavy rock-forming elements, such as aluminum and silicon, to condense. These dense, rocky planets never became large enough to retain an atmosphere of hydrogen and helium, the light, abundant gases in the solar nebula. Beyond 5 AU, the temperatures were cool enough that metals, rocks, and the more volatile gases could condense. These gases are key ingredients for the eventual formation of life: water vapor, methane, ammonia, carbon dioxide, and others. Planets forming in this outer zone grew large enough to retain light atmospheres, so they resemble the Sun in their chemical composition. How did the inner planets acquire their volatile gases? Recent models suggest that these gases were the last additions in the accretion process, coming from bodies that resemble today's meteorites and comets.

Our forming solar system was like a cosmic smelter. The overwhelming majority of the gas in the cloud that formed the solar system collapsed into the newly formed Sun. However, the small amount of material left over contained concentrations of heavy elements enhanced by factors of hundreds or thousands, compared with average values for the Sun (see Table 18–1). At distances of a few AU, planets grew by the accretion of rocks and metals and then ices. We live on an icy cinder that is the residue from the formation process of the star that now warms us. It sounds like an implausible sequence of events. Yet there is a firm physical basis for the ideas of accretion, condensation, and radiation upon which this picture is based. Theory indicates that planet formation should be a general by-product of star formation. What do observations tell us?

As we learned in Chapter 9, the difficult search for extrasolar planets has finally paid off. Astronomers have discovered over two dozen Jupiter-sized planets orbiting stars beyond the Sun. Success has been achieved by looking for the Doppler wobble of a star caused by the tug of a giant planet; no search yet has the sensitivity to detect an Earth-like planet. This dramatic result ends over 2000 years of speculation. We have taken an important new step in the Copernican revolution by showing that planets are scattered through space, and Earth is not the only vantage point from which to view the universe.

The next generation of experiments will reveal planets by direct imaging. Astronomers are already using special techniques to sharpen the images made with ground-based telescopes, compensating for the blurring effect of Earth's atmosphere. They are also planning *interferometers,* linked telescopes with the resolution of a single large telescope equal in size to the separation of the individual telescopes in the array. Interferometers in space could achieve a resolution of 10^{-6} seconds of arc, sufficient to detect Earth-like planets around nearby stars. Beyond imaging, astronomers hope to use large new telescopes to spread the feeble reflected light from extrasolar planets into spectra. We can then learn about the chemistry of the atmospheres of these remote planets. As we saw in Chapter 6, oxygen is highly reactive, so when we see it in a planet atmosphere, it is a strong sign of a metabolism; in other words, oxygen is continually replenished by photosynthesis or another life process. We might be able to infer life on other planets by the presence of oxygen (O_2), along with ozone (O_3) and water vapor (H_2O). The search for planets is an exciting enterprise that will be vigorously pursued in the coming years; it may finally clarify whether or not our own planetary system is commonplace or freakish.

What are the requirements for a planet to be suitable for life? Since we know so little about the diversity of planets, we should try to make as few assumptions as possible. The basic assumption is that the planet's temperature must allow liquid water to exist. Chemical reactions proceed thousands of times more slowly in a gas than in a liquid, and with enormous difficulty in a solid. A liquid is by far the best medium for chemical and biological processes. The normal temperature range for liquid water (0°C to 100°C, or 273 K to 373 K) can be extended if the water is mixed with ammonia or if the water is under pressure below a planetary surface. We do not have to assume that the energy source for life is starlight, since life might be facilitated by geothermal energy within a planet or from tidal flexing of a planet. We assume that both planets and large moons of planets are potential sites for life.

Astronomers have proposed several conditions required to make a planet habitable. The central star should be a main-sequence star; evolutionary stages beyond the main sequence are too short-lived or generate too little energy to shelter life. The central star should be no more than about 1.5 $M_\odot$. This upper boundary allows enough time on the main sequence (at least 2 billion years) for multicellular life to evolve. It also limits stars to about four times the Sun's luminosity; a luminosity beyond this would provide too much ultraviolet radiation, which breaks down organic molecules. The central star should be at least 0.3 $M_\odot$. (Notice that the lower boundary is much more important than the upper bound, since the vast majority of stars are low in mass.) This lower boundary, corresponding to at least $1/100$ of the Sun's luminosity, allows the star to be warm enough for nearby planets to retain water. Even cooler stars have zones where water would be liquid that are so close to the star that a planet at such a small distance would have its atmosphere ripped off. Moreover, planets so close to a star would be tidally locked, such that the same face always points to the star. Figure 18–2 shows the habitable zones around a variety of star types. About 25 percent of the 40 billion stars in the Milky Way are main-sequence stars of spectral types F, G, and K, which satisfy these first two conditions.

We might add to these conditions the requirement that the planet must be large enough for gravity to retain a substantial atmosphere. Also, the planet's orbit must be nearly

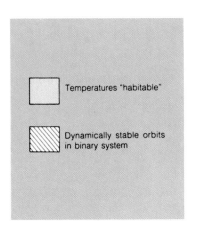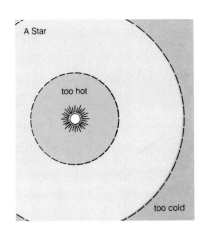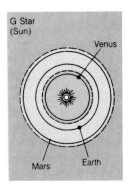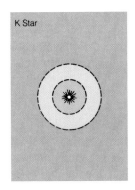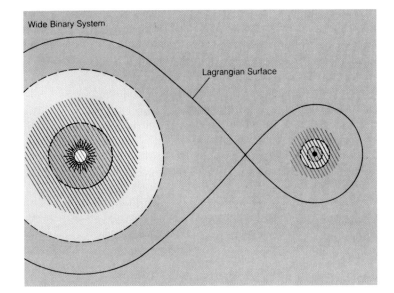

FIGURE 18–2
The green doughnut-shaped regions show the habitable zones around each star, where water neither evaporates nor freezes. These zones are presumed to be "just right" for life. In a binary system (bottom), the situation is complicated because orbits outside the hatched regions experience large perturbations, causing evolution away from circular orbits. This in turn causes large temperature extremes on the planet surface.

(SOURCE: Adapted from A. Huang.)

circular, or at least stable enough to keep it at a proper distance and prevent drastic seasonal changes. Planetary orbits are unlikely to be stable in binary star systems. So we must probably exclude them, except for very close or very widely separated pairs where the planets may be undisturbed by the stellar motions.

Thus, astronomers conclude that sites for life may be found around a wide range of main-sequence stars, on either planets or moons of large planets. Most sites for life may be variations on a familiar theme—Earth-like planets in orbit around a Sun-like star—but nature may have provided an unexpected diversity of habitable places. So far, none of the newly discovered extrasolar planets appears to be suitable for life. While we continue looking, astronomers hope they can learn more about the evolution of life by looking closely at the history of the place we call home.

The Origin of Life on Earth

As we saw in the previous discussion, the basic chemical ingredients for life are distributed widely in the universe. Carbon, nitrogen, and oxygen have been produced in the cores of stars for billions of years and then distributed into the interstellar medium, where they are incorporated into new stars. Planets, the necessary sites for life, are believed to be a natural by-product of the process of star formation. No definitive examples of life have been found beyond our solar system thus far. To understand the origin of life on Earth, we must understand the early history of interactions among the elements essential for life: carbon, nitrogen, oxygen, and hydrogen. How did these elements form complex molecules and then living organisms?

The Production of Complex Molecules

Complex molecules are requirements for life. A living organism must be able to store and transmit large amounts of *information*. As an analogy, consider the way information is stored in a computer. (To review the idea of digital information, see the Science Toolbox in Chapter 10.) A single bit of information in a computer does not convey much complexity; it represents an electric current "high" or "low" or a decision "yes" or "no." However, when millions or billions of bits of information are stored, they can convey great complexity, enabling a computer to solve complicated mathematical problems or play chess better than most humans. From this perspective, molecules are tools for storing information and vehicles for carrying out the many processes of a living organism. Table 18-2 lists the different levels of complexity in biochemical life, the amount of information coded at each level, and information storage analogies in terms of human language.

What are the simplest molecules required for life processes? Increasing in complexity from the life-forming elements are **amino acids.** The 20 amino acids found in living organisms contain between 10 and 27 atoms of carbon, nitrogen, oxygen, and hydrogen. Several hundred amino acids can combine to form **proteins.** Proteins are long molecules with a convoluted structure of chains, loops, and rings; they facilitate, or catalyze, chemical reactions. Proteins have thousands of atoms. In theory, amino acids could combine to form proteins in an enormous number of ways (larger than the number of atoms in the universe), but living systems use only a tiny fraction of them. As with the elements of the periodic table, life operates with a high degree of selectivity.

The next level of molecular complexity is a long and complex molecule called **deoxyribonucleic acid (DNA),** and its close relative *ribonucleic acid (RNA).* DNA is the key to life—the famous double helix, whose structure was determined by James Watson and Francis Crick in 1953. DNA has two strands wound around each other in a graceful spiral. The strands are connected at many points by molecules called *bases.* Out of many possible bases, living systems use only four: adenine (A), cytosine (C), guanine (G), and thymine (T). In terms of complexity and information storage, DNA has prodigious capabilities. The sequence of bases along a strand of DNA represents the message, or the *genetic code.* The alphabet is very small in the language of life; there are only four "letters." However, the message is very long: There are about 30 billion atoms in a typical strand of DNA.

DNA's central function is to transmit an organism's genetic information. The genetic blueprint carried by a single human cell is enormous; the information content is calculated in the next Science Toolbox. Another essential function of DNA is its ability to *replicate.* When the intertwined strands of DNA separate, each strand can form a new double helix. Base pairs cannot link up randomly; adenine (A) pairs only with thymine (T), and cytosine (C) pairs only with gua-

TABLE 18-2
The Information Content of Life

LEVEL	FUNCTION	NUMBER OF ATOMS	AMOUNT OF INFORMATION	LANGUAGE ANALOGY FOR INFORMATION STORAGE
Base	Joins strands of DNA	10 atoms	1 bit	Letter (A, T, C, G)
Codon	Controls genetic code	100 atoms	6 bits	Word (e.g., ATC)
Gene	Determines characteristics of organism	10^3-10^5 atoms	10^2-10^3 bits	Sentence
Bacterium	Simple life form	10^6-10^7 atoms	10^5-10^6 bits	Short book
Human	Complex life form	3×10^{10} atoms	6×10^9 bits	Encyclopedia

SCIENCE TOOLBOX

Life as Digital Information

James Watson, one of the discoverers of the structure of DNA, was once asked by a journalist to summarize the significance of his discovery in a single sentence. Watson thought hard for a moment, and then said, "All life is digital information." We will explore the implications of his insight, but first let us look at a couple of analogies.

Recall that a computer needs two things in order to function. A computer stores information in bits, each of which is a simple distinction between "on" and "off," or "yes" and "no." With a sufficient number of bits, a computer can represent and store large numbers, images, or words in a language. A computer also needs rules for the manipulation of the information. We call these rules the program, or the software.

The English language (or any other language) is a way to convey information. The smallest unit of information is a letter, and the smallest unit of useful information is a word. The English language combines letters in many different combinations to make words, but not all of the possible combinations of letters are allowed, and the average person uses only 3500 to 4000 words to communicate. The rules of English dictate how those words can be combined, and the results can be as beautiful and varied as a haiku poem or a long novel.

The genetic code consists of only four letters: adenine (A), cytosine (C), guanine (G), and thymine (T). Each of these bases is a small molecule consisting of less than a dozen atoms. DNA is constructed from a long string of bases, each one in combination with a sugar molecule and a phosphate molecule; the assemblage of all three is called a *nucleotide*. The bases in a long DNA molecule are like the rungs that join together the two sides of a ladder. The information of the four bases can be represented as two bits:

A = 1 1 *or* on on

T = 1 0 *or* on off

C = 0 1 *or* off on

G = 0 0 *or* off off

The genetic code has only a couple of rules. The shape of the base molecules means that they can only combine in a few ways: Adenine and thymine can combine with each other, and cytosine and guanine can combine with each other, but no other pairings are possible. Therefore, the four possible rungs of the DNA ladder are AT, TA, CG, and GC. The other rule relates to the way that DNA replicates. When DNA splits and reproduces, it is "read" in groups of three bases at a time. Each three-base group is called a *codon*. A codon is analogous to a word in the English language. With four bases, there are $4 \times 4 \times 4 = 64$ ways that bases can combine in groups of three. Each codon therefore has $2 \times 3 = 6$ bits of information.

Now we step up to the next level of complexity. Each codon specifies a particular amino acid. A sequence of codons on the DNA strand (corresponding to a base sequence such as ACCAGGTC . . . and so on) is called a gene, and it corresponds to a specific protein. Proteins vary greatly in size, so a gene may have anywhere from several dozen to several hundred codons. We can think of a gene as analogous to a sentence in the English language. Each gene contains several thousand bits of information. A central feature of living systems is the fact that the sequence of bases in a gene governs a particular chemical process. In a human, that chemical might control hair color, height, the ability to jump, or the tendency to develop a particular disease.

A human has about 100,000 genes. Simple organisms have far fewer. The total number of bases in our DNA is about 3 billion, or 3×10^9. Now we can calculate the information content of a human's genetic material. Each base is two bits, so the information content is $2 \times 3 \times 10^9 = 6$ billion bits. Recall from Chapter 10 that 8 bits equals 1 byte. Thus, the information content is $(6 \times 10^9)/8 = 7.5 \times 10^8$ bits, or 750 megabytes—equal to the information contained in 500 books, or two hours of music on a compact disk. You could easily store the information of the human genetic material on a personal computer or carry around it on a CD-ROM!

Once we have accepted the fact that life is digital information, we might speculate that carbon chemistry is not the only way to store and transmit information. With computers of sufficient complexity, could we mimic or recreate the essential attributes of life? Carbon chemistry is not even the most efficient way to store information. Recall that our biochemistry depends on the shapes of molecules, and the simplest useful molecules in life processes have about a dozen atoms. Physicists are experimenting with using *single* atoms to store information. For example, the spin axis of an atom in a lattice might be used to code information—spin up equals 1, or "on," and spin down equals 0, or "off." The creation of microscopic machines that can store and process information is one of the frontiers of computing.

nine (G). In this way, the genetic blueprint in the parent DNA molecule is precisely reproduced in the two new molecules (Figure 18–3). Each sequence of several hundred bases that codes for an inherited trait is called a **gene**.

How did these complex molecules evolve on Earth? Scientists' models of early Earth are too crude to make detailed calculations of molecular chemistry on the cooling crust. However, biologists have tried to simulate "life in a bottle"—

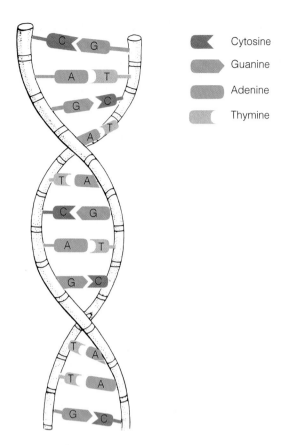

FIGURE 18-3
A section of the DNA molecule. DNA carries the genetic information in a long sequence of base pairs. Adenine (A) pairs only with thymine (T) and cytosine (C) pairs only with guanine (G). When the rungs of the twisted ladder split and DNA reproduces itself, the base-pair sequence is copied exactly.

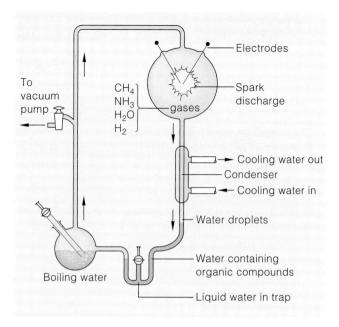

FIGURE 18-4
The apparatus used by Miller and Urey to reproduce the conditions on the primitive Earth. With energy input from an electrical arc, the experiment produced a number of amino acids, the building blocks of proteins.

an attempt to generate complex molecules in an artificial environment. In a famous series of experiments in the 1950s, Stanley Miller and Harold Urey bottled gases such as hydrogen (H_2), methane (CH_4), and ammonia (NH_3), together with water. The gases represented Earth's primitive atmosphere, and water represented the ocean. Energy was added in the form of an electrical discharge, which represented lightning (Figure 18-4). After a few days, the flask contained an organic sludge containing amino acids. This experiment has been successfully repeated many times with different primitive atmospheres and energy sources. Current models suggest that Earth's early atmosphere had much less free hydrogen than was used by Miller and Urey in their experiments. However, even on planets with fewer hydrogen compounds and more volcanic gases, such as carbon dioxide (CO_2), nitrogen (N_2), and carbon monoxide (CO), sunlight tends to initiate photochemical reactions that produce amino acids.

Such laboratory experiments are successful in producing many of the important compounds for living systems. Amino acids and sugars are formed, and chemical pathways are known for all the five critical bases used by DNA and RNA (in RNA, uracil replaces thymine). Note that these experiments are no more than plausible scenarios. Their results are only as reliable as our sketchy knowledge of the conditions on the primitive Earth. Moreover, none of these experiments has ever succeeded in synthesizing DNA, RNA, or any other replicating molecule. Perhaps this failure is not surprising, given that the increase in complexity from an amino acid to the simplest living organisms is a factor of thousands.

For a long time, scientists thought the emergence of complex molecules from simple building blocks must have been an extremely unlikely event. We can imagine small molecules colliding in a liquid and occasionally sticking together. It would take many collisions and reactions to build a large unit. Some calculations have indicated it might take millions of times longer than the age of the universe to build a small strand of DNA! From this perspective, life must be a fluke that may have taken place only on Earth. Recent research shows that carbon molecules in a water solution actually combine rapidly to form complex chemical networks. As molecules grow, their shapes permit a rapidly increasing number of chemical reactions. Complexity leads to more complexity.

However, in solving the great puzzle of how these molecules first appeared on Earth, we are faced with a classic "chicken and egg" problem. DNA and RNA are ideally suited for replication, but to do so, they require protein catalysts. Proteins are ideally suited for life functions, but they cannot be synthesized without the direction of DNA and RNA. How do complex molecules express themselves in the functions of life? Two ideas are being actively explored. English biochemist Leslie Orgel and his colleagues have shown that RNA can catalyze its own reactions, leading to the argument that RNA initiates protein synthesis. This discovery makes it likely that RNA preceded the more complex and stable DNA as the primordial information-carrying molecule. Another biochemist, A. Cairns-Smith, has suggested a primitive replicating system that was independent of DNA, RNA, and proteins: Molecules began to organize themselves on the fine lattices of clay. Two sheets of clay would be like the two strands of a DNA molecule. Many new and strange ideas may have to be explored before we understand how replicating molecules evolved.

Complex Molecules in Space

We have considered planets as the most likely sites for life, but many of the basic molecular building blocks of life form in deep space from material that has been ejected from stars. These molecules can be found in the interstellar medium, so analyzing its chemistry reveals important information about the origins of life. More than 80 molecular types have been identified in the interstellar medium in the past 20 years (Table 18-3). Most of these have been detected by very-low-energy changes in their modes of rotation, which produce spectral features at radio frequencies; a few have been identified by their vibrations, which generate features in the infrared part of the spectrum. The molecules in the interstellar medium include ethyl alcohol (C_2H_5OH), the amino acid glycine ($C_2H_5O_2N$), and a number of other organic molecules.

We can also learn about molecules in cold and deep space by studying meteorites. Occasionally a meteorite "fossil" representing the primordial composition of the solar nebula drops right into our lap. Radioactive dating of such meteorites reveals ages as old as 4.5 billion years, when Earth and Sun were first forming. A small percentage of meteorites are rich in carbon materials; they are called *carbonaceous chondrites* and were described in Chapter 8 (Figure 18-5). Only a few meteorites have been recovered quickly enough to not have been contaminated by molecules from Earth. These rare finds have provided most of our information about complex molecules from space. Dozens of amino acids have been found inside carbonaceous chondrites. In 1983, Cyril Ponnamperuma announced that all five critical bases for coding genetic information in RNA and DNA were found in a single carbonaceous chondrite. Many of life's building blocks arise naturally in extraterrestrial bodies. Most of the amino acids in meteorites do not occur in life on Earth and must have an extraterrestrial origin. Scientists are confident of this conclusion because atoms can link into molecules with either "left-handed" or "right-handed" symmetry. These terms refer to shapes that are mirror images of each other, like left and right hands. Earth uses *only* left-handed molecules, whereas the complex molecules in meteorites occur equally in left- and right-handed varieties.

Comets are icy messengers from the outer solar system that have been shown to contain complex molecules (see Chapter 8). Using the powerful techniques of molecular biochemistry, scientists can analyze meteorites directly, but analysis of cometary material is more indirect and relies on spectroscopy of the reflected light. Since massive molecules have very complex and subtle spectroscopic signatures, identifying amino acids indirectly in comets is beyond the capability of current technology. An upcoming NASA mission hopes to overcome this obstacle by rendezvousing with

TABLE 18-3
Molecules in Interstellar Space

					Number of Atoms						
2	3	4	5	6	7	8	9	10	11	12	13
H_2	N_2H^+	NH_3	C_4H	CH_3OH	CH_3CCH	CH_3COOH	CH_3OCH_3	CH_3C_5N	HC_9N		$HC_{11}N$
CO	HCO^+	H_2CO	CH_2NH	CH_3SH	CH_3CHO	CH_3C_3N	CH_3CH_2OH	CH_3CH_3CO			
CH	HCS^+	H_2CS	NH_2CN	CH_3CN	CH_3NH_2	$HCOOCH^3$	CH_3CH_2CN				
CN	HCN	HNCO	HCOOH	NH_2CHO	CH_2CHCN		CH_3C_4H				
CS	HNC	HNCS	CH_2CO	CH_3NC	HC_5N		HC_7N				
C_2	C_2H	C_3H	HC_3N	C_5H	$CH_2(CN)_2$		C_2H_5OH				
CH^+	H_2O	C_3O	*C_3H_2	HC_2CHO	CH_3CCN		C_2H_5CN				
OH	SO_2	C_3N	SiH_4	C_5H	C_6H						
NO	H_2S	$HOCO^+$	H_2CCO								
NS	HCO	C_2H_2	CH_2CN								
SO	OCS	$NCNH^+$	C_5								
SiO	HNO	HSCC or	SiC_4								
SiS	HOC^+	HSiCC	H_2CCC								
HCl	*SiC_2	H_3O^+	HCCCO								
PN	*H_3^+	CCCN									
NaCl	CCH	CCCO									
KCl	CCO	CCCS									
AlCl	CCS	HCCH									
AlF	C_3										
SiC											
CP											

*Ring molecule.

FIGURE 18-5
A fragment of the Murchison meteorite, a carbonaceous chondrite that fell in 1969 in Australia. Analysis of this meteorite revealed evidence of extraterrestrial amino acids. Similar carbon-based molecules are essential to living cells on Earth. Although no fossils or life-forms have been found in the meteorite, its dark color is due to abundant carbon and carbon compounds. The building blocks of life can form beyond Earth. (SOURCE: Chip Clark, National Museum of Natural History.)

a comet and making a detailed chemical analysis of its nucleus and coma. For example, spacecraft that made close encounters with the nucleus of Halley's comet found a dark surface with a large amount of carbon-rich material. It is indeed surprising that such complex molecules can form in the cold and hostile environment of deep space. Although there has been much speculation about life in comets, *no evidence* has been found of replicating molecules or primitive forms of life in comets or meteorites. However, these primitive bodies may have played a key role in the emergence of living things. They may have deposited organic material during the early bombardment of the solar system, giving a head start to the prebiological chemistry of all the inner planets, including Earth.

We can also learn about complex chemistry from our close planetary neighbors. Outside the inner planets, the interesting satellite Titan may have produced a photochemical smog. With its reddish clouds of carbon compounds and its possible rains of liquid methane, Titan may provide an intriguing natural laboratory for further study of organic chemistry (see Chapter 7). This is not to say that all planets are oozing with organic slime. Sunlight encourages reactions that not only *form* organic molecules but also *break down* organic molecules. When energetic ultraviolet photons strike molecules, the molecules can break apart. On worlds like the Moon and Mars, where there is no ozone layer to block ultraviolet radiation, the rate of destruction is high, and gases are insufficient to provide raw materials to form new organic molecules. As we saw in Chapter 6, sunlight has apparently destroyed any molecules that may have formed on Mars in the past. Indeed, fossil and geochemical evidence suggests that life did not emerge on Earth's land surface until ocean plants, which were shielded from sunlight by seawater, emitted enough oxygen to build up an ozone (O_3) layer in Earth's atmosphere. This ozone layer then shielded land-based life from solar ultraviolet rays.

Although the Viking landers found no organic molecules on Mars, they did reveal interesting clues about the development of such molecules on planets. Unlike on Earth, Martian soil is exposed to strong solar ultraviolet rays, which apparently produce unfamiliar chemical reactive states in minerals. The soil contains material that can synthesize organic molecules from atmospheric carbon dioxide and release gas once nutrients are added. Because the Martian soil in its natural state contains almost no organic molecules, most scientists believe that these processes are not caused by Martian life forms. In fact, laboratory experiments in 1977 duplicated most Viking results using simulated Martian soil (iron oxide minerals exposed to ultraviolet light in carbon dioxide) without any organisms. Nonetheless, the processes may indicate how chemical reactions create material from which life could form on other planets.

From Molecules to Cells

Now we come back down to Earth to describe how the first life-forms on our planet evolved. The point when a complex assemblage of molecules deserves to be called a life form is a matter of conjecture. We do not know how and when those first primitive life-forms emerged, but scientists have come up with a likely scenario. It is generally agreed that the appearance of life depended on the changing conditions on the early Earth. Planet Earth formed about 4.5 billion years ago, with a primitive atmosphere containing gases rich in carbon and nitrogen and deficient in hydrogen and oxygen. During its first 100 million years, Earth was bombarded by many rocky fragments, some of which may have been as large as the Moon. Such impacts would have vaporized the oceans and a large amount of Earth's crust. Material rich in hydrogen, oxygen, and complex molecules arrived somewhat later, in the form of comets and objects from the outer solar system. Early Earth was a strange and almost unrecognizable place, yet life first formed in this environment.

The next step—aggregating complex organic molecules into cell-like structures—must have taken place over the next 500 million years. Although some of the processes can occur in dry environments, water was probably critical to

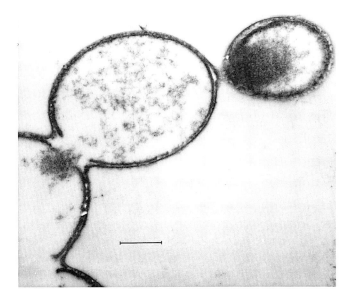

FIGURE 18-6
Protein material in water can form microscopic spheres that have a double-layered boundary similar to a cell membrane. These proteinoids do not contain DNA or genetic information, but they may have been a factor in the development of complex molecules.
(SOURCE: S. Fox and K. Dose.)

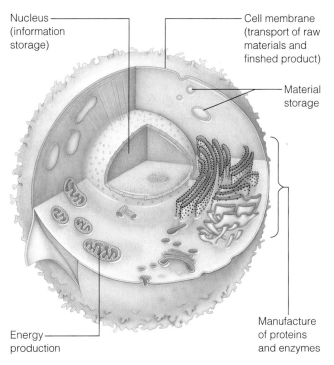

FIGURE 18-7
A cell is a self-contained factory that absorbs raw materials from its surroundings and uses them to maintain its complex functions and reproduce itself. In a complex organism, different types of cells have specialized functions, all contributing to the entire organism.

biochemical evolution because it provided a liquid medium in which materials could move and stick together. One botanist observed that cells of all living organisms are strictly aquatic. Any land-based organism is merely a protective shell filled up with millions of aquatic cells.

Biological processes cannot develop unless they are set apart from the environment and protected from dilution. Some sort of compartment or membrane is required to form a cell. Biologist Sidney Fox has shown that the simple heating of dry amino acids (as might happen on a dry planet) can create protein molecules. Once water is added, these proteins assume the shape of round, cell-like objects called *proteinoids,* which take in material from the surrounding liquid, grow by attaching to each other, and divide (Figure 18–6). Though they are not considered living, they resemble bacteria so much that experts have trouble distinguishing them by appearance. Possibly related to proteinoids are objects discovered in the 1930s by Dutch chemist H. G. Bungenberg de Jong. When proteins are mixed in water solutions with other complex molecules, both sets of substances spontaneously accumulate into cell-sized clusters called *coacervates.* The remaining fluid is almost entirely free of complex organic molecules.

The next step toward recognizable life is even more uncertain. If organic molecules or coacervates are present in a pool of water, they will be left in the pool as the water evaporates. In this way, evaporation of the water in tidewater pools provided high, localized concentrations of amino acids, proteins, and other molecules, allowing cell-like structures to form. The cell-like structures in the primeval pools of "organic broth" could have begun reacting with fluids in the pools and with each other, accumulating more molecules and growing more complex, as suggested by Figure 18–1. This concept was first suggested by Charles Darwin, who speculated on "some warm little pond" where life might have begun. Eventually, these early cells could have evolved into biochemical systems capable of reproducing and increasing in complexity. As shown in Figure 18–7, a cell is a sophisticated chemical factory. It is not surprising that we cannot duplicate this evolution in the laboratory, because the process took half a billion years on Earth.

Earth's Earliest Life-Forms

The earliest biological systems capable of independent life were bacteria. Bacterial cells are **prokaryotes,** cells without nuclei that contain a single long strand of DNA with several thousand genes. Indirect evidence of bacteria has been found in Earth's oldest rocks (Figure 18–8a). The evidence consists of carbon isotopes of possible biological origin found in a 3.8-billion-year-old rock from western Greenland. The earliest "probable" evidence for life is a colony of *stromatolites,* cabbagelike mats of sediment rimmed with bacteria and blue-green algae (Figure 18–8b). These primitive life forms date from 3.5 to 3.6 billion years ago and have been found in Africa and Australia. Fossils of methane-producing bacteria have also been found in 3.4-billion-year-old rocks in South Africa.

The chances of finding recognizable fossils in rocks this old are very small. Therefore, the fact that the oldest fossils are younger than the oldest rocks may not be significant. Life may have originated considerably earlier. However, life probably could not have evolved much before 4.1 billion years ago because of the intense early meteoric bombardment and the possible ocean of liquid lava covering much of Earth's crust. As paleontologist Stephen Jay Gould has observed, life arose "as soon as it could; perhaps it was as inevitable as quartz or feldspar." One set of calculations has been used to argue that life may have originated several times on early Earth, and that all life today would have descended from just one of the origination events. It is also noteworthy that the early Australian stromatolites flourished in strange and hostile environments. Evidence suggests that these organisms lived near shallow hydrothermal vents dominated by island volcanism. In an atmosphere with almost no oxygen, they metabolized the gas hydrogen sulfide (H_2S), which is toxic to most modern life-forms.

Using fossil and chemical evidence, and a little speculation, we can tell the story of life on Earth. For around 2 billion years, prokaryotes ruled Earth's oceans, where water provided a supporting and protective environment. Organisms were mostly soft-bodied and rarely left fossils, so their history is hard to trace. The land was barren. Some areas must have looked like today's deserts or like Mars. Some areas were moist and washed by rains, but instead of luxurious forests, there were only bare acres, eroded gullies, and grand canyons. Brown vistas stretched to the sea.

Gradually, life went through a remarkable transition. Early prokaryotes survived and evolved by using organic compounds in warm ponds and hydrogen sulfide as an energy source. However, as this source of food was extinguished by changes in Earth's atmosphere, some prokaryotes evolved the process of **photosynthesis,** the conversion of sunlight to chemical energy for future use. This fundamental process allowed the proliferation of increasingly complex organisms, with life spans that could endure over a cosmic time scale. Life's destiny became coupled to the fusion energy source deep in the Sun's interior.

One result of photosynthesis is the release of oxygen into Earth's atmosphere. Essentially all the free oxygen on which modern organisms depend (including humans) was dumped into the atmosphere by microscopic life-forms billions of years ago. At first this gas was nothing more than a waste product; oxygen was actually poisonous to the first organisms! Over time, organisms evolved that could use oxygen for metabolism, and the oxygen content began to rise toward the present-day level of 21 percent (Figure 18–9). The atmosphere therefore evolved from more *reducing* conditions (dominated by hydrogen compounds) toward *oxidizing* conditions (dominated by oxygen compounds). As evidence for this change, we find that oxidized sediments are rare before 2 billion years ago and common afterwards. Oxygen production modified the whole environment. Solar ultraviolet radiation broke down some O_2 molecules, and the free oxygen atoms joined with other O_2 molecules to make ozone (O_3) molecules. The result was the formation of an ozone layer high in Earth's atmosphere,

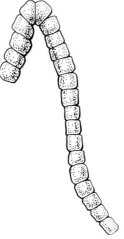

FIGURE 18–8
(a) A microscopic bacterial filament that is one of the oldest known fossils (artist's reconstruction at right). It was found in a 3.5-billion-year-old rock in northwestern Australia. (Source: J. William Schopf.) **(b)** The first fossils were stromatolites, mats of blue-green algae that formed layer by layer in shallow water. This imaginary view shows a patch of stromatolites visible at low tide on the left, at a time when most of the land was still barren. Stromatolites still form today in similar environments. (Source: Painting by Ron Miller.)

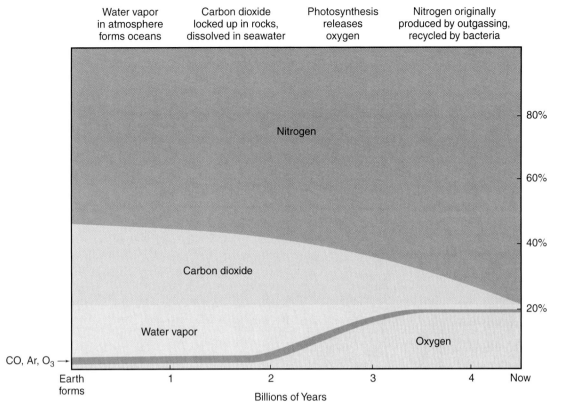

FIGURE 18-9
The changing composition of the atmosphere, from Earth's formation to the present day. All the major changes were caused by the metabolic processes of tiny organisms. The presence of a reactive gas like oxygen in our atmosphere is a clear sign of life.

which absorbs solar ultraviolet radiation and thus protects organisms on the surface.

About 1.4 billion years ago, the complexity of life increased. Cells called **eukaryotes** evolved; a nucleus at the center of eukaryotic cells contains DNA. Eukaryotes contain hundreds of times more genetic material than prokaryotes, with a corresponding increase in the complexity of cell function. New types of organisms appear in the fossil record; they are capable of oxygen metabolism, and, although less resistant to damage from ultraviolet rays, they could flourish because of the new ozone layer. Earth then witnessed the expansion of life from the sea onto the land (Figure 18-10)—a step as momentous as the contemplated colonization of other planets by humans! Sexual reproduction evolved. In this process, an offspring receives half of its genes from each parent. In sexual reproduction gene combinations change from generation to generation, and the new combinations in turn facilitate the experimentation and adaptation that enables organisms to survive in a hostile and changing environment.

FIGURE 18-10
The movement of life onto dry land. Lichens, symbiotic combinations of algae and fungi, were probably some of the first organisms on land. As shown here, they grow on rock surfaces, weathering the rock to produce soil. Lichens are widely adaptable and are found from arctic tundra to hot deserts. Three lichen colonies of red, black, and gray appear in this picture. The actual width of the picture is 6 inches, taken in Tsegi Canyon, Arizona.
(SOURCE: Photograph by WKH)

Evolution, Intelligence, and Technology

So far, we have summarized the first 3.5 to 4 billion years of Earth's history. Primitive life appeared quickly on the cooling crust. But for two-thirds of the entire history of life on Earth, no organism evolved beyond the simplest prokaryotic form. For the better part of another billion years, eukaryotic cells reproduced without any great increase in complexity. After all this time, nothing beyond a single-celled organism existed on the planet. About half a billion years ago, something extraordinary happened. Simple life-forms proliferated, and a couple of them eventually developed intelligence. The last half-billion years of Earth's evolution are thus dramatic because they gave rise to one species—*Homo sapiens,* which has the ability to discover and understand the universe.

From Cells to Intelligence

The evolution of intelligence occurred relatively recently in Earth's long history. To illustrate this, let us consider the history of our planet in terms of the analogy of a single year (Figure 18–11). January 1 marks the origin of Earth, and by the end of February, the first simple cells appear. The atmosphere is enriched with oxygen during the spring and early summer, and by the end of October, cells with nuclei (eukaryotes) exist for the first time. From here on, life exhibits a startling acceleration in complexity. The oceans fill with multicellular life-forms in the Cambrian era, beginning in mid-November. Land animals appear at the end of November. Dinosaurs rule Earth for the first part of December, and primates first appear on December 26. Our hominid ancestors do not arrive on the scene until 10 A.M. on December 31, and *Homo sapiens* does not develop until 3½ minutes before midnight. The entire modern history of astronomy, including our ability to communicate through space with electromagnetic waves, occupies the last tenth of a second of this cosmic year—11:59:59.9 P.M. on December 31.

Now let us focus in on a period when the complexity of life increased dramatically: October and November, in our analogy. About 600 million years ago, Earth's oceans witnessed an explosion of *multicellular organisms*. This evolutionary spurt was probably encouraged by the buildup of oxygen in the atmosphere, which allowed more complex metabolic processes, and by the breakup of the continents, which provided more habitats for living things. Before the surge in evolution, eukaryotes had only three kingdoms: fungi, plants, and animals. After the surge, each of these three kingdoms generated scores of separate evolutionary lines. As with the origin of life itself, there is evidence for several failed experiments in multicellular organisms. All higher-level animals, including humans, evolved from only one of these experiments. The rich diversity of life in the Precambrian oceans has been chronicled by Harvard paleontol-

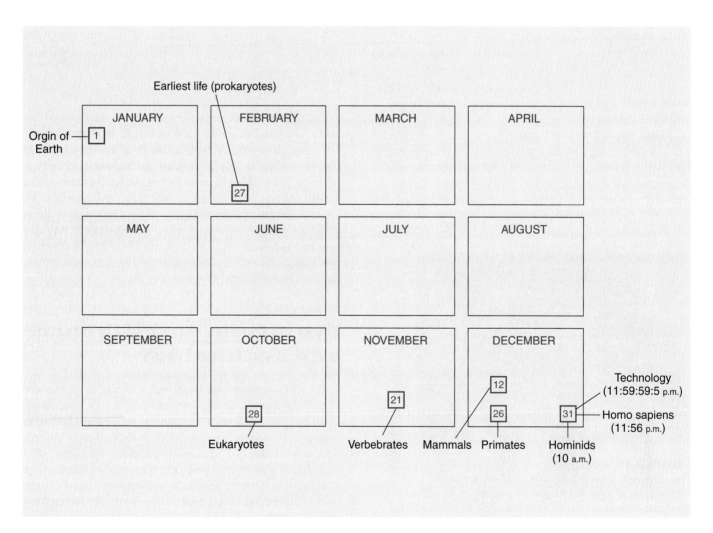

FIGURE 18–11
The Cosmic Year. If Earth formed on January 1, and the present moment is midnight on New Year's Eve, the figure shows the major landmarks in the evolution of life and the development of intelligence and civilization. Life formed very early in the history of Earth but did not attain complexity until relatively recently.

ogist Steven Jay Gould in the story of the Burgess Shale, a fossil site in Canada. Throughout this time, few organisms had "hard parts," leaving the fossil record very sketchy. Therefore, we rely on fortuitous events like those that led to the Burgess Shale, where entire ecosystems were entombed by a mudslide on the ocean floor.

The Cambrian fauna included the first hard-bodied sea creatures, the trilobites (Figure 18–12). Trilobites have two eyes and a complex body structure, but they are extremely primitive by human standards. The vertebrates (animals with backbones) in the oceans of 500 to 600 million years ago can be loosely lumped together as "fishes," but they are strikingly varied in function and form. Most did not survive to live in the oceans of today. About 450 million years ago, plants began to spread across the continents, their spores and seeds carried by the wind. Next came the extraordinary migration of certain types of fish to the land. The transition from a streamlined aquatic animal to one that dragged itself around on poorly formed limbs must have conveyed some adaptive advantage, but it is an unlikely and remarkable transition in retrospect. Equally remarkable is the fact that 350 million years ago, insects first took to the air, followed 200 millions years ago by the descendants of one type of reptile—birds. By this time, the fossil record shows that life had radiated into all possible environments.

It would be a mistake to see the "progression" toward greater sophistication and complexity as predictable or inevitable. About 225 million years ago, after dinosaurs became established on the land, the fossil record in the oceans shows that 95 percent of all marine species became extinct in a short period of time. The cause of this mass extinction is unknown. Then, 65 million years ago, another mass extinction led to the demise of the dinosaurs (see Chapter 5). This second catastrophe helped the evolution of the mammals, which were in competition with the dinosaurs for food. Thereafter, many mammals developed complex central nervous systems and large brains, but for many millions of years, there was no sign of high intelligence. Primates appeared on the scene only 4 million years ago, yet there were many more evolutionary branches and extinct lines before humans arrived, some 300,000 years ago, as modern *Homo sapiens*. It took 99.99 percent of the time since life began on Earth to develop a human level of intelligence. Humans have had the ability to explore and communicate with the cosmos for only a few hundred years—a blink of the eye in the eons since the motor of life first turned over.

Natural Selection

What governs the evolution of complex organisms? At the molecular level, life evolves through slight alterations in the replication of DNA. A change in the sequence of bases in a DNA molecule is called a *mutation*; mutations can result from copying errors or from the influence of an external agent, such as a chemical, a cosmic ray, or a gamma ray. Some mutations are neutral and do not affect the function of an organism. Others can be either helpful or harmful to an organism's ability to survive in its environment. Darwin was unaware of the molecular mechanism of evolution, but he witnessed the diversity and adaptation of species in response to their environments. Mutations occasionally produce offspring with improved survival traits. In turn, these offspring live longer and have more offspring of their own, promoting retention of the new trait. This mechanism is called **natural selection**. "The capacity to blunder slightly is the real marvel of DNA," wrote physician Lewis Thomas. "Without this special attribute, we would still be anaerobic bacteria and there would be no music."

Natural selection has led to life's rich diversity. Nature has rolled the dice many times over billions of years, with results based on the ability of each species to survive in a changing environment. Life on Earth fills a remarkable range of evolutionary habitats. Certain microorganisms can live in Antarctic ponds at 228 K (−49°F) because of dissolved calcium salts; others can live in Yellowstone Hot Springs at temperatures of 363 K (194°F). Some bacteria exist at altitudes where the atmospheric pressure is only 10 percent of the pressure at sea level; others are found several kilometers deep into Earth's crust. Shrunken microbes exist inside rocks

FIGURE 18-12
A fossilized trilobite, a hard-bodied sea animal that first appeared about 600 million years ago and survived for several hundred million years. Trilobites date from the Cambrian era, when life proliferated and took a variety of forms in Earth's oceans. This example is about the size of a human hand. (SOURCE: Grundy Observatory.)

where no sunlight reaches. In 1980, oceanographers discovered entire colonies of sea animals clustered in the darkness around volcanic vents deep on the seafloor. An entire food web is based on bacteria that use volcanic heat and metabolize hydrogen sulfide (H_2S) gas, all occurring at temperatures of up to 523 K (482°F) and pressures of 250 atmospheres—conditions as severe as those on Venus. We consider this environment a bizarre habitat today, but these conditions resemble those of Earth just after the origin of life.

We can understand evolution using the idea of *contingency*, a concept introduced by Alfred Russel Wallace, who developed the theory of natural selection at the same time as Charles Darwin. According to the principle of contingency, the development of human intelligence was the result of many evolutionary branching points, where the progression was heavily influenced by external events such as meteor impacts and climate change. The diversity of life-forms has occasionally been reduced by mass extinctions (Figure 18–13).

The process of evolution is not random, but the details of survival have the character of a lottery. Therefore, it is impossible to predict the nature of higher organisms based on their primitive ancestors. According to Gould, if you "replay life's tape" and watch the development of organisms, you would be unlikely, after 4 billion years, to see humans, primates, or perhaps even any mammals at all. Gould illustrates the expression of contingency with a fine example—the

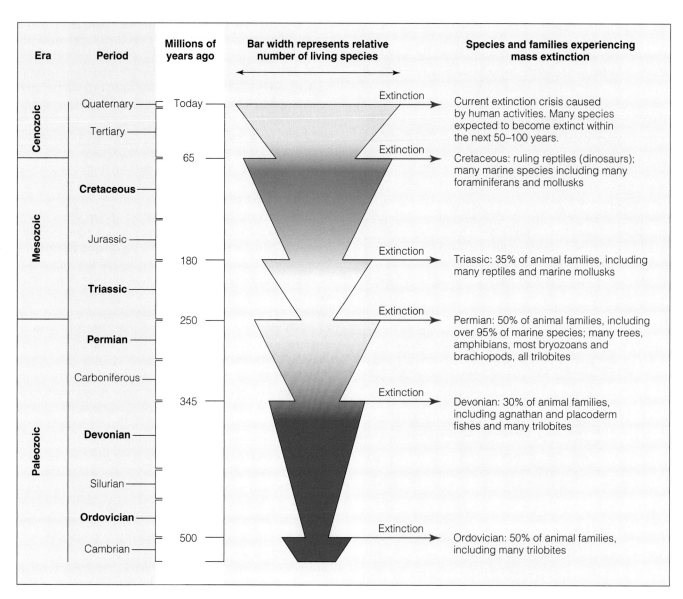

FIGURE 18–13
Over a period of several hundred million years, evolution has consisted of dramatic exits (extinctions) and grand entrances (proliferations of new life-forms). The arrows indicate five major extinctions that have taken place over the past 500 million years; some or all of these may have been caused by impacts from space. The wedge shapes indicate that each mass extinction is followed by a period when new species evolve rapidly to fill new habitats. Many scientists think we are in the midst of a sixth mass extinction, caused by destructive human activities.

Frank Capra movie, *It's a Wonderful Life*. In the movie, a good but poor man (George Bailey, played by Jimmy Stewart) is driven to despair and contemplates suicide. His guardian angel pulls George back from the brink by showing him how the world would have been if he had not existed. In George's absence, his town has been taken over by a robber baron and has slid into grim destitution. Replaying life's tape has yielded a different but sensible outcome, where small changes lead to cascades of accumulating difference.

Life's history on Earth is rich with examples of contingency. Despite a rapid start, it took 3 billion years for the first life-forms to evolve to the first and simplest multicellular organisms. The Sun has only 5 billion years more to live as a main-sequence star; its later giant phase will engulf and destroy all life on Earth. If evolution were only a factor of three slower, the pinnacle and the endpoint of life would be the humble algae or plankton. Given all the errors and uncertainties, the branches and pathways, we can have no confidence that intelligence will necessarily evolve during the stable phase of a solar-type star. The Cambrian oceans witnessed a rapid explosion of fauna. Two dozen quite different "floor plans" for life evolved, yet only a couple survived to propagate their descendants to the present day. Serendipity played a role in the survival of these few lines; no paleontologist could look at the range of Cambrian fauna and safely predict the winners. Move forward to the time of the dinosaurs. Contrary to common perception, dinosaurs were not outwitted by the small and agile mammals. Mammals coexisted with dinosaurs for 100 millions years without any trends toward domination or larger brains. Rather, they adapted to the habitat of the dinosaur's world. Mammals gained ascendancy because of a highly unpredictable event: the impact of an asteroid that caused the demise of large land animals.

Contingency can logically be extended to *Homo sapiens*. There was nothing inevitable about the hominid tribes on the plains of Africa, who stood upright, used tools, and developed language. At each stage in the evolution of life, there were dozens or even hundreds of branching points. The success or failure of each line did not follow a simple notion of "survival of the fittest," but was influenced in large part by sudden and chance events. There have been nearly a billion species of animals on Earth, and a high level of intelligence has arisen in (perhaps) only one. Of course, contingency should not be used to support a purely pessimistic view of the prospects for intelligent life elsewhere in the universe. Despite the apparent obstacles, humans did evolve on Earth. The arrow of evolution does lead in some cases to both complexity and intelligence. On other planets, the proliferation of multicellular organisms and intelligent life may have been rapid, once the basic biochemistry of life was in place.

Highly complex life-forms tend to be fragile. Humans can survive only a 3-percent variation in body temperature, but our skills have nonetheless enabled us to live in equatorial wet jungles and dry deserts, and travel to arctic plains and mountain summits where air pressure is barely half that at sea level. In summary, natural selection on Earth has produced species capable of occupying most environments. Life may have evolved on other planets, but this life may look very strange to us. After all, if mushrooms and corals and woolly mammoths and Venus flytraps all evolved on one planet, imagine how much greater the differences may be between life-forms on two different planets? Feathers and fur, sex and seeds and symphonies—these may very well be the products of Earth only.

Is Intelligence Inevitable?

Once life begins, is the eventual development of intelligence inevitable? This question is central to the likelihood of communication with life beyond the solar system. First, we need to define intelligence. Most people would agree on the relative intelligence of the occupants of this planet. For example, a sequence of species ranked in order of increasing intelligence might be the following: worms, frogs, birds, dogs, chimpanzees, humans. In general, intelligence corresponds to brain size, which correlates with body mass. Intelligence varies continuously among species; for the purposes of this chapter, we can set a high standard. We will define **intelligence** as the capability of abstract thought, coupled with the ability to use tools or technology and to modify the environment.

Life on Earth often evolves toward organisms of greater complexity. You might think that intelligence carries with it an adaptive advantage, and therefore is favored in natural selection. But this optimistic view must be tempered by two facts. First, some life-forms, such as blue-green algae, have remained essentially unchanged for 3.5 billion years. Tropical reefs have also been in existence a long time, and the coral reefs of today's oceans are just the current version of a recurrent ecosystem. Living things can certainly adapt to a changing environment without ever becoming "smart." Second, most species become extinct, and do not reappear. An estimated 500 million species of animal and plant life have existed over the course of Earth's history; 99.8 percent of those are extinct. We can speculate that if humans disappeared, another intelligent species would evolve, but we cannot be sure.

In considering extraterrestrial life, we must avoid arguments that are *anthropocentric,* a term that means derivative of our own human experience. It is not safe to assume that intelligence must be associated with the human form (or primates, or any other specific lineage). It is even less safe to assume that alien life will take a human form (although this is the comfortable premise of much science fiction). In addition, we are interested not only in intelligence, but also in technology and the ability to communicate through space. To believe that aliens would understand the universe with our five senses is highly anthropocentric. Consider that on Earth, there are species that have evolved senses based on infrared radiation (snakes), sonar (bats), and magnetic fields (birds). Many animals communicate by chemical means, using chemicals called pheromones. American scientist Edmund Wilson has estimated the information transfer rate of this method: For a single pheromone, the maximum rate is over 100 bits per second—the equivalent of 20 words of English text per second! In other words, species that humans do not happen to consider highly intelligent have evolved sophisticated modes of communication.

Obviously, intelligence can exist without technology. Dolphins, for example, are land mammals that returned to the oceans millions of years ago. Their brain mass is similar to that of humans, and they have a complex language that is still imperfectly understood by us. Dolphins exhibit highly cooperative social behavior, "farming" their food, playing, and rearing their young for a similar amount of time as humans do. There is no indication that dolphins have higher-level, abstract brain functions, but for us, the only important distinction is that technology is limited in an aquatic environment. Dolphins will never point telescopes at the stars and wonder if they are alone in the universe.

Finally, we must consider the possibility that intelligence might exist independently of organic (carbon) chemistry. Computers can be programmed to carry out many human functions—from flying a plane to playing world-class chess. Of course, the ability to carry out calculations at blinding speed does not make a machine intelligent. Current computers are only as good as their programmers; they have no independent thoughts. However, we are within sight of a time when computers may be able to program themselves and improve their own functioning. Networks of many thousands of processing units will process information in a scaled-down analogy of the human brain. Nobody knows what the limits of this trend will be. Perhaps in a cosmic setting, intelligence can evolve from a biological phase to a computational phase.

We can also describe intelligence in terms of the storage and transmission of information. As shown in Table 18–2, the information content of the human genetic code is about 6×10^9 bits. The electrical activity of our brain is controlled by about 1000 dendrites that project from each brain cell. Each time a brain cell "fires," an electrical contact is made, and information is transmitted. Thus, we can think of each connection as potentially a bit of information (electrical signal on or off, 1 or 0). With a trillion brain cells each connected to a thousand other brain cells, the total number of connections is $10^{12} \times 10^3 = 10^{15}$! This is also the number of bits of information the brain could hold. Even if the brain is inefficient at storing information, this vast capability for information storage and processing exceeds anything in our technology. On Earth, life formed as a chemical network, and intelligence formed as an electrical network. It is an enormous challenge to think of the very different forms that life and intelligence might take across the universe. However, we are not sure enough of the nature and limits of intelligence to estimate the likelihood of intelligent life elsewhere in the universe.

The Influence of the Cosmic Environment

So far, we have looked at the evolution of life on Earth without considering the cosmic environment. Yet Earth does not operate as a closed system. Basic planetary and stellar processes may be involved in facilitating or inhibiting biological evolution. Starting at the geological level, we note that convection within a planet can not only cause plate tectonics, but also change sea levels, ocean currents, wind patterns, and seasonal extremes. This process has been responsible for isolating landmasses such as Australia and allowing different species to evolve there. Similarly, volcanic eruptions may have spewed enough dust into the upper atmosphere to reduce the sunlight reaching the surface. For example, widespread reductions in sunlight of up to 25 percent occurred after the 1883 Krakatoa (Indonesia), 1912 Katmai (Alaska), and 1982 El Chichón (Mexico) eruptions. Rarer, larger events could change temperatures enough to cause the decline of some species and the ascendancy of new ones. Current discoveries of dust layers in active ice packs are beginning to allow scientists to date and reconstruct prehistoric volcanic cataclysms.

Cosmic intruders have been decisive in the history of life. The first wave of impacts early in Earth's history was beneficial to life, bringing in substantial amounts of essential gases and organic materials. Since then, the much rarer large impacts have been catastrophic. The impact of an asteroid or comet could have damaged the ozone layer, exposing organisms to enhanced radiation. The Tunguska explosion of 1908 generated as much as 30 million tons of nitrogen oxide in the upper atmosphere, reacting to deplete up to 45 percent of the ozone in the Northern Hemisphere, consistent with Smithsonian measurements made from 1909 to 1911. As illustrated in Chapter 5, an asteroid impact occurred 65 million years ago, probably causing the dramatic break in the fossil record between the Cretaceous and the Tertiary periods. The ejected dust and soot dwarfed that produced by volcanism and probably caused massive climate change.

Subtle changes in Earth-Sun system can also influence life. Over the course of Earth's history, the Sun has brightened by 30 percent, and even smaller changes in the Sun's radiation may have caused climate change. Astronomers believe that slight changes in Earth's orbit, and the tilt of the planetary axis to the plane of the ecliptic due to gravitational forces, caused major climate changes such as the ice ages. Recent marine studies of traces from the past 700,000 years have shown decisively that cycles of climatic change, including several pulses of continental glaciation, can be tied to the so-called Milankovich cycles of orbital change in Earth-Sun-Moon system. During life's tenure, the solar system has probably been influenced by the gravity of nearby passing field stars and by blasts of radiation and high-energy particles from nearby supernovae. On a larger scale, life's history has encompassed 18 galactic years, and Earth has passed through denser material in the plane of the Milky Way galaxy perhaps 200 times. We are only just beginning to discover the signatures of these cosmic events in our rock and fossil records, and the effects they have had on the evolution of higher forms of life.

Culture and Technology

Our ability to observe and explore space raises an obvious question. What is the role of technology in the evolution of life? We have already seen that technology is not an inevitable consequence of the evolution of intelligence. However, the experience of our planet is that the highest functioning organisms—humans—have developed the tools to modify the environment. However, this ability is very recent. For most of the past 500,000 years, roving tribes of humans left no mark on the planet. It is the same story for most of the 6000 years of city-dwelling civilizations. Some 1000 years ago, the only trace of human activity that would have been visible from a high Earth orbit was the scar of the Great Wall of China. The surge of technology began 200 years ago with the industrial revolution and rapid population growth. It has accelerated over the last 50 years with the development of mass-production techniques, nuclear power, and electronics. We live in a time in which we can decisively change our global environment—for better or worse.

We must consider the possibility that the development of technology could actually end civilization on Earth and potentially elsewhere. As Pulitzer Prize–winning naturalist René Dubois pointed out, we are umbilically connected to Earth. If we alter our planet too much before acquiring an ability to leave it, we are finished. Past wars did not threaten our whole species, because most conflicts were local, and weapons had limited destructive capability. Today's nuclear, biological, and other types of weapons, however, could potentially engulf the whole world. A sufficiently massive nuclear exchange could devastate not only civilization but also future forms of life, whose genetic pool would be exposed to high radiation levels for decades. Just as a large meteor impact can cause climate change, nuclear weapons could put a large enough amount of debris in the upper atmosphere to cause blanketing of the Sun's radiation. For most of the 50 years of the nuclear age, human civilizations have lived in a kind of uneasy peace. Recent years have given cause for optimism, as the nuclear arsenals of the United States and the former Soviet Union have begun to shrink. However, the corresponding proliferation of nuclear weapons in smaller and often less stable countries is a cause for concern. Our planetary culture could wipe itself out by a conscious design of weapons, as irrational as that may seem (Figure 18–14).

Disasters can also happen inadvertently. As our technology reaches planetary scale, our accidents can involve large regions of the planet. Problems as diverse as nuclear power, acid rain, and aerosol cans illustrate the issue. Although humans have been around less than 0.1 percent of the age of our planet, we are already beginning to have brushes with global disaster. One sobering truth stands behind our discussion of the future of humanity: Extinction has been the fate of 99.8 percent of the species on Earth.

By exercising a bit of intelligence and caution, humans can recognize these dangers and avoid them—we hope. But we must overcome a cultural obstacle: the transition from scattered, competing nation-states to stable global or interplanetary societies of intelligence and imagination. We face a race between the good side of our technological abilities, which make civilization possible (plumbing, electricity, stereos, and space travel), and the dark side of technology, which threatens civilization (carcinogenic by-products, pollution, and hydrogen bombs). It is possible to break out of our muddle of competitive strife over resources and ideological space on a finite planet. Perhaps some alien cultures have crossed this cultural hurdle and spread to many planets, ensuring their survival against ecological disaster on any one planet. Such cultures might last and be detectable for millions rather than thousands of years.

The Search for Extraterrestrial Intelligence

In the preceding discussion, we looked at the nature of life, the possible sites for life, and the history of life on Earth. We found that the basic ingredients of life are widespread throughout the universe; they are produced in stars, and simple organic molecules can form even in the rarified

FIGURE 18-14
The haunting image of a nuclear explosion is a reminder that technology can lead to mass destruction as well as to interstellar travel. Even though humans have recently drawn back from the nuclear precipice, we still have enough weapons to destroy ourselves many times over. The lifetime of civilizations with our level of technology is unknown.
(SOURCE: Chris Bjornberg, Photo Researchers, Inc.)

environment of interstellar clouds. We also have indications that sites for life are likely to be widespread. Planets form as a natural by-product of star formation, and extrasolar planets are now being discovered at a rate of about 5 to 6 per year. Possible worlds are illustrated in Figure 18–15. Primitive life formed readily on the early Earth and proliferated into a wide variety of evolutionary environments. There are several places in the solar system where we may yet find microbial life, or at least traces of its past existence. Up to this level of complexity, we might convince ourselves that life is inevitable and should be widespread in the universe.

We must also take into consideration the powerful *principle of mediocrity*. In the history of astronomy, we have seen that our place in the universe is not special. We live on a rocky cinder orbiting a middle-aged, mid-sized star in the outer regions of a typical spiral galaxy, one of billions scattered through expanding space. The same laws of physics govern Earth and the planets in remote parts of the universe. The chemical elements that compose our planet and our bodies are spread throughout the universe. Can we extend this principle to the biological realm as well? The last step in the Copernican revolution would be the demonstration that life in the universe is commonplace.

Statistics alone imply that life should be quite widespread in the universe. With 10^{20} stars in the observable universe, it seems incredibly unlikely that life would have occurred only on this planet. It seems implausible that we are alone in the universe. Yet we must remember that the principle of mediocrity is no more than an assumption. Life may indeed be a natural consequence of the evolution of stars and planets. However, it is also possible that life—and particularly intelligence—was the result of an unlikely series of chemical and biological events that have not taken place anywhere else in the universe. Deciding between these opposing viewpoints is pure speculation until we have more data.

What we do know is that the long road toward intelligence on Earth was not smooth and direct. Adaptation by natural selection does not lead to intelligence in most cases, and the fate of most species is to become extinct. The evolution of life is heavily contingent on changes in the global environment, some of which are catastrophic, and many of which are caused by astronomical processes. In addition, the road from large brains to the use of technology is not preor-

(a)

(b)

(c)

FIGURE 18-15

Too hot, too cold, and just right. Scenes on imaginary planets emphasize that many extrasolar planets may not have conditions suitable for life. **(a)** The surface of this planet, close to a pair of hot, massive stars, is a sun-blasted desert. **(b)** On this world, far from a red and white pair of stars, all water is frozen. **(c)** Some worlds may exist in which liquid water persists and life forms have evolved.

(SOURCE: Paintings by Don Dixon, Ron Miller, and WKH, respectively.)

dained. We share our planet with a number of species, both primates and marine mammals, that have large brains but do not have the ability to communicate through space. Now we must ask some very direct questions about life in the universe. Is there a rationale for searching for intelligent extraterrestrial forms of life? What is the most logical way to communicate? How should we send a message? How would we recognize a message sent to us?

Is There Alien Life in the Solar System?

In the solar system at least, we can be unequivocal. Spacecraft exploration of the planets has virtually proven that there are no advanced forms of life. These same space probes have almost ruled out even simple extraterrestrial life. The terrestrial planets other than Earth are devoid of life, according to space probes, and the outer planets appear too cold to sustain life. The environments that have some potential for fostering life forms are those of Mars, Titan, the middle atmospheres of the giant planets, and the oceans of Europa.

Although ultraviolet radiation has probably destroyed any organic material on the Martian surface, the Martian soil has the ability to synthesize organic molecules from artificial nutrients. Because of this fact, scientists running the Viking missions could not absolutely rule out the possibility of primitive forms of life. At the time, Carl Sagan pointed out that the Viking probes could have landed at a number of arid sites on Earth and failed to detect life! In 1978, American biologist E. I. Freidmann reported the discovery of tiny organisms thriving in the pores of rocks in seemingly lifeless, Mars-like dry valleys in Antarctica, giving some support to the idea of hidden organisms on Mars. Since the Viking probes merely scratched the surface at two places, we cannot rule out the possibility of Martian microbes below the surface. In addition, planetary scientists are convinced that Mars had a wetter, warmer past—which is why many scientists were receptive to the electrifying claim of ancient Martian life described in the opening story of this book. The evidence from ALH 84001 is not yet compelling. It will take the study of more Martian meteorites and samples returned from the red planet to be sure.

Titan's thick atmosphere and ethane-methane chemistry remain an intriguing laboratory for life, especially if "hot springs" of any sort exist on the surface. The atmosphere of giant planets may have hydrogen-dominated layers with temperatures and pressures not too different from the atmosphere of Earth. Sagan and others have speculated about the evolution of organisms that could float in such atmospheres. However, most biologists consider it very unlikely that replicating molecules could ever form in such a diffuse environment.

The discovery of Sun-independent organisms deep within rocks and near deep-sea vents on Earth remind us that entire ecological systems can thrive in extreme environments. Coupled with the discovery of tidal heating of the interiors of Jupiter's satellites, this discovery has raised the possibility of organisms that might have evolved in liquid water zones under the ice of satellites such as Jupiter's moon Europa (a theme developed in Arthur C. Clarke's novel, *2010*). In the future, it might be interesting if we can drill through the icy crust of this moon and study the "buried ocean" believed to exist there. Although we do not expect to meet creatures of any great intellect in our solar system, there are clearly places left to explore that may shed light on biochemical and biological evolution.

Intelligent Life Among the Stars

Is there any way we can possibly estimate the probability of intelligent life in the universe? American astronomer Frank Drake is one of the pioneers in the subject of **SETI**—the search for extraterrestrial intelligence. He introduced a method for combining all the information that we need in order to estimate if there is anyone out there in space that we can talk to. The result has become known as the **Drake equation.** Drake's logic was to combine various factors into an estimate of N, the number of communicating civilizations in our galaxy. The following Science Toolbox gives the Drake equation and discusses the uncertainty in each of its factors.

Caution is required when applying the Drake equation. We need estimates of *all* the factors in order to arrive at the final estimate of the number of civilizations that have the capability for interstellar communication. If any factor in the equation is unknown, then we cannot calculate N. Remember that the error in any product of factors is dominated by the factor with the largest error. In effect, an accurate census of the number of stars in the Milky Way is not helpful as long as the last two factors are so poorly determined. Our astronomical knowledge is swamped by our ignorance of the likelihood that life evolves intelligence and technology. Most importantly of all, empirical justification of the Drake equation is based on only one planet where we know life exists and only one species that we know can use technology to communicate in space. If you recall the discussion in Chapter 1, you will see that we are skating on thin ice with the scientific method. We cannot reliably extrapolate based on one example; the method of induction fails.

As both a metaphor and an analogy for the Drake equation, think of a series of windows lined up. If we look through all the windows, we will see the number N written on a wall beyond the last window. Each of the windows is analogous to a term in the Drake equation. If each of the windows held a perfectly clear pane of glass, we could see the answer clearly. A perfectly clear window corresponds to a perfectly determined factor in the equation. But a poorly determined or unknown factor is like a frosted pane of glass. Even if only one of the windows is frosted, we cannot see the answer at all.

Astronomers tend to think deterministically on the issue of extraterrestrial intelligence. The 40 billion stars in our galaxy represent only a tiny fraction of the possible venues for life. There are about 50 billion other galaxies in the observable universe. Given 10^{20} stars, it seems virtually certain that intelligent life has evolved elsewhere. Yet it is a logical fallacy to think that everything that *can* happen *will* happen. As an example, individual atoms can spontaneously change their energy by small amounts. There is a finite but very

SCIENCE TOOLBOX

The Drake Equation

In 1961, the National Academy of Sciences convened the first conference on the subject of extraterrestrial life. The field was so small that only 10 people were invited. One of them was Frank Drake, a young researcher who had, only 2 years earlier, conducted the first, modest radio search for intelligent signals from planets around nearby stars. Drake had been thinking hard about how to organize the diverse subject matter of the conference. On the first morning, he stepped up to the blackboard and wrote the following equation:

$$N = R f_p n_e f_l f_i f_c L$$

Ever since, the Drake equation has been used as the framework for the discussion of intelligent life in the universe. N is the number of planets in the galaxy that host a civilization intelligent enough to communicate through space. It is equal to the product of a number of factors: R is the average rate of star formation, f_p is the fraction of stars that form planets, n_e is the number of planets in each system with conditions favorable for life, f_l is the fraction of those planets that actually do develop life, f_i is the fraction of planets with life where intelligent beings evolve, f_c is the fraction of those intelligent species that develop the ability for interstellar communication, and L is the average lifetime of such civilizations.

The Drake equation combines a set of factors that range from the purely astronomical to the sociological. We should recognize at the outset that the scientific method is on shaky ground with this calculation, because our entire knowledge of intelligent civilizations is based on just one example: ourselves. Notice also that the equation only gives the number of potential civilizations within communication range in our own galaxy; scientists accept this practical limitation because the light travel times to other galaxies are so large. We would have to multiply the result by roughly 10^{10} to get the number of likely hi-tech civilizations in the universe.

We can scrutinize the Drake equation piece by piece. The first factors are fairly reliable. There are about 40 billion stars in our galaxy, and the age of the galaxy is about 10 billion years. Dividing these numbers gives an average star formation rate of R ≈ 4 stars per year. This might rise to R ≈ 10 if we account for stars that formed in the past but have since exploded or faded from view. The first factor in the Drake equation is the *only one* that we know with an accuracy factor of 2 or better.

The fraction of stars with planets is still uncertain, but it is the subject of one of the most intensive research efforts in the history of astronomy. So far, only Jupiter-sized planets have been detected, but Earth-sized planets will be within reach of the next generation of experiments. Astronomers believe that planets are a natural by-product of star formation, which would imply $f_p = 1$, but a plausible range for this factor might be 0.1 to 1.

The number of habitable planets in each system is highly uncertain, and we are thrown back on the assumption of mediocrity and the hope that our solar system is typical of other planetary systems. Our solar system has a habitable zone—a distance from the star where surface water can exist—that includes Venus, Earth, and Mars. The fact that two out of three of those planets *do not* have surface water tells us that atmospheric conditions are important, too. The number n_e must also take into account the fact that many low-luminosity stars have tiny habitable zones, and the fact that most stars are in binary or multiple systems where stable planet orbits are unlikely. If we also require that the star live long enough for life to develop, estimated at 1 billion years, then this number could be as high as 0.1 or as low as 0.001.

Our knowledge of life's evolution on Earth indicates that life begins quite readily on a suitable site, so the next factor may be close to 1. On the other hand, if life arises as a random outcome of chemical evolution, f_l might be low. Currently

(continued)

TABLE 18-4
Estimates of the Number of Intelligent Civilizations in the Galaxy

PARAMETER	DEFINITION	"PESSIMISTIC" ESTIMATE	"OPTIMISTIC" ESTIMATE	RANGE IN PARAMETER
R	Average rate of star formation in the Milky Way galaxy	4	10	× 3
f_p	Fraction of all stars that form planets of any kind	0.1	1	× 10
n_e	Number of planets in each system with conditions suitable for life	0.001	0.1	× 100
f_l	Fraction of habitable planets that actually develop life	0.01	1	× 100
f_i	Fraction of planets with life where intelligence evolves	0.001	1	Unknown
f_c	Fraction of intelligent species that can communicate through space	0.1	1	Unknown
L	Lifetime in years of an intelligent, communicable civilization	50	10^4	Unknown
N	Number of civilizations we can talk to at any one time	$10^{-9} L$	L	Unknown
t	Light travel time to nearest civilization if civilization is long-lived	10^6 years	10 years	Unknown

SCIENCE TOOLBOX

The Drake Equation (continued)

we have no evidence to decide between these two possibilities. However, if primitive life is discovered on Mars, Europa, or Titan, it would be evidence in support of the idea that $f_l = 1$.

The last three factors are hopelessly uncertain. We do not know if intelligence is a natural or necessary consequence of biological evolution. We have no idea how likely it is that life will develop technology and the ability to communicate into space. In the absence of any evidence, logical arguments can be made for high and low values of f_i and f_c. We are equally in the dark as to how long such a capability will endure. On Earth, $L = 50$ years so far.

Table 18–4 shows "optimistic" and "pessimistic" estimates for the various factors. The table also shows the N that results from the product of the factors and the consequent estimates of the number of potential pen pals in the Galaxy at any particular time. We combine all the low and high estimates of each factor purely to illustrate the range of the product; in practice, the various factors may be independent, in which case a low value for one factor does not imply a low value for the others. We stress that these are little more than educated guesses. The estimate of N is as uncertain as the *most* uncertain factor. Since several of the factors are completely unknown, we must conclude that N cannot be determined.

---

small probability that a brick will raise itself spontaneously off the ground, but to do so would require coincident energy changes of all the many atoms in the brick. The probability of this ever happening is 10^{-100}, an astoundingly small number. It is possible that the contingent set of conditions that led to technological civilizations on Earth were not duplicated on a single other planet in the universe!

Some have argued that SETI is not a scientific enterprise. Without a theory to back up all the hypothesizing, or some empirical evidence to show that the probabilities we have estimated are reasonable, we are left with speculation. Nevertheless, we can follow up the logical consequences of "optimistic" and "pessimistic" estimates, which are shown in Table 18–4. Let us assume that intelligent civilizations are distributed randomly among the stars and planets that can shelter them; in other words, there is no region of the Milky Way that is smarter than any other region. If the number of communicable civilizations is small, then the average distance between them is large. If the number of communicable civilizations is large, then the average distance between them is small. Figure 18–16 shows how the typical distance to a civilization depends of the lifetime of the civilization. Under the optimistic calculation in Table 18–4, the nearest civilization might only be 10 to 15 light-years away—a close neighbor in cosmic terms. Electromagnetic signals could be exchanged within a human generation. In the pessimistic case, our civilization is unique in the Milky Way. Two-way communication would then take millions of years and might be hampered by the limited lifetimes of civilizations.

Where Are They?

The optimists have generally held sway in the debate over extraterrestrial intelligence. In the 1930s, this optimism led the physicist Enrico Fermi to ask the question, Where are they? Fermi was persuaded by the argument that the vast number of stars implies that some of them must host planets with intelligent life. He then wondered why we have not been contacted. Radio astronomers have listened for radio messages and heard none. Our skies are not filled with alien visitors observing us or trying to contact us. There is no concrete evidence of "ancient astronauts" or alien visitations in earlier Earth history (see Chapter 1 for a discussion and critique of these ideas, and Figure 18–17).

Theorists have even argued that any one civilization could colonize all the habitable planets in the galaxy on a time scale short compared to the galaxy's age (but long compared to the time scale of evolution, so that the original civilization might produce many new species as it went). Such a civilization could explore the galaxy without colonization by sending spacecraft to planetary systems to mine asteroids. These probes could then use local materials to replicate themselves and they could rapidly propagate through the galaxy, sending information back at the speed of light to the originators. If this sounds fanciful, remember that it involves only a modest extrapolation of current technology; we are probably less that 100 years away from inventing robots that can construct other robots and space propulsion systems that can reach 10 percent of the speed of light. So why have we not heard from aliens?

We will use scientific criteria to rule out the option that we *have* been contacted. Yet, polls in the United States show that the public has an enormous susceptibility to the idea of alien visitation. Over 50 percent of the population believes that we have already made contact, in the complete absence of any convincing evidence. Perhaps this is not surprising, given that aliens have been in the popular culture for over 100 years, beginning with the influential H. G. Wells book, *War of the Worlds,* and the science fiction of Jules Verne, and progressing to more recent phenomena like *The X-Files* TV show. Long a staple of science fiction, alien life forms took a big step into the public consciousness through the powerful media of TV and film. As examples, consider the rise of the TV show *Star

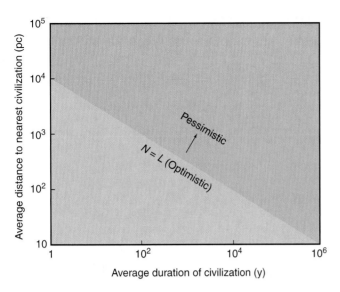

FIGURE 18-16
The relationship between the typical distance to an extraterrestrial civilization and the lifetime of an average civilization. Multiply the number on the vertical axis by about 3 to get the signal travel time in years. The slanting line shows the distance under the most optimistic assumptions that can be put into the Drake equation. The true distances might be orders of magnitude larger, anywhere in the shaded part of the diagram.

Trek and the *Star Wars* movie series, from small cults of popularity to the level of widespread cultural phenomena. Films show us two versions of the alien myth. One is the optimistic view, expressed by *E.T., Close Encounters of the Third Kind,* and, more recently, by *Contact*. This view illustrates our yearning for knowledge and even salvation (Figure 18–18). The other is the pessimistic view, expressed by *Alien, Independence Day,* and many other movies. This view places us in jeopardy in a universe of superior life forms. The belief in aliens is a modern religious metaphor. Perhaps a more interesting question than "Are we alone?" is "Why are we so lonely?"

There are other logical answers to the question posed by Fermi. Maybe there are no aliens; perhaps intelligent life *is* a unique accident. Alternatively, the pessimists could be right, and the nearest civilizations are in distant galaxies. Their radio messages would be million of years old by the time we receive them, and their spaceships would be unlikely to reach Earth if limited to speeds less than the speed of light, as current physics demands. It is entirely consistent with our current knowledge to propose that simple microbial life is quite common in the universe, but that intelligence and technology are extremely rare.

Another possibility is that space exploration and the desire for communication are uniquely human attributes. The vital ingredients for SETI—technology, and the desire to explore and communicate—are unique to humans among the many species on Earth and may be exceedingly rare beyond Earth. Technology is not even universal among human societies. Are humans fated to be explorers, bridge builders, and scientists rather than artists, athletes, or daydreamers? Is the stereotypical aggressive Westerner more representative of the essence of humanity than the stereotypical contemplative Easterner? Our technocracy may be just one type of *cultural* activity, rather than a natural consequence of *biological* evolution. Historically, patterns we once assumed to be biological have turned out to be merely cultural (confusion between these two has led to racist and sexist biases that we are still trying to overcome). We have all been influenced by the appealing image from science fiction of space exploration and extraterrestrial communication as universal

FIGURE 18-17
There is no good evidence of alien encounters at any time in human history. Even if 10,000 alien expeditions had visited Earth at random times during the planet's history, the visits would have been hundreds of thousands of years apart. In such a scenario, even the most recent expedition would not have left any recoverable historical or archaeological traces. (SOURCE: Painting by Jim Nichols.)

cultural activities. In fact, it is absurdly anthropocentric to suppose that beings on other planets would resemble us physically, psychologically, or socially.

This brings us to a last and possibly most significant answer. We may be farther from aliens in evolutionary time than in physical space. Let us imagine a hypothetical experiment. Suppose another planet, a twin of Earth, started evolving at exactly the same time as our Earth. Even with a similar biochemistry, organisms are unlikely to be in a phase of evolution similar to ours. If the evolutionary "clocks" on the two planets were only 1 percent out of synchronization, which could be caused by a slight difference in temperature, they would be 40 million years ahead of us or behind us. This is as far from us evolutionarily as we are from early mammals. If the clocks differ by 10 to 15 percent, then we are talking about the enormous difference between ourselves

(a)

(b)
FIGURE 18–18
The popular culture contains powerful expressions of our yearning for contact with intelligent aliens. We recognize that positive evidence of extraterrestrial civilizations would be a pivotal and dramatic event. **(a)** Contact could come from the discovery of an alien artifact, as dramatized in *2001—A Space Odyssey*. (SOURCE: Copyright 1968, Metro-Goldwyn-Mayer Inc.) **(b)** Contact could come from an alien visitation, as in the arrival of the "mother ship" in *Close Encounters of the Third Kind*. (SOURCE: Copyright 1977, Columbia Pictures Industries Inc.)

and single-celled microorganisms! (Figure 18–17 shows a hypothetical encounter in which the timing mismatch is less than 0.1 percent.) The timing argument carries one important consequence. We have had the capability for interstellar communication for only 50 years, which is an instant in evolutionary and cosmic time. Because of the time it takes signals to travel a significant distance in interstellar space, it is virtually certain that any civilization with which we could make contact will be far more advanced than us, by thousands or perhaps millions of years.

It turns out that the most significant number in the Drake equation is L, the lifetime of a civilization with the ability for, and interest in, interstellar communication. It is also the most uncertain number. With the optimistic calculation from Table 18–4, we have $N \approx L$. Therefore, the number of *currently active* civilizations in our galaxy approximately equals the lifetime of a typical civilization at a level where they can (and do) communicate. Up to this point in history, our number $L = 50$ years, and we have no idea how much longer we will survive as an intelligent species. If L is less than 3000 or 4000 years, intelligent civilizations will be so thinly spread throughout the galaxy that a round-trip message will take longer than the lifetime of the civilization. Communication would then be impossible, and any messages in space would be vestiges of extinct civilizations (see Figure 18–19; also, review space-time diagrams and Figure 17–18). If the pessimists are correct, then we are alone in our galaxy, and communication is virtually futile.

SETI optimists even have an answer for this bleak scenario. Rather than use signals for communication from point to point, alien civilizations might use signals to store information in a vast network. If one civilization succeeded in sending probes across the galaxy, they could use these probes to set up a communication network across the galaxy. These artificial stations could beam information at planets, collect information from planets with intelligent life-forms, and store the information in a growing database. Each newly intelligent civilization would not have to wait thousands of years of light travel time to communicate across the galaxy; they would only have to wait the much shorter time to tap into the nearest node of the network. The information in the network would survive the death of any civilization and act as

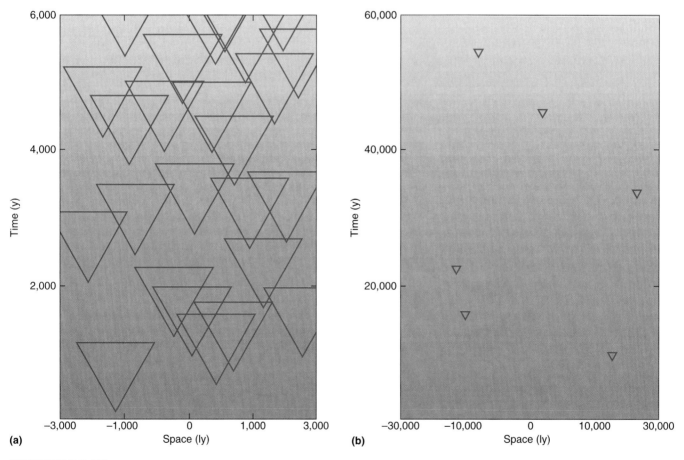

FIGURE 18-19
Interstellar communication depends on the lifetimes of civilizations. These space-time diagrams show the possible isolation of technological civilizations, assuming the civilizations are randomly scattered through space. **(a)** In the optimistic assumption about life in the universe, $N = L$, and the lifetime of a civilization that can communicate in space is 1000 years. The apex of each triangle is the beginning of the time a civilization can communicate, and the top of the triangle shows how far a light signal could travel in the lifetime of the civilization. When the triangles overlap, two-way communication is possible. **(b)** In the second case, $N = L/10,000$, which means that civilizations are more thinly scattered through space. None of the triangles overlap, which means that civilizations become extinct before they can establish communications with each other.

a kind of galactic consciousness. These are hypothetical ideas, but we should recall the words of physicist Freeman Dyson: "Nature always has more imagination than us."

The Best Way to Communicate

SETI arguments are based entirely on speculation. The only way to move beyond this uncertainty is to get some data, to conduct an experiment. We can listen for messages or we can send our own into space, but to do so, we need a strategy. What is the best message to send? What type of communication should we use? Where should we beam our signal, among the many millions of possible targets? It is important to be aware that we cannot justify our search on strictly rational grounds. The case for SETI cannot be made in the same way as the case for building a particle accelerator or for sequencing a gene.

SETI optimists, Carl Sagan among them, have claimed that the results of a search for extraterrestrial intelligence will be significant, no matter what the result is. However, estimates of N are so uncertain that the failure of a SETI experiment will not prove that we are alone. The truth is that SETI is a scientific gamble. The odds are stacked against us, but the stakes are incredibly high. Our role as sentient beings in this vast and inhospitable universe would be profoundly affected by whether or not we are alone. Unless we conduct an experiment, we will never know.

In 1972, the *Pioneer 10* spacecraft was launched toward Jupiter, and it has now become the first human artifact to leave the solar system. A plaque was attached to the spacecraft, carrying a greeting to any civilization that might find it (Figure 18–20a). The plaque shows a naked man and woman next to a silhouette of the Pioneer spacecraft. The top of the drawing shows the spin transition of the hydrogen atom, and the bottom shows the trajectory of the spacecraft within the solar system. The radial pattern represents the position of the solar system within the Milky Way, by triangulation among a set of 14 pulsars. At the time it was launched, the plaque created quite a stir. Many complained about the government-sanctioned nudity, and feminists objected to the fact that the man's hand was raised in greeting, but not the woman's. Also, some people found the pulsar map to be obscure, wondering how aliens would decipher it if *they* could not. The plaque brings home how difficult it is to encapsulate the essence of humans in a single image.

Five years later, a team headed by Carl Sagan and Frank Drake designed a message to be sent into space with the *Voyager* spacecraft. This time they included music as well as images. As Lewis Thomas has said, "I would vote for Bach, all of Bach, streamed out into space, over and over again. We would be bragging, of course, but . . . we can tell the harder truths later." Images and sounds were both encoded onto a 12-inch gold-anodized record, with instructions etched on the cover (Figure 18–20b). It is ironic that the phonograph record technology used to encode the message is already obsolete on Earth! The images tried to represent the diversity of humans and the natural environment. Censorship reared

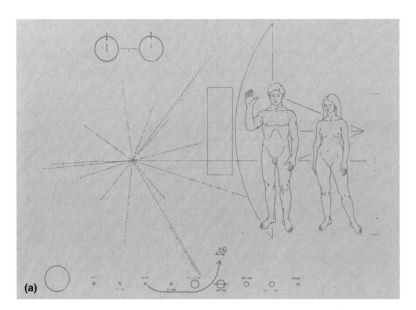

FIGURE 18-20
Our versions of a "message in a bottle" tossed into the interstellar sea. **(a)** The plaque attached to the *Pioneer 10* spacecraft. Two humans are shown with the spacecraft to scale at the right. On the left, the solar system is identified with respect to the sight lines to 14 pulsars, with periods expressed by tick marks in a binary code along each sight line. The graphic of the solar system at the bottom shows the trajectory of the spacecraft. *Pioneer 10* was the first human artifact to leave the solar system, but it will take tens of thousands of years to reach the nearest stars. **(b)** The gold record attached to the *Voyager* spacecraft. It contains selections of music, sounds of life on Earth, digitized pictures, and a greeting spoken in dozens of the world's languages. As we try to communicate with advanced civilizations, it is ironic that the technology used to make the recording—the phonograph record—is obsolete even on Earth. (SOURCE: NASA.)

its ugly head, as NASA vetoed one picture involving nudity. The sound selections did include Bach, jazz, and rock and roll, but also much non-Western music, as well as sounds of life on Earth, such as those made by whales, rain, footsteps, a kiss. Spoken greetings in 55 different languages are followed by a message from the United Nations Secretary General. It is hard not to see the record as a message to ourselves, rather than to aliens.

The messages sent on spacecraft are like bottles tossed onto a vast ocean. *Pioneer 10* is now about 3.5 billion miles away from us, beyond the orbit of Pluto. Even traveling at about 6 mi/s, it will take 100,000 years to reach the nearest star to the Sun. How can we build an interstellar probe that travels at more than $1/30{,}000$ of the speed of light? Unfortunately, the laws of physics work against us. The energy required to accelerate a small payload of 100 kg to one-tenth the speed of light exceeds 1 year's output from all the power plants on Earth! To reach 99 percent of the speed of light using the most efficient energy source imaginable, the annihilation of matter and antimatter, requires a spacecraft with 40,000 times as much mass in fuel as in payload. The energy requirements for transmitting electromagnetic radiation are far less restrictive. The kinetic energy of a radio wave photon is 10^{12} times less than the kinetic energy of an electron travelling at 99 percent of the speed of light.

Electromagnetic waves (or photons) are the preferred carriers of information because they travel at the highest velocity known, and they are easy to transmit, modulate, and receive. However, photons range in frequency and wavelength over a factor of 10^{20} from radio waves to gamma rays. How do we choose an optimum frequency for communication? Nature provides some guidance. First, radio waves contain the least amount of energy per photon, and therefore are the most efficient to produce. Secondly, photons of optical frequencies and higher suffer absorption and scattering by gas and dust in the interstellar medium. The best penetration is achieved by radio waves. Finally, a spectrum of cosmic radio "noise" shows that the quietest region on the dial is the zone around 1000 MHz (a thousand million Hertz, or 10^9 Hz). Figure 18–21 shows that at lower radio frequencies, radiation from high-energy electrons in the Milky Way contaminates the signal. At higher radio frequencies, there is a rising noise source due to the cosmic background radiation. The quiet zone—in other words, the frequency range where cosmic noise is low—also contains the frequency of the spin transition of cold hydrogen, the most abundant element.

Even if we accept the arguments for *radio communication*, we have a classic "needle in a haystack" problem. There is a range of thousands of MHz in the zone in which cosmic noise is low. We do not know in advance what the *bandwidth*, or range of frequencies, of a signal might be. To send a large amount of information, the bandwidth should be large. A single TV channel has a range of 6 MHz, and an FM radio station transmits over a range of 200 kHz (200,000 hertz). On the other hand, to transmit information as efficiently as possible, a narrow bandwidth should be used. However, no matter how narrow the signal is originally, the scattering of radio waves in interstellar space smears the signal to a width of about 0.1 Hz. This smearing is analogous to the blurring of an optical image as it passes through Earth's atmosphere. Therefore, a search strategy for receiving a message involves thousands or perhaps millions of targets, each of which must be searched over billions of separate frequency channels!

Our period of isolation may be nearing an end. For 50 years, we have been inadvertently leaking radio, radar, and television signals into space, creating a bubble of radio energy expanding outward from Earth at the speed of light. Dilute signals of *I Love Lucy* have crossed the paths of approximately 1000 stars, and *The Brady Bunch* has reached several hundred. Only a few dozen star systems

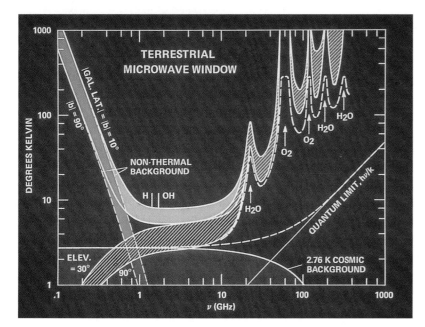

FIGURE 18–21
A radio spectrum of cosmic sources of noise shows a quiet region around a few billion hertz. The vertical axis shows the equivalent temperature of various cosmic sources of noise. At lower frequencies, the signal from high-energy electrons in the Milky Way increases, and at higher frequencies, the cosmic microwave background radiation increases. Our specific Earth environment also makes observations at high radio frequencies difficult (atmospheric absorption from oxygen and water vapor are shown). The fundamental spin transition of the cold hydrogen atom lies in this quiet zone, along with the radio emission from OH; the "water hole" between the H and OH emissions is a favorite search window.
(SOURCE: NASA/SETI Institute.)

have been treated to episodes of *Seinfeld*. The joke goes that aliens are not visiting us because they have received our radio and TV broadcasts and have so far seen no signs of intelligent life. In fact, the spinning Earth sends out radio waves that rise and fall several times per day from the concentration of transmitters in the United States and Europe. Realistically, the content of our weak transmissions could not be deciphered because of the overwhelming radio din from other cosmic sources. Moreover, the two largest sources of our radio leakage have diminished. Powerful early warning radar has been dismantled because of the end of the cold war, and TV transmissions are moving toward fiber and cable technologies.

How to Recognize a Message

As radio signals travel through space, they decrease in intensity by the square of the distance from the transmitter. The biggest obstacle to sending or receiving a signal over large distances is the weakness of the signal in relation to cosmic noise sources. It is like trying to hear a whispering child in a room full of shouting adults. Compared to a natural astronomical signal, a radio message should have a narrow bandwidth and be varied, in order to code information. A narrow bandwidth gives the best detection in the presence of sources of radio noise. Pulsars have a periodic signal, which represents the simplest kind of variation. (The University of Cambridge team that discovered pulsars in 1967 was initially mystified as to the origin of the periodic signals. As they searched for a natural explanation for the radio pulses, they jokingly referred to the pulsars as LGM1, LGM2, etc., for "little green men." Needless to say, when a local paper heard about the joke, it published a splashy story about aliens.) A periodic signal carries very little information—a single frequency or number. Messages carrying a significant amount of information will be more complicated.

A signal that is *pulsed* is optimal for transmission over large distances because the radio energy is concentrated in small bursts. A pulsed form of signal corresponds directly to the digital way that information is stored and transmitted on FM radio stations, CDs, and computers. Information of enormous complexity can be conveyed with a sufficient number of binary elements, or bits. Each bit is a signal, either 1 or 0, on or off. We might hope that any creature with sense receptors is aware of some version of the distinction between bright and dark, hot and cold, loud and quiet, or 1 and 0.

What message should we send? Ideally we might want to send an eloquent statement of our hopes and dreams, but we risk falling into an anthropocentric trap. A universal message should be as free from cultural influences as possible. A language based on mathematics could convey many universal truths. For example, artificial intelligence expert Marvin Minsky has explored the behaviors of all possible processes by creating all possible computers and their programs. This is the computational equivalent of the Miller-Urey experiments mentioned earlier—taking the simplest processes and seeing what complexity arises. Thousands of computational machines were created. Most stopped without accomplishing anything, a few got trapped in circles and senselessly repeated the same steps over and over. However, a few performed a counting operation; essentially, they "invented" arithmetic! Dutch mathematician Hans Freudenthal has developed LINCOS, or *Lingua Cosmica*, a language designed for universal discourse. This language is designed to convey mathematics and physics in a coded form.

Have the messages sent thus far matched the ideal of a universal language? In 1974, a symbolic message was beamed in the direction of the globular star cluster M 13, using the Arecibo radio telescope. The actual message consisted of 1679 on and off pulses, or bits of information (Figure 18–22a). An (alien) mathematician would notice that 1679 is divisible only by 23 and 73, two prime numbers. This suggests forming a two-dimensional pattern. If the bits are arranged as 73 columns of 23 bits each, there is no discernable pattern. The second choice, 23 columns of 73 bits each, gives the pattern shown in Figure 18–22b (where the binary numbers are replaced by their equivalents in a visual field, 0 = dark and 1 = bright). The pattern is clearly nonrandom, but it is instructive to see if you could deduce any parts of the message before reading the caption to the figure (Figure 18–22c). The message is a strange mixture of crude cartoons and binary representations of the chemical constituents of life. It is questionable how clear its true meaning would be to an alien intelligence. We will not find out for a while, since the round-trip light travel time to M 13 is 54,000 years!

It turns out that we may be able to *recognize* a message without being able to *decipher* it. Even if the alien recipients of the Arecibo message cannot interpret the symbolic images, or have no visual sense at all, they could recognize the intelligent intent of the message. For example, the pattern of 0's and 1's is highly nonrandom, with less than a 1 in 10^5 chance of arising by chance. An expert in mathematics could find a number of patterns in a binary-coded picture such as this. If the message were a coded language rather than a picture, pattern recognition techniques would take advantage of the redundancy of natural languages. For example, written English is over 75 percent redundant, in that on average, only one out of every four letters conveys any information at all!

SETI Past, Present, and Future

Speculation about life in the universe has a long and interesting history. Some 2000 thousand years ago, Plutarch wrote about "collections of matter, some of which are other worlds with their own skies and races of men and beasts." In the 11th century, the Church declared the doctrine of the "plurality of worlds" to be heretical, and Giordano Bruno was burned at the stake in 1609 for proposing an infinity of worlds with life on them. Practical SETI began in 1820, when Karl Gauss suggested planting large tracts of Siberian forest in a graphic demonstration of the Pythagorean theorem. In 1840, Joseph von Littrow wanted to ignite kerosene-filled trenches in the Sahara Desert in an assortment of geometric shapes. Neither of these proposals was ever funded. At the turn of the century, the eccentric

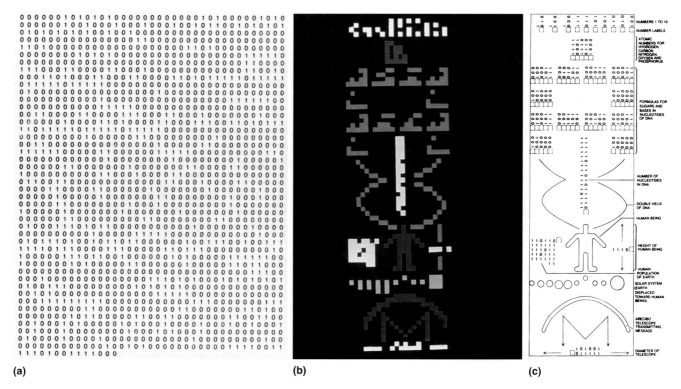

FIGURE 18-22
Astronomers have attempted interstellar communication using radio waves to send a message using a binary code. **(a)** This symbolic message was sent out as a stream of 1679 bits, or on/off signals, beamed in the direction of the globular cluster M 13 in 1974. The arrangement shown requires the recognition that 1679 is the product of only two prime numbers, 23 and 73 (there are, of course, two ways to arrange the message, one of which produces a meaningless pattern). The round-trip light travel time to M 13 is 54,000 years. **(b)** A color representation of the message shows a highly nonrandom pattern. Can you guess the meaning? Cover up the last panel of the figure, and see if you can decode any of the information. **(c)** Running from top to bottom, the coding represents (1) binary representation of the numbers 1 to 10; (2) atomic numbers of the five elements essential to life (H, C, N, O, P); (3) the chemical makeup of the four bases used in the DNA molecule, given by binary representations of the number of H, C, N, O, and P atoms in each one; (4) a cartoon of the DNA double helix; (5) numbers for the human population and human height surrounding a cartoon of a human figure; (6) a schematic view of the solar system, with Earth indicated; and (7) a schematic cartoon of the Arecibo radio telescope, with its size indicated in binary notation. (Source: NASA.)

genius Nikola Tesla built enormous 150-foot electrical coils and strained the power production capacity of Colorado Springs in order to send radio signals to extraterrestrials. Tesla thought he had detected signals from an extraterrestrial civilization, but in fact he discovered the atmospheric phenomenon known as whistlers. Guglielmo Marconi also used his early radio technology to listen for signals from beyond Earth.

Modern SETI began with Project OZMA, which ran from 1959 to 1960. In this experiment, Frank Drake scanned across a 400-kHz band of radiation from the two nearest solar-type stars. Nothing conclusive was found. Since then, dozens of radio SETI projects have been conducted without any convincing sign of extraterrestrial intelligence. Of course, all of these experiments combined have searched only a tiny fraction of the cosmic haystack.

On October 12, 1992—500 years after Columbus' discovery of the New World—NASA began the most ambitious SETI project ever. The prosaically named Microwave Observing Project (MOP) searched more of the cosmic haystack in its first 5 minutes than all previous SETI experiments combined! The long time scales for sending and receiving messages led NASA to a passive strategy: to search for signals that may have been deliberately or accidentally sent our way long ago. The search strategy has two components. The first part consists of the detailed monitoring of the 1000 solar-type stars nearest the Sun. The second part involves a scan of the entire sky, to allow for the possibility of rarer but more powerful signals from across the Galaxy. After funding was repeatedly cut by the United States Congress, SETI activists started to look for private sources of funding. The current plan, called Project Phoenix, is not funded by taxpayers at all.

The new wave of SETI experiments depends on powerful receivers and large telescopes. Previous projects could only search a few hundred different frequency channels simultaneously. The MOP listens in on tens of millions of channels simultaneously. The improvement has been made possible by modern digital techniques and custom integrated circuits. The receivers are also extraordinarily sensitive. Figure 18–23 shows the radio detection of the weak signal of *Pioneer 10* after it left the solar system; the diagonal slope of the detection represents the Doppler shift caused by Earth's rotation. The signal has the equivalent power of a single Christmas tree light at a distance of over 5 billion miles! The 1000-foot-diameter radio dish at Arecibo in Puerto Rico offers even greater sensitivity (Figure 18–24). The dish is so large that it could hold 357 million boxes of corn flakes or all the beer consumed on Earth in an entire year. Detectors at the focus of the Arecibo dish could detect a half-megawatt radio signal, equal to the strength of a modest radio station on the other side of the Milky Way.

More ambitious SETI schemes exist. It would be possible to send a beacon and receiver hundreds of AU from Earth and use the Sun as a gravitational lens to amplify and direct signals to distant targets. Future telescopes in space will allow spectroscopic analysis of the reflected light from planets in other stellar systems. The direct detection of atmospheric chemistry that indicates life would cut through the web of anthropocentric arguments regarding technology and radio communication. In the meantime, researchers must think as broadly as possible about the nature of life (Figure 18–25). The search for life in the universe will become a truly scientific subject, one step at a time.

Few scientific subjects generate as strong an emotional response as SETI. Debates between SETI optimists and pessimists can be acrimonious. Pragmatists argue that the scientific basis for the optimistic calculations is flimsy and that no search strategy can be logically justified. SETI has been unpopular with some politicians, who see it as a frivolous use of taxpayers' money. NASA has had considerable trouble in funding SETI, despite the fact that it accounts for less than 0.1 percent of the agency's science budget. However, popular support for SETI remains strong. "The probability of success is difficult to estimate," wrote physicists Giuseppe Cocconi and Phillip Morrison in 1959 in a paper that started the modern era of SETI. "But if we never search, the chance of success is zero." Few people can resist the excitement of one of the most profound questions humans can ask: What is our place in the universe?

FIGURE 18-24
The 1000-foot-diameter radio dish at Arecibo in Puerto Rico, the world's largest. Radio signals are detected at the focus suspended above the dish by three towers (each the size of the Washington Monument). The dish cannot be steered, and it can observe only a narrow strip of sky close to the zenith. Astronomers use this telescope both to listen and to send signals into space. If alien civilizations had similar technology, we could hear them, and they could hear us, out to a distance that includes hundreds of nearby stars. (SOURCE: Arecibo Observatory, National Astronomy and Ionosphere Center, Cornell University under contract with the NSF.)

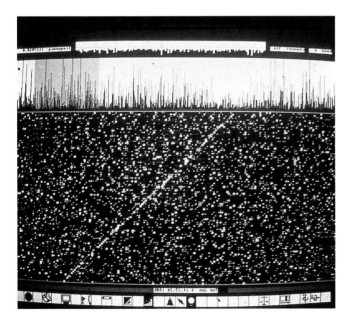

FIGURE 18-23
Detection of the 1-watt signal from the *Pioneer 10* spacecraft, when it was over 5 billion miles from Earth. Each line represents a scan in frequency; the frequency of the *Pioneer 10* signal shifts because of the rotation of Earth. (SOURCE: NASA/SETI Institute.)

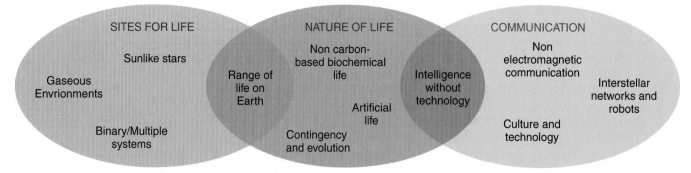

FIGURE 18–25
To move beyond the confines and the anthropocentric assumptions of the Drake equation, we must think as broadly as possible about life in the universe. This graphic indicates some of the questions we might ask about life in the universe, and indicates some of the experiments we might carry out.

The Anthropic Principle

Even though we have not detected life beyond Earth, we are confronted with the mystery of our own existence. Modern cosmology tells us that we live in a universe that is enormously large and old, a universe in which we are insignificant players. Can there be any meaning to life in such a cosmic setting? In response to this question, some scientists have come up with an unusual idea called the **anthropic principle,** stating that the universe is arranged in a particular way that can support life.

Nature exhibits some remarkable coincidences. Miss Marple, the detective hero of Agatha Christie's novels, once said: "A coincidence is always worth noticing. You can always discard it later if it *is* just a coincidence." Consider the fundamental forces of nature that we discussed in the last chapter. If the strong force that binds atomic nuclei together were only slightly stronger, there could be no stable atoms, and stars would fizz through their lives in only a few years. If this force were only slightly weaker, stellar fusion would be impossible. If the weak force in atomic nuclei were much stronger, the big bang would have converted all the hydrogen in the universe into helium, leaving a universe with no long-lived stars and no water. If the weak force were much weaker, the universe would not even contain hydrogen. If the electromagnetic force were much stronger, stars would be reduced to a feeble level of brightness, and no elements heavier than iron could be formed. If the electromagnetic force were much weaker, all stars would be very hot and short-lived.

What do we learn from these hypothetical statements? If the forces of nature had slightly different properties, the universe would be a sensible place governed by physical laws. However, it would be a universe without biological life! Variations in almost any of the physical constants lead to conditions in which life would be impossible: Either there would be no long-lived stars, or there would be no heavy elements. On a cosmological scale, we find ourselves in a universe with a density parameter close to 1. A universe with a density parameter much greater than 1 would collapse into heat death on a time scale much shorter than the age of our universe. A universe with a density parameter much less than 1 would expand so rapidly that stars could not form out of the thinning gas. Once again, these variations represent plausible big bang models, but they are all examples of a universe where biological life could not form. In other words, we should not be surprised that we live in a universe so vast and old—if it were much different, we could not exist!

Is the anthropic principle scientifically useful? Humans observe the universe. One way of stating the anthropic principle is that we can only observe a universe that is capable of allowing observers to exist. This is a tautology, an obvious truth. As a scientific idea, the anthropic principle is flawed because it has no predictive power. Also, it presumes a biological basis for life; we do not know if life *needs* heavy elements and stars. Perhaps the only absolute requirement for life is a departure from thermal equilibrium, since it is almost certain that a uniform and featureless gas cannot carry any coded information. Science is not able to explain why we exist, nor if extraterrestrial life exists. Yet we continue to search—a search that involves how we know, the physical nature of the universe, and our place within it.

Summary

Evidence of alien civilizations, or even alien life-forms, could be a pivotal development in our view of ourselves. Many issues are involved in the calculation of the probability that intelligent life exists beyond Earth. Planets exist beyond the solar system, and astronomers believe that habitable planets exist elsewhere in the universe, although they cannot yet prove it. Laboratory experiments and the early history of life on Earth suggest to us that life should start on other planets if liquid water, energy, and the right chemicals are present. Four elements form the foundation of organic chemistry: hydrogen, carbon, nitrogen, and oxygen. These elements are produced in the cores of stars and are widely distributed across interstellar space. Earth may have been seeded with organic molecules (but not life-forms) by cometary material soon after the solar system formed. We cannot reproduce the progression from organic molecules to simple life forms in the laboratory. Yet we know that life formed on Earth under extreme conditions within the first 10 percent of the planet's history.

The evolution on Earth from single-celled organisms to complex multicellular life-forms was neither smooth nor predictable. Biological evidence shows that life is adaptable and species can evolve to prosper in different environments, from ocean depths to low-pressure atmospheres of different compositions. Life on Earth has evolved toward greater complexity, but we know that complexity is not required for a life-form to endure or adapt to a wide range of environments. Eventually, life on Earth evolved to intelligence in a few species and the ability to manipulate the global environment in one. However, the progress of evolution has often been punctuated by external events. These include the catastrophic effects of meteor and comet impacts and more subtle influences from our solar and galactic environment. We do not know if intelligence is inevitable. We do know that it took nearly 4 billion years to evolve on Earth and has arisen in only a few of the millions of species on the planet.

Despite the enormous uncertainty in our estimate of the number of intelligent civilizations in the galaxy, astronomers have made tentative efforts to communicate across interstellar space. There is no evidence that alien life-forms have either visited or tried to make contact with us. We can only speculate about the reasons. Perhaps they do not exist and we *are* alone. Perhaps they are too far away. Perhaps civilizations destroy themselves before successfully exploring the universe. Perhaps communication across space is a cultural activity not duplicated beyond Earth. Perhaps they are unrecognizable. Arthur C. Clarke has remarked that any technology much advanced beyond our own would look like magic. Perhaps we are too limited by our own concept of civilization. There is no reason to believe that aliens, even if they existed, would resemble us physically or culturally.

Clearly, we have been reduced to speculation by a lack of facts. The search for extraterrestrial intelligence can be criticized on logical grounds, but the emotional appeal is undeniable. If we could demonstrate that we live in a biological universe, it would be another giant step in the Copernican revolution. The only way to reduce the conjecture and increase the proportion of fact is to pursue research in many related fields—astronomy, physics, chemistry, geology, biology, paleontology—and continue to search the skies with our telescopes. There may be surprises out there.

Important Concepts

You should be able to define these concepts and use them in a sentence.

life	prokaryote
cell	photosynthesis
bacteria	eukaryote
organic chemistry	natural selection
amino acids	intelligence
proteins	SETI
deoxyribonucleic acid (DNA)	Drake equation
gene	anthropic principle

How Do We Know?

These questions and answers show how the scientific method is used to learn about the universe.

Q What are the requirements for life?

A Earth is the only planet we know with life on it, so we base our expectations for life on our own home. Life on Earth uses organic chemistry, and carbon has unique advantages for forming complex molecules that can carry information. Biochemical life is based on DNA, the replicating molecule that carries the genetic information of every individual. DNA may not be the only molecule that could form the basis of a living organism. The smallest living units are cells, tiny chemical factories that function in a liquid medium. Water is a solvent for many organic molecules and thus is a perfect medium for complex chemical reactions. Living systems need an energy source. On Earth, the energy source is usually sunlight, as in the enormous food web that is based on photosynthesis. However, the energy source might also be geothermal energy. Finally, we presume that living things need a planet on which to form and evolve. It is very unlikely that biochemical life could evolve on rocky fragments or in the depths of space.

Q How do we know that life must be biological?

A Astronomers believe that life in the universe must be biochemical, but it is no more than an assumption. In general terms, a life-form is something that stores and transmits information, evolves, and adapts to its environment. Carbon chemistry in a liquid medium may not be the only pathway to life. For example, humans have developed computers that mimic many life processes. While no scientist would suggest that computers are alive, we should at least consider the possibility that life might evolve to a mechanical or artificial form.

Q Are there other sites for life?

A Perhaps. Most of the extrasolar planets that have been discovered so far are too massive and hot to be hospitable to life. However, Earth-like planets are not detectable by current experiments. Since astronomers believe that planet formation is a natural by-product of star formation, it is very likely that some fraction of those planets will have the appropriate atmospheric and surface conditions for life. It is significant that life formed under extreme conditions in the first 10 percent of Earth's history. It is also interesting that there are several solar system locations where we cannot rule out current or ancient microbial life—Mars, Europa, and Titan.

Q How do we know the number of intelligent civilizations in the galaxy?

A We do not. The Drake equation combines a number of factors into an estimate of N, the number of intelligent, communicable civilizations in the Milky Way galaxy. The astronomical and chemical factors are uncertain, but there is evidence to support scientific speculation; we know that planets exists around other stars, and we know that the chemical ingredients of life are widely distributed throughout the galaxy. Given the robustness of primitive life on Earth and its variety of environments, astronomers think it plausible that simple life-forms exist beyond Earth. On the other hand, the biological and cultural factors in the Drake equation are completely unknown. We have no idea how likely it is that life will evolve to become intelligent, and we have no idea whether intelligence, technology, and communication go hand in hand. As a consequence, N will remain unknown until we gather more evidence.

Q How do we know the best way to communicate with intelligent life-forms?

A We do not. The modern strategy of communication by radio signals is based on a number of assumptions. Most important is the presumption that intelligent beings will feel the need to communicate across the enormity of interstellar space. Scientists believe that the energy and time requirements for interstellar travel are prohibitive. Given that, it is more energy-efficient to communicate with radio waves than with any other form of electromagnetic radiation. Astronomers are confident that they can distinguish an artificial signal of intelligent origin from the various natural kinds of radio emission in the universe, even though it may be difficult or impossible to decode the message. Despite the optimism that motivates the search, it is possible that SETI will never succeed.

Problems

Use these problems to test your understanding of the information and concepts in this chapter. The * indicates a more advanced or mathematical problem.

1. List the requirements for life, if we assume that the biological basis for life is the same as it is on Earth. Are the chemical ingredients for life found beyond the solar system? How do we know?
2. Have we discovered any plausible sites for life beyond Earth? What is the range of conditions on a planet that might allow life to form? What is the range of stars that could support life? Why do astronomers believe life cannot form without a planetary surface?
3. Summarize the evidence for or against life on the following sites in the solar system: Mars, Europa, and Titan. Why do astronomers think life is unlikely anywhere else in the solar system?
*4. Convert the 4.5-billion-year history of Earth into a day. In terms of this scaled-down 24-hour period, when did the following major events in the history of our planet occur? **(a)** The first primitive forms of life. **(b)** The increase in atmospheric oxygen due to microbial life-forms. **(c)** The emergence of life from the oceans. **(d)** The death of the dinosaurs and many other large species. **(e)** The first appearance of *Homo sapiens*. **(f)** The first human cities and civilizations. **(g)** The invention of radio technology. **(h)** The first manned space flight.
5. Astronomers sent a radio signal from the Arecibo Observatory to M 13, a globular cluster at a distance of 26,000 light-years. Apart from the distance, why is M 13 a poor choice of target if we are trying to inform carbon-based, intelligent organisms like ourselves of our existence?
6. Why would life be unlikely on a planet associated with the following types of star? **(a)** An O-type or B-type main-sequence star. **(b)** A red giant star. **(c)** A white dwarf. **(d)** A pulsar.
7. Describe the ways in which the characteristics of the cosmic environment, ranging from the existence of the solar system and the evolution of stars to our location within the Milky Way, might have affected the evolution of life on Earth or other planets.
*8. Assume that biochemical life is a way of coding digital information. Explain how you could fit the genetic code of a human on the hard disk of an average computer. Computers are designed to copy information perfectly. Explain what causes biochemical life to have slightly flawed copying procedures. Why is this imperfect copying an important feature of the way life works?
*9. Explain why the left-hand side of the Drake equation is a completely unknown number. Taking the factors on the right-hand side one a time, show how the product differs as you multiply in each new factor, taking the two cases of "optimistic" and "pessimistic" estimates. You can plot this on graph paper; take logarithms of the numbers so you can display large and small numbers on the same graph.
10. In your opinion, what would be the consequences for humanity of the following: **(a)** The discovery of microbial life in the icy oceans of Europa. **(b)** Firm evidence that Earth is the only place where life exists in the solar system. **(c)** The discovery of artificial signals of unknown meaning coming from a Sun-like star 10 light-years away. **(d)** Evidence that Earth-like planets are common around a wide range of stars. **(e)** The arrival of an alien spacecraft and visitors at a prominent place, like the United Nations headquarters in New York. **(f)** Strong scientific arguments that intelligent life on Earth was a fluke that was unlikely to have occurred anywhere else in the universe.

Projects

Activities to carry out either individually or in groups.

1. You are project scientist for a mission to send a time capsule into space that will eventually reach other star systems. The payload must fit into a cube 1 meter on a side, and it can contain no living or perishable items. Your goal is to communicate human culture and civilization to any intelligent life-forms that might find the capsule. What would you send, and why?
2. Split a large class into 12 groups according to birth month. Each group is a "civilization" isolated in space. Each group is to design a message using only a binary digital code—that is, a sequence of 1's and 0's, with a pause or a dash to separate the groups of digits and represent punctuation. Use long strips of paper like a cash register roll to write the message in a continuous pattern going from left to right; the information will be sent out into space as a stream of on/off signals. For example, the message might look like this: 1-11-111-1111-11111- . . . (a simple counting signal) or 010001-00001010-0001-1-1-1-000 . . . (a more obscure message). In the first part of the experiment, lasting roughly 1 hour, each civilization must discuss what message to send and how they would code it. The message can convey any aspect of human knowledge or culture. *(Note:* If the message is too simple, it might not be interesting, and if it is too complex, it might be hard to convert into a binary code.) Each civilization should write their message on one or more strips of paper; the instructor then photocopies each message 11 times and distributes them. In the second part of the experiment, also lasting 1 hour, *each* civilization tries to decode the messages of *all other* civilizations. To make it competitive, the instructor can keep track of which group decodes the most messages and which group's message is most easily understood by other groups.

Reading List

Dick, S. 1998. *Life on Other Worlds: A History of the 20th Century Extraterrestrial Life Debate.* Cambridge, Eng.: Cambridge University Press.

Fisher, D., and Fisher, M. 1998. *Strangers in the Night: A Brief History of Life on Other Worlds.* Washington, D.C.: Counterpoint.

Freudenthal, H. 1960. *LINCOS: Design of a Language for Cosmic Intercourse.* Amsterdam: North-Holland.

Goldsmith, D., and Owen, T. 1992. *The Search for Life in the Universe.* Reading, Mass.: Addison-Wesley.

Gould, S. J. 1989. *Wonderful Life.* New York: Norton.

Lewis, J. S. 1998. *Worlds Without End: The Exploration of Planets Known and Unknown.* Reading, Mass.: Perseus Books.

Raup, D. M., and Valentine, J. W. 1983. "Multiple Origins of Life." *Proceedings of the National Academy of Sciences,* vol. 80, p. 2981.

Regis, E., ed. 1985. *Extraterrestrials: Science and Alien Intelligence.* Cambridge, Eng.: Cambridge University Press.

Sagan, C. 1978. *Murmurs of Earth.* New York: Ballantine.

Schidlowski, M. 1988. "A 3800 Million-Year Isotopic Record of Life from Carbon in Sedimentary Rocks." *Nature,* vol. 333, p. 313.

Shostak, S. 1992. "The New Search for Intelligent Life." *Mercury,* vol. 21, p. 115.

Tipler, F. J. 1980. "Extraterrestrial Beings Do Not Exist." *Quarterly Journal of the Royal Astronomical Society,* vol. 21, p. 267.

Appendix A: Scientific and Mathematical Techniques

Appendix A-1
Scientific Notation and Logarithms

In the study of astronomy, you will encounter many extraordinary numbers to describe distances, ages, temperatures, and densities. Astronomers and other scientists have developed a shorthand system for writing very large and very small numbers. It is not convenient to write the distance light travels in a year as 6,000,000,000,000 miles or the density of interstellar space as 0.000000000000000000000002 g/cm^3. The best system for writing large and small numbers is called *scientific notation*.

A number written in scientific notation has two parts, separated by the multiplication sign. The part in front of the multiplication sign is the *coefficient*. The coefficient has only one digit to the left of the decimal point. The power of ten that follows the multiplication sign is the *exponent*. (Scientific notation is also called exponential notation, or powers of ten notation.) For example, the two numbers above can be written in scientific notation as 6×10^{12} miles and 2×10^{-24} g/cm^3—a much more compact way to express large and small numbers.

You can see that the exponent tells you how many places to shift the decimal point to the left or the right to form the number. In other words, 2.51×10^5 means shift the decimal point plus five places, or five places to the right, giving 251,000. On the other hand, 6.8×10^{-7} means shift the decimal point minus seven places, or seven places to the left, giving 0.00000068. If there is a zero in the exponent, the decimal point does not move (which makes sense since $10^0 = 1$). You should practice until you are comfortable going from normal notation to scientific notation and from scientific notation to normal notation. Let us look at some more examples:

$$0.000490372 = 4.90372 \times 10^{-4}$$
$$3001 = 3.001 \times 10^3$$
$$0.000002 = 2 \times 10^{-6} \text{ (or } 2.0 \times 10^{-6})$$
$$100,000,000 = 10^8 \text{ (or } 1.0 \times 10^8)$$
$$0.887 = 8.87 \times 10^{-1}$$
$$148,400 = 1.484 \times 10^5$$

Now we can look at all the powers of ten explicitly. Remember that the exponent is the number of factors of ten that are multiplied together to get the quantity. For example, $10^0 = 1$, $10^1 = 10$, $10^2 = 10 \times 10 = 100$, $10^3 = 10 \times 10 \times 10 = 1000$, and so on. The most often used large numbers are one thousand (10^3), one million (10^6), one billion (10^9), and one trillion (10^{12}). You can see that the exponent in a power of ten just gives the number of zeros that follows the one. On the next page is a list of powers of ten.

Let us see how these forms are used in practice. Rather than write a long decimal version of a large or small number, scientists prefer to use scientific notation. Often they use the prefix (or symbol) in the metric system as shorthand. For example:

0.0000673 m = 6.73×10^{-5} m = 0.0673 mm = 67.3 µm
56,000 J = 5.6×10^4 J = 56 kJ = 0.056 MJ
127,000,000 pc = 1.27×10^8 pc = 127 Mpc = 0.127 Gpc
0.0000000899 s = 8.99×10^{-8} s = 0.00899 µs = 0.899 ns
145 g = 1.45×10^2 g = 145,000 mg = 0.145 kg
51,000,000,000 W = 5.1×10^{10} W = 51 GW = 0.051 TW

Throughout this book, you will see examples of estimation, where we combine large and small numbers to get a rough estimate of an important quantity. Preserving a lot of significant digits is not important in these calculations—estimation is only intended to be accurate to about a factor of 2. To work with large and small numbers, you need to know the rules for multiplying and dividing in scientific notation. To multiply two numbers together, you multiply the coefficients and add the exponents. (Be careful to preserve signs.) Here are some examples.

$$(7.91 \times 10^4) \times (2 \times 10^7) = (7.91 \times 2) \times 10^{4+7}$$
$$= 15.82 \times 10^{11}$$
$$= 1.582 \times 10^{12}$$

$$(6.9 \times 10^8) \times (1.1 \times 10^{-5}) = (6.9 \times 1.1) \times 10^{8+(-5)}$$
$$= 7.59 \times 10^{8-5}$$
$$= 7.59 \times 10^3$$

$$(4 \times 10^{-6}) \times (5.8 \times 10^{-11}) = (4 \times 5.8) \times 10^{-6+(-11)}$$
$$= 23.2 \times 10^{-17}$$
$$= 2.32 \times 10^{-16}$$

To divide two numbers, you *divide* the coefficients and *subtract* the exponents. (Once again, be careful to preserve signs when you subtract exponents.) Here are some examples:

ENGLISH VERSION	DECIMAL VERSION	SCIENTIFIC NOTATION	METRIC PREFIX	SYMBOL
one billionth	0.000000001	1.0×10^{-9}	nano	n
one hundred millionth	0.00000001	1.0×10^{-8}		
one ten millionth	0.0000001	1.0×10^{-7}		
one millionth	0.000001	1.0×10^{-6}	micro	μ
one hundred thousandth	0.00001	1.0×10^{-5}		
one ten thousandth	0.0001	1.0×10^{-4}		
one thousandth	0.001	1.0×10^{-3}	milli	m
one hundredth	0.01	1.0×10^{-2}	centi	c
one tenth	0.1	1.0×10^{-1}	deci	d
one	1	1.0×10^{0}		
ten	10	1.0×10^{1}		
one hundred	100	1.0×10^{2}		
one thousand	1000	1.0×10^{3}	kilo	k
ten thousand	10,000	1.0×10^{4}		
one hundred thousand	100,000	1.0×10^{5}		
one million	1,000,000	1.0×10^{6}	mega	M
ten million	10,000,000	1.0×10^{7}		
one hundred million	100,000,000	1.0×10^{8}		
one billion	1,000,000,000	1.0×10^{9}	giga	G
ten billion	10,000,000,000	1.0×10^{10}		
one hundred billion	100,000,000,000	1.0×10^{11}		
one trillion	1,000,000,000,000	1.0×10^{12}	tera	T

$$(3 \times 10^6) \div (6.3 \times 10^4) = (3 / 6.3) \times 10^{6-4}$$
$$= 0.48 \times 10^2$$
$$= 4.8 \times 10^1 \text{ (or just 48)}$$

$$(8.35 \times 10^6) \div (2.7 \times 10^{-6}) = (8.35 / 2.7) \times 10^{6-(-6)}$$
$$= 3.07 \times 10^{6+6}$$
$$= 3.07 \times 10^{12}$$

$$(7.5 \times 10^{-8}) \div (9 \times 10^{-7}) = (7.5 / 9) \times 10^{-8-7}$$
$$= 0.83 \times 10^{-15}$$
$$= 8.3 \times 10^{-14}$$

Parentheses are used to group items and operations in mathematics. You should always complete the operations within each set of parentheses before moving outside the parentheses. (In other words, do not just work out an equation from left to right.) If a complex problem involves more than two large or small numbers, the principles are the same. You just group all the coefficients and multiply or divide them, and group all the exponents and add or subtract them. Look at the following example:

$$\{(3.4 \times 10^4) \times (6 \times 10^{-8})\} / \{(1.6 \times 10^{-9}) \times (4.7 \times 10^5)\}$$
$$= \{(3.4 \times 6) / (1.6 \times 4.7)\} \times 10^{4+(-8)-(-9)-5}$$
$$= (20.4 / 7.5) \times 10^{4-8+9-5}$$
$$= 2.7 \times 10^0 \text{ (or just 2.7)}$$

The best way to become confident manipulating large and small numbers is to practice. Calculators and computers that cannot show superscripts usually display a number like 5.9×10^6 as 5.9e+06 or 5.9E+06. The usual way to enter a number in scientific notation is to enter the coefficient, then press the button marked EXP or EE, and finally enter the exponent. You should change the sign of the exponent before entering it if the number is a negative power.

In addition to scientific notation, *logarithms* are often used to express and manipulate large and small numbers. The common logarithm, or logarithm to the base 10, of a number is written like this:

If $x = 10^y$, then $y = \log_{10} x$, or $y = \log x$

The logarithm defines a compressed scale that is very handy in astronomy where numbers with an enormous range must often be compared (see also Appendix A–7). On a calculator, you can take the logarithm of a number by entering the number and pressing the button marked 10^x or LOG. The reverse operation, raising a number to a power of ten, is done by entering the power of ten and pressing the button marked 10^x or ALOG. Note that the power of ten does not have to be an integer or a positive number. The table below has the logarithms of integers from 1 to 10, along with the fractional powers of ten from 0.1 to 1. These conversions are quoted with three significant figures, although your calculator can display them with much higher precision. The table shows the very simple form of the logarithms of powers of ten.

x	$\Rightarrow$	log x	log x	$\Rightarrow$ x	x	$\Rightarrow$	log x
1		0.000	0.1	1.259	10^{-4}		-4
2		0.301	0.2	1.585	10^{-3}		-3
3		0.477	0.3	1.995	10^{-2}		-2
4		0.602	0.4	2.512	10^{-1}		-1
5		0.699	0.5	3.162	10^{0}		0
6		0.778	0.6	3.981	10^{1}		1
7		0.845	0.7	5.012	10^{2}		2
8		0.903	0.8	6.310	10^{3}		3
9		0.954	0.9	7.943	10^{4}		4
10		1.000	1.0	10.00	10^{5}		5

The rules for combining and manipulating logarithms are given below:

OPERATION	DEFINITION OF SYMBOLS	RULES
Logarithm of a product	A and B are numbers	$\log(A \times B) = \log A + \log B$
Logarithm of a ratio	A and B are numbers	$\log(A / B) = \log A - \log B$
Logarithm of a power	n is the power of A	$\log(A^n) = n \log A$

In other words, we have an alternative to the multiplication and division of large and small numbers given earlier. To multiply two or more numbers in scientific notation, take the logarithms of the numbers and add them. To divide numbers in scientific notation, take the logarithms of the numbers and subtract them. To square a number in scientific notation, take its logarithm and multiply by two. To raise a number in scientific notation to the $1/4$ power, take its logarithm and multiply by $1/4$. This simple and versatile system has been used for over 300 years.

Appendix A-2
Algebra and Working with Dimensions

The system for manipulating numbers and symbols in an equation is called *algebra*. Scientists use equations to express physical relationships between measurable quantities. Algebra is the tool that scientists use to relate one equation to another, or to convert an equation into a more useful form. In algebra, we use letters of the alphabet (*a, b, c . . . x, y, z*) as symbols to represent numbers or physical quantities. The idea is to manipulate the symbols without inserting numbers until you have the equation in the form you want.

Physical science seems to contain a bewildering array of different things that can be measured and quantified—mass, density, temperature, power, force, acceleration, and so on. Yet almost every quantity in this book is a combination of only three fundamental quantities, or dimensions—*mass, length,* and *time*. We will abbreviate these dimensions as *M, L,* and *T*. The best way to see this is to consider some examples:

Area of a square	$A = d^2$	has dimensions of L^2
Volume of a cube	$V = d^3$	has dimensions of L^3
Density	$\rho = M/V$	has dimensions of M/L^3
Velocity	$v = d/t$	has dimensions of L/T
Acceleration	$a = v/t$	has dimensions of L/T^2
Momentum	$S = mv$	has dimensions of ML/T
Force	$F = ma$	has dimensions of ML/T^2
Pressure	$P = F/A$	has dimensions of M/LT^2
Kinetic energy	$E = \tfrac{1}{2} mv^2$	has dimensions of ML^2/T^2
Power	$P = E/t$	has dimensions of ML^2/T^3

You can always check that an equation in physics makes sense by inserting the dimensions of the quantities, and demonstrating that the dimensions of the left-hand side and the right-hand side are the same. For example, we see that the dimensions of force are mass times length divided by time squared, by starting with the equation $F = ma$. Substituting for acceleration, we get $F = mv/t$. Substituting for velocity, we get $F = md/t^2$, dimensions of ML/T^2. You will notice in algebra that we often omit the times sign ($\times$), since it can be confused with the symbol x. When two numbers or symbols appear next to each other, you can assume that they are multiplied together.

Let us see what the dimensions of the gravitational constant must be. Newton's law of gravity is written

$$F = G m_1 m_2 / r^2$$

In this equation, m_1 and m_2 are the masses of any two objects and r is the distance between them. We can rewrite the equation using the dimensions of each quantity.

$$M L / T^2 = G (M M / L^2)$$

To get G on its own, we follow the rules of algebra. Multiplying each side of the equation by L^2 gives

$$M L^3 / T^2 = G M^2$$

Now, dividing each side of the equation by M^2 gives the result

$$L^3 / T^2 M = G$$

Thus, the dimensions of the gravitational constant G are length cubed divided by time squared times mass. Let us look at another example. The Hubble relation says that the recession velocity of a distant galaxy is proportional to its distance from us. In equation form, $v = H_0 d$ is the Hubble constant. Dividing each side of the equation by d, we get $v/d = H_0$. Therefore, the dimensions of the Hubble constant are $(L/T)/L$ or $1/T$. Taking the reciprocal, the dimensions of $1/H_0$ are T, or time. In the chapter on cosmology, you will learn that the reciprocal of the Hubble constant gives an estimate of the age of the universe.

Now let us review the simple rules of algebra, which are designed to let you manipulate an equation to get the quantity you want on the left-hand side of the equal sign. The equations you will see in this book are simple, but the same rules can be used to solve equations of enormous complexity. In algebra, whatever you do to one side of an equation, you must also do to the other side. For example, if $x = y$, then $x/2 = y/2$ is true and $x - 7 = y - 7$ is true. The rule applies for any operation, so $x^3 = y^3$ is true, $\log x = \log y$ is true, and $\sqrt{x} = \sqrt{y}$ is true. An equality is always true as long as the same operation is performed on both sides of the equation.

RULE FOR	IF THIS IS TRUE	THEN THIS IS TRUE
Addition and subtraction	$a + b = c$	$a = c - b$
Multiplication and division	$a / b = c / d$	$a\,d = b\,c$
Powers and roots	$a = b^2$	$\sqrt{a} = b$

These rules tell us how to manipulate an equation to get the result we want. To "solve" an equation for a certain quantity means removing all other numbers and symbols so that the equation has only the quantity we want on the left-hand side. If the unwanted number or symbol is added, we remove it by subtracting it from both sides of the equation. In the example from the table above, $a + b = c$, we solve for a by subtracting b from both sides of the equation, giving $a + b - b = c - b$, or $a = c - b$. In general, we always perform the inverse operation—remove a quantity that has been multiplied by dividing it, remove a quantity that has been subtracted by adding it, and so on.

Suppose you want to convert a temperature from the Fahrenheit system into the Celsius system (or degrees centigrade). The equation that relates the two is $C = 5/9(F - 32)$. You need to solve this equation for F; that is, you need to get the quantity F alone on the left-hand side of the equation. First, multiply each side by 9, giving $9C = 9 \times 5 / 9(F - 32)$. The nines on the right-hand side cancel to give $9C = 5(F - 32)$. Now divide each side by 5, giving $9C / 5 = 5 / 5(F - 32)$. The fives on the right-hand side cancel to give $9C / 5 = F - 32$. Next, add 32 to each side, giving $9C / 5 + 32 = F - 32 + 32$. The result is $9C / 5 + 32 = F$, and we can just reverse the equation to get the form we want, $F = 9C / 5 + 32$.

Here are some more examples. Newton's law gives the force in terms of the mass acted on and the amount of acceleration caused, $F = ma$. Suppose we want to solve this equation for the acceleration. Simply divide each side of the equation by mass, $F / m = ma / m$. The masses on the right-hand side cancel, giving the result $F / m = a$, or $a = F / m$. Einstein's famous equation that gives the energy locked up in matter is $E = mc^2$. To find the mass equivalent to a certain amount of energy, we need to get the mass on its own, so divide each side of the equation by c^2. The result is $E / c^2 = m$, or $m = E / c^2$. Finally, the kinetic energy of a moving object is given by the equation $E = \frac{1}{2} mv^2$. Suppose we know the energy and mass of the object and want to solve for its velocity. First, we multiply both sides of the equation by two, giving $2E = mv^2$. Then we divide both sides by m, giving $2E / m = v^2$. Finally, we take the square root of both sides of the equation, giving $\sqrt{(2E/m)} = v$. So we have our answer, the velocity in terms of the kinetic energy and mass is $v = \sqrt{2E/m}$.

The equations you will encounter in this book have a variety of simple forms. Remember that these equations are not just abstract pieces of math; each equation expresses a relationship between quantities that we can measure in the real world. Equations are the tools used by scientists to help make sense of the physical universe. The simplest form is a direct relationship between two quantities, which we can write as

$$y = kx$$

In this equation, x and y are two quantities we can measure and k is a number, or constant. In the everyday world, y might be the cost of something and x might be its weight. The *constant of proportionality, k,* is the price per pound, or the price per kilogram. We can also write

$$y \propto x$$

In this form, the math symbol $\propto$ means "proportional to," and we say that y is directly proportional to x. In other words, if the weight doubles, the cost doubles, and if the weight triples, so does the cost. We can get another useful form of this equation by taking two specific situations. In one case, $y_1 = k x_1$, and in the second case, $y_2 = k x_2$. If we divide the two equations, the constant of proportionality cancels out, and we get the result

$$y_1 / y_2 = x_1 / x_2$$

In other words, the ratio of the costs of two items is equal to the ratio of their weights (assuming that k is the same for each item). To take another example, $\propto a$, this means that when an object is subject to a force, the acceleration is proportional to the force. If you double the force, the acceleration will double too.

The table below shows the main types of relationships between quantities that you will encounter in this book.

TYPE OF PROPORTIONALITY	FORM IN SYMBOLS	EXAMPLE
Direct	$y \propto x$	$E \propto m$ (Mass-energy)
Inverse	$y \propto 1/x$	$\lambda \propto 1/T$ (Wien's law)
Square	$y \propto x^2$	$E \propto v^2$ (Kinetic energy)
Inverse square	$y \propto 1/x^2$	$F \propto 1/r^2$ (Newton's gravity)
Cubic	$y \propto x^3$	$V \propto r^3$ Volume of a solid)
Quartic	$y \propto x^4$	$L \propto T^4$ (Stefan-Boltzmann)
General form	$y \propto x^n$	n is the power law index

The first four forms are the most common. Graphic versions of these relationships are shown in Figure A-1. The inverse square relation is particularly important: Gravity and light diminish with distance from their source according to an inverse square law. However, the index n in a power law relationship can have *any* value, positive or negative. Kepler's third law relates the period of a planet's orbit (P) to its distance from the Sun (a) with a form $P^2 \propto a^3$. Taking the square root of both sides, we see that $P \propto a^{3/2}$, or $P \propto a^{1.5}$. Or taking the cube root of both sides, we see that $a \propto P^{2/3}$, or $a \propto P^{0.67}$.

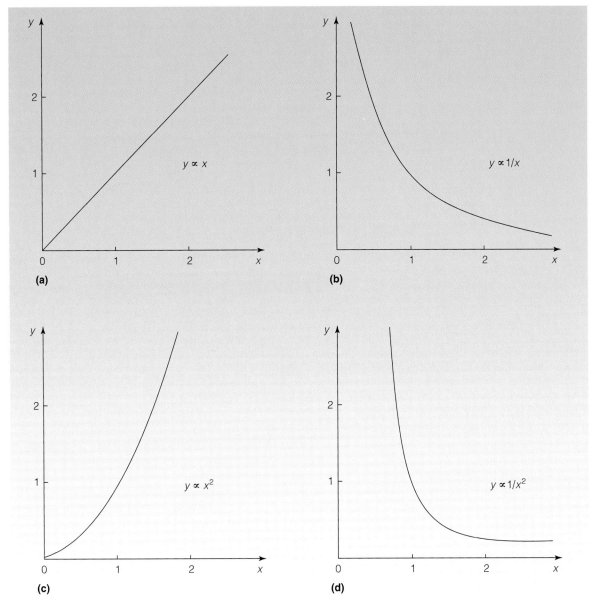

FIGURE A-1
Four common mathematical relationships that can apply to physical quantities. **(a)** $y \propto x$ is a linear proportionality. **(b)** $y \propto 1/x$ is an inverse proportionality. **(c)** $y \propto x^2$ is a nonlinear proportionality. **(d)** $y \propto 1/x^2$ is the familiar inverse square law. In each case, the constant of proportionality $k = 1$, so in these equations "=" can replace "$\propto$".

Appendix A-3
Units and the Metric System

The subject of measurement has a rich and fascinating history. Ever since humans have traded goods, they have needed a standard system of weights and measures. Many of our familiar units are part of the *English system,* which dates back to the Middle Ages. For example, length units were based on the human body: An inch is the length of the end joint of a thumb, a foot is of course the length of a foot, a yard is the distance from the tip of the nose to the end of an outstretched arm, and a fathom is the distance between the fingertips of two arms held straight out. However, these units are variable, since people are different sizes! As early as 1215, King John of England recognized the need for standardization in the Magna Carta: "There shall be standard measures of wine, corn, and ale throughout the kingdom."

Even after standardization, the English system of units is a mess. It consists of a jumble of different units, many of which are divided by different amounts. There are 8 pints in a gallon, 12 inches in a foot, and 16 ounces in a pound.

Many units are accidents of history, but their common use came about because very few people think quantitavely. It is easy to divide up objects when they are contained in groups of 8 (= 2 × 2 × 2) or 12 (× 2 × 2 × 3) or 16 (= 2 × 2 × 2 × 2).

The *metric system* was established by the French Academy of Sciences in 1791. It grew out of the French Revolution and a desire to rationalize many aspects of human affairs. The goal was to create a measurement system based on invariable quantities in nature, rather than on parts of the human body. Thus, the meter was defined as 1 ten-millionth of the distance from Earth's equator to the North Pole, the gram was defined as the mass of a cubic centimeter of water at 4°C, and the second was defined as 1/86,400 of a solar day (1/60 × 1/60 × 1/24). The metric system makes a lot of sense. There is only one basic unit for each quantity and all subdivisions and multiples are powers of ten. Thomas Jefferson was impressed by the metric system and pushed for its adoption in the United States. Finally, the United States and sixteen other countries signed the Treaty of the Meter in 1879. But as you look around, you will see that we are not metric in everyday life. In this regard, the United States is almost a lone holdout—not even the English use the English system any more!

Scientists have embraced the metric system completely. Here are the conversion factors between some familiar quantities in the English system and their counterparts in the metric system:

1 m = 39.37 in	and the reverse	1 in = 0.0254 m = 25.4 mm
1 m = 1.094 yd	and the reverse	1 yd = 0.914 m = 914 mm
1 km = 0.621 mi	and the reverse	1 mi = 1.609 km ≈ 8/5 km
1 g = 0.0353 oz	and the reverse	1 oz = 28.3 g = 0.0283 kg
1 kg = 2.205 lb	and the reverse	1 lb = 0.454 kg = 454 g
1 liter = 1.06 qt	and the reverse	1 qt = 0.94 liters = 940 cm^3
1 liter = 0.264 gal	and the reverse	1 gal = 3.79 liters = 3790 cm^3
1 W = 0.00134 hp	and the reverse	1 horsepower = 745.7 W
1 J = 0.00024 cal	and the reverse	1 calorie = 4186 J

These are just a few of the most common units from the English system. There are many more. You can be glad you do not have to remember the distinctions between cubits and furlongs and pecks and pottles and bushels and jacks and jills! Here are other important quantities that can all be derived from the fundamental units of mass, length, and time:

Frequency	measured in cycles/s	or Hertz (Hz)
Force	measured in kg m/s^2	or Newton (N)
Energy	measured in kg m^2/s^2	or Joules (J)
Power	measured in J/s	or Watts (W)
Temperature	measured in K	or Kelvin (K)

In science, temperature is given on the Kelvin scale, named after the Scottish physicist William Thomson, Lord Kelvin. The Kelvin scale is measured with reference to absolute zero—the temperature at which an object contains no heat and all atomic motions are frozen. Some scientists also use the Celsius scale, named after the Swedish astronomer Anders Celsius. The Celsius scale is also called the centigrade scale, since it is based on the division of the range from the freezing to the boiling points of water into 100 equal degrees. A third temperature scale was named after German physicist Gabriel Fahrenheit, who made the first successful mercury thermometer in 1720. Notice that Kelvin and Celsius degrees are the same size, but Fahrenheit degrees are smaller than either. The Fahrenheit scale is not used by scientists, and it is only used in the United States. The table below lists some important temperature markers and shows how to convert from one temperature scale to another.

TEMPERATURE	KELVIN	CENTIGRADE	FAHRENHEIT
Absolute zero	0 K	−273 °C	−459 °F
Water freezes	273 K	0 °C	32 °F
Water boils	373 K	100 °C	212 °F
Surface of the Sun	5700 K	5427 °C	9797 °F
Any level	°C + 273	5/9(°F − 32)	(9/5) °C + 32

Despite the variety of measurements we can make in the natural world, there are three fundamental properties that cannot be described in any simpler terms. These fundamental properties are *mass, length,* and *time*. Every quantity you see in this book can be expressed as some combination of these three. For example, velocity is distance (or length) divided by time, and density is mass divided by volume (or length cubed).

The metric system has been refined over time. The modern version of the metric system was agreed on in 1960; it is called the *International System of Units (SI)*. We now have fundamental and extremely precise definitions of the units of mass, length, and time. The standard unit for mass is the kilogram (kg). The "true" kilogram is a metal cylinder kept at the International Bureau of Weights and Measures in Paris. The standard unit of length is the meter (m). The meter is defined as the distance that light travels in 1/299,792,458 seconds; this definition is precise because we have measured the speed of light very accurately. The standard unit of time is the second (s). One second is defined as the length of time required for 9,192,631,770 vibrations of the cesium-133 atom. These details of these exact definitions are not important; scales and rulers and clocks work well enough in everyday life. But it is important to know that metric units *do* have precise definitions.

Astronomy covers a vast range of time and space. You will encounter a wide range of scales in this book. The table and figure that follow give rough values of mass (Figure A–2a), length (Figure A–2b), and time (Figure A–2c), ranging from the interior of an atom to the universe of galaxies.

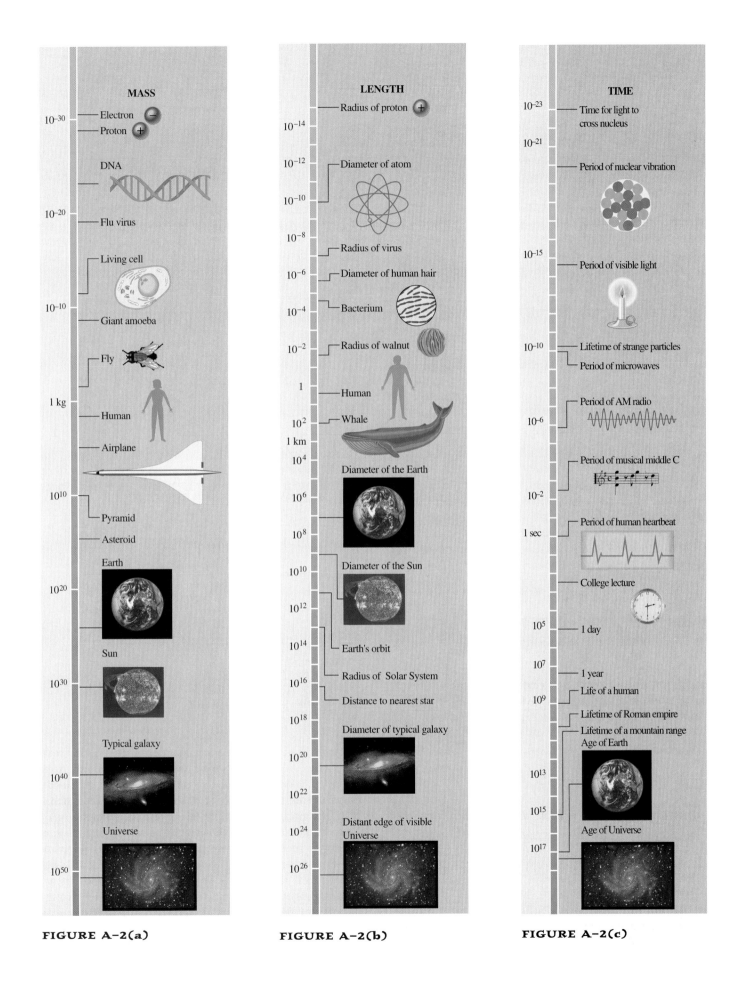

FIGURE A-2(a)

FIGURE A-2(b)

FIGURE A-2(c)

Approximate Values of Mass, Length, and Time

MASS IN KILOGRAMS		LENGTH IN METERS		TIME IN SECONDS	
10^{-30}	Mass of electron	10^{-35}	Planck length	10^{-43}	Planck time
10^{-27}	Mass of proton	10^{-17}	Smallest observable scale	10^{-34}	Inflationary era
10^{-25}	Mass of uranium atom	10^{-15}	Diameter of proton	10^{-23}	Light crosses a proton
10^{-15}	Mass of a bacterium	10^{-14}	Diameter of uranium nucleus	10^{-22}	Mean life of unstable nucleus
10^{-5}	Mass of a mosquito	10^{-10}	Diameter of hydrogen nucleus	10^{-15}	Oscillation of visible light
10^{-2}	Mass of a hummingbird	10^{-8}	Thickness of cell membrane	10^{-13}	Vibration of an atom in solid
1	One liter of water (1 qt)	10^{-6}	Wavelength of visible light	10^{-8}	Oscillation of FM radio wave
10^2	Mass of a person	10^{-3}	Size of a grain of sand	10^{-3}	Duration of nerve impulse
10^3	Mass of a car	1	Height of a young child	1	One heartbeat
10^8	Mass of a large ship	10^2	Length of a football field	10^5	One day
10^{13}	Mass of a small mountain	10^4	Greatest ocean depth	10^7	One year
10^{15}	Mass of a comet nucleus	10^7	Diameter of Earth	10^9	Human life expectancy
10^{23}	Mass of the Moon	10^{11}	Earth-Sun distance (1 AU)	10^{11}	Recorded history
10^{25}	Mass of Earth	10^{16}	One light-year	10^{12}	Light crosses the Milky Way
10^{30}	Mass of the Sun	10^{21}	Diameter of Milky Way galaxy	10^{16}	Age of youngest quasar
10^{42}	Mass of the Milky Way	10^{22}	Distance from Earth to M31	10^{17}	Age of Earth
10^{53}	Mass of known universe	10^{26}	Size of observable universe	10^{18}	Age of universe

Range of scales = largest number divided by smallest number:

$10^{53}/10^{-30} = 10^{80}$ $10^{26}/10^{-35} = 10^{61}$ $10^{18}/10^{-43} = 10^{61}$

Notice the fantastic range of physical quantities we consider in this book. The mass range from an electron to the entire universe is 10^{80}—one with eighty zeros after it! In addition to units based on mass, length, and time, there are two pure numbers that occur frequently in science and mathematics. Each of these numbers can be calculated with arbitrary accuracy—the sequence of digits after the decimal points is random and never repeats. The first dozen or so significant figures are given below; three significant figures are sufficient for most calculations.

$$\pi = 3.141592653589793238... \approx 3.14$$

$$e = 2.718281828459045235... \approx 2.72$$

The fundamental constants of nature that you will encounter most frequently in this book are given in the table below. Many of these numbers are known to a very high precision, but we only quote five significant figures here.

Astronomers mostly use the metric (or SI) system, but certain other units have been retained for convenience or sometimes for historical reasons. The most important examples are given below.

ASTRONOMICAL CONSTANT	SYMBOL	VALUE	UNITS
Astronomical unit	AU =	1.50×10^{11}	m
Parsec	pc =	3.09×10^{16}	m
Light-year	ly =	9.48×10^{15}	m
Solar mass	$M_\odot$ =	1.99×10^{30}	kg
Solar luminosity	$L_\odot$ =	3.90×10^{26}	W
Solar constant	... =	1.39×10^3	W/m²
Earth mass	$M_\oplus$ =	5.98×10^{24}	kg
Earth radius	$R_\oplus$ =	6.37×10^6	m

Because the scales of the universe are so large, we use these units so we do not always have to deal with extremely large numbers. It does not make sense to measure the distance between stars in meters or to measure the mass of a galaxy in kilograms. For example, in this book you will see density expressed in several different units. In a normal situation on Earth, the most appropriate units for density are kilograms per cubic meter (kg/m³). In interstellar space, the most useful units are atoms per cubic meter. However, on the enormous scales defined by a galaxy, the best units to use are solar masses per cubic parsec ($M_\odot/pc^3$). You will become familiar with these and other units as you work through the book.

FUNDAMENTAL CONSTANT	SYMBOL	VALUE	UNITS
Speed of light	c =	2.9979×10^8	m/s
Gravitational constant	G =	6.6726×10^{-11}	N m²/kg²
Planck's constant	h =	6.6261×10^{-33}	N s
Boltzmann's constant	k =	1.3807×10^{-23}	J/K
Stefan-Boltzmann constant	σ =	5.6705×10^{-8}	W/m²K⁴
Mass of an electron	m_e =	9.1094×10^{-31}	kg
Mass of a proton	m_p =	1.6726×10^{-27}	kg
Size of a hydrogen atom	... =	5.2900×10^{-11}	m

Appendix A-4
Precision and Measurement Errors

Scientists record information according to the accuracy of the measurement. The number of digits used to write a measurement indicates the *precision* of the measurement, also called the number of *significant figures*. Suppose we quote the population of Earth as 6,300,000,000 people. This estimate has only two nonzero digits or significant figures and implies a measurement of low precision. Alternatively, we might quote the population of Earth as 6,356,908,417 people. This estimate has ten significant figures and implies that we counted every person!

Clearly, the way we write a measurement carries an implication of our degree of certainty about the measurement. The nonzero digit that is farthest to the left is the *least* significant digit. The nonzero digit that is farthest to the right is the *most* significant digit—it indicates the precision of the measurement. In our population example, the most significant digit of the number 6,300,000,000 is 3, so we infer that the uncertainty in the estimate is one unit in the most significant digit, or 100,000,000. With two significant figures, we see that the accuracy of the estimate is 1 part in 63 or about 1 to 2 percent. The most significant digit in the number 6,356,908,417 is 7, so the implied accuracy of the number is 1 part in 6 billion or about 0.00000001 percent. Since someone is born and dies every second or so, this highly accurate estimate is very implausible! A better estimate of Earth's population might be 6,356,910,000—a precision of six significant figures.

Let us look at a few more examples of measurements and their precision. Remember to count a zero just like any other digit—for example, 9004 has four significant digits.

There are two types of exception where 0 is not counted as a significant digit. The first is in a decimal fraction, where the leading zeros are just placeholders behind the decimal point. The second is the fact that some numbers just happen to be round numbers. For example, a hundred-dollar bill represents exactly $100 and has three significant figures.

65,400 K	has 3 significant figures
0.00002 g/cm^3	has 1 significant figure
980,014 km	has 6 significant figures
7.5 × 10^5 pc	has 2 significant figures
1.007 × 10^{-8} kg	has 4 significant figures
3.3206 × 10^6 W	has 5 significant figures

Now we can see how the precision of a number relates to the implied accuracy of the measurement. In the first table below, we use the diameter of Earth, measured to be 12,756 km, as an example. Notice that it is always possible to quote a measurement with lower precision by rounding the number. Here is how to round off a number: If the most significant digit is 4 or less, you set it to 0, and if the most significant digit is 5 or higher, you set it to 0 and increase the next less significant digit by 1.

In the second table below, we use the example of the speed of light, which is one of the most accurately determined constants in nature. It is a good example of a physical constant that just happens to be nearly a round number in the metric system.

MEASUREMENT	PRECISION	ERROR	ACCURACY	% ERROR
12,756 km	5 significant figures	1 km	1 part in 10,000	0.01
12,760 km	4 significant figures	10 km	1 part in 1000	0.1
12,800 km	3 significant figures	100 km	1 part in 100	1
13,000 km	2 significant figures	1000 km	1 part in 10	10
10,000 km	1 significant figure	10,000 km	1 part in 1	100

MEASUREMENT	PRECISION	ERROR	ACCURACY	% ERROR
299,792.458 km/s	9 significant figures	0.001 km/s	1 part in 10^8	10^{-6}
299,792.460 km/s	8 significant figures	0.01 km/s	1 part in 10^7	10^{-5}
299,792.500 km/s	7 significant figures	0.1 km/s	1 part in 10^6	10^{-4}
299,793.000 km/s	6 significant figures	1 km/s	1 part in 10^5	0.001
299,790.000 km/s	5 significant figures	10 km/s	1 part in 10^4	0.01
299,800.000 km/s	4 significant figures	100 km/s	1 part in 1000	0.1
300,000.000 km/s	3 significant figures	1000 km/s	1 part in 100	1
300,000.000 km/s	2 significant figures	10^4 km/s	1 part in 10	10
300,000.000 km/s	1 significant figure	10^5 km/s	1 part in 1	100

The precision of a number—how many significant figures we quote—has an implication about the accuracy of the measurement. Accuracy reflects how far the measurement deviates from the true value. Therefore, precision has little meaning if we know nothing about the amount of *observational error*. To return to our first example, it is meaningless to quote the population of the Earth with a precision of 9 or 10 significant figures, since no one could possibly count people that accurately. Scientists talk about observational error, but a better term is *uncertainty*. Measurements are not usually wrong or in error, but they are uncertain because of unavoidable limitations in the measuring apparatus.

To be complete, a measurement should have an error attached. The theory of observational errors was worked out by the German mathematician Karl Friedrich Gauss in the 18th century. If we measure the width of a table with a tape measure, we might get the result 89 mm. If the smallest division on the tape measure is 1 mm, that is the probable (or standard) error in our measurement. Therefore, we would quote the measurement as 89 ± 1 mm. We can see some examples:

78.65 ± 0.02 K	implies a range of 78.67 to 78.63 K
0.0022 ± 0.0006 kg	implies a range of 0.0028 to 0.0016 kg
670 ± 50 Mpc	implies a range of 720 to 620 Mpc
0.17893 ± 0.00023 J	implies a range of 0.17916 to 0.17870 J
33 ± 8 μs	implies a range of 41 to 25 μs
91,000 ± 3500 W	implies a range of 94,500 to 87,500 W

We could do even better by trying to read the scale to the nearest half-millimeter, making a number of separate measurements, and taking the average. The result might be 88.2 mm. The standard error is given by the scatter in the individual measurements—usually represented by the Greek letter σ (sigma). Let's say it was 0.7 mm. The new and more accurate result is 88.2 ± 0.7 mm. Notice that our repeated measurements have allowed us to quote the width to a higher precision of three significant figures. Scientists *always* try to increase the reliability of their data by making multiple observations. Often, this is the only way they can reduce the uncertainty in their measurements.

In symbols, we say that if a measurement X has a standard error σ_x, we should quote the result as $X \pm \sigma_x$. What does this actually mean? It means that the true value of the quantity we have measured is probably in the range $X + \sigma_x$ to $X - \sigma_x$. The quantity X can be *any* measurement—the strength of Earth's magnetic field, or the temperature of the core of the Sun, or the number of stars in the Milky Way galaxy. For example, the true value of the width of the table from our previous measurement, 88.2 ± 0.7 mm, is probably in the range 88.2 + 0.7 to 88.2 − 0.7, or 88.9 to 87.5 mm.

In Gauss' theory of errors, if the errors have a random cause, then the measurements will have a well-defined distribution around the true value. This is called the Gaussian or normal distribution, or the "bell curve," which is shown in Figure A–3. Notice that the bell curve falls off rapidly in either direction. This means that random errors are unlikely to throw a measurement very far off the true value. By convention, the standard error σ is defined as the region that encloses 68 percent, or about $2/3$, of the measurements (see the example of measuring a star position in the Science Toolbox in Chapter 1). In other words, if the measured value of a quantity is X, there is a 68 percent chance that the true value lies within the range $X + \sigma_x$ to $X - \sigma_x$. There is a 95 percent chance that the true value lies within the range $X + 2\sigma_x$ to $X - 2\sigma_x$ and a 99.5 percent chance that the true value lies within the range $X + 3\sigma_x$ to $X - 3\sigma_x$. To be cautious, scientists can quote a 3σ error range on a measurement, since the true value is very unlikely to lie outside that range.

Some examples are on the next page. We will give the measurement, its probable error, and the implied range on the true value at three levels of confidence: ± 1σ, ± 2σ, and ± 3σ. In other words, if we make multiple observations, we expect that 68 percent, 95 percent, and 99.5 percent of measurements will lie within these ranges.

Astronomers gather light from the distant universe, and they are often working at the limits of detection. We can see

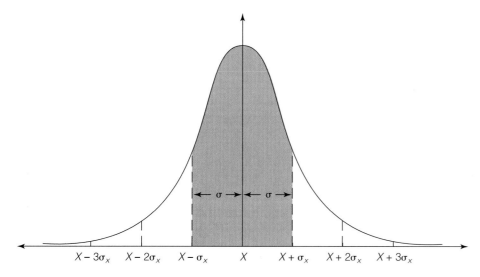

FIGURE A–3
The Gaussian or normal distribution. If the true value of a quantity is X, the height of the curve gives the relative probability of making a measurement with any particular value. The standard error or most probable error is σ. With multiple measurements, there is a $1/3$ probability that any measurement will fall outside the range spanned by plus 1σ and minus 1σ (shaded region), a probability of 0.05 that it will fall outside plus 2σ to minus 2σ, and a probability of 0.005 that it will fall outside plus 3σ to minus σ.

MEASUREMENT	68% WITHIN RANGE	95% WITHIN RANGE	99.5% WITHIN RANGE
0.67 ± 0.05 kg	0.72 to 0.62 kg	0.77 to 0.57 kg	0.82 to 0.52 kg
157.8 ± 6.2 pc	164.0 to 151.6 pc	170.2 to 145.4 pc	176.4 to 139.2 pc
8800 ± 100 N	8900 to 8700 N	9000 to 8600 N	9100 to 8500 N
0.0045 ± 0.0013 s	0.0058 to 0.0032 s	0.0071 to 0.0019 s	0.0084 to 0.0006 s
59.1 ± 0.4 m	59.5 to 58.7 m	59.9 to 58.3 m	60.3 to 57.9 m

what this means in terms of standard error. Suppose we measure the magnetic field of a distant star to be 90 ± 100 Gauss. You can see that the range of the magnetic field within the standard error reaches zero. In other words, there is a greater than $1/3$ chance that the star has no magnetic field at all! We might take more measurements or use a more accurate technique to get a new value of 84 ± 41 Gauss. The new measurement is slightly more than two times the standard error (or 2σ) above zero. There is still a nearly 5 percent chance that the star has no magnetic field. Scientists often demand a measurement of three or four times the standard error (3σ or 4σ) above zero before they are confident that they have *detected* something.

Scientists can use a similar logic to test a theory. Suppose in the previous example we have a theory that predicts that a certain type of star will have a magnetic field of 200 Gauss. Our first measurement, 90 ± 100 Gauss, is low but it is too crude to test the theory. The reason is that the measurement disagrees with the theoretical prediction by only 200 − 90 = 110 Gauss, or 1.1 times the standard error (1.1σ). So there is nearly a $1/3$ chance that the true magnetic field actually agrees with the theory. The second measurement is much more accurate, 84 ± 41 Gauss. Now we disagree with the theoretical prediction by 200 − 84 = 116 Gauss, or 116/41 = 2.9 times the standard error (2.9σ). In other words, there is less than a 1 percent chance that the measurement agrees with the theory. The related ideas of "fitting a model" and "testing a hypothesis" are illustrated in a Science Toolbox in Chapter 11. Scientists often use this type of argument to test theories and advance to a better understanding of nature.

All of our discussion of errors has assumed that their nature is random. A *random error* is equally likely to displace a measurement to the high or to the low side of the true value. If errors are random, we can be confident that more measurements will reduce the standard error and allow us to home in on the true value of a quantity. Unfortunately, scientists often encounter a more difficult problem: systematic errors. A *systematic error* is any kind of error that does not reduce in size when more measurements are made.

To see some examples of systematic error, let us return to our earlier example of measuring the width of a table with a tape measure. Suppose that the machine that printed the scale on the tape measure had been programmed wrong, so that the divisions were all slightly too big. You would then measure all distances to be slightly smaller than they really are. Or, we know that metal expands when it is hot, so if the tape measure had been calibrated when it was still hot, you would measure all distances to be slightly larger than they really are because the tape measure had shrunk since its manufacture. Or, you might systematically misread the scale. None of these problems would be revealed by repeated measurements. The only way scientists can guard against systematic errors is by measuring a quantity using two or more different techniques.

Our discussion of measurement error is not tangential; the correct treatment of errors or uncertainty is at the heart of the scientific method. Most scientists spend as much time trying to understand the error in a measurement as they did making the measurement in the first place! One final caution: Most calculators do their calculations with a large number of significant figures (as many as 16) and can be set to show an arbitrary number of digits following the decimal point. But the numbers you get out of a calculator are only as accurate as the numbers you put in. So if your calculation involves quantities that are only accurate to two significant figures, you should only read off your answer to a precision of two significant figures.

Appendix A-5
Angular Measurement and Geometry

Angular measurement is an important part of astronomy. If you want to describe where an object is in the sky, you must give its position in terms of angles. The basic unit of angular measurement is a degree, given by the symbol 1°. There are 90 degrees in a right angle and 360 degrees in a complete circle.

Why is angular measurement not based on a decimal system? The answer goes back to Babylonian astronomers of 5000 years ago. Early calendars had about 360 days. Since the constellations make a complete cycle of the night sky every year, any particular star must move $1/360$ of an entire circle with respect to the Sun each day. Therefore, the

Babylonians settled on 1/360 of a circle as the unit of angles. They also used 60, which is a factor of 360, as the basis of the counting system. It may seem strange to settle on 60, but for simple calculations it is a good choice: 60 is the lowest number that can be divided into 2, 3, 4, 5, or 6 equal parts.

The subdivisions of a degree are based on factors of 60. There are 60 arc minutes in a degree, and 60 arc seconds in an arc minute. These factors mirror our system of time measurement, with 60 minutes in an hour and 60 seconds in a minute. In other words, our modern systems for measuring time and angle are relics of one of the oldest civilizations on Earth! The divisions of a complete circle into smaller parts are summarized below:

Complete circle	= 360°
Right angle	= 90°
Radian	= 360/2π = 57.3°
1 degree (1°)	= 60 arc minutes = 60'
1 arc minute	= 60 arc seconds = 60"
1 degree (1°)	= 3600 arc seconds = 3600"

What do these angles mean? A degree is roughly the angle made by your thumbnail with your arm held out in front of you. A half-degree is the angle made by the Moon in the sky. (It is also the angle made by the Sun, but you should *never* look directly at the Sun for any length of time!) Astronomers use the word *subtend* to describe the angle made by an object—the angle between imaginary lines drawn from either side of an object that meet at your eye. The subdivisions of a degree represent very tiny angles. One arc minute is the angle subtended by a dime at a distance of about 50 m, and one arc second is the angle subtended by a dime at a distance of about 3 km, or 1.8 miles.

The *angular diameter* (or angular size) of an astronomical object is the angle it covers on the sky as seen by an observer on Earth. The usual symbol for angular diameter is the Greek letter θ (theta). Let us call the distance to an object D and the true (physical) diameter of the object d. Both D and d are measured in distance units of meters or kilometers. Look at the two triangles in Figure A–4a. They show the situation if the same object is viewed from two different distances. Any object will appear smaller when it is far away and larger when it is close. Therefore, the angular diameter is inversely proportional to the distance

$$\theta \propto 1/D \text{ (for a fixed size } d\text{)}$$

Alternatively, we know that different sized objects at the same distance subtend different angles (Figure A–4b). The angular diameter is proportional to the true diameter

$$\theta \propto d \text{ (for a fixed distance } D\text{)}$$

Lastly, we know that a large object far away can appear the same size as a smaller object that is closer. The Sun and the Moon have almost identical angular diameters; this is the

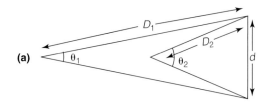
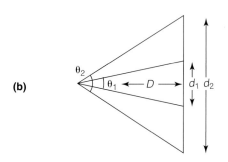
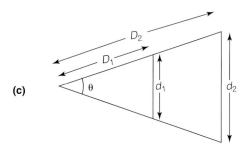

FIGURE A–4
Scaling relationships for a triangle, where D is the distance to an object, d is the diameter of the object, and θ is the angle subtended by the object. **(a)** If the transverse size d is fixed, $\theta \propto 1/D$. **(b)** If the linear distance D is fixed, $\theta \propto d$. **(c)** If the angle θ is fixed, $d \propto D$.

reason for the spectacular phenomena of eclipses. The Sun's much greater size is offset by its much greater distance from Earth. Figure A–4c shows the situation where distance is proportional to physical diameter

$$D \propto d \text{ (for a fixed angle } \theta\text{)}$$

We can combine these results to express the general relation between distance, physical size, and angular size

$$\theta \propto d/D$$

Now we can make an approximation that will allow us to turn this proportionality into an equation. Let us consider angles that are small enough that the diameter of an object at a certain distance is nearly equal to the length of an arc of a circle at that distance (Figure A–5). You can see from the figure that $s \approx d$ as long as the angle θ is not too large. In practice, the approximation is good if $\theta < 10°$, which is true for the angular size of almost all astronomical objects.

In Figure A–5, the ratio of the arc length s to the circumference of the circle is the same as the ratio of the angle θ

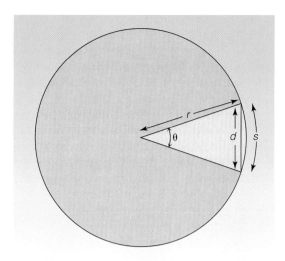

FIGURE A-5
The small-angle approximation. As the angle θ get smaller, the length of the chord d and the arc s become similar. We can set up the proportionality that s (or d) is the same fraction of the circle's circumference as θ is of 360°; this leads to the small-angle equation.

(measured in degrees) to the number of degrees in a complete circle. In equation form

$$s / 2\pi r = \theta / 360$$

Multiplying both sides of the equation by $2\pi r$, we get

$$s = 2\pi r \theta / 360$$

Now we can make three substitutions. We use our approximation to substitute d for s, we substitute the distance D for the radius of the circle, and we note that the numerical factor $2\pi/360$ can be written as 57.3

$$d = \theta D / 57.3$$

This is the *small-angle equation,* and you will find it widely used in this book (see especially the Science Toolbook in Chapter 2). If we can determine any two out of d, D, and θ, the small-angle equation will allow us to calculate the third quantity. You will also see the equation written in the form where θ is measured in units of arc seconds. Since there are 3600 arc seconds in a degree, the denominator becomes 57.3 × 3600 and the small angle equation is $d = \theta D / 206,265$.

Other useful geometric formulas—involving the areas and volumes of spheres, cylinders, and cubes—are given below:

Area of a rectangle of sides a and b	$A = ab$
Volume of a cube of side a	$V = a^3$
Volume of rectangular solid of sides a, b, c	$V = abc$
Circumference of a circle of radius r	$D = 2\pi r$
Area of a circle of radius r	$A = \pi r^2$
Surface area of a sphere of radius r	$A = 4\pi r^2$
Volume of a sphere of radius r	$V = 4/3 \, \pi r^3$
Volume of a cylinder of radius r, height h	$V = \pi r^2 h$

Unusual shapes are rarely encountered in astronomy. Positions on the sky are best described in terms of angles marked off on a circle, and astronomical objects like planets and stars are almost always perfect spheres.

Appendix A-6
Observing the Night Sky

The night sky is the inspiration for astronomy. Even though modern astronomy is carried out using large telescopes and electronic cameras, the beauty of the night sky is the reason most astronomers become hooked on their subject. Everyone can share this experience. Go far from any city or suburb, or travel into a wilderness area. You will see thousands of stars scattered like gems on a velvet backdrop. You will see the pale arch of the Milky Way. You will understand the awe that humans have felt for thousands of years when they contemplate the night sky.

Your unaided eye is good enough to see nearly 6000 stars from a dark site, along with planets and the brightest nebulae. Take a good pair of binoculars and you will also be able to see many nebulae and star clusters. With a small telescope, you can resolve star clusters and see galaxies several million light-years away. You can even try photography; a number of the pictures in this book were taken with a camera, and the captions give the type of film and the exposure time. Use the star charts that follow in this appendix to get started—twelve monthly views of the night sky as seen from a mid-northern latitude. Also refer to Chapter 2 for an explanation of the motions of the night sky and a summary of what you could observe over the course of a year. Your daily newspaper may have a column on what is visible in the night sky. Also, you can get this information from the monthly issues of *Astronomy* magazine and *Sky and Telescope* magazine. No matter what you might read or learn about astronomy, there is no substitute for going out and looking at the sky!

Earth rotates eastward, so the sky appears to rotate above us from east to west. Stars rise in the east and set in the west. All stars appear to rotate in concentric circles around a fixed point in the sky near the bright star Polaris. Star motions appear imperceptible from minute to minute, but over the course of an hour or so, you can see new constellations rise in the east while others set in the west. Planets are bright objects that do not twinkle and change their position among the stars from night to night. You may even see transient events in the upper atmosphere like shooting stars or fireballs. Figure A-6 shows you how to use the star charts that follow.

Even though we know that the stars are all at different distances, it is convenient to imagine them on a *celestial sphere* that rotates overhead. You can orient yourself by finding Polaris, which is in the direction of the *north celestial*

FIGURE A-6
To use the star charts in this book, select the appropriate chart for the date and time. Hold it overhead, and turn it until the direction at the bottom of the chart is the same as the direction you are facing.

pole. The north celestial pole is directly above the direction that defines north on your horizon. Now you can find the *cardinal points*—north, south, east, and west. The *celestial equator* is the projection of Earth's equator onto the sky. If you are in the Northern Hemisphere, you will see it in the southern sky. The point directly above your head is the *zenith* and the point directly below your feet is the *nadir.* You can easily measure approximate angles on the sky. The angle between the zenith and the horizon, or between any two adjacent cardinal points, is 90°. The width of your fist held at arm's length is about 10°. The width of your thumbnail at arm's length is about 1°, and the width of the full Moon is $1/2$°.

We can locate any object on the sky with two angles. This is a direct analogy to the *coordinate system* of Earth, where any place is specified by the two angles of latitude and longitude. One simple coordinate system in astronomy uses *altitude* and *azimuth*. The altitude of an object is its angular distance above the horizon, measured from 0° at the horizon to 90° overhead. The azimuth of an object is its angular distance around the horizon, measured toward the east from the north, from 0° to 360°.

However, astronomers use a different pair of angles defined in the equatorial coordinate system. The equatorial system is tied to Earth as a frame of reference. The angle of an object above or below the celestial equator is called the *declination,* measured in degrees, minutes of arc, and seconds of arc. The declination is positive north of the celestial equator and negative south of the celestial equator. For example, suppose your latitude is +28° N. A star with a declination of +28° would pass directly overhead, a star with a declination of +65° would pass to the north of the zenith, and a star with a declination of 21° would arc through the sky to the south of the zenith. Stars with declinations of less than −62° (in other words, 28° − 90°) are never be visible because they never rise above the horizon. Telescopes need to point very accurately to find faint objects. A full description of declination might be +47° 15′ 58″ or just over $47 1/4$ degrees north of the celestial equator.

The east-west coordinate is more complicated because Earth is rotating. Astronomers use *right ascension,* measured in hours, minutes, and seconds of time. The reference point for right ascension is the time when an object crosses the *meridian,* which is the north-south line that passes overhead. The zero for right ascension is defined as the position of the Sun on the vernal equinox, which is around March 21 each year. Right ascension increases toward the east, because that is the direction in which Earth turns. The range is 0 to 24 hours, and a particular right ascension might be expressed as $6^h\ 47^m\ 56.4^s$. For an object on the celestial equator, we can relate right ascension to angle as follows:

24 hours	Full rotation of the night sky	= 360°
12 hours	Time between rising and setting of a star	= 180°
1 hour	Typical angle between constellations	= 15°
4 minutes	Time for star to cross 2 Moon diameters	= 1°
1 minute	Time for star to cross ½ Moon diameter	= ¼°

Thus, we say that a telescope has to "track" an object; in other words, to remain pointing at a fixed right ascension, it has motors to move it in compensation for Earth's motion. The orbital motion of Earth makes the Sun appear to move eastward among the stars. The path of the Sun in the sky is called the *ecliptic.* Each day the Sun moves about 1° (twice its angular diameter) eastward along the ecliptic. Each night at a particular time, the constellations are seen about 1° farther to the west. In other words, stars rise or set 4 minutes earlier each night. On March 21, the Sun is at a right ascension of 0^h and stars at a right ascension of 12^h are overhead at midnight. On September 21, the Sun is at a right ascension of 12^h and stars at a right ascension of 0^h are overhead at midnight. In this way, you can see different sets of constellations as Earth sweeps around in its orbit of the Sun.

The star charts show the celestial equator with right ascension marked off in hours. The ecliptic is the path followed by the Sun so it defines the plane of the Earth-Sun orbit. Since the solar system is nearly in a plane, all the planets can be found within a few degrees of the ecliptic. Mercury, Venus, Mars, Jupiter, and Saturn are visible by eye. Uranus and Neptune are visible through binoculars. You will need a small telescope (and a dark site) to see Pluto. The celestial equator and the ecliptic intersect with an angle of 23.5° between them, because that is the tilt of Earth's axis as it orbits the Sun. The planets move among the stars from night to night. Mercury and Venus are interior to Earth's orbit, so they never appear far from the Sun. The outer planets may be found anywhere along the ecliptic.

Constellations are shown for a view from a northern latitude; additional constellations are only visible to viewers in the Southern Hemisphere. We have connected the dots—when you look at the sky you will have to use your imagination! The constellations along the ecliptic include the twelve signs of the *zodiac*. The stars have a dot size proportional to their brightness. Astronomers measure brightness using the 2000-year-old system of *magnitudes*. The magnitude system uses a nonlinear scaling with brightness ratio, and it probably has its origin in the fact that the eye is a nonlinear detector of light. If b_1 is the apparent brightness of star 1 and b_2 is the apparent brightness of star 2, the magnitude difference is given by

$$m_1 - m_2 = 2.5 \log (b_2 / b_1)$$

We can also define the relationship in terms of the brightness ratio

$$b_2 / b_1 = 10^x, \text{ where } x = 0.4 \, (m_1 - m_2)$$

In addition to being on a logarithmic scale, the magnitude system is an inverse scale—larger magnitudes correspond to fainter objects. The table to the right gives the relationship between these two measures of brightness (HST is the Hubble Space Telescope).

As you can see, the magnitude system is clumsy and nonintuitive. It is also the most persistent intrusion of nonmetric units into the world of astronomy. There is no reason for you to be burdened by history; throughout this book, we use linear brightness units.

The star charts that follow show the early evening night sky for every month of the year. In a month, the constella-

$m_1 - m_2$	b_2/b_1	EQUIVALENT TO BRIGHTNESS RATIO OF
0	1	Two identical stars at the same distance
1	2.5	Minimum difference easily discernable by eye
2	6.3	Venus at its brightest to Mars at its brightest
3	15	Venus (at its brightest) to brightest star (Sirius)
4	40	Limit for eye to limit for binoculars
5	100	Brightest to faintest star visible by eye
10	10^4	Full Moon to Mars at its brightest
15	10^6	Brightest star to Pluto
20	10^8	Limit for binoculars to limit for HST
25	10^{10}	Sun to the brightest star
30	10^{12}	Brightest to faintest star visible with HST

tions move 30° farther to the west. Equivalently, any star near the celestial equator rises or sets 2 hours earlier each month. (Stars near the celestial pole do not rise and set; they travel on tight concentric circles around Polaris.) Thus, you can use these charts to see what will be in the sky at *any* time of night. Just move forward one month for every two hours later that you want to observe. In other words, the chart that gives the 8 P.M. sky for mid-April also shows the 10 P.M. sky for mid-March, and the midnight sky for mid-February, and the 2 A.M. sky for mid-January, and so on. Hold the chart overhead, as shown in Figure A–6. If you face north, turn the chart until the words Northern Horizon are at the bottom. If you face other directions, turn the chart accordingly. Enjoy the night sky!

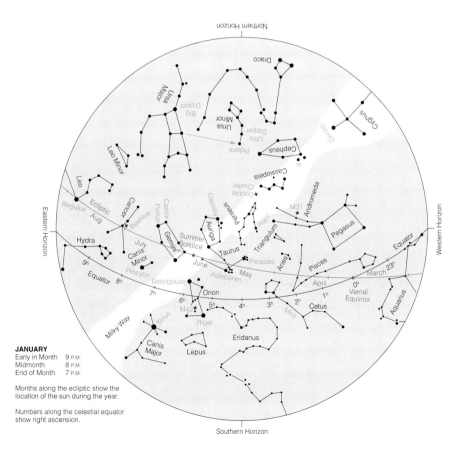

JANUARY
Early in Month   9 P.M.
Midmonth   8 P.M.
End of Month   7 P.M.

Months along the ecliptic show the location of the sun during the year.

Numbers along the celestial equator show right ascension.

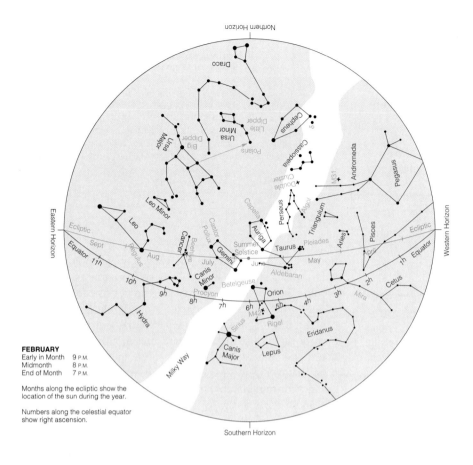

FEBRUARY
Early in Month 9 P.M.
Midmonth 8 P.M.
End of Month 7 P.M.

Months along the ecliptic show the location of the sun during the year.

Numbers along the celestial equator show right ascension.

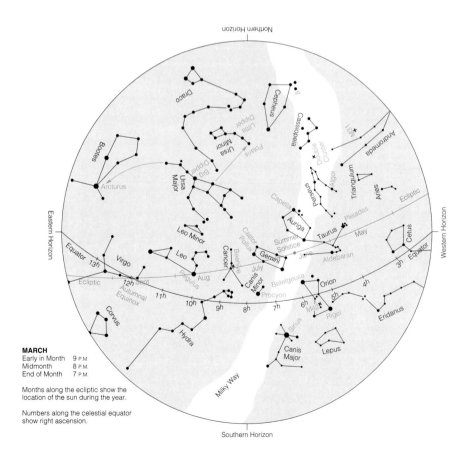

MARCH
Early in Month 9 P.M.
Midmonth 8 P.M.
End of Month 7 P.M.

Months along the ecliptic show the location of the sun during the year.

Numbers along the celestial equator show right ascension.

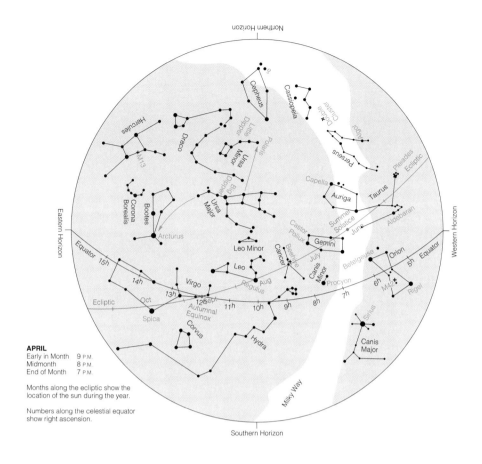

APRIL
Early in Month 9 P.M.
Midmonth 8 P.M.
End of Month 7 P.M.

Months along the ecliptic show the location of the sun during the year.

Numbers along the celestial equator show right ascension.

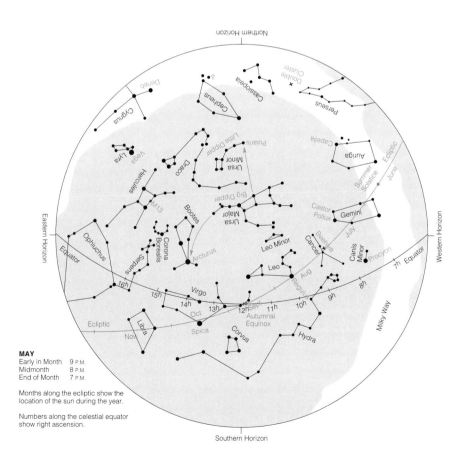

MAY
Early in Month 9 P.M.
Midmonth 8 P.M.
End of Month 7 P.M.

Months along the ecliptic show the location of the sun during the year.

Numbers along the celestial equator show right ascension.

APPENDIX A-6: OBSERVING THE NIGHT SKY 601

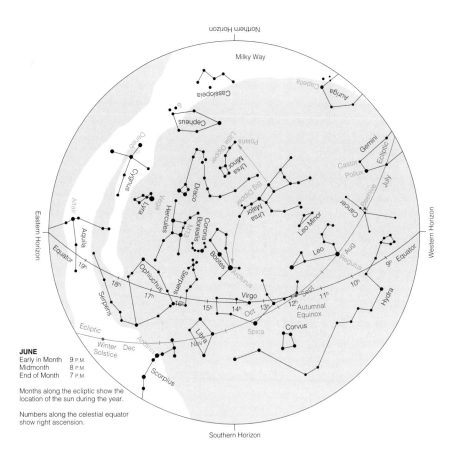

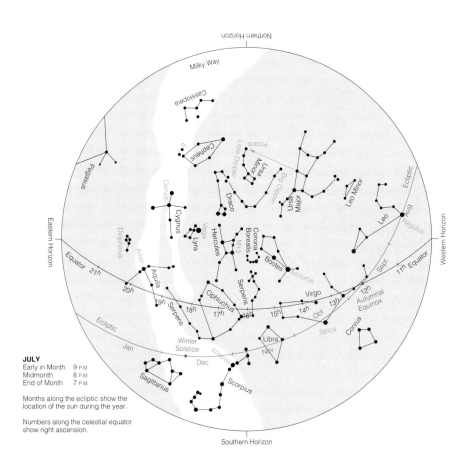

602 APPENDICES

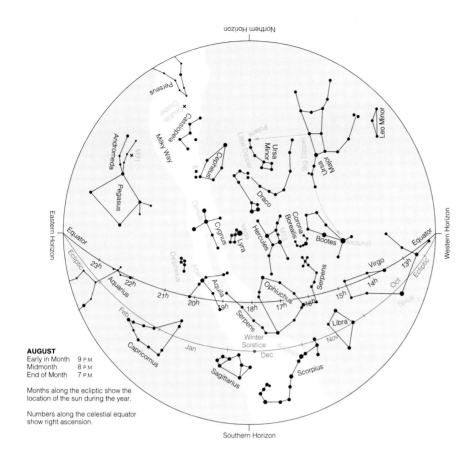

AUGUST
Early in Month 9 P.M.
Midmonth 8 P.M.
End of Month 7 P.M.

Months along the ecliptic show the location of the sun during the year.

Numbers along the celestial equator show right ascension.

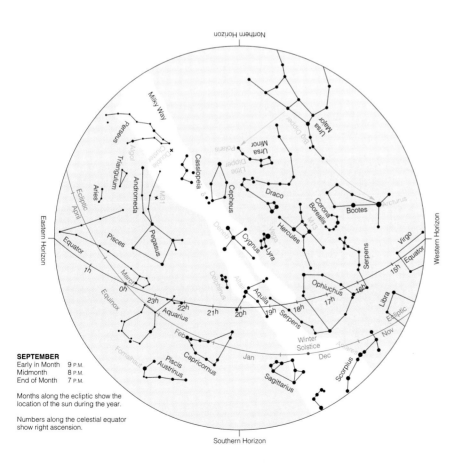

SEPTEMBER
Early in Month 9 P.M.
Midmonth 8 P.M.
End of Month 7 P.M.

Months along the ecliptic show the location of the sun during the year.

Numbers along the celestial equator show right ascension.

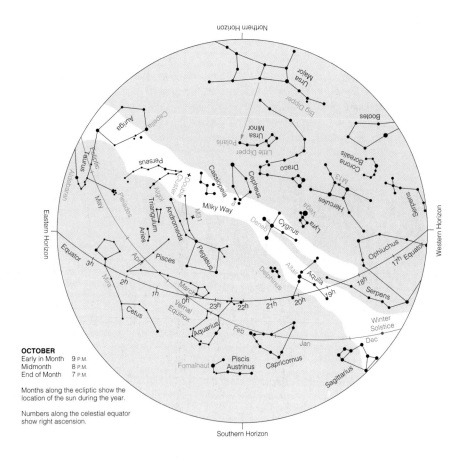

OCTOBER
Early in Month 9 P.M.
Midmonth 8 P.M.
End of Month 7 P.M.

Months along the ecliptic show the location of the sun during the year.

Numbers along the celestial equator show right ascension.

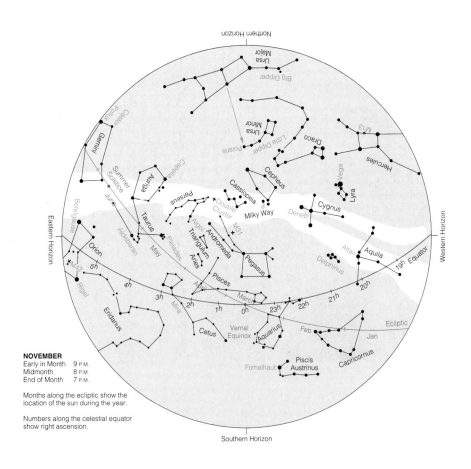

NOVEMBER
Early in Month 9 P.M.
Midmonth 8 P.M.
End of Month 7 P.M.

Months along the ecliptic show the location of the sun during the year.

Numbers along the celestial equator show right ascension.

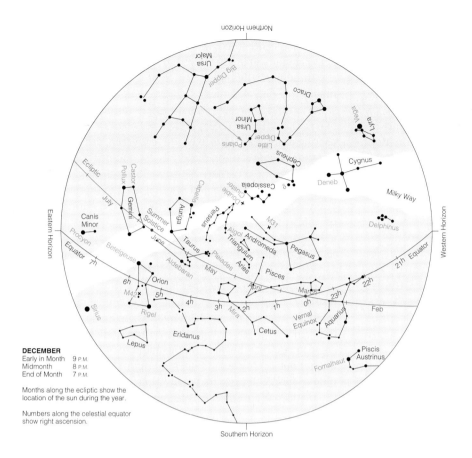

Appendix A-7
Ways of Representing Data

Data are the facts of science. From the earliest days of science, language and mathematics have been used to convey information. In the 2nd century A.D., Ptolemy wrote his masterwork on astronomy, called *al-Magiste,* or the *Almagest.* This encyclopedia contained a description of the geocentric cosmology and a list of over 1000 star positions, making it the first important collection of astronomical data.

By the time of the Renaissance, accurate drawing and painting had become a legitimate way to represent scientific information. Leonardo da Vinci was not only an artist of extraordinary power and sensitivity; he was also an accomplished scientist and engineer. Da Vinci made many drawings of the features and phases of the Moon, and of the trajectories of moving objects. His drawings of cadavers and the human form were so accurate that they formed the basis for modern anatomy. Galileo used pen and ink to record what he saw through the newly invented telescope. Galileo's watercolors of craters and mountains on the Moon helped convince people that the Moon was a world in space, just like Earth. Some 200 years later, Charles Darwin used his meticulous drawings of wildlife to support the theory of evolution. Art has never stopped being a tool of science. You are probably familiar with the pictures by artists who work with paleontologists to recreate the look of dinosaurs and other long-extinct species. This book contains many examples of space art, realistic depictions of imagined worlds.

Scientists have developed ways of extending the visual sense. Dutch craftsmen were the first people to make accurate lenses 400 years ago. As a result, they invented a device for magnifying nearby objects—the microscope—and a device for gathering light from distant objects—the telescope. Scientists also developed ways to render the invisible visible. Michael Faraday used the patterns of iron filings to trace magnetic lines of force. E. E. Chaldni used chalk dust in a similar way to trace the patterns of sound waves coming from a metal plate when he caused it to vibrate. As described in Chapter 5's opening story, scientists 200 years ago were able to record waves from beyond the blue and red ends of the visible spectrum. Since then, we have learned how to record and transmit electromagnetic waves from radio waves to X rays.

Sensors that measure physical quantities and transmit the data to us are a part of our everyday lives. For example, the dashboard of your car might show measurements of speed, distance, engine speed, water temperature, oil pressure, battery electrical charge, and frequency of the waves on your radio receiver. Scientists have become accustomed to recording such an array of data for everything from the tiny world of the atom to the gigantic scales of a distant galaxy.

Since the invention of the telescope, astronomers have benefited from two revolutions in the representation of data. The first came with the development of photography in the mid-19th century. Before photography, astronomers had to make drawings of what they saw. Photographs create a permanent and indisputable record. The second revolution was the development of electronic detectors in the mid-1970s. Charge-coupled devices, or CCDs, enable astronomers to convert light into electrons on a tiny wafer of silicon. The best of these devices have tens of millions of picture elements, or *pixels*. While a photograph can still store more information than a CCD, electronic detectors have the enormous advantage of storing their information in a digital form.

Modern astronomy is based on *digital information*. This means that information content can be represented as an array of bits, or on-off signals. (Review the discussion and the Science Toolbox in Chapter 10). Figure A–7 shows how the information content of a signal depends on the size of the signal and how finely the signal is subdivided. The signal can be any kind of measurement: the strength of a magnetic field, the velocity of a star, or the intensity in a single pixel of a CCD image. A satellite might record an image of one of the moons of Jupiter and send it back to Earth as a stream of bits. Or an astronomer might record an image of a galaxy through a large telescope and store it on a computer hard disk, or send it over the Internet to a colleague. The information revolution has made it easy to store, manipulate, and transmit enormous amounts of information; a high-quality CCD image might contain over 1 billion bits. An all-sky CCD survey is currently under way that will produce over 10^{15} bits of information!

Throughout this book are color images that have been made with digital detectors. CCDs do not sense color in the same way as your eye, which has chemicals that are sensitive to the different colors of light, or your TV screen, which uses beams of electrons to excite red, green, and blue dots of phosphor (Figure A–8). Rather, an astronomer uses a filter in front of the CCD camera to isolate a small range of wavelengths. A single image can only be used to represent black and white, with shades of gray indicating the number of bits of information. Three different exposures through red, green, and blue filters can then be combined to produce a true color image of the night sky.

You should be aware that there are two kinds of *false color* in scientific images. The first is when a single measured intensity is used to represent color. For example, low intensity might be coded blue and high intensity red (you could image the signal strength in Figure A–7 coded that way). A false-color image may be more appealing than a black-and-white image, but it conveys no spectral information. The second type of false color occurs when we are representing a quantity that is not based on visible light. If you see a color image

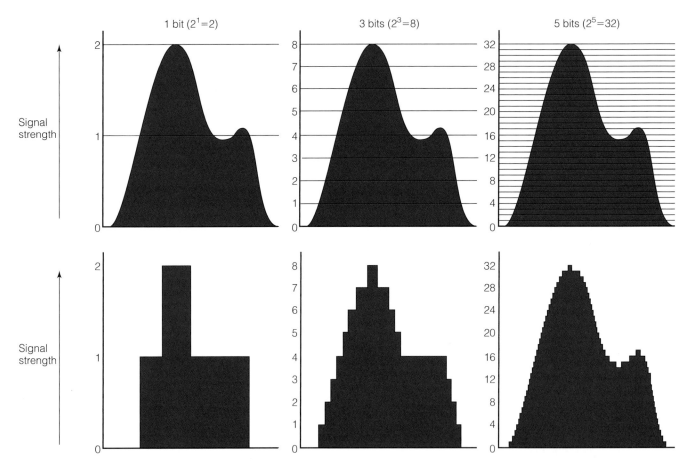

FIGURE A–7
The information content of a signal depends on how finely it is recorded. The top sequence going from left to right shows an increasing number of levels (or bits) used to "digitize" the signal. The lower sequence shows the result—a digitized signal. The vertical axis could be any measurement, such as light intensity or force or magnetic field strength. The horizontal axis could be distance in pixels across a CCD or even a time sequence.

FIGURE A-8
A color TV screen is made up of red, blue, and green regions that can be made visible with a magnifying glass. A handheld lens on the right in the photo shows the three-color structure.

of an X ray, or a color image made with a radio telescope, the colors have no conventional scientific meaning. Most of the information you will see in this book is in the form of images. An image is just a map of intensity in two dimensions. The other main form of presenting information is a *spectrum*, which is a plot of intensity versus wavelength.

Many types of data cannot be represented as images or spectra. The simplest way for scientists to present data is in a *table* (Figure A-9a). A list of percentages or proportions can also be shown graphically as a pie chart (Figure A-9b). This form of representation is more commonly used in business and social science than in physical science. Scientists often use a *histogram* to represent a list of numbers (Figure A-9c). Histograms are useful for showing the relative size of different numbers, but it can be difficult to see the information when the values cover a large range. One solution to this problem is to make a histogram from the logarithm of the quantity, which gives a scale that is highly compressed (Figure A-9d). It is important to remember that equal intervals on a logarithmic plot correspond to equal ratios, whereas equal intervals on a normal histogram correspond to equal differences.

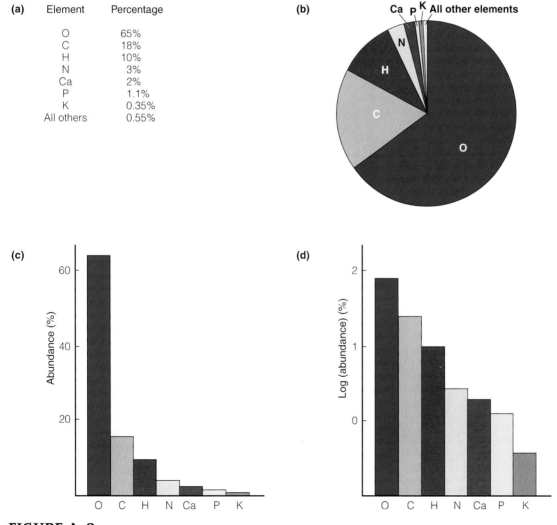

FIGURE A-9
Different ways of representing data. **(a)** A table of values. **(b)** A pie chart. **(c)** A histogram. **(d)** A histogram with a logarithmic scale, which compresses widely separate values.

With more than one quantity, the best graphical representation is a plot, or a *graph*. One quantity is plotted on the horizontal axis (or *x*-axis) and a second quantity is plotted on the vertical axis (or *y*-axis). Scientists often make a graph because they anticipate a relationship between two quantities and they want to know the form of the relationship. Other times, they make a plot without knowing what they will find, as a way of "exploring" their data.

When two quantities are related, we say they are *correlated*. When there is no relationship between the two quantities, we say they are *uncorrelated* (Figure A–10a). The simplest relationship is a linear correlation (Figure A–10b). On a graph, a linear correlation has the algebraic form of the equation for a straight line: $y = ax + b$, where a is the slope of the line and b is the y-intercept. Thus, $y = 2x$ and $y = 0.658x$ are examples of linear correlation. The general form is $y \propto x$. In general, all data should be shown with an associated error bar attached. A correlation is often more complicated than a linear relationship. Thus, $y = 0.112x^2$ and $y = 1.47x^{0.44}$ are examples of nonlinear correlation. Kepler's third law relates the period of a planet's orbit (P) to its distance from the Sun (a) with the equation $P^2 \propto a^3$. The correlation represented by Kepler's third law has a varying slope, since the relationship between period and distance is a power law, $P \propto a^{3/2}$ (Figure A–10c). Alternatively, we can plot Kepler's law on a logarithmic scale (Figure A–10d). Taking the logarithm of Kepler's law, we get $\log P = 3/2 \log a$, which is just the form of a straight line where $y = \log P$, $x = \log a$, and the slope is 3/2. Any power law is therefore a straight line in a logarithmic plot.

We have only considered correlations between two quantities, but nature can be complex. Scientists often extend their analysis to more than two quantities, or *multiple variables*. For example, cancer is difficult to understand because it is controlled by a complex mixture of environmental and genetic factors. In astronomy, the size of a star depends on *both* the mass and the chemical composition. Occasionally, a scientists discovers a new type of star or a new fundamental particle. More often, scientists make progress by sifting through piles of data looking for correlations. The discovery of a significant and unanticipated pattern is one of the thrills of doing science.

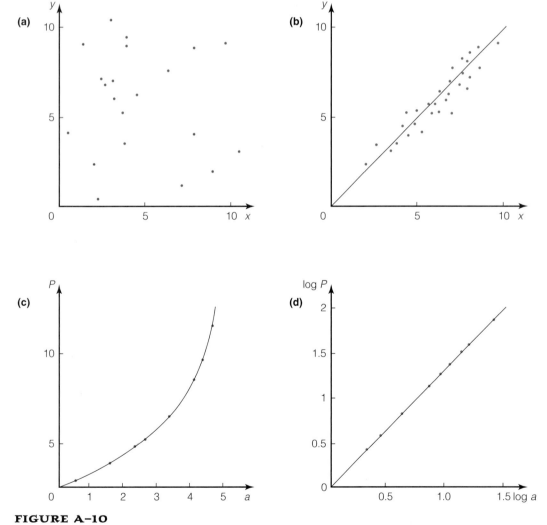

FIGURE A-10
Types of correlation. **(a)** Data with no correlation. **(b)** A linear correlation between two quantities. **(c)** A nonlinear correlation—in this case, Kepler's third law in a hypothetical solar system. **(d)** Same as c, but with logarithms of the two quantities plotted, yielding a straight line with slope 3/2.

Appendix B: Astronomical Data

Appendix B-1
Planetary Data

OBJECT	EQUATORIAL DIAMETER (km)	MASS (kg)[a]	ROTATION PERIOD[b] (d)	ORBITAL PERIOD (DAYS UNLESS MARKED)	DISTANCE FROM PRIMARY (10^3 km UNLESS MARKED)	ORBIT INCLINATION[b,c] (DEGREES)	ORBIT ECCENTRICITY	ESCAPE VELOCITY (km/s UNLESS MARKED)	KNOWN OR PROBABLE SURFACE MATERIAL
Sun	1,391,400	1.99 (30)	25.4	—	0	—	—	617	Ionized gas
Mercury	4878	3.30 (23)	58.6	89	0.387 AU	7.0	0.206	4.2	Basaltic dust and rock
Venus	12,104	4.87 (24)	243R	225	0.723 AU	3.4	0.007	10.4	Basaltic and granite rock
Earth	12,756	5.98 (24)	1.00	365	1.00 AU	0.0	0.017	11.2	Water, granitic soil
Moon	3476	7.35 (22)	S	27	384	18–29	0.055	2.4	Basaltic dust and rock
Mars	6787	6.44 (23)	1.02	687	1.52 AU	1.8	0.093	5.0	Basaltic dust and rock
Phobos	37 × 19	9.6 (15)	S	0.32	9.4	1.0	0.015	11 m/s	Carbonaceous soil
Deimos	15 × 11	1.0 (15)?	S	1.26	23	2.8	0.001	6 m/s	Carbonaceous soil
Asteroids									
1 Ceres	1020	1.2 (21)?	0.38	4.6 y	2.77 AU	10.6	0.08	0.6	Carbonaceous soil
4 Vesta	549	2.4 (20)?	0.22	3.6 y	2.36 AU	7.1	0.09	0.3	Basaltic soil
2 Pallas	538	?	0.33	4.6 y	2.77 AU	34.8	0.24	0.3	Meteoritic soil
10 Hygiea	443	?	0.75	5.6 y	3.15 AU	3.8	0.10	0.2	Carbonaceous soil
511 Davida	341	?	0.21	5.7 y	3.19 AU	15.8	0.17	0.2	Carbonaceous soil
704 Interamnia	338	?	0.36	5.4 y	3.06 AU	17.3	0.15	0.2	Unidentified soil
Jupiter	142,800	1.90 (27)	0.41	11.9 y	5.20 AU	1.3	0.048	60	Liquid hydrogen?
J16 Metis	40	?	?	0.29	128	0.0	0.0	20 m/s	Rock?
J15 Adrastea	25 × 15	?	?	0.30	129	0.0	0.0	10 m/s	Rock?
J5 Amalthea	270 × 150	?	S	0.50	181	0.4	0.003	0.13	Sulfur-coated rock?
J14 Thebe	120? × 90	?	?	0.67	222	0.0	0.0	60 m/s	Rock?
J Io	3630	8.94 (22)	S	1.77	422	0.0	0.000	2.6	Sulfur compounds
J2 Europa	3138	4.80 (22)	S	3.55	671	0.5	0.000	2.0	H_2O ice
J3 Ganymede	5262	1.48 (23)	S	7.16	1070	0.2	0.001	3.6	H_2O ice, dust
J4 Callisto	4800	1.08 (23)	S	16.69	1883	0.2	0.008	2.4	Dust, H_2O ice
J13 Leda	8?	?	?	239	11,094	26.7	0.146	4 m/s?	Carbonaceous rock?
J6 Himalia	180	?	0.4	251	11,480	27.6	0.158	90 m/s?	Carbonaceous rock?
J10 Lysithea	40	?	?	259	11,720	29.0	0.130	20 m/s?	Carbonaceous rock?
J7 Elara	80	?	?	260	11,737	28.0	0.207	40 m/s?	Carbonaceous rock?
J12 Ananke	30	?	?	631	21,200	147R	0.17	16 m/s?	Carbonaceous rock?
J11 Carme	44	?	?	692	22,600	163R	0.21	20 m/s?	Carbonaceous rock?
J8 Pasiphae	70	?	?	735	23,500	148R	0.38	40 m/s?	Carbonaceous rock?
J9 Sinope	40	?	?	758	23,700	153R	0.28	20 m/s?	Carbonaceous rock?
Saturn	120,660	5.69 (26)	0.43	29.5 y	9.54 AU	2.49	0.056	36	Liquid hydrogen?
S15 Atlas	38 × 28	?	?	0.60	138	0.0	0.000	13 m/s?	Ice?
S16 Prometheus	140 × 74	?	?	0.61	139	0.0	0.002	50 m/s?	Ice?

(continued)

(continued)

OBJECT	EQUATORIAL DIAMETER (km)	MASS (kg)[a]	ROTATION PERIOD[b] (d)	ORBITAL PERIOD (DAYS UNLESS MARKED)	DISTANCE FROM PRIMARY (10^3 km UNLESS MARKED)	ORBIT INCLINATION[b,c] (DEGREES)	ORBIT ECCEN- TRICITY	ESCAPE VELOCITY (km/s UNLESS MARKED)	KNOWN OR PROBABLE SURFACE MATERIAL
S17 Pandora	110 × 66	?	?	0.63	142	0.0	0.004	35 m/s?	Ice?
S11 Epimetheus	140 × 100	6 (17)?	S	0.69	151	0.3	0.009	50 m/s?	Ice?
S10 Janus	220 × 160	2 (18)?	S	0.69	151	0.1	0.007	70 m/s?	Ice?
S1 Mimas	394	3.8 (19)	S	0.94	186	1.5	0.02	0.2	Mostly H_2O ice
S2 Enceladus	502	8.4 (19)	S	1.37	234	0.0	0.00	0.2	Mostly H_2O ice
S3 Tethys	1048	7.6 (20)	S	1.89	295	1.1	0.00	0.4	Mostly H_2O ice
S13 Telesto	≈25 × 11	?	?	1.89	295[d]	0	0	7 m/s?	?
S14 Calypso	30 × 16	?	?	1.89	295[d]	0	0	9 m/s?	?
S4 Dione	1118	1.0 (21)	S	2.74	377	0.0	0.00	0.5	Mostly H_2O ice
S12 Helene	36 × 20	?	?	2.74	377[d]	0.2	0.00	11 m/s?	Mostly H2O ice
S5 Rhea	1528	2.5 (21)	S	4.52	527	0.4	0.00	0.7	Mostly H_2O ice
S6 Titan	5150	1.3 (23)	S	15.94	1222	0.3	0.03	2.7	Ices, liquid NH_3 and CH_4
S7 Hyperion	350 × 200	?	chaotic	21.28	1481	0.4	0.10	0.1	Ices?
S8 Iapetus	1436	1.9 (21)	S	79.33	3560	14.7	0.03	0.6	Ice and soil
S9 Phoebe	230 × 210	?	0.4	550.5	12,930	150R	0.16	0.1	Carbonaceous soil
Asteroid/Comet 2060 Chiron	200?	?	0.25	50.7 y	13.70 AU	7.0	0.38	0.1?	Carbonaceous soil (?) and volatile ices
Uranus	50,800	8.76 (25)	0.72R	84.0 y	19.18 AU	0.8	0.05	21	?
U6 Cordelia	26	?	?	0.34	49.3	0	0	14 m/s?	Ice and soil
U7 Ophelia	32	?	?	0.38	53.3	0	0	17 m/s?	Ice and soil
U8 Bianca	44	?	?	0.44	59.1	0	0	23 m/s?	Ice and soil
U9 Cressida	66	?	?	0.46	61.75	0	0	35 m/s?	Ice and soil
U10 Desdemona	58	?	?	0.48	62.7	0	0	31 m/s?	Ice and soil
U11 Juliet	84	?	?	0.49	64.35	0	0	44 m/s?	Ice and soil
U12 Portia	110	?	?	0.52	66.09	0	0	58 m/s?	Ice and soil
U13 Rosalind	58	?	?	0.56	69.92	0	0	31 m/s?	Ice and soil
U14 Belinda	68	?	?	0.62	75.10	0	0	36 m/s?	Ice and soil
U15 Puck	160 × 150	?	?	0.76	85.89	0	0	126 m/s?	Ice and soil
U5 Miranda	484	7 (19)	S	1.41	130	3.4	0.02	0.4	H_2O ice, soil
U1 Ariel	1160	1.4 (21)	S	2.52	192	0	0.00	0.7	H_2O ice, soil
U2 Umbriel	1190	1.2 (21)	S	4.14	267	0	0.00	0.6	H_2O ice, soil
U3 Titania	1600	3.4 (21)?	S	8.71	438	0	0.00	1.1	H_2O ice, soil
U4 Oberon	1550	2.9 (21)	S	13.46	586	0	0.00	1.0	H_2O ice, soil
Neptune	48,600	1.03 (26)	0.67	164.8 y	30.07 AU	1.8	0.01	24	?
N3 Naiad	54	?	?	0.30	48.2	4.5	?	25 m/s?	Ice and soil
N4 Thalassa	80	?	?	0.31	50.0	<1	?	48 m/s?	Ice and soil
N5 Despina	150	?	?	0.33	52.5	<1	?	74 m/s?	Ice and soil
N6 Galatea	180	?	?	0.40	62.0	<1	?	84 m/s?	Ice and soil
N7 Larissa	190	?	?	0.55	73.6	<1	?	106 m/s?	Ice and soil
N8 Proteus	400	?	?	1.12	117.6	<1	?	222 m/s?	Ice and soil
N1 Triton	2705	2.1 (22)	S	5.88	354	159R	0.00	2.5	CH_4 ice
N2 Nereid	400	?	?	360.2	5515	27.6	0.75	0.2?	CH_4 ice
Pluto	2300	1.5 (22)?	6.4	247.7 y	39.44 AU	17.2	0.25	0.9?	CH_4 ice
P1 Charon	1190	2 (21)?	S	6.39	19	0R	0.00	0.6?	CH_4 ice?

Numbers assigned to asteroids and outer planets' satellites indicate order of discovery, except for largest satellites. The tables in this book use a system of symbols to indicate data that is uncertain (?) and not available or not applicable (—).

[a]Numbers in parentheses are powers of 10.

[b]An R in this column indicates retrograde motion; S indicates that synchronous rotation has been confirmed.

[c]To ecliptic for planets; to planet equator for satellites.

[d]S12 in the leading Lagrangian point of Dione's orbit. S13 and S14 are in following and leading Lagrangian points of Tethys' orbit, respectively. Lagrangian points are stable points for small bodies in larger bodies' orbits.

[e]D. Tholen, W. Hartmann, and D. Cruikshank discovered anomalous brightening of this strange object in 1988, probably due to cometary activity. Karen Meech and M. Belton found a coma (a gas and dust cloud emitted by a comet) in 1989. Though it is catalogued as an asteroid, Chiron turns out to be the largest known comet nucleus!

SOURCE: Data from announcements of the International Astronomical Union, Voyager team reports, plus data published by Reitsema, Dunbar, Burns, Thomas, and collaborators.

Appendix B-2
The Nearest Stars

DISTANCE (pc)	STAR NAME	COMPONENT	LUMINOSITY ($L_\odot$)	SPECTRAL TYPE	MASS ($M_\odot$)	RADIUS ($R_\odot$)	SEMIMAJOR AXIS IN MULTIPLE SYSTEMS (AU)
0.0	Sun (and Jupiter)	A	1.0	G2	1.0	1.0	5.2
		B	—		0.001	0.1	
1.3	Alpha Centauri	A	2.5	G2	1.1	1.2	23.6 (AB)
		B	0.4	K0	0.9	0.9	
		C[a]	0.0001	M5	0.1	?	13,000 (AC)
1.8	Barnard's Star	—	0.0005	M5	?	?	—
2.3	Wolf 359	—	2×10^{-5}	M8	?	?	—
2.5	BD + 36°2147	—	0.01	M2	0.35	?	0.07
2.7	L726-8 (= UV Ceti)	A	0.0001	M6	0.11	?	10.9
		B	4×10^{-5}	M6	0.11	?	
2.9	Sirius	A	23	A1	2.3	1.8	19.9
		B	0.004	Wh. dw., A5	1.0	0.02	
2.9	Ross 154 (= Gliese 729)	—	0.0006	M5	?	?	—
3.1	Ross 248	—	0.0001	M6	?	?	—
3.2	L789-6	—	0.0001	M6	?	?	—
3.3	ϵ Eridani	—	0.4	K2	0.9	?	—
3.3	Ross 128	—	0.0003	M5	?	?	—
3.3	61 Cygni	A	0.063	K5	0.63	?	85 (AB)
		B	0.063	K7	0.6	?	
3.4	ϵ Indi	—	0.16	K5	?	?	—
3.5	Procyon	A	6.3	F5	1.8	1.7	15.7
		B	0.0006	Wh. dw.	0.6	0.01	
3.5	Σ 2398 (= Gliese 725)	A	0.004	M4	0.4	?	60
		B	0.0016	M5	0.4	?	
3.5	BD + 43°44 (= Gliese 15)	A	0.01	M1	?	?	156 (AB)
		B	0.0006	M6	?		
		C		K?			
3.6	τ Ceti	—	0.4	G8	?	1.0	—
3.6	CD − 36°15693	—	0.01	M2	?	?	—
3.7	BD + 5°1668 (= Luyten's Star)	A	0.0016	M4	?	?	?
		B		?	?	?	
3.7	G51-15	—	2×10^{-5}	?	?	?	—
3.8	L725-32	—	0.0003	M5	?	?	—
3.8	CD − 39°14192	—	0.025	M0	?	?	—
3.9	Kapteyn's Star	—	0.004	M0	?	?	—
4.0	Krüger 60 (= DO Cep = Gliese 860)	A	0.0016	M4	0.27	0.51	9.5 (AB)
		B	0.0006	M6	0.16	?	
		C		?	0.01	?	
4.0	Ross 614	A	0.0006	M5	0.14	?	3.9
		B	2×10^{-5}	?	0.08	?	
4.0	BD − 12°4523 (= Gliese 628)	—	0.0016	M5	?	?	—

[a]Proxima Centauri, the closest known star beyond the Sun.

SOURCE: Data from van de Kamp and Baum. A companion to Barnard's Star, listed by van de Kamp, is regarded as dubious and deleted here. Luminous stars in the neighborhood of the Sun have all been cataloged. However, the census is incomplete for intrinsically dim stars. This table will change substantially with final results from the *Hipparcos* space mission and the results of ongoing surveys for cool stars.

Appendix B-3
The Brightest Stars

STAR NAME	BOLOMETRIC LUMINOSITY ($L_\odot$)	TYPE[a]	RADIUS[b] ($R_\odot$)	DISTANCE (pc)
Sun	1.0	Main sequence	1.0	4.85×10^{-6}
Sirius (α Canis Majoris)	23	Main sequence (primary)	1.8	2.7
Canopus (α Carinae)	(1400)	Supergiant	30	34
Arcturus (α Boötis)	115	Red giant	(25)	11
Rigel Kent (α Centauri)	1.5	Main sequence (primary)	1.1	1.33
Vega (α Lyrae)	(58)	Main sequence	(3)	8.3
Capella (α Aurigae)	(90)	Red giant (primary)	13	14
Rigel (β Orionis)	(60,000)	Supergiant (primary)	(40)	(280)
Procyon (α Canis Minoris)	6	Main sequence (primary)	2.2	3.5
Achernar (α Eridani)	(650)	Main sequence	(7)	37
Hadar (β Centauri)	(10,000)	Giant (primary)	(10)	150
Betelgeuse (α Orionis)	(10,000)	Supergiant	800	160
Altair (α Aquilae)	(9)	Main sequence	1.5	5
Aldebaran (α Tauri)	125	Red giant (primary)	(40)	21
Acrux (α Crucis)	(2500)	Main sequence (primary)	(3)	(110)
Antares (α Scorpii)	(9000)	Supergiant (primary)	(600)	(160)
Spica (α Virginis)	(2300)	Main sequence (primary)	8	84

[a]The designation *primary* indicates data for the brighter companion in binary pairs.
[b]Parentheses indicate estimates.
SOURCE: Data from Burnham's *Celestial Handbook*.

Appendix B-4
Local Group Galaxies

NAME	CATALOG NUMBER	TYPE[a]	DISTANCE[b] (kpc)	DIAMETER (kpc)	LUMINOSITY ($L_\odot$)	GAS MASS ($M_\odot$)	RADIAL VELOCITY[c] (km/s)
Milky Way		Sb	0	30	2×10^{11}	2×10^{10}	0
Large Magellanic Cloud	LMC	Irr	50	19	3×10^9	2×10^9	12
Small Magellanic Cloud	SMC	Irr	63	8	4×10^8	2×10^9	−30
Draco	DDO 208	Dwarf E2	90	1.8	3×10^5	—	−31
Ursa Minor	DDO 199	Dwarf E4	90	1.4	2×10^5	—	17
Sculptor		Dwarf E3	100	1	3×10^6	—	162
Carina		Dwarf E3	100	2	2×10^4	—	230
Sextans		Dwarf E	100	2	3×10^5	—	230
Fornax		Dwarf E3	150	0.5	4×10^6	—	−51
Leo I	DDO 74	Dwarf E4	180	1	1×10^6	—	285
Leo II	DDO 93	Dwarf E1	180	2	7×10^5	—	70
	NGC 6822	Irr	500	3.5	2×10^8	3×10^8	66
WLM	DDO 221	Irr	610	0.5	—	1×10^6	11
—	IC 5152	Irr	610	7.3	2×10^9	—	74
Andromeda	M 31 (NGC 224)	Sb	670	32	3×10^{10}	4×10^9	−59
	M 32 (NGC 221)	Dwarf E2	670	2.5	2×10^8	—	35
	NGC 205	Dwarf E5	670	3.6	3×10^8	—	−1
	NGC 147	Dwarf E	670	3.7	1×10^8	—	89
	NGC 185	Dwarf E	670	3.9	1×10^8	—	39
Andromeda I		Dwarf E	670	0.7	3×10^6	—	—
Andromeda II		Dwarf E	670	0.7	3×10^6	—	—
Andromeda III		Dwarf E	670	0.9	3×10^6	—	—
Triangulum	M 33 (NGC 598)	Sc	770	11	3×10^9	1×10^9	3
	IC 1613	Irr	770	3.8	8×10^7	6×10^7	−125
Aquarius	DDO 210	Irr	920	0.4	—	1×10^6	11
Pisces	LG 53		920	—	1×10^6	—	

[a] S = spiral, subtypes a, b, c
Irr = irregular
E = elliptical, subtypes 0 through 7
[b] Galaxies clustered in groups, error in distances at least 30%.
[c] Velocity with reference to entire Milky Way galaxy.
SOURCE: Data from Allen, van den Bergh, Hirshfeld, Hodge, and Tully. The census may be incomplete for galaxies under a million times the luminosity of the Sun.

GLOSSARY

absolute brightness The true or intrinsic brightness of an astronomical source, or the number of photons emitted per second by the source. *See also* luminosity.

absorption lines Narrow spectral features that represent a reduction in intensity over a small wavelength range. They are caused by the loss of photons that raise an atom from a particular energy state to a higher energy state. *See also* emission lines.

acceleration A rate of change of velocity with time, in terms of either the size of the velocity or the direction of the motion; measured in units of meters per second per second.

accretion The gradual accumulation of material to make a large astronomical object.

accretion disk A disk of hot gas and dust around a star, usually material that has come from a companion star. Also, a large disk of hot gas at the center of a galaxy that provides fuel for nuclear activity.

accuracy The amount by which a measurement differs from the true value of a quantity. Accuracy can only be determined by multiple, independent measurements.

active galaxy A galaxy whose nucleus emits more energy than other, normal galaxies. Typical signatures of active galaxies are high-velocity gas, variable light intensity, strong radio emission, and strong X-ray emission. Also called active galactic nuclei (AGN).

age of Earth The period since the solar system formed and Earth was built from smaller fragments of rock, estimated at 4.6 billion years.

age of the universe Calculated from the big bang model to be about 13 to 14 billion years. This number is in reasonable agreement with the ages of the oldest stars.

ages of groups of stars Inferred ages by using the H-R diagram and stellar models. Assuming that a group of stars all formed at the same time, the color and luminosity of stars that are just leaving the main sequence are measures of age.

ages of stars Age determinations from using models. The lifetime of stars varies from millions of years for stars more massive than the Sun to many billions of years for stars less massive than the Sun.

altitude The angle of an object above the horizon. Altitude combines with azimuth to specify the position of an object on the sky.

amino acids Simple molecules required for life processes, containing a few dozen atoms. Of the many known amino acids, only 20 are used in living organisms.

Andromeda galaxy Also known as M 31, the nearest spiral galaxy comparable in size to our own, the Milky Way.

angular measurement The angle covered or subtended by something (measured in degrees, minutes of arc, seconds of arc), as opposed to the linear size of something (measured in units of kilometers or parsecs).

angular momentum The amount of rotary motion in a system. Angular momentum increases with the mass, size, and rotation speed of the system.

anisotropic Varying according to direction. In astronomy, this can refer to any physical property that differs depending on the direction of observation.

anthropic principle The idea that certain characteristics of the physical universe are carefully tuned to allow the existence of carbon-based life-forms. It is not clear whether the anthropic principle is a scientifically fruitful idea.

antimatter Material with quantum properties that are the reverse of normal matter. For example, the electrical charges of particles and antiparticles are opposite. Antimatter is rare in the universe. Matter and antimatter annihilate on contact to produce gamma rays.

apparent brightness The number of photons per second collected from an astronomical source. Apparent brightness is no indicator of the true or absolute brightness of a source.

archaeoastronomy The study of the astronomical and cultural practices of ancient cultures.

arrow of time The sense that time moves in only one direction. The flow of time is related to the increase in entropy of all physical systems.

asteroid A rocky or metallic interplanetary body, usually referring to bodies 100 m up to 1000 km in diameter. The largest group of asteroids lies between the orbits of Mars and Jupiter.

astrology The belief that human lives are influenced or controlled by the positions of planets and stars. This superstition has no basis in scientific fact.

astrometry The precise study of the positions and motions of astronomical objects.

astronomical unit (AU) The average distance from the Sun to Earth, equal to about 150 million km.

astronomy The study of all matter and energy in the universe.

atom A microscopic unit of matter consisting of a massive, positively charged nucleus surrounded by light, negatively charged electrons. The number of protons or electrons gives the atomic number; each atomic number corresponds to a substance with unique chemical properties. *See also* element.

atomic nucleus The dense core of every atom. Atomic nuclei are positively charged and contain most of the mass of each atom, in the form of protons and neutrons.

aurora A glowing, moving form of colored light seen at high latitudes on Earth, caused by high-energy particles from the Sun interacting with Earth's atmosphere near the magnetic poles.

average The best estimate of a quantity based on multiple measurements, given by the sum of the independent values divided by the number of measurements made. Also called mean value.

azimuth The angle of an object along the horizon from due north or some other reference point. Azimuth combines with altitude to specify the position of an object on the sky.

bacteria The smallest type of living organisms.

basalt A type of volcanic rock, often formed in lava flows, which is common on the Moon and terrestrial planets.

big bang model The scientific theory for the creation of the universe. In the big bang model, the universe has been expanding from a state of enormous density and temperature for about 13 to 14 billion years. Evidence for the model includes the cosmic background radiation, the universal recession of galaxies, and the abundance of light elements.

binary star Two stars in orbit due to their mutual gravity.

binding energy The amount of energy required to take something apart. Orbiting objects have a binding energy that keeps them in orbit, electrons have a binding energy that keeps them orbiting the atomic nucleus, and the protons and neutrons in a nucleus have a binding energy.

biology The study of living organisms and life processes.

black hole Any object whose surface gravity is so strong that no radiation or matter can escape. A black hole is the end-state of any star with a core more than about three times the Sun's mass. Theoretically, black holes much less massive and much more massive than stars can exist.

bolometric luminosity The total amount of energy emitted per second by an astronomical source, in all forms and at all wavelengths.

brown dwarf A collapsed object with less than 8 percent of the mass of the Sun, so it is not a true star. Brown dwarfs emit thermal radiation and are more than about twice the mass of Jupiter, so they are more massive than planets.

calendar A tool for dividing time; in particular, a way of counting the days in a solar year, or one Earth orbit of the Sun.

carbonaceous material Carbon-rich rocks and soil seen on comets and on the surfaces of some of the moons of the giant planets.

carbon cycle Also called the CNO cycle, a fusion chain that converts hydrogen nuclei into helium nuclei, using carbon as a catalyst. The carbon cycle operates only in stars more than about 1.5 times the mass of the Sun.

cause of the seasons Earth's rotation axis is tilted by 23.5° with respect to the plane of Earth's orbit of the Sun, but the direction of the rotation axis does not change throughout the orbit. As a result, the north pole is titled toward the Sun during one part of the orbit and away from the Sun 6 months later.

CCD detector An electronic detector that can sensitively record light. CCDs (charge-coupled devices) are the most commonly used detectors in astronomy.

cell The smallest unit of life processes. Cells are highly organized chemical factories.

center of gravity A point between two orbiting objects, like a balance point, that represents a position where the combined mass would be concentrated. Alternatively, the place where the entire mass of a large object could be represented (for a spherical object, the center of gravity is the center of the sphere).

Chandrasekhar limit When the core mass of an evolved star is more than 1.4 times the mass of the Sun, the force of gravity causes the star to collapse beyond a white dwarf state to a more compact form.

chaotic orbit Objects that suffer many small gravitational influences have positions that become impossible to predict after many orbital periods. Also, an orbit in which a small change in the initial conditions makes a large change in the eventual outcome. Chaotic orbits are impossible to accurately predict after many cycles.

chemical reaction A process in which elements and compounds combine and separate. Chemical reactions affect electrons but not atomic nuclei.

chemistry The study of the composition, structure, properties, and reactions of atoms and molecules.

chromosphere A thin layer of pink gas, just above the photosphere of the Sun or any other star.

circular velocity The velocity required to keep an object in a circular orbit around a much larger object, such as a satellite orbiting Earth. *See also* escape velocity.

climate change on Mars Long-term evolution of the conditions on Mars. Now-dry river beds on Mars suggest the Martian atmosphere was thicker, wetter, and warmer in the past.

clustering A distribution of any group of objects where the average spacing is less than the average spacing of a randomly distributed population. Gravity causes clustering of both stars and galaxies.

collecting area The area of a telescope that intercepts the radiation from distant objects.

comet An icy and rocky body that moves on a highly elliptical orbit through the solar system. Comets spend most of their time far beyond the orbits of the planets and only show a bright head and diffuse tail when they make their periodic passage close to the Sun.

comet head The bright diffuse part of a comet, caused by glowing gases.

comet nucleus A bright point of light that indicates the solid part of a comet, which cannot be resolved by telescopes on the ground.

comet tail The glowing gases that stream away from the comet nucleus, pointing away from the Sun.

comparative planetology A field of study that brings together astronomy and geology in looking for the principles that unite the properties of different planets.

compound Any substance that consists of molecules or the atoms of more than one element.

condensation sequence The sequence in which chemical compounds condense to form solid grains in a cooling, dense nebula.

conduction Heat flow through atomic collisions. Faster moving atoms or molecules transfer their heat energy to slower moving atoms or molecules.

constellation A grouping or pattern of stars that represents an animal or a mythological figure. The twelve constellations along the ecliptic are the signs of the zodiac.

convection Heat flow through the bulk movement of material. Hot material circulates and mixes with cooler material, thereby transferring heat energy.

Copernican revolution The intellectual impact of the idea of Copernicus (ca. 1543) that the Earth is not the center of the universe. In the past 100 years, we have also learned that Earth has no privileged position in the universe.

core The densest inner region of Earth, of iron-nickel composition. In stars, the core is the hot central region where fusion occurs. In galaxies, the core is the dense and bright central region.

corona The extended outer atmosphere of the Sun or other star, with a temperature of several million degrees.

cosmic abundance The average abundance in the universe of all the stable elements.

cosmic background radiation The nearly uniform thermal radiation that is a relic from the hot big bang. This radiation is observed as microwaves at a temperature of 2.7 K.

cosmic evolution The inevitable fact that distant objects in the universe are observed as they were when they were younger, as a result of the large light-travel time.

cosmic nucleosynthesis The fusion of helium and other light elements in the early universe a few minutes after the big bang. As a result, nearly a quarter of the mass of the universe was converted from hydrogen to helium.

cosmological constant Energy associated with the vacuum of space that acts as a repulsive force that affects the universal expansion. Its physical basis and its status in the big bang model are unclear.

cosmological principle The assumption that the universe is homogeneous and isotropic. The cosmological principle is supported by current observations, as far as it can be tested.

cosmological redshift A shifting of photons to longer wavelengths or lower energies due to the expansion of space. Cosmological redshifts are not the same as redshifts due to the Doppler effect.

cosmology The study of the origin, size, and structure of the universe.

critical density The density that defines a flat universe, at the boundary between open and closed universes; the mass density needed to just halt the universal expansion.

crust The outermost solid layer of Earth or any other planet.

dark matter Mass that makes its presence felt by gravitational forces but that does not emit light or radiation at any wavelength. Dark matter makes up 95 percent of the mass of the universe, and most of it is found on the largest scales of galaxy halos and galaxy clusters.

decimal system The system of counting and arithmetic based on factors of 10 (the binary system is based on factors of 2).

decoupling The time in the early universe, about 300,000 years after the big bang, when the universe became transparent and photons could travel freely through space.

deduction A logical method for combining ideas, observations, measurements, or numbers. In word form, deduction allows well-justified premises to be combined to draw a reliable conclusion. *See also* induction.

density parameter The ratio of the mean mass density of the local universe to the density required to just halt the universal expansion. *See also* critical density.

deoxyribonucleic acid (DNA) A long molecule in the shape of a double helix that is the key to life. DNA carries the genetic code in a four-letter chemical alphabet.

differentiation Any process, including gravity, that separates chemicals from their mixed state and concentrates them in different regions.

digital information Information in the form of (or convertible into) discrete levels suitable for manipulation by a computer. The simplest form of digital information is a binary code consisting of two levels, 0 and 1.

dirty iceberg model The description of comets as large icy bodies with bits of silica or carbon-rich material embedded in them.

distance indicator Any property of a star or galaxy that can be used to measure distance; usually involves a local calibrator whose distance is known via a different technique.

distance scale The set of measurements that define distance from the solar system out to the most remote galaxies. With an overlapping set of techniques, the errors tend to increase with increasing distance from Earth.

Doppler effect The change in the length of a wave due to the relative motion of the source and the observer; applies to light, sound, or any electromagnetic wave. If the source of waves is approaching, the wavelength becomes shorter (or bluer); if the source of waves is receding, the wavelength becomes longer (or redder).

Drake equation The equation that separates out various factors involved in estimating the number of intelligent, communicable civilizations in the Milky Way galaxy. Unfortunately, several of these factors are unknown.

Earth-crossing asteroids Asteroids that have orbits that can cross Earth's orbit of the Sun.

ecliptic The path traveled by the Sun in the sky and, therefore, the plane of Earth's orbit of the Sun. Also, the approximate plane of the planets of the solar system.

electromagnetic radiation Linked electrical and magnetic disturbances that travel through a vacuum at 300,000 km/s. Electromagnetic radiation can also travel through gases and some solids. The radiation has wave properties but can also be thought of as a stream of particles or photons. Light is one familiar example.

electromagnetic spectrum The full range of electromagnetic radiation from long to short wavelengths, or low to high frequencies. Only the narrow wavelength range of the visible spectrum can be detected by the eye. Radio waves, microwaves, and infrared radiation are at longer wavelengths; ultraviolet radiation, X rays, and gamma rays are at shorter wavelengths.

electron The lightest subatomic particle known. Negatively charged electrons define the outer edge of the atom.

element Fundamental substance with a unique set of chemical properties. Each element has as its smallest unit a different type of atom. Ninety-two different elements are found in nature, and several dozen more can be created fleetingly in the laboratory. *See also* atom.

elliptical galaxy A galaxy with a smooth distribution of stars in the shape of a squashed sphere. Elliptical galaxies contain mostly old stars and occur in a wide range of sizes and masses.

elliptical orbit The general shape that describes the motion of one object bound by gravity to another object; a closed figure like a squashed circle.

emission lines Narrow spectral features that result when atoms go from a particular energy state to a lower energy state. *See also* absorption lines.

energy In physical terms, anything with the ability to cause a change or do work. Energy can take many forms, including electromagnetic radiation, heat, motion, and even mass (according to the theory of relativity).

energy levels The discrete values of energy available to electrons in an atom. Atoms of each element have their own unique set of energy levels.

entropy The measure of disorder in a physical system, which always tends to increase with time. Entropy is also related to the number of possible states of a system, since ordered states are less probable. *See also* laws of thermodynamics.

erosion The removal of rock and soil by any natural process, such as wind or rain.

escape velocity The velocity required for an object to escape the gravity force of a much larger object, such as a spacecraft leaving Earth's orbit. The escape velocity is about 40 percent higher than the circular velocity. *See also* circular velocity.

estimation A rough calculation that is accurate to no more than one or two significant figures. Also called an order of magnitude calculation.

eukaryotes Living organisms made of cells with nuclei. The nucleus has a membrane that retains much more genetic material than is found in cells without nuclei. *See also* prokaryotes.

event horizon The imaginary surface of a black hole where the escape velocity is the speed of light. The event horizon is an information barrier, and matter and energy cannot escape from within this region.

evidence Scientific evidence consists of quantitative observations or experimental results that can be confirmed by other investigators.

expanding universe The growth of the universe due to expanding space, which carries all galaxies away from one another. *See also* big bang model.

explosive nucleosynthesis The rapid creation of heavy elements (including atomic nuclei heavier than iron) in the blast wave of a supernova. Heavy elements can also be made by the slower process of neutron capture in evolved stars.

extrasolar planet A planet orbiting a star beyond the solar system.

fireball A large interplanetary body that burns up in Earth's atmosphere. *See also* meteor.

force Anything that causes a change in motion or acceleration. For example, acceleration can be caused by gravity, electric charge, pressure, or radiation.

fossil The remains of a living organism that have been turned to stone over a long period of time.

frequency The number of crests or troughs of a wave that pass each second. For electromagnetic waves, the energy of the wave is proportional to the frequency. *See also* wavelength.

frost line A region between 3 and 4 AU from the Sun in the asteroid belt. Asteroids beyond this region have water that is frozen and inactive. *See also* soot line.

fundamental forces Four forces that account for all the physical behavior in the universe. Gravity and electromagnetism are long-range forces, and the strong and weak nuclear forces operate only within atoms.

galactic bulge The slightly flattened component near the center of a spiral galaxy where stars move on elliptical orbits. Stars in the bulge have a range of ages.

galactic disk The flattened component of a spiral galaxy where stars move on circular orbits. Stars in the disk are mostly young and blue.

galactic halo The spherical component of any galaxy where stars move on elliptical orbits. Stars in the halo are mostly old and red.

galaxy A large system of stars held together by gravity, remote from the Milky Way.

galaxy clusters A concentration of galaxies held in one large region of space by gravity, with anywhere from a few dozen to thousands of members.

galaxy evolution The gradual change in the overall properties of galaxies with time, due to the evolution of the stars they contain.

galaxy formation The process in which enormous clouds of gas collapse by gravity into systems of stars. Most galaxy formation took place 5 to 10 billion years ago.

galaxy group Between a few and a few dozen galaxies held in one region of space by gravity.

galaxy interactions In regions of the universe that are denser than average, the interactions of galaxies at a distance by the force of gravity. Collisions and mergers can also occur.

galaxy supercluster An enormous concentration of galaxies that may contain thousands to tens of thousands of members.

gas A low-density form of matter in which atoms or molecules can move freely and in which collisions are rare. Any substance can be turned into a gas, given a sufficient temperature.

gas giant planet A planet made primarily of gaseous hydrogen and helium; in the solar system, the outer planets Jupiter, Saturn, Uranus, and Neptune.

gas laws The mathematical relationships that govern the temperature, density, and pressure of gases.

gene The minimum amount of genetic material that expresses a characteristic of a living organism; a sequence of several hundred bases along DNA molecule.

general theory of relativity Einstein's theory to describe the behavior of objects in nonuniform motion. It is impossible to distinguish between acceleration due to gravity and acceleration due to any other force. As a consequence, mass bends light, radiation loses energy when leaving an object with strong gravity, and time slows down in a situation of strong gravity. *See also* special theory of relativity.

geocentric cosmology The description of a spherical universe with Earth stationary at the center. As described by Aristotle, this cosmology was used for over 2000 years. *See also* heliocentric cosmology.

geological time scale The sequence of events in the history of Earth.

geology The study of the history, origin, and structure of Earth. This term has been extended to include other planets.

giant star A large evolved star that radiates much more luminosity than a main-sequence star of the same surface temperature.

globular star cluster A dense, spherical group of up to a few million old stars, found in the halo of the Milky Way and in the halos of other galaxies.

granite A pale rock of moderate density and high silica content, formed by differentiation and concentrated in Earth's continents.

gravitational collapse The contraction of a diffuse gas cloud to form a planet, a star, or a galaxy.

gravitational lensing The bending of light by gravity, one of the consequences of general relativity. The light from a distant object can be bent and distorted into arcs, rings, and multiple images.

gravitational microlensing The amplification of light from a distant star due to the passage of a nearby object between us and the star. The bending of light by gravity occurs on a scale too small to measure, so a temporary brightening is the only indication of the event.

gravitational perturbation A small deviation to an orbit that can lead to a very different eventual position.

Great Red Spot A long-lived storm in Jupiter's atmosphere that is bigger than Earth.

greenhouse effect The heating of a planetary atmosphere due to trapped infrared radiation.

greenhouse warming of Venus An extremely hot atmosphere resulting from a high concentration of carbon dioxide.

half-life In radioactivity, the time for half of the original atoms to decay. In general, the time for a variable to change by half its value. Half-life is loosely equivalent to the characteristic time scale of a phenomenon.

heat The energy of motion of atoms and molecules, measured by temperature. Faster atomic or molecular motions mean more heat energy. *See also* temperature.

Heisenberg uncertainty principle The fundamental limit to the precision of all physical measurements. It states that we cannot know with arbitrary accuracy the position and the momentum of a system, or the energy of the system at every instant of time.

heliocentric cosmology The description of the universe with the Sun at the center and the planets in orbit around the Sun; proposed by Copernicus and shown to be correct by Galileo. *See also* geocentric cosmology.

homogeneous Smooth and featureless, or the same at all locations.

horizon The limit of space visible to us. Regions outside the horizon either cannot be seen or are places from which light has not had time to reach us.

H-R diagram An essential tool of stellar astronomy, the plot of photospheric temperature (or color or spectral class) against luminosity; named after astronomers Hertzsprung and Russell.

Hubble constant The slope of the Hubble relation, which gives the current expansion rate of the universe.

Hubble relation A linear correlation between redshift and distance for galaxies. More distant galaxies move away from the Milky Way faster, which is evidence for an expanding universe. The slope of the Hubble relation gives the current expansion rate. Also called the Hubble law.

hydrostatic equilibrium The balance that exists at every point within a stable star between the inward force of gravity and the outward pressure, due to energy released from nuclear reactions.

hypothesis A proposed explanation for an observed phenomenon, or a proposal that a certain observable phenomenon occurs.

icy material Frozen liquids and gases that form much of the material in gas giant planets and comets.

igneous rock Rock crystallized from molten material, such as granite or basalt.

impact crater A roughly circular depression of any size (from microscopic to over 1000 km), caused by an impact from space.

induction A logical method for drawing a broad conclusion from a limited set of observations or experiments. The truth of an inductive argument cannot be guaranteed with certainty. *See also* deduction.

inertia The resistance of an object to a change in its motion; also, a measure of the amount of material in an object, or its mass.

inflationary universe The idea that the universe went through a rapid phase of expansion in the first fraction of a second after the big bang. Inflation can explain the flatness and smoothness of the universe.

inhomogeneous Variable in density and not the same at all locations.

intelligence The capability for abstract thought, coupled with a mastery of tools or technology. In living organisms, intelligence is found only in animals with large and complex brains.

intensity The strength of radiation. In the wave description of radiation, intensity is proportional to the square of the amplitude of the wave.

interferometry The use of widely separated telescopes to increase the angular resolution of an astronomical observation. This technique effectively increases the resolution to that of a single telescope whose size is equal to that of the most widely separated telescopes of the interferometer.

interstellar extinction The absorption, or obscuration, of light by interstellar dust, causing distant objects to appear fainter.

interstellar medium Diffuse gas and dust that occupy the space between stars.

interstellar reddening The preferential absorption, or scattering, of blue light by interstellar dust, causing distant objects to appear redder.

inverse square law The variation of any quantity, such as the force of gravity or the intensity of radiation, that decreases with the square of the distance from the source.

irregular galaxy A small low-mass galaxy with an irregular or amorphous shape. Irregular galaxies usually have young stars and lots of gas and dust. Best-known examples are the Large and Small Magellanic clouds, neighbors of the Milky Way.

isotope A variation of a chemical element with more neutrons in the atomic nucleus; usually radioactive.

isotropic The same in all directions.

Jupiter The largest planet in the solar system, containing just over two-thirds of the mass of all the planets.

Kepler's laws of planetary motion The three laws of planetary motion that describe how the planets move in elliptical orbits of the Sun, and allow accurate prediction of planetary positions.

kiloparsec (kpc) 1000 parsecs, or 3260 light years. *See also* parsec.

kinetic energy The energy of motion. Kinetic energy is proportional to the mass of the object and the square of the velocity. *See also* potential energy.

Kirchhoff's laws of radiation Laws governing radiation from hot objects. Every sufficiently hot gas will emit a continuous spectrum. A hot gas will emit a set of spectral lines characteristic of the elements in the gas. A cool gas in front of a hot continuum source will produce a set of spectral lines in absorption.

Kuiper belt A group of comets lying in the plane of the solar system at a distance of 30 to 100 AU from the Sun, just beyond the orbits of Neptune and Pluto.

large-scale structure The distribution of matter in the universe on the largest scales, from groups and clusters of galaxies to superclusters of galaxies.

law of conservation of angular momentum The fact that the amount of angular momentum in a system is constant. If the system shrinks, the rotation speed will increase; if the system expands, the rotation speed will decrease.

law of conservation of energy The fact that the total amount of energy in a system is constant, even though energy may change forms.

laws of thermodynamics Rules that govern the behavior of heat and other forms of energy. Energy can change forms, but the total amount of energy in a system is constant. As energy changes forms, the proportion of energy in the disordered form of heat increases. It is impossible to remove all the heat from a physical system. *See also* entropy.

length The fundamental measure of size or distance, represented by units of meters in the metric system.

life A self-contained set of chemical reactions involving carbon-based molecules. Living things grow, evolve, reproduce, utilize energy, and respond to their environment.

life on Mars Evidence pointing to fossil microbial life forms in a Martian meteorite that was recovered on Earth.

light curve A plot of the brightness of a star (or any astronomical object) versus time.

light travel time The time it takes light to travel a certain distance in the universe. Typical light travel times are hours within the solar system, years to the nearest stars, millions of years to the nearest galaxies, and billions of years to the most distant galaxies.

light-year The distance light travels in 1 year, 9.5×10^{12} km.

liquid A high-density form of matter in which the atoms or molecules can move freely and take the shape of the container. Matter densities of liquids and solids are similar.

Local Group The concentration of galaxies that contains the Milky Way, Andromeda (M 31), and a few dozen small galaxies.

Local Supercluster The large flattened concentration of galaxies that contains the Milky Way, the Local Group, the Virgo cluster, and thousands of other galaxies.

luminosity The total amount of energy emitted per second by an astronomical source. *See also* absolute brightness.

lunar eclipse The dimming of the Moon as it passes through Earth's shadow cast by the Sun. *See also* solar eclipse.

Magellanic clouds The two galaxies nearest the Milky Way, irregular in form and visible to the naked eye in the Southern Hemisphere.

main asteroid belt The largest group of asteroids, found on circular orbits between Mars and Jupiter.

main-sequence fitting A technique for measuring the relative distance of two groups of stars, based on the vertical offset between the two main sequences on an H-R diagram.

main-sequence stars Stars, like the Sun, that are converting hydrogen into helium by fusion in their interiors. Main-sequence stars form a diagonal band on the H-R diagram.

mantle A region of intermediate density surrounding the core of Earth or any other planet.

Mars The fourth planet from the Sun, most similar to Earth in terms of geology and surface features.

mass The fundamental measure of the amount of material, represented by units of kilograms in the metric system. Also, a measure of the resistance of an object to any change in its motion.

mass-energy conversion The conversion of mass into a much larger amount of equivalent energy, according to Einstein's equation $E = mc^2$. Energy can be converted into mass according to the same equation.

mass extinction A brief interval, compared with the time scale of evolution, when a significant fraction of Earth's species become extinct. Some mass extinctions may be caused by catastrophic impacts from space.

mass-luminosity relation The mathematical relationship between the mass of main-sequence stars and their total emission. More massive stars have far higher values of luminosity.

mass-to-light ratio The ratio of the mass of an astronomical object to its luminosity, in solar units. The mass-to-light ratio of the Sun is one, by definition.

mathematics The study of numbers and their properties, and the symbols and operations that can apply to numbers.

megaparsec One million parsecs, or 3.26 million light-years. *See also* parsec.

measurement A quantitative way to present scientific evidence, in the form of a number, an error or uncertainty, and a unit that represents the type of measurement.

Mercury The small rocky planet closest to the Sun, similar in size to Earth's moon.

meridian An imaginary line in the sky that runs from the direction of the South Pole, through the zenith, to the direction of the North Pole. Objects to the east of the meridian are said to be rising, and objects to the west of the meridian are said to be setting.

metamorphic rock Rock formed when igneous or sedimentary rocks are modified by heat, pressure, or chemical reactions.

meteor A small interplanetary body that burns up in Earth's atmosphere. Also called a shooting star. *See also* meteorite.

meteorite A rocky or metallic interplanetary object large enough to reach Earth's surface. *See also* meteor.

meteor shower A concentrated burst of meteors that occurs at a particular time of year when Earth's orbit intersects debris from a comet.

metric system A system of measurement based on powers of ten, with fundamental units of kilograms for mass, meters for length, and seconds for time.

microwave background fluctuations Tiny features in the cosmic microwave background that represent the seeds of galaxy formation. These fluctuations are observed as variations in the temperature of the radiation, and they represent a view of the universe 300,000 years after the big bang.

Milky Way The band of diffuse light stretching across the sky that represents the plane of our galaxy. Also refers to the entire Milky Way galaxy.

molecular clouds Interstellar clouds of higher-than-average density. These cool regions contain dust grains and molecules and are the sites of much star formation.

molecule A microscopic unit of matter consisting of one or more atoms linked together.

morphology-density relation The relation between galaxy type and large-scale environment. Spiral galaxies tend to be found in low-density regions, and elliptical galaxies tend to be found in high-density regions.

multiple-star system Three or more stars orbiting each other and held in one region of space by their mutual gravity.

natural selection The evolution of species by adaptation to the environment and the near-perfect transmission of genetic material from one generation to the next.

nebula A cloud of gas and dust in interstellar space or surrounding a star.

Neptune The most distant gas giant planet from the Sun.

Newton's laws of motion The physical laws that govern forces and motions. An object does not alter its motion unless a force acts on it. The force is given by the mass times the acceleration. Every force (action) has an equal and opposite force (reaction).

neutrino A subatomic particle with no charge and small mass, which has extremely weak interactions with matter. Neutrinos are produced copiously by nuclear reactions in the Sun.

neutron A massive, neutral particle that occupies the atomic nucleus of every element, except hydrogen. A free neutron decays in roughly 10 minutes. *See also* proton.

neutron star The end state of a massive star, in which the pressure is sufficient for protons and electrons to coalesce into neutrons. Neutron stars have the density of a gigantic atomic nucleus. *See also* pulsar.

nonthermal radiation Radiation that results from the constant excitation or acceleration of high-energy particles. Nonthermal radiation has a much broader spectrum than thermal radiation.

nova The sudden brightening of a star that occurs when mass is transferred from one member of a binary star system to the other.

nuclear fission The spontaneous decay of the atomic nuclei of heavy elements, with the release of particles and radiation. *See also* radioactivity.

nuclear fusion The merging of light atomic nuclei into heavier atomic nuclei, with the release of particles and radiation.

observational error The uncertainty in a measurement, often caused by limitations in the equipment or observation. Every measurement has some degree of error. Random errors can be reduced by multiple measurements, but systematic errors are caused by unrecognized flaws in the measuring apparatus.

Olbers' paradox The problem in an infinite universe that every direction should end in a source of light, which conflicts with the observation that the night sky is dark.

Oort cloud A spherical swarm of comets that extends out to many thousand AU from the solar system.

opacity The ability of a material to obscure light that passes through it. Opacity, or optical depth, is related to the typical number of collisions suffered by a photon in traveling a particular distance through a medium. *See also* photosphere.

open star clusters Groups of 100 to 1000 young stars bound by gravity that are found in the plane of the Milky Way and other galaxies. *See also* stellar associations.

organic chemistry The set of processes based on molecules that contain carbon, including processes unrelated to living organisms.

parallax The angular shift of an object due to the changing position of the observer. Stellar parallax is the small angular shift of a nearby star with respect to more distant stars caused by Earth's orbit of the Sun.

parallax distance A direct and trigonometric measurement of distance for a star, based on the parallax shift observed from Earth.

parent body The larger interplanetary object of which a meteorite is a fragment.

parsec (pc) The distance that produces a parallax, as seen from Earth's orbit, of 1 second of arc, equal to 3.26 light-years, or 3×10^{13} km.

period-luminosity relation A correlation between the luminosity and the period of variation of Cepheid variable stars, which enables astronomers to measure the period of Cepheids in a remote galaxy and infer the distance of the galaxy.

periodic process Any physical process that repeats in a regular way, with a fixed cycle of variations, such as waves, vibrations, and oscillations. *See also* random processes.

periodic table A way of organizing the elements according to the number of outer electrons in each atom. Elements in a particular column of the periodic table share many chemical properties.

phases of the Moon The 29.5-day cycle of changing illumination of the Moon as seen from Earth. The Moon is always half-lit by the Sun, but we see a different fraction of the lit face as the Moon orbits Earth.

Phobos and Diemos The two small, irregularly shaped moons of Mars.

photon The fundamental unit, or quantum, of electromagnetic radiation. Photons are like the particles of radiation, with energy proportional to their frequency.

photometry A measurement of the amount of light of an astronomical object, at all wavelengths or over a range of wavelengths.

photosphere In the Sun, the visible surface where radiation escapes and travels freely through space. In any star or hot gas, the region where the density is low enough that radiation travels without interacting with matter. *See also* opacity.

photosynthesis The conversion of sunlight into stored energy by a living organism. The energy can then be used to run a metabolic process.

physics The study of the forces of nature and the laws that govern the way matter and radiation interact.

planetary nebula A spherical or nearly spherical cloud of hot gas surrounding an evolved star. Nebulae have nothing to do with planets except for a slight resemblance to planetary disks when observed through a small telescope.

planet A rocky or gassy body that orbits a star.

planetesimal A preplanetary body. Planetesimals range in size from several microns to several kilometers.

plate tectonics The theory that Earth's surface features are caused by moving plates of the crust, driven by liquid rock from the mantle below.

plurality of worlds The idea that Earth is just one among many worlds in space, including the idea that there may be applications of geology, chemistry, and biology beyond Earth.

Pluto The ninth planet in the solar system, which many planetary scientists consider not to be a true planet.

polar ice caps of Mars Regions of frozen carbon dioxide at the poles of Mars that grow and shrink with the Martian seasons. A smaller amount of frozen water remains at each pole year round.

potential energy Stored energy, or energy with the potential to do work. Examples are the energy stored in chemical bonds, the energy stored in a raised object, and the mechanical energy in a coiled spring or rubber band. *See also* kinetic energy.

precession The 26,000-year wobble of Earth's rotation axis as it traces out a circle on the sky.

precision A measure of how finely a quantity can be specified, in terms of the number of significant figures in the measurement. Most observations in astronomy are quoted with a precision of no more than three significant figures. *See also* accuracy.

pre-main-sequence star An interstellar cloud that has collapsed to stellar dimensions but has not yet begun to fuse hydrogen into helium. Pre-main-sequence stars are often shrouded in gas and dust.

prograde rotation The normal form of rotation for a planet, in the same direction as its orbit of the Sun.

prokaryotes Simple life-forms that consist of cells without nuclei, and contain a single long strand of DNA. Bacteria are the primary example. *See also* eukaryotes.

proper motion The rate of angular motion of a star across the sky (generally less than 1 second of arc per year). Proper motion can be combined with distance to calculate the transverse velocity of the star, or the velocity perpendicular to the line of sight. *See also* radial velocity.

proteins The chemical workhorses that carry out life functions. Each of the many different kinds of proteins contains thousands of atoms.

proton A massive, positively charged particle that occupies the atomic nucleus. The number of protons in the nucleus is the atomic number and specifies the chemical element. *See also* neutron.

proton-proton chain The main energy source of the Sun and stars like the Sun. A series of nuclear reactions convert hydrogen nuclei into helium nuclei, converting a tiny amount of mass into energy.

protostar A cloud of interstellar gas and dust that is dense and cool enough to gravitationally contract into a star.

pulsar A type of neutron star that emits radio waves resulting from a "hot spot" on its surface. Since neutron stars rotate rapidly, the beam of radio emission creates pulses as the star spins. The period of pulsars can range from a few seconds to tiny fractions of a second. *See also* neutron star.

quantum fluctuations Microscopic violations of the law of conservation of energy, when particles can momentarily come in and out of existence. Quantum fluctuations exist even in the vacuum of space; in the inflationary cosmology, they are magnified into the seeds for galaxy formation.

quantum theory of radiation The idea that energy is not smooth and continuous but comes in discrete packets called photons. Light is made up of these microscopic units of radiation.

quasar An active galaxy where the light from the nucleus swamps the light from the surrounding galaxy. Quasars are at enormous distances, as indicated by their large cosmological redshifts. *Quasar* stands for quasi-stellar radio source, but similar objects called QSOs (quasi-stellar objects) can be selected by their optical emission.

radial velocity The component of a star's motion (or the motion of any astronomical object) along the line of sight. Positive velocities indicate recession; negative velocities indicate approach. *See also* proper motion.

radiation Heat flow through photons. Various forms of energy (including heat energy) can be carried from one place to another by electromagnetic radiation.

radioactive dating A technique for determining the age of a material by measuring the amount of a radioactive isotope and its decay product.

radioactivity The spontaneous decay of a massive atomic nucleus, releasing particles or radiation.

radio galaxy An active galaxy with intense radio emission, often in the form of opposite-directed jets. Radio galaxies are usually elliptical galaxies.

random process A process in which the timing of an individual event is unpredictable, even though the rate of events may be well-determined. Examples are impacts from space and decays of radioactive atoms. *See also* periodic process.

red giant A star that has consumed its hydrogen and is fusing other nuclear fuels. Red giants have compact cores, and they appear red because of their cool outer envelopes.

reflex motion The wobble, or periodic motion, of a star due to the influence of a much smaller planet or companion.

resolution The ability of a telescope or optical device to see fine details of an astronomical object or distinguish small angles on the sky.

resonance An orbit, oscillation, or vibration where two bodies interact strongly. In astronomy, a resonance occurs when the orbital period of one body is a multiple or simple fraction of the orbital period of another body.

retrograde motion The occasional (and temporary) reversal of a planet's motion among the stars to move from east to west; can occur only for planets beyond Earth's orbit and is most noticeable for Mars.

retrograde rotation Rotation from east to west, shown only by Uranus.

revolution The orbital motion of a planet around the Sun or a moon around a planet. *See also* rotation.

ring systems Systems of orbiting particles seen in the equatorial planes of each of the four gas giant planets. The rings are transparent, and the individual particles follow orbits governed by Kepler's laws.

riverbeds on Mars Evidence of erosion on Mars in the form of dry riverbeds, the best evidence that Mars was once wetter than it is today.

Roche's limit Inside this distance from a planet, an orbiting satellite will be disrupted or destroyed by tidal forces.

Roche surface An imaginary surface in three dimensions that contains the gas in a binary star system. Also called a Roche lobe.

rotation The spin of a planet or a moon on its axis. *See also* revolution.

rotation curve A plot of the orbital velocity of material in a spiral disk versus distance from the center of the galaxy. Flat rotation curves indicate substantial amounts of dark matter in the halos of spiral galaxies.

Saturn The second largest planet in the solar system, ten times larger than Earth.

scientific method The method of learning about nature by making observations, formulating hypotheses, and carrying out observational or experimental tests to see if the hypotheses are accurate. The scientific method is based on evidence in the form of repeatable, quantitative measurements.

scientific notation A convenient way of recording and manipulating very large and very small numbers. Also called exponential notation, or powers of ten.

Schwartzschild radius The radius corresponding to the event horizon of a black hole (equal to 3 km for an object with the mass of the Sun), proportional to the mass of the black hole.

sedimentary rocks Rocks formed by the deposition and compression of material that has been dissolved in water, such as limestone or chalk.

SETI The search for extraterrestrial intelligence.

Seyfert galaxy An active galaxy with a bright nucleus containing ionized, high-velocity gas. Seyfert galaxies are usually spiral galaxies.

shepherd satellites Satellites that move near planetary rings and act to confine the ring particles to certain orbits.

solar composition A gas composed of 76 percent hydrogen and 22 percent helium, with much smaller trace amounts of all other elements.

solar core The central region of the Sun, where nuclear reactions release energy under conditions of high density and high temperature.

solar cycle The 22-year cycle of the Sun's magnetic activity, consisting of two 11-year cycles of sunspots numbers.

solar eclipse The partial or total blocking of the Sun's light caused by the passage of the Moon in front of the Sun. *See also* lunar eclipse.

solar nebula The cloud of gas around the Sun during the formation of the solar system.

solar seismology The study of natural vibrations and oscillations in the Sun as a way to probe the structure of the solar interior.

solar wind An stream of charged particles from the Sun that crosses Earth's orbit at speeds of 400 km/s or higher.

solid Material with a rigid structure resulting from the forces between atoms. The atoms in a solid can be arranged in an irregular (or amorphous) structure or in a regular (or crystalline) structure.

soot line A distance of about 2.7 AU in the middle of the asteroid belt, separating stony and metal-rich asteroids closer in from black carbon-rich asteroids farther out. *See also* frost line.

space astronomy Any type of astronomy that is carried out beyond Earth's atmosphere, including high-flying planes and balloons. Essential for detecting types of electromagnetic radiation that are absorbed by Earth's atmosphere: Far-infrared waves, ultraviolet radiation, X rays, and gamma rays.

space curvature The curvature of three-dimensional space; one of the consequences of general relativity. Matter tells space how to curve, and space tells matter how to move.

space-time diagram A graphical way of representing evolution in time and in space. In a space-time diagram, the three dimensions of space are collapsed to one dimension and plotted as the x-axis, while time is plotted as the y-axis.

special theory of relativity Einstein's theory to describe the behavior of objects in constant relative motion. The speed of light is a universal constant. As a consequence, objects moving at a significant fraction of the speed of light become more massive, they contract in the direction of their motion, and their time slows down. *See also* general theory of relativity.

spectral classes Divisions of stellar spectra according to the strength and characteristics of the spectral lines. As currently arranged, spectral classes correspond to a sequence of photosphere temperature.

spectroscopy A measurement of the amount of light at each wavelength from an astronomical object, after the light has been dispersed and recorded.

speed of light A universal constant, written as c, equal to 300,000 km/s in a vacuum. All electromagnetic waves travel at this speed.

sphere of gravitational influence The region within which a planet dominates the motion of any small object. Beyond this region, the gravity of the Sun dictates the motions. All stars and galaxies also have spheres of gravitational influence.

spiral arms Two arms that emerge from the center of a spiral galaxy and trail in the direction of rotation of the disk. The spiral arms contain young blue stars, gas, and dust, although stars move in and out of the spiral features.

spiral galaxy A galaxy with prominent spiral arms containing young stars and a diffuse but massive halo of older stars. The best-studied example is the Milky Way.

standard error A measure of the uncertainty in a single measurement, based on the scatter of multiple measurements of the same quantity. Also called standard deviation.

stars Spheres of gas massive enough to have sufficient temperatures to sustain nuclear reactions.

Stefan-Boltzmann law A law giving the total amount of energy radiated per second from a star, in terms of its surface area and surface (or photospheric) temperature.

stellar associations Loose groups of 10 to 100 young stars found in the plane of the Milky Way and other galaxies. *See also* open star clusters.

stellar evolution The idea that stars must change their composition and structure as they use their nuclear fuel.

stellar populations Large collections of stars that share properties of age, motions, or chemical composition.

sublime The process by which frozen solids in a comet turn directly into gases under the heating influence of the Sun.

sunspots Magnetic disturbances on the Sun's surface that are cooler than the areas that surround them.

supergiant star An extremely large, massive, and luminous star that is short-lived compared to the Sun.

supermassive black hole The hypothetical object at the center of quasars and active galaxies, formed by the gradual accretion of material at the center of a galaxy. A supermassive black hole is like a gravitational engine, accelerating particles and producing large amounts of radiation.

supernova An energetic stellar explosion that creates heavy elements and results in the death of the star. A supernova is the fate of any single star where the core mass exceeds the Chandrasekhar limit, or it can occur in a binary system when mass transfer pushes the mass of one of the stars above the Chandrasekhar limit.

symmetry In physics, the idea that distinct phenomena can share a common basis. High energies or temperatures are usually required to observe the underlying symmetry.

systems of knowledge Ways of dealing with information, such as superstition, appeals to authority, and the advocacy system. The scientific method is the only system that has been successful in leading to an understanding of the natural world.

telescope A device for collecting and magnifying electromagnetic radiation from distant objects.

temperature A quantitative measure of heat energy, or the energy in the motions of atoms and molecules. Scientists use the Kelvin temperature, where 373 K is the boiling point of water and zero corresponds to no atomic motion.

terrestrial planet A planet made primarily of rocky material; in the solar system, the inner planets Mercury, Venus, or Mars.

theory A hypothesis or set of hypotheses that have passed many observational tests and explain a wide range of phenomena.

theory of star formation The theory that describes how stars form from collapsing clouds of interstellar gas and dust.

thermal equilibrium The tendency for heat energy to flow from hotter regions to cooler regions. More generally, the tendency for heat to flow until all regions have the same temperature.

thermal radiation A broad spectrum of electromagnetic radiation that is emitted by any material. The peak wavelength and amount of thermal radiation both increase with increasing temperature, and they do not depend on the chemical composition of the material.

tidal force A situation where the difference between the gravity acting on opposite sides of a planet or satellite causes a stretching (and heating) force.

time The fundamental measure of duration, measured in units of seconds in the metric system.

transformation of energy The process where one form of energy turns into another form. The natural world is full of examples of transforming energy. Mass is also a form of energy.

triple-alpha process A nuclear fusion reaction in which helium is transformed into carbon in the core of a red giant star.

Trojan asteroids Asteroids that lie in two groups, ahead of and behind Jupiter's orbit.

21-cm emission line A low-energy spectral feature that results from a change in the spin state of cold atomic hydrogen.

unification The idea that the four fundamental forces of nature are expressions of a single force. Unification is only manifested at extremely high temperatures or energies.

unit A well-defined entity for representing a measurable quantity.

universal law of gravitation Newton's expression for gravity, which states that the gravity force between two objects is equal to the product of their masses divided by the square of the distance between them, multiplied by the gravitational constant.

universe Everything that exists. Astronomers distinguish between the observable universe, the region from which light has had time to reach us in the age of the universe, and the physical universe, which may be much larger.

Uranus The seventh planet from the Sun, a gas giant with a uniquely large orbital tilt.

variable star A star that varies in brightness on time scales from hours to years. Stars are variable only during brief phases of their evolution.

velocity dispersion The spread or scatter in the motions of any gravitational system, such as a star cluster, a galaxy, or a cluster of galaxies.

Venus The planet most similar in size to Earth and second closest to the Sun.

visible spectrum Light arrayed in wavelength from long to short waves. The range from red to blue light represents the electromagnetic waves that can be perceived by the eye.

void A large empty (or nearly empty) region in the large-scale distribution of galaxies.

volcanism The eruption of molten materials at the surface of a planet or satellite.

volcanism on Venus Evidence for geological activity and resurfacing on Venus.

wavelength The distance between successive crests or troughs of any wave. For electromagnetic radiation, the energy of the wave is inversely proportional to the wavelength. *See also* frequency.

weight The mass of something as indicated by the local gravity force. Weight that has a certain value on Earth's surface would be lower on the Moon, and zero in Earth orbit or in deep space. Weight is distinguished from mass, which is fixed and depends only on the amount of material in an object.

Wien's law A law stating that the wavelength of the peak amount of thermal radiation from any material is inversely proportional to the temperature.

white dwarf star A planet-sized star of very high density that represents an end state of evolution for low-mass stars. White dwarfs are cooling stellar embers where no nuclear reactions are taking place.

zenith The point directly overhead in the sky.

zodiac The band around the sky, centered on the ecliptic, in which the Sun, Moon, and planets move. The zodiac is about 18° wide and is divided into the 12 constellations.

zodiacal light A dim band of light in the night sky, distributed along the ecliptic, caused by sunlight scattering off tiny particle of interplanetary debris.

INDEX

A

Absorption/emission spectra. *See* Stellar spectra
Abundance of elements, 354–355
Accretion
    accretion disc of quasars, 499
    accretion discs near black holes, 387
    of planetesimals, 246–249, 261, 262
Adams, W.S., 373–374
Advocacy, systems of, 18
Age of universe
    big bang model for, 526, 527
    estimating from expansion, 517
    Science Toolbox on, 495
    See also Cosmology
Ages of stars. *See* Stellar evolution
Alcohol, ethyl
    in Sagittarius B-2 molecular cloud, 357
Alexandria library, burning of, 47
Alfonso X (Spain), 55
Alfven, Hannes, 237
Algebra, App. A–2
Alien life. *See* Intelligence, extraterrestrial; Life, extraterrestrial
Alpha Centauri, 400
Altitude, position on the sky, App. A–6
Alvarez, Walter and Louis, 128
Anaxagoras (Greece), 37
Ancient astronomy
    ancient cosmologies, overview of, 508–509
    Arab science, 47
    archaeoastronomy, 32
    Aristarchus, heliocentric cosmology of, 42, 43–44
    astrology, influence of, 50
    Buddhist/Hindu cosmology, 513
    Casa Grande, 33
    Chaco Canyon (Anasazi), 33–34, 377
    Chichen Itza, 32–33
    Chinese astronomy, 48
    distant objects, determining sizes of, 50
    Great Pyramid, 32
    Indian astronomy, 47–48
    Mayan astronomy, 48–49
    monuments, purpose of, 50
    in tropical latitudes, 34
    See also Calendars and time
Andromeda (M31)
    as distance indicator, 478–479
Angular diameter
    and curvature of space, 521–522
    small–angle equation (Toolbox), 38
Angular measurement, App. A–5
Angular momentum
    conservation of, 239, 240–241, 260
    of sun, 250

Anisotropy, 420, 422–424
Antares (supergiant star), 342
Anthropic principle, 580–581
Antimatter/antiparticles
    asymmetry/imbalance of in early universe, 540–541
    and radiation in early universe (Toolbox), 535–536
    and spacecraft propulsion, 576
    See also Particle pairs, Subatomic
Appendices
    algebra and working with dimensions, App. A–2
    angular measurement and geometry, App. A–5
    the brightest stars, App. B–3
    Local Group galaxies, App. B–4
    the nearest stars, App. B–2
    observing the night sky, App. A–6
    planetary data, App. B–1
    precision and measurement errors, App. A–4
    scientific notation and logarithms, App. A–1
    units and the metric system, App. A–3
    ways of representing data, App. A–7
Arab science, 47
Archaeoastronomy. *See* Ancient astronomy
Arecibo radio telescope, 264, 288, 579
Aristarchus, 42, 43–44
Aristotle
    biographical note, 41
    geocentric cosmology of, 42
Arrow of time, 527
Associations, stellar, 412
Asteroids
    4769 Castalia, radar image of, 222
    243 Ida, 222
    asteroid belt, 206, 207, 249
    astronaut landing on, 231
    carbonaceous, 223, 224
    Ceres, 218, 220
    classes of, defined, 223
    as comet nuclei, 223
    composition vs. distance from sun, 223–224
    compound asteroids, 222–223
    defined, 205
    discovery of, 218–220
    distribution, Kirkwood gaps in, 221, 222
    distribution of (by class), 223
    Earth-crossing, 220, 224, 231
    evolution and formation of, 231–232, 233
    frost line, 223, 243
    gravitational perturbation of, 220, 221
    impact frequencies of, 226–227
    impacts, Earth (*See* Impact craters, terrestrial; Interplanetary threat)
    impacts, moons/planets (*See* Impact craters, extraterrestrial)
    metallic asteroids, 223
    named, tabulated data for, App. B–1
    naming of, 220
    Near Earth Asteroid Rendezvous (NASA), 231
    nickel-iron cores in, 224
    orbits of, 206, 207
    as planetesimal survivors, 244
    positions/distribution of, 220–223
    properties of, 206, 207
    soot line, 223, 243
    as sources of raw materials, 231
    stony (rocky) asteroids, 223
    Trojan asteroids, 220, 231–232
Astrology, 18, 35–36, 50
Astrometry
    basic calculations for, 334–335
    and Doppler shift, 334
    See also Stellar motion
Astronomical unit (AU)
    defined, 73, 330
    in measuring stellar parallax, 330
Atmospheres
    of Earth, 170–171, 198, 550
    of giant planets, 178–179, 180–182, 186
    of Mars, 170–171
    planetary gas, retention of, 186
    retention vs. planet size (inner planets), 169–171
    of Titan, 196–197, 198, 558
    of Venus, 151, 170–171
Atomic theory. *See* Fusion, stellar; Matter, structure of; Nuclear energy; Radioactive decay; Radioactive heating
Aurora borealis, 86, 317, 318
Authority, appeal to, 17–18
Azimuth, position on the sky, App. A–6

B

Baade, Walter, 378
Babylon
    Babylonian calendar, accuracy of, 29, 31
    Babylonian systems of measurement, 34
Background radiation, cosmic
    Bell Labs detection of, 508
    and the big bang, 520
    COBE mapping of, 519–520
    COBE spectrum of, 518
    Gamow hypothesis about, 507–508
    and inflationary early universe, 538
Berzelius, Jacob, 87
Bessel, Friedrich, 374
Beta Pictoris, planetary dust around, 362, 363
Bethe, Hans, 364
Big bang

introduction to, 513
and age of the universe, 526, 527
cause of, 545
cosmic background radiation from, 520
cosmological constant, 510, 528
current status of, 528
density parameter (big bang model), 516–517, 525–526
evidence for, 545
Hoyle and big bang theory, 513
oscillating universe, theory of, 528
Planck era physics, 543
See also Cosmology; Universe, density of; Universe, properties of
Big Dipper, Little Dipper, 27
Binary systems
  Alpha Centauri, 400
  astrometric binaries, 396, 398
  binary X-ray sources (with black holes), 387
  contact binary systems, 401
  detection of, 425
  eclipsing binaries, 396, 397
  formation of, 402–403
  in globular clusters, 414, 415
  Herschel study of, 396
  incidence of, 398, 425
  John Mitchell theory about, 394, 396
  mass analysis of, 400
  mass transfer in, 399–401
  Roche surface, 399, 401
  Sirius, 400
  spectroscopic binaries, 396, 397
  supernova/white dwarf binary, 476
  triple binary in Castor system, 399
  visual binaries, 396
  white dwarf/neutron star binary, 401
Biot, Jean, 7
Bit, unit of information, App. A–7, 286–287
Black holes
  accretions discs near, 387
  deflection of light by, 384–386, 512
  detecting, 387, 389
  environment of, 384, 385
  escape velocities of, 384, 385, 386
  event horizon of, 384
  galactic distribution of, 500
  gravitational redshift by, 385–386
  Hawking radiation from, 386
  mass of, 384, 386, 388
  in nearby galaxies, 491–492
  properties of (Toolbox), 386
  and quasars, 500
  rubber sheet analogy for, 385
  Schwarzschild radius of, 384, 492
  space/time curvature from, 384–385
  supermassive, existence of, 504
  supermassive, in Milky Way, 439
  and X-ray sources, 387
Black body radiation, Stefan-Boltzmann law of, 106, 335, 336, 341, 344
Blue sky, explanation of, 408
Blueshift, Doppler
  blueshifts/redshifts in Milky Way, 432, 433
  of nearest galaxies, 469, 470
  and stellar motion, 334
Bode, J., 182
Bode's Law, 249
Bohr, Neils, 274–275
Bolometric luminosity, App. B–3, B–4, 335
Boltzmann, Ludwig
  Boltzmann constant, 96
  Stefan-Boltzmann law, 106, 335, 336, 341, 344

Brahe, Tycho
  biographical sketch of, 59, 61
  observations of comets, 60, 207, 213
  observations of planetary motions, 60
  portrait of, 61
Brightness, stellar
  absolute, 328, 335
  apparent, 326–327, 328, 393–394
  galactic redshift and apparent brightness, 522–523
  intrinsic, 326
  magnitude, App. A–6, 327
Brown dwarfs (substellar objects), 360–361
Buffon, Comte de (Georges-Louis Leclerc), 111–112, 113
Bulge stars. *See* Milky Way

C
Cairns-Smith, A., 557
Calendars and time
  Babylonian calendar, accuracy of, 29, 31
  calendars, historical background to, 28–29
  clocks, development of, 32
  days, naming of, 32
  early Roman calendar, 31
  equinoxes, rituals celebrating, 29
  Gregorian calendar, 31
  hours, division of day into, 32
  Julian calendar, 31
  lunar calendar, 31
  religious influences on, 31
  sidereal day, defined, 30
  solar day, defined, 30
  solstices, marking seasons by, 28, 29
  synodic month, 31
  synodic vs. sidereal period (of Moon), 30
  time, typical values of (table), App. A–3
  See also Ancient astronomy
Calories (in food), 97–98
Cannon, Annie Jump, 323, 331
Carbon dioxide, frozen
  in comets, 211
  on giant planets, 178
  on Mars, 159
Carbon dioxide, gaseous
  in Earth's atmosphere, increase of, 136
  in Earth's primordial atmosphere, 131
  and greenhouse effect (Earth), 135–137, 143
  and greenhouse effect (Venus), 151–152, 153, 172
  on Mars, 159, 160
  as terrestrial planetary atmosphere, 171
  on Venus, 151–152
Carbonaceous materials
  in asteroids, 223, 224
  carbonaceous chondrites, amino acids in, 557–558
  in comets, 211–212
  in meteorites, 244–245
  on moons of giant planets, 193
Casa Grande, 33
Cassen, Pat, 176
Cassini, G.D.
  Cassini's division (in Saturn's rings), 188
  determines Martian rotation period, 155
  discovers Great Red Spot, 180
Castor system, triple binary in, 399
Catastrophes
  solar system perturbation by, 250–251, 261
  See also Interplanetary threat; Mass extinctions
Cat's Eye Planetary Nebula, 371

CCDs (change coupled devices)
  in optical telescopes, App. A–7, 285, 287–288
  and recording data, App. A–7
  and stellar spectra, 331
Cepheid variables
  discovery of, 368–369
  and galaxy distances, 474, 476
  period-luminosity relationships of, 416–417, 443–444
  properties of, 368–370
CFCs in Earth's atmosphere, 134–135, 137
Chaco Canyon (Anasazi), 33–34, 377
Chandrasekhar, Subrahmanyan
  biographical sketch of, 374–375
  Chandrasekhar limit (for white dwarfs), 375
Chaos
  chaos and determinism (Toolbox), 252–253
  chaotic inflation model (of universe), 543
  chaotic orbits, 216, 252–253
  small bodies, chaotic motions of, 253
  See also Orbits
Charon
  composition of, 199
  image of, 200
Chichen Itza, 32–33
Chinese astronomy, 48
Chondrites
  carbonaceous, amino acids in, 557–558
  properties of (typical), 207
  See also Meteorites
Chronometer, navigational, 73
Circular (orbital) velocity, 74–75
Clarke, Arthur C., 194, 219
Classification
  introduction to, 337–338
  and H-R diagram, 339–340
  of main-sequence stars, 347
  of nearest stars, App. B–2
  See also Hertzsprung-Russell (H-R) diagram
Clouds, atmospheric
  Jupiter, cloudscape on (painting), 179
  Uranus, indistinct clouds on, 179, 200
  See also Dust clouds/dust discs, interstellar; Magellanic Cloud; Molecular clouds; Oort Cloud
Clouds, gas (galactic)
  orbits of in galactic plane, 442–443
Clusters, stellar
  associations, 412
  clustered vs. random star distribution, 394–395
  Coma Berenices, 481, 484
  discussion of, 412–415, 418
  Eagle Nebula, 352
  formation from interstellar medium, 357
  globular, distribution in Milky Way, 421
  globular cluster, imaginary view of, 415
  globular cluster 47 Tucanae, 412, 414
  main-sequence turn-off, 417, 418, 419–420
  Jewel Box cluster, 345
  M 92, main-sequence turn-off of, 419–420
  NGC 2264, 362, 364, 417
  NGC 3293 (open cluster), 418
  open, distribution in Milky Way, 421
  open clusters, 413, 418, 439
  Pleiades (as open cluster), 412, 413
  rich vs. poor, 481
  Trapezium, 362
  X rays from globular clusters, 413, 415
  young clusters and young stars, 361–362, 364
  See also Galaxies
COBE (Cosmic Background Explorer) satellite
  background radiation measurements, 519–520

623

COBE satellite *(continued)*
   background radiation spectrum, 518–519
   infrared maps of Milky Way, 431
Cocconi, Giuseppi, 579
Cometary globule CG4, 407
Comets
   asteroids as comet nuclei, 223
   breakup of, 211
   chaotic orbits of, 216
   Comet West, breakup of, 211
   cometary orbits and Kepler's Laws, 214–215
   complex molecules in, 548–558
   composition of, 211, 212–213
   defined, 205
   development of, 208, 209, 249
   dirty iceberg model of, 211
   effect of solar wind on, 208
   evolution of, 231–232
   Giotto space probe (Halley's Comet flyby), 212
   Hale-Bopp, 208, 211–212
   Halley's Comet, 208, 210, 212
   historical note about, 208–210, 228
   Hyukatake, 210
   impending calamity, and, 208, 228
   Kuiper belt, 213, 214–215, 231–232, 260
   life story of, 215–216
   meteor showers as cometary debris, 216
   microscopic dust particles from, 243–244
   naming of, 210
   nucleus, size of (typical), 207
   nucleus, sublimation of, 205, 207–208
   Oort Cloud, 213, 214–215, 231–232, 233, 261
   orbits of, 206, 207, 249
   origins of, 206
   parts of, 207
   as planetesimal survivors, 243–244
   presence of ice in, 232
   properties of (short period/long period), 207
   Shoemaker-Levy, 211, 229–230
   sublimation of, 205
   tails, length of, 207, 208
   visibility of, 208
Comparative planetology. *See* Planetary evolution
Computer simulation. *See* Simulation, computer
Condensation sequence (solar nebula), 242–243, 357–358
Conservation of energy
   in impact crater formation, 95, 97
Constants, physical/astronomical
   Boltzmann constant (k), 96
   conversion, English/metric, App. A–3
   cosmological constant (Λ), 510, 528
   Hubble constant (H_0), 476–479, 516, 517
   Planck's constant *(h)*, 274
   table of, App. A–2
Constellations
   as agricultural guides, 27
   Big Dipper, Little Dipper, 27
   Carina, open cluster NGC 3293 in, 418
   Coma Berenices cluster, 481
   ecliptic, plane of, 27
   historical background to, 26–27
   monthly star charts of (northern latitude), App. A–6
   movement of, 28, 30
   notable constellations (northern sky), 28
   Sagittarius, 420
   Virgo, 481, 486
   visibility of (vs. latitude), 27, 28
   zodiac, 27

See also Galaxies; Milky Way; Nebulae; Night sky, observing
Continental drift
   Wegener's theory of, 120
   See also Plate tectonics
Contingency theory, 564–565
Conversion, English/metric, App. A–3
Copernicus, Nicolaus
   biographical sketch of, 56–57
   Copernican model, parallax in, 58–59, 60
   Copernican model (of solar system), 57–58
Cosmic abundance (of elements), 354–355
Cosmic year, 562
Cosmology
   ancient (*See* Ancient astronomy)
   angular diameter and curvature of space, 521–522
   antimatter asymmetry in early universe, 540–541
   background radiation (*See* Background radiation, cosmic)
   and the big bang (*See* Big bang)
   closed (high density) universe, 516, 520–523
   cosmic background radiation (*See* Background radiation, cosmic)
   cosmic nucleosynthesis, 517–518
   cosmological constant, 510, 528
   cosmological model, observational tests of, 521–523
   cosmos, evolution of (schematic), 506
   critical density (flat) universe, 516, 520–523, 524
   defined, 508
   density of universe, time evolution of, 533, 534
   density parameter and dark matter, 525–526
   density parameter (big bang model), 516–517
   early universe, particles and radiation in, 533–536
   expansion rate vs. time, 516, 539
   fate of the universe, 517, 526–527, 545
   flat universe, evidence for, 536–539
   homogeneous vs. inhomogeneous universe, 514
   inflationary universe (*See* Universe, inflationary)
   isotropic vs. anisotropic universe, 514–515
   Olbers' paradox, 509–510
   open (low-density) universe, 516, 520–523
   oscillating universe model, 528
   particles and radiation in the early universe (Toolbox), 535–536
   Planck era (at big bang), 543
   space-time diagrams, 528–531
   summary discussion of, 544
   temperature of universe, time evolution of, 533, 534
   universe, scale of *(R)*, 515–516
   universe, structure of, 531–533
   universe expansion rate, measurement of, 504, 520
   See also Space/time, curvature of; Universe
Crab Nebula, 376, 377, 379, 380, 381
Cratering. *See* Impact craters, extraterrestrial; Impact craters, terrestrial
Curie, Marie and Pierre, 303
Curved space. *See* Space/time, curvature of
Cycles, observational
   over one-year period, 24, 25
   See also Night sky, observing

D

Da Vinci, Leonardo, 113

Dalton, John, 86–87
Dark matter
   cold, and bottom-up structure formation, 532–533
   and cosmic density parameter, 525–526
   detection through gravitational lensing, 487–489, 490
   detection through gravitational microlensing, 456, 458
   in galaxies, 455–459
   hot, and top-down structure formation, 532–533
   in Milky Way, 437–438, 443, 465
   nature of, 437–438, 465
   in universe, 486–489, 504
Data
   and measurement error, App. A–4
   ways of representing, App. A–7
Davis, Ray, 309
Declination, position on the sky, App. A–7
De Jong, H.G. Bungenberg, 559
Deimos
   atmosphere of, 171
   impact craters on, 169
   irregular shape of, 164–165
   origin of, 164
   photograph of, 165
   physical characteristics of, 164
   temperature of, 169
Democritus (Greece), 36
Density of universe. *See* Universe, density of
Density wave theory (spiral arm formation), 434, 435
Determinism
   chaos and determinism (Toolbox), 252–253
Diameters/radii of stars, App. B–2, B–3, 335, 345
Differentiation, mineral
   in Earth and Moon, 117, 142–143
Diffraction (of light), 270
Digital image processing
   in optical telescopes, App. A–7, 285–288
   true-color vs. false-color in, App. A–7, 288
Digital information
   digital image processing (optical telescopes), App. A–7, 285–288
   life as, 554, 555, 556
   Science Toolbox, 286–287
Dimensional analysis, App. A–2
Dirac, Paul, 528
DIRBE (Diffuse Infrared Background Experiment), 431
Disk stars. *See* Milky Way
Distances of stars. *See* Galaxies; Stellar distances; Universe, expanding
Distribution (normal/Gaussian), App. A–4
DNA (deoxyribonucleic acid), 554–557, 578, 581
Doppler effect
   defined, 257–258
   and planet detection, 257
   and stellar radial velocity, 334–335
   in two-body system, 255–256, 258
   See also Redshift, Doppler
Drake, Frank
   and Voyager message design, 575
Drake equation
   discussion of, 569, 574
   Science Toolbox, 570–571
Draper, Henry, 331
Duhalde, Oscar, 353–354
Dust, cometary, 243–244
Dust clouds/dust discs, interstellar
   Beta Pictoris, planetary dust around, 362, 363

dust extinction and reddening (Toolbox), 411–412
near young stars (cool dust), 361–362, 363
in stellar evolution, 358, 361–362, 363
See also Interstellar medium
Dwarf stars
brown dwarfs (substellar objects), 360–361
giants vs. dwarfs, luminosity of, 343–344
See also White dwarfs
Dyson, Freeman, 575

E

Eagle Nebula, 352
Early universe. *See* Cosmology; Universe, inflationary
Earth
See: Earth, age of
Earth, atmosphere of
Earth, physical properties of
Earth, structure of
Earth-crossing asteroids
Interplanetary threat
Mass extinctions
Plate tectonics
Volcanism, terrestrial
Earth, age of
Buffon's estimate of, 111–112, 113
estimates from fossil record, 125–126
geological time scale, 125–126, 127
historical estimates of, 112–113
from radioactive dating, 113–115, 126, 143
Earth, atmosphere of
carbon dioxide and global warming, 135–137, 143
CFCs in, 134–135, 137
changing composition of, 560–561
common elements in, 550
comparison with Mars, Venus, 170–171
comparison with Titan, 198
ozone layer, damage to, 134–135
primordial composition of, 131
public policy implications, 137–138
ultraviolet radiation, absorption by, 134–135
water vapor in, 131, 135
Earth, physical properties of
atmosphere (*See* Earth, atmosphere of)
axis, inclination of, 45–46
comparison with Mars, Venus, Mercury, 169, 170, 171
crust, composition of, 550
crustal silicates on, 118
erosion on, 117, 118, 119, 122–123
impact craters, 169
magnetic field (and solar wind), 318
oceans, formation of, 131
prograde rotation, 184
size of (Eratosthenes' deduction), 44
structure of (*See* Earth, structure of)
surface mosaic (illus.), 111
surface temperature, 135–137, 169
view from *Galileo*, 110
view from space, 2
Earth, structure of
aluminum and surface iron on, 118
core-mantle-crust structure, 117–118
feldspar and granite on, 117
igneous rocks on, 118
metamorphic rocks on, 118
mountain building, height limits of, 122
nickel-iron core in, 117, 118
radioactive heating in, 116, 119–120, 121, 303

sedimentary rocks on, 118, 134
Earth-crossing asteroids, 220, 224, 231
Earthquakes
effect of lunar tidal forces on, 139
origins of, 121
Eclipses
Anaxagoras deduces nature of, 37
lunar, 40–41, 43–44
solar, 40–41, 42
Ecliptic, plane of
defined, 27
and lunar eclipses, 40–41
Einstein, Albert
biographical sketch of, 380, 384
general theory of relativity, 382–384
proposes cosmological constant, 510, 528
special theory of relativity, 380–381
See also Relativity, theories of; Space/time, curvature of
Einstein rings (in gravitational lensing), 488, 489, 490, 496
Electromagnetic spectrum
described, 267, 268
See also Radiation, electromagnetic; Radiation, thermal; Stellar spectra
Electrons. *See* Matter, structure of
Enceladus, 196, 197
Energy
Boltzmann constant, 96
caloric (in food), 97–98
chemical (Space Shuttle example), 93
conductive heat transfer, 100–101
conservation of (impact crater formation), 95, 97
conservation of (in mechanical systems), 108
convective heat transfer, 101
defined, 91, 92
gravitational, 91
heat, frictional, 92–93
heat transfer, three modes of, 101, 102
heat transfer in stars/planets, 107
Joule's experiment (work into heat), 93, 95
Kelvin temperature scale, 93–94
kinetic energy, 92, 94
kinetic theory of matter, 94
potential energy, 91–92, 94
radiative heat transfer, 101
release, manmade vs. natural, 130–131
Stefan-Boltzmann law, 106, 335, 336, 341, 344
temperature, discussion of, 93–95
temperature scales compared, 95
thermal equilibrium, 99–100
transformation of, 95, 97
units of, App. A–3, 92
velocities of gas atoms/molecules, 96–97, 98
Entropy
and the arrow of time, 527
entropy and time (Toolbox), 374
as a measure of disorder, 372–373, 527
Eratosthenes
deduces size of Earth, 45
determines tilt of Earth's axis, 45
proves Earth is round, 49–50
Erosion
on Earth, 117, 118, 119, 122–123
on Mars, 159, 160, 161
Errors
observational (measurement) errors, App. A–2, 12–14, 20
standard error/standard deviation, App. A–4, 12, 13, 15–16
statistical reduction of, App. A–4, 15–16

systematic vs. random errors, 16
Escape velocity, 75
Estimation, App. A–1, 8, 9, 13
Euclidean vs. non-Euclidean geometry, 511–512
Event horizon (of black holes), 384
EXOSAT orbiting observatory
and white dwarf/neutron star binary, 401
Explosive nucleosynthesis, 376, 389
Exponents, mathematical, App. A–1
Extinction events. *See* Interplanetary threat; Mass extinctions
Extrasolar planets. *See* Planets, extrasolar

F

Faber-Jackson relation (elliptical galaxy distances), 474, 476
Fabricus, David, 368
Faraday, Michael, 266–267
Fireballs, astronomical, 217, 218
Fission, nuclear, 303
Fossils
defined, 119
on Earth, 125–126, 140, 144
possibility in Martian rocks, 163–164
Fraunhofer lines, 299
Freudenthal, Hans, 577
Frost line (in asteroid distribution), 223, 243
Fuller, Buckminster, 549
Fundamental constants, App. A–3
Fusion, cosmic
cosmic nucleosynthesis, 517–518, 525–526
Fusion, stellar
and catalysis (Bethe theory), 364
heavy elements, creation of, 354–355, 368, 369, 375–376, 389
helium flash, 367
hydrogen exhaustion, 367
iron as terminal fusion product, 368, 369
in main-sequence stars, 347, 349–350, 364–365
in protostars, 357
and stellar luminosity, 389
in the sun, 305–306
in supernovae, 476
See also Nuclear energy

G

Galaxies
active galaxy, defined, 491
Andromeda (M31), distance to, 443, 444
barred spirals NGC 4650, NGC 6744, 446
barred spiral M 83 (NGC 5236), 452, 454
black holes, distribution of, 500
black holes in, 491–492
classification scheme for ("tuning fork"), 447, 448
clusters of, 479–482, 483–484
cosmological redshifts and, 521
Cygnus A (3C 405), 493
dark matter in (*See* Dark matter)
density of (and curvature of space), 520–521, 522–523
distance vs. redshift (Toolbox), 475
distances of, measuring, 443–445, 465, 474, 476, 478–479
dwarf elliptical, 446, 449–450
elliptical, evolutionary model of, 523
elliptical, Faber-Jackson relation for, 474, 476
elliptical (spheroidal) galaxies, 447, 455, 456
evolution of, 459, 465
formation, theory of, 462–464
galactic bulge, 420

625

Galaxies (continued)
 galactic disc, 420
 galaxy, defined, 393
 galaxy groups, 479
 gas clouds in, orbits of, 432–433
 giant elliptical galaxy M 87 (NGC 4486), 452, 453
 halos, galactic, 420, 430, 486–487
 Hubble constant, 476–479, 516, 517
 Hubble determines the nature of, 430
 Hubble relation, 471, 472, 473–474, 510, 516, 517
 interaction of (collision/merger), 459–462
 irregular galaxies, 447, 455
 large number of in universe, 5
 Local Group (of galaxies), App. B–4, 447–452, 519
 Magellanic Clouds, 447–449
 many-body gravity, analysis of, 457–458
 maps of (northern sky), 484–486, 487
 mass of, measuring, 455–456
 mass-to-light ratio of, 452–455
 morphology-density relations of, 481–482
 most distant galaxies, 489–490
 NGC 2997, photograph of, 428
 nonthermal radiation from, 493–494
 peculiar galaxies, 447
 poor vs. rich clusters, 481
 protogalaxies, 463
 radio galaxies, 493
 radio jets (quasars), 500, 501, 502
 redshifts of (See Redshift, galactic)
 rotation curves of, 455, 456
 Sc galaxies NGC 253, 2997, 446
 Seyfert galaxies, 492
 Sombrero Galaxy (M 104), 446
 spiral, Tully-Fisher relation for, 474, 476
 spiral arms in (Milky Way), 433–436
 spiral galaxies, 445–447, 455, 456
 star-bursting irregular galaxy NGC 1313, 447
 superclusters of, 482, 484–486
 synchrotron radiation from, 493
 tidal interactions between, 459–460
 velocity dispersion in, 455, 456
 Whirlpool Galaxy (M 51), 446
 See also Clusters, stellar; Cosmology; Milky Way; Nebulae; Universe
Galileo Galilei
 attempts to measure speed of light, 294
 biographical sketch of, 64–65
 concept of inertia, 66
 first use of telescope, 64
 observation of Saturn, 187
 portrait of, 65
 representing data by drawing, App. A–7
 study of mechanics, 65–66
 study of phases of Venus, 65, 66
 trial by Catholic Church, 52–54, 65
Gamov, George, 507–508
Ganymede
 heating of, 194
 surface of, 193–194
Gas giants. See Giant planets
Gas jets (bipolar outflow)
 from T Tauri stars, 362, 363
Gas laws, 183–184
Gauss, Karl Friedrich, App. A–4
Geological time scale
 Earth, 125–126, 127
 Earth-Moon system, 125, 128
Geometry
 Euclidean vs. non-Euclidean, 511–512
Geothermal energy, 319

Giant planets
 appearance of, 177–178
 atmospheres, general properties of, 178–179
 atmospheric gas, retention of, 186
 atmospheric temperature and pressure, 178–179, 180–182
 average density of, 247
 diameters of, 246
 formation of, 186–187, 249
 internal heat, sources of, 185
 internal structures of, 180, 181–182, 201
 mass of, 178, 247
 mean density of, 178, 201
 possibility of life on, 185–186
 rapid rotation of, 178
 rocky cores in, 181, 182
 satellites of, introduction to, 189–193
 satellites of, tabulated data for, App. B–1
 summary discussion of, 200
 temperature of, 180, 185
 thermal radiation from, 185
 water ice on, 178, 181, 201
 See also specific planets & satellites
Giant stars
 Antares (supergiant), 342
 giants, luminosity of, App. B–3
 giants vs. dwarfs, luminosity of, 343–344
 O-type, 333, 362, 412, 417, 419
 red giants, App. B–3, 339, 341, 344–345, 346, 347, 366, 367
 spectra of, 345
Giotto space probe, 212
Global warming, 135–137, 143
Gold, explosive nucleosynthesis of, 376
Goodricke, John, 368–369
Gould, Stephen Jay
 and the Burgess Shale, 562–563
 and contingency in life, 564–565
 on inevitability of life, 560
Graphs, to representing data, App. A–7
Gravitational collapse
 Helmholtz contraction, 241–242
 process of, 239, 241
 See also Solar system
Gravitational influence, sphere of, 191, 193
Gravitational lensing
 analysis of (Toolbox), 498
 Einstein rings in, 488, 489, 490, 496
 and large-scale dark matter, 487–489, 490, 501
 microlensing, detecting dark matter through, 456, 458, 498
 microlensing and stellar brightening, 498
 of quasars, 494–495
 starlight, gravitation deflection of, 383, 511–512
Gravitational perturbation (of orbits), 220, 221
Gravitational redshift. See Redshift, gravitational
Gravity
 in Earth-Moon system, 68, 70
 inverse square law of (Toolbox), 69–70
 many-body, analysis of, 457–458
 mass vs. weight, 68, 71
 Newton's discovery of, 68
 and Sir Isaac Newton, 68–70, 81
 space/time distortion by, 382–383, 384–385
 universal applicability of, 81
Great Pyramid, 32
Great Red Spot (Jupiter), 177, 180, 252
Greece, classical
 Greek thinkers/philosophers, 36–37
 ideas about structure of matter, 86

 Ptolemaic model of solar system, 54–55
 Ptolemy, Claudius, 47
Greenhouse effect
 and global warming, 135–137, 143
 on Venus, 151–152, 153, 172
Guth, Alan, 538

H

H II regions, 409, 410
 See also Nebulae
Hale-Bopp comet, 208, 212
Half-lives, radioactive
 calculations of, 115
 and radioactive dating, 113–115, 126, 143
 See also Radioactive decay
Halley, Edmond
 biographical sketch, 209–210
 Halley's Comet, 207, 208, 210, 212
Halo stars. See Milky Way
Harmonics and resonance (in ring systems), 189, 190, 191
Harrison, John, 73
Hawking, Stephen, 386
Hayashi, Chushiro, 358–359
Heat. See Energy; Entropy
Heavy elements, creation of
 by explosive nucleosynthesis (r-process), 376, 389
 in stellar interiors (s-process), 368, 369, 375–376
Hegel, G.W.F., 219
Heisenberg uncertainty principle, 276, 278
 and virtual pairs in early universe, 536
Heliocentric solar system
 Copernican theory of, 57–58
 evidence for, 81
 observational parallax in, 59, 60
Helium
 atomic structure of, 89, 90
 helium flash, 367
 See also Fusion, stellar
Helix Nebula (NGC 7293), 370, 371
Helmholtz, Hermann von
 estimates energy output of sun, 305
 Helmholtz contraction, 241–242
Herschel, William, 396
 biographical sketch, 392–393
 and Caroline Herschel, 393
 discovery of infrared radiation, 102, 108, 265
 discovery of Uranus, 182, 392
 measures Milky Way, 425–426
Hertz, Heinrich, 293
Hertzsprung, Ejnar. See (H-R) diagram
Hertzsprung-Russell (H-R) diagram
 introduction to, 338–340
 constant stellar radius, lines of, 341, 344–346
 giants vs. dwarfs, luminosity of, 343–344
 interpreting an H-R diagram, 339–340
 Jewel Box star cluster on, 341, 342
 main-sequence fitting (of clusters), 416
 main-sequence stars on, 339–340
 main-sequence turn-off point (clusters), 417, 418, 419–420
 NGC 2264 star cluster on, 362, 364
 non-main-sequence stars, App. B–3, 339, 340–341
 for one hundred nearest stars, 340
 period-luminosity relationships (Cepheid variables), 416–417
 pre-main-sequence tracks vs. stellar mass, 359, 360
 red giants/supergiants on, 339, 341, 344–345, 346, 347, 367

star counts, errors in, 340, 341
stellar masses on, 346
white dwarfs on, 339, 341, 344–345
See also Stellar evolution
Hipparcos satellite, 329
Hoyle, Fred, 513, 528
Hubble, Edwin
   biographical sketch of, 430
   determines nature of galaxies, 430, 445, 450
   Hubble constant, 476–479, 516, 517
   Hubble relation (radial velocity vs. distance), 471, 472, 473–474, 510, 516, 517
   "tuning fork" classification scheme for galaxies, 447, 448
Hubble Space Telescope (HST)
   Charon, image of, 199, 200
   configuration of, 289
   galaxies as seen by, 5
   Mars, digital image of, 158
   Pluto, image of, 199, 200
   resolution of (vs. Schwarzchild radius), 492
   UV image of globular cluster 47 Tucanae, 414
   See also Telescopes, optical; Telescopes, radio; Visible spectrum
Huggins, William, 211
Huxley, Thomas, 297
Huygens, Christian
   identifies Saturn's rings, 187
   maps Mars' markings, 155
Hydrogen, gaseous
   atomic structure of, 89, 90
   energy states of, 331–332
   hydrogen clouds in galactic plane, orbits of, 432–433
   hydrogen-alpha line (in stellar spectra), 331–332
   image of sun in hydrogen light, 302
Hydrogen, liquid metallic, 181
Hydrogen, liquid molecular, 181, 182
Hydrostatic equilibrium (stellar), 337
Hypothesis-testing, App. A–4, 311–312
Hyukatake comet, 210

I

Ice, hydrocarbon
   in comet nuclei, 211
   on giant planets, 178, 181–182
   on Pluto and Charon, 199
   on Titan, 197
   on Triton, 199
Ice, water
   in comet nuclei, 205, 207, 211, 223, 232
   on Enceladus, 196, 197
   on Europa, 194–195
   on Ganymede, 193–194
   on giant planets, 178, 181–182, 201
   on Mars, 155, 159–160
   on Mercury, 150–151
   on the Moon, 141
   in outer moons and planets, 243
   in Saturn's rings, 188
Igneous rocks (on Earth), 118
Impact craters, formation of
   and conservation of energy, 95, 97
   lunar craters, typical, 123, 125
   in Mesozoic extinction event, 129
   on old vs. young surfaces, 124–125, 166
   shape vs. size, 123–124
   water-drop analogy for, 126
Impact craters, extraterrestrial
   counting (on the Moon), 167–168
   cratering throughout solar system, 248
   on Mars, 159, 160, 162, 169

Mercury, 149, 150–151
   on Miranda, 197, 198
   on Moon, 123–125, 167–168
   and Phobos, 165
   on planetary satellites, 193
   Shoemaker-Levy strikes Jupiter, 211, 229–230
   throughout solar system (typical), 248
   Triton impact craters, absence of, 198
   on Venus, 154–155, 156
Impact craters, terrestrial
   impact theory of lunar origin, 141–142, 144
   impermanence of, 123
   Meteor Crater (Arizona), 131
   See also Interplanetary threat; Mass extinctions
Inclination (of Uranus), 184
Indian astronomy, 47–48
Infrared radiation. *See* Radiation, infrared
Intelligence, extraterrestrial
   alien encounters, rarity of, 572
   alien life in solar system, 569
   alien messages, techniques for recognizing, 577–578
   anthropic principle, 580–581
   Arecibo radio dish (and MOP project), 579
   binary-coded messages for, 577–578
   civilizations, lifetime vs. distance, 572
   close encounters of any kind, 571–573
   communication, interstellar, 574
   communication with, 576–577, 582
   distribution of, 573
   Drake equation (discussion), 569, 574, 582
   Drake equation (Toolbox), 570–571
   extrasolar life, introduction to, 569, 571
   extrasolar planets, suitability for life, 568
   lingua cosmica (LINCOS), 577
   mediocrity, principle of, 568
   Pioneer 10 "message in a bottle," 575–576
   Pioneer 10 signal, detection of, 579
   SETI program, 577–579
   Seven of Nine, 709
   Voyager phonograph record, 575–576
   See also Intelligence, terrestrial
Intelligence, terrestrial
   evolution of, 563
   inevitability of, 565–566
   information storage/transmission in, 566
   mediocrity, principle of, 568
Interference (of light), 270–271
Interferometry
   Laser Interferometer Gravitational Observatory (LIGO), 292
   in search for extrasolar planets, 553
   Very Large Array (VLA) radio telescope, 288, 289, 291
Interplanetary bodies
   cometary orbits and Kepler's Laws, 214–215
   mass of, 206
   orbits of, typical, 205–206, 207
   properties of (table), 207
Interplanetary debris
   atmospheric heating of, 205
   collision velocities of (with Earth), 205
   See also Interplanetary bodies; Meteorites; Meteors
Interplanetary threat
   introduction to, 226–228
   asteroid impact defense system (Teller), 229
   asteroid/comet impacts: frequency vs. size, 226–227
   Earth-crossing asteroids, 220, 224, 231
   impact probability, calculation, of, 230, 233
   response strategies to, 228-229

Tunguska Event (1908), 203–205, 225, 227
   See also Impact craters, terrestrial; Mass extinctions
Interstellar extinction (of starlight), 408, 411–412
Interstellar medium
   Beta Pictoris, planetary dust around, 362, 363
   clouds/discs in stellar evolution, 358, 361–362, 363
   complex molecules in, 557
   composition of, 404–405, 425
   cool dust nebulae near young stars, 361–362, 363
   evolution of, 409–410
   gas/dust densities in, 404
   H II regions in, 409, 410
   as inhomogeneous medium, 410
   interaction with starlight, 405–409
   interstellar extinction (of starlight), 408, 411–412
   interstellar grains, 405
   interstellar reddening (of starlight), 406, 408, 409, 411–412
   light scattering (blue appearance), 408–409
   spectral lines in, 404
   superbubbles (supernova shells), 410
   See also Opacity
Io
   iron core in, 196
   sulfur compounds on, 195–196
   tidal heating of, 175, 195–196
   volcanoes on, 175–177, 195–196, 201
Iron
   asteroids, nickel-iron cores in, 224
   creation of, 368, 369
   Earth, nickel-iron core in, 117, 118
   Earth's Moon, iron core in, 117, 118
   Io, iron core in, 196
   iron/stony meteorites, 225–226
   as terminal stellar fusion product, 368, 369, 375
Isotopes, 114
Isotropy
   isotropic vs. anisotropic universe, 514–515
   isotropy vs. anisotropy (Toolbox), 422–423
   in Milky Way, 420
   of universe (from COBE measurements), 519–520

J

Jansky, Karl, 288, 438
Janssen, Pierre, 301–302
Jeans, James, 358
Jewel Box cluster (NGC 4755), 341, 342
Joule, James P.
   experiment (work into heat), 93, 95
   joule, unit of energy defined, 93
Jupiter
   atmospheric temperature/pressure of, 180–181
   belts and zones on, 180
   cloudscape on (painting), 179
   Great Red Spot on, 177, 180, 252
   high winds on, 179
   impact of Shoemaker-Levy on, 211, 229–230
   influence on cometary orbits, 249
   internal structure of, 181–182, 201
   lightning on, 180
   liquid metallic hydrogen on, 181, 182
   moons, tidal heating of, 176
   relative diameter of, 178
   ring system of, 187
   rocky core in, 181, 182

Jupiter (continued)
   satellite orbits of, 193
   satellites of, tabulated data for, App. B–1
   size of (as planetary limit), 259–260

K

Kant, Immanuel, 443
Kelvin, Lord (William Thomson)
   debates age of Sun, 297
   temperature scale, 93–94
Kepler, Johannes
   discovery of planetary orbits, 61, 63
   portrait of, 61
Kepler's Laws
   and binary systems, 400
   and cometary orbits, 214–215
   of planetary motions, 62–64
   and spiral arm formation, 434, 435
   weighing a galaxy, 437
Kirchhoff's laws of radiation, 299, 301
Kirkwood, Daniel (Kirkwood gaps), 221, 222
Knapp, Michelle, 225, 228
Knowledge, creation of, 19
Kuiper, Gerard
   biographical sketch, 213
   Kuiper belt, 213, 214–215, 231–232, 260

L

Lagrangian points (in asteroid orbits), 220
Lambert, Johann, 443
Large Magellenic Cloud
   and supernova SN 1987A, 377–378
Large-scale structures (in universe), 479–486
Lasar Interferometer Gravitational Observatory (LIGO), 292
Leclerc, Georges-Louis (Comte de Buffon)
   estimates age of the Earth, 111–112, 113
Lemaître, Georges
   cosmic singularity hypothesis, 513
Length, physical
   of comet tails, 207, 208
   typical values of (table), App. A–3
   visible spectrum, wavelengths of, 102, 103, 267, 268
   wavelength vs. amplitude, frequency, 267, 268
   X-ray wavelengths, solar image at, 313, 314
Length of day
   on Mercury, 150
   solar day, 30
Levy, David
   discovers Comet Shoemaker-Levy, 229
Life, extraterrestrial
   amino acids in carbonaceous chondrites, 557–558
   anthropocentrism, danger of, 565
   biochemical basis for, 581–582
   extrasolar planets and conditions for life, 568
   on giant planets, 185–186
   habitable zones, circumstellar, 553
   interstellar medium, complex molecules in, 557
   on Mars, 3–4, 162–164, 173
   possible sites for, 582
   requirements for, 581
   SETI program, 577–579
   sites for, 552–553
   UV radiation and destruction of organic molecules, 558
   in solar system, 569
Life, terrestrial
   amino acids, synthesis of, 556
   Burgess Shale, 563
   Cambrian fauna/trilobites, 563
   carbon, importance of, 551–552
   cells, basics of, 559
   common elements in, 550
   complex molecules, aggregating, 558–559
   contingency, importance of, 564–565
   culture and technology, 566–567
   as digital information (discussion/Toolbox), 554, 555, 566
   DNA (deoxyribonucleic acid), 554–556, 578, 581
   earliest terrestrial life forms, 559–561
   effect of on Earth's atmosphere, 560–561
   environmental influences on, 566
   eukaryotes, 561, 562
   extinction events, importance of, 563–565
   genes, 555
   history of (the Cosmic year), 562
   inevitability of (Gould on), 560
   intelligence, evolution of, 563
   intelligence, inevitability of, 565–566
   intelligence as information, 566
   lichens, 561
   membranes, importance of, 559
   multicellular organisms, proliferation of, 562–563
   natural selection, 563–565
   nature of (discussion), 550–552
   organic chemistry, defined, 550
   pheromones, communication by, 565–566
   photosynthesis, 560
   prokaryotes, 559, 560
   proteinoids, 559
   proteins, 554
   requirements for, 581
   RNA (ribonucleic acid), 554–557
   water, importance of, 551–552
   See also Intelligence, extraterrestrial
Light, visible
   diffraction of, 270
   gravitation deflection of, 383, 511–512
   interference of, 270–271
   light-year, defined, 451
   "old" light and stellar distances, 451
   photoelectric effect, 271
   photons, 271, 274, 278
   propagation of (inverse square law), 272, 327, 328
   speed of (c), 267–269, 294
   supernova/white dwarf binary, light curve for, 476
   thermal spectra (stellar), 272–273
   "tired" light and stellar distances, 528
   travel time of, 451
   visible spectrum, 102, 103, 267, 268
   as wave phenomenon, 269–271
   wave/particle duality of, 271, 276, 293
   See also Luminosity; Solar spectrum; Starlight; Visible spectrum
Light-year
   defined, 325
   light travel time (Toolbox), 453
LINCOS (lingua cosmica), 577
Linde, Andrei
   chaotic inflation model (of universe), 543
Lithosphere
   composition of, 118
   crustal plates in, 120–121
   defined, 118
   on Earth and Moon, 118
   over plastic mantle, 118–119
   on Venus, 155
   See also Plate tectonics
Local Group (of galaxies)
   and cosmic background radiation, 519, 536, 537
   data on, App. B–4
   optical resolution of, 447–452
Lockyer, Norman
   discovery of helium in sun, 301-302
   and Stonehenge analysis, 23
Logarithms, App. A–1
Logic
   deduction, 10
   induction, 10–12
   introduction to, 8, 10
   See also Scientific method
Lookback time, 451, 523
Lowell, Percival
   canals on Mars, 156–157, 548–549
Luminosity, stellar
   absolute vs. relative, 328
   bolometric, App. B–3, B–4, 335
   of brightest stars, App. B–3
   of galaxies in large regions of space, 525
   of giant stars, App. B–3
   of giant vs. dwarf stars, 343–344
   of Local Group galaxies, App. B–4
   low-luminosity stars, predominance of, 340, 341
   luminosity class (of a star), 345–346
   of main-sequence stars, App. B–3, 347
   mass-to-light ratio (of galaxies), 452–455
   period-luminosity relationships (Cepheid variables), 416–417
   of quasars, 495, 496, 500
   of brightest stars, App. B–3, 341
   and Stefan-Boltzmann law, 335, 336
   Tully-Fisher relation (for galactic distances), 474, 476
   See also Hertzsprung-Russell (H-R) diagram
Lunar observatory (proposed), 292
Lyell, Charles
   dynamic Earth theory, 119

M

Magellan space probe
   Venus mapping by, 154
Magellanic Clouds
   as irregular galaxies, 447
   Large/Small Magellanic Clouds, 448–449
   Tarantula Nebula in, 449
Magnitude, stellar, App. A–6, 327
Main-sequence stars
   evolution of, 339–340, 362, 364
   fusion in (basics of), 347, 349–350, 364–365
   and galactic mass-to-light ratio, 463
   high-mass stars, nuclear reactions in, 364–365
   on H-R diagram, 339–340
   H-R main-sequence fitting (of clusters), 416
   H-R turn-off point (for clusters), 417, 418, 419–420
   lifetimes of, 365, 366
   low-mass stars, nuclear reactions in, 364
   luminosity/temperature of, App. B–3, 347
   mass of, 347
   period-luminosity relationship (on H-R diagram), 416–417
   spectral types of, 347
   See also Hertzsprung-Russell (H-R) diagram; Stellar evolution
Mantle, Earth/Moon, 117–118
Mariner 10 space probe, 149

Mars
    ancient water/dry riverbeds on, 159–160, 172–173
    atmosphere of (vs. Earth, Venus), 170–171
    canals on (Percival Lowell), 56–157, 548–549
    climate change on, 161–162
    CO_2 ice on, 159, 160
    comparison with Earth, Venus, 169–171
    digital image of (Hubble), 158
    early drawings of, 157
    future exploration of, 164
    historical studies and theories of, 155–158
    human face on (*Viking* orbiter), 158, 159
    impact craters on, 159, 160, 162, 169
    landing on (*Pathfinder*), 160, 161
    landing on (*Viking I*), 147
    life on, 3–4, 162–164, 173
    Martian fiction: *The War of the Worlds*, 157
    meteorites from, 160–161, 162, 164, 173
    moons of, 164–165
    orbit of, 58
    polar ice caps on, 155, 159–160
    retrograde orbital movement of, 58
    surface features of, 146, 158–161
    surface of (*Viking I* lander), 146
    temperature of, 158, 169, 172
    and UFO sightings, 157–158
    ultraviolet radiation on, 558
    volcanoes on, 159, 169
Mass, basics of
    inertial vs. gravitational mass, 382
    mass vs. weight, 68, 71
    mass-energy conversion, 303–304
    mass-energy equivalence, 381
    typical values of (table), App. A–3
Mass, planetary
    of giant planets, 178
    of interplanetary bodies, 206
    limits of, 259–260
    two-bodied systems, center of mass of, 255
Mass, stellar
    introduction to, 335–337
    binary systems, mass transfer in, 399–401
    of black holes, 384, 386
    of brown dwarfs, 360–361
    frequency of occurrence, 345, 346, 347
    and the H-R diagram, 346
    indirect measurement of, 349
    of interplanetary bodies, 206
    of local group galaxies, App. B–4
    of main-sequence stars, 347
    mass limits, large and small, 359–360
    mass loss (in evolved stars), 370, 371
    mass-to-light ratio (of galaxies), 452–455
    Milky Way, mass of, 436–437
    of nearest stars, App. B–2
    of neutron stars/pulsars, 378, 380
    and pre-main-sequence H-R tracks, 359, 360
    of sun vs. other planets, 241
    vs. nuclear reactions (main-sequence stars), 364–365
    of white dwarfs, 375
    See also Hertzsprung-Russell (H-R) diagram
Mass extinctions
    asteroid impacts: frequency vs. energy release, 226–227
    effect on evolution and development of life, 564
    energy release, manmade vs. natural, 130–131
    on geological time scale, 126–127
    Mesozoic extinction (of dinosaurs), 127-130, 143, 144
    Paleozoic extinction (the Great Dying), 130, 143, 144
    periodicity of, 227–228, 564
    random processes, mathematics of, 132–134
    random processes vs. Darwinian evolution, 131
    recurrence of, 130–131
    shocked quartz grains in, 130
    See also Interplanetary threat
Mathematics, basics of
    angular measurement, App. A–5, 34
    average (mean), 12, 13, 15, 16
    decimal system, App. A–1, 34–35
    dimensional analysis, App. A–2
    error, statistical reduction of, App. A–4, 15–16
    exponents/scientific notation, App. A–1, 7–8
    histograms, App. A–7
    logarithms, App. A–1
    mathematics, universal applicability of, 12
    metric system, App. A–3, 8, 35
    ratios/proportionality, App. A–2
    sexigesimal system, 34
    small angles, equation for, App. A–5, 38
    standard deviation (uncertainty), 12, 15
    See also Digital information
Matter, structure of
    alpha particles, 87–88
    atom, molecule, defined, 86
    atoms, basic structure of, 89–90, 273
    atoms, stability of, 88–89
    chemical reactions, introduction to, 90–91
    Dalton atomic theory, 86–87, 107–108
    dimensions of atomic structures, 89
    early Greek ideas about, 86
    electron energy levels, 275–276
    element, compound defined, 86
    emptiness of matter, 89
    gases, structure of, 98
    ionized gas, 98, 99
    kinetic theory of matter, 94
    liquids, structure of, 98
    microscopic structure, evidence for, 91, 92
    molecules, basic structure of, 90
    periodic table of elements, 87, 88, 91
    Planck's quantum theory, 273–276
    proton, neutron, defined, 89
    Rutherford atomic theory, 85–86, 87–88, 89
    solids, structure of, 98
    spectral lines, introduction to, 273, 275–276
    Thomson discovers electron, 87
    transitions (solid to liquid to gas), 99
    See also Energy; Radioactive decay
Maxwell, James Clerk
    kinetic properties of gases, 96
    Maxwell's equations (electromagnetic propagation), 267
Mayan astronomy, 48–49
Measurement systems
    altitude, azimuth, 34, 35
    angles and distances, estimating, 37
    angular measurement, App. A–5, 34
    Babylonian systems of, 34
    decimal system, App. A–1, 34–35
    English system, App. A–3
    field of view (optical), 34, 35
    good measurements, qualities of, 19
    metric system, App. A–3, 8, 35
    sexigesimal system, 34
    small angles, equation for, App. A–5, 38
    See also Mathematics, basics of; Scientific method
Mendeleyev, Dmitri, 87, 88, 91
Mercury
    atmosphere of, 171
    axial tilt of, 151
    day, length of, 150
    impact craters on, 149, 150–151, 169
    perihelion, shift in, 150
    polar caps on, 150–151
    radar image of, 150
    rotation of, 149–150
    size, physical appearance of, 149
    temperature of, 169
    view from *Mariner 10*, 149
Meridian, defined, 25
Messier catalog of nebulae, 403
Metamorphic rocks, 118
Meteor showers
    apparent direction of, 217
    as cometary debris, 216, 232
    Draconid shower, 216
    Leonid shower, 216, 218
    Perseid shower, 216, 218
    prominent showers, table of, 216
    and zodiacal light, 217, 218
    See also Meteorites; Meteors
Meteorites
    Allende, inclusions in, 244, 245
    as asteroid fragments, 233
    brecciated, 225
    carbonaceous chondrites, amino acids in, 557–558
    carbonaceous meteorite, 244–245
    chondrites, properties of (typical), 207
    defined, 205
    discovery of, 224
    formation of, 225–226, 233, 249
    human encounters with, 225, 228
    iron/stony iron types, 225–226
    Jean Biot identification of, 7
    large meteorites, rarity of, 225
    from Mars, 160–161, 162, 164, 173
    meteorite falls, occurrence by type, 225
    meteorite impacts (*See* Impact craters, terrestrial; Interplanetary threat)
    Murchison meteorite, 2558
    origin of, by asteroid collision, 225
    parent body of, 205
    radioactive iodine, puzzling presence of, 245
    stony types, 225–226
    Winona (stony) meteorite, 224
    See also Meteors
Meteors
    defined, 205
    fireballs, 217, 218
    frequency of, 216
    head-on vs. overtaking meteors, 217
    as interplanetary debris, 205
    See also Meteor showers; Meteorites
Metric system, App. A–3, 8, 35
Microlensing, gravitatioal. *See* Gravitational lensing
Milky Way
    introduction to, 420
    bulge stars, population/properties of, 441, 442
    clusters, distribution in, 421
    coordinate system for, 420–421
    cross-section through, 421
    dark matter in, 437–438, 442
    density wave theory (of spiral arm formation), 434, 435

629

Milky Way *(continued)*
   disc stars, population/properties of, 440, 442
   distance of Sun from center, 424
   early studies of (Shapley, et al.), 420
   formation of, 442–443
   galactic bulge in, 430, 431
   galactic center of, 438–439
   galactic plane, view from the north, 432–433
   Grote Reber radio map of, 288
   halo of, 430, 432, 437
   halo stars, panorama of, 439
   halo stars, population/properties of, 440–442
   Herschel measurement of, 425–426
   infrared mapping of, 430–432
   isotropy/anisotropy in, 420, 422–424
   known and estimated features of, 433
   mass of, estimating, 436–437, 464
   name, origin of, 393
   orbit of Sun around, 432
   orbital speeds of stars in, 432–433
   photograph of (region of Sagittarius), 420
   rotation curve for, 436
   shape of (galactic disk, halo, bulge), 420
   size of, 421, 425–426
   spiral arms in, 433–436
   starlight attenuation in, 430, 431, 438
   stellar bar in, 439
   stochastic star-formation theory (of spiral arms), 434–436
   summary discussion of, 464
   supermassive black hole in, 439
   view from above galactic plane, 433–434
   See also Constellations; Dark matter; Galaxies; Nebulae; Night sky, observing
Miller, Stanley, 555
Miranda
   cratered/fractured surface of, 197, 198
   tidal heating on, 197
Mitchell, John
   hypothesizes black holes, 384
   theorizes binary systems, 394, 396
Molecular clouds
   cometary globule CG4, 407
   maser emission from, 356
   Sagittarius, ethyl alcohol in, 357
   and stellar evolution, 355–357, 358
Moon (Earth's)
   age of, 116
   eclipses, lunar, 40–41, 43–44
   feldspar on, 117
   as geologically static, 119–120
   ice on, 141
   impact craters on, 123–125, 167–168
   internal structure of, 117–118
   iron-rich core in, 117, 118
   lack of granite on, 117
   mineral differentiation in, 117, 142–143
   minimal radioactive heating in, 119–120
   origin of (impact hypothesis), 141–142, 144
   phases of, 37, 39–40, 44, 138
   radioactive dating of, 126, 143
   rock samples from (*Apollo 17*), 116
   seismic activity on, 117, 118
   synodic month, 31
   synodic vs. sidereal period of, 30
   temperature of, 169
   tidal effects of, 138–140
Morphology-density relations
   for galaxy formation, 481–482
Morrison, Phillip, 579

Multiple star systems
   formation of, 402–403
   incidence of, 398–399
   See also Binary systems

N

Navigation
   importance of accurate clock, 73
   Polaris, determining latitude using, 25, 26
   by the stars, 25, 26
   by the Sun, 24–25
Nebulae
   catalogs of (Messier, et al.), 403–404
   Cat's Eye Planetary Nebula, 371
   condensation of, 242–243, 357–358
   defined, 242, 403, 424
   Eagle Nebula, star formation in, 352
   Helix Nebula (NGC 7293), 370, 371
   near young stars (cool dust), 361–362, 363
   Omega Nebula (M17), star formation in, 359
   Orion Nebula, in Milky Way, 433
   Orion Nebula, star formation in, 361–362, 403
   planetary nebulae (from stellar mass loss), 370, 371
   scattering/absorption of light by, 424
   solar, appearance of (early), 235, 241
   Tarantula Nebula (in Large Magellanic Cloud), 449
   Trifid Nebula, 407
   See also Galaxies; Milky Way; Stellar evolution
Neptune
   discovery of, 182, 184
   ice on, 181
   internal structure of, 181–182
   liquid ammonia/methane on, 181–182
   ring system of, 188
   satellites of, tabulated data for, App. B–1
Neutrinos
   neutrino astronomy, 292
   solar, detection of, 309, 312
   in solar fusion reaction, 305
Neutron stars/pulsars
   Baade/Zwicky speculation about, 378
   as contact binary systems, 401–402
   in Crab Nebula, 376, 379, 380, 381
   density of, 378
   emission model for, 380
   gamma-ray bursters, 402
   magnetic fields of, 379
   mass requirements for, 378, 380, 388
   millisecond pulsars, 402
   neutron stars, rotation of, 379
   optical pulses from (Crab Nebula), 381
   pulsars, radio emissions from, 379, 380
   pulsars, rotation of, 379–380
   white dwarf/neutron star binary, 401
   X-ray bursters, 402
Newton, Sir Isaac
   and conceptual problems of gravity, 510
   early seminal discoveries of, 66
   law of gravity, 68–70, 81
   Newton's Laws, 66–68
   observes visible spectrum, 102, 103
   personality of, 68
   portrait of, 67
   satellite launching concept, 74
NGC 2264 (star cluster), 362, 364
Night sky, observing
   the celestial sphere, App. A–6
   constellations, App. A–6, 26–28
   declination/right ascension, App. A–6
   the meridian, 25

   navigation by, 24–25, 26
   over one-year period, 24, 25
   Polaris, determining latitude by, 25, 26
   seasons, tracking, 26, 27
   star charts, monthly (northern latitude), App. A–6
   view from Mt. Graham, 391
   the zenith, 25
   See also Milky Way
North star. *See* Polaris
Novae
   difference from supernovae, 401
   intermittent/cyclical flaring of, 401
   and mass transfer in binary pairs, 401
   origin of name, 401
   See also Supernovae
Nuclear energy
   alpha decay, defined, 302
   antimatter/antiparticles, 535–536
   beta decay, defined, 302
   binding energy, atomic, 303
   chain reactions (controlled, natural), 304
   Curie, Marie, 303
   electroweak theory, 541
   fission, basics of, 303
   fusion, energy from (basics), 304–305
   fusion, stellar, 305–306, 347, 349–350, 364–365
   gamma radiation, defined, 302–303
   mass-energy conversion, 303–304
   pair creation/annihilation, 535–536
   quantum fluctuations, 542
   radioactive decay, Rutherford demonstration of, 302
   strong/weak nuclear forces, 541, 542, 580
   unification theory, search for, 541–542
   See also Fusion, stellar
Nucleosynthesis
   and abundance of elements, 354–355
   cosmic, 517–518, 525–526
   explosive, 376, 389
   heavy elements, creation of, 368, 369

O

Observatories, ancient. *See* Ancient astronomy
Olbers' paradox, 509–510
Omega Nebula (M17), star formation in, 359
Oort, Jan
   biographical sketch, 213
   studies of Milky Way, 420
Oort Cloud
   as comet reservoir, 213, 214–215, 231–232, 233, 261
   elongated orbits in, 214–215
   existence of, 233
Opacity and collisions (Toolbox), 312
Orbits, asteroid
   gravitational perturbation of, 220, 221
   Lagrangian points in, 220
   main asteroid belt, 205–206, 297
   properties of (table), 207
Orbits, chaotic, 216, 252–253
Orbits, cometary
   Halley discovery of, 209–210
   Jupiter influence on, 249
   and Kepler's Laws (Toolbox), 214–215
   of short/long period comets, 207
Orbits, meteor
   meteor showers, fireballs, meteorites, 207
Orbits, periodicity of, 75
Orbits, planetary
   elliptical, Kepler's laws of, 61–64
   epicyclic/retrograde (Ptolemaic model), 54–55

of extrasolar planets, 259
of Jovian satellites, 193
Kepler's discovery of, 61, 63
of Mars, 58
of Triton, 197–198
Orbits, satellite
  circular velocity, 74–75
  escape velocity, 75
Orbits, stellar
  of disc stars vs. halo stars (Milky Way), 441
  of gas clouds in galactic plane, 432–433
  of Sun around Milky Way, 432
  See also Binary systems
Orion Nebula
  in Milky Way, 433
  star formation in, 361–362, 403
O-type stars, 333, 362, 412, 417, 419
Outer planets. See Giant planets
Oxygen, gaseous
  ionized oxygen, blue-green color of, 406, 407
  in planetary atmospheres, 171
Ozone layer
  in Earth's atmosphere, 134–135

P

Parallax
  basics of, 58
  Copernican/Ptolemaic models, parallax in, 58–59, 60
  Hipparcos satellite measurement of, 329
  kiloparsec, defined, 424
  megaparsec, defined, 470
  parsec, defined, 325, 330
  and stellar distances, 329–330, 394
Parent body (of meteorites), 205
Parsec
  kiloparsec, defined, 424
  in measuring stellar parallax, 330
  megaparsec, defined, 470
  parsec, defined, 325, 330
Particle pairs, subatomic
  pair annihilation, 535
  pair creation, 535
  virtual particle-antiparticle pairs, 536
  See also Antimatter/antiparticles
Pathfinder, 160, 161
Payne-Gaposhkin, Cecilia, 323–324
Peale, Stanton, 176
Penzias, Arno, 508
Perihelion of Mercury, shift in, 150
Periodic table (of elements), 87, 88, 91
Periodicity
  of mass extinctions, 227–228
  of orbits, 75
  periodic phenomena, 75, 76
  periodic processes, 77–78
  periodic sequences (cards), 75, 76
Perseus
  Perseus arm in Milky Way, 433
Phases of the Moon, 37, 39–40, 44, 138
Phobos
  cratering on, 165, 169
  irregular shape of, 164–165
  origin of, 164
  photographs of, 165
  physical characteristics of, 164
  temperature of, 169
Photoelectric effect, 271
Photometry
  introduction to, 285
  See also Telescopes, optical
Photons
  in black hole environment, 384, 385
  energy loss of (from space expansion), 533
  gravitational redshift of, 383
  movement of (in solar atmosphere), 308–309
  and Olbers' Paradox, 509–510
  photon flux, F (apparent brightness), 393
  in quantum theory, 271, 274, 278
Pioneer 10 spacecraft
  "message in a bottle" on, 575–576
Pixels (digital image processing), App. A-7, 286
Planck, Max
  Planck era, 543
  Planck's constant (h), 274
  proposes quantum theory, 273–274
  See also Matter, structure of; Radiation, electromagnetic
Planetary evolution
  atmospheric retention vs. planet size, 169–171
  Beta Pictoris, planetary dust around, 362, 363
  CO_2 as normal atmosphere, 171
  five rules of, 165–166
  geological activity vs. planet size, 168–169
  impact craters vs. surface age, 166
  impact heating (Earth), 117
  internal heating from radioactive decay, 116, 119–120, 121, 168–169
  oxygen in planetary atmospheres, 171
  planetary growth through collision, 249
  planets as debris after star formation, 262
  Safronov theory of, 236–237, 246
  size limit of, 259–269
  See also Interstellar medium
Planetary nebulae (from stellar mass loss), 370, 371
Planetary orbits. See Orbits, planetary
Planetesimals
  accretion of, 246–249, 261, 262
  defined, 206
Planets. See Planetary evolution; Planetary nebulae; Planets, extrasolar; See also specific planets; Solar system
Planets, extrasolar
  introduction to, 251
  characteristics of, 259
  and conditions for life, 568
  direct detection of (optical), 251, 254
  discussion of, 259–260, 262
  habitability of, parent star conditions for, 553
  indirect detection of (gravitational), 254–256, 262
  orbital data for, 259
  orbital detection of (perturbations), 311–312
  reflex motion (wobble), doppler detection of, 255–256, 258
  search for (interferometric), 552
  size limit of, 259–260
  two-bodied systems, center of mass of, 255
Plate tectonics
  continental drift (Wegener), 120–121
  continents, movement of, 122, 124, 143
  dynamic processes of (illus.), 120
  and mountain building, 122
  and origins of earthquakes, 121
  Pangea, 122, 124
  plate boundaries, map of, 123
  plate boundaries and volcanism, 122–123
  the "ring of fire," 122
  on Venus, 155
Plato (Greece), 36
Pleiades (Seven Sisters)
  as open cluster, 412, 413
  and zodiacal light, 218
Plurality of worlds, 71

Pluto
  differences from satellite Charon, 199–200
  discovery of, 199
  Hubble views of, 199, 200
  planet or interplanetary body, 199
  satellite of, tabulated data for, App. B-1
Polaris
  determining latitude using, 25, 26
Precession, gyroscopic, 32
Pre-main-sequence stars
  contraction/luminosity of (Hayashi), 358–359
  origins of, 358
  place on H-R diagram, 358–359
  T Tauri transitional stars, 361–362, 363
Proton-proton chain (solar nuclear reaction), 305–306
Protons, instability of, 527
Protostars
  formation of, 358–359
  fusion in, 357
Protosun
  evolution of, 239
  See also Solar system; Sun
Ptolemaic model (of solar system)
  epicyclic planetary motions in, 54–55
  Greek (Aristotelian) basis of, 54
  observational parallax in, 59, 60
Ptolemy, Claudius, App. A-7, 47
Pulsars. See Neutron stars/pulsars
Pythagoras, 36–37

Q

Quantum theory
  absorption/emission spectra (See Stellar spectra)
  basics of, 274
  electron behavior, quantum analogy for, 276–277
  Heisenberg uncertainty principle, 276, 278
  photoelectric effect, 271
  photons, 271, 274, 278
  Planck proposes, 273–274
  Planck's constant (h), 274
  quantum, defined, 274
  quantum fluctuations, 536, 542
  See also Fusion, cosmic; Fusion, stellar; Nuclear energy
Quasars
  accretion disc of, 499
  and black holes, 500
  composite spectrum of, 497, 499
  discovery of, 468–469
  distance of, 504
  energy source of, 499–500
  gravitational lensing of, 494–495, 501
  Maarten Schmidt determines nature of, 469
  properties of, 496–497, 499
  quasar, defined, 495
  quasar redshifts vs. galaxy redshifts, 495
  radio jets from, 500, 501, 502

R

Radial/tangential velocity, stellar
  astrometric calculations for, 334–335
  See also Redshift *(various)*; Stellar motion
Radiation, electromagnetic
  introduction to, 266–267
  absorption by Earth's atmosphere, 281
  basic properties of, 274
  characteristics of (wavelength/amplitude/frequency), 267, 268
  cosmic noise sources, spectrum of, 576

631

Radiation, electromagnetic *(continued)*
  detection of, 293–294
  electromagnetic spectrum, described, 267, 268
  synchrotron radiation, 493
  velocity of *(c)*, 267–269
  X rays, discovery of, 265–266
  See also Quantum theory; Quasars; Radiation, infrared; Radio signals; Visible spectrum; X Rays
Radiation, infrared
  from giant planets, 184–185
  incandescent vs. fluorescent lamps, 104–105, 301
  infrared, discovery of (Herschel), 102, 108
  infrared imaging, 104, 105, 108
  Kirchhoff's laws of, 299, 301
  Milky Way, far infrared image of, 431
  Milky Way, near infrared image of, 431–432
  Stefan-Boltzmann law (black body radiation), 106, 335, 336, 341, 344
  thermal spectra (stellar), 272–273
  thermal vs. nonthermal radiation, 493–494
  vs. stellar aging, 523
  vs. temperature, 104, 105–106
  Wien's Law of, 103–106, 108, 272–273
  See also Radiation, electromagnetic; Stellar spectra
Radii/size of stars, App. B–2, B–3, 335, 345
Radio signals
  as electromagnetic waves, 293
  from Milky Way galactic center, 438, 439
  near Milky Way galactic center, 438
  produced by Heinrich Hertz, 293
  from Sagittarius A, 439
  See also Radiation, electromagnetic
Radioactive dating
  of Earth, 113–115, 126, 143
  of Moon, 126, 143
Radioactive decay
  half-lives, radioactive, 115
  and internal planetary heating, 116, 119–120, 121
  and radioactive dating, 113–115, 126, 143
  See also Energy; Matter, structure of
Radioactive heating
  of carbon sphere, 882
  planetary, 116, 119–120, 121, 303
  See also Energy; Matter, structure of; Nuclear energy
Random processes
  mathematics of, 132–134
  random errors, App. A–4
  vs. Darwinian evolution, 131
Ray, John, 113
Reber, Grote, 288, 438
Red giants, App. B–3, 339, 341, 344–345, 346, 347, 366, 367
  hypothetical, in Milky Way, 439
  hypothetical red-giant/white dwarf binary, 398
Reddening of starlight. *See* Starlight
Redshift, Doppler
  defined, 515
  Doppler vs. cosmological redshift, 515–516
  redshifts/blueshifts in Milky Way, 432, 433
  stellar spectra, redshift/blueshift of, 334
  See also Redshift, galactic; Stellar motion
Redshift, galactic
  and angular diameter, 521–522
  and apparent brightness, 522–523
  and cosmic evolution, 523
  cosmological redshift and galactic dimming, 521

cosmological redshift *(z)*, defined, 515
cosmological vs. Doppler redshift, 515–516
discovery of, 469–470
Hubble relation for, 471, 472, 473–474, 510, 516, 517
interpretation of, 419–473
for most distant galaxies, problems with, 490
redshift vs. distance (Toolbox), 475
in testing cosmological model, 521–523
vs. quasar redshifts, 495
See also Redshift, Doppler; Redshift, gravitational; Stellar motion
Redshift, gravitational
  by black holes, 385–386
  of photons, 383
  and Theory of Relativity, 383
Reflex motion (wobble)
  in two-bodied system, 255–256, 258
Relativism, 18
Relativity, theories of
  acceleration, thought experiment about, 382
  general theory of, 382–384, 510
  gravitational mass vs. inertial mass, 382, 510
  gravitational redshift in, 383
  implications for bodies in motion, 510
  light, deflection of (elevator example), 511
  mass-energy equivalence in, 381, 510, 511, 540
  space/time distortion (*See* Space/time, curvature of)
  special theory of, 380–382, 510
  starlight, deflection of, 383, 511, 512
  See also Black holes
Resonance and harmonics
  in Mercury's rotation, 149–150
  in ring systems (Toolbox), 189, 190, 191
Retrograde orbits, planetary
  of Mars, 58
  Ptolemaic model of solar system, 54–55
  of Triton, 197–198
Retrograde rotation
  of Uranus, 184
  See also Retrograde orbits, planetary
Reynolds, Ray, 176
Right ascension, position on the sky, App. A–6
Ring of fire, 122
Ring systems
  Cassini's division in, 188
  general behavior of, 238
  of the giant planets, 187
  Jupiter's, 187
  of Neptune, 188
  origins of, 189
  resonance and harmonics in, 189, 190, 191
  and Roche's limit, 189, 192
  of Saturn, 187–188
  shepherd satellites in, 188
  structure of, 188–189, 201
  tidal forces on, 200
  of Uranus, 188
Ritter, Johann, 265
RNA (ribonucleic acid), 554–557
Roche surface
  in binary systems, 399
  in contact binary systems, 401
Roche's limit (in ring systems), 189, 192
Roemer, Ole, 269
Roentgen, Wilhelm, 265–266
Rotation
  Earth, effect of tidal forces on, 139, 140
  Earth, prograde rotation of, 184
  giant planets, rapid rotation of, 178

Uranus, anomalous rotation/inclination of, 182, 184–185
Rumford, Count Benjamin, 92–93
Russell, Henry Norris
  theorem about stellar evolution, 347, 348
  See also Hertzsprung-Russell (H-R) diagram
Rutherford, Ernest
  biographical note, 86
  demonstrates radioactive decay, 302
  gold foil experiment, 85, 87–88, 89

S

Safronov, Victor S., 236–237, 246
Sagan, Carl
  deduces greenhouse effect on Venus, 151
  and SETI program, 575
Sagittarius (molecular cloud)
  ethyl alcohol in, 357
  in Milky Way, 433
  photograph of, 420
  radio emissions from, 439
Satellites, planetary
  cratered/uncratered regions on, 193
  four categories of, 191, 193
  of giant planets, introduction to, 189–193
  and gravitational sphere of influence, 191, 193
  origins and capture of, 242–250
  shepherd satellites (in ring systems), 188
  See also specific satellites
Saturn
  ice on, 178, 181–182
  internal structure of, 181–182
  ring system of, 187–188
  satellites of, tabulated data for, App. B–1
Schiaparelli, G.V., 216
Schmidt, Maarten, 469
Schwarzschild radius (of black holes), 384, 492
Science, study themes in, 4–5
Science Toolboxes
  Aristarchus and sun-centered cosmology, 43–44
  atom and molecule velocities in a gas, 96–97
  binaries and stellar mass, 400
  black holes, properties of, 386
  chaos and determinism, 252–253
  collisions and uncertainty (photons/random walk), 308–309
  counting craters (on Moon), 167–168
  the critical density of the universe, 524
  digital information, 286–287; *See also* App. A–7
  dividing time, 31
  Doppler effect and planet detection, 257
  the Drake equation, 570–571
  dust extinction and reddening, 411–412
  electromagnetic radiation, properties of, 274
  entropy and time, 374
  errors and statistics, 15–16
  estimation, 9
  galaxy clustering, 483–484
  gas laws, 183–184
  giants and dwarfs, compared, 343–344
  gravitational lensing, 498
  gravitational perturbations, 221
  gravity, inverse square law of, 69–70
  gravity, many-body, 457–458
  greenhouse effect (Venus), 153
  isotropy and anisotropy, 422–424
  Kepler's laws and comet orbits, 214–215
  life as digital information, 555
  lifetimes of stars, 366
  light, travel time of, 451

mass-energy conversion, 304
momentum and angular momentum, 240–241
parallax and stellar distances, 330
particles and radiation in the early universe, 535–536
periodic processes, 77–78
potential/kinetic energy, 94
probability and impacts, 230
random processes, 132–134
resonance and harmonics, 190–191
Roche's limit, 192
scientific (exponential) notation, App. A–1, 7–8
small-angle equation, 38
Stefan-Boltzmann Law, 336
telescopes, collecting area/resolution of, 290–291
testing a hypothesis, 311–312
tidal forces, 139–140
uncertainty and the quantum world, 278
universal applicability of, 12
universe, size and age of, 477
weighing a galaxy, 437
Wien's Law, 105–106
See also Scientific method
Scientific (exponential) notation, App. A–1, 7–8
Scientific method
average (mean), 12, 13, 15, 16
data, ways of representing, App. A–7
deductive reasoning, 10
distribution (normal/Gaussian), App. A–4
error, statistical reduction of, App. A–4, 15–16
error bars, 312
essential steps in, 5–7
estimation, 8, 9, 13
evidence, gathering/evaluating, 7
histograms, App. A–7
hypotheses, testing of, 311–312
inductive reasoning, 10–12
knowledge, creation of, 19
limitations of, 7
logic, introduction to, 8, 10
mathematics, universal applicability of, 12
mean value (average), 12, 13, 15, 16
measurement as basic step in, 7–8
measurements, distribution of, 12–13
metric system, 8, 35
observational (measurement) errors, 12–14, 20
playing card analogy for, 6, 12, 13
precision in, App. A–4, 8
precision vs. accuracy, App. A–4, 12–14
reliability of, 19–20
scatter (in measurements), 15
scientific (exponential) notation, App. A–1, 7–8
standard deviation (uncertainty), 12, 15
standard error, 12, 13, 15–16
systematic vs. random errors, 16
theories, building/testing, 14, 16–17, 19
unanswered questions in, 20
unscientific alternatives to, 18
See also Mathematics, basics of; Measurement systems
Seasons
cause of, 45–46
Eratosthenes deductions regarding, 45
marking by solstices, 28, 29
tracking by observation of night sky, 26–28
on Uranus, 185

Sedimentary rocks, 118
SETI (Search for Extraterrestrial Intelligence)
anthropic principle, 580–581
Arecibo radio telescope, 264, 288, 579
beginnings of (Project OZMA), 577–578
controversy over, 579
future schemes for, 579
Microwave Observing Project (MOP), 578, 579
Pioneer 10 signal, detection of, 579
See also Intelligence, extraterrestrial
Seven Sisters. See Pleiades
Seyfert galaxies, 492
Shapley, Harlow, 420
Shepherd satellites (in ring systems), 188
Shoemaker, Carolyn and Gene, 211, 229
Siberian explosion (Tunguska Event), 203–205, 225, 227
Sidereal day
accuracy of, 30
Significant figures and digits, App. A–1
Simulation, computer
of galaxies merging, 462–463
of gravity of many bodies, 457–458
of stellar spectra, 332
of supernovae, 379
Sirius (as binary system), 400
Sky, observing. See Night sky, observing
Slipher, Vesto, 469
Small angles, equation for, App. A–5, 38
Solar day
accuracy of, 30
defined, 30
Solar spectrum
absorption lines in, 299
color range of (Wien's Law), 103, 104
and discovery of helium in sun, 301–302
and solar surface temperature, 272–273, 300
See also Stars, properties of; Stellar spectra; Sun
Solar system
accretion, process of, 246–249, 261, 262
age of, 261
angular momentum, conservation of, 239, 240–241, 261
angular momentum of sun/solar system, 250
archaeology of: unraveling the clues, 237–238
arrangement of, 72–73
asteroid belt, formation of, 249
astronomical unit (AU), defined, 73, 330
catastrophic events, perturbation by, 250–251, 261
comets, formation of, 249
Copernican (heliocentric) model of, 56
early solar nebula, view of, 235, 241
gravitational collapse, process of, 239, 241
Hannes Alfven comment on, 237
Helmholtz contraction, 241–242
impact craters, typical, 248
largest bodies in (illus.), 148
meteorites, formation of, 249
parallax (Copernican vs. Ptolemaic), 58–59, 60
planetary data, table of, App. B–1
planetary spacing in, 73–74
planets, evolution of (See Planetary evolution)
planets, naming of, 32
planets, number of, 81, 201
properties of, 237, 261
protosun, three stages in evolution of, 239

Ptolemaic model of, 54–55
radioactive isotopes, puzzling presence of, 245
ring systems, general behavior of, 238
satellites, origins and capture of, 242–250
scale of, 72–73, 81
solar disk, early formation/contraction of, 238–239, 241
solar ignition, 242
solar nebula, condensation of, 242–243
terrestrial planets, introduction to, 147–149
terrestrial planets, formation of, 249
theory of origins, 237
See also Solar wind; Sun; *specific planets*
Solar wind
and auroral displays, 318
discussion of, 316–317
effect on comets, 208
interaction with Earth's magnetic field, 319
Solstices
application to architecture, 29–30
marking seasons by, 28, 29
rituals celebrating, 29
summer/winter, defined, 29
See also Stonehenge
Soot line
in asteroid distribution, 223, 243
Sound (as compression wave), 270
Space exploration
future of, 79
history of, 75–76, 78–79
Space/time, curvature of
and angular diameter, 521–522
by black holes, 384–385
closed (positive curvature) universe, 516, 520–523
Euclidean vs. non-Euclidean geometry, 511–512
evidence for, 545
flat universe, evidence for, 536–539
flat (zero curvature) universe, 516, 520–523
by gravity (Solar example), 382–383
measurement of, 520–524
open (negative curvature) universe, 516, 520–523
terrestrial analogy for, 512–513
two-dimensional analogy for, 512
in very early universe, 543
See also Relativity, theories of
Spectral analysis/Spectroscopy. See Electromagnetic spectrum; Solar spectrum; Stellar spectra; Visible spectrum
Standard error/standard deviation, App. A–4, 12, 13, 15–16
Star formation/lifetimes. See Stellar evolution
Starlight
blue light, scattering of, 406, 407, 408–409
blue sky, explanation of, 408
Doppler redshifts/blueshifts (See Redshift, Doppler)
dust extinction and reddening of, 406, 407, 408–409, 411–412, 438
gas atoms, interaction with, 405–406
gravitational deflection of (See Gravitational lensing)
ionized oxygen, blue-green color of, 406, 407
Milky Way dimming of, 430, 431
nebula molecules, interaction with, 406–408
redshift of (See Redshift, galactic *(various)*)
Trifid Nebula, red/blue colors of, 407
See also Luminosity, stellar; Solar spectrum; Stellar spectra

Stars, properties of
  introduction to, 324–325
  ages/lifetimes (*See* Stellar evolution)
  apparent brightness, 326–327, 328, 393–394
  astronomical unit (AU), defined, 73, 330
  brightest stars, tabulated data for, App. B–3
  chemical composition, 334
  classification (*See* Hertzsprung-Russell [H-R] diagram)
  clusters (*See* Clusters, stellar)
  diameters/radii of, App. B–2, B–3, 335, 345
  distances, measuring (*See* Stellar distances)
  giants, luminosity of, App. B–3
  giants vs. dwarfs, luminosity of, 343–344
  habitable planets, conditions for, 553
  habitable zones around, 553
  hydrostatic equilibrium, 337
  intrinsic brightness, 326
  light year, defined, 325
  low-luminance stars, predominance of, 340, 341
  luminosity (*See* Luminosity)
  magnitude, App. A–6, 327
  mass (*See* Mass, stellar)
  megaparsec, defined, 425
  naming, 325
  nearest stars, tabulated data for, App. B–2
  parsec, defined, 325, 330
  propagation of light from (inverse-square law), 327, 328
  proper motion, 334
  radial/tangential velocity, 334–335
  spectra (*See* Stellar spectra)
  spectral classes, 331, 332–333
  star counts, errors in, 340, 341
  stellar parallax, measuring, 329–330
  stellar population, distribution of, 394–395
  stellar population, measuring, 340–341, 350
  stellar radius, measurement along H-R diagram, 341, 344–346
  summary discussion of, 348–349
  temperature classifications, 332–333, 347
  variables/Cepheid variables (*See* Variable stars)
  See also Galaxies; Hertzsprung-Russell (H-R) diagram; Milky Way; Solar spectrum; Starlight; Stellar spectra
Statistics
  counting craters on the Moon, 167–168
  counting statistics (card example), 166
  statistics and errors (Toolbox), 15–16
Statistics and errors, App. A–4, 15–16
Stefan-Boltzmann law, 106, 335, 336, 342, 344
Stellar bar (in Milky Way), 439
Stellar distances
  Andromeda as distance indicator, 478–479
  Astronomical unit (AU), 330
  dist-ance of Sun from center of Milky Way, 424
  distances to (definitions), 325
  distances to (measuring), 327, 329, 330, 349, 425
  Faber-Jackson relation (elliptical galaxy), 474, 476
  galactic distance scale, 444–445, 515–516
  Galactic distance vs. redshift (Toolbox), 475
  galaxies, distance indicators for, 444–445, 451, 478–479
  Hubble relation (galactic radial velocity vs. distance), 471, 472, 473–474, 510, 516, 517
  kiloparsec, defined, 424
  and light travel time, 451–452
  light-year, defined, 452
  Local Group galaxies, optical resolution of, 451–452
  megaparsec, defined, 470
  and "old" light (Toolbox), 451
  parsec, defined, 325, 330
  relating redshift and distance (Toolbox), 475
  stellar parallax, measuring, 329–330
  supernovae as galactic distance indicators, 476
  Tully-Fisher relation (for galactic distances), 474, 476
Stellar evolution
  introduction to, 353, 354–355
  ages, star groups/clusters, 417–420, 425
  black holes, 384–387
  brown dwarfs, 360–361
  dust clouds/dust discs, 358, 361–362, 363
  Eagle Nebula, star formation in, 352
  elements, cosmic abundance of, 354–355
  evolved stars, introduction to, 365, 367
  galactic evolution/interaction (*See* Galaxies)
  heavy elements, creation of, 368, 369, 375–376, 389
  helium flash, 367
  hydrogen exhaustion, 367
  and infrared radiation, 523
  and the Laws of Thermodynamics, 371–373
  lifetimes, main-sequence stars, 365, 366
  main-sequence stars (*See* Main-sequence stars)
  mass limits, large and small, 359–360
  mass loss (in evolved stars), 370, 371
  molecular clouds, 355–357, 358
  neutron stars/pulsars, 378–380, 381
  non-main-sequence stars, 340–341, 342
  Omega Nebula (M17), star formation in, 359
  Orion Nebula, star formation in, 361–362
  precollapse core layering, 368, 369
  pre-main-sequence stars, 358–359, 360, 361–362, 363
  protostars, 358
  protosun, three stages in evolution of, 239
  red giants, 339, 341, 344–345, 346, 347, 366, 367
  Russell-Vogt theorem and, 347–348
  star formation, stages in, 358, 388, 389
  summary of, 388
  supernovae, 375–378
  T Tauri stars, 361–362, 363
  universe, size/age of, 477, 516, 517, 526, 527
  white dwarfs, 339, 366, 373–375
  See also Cosmology; Galaxies; Hertzsprung-Russell (H-R) diagram; Nebulae
Stellar motion
  astrometry, 334–335
  Doppler effect and stellar radial velocity, 334–335
  Doppler effect in two-bodied system, 255–256, 258
  overall motion of galaxies via Hubble relation, 471, 472, 473–474, 510, 516, 517
  proper motion, 334
  radial/tangential velocity, basic calculations for, 334–335
  See also Redshift (*various*); Universe, expanding
Stellar populations
  clustered vs. random distribution of, 394–395
  disk population (Milky Way), 440, 441, 442
  measuring, 340–341, 350
  of Milky Way, 440–442
  nearest stars, tabulated data for, App. B–2
  See also Galaxies; Milky Way
Stellar spectra
  introduction to, 329, 331
  absorption bands, 281, 333
  absorption lines, 280, 299–300, 331–333, 405–406
  blueshift/redshift of (basics), 334
  changes vs. star age, 523
  computer simulation of, 332
  Draper early photographs of, 331
  early stellar spectra (from prism), 322
  emission bands, 279–280, 492
  emission lines, 277, 279, 299–300, 331, 405–406
  Fraunhofer lines, 299
  and free electrons, 300
  and galactic redshift, 473, 523
  and galactic rotation curves, 455, 456
  of halo stars vs. disk stars, 396, 397
  Harvard study of (Annie Cannon), 323, 331
  hydrogen, energy states of, 331–332
  hydrogen-alpha line, 331–332
  in interstellar medium, 404
  and Kirchhoff's laws of radiation, 299, 301
  luminosity class (of a star), 345–346
  molecular clouds, radio-wavelength spectrum of, 355–357, 358
  Payne-Gaposhkin study of, 323–324
  quasar composite spectrum, 497, 499
  in quasar composite spectrum, 497, 499
  remote objects, analysis of, 294
  Seyfert galaxies, emission bands in, 492
  in solar spectrum, 299
  spectral classes, 331, 332–333
  spectroscopic binaries, 396, 397
  thermal spectra, 272–273
  See also Redshift (*various*); Solar spectrum; Starlight; Stars, properties of
Stellar wind
  in formation of white dwarfs, 375
  from mass-shedding stars, 370, 371
  and T Tauri stars, 362
Stern, D.K., 549
Stochastic star-formation theory, 434–436
Stonehenge
  aerial view of, 33
  angular measurements of, 34, 35
  astronomical orientation of, 23–24, 32
  heel stone in, 32
  and summer solstice, 32, 33
  sun rising over, 22, 32
Sublimation of comet nuclei, 205, 207–208, 211
Substellar objects (brown dwarfs), 360–361
Sun
  age of (Kelvin-Huxley debate), 297
  atmosphere of, 308–309, 310, 313
  auroras, 86, 317, 318
  chromosphere, 310, 314
  composition of, 301–302, 319–320, 550
  corona, 307, 312–313, 314
  cross-section of, 307
  discovery of helium in, 301–302
  distance from center of galaxy, 424
  effect on Earth's climate, 317–318, 320
  end-of-life transformation to white dwarf, 375
  energy flow/energy balance in, 307
  fusion reactions in (proton-proton chain), 305–306